Texts and
Monographs
in Physics

Roger G. Newton

Scattering Theory of Waves and Particles

Second Edition

With 35 Illustrations

Springer-Verlag
New York Heidelberg Berlin

Roger G. Newton
Department of Physics
Indiana University
Swain Hall—West 117
Bloomington, IN 47405
U.S.A.

Editors

Wolf Beiglböck
Institut für Angewandte Mathematik
Universität Heidelberg
Im Neuenheimer Feld 5
D-6900 Heidelberg 1
Federal Republic of Germany

Elliott H. Lieb
Department of Physics
Joseph Henry Laboratories
Princeton University
P.O. Box 708
Princeton, NJ 08540
U.S.A.

Walter Thirring
Institut für Theoretische Physik
der Universität Wien
Boltzmanngasse 5
A-1090 Wien
Austria

The first edition was published in 1966 by McGraw-Hill Inc.

Library of Congress Cataloging in Publication Data

Newton, Roger G.
Scattering theory of waves and particles.

(Texts and monographs in physics)
Originally issued in series: International series in pure and applied physics.
Bibliography: p.
Includes index.
1. Scattering (Physics) I. Title. II. Series.
QC794.6.S3N47 1982 530.1 81-23326
AACR2

Printed in the United States of America.
Typeset by Asco Trade Typesetting Ltd., Hong Kong.

9 8 7 6 5 4 3 2 1

ISBN 0-387-10950-1 Springer-Verlag New York Heidelberg Berlin
ISBN 3-540-10950-1 Springer-Verlag Berlin Heidelberg New York

To Ruth, and Julie, Rachel, and Paul

Preface to the Second Edition

Much progress has been made in scattering theory since the publication of the first edition of this book fifteen years ago, and it is time to update it. Needless to say, it was impossible to incorporate all areas of new development. Since among the newer books on scattering theory there are three excellent volumes that treat the subject from a much more abstract mathematical point of view (Lax and Phillips on electromagnetic scattering, Amrein, Jauch and Sinha, and Reed and Simon on quantum scattering), I have refrained from adding material concerning the abundant new mathematical results on time-dependent formulations of scattering theory. The only exception is Dollard's beautiful "scattering into cones" method that connects the physically intuitive and mathematically clean wave-packet description to experimentally accessible scattering rates in a much more satisfactory manner than the older procedure.

Areas that have been substantially augmented are the analysis of the three-dimensional Schrödinger equation for noncentral potentials (in Chapter 10), the general approach to multiparticle reaction theory (in Chapter 16), the specific treatment of three-particle scattering (in Chapter 17), and inverse scattering (in Chapter 20). The additions to Chapter 16 include an introduction to the two-Hilbert space approach, as well as a derivation of general scattering-rate formulas. Chapter 17 now contains a survey of various approaches to the solution of three-particle problems, as well as a discussion of the Efimov effect. I did not include a detailed treatment of n-particle equations for $n > 3$, which is a topic currently under intense development, except for the addition of an extensive bibliography. In Chapter

20 I have added a discussion of the phase problem for the scattering amplitude, the Marchenko equation for solving the inverse scattering problem at fixed angular momentum in the case of central potentials, and a detailed summary of recent results for noncentral potentials. There are many other minor corrections and additions in the text.

The notes to most chapters have been substantially lengthened and the bibliography has been increased by about a thousand entries. In order to save space I have reluctantly abandoned the alphabetized listing of multiply authored papers under each co-author's name.

I am indebted to many people, among whom I should mention particularly Professor H. van Haeringen, for pointing out to me minor mistakes and printing errors in the first edition. It is surely too much to hope that the second edition will be free of such blemishes, but perhaps there will be fewer than in the first.

Bloomington, Indiana
March, 1982

Roger G. Newton

Preface to the First Edition

The importance of observations made in physics by analyzing the scattering of particles or waves can be hardly exaggerated. For the last half century especially, many of the most important discoveries have been made by collision experiments. They range from Rutherford's discovery of the atomic nucleus, over atomic and nuclear spectroscopy, to nuclear fission and the finding of the fundamental particles and their properties. If we add to these phenomena all the observations made by analysis of light scattering, the list becomes truly impressive.

Yet until very recently the theory of scattering has not been a very prominent part of the graduate education of physicists. Although most quantum-mechanics books now contain a chapter on it, there was only one book devoted to collision theory, Mott and Massey's "The Theory of Atomic Collisions," and it was written many years ago. The journal literature on the subject, by contrast, is enormous. It is only lately that the situation has begun to change, and two new volumes on the quantum theory of scattering have appeared.

The present book cuts across the boundary lines of the traditional subdivisions of physics and treats scattering theory in the contexts of classical electromagnetic radiation, classical particle mechanics, and quantum mechanics. The reason for this disregard of the canonical course structure is twofold. One is that it is most instructive to see analogous phenomena occur in apparently dissimilar disciplines. The second is to be able to use similar tools and languages in them. Thirty years ago it used to be standard procedure when possible to explicate a new quantum phenomenon in the more familiar terms of electromagnetic theory. I dare say that nowadays many physicists are more conversant with quantum mechanics than with radiation theory, and in parts of electromagnetic theory it is often helpful to draw on a graduate student's experience in quantum mechanics.

This book is meant to be read from the beginning. That is to say, a student who wants to learn scattering theory should not start with Chapter 6, even if his primary interest is quantum mechanics. The reason is that the physical ideas as well as the concrete pictures and concepts that underlie the more powerful formal considerations of the first two chapters of Part III are introduced and discussed earlier. Without them, Chapters 6 and 7 are too abstract to be useful. Of course, all the relevant details are later discussed again in the quantum-mechanical context, but I want to discourage emphatically the learning (or teaching) of quantum scattering theory by reading Part III only and beginning with Chapter 6. Those interested solely in the quantum-mechanical part need not read all the previous details, but they should at least read those sections explicitly pointed out at the beginning of Chapter 6. In that way also the book can easily be used for part of a quantum-mechanics course.

This book is not meant to be an encyclopedia of scattering theory. There are large areas which have been omitted. The most important is the bulk of relativistic quantum scattering theory. The main reason is that the attitude with which that topic has to be approached at the present time is perforce quite different from that adopted in the theories discussed here. The time is not yet for a book on it that is more than an informal collection of notes (or less than an original and revolutionary treatise). Except for occasional use of relativistic kinematics, a discussion of the Lorentz invariance of cross sections, etc., I therefore left it out altogether.

Another omission, not entirely unrelated to the first, is that of many detailed applications of the theory to known experimental facts and the analysis of experimental data in terms of the concepts discussed. This is a book from which to learn, not the data on proton-proton or electron-nucleus scattering, but the methods, ideas, and principles in terms of which such data can be analyzed. For that reason it often simplifies nature, but, I hope, not beyond recognition. The very use of the Schrödinger equation as a model for the description of particles involves such a simplification. In any case, this is probably the best way to learn theoretical physics.

Each chapter is followed by bibliographical and historical notes. I am not foolhardy enough to claim completeness of the bibliography in any particular area. In some subjects I have made an effort, though, to make more thorough references than are unfortunately often found in the literature. No doubt I am nevertheless occasionally guilty of important omissions, for which I apologize. In other instances the bibliography is meant to be no more than a guide to further reading and to other reference lists.

Each chapter is followed by a set of problems, altogether about 200. They should help the student, even if most of them are not very difficult.

How successful they are remains to be seen. Few have been tested on live subjects.

The preparation I envisage for a reader of this book would fit many second-year graduate students of physics. He should be familiar with classical electromagnetic theory, i.e., with the Maxwell equations. But he need not have completed the standard graduate course in the subject. Part I, in fact, could well be used as a portion of that course. He should know some quantum mechanics. An introductory course to the quantum theory alone is not quite enough. Depending upon the circumstances, the first semester of the usual two- or three-semester sequence of quantum mechanics should be sufficient preparation.

As for mathematics, in addition to a good undergraduate grounding in analysis, the main preparation should include knowledge of the standard parts of the theory of functions of a complex variable and of ordinary differential equations. Some knowledge of linear integral equations and of functional analysis would be useful, but these subjects could be learned concurrently or more thoroughly after using some parts of this book as an introduction.

Now a word about the *notation* used in the quantum-mechanical part. I am not employing Dirac notation, except occasionally for transformation functions. By and large, with some exceptions that should cause no confusion, the capital letters Ψ and Φ are used for abstract state vectors and ψ and φ, etc., for wave functions. The usual physicist's notation of the hermitian inner product is used,* for complex conjugation,† for hermitian conjugation, tr for the trace, and det for the determinant. Sometimes I use a diadic notation for projections, such as $\Psi_a\Psi_b{}^\dagger$. This means the same thing as $|a\rangle\langle b|$ in Dirac's way of writing. In other words, $\Psi_b{}^\dagger$ is the image of Ψ_b in a dual $\mathfrak{H}^\dagger$ of the Hilbert space $\mathfrak{H}$, and the hermitian inner product in $\mathfrak{H}$ is then thought of as a euclidean inner product between members of $\mathfrak{H}$ and $\mathfrak{H}^\dagger$. Re and Im denote the real and imaginary parts of a complex number.

Financial support by the National Science Foundation for some of the research embodied in this book is gratefully acknowledged. I should also like to thank Mr. Michael O. Marcoux and Miss Lucy Tyson for carefully reading parts of the manuscript and pointing out a number of errors.

Roger G. Newton

Contents

PART I

SCATTERING OF ELECTROMAGNETIC WAVES

CHAPTER 1

Formalism and General Results

1.1 The Maxwell Equations

We shall use gaussian units in this book, so that the Maxwell equations for the electric fields $\mathscr{E}$ and $\mathscr{D}$, and the magnetic fields $\mathscr{H}$ and $\mathscr{B}$ read

$$\begin{aligned} \nabla \times \mathscr{H} &= \frac{4\pi}{c}\mathbf{j} + \frac{1}{c}\frac{\partial \mathscr{D}}{\partial t} \\ \nabla \times \mathscr{E} &= -\frac{1}{c}\frac{\partial \mathscr{B}}{\partial t} \\ \nabla \cdot \mathscr{D} &= 4\pi\rho \\ \nabla \cdot \mathscr{B} &= 0. \end{aligned} \tag{1.1}$$

In addition the supplementary equations of the medium are needed, which are

$$\mathscr{D} = \varepsilon\mathscr{E} \qquad \mathscr{B} = \mu\mathscr{H} \qquad \mathbf{j} = \sigma\mathscr{E} \tag{1.2}$$

in terms of the dielectric constant ε, the magnetic permeability μ, and the conductivity σ. All these may be tensors and need not be uniform. The medium, in other words, need not be homogeneous or isotropic.

For fields of (circular) frequency ω we use the complex notation

$$\mathscr{E}(t) = \mathscr{E}e^{-i\omega t} + \mathscr{E}^* e^{i\omega t} \tag{1.3}$$

and similarly for the other fields. The first two Maxwell equations then become

$$\nabla \times \mathscr{H} = -ikn'^2\mathscr{E} \qquad \nabla \times \mathscr{E} = ik\mathscr{B} \tag{1.4}$$

with

$$k = \frac{\omega}{c} = \frac{2\pi}{\lambda_0} \equiv \frac{1}{\lambdabar_0} \qquad n'^2 = \varepsilon + \frac{4\pi i\sigma}{\omega}, \tag{1.5}$$

λ_0 being the wavelength in vacuum. Taking the curl of the second equation in (1.4) leads to

$$\nabla \times (\nabla \times \mathscr{E}) = \mathrm{K}^2\mathscr{E} + \mu^{-1}(\nabla\mu) \times (\nabla \times \mathscr{E}) \tag{1.6}$$

with

$$\mathrm{K} \equiv kn \qquad n^2 \equiv \mu n'^2, \tag{1.7}$$

provided that μ is not a tensor (i.e., it is a multiple of the unit tensor). The second term on the right of (1.6) contributes only if the permeability μ is not uniform.

The last two equations in (1.1) are never needed for our purposes. The last follows from the second equation in (1.4), and the second to the last allows us to calculate the charge density ρ. Since the first equation in (1.4) implies that

$$\nabla \cdot (n'^2\mathscr{E}) = 0$$

we find that

$$\rho = -\frac{i}{\omega}\nabla \cdot (\sigma\mathscr{E}). \tag{1.8}$$

The boundary conditions across the surface of a medium are obtained by integrating $\mathscr{E}$ and $\mathscr{H}$ along a small loop as indicated in Fig. 1.1 and using

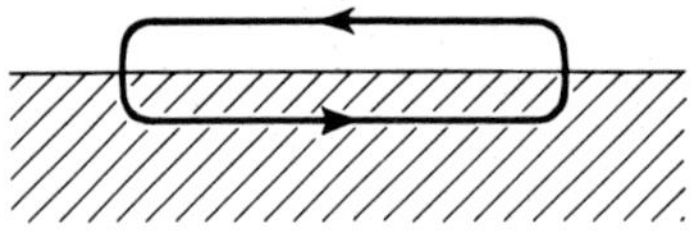

Figure 1.1. Contour of integration across an interface between two media.

Stokes' theorem as well as (1.1). As the area shrinks to zero, with the perpendicular sides small relative to those parallel to the interface, we find that the components of $\mathscr{E}$ and $\mathscr{H}$ parallel to the surface must be continuous. These are all the conditions needed. The divergence equations lead to conditions on the normal components, but these either follow from the others or define a surface charge.

1.2 Stokes Parameters and Polarization

1.2.1 Definition of the Stokes Parameters. A convenient way of parameterizing a plane electromagnetic wave is the following: First an arbitrary plane of reference is chosen through the direction $\hat{\mathbf{k}}$ of wave propagation. In the case of scattering this plane is most conveniently chosen to be that of the incident and scattered beam directions, because then it serves for both. Let $\hat{\mathbf{e}}_{\parallel}$ be the unit vector parallel and $\hat{\mathbf{e}}_{\perp}$, perpendicular to the reference plane, so that

$$\hat{\mathbf{k}} = \hat{\mathbf{e}}_{\parallel} \times \hat{\mathbf{e}}_{\perp}.$$

We now decompose the electric field vector accordingly,

$$= \mathscr{E}_{\parallel}\hat{\mathbf{e}}_{\parallel} + \mathscr{E}_{\perp}\hat{\mathbf{e}}_{\perp}. \tag{1.9}$$

Then the four Stokes parameters are defined by

$$\begin{aligned} I &\equiv |\mathscr{E}_{\parallel}|^2 + |\mathscr{E}_{\perp}|^2 \\ Q &\equiv |\mathscr{E}_{\parallel}|^2 - |\mathscr{E}_{\perp}|^2 \\ U &\equiv -2\,\mathrm{Re}\,(\mathscr{E}_{\parallel}\mathscr{E}_{\perp}^*) \\ V &\equiv -2\,\mathrm{Im}\,(\mathscr{E}_{\parallel}\mathscr{E}_{\perp}^*). \end{aligned} \tag{1.10}$$

Alternatively the wave may be analyzed in terms of right and left circularly polarized components. Set

$$\chi_1 \equiv \frac{1}{\sqrt{2}}(\hat{\mathbf{e}}_{\perp} - i\hat{\mathbf{e}}_{\parallel}) \qquad \chi_{-1} \equiv \chi_1^* \tag{1.11}$$

and write

$$\mathscr{E} = \chi_1\mathscr{E}_r + \chi_{-1}\mathscr{E}_l. \tag{1.12}$$

Then the Stokes parameters (1.10) are expressed in terms of the components $\mathscr{E}_r$ and $\mathscr{E}_l$ by

$$\begin{aligned} I &= |\mathscr{E}_r|^2 + |\mathscr{E}_l|^2 \\ Q &= -2\,\mathrm{Re}\,(\mathscr{E}_l\mathscr{E}_r^*) \\ U &= 2\,\mathrm{Im}\,(\mathscr{E}_l\mathscr{E}_r^*) \\ V &= |\mathscr{E}_r|^2 - |\mathscr{E}_l|^2. \end{aligned} \tag{1.13}$$

The following simply related parameters are sometimes more useful:

$$\begin{aligned} I_{\parallel} &\equiv |\mathscr{E}_{\parallel}|^2 = \tfrac{1}{2}(I + Q) \\ I_{\perp} &\equiv |\mathscr{E}_{\perp}|^2 = \tfrac{1}{2}(I - Q) \\ &U, V \end{aligned} \tag{1.14}$$

or

$$\begin{aligned} I_r &\equiv |\mathscr{E}_r|^2 = \tfrac{1}{2}(I + V) \\ I_l &\equiv |\mathscr{E}_l|^2 = \tfrac{1}{2}(I - V) \\ &Q, U. \end{aligned} \tag{1.15}$$

In neither case are the four parameters independent, since

$$I^2 = Q^2 + U^2 + V^2. \tag{1.16}$$

A simple notation we shall use for plane wave fields is

$$\mathfrak{E}^{(L)}(\mathbf{k},\mathbf{r}) \equiv \begin{pmatrix} \mathscr{E}_{\parallel} \\ \mathscr{E}_{\perp} \end{pmatrix}$$

and

$$\mathfrak{E}^{(C)}(\mathbf{k},\mathbf{r}) \equiv \begin{pmatrix} \mathscr{E}_r \\ \mathscr{E}_l \end{pmatrix},$$

indicating both the observation point $\mathbf{r}$ and the propagation vector $\mathbf{k}$. If we want to refer to both methods of polarization decomposition, without specification, we shall write simply $\mathfrak{E}$. It should be remembered, though, that such two-component vectors are additive only when their propagation vectors are collinear and their reference planes coincide.

1.2.2 Significance of the Parameters. The intensity I is directly related to the average energy density of the wave

$$\bar{U} = \frac{1}{2\pi} I \tag{1.17}$$

and to the average energy flux, or the Poynting vector

$$\mathbf{S} = c\bar{U}\hat{\mathbf{k}} = \frac{c}{2\pi}\hat{\mathbf{k}}I. \tag{1.18}$$

In order to see the significance of the other parameters, let us write the most general elliptically polarized plane wave moving in the z direction,

$$\mathscr{E}(t) = 2\mathscr{E}_0\hat{\mathbf{e}}_1 \sin\beta\cos(\omega t - kz + \alpha) + 2\mathscr{E}_0\hat{\mathbf{e}}_2\cos\beta\sin(\omega t - kz + \alpha) \tag{1.19}$$

in complex notation,

$$\mathscr{E}(t) = \mathscr{E}_0(\hat{\mathbf{e}}_1\sin\beta + \hat{\mathbf{e}}_2 i\cos\beta)e^{-i(\omega t - kz + \alpha)} + \text{c.c.}, \tag{1.20}$$

where $\hat{\mathbf{e}}_1$ and $\hat{\mathbf{e}}_2$ are unit vectors in the directions of the principal axes of the ellipse, as indicated in Fig. 1.2, and c.c. stands for "complex conjugate." The ellipticity is measured by $\tan\beta$. If $\tan\beta = 0$ or $\pm\infty$, then the wave

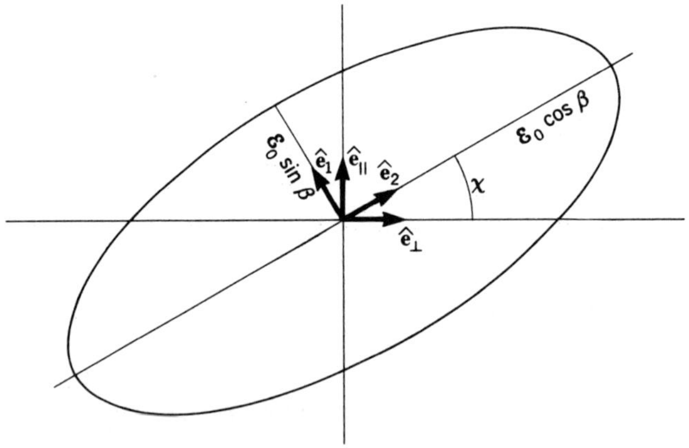

Figure 1.2. The polarization ellipse.

is linearly polarized; if $|\tan\beta| = 1$, it is circularly polarized. If $\tan\beta > 0$ it is right-handedly polarized, the electric vector at fixed z rotating like a right-handed screw relative to the propagation direction; if $\tan\beta < 0$, then it is left-handedly polarized.

Comparison of (1.20) with (1.9) shows that

$$\mathscr{E}_{\|} = \mathscr{E}_0(i\sin\chi\cos\beta + \cos\chi\sin\beta)e^{i(kz-\alpha-\omega t)}$$
$$\mathscr{E}_{\perp} = \mathscr{E}_0(i\cos\chi\cos\beta - \sin\chi\sin\beta)e^{i(kz-\alpha-\omega t)}$$

and therefore

$$I = \mathscr{E}_0^2 \qquad Q = -\mathscr{E}_0^2\cos 2\chi\cos 2\beta$$
$$I_{\|} = \tfrac{1}{2}\mathscr{E}_0^2(1 - \cos 2\chi\cos 2\beta) \tag{1.21}$$
$$I_{\perp} = \tfrac{1}{2}\mathscr{E}_0^2(1 + \cos 2\chi\cos 2\beta)$$
$$U = -\mathscr{E}_0^2\cos 2\beta\sin 2\chi \qquad V = \mathscr{E}_0^2\sin 2\beta, \tag{1.22}$$

the inversion of which is

$$\begin{aligned} \tan 2\chi &= \frac{U}{Q} \\ \tan 2\beta &= V(U^2 + Q^2)^{-1/2} \\ \mathscr{E}_0 &= I^{1/2}. \end{aligned} \tag{1.23}$$

The parameter V is thus most directly related to the ellipticity; its sign determines the "handedness." Q depends, in addition, on the "tilt." A change in reference plane changes only χ; hence I, V, and $Q^2 + U^2$ are invariant under such a rotation. Changing the signs of both $\hat{\mathbf{e}}_{\|}$ and $\hat{\mathbf{e}}_{\perp}$ has no effect on any of the parameters.

Alternatively, we may write

$$\mathscr{E} = \mathscr{E}_0(\boldsymbol{\chi}_{01}\cos\gamma + \boldsymbol{\chi}_{0-1}\sin\gamma)e^{-i(\omega t - kz + \alpha)} + \text{c.c.} \tag{1.24}$$

with

$$\boldsymbol{\chi}_{01} = \frac{1}{\sqrt{2}}(\hat{\mathbf{e}}_1 + i\hat{\mathbf{e}}_2) = \boldsymbol{\chi}_1 e^{i\chi} \qquad \boldsymbol{\chi}_{0-1} = \boldsymbol{\chi}_{01}^* \tag{1.25}$$

and then obtain

$$I = \mathscr{E}_0^2 \qquad V = \mathscr{E}_0^2\cos 2\gamma$$
$$I_r = \mathscr{E}_0^2\cos^2\gamma \qquad I_l = \mathscr{E}_0^2\sin^2\gamma \tag{1.26}$$
$$Q = -\mathscr{E}_0^2\sin 2\gamma\cos 2\chi \qquad U = -\mathscr{E}_0^2\sin 2\gamma\sin 2\chi,$$

the inversion of which is

$$\begin{aligned} &\tan 2\chi = \frac{U}{Q} \qquad \mathscr{E}_0 = I^{1/2} \\ &\tan^2\gamma = \frac{I_l}{I_r} \quad \text{or} \quad \cot 2\gamma = V(Q^2 + U^2)^{-1/2}. \end{aligned} \tag{1.27}$$

1.2.3 Partially Polarized Beams. It is important to notice that the phase α does not appear in the Stokes parameters. Therefore they can be used for the specification of an *incoherent* beam,[1] i.e., of a superposition of many waves of the same propagation directions but with randomly different phases and possibly different polarizations. Labeling each coherent wavelet by a superscript (i) we then have for the intensity of the incoherent beam

$$I = \left|\sum_i \mathscr{E}^{(i)}\right|^2 = \sum_i I^{(i)} \tag{1.28}$$

because the sum of the various phases α averages everything else to zero.

Now let us define for each wavelet a "vector" whose components are the last three Stokes parameters,

$$\mathfrak{P}^{(i)} \equiv (Q^{(i)}, U^{(i)}, V^{(i)}).$$

By (1.16), then, its length is $I^{(i)}$:

$$|\mathfrak{P}^{(i)}| = (\mathfrak{P}^{(i)2})^{1/2} = I^{(i)}.$$

If for the incoherent beam these are all added up,

$$\mathfrak{P} \equiv \sum_i \mathfrak{P}^{(i)} \equiv (Q, U, V)$$

$$Q \equiv \sum_i Q^{(i)} \qquad U \equiv \sum_i U^{(i)} \qquad V \equiv \sum_i V^{(i)},$$

then the familiar triangle inequality says that

$$I^2 \geq \mathfrak{P}^2 = Q^2 + U^2 + V^2. \tag{1.29}$$

The equality sign will hold *if and only if* the χ and β are the same for all the constituent waves. The beam is then fully polarized. If $Q = U = V = 0$, then it is unpolarized. Hence for an incoherent electromagnetic beam the four Stokes parameters are independent, and the degree of polarization may be defined by

$$P \equiv \frac{|\mathfrak{P}|}{I} = \frac{(Q^2 + U^2 + V^2)^{1/2}}{I}. \tag{1.30}$$

Except when $P = 0$, Eqs. (1.23) then also define the ellipticity and the tilt of the ellipse with respect to a given plane, for an incoherent beam. Similarly, Eqs. (1.27) define the relative amount of left- and right-handed circular polarization present, as well as their relative phase. These now amount to the equations

$$\begin{aligned} Q &= -IP\cos 2\chi \cos 2\beta \\ U &= -IP\sin 2\chi \cos 2\beta \\ V &= IP\sin 2\beta \end{aligned} \tag{1.31}$$

1 For a more detailed discussion of incoherent beams, see Sec. 4.3.

and

$$
\begin{aligned}
Q &= -IP\cos 2\chi \sin 2\gamma \\
U &= -IP\sin 2\chi \sin 2\gamma \\
V &= IP\cos 2\gamma.
\end{aligned}
\tag{1.32}
$$

1.2.4 Stokes Vectors. The Stokes parameters give as complete a description of a beam as any "optical" instrument is capable of furnishing (Stokes's principle of optical equivalence). Since for a general incoherent beam they are independent, we may combine them into a four-dimensional "vector" $\mathscr{I}$ whose components are, in three different languages,

$$\mathscr{I}^{(S)} \equiv (I, Q, U, V) \tag{1.33}$$

$$\mathscr{I}^{(L)} \equiv (I_{\parallel}, I_{\perp}, U, V) \tag{1.34}$$

$$\mathscr{I}^{(C)} \equiv (I_r, I_l, Q, U). \tag{1.35}$$

The way in which these vectors transform under a rotation of the reference plane is easily calculated. If the latter is rotated by the angle α in the right-handed-screw sense relative to the propagation vector $\mathbf{k}$, then the tilt angle χ increases by α and the vectors $\mathfrak{E}$ change like

$$\mathfrak{E}^{(L)\prime} = \mathfrak{R}_\alpha^{(L)} \mathfrak{E}^{(L)} \qquad \mathfrak{R}_\alpha^{(L)} = \begin{pmatrix} \cos\alpha & \sin\alpha \\ -\sin\alpha & \cos\alpha \end{pmatrix} \tag{1.36}$$

$$\mathfrak{E}^{(C)\prime} = \mathfrak{R}_\alpha^{(C)} \mathfrak{E}^{(C)} \qquad \mathfrak{R}_\alpha^{(C)} = \begin{pmatrix} e^{i\alpha} & 0 \\ 0 & e^{-i\alpha} \end{pmatrix}, \tag{1.36a}$$

while the Stokes parameters change like

$$
\begin{aligned}
Q' &= Q\cos 2\alpha - U\sin 2\alpha \\
U' &= Q\sin 2\alpha + U\cos 2\alpha \\
I' &= I \\
V' &= V.
\end{aligned}
\tag{1.37}
$$

Hence the vectors $\mathscr{I}'$ with respect to the new plane of reference are

$$\mathscr{I}' = \mathfrak{M}_\alpha \mathscr{I} \tag{1.38}$$

and we obtain

$$\mathfrak{M}_\alpha^{(S)} = \begin{pmatrix} 1 & 0 & 0 & 0 \\ 0 & \cos 2\alpha & -\sin 2\alpha & 0 \\ 0 & \sin 2\alpha & \cos 2\alpha & 0 \\ 0 & 0 & 0 & 1 \end{pmatrix} \tag{1.39}$$

$$\mathfrak{M}_\alpha^{(L)} = \begin{pmatrix} \cos^2\alpha & \sin^2\alpha & -\frac{1}{2}\sin 2\alpha & 0 \\ \sin^2\alpha & \cos^2\alpha & \frac{1}{2}\sin 2\alpha & 0 \\ \sin 2\alpha & -\sin 2\alpha & \cos 2\alpha & 0 \\ 0 & 0 & 0 & 1 \end{pmatrix} \tag{1.40}$$

$$\mathfrak{M}_\alpha^{(C)} = \begin{pmatrix} 1 & 0 & 0 & 0 \\ 0 & 1 & 0 & 0 \\ 0 & 0 & \cos 2\alpha & -\sin 2\alpha \\ 0 & 0 & \sin 2\alpha & \cos 2\alpha \end{pmatrix} \tag{1.41}$$

for the different labelings of $\mathscr{I}$.

1.2.5 Relation to the Density Matrix. A method of handling the polarization which is more customary in quantum mechanics than the Stokes vectors is the use of the density matrix. It is defined by

$$\rho_{ij} = \frac{\mathfrak{E}_i \mathfrak{E}_j^*}{I}. \tag{1.42}$$

In other words, depending on the type of polarization analysis,

$$\begin{aligned} \rho^{(L)} &= \frac{1}{I}\begin{pmatrix} |\mathscr{E}_\parallel|^2 & \mathscr{E}_\parallel \mathscr{E}_\perp^* \\ \mathscr{E}_\perp \mathscr{E}_\parallel^* & |\mathscr{E}_\perp|^2 \end{pmatrix} = \frac{1}{2I}\begin{pmatrix} I+Q & -U-iV \\ -U+iV & I-Q \end{pmatrix} \\ \rho^{(C)} &= \frac{1}{I}\begin{pmatrix} |\mathscr{E}_r|^2 & \mathscr{E}_r \mathscr{E}_l^* \\ \mathscr{E}_l \mathscr{E}_r^* & |\mathscr{E}_l|^2 \end{pmatrix} = \frac{1}{2I}\begin{pmatrix} I+V & -Q-iU \\ -Q+iU & I-V \end{pmatrix}. \end{aligned} \tag{1.43}$$

Because of (1.16) we have for a fully polarized beam

$$\rho^2 = \rho; \tag{1.44}$$

that is, ρ is idempotent.

If the beam is not fully polarized, then the definition (1.30) of the degree of polarization leads to

$$\rho^2 - \rho = \tfrac{1}{4}(P^2 - 1)\,\mathbb{1}. \tag{1.45}$$

In terms of the tilt angle χ and the ellipticity β, the density matrix, by (1.31), is

$$\rho^{(L)} = \frac{1}{2}\begin{pmatrix} 1 - P\cos 2\chi \cos 2\beta & P(\cos 2\beta \sin 2\chi - i \sin 2\beta) \\ P(\cos 2\beta \sin 2\chi + i \sin 2\beta) & 1 + P\cos 2\chi \cos 2\beta \end{pmatrix}, \tag{1.46}$$

so that always

$$\operatorname{tr} \rho = 1$$

and

$$\det \rho = \tfrac{1}{4}(1 - P^2). \tag{1.47}$$

The density matrix contains, of course, precisely as much information as the Stokes vectors. Which is to be used is purely a matter of convenience.

A change in the reference plane for the polarization changes the density matrix in a simple way which is obtained from (1.36) and (1.37),

$$\rho' = \mathfrak{R}_\alpha \rho \mathfrak{R}_{-\alpha}, \tag{1.48}$$

where $\mathfrak{R}_\alpha$ is one of the two matrices given in (1.36) and (1.36a), depending upon whether linear or circular polarization components are employed.

1.3 Scattering

1.3.1 The Scattering Amplitude. The introduction of an obstacle in the path of a plane electromagnetic wave produces scattering. At a distance from the obstacle that is large relative both to the wavelength and to its size (i.e., in the "wave zone") the field consists of a plane wave and a spherical one. Let us form a wave packet out of such fields,

$$\mathscr{E}(\mathbf{r}) = \int d\omega f(\omega)(\mathscr{E}_0 e^{i(\mathbf{k}\cdot\mathbf{r}-\omega t)} + r^{-1}\mathscr{E}_+ e^{i(kr-\omega t)} + r^{-1}\mathscr{E}_- e^{-i(kr+\omega t)})$$

where f is assumed to be sharply peaked around $\omega = \omega_0$. Then, as $t \to -\infty$, the integrand oscillates rapidly and averages to zero, except where the argument of one of the exponentials is stationary as a function of ω. This means there is a contribution for large $|\mathbf{r}|$, from the plane wave at

$$\hat{\mathbf{k}} \cdot \mathbf{r} \simeq -c|t|,$$

and from the incoming spherical wave at

$$r \simeq c|t|.$$

The outgoing spherical wave contributes nothing, since, for $t < 0$, the derivative of $kr - \omega t$ is always positive. The boundary condition appropriate to a situation in which a wave packet is sent in in the remote past in the $\mathbf{k}$ direction is therefore that $\mathscr{E}_- = 0$. At $t \to +\infty$ by the same argument we then get both a plane wave and a contribution from the outgoing spherical wave at $r \simeq ct$. The Maxwell equations being linear, the amplitude of this spherical wave, called the *scattered wave*, is proportional to $\mathscr{E}_0$.

Since at large distances the electric field is *transverse*, also in the spherical wave,[2] we may write both $\mathscr{E}_0$ and $\mathscr{E}_{\text{scatt}}$ in the asymptotic form

$$\mathscr{E}(\mathbf{r}) \underset{r\to\infty}{\simeq} \mathscr{E}_0 e^{i\mathbf{k}\cdot\mathbf{r}} + \mathscr{E}_{\text{scatt}} r^{-1} e^{ikr} \tag{1.49}$$

in the two-component notation. The reference vectors, of course, are not the same, even if the reference plane is chosen to contain both $\hat{\mathbf{k}}$ and $\hat{\mathbf{r}}$. We define the scattering amplitude $\mathfrak{A}$ by writing

$$\mathfrak{E}_{\text{scatt}} = \mathfrak{A}(\mathbf{k}',\mathbf{k})\mathfrak{E}_0, \tag{1.50}$$

where $\mathbf{k}' \equiv k\hat{\mathbf{r}}$. The *scattering plane* is determined by $\mathbf{k}$ and $\mathbf{k}'$, that is, by the incident and the final propagation direction, the scattered spherical wave being locally considered a plane wave. In (1.50) that is chosen as a reference plane. It varies with the observation point, specifically with its angle relative to the incident beam.

2 The meaning of "transverse" in the spherical wave is that $\mathscr{E}$ is perpendicular to $\mathbf{r}$. If a piece of the spherical wave is considered locally plane, then its propagation vector points in the direction of $\mathbf{r}$ and it is transverse in the usual sense, that is, $\mathscr{E} \perp \mathbf{k}$.

1.3.2 Change to a Reference Plane through a Fixed Direction. Alternatively we may adopt a *fixed* direction through which both the initial and the final reference planes are to go; call it the x axis in the coordinate system in which the incident beam moves along the positive z direction. Then the scattering plane must be turned by $-\varphi$ about the incident propagation vector $\mathbf{k}$ in order to coincide with the xz plane, and it must be turned through $-\varphi'$,

$$\cot\varphi' = \cot\varphi\cos\theta, \qquad \frac{\sin\varphi}{\sin\varphi'} > 0, \tag{1.51}$$

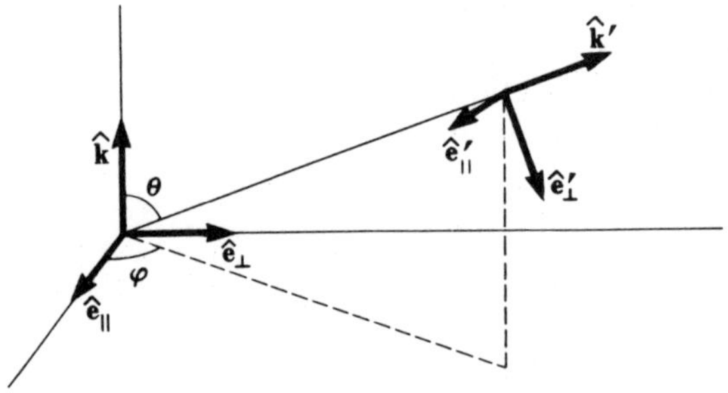

Figure 1.3. Change of reference planes for the polarization.

about the final propagation vector $\mathbf{k}'$ in order to coincide with the xz' plane (the z' axis being in the direction of $\mathbf{k}'$). The components of $\mathfrak{E}_0$ relative to the plane through the fixed direction are then obtained from those with respect to the scattering plane by the linear transformation

$$\mathfrak{E}_{0,\mathrm{fix}} = \mathfrak{R}_{-\varphi}\mathfrak{E}_0,$$

where $\mathfrak{R}_{-\varphi}$ is one of the two matrices defined in (1.36) and (1.36a), while

$$\mathfrak{E}_{\mathrm{scatt,fix}} = \mathfrak{R}_{-\varphi'}\mathfrak{E}_{\mathrm{scatt}}. \tag{1.52}$$

One then gets

$$\mathfrak{E}_{\mathrm{scatt,fix}} = \mathfrak{A}_{\mathrm{fix}}\mathfrak{E}_{0,\mathrm{fix}} \tag{1.53}$$

where

$$\mathfrak{A}_{\mathrm{fix}} = \mathfrak{R}_{-\varphi'}\mathfrak{A}\mathfrak{R}_{\varphi}. \tag{1.54}$$

We shall give here explicitly only the case of circular polarization,

$$\mathfrak{A}_{\mathrm{fix}}^{(C)} = \begin{pmatrix} A_{11}^{C}e^{i(\varphi-\varphi')} & A_{12}^{C}e^{-i(\varphi+\varphi')} \\ A_{21}^{C}e^{i(\varphi+\varphi')} & A_{22}^{C}e^{i(\varphi'-\varphi)} \end{pmatrix}. \tag{1.55}$$

1.3.3 Relation of Circular to Linear Polarization Components in the Scattering Amplitude. The electric field components in the two different languages, those of linearly polarized components and those of circularly polarized ones, are connected by the transformation

$$\mathfrak{E}^{(C)} = \mathfrak{B}^{(CL)}\mathfrak{E}^{(L)} \tag{1.56}$$

with the unitary matrix

$$\mathfrak{B}^{(CL)} = \frac{1}{\sqrt{2}}\begin{pmatrix} i & 1 \\ -i & 1 \end{pmatrix}. \tag{1.57}$$

As a result the scattering amplitudes are related by

$$\mathfrak{A}^{(C)} = \mathfrak{B}^{(CL)}\mathfrak{A}^{(L)}\mathfrak{B}^{(CL)\dagger} \tag{1.58}$$

or, explicitly

$$\mathfrak{A}^{(C)} = \frac{1}{2}\begin{bmatrix} (A_{11}^L + A_{22}^L) - i(A_{21}^L - A_{12}^L) & (A_{22}^L - A_{11}^L) + i(A_{21}^L + A_{12}^L) \\ (A_{22}^L - A_{11}^L) - i(A_{21}^L + A_{12}^L) & (A_{11}^L + A_{22}^L) + i(A_{21}^L - A_{12}^L) \end{bmatrix}. \tag{1.59}$$

1.3.4 Stokes Vectors of the Scattered Wave. At the large distance r the part of the scattered wave which moves in a given direction can "locally" be considered a plane wave. Since the total scattered flux crossing a spherical surface of radius r is

$$\frac{c}{2\pi}\int d\Omega \mathfrak{E}_0^* \mathfrak{A}^\dagger \mathfrak{A}\mathfrak{E}_0,$$

the local intensity as given by

$$I \equiv r^{-2}I_{\text{scatt}} = r^{-2}\mathfrak{E}_0^\dagger \mathfrak{A}^\dagger \mathfrak{A}\mathfrak{E}_0.$$

This is linearly related to the four Stokes parameters of the incident beam. Similarly for the other three Stokes parameters of the scattered beam. We can write

$$\mathscr{I}_{\text{scatt}} = \mathfrak{F}\mathscr{I}_{\text{inc}} \tag{1.60}$$

for the "vectors" defined by (1.33) to (1.35). It is understood that the actual vector $\mathscr{I}$ of the (locally) plane wave observed at a distance r is $r^{-2}\mathscr{I}_{\text{scatt}}$. We list here explicitly

$$\mathfrak{F}^{(L)} = \begin{bmatrix} |A_{11}^L|^2 & |A_{12}^L|^2 & -\operatorname{Re}(A_{11}^L A_{12}^{L*}) & \operatorname{Im}(A_{11}^L A_{12}^{L*}) \\ |A_{21}^L|^2 & |A_{22}^L|^2 & -\operatorname{Re}(A_{21}^L A_{22}^{L*}) & \operatorname{Im}(A_{21}^L A_{22}^{L*}) \\ -2\operatorname{Re}(A_{11}^L A_{21}^{L*}) & -2\operatorname{Re}(A_{12}^L A_{22}^{L*}) & \operatorname{Re}(A_{11}^L A_{22}^{L*} + A_{12}^L A_{21}^{L*}) & \operatorname{Im}(A_{22}^L A_{11}^{L*} - A_{21}^L A_{12}^{L*}) \\ -2\operatorname{Im}(A_{11}^L A_{21}^{L*}) & -2\operatorname{Im}(A_{12}^L A_{22}^{L*}) & \operatorname{Im}(A_{11}^L A_{22}^{L*} + A_{12}^L A_{21}^{L*}) & \operatorname{Re}(A_{22}^L A_{11}^{L*} - A_{21}^L A_{12}^{L*}) \end{bmatrix} \tag{1.61}$$

and

$$\mathfrak{F}^{(C)} = \begin{bmatrix} |A_{11}^C|^2 & |A_{12}^C|^2 & \operatorname{Re}(A_{11}^C A_{12}^{C*}) & -\operatorname{Im}(A_{11}^C A_{12}^{C*}) \\ |A_{21}^C|^2 & |A_{22}^C|^2 & \operatorname{Re}(A_{21}^C A_{22}^{C*}) & -\operatorname{Im}(A_{21}^C A_{22}^{C*}) \\ 2\operatorname{Re}(A_{11}^C A_{21}^{C*}) & 2\operatorname{Re}(A_{12}^C A_{22}^{C*}) & \operatorname{Re}(A_{11}^C A_{22}^{C*} + A_{12}^C A_{21}^{C*}) & \operatorname{Im}(A_{22}^C A_{11}^{C*} - A_{21}^C A_{12}^{C*}) \\ 2\operatorname{Im}(A_{11}^C A_{21}^{C*}) & 2\operatorname{Im}(A_{12}^C A_{22}^{C*}) & \operatorname{Im}(A_{11}^C A_{22}^{C*} + A_{12}^C A_{21}^{C*}) & \operatorname{Re}(A_{22}^C A_{11}^{C*} + A_{21}^C A_{12}^{C*}) \end{bmatrix} \tag{1.62}$$

$$
\begin{aligned}
2F^{(S)}_{\substack{11\\21}} &= |A^L_{11}|^2 \pm |A^L_{12}|^2 \pm |A^L_{21}|^2 + |A^L_{22}|^2 \\
2F^{(S)}_{\substack{12\\22}} &= |A^L_{11}|^2 \mp |A^L_{12}|^2 \pm |A^L_{21}|^2 - |A^L_{22}|^2 \\
F^{(S)}_{\substack{13\\23}} &= -\mathrm{Re}\,(A^L_{11}A^{L*}_{12} \pm A^L_{22}A^{L*}_{21}) \\
F^{(S)}_{\substack{14\\24}} &= -\mathrm{Im}\,(A^L_{11}A^{L*}_{12} \mp A^L_{22}A^{L*}_{21}) \\
F^{(S)}_{\substack{31\\32}} &= -\mathrm{Re}\,(A^L_{11}A^{L*}_{21} \pm A^L_{22}A^{L*}_{12}) \\
F^{(S)}_{\substack{33\\44}} &= \mathrm{Re}\,(A^L_{11}A^{L*}_{22} \pm A^L_{12}A^{L*}_{21}) \\
F^{(S)}_{\substack{34\\43}} &= \mathrm{Im}\,(A^L_{12}A^{L*}_{21} \mp A^L_{11}A^{L*}_{22}) \\
F^{(S)}_{\substack{41\\42}} &= \mathrm{Im}\,(A^L_{21}A^{L*}_{11} \pm A^L_{22}A^{L*}_{12}).
\end{aligned}
\qquad (1.63)
$$

These equations hold both if the vectors $\mathscr{I}$ are expressed with respect to reference planes through a fixed direction and if they are expressed with respect to the scattering plane. The direct transformation from one to the other is obtained by (1.36). Since

$$\mathscr{I}_{\mathrm{inc,fix}} = \mathfrak{M}_{-\varphi}\mathscr{I}_{\mathrm{inc}}, \qquad \mathscr{I}_{\mathrm{scatt,fix}} = \mathfrak{M}_{-\varphi'}\mathscr{I}_{\mathrm{scatt}}, \qquad (1.64)$$

we have

$$\mathfrak{F}_{\mathrm{fix}} = \mathfrak{M}_{-\varphi'}\mathfrak{F}\mathfrak{M}_{\varphi} \qquad (1.65)$$

where $\mathfrak{M}_{\varphi}$ is given by (1.39), (1.40), or (1.41) and φ' by (1.51).

1.3.5 The Differential Cross Section. In a typical scattering experiment we envisage a beam of radiation incident on an obstacle and detectors located at a large distance (compared with both the wavelength and the size of the obstacle) to measure the flux scattered in a given direction. This is possible only if the detectors are located outside the incident beam. The latter must therefore have a finite width, which implies that its wave vector is not uniquely defined. The uncertainty Δk in its transverse-wave vector component is related to the beamwidth W by

$$\Delta k W \simeq 1.$$

This produces a "smearing out" of the incident direction, and hence of the scattering angle, by

$$\frac{\Delta k}{k} \simeq \frac{1}{Wk}.$$

In order to define the scattering angle accurately, we must therefore make $W \gg \lambda$. In addition, of course, it is clear that the beamwidth must be larger than the size D of the obstacle. The location of the detectors at a distance large compared with both W and D is compatible with these demands on the beam and with the requirement that they be situated outside it.

The differential scattering cross section is defined as the ratio of the scattered flux per unit solid angle (at large distance from the scattering center) to the incident flux. This means

$$\frac{d\sigma}{d\Omega} = \frac{I_{\text{scatt}}}{I_{\text{inc}}}. \tag{1.66}$$

It can be calculated immediately from $\mathfrak{F}$ for any polarization of the incident beam. Similarly, the polarization parameters of the scattered radiation can be calculated from $\mathfrak{F}$ in terms of those of the incoming wave. Using (1.31) and (1.63), we get the following general expression for the differential cross section in terms of the parameters P, χ, and β of the incident wave,

$$\begin{aligned} \frac{d\sigma}{d\Omega} = {} & \tfrac{1}{2}(|A_{11}^L|^2 + |A_{12}^L|^2 + |A_{21}^L|^2 + |A_{22}^L|^2) \\ & + \tfrac{1}{2}P\cos 2\chi\cos 2\beta\,(|A_{12}^L|^2 - |A_{11}^L|^2 - |A_{21}^L|^2 + |A_{22}^L|^2) \\ & + P\sin 2\chi\cos 2\beta\,\mathrm{Re}\,(A_{12}^L A_{11}^{L*} + A_{22}^L A_{21}^{L*}) \\ & + P\sin 2\beta\,\mathrm{Im}\,(A_{12}^L A_{11}^{L*} + A_{22}^L A_{21}^{L*}). \end{aligned} \tag{1.67}$$

If the incident beam is unpolarized, the polarization of the scattered radiation is found to be

$$P^2 = 1 - \frac{|\det \mathfrak{A}|^2}{(d\sigma/d\Omega)^2} \tag{1.68}$$

and its ellipticity

$$\sin 2\beta = \frac{\mathrm{Im}\,(A_{21}^L A_{11}^{L*} + A_{22}^L A_{12}^{L*})}{[(d\sigma/d\Omega)^2 - |\det \mathfrak{A}|^2]^{1/2}}. \tag{1.69}$$

Two examples: For an unpolarized incident beam the intensity of right-handedly polarized scattered radiation per unit incident total intensity is given by

$$\tfrac{1}{2}|A_{11}^C|^2 + \tfrac{1}{2}|A_{12}^C|^2 = |A_{11}^L - iA_{21}^L|^2 + |A_{22}^L + iA_{12}^L|^2.$$

If the incident radiation is 100 percent right-handedly circularly polarized, then

$$\mathscr{I}_{\text{scatt}}^{(C)} = [|A_{11}^C|^2, |A_{21}^C|^2, -2\,\mathrm{Re}\,(A_{11}^C A_{21}^{C*}), 2\,\mathrm{Im}\,(A_{21}^C A_{11}^{C*})],$$

so that the differential cross section is

$$d\sigma/d\Omega = |A_{11}^C|^2 + |A_{21}^C|^2.$$

The scattered beam is 100 percent polarized, and the ellipticity is given by

$$2\tan 2\beta = \frac{|A_{11}^C|}{|A_{21}^C|} - \frac{|A_{21}^C|}{|A_{11}^C|}.$$

1.3.6 The Density Matrix of the Scattered Wave. If the incident beam is described by means of the density matrix ρ_{inc}, then

$$
\begin{aligned}
I_{\text{scatt}}(\rho_{\text{scatt}})_{ij} &= (\mathfrak{E}_{\text{scatt}})_i(\mathfrak{E}_{\text{scatt}})_j^* \\
&= \sum_{kl} (\mathfrak{E}_{\text{inc}})_k(\mathfrak{E}_{\text{inc}})_l^* A_{ik} A_{jl}^* \\
&= \sum_{kl} (\rho_{\text{inc}})_{kl} A_{ik} A_{jl}^* I_{\text{inc}}
\end{aligned}
$$

or, in matrix notation,

$$\frac{d\sigma}{d\Omega}\rho_{\text{scatt}} = \mathfrak{A}\rho_{\text{inc}}\mathfrak{A}^\dagger \tag{1.70}$$

where

$$\frac{d\sigma}{d\Omega} = \frac{I_{\text{scatt}}}{I_{\text{inc}}} = \operatorname{tr} \mathfrak{A}\rho_{\text{inc}}\mathfrak{A}^\dagger$$

is the differential cross section when both polarization components of the scattered wave are included. Use of (1.46) then allows us directly to calculate all the polarization parameters of the scattered wave in terms of those of the incident beam. For initially unpolarized radiation, for instance, we have

$$\frac{d\sigma}{d\Omega} = \tfrac{1}{2}\operatorname{tr} \mathfrak{A}^\dagger \mathfrak{A} \tag{1.71}$$

and

$$\rho_{\text{scatt}} = \frac{\mathfrak{A}\mathfrak{A}^\dagger}{\operatorname{tr} \mathfrak{A}^\dagger \mathfrak{A}}. \tag{1.72}$$

1.3.7 Azimuthal Dependence of Forward and Backward Scattering. A word should be said about the forward- and backward-scattering amplitudes. When $\theta = 0$ or $180°$, the amplitudes $\mathfrak{A}_{\text{fix}}$ must be φ-independent since the azimuthal angle is then undefined. But the amplitudes $\mathfrak{A}$ are *not* φ-independent, because φ still determines the orientation of the reference plane for the components of $\mathfrak{E}$. Inversion of (1.55) shows that the φ dependence of $\mathfrak{A}^{(C)}$ is of a very simple nature. For *forward* scattering $\varphi' = \varphi$ and ${A_{11}}^C$ and ${A_{22}}^C$ are φ-independent, whereas ${A_{12}}^C \propto e^{2i\varphi}$, ${A_{21}}^C \propto e^{-2i\varphi}$. For backward scattering $\varphi' = \pi - \varphi$, and ${A_{12}}^C$ and ${A_{21}}^C$ are φ-independent, while ${A_{11}}^C \propto e^{-2i\varphi}$, ${A_{22}}^C \propto e^{2i\varphi}$. The φ dependence of $\mathfrak{A}^{(L)}$ is more complicated.

1.3.8 Effects of Rotational or Reflectional Symmetry. If the scatterer has complete rotational-reflectional symmetry, then the scattering amplitude $\mathfrak{A}^{(L)}$ must be *diagonal*. This can be seen simply as follows.

The result of reflecting the scatterer on the scattering plane must be the same as that of reflecting all but the scatterer. In the latter case $\mathscr{E}_{\perp\text{inc}}$ and $\mathscr{E}_{\perp\text{scatt}}$ change sign, while $\mathscr{E}_{\parallel\text{inc}}$ and $\mathscr{E}_{\parallel\text{scatt}}$ do not. That means that ${A_{12}}^L$ and ${A_{21}}^L$ change sign. If the scatterer is symmetric under such a reflection, then ${A_{12}}^L$ and ${A_{21}}^L$ must not change; hence, for this particular direction of incidence and scattering, they must be *zero*. If the scatterer is rotationally symmetric about the direction of incidence and has mirror symmetry on any plane through it, then for that particular direction of incidence $\mathfrak{A}^{(L)}$ must be diagonal no matter what the direction of scattering. Finally, for a scatterer with complete rotational-reflectional symmetry, $\mathfrak{A}^{(L)}$ is a diagonal matrix for all initial and all final directions.

That $\mathfrak{A}^{(L)}$ is diagonal does not imply that either $\mathfrak{A}_{\text{fix}}{}^{(L)}$ or $\mathfrak{A}^{(C)}$ is diagonal. The argument cannot be made with circular components. But (1.59) shows that $\mathfrak{A}^{(C)}$ is *symmetric* and $A_{11}{}^C = A_{22}{}^C$ if $\mathfrak{A}^{(L)}$ is diagonal. Rotational symmetry alone about the axis of incidence does not rule out a sense of rotation in the scatterer, and hence this is not sufficient to make $A_{11}{}^C = A_{22}{}^C$, that is, $\mathfrak{A}^{(L)}$ diagonal.

If the scatterer is rotationally symmetric about the axis of incidence, then $\mathfrak{A}$ cannot have any φ dependence. Owing to the fixed initial reference plane, on the other hand, the amplitude $\mathfrak{A}_{\text{fix}}$ has the simple φ dependence given for $\mathfrak{A}_{\text{fix}}{}^{(C)}$ by (1.55).

Assuming now rotation-reflection symmetry, we may write

$$A_{11}^L A_{22}^{L*} \equiv \alpha + i\beta.$$

Then (1.61) shows that $\mathfrak{F}^{(L)}$ simplifies to

$$\mathfrak{F}^{(L)} = \begin{pmatrix} |A_{11}^L|^2 & 0 & 0 & 0 \\ 0 & |A_{22}^L|^2 & 0 & 0 \\ 0 & 0 & \alpha & -\beta \\ 0 & 0 & \beta & \alpha \end{pmatrix}. \tag{1.73}$$

Furthermore $A_{11}{}^C A_{12}{}^{C*} = \frac{1}{2}|A_{22}{}^L|^2 - \frac{1}{2}|A_{11}{}^L|^2 + i\beta$ and

$$\mathfrak{F}^{(C)} =$$
$$\begin{bmatrix} \frac{1}{2}|A_{11}^L + A_{22}^L|^2 & \frac{1}{2}|A_{22}^L - A_{11}^L|^2 & \frac{1}{2}|A_{11}^L|^2 - \frac{1}{2}|A_{22}^L|^2 & \beta \\ \frac{1}{2}|A_{22}^L - A_{11}^L|^2 & \frac{1}{2}|A_{11}^L + A_{22}^L|^2 & \frac{1}{2}|A_{11}^L|^2 - \frac{1}{2}|A_{22}^L|^2 & -\beta \\ |A_{22}^L|^2 - |A_{11}^L|^2 & -|A_{22}^L|^2 - |A_{11}^L|^2 & |A_{11}^L|^2 + |A_{22}^L|^2 & 0 \\ -2\beta & 2\beta & 0 & |A_{11}^L|^2 + |A_{22}^L|^2 \end{bmatrix} \tag{1.74}$$

The polarization of the scattered radiation for an unpolarized incident beam then comes out simply,

$$P = \frac{|A_{11}^L|^2 - |A_{22}^L|^2}{|A_{11}^L|^2 + |A_{22}^L|^2}. \tag{1.75}$$

Half of it is left-handed, half right-handed, and the ellipticity is zero.

Consider now forward and backward scattering, for which we already know that $\mathfrak{A}_{\text{fix}}$ is independent of φ. If:

1. The scatterer is axially (rotationally) symmetric about the line of incidence, then $\mathfrak{A}$ must also be independent of φ. Equation (1.55) then shows that for *forward* scattering ($\varphi' = \varphi$) $\mathfrak{A}^{(C)}$ is diagonal; (1.59) therefore implies

that $A_{11}{}^L = A_{22}{}^L$, $A_{12}{}^L = -A_{21}{}^L$. In the *backward* direction ($\varphi' = \pi - \varphi$) we find similarly that $A_{11}{}^C = A_{22}{}^C = 0$, and hence $A_{11}{}^L = -A_{22}{}^L$ and $A_{12}{}^L = A_{21}{}^L$. Such a scatterer may still have a sense of rotation about the axis of incidence. That is the reason why right- and left-handed radiation need not be scattered equally.

Suppose that

2. The scatterer has, *in addition*, reflection symmetry on a plane through the axis of incidence. Then we know that $\mathfrak{A}^{(L)}$ must be diagonal. It follows that for *forward* scattering $A_{11}{}^C = A_{22}{}^C = A_{11}{}^L = A_{22}{}^L \equiv A$, while all others are zero. For *backward* scattering we get $-A_{11}{}^L = A_{22}{}^L = A_{21}{}^C = A_{12}{}^C$ and all others zero. The additional reflection symmetry means that it is impossible for the scatterer to have a rotation sense attached, and therefore it scatters right- and left-handed radiation equally in the forward direction. In the backward direction it also scatters them equally, but the rotation sense (angular momentum) being conserved, the right-handed wave becomes left-handed, and vice versa.

1.3.9 Forward Scattering; the Optical Theorem. In the forward direction or almost forward direction, the electric field at large distance from the scattering center is given by

$$\mathfrak{E} \simeq [1 + r^{-1} e^{ik(r-z)} \mathfrak{A}(0)] \mathfrak{E}_{\text{inc}}$$

according to (1.49). Suppose that this radiation is observed on a screen whose dimensions are small compared with the distance from the scatterer. This means that

$$x^2 + y^2 \ll z^2,$$

and we may expand

$$r = [z^2 + (x^2 + y^2)]^{1/2} \simeq z + \tfrac{1}{2} z^{-1} (x^2 + y^2)$$

so that

$$r^{-1} e^{ik(r-z)} \simeq z^{-1} e^{ik(x^2+y^2)/2z}.$$

The intensity at a point (x, y) on the screen at distance $z \gg \lambda$ in the forward direction is therefore

$$\begin{aligned} \mathfrak{E}^\dagger \mathfrak{E} \simeq \mathfrak{E}_{\text{inc}}^\dagger \left\{1 + z^{-1} \exp\left[\frac{-ik(x^2 + y^2)}{2z}\right] \mathfrak{A}^\dagger(0)\right\} \\ \left\{1 + z^{-1} \exp\left[\frac{ik(x^2 + y^2)}{2z}\right] \mathfrak{A}(0)\right\} \mathfrak{E}_{\text{inc}} \\ \simeq \mathfrak{E}_{\text{inc}}^\dagger \mathfrak{E}_{\text{inc}} + 2z^{-1} \operatorname{Re}\left\{\exp\left[\frac{ik(x^2 + y^2)}{2z}\right] \mathfrak{E}_{\text{inc}}^\dagger \mathfrak{A}(0) \mathfrak{E}_{\text{inc}}\right\}. \end{aligned}$$

We calculate the total energy received on the screen by integrating over its area. If the dimensions of the screen are "large," but not too large, i.e., so that on its edge

$$\frac{k(x^2 + y^2)}{z} \gg 2\pi$$

while still everywhere on it

$$x^2 + y^2 \ll z^2,$$

then the resulting integral

$$\iint dx\,dy \exp\left[\frac{ik(x^2 + y^2)}{2z}\right]$$

may be taken all the way to $x, y = \pm\infty$ without appreciably altering its value. The conditions on the dimensions D of the receiving screen for this to be true are

$$(z\lambda)^{1/2} \ll D \ll z. \tag{1.76}$$

The gaussian integral can then be evaluated and yields

$$\int_{-\infty}^{\infty} dx \int_{-\infty}^{\infty} dy \exp\left[\frac{ik(x^2 + y^2)}{2z}\right] = 2\pi i z k^{-1}.$$

If a is the area of the screen, the intensity integrated over it is therefore

$$\mathfrak{E}_{\text{inc}}^{\dagger}\mathfrak{E}_{\text{inc}}a - 4\pi k^{-1}\,\text{Im}\,[\mathfrak{E}_{\text{inc}}^{\dagger}\mathfrak{A}(0)\mathfrak{E}_{\text{inc}}].$$

The second term represents the diminution owing to the presence of the scatterer. Depending on the polarization direction of the incident beam, this diminution per unit incident intensity is consequently

$$4\pi k^{-1}\,\text{Im}\,A_{11}(0) \quad \text{or}\ 4\pi k^{-1}\,\text{Im}\,A_{22}(0).$$

It must be remembered that for this to be observed as a blockage of radiation, the receiving screen must be *large* in the sense of (1.76).

This blockage, now, is the energy removed from the original beam in the forward direction per unit time by interference with the scattered radiation, the so-called *shadow scattering*. By conservation of the total electromagnetic energy, this must equal the sum of the total energy scattered (per unit time) and absorbed,[3]

$$\sigma_{\text{total}}^{(i)} \equiv \sigma_{\text{scatt}}^{(i)} + \sigma_{\text{abs}}^{(i)} = 4\pi k^{-1}\,\text{Im}\,A_{ii}(0). \tag{1.77}$$

The index i refers to the polarization direction of the incident beam, either linear or circular. If the incident beam is partially polarized, then we get

$$\begin{aligned}\sigma_{\text{total}} = 2\pi k^{-1}[&\text{Im}\,(A_{11}^L + A_{22}^L) + Q\,\text{Im}\,(A_{11}^L - A_{22}^L)\\ &- U\,\text{Im}(A_{12}^L + A_{21}^L) + V\,\text{Re}(A_{12}^L - A_{21}^L)],\end{aligned} \tag{1.78}$$

all the amplitudes being evaluated in the forward direction. For an unpolarized beam this reduces, of course, to

3 In view of the significance of this cross section as measuring the removal of radiation from a beam, it is also referred to as the *extinction cross section*.

$$\begin{aligned}\sigma_{\text{total}} &= 2\pi k^{-1}\,\mathrm{Im}\,[A^L_{11}(0) + A^L_{22}(0)] \\ &= 2\pi k^{-1}\,\mathrm{Im}\,[A^C_{11}(0) + A^C_{22}(0)].\end{aligned} \tag{1.79}$$

If the beam is incident along an axis of symmetry, then according to the results of the discussion in Sec. 1.3.8,

$$\sigma_{\text{total}} = 4\pi k^{-1}[\mathrm{Im}\,A^L_{11}(0) + P\sin 2\beta\,\mathrm{Re}\,A^L_{12}(0)].$$

Finally, if the scatterer has complete rotational-reflectional symmetry, then simply

$$\sigma_{\text{total}} = 4\pi k^{-1}\,\mathrm{Im}\,A(0). \tag{1.80}$$

The analog of these equations in quantum scattering theory is usually referred to as the *optical theorem*.[4] We shall use that name here.

1.4 Double Scattering

Radiation once scattered by an obstacle may be rescattered by another. If the two centers are separated by a distance comparable with the wavelength of the radiation, or smaller, then they must be treated as composing a single scatterer. But if their separation is large compared both with the wavelength and with their size, they are in one another's wave zone and the flux at the second can be calculated by the scattering approximation (1.49) from the scattering amplitude of the first.

If we call the positions of the two scatterers 1 and 2 and that of the detector 3, then the electric field at 3, in simplified notation, is

$$\begin{aligned}&\mathscr{E}_0 e^{i\mathbf{k}\cdot\mathbf{r}} + r_{13}^{-1}e^{ikr_{13}}\mathscr{E}_{31} + r_{23}^{-1}e^{ikr_{23}}\mathscr{E}_{32} \\ &\qquad + r_{23}^{-1}e^{ikr_{23}}r_{12}^{-1}e^{ikr_{12}}\mathscr{E}_{321} + r_{13}^{-1}e^{ikr_{13}}r_{12}^{-1}e^{ikr_{12}}\mathscr{E}_{312}.\end{aligned}$$

$\mathscr{E}_{31}$ here means the field scattered by 1 toward 3, $\mathscr{E}_{321}$ that scattered by 1 toward 2 and, subsequently, toward 3, and r_{ij} is the distance from point i to point j.

To find the intensity and the density matrix of the scattered radiation, we must form the product of the scattered part of this with its complex conjugate. This will yield cross terms containing $\exp(ikr_{12})$ or $\exp[ik(r_{13} - r_{23})]$, etc. These can contribute to the measured flux only if the positions of the scatterers are fixed to within a fraction of a wavelength and the scattering angle is defined much better than is practically possible. Hence these terms do not contribute to the measured intensity at 3, and we get

$$I_3 = r_{13}^{-2}I_{31} + r_{23}^{-1}I_{32} + r_{23}^{-1}r_{12}^{-1}I_{321} + r_{13}^{-1}r_{12}^{-1}I_{312}. \tag{1.81}$$

In order to convert this statement into a conventional cross section and to write its analog for the density matrix, we have to assume one of two

4 It is also known as the *Bohr-Peierls-Placzek relation*.

things. Either the distances r_{13} and r_{23} are large compared with r_{12} so that the directions from the scatterers to the detector are essentially parallel, or else the detector is set up to register radiation from only one of the scatterers by pointing in its direction and excluding flux coming from the other. In the latter case the second scatterer is likely not to receive any incoming flux directly, but only via a scattering by the first. We take up the former case first.

Assuming that $r_{13} \gg r_{12}, r_{23} \gg r_{12}$, we can convert (1.81) directly into a differential cross section. The scattering direction is well defined. If we call $d\sigma(3)/d\Omega$ that cross section, both polarizations being counted, $d\sigma(31)/d\Omega$ the cross section for scatterer 1 only (toward 3), and $d\sigma(321)/d\Omega$ the cross section for scattering from 1 to 2 to 3, then

$$\frac{d\sigma(3)}{d\Omega} = \frac{d\sigma(31)}{d\Omega} + \frac{d\sigma(32)}{d\Omega} + \frac{d\sigma(321)}{d\Omega} + \frac{d\sigma(312)}{d\Omega}, \tag{1.82}$$

and we find that

$$\begin{aligned} \frac{d\sigma(321)}{d\Omega} &= r_{12}^{-2}\tfrac{1}{2}\mathrm{tr}\,[\rho_{\mathrm{inc}}\mathfrak{A}^\dagger(21)\mathfrak{A}^\dagger(32)\mathfrak{A}(32)\mathfrak{A}(21)] \\ \frac{d\sigma(312)}{d\Omega} &= r_{12}^{-2}\tfrac{1}{2}\mathrm{tr}\,[\rho_{\mathrm{inc}}\mathfrak{A}^\dagger(12)\mathfrak{A}^\dagger(31)\mathfrak{A}(31)\mathfrak{A}(12)] \end{aligned} \tag{1.83}$$

because

$$\mathfrak{E}_{321} = \mathfrak{A}(32)\mathfrak{A}(21)\mathfrak{E}_0. \tag{1.84}$$

The amplitudes here must be chosen to refer to reference planes through a fixed direction. One may choose that direction conveniently to be perpendicular to the plane formed by the incident beam and by the line joining the two scatterers. Then it can be used for all four amplitudes. Or else one may choose the direction *in* the first scattering plane, perpendicular to the line joining 1 and 2. Notice that the ratio of the (321) cross section to the (21) cross section is essentially equal to that of the (32) cross section to the square of the distance r_{12}. Under most circumstances this, of course, is quite small.

The density matrix of the scattered radiation is similarly found to be

$$\begin{aligned} \frac{d\sigma(3)}{d\Omega}\rho_{\mathrm{scatt}}(3) = {} & \frac{d\sigma(31)}{d\Omega}\rho_{\mathrm{scatt}}(31) + \frac{d\sigma(32)}{d\Omega}\rho_{\mathrm{scatt}}(32) \\ & + \frac{d\sigma(321)}{d\Omega}\rho_{\mathrm{scatt}}(321) + \frac{d\sigma(312)}{d\Omega}\rho_{\mathrm{scatt}}(312) \end{aligned} \tag{1.85}$$

$$\begin{aligned} \frac{d\sigma(321)}{d\Omega}\rho_{\mathrm{scatt}}(321) &= r_{12}^{-2}\mathfrak{A}(32)\mathfrak{A}(21)\,\rho_{\mathrm{inc}}\mathfrak{A}^\dagger(21)\mathfrak{A}^\dagger(32) \\ \frac{d\sigma(312)}{d\Omega}\rho_{\mathrm{scatt}}(312) &= r_{12}^{-2}\mathfrak{A}(31)\mathfrak{A}(12)\,\rho_{\mathrm{inc}}\mathfrak{A}^\dagger(12)\mathfrak{A}^\dagger(31). \end{aligned} \tag{1.86}$$

Of course, this can all be done also in terms of the Stokes vectors, with

$$F_{321} = r_{12}^{-1}F_{32}F_{21}, \text{ etc.}$$

We now have to relate the angles of the two scatterings. Let us choose our coordinate system so that the vector $\hat{\mathbf{k}}''$ pointing from the first to the second scatterer lies in the xz plane, and let us choose the first scattering plane as a reference plane for the polarization in the first scattering. Since this contains the x' axis (see Fig. 1.4), we may use the x' axis directly as a

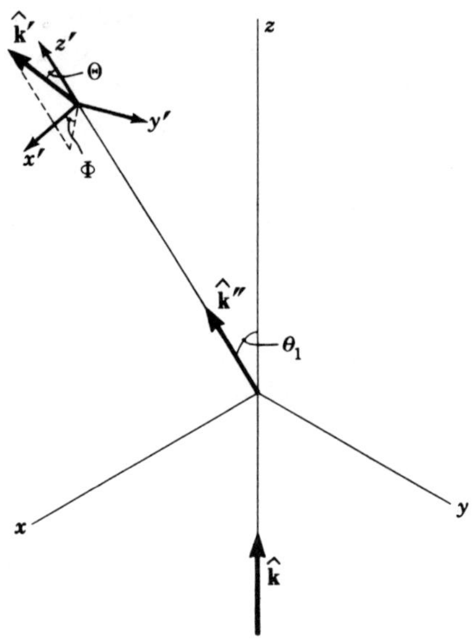

Figure 1.4. The reference planes in double scattering.

fixed direction through which the initial and final reference planes for the second scattering are to go. The angles Θ and Φ in the (x', y', z') system are then related to the angles θ and φ measured in the first system (x, y, z) by

$$\begin{aligned}\cos\Theta &= \cos\theta\cos\theta_1 + \cos\varphi\sin\theta\sin\theta_1\\ \cot\Phi &= \cot\varphi\cos\theta_1 - \csc\varphi\cot\theta\sin\theta_1\end{aligned} \tag{1.87}$$

if θ_1 is the (fixed) angle of the line from the first to the second scatterer relative to the z axis, the direction of incidence. The angles Φ and Θ must be used in place of φ and θ for the computation of φ' by (1.51) and in (1.54) to convert from the scattering plane as a reference plane to those through x' in the second scattering. In the (312) cross section θ_1 has to be replaced by $\pi - \theta_1$.

If an experimental setup in used in which radiation from scatterer 2 only is detected and all the flux incident on 2 comes from 1, then one measures directly $d\sigma(321)/d\Omega$ and, if polarization is measured too, $\rho(321)$. The purpose of such an experiment could be to measure the polarization produced by a single scattering without having to detect polarizations.

Suppose that both scatterers have rotational reflectional symmetry but that the two components of linear polarization are scattered differently. Then initially unpolarized radiation is polarized by the first scattering and the second scattering produces an azimuthal dependence which serves to analyze this polarization. Call the cross sections of the first and second scatterer for unpolarized radiation $d\sigma^{(0)}(21)/d\Omega$ and $d\sigma^{(0)}(32)/d\Omega$, and call the difference of the respective cross sections for radiation linearly polarized parallel and perpendicular to the scattering planes $\Delta d\sigma/d\Omega$, that is,

$$\Delta\frac{d\sigma}{d\Omega} \equiv \left||A_\perp^L|^2 - |A_\parallel^L|^2\right| = 2P\frac{d\sigma^{(0)}}{d\Omega},$$

P being the polarization produced if the initial beam is unpolarized. Then a simple computation using (1.55) and (1.62) in (1.83) shows that

$$\begin{aligned}\frac{d\sigma(321)}{d\Omega} &= r_{12}^{-2}\left[\frac{d\sigma^{(0)}(32)}{d\Omega}\frac{d\sigma^{(0)}(21)}{d\Omega} + \tfrac{1}{4}\Delta\frac{d\sigma(32)}{d\Omega}\Delta\frac{d\sigma(21)}{d\Omega}\cos 2\Phi\right] \\ &= r_{12}^{-2}\frac{d\sigma^{(0)}(32)}{d\Omega}\frac{d\sigma^{(0)}(21)}{d\Omega}[1 + P(32)P(21)\cos 2\Phi].\end{aligned} \tag{1.88}$$

The difference between the double scattering in the plane of the first single scattering and that perpendicular to it is therefore

$$\begin{aligned}\Delta\frac{d\sigma(321)}{d\Omega} &= \tfrac{1}{2}r_{12}^{-2}\Delta\frac{d\sigma(32)}{d\Omega}\Delta\frac{d\sigma(21)}{d\Omega} \\ &= 2r_{12}^{-2}P(32)P(21)\frac{d\sigma^{(0)}(32)}{d\Omega}\frac{d\sigma^{(0)}(21)}{d\Omega}.\end{aligned} \tag{1.89}$$

The two polarizations can thus be measured without the use of polarized beams or polarization filters.

1.5 Scattering by a Cloud of Many Particles

1.5.1 Addition of Cross Sections. If an electromagnetic wave passes through a medium of many randomly distributed particles (not a crystal!), then the phases of the waves scattered in a given direction, other than forward, coming as they do from randomly distributed centers, are incoherent. As a result, the total intensity is the sum of those coming from all the individual scattering centers,

$$I = \sum_i I^{(i)}$$

and the scattering cross sections are additive,

$$\frac{d\sigma}{d\Omega} = \sum_i \frac{d\sigma^{(i)}}{d\Omega}. \tag{1.90}$$

Naturally, in order to add the differential cross sections all for the same direction, the detection must take place at a distance, not only large relative to the wavelength, but large also relative to the size of the cloud.

In the forward direction no phase difference is introduced by changing the position of the scatterer, and so we must add *amplitudes*,

$$\mathfrak{A}(0) = \sum_i \mathfrak{A}^{(i)}(0).$$

It then follows from the optical theorem of Sec. 1.3.9 that the total cross sections are additive,

$$\sigma_{\text{total}} = \sum_i \sigma_{\text{total}}^{(i)}. \tag{1.91}$$

Notice, though, that the reasons for (1.90) and (1.91) are quite different. As a consequence of (1.90) and (1.91), the absorption cross sections are also additive.

1.5.2 Index of Refraction. If we want to correlate the "optical" properties of a medium with those of its constituent particles, two extreme cases can be most readily approached. One is that in which the wavelength is large compared with the interparticle distance. This is essentially the static limit, and its result, under certain additional assumptions, is the Lorentz–Lorenz formula. We are not interested in this here.

The other limit is the one in which the wavelength is small relative to the interparticle distance. We then get a simple result only if the medium is of *low density*. That means that the average interparticle distance D is large compared with the particle size R,

$$D \gg R. \tag{1.92}$$

(The medium is for the moment assumed to consist of only one kind of particle.) In a column of unit cross section and length x, the area blocked out is roughly

$$xNR^2 = \frac{xR^2}{D^3}$$

if there are N particles per unit volume. The characteristic distance in the medium over which the beam is appreciably altered is therefore

$$x \simeq \frac{D^3}{R^2} = D\left(\frac{D}{R}\right)^2 \gg D. \tag{1.93}$$

It is consequently possible to choose a thickness dz of a slab of material which is "infinitesimal" in the sense that the beam can be considered constant and unattenuated throughout it, and yet which contains many particle "layers,"

$$D \ll dz \ll D\left(\frac{D}{R}\right)^2. \tag{1.94}$$

We position ourselves at a point P, a distance d above the surface of this slab, which is small compared with dz and yet also large compared with the interparticle distance

$$D \ll d \ll dz. \tag{1.95}$$

Then d is such that an appropriately chosen narrow cone C from P perpendicularly into the slab, though narrow, may contain many particles laterally. Under a microscope of "intermediate" power, then, the medium looks continuous, and P lies on the surface of the infinitesimal slab dz.

We want to know the electric field at P, given its value before entering the slab of thickness dz. If we assume that the wavelength λ is small compared with the interparticle distance,

$$\lambda \ll D, \tag{1.96}$$

then we are automatically in the wave zone at P and the electric field may be obtained from the scattering amplitude. We are interested only in the coherent part, i.e., in the plane wave which propagates in the forward direction. By the same approximation used in Sec. 1.3.9 this is

$$\mathfrak{E} \simeq \mathfrak{E}_{\text{inc}} + \sum_j z_j^{-1} \exp\left[ik\tfrac{1}{2}z_j^{-1}(x_j^2 + y_j^2)\right]\mathfrak{A}(0)\mathfrak{E}_{\text{inc}}.$$

The summation here runs over all particles in the intersection of the slab with the narrow cone C. Because of (1.92) the multiple scattering contribution can be neglected. By virtue of the assumptions made, there are many terms in the sum, even at practically fixed z, and we may replace the sum by an integral. What is more, the integral may be laterally extended to infinity, for the opening of the cone may be chosen wide enough so that on its edge

$$k(x^2 + y^2)z^{-1} \gg 2\pi,$$

i.e.,

$$x^2 + y^2 \gg \lambda d$$

while it is still narrow, i.e.,

$$x^2 + y^2 \ll d^2.$$

The result of carrying out the gaussian integral is

$$\begin{aligned}\mathfrak{E} &\simeq \mathfrak{E}_{\text{inc}} + N \int_{-\infty}^{\infty} dx \int_{-\infty}^{\infty} dy\, dz\, z^{-1} \exp\left[ik\tfrac{1}{2}z^{-1}(x^2 + y^2)\right]\mathfrak{A}(0)\mathfrak{E}_{\text{inc}} \\ &= \mathfrak{E}_{\text{inc}} + 2\pi i N k^{-1}\, dz\, \mathfrak{A}(0)\mathfrak{E}_{\text{inc}}.\end{aligned}$$

In the slab of thickness dz the wave has therefore changed by

$$d(\mathfrak{E}e^{-ikz}) = 2\pi i N k^{-1}\mathfrak{A}(0)(\mathfrak{E}e^{-ikz})\, dz.$$

This implies that

$$\mathfrak{E} = e^{iknz}\mathfrak{E}_{00},$$

where $\mathfrak{E}_{00}$ is constant (i.e., independent of z) and the index of refraction is given by

$$n = 1 + 2\pi N k^{-2}\mathfrak{A}(0). \tag{1.97}$$

Notice that n is, first of all, *complex* and, second, a *matrix*. The diagonal components give the complex index of refraction for differently polarized radiation. For linear polarization one uses $\mathfrak{A}^{(L)}$ and for circular polarization $\mathfrak{A}^{(C)}$. The off-diagonal components give a distance-dependent change in the polarization, i.e., a rotation of the plane of polarization. The diagonal elements of the real part,

$$\mathrm{Re}\, n_{ii} = 1 + 2\pi N k^{-2}\, \mathrm{Re}\, \mathfrak{A}_{ii}(0) \tag{1.98}$$

describe the change in wavelength in the medium for the two polarization directions (double refraction or birefringence). The fact that the phase velocity c/n depends upon the frequency is called *dispersion*. The diagonal elements of the imaginary part

$$\mathrm{Im}\, n_{ii} = 2\pi N k^{-2}\, \mathrm{Im}\, \mathfrak{A}_{ii}(0) = \frac{N}{2k}\sigma_{\text{total}}^{(i)} \tag{1.99}$$

describe the (possibly different) absorption of the two polarization components in the medium (dichroism). The absorption coefficient γ being defined by

$$I = I_0 e^{-\gamma z} \tag{1.100}$$

we get

$$\gamma^{(i)} = N\sigma_{\text{total}}^{(i)}. \tag{1.101}$$

For the medium to be absorptive it is not necessary for the individual scatterers to be so. Even if the latter only scatter, they thereby remove energy from the forward direction and thus from the coherent part of the wave.

It should be remarked that the index of refraction given by (1.97), by the nature of the assumptions made to derive it, must always be close to 1. What was there called R, the "radius" of the scattering particle, is to be interpreted essentially as $|\mathfrak{A}|$, and therefore

$$|n - 1| \simeq \lambda^2 N R = \left(\frac{\lambda}{D}\right)^2 \frac{R}{D} \ll 1. \tag{1.102}$$

1.5.3 More than One Kind of Particle. If there are various kinds of particles in the medium, which includes the possibility of particles of different *orientation*, then the foregoing results are readily generalized to

$$n = 1 + 2\pi k^{-2} \sum_{\alpha} N_\alpha \mathfrak{A}_\alpha(0) \tag{1.103}$$

where N_α and $\mathfrak{A}_\alpha$ are the number of particles of types α per unit volume and their scattering amplitude, respectively. Consider, specifically, a medium of equal *kinds* of particles but of different orientation.

First let them all be oriented in the same way along the axis of the beam, but with randomly different azimuthal angles about it. We must then average over φ, and (1.55) tells us, $\mathfrak{A}_{\text{fix}}$ being independent of φ for forward scattering, that the off-diagonal elements of $\mathfrak{A}^{(C)}$ average to zero. It is as though *each* particle had rotational symmetry about that axis. Such a medium can still have circular birefringence and dichroism, but there can be no linear birefringence or dichroism. There may still be a rotation in the polarization direction.

Next, let each particle have axial symmetry, and let them be lined up along the beam, but not "polarized"; i.e., for each particle of "forward" orientation there is one of "backward" orientation. If the particles look the same fore and aft, in spite of not having reflection symmetry on a plane through the beam axis, they must each have a *screw sense*. Then circular birefringence and dichroism are still possible. If turning a particle around makes it look optically like its mirror image, then ${A_{12}}^L$ and ${A_{21}}^L$ must change sign, ${A_{11}}^C$ and ${A_{22}}^C$ are interchanged, and the averaging makes ${n_{11}}^C = {n_{22}}^C$. For such particles, which carry a rotation sense but no screw sense, lining up can produce no birefringence, dichroism, or rotation of polarization. It is necessary to *polarize* them in order to produce such effects, i.e., to make them "point" the same way.

Among many other possibilities we may finally consider an altogether randomly oriented mixture of equal particles. If each particle is equal to its own (reoriented) mirror image, then the averaging must make ${n_{11}}^C = {n_{22}}^C = {n_{11}}^L = {n_{22}}^L$ and no birefringence or dichroism can exist. But if the particles differ from their own (reoriented) mirror image, then circular birefringence and dichroism may persist even in the random mixture.

Notes and References

1.2 The Stokes parameters and the principle of optical equivalence were introduced by G. C. Stokes (1852). The use of Stokes vectors here follows essentially H. C. van de Hulst (1957). For other references see also M. Born and E. Wolf (1959), p. 550.

The density matrix is a common quantum-mechanical tool, but it does not appear to have seen much use in classical electromagnetic theory. However, see Born and Wolf (1959), sec. 1081, for what there is called the *coherence matrix*.

1.3.1 For more rigorous mathematical treatments of the argument of stationary phase, see, for example, J. G. van der Corput (1934 and 1936); A. Erdelyi (1955); G. Braun (1956).

1.3.4 The use of the matrix $\mathfrak{F}$ follows H. C. van de Hulst (1959).

1.3.9 The origins of the optical theorem probably go back to Lord Rayleigh. It was certainly known to Kramers; see, for example, H. A. Kramers (1927), eq. (12). I have not tried to trace it further. In quantum mechanics it was independently discovered by E. Feenberg (1932). An amusing interlude is provided by H. C. van de Hulst (1949) and the remark on p. 39 of his book (1959). The physical argument given in Sec. 1.3.9 is van de Hulst's. For a further account of its history, see R. G. Newton (1976a). The physical argument given in Sec. 1.3.9 is van de Hulst's. See also E. Wolf (1973).

1.3 For a discussion of scattering in a polarizable medium, see Vezzetti and Keller (1967), and from optically *active* (as opposed to absorbing) objects, see M. Kerker

(1978). For other general discussions, see O. R. Cruzon (1962); J. J. Bowman et al. (1969); P. C. Waterman (1969); Hizal and Marincic (1970); Hizal and Yasa (1973); Barber and Yeh (1975); Uzunoglu and Holt (1977); N. K. Uzunoglu et al. (1978); Wang and Barber (1979).

1.4 An account of *multiple* scattering is much beyond the scope of this book. One then gets involved in transport problems. For treatments and references, see, for example, S. Chandrasekhar (1950); M. Kerker (1963), part 6; B. Peterson and S. Ström (1973); and A. J. Devaney (1980).

1.5 For a variety of approaches to the scattering by many scatterers, see, for example, Zimm and Dandliker (1954); F. C. Karal and J. B. Keller (1964); C. Mavroyannis (1967); K. M. Watson (1969); Peterson and Ström (1974); S. Ström (1974); A. Ishimaru (1978); V. Twersky (1978); V. K. Varadan et al. (1979); Andreo and Krill (1980); and V. K. Varadan (1980).

Problems

1 Prove the statements in the paragraph below Eq. (1.8).

2 Suppose that the relative difference between the intensities of left and right circularly polarized radiation is $\frac{1}{4}$. What is the minimal value of the polarization? If the polarization is 50 percent, what is the ellipticity?

3 In a given beam the relative difference between the intensities of right and left circularly polarized radiation is $\frac{1}{2}$, between those of plane polarizations parallel and perpendicular to a given reference plane it is also $\frac{1}{2}$. What is the minimal value of the polarization? Suppose that the polarization is 75 percent. What is the tilt angle χ of the ellipse relative to the reference plane? What is the ellipticity?

4 If a beam is 30 percent circularly polarized, what is the value of its plane polarization?

5 Express the density matrix for circular polarization in terms of the angles γ and χ of (1.24) and (1.25).

6 Express the polarization in terms of the trace of ρ^2.

7 Find the unitary transformation that transforms the matrix $\rho^{(L)}$ into $\rho^{(C)}$.

8 Prove Eq. (1.51).

9 Calculate the analog of (1.55) for plane polarization.

10 Prove Eqs. (1.68) and (1.69).

11 Suppose that at a given angle the scattering amplitudes are such that $A_{12}{}^L = A_{21}{}^{L*}$ and $A_{11}{}^L = -A_{22}{}^{L*}$. How does the differential cross section depend on the polarization of the incident beam?

12 Give the general expression for the polarization of the scattered wave in terms of the scattering amplitude and the polarization parameters of the incident beam.

13 Find the expression for the total cross section analogous to (1.78) but in terms of the scattering amplitude components for *circular* polarization.

14 Prove Eq. (1.88).

15 Suppose that a double-scattering cross section for an unpolarized beam is measured as a function of the angle about the line connecting the two scatterers (at a fixed angle of elevation from that line). Let R be the ratio of the minimum to the maximum so found. What is the product of the polarizations produced in the two single scatterings for unpolarized beams? At what angle between the two scattering planes is the double-scattering cross section minimal? Suppose that that minimum is zero. What can you say about the two polarizations?

16 Visible light is scattered by droplets of (*a*) 10^{-2} cm; (*b*) 10^{-3} cm; (*c*) 10^{-4} cm diameter. Assume that, except for the sharp diffraction peak in the forward direction, the scattering cross section is essentially isotropic. Estimate by what factor (at least) the forward-scattering cross section must dominate over the average in cases *a*, *b*, and *c*.

CHAPTER 2

Spherically Symmetric Scatterers

2.1 Spherical Harmonics

An important tool in the calculation of scattering amplitudes is expansion on the basis of spherical harmonics or vector spherical harmonics. In this section we define these and list some of their properties.

2.1.1 Legendre Polynomials. The Legendre polynomials are defined by Rodrigues's formula

$$P_l(\cos\theta) = \frac{(-)^l}{2^l l!}\left(\frac{d}{d\cos\theta}\right)^l (\sin\theta)^{2l} \tag{2.1}$$

or by the generating function

$$(1 - 2r\cos\theta + r^2)^{-1/2} = \sum_{l=0}^{\infty} r^l P_l(\cos\theta) \qquad r < 1. \tag{2.2}$$

They satisfy the recurrence relations

$$\begin{aligned} lP_{l-1} &= l\cos\theta P_l + \sin^2\theta P_l' \\ (l+1)P_{l+1} &= (l+1)\cos\theta P_l - \sin^2\theta P_l' \end{aligned} \tag{2.3}$$

and have the parity of l,

$$P_l(-x) = (-)^l P_l(x). \tag{2.4}$$

They form a complete orthogonal set on the interval $-1 \leq x \leq 1$,

$$\int_{-1}^{1} dx\, P_l(x) P_{l'}(x) = \frac{\delta_{ll'}}{l + \frac{1}{2}}$$
$$\sum_l (l + \tfrac{1}{2}) P_l(x) P_l(x') = \delta(x - x'). \tag{2.5}$$

2.1.2 Associated Legendre Functions. The associated Legendre functions (of the first kind) are defined in terms of the Legendre polynomials by

$$P_l^m(\cos\theta) \equiv (\sin\theta)^m \left(\frac{d}{d\cos\theta}\right)^m P_l(\cos\theta) \qquad 0 \le m \le l \tag{2.6}$$

but they can also be written

$$P_l^m(\cos\theta) = \frac{(-)^{l+m}}{2^l l!} \frac{(l+m)!}{(l-m)!} (\sin\theta)^{-m} \left(\frac{d}{d\cos\theta}\right)^{l-m} (\sin\theta)^{2l}. \tag{2.7}$$

This may be used for either sign of m, so that

$$P_l^{-m} = (-)^m \frac{(l-m)!}{(l+m)!} P_l^m. \tag{2.8}$$

Their parity is $l + m$,

$$P_l^m(-x) = (-)^{l+m} P_l^m(x). \tag{2.9}$$

All the Legendre functions are solutions of Legendre's differential equation,

$$\frac{d}{dx}(1 - x^2)\frac{d}{dx} P_l^m + \left(l(l+1) - \frac{m^2}{1 - x^2}\right) P_l^m = 0. \tag{2.10}$$

2.1.3 Spherical Harmonics. The spherical harmonics used in this book are defined by

$$Y_l^m(\hat{\mathbf{r}}) \equiv (-)^m i^l \left[\frac{2l+1}{4\pi} \frac{(l-m)!}{(l+m)!}\right]^{1/2} e^{im\varphi} P_l^m(\cos\theta), \tag{2.11}$$

θ and φ being the polar angles of the unit vector $\hat{\mathbf{r}}$. By (2.8) their complex conjugates are given by

$$Y_l^{m*} = (-)^{l+m} Y_l^{-m}, \tag{2.12}$$

and they have the parity l,

$$Y_l^m(-\hat{\mathbf{r}}) = (-)^l Y_l^m(\hat{\mathbf{r}}). \tag{2.13}$$

They form a complete orthonormal set,

$$\int d\Omega_r\, Y_l^{m*}(\hat{\mathbf{r}})\, Y_{l'}^{m'}(\hat{\mathbf{r}}) = \delta_{ll'}\delta_{mm'}$$
$$\sum_{l=0}^{\infty} \sum_{m=-l}^{l} Y_l^m(\hat{\mathbf{r}})\, Y_l^{m*}(\hat{\mathbf{r}}') = \delta_\Omega(\hat{\mathbf{r}}', \hat{\mathbf{r}}), \tag{2.14}$$

where δ_Ω is the solid-angle Dirac δ function

$$\int d\Omega_r \delta_\Omega(\hat{\mathbf{r}},\hat{\mathbf{r}}')f(\hat{\mathbf{r}}) = f(\hat{\mathbf{r}}'). \tag{2.15}$$

They solve the partial differential equation

$$-\left[\csc\theta\frac{\partial}{\partial\theta}\sin\theta\frac{\partial}{\partial\theta} + \csc^2\theta\frac{\partial^2}{\partial\varphi^2}\right] Y_l^m = l(l+1)\,Y_l^m \tag{2.16}$$

or

$$\mathbf{L}^2 Y_l^m \equiv [\mathbf{r} \times (-i\boldsymbol{\nabla})]^2 Y_l^m = l(l+1)\,Y_l^m \tag{2.16a}$$

and

$$L_z Y_l^m \equiv [\mathbf{r} \times (-i\boldsymbol{\nabla})]_z Y_l^m = -i\frac{\partial}{\partial\varphi} Y_l^m = mY_l^m. \tag{2.17}$$

Since

$$4\pi \sum_{m=-l}^{l} Y_l^m(\hat{\mathbf{r}}_1)\,Y_l^{m*}(\hat{\mathbf{r}}_2) = (2l+1)P_l(\hat{\mathbf{r}}_1 \cdot \hat{\mathbf{r}}_2), \tag{2.18}$$

the most general rotationally invariant function of $\hat{\mathbf{r}}_1$ and $\hat{\mathbf{r}}_2$ can be expanded in the form

$$f(\hat{\mathbf{r}}_1,\hat{\mathbf{r}}_2) = \sum_{l,m} c_l\, Y_l^m(\hat{\mathbf{r}}_1)\,Y_l^{m*}(\hat{\mathbf{r}}_2). \tag{2.19}$$

2.1.4 Vector Spherical Harmonics. For the expansion of vector functions such as the radiation field we require vector spherical harmonics. First one defines

$$\boldsymbol{\chi}_1 \equiv \frac{1}{\sqrt{2}}(\hat{\mathbf{e}}^{(y)} - i\hat{\mathbf{e}}^{(x)}) \qquad \boldsymbol{\chi}_0 \equiv i\hat{\mathbf{e}}^{(z)} \qquad \boldsymbol{\chi}_{-1} \equiv \frac{1}{\sqrt{2}}(\hat{\mathbf{e}}^{(y)} + i\hat{\mathbf{e}}^{(x)}) \tag{2.20}$$

so that

$$\mathbb{1} = \sum_u \boldsymbol{\chi}_\mu \boldsymbol{\chi}_\mu^* \qquad \boldsymbol{\chi}_\mu^* \cdot \boldsymbol{\chi}_{\mu'} = \delta_{\mu\mu'}, \tag{2.21}$$

$\mathbb{1}$ being the unit tensor in dyadic notation. They are defined in such a way that

$$\begin{aligned} S_z \boldsymbol{\chi}_\mu &= \mu \boldsymbol{\chi}_\mu \\ \mathbf{S}^2 \boldsymbol{\chi}_\mu &= 2\boldsymbol{\chi}_\mu \end{aligned} \tag{2.22}$$

where $\mathbf{S}$ is the "spin operator" whose components are defined by

$$S_i \mathbf{f} \equiv i\hat{\mathbf{e}}^{(i)} \times \mathbf{f}. \tag{2.23}$$

Furthermore, their phases are such that

$$\boldsymbol{\chi}_\mu^* = (-)^{1+\mu}\boldsymbol{\chi}_{-\mu}. \tag{2.24}$$

Now

$$\mathbf{Y}_{lm\mu}(\hat{\mathbf{r}}) \equiv Y_l^m(\hat{\mathbf{r}})\boldsymbol{\chi}_\mu \tag{2.25}$$

are the eigenfunctions of the angular-momentum operators defined by (2.16a), (2.17), and (2.23),

$$L_z\mathbf{Y}_{lm\mu} = m\mathbf{Y}_{lm\mu} \qquad S_z\mathbf{Y}_{lm\mu} = \mu\mathbf{Y}_{lm\mu} \qquad \mathbf{L}^2\mathbf{Y}_{lm\mu} = l(l+1)\mathbf{Y}_{lm\mu}. \tag{2.26}$$

We combine them by means of Clebsch–Gordan coefficients to form eigenfunctions of the total angular momentum. That is, for $J = l-1, l, l+1$, the functions

$$\begin{aligned}\mathbf{Y}_{Jl}^M(\hat{\mathbf{r}}) &\equiv \sum_{m\mu} C(l1J, m\mu M)\mathbf{Y}_{lm\mu}(\hat{\mathbf{r}})\\ &= \sum_{\mu} C(l1J, M-\mu\mu M)\mathbf{Y}_{lM-\mu\mu}(\hat{\mathbf{r}})\end{aligned} \tag{2.27}$$

satisfy

$$\begin{aligned}(L_z + S_z)\mathbf{Y}_{Jl}^M &= M\mathbf{Y}_{Jl}^M\\ (\mathbf{L}+\mathbf{S})^2\mathbf{Y}_{Jl}^M &= J(J+1)\mathbf{Y}_{Jl}^M\\ \mathbf{L}^2\mathbf{Y}_{Jl}^M &= l(l+1)\mathbf{Y}_{Jl}^M.\end{aligned} \tag{2.28}$$

Explicitly written out, they are

$$\begin{aligned}\mathbf{Y}_{JJ+1}^M = &\left[\frac{(J+1+M)(J+2+M)}{2(J+1)(2J+3)}\right]^{1/2} Y_{J+1}^{M+1}\boldsymbol{\chi}_{-1}\\ &-\left[\frac{(J+1+M)(J+1-M)}{(J+1)(2J+3)}\right]^{1/2} Y_{J+1}^{M}\boldsymbol{\chi}_{0}\\ &+\left[\frac{(J+1-M)(J+2-M)}{2(J+1)(2J+3)}\right]^{1/2} Y_{J+1}^{M-1}\boldsymbol{\chi}_{1}\end{aligned} \tag{2.29}$$

$$\begin{aligned}\mathbf{Y}_{JJ}^M = &\left[\frac{(J+1+M)(J-M)}{2J(J+1)}\right]^{1/2} Y_{J}^{M+1}\boldsymbol{\chi}_{-1}\\ &+\frac{M}{[J(J+1)]^{1/2}} Y_{J}^{M}\boldsymbol{\chi}_{0}\\ &-\left[\frac{(J+M)(J+1-M)}{2J(J+1)}\right]^{1/2} Y_{J}^{M-1}\boldsymbol{\chi}_{1}\end{aligned} \tag{2.29a}$$

$$\begin{aligned}\mathbf{Y}_{JJ-1}^M = &\left[\frac{(J-1-M)(J-M)}{2J(2J-1)}\right]^{1/2} Y_{J-1}^{M+1}\boldsymbol{\chi}_{-1}\\ &+\left[\frac{(J+M)(J-M)}{J(2J-1)}\right]^{1/2} Y_{J-1}^{M}\boldsymbol{\chi}_{0}\\ &+\left[\frac{(J-1+M)(J+M)}{2J(2J-1)}\right]^{1/2} Y_{J-1}^{M-1}\boldsymbol{\chi}_{1}.\end{aligned} \tag{2.29b}$$

These vector functions are complete and orthogonal in the sense that

$$\int d\Omega_r \, \mathbf{Y}_{Jl}^{M*}(\hat{\mathbf{r}}) \cdot \mathbf{Y}_{J'l'}^{M'}(\hat{\mathbf{r}}) = \delta_{JJ'}\delta_{MM'}\delta_{ll'}$$
$$\sum_{JlM} \mathbf{Y}_{Jl}^{M}(\hat{\mathbf{r}})\mathbf{Y}_{Jl}^{M*}(\hat{\mathbf{r}}') = \mathbb{1}\delta_\Omega(\hat{\mathbf{r}},\hat{\mathbf{r}}'), \tag{2.30}$$

the last equation being written in dyadic notation, with $\mathbb{1}$ the unit dyadic. For $J = 0$ there is only one nonvanishing vector function, $\mathbf{Y}_{01}{}^0$.

2.1.5 Transverse and Longitudinal Vector Spherical Harmonics. Out of these vector functions new vector spherical harmonics are formed,

$$\mathbf{Y}_{JM}^{(e)} \equiv \left(\frac{J+1}{2J+1}\right)^{1/2} \mathbf{Y}_{JJ-1}^{M} - \left(\frac{J}{2J+1}\right)^{1/2} \mathbf{Y}_{JJ+1}^{M}$$
$$\mathbf{Y}_{JM}^{(m)} \equiv \mathbf{Y}_{JJ}^{M} \tag{2.31}$$
$$\mathbf{Y}_{JM}^{(0)} \equiv \left(\frac{J}{2J+1}\right)^{1/2} \mathbf{Y}_{JJ-1}^{M} + \left(\frac{J+1}{2J+1}\right)^{1/2} \mathbf{Y}_{JJ+1}^{M},$$

the first two of which are transverse,

$$\hat{\mathbf{r}} \cdot \mathbf{Y}_{JM}^{(e)}(\hat{\mathbf{r}}) = \hat{\mathbf{r}} \cdot \mathbf{Y}_{JM}^{(m)}(\hat{\mathbf{r}}) = 0 \tag{2.32}$$

and the third longitudinal,

$$\hat{\mathbf{r}} \times \mathbf{Y}_{JM}^{(0)}(\hat{\mathbf{r}}) = 0. \tag{2.33}$$

All are orthogonal to one another in the ordinary vector sense. For $\lambda, \lambda' = e, m, 0$,

$$\mathbf{Y}_{JM}^{(\lambda)*}(\hat{\mathbf{r}}) \cdot \mathbf{Y}_{JM}^{(\lambda')}(\hat{\mathbf{r}}) = 0 \qquad \text{unless } \lambda = \lambda'. \tag{2.34}$$

Equations (2.31) are easily inverted to read

$$\mathbf{Y}_{JJ-1}^{M} = \left(\frac{J+1}{2J+1}\right)^{1/2} \mathbf{Y}_{JM}^{(e)} + \left(\frac{J}{2J+1}\right)^{1/2} \mathbf{Y}_{JM}^{(0)}$$
$$\mathbf{Y}_{JJ+1}^{M} = -\left(\frac{J}{2J+1}\right)^{1/2} \mathbf{Y}_{JM}^{(e)} + \left(\frac{J+1}{2J+1}\right)^{1/2} \mathbf{Y}_{JM}^{(0)}. \tag{2.35}$$

For $J = 0$, there is only the longitudinal function $\mathbf{Y}_{00}{}^{(0)}$. Furthermore, it follows from (2.13) that

$$\mathbf{Y}_{JM}^{(e)}(-\hat{\mathbf{r}}) = (-)^{J+1}\mathbf{Y}_{JM}^{(e)}(\hat{\mathbf{r}})$$
$$\mathbf{Y}_{JM}^{(m)}(-\hat{\mathbf{r}}) = (-)^{J}\mathbf{Y}_{JM}^{(m)}(\hat{\mathbf{r}}) \tag{2.36}$$
$$\mathbf{Y}_{JM}^{(0)}(-\hat{\mathbf{r}}) = (-)^{J+1}\mathbf{Y}_{JM}^{(0)}(\hat{\mathbf{r}}).$$

Because under a complete coordinate reflection

$$P\mathbf{Y}(\hat{\mathbf{r}}) \equiv \mathbf{Y}'(\hat{\mathbf{r}}') = -\mathbf{Y}(-\hat{\mathbf{r}}),$$

their parities are

$$\begin{aligned} P\mathbf{Y}_{JM}^{(e)} &= (-)^J \mathbf{Y}_{JM}^{(e)} \\ P\mathbf{Y}_{JM}^{(m)} &= (-)^{J+1} \mathbf{Y}_{JM}^{(m)} \\ P\mathbf{Y}_{JM}^{(0)} &= (-)^J \mathbf{Y}_{JM}^{(0)}. \end{aligned} \tag{2.37}$$

Alternative ways of writing these vector functions are

$$\begin{aligned} \mathbf{Y}_{JM}^{(e)}(\hat{\mathbf{r}}) &= [J(J+1)]^{-1/2} r \nabla Y_J^M(\hat{\mathbf{r}}) \\ \mathbf{Y}_{JM}^{(m)}(\hat{\mathbf{r}}) &= \hat{\mathbf{r}} \times \mathbf{Y}_{JM}^{(e)}(\hat{\mathbf{r}}) = r\nabla \times \mathbf{Y}_{JM}^{(e)}(\hat{\mathbf{r}}) \\ \mathbf{Y}_{JM}^{(0)}(\hat{\mathbf{r}}) &= \hat{\mathbf{r}} Y_J^M(\hat{\mathbf{r}}) = -r^2 [J(J+1)]^{-1/2} \nabla \times (\nabla \times \mathbf{Y}_{JM}^{(e)}(\hat{\mathbf{r}})). \end{aligned} \tag{2.38}$$

They are complete and orthogonal in the sense

$$\begin{aligned} \sum_{\lambda JM} \mathbf{Y}_{JM}^{(\lambda)}(\hat{\mathbf{r}}) \mathbf{Y}_{JM}^{(\lambda)*}(\hat{\mathbf{r}}') &= \mathbb{1}\delta_\Omega(\hat{\mathbf{r}}, \hat{\mathbf{r}}') \\ \int d\Omega_r \mathbf{Y}_{JM}^{(\lambda)*}(\hat{\mathbf{r}}) \cdot \mathbf{Y}_{J'M'}^{(\lambda')}(\hat{\mathbf{r}}) &= \delta_{JJ'} \delta_{MM'} \delta_{\lambda\lambda'}. \end{aligned} \tag{2.39}$$

The first is written in dyadic notation, and λ is meant to assume the values e, m, and 0. The notation is such that in the expansion of an *electric* field vector

$$\mathscr{E}(\mathbf{r}) = \sum_{JM} [\mathscr{E}_{JM}^{(e)}(r)\mathbf{Y}_{JM}^{(e)}(\hat{\mathbf{r}}) + \mathscr{E}_{JM}^{(m)}(r)\mathbf{Y}_{JM}^{(m)}(\hat{\mathbf{r}}) + \mathscr{E}_{JM}^{(0)}(r)\mathbf{Y}_{JM}^{(0)}(\hat{\mathbf{r}})]$$

the terms labeled by e, J, M identify the *electric* 2^J-pole radiation and the terms labeled by m, J, M identify the *magnetic* 2^J-pole radiation.

2.1.6 Rotationally Invariant Tensor Functions. By (2.21) Eq. (2.18) can also be expressed in terms of the vector functions,

$$(2l+1)\mathbb{1}P_l(\hat{\mathbf{r}}_1 \cdot \hat{\mathbf{r}}_2) = 4\pi \sum_{JM} \mathbf{Y}_{Jl}^M(\hat{\mathbf{r}}_1)\mathbf{Y}_{Jl}^{M*}(\hat{\mathbf{r}}_2) \tag{2.40}$$

in the proof of which the inversion of (2.27) is used,

$$Y_l^m \chi_\mu = \sum_{JM} C(l1J, m\mu M)\mathbf{Y}_{Jl}^M, \tag{2.41}$$

i.e.,

$$\sum_{JM} C(lsJ, m\mu M)C(lsJ, m'\mu' M) = \delta_{mm'}\delta_{\mu\mu'} \tag{2.42}$$

as well as

$$\sum_{m\mu} C(lsJ, m\mu M)C(lsJ', m\mu M') = \delta_{JJ'}\delta_{MM'}. \tag{2.43}$$

However, the most general rotationally invariant tensor function of $\hat{\mathbf{r}}_1$ and $\hat{\mathbf{r}}_2$ is not necessarily obtained by summing (2.40). It is instead of the form

$$\mathfrak{F}(\hat{\mathbf{r}}_1, \hat{\mathbf{r}}_2) = \sum_{JMll'} \mathbf{Y}_{Jl}^M(\hat{\mathbf{r}}_1) c_{ll'}^J \mathbf{Y}_{Jl'}^{M*}(\hat{\mathbf{r}}_2), \tag{2.44}$$

the proof of which follows from the transformation properties of the spherical harmonics under rotation, as given, for example, by Wigner (1959).

2.1.7 Complex Conjugation Properties. The complex conjugation properties of the vector functions are obtained from (2.12), (2.24), and the property of the (real) Clebsch-Gordan coefficients

$$C(lsJ, -m - \mu - M) = (-)^{l+s-J} C(lsJ, m\mu M). \tag{2.45}$$

We obtain

$$\mathbf{Y}_{Jl}^{M*} = (-)^{J+M} \mathbf{Y}_{Jl}^{-M} \tag{2.46}$$

and consequently, by (2.31), also

$$\mathbf{Y}_{JM}^{(\lambda)*} = (-)^{J+M} \mathbf{Y}_{J-M}^{(\lambda)} \qquad \lambda = e, m, 0. \tag{2.47}$$

2.1.8 θ and φ Components. It is often useful to know the comonents of the vector spherical harmonics in the directions of increasing polar angles θ and φ as shown in Fig. 2.1. The three unit vectors $\hat{\boldsymbol{\varphi}}$, $\hat{\boldsymbol{\theta}}$, and

$$\hat{\mathbf{r}} = \hat{\boldsymbol{\theta}} \times \hat{\boldsymbol{\varphi}} \tag{2.48}$$

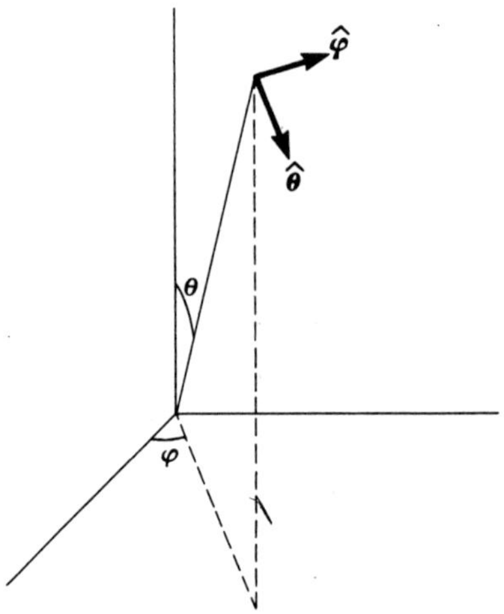

Figure 2.1. The unit vectors in the directions of increasing polar angles.

form an orthogonal basis in which the line element is

$$d\mathbf{r} = \hat{\mathbf{r}}\, dr + \hat{\boldsymbol{\theta}} r\, d\theta + \hat{\boldsymbol{\varphi}} r \sin\theta d\varphi,$$

and hence the gradient vector is given by

$$\boldsymbol{\nabla} = \hat{\mathbf{r}} \frac{\partial}{\partial r} + \hat{\boldsymbol{\theta}} r^{-1} \frac{\partial}{\partial \theta} + \hat{\boldsymbol{\varphi}} r^{-1} \csc\theta \frac{\partial}{\partial \varphi}. \tag{2.49}$$

As a result we get from (2.38)

$$
\begin{aligned}
[\mathbf{Y}_{JM}^{(e)}(\hat{\mathbf{r}})]_\theta &= C_{JM}e^{iM\varphi}\frac{d}{d\theta}P_J^M \\
[\mathbf{Y}_{JM}^{(e)}(\hat{\mathbf{r}})]_\varphi &= iC_{JM}e^{iM\varphi}M\csc\theta\, P_J^M \\
[\mathbf{Y}_{JM}^{(m)}(\hat{\mathbf{r}})]_\theta &= -iC_{JM}e^{iM\varphi}M\csc\theta\, P_J^M \\
[\mathbf{Y}_{JM}^{(m)}(\hat{\mathbf{r}})]_\varphi &= C_{JM}e^{iM\varphi}\frac{d}{d\theta}P_J^M
\end{aligned}
\tag{2.50}
$$

where

$$
C_{JM} = (-)^M i^J\left[\frac{2J+1}{4\pi J(J+1)}\frac{(J-M)!}{(J+M)!}\right]^{1/2}. \tag{2.51}
$$

Out of the θ and φ components we may form "circular" components by defining $\boldsymbol{\chi}$ as in (2.20), but relative to the $(\hat{\mathbf{r}}, \hat{\boldsymbol{\theta}}, \hat{\boldsymbol{\varphi}})$ system,

$$
\boldsymbol{\chi}_1' = \frac{1}{\sqrt{2}}(\hat{\boldsymbol{\varphi}} - i\hat{\boldsymbol{\theta}}) \qquad \boldsymbol{\chi}_0' = i\hat{\mathbf{r}} \qquad \boldsymbol{\chi}_{-1}' = \frac{1}{\sqrt{2}}(\hat{\boldsymbol{\varphi}} + i\hat{\boldsymbol{\theta}}). \tag{2.52}
$$

The result ($\lambda = 1$ meaning $\lambda = e$, $\lambda = 2$ meaning m, and $\mu = 0, \pm 1$) is

$$
\begin{aligned}
[\mathbf{Y}_{JM}^{(\lambda)}]_\mu &= -\frac{1}{\sqrt{2}}(-i\mu)^\lambda C_{JM}e^{iM\varphi}\left(\mu M\csc\theta + \frac{d}{d\theta}\right)P_J^M \\
[\mathbf{Y}_{JM}^{(0)}]_0 &= -iC_{JM}e^{iM\varphi}P_J^M.
\end{aligned}
\tag{2.53}
$$

2.1.9 The z **Axis along r.** Of special interest is the case when $\hat{\mathbf{r}}$ points along the positive z axis. Then

$$
\begin{aligned}
\mathbf{Y}_{JM}^{(e)}(\hat{\mathbf{r}}) &= i^{J-1}\left(\frac{2J+1}{8\pi}\right)^{1/2}\boldsymbol{\chi}_M & M &= \pm 1 \\
\mathbf{Y}_{JM}^{(m)}(\hat{\mathbf{r}}) &= -i^J\left(\frac{2J+1}{8\pi}\right)^{1/2}\boldsymbol{\chi}_M M & M &= \pm 1 \\
\mathbf{Y}_{JM}^{(0)}(\hat{\mathbf{r}}) &= i^{J-1}\left(\frac{2J+1}{8\pi}\right)^{1/2}\boldsymbol{\chi}_0 & M &= 0
\end{aligned}
\tag{2.54}
$$

and zero for all other values of M. Consequently in an expression such as (2.44), evaluated in a coordinate system in which $\hat{\mathbf{r}}_2$ points along the z axis, we need only $M = 0, \pm 1$. These special cases of (2.53) for $M = 0, \pm 1$, are

$$
\begin{aligned}
[\mathbf{Y}_{JM}^{(e)}]_{\pm 1} &= i^{J-1}\left(\frac{2J+1}{8\pi}\right)^{1/2} e^{iM\varphi}(\pi_J \pm M\tau_J) & M &= \pm 1 \\
[\mathbf{Y}_{J0}^{(e)}]_{\pm 1} &= \pm i^{J-1}\left(\frac{2J+1}{8\pi}\right)^{1/2} P_J^1 \\
[\mathbf{Y}_{JM}^{(m)}]_{\pm 1} &= -i^J\left(\frac{2J+1}{8\pi}\right)^{1/2} e^{iM\varphi}(M\tau_J \pm \pi_J) & M &= \pm 1 \\
[\mathbf{Y}_{J0}^{(m)}]_{\pm 1} &= -i^J\left(\frac{2J+1}{8\pi}\right)^{1/2} P_J^1
\end{aligned}
\tag{2.55}
$$

$$
\begin{aligned}
[\mathbf{Y}_{JM}^{(0)}]_0 &= \pm\frac{i^{J-1}}{J(J+1)}\left(\frac{2J+1}{8\pi}\right)^{1/2} e^{iM\varphi}P_J^1 \qquad M = \pm 1 \\
[\mathbf{Y}_{J0}^{(0)}]_0 &= i^{J-1}\left(\frac{2J+1}{8\pi}\right)^{1/2} P_J,
\end{aligned}
\tag{2.55}
$$

where τ_J and π_J are defined by

$$
\begin{aligned}
\pi_J(\cos\theta) &\equiv \frac{1}{J(J+1)}\frac{d}{d\cos\theta}P_J(\cos\theta) \\
\tau_J(\cos\theta) &\equiv \frac{1}{J(J+1)}\frac{d}{d\theta}P_J^1(\cos\theta) = \cos\theta\,\pi_J - \sin^2\theta\,\pi_J'.
\end{aligned}
\tag{2.56}
$$

2.2 Multipole Expansions

2.2.1 Expansion of a Plane Wave; Spherical Bessel Functions. A scalar plane wave can be expanded on the basis of the spherical harmonics,

$$
\exp(i\mathbf{k}\cdot\mathbf{r}) = \frac{4\pi}{kr}\sum_{lm} i^l u_l(kr)\,Y_l^m(\hat{\mathbf{r}})\,Y_l^{m*}(\hat{\mathbf{k}}). \tag{2.57}
$$

The coefficients are the Riccati–Bessel functions

$$
u_l(z) \equiv \sqrt{\tfrac{1}{2}\pi z}\,J_{l+1/2}(z) = zj_l(z) \qquad [=z\psi_l(z)] \tag{2.58}
$$

whose asymptotic values are

$$
u_l(z) \underset{z\to\infty}{\simeq} \sin(z - \tfrac{1}{2}\pi l). \tag{2.59}
$$

They are decomposed into outgoing and incoming wave parts by writing them in terms of the Riccati–Hankel functions

$$
\begin{aligned}
w_l^{(+)}(z) &\equiv ie^{i\pi l}\sqrt{\tfrac{1}{2}\pi z}\,H_{l+1/2}^{(1)}(z) = ie^{i\pi l}zh_l^{(1)}(z) \qquad [= ie^{i\pi l}z\zeta_l^{(1)}(z)] \\
w_l^{(-)}(z) &\equiv w_l^{(+)}(-z) = e^{i\pi l}w_l^{(+)}(z)^* = -i\sqrt{\tfrac{1}{2}\pi z}\,H_{l+1/2}^{(2)}(z) \\
&= -izh_l^{(2)}(z) \qquad [= -iz\zeta_l^{(2)}(z)]
\end{aligned}
\tag{2.60}
$$

whose asymptotic values are

$$
w_l^{(\pm)}(z) \underset{z\to\infty}{\simeq} e^{\pm iz + 1/2 i\pi l} \tag{2.61}
$$

as

$$
u_l = \tfrac{1}{2}i(w_l^{(-)} - e^{-i\pi l}w_l^{(+)}). \tag{2.62}
$$

Since these functions will frequently be needed later, we shall list some of their most important properties here.

They all solve the differential equation

$$
y_l'' - l(l+1)z^{-2}y_l + y_l = 0. \tag{2.63}
$$

The function u_l is the solution regular at $z = 0$. It has the small-z behavior

$$u_l(z) = \frac{z^{l+1}}{(2l+1)!!} - \frac{1}{2}\frac{z^{l+3}}{(2l+3)!!} + O(z^{l+5}). \tag{2.64}$$

The Riccati–Neumann function

$$v_l(z) \equiv \sqrt{\tfrac{1}{2}\pi z}\, N_{l+1/2}(z) = z n_l(z) \tag{2.65}$$

is a solution irregular at $z = 0$, in terms of which

$$w_l^{(+)}(z) = e^{i\pi(l+1)}[v_l(z) - iu_l(z)]. \tag{2.66}$$

Its small-z behavior is

$$v_l(z) = -z^{-l}(2l-1)!! - \tfrac{1}{2}z^{-l+2}(2l-3)!! + O(z^{-l+4}) \tag{2.67}$$

and its asymptotic value

$$v_l(z) \underset{z\to\infty}{\simeq} -\cos(z - \tfrac{1}{2}\pi l). \tag{2.68}$$

The solutions u_l, v_l, and $w_l^{(-)}$ of (2.63) obey the recurrence relations

$$\begin{aligned} y_{l-1}(z) &= y_l'(z) + lz^{-1}y_l(z) \\ y_{l+1}(z) &= -y_l'(z) + (l+1)z^{-1}y_l(z), \end{aligned} \tag{2.69}$$

and their wronskians are

$$\begin{aligned} W(u_l, v_l) &\equiv u_l v_l' - u_l' v_l = 1 \\ W(w_l^{(-)}, w_l^{(+)}) &= 2ie^{i\pi l}. \end{aligned} \tag{2.70}$$

The first few of the Riccati–Bessel functions are listed here explicitly,

$$\begin{aligned} u_0 &= \sin z \\ u_1 &= -\cos z + z^{-1}\sin z \\ u_2 &= -3z^{-1}\cos z + (3z^{-2} - 1)\sin z \\ v_0 &= -\cos z \\ v_1 &= -\sin z - z^{-1}\cos z \\ v_2 &= -3z^{-1}\sin z + (1 - 3z^{-2})\cos z \\ w_0^{(+)} &= e^{iz} \\ w_1^{(+)} &= (i - z^{-1})e^{iz} \\ w_2^{(+)} &= (3z^{-2} - 3iz^{-1} - 1)e^{iz}. \end{aligned} \tag{2.71}$$

The next few are easily generated by means of (2.69).

Because of (2.18) and (2.40) the plane-wave unit tensor is expanded in the series

$$\mathbb{1}\exp(i\mathbf{k}\cdot\mathbf{r}) = \frac{4\pi}{kr}\sum_{JlM} i^l \mathbf{Y}_{Jl}^M(\hat{\mathbf{r}})u_l(kr)\mathbf{Y}_{Jl}^{M*}(\hat{\mathbf{k}}) \tag{2.57a}$$

in dyadic notation.

2.2.2 Expansion of the Electric Field. The most general rotationally invariant tensor function of $\hat{\mathbf{k}}$ and $\hat{\mathbf{r}}$ is of the form (2.44), i.e.,

$$\mathfrak{F} = \frac{4\pi}{kr} \sum_{\substack{JM \\ ll'}} i^{l'} \mathbf{Y}_{Jl}^{M}(\hat{\mathbf{r}}) f_{ll'}^{J}(r) \mathbf{Y}_{Jl'}^{M*}(\hat{\mathbf{k}}). \tag{2.72}$$

If $\mathscr{E}(\mathbf{r}) = \mathfrak{F} \cdot \mathscr{E}_0$, with

$$\mathscr{E}_0 \cdot \mathbf{k} = 0$$

is to satisfy the Maxwell equation

$$\nabla \times (\nabla \times \mathscr{E}) = k^2 \mathscr{E}$$

and the scattering boundary condition, then the functions $f_{ll'}^{J}$ must have the form

$$f_{ll'}^{J} = \tfrac{1}{2} i [w_l^{(-)}(kr)\delta_{ll'} - (-)^l S_{ll'}^{J} w_l^{(+)}(kr)] \tag{2.73}$$

and furthermore

$$\begin{aligned} S_{J-1J-1}^{J} &= S_{J+1J+1}^{J} \\ J S_{J-1J+1}^{J} &= (J+1) S_{J+1J-1}^{J} \\ \sqrt{J} S_{J-1J}^{J} &= \sqrt{J+1} S_{J+1J}^{J}. \end{aligned} \tag{2.74}$$

The incoming wave part of $\mathfrak{F} \cdot \mathscr{E}_0$ is then equal to that of $\mathscr{E}_0 e^{i\mathbf{k}\cdot\mathbf{r}}$.

Rewriting the expansion in terms of the transverse and longitudinal vector functions, we find for the electric field vector in the presence of a rotationally invariant scatterer

$$\mathscr{E}(\mathbf{k}\mu, \mathbf{r}) = \frac{2\pi}{kr} \sum_{\substack{JM \\ \lambda\lambda'}} i^{J+1} \mathbf{Y}_{JM}^{(\lambda')}(\hat{\mathbf{r}}) f_{\lambda'\lambda}^{J}(r) \mathbf{Y}_{JM}^{(\lambda)*}(\hat{\mathbf{k}}) \cdot \boldsymbol{\chi}_{\mu}'. \tag{2.75}$$

The field has been labeled by the wave vector $\hat{\mathbf{k}}$ and the polarization of the incident beam,

$$\mathscr{E}_0(\mathbf{k}\mu, \mathbf{r}) = \exp(i\mathbf{k} \cdot \mathbf{r}) \boldsymbol{\chi}_{\mu}' \tag{2.76}$$

where $\boldsymbol{\chi}_{\mu}'$ is a vector like $\boldsymbol{\chi}_{\mu}$ of (2.20) but relative to a z axis along $\mathbf{k}$. $\mathscr{E}_0$ being transverse, μ runs only over ± 1, and λ in (2.75) only over e and m. The minimum value of J is 1, since, for $J = 0$, the only nonvanishing vector harmonic is longitudinal. The functions $f_{\lambda'\lambda}{}^{J}$ are given by

$$\begin{aligned} f_{ee}^{J} &= -i[w_J^{(-)\prime} - (-)^J S_{ee}^{J} w_J^{(+)\prime}] \\ f_{mm}^{J} &= w_J^{(-)} - (-)^J S_{mm}^{J} w_J^{(+)} \\ f_{me}^{J} &= -(-)^J S_{me}^{J} w_J^{(+)} \\ f_{em}^{J} &= i(-)^J S_{em}^{J} w_J^{(+)\prime} \\ f_{0e}^{J} &= -i\sqrt{J(J+1)}[w_J^{(-)} - (-)^J S_{ee}^{J} w_J^{(+)}] \\ f_{0m}^{J} &= i(-)^J \sqrt{J(J+1)} S_{em}^{J} w_J^{(+)} \end{aligned} \tag{2.77}$$

with

$$
\begin{aligned}
S_{mm}^{J} &= S_{JJ}^{J} \\
S_{ee}^{J} &= S_{J-1J-1}^{J} + \left(\frac{J}{J+1}\right)^{1/2} S_{J-1J+1}^{J} \\
S_{me}^{J} &= -i\left(\frac{J+1}{2J+1}\right)^{1/2} S_{JJ-1}^{J} - i\left(\frac{J}{2J+1}\right)^{1/2} S_{JJ+1}^{J} \\
S_{em}^{J} &= i\left(\frac{2J+1}{J}\right)^{1/2} S_{J+1J}^{J}.
\end{aligned}
\tag{2.78}
$$

In the derivation of these equations the recurrence formulas (2.69) have been used.

Observe that the scattering matrix $\mathscr{S}^J$, whose elements are $S_{\lambda\lambda'}{}^J$, is *not* necessarily diagonal. The off-diagonal elements in the (e,m) language represent parity nonconserving "transitions," the electric and magnetic modes of a given J having opposite parity. Such off-diagonal elements are possible because no assumption has been made that the scatterer has *reflection* symmetry. Only rotation invariance has been assumed.

It is easily verified that the electric field (2.75) satisfies the equation

$$\nabla \cdot \mathscr{E} = 0.$$

Therefore it may be used inside a medium, where $k \neq \omega/c$, only if the charge density vanishes.

2.2.3 The Magnetic Field. The magnetic field is obtained from (2.75) by means of the equations

$$
\begin{aligned}
\nabla \times (f\mathbf{Y}_{JM}^{(e)}) &= r^{-1}(rf)'\mathbf{Y}_{JM}^{(m)} \\
\nabla \times (f\mathbf{Y}_{JM}^{(m)}) &= -r^{-1}(rf)'\mathbf{Y}_{JM}^{(e)} - r^{-1}f[J(J+1)]^{1/2}\mathbf{Y}_{JM}^{(0)} \\
\nabla \times (f\mathbf{Y}_{JM}^{(0)}) &= -r^{-1}f[J(J+1)]^{1/2}\mathbf{Y}_{JM}^{(m)}
\end{aligned}
\tag{2.79}
$$

and

$$\mathscr{B} = -ik^{-1}\nabla \times \mathscr{E}.$$

We find

$$\mathscr{B}(\mathbf{k}\mu,\mathbf{r}) = \frac{2\pi}{kr}\sum_{\substack{JM\\ \lambda\lambda'}} i^{J+1}\mathbf{Y}_{JM}^{(\lambda')}(\hat{\mathbf{r}}')g_{\lambda'\lambda}^{J}(r)\mathbf{Y}_{JM}^{(\lambda)*}(\hat{\mathbf{k}}')\cdot\boldsymbol{\chi}_{\mu}' \tag{2.80}$$

with

$$
\begin{aligned}
g_{ee}^{J} &= i(-)^{J+1}S_{me}^{J}w_J^{(+)\prime} \\
g_{mm}^{J} &= (-)^{J+1}S_{em}^{J}w_J^{(+)} \\
g_{me}^{J} &= w_J^{(-)} - (-)^J S_{ee}^{J}w_J^{(+)} \\
g_{em}^{J} &= i[w_J^{(-)\prime} - (-)^J S_{mm}^{J}w_J^{(+)\prime}]
\end{aligned}
\tag{2.81}
$$

$$\begin{aligned} g_{0e}^J &= i(-)^{J+1}[J(J+1)]^{1/2} S_{me}^J w_J^{(+)} \\ g_{0m}^J &= i[J(J+1)]^{1/2}[w_J^{(-)} - (-)^J S_{mm}^J w_J^{(+)}]. \end{aligned} \tag{2.81}$$

The label μ on $\mathscr{B}$ still refers to the polarization of the incident *electric* field.

2.2.4 The $\mathscr{K}$ Matrix. The coefficients $S_{\lambda\lambda'}{}^J$, are determined from the tangential components of $\mathscr{E}$ and $\mathscr{B}$ on the surface of the scatterer. It is then convenient to use the real Riccati-Neumann functions (2.65) and the matrix $\mathscr{K}$ in terms of which

$$\mathscr{S}^J = (\mathbb{1} + i\mathscr{K}^J)(\mathbb{1} - i\mathscr{K}^J)^{-1}. \tag{2.82}$$

If we write

$$\begin{aligned} f_{ee}^J &= -iF_{ee}^{J\prime} \qquad & f_{em}^J &= -iF_{em}^{J\prime} \\ f_{me}^J &= F_{me}^J \qquad & f_{mm}^J &= F_{mm}^J \end{aligned} \tag{2.83}$$

and

$$F_{\lambda'\lambda}^J = w_J^{(-)} - (-)^J S_{\lambda'\lambda}^J w_J^{(+)} \qquad \lambda', \lambda = e, m \tag{2.84}$$

then

$$\mathfrak{F}^J = 2(\mathbb{1} - i\mathscr{K}^J)^{-1}(u_J \mathbb{1} - v_J \mathscr{K}^J) \tag{2.85}$$

is the matrix whose elements are $F_{\lambda'\lambda}^J$.

2.2.5 The Scattering Amplitude. The asymptotic value of the electric field is directly obtained from (2.61) and (2.75). The longitudinal components tend to zero faster than r^{-1}, and the field is purely transverse, as it must be,

$$\begin{aligned} \mathscr{E}(\mathbf{k}\mu,\mathbf{r}) \underset{kr\gg 1}{\simeq} & \exp(i\mathbf{k}\cdot\mathbf{r})\chi'_\mu \\ & + \frac{2\pi}{ikr}\exp(ikr) \sum_{JM\lambda\lambda'} \mathbf{Y}_{JM}^{(\lambda')}(\hat{\mathbf{r}})(S_{\lambda\lambda}^J - \delta_{\lambda'\lambda})\mathbf{Y}_{JM}^{(\lambda)*}(\hat{\mathbf{k}})\cdot\chi'_\mu \end{aligned} \tag{2.86}$$

where λ and λ' run over e and m. The scattering-amplitude tensor is therefore

$$\mathscr{A}(\mathbf{k}',\mathbf{k}) = \frac{2\pi}{ik} \sum_{JM\lambda\lambda'} \mathbf{Y}_{JM}^{(\lambda')}(\hat{\mathbf{k}}')(S_{\lambda'\lambda}^J - \delta_{\lambda'\lambda})\mathbf{Y}_{JM}^{(\lambda)*}(\hat{\mathbf{k}}) \tag{2.87}$$

with $J \geq 1$, so that

$$\mathscr{E}_{\text{scatt}} = \mathscr{A}(\mathbf{k}',\mathbf{k})\cdot\mathscr{E}_{\text{inc}}. \tag{2.88}$$

2.2.6 The z Axis along k. We now write things out more explicitly. If the z axis is chosen along the incident beam, then the $\mathbf{Y}_{JM}{}^{(\lambda)}(\hat{\mathbf{k}})$ are given by (2.54). Furthermore, if the scattering plane is the reference plane for the polarization, then the basis vectors for the right and left circular polarization of the incident beam, by (1.11), (2.21), and (1.53), are

$$\chi''_1 = e^{i\varphi}\chi_1 \qquad \chi''_{-1} = e^{-i\varphi}\chi_{-1}. \tag{2.89}$$

For the scattered wave they are directly χ'_μ, and we use (2.55). We then get for the scattering-amplitude matrix for circular-polarization components, using the scattering plane as reference plane,

$$A^C_{\substack{rr\\ll}} = \frac{1}{4ik}\sum_{J=1}^{\infty}(2J+1)(\mp S^J_{em} \mp S^J_{me} + S^J_{ee} + S^J_{mm} - 2)(\tau_J + \pi_J) \tag{2.90}$$

$$A^C_{\substack{rl\\lr}} = \frac{1}{4ik}\sum_{J=1}^{\infty}(SJ+1)(\pm S^J_{em} \mp S^J_{me} + S^J_{ee} - S^J_{mm})(\tau_J - \pi_J) \tag{2.91}$$

or

$$\mathfrak{A}^{(C)} = \frac{1}{4ik}\sum_{J=1}^{\infty}(2J+1)[(\mathfrak{S}^C_J - \mathbb{1})\tau_J + \varepsilon(\mathfrak{S}^C_J - \mathbb{1})\varepsilon\pi_J] \tag{2.92}$$

where

$$\mathfrak{S}^C_J = \frac{1}{\sqrt{2}}\begin{pmatrix}1 & -1\\ 1 & 1\end{pmatrix}\begin{pmatrix}S^J_{ee} & S^J_{em}\\ S^J_{me} & S^J_{mm}\end{pmatrix}\frac{1}{\sqrt{2}}\begin{pmatrix}1 & 1\\ -1 & 1\end{pmatrix} \tag{2.93}$$

and

$$\varepsilon \equiv \begin{pmatrix}-1 & 0\\ 0 & 1\end{pmatrix}. \tag{2.94}$$

For the amplitude for linear polarization we get from (1.58)

$$\mathfrak{A}^{(L)} = \frac{1}{2ik}\sum_{J=1}^{\infty}(2J+1)\left[\begin{pmatrix}S^J_{ee} - 1 & iS^J_{em}\\ -iS^J_{me} & S^J_{mm} - 1\end{pmatrix}\tau_J + \begin{pmatrix}S^J_{mm} - 1 & iS^J_{me}\\ -iS^J_{em} & S^J_{ee} - 1\end{pmatrix}\pi_J\right]. \tag{2.95}$$

The forward-scattering amplitude is obtained from this by setting $\cos\theta = 1$; then

$$\tau_J(1) = \pi_J(1) = \tfrac{1}{2}. \tag{2.96}$$

Therefore

$$\begin{aligned} A^C_{rl}(0) &= A^C_{lr}(0) = 0\\ A^C_{\substack{rr\\ll}}(0) &= \frac{1}{4ik}\sum_{1}^{\infty}(2J+1)(\mp S^J_{em} \mp S^J_{me} + S^J_{ee} + S^J_{mm} - 2) \end{aligned} \tag{2.97}$$

$$\mathfrak{A}^{(L)} = \frac{1}{4ik}\sum_{1}^{\infty}(2J+1)\begin{pmatrix}S^J_{ee} + S^J_{mm} - 2 & i(S^J_{em} + S^J_{me})\\ -i(S^J_{em} + S^J_{me}) & S^J_{ee} + S^J_{mm} - 2\end{pmatrix}. \tag{2.98}$$

The total cross section is obtained from this by the optical theorem (1.77). For circularly polarized radiation

$$\sigma^{(r)}_{(l)\,\text{total}} = \pi k^{-2}\sum_{J=1}^{\infty}(2J+1)\operatorname{Re}(2 - S^J_{ee} - S^J_{mm} \pm S^J_{em} \pm S^J_{me}), \tag{2.99}$$

and for 100 percent linearly polarized radiation

$$\sigma_{\text{total}}^{(\|)} = \sigma_{\text{total}}^{(\perp)} = \pi k^{-2} \sum_{J=1}^{\infty} (2J+1) \operatorname{Re}(2 - S_{ee}^J - S_{mm}^J). \tag{2.100}$$

The total scattering cross sections, on the other hand, are obtained by integrating the squared magnitudes of (2.90),

$$\begin{aligned} \sigma_{\text{scatt}}^{(i)} &= \sum_j \sigma_{\text{scatt}}^{(ji)} = \sum_j \int d\Omega |A_{ji}|^2 \\ &= 2\pi \sum_j \int_{-1}^{1} d\cos\theta \, |A_{ji}|^2. \end{aligned} \tag{2.101}$$

These are easily calculated from

$$\begin{aligned} \int_{-1}^{1} d\cos\theta (\pi_J \pi_{J'} + \tau_J \tau_{J'}) &= \frac{\delta_{JJ'}}{J + \frac{1}{2}} \\ \int_{-1}^{1} d\cos\theta (\pi_J \tau_{J'} + \pi_{J'} \tau_J) &= 0 \end{aligned} \tag{2.102}$$

which follow from (2.5) and definition (2.56) by partial integrations and the use of the Legendre equation (2.10). We find, for example,

$$\begin{aligned} \sigma_{\text{scatt}}^{\binom{r}{l}} = \tfrac{1}{2}\pi k^{-2} \sum_{J=1}^{\infty} (2J+1) \{ &|S_{ee}^J - 1|^2 + |S_{mm}^J - 1|^2 + |S_{em}^J|^2 + |S_{me}^J|^2 \\ &\mp 2 \operatorname{Re}[(S_{ee}^J - 1) S_{em}^{J*} + (S_{mm}^J - 1) S_{me}^{J*}] \}. \end{aligned} \tag{2.103}$$

The absorption cross section is then calculated by (1.77) from (2.99) and (2.103). The result is

$$\sigma_{\text{abs}}^{\binom{r}{l}} = \pi k^{-2} \sum_{J=1}^{\infty} (2J+1)(1 - \tfrac{1}{2}|S_{ee}^J \mp S_{em}^J|^2 - \tfrac{1}{2}|S_{mm}^J \mp S_{me}^J|^2). \tag{2.104}$$

2.3 Unitarity and Reciprocity

2.3.1 Energy Conservation and Unitarity. According to (2.75) an individual spherical wave of a given value of J and M is of the form

$$\mathscr{E}_{JM}(\mathbf{r}) = \sum_{\lambda\lambda'} \mathbf{Y}_{JM}^{(\lambda')}(\hat{\mathbf{r}}) f_{\lambda'\lambda}^J(r) a^\lambda. \tag{2.105}$$

Its asymptotic value is therefore

$$\mathscr{E}_{JM}(\mathbf{r}) \simeq i^J (kr)^{-1} \sum_{\lambda\lambda'} \mathbf{Y}_{JM}^{(\lambda')}(\hat{\mathbf{r}}) [\varepsilon_{\lambda'\lambda} e^{-ikr} - (-)^J S_{\lambda'\lambda}^J e^{ikr}] a^\lambda \tag{2.106}$$

where ε was defined in (2.94) and λ and λ' run over e and m. The total incoming energy flux, integrated over the surface of a large sphere, is obtained from this as

$$\frac{c}{2\pi k^2} \sum_\lambda |a^\lambda|^2,$$

while the total outgoing energy flux is

$$\frac{c}{2\pi k^2}\sum_{\lambda\lambda'} a^{\lambda *}(\mathscr{S}^{J\dagger}\mathscr{S}^J)_{\lambda\lambda'}a^{\lambda'}.$$

Assuming that there is no absorption of energy in the scatterer, the outgoing flux must equal the incoming flux, whatever a^λ. Therefore the matrix $\mathscr{S}^J$ must be unitary,

$$\mathscr{S}^{J\dagger}\mathscr{S}^J = \mathbb{1} \tag{2.107}$$

or explicitly

$$\sum_{\lambda''=e,m} S^{J*}_{\lambda''\lambda}S^J_{\lambda''\lambda'} = \delta_{\lambda\lambda'} \qquad \lambda, \lambda' = e, m.$$

If there is absorption in the scatterer, for example, if the scattering sphere is conducting, then we cannot draw this conclusion. However, since the outgoing flux must be less than the incoming, we conclude that

$$\mathscr{D}^J \equiv \mathbb{1} - \mathscr{S}^{J\dagger}\mathscr{S}^J$$

must be a positive definite matrix,

$$\sum_{\lambda\lambda'} a^{\lambda *}\mathscr{D}^J_{\lambda\lambda'}a^{\lambda'} > 0. \tag{2.108}$$

It should be noted that, $\mathscr{S}^J$ being a finite dimensional matrix, (2.107) implies that also

$$\mathscr{S}^J\mathscr{S}^{J\dagger} = \mathbb{1}. \tag{2.107a}$$

However, if there is no unitarity, then we cannot necessarily conclude from (2.108) that $\mathbb{1} - \mathscr{S}^J\mathscr{S}^{J\dagger}$ is also positive definite.

If we set

$$a^\lambda = \begin{cases} 1 & \lambda = e \\ \pm 1 & \lambda = m \end{cases}$$

then (2.108) tells us that

$$2 - |S^J_{ee} \mp S^J_{em}|^2 - |S^J_{mm} \mp S^J_{me}|^2 > 0.$$

Comparing with (2.104), we see that this is the condition for the total cross section for 100 percent circularly polarized radiation to be greater than the scattering cross section. As it should be, σ_{abs} is positive.

2.3.2 Phase Shifts. Suppose for the moment that $\mathscr{S}^J$ is diagonal as well as unitary. Then it can be written

$$S^J_{ee} = e^{2i\alpha_J} \qquad S^J_{mm} = e^{2i\beta_J} \tag{2.109}$$

where α_J and β_J are real. The significance of these angles is seen by inserting (2.109) in the asymptotic form of (2.77) and comparing the result with (2.59). In the absence of the scatterer, the asymptotic form of the electric field of an electric 2^J-pole wave is

$$r^{-1}\mathbf{Y}^{(e)}_{JM}(\hat{\mathbf{r}})\cos(kr - \tfrac{1}{2}\pi J),$$

and that of a magnetic 2^J-pole wave is

$$r^{-1}\mathbf{Y}^{(m)}_{JM}(\hat{\mathbf{r}})\sin(kr - \tfrac{1}{2}\pi J).$$

In the presence of the scatterer, these become instead

$$r^{-1}\mathbf{Y}^{(e)}_{JM}(\hat{\mathbf{r}})\cos(kr - \tfrac{1}{2}\pi J + \alpha_J)e^{i\alpha_J}$$

and (2.110)

$$r^{-1}\mathbf{Y}^{(m)}_{JM}(\hat{\mathbf{r}})\sin(kr - \tfrac{1}{2}\pi J + \beta_J)e^{i\beta_J},$$

respectively. The α_J and β_J are therefore called the *phase shifts* of the scattered wave.

Suppose now that $\mathscr{S}^J$, though unitary, is not diagonal. Then $\mathscr{S}^J\varepsilon$ is also unitary [ε being given by (2.94)], and it can be diagonalized by a unitary matrix,

$$\mathscr{S}^J\varepsilon = U^J\begin{pmatrix}-e^{2i\alpha_J} & 0\\ 0 & e^{2i\beta_J}\end{pmatrix}U^{J\dagger} \tag{2.111}$$

and the eigenvalues are of magnitude 1. We are thus led to consider waves which are mixtures of electric and magnetic radiation, in the ratios of the elements of the columns of U^J. Their asymptotic values are

$$\begin{aligned} &r^{-1}[\mathbf{Y}^{(e)}_{JM}(\hat{\mathbf{r}})U^J_{e1} + \mathbf{Y}^{(m)}_{JM}(\hat{\mathbf{r}})U^J_{m1}]\cos(kr - \tfrac{1}{2}\pi J + \alpha_J)\\ &r^{-1}[\mathbf{Y}^{(e)}_{JM}(\hat{\mathbf{r}})U^J_{e2} + \mathbf{Y}^{(m)}_{JM}(\hat{\mathbf{r}})U^J_{m2}]\sin(kr - \tfrac{1}{2}\pi J + \beta_J). \end{aligned} \tag{2.112}$$

The angles α_J and β_J are therefore called the *eigenphase shifts* of the wave.

2.3.3 Time Reversal and Reciprocity. The Maxwell equations (1.1) are invariant under the simultaneous replacement of $\mathbf{r}$ by $-\mathbf{r}$ and of t by $-t$ if there are no charges or currents. For fields of a specific frequency $\omega = ck$, this expresses itself in the fact that if $\mathscr{E}(\mathbf{r})$ is a solution of Eqs. (1.4) then $\mathscr{E}^*(-\mathbf{r})$ is also a solution, provided that n is real, that is, the conductivity vanishes everywhere. Now a solution of a specific value of J and M is necessarily of the form (2.105). Using (2.77), (2.74), and the unitarity (2.107), we get

$$f^J_{\lambda\lambda'}(r)^* = \sum_{\lambda''}\varepsilon_{\lambda\lambda}h^J_{\lambda\lambda''}(r)S^*_{\lambda''\lambda'} \tag{2.113}$$

where $\varepsilon_{\lambda\lambda}$ is given by (2.94) and $h_{\lambda\lambda'}{}^J$ is the same as $f_{\lambda\lambda'}{}^J$ except that $S_{\lambda\lambda'}{}^J$ is replaced by $S_{\lambda'\lambda}{}^J$. Because of (2.36) and (2.47) we therefore find that

$$\mathscr{E}_{JM}(-\mathbf{r})^* = \sum_{\lambda\lambda'\lambda''}\mathbf{Y}^{(\lambda)}_{\lambda-M}(\hat{\mathbf{r}})h^J_{\lambda\lambda''}(r)\,S^{J*}_{\lambda''\lambda'}a^{\lambda'}{}^*(-)^M.$$

This is again a solution provided that $h_{\lambda\lambda'}{}^J = f_{\lambda\lambda'}{}^J$, i.e.,

$$\tilde{\mathscr{S}}^J = \mathscr{S}^J. \tag{2.114}$$

In other words, time-reversal invariance implies that the $\mathscr{S}$ matrix is *symmetric*. The symmetry condition (2.114) is also called *reciprocity*. If the scatterer is absorptive, owing to conductivity, then there is no time-reversal invariance and the reciprocity may not hold.

A complex index of refraction implies an absence of time-reversal invariance in the scattering only if it is due to conductivity, but not necessarily if it is due to the presence of many nonabsorbing particles in the scatterer. According to (1.97) this also leads to a complex index of refraction. But the derivation of (1.97) shows the meaning of n there. It refers to the coherent wave propagation only, and it cannot be used in general for the calculation of the scattering by the cloud. The incoherent parts of the wave will, after all, emerge from the cloud, too, and contribute to the scattering.

Under conditions of time-reversal invariance, then, (2.90) and (2.91) simplify to

$$\begin{aligned} A^C_{\substack{rr\\ll}} &= \frac{1}{2ik}\sum_{J=1}^{\infty}(2J+1)(s^J \pm \Delta^J - 1)(\tau_J + \pi_J) \\ A^C_{rl} = A^C_{lr} &= \frac{1}{2ik}\sum_{J=1}^{\infty}(2J+1)\delta^J(\tau_J - \pi_J) \end{aligned} \qquad (2.115)$$

where

$$\begin{aligned} s^J &= \tfrac{1}{2}(S^J_{ee} + S^J_{mm}) \\ \Delta^J &= S^J_{me} = S^J_{em} \\ \delta^J &= \tfrac{1}{2}(S^J_{ee} - S^J_{mm}). \end{aligned} \qquad (2.116)$$

The expression (2.87) for the amplitude tensor in dyadic notation then has the reciprocity property

$$\tilde{\mathscr{A}}(\mathbf{k}',\mathbf{k}) = \mathscr{A}(\mathbf{k},\mathbf{k}'). \qquad (2.117)$$

2.3.4 The Generalized Optical Theorem. We may now also insert the unitarity (2.107) in (2.87). This leads to

$$\int d\Omega_k \mathscr{A}(\mathbf{k}',\mathbf{k})\cdot\mathscr{A}^\dagger(\mathbf{k}'',\mathbf{k}) = \frac{2\pi}{ik}[\mathscr{A}(\mathbf{k}',\mathbf{k}'') - \mathscr{A}^\dagger(\mathbf{k}'',\mathbf{k}')] \qquad (2.118)$$

or, together with the symmetry (2.117),

$$\begin{aligned} \int d\Omega_k \mathscr{A}(\mathbf{k}',\mathbf{k})\cdot\mathscr{A}^*(\mathbf{k},\mathbf{k}'') &= \frac{2\pi}{ik}[\mathscr{A}(\mathbf{k}',\mathbf{k}'') - \mathscr{A}^*(\mathbf{k}',\mathbf{k}'')] \\ &= 4\pi k^{-1}\,\mathrm{Im}\,\mathscr{A}(\mathbf{k}',\mathbf{k}''). \end{aligned} \qquad (2.119)$$

A special case of this is obtained by setting $\mathbf{k}' = \mathbf{k}''$. Then explicitly,

$$4\pi k^{-1}\,\mathrm{Im}\,A_{\mu\mu}(\mathbf{k}',\mathbf{k}') = \sum_\nu \int d\Omega_k |A_{\nu\mu}(\mathbf{k},\mathbf{k}')|^2.$$

But the right-hand side is the total scattering cross section for an incident beam of polarization μ and incident direction $\mathbf{k}'$. This is the optical theorem (1.77) in this special case.

2.3.5 Generalization to Absence of Spherical Symmetry. The arguments leading to unitarity [(2.107)] and symmetry [(2.114)] are generalizable to the case without spherical symmetry. Of course, in that case $\mathscr{S}$ is not diagonal in J or necessarily a multiple of the unity in M. Nevertheless analogous results hold under the same conditions as for the rotationally invariant case.

2.4 Scattering by a Uniform Sphere (Mie Theory)

2.4.1 Calculation of the $\mathscr{K}$ Matrix. Supposing the scatterer to be a uniform sphere of isotropic (nontensorial) dielectric constant ε, conductivity σ, and magnetic permeability μ, we can easily calculate the $\mathscr{S}$ matrix by the methods of Sec. 2.2. The Maxwell equations inside the scatterer then read

$$\begin{aligned} \nabla \times (\nabla \times \mathscr{E}) &= \kappa^2 \mathscr{E} \\ \nabla \times (\nabla \times \mathscr{B}) &= \kappa^2 \mathscr{B} \end{aligned} \tag{2.120}$$

where

$$\kappa^2 = k^2 \mu \left(\varepsilon + \frac{4\pi i \sigma}{\omega} \right) = n^2 k^2 \tag{2.121}$$

and the tangential components of $\mathscr{H}$ and $\mathscr{E}$ must be continuous on the surface.

Inside the scattering sphere the electric field of fixed value of J and M must be of a form analogous to (2.75), namely,

$$\mathscr{E}_{JM}(\mathbf{r}) = \sum_{\lambda'} \frac{\mathbf{Y}_{JM}^{(\lambda')}(\hat{\mathbf{r}}) \bar{f}_{\lambda'\lambda}^{J}(r)}{\kappa r}. \tag{2.122}$$

Here $\bar{f}$ is formed like f of (2.77) but is made up of *regular* Bessel functions. We write the transverse part ($\lambda, \lambda' = e, m$) in matrix notation,

$$\bar{f} = \begin{pmatrix} -i\bar{u}_J' & 0 \\ 0 & \bar{u}_J \end{pmatrix} \mathscr{C}$$

with $\bar{u}_J \equiv u_J(\kappa r)$, and $\mathscr{C}$ an arbitrary constant matrix. The magnetic field is obtained by the analog of (2.81)

$$\mathscr{B}_{JM}(\mathbf{r}) = \sum_{\lambda'} \frac{\mathbf{Y}_{JM}^{(\lambda')}(\hat{\mathbf{r}}) \bar{g}_{\lambda'\lambda}^{J}(r)}{\kappa r} \tag{2.123}$$

with the transverse part of $\bar{g}$ in matrix notation given by

$$\bar{g} = \frac{\kappa}{k} \begin{pmatrix} 0 & i\bar{u}_J' \\ \bar{u}_J & 0 \end{pmatrix} \mathscr{C}.$$

Outside the sphere by (2.75) and (2.80), we have

$$\mathscr{E}_{JM}(\mathbf{r}) = \sum_{\lambda'} \frac{\mathbf{Y}_{JM}^{(\lambda')}(\hat{\mathbf{r}}) f_{\lambda'\lambda}^{J}(r)}{kr} \tag{2.124}$$

$$\mathscr{B}_{JM}(\mathbf{r}) = \sum_{\lambda'} \frac{\mathbf{Y}_{JM}^{(\lambda')}(\hat{\mathbf{r}}) g_{\lambda'\lambda}^{J}(r)}{kr}. \tag{2.125}$$

According to (2.85) and (2.83) the transverse parts of f and g can be chosen to be

$$f = \begin{pmatrix} -iu_J' & 0 \\ 0 & u_J \end{pmatrix} - \begin{pmatrix} -iv_J' & 0 \\ 0 & v_J \end{pmatrix} \mathscr{K}^J$$

$$g = \begin{pmatrix} 0 & iu_J' \\ u_J & 0 \end{pmatrix} - \begin{pmatrix} 0 & iv_J' \\ v_J & 0 \end{pmatrix} \mathscr{K}^J$$

where all functions are evaluated at kr. On the surface of the sphere of radius R the tangential components of $\mathscr{E}$ and $\mathscr{H}$ must be continuous. Hence only the transverse components of the fields are involved in this, and we have, at $r = R$,

$$f = \bar{f} \qquad g = \frac{\bar{g}}{\mu}$$

or

$$\begin{pmatrix} u_J' & 0 \\ 0 & u_J \end{pmatrix} - \begin{pmatrix} v_J' & 0 \\ 0 & v_J \end{pmatrix} \mathscr{K}^J = \begin{pmatrix} \bar{u}_J' & 0 \\ 0 & \bar{u}_J \end{pmatrix} \mathscr{C}$$

$$\begin{pmatrix} 0 & u_J' \\ u_J & 0 \end{pmatrix} - \begin{pmatrix} 0 & v_J' \\ v_J & 0 \end{pmatrix} \mathscr{K}^J = \frac{\kappa}{\mu k} \begin{pmatrix} 0 & \bar{u}_J' \\ \bar{u}_J & 0 \end{pmatrix} \mathscr{C}.$$

These matrix equations are easily solved for $\mathscr{K}$. Owing to the reflection symmetry it comes out diagonal. Writing

$$\begin{aligned} x &\equiv kR = \frac{2\pi R}{\lambda} \\ \bar{X} &\equiv \kappa R = nx \end{aligned} \tag{2.126}$$

we get

$$\begin{aligned} K_e^J \equiv K_{ee}^J &= \frac{nu_J'(x)u_J(\bar{X}) - \mu u_J'(\bar{X})u_J(x)}{nv_J'(x)u_J(\bar{X}) - \mu u_J'(\bar{X})v_J(x)} \\ K_m^J \equiv K_{mm}^J &= \frac{\mu u_J'(x)u_J(\bar{X}) - nu_J'(\bar{X})u_J(x)}{\mu v_J'(x)u_J(\bar{X}) - nu_J'(\bar{X})v_J(x)}. \end{aligned} \tag{2.127}$$

The $\mathscr{S}$ matrix elements are obtained from (2.82) as

$$S_\lambda^J \equiv S_{\lambda\lambda}^J = \frac{1 + iK_\lambda^J}{1 - iK_\lambda^J} \qquad \lambda = e, m \tag{2.128}$$

and therefore, by (2.66)

$$S_e^J = e^{i\pi J} \frac{nw_J^{(-)\prime} u_J - \mu w_J^{(-)} u_J'}{nw_J^{(+)\prime} u_J - \mu w_J^{(+)} u_J'}$$
$$S_m^J = e^{i\pi J} \frac{\mu w_J^{(-)\prime} u_J - n w_J^{(-)} u_J'}{\mu w_J^{(+)\prime} u_J - n w_J^{(+)} u_J'} \tag{2.127a}$$

every w_J being evaluated at x and every u_J, at $\overline{X}$.

Notice that in the absence of conductivity n is real and hence so is $\overline{X}$. As a result $\mathscr{K}$ is *real* and $\mathscr{S}$ unitary. That is to say,

$$|S_e^J|^2 = |S_m^J|^2 = 1.$$

In terms of the phase shifts (2.109) we have

$$K_e^J = \tan \alpha_J$$
$$K_m^J = \tan \beta_J. \tag{2.129}$$

If the sphere is conducting, then there is absorption and

$$|S_e^J|^2 < 1 \qquad |S_m^J|^2 < 1. \tag{2.130}$$

We can still define α_J and β_J by (2.129), but the phase shifts are not real.

The difference between $S_e{}^J$ and $S_m{}^J$ vanishes in the special case in which $\varepsilon = \mu$. Then we have, not only $A_{rr}{}^C = A_{ll}{}^C$, but also $A_{rl}{}^C = A_{lr}{}^C = 0$.

2.4.2 The Scattering Amplitude. The scattering amplitudes (2.90) and (2.91) now simplify to

$$A_{rr}^C = A_{ll}^C = \frac{1}{4ik} \sum_{J=1}^{\infty} (2J+1)(e^{2i\alpha_J} + e^{2i\beta_J} - 2)(\tau_J + \pi_J)$$
$$A_{rl}^C = A_{lr}^C = \frac{1}{4ik} \sum_{J=1}^{\infty} (2J+1)(e^{2i\alpha_J} - e^{2i\beta_J})(\tau_J - \pi_J) \tag{2.131}$$

and by (2.95)

$$A_{12}^L = A_{21}^L = 0$$
$$A_{\parallel}^L \equiv A_{11}^L = \frac{1}{2ik} \sum_{J=1}^{\infty} (2J+1)[(e^{2i\alpha_J} - 1)\tau_J + (e^{2i\beta_J} - 1)\pi_J] \tag{2.132}$$
$$A_{\perp}^L \equiv A_{22}^L = \frac{1}{2ik} \sum_{J=1}^{\infty} (2J+1)[(e^{2i\beta_J} - 1)\tau_J + (e^{2i\alpha_J} - 1)\pi_J].$$

The amplitudes can be written in terms of two simpler series,

$$A_e(\cos\theta) = \frac{1}{2ik} \sum_{J=1}^{\infty} (2J+1)(e^{2i\alpha_J} - 1)\pi_J(\cos\theta)$$
$$A_m(\cos\theta) = \frac{1}{2ik} \sum_{J=1}^{\infty} (2J+1)(e^{2i\beta_J} - 1)\pi_J(\cos\theta) \tag{2.133}$$

namely,

$$\begin{aligned}
A_{\|}^{L} &= -\sin\theta \frac{d}{d\cos\theta} \sin\theta\, A_e + A_m \\
A_{\perp}^{L} &= A_e - \sin\theta \frac{d}{d\cos\theta} \sin\theta\, A_m \\
A_{rr}^{C} &= A_{ll}^{C} = \left(1 - \sin\theta \frac{d}{d\cos\theta} \sin\theta\right)\frac{1}{2}(A_e + A_m) \\
A_{rl}^{C} &= A_{lr}^{C} = \left(1 + \sin\theta \frac{d}{d\cos\theta} \sin\theta\right)\frac{1}{2}(A_m - A_e).
\end{aligned} \tag{2.134}$$

The total cross section is obtained from (2.99),

$$\sigma_{\text{total}} = \left[2x^{-2}\sum_{J=1}^{\infty}(2J+1)(\sin^2\alpha_J + \sin^2\beta_J)\right]\pi R^2 \tag{2.135}$$

if α_J and β_J are real. It does not depend on the polarization. The bracket is the ratio of the total cross section to the geometrical cross section of the scattering sphere. If α_J and β_J are not real, then

$$\sigma_{\text{total}} = \left[x^{-2}\sum_{J=1}^{\infty}(2J+1)(2 - \operatorname{Re} S_e^J - \operatorname{Re} S_m^J)\right]\pi R^2. \tag{2.136}$$

The parameter x is defined by (2.126). It is a measure of the size of the scatterer relative to the wavelength of the radiation.

The scattering cross section is obtained from (2.103) as

$$\sigma_{\text{scatt}} = \left[\tfrac{1}{2}x^{-2}\sum_{J=1}^{\infty}(2J+1)(|S_e^J - 1|^2 + |S_m^J - 1|^2)\right]\pi R^2. \tag{2.137}$$

As it must, this reduces to (2.135) if there is no absorption. If there is, we get, from (2.104),

$$\sigma_{\text{abs}} = \left[x^{-2}\sum_{J=1}^{\infty}(2J+1)(1 - \tfrac{1}{2}|S_e^J|^2 - \tfrac{1}{2}|S_m^J|^2)\right]\pi R^2. \tag{2.138}$$

The ratio of the scattering to the total cross section is called the *albedo*,

$$\text{Albedo} \equiv \frac{\sigma_{\text{scatt}}}{\sigma_{\text{total}}} = 1 - \frac{\sigma_{\text{abs}}}{\sigma_{\text{total}}}. \tag{2.139}$$

Notes and References

2.1.1 and **2.1.2** For detailed treatments of the properties of Legendre functions, see Whittaker and Watson (1948); A. Erdelyi (1953), vol. 1.

2.1.3 The spherical harmonics here used differ from those of Blatt and Weisskopf (1952) and from those of M. E. Rose (1955), for example, by a factor of i^l; they differ from those of J. D. Jackson (1962) by a factor of $(-)^m i^l$. For a comparison with the phases used in other works, see M. E. Rose (1955), appendix A. The definition (2.11) makes for more convenient properties under complex conjugation, i.e., time reversal.

2.1.4 For more detailed treatments of vector spherical harmonics, see, for example, M. E. Rose (1955). The purpose of the phase of the vectors $\boldsymbol{\chi}$ adopted in (2.20) is the same as for that of the spherical harmonics.

For detailed treatments of Clebsch–Gordan (or vector addition) coefficients, see M. E. Rose (1952), A. R. Edmonds (1957), and E. P. Wigner (1959).

2.1.5 The present notation for the transverse and longitudinal vector spherical harmonics differs from that of others works. For example, in the notation of Blatt and Weisskopf (1952), $\mathbf{Y}_{JM}{}^{(m)} = i^{1+l}X_{JM}$, and in that of J. D. Jackson (1962), $\mathbf{Y}_{JM}{}^{(m)} = i^{1+l}(-)^m X_{JM}$.

2.1.9 The functions π_J and τ_J differ from those denoted by these symbols by H. C. van de Hulst (1957) by a factor of $1/J(J + 1)$.

2.2.1 The notation j_l, n_l, and h_l used in (2.58), (2.60), and (2.65) is the one now used most commonly in quantum mechanics; the symbols ψ_l and ζ_l indicated in brackets are those employed by Watson and Sommerfeld and in much of the electromagnetic literature. For the purpose of the radial differential equations, the functions u_l, v_l, and w_l, which differ from those mentioned by a factor of z or kr, are much more convenient.

For detailed discussions of the properties of Bessel functions, see A. Erdelyi (1953), vol. 2, and G. N. Watson (1958).

2.2.2 The S matrix was first independently introduced in quantum scattering theory by J. A. Wheeler (1937a) and by W. Heisenberg (1943).

2.2.5 The partial wave expansion of the scattering amplitude in a form analogous to (2.87) was first performed in a quantum-mechanical context by Faxen and Holtsmark (1927).

2.4 The exact solution to the electromagnetic-wave problem with a uniform dielectric sphere, using partial waves, was first given by G. Mie (1908); see also A. Clebsch (1863); and L. Lorenz (1890). The early history of the Mie solution is traced by N. A. Logan (1962). For Mie scattering by optically active particles, see Alexopoulos and Uzunoglu (1978).

A survey of early studies of scattering by a sphere can be found in N. A. Logan (1965), and a summary of more recent work in R. F. Goodrich et al. (1961); by a cone, in Kleinman and Senior (1963). These papers contain extensive bibliographies. Scattering by spheroidal particles is treated in Warner and Hizal (1976); A. R. Holt et al. (1978); S. Asano (1979); Holt and Shepherd (1979); and Asano and Sato (1980); and by axisymmetric particles, in Morrison and Cross (1974); Medgyest–Mitschang and Eftimiu (1979).

Problems

1 Prove Eqs. (2.73) and (2.74).

2 Prove Eqs. (2.77) and (2.78).

3 Prove Eqs. (2.80) and (2.81).

4 Prove Eqs. (2.102).

5 Is Eq. (2.137) valid if the incident radiation is polarized? If not, what is the total scattering cross section in that case?

6 Express the density matrix of the wave scattered in a given direction by a uniform dielectric sphere in terms of the density matrix of the incoming beam and in terms of the phases α_J and β_J defined by (2.129) and (2.127). Do this for plane and for circular polarization. Can the scattered wave be polarized if the incident beam is not?

7 What is the reference plane for the polarization in Eq. (2.132)?

8 Equation (2.131) implies that a sphere on which only left circularly polarized radiation falls produces some right circularly polarized radiation, and vice versa. Discuss this possibility in angular-momentum terms.

9 Give the fundamental reasons why the amplitudes A^C_{rr} and A^C_{ll} in (2.131) are equal, and why $A^C_{rl} = A^C_{lr}$.

CHAPTER 3

Limiting Cases and Approximations

According to Eq. (2.127) the ratio of the scattering cross section of a uniform sphere to its geometrical cross section depends upon three independent parameters

$$x = kR = \frac{2\pi R}{\lambda}$$

$$n \equiv \mu^{1/2} n' = \left[\mu\left(\varepsilon + \frac{4\pi i \sigma}{\omega}\right)\right]^{1/2}$$

$$\frac{n}{\mu} = \left[\frac{1}{\mu}\left(\varepsilon + \frac{4\pi i \sigma}{\omega}\right)\right]^{1/2}.$$

The expression for the cross section simplifies when one or more of these are large or small, in the case of x, or close to 1, in the case of n and μ. We now want to look at these limiting cases. They will each be discussed in the context of the Mie theory and then in that of a more general approximation scheme.

3.1 Small Spheres, Not Too Dense (Rayleigh Scattering)

If the radius of the sphere is small relative to the wavelength, $R \ll \lambda$, then

$$x \ll 1 \tag{3.1}$$

and we may use the first terms of the expansions (2.64) and (2.67) of u_J

and v_J about $x = 0$ in (2.127). In order to be able to use these first terms also for $u_J(\overline{X})$ and $v_J(\overline{X})$, we must assume that n is not too large, i.e.,

$$xn \ll 1. \tag{3.2}$$

It is to be expected on physical grounds that the limitation (3.2) plays an important role because

$$2x(n - 1) = 2Rkn - 2Rk$$

is the shift in the phase of a wave that passes straight through the sphere. We must suppose that this is small in order for little scattering to occur.

Insertion of (2.64) and (2.67) in (2.127) leads to

$$\begin{aligned} \tan \alpha_J &\simeq \frac{J+1}{(2J+1)!!(2J-1)!!} \frac{n^2 - \mu}{Jn^2 + \mu(J+1)} x^{2J+1} \\ \tan \beta_J &\simeq \frac{J+1}{(2J+1)!!(2J-1)!!} \frac{\mu - 1}{\mu J + J + 1} x^{2J+1}. \end{aligned} \tag{3.3}$$

Because $x \ll 1$, this means that only the term with $J = 1$ need be retained in the series (2.131) or (2.132). For $J = 1$,

$$\alpha_1 \simeq \tfrac{2}{3}\mathfrak{N}x^3 \qquad \beta_1 \simeq \tfrac{2}{3}\mathfrak{M}x^3$$

where

$$\begin{aligned} \mathfrak{N} &= \frac{n'^2 - 1}{n'^2 + 2} \\ \mathfrak{M} &= \frac{\mu - 1}{\mu + 2}. \end{aligned} \tag{3.4}$$

Notice that the electric and magnetic contributions separate (in this lowest approximation), the electric phase shift depending solely on the dielectric constant and on the conductivity and the magnetic phase shift depending in the same manner on the magnetic permeability.

From (2.131) and (2.56) we get the amplitudes for circular polarization,

$$\begin{aligned} A_{rr}^C = A_{ll}^C &\simeq x^2 R(\mathfrak{N} + \mathfrak{M}) \cos^2 \tfrac{1}{2}\theta \\ A_{rl}^C = A_{lr}^C &\simeq x^2 R(\mathfrak{M} - \mathfrak{N}) \sin^2 \tfrac{1}{2}\theta \end{aligned} \tag{3.5}$$

and from (2.132) those for plane polarization,

$$\begin{aligned} A_{\parallel}^L &\simeq x^2 R(\mathfrak{N} \cos\theta + \mathfrak{M}) \\ A_{\perp}^L &\simeq x^2 R(\mathfrak{N} + \mathfrak{M} \cos\theta). \end{aligned} \tag{3.6}$$

If there is absorption, the total cross section, according to (2.136), is

$$\sigma_{\text{total}} \simeq 4x \operatorname{Im} \mathfrak{N} \pi R^2. \tag{3.7}$$

The extinction in this case is practically all caused by absorption, since the scattering cross section is found by (2.137) to be much smaller than the total cross section,

$$\sigma_{\text{scatt}} \simeq \tfrac{8}{3}x^4(|\mathfrak{N}|^2 + |\mathfrak{M}|^2)\pi R^2. \tag{3.8}$$

The polarization of an initially unpolarized beam is obtained by inserting (3.6) in (1.68),

$$P = \frac{(|\mathfrak{N}|^2 - |\mathfrak{M}|^2)\sin^2\theta}{(|\mathfrak{N}|^2 + |\mathfrak{M}|^2)(1 + \cos^2\theta) + 4\mathfrak{M}\,\mathrm{Re}\,\mathfrak{N}\cos\theta}. \tag{3.9}$$

When the magnetic permeability is unity, the polarization is independent of n,

$$P = \frac{\sin^2\theta}{1 + \cos^2\theta}. \tag{3.10}$$

The radiation scattered at right angles is therefore 100 percent polarized.

These are the characteristic results of *Rayleigh* scattering. They can also be derived as the scattering by a point dipole with the polarizability of a sphere. The strong λ^{-4} wavelength dependence of the scattering cross section visible in (3.8) is responsible for the blue color of the sky and for the effectiveness of infrared light for penetrating haze consisting of very small particles. The strong polarization of the sky's light at right angles to the sun is also easily observed. If the next powers of x are taken into account, P is no longer 100 percent but differs from it by a term of order x^4. As a consequence the polarization is better for long wavelengths (*dispersion of polarization*). The resulting increase of the blueness of the sky at right angles to the sun, when observed through a polarization filter, is a familiar effect in color photography. In addition, the angular position of the polarization maximum depends on the wavelength (*polychroism*).

The next term in powers of x^2 also removes the forward-backward symmetry of the Rayleigh differential cross section and increases it in the forward direction.

3.2 Low Optical Density, Not Too Large (Rayleigh–Gans; Born Approximation)

We now consider the limit in which

$$|n - 1| \ll 1 \qquad |\mu - 1| \ll 1. \tag{3.11}$$

The size of the scattering sphere is arbitrary, but it should not be too large. In other words, we also want

$$|n - 1|x \ll 1. \tag{3.12}$$

In the evaluation of the phase shifts by (2.127) we may then replace $u_J(\overline{X})$ by

$$u_J(nx) \simeq u_J(x) + (n - 1)xu'_J(x)$$

and similarly for v_J. From the differential equation (2.63) for $u_J(x)$ and $v_J(x)$ and the boundary conditions we obtain the wronskian (2.70) and similarly

$$u_J(nx)u_J'(x) - nu_J(x)u_J'(nx) = (n^2 - 1)\int_0^x dx'\, u_J(x')u_J(nx'). \tag{3.13}$$

Insertion in (2.127) yields the first approximation,

$$\alpha_J \simeq (n^2 - 1)\int_0^x dx'\,[u_J(x')]^2 + (n^2 - \mu)\mu_J(x)u_J'(x)$$

$$\beta_J \simeq (n^2 - 1)\int_0^x dx'\,[u_J(x')]^2 + (\mu - 1)u_J(x)u_J'(x).$$

By (2.132) this gives us the amplitude

$$A_{\parallel}^L = (n^2 - 1)k^{-1}\sum_{J=1}^{\infty}(2J + 1)$$
$$\left[\left(\int_0^x dx'\, u_J^2(x') + u_J u_J'\right)\tau_J + \int_0^x dx'\, u_J^2(x')\pi_J\right]$$
$$+ (\mu - 1)k^{-1}\sum_{J=1}^{\infty}(2J + 1)u_J(x)u_J'(x)(\pi_J - \tau_J)$$

and $A_{\perp}{}^L$, by exchanging π_J and τ_J. These expressions are simplified by writing

$$u_J u_J' = \int_0^x dx'(u_J'^2 + u_J u_J''),$$

and using the differential equation (2.63) as well as the recursion relations (2.69),

$$(2J + 1)u_J u_J' = \int_0^x dx'[(J + 1)u_{J-1}^2 - (2J + 1)u_J^2 + Ju_{J+1}^2].$$

At this point the recursion relations for the Legendre polynomials (2.3) are brought into play, with the result that

$$\sum_J (2J + 1)u_J u_J'\pi_J = (\cos\theta - 1)\sum_J \frac{2J + 1}{J(J + 1)}P_J'\int_0^x dx'\, u_J^2$$

$$\sum_J (2J + 1)u_J u_J'\tau_J = (\cos\theta - 1)\sum_J \frac{2J + 1}{J(J + 1)}[J(J + 1)P_J + P_J']\int_0^x dx'\, u_J^2.$$

A simple calculation then gives

$$A_{\parallel}^L = k^2\int_0^R dr\, r^2\sum_{l=0}^{\infty}(2l + 1)[j_l(kr)]^2 P_l(\cos\theta)[(n'^2 - 1)\cos\theta + \mu - 1]$$

$$A_{\perp}^L = k^2\int_0^R dr\, r^2\sum_{l=0}^{\infty}(2l + 1)[j_l(kr)]^2 P_l(\cos\theta)[(\mu - 1)\cos\theta + n'^2 - 1].$$

The infinite series involved can be expressed in closed form. Expansion of the two exponentials according to (2.57) and use of the orthogonality (2.14) and of (2.18) gives

$$
\begin{aligned}
F(\theta) &\equiv \int_{|\mathbf{r}|\le R} d\mathbf{r} \exp[i(\mathbf{k}-\mathbf{k}')\cdot\mathbf{r}] \\
&= 4\pi \sum_{l=0}^{\infty} (2l+1)P_l(\cos\theta) \int_0^R dr\, r^2 [j_l(kr)]^2.
\end{aligned}
\tag{3.14}
$$

Therefore

$$
\begin{aligned}
A_{\|}^L &= \frac{k^2}{4\pi} F(\theta)[(n'^2-1)\cos\theta + \mu - 1] \\
A_{\perp}^L &= \frac{k^2}{4\pi} F(\theta)[(\mu-1)\cos\theta + n'^2 - 1].
\end{aligned}
\tag{3.15}
$$

This result can be obtained also in an entirely different way, known as the *Rayleigh–Gans approximation*. The scattering by the sphere is represented as the sum of Rayleigh scatterings by each volume element. Because of (3.11) the results (3.6) simplify in that

$$
\begin{aligned}
\mathfrak{M} &\simeq \tfrac{1}{3}(\mu - 1) \\
\mathfrak{N} &= \tfrac{1}{3}(n'^2 - 1).
\end{aligned}
$$

We then write the Rayleigh amplitudes of a particle of volume $d\mathbf{r}$ as

$$
\begin{aligned}
dA_{\|}^L &= \frac{k^2}{4\pi} d\mathbf{r}[(n'^2-1)\cos\theta + \mu - 1] \\
dA_{\perp}^L &= \frac{k^2}{4\pi} d\mathbf{r}[(\mu-1)\cos\theta + n'^2 - 1].
\end{aligned}
$$

Before summing, however, we must take into account relative phase changes which are due to the fact that the scattering volume $d\mathbf{r}$ is situated, not at the origin, but at $\mathbf{r}$ (see Fig. 3.1). That means that $dA_{\|}{}^L$ must be multiplied by $\exp(i\delta)$, where

$$
\delta = \mathbf{r}\cdot(\mathbf{k}-\mathbf{k}').
$$

Integration then immediately gives

$$
\begin{aligned}
A_{\|}^L &= \frac{k^2}{4\pi}(F_e\cos\theta + F_m) \\
A_{\perp}^L &= \frac{k^2}{4\pi}(F_m\cos\theta + F_e)
\end{aligned}
\tag{3.16}
$$

which may be used even in the case in which n' and μ are not constant, i.e., the scatterer is not homogeneous. The functions

$$
\begin{aligned}
F_e &\equiv \int d\mathbf{r}(n'^2-1)\exp[i(\mathbf{k}-\mathbf{k}')\cdot\mathbf{r}] \\
F_m &\equiv \int d\mathbf{r}(\mu-1)\exp[i(\mathbf{k}-\mathbf{k}')\cdot\mathbf{r}]
\end{aligned}
\tag{3.17}
$$

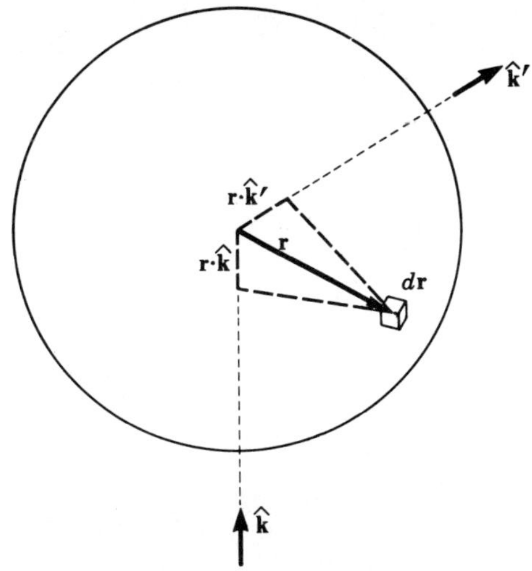

Figure 3.1 The optical-path length for scattering by $d\mathbf{r}$.

are the *form factors* of the scatterer, and the integrals extend over its volume. If n' and μ are constant and the scatterer is a sphere, (3.16) reduces to (3.15). The polarization is then the same as in the Rayleigh case, namely, (3.9).

The Rayleigh–Gans approximation is the exact analog of the Born approximation in quantum mechanics, as can be seen by comparing (3.17) with (10.97) and replacing the potential by a constant index of refraction in the region of the scatterer.

If the scatterer is spherically symmetric, (3.17) reduces to

$$\begin{aligned} F_e &= 2\pi k^{-1} \int_0^R dr\, r(n'^2 - 1) \frac{\sin(2kr \sin\frac{1}{2}\theta)}{\sin\frac{1}{2}\theta} \\ F_m &= 2\pi k^{-1} \int_0^R dr\, r(\mu - 1) \frac{\sin(2kr \sin\frac{1}{2}\theta)}{\sin\frac{1}{2}\theta} \end{aligned} \tag{3.18}$$

and if n' and μ are constant, the single function F of (3.15) becomes

$$F = 2\pi R^2 \frac{j_1(2kR\sin\frac{1}{2}\theta)}{k\sin\frac{1}{2}\theta}. \tag{3.19}$$

Consequently

$$\begin{aligned} A_{\parallel}^L &= \tfrac{1}{2} Rx \frac{j_1(2x\sin\frac{1}{2}\theta)}{\sin\frac{1}{2}\theta}[(n'^2 - 1)\cos\theta + \mu - 1] \\ A_{\perp}^L &= \tfrac{1}{2} Rx \frac{j_1(2x\sin\frac{1}{2}\theta)}{\sin\frac{1}{2}\theta}[(\mu - 1)\cos\theta + n'^2 - 1]. \end{aligned} \tag{3.20}$$

If the permeability is unity, then the resulting cross sections are independent of whether $n > 1$ or $n < 1$. The latter situation may occur if the outside medium is not a vacuum, as, for example, in the scattering by bubbles in a fluid.

The total scattering cross section for unpolarized radiation is obtained from (3.20) by integrating

$$\begin{aligned}\sigma_{\text{scatt}} &= \pi\int_{-1}^{1} d\cos\theta(|A_{\|}^{L}|^2 + |A_{\perp}^{L}|^2)\\ &= \pi R^2\{\tfrac{1}{4}[|n'^2 - 1|^2 + (\mu - 1)^2]I_1(x)\\ &\qquad + \tfrac{1}{2}(\mu - 1)\operatorname{Re}(n'^2 - 1)I_2(x)\}\end{aligned} \tag{3.21}$$

where

$$\begin{aligned}I_1(x) &= 8x^2\int_0^{2x}\frac{dz}{z}[j_1(z)]^2\left(1 - \frac{z^2}{2x^2} + \frac{z^4}{8x^4}\right)\\ &= \frac{5}{2} + 2x^2 - \frac{\sin 4x}{4x} - 7\frac{1 - \cos 4x}{(4x)^2} + \left(\frac{1}{4x^2} - 1\right)4\int_0^{2x}\frac{dz}{z}\sin^2 z\\ I_2(x) &= 8x^2\int_0^{2x}\frac{dz}{z}[j_1(z)]^2\left(1 - \frac{z^2}{2x^2}\right)\\ &= \frac{3}{2} + 2x^2 - 3\frac{\sin 4x}{4x} + 3\frac{1 - \cos 4x}{(4x)^2} - 4\int_0^{2x}\frac{dz}{z}\sin^2 z.\end{aligned}$$

The remaining integral can be expressed as

$$2\int_0^{2x} dz\, z^{-1}\sin^2 z = \gamma + \ln 4x - \operatorname{Ci}(4x),$$

γ being Euler's constant,

$$\gamma = 0.5771\cdots$$

and the cosine integral

$$\operatorname{Ci}(x) = -\int_x^{\infty} dy\, y^{-1}\cos y.$$

The forward amplitude obtained from (3.18) agrees with the Rayleigh forward amplitude. That must be so because in that case no phase differences arise from the different locations of the scattering elements. The total cross section consequently also agrees with the Rayleigh case.

The expressions (3.20) may be further simplified by assuming either $x \ll 1$ or $x \gg 1$. In the first instance the result, of course, reduces to the Rayleigh amplitudes. In the second case, that of *large* spheres relative to the wavelength [but not too large, i.e., still retaining (3.12)], (3.20) becomes

$$\begin{aligned}A_{\|}^{L} &= -\tfrac{1}{4}R\frac{\cos(2x\sin\frac{1}{2}\theta)}{\sin^2\frac{1}{2}\theta}[(n'^2 - 1)\cos\theta + \mu - 1]\\ A_{\perp}^{L} &= -\tfrac{1}{4}R\frac{\cos(2x\sin\frac{1}{2}\theta)}{\sin^2\frac{1}{2}\theta}[(\mu - 1)\cos\theta + n'^2 - 1]\end{aligned} \tag{3.22}$$

provided that the scattering angle is not too small. In other words, we must assume that

$$x \sin \tfrac{1}{2}\theta \gg 1.$$

The total scattering cross section reduces to

$$\sigma_{\text{scatt}} = \tfrac{1}{2}|n'^2 + \mu - 2|^2(x^2 - \ln x)\pi R^2 \tag{3.23}$$

which, owing to the assumption (3.12), must still be much smaller than the geometrical cross section.

Observe that the amplitudes (3.20) have zeros at the angles and wavelengths determined by

$$J_{1/2}(2kR \sin \tfrac{1}{2}\theta) = 0.$$

These are characteristic diffraction phenomena. The exact amplitude has similar zeros, but not precisely where (3.20) vanishes. If the optical density drops down gradually instead of having a sharp edge, the oscillations become smoother.

3.3 Small Dense Spheres

3.3.1 Resonance Scattering. We now want to consider the limiting case of

$$\frac{n}{\mu} \gg 1. \tag{3.24}$$

Then (2.127) becomes

$$\begin{aligned} \tan \alpha_J &\simeq \frac{u_J'(x)/v_J'(x)}{1 - (\mu/n)[v_J(x)/v_J'(x)][u_J'(\bar{X})/u_J(\bar{X})]} \\ \tan \beta_J &\simeq \frac{u_J(x)/v_J(x)}{1 - (\mu/n)[v_J'(x)/v_J(x)][u_J(\bar{X})/u_J'(\bar{X})]}. \end{aligned} \tag{3.25}$$

If, in addition, the scatterer is *small*, i.e.,

$$x \ll 1, \tag{3.26}$$

then the phase shifts are simply

$$\begin{aligned} \tan \alpha_J &\simeq \frac{(J+1)x^{2J+1}}{J(2J+1)!!(2J-1)!!}\left[1 + \frac{\mu}{n}\frac{x\,u_J'(\bar{X})}{J\,u_J(\bar{X})}\right]^{-1} \\ \tan \beta_J &\simeq -\frac{x^{2J+1}}{(2J+1)!!(2J-1)!!}\left[1 + \frac{\mu}{n}\frac{J\,u_J(\bar{X})}{x\,u_J'(\bar{X})}\right]^{-1}. \end{aligned} \tag{3.27}$$

As they should be for small scatterers, these are generally quite small, and the larger J, the smaller they are. There are, however, special values of x for which a denominator in (3.27) vanishes and therefore the corresponding phase shift goes through $\tfrac{1}{2}\pi$. This causes that specific partial amplitude (term in the amplitude expansion) to be *maximal* in magnitude, while all others are small. The scattering cross section in the vicinity of the frequency at which a denominator in (3.27) vanishes consequently has a sharp peak. There is a *resonance*.

The resonance frequencies can be easily obtained from (3.27). The electric phase shift α_J goes through $\frac{1}{2}\pi$ (modulo π) when

$$kR = x = x_{eJ} \equiv \frac{\omega_{eJ} R}{c} \simeq \frac{z_J}{n} \tag{3.28}$$

if z_J is a root of $u_J(z)$,

$$u_J(z_J) = 0. \tag{3.29}$$

The amplitudes at this specific frequency are well approximated by

$$\begin{aligned} A_{rr}^C = A_{ll}^C &\simeq -\frac{1}{2ik}(2J+1)(\tau_J + \pi_J) \\ A_{rl}^C = A_{lr}^C &\simeq -\frac{1}{2ik}(2J+1)(\tau_J - \pi_J). \end{aligned} \tag{3.30}$$

In order to obtain the frequency at which the magnetic phase shift goes through $\frac{1}{2}\pi$, the recurrence relations (2.69) ought to be used. One then finds that it happens when

$$x = x_{mJ} \equiv \frac{\omega_{mJ} R}{c} \simeq \frac{z'_{J-1}}{n}. \tag{3.31}$$

When the magnetic permeability is unity, the number z'_{J-1} equals z_{J-1}; otherwise it is determined by the equation

$$u_{J-1}(z'_{J-1}) + \frac{(\mu - 1) J u_J(z'_{J-1})}{z'_{J-1}} = 0. \tag{3.32}$$

The amplitudes at such a magnetic resonance are approximately (3.30), except that $A_{rl}{}^C$ and $A_{lr}{}^C$ have the opposite sign from that in (3.30). The first few solutions of (3.29) are listed in Table 3.1.

Each frequency at which $\tan \alpha_J$ or $\tan \beta_J = \infty$ is closely followed or preceded by a frequency at which $\tan \alpha_J$ or $\tan \beta_J = 0$. At such a point the phase shift is an integral multiple of π, and the corresponding partial cross section *vanishes*. But since the other partial amplitudes do not, this usually does not produce an easily observable phenomenon.

Table 3.1 Zeros of spherical Bessel functions, $u_J(z_J{}^{(n)}) = 0$‡

n	$J = 1$	$J = 2$
1	4.4934	5.7635
2	7.7253	9.0950
3	10.9041	12.3229
4	14.0662	15.5146

‡From Jahnke, Emde, and Lösch, "Tables of Higher Functions," 6th ed., p. 198, McGraw-Hill Book Company, New York. 1960.

The dependence of a partial amplitude on the frequency near a resonance is seen most clearly by writing there

$$\frac{1}{2i}(e^{2i\alpha_J} - 1) = \frac{\tan\alpha_J}{1 - i\tan\alpha_J} \simeq \frac{\frac{1}{2}\Gamma_{eJ}}{\omega - \omega_{eJ} + i\frac{1}{2}\Gamma_{eJ}}$$
$$\frac{1}{2i}(e^{2i\beta_J} - 1) \simeq \frac{\frac{1}{2}\Gamma_{mJ}}{\omega - \omega_{mJ} + i\frac{1}{2}\Gamma_{mJ}} \tag{3.33}$$

with

$$\Gamma_{eJ} = \frac{2x_{eJ}^{2J+2}}{J^2[(2J-1)!!]^2}\frac{c\mu}{Rn^2}$$
$$\Gamma_{mJ} = \frac{2x_{mJ}^{2J}}{[(2J-1)!!]^2}\frac{c\mu/R}{(nx_{mJ})^2 + (\mu - 1)J + (\mu^2 - 1)J^2}. \tag{3.34}$$

These equations show that the partial amplitudes have poles in the complex x plane at

$$\omega = \omega_{eJ} - i\tfrac{1}{2}\Gamma_{eJ} \quad \text{and} \quad \omega = \omega_{mJ} - i\tfrac{1}{2}\Gamma_{mJ}.$$

For real n, they are very close to the real axis. As k passes k_{eJ} or k_{mJ}, the corresponding phase shift quickly changes by π, thus going in rapid succession through both $\frac{1}{2}\pi$ and π (modulo π). This is how both a peak and a zero are produced in the partial amplitude. The partial cross section near the resonance for real n being proportional to

$$\frac{\frac{1}{4}\Gamma^2}{(\omega - \omega_0)^2 + \frac{1}{4}\Gamma^2}$$

the significance of Γ is evidently that of the *width* of the resonance line. At $\omega = \omega_0 \pm \frac{1}{2}\Gamma$ the curve has fallen to half its value at the peak. The typical resonance-line shape is illustrated in Fig. 3.2.

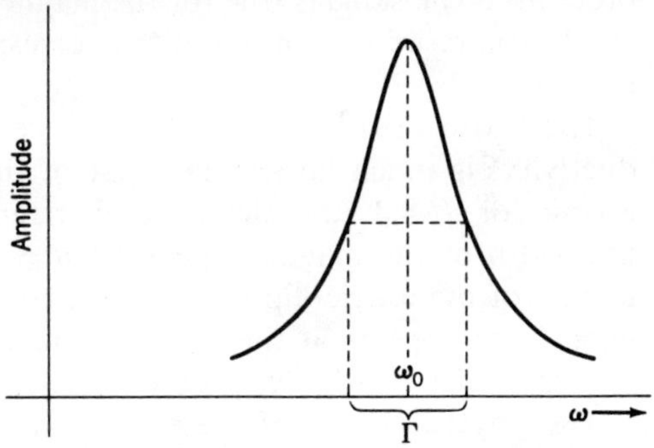

Figure 3.2. The shape of a resonance line (*Lorentz shape*).

If n is complex, resonances are still possible but then there is an additional contribution to the width coming from the conductivity. Since the zeros z_J of the Bessel functions are real, x_{eJ} and x_{mJ} have imaginary parts equal to

$$\operatorname{Im} x_{eJ} \simeq -\frac{z_J}{|n|^2}\operatorname{Im} n$$

and similarly for x_{mJ}. These are not necessarily small (compared with the real parts), and the resonances are not necessarily sharp. If the conductivity is small, i.e.,

$$\frac{\sigma}{\varepsilon|\omega|} \ll 1$$

then we find that

$$\operatorname{Im} n \simeq \frac{2\pi\sigma}{\sqrt{\varepsilon}c|k|}$$

and the contribution to the width is

$$\Delta\Gamma \simeq \frac{4\pi\sigma}{\varepsilon}. \tag{3.35}$$

This may or may not dominate over the contributions (3.34), depending on the conductivity, but in any event it has a quite different J dependence. If the conductivity is large, then the width of the ensuing resonances is usually not small compared with their position. In fact, in this case the poles appear essentially on the imaginary axis, at

$$\omega \simeq -i\frac{c^2}{4\pi\sigma R^2}z^2, \tag{3.36}$$

where z is z_J for the electric case and z'_{J-1} for the magnetic.

In the absence of magnetic permeability, it follows from (3.28) and (3.31) that the "electric" peaks of successive J values interlace one another, i.e., between two successive peaks of the same J there lies always exactly one of $J + 1$. The distance Δx between successive resonances of equal J is of order n^{-1}. The same is true for the magnetic ones. But the distance between an electric resonance of J and the corresponding magnetic one of $J + 1$ is of order n^{-3}.

The width of the magnetic resonance at ω_{mJ+1} (in the absence of conductivity) is much larger than that of the electric resonance at ω_{eJ}, and, except for $J = 1$, the electric and magnetic resonances both fall within the width of the magnetic peak (if $\mu = 1$). Thus they are separable only in case of the elctric-dipole and the magnetic-quadrupole peaks. The resonance that appears at the lowest frequency is a magnetic-dipole one. The next is an electric dipole, closely followed by a magnetic quadrupole, etc.

The physical cause of these resonances is best understood by realizing the significance of a solution of the Maxwell equations at a frequency for which S^J has a pole. If this were to occur at a real value of the frequency, the electromagnetic field would consist of an outgoing spherical wave only and the scatterer would be capable of radiating energy continually without receiving any. Such a solution, of course, cannot exist. Its frequency must be complex, with a small *negative* imaginary part, so that the electromagnetic energy, if originally confined to a finite region, spreads out slowly and the energy density everywhere decreases, whether there is absorption or not.

As they should, the Γ of (3.34) are always positive, and so is the contribution to Γ from the complex index of refraction. The electromagnetic energy in a given region being a quadratic function of the field, it decreases with the characteristic exponential time dependence $\exp(-\Gamma t)$. The time interval τ within which it falls off by a factor of e is therefore $\tau = 1/\Gamma$. If the index of refraction is real, then this decrease is due solely to the fact that energy has been radiated away. If it is complex, some of it has been absorbed.

The scattering process in the presence of a resonance can be pictured in terms of wave packets, if the latter are made up primarily of frequencies in the resonance region. The packet passes the scatterer with essentially the speed of light, and some of it is scattered into the usual spherical wave that spreads out immediately. An appreciable fraction of the energy, however, is retained in the scatterer and its neighborhood, and it leaks out much more slowly. The scatterer will then radiate energy long after the exciting wave has passed, and it will continue to do so with a characteristic half-life $\tau = 1/\Gamma$. The scattering cross section counts *all* the scattered radiation, delayed or not, and it is enhanced by the temporary capture of radiation and its subsequent release. Both the peak in the cross section and the time delay are characteristic of the resonance process. The more detailed mathematical description of the delay can be found in Chap. 19.

3.3.2 Totally Reflecting Spheres. At nonresonant frequencies we may go to the limit $n \to \infty$ and disregard the denominators in (3.27). This is the limit of the *perfect conductor* ($\sigma \to \infty$), or *total reflection.* Because $x \ll 1$, only the $J = 1$ term in the amplitude matters and one obtains

$$A^C_{rr} \simeq A^C_{ll} \simeq \tfrac{1}{2}x^2 R\cos^2 \tfrac{1}{2}\theta \qquad A^C_{rl} \simeq A^C_{lr} \simeq -\tfrac{3}{2}x^2 R\sin^2 \tfrac{1}{2}\theta \tag{3.37}$$

$$A^L_{\parallel} \simeq x^2 R(\cos\theta - \tfrac{1}{2}) \qquad A^L_{\perp} \simeq x^2 R(1 - \tfrac{1}{2}\cos\theta). \tag{3.38}$$

The total cross section is now all due to scatering, and for unpolarized incident radiation it equals

$$\sigma_{\text{total}} = \tfrac{10}{3}x^4(\pi R^2). \tag{3.39}$$

The polarization of an initially unpolarized beam is given by

$$P = \frac{3\sin^2\theta}{5\cos^2\theta + 5 - 8\cos\theta}. \tag{3.40}$$

It should be noted that these results differ from those obtained from Rayleigh scattering by simply letting $n \to \infty$. The assumption $x|n - 1| \ll 1$ is essential for the Rayleigh case, and one cannot let $n \to \infty$ there. The physical reason for the difference is that Rayleigh scattering is due entirely to an *electric* dipole field. But as $n \to \infty$, the skin depth in the particle becomes much less than its radius and the resulting surface current gives rise to a *magnetic dipole* field. The sum of these two results in (3.37) and (3.38). Nevertheless, the order of magnitude of the cross section is the same as in the Rayleigh case, and the polarization again vanishes in the forward and backward directions. The main difference from the Rayleigh case for $\mu = 1$ is the forward-backward asymmetry. The polarization at right angles is no longer 1, but $\frac{3}{5}$, the maximal polarization $P = 1$ being obtained at 60°.

Mathematically speaking, the limit $n \to \infty$ does not exist unless n is complex. This is because when x is held fixed the denominators in (3.27) keep on going through zeros as n increases. When n is complex, however, and it approaches infinity by virtue of $\sigma \to \infty$, then this is not the case and the limit is the one indicated.

3.4 Large Diffuse Spheres (van de Hulst Scattering)

When the scatterer is large relative to the wavelength, we are in the region near geometrical optics. But before going into this general area of transition, we first want to look at the special physical approximations possible when $|n - 1| \ll 1$ at the same time. Since for short wavelengths (or large obstacles) the scattering is mostly in the forward direction, we are mainly concerned with small angles. The idea is to improve somewhat on Fraunhofer diffraction. The limiting procedure that starts from the exact Mie solution being treated in the next section, we now want to look at the approximation method introduced by Van de Hulst.

3.4.1 Forward Scattering. We first consider forward scattering. The path length of the light ray that passes through the scattering sphere can be written

$$l = 2R\cos\gamma,$$

γ being the angle indicated in Fig. 3.3. The phase of the field in this pencil of radiation is therefore shifted from $\exp(ikz)$ to

$$\exp[ikz + ik(n - 1)2R\cos\gamma].$$

At a radial distance $R\sin\gamma$ from the center line, in the geometrical shadow area, the beam in the forward direction is thus the incident beam plus

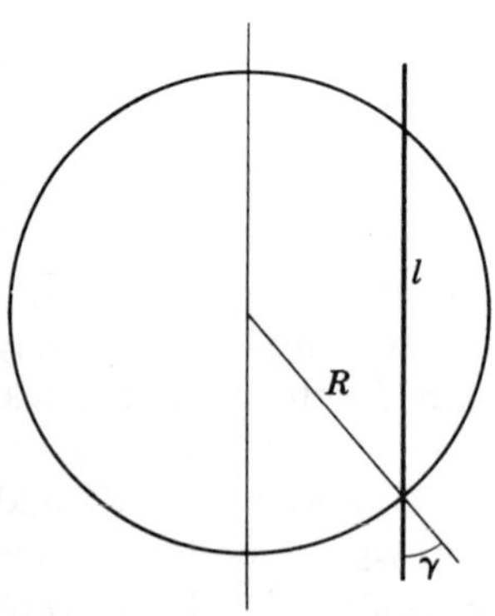

Figure 3.3. The angle γ.

$$(e^{2ix(n-1)\cos\gamma} - 1)\mathscr{E}_0 e^{ikz}.$$

At a large distance (compared with R) one observes a superposition of all the "rays" that passed through the scatterer in the (almost) forward direction. The observed field therefore differs from the incident by

$$\int dS(e^{2ix(n-1)\cos\gamma} - 1)\mathscr{E}_0 e^{ikz},$$

the integral extending over the cross section of the scatterer. This should equal

$$\mathscr{A}(0) \cdot \mathscr{E}_0 e^{ikr} r^{-1} \simeq \mathscr{A}(0) \cdot \mathscr{E}_0 z^{-1} e^{ikz + ik(x^2+y^2)/2z}.$$

If the receiving screen is large compared with $(\lambda r)^{1/2}$, the same approximation as in Sec. 1.3.9 can be made and the total intensity observed on it is that obtained from the incoming beam, plus

$$\mathscr{A}(0) \cdot \mathscr{E}_0 e^{ikz} \frac{2\pi i}{k}.$$

We conclude that for the purpose of collecting the total radiation in the forward direction on a large screen, and hence for that of calculating the total cross section, we must have

$$\mathscr{A}(0) \simeq \frac{k}{2\pi i} \int dS\,(e^{2ix(n-1)\cos\gamma} - 1) \tag{3.41}$$

integrated over the cross-sectional area of the scatterer. According to the optical theorem the total cross section is therefore

$$\sigma_{\text{total}} \simeq 2\,\text{Re} \int dS\,(1 - e^{2ix(n-1)\cos\gamma}). \tag{3.42}$$

For a circular geometrical cross section this integral can be carried out by setting $\cos\theta = u$,

$$\mathscr{A}(0) = ixRK[-2ix(n-1)], \tag{3.43}$$

where

$$K(y) = \int_0^1 du\,u(1 - e^{-yu}) = \tfrac{1}{2} + y^{-1}e^{-y} + y^{-2}(e^{-y} - 1)$$

and

$$\sigma_{\text{total}} \simeq 4\,\text{Re}\,K[-2ix(n-1)]\pi R^2. \tag{3.44}$$

For real n this becomes

$$\sigma_{\text{total}} \simeq 2(1 - \rho^{-1}\sin 2\rho + \rho^{-2}\sin^2\rho)\pi R^2 \tag{3.45}$$

with

$$\rho = x(n-1). \tag{3.46}$$

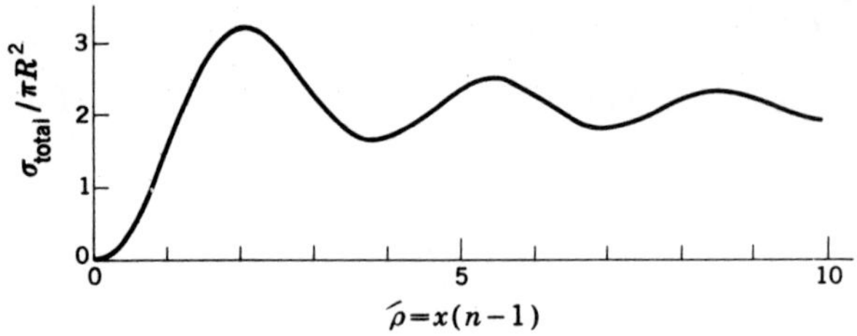

Figure 3.4. The total cross section according to Eq. (3.45). [*From H. C. Van de Hulst* (1957), *p.* 177.]

The functional dependence of σ on ρ is shown in Fig. 3.4. The physical cause of the maxima and minima is interference between refracted and diffracted waves.

When ρ is small, $\rho \ll 1$, then (3.45) reduces to the large x limit of the Rayleigh-Gans theory [(3.23)] for $\mu = 1$. When $\rho \gg 1$, on the other hand, then

$$\sigma_{\text{total}} \simeq 2\pi R^2. \tag{3.47}$$

A large sphere thus blocks out *twice* its geometrical cross-sectional area. This is known as the *extinction paradox*. Remember, though, that in order to observe this one must be beyond the region of sharp shadow.

The frequency independence of the limiting result (3.47) should be noted. In a physically realistic case it tells us, for example, that, for penetrating a real fog of large droplets, infrared light is no better than visible light.

When n is complex, the oscillations in Fig. 3.4 are smoothed out. By an argument similar to that leading to (3.41) one finds for the absorption

$$\sigma_{\text{abs}} \simeq \int dS\,(1 - e^{-4x\,\text{Im}\,n\cos\gamma}) \tag{3.48}$$

integrated over the geometrical cross section of the scatterer. For a circular cross section this yields

$$\sigma_{\text{abs}} \simeq 2K(4x\,\text{Im}\,n)\pi R^2 \tag{3.49}$$

which for small $x\,\text{Im}\,n$ reduces to

$$\sigma_{\text{abs}} \simeq \tfrac{8}{3}x\,\text{Im}\,n\,\pi R^2 \tag{3.50}$$

since

$$K(y) = \tfrac{1}{3}y - \tfrac{1}{8}y^2 + \cdots.$$

The total cross section is then all due to absorption, and the limit of (3.44) for $x|n-1| \ll 1$ agrees with (3.50). In order to calculate the total scattering cross section, the next term in the expansion of K must be kept. If the absorption cross section is subtracted from the total, the result

$$\sigma_{\text{scatt}} \simeq 2x^2|n-1|^2\pi R^2 \tag{3.51}$$

again agrees with the Rayleigh–Gans limit.

For $x\,\text{Im}\,n \gg 1$ one obtains simply

$$\sigma_{\text{abs}} \simeq \pi R^2, \tag{3.52}$$

and therefore (when $x|n - 1| \gg 1$)

$$\sigma_{\text{scatt}} = \sigma_{\text{scatt}} - \sigma_{\text{abs}} \simeq \pi R^2 \tag{3.53}$$

too. This means that then

$$\text{Albedo} \equiv \frac{\sigma_{\text{scatt}}}{\sigma_{\text{total}}} = \frac{1}{2}. \tag{3.54}$$

The fact that the absorption cross section equals the geometrical means that such a scatterer, with $x|n - 1| \gg 1$, $x \operatorname{Im} n \gg 1$, $|n - 1| \ll 1$, constitutes a *blackbody*. We need $x \gg 1$ in order to be able to talk about geometrically incident radiation; $|n - 1| \ll 1$, so that there is no reflection, and $x \operatorname{Im} n \gg 1$, so that all radiation is absorbed. It is thus the only real blackbody.

3.4.2 Small-Angle Scattering. The small-angle scattering under conditions $x \gg 1$ and $|n - 1| \ll 1$ is pictured as similar to that in the forward direction, i.e., as made up of scatterings by infinitesimal disks. The only distinction from (3.41) is that there is an additional phase factor $e^{i\delta}$ as in the Rayleigh-Gans case, coming from the difference in optical path length of the rays scattered by disks at various positions. This phase difference is found from Fig. 3.5, for small angles θ,

$$\delta = -k\theta R \sin\gamma \cos\varphi$$

since $|n - 1| \gg 1$. Therefore

$$\mathscr{A}(\theta) \simeq \frac{k}{2\pi i} \int dS (1 - e^{2ix(n-1)\cos\gamma}) e^{-ix\theta \sin\gamma\cos\varphi} \tag{3.55}$$

integrated over the cross section of the sphere. The φ integral is Bessel's integral

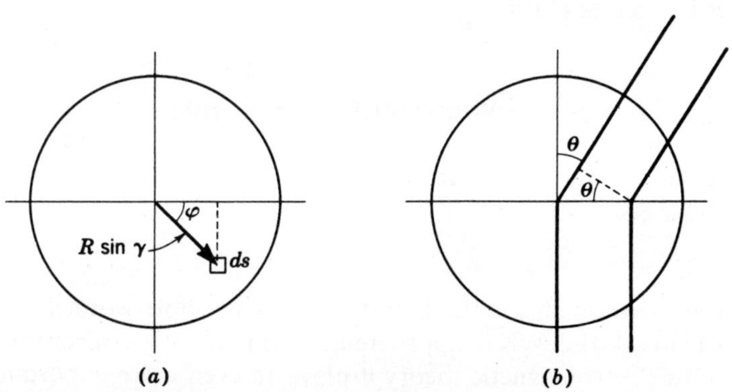

Figure 3.5. (*a*) Looking in the direction of the beam; (*b*) looking down on the scattering plane.

$$J_0(z) = \frac{1}{2\pi}\int_0^{2\pi} d\varphi\, e^{iz\cos\varphi}$$

and we obtain

$$\mathscr{A}(\theta) = ixRA[2x(n-1), x\theta] \tag{3.56}$$

where the function A is defined by

$$A(\rho, z) \equiv \int_0^{1/2\pi} d\gamma \sin\gamma\cos\gamma(1 - e^{i\rho\cos\gamma})J_0(z\sin\gamma). \tag{3.57}$$

The integrals involved in (3.57) are Sonine's. The imaginary part (for real n) can be integrated,

$$\operatorname{Im} A = \frac{\rho}{\rho^2 + z^2} u_1[(\rho^2 + z^2)^{1/2}].$$

The real part has to be expanded or integrated numerically. For details of such evaluations the reader is referred to van de Hulst (1957).

It should be noted that $x^{-1}\mathscr{A}(\theta)$ depends only on the combinations $\rho = (n-1)x$ and $x\theta$. The scattering curves for different values of n therefore look the same in shape and differ only in scale. The result (3.56) can be used for real or complex n and for tensor indices of refraction as well. In the latter case (3.57) has to be evaluated by expansion.

When $\rho \ll 1$, the scattering amplitude reduces to the small-angle limit of the Rayleigh–Gans result (3.20) (for $\mu = 1$). When $\rho \gg 1$, the leading term in (3.57) is the 1 term. That integral can be carried out, and we obtain, for $x \gg 1$, $|n - 1| \ll 1$, $x|n-1| \gg 1$,

$$\mathscr{A}(\theta) \simeq \frac{iRJ_1(x\theta)}{\theta}. \tag{3.58}$$

This is the Fraunhofer diffraction amplitude for small angles. It has the characteristic zeros and maxima of the Bessel function J_1. The more general result (3.56) differs from the Fraunhofer case, and it is called *anomalous diffraction* by van de Hulst.

3.5 Large Spheres (Geometrical-Optics Limit)

We now return to the exact Mie solution (2,127) and go the *geometrical-optics* limit,

$$x \gg 1.$$

For a given scatterer this is the high-frequency or short-wavelength limit, the equivalent of the WKB approximation in quantum mechanics (see Chap. 18). In electromagnetic theory it plays an even more important role, for here the analog of the potential in quantum mechanics is $k^2(n^2 - 1)$, and this tends to infinity with k^2. As a result the phase shifts α_J and β_J

do not tend to zero as $k \to \infty$, and the Born approximation is not available as an exact high-energy limit.

Mathematically speaking, the special methods and results in this region rest on the property of the Bessel functions to drop to a very small value when the order much exceeds the argument. That this is so can be seen quite easily in Bessel's integral,

$$J_\lambda(z) = \frac{1}{2\pi}\int_{-\pi}^{\pi} d\theta\, e^{i(\lambda\theta - z\sin\theta)}.$$

For large λ the integrand oscillates rapidly, and its average value is small, except where the phase is stationary. This means that for large λ most of the integral comes from

$$\cos\theta = \frac{\lambda}{z}$$

and if λ/z much exceeds unity, the integral is small. The larger the variable z, the sharper the drop when $\lambda > z$. On the other hand, when $z \ll \lambda$, then the first term in the expansion about $z = 0$ is a good approximation; for the Riccati-Bessel functions this is (2.64). This means that the Bessel function is of appreciable size only in the vicinity of $z \simeq \lambda$.

As a consequence of this property of the Bessel function, the tangents of the phase shifts α_J and β_J [(2.127)] are negligible when $J \gg x$, as well as when $x \gg J$, and they drop the more sharply as functions of J the larger z becomes. When $x \gg 1$, they are negligible as soon as $J + \frac{1}{2} > x$.

Physically this cutoff for $J + \frac{1}{2} > x$ is understood in terms of the *optical-localization principle*, which asserts that, vaguely, the term of order J in the partial-wave expansion corresponds to a "ray" of distance $(J + \frac{1}{2})/k$ from the center. It is the analog of the well-known quantum-mechanical fact that the partial wave of angular momentum J "corresponds" to a particle of impact parameter $(J + \frac{1}{2})/k$, and the correspondence is better for larger k. On this basis one expects to be able to neglect those values of J for which $(J + \frac{1}{2})/k > R$, for the corresponding rays never meet the sphere of radius R. This is exactly the cutoff obtained from the mathematical discussion.

When $x \gg 1$, to summarize, we may truncate the sums in (2.131) and (2.132) at $J + \frac{1}{2} \simeq x$. In the remaining finite sum we treat first the 1 terms in (2.132), that is, those independent of the nature of the scatterer, and then, separately, the other terms containing $e^{2i\alpha_J}$ and $e^{2i\beta_J}$.

3.5.1 Fraunhofer Diffraction. The 1 term is sharply peaked in the forward direction. If it were not for the cutoff at $J + \frac{1}{2} \simeq x$, it would give rise to a δ-function contribution in the forward direction; hence, when $x \gg 1$, it must be negligible for all but small θ. The Legendre polynomials may therefore be replaced by their asymptotic values for large J,

$$P_J(\cos\theta) \simeq (\cos\theta)^J J_0[(J + \tfrac{1}{2})\sin\theta] \simeq (\cos\theta)^J J_0[(J + \tfrac{1}{2})\theta] \quad (3.59)$$

which are good provided that $\sin\theta \leq 1/J$. This, together with the recurrence relations, implies that the functions π_J and τ_J are approximately given by

$$\begin{aligned}\pi_J &\simeq \tfrac{1}{2}J_0(z) + \tfrac{1}{2}J_2(z)\\ \tau_J &\simeq \tfrac{1}{2}J_0(z) - \tfrac{1}{2}J_2(z)\end{aligned} \qquad z = (J + \tfrac{1}{2})\theta. \tag{3.60}$$

Consequently that part of the amplitude is well approximated by

$$A^{\mathrm{Fr}} \equiv A_{\parallel}^{\mathrm{Fr}} = A_{\perp}^{\mathrm{Fr}} \simeq -\frac{1}{2ik}\sum_{J=1}^{x}(2J+1)J_0[(J+\tfrac{1}{2})\theta].$$

For large x there are many terms in this sum, and they differ by little. It may therefore with little error be replaced by an integral

$$-2ikA^{\mathrm{Fr}} \simeq \frac{1}{2}\int_0^{2x} dy\, y J_0(\tfrac{1}{2}\theta y)$$

which can be carried out. The result is

$$A^{\mathrm{Fr}}(\theta) \simeq \frac{iRJ_1(x\theta)}{\theta}. \tag{3.61}$$

This is nothing but the small-angle expression for the Fraunhofer diffraction, seen already in (3.58). It is limited to the region $\theta \lesssim x^{-1}$, for only there is the approximation (3.59) valid.

3.5.2 Nonforward and Nonbackward Scattering; Real Index of Refraction. We must now treat the terms containing the phase shifts. For the more elaborate approximations necessary here it is convenient to introduce the parameter γ by

$$\sin\gamma \equiv (J + \tfrac{1}{2})x^{-1} \tag{3.62}$$

whose significance for a "localized" ray is shown in Fig. 3.3. It is the angle between the ray and the normal to the surface of the scattering sphere. Similarly, the angle γ' defined by

$$\sin\gamma' = (J + \tfrac{1}{2})(nx)^{-1} = n^{-1}\sin\gamma \tag{3.63}$$

is the refracted angle of incidence. The index of refraction n is here assumed to be real.

The asymptotic value of the Hankel function for large order is

$$H_{J+1/2}^{(1)}(x) \simeq \frac{1}{\sqrt{3}}\cot\gamma\, e^{i(\varphi+\pi/6)}H_{1/3}^{(1)}(Z) \tag{3.64}$$

with

$$\begin{aligned}\varphi &= x\cos\gamma - Z - (J + \tfrac{1}{2})(\tfrac{1}{2}\pi - \gamma)\\ Z &= \frac{1}{3}(J + \tfrac{1}{2})\cot^3\gamma = \frac{x^3\cos^3\gamma}{3(J + \tfrac{1}{2})^2}.\end{aligned}$$

When x is very large, then the edge effects due to rays of grazing incidence are relatively small. We shall treat them separately in Sec. 3.8. Here we shall not take into account values of γ close to 90°, and Z may be assumed large. ${H_{1/3}}^{(1)}$ may therefore be replaced by its asymptotic value for large variable, and

$$H_{J+1/2}^{(1)}(x) \simeq \left(\frac{1}{\pi x \cos\gamma}\right)^{1/2} e^{i(\Gamma x - 1/4\pi)} \tag{3.65}$$

$$\begin{aligned} w_J^{(+)}(x) &\simeq (-)^J(\cos\gamma)^{-1/2} e^{i(\Gamma x + 1/4\pi)} \\ v_J(x) &\simeq -(\cos\gamma)^{-1/2}\cos(\Gamma x + \tfrac{1}{4}\pi) \\ u_J(x) &\simeq (\cos\gamma)^{-1/2}\sin(\Gamma x + \tfrac{1}{4}\pi) \end{aligned} \tag{3.66}$$

with

$$\Gamma = \cos\gamma + (\gamma - \tfrac{1}{2}\pi)\sin\gamma. \tag{3.67}$$

Similarly for the functions of $nx = \overline{X}$.

Insertion of these values in (2.127) leads to

$$\begin{aligned} \tan\alpha_J &\simeq \frac{\rho_1\cos(\Gamma x + \Gamma'\overline{X}) - \sin(\Gamma x - \Gamma'\overline{X})}{\rho_1\sin(\Gamma x + \Gamma'\overline{X}) + \cos(\Gamma x - \Gamma'\overline{X})} \\ \tan\beta_J &\simeq \frac{\rho_2\cos(\Gamma x + \Gamma'\overline{X}) - \sin(\Gamma x - \Gamma'\overline{X})}{\rho_2\sin(\Gamma x + \Gamma'\overline{X}) + \cos(\Gamma x - \Gamma'\overline{X})} \end{aligned} \tag{3.68}$$

with

$$\Gamma' = \cos\gamma' + (\gamma' - \tfrac{1}{2}\pi)\sin\gamma' \tag{3.67a}$$

and ρ_1 and ρ_2 are the Fresnel reflection coefficients

$$\begin{aligned} \rho_1 &= \frac{n\cos\gamma - \mu\cos\gamma'}{n\cos\gamma + \mu\cos\gamma'} \\ \rho_2 &= \frac{\mu\cos\gamma - n\cos\gamma'}{\mu\cos\gamma + n\cos\gamma'}. \end{aligned} \tag{3.69}$$

Consequently the elements of the $\mathcal{S}$ matrix are

$$\begin{aligned} e^{2i\alpha_J} &\simeq e^{-2i(\Gamma x - \Gamma'\overline{X})}\frac{1 + i\rho_1 e^{-2i\Gamma'\overline{X}}}{1 - i\rho_1 e^{2i\Gamma'\overline{X}}} \\ e^{2i\beta_J} &\simeq e^{-2i(\Gamma x - \Gamma'\overline{X})}\frac{1 + i\rho_2 e^{-2i\Gamma'\overline{X}}}{1 - i\rho_2 e^{2i\Gamma'\overline{X}}}. \end{aligned} \tag{3.70}$$

The next step is to expand these expressions in series in powers of $\rho_1 e^{(2i\Gamma'\overline{X})}$ and $\rho_2 e^{(2i\Gamma'\overline{X})}$, respectively,

$$\begin{aligned} e^{2i\alpha_J} &= ie^{-2i\Gamma\overline{X}}\sum_{p=0}^{\infty}\varepsilon_p^{(1)}(-ie^{2i\Gamma'\overline{X}})^p \\ e^{2i\beta_J} &= ie^{-2i\Gamma\overline{X}}\sum_{p=0}^{\infty}\varepsilon_p^{(2)}(-ie^{2i\Gamma'\overline{X}})^p \end{aligned} \tag{3.71}$$

with the coefficients

$$\varepsilon_p^{(1,2)} = \begin{cases} \rho_{1,2} & p = 0 \\ (1 - \rho_{1,2}^2)(-\rho_{1,2})^{p-1} & p = 1, 2, 3, \cdots. \end{cases} \tag{3.72}$$

Furthermore, the Legendre polynomials are approximated by

$$P_J(\cos\theta) \simeq \left[\frac{2}{\pi(J + \frac{1}{2})\sin\theta}\right]^{1/2} \cos[(J + \tfrac{1}{2})\theta - \tfrac{1}{4}\pi] \tag{3.73}$$

an approximation which is good for $J \gg 1$ and for angles which are neither too small nor too large: $\sin\theta \gtrsim 1/J$. For very small angles the Fraunhofer diffraction, trated before, dominates; angles close to π will be discussed in Sec. 3.7. The replacement of P_J by (3.73) means that

$$\begin{aligned} \pi_J &\simeq \frac{1}{J+1}\left(\frac{2}{\pi J \sin^3\theta}\right)^{1/2} \sin\left[\left(J + \frac{1}{2}\right)\theta - \tfrac{1}{4}\pi\right] \\ \tau_J &\simeq \frac{1}{J+1}\left(\frac{2J}{\pi \sin\theta}\right)^{1/2} \cos\left[\left(J + \frac{1}{2}\right)\theta - \tfrac{1}{4}\pi\right] \\ &= \frac{1}{J+1}\left(\frac{J}{2\pi \sin\theta}\right)^{1/2} \sum_{q=-1,1} e^{iq[(J+1/2)\theta - 1/4\pi]}. \end{aligned} \tag{3.74}$$

Since small values of J contribute little to the sum, we have for all relevant values of J $|\tau_J| \gg |\pi_J|$, and we retain only τ_J in (2.132). As a consequence $A_{\|}{}^L$ contains only α_J and $A_{\perp}{}^L$ only β_J. Otherwise the series are identical, and we write simply A and ε_p. For $A_{\|}{}^L$, then, $\varepsilon_p{}^{(1)}$ has to be used, and for $A_{\perp}{}^L$, $\varepsilon_p{}^{(2)}$. Thus the amplitude is written

$$A(\theta) \simeq \frac{1}{k}\left(\frac{x}{2\pi\sin\theta}\right)^{1/2} \sum_{p=0}^{\infty} \sum_{J=f}^{x} \sum_{q=-1,1} \varepsilon_p (\sin\gamma)^{1/2} e^{i\xi} \tag{3.75}$$

$$\begin{aligned} \xi &= 2x(np\Gamma' - \Gamma) - \tfrac{1}{2}\pi p - \tfrac{1}{4}\pi q + q\theta x \sin\gamma + 2\pi m(x\sin\gamma - \tfrac{1}{2}) \\ &= 2x\{np\cos\gamma' - \cos\gamma + [\tfrac{1}{2}\pi(1 - p + 2m) - \gamma + p\gamma' + \tfrac{1}{2}q\theta]\sin\gamma\} \\ &\qquad - \tfrac{1}{2}\pi(p + \tfrac{1}{2}q + 2m). \end{aligned}$$

The last term in the first expression for ξ arises because an integral multiple of $2\pi J = 2\pi(x\sin\gamma - \frac{1}{2})$ may be added to ξ with impunity; m is meant to be an arbitrary integer.

At this point the summation over J is replaced by an integral, since there are many relatively slowly varying terms in it. As integration variable one conveniently adopts γ so that

$$1 \rightarrow dJ = x\, d\sin\gamma$$

and hence

$$A(\theta) \simeq \frac{1}{k}\left(\frac{x}{2\pi\sin\theta}\right)^{1/2} \sum_{p=0}^{\infty} \sum_{q=-1,1} x \int_0^1 d\sin\gamma (\sin\gamma)^{1/2} \varepsilon_p e^{i\xi}. \tag{3.76}$$

The integral is now evaluated by the method of stationary phase. The larger x, the more rapidly the integrand oscillates, except where the factor of x in ξ is stationary as a function of γ. This means that for $x \gg 1$ most of the contributions to the integral come from that value γ_0 of of γ where

$$\frac{d\xi}{d\gamma} = 2x[\tfrac{1}{2}\pi(1 - p + 2m) - \gamma + p\gamma' + \tfrac{1}{2}q\theta]\cos\gamma = 0,$$

i.e.,

$$\Theta + q\theta - 2\pi m = 0 \tag{3.77}$$

with

$$\Theta \equiv 2p(\gamma_0' - \tfrac{1}{2}\pi) + 2(\tfrac{1}{2}\pi - \gamma_0). \tag{3.78}$$

The value of ξ at this point is

$$\xi_0 = 2x(np\cos\gamma_0' - \cos\gamma_0) - \tfrac{1}{2}\pi(p + \tfrac{1}{2}q) -- \pi m.$$

In its vicinity

$$\xi = \xi_0 + a(J - J_0)^2 + \cdots$$

with γ_0 and J_0 connected by (3.62), and

$$a = \frac{1}{2}\frac{\partial^2\xi}{\partial J^2}\bigg|_{J=J_0} = \frac{1}{2x\cos\gamma_0}\frac{d\Theta}{d\gamma_0}. \tag{3.79}$$

We then have

$$A(\theta) \simeq \frac{1}{k}\left(\frac{x\sin\gamma_0}{2\pi\sin\theta}\right)^{1/2} \sum_{p,q} \varepsilon_p e^{i\xi_0} \int dJ e^{ia(J-J_0)^2}.$$

For large x this integrand oscillates extremely rapidly at the end points $J = 0$ and $J = x$ of the integral. It may therefore be extended from $-\infty$ to $+\infty$ with very little error and then evaluated explicitly,

$$\int_{-\infty}^{\infty} dJ\, e^{ia(J-J_0)^2} = \left(\frac{\pi}{|a|}\right)^{1/2} e^{i1/4\pi \operatorname{sgn} a}.$$

As a result we get (writing simply γ for γ_0)

$$A(\theta) \simeq \sum_{p,q} R\varepsilon_p\left(\frac{\sin 2\gamma}{2\sin\theta|d\Theta/d\gamma|}\right)^{1/2} e^{i\pi(1/4s - 1/2p - 1/4q - m) + i\delta} \tag{3.80}$$

where

$$s \simeq \operatorname{sgn}\frac{d\Theta}{d\gamma}$$
$$\delta \simeq 2x(pn\cos\gamma' - \cos\gamma). \tag{3.81}$$

The result (3.80) is essentially that of geometrical optics. In that region each incident ray is partially refracted and infinitely many times internally reflected. At each reflection a part of it emerges from the sphere. If p counts the number of internal reflections of a ray which initially makes the angle γ with the surface normal, then ε_p gives the appropriate factor, two factors of $(1 - \rho^2)^{1/2}$ being due to the two refractions the ray suffers before emerging, and each factor of $-\rho$ from an internal reflection. The deflection angle after no internal reflections is $\pi - 2\gamma$; after one, $2(\gamma - \gamma')$; etc. The angle Θ defined by (3.78) is therefore just this deflection angle. It can have any value. The angle θ is related to Θ by (3.77) or, since $q = \pm 1$, by

$$\theta = -q\Theta + 2\pi m q.$$

The integer m must be chosen so that θ lies between 0 and π. If there is a uniform incident beam, the outgoing intensity in the solid angle element $d\Omega$ is, then, for $p - 1$ internal reflections,

$$I_{\text{inc}} \frac{d\sigma}{d\Omega} \sin\theta \, d\theta \, d\varphi = I_{\text{inc}} \varepsilon_p^2 (R \sin\gamma \, d\varphi)(R \, d \sin\gamma)$$

and therefore the contribution to the differential cross section from $p - 1$ internal reflections is

$$\left.\frac{d\sigma}{d\Omega}\right|_p = \varepsilon_p^2 R^2 \frac{\sin 2\gamma}{2 \sin\theta |d\Theta/d\gamma|}$$

with ε_p chosen $\varepsilon_p^{(1)}$ or $\varepsilon_p^{(2)}$ according to the direction of polarization. This geometrical-optics result is identical with the cross-section contribution from a single p calculated by means of (3.80).

It must be remembered, however, that the rays of various numbers of internal reflections are coherent. One has to add *amplitudes* and not cross sections. In contrast to a ray going straight through the vacuum path of length $2R\cos\gamma$, a ray reflected $p - 1$ times has an optical path length $2Rpn\cos\gamma'$. Hence there is a phase shift δ given by (3.81). The remaining phase shift in (3.80) is due to the passage of focal lines. The entire expression (3.80) is thus accounted for on the basis of ray optics, together with a simple phase-shift consideration. Its derivation is consequently the demonstration of the transition from physical to geometrical optics, the exact analog of the demonstration of the transition from quantum mechanics to classical mechanics via the WKB approximation. The amplitude (3.80) is a good approximation when the radius of the scatterer is about three times the wavelength, or greater, i.e., when $x \lesssim 15$, except in special cases to be treated separately in Secs. 3.6, 3.7, and 3.8.

3.5.3 Large Diffuse Spheres. We may now consider (3.80) in the further specialized case in which $|n - 1| \ll 1$ and $|\mu - 1| \ll 1$. The reflection coefficients (3.69) are then small, and according to (3.72) all ε_p are negligible, except for $p = 1$,

$$\varepsilon_1^{(1)} \simeq \varepsilon_1^{(2)} \simeq 1.$$

Furthermore

$$\gamma' \simeq \gamma - (n - 1)\tan\gamma$$

$$\Theta = 2(\gamma' - \gamma) = 2(n - 1)\tan\gamma$$

$$\frac{d\Theta}{d\gamma} = -\frac{2(n - 1)}{\cos^2\gamma}$$

and $\theta = -\Theta$. Hence $q = +1$, $m = 0$, $s = -1$ (except when the scatterer is more diffuse than the surrounding medium; then $s = +1$), and

$$\delta \simeq \frac{2x(n-1)}{\cos\gamma}.$$

As a result

$$A_{\parallel}^{L}(\theta) \simeq A_{\perp}^{L}(\theta) \simeq -R\frac{\cos^2\gamma}{2(n-1)}e^{2ix(n-1)\sec\gamma}$$

$$= -R\frac{2(n-1)}{\theta^2 + 4(n-1)^2}e^{ix[\theta^2+4(n-1)^2]1/2}$$

and

$$\frac{d\sigma}{d\Omega} \simeq R^2\frac{4(n-1)^2}{[\theta^2 + 4(n-1)^2]^2}. \tag{3.82}$$

It must be remembered that, for (3.80) to hold, the angle θ must not be too small. Equation (3.82) therefore cannot be used for the forward direction. As far as it holds, it is peaked forward and, as expected, much smaller than the forward result (3.61).

3.5.4 Large Dense Spheres. When $n \gg 1$, then $\rho_1 \simeq 1$, $\rho_2 \simeq -1$, all ε_p are negligible except $\varepsilon_0{}^{(1)} = 1$, $\varepsilon_0{}^{(2)} = -1$, and

$$\delta \simeq -2x\cos\gamma \qquad \Theta \simeq \pi - 2\gamma \qquad \frac{d\Theta}{d\gamma} = -2$$

$$\theta = \Theta \qquad q = -1 \qquad m = 0 \qquad s = -1.$$

As a result

$$A_{\parallel}^{L} \simeq A_{\perp}^{L} \simeq \tfrac{1}{2}Re^{-2ix\sin 1/2\theta}$$

$$\frac{d\sigma}{d\Omega} \simeq \tfrac{1}{4}R^2. \tag{3.83}$$

A large, totally reflecting sphere scatters isotropically, except for the diffraction in the forward direction.

3.5.5 Complex Index of Refraction. If n is complex, then everything in Sec. 3.5.2 applies as before until we come to the evaluation of the integral in (3.76). This must now be done by saddle-point integration (see the Appendix to this chapter). The contour leading from $\sin\gamma = 0$ to $\sin\gamma = 1$ is distorted so as to lead through the complex value of $\sin\gamma_0$ where (3.77) holds, with the real Θ given by (3.78). This means by (3.63) that

$$\operatorname{Im}\gamma_0 = p\operatorname{Im}\gamma_0' = p\operatorname{Im}\sin^{-1}(n^{-1}\sin\gamma_0)$$

and, for $n_2 \equiv \operatorname{Im} n \ll \operatorname{Re} n \equiv n_1$,

$$\operatorname{Im}\gamma_0 \simeq \frac{pn_2\sin\gamma_0}{n_1(p\cos\gamma_0 - n_1\cos\gamma_0')}.$$

γ_0 on the right-hand side is the real value obtained from (3.77) by taking $n_2 = 0$. Then

$$\operatorname{Im}\delta = \operatorname{Im}\xi_0 \simeq 2xn_2 p\cos\gamma_0' \tag{3.84}$$

and there is an additional phase factor $e^{i\varphi}$ from the imaginary part of the square root in (3.80),

$$\varphi = \frac{n_2}{n_1}\frac{2 + d\Theta/d\gamma}{|d\Theta/d\gamma|^2}\left[2(n-1) + n\frac{d\Theta}{d\gamma} + \frac{n^2(n^2-1)\sin^2\gamma}{p^2\cos^4\gamma}\left(1 + \frac{1}{2}\frac{d\Theta}{d\gamma}\right)^2\right]. \tag{3.85}$$

For small n_2, the magnitude of $A(\theta)$ is reduced by the factor $1 - \operatorname{Im}\delta$, and its phase is changed by φ.

The more interesting modification when n is complex is that due to the *grazing rays*. It will be treated in Sec. 3.8.

3.6 The Rainbow

For ages the rainbow has enraptured the feelings of human beings:

First the flaming Red
Sprung vivid forth; the tawny Orange next;
And next delicious Yellow; by whose side
Fell the kind beams of all-refreshing Green;
Then the pure Blue, that swells autumnal skies,
Ethereal played; and then, of sadder hue
Emerged the deepened Indigo, as when
The heavy-skirted evening droops with frost;
While the last gleamings of refracted light
Died in the fainting Violet away.
These, when the clouds distil the rosy shower,
Shine out distinct adown the watery bow;
While o'er our heads the dewy vision bends
Delightful melting on the fields beneath.
Myriads of mingling dyes from these result,
And myriads still remain—infinite source
Of beauty, ever blushing, ever new! . . .

And befuddled their brains:[1]

Even the rainbow has a body
Made of drizzling rain
And is an architecture of glistening atoms
Built up, built up
Yet you can't lay your hand on it,
Nay, nor even your mind.

1 Acknowledgment is made to the Viking Press, Inc., for permission to reprint "The Rainbow" by D. H. Lawrence from *The Complete Poems of D. H. Lawrence*, Copyright 1933 by Frieda Lawrence, All Rights Reserved.

It is a thing of beauty but it is also well understood.

When the deflection angle is stationary as a function of the surface angle γ of incidence on a sphere, the scattered radiation must be particularly intense. For sunlight scattered by water droplets large compared with the wavelength of visible light, the resulting phenomenon is the rainbow. The scattering angle at which

$$\frac{d\Theta}{d\gamma} = 0 \tag{3.86}$$

is therefore referred to as the *rainbow angle*.

A look at the limiting result (3.80) shows that for $x \gg 1$ the amplitude is indeed infinite. Using (3.78) as well as Snell's law, (3.63), we find that

$$\cos\gamma_R = \left(\frac{n^2 - 1}{p^2 - 1}\right)^{1/2} \tag{3.87}$$

is the cosine of the surface angle of incidence which produces the rainbow for $p - 1$ internal reflection.

In order to find the amplitude near the rainbow angle, we must return to the derivation of (3.80). Instead of the expansion of ξ used in the evaluation of (3.76), we now have

$$\xi = \xi_0 + (J - J_R)\frac{\partial\xi}{\partial J} + \frac{1}{6}(J - J_R)^3\frac{\partial^3\xi}{\partial J^3} + \cdots$$

the derivatives on the right being evaluated at $J = J_R$, the value of J that corresponds to γ_R according to (3.62). The second derivative term is now missing because (3.86) means that $\partial^2\xi/\partial J^2 = 0$ at $J = J_R$. The gaussian integral which leads to (3.80) must therefore be replaced by

$$\int dJ \exp\left[ib(J - J_R) + \tfrac{1}{3}ic(J - J_R)^3\right]$$

in which the constants b and c are easily calculated to be

$$b = \Theta_R + q\theta - 2\pi m$$

$$c = -x^{-2}\frac{(p^2 - 1)^2}{p^2}\frac{(p^2 - n^2)^{1/2}}{(n^2 - 1)^{3/2}}.$$

If the integral is extended from $-\infty$ to $+\infty$, it becomes

$$2\pi c^{-1/3}\,\mathrm{Ai}\,(bc^{-1/3}),$$

$\mathrm{Ai}\,(z)$ being Airy's rainbow integral,

$$\mathrm{Ai}\,(z) = \frac{1}{2\pi}\int_{-\infty}^{\infty} du\, e^{i(zu + 1/3u^3)} = \frac{1}{\pi}\int_0^{\infty} du \cos\left(zu + \tfrac{1}{3}u^3\right). \tag{3.88}$$

The justification for extending the limits of the integral to $\pm\infty$ is that, for large x, J_R is large and the value of the integral comes mainly from the vicinity of $J = J_R$ in any case. The amplitude now becomes

$$A_p(\theta) \simeq -R\varepsilon_p \left(\frac{2\pi}{\sin\theta_R}\right)^{1/2} x^{1/6} \frac{p^{2/3}}{(p^2-1)^{11/12}} (p^2-n^2)^{1/12}(n^2-1)^{1/2} \times e^{-i\pi(1/2p+1/4q+m)+i\delta}\,\mathrm{Ai}\,(z) \tag{3.89}$$

where

$$\delta = 2x[(p^2-1)(n^2-1)]^{1/2}$$
$$z = q(\theta_R-\theta)x^{2/3}\frac{p^{2/3}}{(p^2-1)^{2/3}}\frac{(n^2-1)^{1/2}}{(p^2-n^2)^{1/6}}. \tag{3.90}$$

The rainbow angle θ_R is determined by

$$\theta_R = (2\pi m - \Theta_R)q,$$

q and m being chosen so that $0 \le \theta_R \le \pi$ and Θ_R being calculated by inserting (3.87) in (3.78). The coefficients ε_p are calculated by (3.72) from the reflection coefficients

$$\rho_1 = \frac{n^2-p}{n^2+p}$$
$$\rho_2 = \frac{1-p}{1+p} \tag{3.91}$$

(for $\mu = 1$), which yield

$$\varepsilon_p^{(1)} = \frac{4pn^2}{(p+n^2)^2}\left(\frac{p-n^2}{p+n^2}\right)^{p-1}$$
$$\varepsilon_p^{(2)} = \frac{4p}{(p+1)^2}\left(\frac{p-1}{p+1}\right)^{p-1} \tag{3.92}$$

for $p \ge 2$.

The Airy integral is the regular solution of the differential equation

$$\mathrm{Ai}\,(z)'' = z\,\mathrm{Ai}\,(z), \tag{3.93}$$

and it is tabulated in a number of tables. We plot it here in Fig. 3.6. It can also be expressed in terms of Bessel functions,

$$\mathrm{Ai}\,(z) = \begin{cases} \dfrac{1}{\pi}\left(\dfrac{z}{3}\right)^{1/2} K_{1/3}\left(\dfrac{2z^{3/2}}{3}\right) & z > 0 \\ -\frac{1}{3}(-z)^{1/2}\{J_{1/3}[\frac{2}{3}(-z)^{3/2}] + J_{-1/3}[\frac{2}{3}(-z)^{3/2}] & z < 0. \end{cases} \tag{3.94}$$

Equation (3.89) is easily seen to have a number of characteristic properties. The largest amplitude is obtained from the smallest possible p, which for $n > 1$ is $p = 2$. This is the main rainbow, due to the once internally reflected ray. A little geometrical consideration shows that for this

$$\theta = \pi - 4\gamma' + 2\gamma$$

and hence $m = 0$, $q = 1$. The angular distribution near $\theta = \theta_R$ is determined by the Airy integral (Fig. 3.6). When $\theta < \theta_R$, the cross section rapidly drops to zero. The main rainbow angle is approached from the direction of *larger* angles of scattering when γ approaches γ_R (from either side). Hence it is the *inner* side of the rainbow which is bright. For the twice internally reflected rays ($p = 3$) one finds $q = -1$. Hence the situation is reversed, and the outer side is bright. Notice the shift of the maximum intensity. It occurs, not at $\theta = \theta_R$, but at a somewhat larger angle (at $z = -1.0188$). The secondary maxima produce the "supernumerary bows." The cross section at maximum is proportional to $x^{1/3}$. The scatterer must therefore be quite large compared with the wavelength for the phenomenon to be clearly visible; that is, $x \gtrsim 2{,}000$. The rainbow angle itself is independent of the size of the scatterer, but since the maximum intensity occurs at $z = -1.0188$ and not at $z = 0$, there is a slight dependence on the radius. The width, as well as the separation of the secondary peaks, is proportional to $x^{-2/3}$. The smaller the scatterer, the wider the peaks and the farther they are separated. Furthermore, for the main rainbow

$$\frac{\varepsilon_p^{(2)}}{\varepsilon_p^{(1)}} = \frac{27n^2(2 - n^2)}{(2 + n^2)^3}$$

whose square for water ($n \simeq \frac{4}{3}$) is much less than unity. Hence the rainbow is strongly polarized.

In the case of water droplets the main rainbow angle is $\theta \simeq 138°$. Since, according to (3.87), it depends on the index of refraction, which in turn varies somewhat with the wavelength of the light, the maximum depends slightly on the wavelength and the colors are separated. It is because of the slow dependence of the angle of maximum intensity on the droplet size that the rainbow is so clear and so common a phenomenon. When the droplets become too small, then the maximum becomes too broad and the colors overlap. The result is a *white* rainbow.

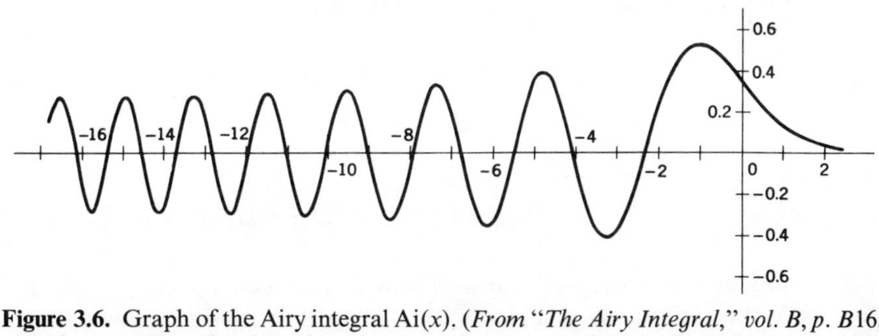

Figure 3.6. Graph of the Airy integral Ai(x). (*From "The Airy Integral," vol. B, p. B*16, *British Association for the Advancement of Science, Mathematical Tables, Cambridge University Press, New York*, 1946.)

3.7 The Glory

The scattering amplitude (3.80) for $x \gg 1$ contains the factor

$$\left(\frac{\sin 2\gamma}{\sin \theta}\right)^{1/2}$$

which is infinite in the forward and backward directions unless $\gamma = 0$ or $\gamma = \frac{1}{2}\pi$. This means that the forward- or backward-scattering cross sections are infinite whenever it is possible for a ray of neither normal nor glancing incidence to emerge in the forward or backward direction. The resulting phenomenon is called the *glory*. In the forward direction it is usually masked by diffraction scattering and therefore is difficult to observe. But in the backward direction it is a striking phenomenon observed in the back scattering of sunlight from clouds, albeit much more rarely than the rainbow.[2] We take up back scattering only.

We return to the series (2.132) for the amplitudes, truncated for $x \gg 1$ to run to $J \simeq x$ only, as discussed in Sec. 3.5. For small $\theta' = \pi - \theta$ we have the analog of (3.60),

$$\begin{aligned} \pi_J &\simeq \tfrac{1}{2}(-)^{J+1}[J_0(z) + J_2(z)] \\ \tau_J &\simeq \tfrac{1}{2}(-)^{J}[J_0(z) - J_2(z)] \end{aligned} \tag{3.95}$$

with

$$z = (J + \tfrac{1}{2}) \sin \theta \simeq (J + \tfrac{1}{2})\theta' = \theta' x \sin \gamma. \tag{3.96}$$

Keeping the exponential terms in (2.132) only, because the 1 terms contribute only to the forward diffraction, we get

$$A_{\parallel}^{L} \simeq \frac{i}{4k} \sum_{J=1}^{x} (2J + 1)(-)^{J}[e^{2i\alpha_J}(J_2 - J_0) + e^{2i\beta_J}(J_0 + J_2)]$$

whereas $A_{\perp}{}^{L}$ is obtained by exchanging α_J and β_J. Substitution of the phase-shift approximations (3.71) gives

$$A_{\parallel}^{L} \simeq -\frac{1}{4k} \sum_{J,p} (2J + 1)(-)^{J} e^{-2i\Gamma x}(-ie^{2i\Gamma' \bar{X}})^{p}$$
$$\times\ [\varepsilon_p^{(1)}(J_2 - J_0) + \varepsilon_p^{(2)}(J_2 + J_0)]$$

and $A_{\perp}{}^{L}$ by interchanging $\varepsilon_p{}^{(1)}$ and $\varepsilon_p{}^{(2)}$. Insertion of (3.67) and (3.67a) and replacement of the sum over J by an integral yields

$$A_{\parallel}^{L} \simeq \frac{i}{4k} \sum_{p} [(\varepsilon_p^{(2)} - \varepsilon_p^{(1)})J_0 + (\varepsilon_p^{(2)} + \varepsilon_p^{(1)})J_2] \int dJ (2J + 1) e^{i\xi}$$

where ξ is given by (3.75) with $q = 0$ and Θ replaced by $\Theta + \pi$. As in Sec. 3.5.2, the integral is evaluated by the method of stationary phase.

2 A practical instance in which the glory is a commonly observed phenomenon is that of special reflecting paint used on traffic signs.

The main contribution comes from the analog of (3.77),

$$\Theta \simeq -(2m + 1)\pi, \tag{3.97}$$

Θ being defined by (3.78). Then ξ is expanded about $J = J_g$, the value of J for which (3.97) holds, and the gaussian integral is carried out as in the derivation of (3.80). The result is

$$A^L \simeq iR \sum_p \left(\frac{\pi x \sin^2 \gamma \cos \gamma}{2|d\Theta/d\gamma|} \right)^{1/2} e^{i\pi(1/4s - 1/2p - m) + i\delta} \times \left[(\varepsilon_p^{(2)} + \varepsilon_p^{(1)}) J_2(z) \pm (\varepsilon_p^{(2)} - \varepsilon_p^{(1)}) J_0(z) \right]. \tag{3.98}$$

The parameters s and δ are those defined in (3.81), z is defined in (3.96), and γ is everywhere the surface angle of incidence responsible for the glory effect, namely, that determined by (3.97). The upper sign in (3.98) holds for $A_\parallel{}^L$, and the lower for $A_\perp{}^L$.

In the series over p practically only the first term for which back scattering for an angle $\gamma \neq 0$ of incidence is possible need be kept. The maximum cross section is proportional to x, and the angular distribution is determined by the two Bessel functions. Notice that in the exact backward direction the cross section is finite. Since, according to (3.96), z is proportional to $x\theta'$, for $x \gg 1$ there is a maximum at an angle which differs by very little from π. Notice also that both kinds of reflection coefficient occur in each amplitude.

The foregoing simple theory is, unfortunately, not quite capable of quantitatively explaining the glory or halo effect sometimes seen in the back scattering of sunlight from clouds. For water droplets, with $n \simeq \frac{4}{3}$, the smallest value of p for which (3.97) has a solution is 5. At least four internal reflections are necessary for back scattering to be possible. But for that high a value of p the cross section is already rather small. It is therefore assumed that edge effects make a glory possible for one or two internal reflections.[3]

3.8 Grazing Rays (The Watson Method)

When there is appreciable absorption in the scatterer, then it is clear that for $kr \gg 1$ the main contributions to the scattering come from the rays which are immediately reflected and from those which, though entering the scattering sphere, arrive at a glancing angle with the surface. The more grazing a refracted incident ray, the less the distance it travels inside. Therefore, for large absorption we treat here separately the contribution due to the grazing rays for which the approximations of Sec. 3.5.2 are invalid. Physically the waves to be considered differ from those already encountered in the geometrical-optics limit by modifications induced by surface "creep" waves.

3 See H. C. van de Hulst (1957), sec. 17.42.

In the language of Sec. 3.5.2 the immediately reflected ray has $q = -1$, but the grazing refracted ones have $q = +1$, no matter how many times they are internally reflected. (The greater the number of internal reflections, the closer γ must be to $\frac{1}{2}\pi$ for this to be true.) We can therefore recognize the contribution to the scattering we seek by the sign of q.

Mathematically speaking the treatment of the grazing rays gives us an opportunity to explore a fruitful method of replacing the partial-wave series by another which in this (and some other) case converges much more rapidly. It is based on a procedure of replacing an infinite series by a contour integral.

3.8.1 The Watson Transform. Suppose that we write the amplitudes (2.133) in the form (with $z = \cos\theta$)

$$A(z) = \frac{dB(z)}{dz}$$

$$B(z) = \frac{1}{2ik}\sum_{J=1}^{\infty}\frac{2J+1}{J(J+1)}(S^J - 1)(-)^J P_J(-z)$$

using the circumstance that $P_J(-z) = (-)^J P_J(z)$. In order to obtain A_e, $S^J = \exp 2i\alpha_J$ must be inserted and, for A_m, $S^J = \exp 2i\beta_J$.

We now replace P_J and S^J by analytic functions of a variable λ, whose values at $\lambda = J + \frac{1}{2}$ are just the given quantities S^J and P_J. If the resulting integral

$$\int_C d\lambda\lambda(\lambda^2 - \tfrac{1}{4})^{-1}[S(\lambda) - 1]P_{\lambda-1/2}(-z)\sec\pi\lambda$$

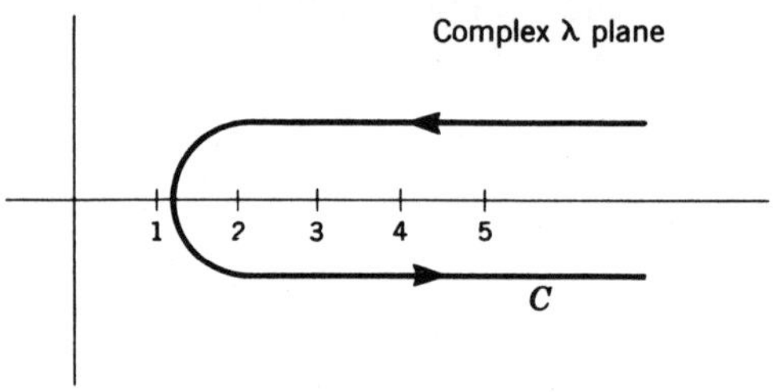

Figure 3.7. Contour in the complex λ plane.

over the contour C shown in Fig 3.7 converges, it can be carried out by Cauchy's residue theorem, and its value is

$$-i\sum_{1}^{\infty}\frac{2J+1}{J(J+1)}(-)^J(S^J - 1)P_J(-z) = 2kB(z).$$

Consequently

$$B(z) = \frac{1}{2k}\int_C \frac{d\lambda\lambda}{\lambda^2 - \frac{1}{4}}[S(\lambda) - 1]P_{\lambda-1/2}(-z)\sec\pi\lambda. \qquad (3.99)$$

The interpolation of the numbers P_J and S^J by analytic functions offers no difficulty. The Legendre polynomials $P_\nu(z)$ have a standard analytic extension which is obtained from Legendre's differential equation by merely letting the parameter there take on arbitrary values. The regular solution then is the hypergeometric function[4]

$$P_\nu(z) = F(-\nu, \nu + 1; 1; \tfrac{1}{2}(1 - z)) \tag{3.100}$$

which is an analytic function of ν.

If we are dealing with the case of scattering by a uniform sphere, the $\mathcal{S}$-matrix elements S^J are also readily extended to nonintegral values of J. The functions that appear in (2.127) are all directly defined in terms of Bessel and Hankel functions, which are analytic functions of their order. According to the definitions (2.58) and (2.60) of the Riccati functions, we have

$$\begin{aligned} S_e(\lambda) &= -\frac{nH_\lambda^{(2)\prime} - \mu H_\lambda^{(2)}\xi_\lambda + \frac{1}{2}\overline{X}^{-1}(n^2 - \mu)H_\lambda^{(2)}}{nH_\lambda^{(1)\prime} - \mu H_\lambda^{(1)}\xi_\lambda + \frac{1}{2}\overline{X}^{-1}(n^2 - \mu)H_\lambda^{(1)}} \\ S_m(\lambda) &= -\frac{\mu H_\lambda^{(2)\prime} - nH_\lambda^{(2)}\xi_\lambda + \frac{1}{2}x^{-1}(\mu - 1)H_\lambda^{(2)}}{\mu H_\lambda^{(1)\prime} - nH_\lambda^{(1)}\xi_\lambda + \frac{1}{2}x^{-1}(\mu - 1)H_\lambda^{(1)}} \end{aligned} \tag{3.101}$$

with

$$\xi_\lambda \equiv \frac{J_\lambda'(\overline{X})}{J_\lambda(\overline{X})}$$

The Hankel functions are evaluated at x.

It should be recognized that these interpolations are, of course, not unique. There exist infinitely many other analytic functions which for $\lambda = J + \frac{1}{2}$ agree with the given quantities S^J and P_J. One need only add to those given any other analytic function multiplied by $\cos \pi\lambda$, and one obtains another valid interpolation. The particular choice made is justified by the fact that it permits the further steps to be carried out.

Contour Change. The integral over the contour C in (3.99) is now transformed in a number of steps as follows. First, the turning point of the loop is pushed to the left, to $\lambda = 0$. Call the new contour C'. There is then a residue contribution from the double pole at $\lambda = \frac{1}{2}$. Since $A(z)$ is obtained from $B(z)$ by differentiation, we are interested only in that part of the residue which does not vanish when differentiated with respect to z. To within a constant we have

$$B(z) = \frac{i}{2k}(S^0 - 1) \lim_{\nu\to 0} \frac{\partial}{\partial \nu} P_\nu(-z) + \frac{1}{2k}\int_{C'} \cdots .$$

The limit is easily calculated from (3.100) and the hypergeometric series,

$$\lim_{\nu\to 0} \frac{\partial}{\partial \nu} P_\nu(-z) = \ln\frac{1}{2}(1 - z). \tag{3.102}$$

4 A. Erdelyi (1953), vol. 1, p. 122, eq. (3).

To within a constant, then

$$B(z) = \frac{i}{2k}(S^0 - 1)\ln\frac{1}{2}(1 - z) + \frac{1}{2k}\int_{C'} \cdots. \tag{3.103}$$

The quantities $S_e{}^0$ and $S_m{}^0$ are easily obtained from (2.127a).

We must now consider separately the integrals over the upper branch $C_1{}'$ and the lower branch $C_2{}'$ of C' in (3.103). To $C_1{}'$ we add a large quarter circle in the first quadrant. Owing to the strong decrease of the integrand as $\operatorname{Im}\lambda \to \infty$ and as $\operatorname{Re}\lambda \to \infty$, to be shown in Sec. 3.8.1, this contributes nothing. Then we add and subtract the integral over the positive imaginary axis. The result is

$$\int_{C_1'} = -\int_0^{i\infty} - \oint_{\mathfrak{C}_1},$$

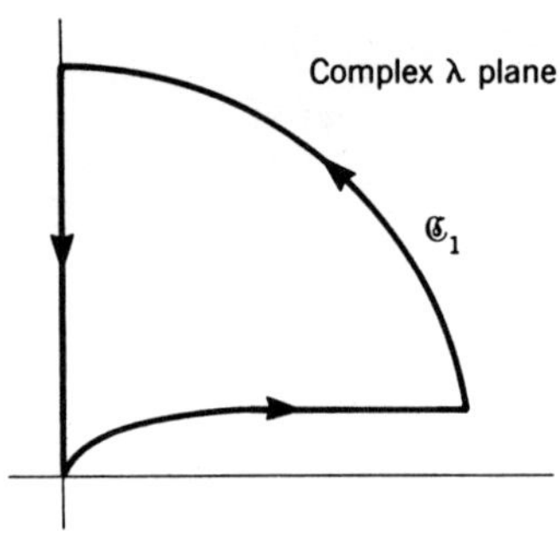

Figure 3.8. Contour $\mathfrak{C}_1$.

where the closed contour $\mathfrak{C}_1$ is shown in Fig. 3.8. The integral over $\mathfrak{C}_1$ is evaluated by Cauchy's residue theorem. If S has poles at $\lambda = \lambda^{(n)}$, with residues $R^{(n)}$, we get

$$\int_{C_1'} = -\int_0^{i\infty} - 2\pi i \sum_n \frac{\lambda^{(n)}}{\lambda^{(n)2} - \frac{1}{4}} R^{(n)} P_{\lambda^{(n)}-1/2}(-z)\sec\pi\lambda^{(n)}. \tag{3.104}$$

Next we transform the integral over $C_2{}'$. Here the S part and the 1 part must be separated. The 1 part is treated analogously to the integral over $C_1{}'$. A large quarter circle in the fourth quadrant is added. Since P_ν behaves as in (3.73) when $|\nu| \to \infty$, this changes nothing. Then the integral over the negative imaginary axis is added and subtracted.

$$\int_{C_2'} = -\int_{-i\infty}^{0} - \oint_{\mathfrak{C}_2},$$

the closed countour $\mathfrak{C}_2$ being shown in Fig. 3.9. There is no singularity inside $\mathfrak{C}_2$ (as will be shown in Sec. 3.8.2); therefore

$$\int_{C_2'} = -\int_{-i\infty}^{0}.$$

The S part of the integral over $C_2{}'$ is split into two parts, necessitated

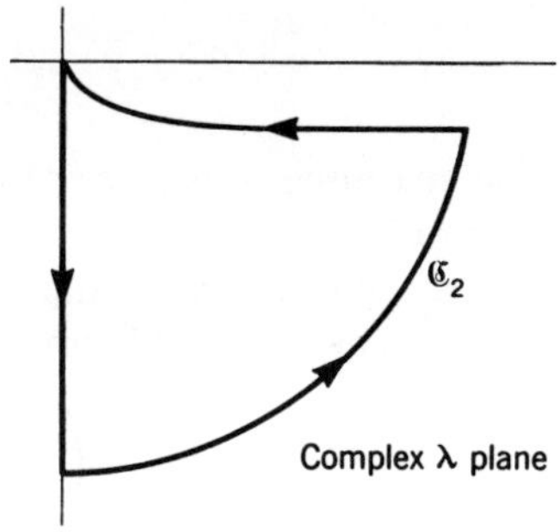

Figure 3.9. Contour $\mathfrak{C}_2$.

by the increase of $S(\lambda)$ as $\operatorname{Im}\lambda \to -\infty$ (to be shown in Sec. 3.8.2). We use the fact that[5]

$$P_\nu(-z) = e^{-i\pi\nu}P_\nu(z) - \frac{2}{\pi}\sin\pi\nu\, Q_\nu(z + i0). \tag{3.105}$$

$Q_\nu(z)$ is the Legendre function of the second kind. It has a branch cut in the z plane from $z = -\infty$ to $z = +1$, and for real z

$$Q_\nu(z + i0) \equiv \lim_{\varepsilon\to 0+} Q_\nu(z + i\varepsilon).$$

The Q_ν part of the S integral over C_2' is left as it is. The P_ν part, armed as it is with the exponentially decreasing factor $e^{-i\pi\lambda}$, is treated just like the 1 term. It will be shown in Sec. 3.8.1 that $S(\lambda)$ has no poles in the fourth quadrant. Consequently

$$\int_{C_2'} = -\int_{-i\infty}^{0} d\lambda\lambda(\lambda^2 - \tfrac{1}{4})^{-1}\sec\pi\lambda[Se^{-i\pi(\lambda-1/2)}P_{\lambda-1/2}(z) - P_{\lambda-1/2}(-z)]$$
$$+ \frac{2}{\pi}\int_{C_2'} d\lambda\lambda(\lambda^2 - \tfrac{1}{4})^{-1}S(\lambda)Q_{\lambda-1/2}(z + i0). \tag{3.106}$$

The first integral here is finally changed into one from 0 to $+i\infty$ by changing λ into $-\lambda$. It follows from the symmetry of the hypergeometric function in its first two variables that

$$P_{-\lambda-1/2} = P_{\lambda-1/2}. \tag{3.107}$$

Hence, combining (3.103), (3.104), and (3.106) and using (3.107) again, we get

$$B(z) = B_{cr} + B_{geo}$$
$$B_{cr}(z) = \frac{i}{2k}(S^0 - 1)\ln\frac{1}{2}(1 - z)$$
$$- \frac{\pi i}{k}\sum_n \frac{\lambda^{(n)}}{\lambda^{(n)2} - \frac{1}{4}} R^{(n)}P_{\lambda^{(n)}-1/2}(-z)\sec\pi\lambda^{(n)} \tag{3.108}$$
$$+ \frac{i}{2k}\int_0^{i\infty}\frac{d\lambda\lambda}{\lambda^2 - \frac{1}{4}}\sec\pi\lambda\, P_{\lambda-1/2}(z)[S(-\lambda)e^{i\pi\lambda} - S(\lambda)e^{-i\pi\lambda}]$$

5 *Ibid.*, p. 140, eq. (10).

$$B_{geo}(z) = \frac{1}{\pi k} \int_{i\infty}^{\infty} \frac{d\lambda \lambda}{\lambda^2 - \frac{1}{4}} S(\lambda) Q_{\lambda-1/2}(z + i0). \tag{3.109}$$

The last integral is meant to run down the positive imaginary axis and then along C_2'.

The significance of the split-up of $B(z)$ into B_{cr} and B_{geo} may be seen from the asymptotic form as $|v| \to \infty$,[6]

$$Q_v(z + i0) \simeq \left(\frac{\pi}{2v \sin\theta}\right)^{1/2} \exp\left\{\mp i \left[\left(v + \frac{1}{2}\right)\theta + \tfrac{1}{4}\pi\right]\right\}. \tag{3.110}$$

When $x \gg 1$, B_{geo} may be treated by the stationary-phase method, as in Sec. 3.5.2. It then gives rise to the $q = -1$ part of the geometrical-optics limit of the amplitude. If there is absorption, the leading term there is that from the immediately reflected ray. In the high-absorption limit the split-off part B_{cr} is therefore just that due to the grazing rays, or the creep-wave part.

The Creep Wave. Consider now B_{cr}. As $|n| \gg 1$, (3.101) shows that

$$\begin{aligned} S_e(\lambda) &\simeq -\frac{2xH_\lambda^{(2)\prime}(x) + H_\lambda^{(2)}(x)}{2xH_\lambda^{(1)\prime}(x) + H_\lambda^{(1)}(x)} \\ S_m(\lambda) &\simeq -\frac{H_\lambda^{(2)}(x)}{H_\lambda^{(1)}(x)}. \end{aligned} \tag{3.111}$$

But the Hankel functions have the property[7]

$$\begin{aligned} H_{-\lambda}^{(2)}(x) &= H_\lambda^{(2)}(x) e^{-i\pi\lambda} \\ H_{-\lambda}^{(1)}(x) &= H_\lambda^{(1)}(x) e^{i\pi\lambda}. \end{aligned} \tag{3.112}$$

Therefore in the limit of large absorption both S_e and S_m satisfy the relation

$$S(-\lambda) \simeq S(\lambda) e^{-2\pi i\lambda}, \tag{3.113}$$

and as a consequence the integral in (3.108) vanishes.

The first term in (3.108) is of order $\lambda = Rx^{-1}$. B_{geo}, on the other hand, is, for large x, of the order R. Thus we may neglect the first term in (3.108) too, and we obtain for the creep-wave contribution to the amplitude

$$A_{cr} \simeq i\pi R x^{-1} \sum_n \lambda^{(n)} R^{(n)} \sec \pi\lambda^{(n)} \pi_{\lambda^{(n)}-1/2}(-z). \tag{3.114}$$

In order to estimate the size of (3.114), we must estimate $\lambda^{(n)}$ and $R^{(n)}$ for large x. The first poles occur for $\lambda \simeq x$, where we may use the approximation (3.64) for the Hankel function. This shows that, when $|\lambda - x| \ll x$, then

$$H_\lambda^{(1)}(x) \simeq \left[\frac{2}{3}\left(1 - \frac{\lambda}{x}\right)\right]^{1/2} e^{i1/6\pi} H_{1/3}^{(1)}(Z), \tag{3.115}$$

6 *Ibid.*, p. 162, eqs. (1) and (2), and p. 143, eq. (2).

7 *Ibid.* (1953), vol. 2, p. 4, eq. (9).

with the abbreviation

$$Z = \tfrac{1}{3}\lambda^{-2}(x^2 - \lambda^2)^{3/2} \simeq \tfrac{2}{3}(\tfrac{1}{2}x)^{-1/2}(x - \lambda)^{3/2}$$

and therefore

$$2xH_\lambda^{(1)\prime}(x) + H_\lambda^{(1)}(x) \simeq \left[\frac{2}{3}\left(1 - \frac{\lambda}{x}\right)\right]^{1/2} e^{i1/6\pi} Z H_{1/3}^{(1)\prime}(Z). \tag{3.116}$$

Consequently the first poles of S_m are determined by the roots of

$$H_{1/3}^{(1)}(Z) = 0$$

and those of S_e by

$$H_{1/3}^{(1)\prime}(Z) = 0.$$

But since[8]

$$H_\lambda^{(1)} = -i \csc \pi\lambda (J_{-\lambda} - e^{-i\pi\lambda} J_\lambda) \tag{3.117}$$

and[9]

$$J_\lambda(ze^{i\pi}) = e^{i\pi\lambda} J_\lambda(z), \tag{3.118}$$

it follows for $\lambda = \frac{1}{3}$, by (3.94), that

$$\begin{aligned} H_{1/3}^{(1)}(Z) &= 2\sqrt{3} e^{i5\pi/6} q^{-1/2} \operatorname{Ai}(-q) \\ q &= (\tfrac{3}{2}Z)^{2/3} e^{i2\pi/3} \simeq (\tfrac{1}{2}x)^{-1/3}(\lambda - x)e^{-i\pi/3}. \end{aligned} \tag{3.119}$$

The (positive) zeros q_n of the Airy function $\operatorname{Ai}(-q)$ therefore determine the relevant roots of the Hankel function, and hence the complex values $\lambda_m{}^{(n)}$ at which S_m has its first poles, as

$$\lambda_m^{(n)} \simeq x + 2^{-1/3} x^{1/3} q_n e^{i\pi/3}. \tag{3.120}$$

The phase is determined by the requirement that λ lie in the first quadrant.

Similarly, the first poles $\lambda_e{}^{(n)}$ of S_e are obtained by

$$\lambda_e^{(n)} \simeq x + 2^{-1/3} x^{1/3} q_n' e^{i\pi/3} \tag{3.121}$$

from the zeros of the derivative

$$[q_n'^{-1/2} \operatorname{Ai}(-q_n')]' = 0.$$

The residues at these poles are obtained from the wronskian (easily calculated from Bessel's equation and the boundary conditions)

$$H_\lambda^{(1)}(x) H_\lambda^{(2)\prime}(x) - H_\lambda^{(1)\prime}(x) H_\lambda^{(2)}(x) = -4i(\pi x)^{-1} \tag{3.122}$$

which implies that, when $H_\lambda{}^{(1)}(x) = 0$, then

$$H_\lambda^{(2)}(x) = \frac{4i}{\pi x H_\lambda^{(1)\prime}(x)}.$$

8 *Ibid.*, p. 4, eq. (5).
9 *Ibid.*, p. 12, eq. (61).

The residue at a pole of S_m is found to be

$$R_m^{(n)} \simeq \frac{3ix}{2\pi q_n^2}[\mathrm{Ai}'(-q_n)]^{-2} \tag{3.123}$$

and that, at a pole of S_e,

$$R_e^{(n)} \simeq \frac{3ix}{2\pi(q_n'^2 - \frac{1}{4}q_n'^{-1})}[\mathrm{Ai}(-q_n')]^{-2}. \tag{3.124}$$

We now insert these residues and (3.73) in (3.114). The first terms in the series of A_{cr} then become

$$A_{cr} \simeq 3iR(2\pi x \sin^3\theta)^{-1/2} e^{i(1/4\pi + x\theta)}\Sigma$$

where in the electric case

$$\Sigma_e \simeq \sum_n (q_n'^2 - \tfrac{1}{4}q_n'^{-1})^{-1}[\mathrm{Ai}(-q_n')]^{-2}\exp[i\tfrac{1}{2}(\tfrac{1}{2}x)^{1/3}q_n'\theta - \tfrac{1}{2}\sqrt{3}(\tfrac{1}{2}x)^{1/3}q_n'\theta] \tag{3.125}$$

and in the magnetic case

$$\Sigma_m \simeq \sum_n q_n^{-2}[\mathrm{Ai}'(-q_n)]^{-2}\exp[i\tfrac{1}{2}(\tfrac{1}{2}x)^{1/3}q_n\theta - \tfrac{1}{2}\sqrt{3}(\tfrac{1}{2}x)^{1/3}q_n\theta]. \tag{3.126}$$

According to (2.134), the actual amplitudes are obtained from this as

$$A_{\parallel}^L \simeq -\frac{3Rx^{1/2}}{(2\pi\sin\theta)^{1/2}}e^{i(1/4\pi + x\theta)}\Sigma_e \qquad A_{\perp}^L \simeq -\frac{3Rx^{1/2}}{(2\pi\sin\theta)^{1/2}}e^{i(1/4\pi + x\theta)}\Sigma_m. \tag{3.127}$$

These results show that the effect of the grazing rays, although negligible in the strict geometrical-optics limit of $x \to \infty$, can be sizable for reasonably large values of x. The amplitudes are of order $Rx^{1/2}\exp[-\frac{1}{2}\sqrt{3}(\frac{1}{2}x)^{1/3}q_n\theta]$, whereas those of (3.80) are of order R. Furthermore, the series converge quite rapidly. (That they actually converge will be shown in the next section. Here we are interested only in the fact that the first few terms suffice to evaluate them approximately.) The smaller the angle, the larger the contribution of the grazing rays, except, of course, when $\theta \lesssim x^{-1}$. Then the diffraction peak dominates.

3.8.2 Convergence Questions. There are a number of questions in connection with the Watson method which still have to be answered before the result can be accepted. The behavior of S must be investigated in the limit of large $|\lambda|$ to see whether or not the transformations of the contours C_1 and C_2 are allowed and whether or not the series (3.114) converges. The results (3.120), (3.121), (3.123), and (3.124) hold only when $|\lambda^{(n)}|/x \simeq 1$. Finally, a general proof is to be supplied that S has no poles in the fourth quadrant of the λ plane.

When $|\lambda| \gg z$, the Bessel function has the asymptotic form

$$J_\lambda(z) \simeq \frac{(\frac{1}{2}z)^\lambda}{\Gamma(1+\lambda)}.$$

Using Stirling's formula in the approximation of

$$\frac{\Gamma(1+\lambda)}{\Gamma(1-\lambda)} = \frac{1}{\pi}\Gamma(1+\lambda)\Gamma(\lambda)\sin\pi\lambda \simeq 2\sin\pi\lambda\, e^{2\lambda\ln(\lambda/e)}$$

we get from (3.117)

$$e^{i\pi\lambda}\sin\pi\lambda\,\Gamma(1+\lambda)(\tfrac{1}{2}x)^{-\lambda}H_\lambda^{(1)}(x) \simeq i + \xi$$
$$\xi = -2i\sin\pi\lambda\, e^{i\pi\lambda + 2\lambda\ln(2\lambda/ex)}.$$

Similarly, from the fact that

$$H_\lambda^{(2)}(x) = -i\csc\pi\lambda(e^{i\pi\lambda}J_\lambda - J_{-\lambda})$$
$$-e^{i\pi\lambda}\sin\pi\lambda\,\Gamma(1+\lambda)(\tfrac{1}{2}x)^{-\lambda}H_\lambda^{(2)}(x) \simeq ie^{2i\pi\lambda} + \xi \tag{3.128}$$

while

$$\frac{J_\lambda'(\bar{X})}{J_\lambda(\bar{X})} \simeq \frac{\lambda}{\bar{X}}.$$

Notice that when $\operatorname{Im}\lambda \to \pm\infty$,

$$e^{2\lambda\ln\lambda} \simeq e^{-\pi|\operatorname{Im}\lambda|}|\lambda|^{2\operatorname{Re}\lambda}$$

so that, when $\operatorname{Re}\lambda \to \infty$, ξ is unbounded and, when $\operatorname{Im}\xi \to +\infty$,

$$|\xi| \simeq e^{2\lambda\ln(2\lambda/ex)} \simeq \left|\frac{2\lambda}{ex}\right|^{2\operatorname{Re}\lambda} e^{-\pi\operatorname{Im}\lambda}.$$

But when $\operatorname{Im}\lambda \to -\infty$,

$$|\xi| \simeq |e^{2i\pi\lambda}|e^{2\lambda\ln(2\lambda/ex)} \simeq \left|\frac{2\lambda}{ex}\right|^{2\operatorname{Re}\lambda} e^{\pi|\operatorname{Im}\lambda|}.$$

A look at (3.101) therefore shows that, as $\operatorname{Re}\lambda \to \infty$,

$$S_m \simeq S_e \simeq 1$$

and, as $\operatorname{Im}\lambda \to +\infty$, apart from unimportant factors,

$$|S| \simeq |\xi| \simeq \left|\frac{2\lambda}{ex}\right|^{2\operatorname{Re}\lambda}\exp(-\pi\operatorname{Im}\lambda) \tag{3.129}$$

(both for S_e and S_m), except along the direction where the poles lie; however, as $\operatorname{Im}\lambda \to -\infty$,

$$|S| \simeq \frac{|e^{2\pi i\lambda}|}{|\xi|} \simeq \left|\frac{ex}{2\lambda}\right|^{2\operatorname{Re}\lambda} e^{\pi|\operatorname{Im}\lambda|} \tag{3.130}$$

both for S_e and for S_m.

The foregoing estimates when used in (3.99) and in the steps leading to (3.106) together with (3.105) and the asymptotic behavior (3.73) of P_ν show that the integrands vanish sufficiently rapidly at infinity to allow the contour transformations there performed. They also show that the use of the split-up (3.105) was essential.

The estimates also allow us to find the asymptotic distribution of poles. Since they lie in the upper half plane, we have

$$S \simeq 1 - i\frac{(e^{2\pi i\lambda} - 1)a}{\xi - ia}$$

where for S_e

$$a = a_e = \frac{n^2 - \mu}{n^2 + \mu}$$

and for S_m

$$a = a_m = \frac{\mu - 1}{\mu + 1}.$$

The faraway poles of S are therefore determined by the roots of

$$\xi - ia = 0$$

or

$$2\lambda \ln \frac{2\lambda}{ex} = \ln a + 2\pi i\left(N + \frac{1}{4}\right),$$

N being an integer. This means that

$$\begin{aligned} \operatorname{Re}\lambda_N &\simeq \tfrac{1}{2}\pi^2\left(N + \frac{1}{4}\right)\left[\ln\frac{2\pi(N + \frac{1}{4})}{ex}\right]^{-2}\left[1 + O\left(\frac{\ln\ln N}{\ln N}\right)\right] \\ \operatorname{Im}\lambda_N &\simeq \pi\left(N + \frac{1}{4}\right)\left[\ln\frac{2\pi(N + \frac{1}{4})}{ex}\right]^{-1}\left[1 + O\left(\frac{\ln\ln N}{\ln N}\right)\right] \end{aligned} \tag{3.131}$$

for the poles of S_e as well as for those of S_m. As $N \to \infty$, both $\operatorname{Re}\lambda_N$ and $\operatorname{Im}\lambda_N$ tend to infinity, but the phase of λ_N tends to $\frac{1}{2}\pi$.

The residues are easily calculated too,

$$R_N \simeq \frac{1}{2}\left[\ln\frac{2|\lambda|}{x}\right]^{-1}. \tag{3.132}$$

Insertion of this and (3.73) in (3.114) shows that the series due to the poles of S converges.

There remains the demonstration that there are no poles of S in the fourth quadrant of the λ plane. That is proved by considering a function f which satisfies the differential equation

$$-\frac{d^2}{dr^2}f + \left(\lambda^2 - \frac{1}{4}\right)r^{-2}f = n^2k^2f$$

when $0 \leq r \leq R$, and

$$-\frac{d^2}{dr^2}f + \left(\lambda^2 - \frac{1}{4}\right)r^{-2}f = k^2f$$

when $r > R$. At $r = E$, let f be continuous, but let its first derivative have a discontinuity so that

$$\lim_{r\to R+} \frac{d}{dr} f = \varepsilon \lim_{r \to R-} \frac{d}{dr} f \qquad \varepsilon > 0.$$

The function

$$f = \begin{cases} -2u_J(nkr)e^{i\pi\lambda} & r < R \\ Dw_J^{(-)}(kr) - Nw_J^{(+)}(kr) & r > R. \end{cases}$$

$(J = \lambda - \frac{1}{2})$ is a solution of this problem, if

$$D = \varepsilon n u_J' w_J^{(+)} - u_J w_J^{(+)\prime}$$
$$N = \varepsilon n u_J' w_J^{(-)} - u_J w_J^{(-)\prime}$$

in which u_J is evaluated at $\overline{X} = nkR$ and w_J at $x = kR$.

Now use the standard wronskian technique. Multiply the differential equation for f by f^*, that for f^* by f, and subtract,

$$\frac{d}{dr} W(f^*,f) \equiv \frac{d}{dr}(f^*f' - f^{*\prime}f) = 2i \operatorname{Im} \lambda^2 r^{-2}|f|^2.$$

Integrate this equation from R to ∞, and use the asymptotic form of w_J in order to evaluate the limit of the left-hand side,

$$\lim_{r\to\infty} W(f^*,f) = 2i(|N|^2 - |D|^2).$$

At $r = R$ we use the boundary condition on f,

$$\lim_{r\to R+} W(f^*,f) = 4\varepsilon \exp(-\pi \operatorname{Im} \lambda) W(u_J^*, u_J),$$

the right-hand wronskian being evaluated at $\overline{X}$. Hence

$$|N|^2 - |D|^2 = \operatorname{Im} \lambda^2 \int_R^\infty dr r^{-2}|f|^2 - 2i \exp(-\pi \operatorname{Im} \lambda) \varepsilon W(u_J^*, u_J).$$

The same technique applied to $r < R$ shows that, if $\operatorname{Re} \lambda > 0$, then

$$2 \exp(-\pi \operatorname{Im} \lambda) W(u_J^*, u_J) = i \operatorname{Im} \lambda^2 \int_0^R dr r^{-2}|f|^2$$

and consequently

$$|N|^2 - |D|^2 = \operatorname{Im} \lambda^2 (\varepsilon \int_0^R dr r^{-2}|f|^2 + \int_R^\infty dr r^{-2}|f|^2).$$

This is the desired result. It shows that $D = 0$ is possible only if $\operatorname{Im} \lambda^2 > 0$. In other words, for $\operatorname{Re} J > -\frac{1}{2}$, real x, $\overline{X}$, and $\sigma \geq 0$,

$$D = \sigma u_J'(\overline{X}) w_J^{(+)}(x) - u_J(\overline{X}) w_J^{(+)\prime}(x) = 0$$

implies that $\operatorname{Im} J > 0$.

According to (2.127a) the denominators of both S_e and S_m are of the form D. It has therefore been demonstrated that all the poles of both elements of the $\mathscr{S}$ matrix in the right-hand half of the λ plane occur for $\operatorname{Im} \lambda > 0$.

Appendix: Saddle-Point Intergration (The Method of Steepest Descent)

The aim of the method is to find the asymptotic value of an integral of the form

$$I(\omega) = \int_C dz\, g(z) e^{\omega f(z)}$$

for large values of ω. The contour C of integration is some path inside the region of analyticity of $f(z)$ and $g(z)$.

There are two things to watch in the integral: the magnitude of the integrand and its phase. If f has a nonvanishing imaginary part, large values of ω will make the integrand oscillate rapidly and hence will tend to make the integral small. But if f has a nonvanishing real part, then large values of ω will also tend to make it larger, the larger the $\mathrm{Re} f$. The first effect favors those regions of the integrand where $v = \mathrm{Im}\, f$ is stationary; the second effect favors those regions where $u = \mathrm{Re}\, f$ has a maximum. Hence it is clear that, if the path of integration can be distorted so as to run through a point z_0 in such a direction that along it $u(z)$ has a maximum and at the same time $v(z)$ is stationary, then in the limit of large ω the entire value of $I(\omega)$ comes from the immediate vicinity of z_0.

Let us look at these conditions on z_0 and on the chosen direction. At $z = z_0$

$$\frac{\partial u}{\partial x} dx + \frac{\partial u}{\partial y} dy = 0 \qquad \frac{\partial v}{\partial x} dx + \frac{\partial v}{\partial y} dy = 0.$$

But the Cauchy–Riemann equations expressing the analyticity of f are

$$\frac{\partial v}{\partial x} = -\frac{\partial u}{\partial y} \qquad \frac{\partial v}{\partial y} = \frac{\partial u}{\partial x}$$

so that the last equation becomes

$$\frac{\partial u}{\partial y} dx - \frac{\partial u}{\partial x} dy = 0.$$

This, together with the first equation, implies that there exists a direction such as we are looking for if and only if

$$\frac{\partial u}{\partial x} = \frac{\partial u}{\partial y} = 0 \quad \text{and} \quad \frac{\partial v}{\partial x} = \frac{\partial v}{\partial y} = 0.$$

This in turn implies that u and v are stationary at z_0 along *all* direction. But since they cannot have maxima or minima, they must have *saddle points* at z_0. The direction through z_0 must of course be chosen to that u has a maximum along it. Furthermore, the value of I will for a given large ω come most nearly from the integrand at z_0 if the path is chosen along that direction in which the maximum of u is most narrow. This is the direction of *steepest descent.*

These statements do not make the method entirely foolproof, especially if the contour C has finite end points. In that instance u may have an absolute maximum at an end point without being stationary.

Suppose that f is real for real z. It then follows from Schwarz's reflection principle that $f(z)^* = f^*(z)$ and hence $u(x,y) = u(x,-y)$. Consequently the y derivative of u vanishes on the real axis. If it has an extremum along the real axis at x_0, it must have a saddle point there. If x_0 is a *maximum* along the real line, then the x axis is also the direction of steepest descent. If it is a *minimum*, on the other hand, then u has a maximum with the direction of steepest descent along the direction parallel to the imaginary axis. But, of course, if C runs along the real axis, then a minimum at x_0 cannot be exploited by distorting C to run through x_0 at right angles, because then there are larger values of u along the real axis and v is everywhere stationary there. We expand f about $z = z_0 = x_0$,

$$f(z) = f(x_0) + \tfrac{1}{2}(z - z_0)^2 f''(x_0) + \cdots.$$

Let x_0 be a *maximum* for real z. If C runs along the real axis or can be distorted to to so without running through points where u is larger, then we integrate in the real direction, where $(z - z_0)^2$ is positive while $f''(x_0)$ is negative. The integral may be extended to $\pm\infty$ since almost all of it comes from the vicinity of x_0 in any case, and we get, depending on the original direction of C,

$$\begin{aligned} I(\omega) &\simeq \pm g(x_0)e^{\omega f(x_0)} \int_{-\infty}^{\infty} dx\, e^{-1/2(x-x_0)^2\omega|f''(x_0)|} \\ &= \pm\left(\frac{2\pi}{\omega|f''(x_0)|}\right)^{1/2} g(x_0)e^{\omega f(x_0)}. \end{aligned}$$

On the other hand, if f has a *minimum* at x_0 (in the real direction) and C can be distorted to run through it without passing through larger values of u (which rules out a C that leads from a real point to real point), then we integrate in the imaginary direction. Now $f''(x_0)$ is positive and $(z - z_0)^2$ is negative, and we get, depending on the original direction of C,

$$\begin{aligned} I(\omega) &\simeq \pm g(x_0)e^{\omega f(x_0)}\, i \int_{-\infty}^{\infty} dy\, e^{-1/2(y-y_0)^2\omega|f''(x_0)|} \\ &= \pm i\left(\frac{2\pi}{\omega|f''(x_0)|}\right)^{1/2} g(x_0)e^{\omega f(x_0)}. \end{aligned}$$

A situation that often arises in practice is closely related to the saddle-point method. In order to put it in the form in which it usually occurs, let us write

$$I(\omega) = \int_C dz\, g(z)e^{i\omega h(z)}.$$

Suppose that C runs along the real axis, and let h be real there. Furthermore assume that $h(z)$ is stationary along the real axis at $x = x_0$,

$$\left.\frac{\partial h}{\partial x}\right|_{x=x_0} = 0.$$

Then it follows, as in the previous case, that x_0 is a saddle point of $\operatorname{Re} h$. At the same time, since $\operatorname{Im} h$ is an odd function of y, x_0 is a *point of inflection* of $\operatorname{Im} h$ along the direction parallel to the imaginary axis. One might therefore be tempted to suppose that the integral can be made *smaller* by distorting the contour in the direction toward larger $\operatorname{Im} h$. The fallacy of that supposition lies in the fact that, although along the imaginary direction $\operatorname{Im} h$ has a point of inflection at x_0, it is nevertheless a saddle point of $\operatorname{Im} h$. The contour cannot be shifted away from x_0 without bringing in smaller values of $\operatorname{Im} h$. Now in the method of stationary phase the integral is assumed to come from the vicinity of x_0 only, but one does not search for the direction of steepest descent. The ensuing gaussian integral is usually carried out as it stands. However, the argument of steepest descent is merely hidden in the evaluation of the gaussian integral. In order to carry it out properly, the direction must be chosen so that the exponential is *real*, and this is the steepest descent.

Notes and References

A more detailed survey of the (n, x) domain and the approximations there can be found in H. C. van de Hulst (1957). Further references are to be found in this book. In contrast to his, most of the present results do not assume $\mu = 1$. See also Born and Wolf (1959), pp. 630ff, and N. A. Logan (1965). More recent contributions and references can be found in M. Kerker (1963 and 1977).

3.1 The cross section for $x \ll 1$ was first derived by Lord Rayleigh (1871); see also Lord Rayleigh (1899); and for recent work, A. Ramm (1980a).

3.2 The name Rayleigh–Gans is based on the following papers: Lord Rayleigh (1881); R. Gans (1925); for more recent work, see, for example, M. Kerker et al. (1963); and for Rayleigh–Gans scattering by nonspherical particles, see Barber and Wang (1979).

3.3 The original work on resonances scattering is that by P. Debye (1909a).

3.4 For more details, see H. C. van de Hulst (1957), chap. 11. See also L. I. Schiff (1956b).

3.5 The method of obtaining the geometrical-optics limit from physical optics goes back to P. Debye (1909a). The outline given follows essentially H. C. van de Hulst (1957).

3.6 and **3.7** The poetic description of the rainbow is from *Ode to Sir Isaac Newton* by James Thomson. The second poem is *The Rainbow*, by D. H. Lawrence. I am indebted to Prof. John Hollander of Yale University for these references.

For more details on the rainbow, see T. Ljunggrén (1948); H. C. van de Hulst (1957). Relevant numerical tables can be found in J. C. P. Miller (1946); *Integrals* (1958). For recent work see H. M. Nussenzveig (1969b); and Khare and Nussenzveig (1974).

The single fact which is perhaps most notable about the glory from the point of view of the history of physics is that in the attempt to study it experimentally C. T. R. Wilson discovered the cloud chamber. See the quotation on pp. 14 and 15 of C. N. Yang (1961).

3.8 The procedure on which the Watson method is based is due to Poincaré and Nicholson. It was first used in the context of electromagnetic theory by G. N. Watson (1918). See also E. P. White (1922); O. Laporte (1923); Van der Pol and Bremmer (1937); A. Sommerfeld (1949), pp. 282ff.

The split-up into the geometrical-optics part and the creep-wave part is due to W. Franz (1954).

The zeros of Hankel functions and of their derivatives, in the complex λ plane, are studied by J. B. Keller et al. (1963), where other references may be found.

Other recent references are Kazarinoff and Ritt (1959); Goodrich and Kazarinoff (1963); Senior and Goodrich (1964); H. M. Nussenzveig (1965 and 1970).

Appendix The method of saddle-point integration goes back to F. Debye (1909b and 1910). For examples of more recent discussions see A. Sommerfeld (1949), pp. 98ff. and 116ff.; B. L. van der Waerden (1951); Morse and Feshbach (1953), vol. 1, pp. 434ff; A. Erdelyi (1955).

Problems

1 Show that, when (3.1) holds, the ratio of forward- to backward-scattering intensity (for a sphere of $\mu = 1$) is given by

$$1 + \frac{4}{15}x^2\frac{(n^2 + 4)(n^2 + 2)}{2n^2 + 3} + O(x^4).$$

2 Show that, under the same conditions as in Prob. 1, the polarization at 90° is given by

$$P = 1 - 2x^4\left[\frac{(n^2 - 1)(n^2 + 2)}{15(2n^2 + 3)}\right]^2 + \cdots.$$

3 Show that under the conditions of Prob. 1 the angle at which the polarization is maximal (polychroism) is given by

$$\cos\theta = -x^2\frac{(n^2 - 1)(n^2 + 2)}{15(2n^2 + 3)}.$$

4 Calculate the density matrix of the scattered wave for Rayleigh scattering in terms of the density matrix of the incident beam.

5 Check the two equations preceding (3.14).

6 Derive the differential and total scattering cross sections for a diffuse sphere (that is, $|n - 1| \ll 1$) of unit permeability ($\mu = 1$) whose radius is large compared with the wavelength ($x \gg 1$) but still not too large: $x|n - 1| \ll 1$. Assume that the scattering angle is not too small ($x \sin\frac{1}{2}\theta \gg 1$).

7 Calculate the polarization (as a function of the scattering angle) of the wave scattered by a uniform dielectric sphere when $|n - 1| \ll 1$, $|\mu - 1| \ll 1$, $x|n - 1| \ll 1$, and the incident radiation is unpolarized.

8 What is the polarization (as a function of angle) of the scattered wave at an electric resonance of order J (that is, $\alpha_J = \frac{1}{2}\pi$) if the incident radiation is unpolarized?

9 Prove the last sentence in the paragraph following Eq. (3.36).

10 Suppose a given dense sphere of radius 0.1 cm is known to have a resonance at a wavelength of 6 cm. Its scattering cross section there is σ. Another sphere of radius 0.15 cm is made of the same material. At what wavelength will it have a resonance, and what is its approximate cross section there?

11 What are the degree and direction of polarization of the radiation scattered at 60° by a totally reflecting sphere whose radius is small compared with the wavelength?

12 Let light of wavelength 5×10^{-5} cm shine on a sphere or radius 0.01 cm. What are the radii of the two innermost dark diffraction rings when seen on a screen at a distance of 5 m?

13 What is the polarization of the light in the region of Fraunhofer diffraction if the incident beam is unpolarized?

14 Calculate the relative intensities of the rays directly reflected from the surface of a sphere ($p = 0$), those twice refracted ($p = 1$), and those once internally reflected ($p = 2$), for both polarization directions. To get an idea of the relative importance of terms in the series in (3.80), make a table of these values for $n = 4/3$, 3, and 5. You may compare your results with van de Hulst (1957), pp. 219–220.

15 Calculate and plot the differential scattering cross sections as functions of the angle for a sphere of 1 mm radius made of a material of refractive index $n = 1.1$, if the scattered radiation has a wavelength of (*a*) 1 cm, (*b*) 10 cm, (*c*) 0.01 cm. Assume that the index of refraction is independent of the wavelength.

CHAPTER 4

Miscellaneous

4.1 Other Methods

4.1.1 Debye Potentials. The Maxwell equations (1.4) for a uniform isotropic nonmagnetic medium of refractive index n can be simplified by setting

$$\mathfrak{F}_{\pm} = \mathscr{B} \mp in\mathscr{E}. \tag{4.1}$$

The functions $\mathfrak{F}_{\pm}$ then satisfy the equations

$$\nabla \times \mathfrak{F}_{\pm} = \pm kn\mathfrak{F}_{\pm} \tag{4.2}$$

inside and outside the medium, but not on the boundary. These equations are solved by the introduction of two scalar potential functions

$$u_{\pm} = \phi \pm n\psi \tag{4.3}$$

which satisfy the wave equation

$$\nabla^2\psi + (kn)^2\psi = 0 \qquad \nabla^2\phi + (kn)^2\phi = 0. \tag{4.4}$$

It is easily verified by direct computation that

$$\mathfrak{F}_{\pm} = k\nabla \times (\mathbf{r}u_{\pm}) \pm \frac{1}{n}\nabla \times [\nabla \times (\mathbf{r}u_{\pm})] \tag{4.5}$$

then satisfies (4.2). The functions $u_{\pm}$ are called the *Debye potentials.* The electric and magnetic fields are expressed in terms of these by

$$\begin{aligned} -i\mathscr{E} &= k\nabla \times (\mathbf{r}\psi) + \frac{1}{n^2}\nabla \times [\nabla \times (\mathbf{r}\phi)] \\ \mathscr{B} &= k\nabla \times (\mathbf{r}\phi) + \nabla \times [\nabla \times (\mathbf{r}\psi)] \end{aligned} \tag{4.6}$$

or in terms of the r, θ, and φ components (see Sec. 2.1.8)

$$
\begin{aligned}
-i\mathscr{E}_r &= k^2 r\phi + \frac{1}{n^2}\frac{\partial^2}{\partial r^2}(r\phi) \\
-i\mathscr{E}_\theta &= \frac{k}{r\sin\theta}\frac{\partial}{\partial\varphi}(r\psi) + \frac{1}{n^2}\frac{1}{r}\frac{\partial^2}{\partial r\partial\theta}(r\phi) \\
-i\mathscr{E}_\varphi &= -\frac{k}{r}\frac{\partial}{\partial\theta}(r\psi) + \frac{1}{n^2 r\sin\theta}\frac{\partial^2}{\partial r\partial\varphi}(r\phi)
\end{aligned}
\tag{4.7}
$$

$$
\begin{aligned}
\mathscr{B}_r &= (nk)^2 r\psi + \frac{\partial^2}{\partial r^2}(r\psi) \\
\mathscr{B}_\theta &= \frac{k}{r\sin\theta}\frac{\partial}{\partial\varphi}(r\phi) + \frac{1}{r}\frac{\partial^2}{\partial r\partial\theta}(r\psi) \\
\mathscr{B}_\varphi &= -\frac{k}{r}\frac{\partial}{\partial\theta}(r\phi) + \frac{1}{r\sin\theta}\frac{\partial^2}{\partial r\partial\varphi}(r\psi).
\end{aligned}
\tag{4.8}
$$

If the boundary of the medium is spherical, then the boundary conditions on ψ and ϕ are simple. Since the tangential components $\mathscr{E}_\theta$, $\mathscr{E}_\varphi$, $\mathscr{B}_\theta$, and $\mathscr{B}_\varphi$ of $\mathscr{E}$ and $\mathscr{B}$ must be continuous, we must make ϕ and ψ, their θ and φ derivatives, and $\partial(r\psi)/\partial r$ and $(1/n^2)\ \partial(r\phi)/\partial r$ continuous. The scattering problem of a sphere is therefore reduced to two scalar scattering problems for Eq. (4.4) with the boundary conditions just given.

The scalar functions ϕ and ψ can be expanded on the basis of the Legendre functions. The plane wave coming in along the positive z direction and polarized along the x axis is described by the potentials

$$
\begin{aligned}
r\phi_{\text{inc}} &= -k^{-2}\cos\varphi \sum_{l=1}^{\infty} i^l \frac{2l+1}{l(l+1)} u_l(kr) P_l^1(\cos\theta) \\
r\psi_{\text{inc}} &= -ik^{-2}\sin\varphi \sum_{l=1}^{\infty} i^l \frac{2l+1}{l(l+1)} u_l(kr) P_l^1(\cos\theta).
\end{aligned}
\tag{4.9}
$$

In the derivation of this result (2.57) has been used, as well as the recurrence relations for the Legendre polynomials and for the Bessel functions. Because of the boundary conditions on ϕ and ψ, the outgoing wave must contain the same φ dependence and

$$
\begin{aligned}
r\phi_{\text{scatt}} &= -k^{-2}\cos\varphi \sum_{l=1}^{\infty} i^l \frac{2l+1}{l(l+1)} f_l(r) P_l^1(\cos\theta) \\
r\psi_{\text{scatt}} &= -ik^{-2}\sin\varphi \sum_{l=1}^{\infty} i^l \frac{2l+1}{l(l+1)} g_l(r) P_l^1(\cos\theta)
\end{aligned}
$$

where f_l and g_l must satisfy the equations

$$
-f_l'' + l(l+1)r^{-2}f_l = (kn)^2 f_l \qquad -g_l'' + l(l+1)r^{-2}g_l = (kn)^2 g_l
$$

inside the sphere, and the same with $n = 1$, outside. They are therefore multiples of Riccati–Hankel functions outside, multiples of Riccati–Bessel functions inside, and the constants are found from the boundary conditions.

The results are the same as those of Sec. 2.4 for $\mu = 1$, and we need not repeat them here.

4.1.2 The Green's-Function Method. The introduction of Green's functions appropriate to the boundary condition at hand is a common procedure in solving electromagnetic problems. The Green's functions used, though, are usually adapted to potential or diffraction problems with perfectly conducting boundaries, or the like. We shall here use a method which is customary in quantum scattering theory and shall adapt it to the electromagnetic case.

The differentiated Maxwell equations (1.4) may be written

$$\nabla \times (\nabla \times \mathscr{E}) - k^2\mathscr{E} = k^2(n'^2 - 1)\mathscr{E} + \nabla \times [(1 - \mu^{-1})\nabla \times \mathscr{E}] \tag{4.10}$$

where the index of refraction n'

$$n'^2 = \varepsilon + \frac{4\pi i\sigma}{\omega}$$

and the permeability μ may both be tensors. If μ is not a tensor, (4.10) may also be rewritten as in (1.6),

$$\nabla \times (\nabla \times \mathscr{E}) - k^2\mathscr{E} = k^2(n^2 - 1)\mathscr{E} + \mu^{-1}(\nabla\mu) \times (\nabla \times \mathscr{E}) \tag{4.11}$$

with

$$n^2 = \mu n'^2.$$

The tensor Green's function

$$\begin{aligned}\mathbf{\Gamma}(k;\mathbf{r},\mathbf{r}') &= (\mathbb{1} + k^{-2}\nabla\nabla)\frac{\exp(ik|\mathbf{r} - \mathbf{r}'|)}{4\pi|\mathbf{r} - \mathbf{r}'|} \\ &= \lim_{\varepsilon\to 0+}\int \frac{d\mathbf{k}'}{(2\pi)^3}\frac{\mathbb{1}k^2 - \mathbf{k}'\mathbf{k}'}{k^2(k'^2 - k^2 - i\varepsilon)}\exp[i\mathbf{k}'\cdot(\mathbf{r} - \mathbf{r}')],\end{aligned} \tag{4.12}$$

written in dyadic notation, allows us to combine the differential equation (4.10) and the scattering boundary condition into an integral equation. It satisfies the differential equation

$$\nabla \times (\nabla \times \mathbf{\Gamma}) - k^2\mathbf{\Gamma} = \mathbb{1}\delta(\mathbf{r} - \mathbf{r}') \tag{4.13}$$

as well as

$$\nabla \cdot \mathbf{\Gamma} = 0 \quad \text{for } \mathbf{r} \neq \mathbf{r}'. \tag{4.14}$$

In addition, at large distance r it contains *outgoing* spherical waves only and is thus appropriate to the scattering boundary condition.

The integral equation for $\mathscr{E}(\mathbf{k}\nu,\mathbf{r})$, in the notation explained after (2.75) (but with ν for the index of $\boldsymbol{\chi}_\nu$ in order not to confuse it with the permeability), is

$$\begin{aligned}\mathscr{E}(\mathbf{k}\nu,\mathbf{r}) = \mathscr{E}_0(\mathbf{k}\nu,\mathbf{r}) + \int d\mathbf{r}'\,\mathbf{\Gamma}(k;\mathbf{r},\mathbf{r}') \\ \cdot\{k^2(n'^2 - 1)\cdot\mathscr{E}(\mathbf{k}\nu,\mathbf{r}') + \nabla' \times [(1 - \mu^{-1})\nabla' \times \mathscr{E}(\mathbf{k}\nu,\mathbf{r}')]\}.\end{aligned} \tag{4.15}$$

The integral extends over the volume of the scatterer only.

The asymptotic behavior of the field as $r \to \infty$ is obtained from that of the Green's function. Since

$$|\mathbf{r} - \mathbf{r}'| = r(1 - 2\mathbf{r}' \cdot \hat{\mathbf{r}} r^{-1} + r'^2 r^{-2})^{1/2} \simeq r - \mathbf{r}' \cdot \hat{\mathbf{r}}$$

when $r \gg r'$, we have

$$\boldsymbol{\Gamma}(k;\mathbf{r},\mathbf{r}') \underset{r\to\infty}{\simeq} (\mathbb{1} - \hat{\mathbf{r}}\hat{\mathbf{r}}) \frac{e^{ikr}}{4\pi r} e^{-ik\hat{\mathbf{r}}\cdot\mathbf{r}'} . \tag{4.16}$$

Therefore, when $kr \gg 1$ and r is much larger than the dimensions of the scatterer,

$$\begin{aligned} \mathscr{E}(\mathbf{k}\nu,\mathbf{r}) \simeq \mathscr{E}_0(\mathbf{k}\nu,\mathbf{r}) + \frac{e^{ikr}}{4\pi r} \int d\mathbf{r}' \, (\mathbb{1} - \hat{\mathbf{r}}\hat{\mathbf{r}}) e^{-ik\hat{\mathbf{r}}\cdot\mathbf{r}'} \\ \times \left[k^2(n'^2 - 1) + \boldsymbol{\nabla}' \times (1 - \mu^{-1}) \boldsymbol{\nabla}' \times\right] \mathscr{E}(\mathbf{k}\nu,\mathbf{r}'). \end{aligned} \tag{4.17}$$

The incoming plane wave is that of (2.76),

$$\mathscr{E}_0(\mathbf{k}\nu,\mathbf{r}) = \exp(i\mathbf{k} \cdot \mathbf{r}) \boldsymbol{\chi}'_\nu$$

where $\boldsymbol{\chi}'_\nu$ is the circular-polarization vector relative to $\mathbf{k}$ as the z axis, with $\nu = \pm 1$. We may then write

$$(\mathbb{1} - \hat{\mathbf{r}}\hat{\mathbf{r}}) \exp(-ik\hat{\mathbf{r}} \cdot \mathbf{r}') = \sum_{\nu'=\pm 1} \boldsymbol{\chi}''_{\nu'} \mathscr{E}_0^*(\mathbf{k}'\nu',\mathbf{r}')$$

with $\mathbf{k}' = \hat{\mathbf{r}}k$ and $\boldsymbol{\chi}''_\nu$ the circular-polarization vector relative to $\mathbf{k}'$ as the z axis. Then

$$\mathscr{E}(\mathbf{k}\nu,\mathbf{r}) \underset{r\to\infty}{\simeq} \mathscr{E}_0(\mathbf{k}\nu,\mathbf{r}) + r^{-1} e^{ikr} \sum_{\nu'=\pm 1} \boldsymbol{\chi}''_{\nu'} A(\mathbf{k}'\nu',\mathbf{k}\nu) \tag{4.18}$$

and the scattering amplitiude is given by

$$\begin{aligned} A(\mathbf{k}'\nu',\mathbf{k}\nu) = \frac{1}{4\pi} \int_{\text{scatterer}} d\mathbf{r} \, \mathscr{E}_0^*(\mathbf{k}'\nu',\mathbf{r}) \cdot \{ k^2(n'^2 - 1) \mathscr{E}(\mathbf{k}\nu,\mathbf{r}) \\ + \boldsymbol{\nabla} \times [(1 - \mu^{-1}) \boldsymbol{\nabla} \times \mathscr{E}(\mathbf{k}\nu,\mathbf{r})]\}. \end{aligned} \tag{4.19}$$

An integration by parts yields the somewhat simpler expression

$$\begin{aligned} A(\mathbf{k}'\nu',\mathbf{k}\nu) = \frac{k^2}{4\pi} \int_{\text{scatterer}} d\mathbf{r} \, [\mathscr{E}_0^*(\mathbf{k}'\nu',\mathbf{r}) \cdot (n'^2 - 1) \cdot \mathscr{E}(\mathbf{k}\nu,\mathbf{r}) \\ + \mathscr{H}_0^*(\mathbf{k}'\nu',\mathbf{r}) \cdot (\mu - 1) \cdot \mathscr{H}(\mathbf{k}\nu,\mathbf{r})] \end{aligned} \tag{4.20}$$

where the labels on the magnetic field refer to the polarization of the *electric* field. That is,

$$\mathscr{H}_0(\mathbf{k}\nu,\mathbf{r}) = -ik^{-1} \boldsymbol{\nabla} \times \mathscr{E}_0(\mathbf{k}\nu,\mathbf{r})$$

and

$$\mu \mathscr{H}(\mathbf{k}\nu,\mathbf{r}) = -ik^{-1} \boldsymbol{\nabla} \times \mathscr{E}(\mathbf{k}\nu,\mathbf{r}).$$

We expand the Green's function on the basis of the vector spherical harmonics. Since

$$-\frac{\exp(ik|\mathbf{r}-\mathbf{r}'|)}{4\pi|\mathbf{r}-\mathbf{r}'|} = \frac{1}{krr'}\sum_{l,m} Y_l^m(\hat{\mathbf{r}})\Gamma_l(k;r,r')Y_l^{m*}(\hat{\mathbf{r}}')(-)^l \tag{4.21}$$

with[1]

$$\Gamma_l(k;r,r') = u_l(kr_<)w_l^{(+)}(kr_>) \tag{4.22}$$

(u_l and $w_l^{(+)}$ being the spherical Riccati–Bessel functions defined in Sec. 2.2.1), we obtained by the use of the recurrence relations (2.69)

$$\boldsymbol{\Gamma}(k;\mathbf{r},\mathbf{r}') = (krr')^{-1}\sum_{JM\lambda\lambda'} \mathbf{Y}_{JM}^{(\lambda)}(\hat{\mathbf{r}})\Gamma_{\lambda\lambda'}^{J}(k;r,r')\mathbf{Y}_{JM}^{(\lambda')*}(\hat{\mathbf{r}}')(-)^J \tag{4.23}$$

with

$$\begin{aligned}
\Gamma_{ee}^J &= u_J'(kr_<)w_J^{(+)\prime}(kr_>)\\
\Gamma_{mm}^J &= -u_J(kr_<)w_J^{(+)}(kr_>)\\
\Gamma_{00}^J &= -\frac{J+1}{k^2rr'}u_J(kr_<)w_J^{(+)}(kr_>)
\end{aligned} \tag{4.24}$$

$$\Gamma_{e0}^J(k;r,r') = \Gamma_{0e}^J(k;r',r) = -\frac{[J(J+1)]^{1/2}}{k^2r'}\frac{\partial}{\partial r}[u_J(kr_<)w_J^{(+)}(kr_>)]$$

and all others, zero. The integral equation (4.15) may now be projected onto specific states of $JM\lambda$,

$$\mathscr{E}(\mathbf{k}\nu,\mathbf{r}) = \frac{4\pi}{kr}\sum_{\substack{JM\lambda\\J'M'\lambda'}} \mathbf{Y}_{JM}^{(\lambda)}(\hat{\mathbf{r}})\mathscr{E}_{JM,J'M'}^{\lambda\lambda'}(r)i^J\mathbf{Y}_{J'M'}^{(\lambda')*}(\hat{\mathbf{k}})\cdot\boldsymbol{\chi}_\nu'. \tag{4.25}$$

Then

$$\begin{aligned}
\mathscr{E}_{JM,J'M'}^{\lambda\lambda'}(r) &= u_J(kr)\delta_{JJ'}\delta_{MM'}\delta_{\lambda\lambda'}(\delta_{\lambda0}-1)\\
&\quad + k^2\sum_{\substack{J''M''\\\lambda'''\lambda''}}\int_0^\infty dr'\,\Gamma_{\lambda\lambda'''}^J(r,r')\mathfrak{N}_{JM,J''M''}^{\lambda'''\lambda''}(r')\mathscr{E}_{J''M'',J'M'}^{\lambda''\lambda'}(r')
\end{aligned} \tag{4.26}$$

with

$$\mathfrak{N}_{JM,J''M''}^{\lambda'''\lambda''}(r) = i^{J+J''}\int d\Omega\,\mathbf{Y}_{JM}^{(\lambda''')*}(\hat{\mathbf{r}})\cdot(n^2-1)\cdot\mathbf{Y}_{J''M''}^{(\lambda'')}(\hat{\mathbf{r}}). \tag{4.27}$$

If the scatterer is spherically symmetric, then n can be taken out of the integral in (4.27) and

$$\mathfrak{N}_{JM,J''M''}^{\lambda'''\lambda''}(r) = (-)^J\delta_{JJ''}\delta_{MM''}\delta_{\lambda'''\lambda''}[n^2(r)-1].$$

Equation (4.26) then reduces to the simpler set of three coupled equations

[1] The subscripts < and > indicate the lesser and the larger of the two quantities r and r'.

$$\mathscr{E}^{\lambda\lambda'}_{JM}(r) = u_J(kr)\delta_{\lambda\lambda'}(\delta_{\lambda 0} - 1) + (-)^J k^2 \sum_{\lambda''} \int_0^R dr'\, \Gamma^J_{\lambda\lambda''}(r,r')[n^2(r') - 1]\mathscr{E}^{\lambda''\lambda'}_{JM}(r'). \quad (4.28)$$

However, if n is a tensor, then even a homogeneous spherical scatterer produces couplings between J and $J \pm 2$ and Eqs. (4.26) do not separate.

If the scatterer is small enough and not too dense, the integral equation (4.15) or (4.26) can be solved by iteration. The exact conditions under which this series of successive approximations converges will not be discussed here. The integral equation (4.15) and its partial-wave analog (4.26) are exactly of the same form as the corresponding integral equations for the scattering of a particle by a potential of finite range in quantum mechanics. We therefore refer to Chap. 9 for the relevant discussion.

In case $|n^2 - 1| \ll 1$ and also $x|n^2 - 1| \ll 1$ it suffices to insert the incoming plane wave $\mathscr{E}_0(k\nu,\mathbf{r})$, that is, the "zeroth" approximation to (4.15), in the expression (4.19) or (4.20) for the scattering amplitude. We then get what is called the *Born approximation* in quantum mechanics,

$$A(\mathbf{k}'\nu',\mathbf{k}\nu) \simeq \frac{k^2}{4\pi}\int d\mathbf{r}\exp[i\mathbf{r}\cdot(\mathbf{k} - \mathbf{k}')]\boldsymbol{\chi}''^{*}_{\nu'}\cdot[n'^2 - 1 + \nu\nu'(\mu - 1)]\cdot\boldsymbol{\chi}'_{\nu} \quad (4.29)$$

because the definitions (2.20) show that

$$\hat{\mathbf{k}} \times \boldsymbol{\chi}'_{\nu} = -i\nu\boldsymbol{\chi}'_{\nu}.$$

If the scattering plane is used as reference plane, then

$$\hat{\mathbf{e}}''_y = \hat{\mathbf{e}}'_y = \hat{\mathbf{e}}_{\perp}$$

$$\hat{\mathbf{e}}''_x = \hat{\mathbf{e}}'_x\cos\theta - \hat{\mathbf{k}}\sin\theta = \hat{\mathbf{e}}_{\parallel}\cos\theta - \hat{\mathbf{k}}\sin\theta.$$

Therefore, if n' and μ are not tensors, then (4.29) is seen to reduce to the Rayleigh–Gans result (3.16). In the more general case it is best to keep it in the form (4.29) and to remember that $\boldsymbol{\chi}_1$ is the basis vector for right-handed circular polarization.

4.2 Causality and Dispersion Relations

4.2.1 Introduction. Let us look again at the description of the scattering process in time. We consider a packet of radiation sent toward an obstacle. After it gets there, the scattered radiation moves outward in all directions. First we look specifically in the forward direction.

Since electromagnetic energy travels with the finite velocity c, "causality" requires that no scattered radiation be observed at any point P beyond the scatterer until sufficient time has elapsed for the incident radiation to have reached the scatterer and for the scattered to have subsequently traveled to P. That requirement implies an important restriction on the behavior of the scattering amplitude as a function of the frequency.

We Fourier-analyze the incident wave packet, which travels in the direction $\hat{\mathbf{k}}$,

$$\mathscr{E}_{\rm inc}(t,\mathbf{r}) = \int_{-\infty}^{\infty} d\omega\, \mathscr{E}_{\rm inc}(\omega) \exp\left[i\omega\left(\frac{1}{c}\hat{\mathbf{k}}\cdot\mathbf{r} - t\right)\right]$$

and therefore

$$\mathscr{E}_{\rm inc}(\omega) = \frac{1}{2\pi}\int_{-\infty}^{\infty} dt\, \mathscr{E}_{\rm inc}(t,\mathbf{r}) \exp\left[-i\omega\left(\frac{1}{c}\hat{\mathbf{k}}\cdot\mathbf{r} - t\right)\right].$$

If there is a *sharp wave front* which travels toward the origin and arrives there at time $t = 0$, then the last integral runs only from $t = t_0 \equiv \hat{\mathbf{k}}\cdot\mathbf{r}/c$,

$$\mathscr{E}_{\rm inc}(\omega) = \frac{1}{2\pi}\int_{t_0}^{\infty} dt\, \mathscr{E}_{\rm inc}(t,\mathbf{r}) e^{i\omega(t-t_0)}.$$

But in such an integral it is obviously possible to let ω be complex, with a positive imaginary part. Since the derivative with respect to ω also exists when $\operatorname{Im}\omega > 0$, each component of $\mathscr{E}_{\rm inc}(\omega)$ is the boundary value of an analytic function of ω regular in the upper half of the complex ω plane.[2] What is more, when $|\omega| \to \infty$ in $\operatorname{Im}\omega > 0$, then $\mathscr{E}_{\rm inc}(\omega)$ vanishes. Conversly, these two conditions, analyticity and asymptotic vanishing, are sufficient to assure that, when $t < t_0$, the Fourier integral for $\mathscr{E}_{\rm inc}(t,\mathbf{r})$ may be carried out by adding a large semicircle in the upper half of the ω plane; it then vanishes by Cauchy's theorem.

4.2.2 Forward-Dispersion Relations. We now switch to the two-component notation introduced in Sec. 1.2.1. The foregoing discussion, of course, applies then too. The forward-scattered wave components are determined from the incident for each frequency by the linear relation

$$\mathfrak{E}_{\rm scatt}(\omega) = \mathfrak{A}(\omega)\cdot\mathfrak{E}_{\rm inc}(\omega),$$

$\mathfrak{A}(\omega)$ being the forward-scattering amplitude. The requirement of causality is that *whenever* the incident wave vanishes until $t \geq t_0 = \hat{\mathbf{k}}\cdot\mathbf{r}/c$, then the forward-scattered wave must vanish also until $t \geq t_0$ (for $t_0 > 0$). Translated into the Fourier-transformed language, this means that, whenever $\mathfrak{E}_{\rm inc}(\omega)$ is chosen to be the boundary value of a function analytic in the upper half of the complex ω plane and to vanish there as $|\omega| \to \infty$, $\mathfrak{E}_{\rm scatt}(\omega)$ must have the same property. Clearly this can be the case only if the forward-scattering amplitude $\mathfrak{A}(\omega)$ itself *is the boundary value of an analytic function regular in the upper half of the ω plane, and it is bounded there when* $|\omega| \to \infty$. This is thus the consequence of the causality requirement.

2 We are here more interested in an intuitive feeling for why these Fourier transforms are boundary values of analytic functions than in the exact mathematical conditions under which they are. This is why nothing specific is said about square integrability, etc. We take it for granted that the functions discussed tend to zero at infinity, along the real line, sufficiently rapidly to be square-integrable. For more detailed discussions of the mathematical points involved here, see E. C. Titchmarsh (1937).

The analyticity of a function $f(\omega)$ can be expressed simply in the form of an integral. Suppose that $f(\omega)$ is the boundary value of an analytic function regular in the upper half plane and that, when $|\omega| \to \infty$ there, $f(\omega) \to 0$. Then

$$\int_{-\infty}^{\infty} d\omega' \frac{f(\omega')}{\omega' - \omega}$$

can be evaluated by adding a large semicircle in the upper half plane and then using Cauchy's residue theorem. If $\operatorname{Im}\omega > 0$, the pole at $\omega = \omega'$ gives

$$\int_{-\infty}^{\infty} d\omega' \frac{f(\omega')}{\omega' - \omega} = 2\pi i f(\omega)$$

but if $\operatorname{Im}\omega < 0$, the result is zero. For real ω we therefore obtain

$$f(\omega) = \frac{1}{i\pi} \mathscr{P} \int_{-\infty}^{\infty} d\omega' \frac{f(\omega')}{\omega' - \omega}, \tag{4.30}$$

the Cauchy principal value being defined by

$$\mathscr{P} \int_{-\infty}^{\infty} d\omega' \frac{f(\omega')}{\omega' - \omega} = \lim_{\varepsilon \to 0} \frac{1}{2} \int_{-\infty}^{\infty} d\omega' f(\omega') \left(\frac{1}{\omega' - \omega + i\varepsilon} + \frac{1}{\omega' - \omega - i\varepsilon} \right).$$

Taking real and imaginary parts, we find that the real and imaginary parts of f are one another's *Hilbert transforms*,

$$\begin{aligned} \operatorname{Re} f(\omega) &= \frac{1}{\pi} \mathscr{P} \int_{-\infty}^{\infty} d\omega' \frac{\operatorname{Im} f(\omega')}{\omega' - \omega} \\ \operatorname{Im} f(\omega) &= -\frac{1}{\pi} \mathscr{P} \int_{-\infty}^{\infty} d\omega' \frac{\operatorname{Re} f(\omega')}{\omega' - \omega}. \end{aligned} \tag{4.31}$$

Since they imply one another, we need to keep only the first.

If the function $f(\omega)$ does not vanish at $|\omega| \to \infty$ when $\operatorname{Im}\omega \geq 0$ but is bounded there, we may take, instead,

$$g(\omega) = \frac{f(\omega) - f(\omega_0)}{\omega - \omega_0}.$$

Then it follows as before that

$$\operatorname{Re} \frac{f(\omega) - f(\omega_0)}{\omega - \omega_0} = \frac{1}{\pi} \mathscr{P} \int_{-\infty}^{\infty} \frac{d\omega'}{\omega' - \omega} \operatorname{Im} \frac{f(\omega') - f(\omega_0)}{\omega' - \omega_0}$$

or, for real ω_0,

$$\operatorname{Re} f(\omega) = \operatorname{Re} f(\omega_0) + \frac{1}{\pi} \mathscr{P} \int_{-\infty}^{\infty} d\omega' \operatorname{Im} [f(\omega') - f(\omega_0)] \frac{\omega - \omega_0}{(\omega' - \omega)(\omega' - \omega_0)}. \tag{4.32}$$

Let us assume first in addition to causality that the forward-scattering amplitude vanishes in the high-frequency limit. Then it must satisfy the equation

$$\operatorname{Re} \mathfrak{A}(\omega) = \frac{1}{\pi} \mathscr{P} \int_{-\infty}^{\infty} d\omega' \frac{\operatorname{Im} \mathfrak{A}(\omega')}{\omega' - \omega}. \tag{4.33}$$

This is simplified by the reality requirement on the field. If the time-dependent electric field is to be real, its Fourier transform must satisfy

$$\mathscr{E}(-\omega) = \mathscr{E}^*(\omega)$$

and therefore

$$\mathfrak{E}^{(L)}(-\omega) = \mathfrak{E}^{(L)*}(\omega)$$
$$\mathfrak{E}^{(C)}(-\omega) = \Upsilon \mathfrak{E}^{(C)*}(\omega)$$

with

$$\Upsilon = \begin{pmatrix} 0 & 1 \\ 1 & 0 \end{pmatrix}.$$

The scattering amplitude must be such that every real incident wave produces a real scattered wave. Therefore it must be such that

$$\mathfrak{A}^{(L)}(-\omega) = \mathfrak{A}^{(L)*}(\omega) \qquad \mathfrak{A}^{(C)}(-\omega) = \Upsilon \mathfrak{A}^{(C)*}(\omega) \Upsilon \tag{4.34}$$

for real ω. That means specifically that $\operatorname{Re} \mathfrak{A}^{(L)}(\omega)$ is an even function of ω and that $\operatorname{Im} \mathfrak{A}^{(L)}(\omega)$ is odd. The Hilbert transform thus becomes

$$\operatorname{Re} \mathfrak{A}^{(L)}(\omega) = \frac{2}{\pi} \mathscr{P} \int_0^{\infty} \frac{d\omega' \, \omega'}{\omega'^2 - \omega^2} \operatorname{Im} \mathfrak{A}^{(L)}(\omega'). \tag{4.33a}$$

This equation becomes of special interest when used in conjunction with the optical theorem (1.77). We then get for the diagonal elements

$$\operatorname{Re} A_{ii}^{L}(\omega) = \frac{1}{2\pi^2 c} \mathscr{P} \int_0^{\infty} \frac{d\omega' \, \omega'^2}{\omega'^2 - \omega^2} \sigma_{\text{total}}^{L(i)}(\omega'). \tag{4.35}$$

In other words, the causality requirement allows us to calculate the forward-scattering amplitude from a knowledge of the total cross section as a function of the frequency. The analogous equations for circular-polarization components are somewhat more complicated.

An application of this result which is of specific importance is its use for the index of refraction n and the absorption coefficient γ of a medium. According to (1.98) and (1.101) these are expressible in terms of the real part of the forward scattering amplitude and the total cross section. We then obtain

$$n(\omega) = 1 + \frac{c}{\pi} \mathscr{P} \int_0^{\infty} \frac{d\omega' \, (\omega'/\omega)^2}{\omega'^2 - \omega^2} \gamma(\omega'). \tag{4.36}$$

n is here meant to be the *geometrical* (real) index of refraction, and, of course, n and γ must refer to the same polarization. This equation expresses the frequency dependence of the index of refraction in terms of the absorption coefficient, and hence it determines the amount of *dispersion*. It is therefore usually referred to as a *dispersion relation*, and so is the corresponding equation (4.35) for the amplitude.

The dispersion formula (4.36), taken literally, does not really follow from (4.35). The connection between the forward-scattering amplitude and index of refraction holds only in the limit of short wavelengths, i.e., large frequencies. But the argument for the dispersion formula for n can be made directly by means of the Maxwell equation (1.4),

$$\nabla \times \mathscr{H} = -ikn^2\mathscr{E}.$$

If $\mathscr{E}(\omega)$ is analytic in the upper half plane and vanishes there at infinity, then $\mathscr{H}(\omega)$ will have that same property only if $n(\omega)$ is analytic in the upper half plane and is bounded there at infinity. Unless both $\mathscr{E}(\omega)$ and $\mathscr{H}(\omega)$ have this property, the medium does not allow a sharp wave front to progress with the velocity of light (or less). Instead it produces an advance leakage of energy at infinite speed. If this is to be ruled out, the dispersion relation

$$n(\omega) = 1 + \frac{c}{\pi}\mathscr{P}\int_0^\infty \frac{d\omega'\,\gamma(\omega')}{\omega'^2 - \omega^2} \tag{4.37}$$

follows under the additional assumption that the complex index of refraction approaches 1 as $|\omega| \to \infty$, and that it satisfies the "reality condition" $n(-\omega) = n^*(\omega)$. If one makes the stronger assumption that

$$(n - 1)\omega^2 \to 0$$

when $|\omega| \to \infty$, then (4.36) follows.

There is no contradiction between these two dispersion formulas. In fact, many others can be written down if n approaches unity very strongly at infinity. The "weaker" dispersion relation (4.37), which assumes only that $|n| \to 1$ as $|\omega| \to \infty$ and hence $|A(\omega)| = o(\omega^2)$ by (1.97), is equivalent to

$$\operatorname{Re} A_{ii}^L(\omega) = A_{ii}^L(0) + \frac{\omega^2}{2\pi^2 c}\mathscr{P}\int_0^\infty \frac{d\omega'}{\omega'^2 - \omega^2}\,\sigma_{\text{total}}^{L(i)}(\omega'). \tag{4.38}$$

This is weaker than (4.35) in the sense that, in addition to analyticity, it implies only that $\operatorname{Re} A_{ii}(\omega) = o(\omega^2)$ as $|\omega| \to \infty$, whereas (4.35) implies that $\operatorname{Re} A_{ii}(\omega) = o(1)$. Every amplitude which satisfies (4.35) also satisfies (4.38), but not vice versa.

4.2.3 Nonforward-Dispersion Relations. The argument leading from causality to a forward-dispersion relation can be generalized to nonforward-scattering amplitudes. In this case, however, the time t_0 prior to which the scattered wave vanishes is not the same as that prior to which the incident wave is zero. As seen in Fig. 4.1, if the radius of the scattering sphere is R and the scattering angle is θ, a ray that makes the angle α with the surface may save itself a distance

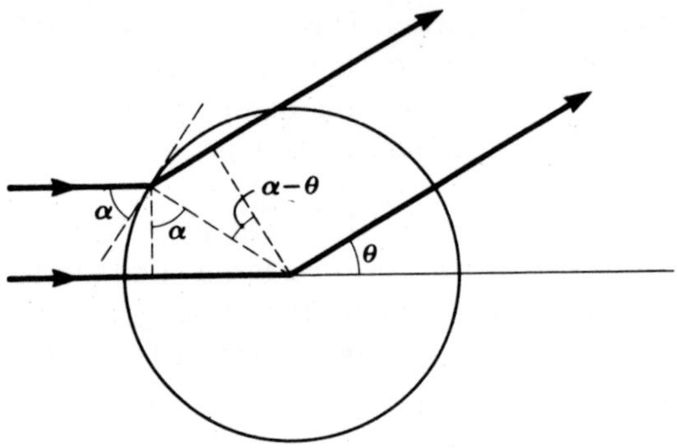

Figure 4.1. Path lengths in scattering by a sphere.

$$\Delta = R[\sin\alpha - \sin(\alpha - \theta)]$$

by being deflected immediately on the surface. This saving has the maximal value $2R\sin\frac{1}{2}\theta = R|\hat{\mathbf{k}}' - \hat{\mathbf{k}}|$ when $\alpha = \frac{1}{2}\theta$. Thus, if $\mathscr{E}_{\text{inc}}(t,\mathbf{r}) = 0$ for $ct < \hat{\mathbf{k}} \cdot \mathbf{r}$, then $\mathscr{E}_{\text{scatt}}(t,\mathbf{r}) = 0$ for $ct < \hat{\mathbf{k}} \cdot \mathbf{r} - 2R\sin\frac{1}{2}\theta \equiv ct_1$. Therefore, whenever $\mathscr{E}(\omega)$ is the boundary value of an analytic function regular in the upper half of the complex plane and vanishes there at infinity, then so is and does

$$\begin{aligned} e^{2ikR\sin 1/2\theta}\mathscr{E}_{\text{scatt}}(\omega) &= \frac{1}{2\pi}\int_t^\infty dt\,\mathscr{E}_{\text{scatt}}(t,\mathbf{r})e^{i\omega(t-t_0)}\,e^{i\omega(t_0-t_1)} \\ &= \frac{1}{2\pi}\int_{t_1}^\infty dt\,\mathscr{E}_{\text{scatt}}(t,\mathbf{r})e^{i\omega(t-t_1)}. \end{aligned}$$

Consequently the function

$$e^{iR|\mathbf{k}'-\mathbf{k}|}\mathfrak{A}(\mathbf{k}',\mathbf{k}) = e^{2i\omega(R/c)\sin 1/2\theta}\mathfrak{A}(\omega;\theta,\varphi)$$

satisfies a dispersion relation, but not necessarily $\mathfrak{A}(\mathbf{k}',\mathbf{k})$ itself. Under the additional assumption that $\mathfrak{A}(\omega;\theta,\varphi)$ vanishes when $|\omega| \to \infty$ in $\operatorname{Im}\omega \geq 0$, this means that

$$\begin{aligned} &\operatorname{Re}\left\{\exp\left[2i\omega\left(\frac{R}{c}\right)\sin\tfrac{1}{2}\theta\right]\mathfrak{A}^{(L)}(\omega;\theta,\varphi)\right\} \\ &\quad = \frac{2}{\pi}\mathscr{P}\int_0^\infty \frac{d\omega'\omega'}{\omega'^2 - \omega^2}\operatorname{Im}\left\{\exp\left[2i\omega'\left(\frac{R}{c}\right)\sin\tfrac{1}{2}\theta\right]\mathfrak{A}^{(L)}(\omega';\theta,\varphi)\right\}. \end{aligned} \tag{4.39}$$

This dispersion relation for nonforward angles depends on the radius R of the scatterer and is considerably less useful than (4.33a). The real and imaginary parts of $\mathfrak{A}^{(L)}$ are both contained on both sides of (4.39). Moreover, we do not have an optical theorem to give the right-hand side additional experimental significance.

We are tempted at this point to consider the scattering amplitude, not as a function of ω and the angle θ (and φ), but as a function of ω and[3] $\tau \equiv 2\omega\sin\frac{1}{2}\theta$ (and φ), because then the troublesome factor

3 In quantum-mechanical language $\hbar\tau/c$ is the momentum transfer of the scattered photon, i.e., the magnitude of the difference between its final and initial 3-momenta. It is a natural variable to consider in place of θ.

$\exp[2i\omega(R/c)\sin\frac{1}{2}\theta] = \exp[i\tau(R/c)]$ can be held *fixed* when $|\omega| \to \infty$. This, however, implies that, when we analytically continue the amplitude into the complex ω plane, then the scattering angle θ, and hence

$$\cos\theta = 1 - \frac{1}{2}\left(\frac{\tau}{\omega}\right)^2$$

becomes complex. The simple causality argument given no longer suffices now to prove that the amplitude is an analytic formation of ω. What is more, even if the necessary regularity has been demonstrated, the concomitant dispersion relation for fixed τ contains an "unphysical region" in which $\omega < \tau/2$ and hence $|\cos\theta| > 1$. We shall therefore not pursue this possibility here any further.

4.2.4 Partial-Wave-Dispersion Relations. The causality argument can be applied just as well to spherical waves of a fixed value of J and M. If a spherical wave with a sharp front at $ct = r$ is sent toward a scattering sphere of radius R, then the outgoing spherical wave must be zero until $ct = r - 2R$. This implies that whenever the incoming wave amplitude is the boundary value of an analytic function of ω regular in the upper half plane and it vanishes there at infinity, then the outgoing wave multiplied by e^{2ikR} must have the same properties. Consequently $e^{2ikR}\mathscr{S}^J$ must be the boundary value of an analytic function of ω or of k, regular in the upper half of the complex plane, and bounded there when $|k| \to \infty$. In this case $e^{2ikR}\mathscr{S}^J$ will surely not vanish when $k \to \pm\infty$; so we must write a dispersion formula like (4.32), i.e., "once subtracted." Since $\mathscr{S}(0) = 1$, we choose $k_0 = \omega_0/c = 0$ and get

$$\mathrm{Re}[e^{2ikR}\mathscr{S}^J(k)] = 1 + \frac{1}{\pi}\mathscr{P}\int_{-\infty}^{\infty} \frac{dk'\,(k/k')}{k' - k}\mathrm{Im}[e^{2ik'R}\mathscr{S}^J(k')].$$

The property

$$\mathscr{S}^J(-k) = \mathscr{S}^{J*}(k) \tag{4.40}$$

for real k follows from the corresponding relation (4.34) for the amplitude. Hence we get

$$\mathrm{Re}[e^{2ikR}\mathscr{S}^J(k)] = 1 + \frac{2}{\pi}\mathscr{P}\int_0^{\infty} \frac{dk'\,k^2/k'}{k'^2 - k^2}\mathrm{Im}\,[e^{2ik'R}\mathscr{S}^J(k')]. \tag{4.41}$$

If $\mathscr{S}^J$ is diagonal and there is no absorption, then unitarity implies that the dispersion relations read

$$\cos[2\alpha_J(k) + 2kR] = 1 + \frac{2}{\pi}\mathscr{P}\int_0^{\infty} \frac{dk'\,k^2/k'}{k'^2 - k^2}\sin[2\alpha_J(k') + 2k'R] \tag{4.42}$$

and the same for the other phase shift β_J. One may then, however, more usefully try to write down dispersion relations for $\ln\mathscr{S}^J$, that is, for $\alpha_J(k)$ and $\beta_J(k)$. Before this can be done, the zeros of $\mathscr{S}^J$ must be removed, of which there are generally infinitely many, and the dispersion relation must contain them explicitly. It is therefore not very useful.

4.3 Intensity-Fluctuation Correlations (Hanbury Brown and Twiss Effect)

We shall now discuss a method of experimentally measuring the *phase* of a scattering amplitude as a function of the scattering angle. It involves the fluctuations in incoherent beams. For a clear understanding of the principles we first go back to a conventional scattering experiment described in terms of its development in time.

An incoherent beam of radiation may be thought of as consisting of a random sequence of wave packets of random phases. At a fixed location such a wave, if essentially monochromatic, may be represented by

$$\mathfrak{E}(t) = \sum_n \eta(t - t_n)(\mathfrak{E}^{(n)} e^{-i\omega t} + \text{c.c.}), \tag{4.43}$$

$\eta(t)$ being the packet shape. The length τ of each packet is large compared with $1/\omega$. If the beam is intense enough so that there is appreciable overlap between successive packets, the result is a field whose amplitude and phase fluctuate "randomly" but so that most of the frequencies contained in these fluctuations are much smaller than ω. We may write

$$\mathfrak{E}(t) = \mathrm{E}(t) e^{-i\omega t} + \text{c.c.} \tag{4.44}$$

The energy flux of such a wave is

$$S(t) = \frac{c}{4\pi}\{[\mathrm{E}(t)^2 e^{-2i\omega t} + \text{c.c.}] + 2\mathrm{E}^*(t) \cdot \mathrm{E}(t)\}. \tag{4.45}$$

The first two terms oscillate with frequencies near 2ω, while the third term has only the low frequences in $\mathrm{E}(t)$. Every practical detector will average over the former, but not necessarily over the latter. Thus the effective flux is

$$S_{\text{eff}}(t) = \frac{c}{2\pi}\mathrm{E}^*(t) \cdot \mathrm{E}(t) = \frac{c}{2\pi} I(t) \tag{4.46}$$

and the average flux measured is

$$S = \langle S(t) \rangle = \langle S_{\text{eff}}(t) \rangle = \frac{c}{2\pi}\langle I(t) \rangle. \tag{4.47}$$

The other Stokes parameters of the beam are similarly treated. For example, according to (1.10) the "instantaneous" Stokes parameter $Q(t)$ of the wave (4.44) is

$$Q(t) = |\mathscr{E}_{\parallel}(t)|^2 - |\mathscr{E}_{\perp}(t)|^2$$

with respect to a given reference plane. This is automatically the "effective" value, since the Stokes parameters are defined so as not to refer to the oscillations of frequency 2ω. The average value of $Q(t)$ then can be written in terms of the polarization, the tilt angle χ, and the ellipticity β. According to (1.31)

$$\langle Q(t)\rangle = -P\langle I(t)\rangle \cos 2\chi \cos 2\beta. \tag{4.48}$$

Similarly, $U(t)$ and $V(t)$ are defined instantaneously, and then

$$\begin{aligned}\langle U(t)\rangle &= -P\langle I(t)\rangle \sin 2\chi \cos 2\beta \\ \langle V(t)\rangle &= P\langle I(t)\rangle \sin 2\beta.\end{aligned} \tag{4.49}$$

These equation *define* P, χ, and β for the beam.

The fluctuations of the intensity away from its average value depend on the density of the beam. If the average spacing in time of the packets in (4.43) is larger than the packet length τ, then they will rarely overlap and the intensity has essentially only two values, zero or that of a single packet (which may be taken as constant). But if there is a large amount of overlap so that usually many packets are present simultaneously, then most of the time the random phases make the field quite small. But this is compensated for by the rare occurrence of very high peaks owing to occasional *constructive* interference of the fields. The denser the beam, the rarer and the higher are these peaks in the intensity, an effect also known as *photon bunching*. It is quantum mechanically ascribed to their boson nature and consequent effective attraction.

We now use the beam (4.44) in a scattering experiment. Then for a monochromatic wave

$$\mathfrak{E}(\omega)_{\mathrm{scatt}} = \mathfrak{A}(\omega) \cdot \mathfrak{E}(\omega)_{\mathrm{inc}}.$$

Since the frequencies in the fluctuation are small compared with ω, we get also for the fluctuating beam [assuming plane-polarization components, though we drop the superscript (L)]

$$\begin{aligned}\mathfrak{E}_{\mathrm{scatt}}(t) &= \mathfrak{A}(\omega) \cdot \mathrm{E}_{\mathrm{inc}}(t)e^{-i\omega t} + \mathrm{c.c.} \\ &= \mathrm{E}_{\mathrm{scatt}}(t)e^{-i\omega t} + \mathrm{c.c.}\end{aligned} \tag{4.50}$$

because of the "reality condition" (4.34). Equation (4.50) will be correct provided that $\mathfrak{A}$ is a smooth function of the frequency near ω, that is, essentially constant over the bandwidth of the beam. Hence the effective scattered flux is

$$\begin{aligned}S_{\mathrm{eff\ scatt}}(t) &= \frac{c}{2\pi} \mathrm{E}^*_{\mathrm{scatt}}(t) \cdot \mathrm{E}_{\mathrm{scatt}}(t) \\ &= \frac{c}{2\pi} \mathrm{E}^*_{\mathrm{inc}}(t) \cdot \mathfrak{A}^\dagger \cdot \mathfrak{A} \cdot \mathrm{E}_{\mathrm{inc}}(t).\end{aligned} \tag{4.51}$$

For each of the components of the Stokes vector we then get [as in (1.60)]

$$\mathscr{I}_{\mathrm{scatt}}(t) = \mathfrak{F} \cdot \mathscr{I}_{\mathrm{inc}}(t)$$

and both sides can be averaged,

$$\langle \mathscr{I}_{\mathrm{scatt}}(t)\rangle = \mathfrak{F} \cdot \langle \mathscr{I}_{\mathrm{inc}}(t)\rangle. \tag{4.52}$$

Specifically, if the incident beam is unpolarized, then

$$\langle I'(t)\rangle = \frac{d\sigma}{d\Omega}\bigg|_{\omega}\langle I(t)\rangle$$
$$\frac{d\sigma}{d\Omega} = \frac{1}{2}(|A_{11}|^2 + |A_{12}|^2 + |A_{21}|^2 + |A_{22}|^2). \tag{4.53}$$

Notice that this type of measurement depends not at all on the phases of the scattering amplitudes, but only on their absolute magnitudes. Such an experiment therefore cannot determine this phase. A look at (1.61), (1.62), and (1.63) or at (1.70) shows, moreover, that, even if the incident beam is polarized and the polarization parameters of the scattered radiation are measured as well, the *overall* phase of the scattering amplitude matrix is not contained in $\mathfrak{F}$ and in the density matrix, and hence it cannot be measured in this way.

Suppose, now, that two scattering experiments are performed simultaneously, with two beams and two detectors, as shown in Fig. 4.2. Let beam 1 be represented by

$$\mathfrak{E}_1(t) = \mathrm{E}_1(t)e^{-i\omega t} + \text{c.c.}$$

and beam 2 by

$$\mathfrak{E}_2(t) = \mathrm{E}_2(t)e^{-i\omega t} + \text{c.c.}$$

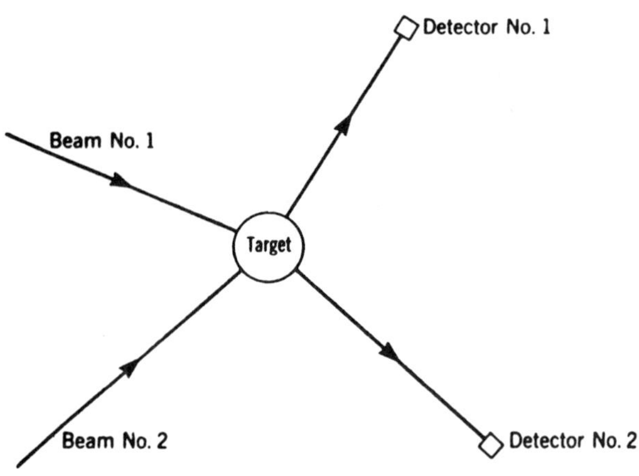

Figure 4.2. Scattering of two beams in two directions.

The two beams are assumed to be independent, i.e., incoherent. The magnitudes and phases of E_1 and E_2 then fluctuate independently. For simplicity we assume first that all four beams lie in a plane, that both incident beams are 100 percent polarized perpendicular to that plane, and that each detector measures only that polarization direction. If we call the scattering amplitude at frequency ω of beam i to detector j $A(ji)$, then the scattered field component to which the detector j responds is

$$\mathfrak{E}_j(t) = [A(j1)\mathrm{E}_1(t) + A(j2)\mathrm{E}_2(t)]e^{-i\omega t} + \text{c.c.}$$

The flux at detector j, averaged over times long compared with $1/\omega$ but short compared with the fluctuations, is given by

$$S_j(t) = \frac{c}{4\pi}\{[A^*(j1)\mathrm{E}_1^*(t) + A^*(j2)\mathrm{E}_2^*(t)][A(j1)\mathrm{E}_1(t) + A(j2)\mathrm{E}_2(t)] + \text{c.c.}\}$$

$$= \frac{c}{2\pi}[|A(j1)|^2 I_1(t) + |A(j2)|^2 I_2(t) + A^*(j1)A(j2)\mathrm{E}_1^*(t)\mathrm{E}_2(t) + A(j1)A^*(j2)\mathrm{E}_1(t)\mathrm{E}_2^*(t)]. \qquad (4.54)$$

If this is averaged over the fluctuations, the independence of the phases of beams 1 and 2 makes the last two terms vanish and we get

$$\langle S_j(t)\rangle = |A(j1)|^2\langle S_{1in}(t)\rangle + |A(j2)|^2\langle S_{2in}(t)\rangle. \qquad (4.55)$$

But suppose that the detectors are arranged to register only if both are irradiated simultaneously, so as to measure the instantaneous product of the two fluxes. Then we have

$$S_1(t)S_2(t) = \left(\frac{c}{2\pi}\right)^2\{[|A(11)|^2 I_1 + |A(12)|^2 I_2][|A(21)|^2 I_1 + |A(22)|^2 I_2] + [A(11)A^*(12)A^*(21)A(22) + \text{c.c.}]I_1 I_2\} \qquad (4.56)$$

plus terms which, on account of the phase independence of $\mathfrak{E}_1$ and $\mathfrak{E}_2$, vanish when averaged. Thus we have

$$\langle S_1(t)S_2(t)\rangle = \left(\frac{c}{2\pi}\right)^2[|A(11)|^2|A(21)|^2\langle I_{1in}^2\rangle + |A(22)|^2|A(12)|^2\langle I_{2in}^2\rangle + |A(11)A(22) + A(12)A(21)|^2\langle I_{1\,in}I_{2\,in}\rangle] \qquad (4.57)$$

and the result depends, not only upon the magnitudes of the four scattering amplitudes, but also upon their relative phases. By varying the angles of the detectors we may thus measure this phase as a function of the scattering angle. The only remaining phase ambiguity is a constant one, independent of the angle. This, finally, may be determined by measuring the total cross section and using the optical theorem.

The expression (4.57) may be simplified by realizing that the average of the square of the fluctuation is

$$\langle(I - \langle I\rangle)^2\rangle \equiv \Delta I^2 = \langle I^2\rangle - \langle I\rangle^2$$

and because of the independence of the two incoming beams,

$$\langle I_{1in}I_{2in}\rangle = \langle I_{1in}\rangle\langle I_{2in}\rangle.$$

Hence by (4.55)

$$\langle S_1 S_2\rangle - \langle S_1\rangle\langle S_2\rangle = 2\,\mathrm{Re}[A(11)A(22)A^*(12)A^*(21)]\langle S_{1in}\rangle\langle S_{2in}\rangle + |A(11)|^2|A(21)|^2\Delta S_{1in}^2 + |A(22)|^2|A(12)|^2\Delta S_{2in}^2. \qquad (4.58)$$

The second and third terms are much smaller than the first, and we get to a very good approximation

$$\langle S_1 S_2\rangle - \langle S_1\rangle\langle S_2\rangle \simeq 2\,\mathrm{Re}[A(11)A(22)A^*(12)A^*(21)]\langle S_{1in}\rangle\langle S_{2in}\rangle. \tag{4.59}$$

As it stands, the average of the product of the simultaneous fluxes is not a realistic representation of what two detectors can measure. There is always a certain delay time T between the two, partly because of the electronics and partly because the travel distances to the two detectors cannot be made *exactly* the same. Therefore we must really evaluate

$$\begin{aligned} S_1(t)S_2(t-T) &= \left(\frac{c}{2\pi}\right)^2 \{[|A(11)|^2 I_1(t) + |A(12)|^2 I_2(t)][|A(21)|^2 I_1(t-T) \\ &+ |A(22)|^2 I_2(t-T)] \\ &+ [A^*(11)A(12)A(21)A^*(22)\mathrm{E}_1^*(t)\mathrm{E}_1(t-T)\mathrm{E}_2(t)\mathrm{E}_2^*(t-T) + \text{c.c.}]\} \end{aligned} \tag{4.60}$$

plus terms which vanish when averaged. We must distinguish between two extreme cases. If the delay time T is small compared with the characteristic time T_0 of the fluctuations, i.e., small compared with the inverse of the typical noise frequencies, then $\mathrm{E}(t) \simeq \mathrm{E}(t-T)$ and we get the same results as (4.56) to (4.59). But if T is larger than the *coherence time* T_c, then $\mathrm{E}(t)$ and $\mathrm{E}(t-T)$ are *independent* and the last term in (4.60) vanishes when averaged. The effect of interest then disappears. The reason is, of course, that if $T > T_c$ then the independence of $S(t)$ and $S(t-T)$ results immediately in

$$\langle S_1(t)S_2(t-T)\rangle = \langle S_1\rangle\langle S_2\rangle.$$

The intermediate situation is that in which T is neither small compared with the typical noise period T_0 nor larger than the coherence time T_c. Then

$$\langle \mathrm{E}_i^*(t)E_i(t-T)\rangle \equiv \langle I_i\rangle\gamma_i(T)$$

where γ_i is called the *degree of self-coherence* of beam i. Neglecting the fluctuation terms again, we get in place of (4.59)

$$\begin{aligned} &\langle S_1(t)S_2(t-T)\rangle - \langle S_1\rangle\langle S_2\rangle \\ &\quad = 2\,\mathrm{Re}[A(11)A(22)A^*(12)A^*(21)]\langle S_{1in}\rangle\langle S_{2in}\rangle\gamma_1(T)\gamma_2(T). \end{aligned} \tag{4.59a}$$

Similar modifications can be made if the two incident beams are partially coherent. One then introduces an analogous *degree of mutual coherence*.

We generalize these results first to the case in which the detectors respond to both polarization components (but the incident beams are still 100 percent polarized and all four beams lie in a plane). Then the total flux at the detector is not (4.54) but

$$S_j(t) = \frac{c}{2\pi}\{[\mathfrak{A}^\dagger(j1)\mathfrak{A}(j1)]_{11}I_{1in} + [\mathfrak{A}^\dagger(j2)\mathfrak{A}(j2)]_{22}I_{2in} + [\mathfrak{A}^\dagger(j1)\mathfrak{A}(j2)]_{12}\mathrm{E}_1^*\mathrm{E}_2 + [\mathfrak{A}^\dagger(j2)\mathfrak{A}(j1)]_{21}\mathrm{E}_2^*\mathrm{E}_1\}. \tag{4.61}$$

The scattering-amplitude matrices are meant to be taken relative to the scattering plane. For the average instantaneous product of the two fluxes we get, instead of (4.57),

$$\begin{aligned}\langle S_1(t)S_2(t)\rangle = \left(\frac{c}{2\pi}\right)^2 &\{[\mathfrak{A}^\dagger(11)\mathfrak{A}(11)]_{11}[\mathfrak{A}^\dagger(21)\mathfrak{A}(21)]_{11}\langle I_{1in}^2\rangle \\ &+ [\mathfrak{A}^\dagger(12)\mathfrak{A}(12)]_{22}[\mathfrak{A}^\dagger(22)\mathfrak{A}(22)]_{22}\langle I_{2in}^2\rangle \\ &+ [[\mathfrak{A}^\dagger(11)\mathfrak{A}(11)]_{11}[\mathfrak{A}^\dagger(22)\mathfrak{A}(22)]_{22} \\ &+ [\mathfrak{A}^\dagger(12)\mathfrak{A}(12)]_{22}[\mathfrak{A}^\dagger(21)\mathfrak{A}(21)]_{11}]\langle I_{1in}\rangle\langle I_{2in}\rangle \\ &+ 2\,\mathrm{Re}\,[[\mathfrak{A}^\dagger(11)\mathfrak{A}(12)]_{12}[\mathfrak{A}^\dagger(22)\mathfrak{A}(21)]_{21}]\langle I_{1in}\rangle\langle I_{2in}\rangle\}\end{aligned}$$

and therefore

$$\begin{aligned}\langle S_1S_2\rangle - \langle S_1\rangle\langle S_2\rangle \\ = 2\,\mathrm{Re}\,&\{[\mathfrak{A}^\dagger(11)\mathfrak{A}(12)]_{12}[\mathfrak{A}^\dagger(22)\mathfrak{A}(21)]_{21}\}\langle S_{1in}\rangle\langle S_{2in}\rangle \\ &+ [\mathfrak{A}^\dagger(11)\mathfrak{A}(11)]_{11}[\mathfrak{A}^\dagger(21)\mathfrak{A}(21)]_{11}\Delta S_{1in}^2 \\ &+ [\mathfrak{A}^\dagger(22)\mathfrak{A}(22)]_{22}[\mathfrak{A}^\dagger(12)\mathfrak{A}(12)]_{22}\Delta S_{2in}^2\end{aligned} \tag{4.62}$$

in place of (4.58). The last two terms may again be neglected, and the first term is the analog of (4.59) for no polarization detection. The subscripts on the matrix products refer to the polarization directions of the two incident beams.

Suppose now that both beams are unpolarized. We may then think of each as made up of two incoherent components, one for each polarization direction, each with independently varying random phases. In the course of the averaging process terms such as

$$\mathfrak{A}(11)\cdot\mathrm{E}_1(t)\mathrm{E}_1^*(t)\cdot\mathfrak{A}^\dagger(21)$$

become

$$\tfrac{1}{2}\mathfrak{A}(11)\mathfrak{A}^\dagger(21)I_1(t)$$

because the cross terms containing one component of E_1 multiplied by another of E_1^* average to zero. One therefore finds

$$\langle S_j\rangle = \tfrac{1}{2}\mathrm{tr}\,[\mathfrak{A}^\dagger(j1)\mathfrak{A}(j1)]\langle S_{1in}\rangle + \tfrac{1}{2}\mathrm{tr}\,[\mathfrak{A}^\dagger(j2)\mathfrak{A}(j2)]\langle S_{2in}\rangle$$

and

$$\begin{aligned}\langle S_1S_2\rangle - \langle S_1\rangle\langle S_2\rangle = \tfrac{1}{4}\sum_j &\{[\mathfrak{A}^\dagger(1j)\mathfrak{A}(1j)]_{12}[\mathfrak{A}^\dagger(2j)\mathfrak{A}(2j)]_{21} \\ &+ [\mathfrak{A}^\dagger(1j)\mathfrak{A}(1j)]_{21}[\mathfrak{A}^\dagger(2j)\mathfrak{A}(2j)]_{12}\}\langle S_{jin}\rangle^2 \\ &+ \tfrac{1}{2}\mathrm{Re\,tr}\,[\mathfrak{A}^\dagger(11)\mathfrak{A}(12)\mathfrak{A}^\dagger(22)\mathfrak{A}(21)]\langle S_{1in}\rangle\langle S_{2in}\rangle\end{aligned} \tag{4.63}$$

plus negligible terms proportional to the average fluctuations. The first sum is a new contribution which owes its existence to the fact that the detectors do not pay attention to the polarization. The subscripts there refer to the two polarization components of the *product* matrix.

The experimental setup discussed may be modified so as to replace the two incident beams by a single unpolarized one. Each of the two polarization components then takes the place of one of the two beams in the previous arrangement. Of course, the scatterer must be such as to produce both polarizations from each incident polarization; that is to say, it must rotate the plane of polarization. The result can then be obtained either from (4.63), by setting $S_{2in} = 0$, or from (4.62), by letting S_{2in} there come from the same direction as S_{1in}, but with a different polarization,

$$\langle S_1 S_2\rangle - \langle S_1\rangle\langle S_2\rangle = \tfrac{1}{2}\mathrm{Re}\,\{[\mathfrak{A}^\dagger(1)\mathfrak{A}(1)]_{12}[\mathfrak{A}^\dagger(2)\mathfrak{A}(2)]_{21}\}\,\langle S_{in}\rangle^2. \tag{4.64}$$

This time it is sufficient to indicate the outgoing direction on the amplitude. In this way it is possible to measure the polarization components of the amplitude with an unpolarized beam and without measuring the outgoing polarization. For example, if the scatterer has complete rotation-reflection symmetry, then according to the discussion in Sec. 1.3.8, $A_{11}{}^C = A_{22}{}^C$ and $A_{12}{}^C = A_{21}{}^C$. Since we must use amplitudes relative to reference planes through a fixed direction, the right-hand side of (4.64) becomes therefore, by (1.55),

$$2\,\mathrm{Re}\,[A_{11}^{C*}(1)A_{12}^C(1)e^{-2i\varphi_1}]\,\mathrm{Re}\,[A_{11}^{C*}(2)A_{12}^C(2)e^{-2i\varphi_2}]$$

for unit incident flux. The overall phase of the amplitude, of course, is not measured in this way.

A further possibility is to use a single unpolarized incident beam and two detectors in almost the same position, but sensitive to different polarization. In this case (4.59) gives

$$\langle S_1 S_2\rangle - \langle S_1\rangle\langle S_2\rangle = \tfrac{1}{2}\mathrm{Re}\,(A_{11}A_{22}A_{12}^*A_{21}^*)\langle S_{in}\rangle^2$$

the amplitudes being the components of $\mathfrak{A}_{\mathrm{fix}}{}^{(L)}$ or of $\mathfrak{A}_{\mathrm{ix}}{}^{(C)}$ depending on whether the detectors discriminate between linear or circular polarizations.

Other effects are possible if the incident beams are only *partially* polarized. The results are then considerably more complicated, and we shall not go into them.

If the two detectors do not lie in the same plane as the incident beams, then the amplitudes used must refer to reference planes through a fixed direction. This direction is most conveniently chosen to be perpendicular to the plane of both incident beams. Then everything done so far holds again, except for the altered meaning of the amplitudes in (4.63).

Notes and References

4.1.1 The two scalar potentials were introduced by P. Debye (1909a). For more recent uses, see, for example, J. A. Stratton (1941); H. C. van de Hulst (1957); J. D. Jackson (1962).

4.1.2 For more detailed expositions of the method of Green's functions in electromagnetic theory, see A. Sommerfeld (1949); Morse and Feshbach (1953), vol. 1; J. D. Jackson (1962); and, specifically, Levine and Schwinger (1951). For earlier use of Green's-function techniques, see A. Sommerfeld (1910 and 1912), for example.

Another method, transferred from quantum mechanics, that has been found useful is called the *T matrix*. It can be found in P. C. Waterman (1965, 1971, 1979, and 1980); Peterson and Ström (1973 and 1974); S. Ström (1974), G. S. Agarwal (1976); N. Morita (1979); P. W. Barber (1980); R. H. T. Bates (1980); and A. Hizal (1980). The Fredholm method has also been used; see Uzunoglu et al. (1976); and A. R. Holt (1980).

4.2.1 For a more detailed discussion of Hilbert transforms, see E. C. Titchmarsh (1937). The disperison relation (4.36) was first derived by R. Kronig (1926); H. A. Kramers (1927). Both obtained it as a limit of the dispersion due to atomic resonance-absorption lines, but Kramers connected the result to the causality argument. More recent and detailed discussions can be found in the papers by N. G. van Kampen (1953a) and J. S. Toll (1956), which also contain further references. See also J. Hilgevoord (1960) and E. Gerjuoy (1965a).

4.3 The usefulness of measuring intensity fluctuation correlations was discovered by Hanbury Brown and Twiss (1954, 1956, and 1957). It was discussed also by E. M. Purcell (1956) and by Goldberger, Lewis, and Watson (1963). This paper also treats the analogous method in quantum mechanics, which we shall not go into in this book. The principle is, of course, the same.

Further treatments of fluctuation and coherence phenomena are by Mandel and Wolf (1965) and J. M. Conway (1967) in the electromagnetic case, and for quantum mechanics Goldberger and Watson (1964c and 1965). See also E. Wolf (1972 and 1970); Kano and Wolf (1962); Dialetis and Wolf (1967); C. L. Mehta (1968); Kohler and Mandel (1970); J.-M. Lévy-Leblond (1977).

For detailed discussions of noise and radiation-fluctuation phenomena in general see Lawson and Uhlenbeck (1950); Born and Wolf (1959), chap. X.

The following papers treat orders of coherence of the quantum radiation field: R. J. Glauber (1963, 1965, and 1966a); Mandel and Wolf (1963); E. C. G. Sudarshan (1963); Titulaer and Glauber (1965 and 1966).

For a review of the utilization of polarization information for inverse scattering, see W.-M. Boerner (1979).

Problems

1 Prove Eqs. (4.23) and (4.24).

2 Assume that $A(\omega)$ is analytic in the upper half of the complex ω plane and that it tends to infinity as $|\omega|$ when $|\omega| \to \infty$ there. Write down a dispersion relation.

3 The Maxwell equations are real. What then is the origin of the distinction between $+i$ and $-i$ implied by the analyticity of the scattering amplitude in the *upper* half of the complex ω plane but not in the lower?

4 In quantum mechanics the scattering amplitude may have poles in the upper half of the k plane. Why can it have no poles there in the electromagnetic case?

5 Consider the scattering of radiation by air bubbles in an infinite refractive medium. Are the arguments for dispersion relations changed? If so, in what way?

Additional References for Part I

In addition to the literature cited at the end of each chapter, the following books should be mentioned for general reference: King and Wu (1959); J. M. Stone (1963); M. Kerker (1969), and for mathematical reference Lax and Phillips (1964 and 1967).

PART II

SCATTERING OF CLASSICAL PARTICLES

CHAPTER 5

Particle Scattering in Classical Mechanics

5.1 The Orbit Equation and the Deflection Angle

5.1.1 The Nonrelativistic Case. The scattering of classical particles by a central field of force or by one another is described in terms of the particle's orbit of motion. The latter is most simply obtained from the hamiltonian via the Hamilton–Jacobi equation.

Suppose that a particle of mass m is subject to a conservative force field whose potential function is V. We shall assume that V is rotationally invariant, so that it depends only on the magnitude of the distance r of the particle from the force center. Alternatively we may consider two particles in their center-of-mass coordinate system, with an interaction force between them described by the potential V. The symbols m and r then have the meaning of the reduced mass of the two particles and the distance between them. The second situation goes over into the first in the limit in which the mass of one of the particles tends to infinity, so that it suffers no recoil.

The hamiltonian of the system is

$$H = \frac{p^2}{2m} + V(r) \tag{5.1}$$

and the momentum vector may be decomposed into its r, ϑ, and φ components,

$$\mathbf{p}^2 = p_r^2 + r^{-2}p_\vartheta^2 + r^{-2}\csc^2\vartheta\, p_\varphi^2 = p_r^2 + r^{-2}J^2 \tag{5.2}$$

where J is the magnitude of the conserved total angular momentum. The Hamilton–Jacobi equation for the radial contribution to Hamilton's characteristic function is therefore

$$\frac{1}{2m}\left(\frac{dS_r}{dr}\right)^2 + \frac{1}{2mr^2}J^2 + V(r) = E$$

and hence

$$S(r,\vartheta) = \int dr\,[2m(E - V) - r^{-2}J^2]^{1/2} + J\vartheta.$$

The orbit equation being obtained from this by differentiation with respect to J, we have

$$\vartheta = J\int_r^\infty dr\,r^{-2}(p^2 - \mathscr{V} - r^{-2}J^2)^{-1/2} \tag{5.3}$$

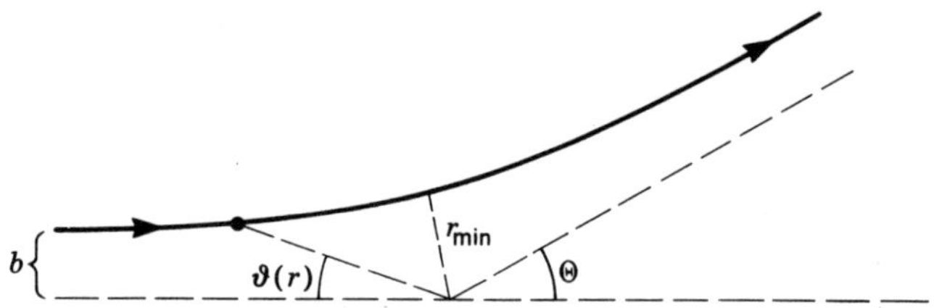

Figure 5.1. The trajectory of a particle.

if the angle ϑ is defined as shown in Fig. 5.1, p is the asymptotic momentum of the particle, and $\mathscr{V} = 2mV$. The motion is symmetric with respect to the minimal distance $r_{\min}$. The deflection angle Θ is consequently given by

$$\Theta = \pi - 2\int_{r_{\min}}^\infty dr\,r^{-2}[(p^2 - \mathscr{V})J^{-2} - r^{-2}]^{-1/2} \tag{5.4}$$

where $r_{\min}$ is the largest value of r at which the radical vanishes.

The *impact parameter* b is defined as shown in Fig. 5.1. It is the distance of the asymptotic path of the particle from the line of head-on collision. It is therefore connected with its angular momentum and with its asymptotic momentum and energy by

$$J = pb = \sqrt{2mE}b. \tag{5.5}$$

We may thus write

$$\begin{aligned}\Theta &= \pi - 2\int_{r_{\min}}^\infty dr\,r^{-2}\left[b^{-2}\left(1 - \frac{V}{E}\right) - r^{-2}\right]^{-1/2} \\ &= \pi - 2\int_{\xi_0}^\infty \frac{d\xi}{\xi}\{\xi^2[1 - f(\xi)] - 1\}^{-1/2}\end{aligned} \tag{5.6}$$

where

$$\xi = \frac{r}{b}$$
$$f(\xi) = \frac{V(b\xi)}{E} \tag{5.7}$$

with ξ_0 determined by the equation

$$f(\xi_0) = 1 - \xi_0^{-2}. \tag{5.8}$$

Notice that, so far as the deflection angle is concerned, changes in energy or impact parameter are equivalent to simple changes in the shape of the potential curve. A variation of E is equivalent to a change in the vertical scale of the potential and a variation of b to a change in its horizontal scale.

5.1.2 The Relativistic Case. The relativistic energy-momentum relation of a free particle is

$$W^2 = m^2c^4 + \mathbf{p}^2c^2.$$

In this equation the energy W contains both the kinetic and the rest contribution. In the nonrelativistic limit, in which

$$\mathbf{p}^2 \ll m^2c^2$$

it reduces to

$$W \simeq mc^2 + \frac{\mathbf{p}^2}{2m}$$

and only the second part is retained in the nonrelativistic definition of energy.

If the particle is subject to a potential $V(r)$, its relativistic hamiltonian may be given by

$$H = (\mathbf{p}^2 + m^2c^2)^{1/2}c + V = (p_r^2 + m^2c^2 + J^2r^{-2})^{1/2}c + V \tag{5.9}$$

or

$$(H - V)^2c^{-2} = p_r^2 + m^2c^2 + J^2r^{-2}. \tag{5.9a}$$

Therefore the Hamilton–Jacobi equation for Hamilton's characteristic function becomes

$$(W - V)^2c^{-2} = \left(\frac{dS_r}{dr}\right)^2 + J^2r^{-2} + m^2c^2.$$

The orbit equation is consequently given by

$$\vartheta = J\int_r^\infty dr\, r^{-2}[(W - V)^2c^{-2} - m^2c^2 - J^2r^{-2}]^{-1/2}$$

instead of (5.3). Using again the letter p for the asymptotic momentum of the particle, we can write the deflection angle Θ in the form

$$\Theta = \pi - 2J \int_{r_{\min}}^{\infty} dr\, r^{-2}[p^2 - V(2W - V)c^{-2} - J^2 r^{-2}]^{-1/2} \quad (5.10)$$

in place of (5.4). Comparison shows that the expression is the same, except that $\mathscr{V}$ is not equal to $2mV$, but

$$\mathscr{V} = V(2W - V)c^{-2}. \quad (5.11)$$

In the nonrelativistic limit the rest energy mc^2 dominates the parenthesis, and the right-hand side goes over into $2mV$, as it should.

The most striking feature of the effective potential $\mathscr{V}$ is that when it is singular it is *always negative*, no matter what the sign of V is. However, since the region in which $2W < V$ is energetically inaccessible to the particle, the sign change is physically irrelevant.

The second thing to notice about (5.11) is that when V is singular then $\mathscr{V}$ is much more singular in the relativistic case than in the nonrelativistic, in fact, as the square of the latter. We shall return to this point in Sec. 5.6.

The third property of (5.11) is that the effective relativistic potential V is *energy-dependent*. The larger the energy, the stronger is the effective force.

It should be remarked that relativistic potential scattering is a rather artificial construction. Although mathematically well defined, because retardation effects are neglected it is physically not consistent. A physically acceptable theory (quantum mechanics apart) would have to be constructed separately for each particular type of interaction. Electrodynamics is one such example, general relativity another. One then has very little freedom, if any, of choosing the potential.

5.2 The Scattering Cross Section

In a scattering experiment one does not observe the path of a single particle. Instead, a "beam" of particles, ideally all of the same energy, is sent toward the target, and the flux scattered in a specific direction is counted. This means that the incident missiles do not have a fixed impact parameter nor a fixed azimuthal angle, but they are uniformly distributed over φ (about the incident direction as a z axis) and over b. All the missiles in the incoming "tube" element of area $b\, db\, d\varphi$, however, go into the specific infinitesimal outgoing cone of solid angle $d\varphi\, d\theta \sin\theta$, θ being connected to b by Eq. (5.6). Thus, if N is the number of particles per unit time and area in the incident beam, then the number of particles per unit time and solid angle going in the given direction (θ, φ) is

$$N \frac{d\sigma}{d\Omega} d\Omega = N \frac{d\sigma}{d\Omega} d\varphi\, d\theta \sin\theta = Nb\, db\, d\varphi$$

or

$$\frac{d\sigma}{d\Omega} = \frac{b \csc\theta}{|d\Theta/db|}. \quad (5.12)$$

The deflection angle Θ must be evaluated by (5.6) as a function of b, and the functional relation must then be inverted to find $b(\Theta)$.

Notice immediately from (5.12) that if the scattering potential has an infinite range so that it deflects particles of *all* impact parameters then the differential cross section is always infinite in the forward direction, and in fact more so than $1/\theta$. This is because $\Theta \to 0$ when $b \to \infty$ and furthermore, $d\Theta/db \to 0$ when $b \to \infty$, that is, when $\Theta \to 0$. Hence $(d\sigma/d\Omega)\sin\theta$ tends to infinity when $\Theta \to 0$ on account of both the numerator and the denominator in (5.12).

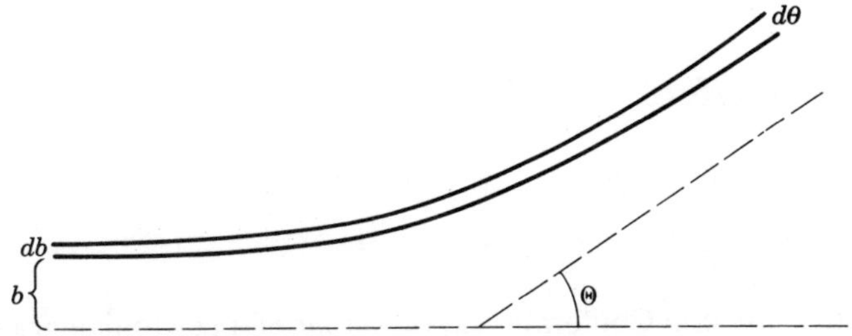

Figure 5.2. The deflection "tube."

A word should be said about the relation between the deflection angle Θ, which may have any value between $-\infty$ and π, and the scattering angle θ, which, by definition, always lies between zero and π. The two angles have to be connected by

$$\Theta = q\theta - 2\pi m \tag{5.13}$$

where $q = \pm 1$ and m is a positive integer (or zero). For a given Θ, the numbers q and m are chosen so that θ lies between zero and π.

An alternative way of thinking about the differential-scattering cross section is in terms of a "bunch" of particles being sent toward the target, rather than in terms of a steady stream. In the distant past a monoenergetic cloud of missiles was prepared and sent toward the scatterer, with a certain given distribution of impact parameters or angular momenta $f(J)$. As they approach the target, they are deflected in various directions, and in the distant future they fly away radially. The flux of outgoing particles in the solid angle element $d\Omega$ is then

$$\frac{d\rho}{d\Omega}\,d\Omega f(J)$$

where J is connected with θ by (5.4) and (5.13).

The total scattering cross section is obtained from (5.12) by integrating over all solid angles,

$$\begin{aligned}\sigma &= \int d\Omega \frac{d\sigma}{d\Omega} = \int d\varphi\, d\theta\, b \left|\frac{db}{d\Theta}\right| \\ &= 2\pi \int_0^{b_{\max}} db\, b = \pi b_{\max}^2,\end{aligned} \tag{5.14}$$

b_{max} being the maximal impact parameter that is deflected at all. If the "range" of the force is R, so that for $r > R$ the potential vanishes, then missiles with an impact parameter greater than R are not deflected. The scattering cross section is then πR^2, the geometrical cross section of the scattering region. For potentials without such a finite cutoff and which, however weakly, deflect missiles with *all* impact parameters, the total scattering cross section for classical particles is *always infinite.* Because of the remarks made below (5.12), the origin of this infinity is always the sharp forward peaking of the differential cross section.

5.3 The Rutherford Cross Section

If the potential is of the form

$$V = \alpha r^{-1}$$

as it is in the case of a Coulomb field or in that of a newtonian gravitational field, the integral in (5.6) can be carried out in closed form. The result is

$$\Theta = \pi - 2\cot^{-1}\frac{\alpha}{2Eb}$$

or

$$b = \frac{\alpha}{2E}\cot\tfrac{1}{2}\theta. \tag{5.15}$$

The differential cross section is therefore

$$\frac{d\sigma}{d\Omega} = \left(\frac{\alpha}{4E\sin^2\frac{1}{2}\theta}\right)^2. \tag{5.16}$$

In the case of the scattering of a charge Z_1e by another charge Z_2e we have $\alpha = Z_1Z_2e^2$, and the result is the famous Rutherford scattering formula. It is characterized by a very sharp rise to infinity in the forward direction. Notice also that it does not depend on whether the charges attract or repel one another, in spite of the fact that the individual particle trajectories, of course, are different in the two cases.

In the scattering point charge is replaced by an extended charge distribution, the scattering will be changed in a manner that depends on whether the charges attract or repel one another. In the repulsive case, the distance of closest approach is $r_{min} = Z_1Z_2e^2/E$, and hence, if the radius of the charge distribution is smaller than this, the cross section in unaltered. If there is attraction, the head-on-pointing ($b = 0$) particles go right through the center. Hence they *always* "feel out" the charge distribution.

In the relativistic case the deflection angle is also easily calculated. For $J > \alpha/c$ one finds

$$\Theta = \pi - \frac{2}{\eta}\cot^{-1}\frac{\alpha W}{c^2 p^2 b\eta} = \pi - \frac{2}{\eta}\cot^{-1}\frac{\alpha}{vJ\eta}$$
$$\eta = \left(1 - \frac{\alpha^2}{c^2p^2b^2}\right)^{1/2} = \left[1 - \left(\frac{\alpha}{cJ}\right)^2\right]^{1/2} \tag{5.17}$$

if v is the asymptotic velocity of the particle. In the attractive case this expression is not valid for $J < \alpha/c$, because then the distance of closest approach is zero, and the integral in (5.10) diverges; the deflection angle is (negatively) infinite. (We shall return to this point in Sec. 5.6.)

Equation (5.17) cannot be solved in closed form for b in terms of θ, as could (5.15). Consequently, we cannot write down explicitly the relativistic Coulomb-scattering cross section. Nevertheless, (5.17) and (5.12) allow us to calculate the cross section exactly.

5.4 Orbiting (Spiral Scattering)

If the force is everywhere repulsive, i.e., the potential monotonely decreases to zero, then the function of (5.7) is such that $f(\xi) < f(\xi_0)$ for $\xi > \xi_0$ and

$$\int_{\xi_0}^{\infty} d\xi\,\xi^{-1}\{\xi^2[1 - f(\xi)] - 1\}^{-1/2} < \int_{\xi_0}^{\infty} d\xi\,\xi^{-1}\{\xi^2[1 - f(\xi_0)] - 1\}^{-1/2}$$
$$= \int_{1}^{\infty} d\xi\,\xi^{-1}(\xi^2 - 1)^{-1/2} = \tfrac{1}{2}\pi.$$

Therefore, as is physically obvious,

$$0 \le \Theta \le \pi.$$

On the other hand, if the potential is not monotonely decreasing, then the integral in (5.6) can have any positive value you please. If the force is attractive, the scattered particle can go around the center many times.

Suppose that V has a maximum V_0 at $r = r_0$. Then f has a maximal value V_0/E at $\xi = r_0/b$, and there exists a family of values E and b for which a point near the maximum gets quite close to the curve $1 - \xi^{-2}$. The integral in (5.6) then becomes large and, when the curve $f(\xi)$ touches $1 - \xi^{-2}$, logarithmically *infinite*. This does not happen exactly at the maximum of $f(\xi)$, but for every $V_0 < E < \max[V + \frac{1}{2}rV']$ it will happen[1] at some value b_0 of b. When $b < b_0$, then the peak of $f(\xi)$ almost touches $1 - \xi^{-2}$; when $b > b_0$, then the peak protrudes above $1 - \xi^{-2}$ and suddenly a large inner region of attractive force is no longer accessible to the particles, because the minimal distance of approach is determined by the crossing of $f(\xi)$ and $1 - \xi^{-2}$. At the impact parameter for which the two curves touch, the particle spirals around the center and asymptotically approaches

1 W. H. Miller (1969b).

the minimal radius.[1a] This phenomenon is called *orbiting*, or *spiral scattering*. The functional dependence of $\pi - \Theta$ upon b is then as schematically indicated in Fig. 5.3. At $b = b_0$ it approaches infinity logarithmically. In order

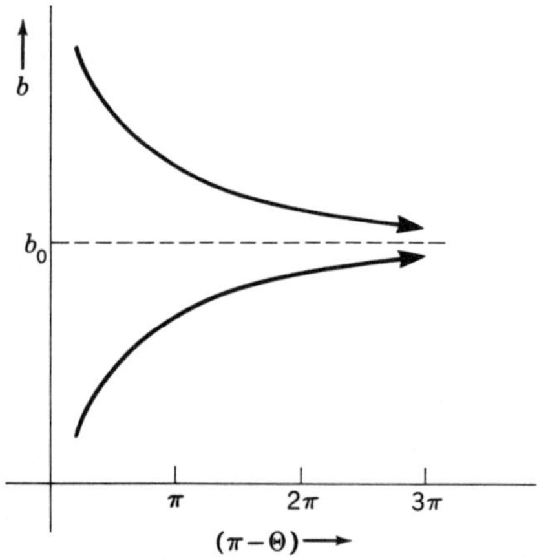

Figure 5.3. The functional dependence of the impact parameter on the deflection angle when orbiting occurs.

to obtain $b(\theta)$, this curve must be "folded" back and forth at each multiple of π. As a result, $b(\theta)$ is thus an *infinitely many-valued function* of θ. For each value of θ there are infinitely many values of b, and each contributes to the cross section,

$$\frac{d\sigma}{d\Omega} = \frac{1}{2} \sum_{m=1}^{\infty} \left| \frac{db_m^2}{d\theta} \right| \csc\theta. \tag{5.18}$$

Because Θ diverges logarithmically as b approaches the critical value b_0,

$$-\Theta \simeq \begin{cases} \text{const} + c \ln(b - b_0) & \text{for } b > b_0 \\ \text{const} + 2c \ln(b_0 - b) & \text{for } b < b_0. \end{cases} \tag{5.19}$$

For $b > b_0$ we have

$$b_m \simeq b_0 + e^{qm\theta/c} e^{-2m\pi/c}$$

and for each m there is one branch with $q = +1$ and one with $q = -1$. The series in (5.18) therefore converges as

$$\sum_m e^{-2\pi m/c} = (1 - e^{-2\pi/c})^{-1}. \tag{5.20}$$

If the energy of the incoming particles is varied, orbiting cannot take place so long as the energy is less than the peak in the potential. When $E = V_0$, then, it happens for the particles of zero impact parameter or zero angular momentum.

1a This implies that the particles of this impact parameter spend an infinite length of time near the scatterer and thus never emerge. Since this happens for a specific impact parameter only, it does not destroy the possibility of defining a cross section.

If the potential is everywhere negative and monotonely increasing (i.e., the force is everywhere attractive), then for a sufficiently large impact parameter there may exist two values of the energy at which $f(\xi)$ is tangent to $1 - \xi^{-2}$. At the smaller of these, orbiting occurs. As b decreases, these two values of the energy move toward one another, and at some value of b they must coincide. At that special value of E and b (the critical point) the curves $y = f(\xi)$ and $y = 1 - \xi^{-2}$ cross in an osculating manner, i.e., $1 - \xi^{-2} - f$ has a triple zero. The integral in (5.6) then diverges more strongly $[-\Theta \simeq \text{const} + c(b - b_0)^{-1/4}]$ but the cross section is still finite. For larger values of E no orbiting takes place, no matter what b is.

5.5 Glory and Rainbow Scattering

A look at (5.12) shows that the differential cross section is infinite if

$$\frac{d\Theta}{db} = 0. \tag{5.21}$$

The deflection angle being stationary there as a function of the impact parameter, it is clear that many particles are concentrated near this particular angle. The effect is perfectly analogous to the rainbow in electromagnetic scattering.

That it is indeed possible for (5.21) to be satisfied can be easily seen. Equation (5.6) shows that, when b increases, the radical decreases and hence the integrand, at fixed r, increases. At the same time, however, the value of $r_{\min}$ will usually increase too, and hence the region of integration will decrease. As a result $\Theta\ (b)$ need not be monotone.

If the potential is everywhere attractive (i.e., negative and monotonely increasing), then it is clear on physical grounds that, at zero impact parameter, the deflection angle is *zero*. This can also be verified directly from (5.6). As $b \to 0$,

$$f(\xi) \to \frac{V(0)}{E}$$

and the integral approaches $\frac{1}{2}\pi$. The same is true if the potential is positive but $E > V(0)$, so that at zero-impact parameter the particle goes right through the origin. Since the deflection angle must also go to zero as $b \to \infty$, it must have either an infinity at some finite value of b or an extremum. Consequently, for such potentials (and energies) either orbiting must occur or a rainbow.

Equation (5.12) also shows that the cross section in the backward direction is infinite if back scattering is possible for impact parameters other than zero. This effect is the analog of the *glory* in the electromagnetic case. It will surely happen when the deflection angle for some impact parameters a less than $-\pi$. Consequently it occurs whenever the potential has a maximum V_0 which is less than the energy. In fact, whenever orbiting takes place, then the glory effect occurs, too. On the other hand, if the potential is everywhere positive and monotonically decreasing, then a glory cannot happen.

5.6 Singular Potentials

If the effective potential $\mathscr{V}$ behaves less singularly near $r = 0$ than r^{-2}, then (5.4) and (5.10) show that the radicand in the integrand always becomes negative at a finite distance r. There is, then, at fixed E and b, always a distance $r_{\min}$ of closest approach. The same is true if the potential is as singular as r^{-2}, or more so, but repulsive (positive). Difficulties arise only if it is singular and *attractive* (negative).[2]

If $r^2\mathscr{V}$ tends to $-\infty$ as $r \to 0$, then there exists no distance of closest approach. For all energies and impact parameters the particle goes right through the origin and the integral in (5.6) extends down to zero. Under the stronger assumption that there exists an $\varepsilon > 0$ (or $\varepsilon = \infty$) so that

$$r^4\mathscr{V}(r) \to -\varepsilon \quad \text{as } r \to 0 \tag{5.22}$$

the absence of a distance of closest approach has no undesirable consequences.[3] The particle then races through the center and emerges at a finite, well-defined angle. But if the singularity of $\mathscr{V}$ lies between r^{-2} and r^{-4} every trajectory has an infinite curvature at the origin. Its continuation and the definition of a deflection angle are then possible only by means of additional assumptions.

If the potential is such that

$$\lim_{r\to 0} r^2\mathscr{V}(r) = -|a| \tag{5.23}$$

then the integral in (5.6) diverges so long as

$$b < \left(\frac{|a|}{E}\right)^{1/2} = b_0 \quad \text{or } J < (2m|a|)^{1/2} = J_0. \tag{5.24}$$

For small enough impact parameters the particle then spirals in toward the center, and the deflection angle is infinite. A scattering angle in that case is undefined. The physical problem of where the particle emerges in this event

2 Note that in the relativistic case the dividing line between regular and singular potentials $V(r)$ is r^{-1}. According to (5.11) that produces an r^{-2} behavior in the effective potential $\mathscr{V}$. Furthermore, the singular *repulsive* case never arises for relativistic particles. If V is singular and repulsive, then $\mathscr{V}$ still goes to $-\infty$. However, as we have already remarked, there still exists a distance of closest approach, so that no difficulties arise. It is therefore true relativistically as well as nonrelativistically that repulsive singular potentials cause no trouble.

3 If the potential decreases faster than r^{-2} at infinity then (5.22) implies that the potential, together with the centrifugal term, has a maximum. Hence orbiting must occur at some impact parameter. As a result, the cross section *may* be infinite. But even if it is, that infinity cannot be properly ascribed to the singularity of the potential. It is therefore not correct to say that an attractive singular potential necessarily leads to difficulties in classical particle scattering. It does so only if its singularity lies between r^{-2} and r^{-4}.

is again unanswerable. A slight alteration of the potential which cuts off its r^{-2} behavior near the origin (such as a finite particle size, for example) produces a large change in the deflection angle. Under such circumstances a scattering cross section is extremely sensitive to the cutoff.

For angular momenta larger than J_0 there are a well-defined distance of closest approach and a finite deflection angle. How Θ behaves when J decreases toward its critical value depends upon the more detailed nature of the potential.

As an example we consider the nonrelativistic potential

$$V = ar^{-2} \qquad a > 0.$$

A simple computation from (5.6) then yields for $b^2 > a/E$

$$\Theta = \pi - \pi\left(1 + \frac{a}{Eb^2}\right)^{-1/2}$$

or

$$b^2 = \frac{a}{E}\frac{(\pi - \theta)^2}{(2\pi - \theta)\theta}$$

and the scattering cross section is

$$\frac{d\sigma}{d\Omega} = \frac{a\pi^2(\pi - \theta)}{E\sin\theta(2\pi - \theta)^2\theta^2}.$$

It diverges strongly in the forward direction (although less so than the Rutherford cross section), and it has the finite value $a/E\pi^2$ in the backward direction.

If the potential is attractive, then $a < 0$ and the deflection angle Θ tends to $-\infty$ when b approaches its critical value b_0 from above. If the beam is arranged to contain no impact parameters less than b_0, then a cross section of sorts can be defined. The scattering angle, when $b > b_0$, is obtained from (5.13). For every θ there are infinitely many impact parameters,

$$b_{m,q}^2 = \frac{|a|}{E}\frac{(2m + 1 - q\theta/\pi)^2}{(2m - q\theta/\pi)(2m + 2 - q\theta/\pi)} \qquad \begin{array}{l} q = \pm 1,\, m = 0, 1, 2, \ldots \\ \text{but not } q = 1,\, m = 0 \end{array}$$

and the "differential cross section" is the sum of the contributions from all, as in (5.18). The series is easily summed, with the result

$$\frac{\text{“}d\sigma\text{”}}{d\Omega} = \frac{|a|\pi}{2E\theta^2\sin\theta}.$$

The quotation marks are to indicate the special nature of this cross section, defined only for a "hollow" beam. Notice that this is infinite in the backward (as well as the forward) direction.

A similar situation occurs in the case of relativistic Coulomb scattering. The expression (5.11) shows that when $V = \pm\alpha r^{-1}$ then the effective relativistic potential goes as $-\alpha^2c^{-2}r^{-2}$ near the origin. This will matter only if the potential is *attractive*, though, because otherwise the particle does not penetrate to the troublesome region. In contrast to the nonrelativistic situation there is therefore an essential difference in the scattering between attraction and repulsion. In the attractive case there is a minimal angular momentum $J_0 = \alpha/c$ below which the deflection angle is infinite and a cross section is undefined.

5.7 Transformation Between Laboratory and Center-of-Mass Coordinate Systems

If the scattering is, not that of a particle by a fixed potential, but that of two particles by one another, then everything said so far holds in the center-of-mass coordinate frame of reference, except that the mass m has to be replaced by the reduced mass $\mu = m_1 m_2/M$, with $M = m_1 + m_2$, of the two particles. Experiments, however, are most frequently done in other systems. We must therefore find the transformation that relates the differential scattering cross section and the angles in the barycentric system to those in another frame. In many experimental situations one of the particles is initially at rest, and the reference frame in which this is the case is usually called the *laboratory system*. We shall be more general and specialize to this case only later.

Let us call the initial and final momenta of particle 1 in the general system $\mathbf{P}_1$ and $\mathbf{P}_1'$, respectively, and let E and $\mathbf{P}$ be the total energy and momentum there. The direction of $\mathbf{P}_1'$ will be specified by its angle ϑ relative to $\mathbf{P}_1$ and its angle φ of rotation around $\mathbf{P}_1$ out of the initial plane of $\mathbf{P}_1$ and $\mathbf{P}_2$:

$$\hat{\mathbf{P}}_1 \cdot \hat{\mathbf{P}}_1' = \cos\vartheta, \qquad (\hat{\mathbf{P}}_1 \times \hat{\mathbf{P}}) \cdot (\hat{\mathbf{P}}_1 \times \hat{\mathbf{P}}_1') = \sin\alpha \sin\vartheta \cos\varphi,$$

where $\hat{\mathbf{P}} \cdot \hat{\mathbf{P}}_1 = \cos\alpha$. If we set $\hat{\mathbf{P}} \cdot \hat{\mathbf{P}}_1' = \cos\beta$ it follows that

$$\cos\beta = \cos\vartheta \cos\alpha + \sin\vartheta \sin\alpha \cos\varphi. \tag{5.25}$$

The total energy, being conserved, is related to P, P_1, and α by

$$2E = \frac{P_1^2}{\mu} + \frac{P^2}{m_2} - 2\frac{PP_1}{m_2}\cos\alpha$$

and to P, P_1', and β by

$$2E = \frac{P_1'^2}{\mu} + \frac{P^2}{m_2} - 2\frac{PP_1'}{m_2}\cos\beta.$$

This equation determines P_1' in terms of P, P_1, E, and β:

$$\frac{P_1}{P}\frac{M}{m_1} = \cos\beta + \left[\cos^2\beta + \frac{M}{m_1}\left(\frac{2Em_2}{P^2} - 1\right)\right]^{1/2}.$$

It follows that β must always be such that

$$\cos^2\beta \geq \frac{M}{m_1}\left(1 - \frac{2Em_2}{P^2}\right).$$

In the barycentric system the initial and final momenta of particle 1 are $\mathbf{p}_1$ and $\mathbf{p}_1'$, respectively, which are of equal magnitude, and they are related to $\mathbf{P}_1$ and $\mathbf{P}_1'$ by

$$\mathbf{p}_1 = \mathbf{P}_1 - \frac{m_1}{M}\mathbf{P}, \qquad \mathbf{p}_1' = \mathbf{P}_1' - \frac{m_1}{M}\mathbf{P}.$$

The barycentric angles θ and ϕ are defined by

$$\hat{\mathbf{p}}_1 \cdot \hat{\mathbf{p}}_1' = \cos\theta, \qquad (\hat{\mathbf{p}}_1 \times \hat{\mathbf{P}}) \cdot (\hat{\mathbf{p}}_1 \times \hat{\mathbf{p}}_1') = (P_1/p)\sin\alpha\sin\theta\cos\phi,$$

where we have written $p = p_1 = p_2$. The first equation leads to

$$p_1^2\cos\theta = P_1 P_1' \cos\vartheta + (m_1/M)P[(m_1/M)P - P_1\cos\alpha - P_1'\cos\beta] \tag{5.26}$$

and the second to

$$p\sin\theta\sin\phi = P_1'\sin\vartheta\sin\varphi. \tag{5.27}$$

Equations (5.25), (5.26), and (5.27) determine the angles θ and ϕ from ϑ and φ.

The Jacobian of the transformation from ϑ and φ to θ and ϕ can now be calculated and one finds that

$$\frac{\partial(\cos\theta, \phi)}{\partial(\cos\vartheta, \varphi)} = \frac{P_1'^3}{|\mathbf{P}_1' \cdot \mathbf{p}_1'|p}. \tag{5.28}$$

Therefore the differential cross section for the scattering of particle 1 in the general frame is given in terms of that in the barycentric system

$$\frac{d\sigma^{\text{gen}}}{d\Omega_1} = \frac{P_1'^3}{p|\mathbf{p}_1' \cdot \mathbf{P}_1'|}\frac{d\sigma^{\text{CM}}}{d\Omega}. \tag{5.29}$$

We note that this depends on φ even if the C.M. cross section depends only on θ. Note also that the cross section diverges at an angle for which $\mathbf{P}_1'$ is orthogonal to $\mathbf{p}_1'$.

In terms of $a = v_2'/v_1'$, where v_1' and v_2' are the final velocities of particles 1 and 2 in the general system, respectively, and $\cos\delta = \hat{\mathbf{P}}_1' \cdot \hat{\mathbf{P}}_2'$, (5.29) may also be expressed as

$$\frac{d\sigma^{\text{gen}}}{d\Omega_1} = \frac{[1 + (m_1/m_2)]^2}{(1 + a^2 - 2a\cos\delta)^{1/2}|1 - a\cos\delta|}\frac{d\sigma^{\text{CM}}}{d\Omega}. \tag{5.29a}$$

If the potential is symmetric under reflection, so that $V(\mathbf{r}_1 - \mathbf{r}_2) = V(\mathbf{r}_2 - \mathbf{r}_1)$, then the barycentric cross section is symmetric under an exchange of particles 1 and 2, and hence

$$\frac{d\sigma^{\text{gen}}}{d\Omega_1} = \left(\frac{m_1}{m_1}\right)^2 \frac{1}{a^2}\left|\frac{a - \cos\delta}{1 - a\cos\delta}\right|\frac{d\sigma^{\text{gen}}}{d\Omega_2}. \tag{5.30}$$

We also conclude from (5.29) that $|\mathbf{p}_1' \cdot \mathbf{P}_1'|P_1'^{-3}\,d\sigma^{\text{gen}}/d\Omega_1$ is a Galilean invariant.

If particle 2 is initially at rest, so that $\mathbf{P} = \mathbf{P}_1$, $\alpha = 0$, $\beta = \vartheta$, then $\varphi = \phi$ and (5.25) and (5.26) lead to the relation for the laboratory angle

$$\tan\vartheta = \frac{\sin\theta}{(m_1/m_2) + \cos\theta}. \tag{5.27a}$$

The relation between the cross sections then becomes

$$\frac{d\sigma^{\text{lab}}}{d\Omega_1} = \frac{[1 + (m_1/m_2)^2 + 2(m_1/m_2)\cos\theta]^{3/2}}{|1 + (m_1/m_2)\cos\theta|}\frac{d\sigma^{\text{CM}}}{d\Omega}. \tag{5.29b}$$

These equations are easily generalized to the relativistic case. One finds that if w_1 and w_2 are the energies of the two particles in the barycentric system, and c the velocity of light, then the relativistic version of (5.27a) is

$$\tan\vartheta = \frac{m_2c^2\sin\theta}{w_1 + w_2\cos\theta} \tag{5.27b}$$

and that of (5.29a),

$$\frac{d\sigma^{\text{lab}}}{d\Omega_1} = \frac{[(m_1^2 + m_2^2)c^4 + p^2c^2(1 + \cos^2\theta) + 2w_1w_2\cos\theta]^{3/2}}{m_2^2c^4|w_2 + w_1\cos\theta|}\frac{d\sigma^{\text{CM}}}{d\Omega}. \tag{5.29c}$$

When $m_2 < m_1$ the denominator vanishes at an angle θ where $\cos\theta = -w_2/w_1$, i.e., $\sin\vartheta = m_2/m_1$. At this angle the laboratory cross section diverges.

When the two masses are equal (5.27b) becomes

$$\tan\vartheta = 2\frac{mc^2}{w}\tan\tfrac{1}{2}\theta \tag{5.27c}$$

where w is the total C.M. energy. The backward angle $\theta = 180°$ in the barycentric system then corresponds to right-angle scattering $\vartheta = 90°$ in the laboratory, and scattering with $\vartheta > 90°$ is impossible. In the nonrelativistic limit (5.27c) becomes $\vartheta = \frac{1}{2}\theta$, whereas relativistically we have $\vartheta \leq \frac{1}{2}\theta$. The laboratory cross section in the equal-mass case is given by

$$\frac{d\sigma^{\text{lab}}}{d\Omega} = \frac{w^2}{m^2c^4}\frac{\cos^4\frac{1}{2}\theta}{\cos^3\vartheta}\frac{d\sigma^{\text{CM}}}{d\Omega} \tag{5.29d}$$

$$\underset{NR}{\rightarrow} 4\cos\vartheta\frac{d\sigma^{\text{CM}}}{d\Omega}.$$

Unless the barycentric cross section is infinite at $\theta = 180°$, the laboratory cross section for two equally massive particles vanishes at right angles (and beyond).

Equation (5.27c) is at the same time the limit of (5.27b) for very large center-of-mass energy, except near the backward direction. All particles, except those in a small backward cone in the barycentric system, are confined to a small forward cone in the laboratory. The angular openings of both cones are inversely proportional to the total center-of-mass energy.

Notice that if $m_2 < m_1$ then $\tan \vartheta > 0$ and the scattering is always in the forward cone, that is, $\vartheta < \frac{1}{2}\pi$. A particle cannot be turned back by a collision with a stationary lighter particle. The maximal angle of scattering in the laboratory is given by

$$\sin \vartheta_{\max} = \frac{m_2}{m_1}$$

both relativistically and nonrelativistically. When $m_1 < m_2$, then the forward-scattering cone ($\theta < \frac{1}{2}\pi$) in the center-of-mass frame corresponds to

$$\vartheta < \tan^{-1} \frac{2wm_2c^2}{w^2 + (m_1^2 - m_2^2)c^4} \underset{NR}{\rightarrow} \tan^{-1} \frac{m_2}{m_1}$$

the forward cone in the laboratory, to

$$\tan \tfrac{1}{2}\theta < \frac{w}{(m_2^2 - m_1^2)^{1/2}c^2}$$

or, nonrelativistically, to

$$\cos \theta > -\frac{m_1}{m_2}.$$

$\theta = 0$ corresponds to $\vartheta = 0$, and $\theta = \pi$ to $\vartheta = \pi$. As $m_2 \to \infty$, the two angles approach one another, as they should, since a very massive particle suffers little recoil.

In the center-of-mass frame the two particles emerge in opposite directions, $\theta_2 = \pi - \theta_1$. In the laboratory we get

$$\tan \vartheta_2 = \frac{2m_2c^2w}{w^2 + (m_2^2 - m_1^2)c^4} \cot \tfrac{1}{2}\theta \underset{NR}{\rightarrow} \cot \tfrac{1}{2}\theta. \tag{5.31}$$

In the nonrelativistic limit, therefore, simply

$$\vartheta_2 = \tfrac{1}{2}\pi - \tfrac{1}{2}\theta = \tfrac{1}{2}\theta_2.$$

Thus the target particle reciols in the laboratory at half the angle of emergence in the center-of-mass system. If the two masses are equal, (5.31) becomes

$$\tan \vartheta_2 = \frac{2mc^2}{w} \cot \tfrac{1}{2}\theta = \left(\frac{2mc^2}{w}\right)^2 \cot \vartheta_1. \tag{5.31a}$$

The angle between the two particles of equal mass is therefore given by

$$\tan(\vartheta_1 + \vartheta_2) = \frac{4mc^2w}{w^2 - 4m^2c^4} \csc \theta. \tag{5.32}$$

In the nonrelativistic limit this goes over into the elementary fact that the particles emerge at right angles.

In the laboratory the scattered particle suffers a change in energy, some of it being transferred to the target. If $\Delta W_1 = \Delta E_1$ is the difference between the initial and final energies of the missile and $E_1 = W_1 - m_1c^2$ is its initial kinetic energy, then we easily find that

$$\frac{\Delta E_1}{E_1} = \left[1 - \frac{(m_1 - m_2)^2 c^4}{w^2}\right] \sin^2 \tfrac{1}{2}\theta \underset{NR}{\rightarrow} \frac{4m_1 m_2}{(m_1 + m_2)^2} \sin^2 \tfrac{1}{2}\theta. \quad (5.33)$$

As it should be, the energy loss is maximal for backward scattering in the center-of-mass system. The maximal relative kinetic-energy loss is

$$\left(\frac{\Delta E_1}{E_1}\right)_{\max} = 1 - \frac{(m_1 - m_2)^2 c^4}{w^2} \underset{NR}{\rightarrow} \frac{4m_1 m_2}{(m_1 + m_2)^2}. \quad (5.34)$$

Only if the two masses are equal can the missile lose all its energy to the target. In that case we have simply

$$\frac{\Delta E_1}{E_1} = \sin^2 \tfrac{1}{2}\theta. \quad (5.33a)$$

This simple expression is at the same time the limit of (5.33) when the total barycentric energy is large compared with the difference between the two rest energies.

The square of the momentum transferred by the missile to the target in the center-of-mass system is

$$\tau^2 = (\mathbf{p}_1' - \mathbf{p}_1)^2 = 2\mathbf{p}_1^2(1 - \cos\theta) = 4\mathbf{p}_1^2 \sin^2 \tfrac{1}{2}\theta$$

or

$$\tau = 2|\mathbf{p}_1| \sin \tfrac{1}{2}\theta. \quad (5.35)$$

It is therefore possible to use, in place of the angle of scattering in the center-of-mass system, either the momentum transfer or else the energy transfer in the laboratory system. The former varies between zero and twice the initial momentum (of one particle), the latter between zero and the maximal value given by (5.34).

The energy loss ΔE_1 of the missile equals the recoil kinetic energy E_2 of the target in the laboratory. Hence we have the simple relation

$$\cos\theta = 1 - 2(E_2/E_{2\max}) \quad (5.36)$$

which permits the measurement of the differential scattering cross section in the center-of-mass system directly by measuring the spectrum of recoil energies in the laboratory. Equation (5.36) is valid both relativistically and nonrelativistically.

5.8 Identical Particles

If the two colliding particles are indistinguishable, then the flux in a given direction, which consists both of missiles and of recoil particles associated with missiles flying off in the opposite direction (in the center-of-mass system), has to be measured as a whole. Consequently the center-of-mass scattering cross section is the sum of the cross sections for the two diametrically opposed directions,

$$\frac{d\sigma^{\mathrm{CM}}}{d\Omega}(\theta,\varphi) = \frac{d\sigma_1}{d\Omega}(\theta,\varphi) + \frac{d\sigma_1}{d\Omega}(\pi - \theta, \varphi + \pi), \tag{5.37}$$

$d\sigma_1/d\Omega$ being calculated according to (5.12). For the laboratory cross section we have

$$\frac{d\sigma^{\mathrm{lab}}}{d\Omega}(\vartheta) = \frac{d\sigma_1^{\mathrm{lab}}}{d\Omega}(\vartheta_1) + \frac{d\sigma_2^{\mathrm{lab}}}{d\Omega}(\vartheta_2) \tag{5.38}$$

where $\vartheta = \vartheta_1$ and ϑ_2 are related by (5.31a), and $d\sigma_1{}^{\mathrm{lab}}/d\Omega$ must be calculated by means of (5.29d),

$$\frac{d\sigma^{\mathrm{lab}}}{d\Omega}(\vartheta) = \frac{d\sigma_1}{d\Omega}(\theta)\frac{(w^2 + 4m^2c^4\tan^2\frac{1}{2}\theta)^{3/2}\cos^4\frac{1}{2}\theta}{wm^2c^4}$$
$$+ \frac{d\sigma_1}{d\Omega}(\pi - \theta)\frac{(w^2 + 4m^2c^4\cot^2\frac{1}{2}\theta)^{3/2}\sin^4\frac{1}{2}\theta}{wm^2c^4}, \tag{5.39}$$

θ and ϑ being connected by (5.27*c*).

The center-of-mass cross section for identical particles is necessarily symmetric about 90° (if it is φ independent). The total scattering cross section is obtained from (5.37) by integrating from $\theta = 0$ to $\theta = \frac{1}{2}\pi$ only, so that it has the same value as before. Otherwise each particle would be counted twice, once as a missile and once as a recoil particle.

5.9 The Inverse Problem

In the customary formulation of the scattering problem the potential is assumed given, and the scattering cross section is calculated from it. In the inverse scattering problem one assumes that the cross section is experimentally determined and the potential is to be found.

Suppose that the cross section, at a fixed energy, is given as a function of the angle θ. If to each θ there corresponds exactly one impact parameter, that is, θ is a monotone function of b, then the latter may be determined as a function of the angle from (5.12) by

$$b^2(\theta) = 2\int_\theta^\pi d\theta \sin\theta \frac{d\sigma}{d\Omega}(\theta) \tag{5.40}$$

since in such a case the deflection angle must be equal to π when the impact parameter is zero. (If $\theta = 0$ for $b = 0$, θ cannot be a monotone function of b, since $\theta = 0$ for $b \to \infty$.)

The problem now is to determine the potential from $b(\theta)$ or from $\theta(b)$. Let us write

$$v(r) = 1 - \frac{V(r)}{E}$$

in the first line of (5.6) and introduce

$$x = \frac{1}{r^2 v}$$

as a new variable of integration. Then we have

$$\tfrac{1}{2}[\pi - \Theta(y)] = \int_0^y dx\, g(x)(y - x)^{-1/2} \tag{5.41}$$

where

$$g(x) = -\frac{\sqrt{x}}{r}\frac{dr}{dx} = \frac{rv^{3/2}}{2v + r\,dv/dr}$$

and

$$y = b^{-2}.$$

According to (5.41), the function g is an Euler transform of $\tfrac{1}{2}(\pi - \Theta)$. Its inversion is equivalent to solving an integral equation of the Abel type,

$$\begin{aligned} g(x) &= \tfrac{1}{2}x^{-1/2} - \frac{1}{2\pi}\int_0^x dy \frac{\Theta'(y)}{\sqrt{x - y}} \\ &= \tfrac{1}{2}x^{-1/2} - \frac{1}{2\pi}\int_0^{\theta(x)} \frac{d\theta}{[x - y(\theta)]^{1/2}} \end{aligned} \tag{5.42}$$

because $\Theta\,(0) = 0$. The upper limit $\theta(x)$ of the integral in (5.42) is determined by $x = y(\theta)$. Now the function $v(r)$ is obtained by

$$-\frac{\sqrt{x}}{g(x)} = r\frac{dx}{dr} = -x\left(2 + r\frac{dx}{dr}\frac{d\ln v}{dx}\right)$$

so that

$$\frac{d\ln v}{dx} = \frac{2g(x)}{\sqrt{x}} - \frac{1}{x} = -\frac{1}{\pi\sqrt{x}}\int_0^{\theta(x)} \frac{d\theta}{[x - y(\theta)]^{1/2}}$$

or because $v \to 1$ as $r \to \infty$, that is, $x \to 0$,

$$v = \exp\left\{-\frac{1}{\pi}\int_0^x \frac{dx'}{\sqrt{x'}}\int_0^{\theta(x')} \frac{d\theta}{[x' - y(\theta)]^{1/2}}\right\}.$$

If we invert the order of integrations, the x' integral can be carried out,

$$\begin{aligned} \int_0^x \frac{dx'}{\sqrt{x'}}\int_0^{\theta(x')} \frac{d\theta}{[x' - y(\theta)]^{1/2}} &= \int_0^{\theta(x)} d\theta \int_{y(\theta)}^x \frac{dx'}{\sqrt{x'}[x' - y(\theta)]^{1/2}} \\ &= 2\int_0^{\theta(x)} d\theta \cosh^{-1}\left[\frac{x}{y(\theta)}\right]^{1/2}, \end{aligned}$$

$\theta(x)$ being determined by

$$xb^2(\theta) = 1.$$

Therefore

$$v = \exp\left\{-\frac{2}{\pi}\int_0^{\theta(x)} d\theta \cosh^{-1}[\sqrt{x}b(\theta)]\right\}. \tag{5.43}$$

The procedure is thus to find $b(\theta)$ by (5.40) from the differential cross section and then to compute v by means of (5.43) in terms of x. Then r is determined as a function of x, or x as a function of r, by

$$r = \frac{1}{\sqrt{xv}}. \tag{5.44}$$

This is inserted in $v(x)$, and $V(r)$ is given by

$$V = E(1 - v).$$

For this inversion procedure to work, we had to assume that the deflection angle Θ is a monotone function of the impact parameter b, so that $b(\theta)$ can be obtained from the cross section by (5.40). Then, for $x = 1/r^2v$ to be usable as a variable of integration, $r^2(E - V)$ must be monotone. The latter requirement will be satisfied if V is everywhere positive and decreasing and $E > V(0)$. But this is not a necessary requirement. The monotonicity of $\Theta(b)$ is, in any case, an additional necessary assumption. At a given energy, the potential can, of course, be obtained only down to the distance of closest approach. The inner region beyond this is not accessible to the particle, and its scattering cross section is independent of it. In the inversion procedure this manifests itself in the fact that, when $x \to \infty$, then the integral in (5.43) tends to infinity and hence $v \to 0$. The variable r in (5.44) then approaches a value different from zero.

Notes and References

5.1 For discussions of the Hamilton-Jacobi equation see, for example, H. Goldstein, (1980), where further references can be found at the end of chaps. 9 and 10.

5.3 The cross section for the scattering of charged particles in a Coulomb field was first calculated by E. Rutherford (1911), and it was instrumental in his discovery of the atomic nucleus from the atomic scattering of α particles. It is one of the striking facts of quantum mechanics that, if effects of relativity are neglected, this cross section has exactly the same value when calculated classically and quantum mechanically. We can only speculate for how long the discovery of the nucleus would have been delayed if it had not been for this coincidence which made the classical argument used by Rutherford correct.

For a detailed explanation of a general contour integration technique of carrying out an integral such as (5.6), say, in the Coulomb case, see A. Sommerfeld (1933), pp. 645ff.

See I. Herbst (1974a) for trajectories of particles in a Coulomb field.

5.4 For discussions of orbiting, or spiral scattering, see Hirschfelder, Curtiss, and Bird, (1954) and Ford and Wheeler (1959).

5.5 For discussions of glory and rainbow scattering in electromagnetic scattering, see Chap. 3; for their discussion in classical particle scattering, see Ford and Wheeler (1959).

5.6 A discussion of classical particle scattering by an r^{-4} potential can be found in Vogt and Wannier (1954).

5.7 The derivation of the relation between the c.m. cross section and the cross section in a general coordinate system follows R. G. Newton (1972). For a detailed discussion of classical scattering in the laboratory system, see M. Gryzinski (1965a).

5.9 The content of this section is based on a paper by Keller, Kay, and Shmoys (1956), in which the examples of Rutherford scattering and scattering by an r^{-2} potential are also worked out explicitly. See also F. C. Hoyt (1939); Minerbo and Levy (1969); and W. H. Miller (1969b).

Integral equations of the Abel type are given in G. Doetsch (1956), pp. 157ff. Integrals of the Euler-transform type (5.41) are also called *Riemann-Liouville fractional integrals*, and they are tabulated in A. Erdelyi (1954), vol. 2, pp. 185ff.

General Note. For formulations of classical particle scattering in a language related to that used in quantum mechanics, see Prigogine and Henin (1957 and 1959); P. Résibois (1959); Miles and Dahler (1970); B. C. Eu (1971a); W. Hunziker (1974); T. A. Osborn et al. (1980); Narnhofer and Thirring (1981); Bollé and Osborn (1981); K. Yajima (1981e).

For classical treatments of the three-body scattering problem, in which particle A impinges upon a bound system of particles B and C and either A and the bound system (B, C) emerge unscathed or else B and the bound system (A, C), or C and (A, B), emerge, see Karplus, Porter, and Sharma (1964 and 1965), Karplus and Raff (1964), and Blais and Bunker (1964). See also M. Gryzinski (1965b); B. C. Eu (1971).

Problems

1 Use the equation preceding (5.10) to calculate the advance of the perihelion of a particle in a Coulomb bound state.

2 Calculate the first relativistic correction (i.e., the term of lowest order in an expansion in powers of v/c) to the classical Rutherford cross section.

3 Check Eq. (5.19).

4 Show that if a particle goes through the origin, $r = 0$, its velocity there must be infinite.

5 Calculate the impact parameter at which orbiting occurs for the potential $V = -ar^{-4}$, $a > 0$. What is the orbit equation for this impact parameter? Is the scattering cross section finite for such a potential? What if the potential is cut off at some small value of r?

6 Calculate relativistically the relation between the laboratory and center-of-mass scattering angles.

7 Carry out the inversion procedure of Sec. 5.9 by starting with the Rutherford cross section.

PART III

QUANTUM SCATTERING THEORY

CHAPTER 6

Time-Dependent Formal Scattering Theory

Having discussed the scattering of classical electromagnetic waves in Part I and that of classical particles in Part II, we now turn to the treatment of particle scattering from the quantum-mechanical point of view. The language, the general approach, and many of the tools here resemble those used in the previous chapters. That is in the nature of quantum mechanics. The description is most easily "visualized" in terms of waves. It therefore bears a great deal of resemblance to that of classical electromagnetic theory. Yet the wave formalism is tied intimately to a particle interpretation, in which language it is related to the discussion of Chap. 5.

At the same time the quantum scattering theory is considerably more elaborate than either of the other two. This is simply because it is very much richer in the number and variety of natural phenomena it describes. Because of this it is important to introduce an element of economy into the description so that the essentials common to all are not lost sight of in the details of special instances. The price, as always in physics, is a more abstract mathematical language. In this case the relevant new mathematical tools are those of Hilbert space theory and functional analysis.

The intimate relation of the ideas and approaches of quantum scattering theory to those treated in the first two parts of the book makes it unnecessary for us to give an introduction on a concrete physical level before turning to an abstract and general formulation of the scattering problem. The reader is urged at this point to remember those parts of the first five chapters which describe the scattering process in general. They are primarily to be found in Secs. 1.3 (especially 1.3.1 and 1.3.5), 2.2, 2.3, and 5.2. The details of the expressions for the scattering amplitudes, for example, are not relevant in the following, and they will, of course, be rederived in the new context. But the physical ideas involved should be familiar. We briefly summarize them as follows:

A *cloud* of particles is sent toward the target in the form of a *wave packet*. In the remote past, when the packet was prepared, the particles were free, i.e., did not interact with the scatterer. Their spins, etc., are fixed, and the definition of their momenta is limited only by the finite size of the cloud. In the course of time the missiles interact with the target, and they are scattered. In the far future they move outward in all directions, where they are counted by detectors arranged far from the scattering center. The differential cross section expresses the counting rate as a function of the angles from the direction of incidence. It is based, however, on the idealization that the momentum of the incoming particles is prescribed. This contradicts the assumption of a finite incident cloud and implies that instead the missiles come in as a steady *beam*. The theory has to reflect this idealization, which implies an infinite number of particles. Before the differential cross section is defined, a transition from the idea of a moving wave packet will have to be made to that of a steady, constant stream of infinite length and width. The previously mentioned sections embody many of the physical details of these concepts. The *formal* theory is their general, abstract mathematical expression, and without physical ideas the formalism remains empty.

Before turning to the abstract scattering theory itself, we recall the different ways in which time dependence can be handled in quantum mechanics. One is the *Schrödinger picture*, in which the state vectors describing a physical system depend on the time. The operators corresponding to the dynamical variables of the system are constant, i.e., time-independent, except in so far as they depend *explicitly, intrinsically*, on time. This is the picture most readily interpreted in the language of waves. The time dependence of the states is described by the Schrödinger equation.

Then there is the *Heisenberg picture*, in which all state vectors are constant. The time dependence is carried by the operators, which correspond to the dynamical variables of the system. It is described by Heisenberg's equation of motion. This picture corresponds most directly to the mode of thought of classical particle mechanics. From the point of view of relativistic field theory it has the advantage over the Schrödinger picture, of not treating time and space dependence (of field operators) on different footings.

The *interaction picture*, finally, stands partway between the other two. In it both the state vectors and the dynamical variables depend on the time. The time development of the former is described by a Schrödinger equation in which only the interaction appears, that of the latter by a Heisenberg equation with the *free hamiltonian* only. This picture has certain calculational advantages. In their observable results, all these pictures are, of course, equivalent.

6.1 The Schrödinger Equation

We shall be concerned primarily with the nonrelativistic region. However, the formalism set up is such that, provided that there exists a consistent relativistic quantum theory, the transition to the relativistic domain is relatively simple.

The state vector $\Psi(t)$ of the given physical system is assumed to satisfy the Schrödinger equation.

$$i\hbar\frac{\partial}{\partial t}\Psi(t) = H\Psi(t), \tag{6.1}$$

H being the hamiltonian operator. For the purpose of associating a concrete physical picture with the formal developments below, it is perhaps most natural to think of $\Psi(t)$ as a function of the particle coordinates, as a wave in configuration space. But it is the virtue of the abstract formulation of quantum mechanics that it is not tied in any way to the coordinate representation. The assertions below will hold equally well in the momentum representation or any other. This is the meaning of the statement that $\Psi(t)$ is a vector in an abstract Hilbert space. A representation is merely one of its specific realizations.

A word should be said about units and the fact that the Schrödinger equation is complex. In order to save ourselves some writing it is convenient to adopt units of time such that

$$\hbar = 1.$$

If the unit of energy is the electron volt (ev), then this amounts to a unit of time of about 0.7×10^{-15} sec. Such units are usually referred to as *natural*. The Schrödinger equation now reads

$$i\frac{\partial}{\partial t}\Psi(t) = H\Psi(t). \tag{6.1a}$$

The fact that the imaginary $+i$ appears in the Schrödinger equation entails a fundamental distinction in the solutions between $+i$ and $-i$, that is, between upper and lower complex half planes of energy and other variables. This distinction is, of course, entirely conventional and has its origin in the custom of writing the time dependence of a wave of frequency ω in the form $e^{-i\omega t}$ rather than $e^{+i\omega t}$. As a consequence e^{ikr} represents an *outgoing* spherical wave and e^{-ikr} an *incoming* one. It is well to remember this conventional origin of the distinction between $+i$ and $-i$.

In Secs. 6.2 to 6.6 we shall develop the theory in a heuristic manner without worrying about finer mathematical points, such as the meaning of the necessary limiting processes. Section 6.7 is devoted to these questions of a mathematical nature.

6.2 Time Development of State Vectors in the Schrödinger Picture

In order to solve Eq. (6.1a) we assume that the hamiltonian operator H can be split into two parts,

$$H = H_0 + H' \tag{6.2}$$

so that H_0 represents the hamiltonian of the particles in the absence of interaction between them. In other words, H_0 represents the *kinetic* energy alone. There are important areas, relativistic field theory, for example, in which this split-up is highly problematical. What is important for our purpose is that H_0 represent the energy of the particles in a scattering experiment when they are far apart. We shall return to the discussion of the problematical nature of (6.2) in rearrangement collisions in Chap. 16. For the present we have primarily elastic scattering in mind, and we may think of H' as the potential energy of interaction between two particles or else of interaction of a particle with a fixed force center. We assume that H is independent of the time.

The Green's Functions. In order to solve Eq. (6.1a), we define four kinds of propagators, or Green's functions, by the equations

$$\begin{aligned} \left(i\frac{\partial}{\partial t} - H_0\right) G^{\pm}(t) &= \mathbb{1}\delta(t) \\ \left(i\frac{\partial}{\partial t} - H\right) \mathcal{G}^{\pm}(t) &= \mathbb{1}\delta(t) \end{aligned} \tag{6.3}$$

and the initial conditions

$$\begin{aligned} G^{+}(t) = \mathcal{G}^{+}(t) &= 0 \quad \text{for } t < 0 \\ G^{-}(t) = \mathcal{G}^{-}(t) &= 0 \quad \text{for } t > 0. \end{aligned} \tag{6.4}$$

Thus G^+ and $\mathcal{G}^+$ are *retarded* Green's functions and G^- and $\mathcal{G}^-$ *advanced* ones. In (6.3), $\mathbb{1}$ stands for the unit operator and $\delta(t)$ for Dirac's delta function. These equations can be solved in a symbolic fashion by writing

$$\begin{aligned} G^{+}(t) &= \begin{cases} -ie^{-iH_0 t} & t > 0 \\ 0 & t < 0 \end{cases} \\ G^{-}(t) &= \begin{cases} 0 & t > 0 \\ ie^{-iH_0 t} & t < 0 \end{cases} \end{aligned} \tag{6.5}$$

$$\begin{aligned} \mathcal{G}^{+}(t) &= \begin{cases} -ie^{-iHt} & t > 0 \\ 0 & t < 0 \end{cases} \\ \mathcal{G}^{-}(t) &= \begin{cases} 0 & t > 0 \\ ie^{-iHt} & t < 0. \end{cases} \end{aligned} \tag{6.6}$$

These symbolic solutions, however, are not necessarily more useful or more meaningful than the defining equations (6.3) and (6.4).

The circumstance that the defining equations (6.3) and (6.4) for G contain no operators which fail to commute with H_0 implies that $G^\pm$ must commute with H_0 and similarly $\mathcal{G}^\pm$ with H. These properties are, of course, explicit in (6.5) and (6.6). It then follows from the fact that the operators H_0 and H must, for physical reasons, be hermitian that

$$\begin{aligned} G^+(t)^\dagger &= G^-(-t) \\ \mathcal{G}^+(t)^\dagger &= \mathcal{G}^-(-t) \end{aligned} \tag{6.7}$$

which, again are obvious in the notations (6.5) and (6.6).

It is convenient to convert the differential equations (6.3) and initial conditions (6.4) into integral equations:

$$\mathcal{G}^\pm(t - t') = G^\pm(t - t') + \int_{-\infty}^{\infty} dt''\, G^\pm(t - t'')H'\mathcal{G}^\pm(t'' - t') \tag{6.8}$$

or

$$\mathcal{G}^\pm(t - t') = G^\pm(t - t') + \int_{-\infty}^{\infty} dt''\, \mathcal{G}^\pm(t - t'')H'G^\pm(t'' - t') \tag{6.8a}$$

which may be construed as equations for $\mathcal{G}$ in terms of G, or vice versa. Usually the former is more convenient, since often G can be explicitly written down in a specific representation. The integrals in (6.8) and (6.8a) really extend only from $t'' = t$ to $t'' = t'$, or vice versa, depending upon their use for G^+ or G^-, so that convergence questions in them do not arise. It is easily verified that a solution $\mathcal{G}^+$ or $\mathcal{G}^-$ of (6.8) or (6.8a) satisfies (6.3) and (6.4) if G^+ and G^- satisfies them. Furthermore, since, for example, for $t > 0$,

$$\mathcal{G}^+(t) = G^+(t) + \int_0^t dt''\, G^+(t - t'')H'\mathcal{G}^+(t'')$$

the integral equations are of the Volterra type (though with operator-valued kernels and inhomogeneities) and always have unique solutions. Thus they are equivalent to Eqs. (6.3) and (6.4).

Because the functions G and $\mathcal{G}$ describe the propagation characteristics of the state vectors in time, they are called *propagators*. Let $\Psi_0(t)$ be a state vector which satisfies the free Schrödinger equation

$$i\frac{\partial}{\partial t}\Psi_0(t) = H_0\Psi_0(t). \tag{6.9}$$

Then the operator G^+ allows us to express the state vector $\Psi_0(t')$ for any time t' later than t, $t' > t$, in terms of its value at $t' = t$,

$$\Psi_0(t') = iG^+(t' - t)\Psi_0(t). \tag{6.10}$$

It is readily verified explicitly that $\Psi_0(t')$ then satisfies (6.9) for $t' > t$ and that the vector $\Psi_0(t')$ on the left approaches the vector $\Psi_0(t)$ on the right when $t' \to t^+$. That is so because

$$\lim_{t \to 0+} G^+(t) = -i\mathbb{1} \tag{6.11}$$

as well as

$$\lim_{t \to 0+} \mathscr{G}^+(t) = -i\mathbb{1} \tag{6.11a}$$

and

$$\lim_{t \to 0-} G^-(t) = \lim_{t \to 0-} \mathscr{G}^-(t) = i\mathbb{1}. \tag{6.11b}$$

Similarly we have for $t' > t$

$$\Psi(t') = i\mathscr{G}^+(t' - t)\Psi(t) \tag{6.10a}$$

and for $t' < t$

$$\begin{aligned} \Psi_0(t') &= -iG^-(t' - t)\Psi_0(t) \\ \Psi(t') &= -i\mathscr{G}^-(t' - t)\Psi(t). \end{aligned} \tag{6.10b}$$

The operators G^+ and $\mathscr{G}^+$ thus describe the propagation of waves subject to the hamiltonians H_0 and H, respectively, in the future and G^- and $\mathscr{G}^-$ in the past. Comparison of (6.10) and (6.10b) shows immediately that for $t' > t$

$$G^-(t - t')G^+(t' - t) = G^+(t' - t)G^-(t - t') = \mathbb{1},$$

i.e., in view of (6.7) for $t > 0$,

$$G^-(-t) = G^+(t)^\dagger = G^+(t)^{-1} \tag{6.12}$$

and, similarly,

$$\mathscr{G}^-(-t) = \mathscr{G}^+(t)^\dagger = \mathscr{G}^+(t)^{-1} \tag{6.12a}$$

while, for $t < 0$,

$$\begin{aligned} G^+(-t) &= G^-(t)^\dagger = G^-(t)^{-1} \\ \mathscr{G}^+(-t) &= \mathscr{G}^-(t)^\dagger = \mathscr{G}^-(t)^{-1}. \end{aligned} \tag{6.12b}$$

The propagators $G^+(t)$ and $\mathscr{G}^+(t)$ are unitary for $t > 0$, and $G^-(t)$ and $\mathscr{G}^-(t)$ are unitary for $t < 0$. These properties are at once apparent in Eqs. (6.5) and (6.6).

In and Out States. Let us define

$$\Psi_0(t) \equiv iG^+(t - t')\Psi(t'). \tag{6.13}$$

This is a state vector whose time development for $t > t'$ is governed by the free hamiltonian H_0 but which at the time t_0 was equal to $\Psi(t_0)$. Let us now allow $t' \to -\infty$. This defines the state

$$\Psi_{\text{in}}(t) \equiv \lim_{t' \to -\infty} iG^+(t - t')\Psi(t'). \tag{6.14}$$

Then Ψ_{in} is a free state vector, i.e., one which at all times develops according to H_0, in which the particles never interact, but which in the infinite past was equal to the exact state vector of the complete interacting system, $\Psi(t)$. That such a state Ψ_{in} exists is a special assumption concerning the physical nature of the system under consideration, which makes it a *scattering system*. In the remote past (as well as in the distant future) all particles are assumed to be sufficiently far apart from one another so that their interactions can be neglected.

Now notice that

$$\begin{aligned} i\frac{\partial}{\partial t'}[G^+(t-t')\Psi(t')] &= -i\frac{\partial}{\partial t}G^+(t-t')\Psi(t') + G^+(t-t')i\frac{\partial}{\partial t'}\Psi(t') \\ &= -\delta(t-t')\Psi(t') + G^+(t-t')H'\Psi(t') \end{aligned}$$

because of (6.3) and since $G^+(t)$ and H_0 commute. Integration from $t' = -\infty$ to $+\infty$ yields, according to (6.14),

$$\Psi(t) = \Psi_{\text{in}}(t) + \int_{-\infty}^{\infty} dt'\, G^+(t-t')H'\Psi(t') \tag{6.15}$$

as an integral equation satisfied by $\Psi(t)$. The upper limit of the integral is in reality equal to t, of course, and we may write it, by (6.5),

$$\Psi(t) = \Psi_{\text{in}}(t) - i\int_{-\infty}^{t} dt'\, e^{-iH_0(t-t')}H'\Psi(t'). \tag{6.15a}$$

It aids our physical insight to solve Eq. (6.15) by iteration (without worrying for the moment about convergence),

$$\begin{aligned} \Psi(t) = \Psi_{\text{in}}(t) &+ \int_{-\infty}^{t} dt'\, G^+(t-t')H'\Psi_{\text{in}}(t') \\ &+ \int_{-\infty}^{t} dt' \int_{-\infty}^{t'} dt''\, G^+(t-t')H'G^+(t'-t'')H'\Psi_{\text{in}}(t'') + \cdots. \end{aligned}$$

This is reasonably interpreted by saying that the complete state Ψ differs from its corresponding free state Ψ_{in} by a superposition of terms which refer to a free state interacting once at t' and subsequently propagating freely up to t; a free state interacting at t'', propagating freely to t', and then interacting again and going along freely to t; etc. Thus the nth term in the iteration series is interpreted as referring to n elementary interactions at different times.

In a similar vein one defines

$$\Psi_{\text{out}}(t) = \lim_{t'\to\infty} -iG^-(t-t')\Psi(t') \tag{6.16}$$

which is a free state vector whose time development is governed by H_0, equal to the complete state Ψ in the infinite future. We then get

$$\Psi(t) = \Psi_{\text{out}}(t) + \int_{-\infty}^{\infty} dt'\, G^-(t-t')H'\Psi(t') \tag{6.17}$$

or symbolically

$$\Psi(t) = \Psi_{\text{out}}(t) + i \int_t^{\infty} dt' \, e^{-iH_0(t-t')} H' \Psi(t'). \tag{6.17a}$$

The procedure may just as easily be reversed. The state vector

$$\Psi(t) = i\mathscr{G}^+(t - t')\Psi_0(t') \tag{6.18}$$

develops in time according to the full hamiltonian, for all times t later than t', and at $t = t'$ it was equal to Ψ_0. Hence, if we let $t' \to \pm\infty$, we must obtain

$$\begin{aligned} \Psi(t) &= \lim_{t' \to -\infty} i\mathscr{G}^+(t - t')\Psi_{\text{in}}(t') \\ &= \lim_{t' \to +\infty} -i\mathscr{G}^-(t - t')\Psi_{\text{out}}(t') \end{aligned} \tag{6.19}$$

and from this, as (6.15) was derived,

$$\Psi(t) = \Psi_{\text{in}}(t) + \int_{-\infty}^{\infty} dt' \, \mathscr{G}^+(t - t') H' \Psi_{\text{in}}(t') \tag{6.20}$$

$$\Psi(t) = \Psi_{\text{out}} + \int_{-\infty}^{\infty} dt' \, \mathscr{G}^-(t - t') H' \Psi_{\text{out}}(t') \tag{6.20a}$$

or symbolically

$$\begin{aligned} \Psi(t) &= \Psi_{\text{in}}(t) - i \int_{-\infty}^{t} dt' \, e^{-iH(t-t')} H' \Psi_{\text{in}}(t') \\ &= \Psi_{\text{out}}(t) + i \int_t^{\infty} dt' \, e^{-iH(t-t')} H' \Psi_{\text{out}}(t'). \end{aligned} \tag{6.20b}$$

These equations, which express the full state $\Psi(t)$ in terms of Ψ_{in} or Ψ_{out}, may be regarded as "solving" the integral equations (6.15) and (6.17). They do so at the expense of containing the complete Green's function $\mathscr{G}^{\pm}$, which is usually no better known than $\Psi(t)$.

It is important to have a full understanding of the meaning of the states Ψ_{in} and Ψ_{out} for a given state Ψ. In an experimental scattering situation we expect $\Psi_{\text{in}}(t)$ to describe a more or less collimated beam. It is true that we must not, for mathematical reasons, take it to be *monochromatic*. This would produce convergence difficulties. We do want it to be a *wave packet*. Nevertheless, it contains all the information of how that wave packet was prepared in the remote past, that the particles were sent toward the target along a given direction, say, with (approximately) the given momentum $\mathbf{p}$, spin along the x axis; or else that a converging spherical wave of angular momentum j was arranged, etc. All these specifications must refer to quantum numbers, or eigenvalues of dynamical variables, which commute with H_0 and can thus be specified in a free state Ψ_0. A complete state thus determined is denoted by $\Psi^{(+)}(t)$ and labeled by the same quantum numbers as Ψ_{in}. In other words, $\Psi^{(+)}(\alpha, t)$ is a complete state which in the remote past was equal to the free state $\Psi_{\text{in}}(\alpha, t)$, α being a complete set of variables that commute with H_0 (but not necessarily with H). The state $\Psi^{(+)}(\alpha, t)$ is the solution of

$$\Psi^{(+)}(\alpha, t) = \Psi_{\text{in}}(\alpha, t) + \int_{-\infty}^{\infty} dt'\, G^+(t - t')H'\Psi^{(+)}(\alpha, t'). \tag{6.21}$$

It was *controlled*, so to speak, in the remote past, and its labels refer to that control.

In the far future $\Psi^{(+)}(\alpha, t)$ will again be essentially a free state, namely, Ψ_{out}, and it satisfies (6.17) in terms of this. But Ψ_{out} is *not* controlled. It contains, in addition to the controlled "beam" part of Ψ_{in}, say, an unknown amount of outgoing scattered wave. That is to say, the fact that $\Psi^{(+)}$ satisfies (6.17) and (6.20a) is of no practical use, since Ψ_{out} there is uncontrolled, unknown. The important equations are (6.15) and (6.20) or, explicitly, (6.21).

It is equally possible, although physically not quite so experimentally meaningful, to define a complete state by its controlled behavior in the far future. Such a state is called $\Psi^{(-)}(\alpha, t)$. Its label α refers to the quantum numbers of $\Psi_{\text{out}}(\alpha, t)$. They must be of the same kind as before, i.e., must commute with H_0, but now the important integral equation is (6.17) or (6.20a) or, explicitly,

$$\Psi^{(-)}(\alpha, t) = \Psi_{\text{out}}(\alpha, t) + \int_{-\infty}^{\infty} dt'\, G^-(t - t')H'\Psi^{(-)}(\alpha, t'). \tag{6.22}$$

The fact that $\Psi^{(-)}$ also satisfies (6.15) and (6.20) is of no practical use, since now Ψ_{in} is uncontrolled and differs from $\Psi_{\text{out}}(\alpha, t)$, say, by an unknown amount of incoming spherical waves. The meaning of the labels α in $\Psi^{(+)}(\alpha, t)$ and $\Psi^{(-)}(\alpha, t)$ is therefore quite different. The state $\Psi^{(+)}(\alpha, t)$ is such that in the remote past it was a controlled "beam" completely specified by α; in the far future it will be whatever the hamiltonian decrees. The state $\Psi^{(-)}(\alpha, t)$ is such that in the distant future it will be a controlled beam specified by α; in the remote past it must therefore have been a complicated free state whose exact nature depends on H.

6.3 The Møller Wave Operator in the Schrödinger Picture

Let us now insert (6.10b) in (6.20). This yields

$$\Psi(t) = \Omega^{(+)}\Psi_{\text{in}}(t) \tag{6.23}$$

if we use the abbreviation

$$\begin{aligned} \Omega^{(+)} &= \mathbb{1} - i\int_{-\infty}^{\infty} dt'\, \mathcal{G}^+(t - t')H'G^-(t' - t) \\ &= \mathbb{1} - i\int_{-\infty}^{\infty} dt\, \mathcal{G}^+(-t)H'G^-(t). \end{aligned} \tag{6.24}$$

$\Omega^{(+)}$ is called the *wave operator*. It is a time-independent operator which converts the free state $\Psi_{\text{in}}(t)$ directly into that complete state $\Psi(t)$ which corresponds to it in the sense that it was essentially equal to it in the remote past.

It is convenient to introduce another operator $K^{(+)}$ by the definition

$$\Omega^{(+)} = \mathbb{1} + K^{(+)} \tag{6.25}$$

so that

$$\begin{aligned} K^{(+)} &= -i \int_{-\infty}^{\infty} dt\, \mathcal{G}^{+}(-t) H' G^{-}(t) \\ &= -i \int_{-\infty}^{0} dt\, e^{iHt} H' e^{-iH_0 t}. \end{aligned} \tag{6.26}$$

Since the last equation can be written

$$K^{(+)} = -\int_{-\infty}^{0} dt \frac{d}{dt} (e^{iHt} e^{-iH_0 t}) \tag{6.26a}$$

we have equally well

$$\begin{aligned} \Omega^{(+)} &= \lim_{t \to -\infty} \mathcal{G}^{+}(-t) G^{-}(t) \\ &= \lim_{t \to -\infty} e^{iHt} e^{-iH_0 t}. \end{aligned} \tag{6.27}$$

Similarly one may insert (6.10) in (6.20a) and get

$$\Psi(t) = \Omega^{(-)} \Psi_{\text{out}}(t) \tag{6.28}$$

where

$$\Omega^{(-)} = \mathbb{1} + K^{(-)} = \mathbb{1} + i \int_{-\infty}^{\infty} dt\, \mathcal{G}^{-}(-t) H' G^{+}(t) \tag{6.29}$$

or

$$K^{(-)} = i \int_{0}^{\infty} dt\, e^{iHt} H' e^{-iH_0 t}. \tag{6.30}$$

Because

$$K^{(-)} = \int_{0}^{\infty} dt \frac{d}{dt} (e^{iHt} e^{-iH_0 t}) \tag{6.30a}$$

we can therefore write

$$\begin{aligned} \Omega^{(-)} &= \lim_{t \to \infty} \mathcal{G}^{-}(-t) G^{+}(t) \\ &= \lim_{t \to \infty} e^{iHt} e^{-iH_0 t}. \end{aligned} \tag{6.31}$$

Other useful representations of the wave operators are

$$\begin{aligned} \Omega^{(+)} &= \lim_{\varepsilon \to 0+} \varepsilon \int_{-\infty}^{0} dt\, e^{\varepsilon t} e^{iHt} e^{-iH_0 t} \\ \Omega^{(-)} &= \lim_{\varepsilon \to 0+} \varepsilon \int_{0}^{\infty} dt\, e^{-\varepsilon t} e^{iHt} e^{-iH_0 t}. \end{aligned} \tag{6.32}$$

For the proof of the second equation, for example, we divide the integral into two parts,

$$\int_0^\infty dt \cdots = \int_0^T dt \cdots + \int_T^\infty dt \cdots.$$

Then, no matter how large T, the first part will vanish as $\varepsilon \to 0$. The second differs for large T arbitrarily little from

$$\Omega^{(-)}\varepsilon \int_T^\infty dt\, e^{-\varepsilon t} \underset{\varepsilon \to 0}{\to} \Omega^{(-)}.$$

All these relations can be easily inverted. Inserting (6.10b) in (6.15) we get

$$\Psi_{\text{in}}(t) = \left[\mathbb{1} + i \int_{-\infty}^{\infty} dt'\, G^{+}(t')H'\mathscr{G}^{-}(-t')\right]\Psi(t).$$

Because of (6.7), (6.24), and the hermiticity of H', this means that

$$\Psi_{\text{in}}(t) = \Omega^{(+)\dagger}\Psi(t) \tag{6.33}$$

and similarly

$$\Psi_{\text{out}}(t) = \Omega^{(-)\dagger}\Psi(t). \tag{6.34}$$

General Properties of the Wave Operators. Comparison of (6.33) and (6.34) with (6.23) and (6.28) reveals that

$$\Psi_{\text{in}}(t) = \Omega^{(+)\dagger}\Omega^{(+)}\Psi_{\text{in}}(t)$$

and

$$\Psi_{\text{out}}(t) = \Omega^{(-)\dagger}\Omega^{(-)}\Psi_{\text{out}}(t).$$

It must be counted as among our fundamental assumptions that the free states $\Psi_{\text{in}}(t)$ and $\Psi_{\text{out}}(t)$ are wave packets which span the entire Hilbert space. That is, at a fixed time, every state can be arbitrarily closely approximated by a superposition of states Ψ_{in} or Ψ_{out}. On the basis of this assumption we may therefore conclude that $\Omega^{(+)}$ and $\Omega^{(-)}$ are *isometric*,

$$\Omega^{(+)\dagger}\Omega^{(+)} = \Omega^{(-)\dagger}\Omega^{(-)} = \mathbb{1}. \tag{6.35}$$

This means that the operators $\Omega^{(\pm)}$ do not change the length of a vector,

$$\begin{aligned}\|\Omega\Psi\|^2 &= (\Omega\Psi, \Omega\Psi)\\ &= (\Psi, \Omega^\dagger\Omega\Psi)\\ &= (\Psi, \Psi) = \|\Psi\|^2.\end{aligned}$$

It does *not* imply, though, that $\Omega^{(+)}$ or $\Omega^{(-)}$ is necessarily unitary. We cannot conclude from

$$\Psi(t) = \Omega^{(+)}\Omega^{(+)\dagger}\Psi(t)$$

that (6.35) also holds with the order of its factors reversed. The scattering states $\Psi(t)$ do not necessarily span the Hilbert space. If H has bound states, these cannot be approximated by packets $\Psi(t)$ which evolve in time from free wave packets sent in from far away. All the scattering states are orthogonal to the bound states.

Let us insert (6.23) or (6.28) into the Schrödinger equations (6.1a) and (6.9). Because of the assumption that the states Ψ_{in} and the states Ψ_{out} each form complete sets, we may conclude that

$$H\Omega = \Omega H_0, \tag{6.36}$$

Ω being either $\Omega^{(+)}$ or $\Omega^{(-)}$. It can be recognized from this relation alone that Ω cannot, in general, be unitary. If it were, the spectra of H_0 and H would necessarily be the same and bound states of H would be ruled out. As it stands, (6.36) implies only that every eigenvalue of H_0 is also an eigenvalue of H, but not vice versa. Indeed, we can reduce the question of unitarity of $\Omega^{(+)}$ and $\Omega^{(-)}$ to the question of bound states of H as follows.

Let $\Psi_0(E,\alpha)$ be the set of eigenstates of H_0 with eigenvalue E.[1] The set of variables α is any other set of quantum numbers necessary in order to remove the degeneracy. The completeness of these states is expressible by a "resolution" of the identity,[2]

$$\mathbb{1} = \sum_\alpha \int_0^\infty dE\, \Psi_0(E,\alpha)\Psi_0^\dagger(E,\alpha). \tag{6.37}$$

Now, (6.36) implies that $\Omega\Psi_0(E,\alpha)$ is an eigenstate of H with the same eigenvalue E,

$$H\Omega\Psi_0(E,\alpha) = \Omega H_0\Psi_0(E,\alpha) = E\Omega\Psi_0(E,\alpha)$$

i.e.,

$$\Omega\Psi_0(E,\alpha) = \Psi(E,\alpha).$$

Left multiplication of (6.37) by Ω therefore shows that the wave operators have expansions of the form

$$\Omega = \sum_\alpha \int_0^\infty dE\, \Psi(E,\alpha)\Psi_0^\dagger(E,\alpha). \tag{6.37a}$$

Furthermore, if

$$(\Psi_0(E,\alpha), \Psi_0(E',\alpha')) = \delta(E - E')\delta_{\alpha\alpha'},$$

1 We use the physicist's terminology and refer to Ψ as an eigenstate of H even when the eigenvalue is in the continuum and Ψ not normalizable. Only in the mathematical sections shall we be more circumspect.

2 We use a notation in which products such as $\Phi\Psi^\dagger$ are always meant as *dyadics*. If Φ lies in a Hilbert space $\mathscr{H}$, then $\Phi^\dagger$ is the corresponding state in the dual space $\mathscr{H}^\dagger$. The correspondence is arbitrary, of course, but fixed once and for all, and such that, if $\Phi^\dagger$ corresponds to Φ, then $C^*\Phi^\dagger$ corresponds to $C\Phi$. The hermitian inner product in $\mathscr{H}$ can then be thought of as a euclidean inner product between vectors in $\mathscr{H}$ and $\mathscr{H}^\dagger$. In the Dirac notation, vectors Φ are *kets*,

$$\Phi = |\cdot\rangle$$

and vectors $\Phi^\dagger$ *bras*,

$$\Phi^\dagger = \langle\cdot|.$$

then according to (6.35)

$$(\Psi(E,\alpha),\Psi(E',\alpha')) = (\Psi_0(E,\alpha),\Omega^\dagger\Omega\Psi_0(E',\alpha')) = \delta(E - E')\delta_{\alpha\alpha'}.$$

Thus the states $\Psi(E,\alpha)$ have the right normalization. (See Sec. 7.2.1 for further details.)

We now insert (6.37) in the center of the product $\Omega\Omega^\dagger$ and get

$$\begin{aligned}\Omega\Omega^\dagger &= \Omega\int_0^\infty dE\sum_\alpha \Psi_0(E,\alpha)\Psi_0^\dagger(E,\alpha)\Omega^\dagger \\ &= \int_0^\infty dE\sum_\alpha \Psi(E,\alpha)\Psi^\dagger(E,\alpha) \\ &= \mathbb{1} - \Lambda.\end{aligned} \tag{6.38}$$

Λ is called the *unitary deficiency* of Ω. From the completeness of the set of bound *and* scattering states of H it follows that

$$\Lambda = \sum_n \Psi_{\text{bd}}^{(n)}\Psi_{\text{bd}}^{(n)\dagger}, \tag{6.39}$$

that is, Λ is the projection onto the space spanned by the bound states of H. If H has no bound states, then $\Omega^{(+)}$ and $\Omega^{(-)}$ are *unitary*.

The hermiticity of both H and H_0 implies that the hermitian conjugate of (6.36) reads

$$H_0\Omega^\dagger = \Omega^\dagger H. \tag{6.36a}$$

Consequently

$$H_0\Omega^\dagger\Psi(E,\alpha) = E\Omega^\dagger\Psi(E,\alpha)$$

which implies that, if E is in the spectrum of H but not in that of H_0, then

$$\Omega^\dagger\Psi(E,\alpha) = 0. \tag{6.40}$$

As a result

$$\Omega^\dagger\Lambda = 0 \tag{6.40a}$$

and left multiplication of (6.38) by $\Omega^\dagger$ leads to the necessary consistency with (6.35).

The range[3] of the operators $\Omega^{(\pm)}$ is not the entire Hilbert space. They map the whole space onto the subspace spanned by the continuum states of H. The subspace spanned by the bound states of H is not reached by these operators. This implies that an inverse cannot be defined for the whole space. The operators $\Omega^{(\pm)\dagger}$ are the nearest thing possible. On the subspace spanned by the scattering states of H, they are the inverses of $\Omega^{(\pm)}$, respectively, and the subspace spanned by the bound states is annihilated by them.

3 The *range* of an operator A is the space of all vectors Φ for which a vector Ψ exists such that $\Phi = A\Psi$. It is the space onto which A maps its domain of definition.

6.4 The S Matrix

We can use (6.34) and (6.23) in order to express the out state of a given state vector in terms of its in state,

$$\Psi_{\text{out}}(t) = \Omega^{(-)\dagger}\Omega^{(+)}\Psi_{\text{in}}(t). \tag{6.41}$$

The principal question asked in scattering theory is this: In the remote past the system was in the controlled state $\Psi_{\text{in}}(\alpha, t) = \Psi_0(\alpha, t)$ so that now it is in the state $\Psi^{(+)}(\alpha, t)$. What is the probability for finding it in the state $\Psi_0(\beta, t)$ in the distant future? The amplitude for this probability is evidently

$$\begin{aligned}\lim_{t\to\infty} (\Psi_0(\beta, t), \Psi^{(+)}(\alpha, t)) &= (\Psi_0(\beta, t), \Psi_{\text{out}}(t)) \\ &= (\Psi_0(\beta, t), \Omega^{(-)\dagger}\Omega^{(+)}\Psi_{\text{in}}(t)) \\ &= (\Psi_0(\beta, t), S\Psi_0(\alpha, t))\end{aligned} \tag{6.42}$$

according to (6.41), if we define the scattering operator

$$S \equiv \Omega^{(-)\dagger}\Omega^{(+)}. \tag{6.43}$$

The matrix of S on the basis of such free states is called the *S matrix*. Since (6.36) and (6.36a) imply that

$$\Omega^{(-)\dagger}\Omega^{(+)}H_0 = \Omega^{(-)\dagger}H\Omega^{(+)} = H_0\Omega^{(-)\dagger}\Omega^{(+)}$$

the scattering operator commutes with H_0,

$$[S, H_0] = 0. \tag{6.44}$$

Therefore the inner product on the right of (6.42), the matrix $S_{\beta\alpha}$, is *independent of the time*.

The fundamental question of scattering theory can also be asked in a slightly different way which does not involve a limiting process directly. If the system is described by the state vector $\Psi^{(+)}(\alpha, t)$, which is known to have been in the controlled state $\Psi_{\text{in}}(\alpha, t)$ in the remote past, what is the probability of finding it in the state $\Psi^{(-)}(\beta, t)$, which is known to go over into the controlled state $\Psi_{\text{out}}(\beta, t)$ in the distant future? Its probability amplitude is

$$\begin{aligned}(\Psi^{(-)}(\beta, t), \Psi^{(+)}(\alpha, t)) &= (\Omega^{(-)}\Psi_{\text{out}}(\beta, t), \Omega^{(+)}\Psi_{\text{in}}(\alpha, t)) \\ &= (\Psi_{\text{out}}(\beta, t), \Omega^{(-)\dagger}\Omega^{(+)}\Psi_{\text{in}}(\alpha, t)) \\ &= (\Psi_{\text{out}}(\beta, t) S\Psi_{\text{in}}(\alpha, t)) \\ &= (\Psi_0(\beta, 0) S\Psi_0(\alpha, 0)).\end{aligned} \tag{6.45}$$

The answer, of course, is the same.

Instead of inserting (6.23) in (6.34), we may equally well insert (6.28) in (6.33) to get

$$\begin{aligned}\Psi_{\text{in}}(t) &= \Omega^{(+)\dagger}\Omega^{(-)}\Psi_{\text{out}}(t) \\ &= S^\dagger\Psi_{\text{out}}(t).\end{aligned}$$

Since the sets $\{\Psi_{\text{in}}\}$ and $\{\Psi_{\text{out}}\}$ are assumed to be complete, we conclude that S is unitary,

$$S^\dagger S = SS^\dagger = \mathbb{1}. \tag{6.46}$$

Notice that this conclusion may be drawn irrespective of whether the $\Omega^{(\pm)}$ are unitary or not.[4]

The Reactance Operator. In order to represent the unitary operator S, it is often convenient to introduce a Caley transformation by defining the *reactance operator* K by

$$K \equiv i(\mathbb{1} - S)(\mathbb{1} + S)^{-1}$$

or

$$S = (\mathbb{1} + iK)(\mathbb{1} - iK)^{-1}. \tag{6.47}$$

The unitarity of S then implies that K is *hermitian*. Equation (6.47) can also be written in the form of an equation for S in terms of the hermitian operator K,

$$S - \mathbb{1} = 2iK + iK(S - \mathbb{1}) \tag{6.47a}$$

which is known as *Heitler's integral equation.*[5]

The Operator S'. A modified scattering operator is introduced in a natural fashion by relating $\Psi^{(+)}$ and $\Psi^{(-)}$ for the same quantum numbers; i.e., the states which go over into $\Psi_0(\alpha,t)$ in the distant past or future. According to (6.23), (6.28), (6.21), and (6.22),

$$\begin{aligned} \Psi^{(+)}(\alpha,t) &= S'\Psi^{(-)}(\alpha,t) \\ \Psi^{(-)}(\alpha,t) &= S'^\dagger\Psi^{(+)}(\alpha,t) \end{aligned} \tag{6.48}$$

where

$$S' = \Omega^{(+)}\Omega^{(-)\dagger}. \tag{6.49}$$

The operator S' is generally *not* unitary. Using (6.35) and (6.38), we find that

$$S'^\dagger S' = S'S'^\dagger = \mathbb{1} - \Lambda, \tag{6.50}$$

Λ being the projection onto the space spanned by the bound states of H, (6.39). In contrast to (6.44), S' commutes with H. Use of (6.36) shows that

$$[S', H] = 0. \tag{6.51}$$

The operator S' can be used instead of S to answer the fundamental scattering question; by (6.45) and (6.51),

4 This argument is not foolproof. It was shown by Kato and Kuroda (1959) that there are instances in which the wave operators exist, yet the S operator is not unitary. The exceptional cases, however, can arise only when the ranges of $\Omega^{(+)}$ and $\Omega^{(-)}$ are not equal. See particularly the corollary to theorem 3.1 on p. 439 of S. T. Kuroda (1959a), and the discussion in B. Simon (1971), pp. 98ff.

5 W. Heitler (1941).

$$\begin{aligned}(\Psi^{(-)}(\beta,t),\Psi^{(+)}(\alpha,t)) &= (\Psi^{(-)}(\beta,0),S'\Psi^{(-)}(\alpha,0)) \\ &= (\Psi^{(+)}(\beta,0),S'\Psi^{(+)}(\alpha,0)).\end{aligned} \tag{6.52}$$

The difference is that the relevant matrix elements of S' must be evaluated on the basis of *complete* states rather than free ones, which is considerably less convenient.

6.5 The Interaction Picture

Let us define a time-dependent wave operator by means of the unitary transformation

$$\mathbf{\Omega}(t) \equiv e^{iH_0t}\Omega e^{-iH_0t} \tag{6.53}$$

which may be taken to be either $\Omega^{(-)}$ or $\Omega^{(+)}$. Because of (6.36) $\mathbf{\Omega}(t)$ can also be written

$$\mathbf{\Omega}(t) = e^{iH_0t}e^{-iHt}\Omega. \tag{6.54}$$

It satisfies the differential equation

$$\begin{aligned}i\frac{d}{dt}\mathbf{\Omega}(t) &= e^{iH_0t}[\Omega,H_0]e^{-iH_0t} \\ &= e^{iH_0t}H'\Omega e^{-iH_0t}\end{aligned}$$

on account of (6.36). If we now define

$$\mathbf{H}'(t) \equiv e^{iH_0t}H'e^{-iH_0t} \tag{6.55}$$

then $\mathbf{\Omega}(t)$ obeys the equation

$$i\frac{d}{dt}\mathbf{\Omega}(t) = \mathbf{H}'(t)\mathbf{\Omega}(t). \tag{6.56}$$

The operators $\mathbf{H}'$ and $\mathbf{\Omega}$ are specific examples of operators in the *interaction picture*, which is intermediate between the Schrödinger and the Heisenberg pictures. We started off in the Schrödinger picture, in which all (but the explicit) time dependences are carried by the state vectors and none by the operators. If O is a Schrödinger operator, then the corresponding interaction-picture operator is defined by

$$\mathbf{O}(t) \equiv e^{iH_0t}Oe^{-iH_0t}. \tag{6.57}$$

The interaction-picture *state vector* is defined in terms of the Schrödinger-picture state vector by

$$\mathbf{\Psi}(t) \equiv e^{iH_0t}\Psi(t) \tag{6.58}$$

so that

$$(\Psi_1,O\Psi_2) = (\mathbf{\Psi}_1,\mathbf{O}\mathbf{\Psi}_2).$$

The operators then satisfy the Heisenberg equation of motion with the *free* hamiltonian

$$i\frac{d}{dt}\boldsymbol{O} = [\boldsymbol{O}, \mathbf{H}_0] \tag{6.59}$$

in which, of course, $\mathbf{H}_0 = H_0$. The state vectors obey a Schrödinger equation

$$i\frac{\partial}{\partial t}\boldsymbol{\Psi} = e^{iH_0t}(H - H_0)\Psi = \mathbf{H}'\boldsymbol{\Psi} \tag{6.60}$$

in which only the interaction appears (in the form of an operator in the interaction picture).

We want to solve the differential equation (6.56) for $\boldsymbol{\Omega}$ in a symbolic fashion. In order to do this, we need a boundary condition. Inserting (6.27) in (6.35) and using (6.54), we find that (6.56) is subject to the boundary condition

$$\boldsymbol{\Omega}^{(+)}(-\infty) = \mathbb{1}. \tag{6.61}$$

It can therefore easily be solved. The only special point to remember is that $\mathbf{H}'(t_1)$ and $\mathbf{H}'(t_2)$ do not commute when $t_1 \neq t_2$. Formally, the solution can be written simply,

$$\boldsymbol{\Omega}^{(+)}(t) = \{\exp[-i\int_{-\infty}^{t} dt'\,\mathbf{H}'(t')]\}_+ \tag{6.62}$$

in which the symbol + indicates that the operators are to be *time-ordered*; i.e., their order is to be determined by the time to which they refer, with operators referring to earlier times always on the right. The meaning of the right-hand side of (6.62) is clear only if the exponential is expanded in a power series. Then

$$\begin{aligned}\boldsymbol{\Omega}^{(+)}(t) &= \mathbb{1} + \sum_{n=1}^{\infty}\frac{(-i)^n}{n!}\int_{-\infty}^{t} dt_n \int_{-\infty}^{t} dt_{n-1}\cdots\int_{-\infty}^{t} dt_1\,[\mathbf{H}'(t_n)\cdots\mathbf{H}'(t_1)]_+ \\ &= \mathbb{1} + \sum_{n=1}^{\infty}(-i)^n\int_{-\infty}^{t} dt_n \int_{-\infty}^{t_n} dt_{n-1}\cdots\int_{-\infty}^{t_2} dt_1\,\mathbf{H}'(t_n)\cdots\mathbf{H}'(t_1).\end{aligned} \tag{6.63}$$

Equation (6.62) is merely shorthand for (6.63).

Now let us insert (6.31) and (6.43) in (6.54) again. Then we find that whereas the interaction-picture wave operator $\boldsymbol{\Omega}^{(+)}$ starts out, at $t = -\infty$, as the unit operator, at $t = +\infty$ it is the scattering operator,

$$\boldsymbol{\Omega}^{(+)}(\infty) = S. \tag{6.64}$$

As a result, (6.62) gives us a symbolic expression for the S operator,

$$S = \{\exp[-i\int_{-\infty}^{\infty} dt\,\mathbf{H}'(t)]\}_+ \tag{6.65}$$

or, in terms of the Schrödinger operators,

$$S = \{\exp[-i\int_{-\infty}^{\infty} dt\, e^{iH_0t}H'e^{-iH_0t}]\}_+. \tag{6.66}$$

This is meant as a shorthand for the infinite series obtained from (6.63)

$$S = \mathbb{1} + \sum_1^\infty (-i)^n \int_{-\infty}^{\infty} dt_n \int_{-\infty}^{t_n} dt_{n-1} \cdots \int_{-\infty}^{t_2} dt_1 \\ e^{iH_0t_n}H'e^{iH_0(t_{n-1}-t_n)}H' \cdots H'e^{-iH_0t_1}. \tag{6.67}$$

Remembering (6.5), we may also write the S operator,

$$S = \mathbb{1} - i\sum_1^\infty \int_{-\infty}^{\infty} dt_n \cdots \int_{-\infty}^{\infty} dt_1 \\ e^{iH_0t_n}H'G^+(t_n - t_{n-1})H' \cdots G^+(t_2 - t_1)H'e^{-iH_0t_1}. \tag{6.67a}$$

This expression facilitates the physical interpretation analogous to that of the iterative solution of (6.15). The scattering operator is a sum of terms in which n successive scatterings take place, with free propagation between them. Comparing this with the solution of (6.8) obtained by iteration, we find that

$$\begin{aligned} S &= \mathbb{1} - i\int_{-\infty}^{\infty} dt\, e^{iH_0t}H'e^{-iH_0t} - i\int_{-\infty}^{\infty} dt \int_{-\infty}^{\infty} dt'\, e^{iH_0t}H'\mathscr{G}^+(t-t')H'e^{-iH_0t'} \\ &= \quad - i\int_{-\infty}^{\infty} dt\, e^{iH_0t}H'e^{-iH_0t} \\ &\qquad - \int_{-\infty}^{\infty} dt \int_{-\infty}^{t} dt'\, e^{iH_0t}H'e^{-iH(t-t')}H'e^{-iH_0t'}. \end{aligned} \tag{6.67b}$$

6.6 The Heisenberg Picture

In the Heisenberg picture all the time dependence is carried by the dynamical variables, and the state vectors are independent of the time. The equation of motion of the operators A is that of Heisenberg,

$$i\frac{d}{dt}\mathsf{A}(t) = [\mathsf{A}(t),\mathsf{H}] \qquad H = \mathsf{H}. \tag{6.68}$$

We may use the propagators $\mathscr{G}$ directly to solve it,

$$\begin{aligned} \mathsf{A}(t) &= \mathscr{G}^{+\dagger}(t-t')\mathsf{A}(t')\mathscr{G}^+(t-t') \\ &= e^{i\mathsf{H}(t-t')}\mathsf{A}(t')e^{-i\mathsf{H}(t-t')} \qquad t \geq t'. \end{aligned} \tag{6.69}$$

This expresses the fact that the operator develops in time according to a continuously unfolding similarity transformation, in perfect analogy with the view of the time change of dynamical variables in the classical theory of contact transformations.

Now if we ask for the probability of finding the value α upon measurement of the dynamical variable A at the time t, then we need the corresponding eigenstate $\psi(\alpha)$. This vector *depends on the time* by virtue of the fact that it is an eigenstate of the operator $\mathsf{A}(t)$, which depends on the time. Its time dependence can therefore be obtained directly from the operator $\mathsf{A}(t)$. Because of (6.69), (6.12a), and (6.11a) we find that

$$\psi(\alpha, t) = i\mathcal{G}^{+}(t - t')\psi(\alpha, t'). \tag{6.70}$$

Notice that although this equation looks just like (6.10a) its meaning is different. Equation (6.10a) refers to the time development of the state vector of the system in the Schrödinger picture; in the Heisenberg picture this state vector is constant. Equation (6.70), on the other hand, describes the time development of an eigenstate of the changing operator $\mathsf{A}(t)$ in the Heisenberg picture; in the Schrödinger picture this operator and its eigenvectors are constant.

Now the operator[6]

$$\mathsf{A}_0(t) = e^{iH_0(t-t')}e^{-iH(t-t')}\mathsf{A}(t)e^{iH(t-t')}e^{-iH_0(t-t')} \tag{6.71}$$

develops in time according to the *free* hamiltonian,

$$i\frac{d}{dt}\mathsf{A}_0(t) = [\mathsf{A}_0(t), H_0] \tag{6.72}$$

and at the time t' it was equal to the full Heisenberg operator $\mathsf{A}(t')$. If we let $t \to \pm\infty$, (6.71) becomes[7]

$$\mathsf{A}_0(t) \xrightarrow[t\to-\infty]{} \Omega^{(+)\dagger}\mathsf{A}(t)\Omega^{(+)} \equiv \mathsf{A}_{\text{in}}(t) \tag{6.73}$$

$$\mathsf{A}_0(t) \xrightarrow[t\to+\infty]{} \Omega^{(-)\dagger}\mathsf{A}(t)\Omega^{(-)} \equiv \mathsf{A}_{\text{out}}(t). \tag{6.74}$$

Also the other way around,

$$\mathsf{A}(t) \xrightarrow[t\to-\infty]{} \Omega^{(+)}\mathsf{A}_0(t)\Omega^{(+)\dagger} \tag{6.75}$$

and

$$\mathsf{A}(t) \xrightarrow[t\to+\infty]{} \Omega^{(-)}\mathsf{A}_0(t)\Omega^{(-)\dagger} \tag{6.76}$$

If the operator A_0 commutes with H_0, then it is constant and (6.75) and (6.76) read simply[8]

6 The operator H_0 here is in the Schrödinger picture. In the Heisenberg picture H_0 generally depends on the time. The complete Hamiltonian H is the same in both pictures. We may fix the correspondence between the Schrödinger and Heisenberg pictures so that at $t = t'$ all operators in the two pictures are equal. Then (6.71) is the same as (6.57). $\mathsf{A}_0(t)$ is, of course, an interaction-picture operator, as (6.72) shows.

7 According to (6.59) and (6.36) A_{in} and A_{out} are also interaction-picture operators, but they "correspond" to the Heisenberg operator A in a sense different from that of A_0. Note that A_{in} and A_{out} are generally not unitary transforms of A. The parts of A that are defined on, or map on, the bound states of H are missing from A_{in} and A_{out}. According to (6.38) and (6.73) and (6.74), $\Omega^{(+)}\mathsf{A}_{\text{in}}\Omega^{(+)\dagger} = \Omega^{(-)}\mathsf{A}_{\text{out}}\Omega^{(-)\dagger} = (\mathbb{1} - \Lambda)\mathsf{A}(\mathbb{1} - \Lambda)$.

8 Let us explicitly indicate the t' dependence of A_0 for a moment, and call it $\mathsf{A}_0(t, t')$. Now $[\mathsf{A}_0(t, t'), H_0] = 0$ does not imply that $[\mathsf{A}_0(t, t''), H_0] = 0$ for $t' \neq t''$. But with the convention of footnote 6 above for agreement between Schrödinger and Heisenberg pictures at t', H_0 is the Heisenberg operator $\mathsf{H}_0(t')$, and $[\mathsf{A}_0(t, t'), \mathsf{H}_0(t')] = 0$ is equivalent to $[\mathsf{A}_0(t, t''), \mathsf{H}_0(t'')] = 0$ for $t'' \neq t'$.

$$\mathsf{A}(t) \xrightarrow[t \to -\infty]{} \Omega^{(+)} \mathsf{A}_0 \Omega^{(+)\dagger} \tag{6.77}$$

$$\mathsf{A}(t) \xrightarrow[t \to +\infty]{} \Omega^{(-)} \mathsf{A}_0 \Omega^{(-)\dagger}. \tag{6.78}$$

For the description of an idealized scattering experiment, in which all initial and final quantum numbers must be constants of the motion for *free* particles, only such operators can be used. An example is $\mathsf{A}(t) = \mathsf{H}$. Then (6.73), (6.74), (6.35), and (6.36) show that

$$\mathsf{H}_{\text{in}} = \mathsf{H}_{\text{out}} = H_0. \tag{6.79}$$

As is physically required, the in and out operator of the full hamiltonian is the free hamiltonian (in the Schrödinger picture) which expresses the fact that at $t \to \pm\infty$ the particles are free.

The fundamental question in scattering theory now is: What is the probability amplitude in the far future for finding the system in an eigenstate $\boldsymbol{\psi}^{(-)}(\beta)$ of the operator $\mathsf{B}(\infty)$ if it was prepared in the remote past in an eigenstate $\boldsymbol{\psi}^{(+)}(\alpha)$ of the operator $\mathsf{A}(-\infty)$?

Since the former is given by

$$\boldsymbol{\psi}^{(-)}(\beta) = \Omega^{(-)} \boldsymbol{\psi}_0(\beta) \tag{6.80}$$

in terms of the corresponding eigenstate $\boldsymbol{\psi}_0$ of B_0 and the latter by

$$\boldsymbol{\psi}^{(+)}(\alpha) = \Omega^{(+)} \boldsymbol{\psi}_0(\alpha) \tag{6.80a}$$

in terms of that of A_0, the amplitude is evidently

$$(\boldsymbol{\psi}^{(-)}(\beta), \boldsymbol{\psi}^{(+)}(\alpha)) = (\boldsymbol{\psi}_0(\beta), S\, \boldsymbol{\psi}_0(\alpha)) \tag{6.81}$$

just as in the Schrödinger picture.

6.7 Scattering into Cones

A convenient and intuitive way of thinking about the scattering of particles is to relate their motion in configuration space to their momenta. Suppose a free n-particle system, subject to the hamiltonian $H_0(p^2)$, is described by the configuration-space wave function $\psi(x, t)$, where x stands for all the $m = 3n$ particle coordinates. Then the probability that the particle will be located in an infinite cone C with apex at the origin is given by

$$P^C(\infty) = \lim_{t \to \infty} P^C(t) = \lim_{t \to \infty} \int_C d^m x |\psi(x, t)|^2.$$

Dollard's[8a] "scattering-into-cones theorem" asserts the intuitive fact that this probability equals

$$P^C(\infty) = \int_C d^m k |\hat{\psi}(k)|^2, \tag{6.82}$$

8a J. D. Dollard (1969 and 1973).

where $\hat{\psi}(k)$ is the Fourier transform of $\psi(x,t)$, i.e., the momentum-space wave function. In other words, the probability that in the infinite future the particles will be found in C is equal to the probability that their momenta lie in the same cone[8b]. This theorem has been proved for general forms of $H_0(p^2)$, provided only the derivative of H_0 with respect to p^2 is positive definite.[8c]

The wave function of interacting particles, on the other hand, approaches (in a sense to be discussed in section 6.8) the free wave function ψ_{out} in the distant future. Therefore we may write more generally,

$$P^C(\infty) = \int_C d^m k |\hat{\psi}_{\text{out}}(k)|^2. \tag{6.82$'$}$$

Now let A be the intersection of C with a spherical surface of radius R centered at the vertex; it divides C into C', between A and the vertex, and C''. Let B be the surface of C, separated by the circumference of A into B' and B'', the surfaces of C' and C'', respectively. Then the Riemann–Lebesgue lemma implies

$$P^{C'}(\infty) = 0,$$

which expresses the physical fact that free particles will eventually escape from any finite region. If C does not contain the initial direction of the incoming particles then we have

$$P^C(-\infty) = 0$$

(which is compatible with $P^C(\infty) \neq 0$ because the interacting state does not, in fact, develop according to H_0). Furthermore, $P^C = P^{C'} + P^{C''}$; therefore

$$P^C(\infty) = P^{C''}(\infty) = \int_{-\infty}^{\infty} dt \frac{d}{dt} P^{C''}(t).$$

But the conservation of flux implies that

$$\frac{d}{dt} P^V(t) = -\int_S d\mathbf{s} \cdot \mathbf{f},$$

if S is the surface of the volume V, and f is the flux density. If we write

$$\eta(S) = \int_{-\infty}^{\infty} dt \int_S d\mathbf{s} \cdot \mathbf{f}$$

with the understanding that the surface normal points away from the vertex, then it follows that

$$P^C(\infty) = \eta(A) - \eta(B'').$$

8b The cones are the same, provided that a metric is used in which $|x|^2 = \Sigma m_i |\mathbf{r}_i|^2/M$, and $|k|^2 = M\Sigma |\mathbf{k}_i|^2/m_i$.

8c Jauch, Lavine, and Newton (1972).

Let us finally allow R to grow to infinity. Then we expect that $\eta(B'')$ tends to naught, and for free hamiltonians this has been proved[8d]. As a result, using (6.82),

$$\lim_{R\to\infty} \eta(A) = \int_C d^m k |\hat{\psi}_{\text{out}}(k)|^2. \tag{6.83}$$

The content of this formula forms the basis of the experimental interpretation of theoretically calculated scattering cross sections. As we shall see later the state vector $\psi^{(+)}$ describes particles which at large distances move *outwards*. Therefore the left-hand side can be interpreted as the probability that the particle(s) will, at some time, once cross the distant surface A, and thus trigger a counter. The righthand side, on the other hand, is the quantity most readily calculated in terms of momentum-space wave functions.

6.8 Mathematical Questions

We must now look a little more carefully at some mathematical points so far treated very cavalierly. In the preceding sections we have paid no attention whatever to the meaning of the limits of operators and vectors in Hilbert space, say, as $t \to \pm\infty$, and to the sense in which such limits may or may not exist. The rigorous formulation of these questions is intimately related to the realization of what kinds of states one has to have in mind when going through the formal developments of the theory.

Consider, for example, Eq. (6.14). If the state vector $\Psi(t')$ is an exact eigenstate[9] of the hamiltonian then its time dependence is as $\exp(-iEt')$. The inner product of the right-hand side of (6.14) with an eigenstate of H_0, of eigenvalue E', is consequently

$$e^{-iE't'(t-t')-iEt'}(\Psi_0(E',0), \Psi(E,0))$$

and the limit at $t' \to -\infty$ surely does not exist. That neither of these vectors is normalizable is part of the trouble, but not all.

6.8.1 Convergence of Vectors. The first step is the recognition of two kinds of convergence of state vectors. For example, consider the square-integrable function of x

$$f_k(x) = f(x)e^{ikx}$$

as a function of the parameter k, when $k \to \infty$. It certainly does not tend to zero, and

$$\|f_k\|^2 = \int dx\, f_k^*(x) f_k(x) = \int dx\, |f_0(x)|^2$$

8d Combes, Newton and Shtokhamer (1975)

9 In the physicist's loose sense; if E is in the continuous spectrum then Ψ is not normalizable.

does not tend to zero either. But if we take any other square-integrable function $g(x)$, then according to the Riemann–Lebegue lemma

$$\lim_{k\to\infty} \int dx\, g^*(x) f_k(x) = \lim_{k\to\infty} (g, f_k) = 0.$$

As a vector in the Hilbert space[10] L^2, f_k is then said to converge to zero *weakly*,

$$f_k \to 0 \quad \text{as } k \to \infty$$

if for all fixed vectors g

$$\lim_{k\to\infty} (g, f_k) = 0.$$

On the other hand, f_k is said to converge *strongly* to zero, written

$$f_k \Rightarrow 0 \quad \text{as } k \to \infty$$

if

$$\lim_{k\to\infty} \|f_k\|^2 = \lim_{k\to\infty} (f_k, f_k) = 0.$$

It follows immediately from Schwarz's inequality that strong convergence implies weak convergence. That the converse is not true was seen in the example above.[11]

As for convergence of one vector to another, one says

$$f_k \to f \quad \text{as } k \to \infty$$

if

$$f_k - f \to 0$$

and

$$f_k \Rightarrow f \quad \text{as } k \to \infty$$

if

$$f_k - f \Rightarrow 0.$$

We may immediately prove a useful *lemma*[12]: If $f_k \to f$ and $\|f_k\|^2 \to \|f\|^2$, then $f_k \Rightarrow f$. The proof is very simple.

$$\begin{aligned} \|f_k - f\|^2 &= \|f_k\|^2 + \|f\|^2 - (f_k, f) - (f, f_k) \\ &\to \|f_k\|^2 - \|f\|^2 \to 0. \end{aligned}$$

A corollary of this lemma is that, if $\|f_k\|^2 \equiv 1$ and if $f_k \to f$, then $f_k \Rightarrow f$ if and only if $\|f\|^2 = 1$.

10 L^2 is the space of square-integrable functions.

11 In view of the fact that $\|\Phi\| = \sup|(\Psi, \Phi)|/\|\Psi\|$ strong convergence may also be called *uniform* weak convergence ("sup" means least upper bound).

12 In order not to complicate the notation unnecessarily, we use the simple arrow also for ordinary numerical limits.

At this point it is appropriate to recall that by definition, a Hilbert space $\mathscr{H}$ is *complete*. This means that all sequences $\{f_n\}$ with the strong Cauchy property, that is, $\|f_n - f_m\| \to 0$, converge strongly. However, $\mathscr{H}$ is also complete in the weak sense.[13] A *compact* set of vectors is one in which every sequence has a strongly convergent subsequence; a *weakly compact* set is one in which it has a weakly convergent subsequence. It is an often useful fact that every (strongly) bounded set in $\mathscr{H}$ is weakly compact.[14] This is the weak analog of the Bolzano–Weierstrass property, which $\mathscr{H}$ does not have in the strong sense. What amounts to the converse is also true: every weakly convergent sequence is (strongly) bounded.[15]

Differentiability and Analyticity. Once convergence has been defined we can introduce the notion of the *derivative* of a vector-valued function of a variable k. One would, to start with, define weak and strong derivatives as the weak and strong limits, respectively, of $[f(k + \Delta) - f(k)]/\Delta$ as Δ approaches zero. By means of this notion one then defines *analyticity* of a vector-valued function. It is a remarkable fact that there is only one kind of analyticity[16], i.e., that weak differentiability in an open set of the complex plane implies strong differentiability. (The requisite *continuity* is always meant in the strong sense.)

6.8.2 Operator Convergence. The notion of convergence of operator sequences or of operator-valued functions is based on that of vectors. The operator $A(t)$ is said to converge *weakly* to zero as $t \to \infty$, written

$$A(t) \to 0$$

if for all vectors in the Hilbert space

$$A(t)\Psi \to 0,$$

i.e., if for any two vectors Ψ and Φ

$$\lim_{t\to\infty} (\Phi, A(t)\Psi) = 0.$$

This means that all matrix elements[17] of the operator vanish in the limit. On the other hand, it converges *strongly* to zero,

$$A(t) \Rightarrow 0$$

if for all vectors

$$A(t)\Psi \Rightarrow 0,$$

13 N. I. Akhiezer and I. M. Glazman (1961), vol. 1, p. 45.

14 *Ibid.* p. 46.

15 *Ibid.* p. 45.

16 A. E. Taylor (1958), p. 205.

17 "Matrix elements" here implies "between elements of the Hilbert space," in other words, between normalizable states.

i.e., if for all vectors

$$\lim_{t\to\infty}(A(t)\Psi, A(t)\Psi) = \lim_{t\to\infty}\langle A^\dagger(t)A(t)\rangle = 0.$$

In order to get some feeling for the difference between strong and weak operator convergence, let us discuss this a little further. It is clear that $A(t) \rightarrow 0$ implies $A^\dagger(t) \rightarrow 0$, since the matrix elements of $A^\dagger(t)$ are the complex conjugate of elements of $A(t)$. But $A(t) \Rightarrow 0$ does *not* imply $A^\dagger(t) \Rightarrow 0$.

It follows from Schwarz's inequality that

$$|(\Phi, A^\dagger A\Psi)|^2 \leq (\Phi, A^\dagger A\Phi)(\Psi, A^\dagger A\Psi)$$

and hence $A(t) \Rightarrow 0$ implies that $A^\dagger(t)A(t) \rightarrow 0$. In fact, these two statements are equivalent to one another, since the demand that *all* matrix elements of $A^\dagger(t)A(t)$ should approach zero includes the diagonal ones. But notice that $A(t) \Rightarrow 0$ does *not* imply that $A^\dagger(t)A(t) \Rightarrow 0$.

One must be careful with the limit of a product, as compared with the product of the limits. The weak convergence of A and B does not imply the weak convergence of AB. A simple counterexample is furnished by setting $A = B^\dagger$ and assuming that A and B do not converge strongly. Nor does strong convergence of A and B imply strong convergence of AB; it does not even imply the weak convergence of AB. On the other hand, the use of Schwarz's inequality shows that the strong convergence of $A^\dagger$ and B implies the weak convergence of AB.

The reason why the convergence of two operators does not imply that of the product is that these convergences may not be uniform with respect to the choice of different vectors in the space. For instance, strong convergence of $A(t)$ means that for any $\varepsilon > 0$ and Ψ there exists a T so that

$$\|A(t)\Psi\| < \varepsilon$$

for all $t > T$. But the value of T depends on Ψ, and there may exist a sequence of (normalized) vectors Ψ_n so that, as $n \rightarrow \infty$, the necessary value of T tends to ∞. This shows that even strong convergence of an operator sequence is not so strong a statement as one might sometimes wish. There is, in fact, a third kind of convergence, called *convergence in the norm*, which we may write

$$A(t) \rightarrow 0.$$

This means that

$$\lim \|A(t)\| = 0,$$

the norm of an operator being defined by [18]

$$\|A\| = \sup \frac{\|A\Psi\|}{\|\Psi\|}$$

18 The expression "sup" means "least upper bound."

for any choice of normalizable Ψ.[19] The convergence in the norm is therefore a strong *uniform* convergence.[20] Since

$$\|AB\Psi\| \leq \|A\|\,\|B\Psi\| \leq \|A\|\,\|B\|\,\|\Psi\|$$

the operator norm obeys Schwarz's inequality,

$$\|AB\| \leq \|A\|\,\|B\|.$$

Hence it follows that convergence in the norm of two operators implies the same for their product. Furthermore, clearly, $\|A(t)\| \to 0$ implies $\|A^\dagger(t)\| \to 0$.

In a similar manner we define the weak approach of $A(t)$ to A,

$$A(t) \rightarrow A$$

by the weak approach of $A(t) - A$ to zero,

$$A(t) - A \rightarrow 0,$$

the strong approach

$$A(t) \Rightarrow A$$

by the strong vanishing of their difference,

$$A(t) - A \Rightarrow 0$$

and the uniform approach

$$A(t) \boldsymbol{\rightarrow} A$$

by

$$A(t) - A \boldsymbol{\rightarrow} 0.$$

We may immediately state a useful *lemma* analogous to that of Sec. 6.7.1:

If $U^\dagger(t)U(t) = \mathbb{1}$ for all finite t and $U(t) \rightarrow U$, then $U(t) \Rightarrow U$ if and only if $U^\dagger U = \mathbb{1}$.

Notice that only *isometry* of $U(t)$ and U is required, not unitarity. Even if $U(t)$ is unitary for all finite t, U need be isometric only in order for $U(t) \Rightarrow U$ to follow.

The proof depends on the fact that $U(t) \rightarrow U$ implies $U^\dagger(t) \rightarrow U^\dagger$. Therefore

19 Since by Schwarz's inequality

$$|(\Phi, A\Psi)| \leq \|\Phi\|\,\|A\Psi\|$$

and the equality sign holds if $\Phi = A\Psi$, the norm is also given by

$$\|A\| = \sup \frac{|(\Phi, A\Psi)|}{\|\Phi\|\,\|\Psi\|}$$

for any choice of normalizable Φ and Ψ.

20 Weak operator convergence refers to *two* fixed vectors. Strong convergence of $A(t)$ means uniformity in the choice of left-hand vectors, strong convergence of $A^\dagger(t)$ in that of the right-hand ones, and convergence in the norm means uniformity in both.

$$(\Psi, [U(t) - U]^\dagger [U(t) - U]\Psi) = (\Psi, \Psi) + (\Psi, U^\dagger U \Psi)$$
$$- (U\Psi, U(t)\Psi) - (\Psi, U^\dagger(t) U \Psi) \to (\Psi, \Psi) - (\Psi, U^\dagger U \Psi) \to 0.$$

As a corollary we conclude that, if $U(t)$ is *unitary* for all finite t and its strong limit U is isometric but *not* unitary, then $U^\dagger(t) \not\Rightarrow U^\dagger$. To put this in another way: If $U(t)$ is unitary and $U(t) \Rightarrow U$ *and* $U^\dagger(t) \Rightarrow U^\dagger$, then U must be unitary. These conclusions follow from the realization that unitarity of an operator U is equivalent to the isometry of both U and $U^\dagger$.

Convergence in the norm,

$$U(t) \to U$$

of course implies both that $U^\dagger(t) \to U^\dagger$ and that if $U(t)$ is isometric or unitary then so is U.

Differentiability and Analyticity. The derivative of an operator-valued function $A(t)$ is defined in terms of that of the vector-valued function $A(t)\Psi$. Similarly for *analyticity*. Strong and weak differentiability in an open set of the complex plane are therefore equivalent (see Sec. 6.8.1). What is more, they are equivalent to "differentiability in the norm," which requires the existence of the limit-in-the-norm of $[A(t + \Delta) - A(t)]/\Delta$ as $\Delta \to 0$[21]. The notion of an analytic operator-valued function of a complex variable is therefore an unambiguous generalization of that of a complex-number-valued function. The required continuity is always meant in the sense of continuity *in the norm*, i.e., *uniform* in the choice of vectors on which the operator acts.

6.8.3 Convergences in the Schrödinger Picture. Let us consider now the convergence of Eq. (6.14) or (6.19). We have already seen that there is *no* convergence if $\Psi_{\text{in}}(t')$ is monoenergetic. One way to circumvent this inconvenient fact is to imagine that the interaction H' carries with it a factor $e^{-\varepsilon|t|}$, $\varepsilon > 0$, so that it vanishes as $t \to \pm\infty$. Then, of course, all the limits in question exist. Subsequently ε is allowed to vanish. This process of slowly turning the interaction on and off is called *adiabatic switching*. The fact that for sufficiently small ε the eigenstates of H_0 are in this manner turned into eigenstates of H with the same eigenvalue is called the *adiabatic theorem*. The entire device is a sometimes convenient method for being able to "get away with" the use of plane waves, or stationary states. But it does not contain much physical reality.

In the absence of adiabatic switching we *must* use wave packets. If, for example, the wave function $\psi(\mathbf{r}, t)$ differs from the free function

$$\psi_{\text{in}}(\mathbf{r}, t) = \int dE\, g(E) \exp[i(\mathbf{k} \cdot \mathbf{r} - Et)]$$

by

$$\int dE\, g(E) e^{i(kr - Et)} f(E, \mathbf{r})$$

21 See A. E. Taylor (1958), p. 206.

then the difference tends to zero as $t \to -\infty$. This, however, is not sufficient. After all, $\psi_{\text{in}}(\mathbf{r}, t)$ also tends to zero, pointwise, in that limit, i.e., when $\mathbf{r}$ is held fixed. At the same time, if $g(E)$ is peaked at $E = E_0$, then the stationary-phase argument[22] shows that $|\psi_{\text{in}}(\mathbf{r}, t)|$ is maximal at a point $\mathbf{r}$ which changes with t as

$$\hat{\mathbf{k}} \cdot \mathbf{r} \left.\frac{dk}{dE}\right|_{E=E_0} = t.$$

What is more,

$$\int d\mathbf{r}\, F(\mathbf{r}) \int dE\, g(E) \exp\left[i(\mathbf{k} \cdot \mathbf{r} - Et)\right]$$

tends to zero when $t \to \pm\infty$, if $F(\mathbf{r})$ is square-integrable. In other words, Ψ_{in} itself tends to zero weakly. Hence it is not very interesting to say that

$$\Psi(t) - \Psi_{\text{in}}(t) \to 0 \quad \text{as } t \to -\infty.$$

On the other hand $\|\Psi_{\text{in}}\|^2$ is independent of the time; so it does *not* tend to zero strongly. We must therefore demand that

$$\Psi(t) - \Psi_{\text{in}}(t) \Rightarrow 0 \quad \text{as } t \to -\infty$$

and, similarly,

$$\Psi(t) - \Psi_{\text{out}}(t) \Rightarrow 0 \quad \text{as } t \to +\infty.$$

As a result

$$\begin{aligned} e^{iH_0t'}\Psi(t') &= e^{iH_0t'}[\Psi(t') - \Psi_{\text{in}}(t')] + \Psi_{\text{in}}(0) \\ &\Rightarrow \Psi_{\text{in}}(0) \quad \text{as } t' \to -\infty \end{aligned}$$

and the limits in (6.14), (6.16), and (6.19) are all required to be *strong* ones.

In the same vein the operator convergence of $e^{-iHt}e^{iH_0t}$ is examined. Equation (6.19) says that according to (6.27) and (6.23)

$$e^{iHt'}e^{-iH_0t'}\Psi_{\text{in}}(0) \underset{t' \to -\infty}{\Longrightarrow} \Psi(0) \equiv \Omega^{(+)}\Psi_{\text{in}}(0).$$

Since the vectors $\Psi_{\text{in}}(0)$ are assumed to form a complete set, this means that the limits in (6.27) and, similarly, in (6.31) are required to be *strong* operator limits. The same conclusion also follows from the fact that the $\Omega^{(\pm)}$ are isometric, and so are $e^{-iHt}e^{iH_0t}$. However, $e^{iH_0t}e^{-iHt}$ is also isometric, and

$$e^{-iH_0t}e^{iHt} \to \Omega^{(\pm)\dagger} \quad \text{as } t \to \pm\infty$$

follows from

$$e^{-iHt}e^{iH_0t} \to \Omega^{(\pm)} \quad \text{as } t \to \pm\infty.$$

But if H has bound states, then the $\Omega^{(\pm)}$ are not unitary. Hence by the lemma

22 See Secs. 1.3.1 and 11.2.2.

of Sec. 6.8.2 we have in general

$$e^{-iHt}e^{iH_0t} \Rightarrow \Omega^{(\pm)} \quad \text{as } t \to \pm\infty \tag{6.84}$$

and

$$e^{-iH_0t}e^{iHt} \to \Omega^{(\pm)\dagger} \quad \text{as } t \to \pm\infty \tag{6.85}$$

whereas

$$e^{-iH_0t}e^{iHt} \Rightarrow \Omega^{(\pm)\dagger} \quad \text{as } t \to \pm\infty$$

is true *if and only if H has no bound states.*

Whether, in fact, free in and out states exist which are the strong limits of $\Psi(t)$ or, equivalently, whether (6.84) and (6.85) are true depends on the physical system, of course, i.e., on the hamiltonian. Such systems may be called *scattering systems.* We shall not go into the mathematical question here of exactly under what conditions H is the hamiltonian of a scattering system.

It is well to realize that if the vector $\Psi(t')$ in (6.14) or (6.16) is a bound state of H then, according to (6.40), the strong limits are *zero*. The limits in (6.19) are never zero.

6.8.4 The Limits in the Interaction Picture. The situation for the operator

$$\mathbf{\Omega}(t) = e^{iH_0t}e^{-iHt}\Omega^{(+)} = e^{iH_0t}\Omega^{(+)}e^{-iH_0t}$$

is somewhat different from that of (6.85). Here the right-hand factor of $\Omega^{(+)}$ eliminates the subspace spanned by the bound states of H. The operator $\mathbf{\Omega}(t)$ approaches unity as $t \to -\infty$, and S as $t \to +\infty$. Since $\mathbf{\Omega}(t)$ is isometric for all finite t, and so are both of its limiting values as $t \to \pm\infty$, it follows from the lemma of Sec. 6.7.2 that both (6.61) and (6.64) are *strong* operator limits,

$$\mathbf{\Omega}^{(+)}(t) \Rightarrow \begin{cases} \mathbb{1} & \text{as } t \to -\infty \\ S & \text{as } t \to +\infty \end{cases} \tag{6.86}$$

The question of the convergence of the series in (6.63), (6.67), and (6.67a), is a somewhat more ticklish one. The corresponding series obtained by iterating (6.8) converges in the norm if H' is a bounded operator:[23] $\|H'\| < \infty$. This can easily be seen from the facts that the integrals run only from t' to t and that G is unitary in the region of integration.[24] In (6.63) we have to worry about divergences at $t \to -\infty$ and in (6.67) and (6.67a), as well as in (6.67b) at $t \to +\infty$ in addition. Since, for example,

23 An operator is called *bounded* if its norm is finite. We shall leave it as a problem to show that if (and only if) A is bounded, then it is *continuous*. This means that it defines a continuous mapping, or, formally: $\Psi_n \Rightarrow \Psi$ implies $A\Psi_n \Rightarrow A\Psi$.

24 We leave the proof as a problem.

$$\begin{aligned} e^{iH_0t}H'e^{-iHt} &= e^{iH_0t}e^{-iHt}H - H_0e^{iH_0t}e^{-iHt} \\ &= (e^{iH_0t}e^{-iHt} - \Omega^{(-)\dagger})H - H_0(e^{iH_0t}e^{-iHt} - \Omega^{(-)\dagger}) \\ &\to 0 \quad \text{as } t \to \infty \end{aligned}$$

the convergence of the second integral in (6.67b) depends upon *how fast* the limits (6.82) and (6.83) are approached. The simplest thing to do is to insert a convergence factor $e^{-\varepsilon|t|}$ in the integral in (6.67b), which amounts to using *adiabatic switching*. Equations (6.32) then show that the convergence of these equations depends on exactly *how* the limits in (6.32) are approached. But even if (6.67b) converges, the series in (6.67a) and (6.67) may not approach that value. We shall look at the question later from a different point of view. But no matter how the question is asked, (6.67) amounts to the *Born series* for S, that is, a series in powers of the interaction, so to speak. We may multiply H' by γ; then the series is a power series in γ. It converges, as we shall see, in general at best for *sufficiently small* γ.

6.8.5 The Limits in the Heisenberg Picture. If the limits in (6.77) and (6.78) hold in the weak sense, then we get for arbitrary matrix elements in the limit as $t \to \pm\infty$

$$\begin{aligned} &\langle[\mathsf{A}(t) - \Omega\mathsf{A}_0\Omega^\dagger]^\dagger[\mathsf{A}(t) - \Omega\mathsf{A}_0\Omega^\dagger]\rangle \\ &\qquad \to \langle \mathsf{A}(t)^\dagger\mathsf{A}(t) - \Omega\mathsf{A}_0^\dagger\mathsf{A}_0\Omega^\dagger\rangle \\ &\qquad \to \langle (e^{iHt}e^{-iH_0t} - \Omega)\mathsf{A}_0^\dagger\mathsf{A}_0(e^{iHt}e^{-iH_0t} - \Omega)^\dagger\rangle \end{aligned}$$

which generally does not vanish in the limit because the approach of (6.85) is *not* strong. Consequently the limits in (6.75) to (6.78) are all generally *weak*, except when H has no bound states. In (6.73) and (6.74), on the other hand, the order of Ω and $\Omega^\dagger$ is reversed and therefore, according to (6.84), the limits are approached in the *strong* sense,

$$\mathsf{A}_{\text{in}}(t) \underset{t\to-\infty}{\Longrightarrow} \mathsf{A}_0 \tag{6.87}$$

$$\mathsf{A}_{\text{out}}(t) \underset{t\to+\infty}{\Longrightarrow} \mathsf{A}_0 \tag{6.88}$$

if A_0 is a bounded operator.

Notes and References

The correspondence between the motion of classical particles and that of quantum-mechanical wave packets is most directly contained in *Ehrenfest's theorem*, P. Ehrenfest (1927), which may be found in most standard books on quantum mechanics, e.g., L. I. Schiff (1968), p. 28. For some generalizations, see B. A. Lippmann (1965); E. Gerjuoy (1965c).

The "classical" papers on time-dependent scattering theory are Lippmann and Schwinger (1950); Gell-Mann and Goldberger (1953). For an excellent general review see Brenig and Haag (1959), a translation of which can be found in M. Ross (1963a). This book also contains reprints of the papers by Lippmann and Schwinger and Gell-

Mann and Goldberger. Other general reviews are B. S. DeWitt (1955a); F. E. Low (1959) and R. Haag (1961).

Other general references are Jauch and Rohrlich (1955), chaps. 7 and 8; S. S. Schweber (1961), chap. 11; Wu and Ohmura (1962), chap. 4; LaVita and Marchand (1974); J. A. DeSanto et al. (1980).

The following papers may also be consulted: S. T. Ma (1953); Coester, Hamermesh, and Tanaka (1954); S. Sunakawa (1955); H. Eckstein (1956a and b); L. van Hove (1955 and 1956); Jauch and Zinnes (1959).

A specific discussion of the scattering of wave packets is given by Dodd and McCarthy (1964).

For extensions to the relativistic region, see Fong and Sucher (1964); Jordan, Macfarlane, and Sudarshan (1964); F. Coester (1965 and 1980); L. B. Redei (1965); K. Yajima (1976).

On the connection between time-dependent and time-independent theory, see Belinfante and Møller (1953).

6.3 The original references to the Møller wave operator are C. Møller (1945 and 1946), the first of which is reprinted in M. Ross (1963a). These papers are important early references on modern scattering theory. For other work on the wave operator, see S. Fubini (1952a); H. Eckstein (1954); M. N. Hack (1958); T. Kato (1967b); K. Yajima (1977); V. Enss (1978). For definitions of the wave operator in the presence of static electric and magnetic fields see Avron and Herbst (1977 and 1978); K. Yajima (1979b and 1981d).

6.4 The **S** matrix was independently invented by Wheeler and Heisenberg, in J. A. Wheeler (1937a); W. Heisenberg (1943 and 1944). For its early use in field theory see E. C. G. Stueckelberg (1943, 1945, and 1946); J. Schwinger (1948); R. P. Feynman (1949); F. J. Dyson (1949). See also the papers by C. Møller (1945 and 1946) and, for a demonstration of the equivalence of various **S** matrix definitions, S. Fubini (1952).

6.5 The interaction picture is closely related to Dirac's time-dependent amplitudes in the method of variation of constants, introduced by P. A. M. Dirac (1926 and 1927). It has become extremely important in the formulation of relativistically covariant quantum field theory and was introduced for that purpose by S. Tomonaga (1946); J. Schwinger (1948b). For the time-ordering device used in formal expressions for the wave operator and the S operator, see F. J. Dyson (1949); R. P. Feynman (1951); I. Fujiwara (1952).

E. A. Remler (1975) discusses scattering in the Wigner representation.

6.7 The scattering-into-cones theorem is due to J. D. Dollard (1969 and 1973); see also J. M. Jauch et al. (1972); J.–M. Combes et al. (1975).

6.8 Our attitude in this section, it will be noticed, is not to prove the results of the previous sections rigorously, starting from a given class of hamiltonians, but to point out what notions and distinctions must be introduced for a rigorous treatment and what, exactly, must be shown for the general theory to work. For rigorous demonstrations the reader is referred especially to J. Cook (1957). Other articles using rigorous mathematical treatments of scattering theory from a time-dependent point of view are K. O. Friedrichs (1952); J. M. Jauch (1958a); Kato and Kuroda (1959); A. Galindo Tixaire (1959); S. T. Kuroda (1959a and 1962); Green and Lanford (1960); T. F. Jordan (1962); I. V. Stankevich (1962); F. H. Brownell (1962); J. D. Dollard (1964); R. T. Prosser (1964); F. Rhys (1965); C. H. Wilcox (1965); J.–M. Combes (1980a). The books by B. Simon (1971), W. O. Amrein et al. (1977), and Reed and Simon (1979) contain a wealth of information.

For further references concerning mathematical approaches, see Sec. 7.3.

6.8.3 The following papers may be consulted concerning the use of the adiabatic theorem in scattering theory: H. S. Snyder (1951); B. Ferretti (1951); Suura, Mimura, and Kimura (1952); H. E. Moses (1955).

For some rigorous treatments of the stationary-phase method, see J. G. van der Corput (1934 and 1936); A. Erdelyi (1955); G. Braun (1956).

Problems

1 Show the unitarity of S directly from (6.43) and the properties of the wave operators.

2 Suppose that a sequence of vectors in a Hilbert space converges weakly but *not* strongly to zero. Show that then it cannot converge strongly.

3 Is the weak limit of a sequence of hermitian operators necessarily hermitian?

4 Show that $A(t) \to A$ implies $\|A(t)\| \to \|A\|$.

5 Can a sequence of unbounded operators converge in the norm to a bounded operator?

6 Is the convergence of $\Omega^{(+)}(t)^\dagger$, as $t \to \pm\infty$, strong or weak?

7 Check Eqs. (6.87) and (6.88).

8 Prove that if H' is a bounded operator, then the Volterra equation (6.8) has a solution obtainable by iteration.

9 Prove that an operator is continuous if and only if it is bounded.

10 Show that if $\{\Psi_n\}$ is an infinite set of orthonormal vectors, then $\Psi_n \to 0$ as $n \to \infty$.

CHAPTER 7

Time-Independent Formal Scattering Theory

While the time-dependent scattering theory has great mathematical, physical, and conceptual advantages over the theory based on the time-independent Schrödinger equation, for the purpose of actual calculation, at least in the nonrelativistic domain, it is not very useful. Such calculations are most conveniently performed at a fixed energy. It has already been pointed out, however, that many of the mathematical steps involved require the use of *wave packets* for convergence. In the stationary-state theory wave packets cannot be used without a constant presence of cumbersome weight functions and integrations over the energy. Without them certain convergence difficulties are inevitable.

One of the consequences of the use of wave functions of monoenergetic beams is the necessity of introducing improper functions such as the Dirac δ function.[1] Most of the time there is, of course, nothing wrong with this. The fact that statements containing such "distributions" are meaningful only when integrated over with an appropriate weight or test function is a simple expression of the circumstance that, in order to attach unambiguous significance to them, wave packets must be formed. There are, nevertheless, occasions when the occurrence of δ functions in the time-independent theory causes serious difficulties which are more troublesome to circumvent.

1 For a simple and concise introduction to the use of Dirac's δ function from the mathematical point of view, see M. J. Lighthill (1958).

7.1 Green's Functions and State Vectors

7.1.1 The Green's Functions. We Fourier-analyze the Green's functions of Sec. 6.2 and define

$$\begin{aligned} G^{\pm}(E) &= \int_{-\infty}^{\infty} dt\, e^{iEt} G^{\pm}(t) \\ \mathscr{G}^{\pm}(E) &= \int_{-\infty}^{\infty} dt\, e^{iEt} \mathscr{G}^{\pm}(t). \end{aligned} \tag{7.1}$$

These integrals, however, are not well defined as they stand. Since for G^+ and $\mathscr{G}^+$ the integrals run from $t = 0$ to $t = +\infty$, a convergence factor $e^{-\varepsilon t}$, $\varepsilon > 0$, is required. If this is inserted then the formal equations (6.5) and (6.6) immediately lead to the formal solutions

$$\begin{aligned} G^+(E) &= (E + i\varepsilon - H_0)^{-1} \\ \mathscr{G}^+(E) &= (E + i\varepsilon - H)^{-1}. \end{aligned} \tag{7.2}$$

Similarly, the integrals in G^- and $\mathscr{G}^-$ require a convergence factor $e^{\varepsilon t}$, since they run to $t = -\infty$. This implies

$$\begin{aligned} G^-(E) &= (E - i\varepsilon - H_0)^{-1} \\ \mathscr{G}^-(E) &= (E - i\varepsilon - H)^{-1}. \end{aligned} \tag{7.3}$$

The ε in these equations is meant to be allowed to approach zero. It can be regarded as a part of the variable E. We may think of the operators

$$G(E) = (E - H_0)^{-1} \tag{7.4}$$

and

$$\mathscr{G}(E) = (E - H)^{-1} \tag{7.5}$$

as functions of the complex variable. E. They are then called the *resolvents* of the operators H_0 and H. Their singularities determine the spectra.

Where $\mathscr{G}(E)$ has a simple pole, H has a discrete eigenvalue there (see Sec. 7.3). The circumstance that both H and H_0 have their continuous spectra from $E = 0$ to $E = \infty$ is responsible for the fact that $G(E)$ and $\mathscr{G}(E)$ have *branch cuts* there (see Sec. 9.4). In other words, because H_0 has no eigenvalues other than the continuum from $E = 0$ to $E = \infty$, $G(E)$ is an analytic function of E regular everywhere in the finite E plane cut from $E = 0$ to $E = \infty$. If H has no bound states, then $\mathscr{G}(E)$ has the same properties; if it has bound states with energies E_n, then $\mathscr{G}(E)$ has simple poles at $E = E_n$.[2] Both $G(E)$ and $\mathscr{G}(E)$ having branch cuts from $E = 0$ to $E = \infty$, their values are different depending upon whether we approach the positive real E axis from above or below. In the first case we are on the upper, in the second on the lower, rim of the cut. The two different limiting values are $G^+(E)$ and $G^-(E)$, or $\mathscr{G}^+(E)$ and $\mathscr{G}^-(E)$. For $\operatorname{Im} E = 0$, $\operatorname{Re} E > 0$,

2 See Sec. 7.3.

$$\begin{aligned} G^+(E) &= \lim G(E) \quad \text{as } \operatorname{Im} E \to 0+ \\ G^-(E) &= \lim G(E) \quad \text{as } \operatorname{Im} E \to 0- \\ \mathscr{G}^+(E) &= \lim \mathscr{G}(E) \quad \text{as } \operatorname{Im} E \to 0+ \\ \mathscr{G}^-(E) &= \lim \mathscr{G}(E) \quad \text{as } \operatorname{Im} E \to 0-. \end{aligned} \tag{7.6}$$

This is the meaning of the ε in (7.2) and (7.3).

The difference between $G^-(E)$ and $G^+(E)$ can be symbolically represented by writing in general

$$\frac{1}{x - x_0 \pm i\varepsilon} = \frac{\mathscr{P}}{x - x_0} \mp i\pi\delta(x - x_0). \tag{7.7}$$

Here $\mathscr{P}$ stands for the Cauchy principal value, meaning that

$$\lim_{\varepsilon \to 0+} \int dx f(x)(x - x_0 \pm i\varepsilon)^{-1} = \mathscr{P} \int dx f(x)(x - x_0)^{-1} \mp i\pi f(x_0)$$

if the range of the integral includes x_0. Equation (7.7) is easily demonstrated after realizing that

$$\lim_{\varepsilon \to 0} \frac{1}{2}\left(\frac{1}{x - x_0 + i\varepsilon} + \frac{1}{x - x_0 - i\varepsilon}\right)$$

is one of the definitions of the principal value, and

$$\lim_{\varepsilon \to 0} \frac{\varepsilon/\pi}{x^2 + \varepsilon^2}$$

is one of the standard representations of $\delta(x)$.

We may consequently write

$$\begin{aligned} G^\pm(E) &= G^P(E) \mp i\pi\delta(E - H_0) \\ \mathscr{G}^\pm(E) &= \mathscr{G}^P(E) \mp i\pi\delta(E - H). \end{aligned} \tag{7.8}$$

In this symbolism G^P and $\mathscr{G}^P$ are the *principal-value Green's functions*

$$G^P(E) = \frac{\mathscr{P}}{E - H_0}$$

$$\mathscr{G}^P(E) = \frac{\mathscr{P}}{E - H}$$

and the operator δ functions have the obvious meaning

$$\begin{aligned} \delta(E - H_0)\Psi_0(E') &= \delta(E - E')\Psi_0(E') \\ \delta(E - H)\Psi(E') &= \delta(E - E')\Psi(E') \end{aligned}$$

if $\Psi_0(E')$ and $\Psi(E')$ are eigenstates of H_0 and H, respectively, with eigenvalues E'.

The operators H and H_0 being hermitian, it is clear that

$$\begin{aligned} G^-(E) &= G^{+\dagger}(E) \\ \mathscr{G}^-(E) &= \mathscr{G}^{+\dagger}(E). \end{aligned} \tag{7.9}$$

The principal-value functions $G^P(E)$ and $\mathcal{G}^P$ are the *hermitian parts* of $G^{\pm}$ and $\mathcal{G}^{\pm}$, whereas $\mp \pi i \delta(E - H_0)$ and $\mp \pi i \delta(E - H)$ are their *skew-hermitian parts*,

$$G^{P\dagger}(E) = G^P(E) \qquad \mathcal{G}^{P\dagger}(E) = \mathcal{G}^P(E). \tag{7.10}$$

It should be noticed that whereas the integral equations for $\mathcal{G}^{\pm}$, obtained from the Fourier analysis of (6.8) and (6.8a), are[3]

$$\begin{aligned} \mathcal{G}^{\pm}(E) &= G^{\pm}(E) + \mathcal{G}^{\pm}(E)H'G^{\pm}(E) \\ &= G^{\pm}(E) + G^{\pm}(E)H'\mathcal{G}^{\pm}(E) \end{aligned} \tag{7.11}$$

the principal-value Green's function does *not* obey an analogous equation,

$$\begin{aligned} \mathcal{G}^{P}(E) &\neq G^{P}(E) + \mathcal{G}^{P}(E)H'G^{P}(E) \\ &\neq G^{P}(E) + G^{P}(E)H'\mathcal{G}^{P}(E). \end{aligned} \tag{7.12}$$

The Green's functions satisfy the operator equations

$$\begin{aligned} (E - H_0)G^{\pm}(E) &= \mathbb{1} \\ (E - H)\mathcal{G}^{\pm}(E) &= \mathbb{1} \end{aligned} \tag{7.13}$$

which are, of course, nothing but (7.2) and (7.3) rewritten. From the point of view of these equations, the $(\pm)$ or $\pm i\varepsilon$ ambiguity is an expression of the fact that, for real positive E, the homogeneous equations have non-trivial solutions. Consequently the solution of the corresponding inhomogeneous equation cannot be unique. Since the solutions of the homogeneous equation are not normalizable, though, the inverse nevertheless *exists*; i.e., the inhomogeneous equations *have* solutions (see Sec. 7.3 for a more rigorous discussion).

7.1.2 The State Vectors. Next we define the Fourier transform of the Schrödinger state vectors,

$$\begin{aligned} \Psi(E) &= \int_{-\infty}^{\infty} dt\, e^{iEt}\Psi(t) \\ \Psi_{\text{in,out}}(E) &= \int_{-\infty}^{\infty} dt\, e^{iEt}\Psi_{\text{in,out}}(t). \end{aligned} \tag{7.14}$$

Multiplication of (6.15) and (6.17) by e^{iEt} and integration over t then yields the integral equation for $\Psi(E)$,

$$\begin{aligned} \Psi(E) &= \Psi_{\text{in}}(E) + G^{+}(E)H'\Psi(E) \\ &= \Psi_{\text{out}}(E) + G^{-}(E)H'\Psi(E) \end{aligned}$$

whereas their solution are obtained similarly from (6.20) and (6.20a),

3 These equations are usually referred to as the *Lippmann-Schwinger equations*. See Lippmann and Schwinger (1950).

$$\begin{aligned}\Psi(E) &= \Psi_{\text{in}}(E) + \mathscr{G}^+(E)H'\Psi_{\text{in}}(E)\\ &= \Psi_{\text{out}}(E) + \mathscr{G}^-(E)H'\Psi_{\text{out}}(E).\end{aligned}$$

As explained in detail in Sec. 6.2, there are two kinds of states Ψ. One is $\Psi^{(+)}$; its in state is "controlled," prepared as an eigenstate of some operator A which commutes with H_0, with the eigenvalue α. According to (6.21),[4]

$$\begin{aligned}\Psi^{(+)}(E,\alpha) &= \Psi_0(E,\alpha) + G^+(E)H'\Psi^{(+)}(E,\alpha)\\ &= \Psi_0(E,\alpha) + \mathscr{G}^+(E)H'\Psi_0(E,\alpha).\end{aligned} \tag{7.15}$$

The label "in" has been replaced by 0 here, since $\Psi_{\text{in}}(E,\alpha)$ is simply a free state with quantum numbers E and α. Its relation to $\Psi^{(+)}$ makes it an in state. The state $\Psi^{(+)}(E,\alpha)$ also satisfies similar equations with G^- and Ψ_{out}, but the latter cannot be specified; it is unknown until the scattering problem has been solved.

Another state of interest to us is $\Psi^{(-)}$, whose *out state* is controlled. According to (6.22) and (6.20a),

$$\begin{aligned}\Psi^{(-)}(E,\alpha) &= \Psi_0(E,\alpha) + G^-(E)H'\Psi^{(-)}(E,\alpha)\\ &= \Psi_0(E,\alpha) + \mathscr{G}^-(E)H'\Psi_0(E,\alpha).\end{aligned} \tag{7.15a}$$

The state $\Psi_0(E,\alpha)$ here is the same as that in (7.15), but in relation to $\Psi^{(-)}$ it is an out state. Notice that the α in $\Psi^{(\pm)}(E,\alpha)$ does *not* refer to an eigenvalue of an operator A which commutes with H and hence is a constant of the motion; in $\Psi^{(+)}$, α refers to a set of quantum numbers of the in state, in $\Psi^{(-)}$ to a set of quantum numbers of the out state. In both cases α refers to a label on the inhomogeneity in the integral equation, whereas $(\pm)$ refers to a property of the Green's function.

The "differential" equations satisfied by the time-independent state vectors are obtained by inserting the Fourier transformations (7.14) in the time-dependent Schrödinger equations (6.1) and (6.9). They are the time-independent Schrödinger equations

$$H_0\Psi_0(E) = E\Psi_0(E) \tag{7.16}$$

$$H\Psi(E) = E\Psi(E) \tag{7.17}$$

which say that $\Psi(E)$ is an eigenstate of the full hamiltonian H (with the eigenvalue E) and that $\Psi_0(E)$ is an eigenstate of the free hamiltonian H_0 (with the eigenvalue E).

Boundary Conditions. In order to specify the solutions of these equations, they have to be supplemented by boundary conditions. The first requirement ought to be that the solutions be normalizable. This, however, is impossible if E lies in the continuous spectrum. In view of the hermiticity of H and H_0, the states are automatically orthogonal in the sense that

$$\begin{aligned}(\Psi_0(E),\Psi_0(E')) &= \text{const}\,\delta(E-E')\\ (\Psi(E),\Psi(E')) &= \text{const}\,\delta(E-E').\end{aligned}$$

4 These equations and (7.15a) are also referred to as the *Lippmann-Schwinger equations.*

But fixing the constants does not make the solutions unique. When the energy lies in the scattering range, then there are, in fact, infinitely many different solutions for each value of E and the boundary conditions have to pick out a single one of them.

Among the solutions to the free equation (7.16) we are interested primarily in one particular class, those which behave like collimated beams going in specific directions. This means that Ψ_0 is an eigenstate of the momenta of the incident particles. This requirement may still not remove the degeneracy completely. There may be additional degrees of freedom, such as spins and isotopic spins, and their components, or other quantum numbers. If so, a sufficiently large set of commuting observables has to be introduced which removes the degeneracy. All their eigenvalues, including those of the momenta, are meant to be combined in the label α. Another class of solutions of (7.16) which we need later is that of eigenstates of the angular momenta of the incident particles, rather than the linear momenta. These must then also be combined with additional operators, if necessary, to form a complete set.

The complete state Ψ is labeled by the same quantum numbers as Ψ_0, but these are *not* necessarily eigenvalues of operators which commute with H. Instead their meaning for Ψ resides in the integral equations (7.15) or (7.15a), i.e., in the boundary conditions. Apparently, then, there is an additional degree of freedom in Ψ which Ψ_0 does not have. That is, $\Psi(E,\alpha)$ can have $\Psi_0(E,\alpha)$ as an in or as an out state. In the first instance it is called $\Psi^{(+)}(E,\alpha)$, in the second $\Psi^{(-)}(E,\alpha)$. This does not imply, though, that there are twice as many full solutions for a given energy as there are free ones. Each $\Psi^{(-)}(E,\alpha)$ is a linear combination of the various $\Psi^{(+)}(E,\alpha')$ of the same energy, and vice versa. This follows from (6.48), together with the fact that S' commutes with H [(6.51)]. It will be explicity visible below in (7.65).

Finally we fix the normalizations. For this purpose we need fix only the normalization of the Ψ_0. The integral equations then determine $\Psi^{(\pm)}$ uniquely. We shall assume that Ψ_0 is such that

$$(\Psi_0(E,\alpha), \Psi_0(E',\alpha')) = \delta(E - E')\delta_{\alpha\alpha'} \tag{7.18}$$

where the Kronecker δ may be replaced by a Dirac δ function for continuous quantum numbers. The normalization of the complete states defined by the integral equation (7.15) or (7.15a) is then automatically the same; i.e., when E and E' lie in the continuum,

$$(\Psi^{(+)}(E,\alpha), \Psi^{(+)}(E',\alpha')) = \delta(E - E')\delta_{\alpha\alpha'} \tag{7.19}$$

and

$$(\Psi^{(-)}(E,\alpha), \Psi^{(-)}(E',\alpha')) = \delta(E - E')\delta_{\alpha\alpha'}. \tag{7.19a}$$

This has already been verified in Sec. 6.3. Of course, the states so normalized do not represent the Fourier transforms of normalizable wave packets as in (7.14). Before we can form the latter, suitable weight functions have to be introduced, so that

$$\Psi_0(t,\alpha) = \frac{1}{2\pi}\int_0^\infty dE\, f(E) e^{-iEt}\Psi_0(E,\alpha).$$

Standing Waves or Principal-Value States. Another state vector of interest is the solution of the integral equation

$$\Psi^{(P)}(E,\alpha) = \Psi_0(E,\alpha) + G^P(E)H'\Psi^{(P)}(E,\alpha) \tag{7.20}$$

G^P being the principal-value Green's function of (7.8). Using the integral equations (7.15) and (7.18), we find

$$\Delta\Psi(E,\alpha) = i\pi\delta(E - H_0)H'\Psi^{(+)}(E,\alpha) + G^{(P)}(E)H'\Delta\Psi(E,\alpha)$$

if we set

$$\Psi^{(P)}(E,\alpha) = \Psi^{(+)}(E,\alpha) + \Delta\Psi(E,\alpha).$$

We expand the δ-function operator

$$\delta(E - H_0) = \sum_\beta \Psi_0(E,\beta)\Psi_0^\dagger(E,\beta)$$

so that

$$\Delta\Psi(E,\alpha) = i\pi\sum_\beta \Psi_0(E,\beta)\mathfrak{T}_{\beta\alpha}(E) + G^P(E)H'\Delta\Psi(E,\alpha)$$

where

$$\mathfrak{T}_{\beta\alpha}(E) = (\Psi_0(E,\beta), H'\Psi^{(+)}(E,\alpha)).$$

From this we may condlude that

$$\Delta\Psi(E,\alpha) = i\pi\sum_\beta \Psi^{(P)}(E,\beta)\mathfrak{T}_{\beta\alpha}(E),$$

i.e.,

$$\Psi^{(+)}(E,\alpha) = \Psi^{(P)}(E,\alpha) - i\pi\sum_\beta \Psi^{(P)}(E,\beta)\mathfrak{T}_{\beta\alpha}(E). \tag{7.21}$$

For the significance of $\mathfrak{T}_{\beta\alpha}$, see (7.40) below. Similarly we find

$$\Psi^{(-)}(E,\alpha) = \Psi^{(P)}(E,\alpha) + i\pi\sum_\beta \mathfrak{T}_{\alpha\beta}^*(E)\Psi^{(P)}(E,\beta). \tag{7.22}$$

7.1.3 Expansion of the Green's Functions. The assumption that the eigenstates of H_0 form a complete set immediately leads to an expansion of the Green's functions. If we let $G(E)$ act on (6.37), then we obtain for the resolvent

$$G(E) = \int_0^\infty dE' \frac{\sum_\alpha \Psi_0(E',\alpha)\Psi_0^\dagger(E',\alpha)}{E - E'}. \tag{7.23}$$

From this one gets $G^\pm(E)$ by letting E approach the real axis from above or from below,

$$G^{\pm}(E) = \int_0^{\infty} dE' \frac{\sum_{\alpha} \Psi_0(E',\alpha)\Psi_0^{\dagger}(E',\alpha)}{E - E' \pm i\varepsilon} \tag{7.24}$$

and similarly

$$G^{P}(E) = \mathscr{P} \int_0^{\infty} dE' \frac{\sum_{\alpha} \Psi_0(E',\alpha)\Psi_0^{\dagger}(E',\alpha)}{E - E'}. \tag{7.24a}$$

In the same way we may expand the complete Green's functions on the basis of the complete set of eigenstates of H. Suppose that the states $\Psi^{(+)}(E,\alpha)$, together with the bound states, form a complete set,

$$\mathbb{1} = \sum_n \Psi_n \Psi_n^* + \int_0^{\infty} dE' \sum_{\alpha} \Psi^{(+)}(E',\alpha)\Psi^{(+)\dagger}(E',\alpha). \tag{7.25}$$

We get

$$\mathscr{G}^{\pm}(E) = \sum_n \frac{\Psi_n \Psi_n^{\dagger}}{E - E_n} + \int_0^{\infty} dE' \frac{\sum_{\alpha} \Psi^{(+)}(E',\alpha)\Psi^{(+)\dagger}(E',\alpha)}{E - E' \pm i\varepsilon} \tag{7.26}$$

$$\mathscr{G}^{P}(E) = \sum_n \frac{\Psi_n \Psi_n^{\dagger}}{E - E_n} + \mathscr{P} \int_0^{\infty} dE' \frac{\sum_{\alpha} \Psi^{(+)}(E',\alpha)\Psi^{(+)\dagger}(E',\alpha)}{E - E'}. \tag{7.26a}$$

That the normalization of the $\Psi^{(+)}$ defined by the integral equation (7.15) is correct for this representation has already been verified in (7.19). Alternatively $\Psi^{(-)}$ may be used. Because of Eq. (7.65) below and the unitarity of the **S** matrix we have

$$\sum_{\alpha} \Psi^{(+)}(E,\alpha)\Psi^{(+)\dagger}(E,\alpha) = \sum_{\alpha} \Psi^{(-)}(E,\alpha)\Psi^{(-)\dagger}(E,\alpha). \tag{7.27}$$

On the other hand, we may *not* use $\Psi^{(P)}$ so simply. According to (7.64) below, one gets

$$\sum_{\alpha} \Psi^{(+)}(E,\alpha)\Psi^{(+)\dagger}(E,\alpha) = \sum_{\alpha,\beta} \Psi^{(P)}(E,\alpha)[(\mathbb{1} + \mathbf{K}^2)^{-1}]_{\alpha\beta}\Psi^{(P)\dagger}(E,\beta) \tag{7.28}$$

which shows that the $\Psi^{(P)}$ are not correctly normalized. The right normalization is obtained by setting

$$\Psi^{(P)\prime}(E,\alpha) \equiv \sum_{\beta} [(\mathbb{1} + \mathbf{K}^2)^{-1/2}]_{\beta\alpha} \Psi^{(P)}(E,\beta) \tag{7.29}$$

which is well defined because $\mathbf{K}$ is a hermitian matrix and therefore $\mathbf{K}^2$ is positive semidefinite.

7.2 The Wave Operator and the S Matrix

7.2.1 The Operators Ω, S, and S'. Since the wave operators $\Omega^{(\pm)}$ are independent of the time, we can insert (6.23) and (6.28) directly into the Fourier transforms (7.14) and obtain

$$\Psi^{(\pm)}(E,\alpha) = \Omega^{(\pm)}\Psi_0(E,\alpha) \tag{7.30}$$

with $\Omega^{(\pm)}$ independent of E (as well as of α). These operators, then, take us directly from the free states to the corresponding full states, as noted below (6.37). The isometry of $\Omega^{(\pm)}$ gives us also

$$\Psi_0(E,\alpha) = \Omega^{(\pm)\dagger}\Psi^{(\pm)}(E,\alpha) \tag{7.31}$$

and therefore the Fourier transforms of (6.41) and (6.48),

$$\Psi^{(+)}(E,\alpha) = S'\Psi^{(-)}(E,\alpha) \tag{7.32}$$

$$\Psi_{\text{out}}(E) = S\Psi_{\text{in}}(E). \tag{7.33}$$

The first relates the state which in the far *future* will be controlled and found to have the quantum numbers α to the state which in the distant past was controlled and prepared will the same quantum numbers. The second equation relates the (free) state into which a given exact one will develop in the future to the (free) state from which it evolved in the past. We cannot put labels on these without risking confusion. If the full state is $\Psi^{(+)}(E,\alpha)$, then the Ψ_{in} in (7.33) is $\Psi_0(E,\alpha)$, but the left-hand Ψ_{out} is not simple; if the full state is $\Psi^{(-)}(E,\alpha)$, then the Ψ_{out} in (7.33) is $\Psi_0(E,\alpha)$, but the right-hand Ψ_{in} is not simple.

Because of (7.30) we can write the expansion (6.37a) in more detail,

$$\Omega^{(\pm)} = \sum_\alpha \int_0^\infty dE\, \Psi^{(\pm)}(E,\alpha)\Psi_0^\dagger(E,\alpha). \tag{7.34}$$

If this is substituted in (6.49), we get the expansion

$$S' = \sum_\alpha \int_0^\infty dE\, \Psi^{(+)}(E,\alpha)\Psi^{(-)\dagger}(E,\alpha). \tag{7.35}$$

There is no similarly simple expansion of the more useful operator S.

7.2.2 The T Matrix. We saw in (6.42) and (6.45) that the fundamental scattering question is answered by the elements of the **S** matrix. If we choose the states $\Psi^{(-)}$ and $\Psi^{(+)}$ to be monoenergetic, then we get

$$(\Psi^{(-)}(E_\beta,\beta),\Psi^{(+)}(E_\alpha,\alpha)) = (\Psi_0(E_\beta,\beta),S\Psi_0(E_\beta,\alpha)) \tag{7.36}$$

on the basis of *free* states. But since S commutes with H_0, this is a multiple of $\delta(E_\beta - E_\alpha)$. Furthermore, since $S = \mathbb{1}$ when $H = H_0$ and no scattering takes place, we conveniently define[5]

$$(\Psi_0(E_\beta,\beta),S\Psi_0(E_\alpha,\alpha)) = (\Psi_0(E_\beta,\beta),\Psi_0(E_\alpha,\alpha)) - 2\pi i\delta(E_\beta - E_\alpha)\mathfrak{T}_{\beta\alpha}(E). \tag{7.37}$$

We can then immediately obtain a practically useful formula for $\mathfrak{T}_{\beta\alpha}(E)$ by using (7.15) and (7.15a) to find the difference between $\Psi^{(+)}$ and $\Psi^{(-)}$.

5 There is considerable confusion in the customary notation for the **T** matrix. Some authors use **T** matrices that differ from ours by a sign or by factors of 2π or π.

According to (7.8),

$$\Psi^{(+)}(E,\alpha) - \Psi^{(-)}(E,\alpha) = -2\pi i\delta(E - H)H'\Psi_0(E,\alpha). \tag{7.38}$$

Insertion in (7.36) yields

$$\begin{aligned}(\Psi_0(E_\beta,\beta), S\Psi_0(E_\alpha,\alpha)) &= (\Psi^{(+)}(E_\beta,\beta), \Psi^{(+)}(E_\alpha,\alpha)) \\ &\quad -2\pi i\delta(E_\beta - E_\alpha)(\Psi_0(E_\alpha,\beta), H'\Psi^{(+)}(E_\alpha,\alpha)) \\ &= (\Psi^{(-)}(E_\beta,\beta), \Psi^{(-)}(E_\alpha,\alpha)) \\ &\quad -2\pi i\delta(E_\beta - E_\alpha)(\Psi^{(-)}(E_\alpha,\beta), H'\Psi_0(E_\alpha,\alpha)).\end{aligned} \tag{7.39}$$

Because of (7.30) and the isometry (6.35) of $\Omega^{(\pm)}$, the first terms on both right-hand sides of (7.39) are equal to $(\Psi_0(E_\beta,\beta), \Psi_0(E_\alpha,\alpha))$. Comparison with (7.37) therefore tells us that

$$\begin{aligned}\mathfrak{T}_{\beta\alpha}(E) &= (\Psi_0(E,\beta), H'\Psi^{(+)}(E,\alpha)) \\ &= (\Psi^{(-)}(E,\beta), H'\Psi_0(E,\alpha)).\end{aligned} \tag{7.40}$$

These are the formulas most commonly used in the calculation of **S**-matrix elements and scattering amplitudes. Insertion of Eqs. (7.15) and (7.15a) and use of (7.9) allows us to verify that both right-hand sides of (7.40) are equal and that

$$\mathfrak{T}_{\beta\alpha}(E) = (\Psi_0(E,\beta), [H' + H'\mathcal{G}^+(E)H']\Psi_0(E,\alpha)). \tag{7.41}$$

The same result also follows immediately from (6.67b), which represents its derivation from the point of view of the interaction picture. That the answer is the same in the Heisenberg picture follows from (6.81).

We now have to distinguish explicitly between the case in which a particle is scattered by a fixed target and that in which two particles are scattered by one another (whether or not a reaction takes place or particles are produced in the collision). In the former instance we can use (7.40) or (7.41) directly. But in the latter, the hamiltonian is invariant under translations, and as a consequence the total particle momentum **P** is conserved. Let us use **P** explicitly as one of the labels on the states, in addition to E and the remaining quantum number α. It is then convenient to choose the normalization explicitly as

$$(\Psi_0(E,\mathbf{P},\alpha), \Psi_0(E',\mathbf{P}',\alpha')) = \delta(E - E')\delta(\mathbf{P} - \mathbf{P}')\delta(\alpha - \alpha') \tag{7.42}$$

in place of (7.18), where the last δ function may be partly a Kronecker δ for the discrete quantum numbers among the α. The normalizations (7.19) and (7.19a) are, of course, changed accordingly. The **T** matrix must then contain a three-dimensional momentum δ function that expresses the conservation of the total momentum,

$$\mathfrak{T}_{\beta\alpha}(E) = \delta(\mathbf{P} - \mathbf{P}')T_{\beta\alpha}(E). \tag{7.43}$$

The state vectors $\Psi_0(E,\mathbf{P},\alpha)$ are direct products of eigenvectors of **P** and a remainder which, though depending on $|\mathbf{P}|$, can, because of rotational invariance, be chosen to be independent of the direction of **P**,

$$\Psi_0(E,\mathbf{P},\alpha) = \Psi_0(\mathbf{P}) \times \Psi_0(E,|\mathbf{P}|,\alpha).$$

The same is true of $\Psi^{(\pm)}$. We thereupon get from (7.40)

$$\begin{aligned} T_{\beta\alpha}(E,|\mathbf{P}|) &= (\Psi_0(E,|\mathbf{P}|,\beta), H'\Psi^{(+)}(E,|\mathbf{P}|,\alpha)) \\ &= (\Psi^{(-)}(E,|\mathbf{P}|,\beta), H'\Psi_0(E,|\mathbf{P}|,\alpha)) \end{aligned} \tag{7.44}$$

and a corresponding formula from (7.41). The dependence of the **T** matrix upon the total momentum, after removal of the δ function, is usually not explicitly indicated. It is implicit in the choice of reference frame. In the center of mass system, of course, $\mathbf{P} = 0$. When referring to the **T** matrix we shall in the future mean (7.44) for the collision of two particles and (7.40) for that of one with a fixed target, and the German letter $\mathfrak{T}$ will no longer be used.

The two alternative ways of calculating the **T** matrix, as written in (7.44), are associated with two different but equally valid physical pictures accompanying the scattering process. In the first, we think of the full state vector as having been prepared in the distant past as a free state with the quantum numbers α. This is $\Psi^{(+)}(E,\alpha)$. As we shall see in the discussion of the coordinate representation, this implies that $\Psi^{(+)}$ is a state whose *incoming* wave part is *controlled* or prepared. Its outgoing wave part, or its behavior in the distant future, is uncontrolled and subject to the scattering interaction. This state is now subjected to a measurement to see how much of it is in the free β state. That is, counters are arranged to find out how many particles are going in a given direction. In the second version of (7.44) we must think of the full physical state as being known to become a controlled beam in the far future. As we shall see later, this implies that its *outgoing* wave part is controlled or known, while its incoming wave part, or its behavior in the distant past, is unknown; it must have been such a complicated state that in the far future it turns out to be simple. Such a state is now subjected to the measurement of how much of it was in fact in the free state $\Psi_0(E,\alpha)$ in the past. The first version may appeal more to our physical intuition than the second, especially with regard to the notion of causality, but both are equally valid descriptions.

Off-the-energy-shell Extensions. According to the definition (7.37) of $T_{\beta\alpha}(E)$ it refers only to a single energy. This is simply an expression of the fact that energy is conserved. Nevertheless, the expressions on the right-hand sides of (7.40) or (7.44) can be evaluated with two different energies:

$$\begin{aligned} T_{\beta\alpha}^{(+)} &\equiv (\Psi_0(E_\beta,\beta), H'\Psi^{(+)}(E_\alpha,\alpha)) \\ T_{\beta\alpha}^{(-)} &\equiv (\Psi^{(-)}(E_\beta,\beta), H'\Psi_0(E_\alpha,\alpha)). \end{aligned} \tag{7.45}$$

These are called (*half*) *off-the-energy-shell extensions* of the matrix **T**, in contrast to (7.40) or (7.44), which is *on the energy shell*. We then have

$$\begin{aligned} T_{\beta\alpha}^{(+)} &\to T_{\beta\alpha} \quad \text{as } E_\beta \to E_\alpha \\ T_{\beta\alpha}^{(-)} &\to T_{\beta\alpha} \quad \text{as } E_\beta \to E_\alpha. \end{aligned}$$

There are, of course, infinitely many other off-the-energy-shell extensions of **T**, but these two are usually the most convenient. The only function of physical significance is $T_{\beta\alpha}(E)$ itself, *on* the energy shell.

Equation (7.41) shows that $T_{\beta\alpha}(E)$ is the matrix element, on the basis of free states of energy E, of the operator

$$T(E) = H' + H'\mathscr{G}^+(E)H'. \tag{7.46}$$

Insertion of the integral equation (7.11) for $\mathscr{G}^+$ gives us an integral equation for the operator T,

$$\begin{aligned} T(E) &= H' + T(E)G^+(E)H' \\ &= H' + H'G^+(E)T(E). \end{aligned} \tag{7.47}$$

Since

$$\begin{aligned} T_{\beta\alpha}^{(+)} &= (\Psi_0(E_\beta,\beta), T(E_\alpha)\Psi_0(E_\alpha,\alpha)) \\ T_{\beta\alpha}^{(-)} &= (\Psi_0(E_\beta,\beta), T(E_\beta)\Psi_0(E_\alpha,\alpha)) \\ &= [T^\dagger]_{\alpha\beta}^{(+)*} \end{aligned} \tag{7.48}$$

the operator equations for $T(E)$ can be replaced by equations for the matrices $T_{\beta\alpha}^{(+)}$ and $T_{\beta\alpha}^{(-)}$,

$$\begin{aligned} T_{\beta\alpha}^{(+)} &= H'_{\beta\alpha} + \sum_\gamma \int \frac{dE_\gamma}{E_\alpha - E_\gamma + i\varepsilon} H'_{\beta\gamma} T_{\gamma\alpha}^{(+)} \\ T_{\beta\alpha}^{(-)} &= H'_{\beta\alpha} + \sum_\gamma \int \frac{dE_\gamma}{E_\beta - E_\gamma + i\varepsilon} T_{\beta\gamma}^{(-)} H'_{\gamma\alpha} \end{aligned} \tag{7.49}$$

which are obtained by inserting the complete set of free states between T and G in (7.44). Because of the necessity for inserting a complete set of states there are no analogous equations for the on-the-energy-shell **T** matrix. The latter has to be obtained from either version of (7.49) by setting $E_\alpha = E_\beta$ *after* solving it.

Sometimes it is convenient to use a more general (*fully*) off-the-energy-shell extension of **T** by taking matrix elements of $T(E)$ between states neither of which have the energy E,

$$T_{\beta\alpha}(E_\beta, E_\alpha; E) = (\Psi_0(E_\beta,\beta), T(E)\Psi_0(E_\alpha,\alpha)). \tag{7.48a}$$

Then, of course,

$$\begin{aligned} T_{\beta\alpha}^{(+)}(E_\beta, E_\alpha) &= T_{\beta\alpha}(E_\beta, E_\alpha; E_\alpha) \\ T_{\beta\alpha}^{(-)}(E_\beta, E_\alpha) &= T_{\beta\alpha}(E_\beta, E_\alpha; E_\beta). \end{aligned}$$

Note that according to (7.46)

$$[T(E + i\varepsilon)]^\dagger = T(E - i\varepsilon) \tag{7.50}$$

so that

$$T_{\beta\alpha}^{(-)}(E_\beta + i\varepsilon, E_\alpha) = T_{\alpha\beta}^{(+)*}(E_\alpha, E_\beta - i\varepsilon). \tag{7.51}$$

The Generalized Optical Theorem. Another set of equations can be obtained by taking matrix elements of (7.46) between free states $\Psi_0(E_\alpha,\alpha)$ and $\Psi_0(E_\beta,\beta)$, with $E_\beta = E$, and using the expansion of $\mathscr{G}$ on a complete set of eigenstates of H, including the bound states. This yields

$$T^{(+)}_{\alpha\beta} = H'_{\alpha\beta} + \sum_\gamma \int_0^\infty dE_\gamma \frac{T^{(+)}_{\alpha\gamma} T^{(+)*}_{\beta\gamma}}{E_\beta - E_\gamma + i\varepsilon} + \sum_n (E_\alpha - E_n)(\Psi_0(E_\alpha,\alpha),\Psi_n)(\Psi_n,\Psi_0(E_\beta,\beta)) \tag{7.52}$$

because, for example,

$$(\Psi_0(E_\alpha,\alpha), H'\Psi_n) = (E_n - E_\alpha)(\Psi_0(E_\alpha,\alpha),\Psi_n).$$

Similarly,

$$T^{(-)}_{\alpha\beta} = H'_{\alpha\beta} + \sum_\gamma \int_0^\infty dE_\gamma \frac{T^{(-)*}_{\alpha\gamma} T^{(-)}_{\beta\gamma}}{E_\alpha - E_\gamma + i\varepsilon} + \sum_n (E_\beta - E_n)(\Psi_0(E_\alpha,\alpha),\Psi_n)(\Psi_n,\Psi_0(E_\beta,\beta)). \tag{7.52a}$$

Exchanging α and β, taking complex conjugates, subtracting, and using the hermitian character of H', we get

$$T^{(+)}_{\alpha\beta} - T^{(+)*}_{\beta\alpha} = \sum_\gamma \int_0^\infty dE_\gamma\, T^{(+)}_{\alpha\gamma} T^{(+)*}_{\beta\gamma} \left(\frac{1}{E_\beta - E_\gamma + i\varepsilon} - \frac{1}{E_\alpha - E_\gamma - i\varepsilon} \right) + (E_\alpha - E_\beta) \sum_n (\Psi_0(E_\alpha,\alpha),\Psi_n)(\Psi_n,\Psi_0(E_\beta,\beta) \tag{7.53}$$

$$T^{(-)}_{\alpha\beta} - T^{(-)*}_{\beta\alpha} = \sum_\gamma \int_0^\infty dE_\gamma\, T^{(-)*}_{\alpha\gamma} T^{(-)}_{\beta\gamma} \left(\frac{1}{E_\alpha - E_\gamma + i\varepsilon} - \frac{1}{E_\beta - E_\gamma - i\varepsilon} \right) + (E_\beta - E_\alpha) \sum_n (\Psi_0(E_\alpha,\alpha),\Psi_n)(\Psi_n,\Psi_0(E_\beta,\beta)). \tag{7.53a}$$

If we finally let $E_\alpha = E_\beta$, then all the **T**-matrix elements move onto the energy shell and we obtain

$$T_{\alpha\beta} - T^*_{\beta\alpha} = -2\pi i \sum_\gamma T_{\alpha\gamma} T^*_{\beta\gamma}. \tag{7.54}$$

This is the formal statement of the *generalized optical theorem* derived in Sec. 2.3.4 for electromagnetic scattering. The optical theorem (derived in Sec. 1.3.9 for the electromagnetic case) follows by setting $\alpha = \beta$,

$$\operatorname{Im} T_{\alpha\alpha} = -\pi \sum_\gamma |T_{\alpha\gamma}|^2. \tag{7.55}$$

Equations (7.53) and (7.53a), which are the generalization of the optical theorem to the off-the-energy-shell **T** matrices, are also known as the *Low equations*.

7.2.3 The K Matrix. In analogy with (7.40) we may define a *reactance matrix*,

$$K_{\beta\alpha}(E) = -\pi(\Psi_0(E,\beta), H'\Psi^{(P)}(E,\alpha)) \tag{7.56}$$

in terms of the principal-value state vector of (7.20). From (7.40) and (7.21) the following relation between the **T** and **K** matrices is obtained,

$$\pi T_{\beta\alpha}(E) = -K_{\beta\alpha}(E) + i\pi \sum_{\gamma} K_{\beta\gamma}(E) T_{\gamma\alpha}(E)$$

which is the on-the-energy-shell statement of Heitler's integral equation (6.47a). In matrix form it reads

$$\pi \mathbf{T} = -\mathbf{K}(\mathbb{1} - i\mathbf{K})^{-1} \tag{7.57}$$

or, if we write the on-the-energy-shell **S** matrix,

$$\mathbf{S} = \mathbb{1} - 2\pi i \mathbf{T} \tag{7.58}$$

$$\mathbf{S} = (\mathbb{1} + i\mathbf{K})(\mathbb{1} - i\mathbf{K})^{-1}. \tag{7.59}$$

It should be noted that in contrast to (7.49) these are all on-the-energy-shell equations.

It is easily verified by the use of the second line in Eq. (7.40) and of Eq. (7.20) that **K** can also be written

$$K_{\beta\alpha}(E) = -\pi(\Psi^{(P)}(E,\beta), H'\Psi_0(E,\alpha)) \tag{7.56a}$$

which expresses the hermitian character of the reactance matrix,

$$\mathbf{K}^{\dagger} = \mathbf{K}. \tag{7.60}$$

Equations (7.56) and (7.56a) lead to "natural" extensions of the **K** matrix off the energy shell,

$$\begin{aligned} K_{\beta\alpha}^{(+)} &\equiv -\pi(\Psi_0(E_\beta,\beta), H'\Psi^{(P)}(E_\alpha,\alpha)) \\ K_{\beta\alpha}^{(-)} &\equiv -\pi(\Psi^{(P)}(E_\beta,\beta), H'\Psi_0(E_\alpha,\alpha)). \end{aligned} \tag{7.61}$$

Inserting the integral equation (7.20) for $\Psi^{(P)}$ in this, we get the analog of (7.49),

$$\begin{aligned} K_{\beta\alpha}^{(+)} &= -\pi H'_{\beta\alpha} + \sum_{\gamma} \mathscr{P}\int \frac{dE_\gamma}{E_\alpha - E_\gamma} H'_{\beta\gamma} K_{\gamma\alpha}^{(+)} \\ K_{\beta\alpha}^{(-)} &= -\pi H'_{\beta\alpha} + \sum_{\gamma} \mathscr{P}\int \frac{dE_\gamma}{E_\beta - E_\gamma} K_{\beta\gamma}^{(-)} H'_{\gamma\alpha}. \end{aligned} \tag{7.62}$$

The two off-the-energy-shell extensions of the reactance matrix are related by

$$\mathbf{K}^{(+)\dagger} = \mathbf{K}^{(-)} \tag{7.63}$$

which goes over into (7.60) on the energy shell.

We may now express (7.21) and (7.22) in matrix form and invert the matrices according to (7.57). Then

$$\begin{aligned} \Psi^{(P)}(E,\alpha) &= \sum_{\beta} \Psi^{(+)}(E,\beta)(\mathbb{1} - i\mathbf{K})_{\beta\alpha} \\ &= \sum_{\beta} \Psi^{(-)}(E,\beta)(\mathbb{1} + i\mathbf{K})_{\beta\alpha} \end{aligned} \tag{7.64}$$

and therefore by (7.59) and (7.60)

$$\Psi^{(+)}(E,\alpha) = \sum_{\beta} \Psi^{(-)}(E,\beta) S_{\beta\alpha}(E) \tag{7.65}$$

$$\Psi^{(-)}(E,\alpha) = \sum_{\beta} \Psi^{(+)}(E,\beta) S^*_{\alpha\beta}(E). \tag{7.65a}$$

This must not be confused with Eq. (7.32); S' there is an *operator* which takes us from $\Psi^{(-)}(E,\alpha)$ to $\Psi^{(+)}(E,\alpha)$. In (7.65), $S_{\beta\alpha}$ is a set of numerical coefficients which express $\Psi^{(+)}(E,\alpha)$ linearly in terms of the various $\Psi^{(-)}(E,\beta)$.

7.2.4 Unitarity and Reciprocity. Equations (7.65) and (7.65a), which are expressions of (7.59) together with the hermiticity (7.60) of $\mathbf{K}$, imply that the $\mathbf{S}$ matrix is unitary,

$$\mathbf{S}(E)\mathbf{S}^\dagger(E) = \mathbf{S}^\dagger(E)\mathbf{S}(E) = \mathbb{1}. \tag{7.66}$$

This is of course a consequence of the unitarity of the S operator. In terms of the $\mathbf{T}$ matrix, related to $\mathbf{S}$ by (7.58), the unitarity reads

$$\tfrac{1}{2}i(\mathbf{T}^\dagger - \mathbf{T}) = -\pi\mathbf{T}\mathbf{T}^\dagger = -\pi\mathbf{T}^\dagger\mathbf{T}. \tag{7.67}$$

The skew-hermitian part of the $\mathbf{T}$ matrix is thus negative semidefinite. This is the formal statement of the generalized optical theorem (7.54). The essential ingredient of the proof is the hermitian character of the hamiltonian. If for some reason this is sacrificed, then (7.66) and (7.67) no longer hold. We would then call the system *absorptive* if the skew-hermitian part of $\mathbf{T}$ is *less* than $-\pi\mathbf{T}\mathbf{T}^\dagger$ (in the sense that the matrix

$$\tfrac{1}{2}i(\mathbf{T} - \mathbf{T}^\dagger) - \pi\mathbf{T}\mathbf{T}^\dagger$$

is positive semidefinite). As was shown in Sec. 2.3.1 for the electromagnetic case, this assures that the flux going out from the scattering center is less than that coming in.

Note that (7.67) is meant as an equation on the energy shell. If we do not restrict the matrix elemenets to this, it must be written more explicitly

$$\begin{aligned}\tfrac{1}{2}i[T^*_{\beta\alpha}(E) - T_{\alpha\beta}(E)] &= -\pi\sum_{\gamma}\delta(E - E_\gamma)T_{\alpha\gamma}(E)T^*_{\beta\gamma}(E)\\ &= -\pi\sum_{\gamma}\delta(E - E_\gamma)T^*_{\gamma\alpha}(E)T_{\gamma\beta}(E).\end{aligned} \tag{7.67a}$$

In this form it can be extended off the energy shell.

The unitarity may also be incorporated in the $\mathbf{T}$ matrix by expressing it in terms of the $\mathbf{K}$ matrix, as in (7.57), which may be written as

$$\pi\mathbf{T} = -\mathbf{K} - \mathbf{K}(\pi\mathbf{T} + i)\mathbf{K}. \tag{7.68}$$

If this is to be extended off the energy shell, one must, of course, insert the appropriate δ functions as in (7.67a),

$$\pi T_{\alpha\beta}(E) =$$
$$-K_{\alpha\beta}(E) - \sum_{\gamma\sigma} K_{\alpha\gamma}(E)\delta(E - E_\gamma)[\pi T_{\gamma\sigma}(E) + i\delta_{\gamma\sigma}]\delta(E - E_\sigma)K_{\sigma\beta}(E).$$

In this form it may be taken off the energy shell. But because of the δ functions, the **T** matrix on the right remains on the shell. We may express it in terms of the on-shell **K** matrix there and get

$$\pi T_{\alpha\beta}(E_\alpha, E_\beta; E) = \\ -K_{\alpha\beta}(E_\alpha, E_\beta; E) - i\sum_{\gamma\sigma} K_{\alpha\gamma}(E_\alpha, E; E)\{[\mathbb{1} - i\mathbf{K}(E)]^{-1}\}_{\gamma\sigma} K_{\sigma\beta}(E, E_\beta; E) \tag{7.69}$$

the **K** matrix in the center being *on the energy shell.*

A further condition on the **S** matrix and the **T** matrix follows if the hamiltonian is invariant under *time* reversal. Let ϑ be the antiunitary[6] time reversal operator, which has the property that

$$(\vartheta\Psi_1, \vartheta\Psi_2) = (\Psi_2, \Psi_1). \tag{7.70}$$

Since H_0 is time-reversal-invariant, i.e.,

$$[\vartheta, H_0] = 0$$

the free states of a given energy are mapped into linear combinations of each other by ϑ. Let us assume that the quantum numbers α are chosen so that

$$\vartheta\Psi_0(E, \alpha) = \varphi_\alpha \Psi_0(E, \alpha') \tag{7.71}$$

where φ_α is a phase factor only. (This equation defines the "time-reversed" quantum numbers α'.) It follows that, if H is also invariant,

$$[\vartheta, H] = 0$$

then

$$\vartheta\Psi^{(+)}(E, \alpha) = \varphi_\alpha \Psi^{(-)}(E, \alpha') \tag{7.71a}$$

because time reversal changes a boundary condition at $t \to -\infty$ to one at $t \to +\infty$, and vice versa. Consequently

$$\begin{aligned} T_{\alpha\beta}(E) &= (\Psi_0(E, \alpha), H'\Psi^{(+)}(E, \beta)) \\ &= (\vartheta H'\Psi^{(+)}(E, \beta), \vartheta\Psi_0(E, \alpha)) \\ &= (\Psi^{(-)}(E, \beta'), H'\Psi_0(E, \alpha'))\varphi_\alpha\varphi_\beta^* \\ &= \varphi_\alpha\varphi_\beta^* T_{\beta'\alpha'} \end{aligned} \tag{7.72}$$

which implies specifically that

$$|T_{\alpha\beta}|^2 = |T_{\beta'\alpha'}|^2. \tag{7.73}$$

This is called the *reciprocity* property of the **T** matrix or of the **S** matrix. In a specific representation the reciprocity may express itself in the fact that the **S** matrix is symmetric. But this depends on a special choice of phase conventions for the wave functions (see Sec. 15.1.3).

6 An antiunitary operator is a special instance of an antilinear operator, defined by the property that $AC\Phi = C^*A\Phi$, if C is a scalar. For a general discussion of such operators, see E. P. Wigner (1960).

7.2.5 Additive Interactions. The calculation of state vectors or scattering amplitudes need not start with the kinetic energy as a zero-order hamiltonian. Suppose that the hamiltonian is written

$$H = H_0 + H_1' + H_2'.$$

Then one may first calculate the states, Green's functions, etc., of

$$H_1 = H_0 + H_1'.$$

We shall give all the corresponding quantities the subscript 1. For example,

$$\mathcal{G}_1^{\pm}(E) = (E - H_1 \pm i\varepsilon)^{-1}.$$

The calculation of the exact quantities corresponding to the full hamiltonian H can then proceed from those corresponding to H_1 as a starting point. We have

$$\begin{aligned} \mathcal{G}^{\pm}(E) &= \mathcal{G}_1^{\pm}(E) + \mathcal{G}_1^{\pm}(E)H_2'\mathcal{G}^{\pm}(E) \\ &= \mathcal{G}_1^{\pm}(E) + \mathcal{G}^{\pm}(E)H_2'\mathcal{G}_1^{\pm}(E) \end{aligned} \tag{7.74}$$

and

$$\begin{aligned} \Psi^{(\pm)}(E,\alpha) &= \Psi_1^{(\pm)}(E,\alpha) + \mathcal{G}_1^{\pm}(E)H_2'\Psi^{(\pm)}(E,\alpha) \\ &= \Psi_1^{(\pm)}(E,\alpha) + \mathcal{G}^{\pm}(E)H_2'\Psi_1^{(\pm)}(E,\alpha). \end{aligned} \tag{7.75}$$

Notice that in this case the inhomogeneous term carries the label $\pm$; it already contains outgoing or incoming waves, in addition to the plane waves (if this is the choice of α). We now insert (7.75) in (7.36),

$$\begin{aligned} (\Psi^{(-)}(E_\beta,\beta),\Psi^{(+)}(E_\alpha,\alpha)) &= (\Psi_1^{(-)}(E_\beta,\beta),\Psi^{(+)}(E_\alpha,\alpha)) \\ &\qquad + (\Psi_1^{(-)}(E_\beta,\beta),H_2'\mathcal{G}^{\pm}(E_\beta)\Psi^{(+)}(E_\alpha,\alpha)) \\ &= (\Psi_1^{(-)}(E_\beta,\beta),\Psi_1^{(+)}(E_\alpha,\alpha)) \\ &\qquad + (\Psi_1^{(-)}(E_\beta,\beta),\mathcal{G}_1^{+}(E_\alpha)H_2'\Psi^{(+)}(E_\alpha,\alpha)) \\ &\qquad + (\Psi_1^{(-)}(E_\beta,\beta),H_2'\mathcal{G}^{+}(E_\beta)\Psi^{(+)}(E_\alpha,\alpha)). \end{aligned}$$

In the second term $\mathcal{G}^{+}(E_\alpha)$ may be replaced by $1/(E_\alpha - E_\beta + i\varepsilon)$; in the third $\mathcal{G}^{+}(E_\beta)$ may be replaced by $1/(E_\beta - E_\alpha + i\varepsilon)$. Since by (7.7)

$$\frac{1}{E_\alpha - E_\beta + i\varepsilon} + \frac{1}{E_\beta - E_\alpha + i\varepsilon} = -2\pi i\delta(E_\beta - E_\alpha)$$

we obtain

$$\begin{aligned} (\Psi^{(-)}(E_\beta,\beta),\Psi^{(+)}(E_\alpha,\alpha)) &= (\Psi_1^{(-)}(E_\beta,\beta),\Psi_1^{(+)}(E_\alpha,\alpha)) \\ &\qquad - 2\pi i\delta(E_\beta - E_\alpha)(\Psi_1^{(-)}(E_\alpha,\beta),H_2'\Psi^{(+)}(E_\alpha,\alpha)) \end{aligned}$$

or

$$T_{\beta\alpha}(E) = T_{1\beta\alpha}(E) + (\Psi_1^{(-)}(E,\beta),H_2'\Psi^{(+)}(E,\alpha)) \tag{7.76}$$

and similarly

$$T_{\beta\alpha}(E) = T_{1\beta\alpha}(E) + (\Psi^{(-)}(E,\beta), H_2' \Psi_1^{(+)}(E,\alpha)). \tag{7.76a}$$

These equations can also be written in a symmetrical form by inserting (7.75) in them,

$$T_{\beta\alpha}(E) = T_{1\beta\alpha}(E) + (\Psi_1^{(-)}(E,\beta), [H_2' + H_2'\mathscr{G}^+(E)H_2']\Psi_1^{(+)}(E,\alpha)). \tag{7.76b}$$

A simple case in which these expressions are of use is that of the scattering of charged particles subject to additional interactions, such as protons, for example. If H_1' is taken to be the Coulomb interaction and H_2' the nuclear force, then the matrix $\mathbf{T}_1$ can be calculated exactly and so can the Coulomb wave functions.

As another example of the utility of these expressions we consider two particles subject to the external potentials V_a and V_b and also interacting with one another, via V_{ab}. Let particle b initially impinge upon a bound state of particle a (with the external potential V_a), and finally let b be bound (with V_b) and a emerge freely. This is called a *rearrangement collision* (treated in more detail in Chaps. 16 and 17). We set

$$H_1' = V_a + V_b$$
$$H_2' = V_{ab}.$$

Then the eigenstates of H_1 are *direct products* of eigenstates of $H_{0a} + V_a$ and of $H_{0b} + V_b$. The initial outgoing-wave eigenstate of H_1, in an obvious notation, is

$$\Psi_1^{(+)}(E_\alpha, \alpha) = \Psi_a^{(\mathrm{bd})} \times \Psi_b^{(+)}(E_\alpha - E_a^{(\mathrm{bd})}, \alpha) \tag{7.77}$$

and the final incoming-wave eigenstate of H_1 is

$$\Psi_1^{(-)}(E_\beta, \beta) = \Psi_a^{(-)}(E_\beta - E_b^{(\mathrm{bd})}, \beta) \times \Psi_b^{(\mathrm{bd})}. \tag{7.77a}$$

The bound-state vectors being orthogonal to the scattering states, the zero-order $\mathbf{T}$-matrix element vanishes,

$$T_{1\beta\alpha} = 0. \tag{7.78}$$

We are left with two apparently different versions of the relevant $\mathbf{T}$-matrix element,

$$\begin{aligned} T_{\beta\alpha}(E) &= (\Psi_a^{(-)}(E - E_b^{(\mathrm{bd})}, \beta) \times \Psi_b^{(\mathrm{bd})}, V_{ab}\Psi^{(+)}(E,\alpha)) \\ &= (\Psi^{(-)}(E,\beta), V_{ab}\Psi_a^{(\mathrm{bd})} \times \Psi_b^{(+)}(E - E_a^{(\mathrm{bd})}, \alpha)). \end{aligned} \tag{7.79}$$

These two results give rise in a natural way to the *distorted-wave Born approximation* (see Sec. 9.1.2). It should be noted that the first terms in (7.76) and (7.76a) are absent, i.e., (7.78) is true only if the interactions V_a and V_b are *external*, i.e., with an infinitely massive third particle. Otherwise the bound and scattering states in (7.77) and (7.77a) are in different coordinate systems and there is a recoil effect.

Green's-Function Approach. Another procedure is applicable if the interaction hamiltonian consists of a sum,

$$H' = \sum_i H_i' \tag{7.80}$$

and the solutions, Green's functions, etc., corresponding to the individual pieces are known,

$$\mathcal{G}_i^{\pm}(E) = G^{\pm}(E) + \mathcal{G}_i^{\pm}(E)H_i'G^{\pm}(E) \tag{7.81}$$

$$T_i(E) = H_i' + H_i'\mathcal{G}_i^{+}(E)H_i'. \tag{7.82}$$

Notice that each $\mathcal{G}_i$ contains the *entire* kinetic energy, but only one of the interactions. We want to calculate the total **T** matrix, which according to (7.47) satisfies the equation

$$T(E) = \sum_i H_i' + \sum_i H_i'G^{+}(E)T(E). \tag{7.83}$$

We define the auxiliary operator

$$T_i'(E) = H_i' + H_i'G^{+}(E)T(E) \tag{7.84}$$

so that

$$T(E) = \sum_i T_i'(E). \tag{7.85}$$

Insertion of (7.81) in (7.84) and subsequent use of (7.83) yields

$$\begin{aligned} T_i' &= H_i' + H_i'\mathcal{G}_i^{+}T - H_i'\mathcal{G}_i^{+}H_i'G^{+}T \\ &= H_i' + \sum_j H_i'\mathcal{G}_i^{+}H_j' + \sum_{j\neq i} H_i'\mathcal{G}_i^{+}H_j'G^{+}T \\ &= T_i + \sum_{j\neq i} H_i'\mathcal{G}_i^{+}T_j'. \end{aligned}$$

But comparison of (7.82) with the integral equation

$$T_i = H_i' + T_iG^{+}H_i'$$

shows that

$$H_i'\mathcal{G}_i^{+}(E) = T_i(E)G^{+}(E). \tag{7.86}$$

We therefore get the coupled equations

$$T_i'(E) = T_i(E) + \sum_{j\neq i} T_i(E)G^{+}(E)T_j'(E) \tag{7.87}$$

for the auxiliary operators $T_i'(E)$.

An alternative procedure is to work with the Green's function instead of the T operator. We may define

$$L_i(E) = \mathcal{G}_i^{+}(E) - G^{+}(E) \tag{7.88}$$

so that according to (7.81) and (7.86)

$$\begin{aligned} L_i(E) &= \mathcal{G}_i^{+}(E)H_i'G^{+}(E) \\ &= G^{+}(E)H_i'\mathcal{G}_i^{+}(E) \\ &= G^{+}(E)T_i(E)G^{+}(E). \end{aligned} \tag{7.89}$$

Then the complete Green's function is written in the form

$$\mathscr{G}^+(E) = G^+(E) + \sum_i L_i(E) + C(E) \tag{7.90}$$

which defines $C(E)$. In order to find an expression for $C(E)$, we substitute the equation for the complete Green's function $\mathscr{G}^+(E)$ in terms of the partial ones,

$$\mathscr{G}^+ = \mathscr{G}_i^+ + \sum_{j \neq i} \mathscr{G}_i^+ H_j' \mathscr{G}^+$$

in the equation for $\mathscr{G}^+$ in terms of G^+,

$$\begin{aligned}\mathscr{G}^+ &= G^+ + \sum_i G^+ H_i' \mathscr{G}^+ \\ &= G^+ + \sum_i G^+ H_i' \mathscr{G}_i^+ + \sum_{i \neq j} G^+ H_i' \mathscr{G}_i^+ H_j' \mathscr{G}^+ .\end{aligned}$$

Comparison with (7.89) and (7.90) shows that therefore

$$C(E) = \sum_{i \neq j} L_i H_j' \mathscr{G}^+ . \tag{7.91}$$

We finally insert (7.90) in this in order to get an equation for $C(E)$,

$$C(E) = B(E) + \sum_{i \neq j} L_i(E) H_j' C(E) \tag{7.92}$$

$$\begin{aligned}B(E) &= \sum_{i \neq j} L_i(E) H_j' [G^+(E) + \sum_k L_k(E)] \\ &= \sum_{i \neq j} L_i H_j' \mathscr{G}_j^+ + \sum_{\substack{i \neq j \\ k \neq j}} L_i H_j' L_k \\ &= \sum_{i \neq j} L_i T_j G^+ + \sum_{\substack{i \neq j \\ k \neq j}} L_i H_j' L_k \\ &= \sum_{i \neq j} G^+ T_i G^+ T_j G^+ + \sum_{\substack{i \neq j \\ k \neq j}} G^+ T_i G^+ H_j' G^+ T_k G^+ .\end{aligned} \tag{7.93}$$

The great advantages of Eqs. (7.87) and (7.92) in problems involving more than two particles will be discussed in Sec. 17.4.

7.3 Mathematical Questions

We wish to discuss here the mathematical structure of the resolvent of an operator K, defined by[7]

$$(\alpha - K)^{-1}.$$

The discussion is applicable to the Green's functions $G(E)$ and $\mathscr{G}(E)$ if K

7 Strictly speaking we should write $\alpha - K$ in the form $\alpha \mathbb{1} - K$, but we shall forgo this bit of pedantry. Note that in the theory of integral equations $\alpha K(\alpha - K)^{-1}$ is also called the resolvent of K.

is taken to be H_0 or H, respectively, and $E = \alpha$ and to the Lippmann–Schwinger equation if

$$K = GH'$$

at a fixed energy, so that (7.15) reads $(\mathbb{1} - K)\Psi = \Psi_0$.

First a remark about the domain of definition (or simply, the *domain*) of our operators. We cannot generally restrict our attention to bounded,[8] i.e., continuous, operators. But we shall assume that all the operators we are dealing with are the next best thing, namely, *closed*,[9] and that their domains are dense in the Hilbert space.[10] The last point is not trivial because an unbounded closed operator cannot be everywhere defined.[11] For example, the kinetic energy is a differential operator in the coordinate representation, or a multiplicative operator (multiplication by p^2) in the momentum representation, neither of which take all L^2 functions [12] into L^2 functions. But every L^2 function can be arbitrarily closely approximated (in the norm) by a function $f(\mathbf{r})$ for which $\nabla^2 f(\mathbf{r})$ [or by an $h(\mathbf{p})$ for which $p^2 h(\mathbf{p})$] is in L^2. Hence the domain of the kinetic-energy operator, though not all of L^2, is everywhere dense in it.

7.3.1 The Spectrum. There are four distinct possibilities for the properties of the operator $\alpha - K$ at a given value of α:

1. $\alpha - K$ maps the Hilbert space $\mathscr{H}$, one to one, onto the whole space or onto an everywhere dense subset. What is more, there exists no sequence of vectors $\{\Psi_n\}$, each normalized to unity,[13] such that

$$(\alpha - K)\Psi_n \Rightarrow 0. \tag{7.94}$$

In this event there exists a well-defined *bounded* inverse $(\alpha - K)^{-1}$, and α is said to be in the *resolvent set* of K. If the range[14] of $\alpha - K$ is not all of $\mathscr{H}$ but an everywhere dense subset, then $(\alpha - K)^{-1}$ is a priori defined only on that subset, though (for a closed K) it can be immediately extended to include its limit points by defining

$$(\alpha - K)^{-1}\Psi = \Phi$$

if

$$(\alpha - K)^{-1}\Psi_n \Rightarrow \Phi \quad \text{as } \Psi_n \Rightarrow \Psi.$$

8 A bounded operator is an operator of finite norm. See footnote 23 of Chap 6.

9 An operator C is called *closed* if $\Psi_n \Rightarrow \Psi$ *together with strong convergence of* $C\Psi_n$ implies $C\Psi_n \Rightarrow C\Psi$.

10 We also assume that the Hilbert space of interest is *separable*. This means that there exists an enumerable set of vectors in it that forms a basis.

11 This is known as the "closed graph theorem"; see Taylor, *op. cit.*, p. 181.

12 See footnote 10 of Chap 6 for the definition.

13 In this section we shall be careful in our usage. All members of the Hilbert space are of finite norm. So whenever we talk of a vector, it *must* be normalizable. In order to remind the reader, we shall usually put *normalizable* in parentheses.

14 The range of an operator is defined in footnote 3 of Chap. 6.

2. The mapping of $\alpha - K$ is not one to one, i.e., for some Ψ_0 there exists more than one Ψ with

$$(\alpha - K)\Psi = \Psi_0. \tag{7.95}$$

Then $\alpha - K$ must annihilate the difference between two such vectors, and there exists a (normalizable) vector Φ such that

$$K\Phi = \alpha\Phi. \tag{7.95a}$$

In this case $(\alpha - K)^{-1}$ does not exist, and α is in the *point spectrum* of K.

3. The mapping of $\alpha - K$ is one to one, its range is $\mathscr{H}$ (or everywhere dense in it), but there exists a sequence of normalized vectors Ψ_n for which (7.94) is true. That is to say, for every $\varepsilon > 0$ there exists a Φ_ε such that

$$\|(\alpha - K)\Phi_\varepsilon\| < \varepsilon\|\Phi_\varepsilon\|.$$

Then $(\alpha - K)^{-1}$ is well defined, but not bounded, and α is in the *continuous* or *essential spectrum* of K.[15]

4. The mapping of $\alpha - K$ is one to one, but the range of $\alpha - K$ is a proper subset of $\mathscr{H}$ which is not everywhere dense in it. In this event $(\alpha - K)^{-1}$ is not everywhere defined; it maps a subset of $\mathscr{H}$, not dense in it, onto all of $\mathscr{H}$. Then α is in the *residual spectrum* of K. There must, then, exist a (normalizable) vector Φ that is orthogonal to the range of $\alpha - K$, so that

$$(\Phi, (\alpha - K)\Psi) = 0$$

for all Ψ. Consequently

$$K^\dagger\Phi = \alpha^*\Phi$$

that is, α is in the residual spectrum of K if and only if it is not in its point spectrum but α^* is in the point spectrum of $K^\dagger$.

It should be remarked parenthetically that the "continuous spectrum" need not form a continuous set, nor need the "point spectrum" consist of isolated points. It is possible for isolated points to be in the continuous spectrum, and for the point spectrum to be continuous. (See Prob. 5.)

The set of all points in one of the three types of spectrum is called the *spectrum* of K. Every complex number α, then, is either in the resolvent set or in the spectrum. The resolvent set is *open*, and hence the spectrum is *closed*[16], so that it contains all its finite limit points. In the mathematical usage, to which we adhere in this section, only the members of the point spectrum are called *eigenvalues*.

15 In some of the mathematical literature, the point and continuous spectra are combined into the *approximate spectrum* [see Friedman (1957)]. It consists of all points α for which (7.94) is true. Furthermore, in some of the literature the subdivisions of the spectrum are defined in such a way that a point may belong both to the continuous and to the point spectrum. This can be done simply by including in the latter those points in the former for which a sequence Ψ_n exists, each member of which is orthogonal to all (normalizable) eigenvectors of K belonging to the eigenvalue α, and for which (7.94) is true.

16 See A. E. Taylor (1958), p. 257.

Left and Right Inverses. A word should be said about left and right inverses. To start with, we are interested only in the *left* inverse

$$L_\alpha(\alpha - K) = \mathbb{1}$$

which allows us to solve Eq. (7.95) in the form

$$\Psi = L_\alpha \Psi_0.$$

If the range of $\alpha - K$ is all of $\mathscr{H}$ (or is dense in it), then it follows that also

$$(\alpha - K)L_\alpha = \mathbb{1}.$$

In case 1, therefore, $(\alpha - K)^{-1}$ is a two-sided inverse. In case 2, a left inverse cannot exist. However, it is not impossible that there is a right-hand one,

$$(\alpha - K)R_\alpha = \mathbb{1}.$$

This would imply that

$$R_\alpha^\dagger(\alpha^* - K^\dagger) = \mathbb{1}$$

and hence α^* cannot be in the point spectrum of $K^\dagger$; α^* thus must lie in the residual spectrum of $K^\dagger$. Thus if α is in the point spectrum of K and α^* is in the point spectrum of $K^\dagger$, then neither left nor right inverses of $\alpha - K$ exist. If α^* is in the residual spectrum of $K^\dagger$, then a right inverse of $\alpha - K$ exists but its range cannt be all of $\mathscr{H}$.

In case 3 the left and right inverses exist and are equal. In case 4 the left-hand inverse L_α exists, but it is defined only in the range of $\alpha - K$. A right-hand inverse of $\alpha - K$ cannot exist in this case because its existence would imply that the range of $\alpha - K$ is all of $\mathscr{H}$.

It is useful to discuss these things also from the point of view of the *existence* of a solution Ψ to Eq. (7.95). In case 2, when α is in the point spectrum of K, it is not impossible that (7.95) has a solution, even though it is not unique. It would then be necessary and sufficient that Ψ_0 lie in the range of $\alpha - K$. But the range of an operator is orthogonal to the null space[17] of its adjoint, and any vector orthogonal to the range is in the null space of the adjoint. Consequently, if the range of $\alpha - K$ is closed, it is the orthogonal complement[18] of the null space of $\alpha^* - K^\dagger$. Hence in that case a necessary and sufficient condition for (7.95) to have a solution is that Ψ_0 be orthogonal to all eigenvectors of $K^\dagger$ with the eigenvalue α^*, if there are any. Furthermore the solution can be made unique by demanding that it be orthogonal to all eigenvectors of K with the eigenvalue α.

A *normal* operator N is defined by the demand that

$$[N, N^\dagger] = 0.$$

It follows that a *normal operator cannot have a residual spectrum*. This is

17 The null space of an operator is the space of all vectors annihilated by it.

18 The *orthogonal complement* of a set $\mathscr{S}$ of vectors is the set of all vectors which are orthogonal to all members of $\mathscr{S}$; it is a *closed* subspace.

because if

$$(N^\dagger - \alpha^*)\Phi = 0$$

then

$$\|(N - \alpha)\Phi\|^2 = (\Phi, (N^\dagger - \alpha^*)(N - \alpha)\Phi) = 0$$

and hence

$$(N - \alpha)\Phi = 0.$$

Thus, if α^* is in the point spectrum of $N^\dagger$, then α is in the point spectrum of N (with the same eigenvectors). Of course, if K is *self-adjoint*, then it is necessarily normal, and a fortiori a self-adjoint operator has no residual spectrum. What is more, if N is normal, then $(\alpha - N)^{-1}$ is a two-sided inverse if it exists, and it fails to exist as a left inverse if and only if it fails to exist as a right one. As for solving (7.95) if K is normal (and if the range of $\alpha - K$ is closed), a solution exists now if and only if Ψ_0 is orthogonal to all the eigenvectors of K with eigenvalue α.

7.3.2 Compact Operators. The operators with properties most directly analogous to those of operators on a finite dimensional vector space are the *completely continuous*, or *compact*, ones. An operator C is called compact if, for every bounded sequence of vectors Ψ_n, the sequence $C\Psi_n$ contains a strongly convergent subsequence. Such an operator is necessarily *continuous* or *bounded*. For if it were not bounded, then there would have to exist a sequence of normalized vectors Ψ_N such that

$$\|C\Psi_N\| > N,$$

and this sequence could not contain a convergent subsequence. A sufficient condition for C to be compact is that it is in the *Hilbert–Schmidt* class,[19] i.e.,

$$\operatorname{tr} CC^\dagger < \infty.$$

Here are some general facts about compact operators, stated without proof:

1. An operator is compact if and only if it maps every bounded set of vectors into a compact one.[20]
2. An operator is compact if and only if it transforms every weakly convergent sequence of vectors into a strongly convergent one; that is, $\{C\Psi_n\}$ converges strongly whenever $\{\Psi_n\}$ converges weakly.[21]
3. An operator is compact if and only if $(\Psi_n, C\Phi_n)$ converges to $(\Psi, C\Phi)$ whenever $\Psi_n \rightharpoonup \Psi$ and $\Phi_n \rightharpoonup \Phi$.[22]

19 This class is also known as the L^2, or *Fredholm*, class.

20 Akhiezer and Glazman (1961), vol. 1, p. 56.

21 Riesz and Nagy (1955), p. 206.

22 *Ibid.*

4. The adjoint of a compact operator is compact,[23] and so are the products of compact operators and bounded ones.[24]
5. The limit in the norm of a sequence of compact operators is compact.[25]
6. A compact operator may be arbitrarily closely approximated (in the norm) by an operator of finite rank.[26] Specifically, if $\{\Psi_n\}$ is an orthonormal basis in $\mathscr{H}$, then

$$C_N \equiv \sum_{n,m=1}^{N} (\Psi_n, C\Psi_m)\Psi_n\Psi_m^\dagger \to C \quad \text{as } N \to \infty.$$

It follows from 5 that this is true of compact operators *only*.

7. If C is bounded and $C^\dagger C$ is compact, then C is compact.[27]
8. If C is compact and α is an eigenvalue other than zero, then α^* is an eigenvalue of $C^\dagger$ of the same degeneracy as α.[28]

Now the important spectral property of compact operators is that, except possibly for the point $\alpha = 0$, they have a *point spectrum only*. Furthermore, the spectrum is a countable set with zero as the only possible limit point,[29] and the degeneracy of every eigenvalue, except zero, is finite.[30] If $C \neq 0$ is self-adjoint as well as compact, then the spectrum must contain at least one point other than zero.[31]

That a compact operator cannot have a residual spectrum follows immediately from property 8. The absence of a continuous spectrum follows from the fact that if to each n there exists a normalized Ψ_n so that

$$\|(C - \alpha)\Psi_n\| < 1/n \tag{7.94a}$$

and C is compact, then there must exist a subsequence $\{\Psi_m\}$ such that $C\Psi_m \Rightarrow \Psi$, and hence, by (7.94a), $\Psi_n \Rightarrow \Psi' = \alpha^{-1}\Psi$. But since C is continuous, $C\Psi' = \Psi$. Thus α is in the point spectrum.

It follows that a compact operator either has a finite number of eigenvalues or else zero is an accumulation point of them. If the domain of C is infinite dimensional then zero *must* belong to its spectrum.[32] This implies that C^{-1} cannot be everywhere defined, and the range of C cannot be the whole Hilbert space.[33] In other words, the equation

$$C\Psi = \Psi_0$$

23 Taylor, *op. cit.*, p. 275.

24 *Ibid.*, p. 285.

25 Akhiezer and Glazman, *op. cit.*, vol. 1, p. 57.

26 An operator of finite rank is one whose range is finite-dimensional; Riesz and Nagy, *op. cit.*, p. 204.

27 Akhiezer and Glazman, *op. cit.*, vol. 1, p. 57.

28 *Ibid.*, pp. 122 and 124.

29 Taylor, *op. cit.*, p. 281.

30 *Ibid.*, p. 278.

31 Akhiezer and Glazman, *op. cit.*, vol. 1, p. 124.

32 Taylor, *op. cit.*, p. 286, Prob. 9.

33 Note that C being closed C^{-1} is also closed; see Taylor, *op. cit.*, p. 177.

which in the theory of integral equations, is called an *integral equation of the first kind*, cannot have a solution for all Ψ_0.

For a compact operator C the range of $\alpha - C$ is *closed*.[34] Hence the range of $\alpha - C$ and the null space of $\alpha^* - C^\dagger$ are orthogonal complements. This means that for a compact operator K and every given number α either (7.95) has a unique solution, or the associated homogeneous equation (7.95a) has a solution.[35] Furthermore, if (7.95a) has n linearly independent solutions, then so has its adjoint equation, and the necessary and sufficient condition for the existence of solutions to (7.95) is that Ψ_0 be orthogonal to all solutions of the adjoint of (7.95a). If Ψ is required to be orthogonal to all solutions of (7.95a), then it is unique.

The following simple lemma is often useful in determining the compactness of a given operator. It is intuitively plausible on the basis of the simultaneous diagonalizability of commuting operators (if they are diagonalizable).

Lemma. *Suppose that A is unitary (or self-adjoint), C is compact, and they commute, $[A, C] = 0$. Then the null space of C must include the space spanned by the continuum part of the spectrum of A.*

Note that the lemma refers to a "continuum part" of the spectrum of A, and not just to the continuous spectrum. It asserts that C may fail to vanish at most on the "discrete" points of the spectrum of A. If the entire spectrum of A forms a continuum, C must be zero. In other words: *No nonzero compact operator can commute with a unitary operator whose spectrum consists of a continuous set of points only.*

We shall give the proof in an appendix to this chapter because it uses the tools of the next sections.

7.3.3 Hermitian and Unitary Operators. Hermitian operators are defined by the property that

$$(\Psi, A\Phi) = (A\Psi, \Phi)$$

for all vectors Ψ and Φ in the domain of definition of A, which is assumed to be dense in $\mathscr{H}$. They are sufficiently familiar to anyone acquainted with quantum mechanics to make anything but a brief discussion here unnecessary. If the domains of definition of A and $A^\dagger$ are equal, then A is called *self-adjoint*, and we may write simply

$$A^\dagger = A.$$

For most of our purposes we need not pay attention to the distinction between self-adjointness and hermiticity.

34 *Ibid.*, p. 279. We suppose that $\alpha \neq 0$ in this paragraph.

35 In the theory of integral equations with Hilbert–Schmidt kernels this is called the *Fredholm alternative*.

The eigenvalues λ of A are real and lie in the region[36]

$$\inf_{\|\Psi\|=1} (\Psi, A\Psi) = m(A) \leq \lambda \leq M(A) = \sup_{\|\Psi\|=1} (\Psi, A\Psi).$$

The larger of the two, $|M|$ or $|m|$, is the norm of A. The eigenvectors corresponding to different eigenvalues are *orthogonal*,

$$(\Psi_\lambda, \Psi_\beta) = 0$$

if

$$A\Psi_\lambda = \lambda\Psi_\lambda \qquad A\Psi_\beta = \beta\Psi_\beta \qquad \lambda \neq \beta.$$

These statements are too familiar and easy to prove to require demonstration here. All expectation values, or diagonal elements, of a hermitian operator are, or course, real.

The spectrum of a bounded self-adjoint operator lies on the interval of the real axis from $m(A)$ to $M(A)$, and it includes the end points m and M.[37] We already know that it consists of a point spectrum and a continuous spectrum only (one of which may be empty.)

The mathematical statement of Hilbert's centrally important *spectral theorem* for hermitian operators makes use of an orthogonal-projection-valued[38] function $P(\lambda)$ defined for all real λ such that for all λ_1 and λ_2

$$[P(\lambda_1), P(\lambda_2)] = 0$$

and

$$[P(\lambda), A] = 0.$$

It is monotonely *increasing* from 0 to $\mathbb{1}$, in the sense that the space on which $P(\lambda)$ projects gets larger and larger as λ increases, starting from the zero vector and ending with all of $\mathscr{H}$. That means that, if $\lambda_1 \leq \lambda_2$, then

$$P(\lambda_1)P(\lambda_2) = P(\lambda_1)$$

and $P(\lambda) = 0$ if $\lambda \leq m(A)$, $P(\lambda) = \mathbb{1}$ if $\lambda > M(A)$. Furthermore it is strongly continuous from the right,

$$P(\lambda_1) \Rightarrow P(\lambda) \quad \text{as } \lambda_1 \to \lambda+$$

but not necessarily from the left. Such a projection function $P(\lambda)$ is called a *resolution of the identity* or a *spectral family*. The spectral theorem[39] asserts that such a function exists with the property that for every continuous function f we can write

$$f(A) = \int_\alpha^\beta f(\lambda)\, dP(\lambda) \tag{7.96}$$

36 The abbreviation "inf" stands for "greatest lower bound."

37 Taylor, *op. cit.*, p. 330.

38 A reminder: An operator P is a *projection* if it is *idempotent*: $P^2 = P$. Its range is the space onto which it projects. It is called *orthogonal* if it is hermitian.

39 Taylor, *op. cit.*, p. 349.

the right-hand side being a Stieltjes integral and $\alpha < m(A)$, $\beta > M(A)$. If f is the constant function, this reads specifically

$$\mathbb{1} = \int_{\alpha}^{\beta} dP(\lambda). \tag{7.96a}$$

The real part of the resolvent set of A consists of those points in a two-sided neighborhood of which $P(\lambda)$ is *constant*; the point spectrum, of those points where $P(\lambda)$ is discontinuous; and the continuous spectrum, of those points not in the resolvent set where $P(\lambda)$ is continuous.[40] It follows from this that the isolated points of the spectrum of a self-adjoint operator are eigenvalues, i.e., members of the point spectrum. Those members of a continuum which are eigenvalues are isolated from each other; they form the part of the point spectrum that is "embedded in the continuum." The remaining members of a continuum are in the continuous spectrum. The finite accumulation points of eigenvalues may be in either part of the spectrum.

The Stieltjes integral is the limit of the corresponding sum; when A acts on a vector Ψ,

$$A\Psi \simeq \sum_n \lambda_n(P_n\Psi - P_{n-1}\Psi).$$

Now suppose that Ψ is chosen so that

$$P_n\Psi = \begin{cases} 0 & \text{for } n < N \\ \Psi & \text{for } n \geq N. \end{cases}$$

Then the sum gives

$$A\Psi \simeq \lambda_N\Psi$$

which means that Ψ is approximately an eigenvector of A with the eigenvalue $\lambda = \lambda_N$. If P has a discontinuity at λ_N, then we may make the subdivisions smaller and smaller and there will exist a vector such that $A\Psi = \lambda\Psi$. But if P is continuous, then, as the differences $\lambda_n - \lambda_{n-1}$ decrease, the approximate eigenvectors have to be shifted. There is no vector in the space such that $A\Psi = \lambda\Psi$. Then λ is in the continuous spectrum. Only by including non-normalizable vectors χ in our considerations may we speak of the limit of such a sequence. Then one may write

$$\begin{aligned}\Psi &= \int_{\alpha}^{\beta} d\lambda \frac{dP}{d\lambda}\Psi + \sum_n \Psi_n(\Psi_n, \Psi) \\ &= \int_{\alpha}^{\beta} d\lambda \sum_m \chi_m(\lambda)(\chi_m(\lambda), \Psi) + \sum_n \Psi_n(\Psi_n, \Psi)\end{aligned}$$

if we represent

$$\frac{dP}{d\lambda} = \sum_m \chi_m(\lambda)\chi_m(\lambda)^{\dagger}.$$

40 *Ibid.*, pp. 352, 353.

This is the way the resolution of the identity is usually written in the physics literature, or, more symbolically,[40a]

$$\mathbb{1} = \int_\alpha^\beta d\lambda \sum_m \chi_m(\lambda)\chi_m(\lambda)^\dagger + \sum_n \Psi_n\Psi_n^\dagger. \tag{7.97}$$

This split of the resolution of the identity into a continuum and a discrete part corresponds to defining an everywhere continuous function $P_c(\lambda)$ by subtracting from $P(\lambda)$ its jumps. Then (7.97) reads

$$\mathbb{1} = P_c + \sum_n P_n \qquad P_c = \int_\alpha^\beta dP_c(\lambda)$$

if P_n is the orthogonal projection on the nth eigenspace of A. The operator P_c is the projection on the subspace of the continuum part of the spectrum.

The resolvent of a hermitian operator can be written in terms of the $P(\lambda)$,

$$(\lambda - A)^{-1} = \int_\alpha^\beta \frac{dP(\lambda')}{\lambda - \lambda'}. \tag{7.98}$$

At those points λ_n where $P(\lambda')$ is discontinuous, we get contributions of the form $Q_n/(\lambda_n - \lambda)$. Thus the resolvent has *simple poles* at the point spectrum. Where $P(\lambda')$ is continuous (but not constant), the integral is at best writable in the form

$$\int \frac{dP}{d\lambda'}(\lambda' - \lambda)^{-1}\, d\lambda'$$

which has one or more *branch lines* along the continuous spectrum.

If the self-adjoint operator A is unbounded, as the hamiltonian of physical systems almost always is, then the integrals over λ are improper. The lower limit is usually finite, but the upper limit is infinite. This produces additional convergence problems that may be circumvented by considering the Cayley transform

$$U = (\mathbb{1} + iA)(\mathbb{1} - iA)^{-1}$$

which is *unitary*.

The spectral theory of unitary operators is essentially identical to that of self-adjoint operators, except that the spectrum is confined to the unit circle. Again it consists of a point spectrum and a continuous spectrum only (one of which may be empty), because a unitary operator is normal. There exists a resolution of the identity $P(\lambda)$ such that for every continuous function f defined on the unit circle one can write[41]

$$f(U) = \int_0^{2\pi} dP(\lambda) f(e^{i\lambda}).$$

40a Eq. (7.97) and the argument leading to it are valid provided that the spectrum has the property of being "absolutely continuous." [See Reed and Simon (1972), vol. I, p. 230.]

41 *Ibid.*, p. 358.

If we choose

$$f(z) = -i \ln z$$

then

$$A = \int_0^{2\pi} dP(\lambda)\lambda$$

defines a bounded hermitian operator with $m = 0$ and $M = 2\pi$ which is such that

$$U = e^{iA}.$$

What is more, every continuous one-parameter group $U(t)$ of unitary transformation, with

$$U(0) = \mathbb{1} \qquad U(s + t) = U(s)U(t)$$

admits a spectral representation of the form

$$U(t) = \int_{-\infty}^{\infty} dP(\lambda)e^{i\lambda t}.$$

This is *Stone's theorem*.[42]

7.3.4 Analyticity of the Resolvent. The resolvent operator $(\alpha - K)^{-1}$ of a closed operator K is an analytic (operator-valued) function of the complex variable α, regular at all points α in the resolvent set.[43] The notion of an analytic operator-valued function was defined in Sec. 6.7.2. If K is compact, then $(\alpha - K)^{-1}$ is everywhere analytic in the complex α plane, save for isolated poles of finite order, which are all contained in a finite region and which cannot have any accumulation points, except at $\alpha = 0$.

Suppose that, at $\alpha = \alpha_0$, the resolvent operator $(\alpha - K)^{-1}$ has a pole of finite order $M \neq 0$. Then we expand

$$(\alpha - K)^{-1} = \sum_{n=-M}^{\infty} A_n(\alpha - \alpha_0)^n. \tag{7.99}$$

Multiplying both sides by $\alpha - K$ must yield the unit operator. Term-by-term comparison therefore gives us the equations

$$\begin{aligned} A_{n-1} &= (K - \alpha_0)A_n = A_n(K - \alpha_0) \quad \text{for } n \neq 0 \\ A_{-1} - \mathbb{1} &= (K - \alpha_0)A_0 = A_0(K - \alpha_0) \\ (K - \alpha_0)A_{-M} &= A_{-M}(K - \alpha_0) = 0. \end{aligned} \tag{7.100}$$

The first set implies that for $m > 0$

$$A_{-(m+1)} = (K - \alpha_0)^m A_{-1} = A_{-1}(K - \alpha_0)^m \tag{7.101}$$

$$A_0 = (K - \alpha_0)A_1 = (K - \alpha_0)^m A_m = A_m(K - \alpha_0)^m \tag{7.102}$$

42 Riesz and Nagy, *op. cit.*, p. 383.
43 K. Yosida (1965), p. 211.

insertion of which in the second line of (7.100) yields

$$A_{-1} - \mathbb{1} = A_{m-1}(K - \alpha_0)^m = (K - \alpha_0)^m A_{m-1}. \tag{7.103}$$

But since $A_{-M-1} = 0$, (7.101) tells us that

$$A_{-1}(K - \alpha_0)^M = (K - \alpha_0)^M A_{-1} = 0 \tag{7.104}$$

and therefore by (7.103) for $m = M$

$$(A_{-1} - \mathbb{1})A_{-1} = 0. \tag{7.105}$$

Consequently α_0 is in the point spectrum of K. The residue operator A_{-1} is a projection whose range is equal[44] to the null space of $(K - \alpha_0)^M$ and whose null space (or the range of $\mathbb{1} - A_{-1}$) is equal to the range of $(K - \alpha_0)^M$. Now $A_m \neq 0$ for all $0 \neq m \geq -M$, because (7.100) shows that $A_m = 0$ implies $A_n = 0$ for all $n \leq m$. Hence, according to (7.101), for $m < M$ the null space of $(K - \alpha_0)^m$ is strictly *smaller* than the range of A_{-1}, and its range is strictly *larger* than the null space of A_{-1}. For $m \geq M$ the null spaces of $(K - \alpha_0)^m$ remain the same, and so do their ranges. This is expressed by saying that the *ascent* and the *descent* of $K - \alpha_0$ are both equal to M.[45] Since the range and null space of a projection are complementary, so are the range and null space of $(K - \alpha_0)^M$.

The assumption (7.99) implies that

$$(\alpha - K^\dagger)^{-1} = \sum_{n=-M}^{\infty} A_n^\dagger(\alpha - \alpha_0^*)^n$$

and hence $\alpha_0{}^*$ is in the point spectrum of $K^\dagger$, $A_{-1}{}^\dagger$ being the projection onto the eigenspace of $K^\dagger$ at $\alpha_0{}^*$. Now, if K is *normal*, then the eigenspaces of K and $K^\dagger$ at α_0 and $\alpha_0{}^*$, respectively, are equal, and hence so are the ranges of A_{-1} and $A_{-1}{}^\dagger$. But the range of the projection A_{-1} is complementary to its null space, and the range of $A_{-1}{}^\dagger$ is orthogonal to the null space of A_{-1}. Consequently, if K is normal, then the range and the null space of A_{-1}, and consequently also of $(\alpha_0 - K)^M$, are *orthogonal complements*. That means that the equation

$$(\alpha_0 - K)^M \Psi = \Psi_0$$

has a solution if and only if Ψ_0 is orthogonal to the null space of $(\alpha_0 - K)^M$. The residue operator A_{-1} is now an *orthogonal projection*,

$$A_{-1} = A_{-1}^\dagger.$$

If the orthonormal set $\{\Psi_n\}$ spans the range of A_{-1}, that is, the null space of $(\alpha_0 - K)^M$, then we can write

44 It follows from (7.104) that its range is inside the null space of $(K - \alpha_0)^M$ and from (7.103) that the null space of $(K - \alpha_0)^M$ is inside its range.

45 The ascent of B is the smallest integer M such that the null spaces of B^M and B^{M+1} are equal; its descent is the smallest integer N such that the ranges of B^N and B^{N+1} are equal. If B is everywhere defined and both ascent and descent are finite, they are equal (Taylor, *op. cit.*, p. 273). If furthermore B is compact and $\alpha \neq 0$, then the ascent and descent of $\alpha - B$ are both finite and hence equal (Taylor, *op. cit.*, p. 279).

$$A_{-1} = \sum_n \Psi_n \Psi_n^\dagger.$$

If K is normal, then the ascent of $\alpha_0 - K$ is either 0 or 1. This is because then the null spaces of $\alpha_0 - K$ and $\alpha_0{}^* - K^\dagger$ are equal, and hence the range and null space of $\alpha_0 - K$ are orthogonal. Consequently $\alpha_0 - K$ cannot map anything into its own the null space, and the null spaces of $\alpha_0 - K$ and $(\alpha_0 - K)^2$ must be equal. *For a normal* K, therefore, $(\alpha - K)^{-1}$ *cannot have a pole of finite order greater than* 1.

We return now to the general theory of $(\alpha - K)^{-1}$. It is possible for $(\alpha - K)^{-1}$ to have a "pole of infinite order"; i.e., the Laurent expansion (7.99) need not terminate at a finite negative n. Then $\alpha - K$ has an *isolated essential singularity* at $\alpha = \alpha_0$.

The converse of the foregoing discussion will not be proved here.[46] Suppose that there exists an integer n such that the range and null space of $(\alpha_0 - K)^n$ are complementary, that the range of $(\alpha_0 - K)^n$ is closed, and that $m \geq 1$ is the smallest such integer. Then α_0 is a pole of the resolvent $(\alpha - K)^{-1}$ of order m. If the domain of K is all of $\mathscr{H}$ and K is compact, then each nonzero point of the spectrum is a pole.[47] This follows essentially from the fact that the ascent and descent of a compact operator are finite.

It is useful to correlate the notion of ascent and the occurrence of higher-order poles of the resolvent with another property the possibility of whose existence is sometimes overlooked. Suppose that the ascent of $K - \alpha$ is greater than 1. Then there exists a vector Φ_α which lies both in the range and in the null space of $K - \alpha$,

$$\Phi_\alpha = (K - \alpha)\Psi_\alpha \qquad (K - \alpha)\Phi_\alpha = 0.$$

In other words, Ψ_α is not annihilated by $K - \alpha$, but it is by $(K - \alpha)^2$. Let $\overline{\Phi}_{\alpha^*}$ be any eigenvector of $K^\dagger$ with the eigenvalue α^*. Then

$$(\overline{\Phi}_{\alpha^*}, \Phi_\alpha) = (\overline{\Phi}_{\alpha^*}, (K - \alpha)\Psi_\alpha) = 0.$$

Thus an eigenvector Φ_α of K with the eigenvalue α is orthogonal to all eigenvectors $\overline{\Phi}_{\alpha^*}$ of $K^\dagger$ with the eigenvalue α^*. Similarly there exists an eigenvector $\overline{\Phi}_{\alpha^*}$ of $K^\dagger$ with the eigenvalue α^* which is orthogonal to all eigenvectors Φ_α of K with the eigenvalue α. This situation being associated with an ascent of $K - \alpha$ greater than unity, it is also associated with a pole of the resolvent of K of order greater than 1. If K is normal, of course, this cannot happen.

The converse is also true. Suppose that $(\alpha - K)^{-1}$ has a simple pole at $\alpha = \beta$. Then we know that the null space and range of $K - \beta$ are complementary. That is, every vector Ψ can be written as a linear combination of a vector Φ_β in the null space and a vector in the range,

$$\Psi = \Phi_\beta + (K - \beta)\Phi.$$

46 Taylor, *op. cit.*, p. 310.
47 *Ibid.*, p. 311.

Now take the inner product of Ψ and an eigenvector $\overline{\Phi}_\beta$ of $K^\dagger$, with the eigenvalue β^*,

$$(\overline{\Phi}_\beta, \Psi) = (\overline{\Phi}_\beta, \Phi_\beta).$$

Since $\overline{\Phi}_{\beta^*}$ cannot be orthogonal to *all* vectors Ψ, it follows that the inner product on the right cannot vanish for all Φ_β. Thus there exists an eigenvector $\overline{\Phi}_{\beta^*}$ of $K^\dagger$ with the eigenvalue β^* which is orthogonal to the entire eigenspace of K at the eigenvalue β, and there is an eigenvector Φ_β of K with the eigenvalue β which is orthogonal to the eigenspace of $K^\dagger$ at β^* *if and only if* the resolvent $(\alpha - K)^{-1}$ has a pole of order greater than 1 at $\alpha = \beta$.

Connection with Completeness. Suppose that the bounded operator K has a point spectrum $\{\alpha_n\}$ only, and let the ascent and descent of $K - \alpha_n$ be M_n. Suppose further that the resolvent $(\alpha - K)^{-1}$ can be expanded in a Mittag-Leffler series

$$(\alpha - K)^{-1} = \sum_{n=1}^{\infty} \sum_{m=-M_n}^{-1} A_m^{(n)}(\alpha - \alpha_n)^m. \tag{7.106}$$

Then letting $|\alpha| \to \infty$ tells us immediately that

$$\mathbb{1} = \sum_{n=1}^{\infty} A_{-1}^{(n)} \tag{7.107}$$

and hence the spectrum is complete in the extended sense that the direct sum of the null spaces of $(K - \alpha_n)^{M_n}$ is the entire Hilbert space. In other words, in order to get a complete basis one must, for each α_n, take a set of vectors that spans the null space of $(K - \alpha_n)^{M_n}$, and not only that of $K - \alpha_n$. Only if each pole of the resolvent is simple can the set of eigenvectors of K form a complete set.

Of course the residues $A_{-1}{}^{(n)}$ are in general not *orthogonal* projections, and neither are they *mutually* orthogonal. But if for each n $\{\Psi_i^{(n)}\}$ is a set of linearly independent vectors that spans the null space of $(K - \alpha_n)^{M_n}$, then we may form for each n a set of linearly independent vectors $\{\Phi_i^{(n)}\}$ that spans the null space of $(K^\dagger - \alpha_n{}^*)^{M_n}$ and which is such that

$$(\Phi_i^{(n)}, \Psi_j^{(m)}) = \delta_{ij}\delta_{nm}.$$

These two bases are called *biorthogonal* sets, and we have in the dyadic notation

$$A_{-1}^{(n)} = \sum_i \Psi_i^{(n)}\Phi_i^{(n)\dagger}.$$

The completeness statement (7.107) then reads

$$\Psi = \sum_{in} a_{in}\Psi_i^{(n)} \qquad a_{in} = (\Phi_i^{(n)}, \Psi)$$

for each Ψ in the Hilbert space.[48]

48 It is to be noted that these arguments are of a heuristic nature. We have paid no attention to convergence here. For more details, see, for example, Riesz and Nagy, *op. cit.*, p. 208.

Appendix

Proof of the Lemma of Sec. 7.3.2. Let $P_c(\lambda)$ be the spectral family of the continuum part of A, in the sense of Eq. (7.97a), so that $P_c(\lambda)$ continuously increases from $P_c(\lambda_1) = 0$. Let $\alpha \neq 0$ be an eigenvalue of $C^\dagger C$; and assume that Q is the orthogonal projection onto the eigenspace of $C^\dagger C$ at α. Then $[Q, P_c(\lambda)] = 0$ for all λ and hence $R(\lambda) \equiv QP_c(\lambda)$ is an orthogonal projection too. Since $C^\dagger C$ is compact, the range of Q is finite dimensional, and so is that of $R(\lambda)$. Furthermore $R(\lambda)$ is a continuously increasing function of λ that is zero for $\lambda = \lambda_1$. We now introduce a basis in the range of $R(\lambda_2)$ and calculate $\text{tr } R(\lambda) = n(\lambda) < \infty$ for $\lambda \leq \lambda_2$. Then $n(\lambda)$ must be continuous, and an integer. Since $R(\lambda_1) = 0$, n vanishes for all λ.

This shows that on the range of P_c, the operator $C^\dagger C$ cannot have any nonzero eigenvalues. But $C^\dagger C$ is self-adjoint and compact. Hence it must vanish on P_c.

Notes and References

For general treatments of scattering theory from a time-independent point of view we give the following references: Mott and Massey (1949); W. Heitler (1954); Jauch and Rohrlich (1955); Wu and Ohmura (1962); W. O. Amrein et al. (1977); Reed and Simon (1979); as well as most modern books on quantum mechanics. Other general surveys of the formal theory are B. S. DeWitt (1955a); Brenig and Haag (1959); F. E. Low (1959); and R. Haag (1961). See also the important papers C. Møller (1945 and 1946).

Other articles on formal scattering theory from a time-independent point of view are E. Feenberg (1948); M. L. Goldberger (1951a and b); M. N. Hack (1954); S. Epstein (1955); B. S. DeWitt (1955b); H. Rollnik (1956); Foldy and Tobocman (1957); E. Gerjuoy (1958a and b); T. Sasakawa (1963); K. L. Kowalski (1972b); S. K. Adhikari (1979b); as well as the papers by Wheeler, Heisenberg, Lippmann and Schwinger, and Gell-Mann and Goldberger mentioned in Chap. 6. For a discussion of the relation between the time-dependent and time-independent points of view, see Belinfante and Møller (1953).

7.1.1 For early use of Green's-function techniques in scattering problems, see A. Sommerfeld (1910, 1912, and 1931); J. Meixner (1933, 1934, and 1937). The following paper makes particular use of the resolvent, or Green's function: M. Schönberg (1951).

7.2.1 See the Notes and References for Secs. 6.3 and 6.4.

7.2.2 For a specific discussion of the occurrence of the incoming wave state as a final state and the outgoing wave state as the initial state, see Breit and Bethe (1954) and also S. Altshuler (1956). For studies of the off-energy-shell T matrix, see J. Nuttall (1967c); R. D. Amado (1970); M. G. Fuda (1971c); and B. R. Karlsson (1972).

The so-called *optical theorem* was discovered in quantum mechanics by E. Feenberg (1932). It is sometimes unjustifiably referred to as the *Bohr–Peierls–Placzek relation*, after Bohr, Peierls, and Placzek (1939). For some remarks concerning its history in electromagnetic theory, where it was first known, see the Notes and References to Sec. 1.3.9. Its history is traced in detail in R. G. Newton (1976a).

The utility and validity of expanding the T matrix in terms of operators of finite rank is discussed by T. A. Osborn (1969, 1973) and J. S. Levinger (1973). For a specific discussion of the effects of exchange, see R. Mapleton (1954). The *Low equation* was introduced by F. E. Low (1955). Its solvability is discussed by R. L. Warnock (1968).

7.2.3 For a standing-wave form of scattering theory, see Kouri and Levin (1974b).

7.2.4 Equation (7.69) was formulated originally by C. Lovelace (1964). For an extensive discussion of the time-reversal operation, see E. P. Wigner (1959), chap. 26. For discussions of reciprocity and its relation to *detailed balance*, see, for example, Wigner and Eisenbud (1947); Blatt and Weisskopf (1952), chap. X, 2D and E. See also F. Coester (1953); S. Watanabe (1955).

7.2.5 For discussion and derivations of (7.76), (7.76a), see Day, Rodberg, Snow, and Sucher (1961); W. R. Gibbs (1974); and J. Zorbas (1976a).

Equations (7.87) are due to L. D. Faddeev (1960, 1961, and 1962). Equation (7.92) is due to S. Weinberg (1964), sec. III. See also W. Hunziker (1964).

7.3 The following books, among others, may be consulted on the mathematical questions that arise in this chapter: J. von Neumann (1955); Riesz and Sz-Nagy (1955); B. Friedman (1957); Hille and Phillips (1957); A. E. Taylor (1958); Dunford and Schwartz (1958 and 1963); K. O. Friedrichs (1960); Akhiezer and Glazman (1961); K. Yosida (1965); Reed and Simon (1972). Among these, perhaps the books by Taylor and Reed and Simon are the most accessible to nonmathematicians. The following article may also usefully be consulted: S. Weinberg (1964a), especially appendix A.

To a reader unfamiliar with vector-space theory the following book is highly recommended as a first approach before proceeding to Hilbert-space theory: P. R. Halmos (1942).

An important paper on the self-adjoint nature of the hamiltonian is T. Kato (1951a); see also N. Limic (1966).

The following mathematical papers concerned with the change of continuous spectra under perturbations are also relevant: K. O. Friedrichs (1948); T. Kato (1957a and b); M. Rosenblum (1957); N. Aronszajn (1957); Ladyzhenskaya and Faddeev (1958); S. T. Kuroda (1959 and 1960); L. Thomas (1972). See also A. Ya. Povsner (1953 and 1955). For other rigorous treatments of scattering theory from the time-independent point of view see K. Kodaira (1949 and 1950); T. Ikebe (1960); K. Meetz (1962); T. F. Jordan (1962); L. D. Faddeev (1963); J. G. Belinfante (1964); B. Simon (1971); S. Agmon (1975); W. O. Amrein et al. (1977); Reed and Simon (1979); J.–M. Combes (1980b). See also Zemach and Odeh (1960); and B. Misra et al. (1963).

The remarks connecting a Mittag–Leffler expansion of the resolvent to completeness were stimulated by Fonda et al. (1966).

Problems

1 The coordinate representation of the state vectors $\Psi^{(+)}(E,\alpha)$ and $\Psi^{(-)}(E,\alpha)$ describe traveling waves in the asymptotic region (when the particles are far apart), the first, outgoing scattered waves, the second, incoming. How must Ψ_0 be chosen in order that $\Psi^{(P)}$, defined by (7.20), describe a *standing wave* in the asymptotic region? Does the requirement that the scattered wave part of the asymptotic region be purely outgoing, incoming, or standing define $\Psi^{(+)}$, $\Psi^{(-)}$, or $\Psi^{(P)}$, respectively, uniquely (apart from a normalization factor)?

2 Write down three off-the-energy-shell extensions of $T_{\beta\alpha}$ in addition to (7.45).

3 Suppose that only a relatively weak part H_2' of the interaction term in the hamiltonian violates time-reversal invariance. Calculate the **T** matrix in perturbation theory with respect to H_2'. Will the violation be detectable in first order? Discuss.

4 It was pointed out below Eq. (7.79) that Eq. (7.78) is valid only when the third particle is infinitely massive. Assuming that it is very massive, but not infinitely so, calculate the first-order recoil correction.

5 Let $\{\Psi_n\}$, $n = 0, 1, 2, \ldots, \infty$, be a complete set of orthonormal vectors, and define the operator a by the equations: $a\Psi_0 = 0$, $a\Psi_n = \Psi_{n-1}$, $n = 1, 2, \ldots$. (*a*) Give the domain of definition of a and its range. Give the spectrum of a; that is, give separately the point spectrum, the continuous spectrum, and the residual spectrum. Is a isometric? Does it have a left or right inverse? If so, what are they? Is a bounded? Is it compact? Is it a normal operator? What are its ascent and its descent? Give the region of analyticity of the resolvent operator $(z - a)^{-1}$. For what values of γ and what Ψ' does the equation $\Psi = \Psi' + \gamma a\Psi$ have a solution? When is it unique? (*b*) Answer the same questions for the operator $b = a^\dagger$.

6 Let $\{\Psi_n\}$, $n = 0, 1, 2, \ldots, \infty$, be a complete set of orthonormal vectors, and define the operator a by the equations $a\Psi_0 = 0$, $a\Psi_n = n^{-1/2}\Psi_{n-1}$, $n = 1, 2, \ldots$. (*a*) Answer the same questions as for Prob. 5*a*. (*b*) Answer the same questions for the operator $b \equiv a^\dagger$. (*c*) Answer the same questions for the hermitian operators $aa^\dagger$ and $a^\dagger a$.

7 Can a compact operator C in a Hilbert space have a finite number of different eigenvalues? If so, give an example. Suppose that C is hermitian, what is the dimensionality of its null space?

8 Show that, if the domain of A is dense in the Hilbert space, then the null space of its adjoint is the orthogonal complement of its range.

9 Show that an operator all of whose expectation values are real is hermitian.

10 Let N be a normal operator. Show that any two of its eigenvectors belonging to different eigenvalues are orthogonal. Show that its "spectrum is complete" in the same sense as that of a hermitian operator (by reducing the proof to that for the latter).

11 Show that an operator of finite rank is compact.

12 Using property 6 listed in Sec. 7.3.2, show that every nonzero number in the spectrum of a compact operator is an eigenvalue (i.e., in the point spectrum) and has finite degeneracy. Also show that zero is the only possible limit point of its spectrum.

13 Using properties 5 and 6 listed in Sec. 7.3.2, show that if tr $CC^\dagger < \infty$, then C is compact.

14 Let $\{\Psi_n\}$ be a complete orthononormal set, and define $C = \sum_1^\infty n^{-2}\Psi_n\Psi_n^\dagger$. Prove that C is compact, that its range is dense in the Hilbert space, and that zero is in the continuous spectrum.

CHAPTER 8

Cross Sections

8.1 General Definition of Differential Cross Sections

Let us consider the relation of the state vector $\Psi^{(+)}(\alpha, t)$ to an experimentally measured counting rate. We first take the case of a particle of mass m in interaction with a center.

If we wish to describe a wave packet with initial momenta distributed according to an amplitude weight function $g(\mathbf{p}) = g_0(\mathbf{p} - \mathbf{p}_i)$ that is peaked at $\mathbf{p} = \mathbf{p}_i$ with a width $\Delta \ll |\mathbf{p}_i|$ and centered at an impact parameter $\mathbf{b} \perp \mathbf{p}_i$ (see Chapter 5) then the momentum-space wave function is given by

$$\cdot \hat{\psi}_{\text{in}}(\mathbf{b}, \alpha', \mathbf{p}, t) = g(p)\delta_{\alpha\alpha'} \exp[-i(\mathbf{p} \cdot \mathbf{b} + tp^2/2m)]. \tag{8.1}$$

The label α is used here for all other needed quantum numbers, and we assume that

$$\int d\mathbf{p}\, |g(\mathbf{p})|^2 = 1. \tag{8.2}$$

According to (6.41) the state vector in the distant future will be

$$\hat{\psi}_{\text{out}}(\beta, \mathbf{p}, t) = \sum_{\alpha'} \int d\mathbf{p}'\, S(\beta, \mathbf{p}; \alpha', \mathbf{p}')\psi_{\text{in}}(\mathbf{b}, \alpha', \mathbf{p}', t),$$

or, if $|\mathbf{p} - \mathbf{p}_0| > \Delta$, by (7.37)

$$\begin{aligned} &\hat{\psi}_{\text{out}}(\beta, \mathbf{p}, t) \\ &= -2\pi i \int d\mathbf{p}'\, \delta(E - E')g(\mathbf{p}')T(E; \beta, \hat{\mathbf{p}}; \alpha, \hat{\mathbf{p}}') \exp[-i(\mathbf{p}' \cdot \mathbf{b} + tE)]. \end{aligned} \tag{8.3}$$

According to (6.83) the probability that a particle with quantum number β will, at some time, trigger a distant counter subtending a cone C (from the center) that lies outside the peak of g at $\mathbf{p}_i$ is

$$\begin{aligned} P^b &= \int_C d\mathbf{p}\,|\hat{\psi}_{\text{out}}(\beta,\mathbf{p},t)|^2 \\ &= (2\pi)^2 \int_C d\mathbf{p} \int d\mathbf{p}'\,d\mathbf{p}''\,T(E;\beta,\hat{\mathbf{p}};\alpha,\hat{\mathbf{p}}')T^*(E;\beta,\hat{\mathbf{p}};\alpha,\hat{\mathbf{p}}'') \\ &\qquad \delta(E-E')\delta(E-E'')g(\mathbf{p}')g^*(\mathbf{p}'')e^{i\mathbf{b}\cdot(\mathbf{p}''-\mathbf{p}')}. \end{aligned} \tag{8.4}$$

It now becomes important to assume that the "beam" of incoming particles forms a mixed state made up incoherently of packets of various impact parameters $\mathbf{b}$. This is entirely analogous to the classical procedure given in Sec. 5.2, and it is justified in most experimental situations. Let $a = \pi r^2$ be the cross-sectional area of the beam of radius r and assume a uniform distribution of impact parameters. Then the probability becomes

$$P = \frac{1}{a}\int_{|\mathbf{b}|\leq r} d\mathbf{p}\ \ P^b. \tag{8.5}$$

It is physically clear (and has been shown[1] mathematically for reasonable potentials) that the diameter of the set B of impact parameters $\mathbf{b}$ for which P^b differs significantly from zero is of the order of magnitude of the (microscopic) range of the interaction. When $\mathbf{b}$ lies outside B the incoming particles miss the scattering center and ψ_{out} vanishes unless $\beta = \alpha$ and $\mathbf{p} = \mathbf{p}_i$. On the other hand, r is macroscopic and hence much larger than the diameter of B. We may therefore with little error extend the $\mathbf{b}$-integral to the whole plane orthogonal to $\mathbf{p}_i$ and use

$$\int d\mathbf{b}\,e^{i\mathbf{b}\cdot(\mathbf{p}''-\mathbf{p}')} = (2\pi)^2\delta(\mathbf{p}_\perp'' - \mathbf{p}_\perp'),$$

in which $\mathbf{p}_\perp'$ and $\mathbf{p}_\perp''$ indicate the components orthogonal to $\mathbf{p}_i$.

We now shift the variables of integration $\mathbf{p}'$ and $\mathbf{p}''$ by $\mathbf{p}_i$. Then

$$\mathbf{p}'^2 - \mathbf{p}''^2 \to (\mathbf{p}' + \mathbf{p}_i)^2 - (\mathbf{p}'' + \mathbf{p}_i)^2 = \mathbf{p}'^2 - \mathbf{p}''^2 + 2\mathbf{p}_i\cdot(\mathbf{p}' - \mathbf{p}''),$$

and since g_0 is sharply peaked at zero with a width $\Delta \ll |\mathbf{p}_i|$ we may replace

$$\begin{aligned} \delta(E' - E'') &= 2m_i\delta(\mathbf{p}'^2 - \mathbf{p}''^2) \to 2m_i\delta[2\mathbf{p}_i\cdot(\mathbf{p}' - \mathbf{p}'')] \\ &= (m_i/p_i)\delta[\hat{\mathbf{p}}_i\cdot(\mathbf{p}' - \mathbf{p}'')] \end{aligned} \tag{8.6}$$

and combine the δ-functions into $\delta(\mathbf{p}' - \mathbf{p}'')$. Furthermore we use the sharp peaking of g_0 to evaluate T and $\delta(\mathbf{p}^2 - \mathbf{p}'')$ at $\mathbf{p}' = \mathbf{p}_i$, and then utilize the normalization of g. The result is the probability

$$P = (2\pi)^4(m_i/ap_i)\int_C d\mathbf{p}\,\delta(E - E_0)|T(E_0;\beta,\hat{\mathbf{p}};\alpha,\hat{\mathbf{p}}_0)|^2$$

1 Campesino-Romeo and Taylor (1975).

$$= (2\pi)^4 \frac{m_f m_i p_f}{a p_i} \int_C d\hat{\mathbf{p}}\,|T_{fi}|^2. \tag{8.7}$$

Note that in (8.7) we have allowed for the possibility that the initial and final masses and momenta are unequal. For elastic scattering of course $m_i = m_f$ and $p_i = p_f$.

Suppose that the "beam" has a duration t (i.e., a spatial length tp_i/m) so that it carries a flux of $F = (1/at)$ particles per unit time and area. Then the probability of finding the (elastically) scattered particle in the cone C is

$$P = (2\pi)^4 (m_f m_i p_f/p_i) F t \int_C d\hat{\mathbf{p}}\,|T_{fi}|^2$$

and hence by (6.83) the number of particles per unit time that cross a distant surface intersecting C is

$$N_C = P/t = F \int_C d\hat{\mathbf{p}}\,[(2\pi)^4 (m_f m_i p_f/p_i)|T_{fi}|^2]. \tag{8.8}$$

The *differential cross section* is this number per unit incident flux per unit sold angle; therefore,

$$d\sigma/d\Omega = (2\pi)^4 (m_f m_i p_f/p_i)|T_{fi}|^2. \tag{8.9}$$

In this expression the T-matrix is to be calculated by means of (7.40) or (7.41) with a free wave function that in the coordinate representation reads

$$\psi_0(\mathbf{p},\alpha;\mathbf{r}) = (2\pi)^{-3/2} \chi_\alpha e^{i\mathbf{p}\cdot\mathbf{r}} \tag{8.10}$$

where χ_α is the normalized eigenstate appropriate to the quantum numbers α, so that

$$\int d\mathbf{r}\,\psi_0^*(\mathbf{p},\alpha;\mathbf{r})\psi_0(\mathbf{p}',\alpha;\mathbf{r}) = \delta_{\alpha\alpha'}\delta(\mathbf{p}-\mathbf{p}'). \tag{8.11}$$

On the other hand, if an *energy normalization* is adopted for the calculation of T, so that

$$\int d\mathbf{r}\,\psi_0^*(E,\hat{\mathbf{p}},\alpha;\mathbf{r})\psi_0(E',\hat{\mathbf{p}}',\alpha';\mathbf{r}) = \delta_{\alpha\alpha'}\delta(E-E')\delta_\Omega(\hat{\mathbf{p}},\hat{\mathbf{p}}'), \tag{8.12}$$

where δ_Ω is the solid-angle δ function defined in (2.15), then

$$\psi_0(E,\hat{\mathbf{p}},\alpha;\mathbf{r}) = \frac{(mp)^{1/2}}{(2\pi)^{3/2}} \chi_\alpha e^{i\mathbf{p}\cdot\mathbf{r}}, \tag{8.13}$$

and the cross section is expressed as

$$d\sigma/d\Omega = (2\pi)^4 p_i^{-2} |T_{fi}|^2. \tag{8.14}$$

The corresponding formulas for the scattering of two particles by one another are obtained by passing from the individual momenta $\mathbf{p}_1$ and $\mathbf{p}_2$ to the total momentum $\mathbf{P}$ and the momentum of relative motion $\mathbf{p} = (\mathbf{p}_2 m_1 - \mathbf{p}_1 m_2)/(m_1 + m_2)$. According to (7.43) the integral in P^b then contains the product $\delta(\mathbf{P} - \mathbf{P}')\delta(\mathbf{P} - \mathbf{P}'')$ as well as integrations over $\mathbf{P}'$ and $\mathbf{P}''$. If we wish to calculate the differential cross section $d\sigma/d\Omega$ in the center-of-mass system, where Ω is the solid angle of the momentum of relative motion, then the probability also contains an integral over all $\mathbf{P}$, since we don't care what the total final momentum is. The result is then again (8.8), except that m is replaced by the reduced mass $\mu = m_1 m_2/(m_1 + m_2)$. In that case, however, the number N_C is not a real particle number and it does not correspond directly to experimental counts. The differential cross section for the realistic case, in which one of the particles is detected and the momentum of the other is determined by momentum conservation, is obtained from

$$\int_C d\mathbf{p}_1 \int d\mathbf{p}_2\, \delta(\mathbf{P} - \mathbf{P}_0)\delta(E - E_0) \cdots = \int_C d\mathbf{p}_1\, \delta(E - E_0)|T_{fi}|^2.$$

In order to convert cross sections to particle counts the flux F in (8.8) must now be replaced by the product flux $\times$ density, i.e., the product of the two densities multiplied by the magnitude of the relative velocity of the two particles. (This is also called the *luminosity*).

The most common experimental situation is one in which a particle is scattered by many identical centers, as in a thin foil. We will see later (sec. 10.3.3) that the T-matrix for a scatterer shifted by $\mathbf{r}$ is related to that for the unshifted one by

$$T(E;\beta,\hat{\mathbf{p}};\alpha,\hat{\mathbf{p}}')\exp[i(\mathbf{p}' - \mathbf{p})\cdot\mathbf{r}].$$

Hence, if the distance between scatterers is large both compared to the range of the interaction and to the "scattering diameter" so that coherent and multiple-scattering effects can be neglected, the effective T-matrix of the collection of scatterers located at r_γ is

$$T_{\text{coll}}(E;\beta,\hat{\mathbf{p}};\alpha,\hat{\mathbf{p}}') = T(E;\beta,\hat{\mathbf{p}};\alpha,\hat{\mathbf{p}}')\sum_\gamma \exp[i(\mathbf{p}' - \mathbf{p})\cdot\mathbf{r}_\gamma]$$

if T is that of the individual centers. This leads to the appearance of the factor

$$\sum_{\gamma,\beta} \exp[i(\mathbf{p}' - \mathbf{p})\cdot\mathbf{r}_{\gamma\beta}],$$

where $\mathbf{r}_{\gamma\beta} = \mathbf{r}_\gamma - \mathbf{r}_\beta$, in the integral over $\mathbf{p}$ and $\mathbf{p}'$. If $|\mathbf{p}_0|\,|\mathbf{r}_{\gamma\beta}| \gg 1$ then this factor is a rapidly varying function of $\hat{\mathbf{p}}$ and $\hat{\mathbf{p}}'$, which leads to closely spaced interference fringes. If either the acceptance angle of the detector is large enough or the definition of the beam poor enough to contain many fringes the only terms that will not average to zero are those for which $\gamma = \beta$. We therefore obtain simply a factor of $\overline{N}$, the number of scatterers crossed by the beam. As a result we obtain (8.8) again, except that F is the product of flux $\times$ density (if the left-hand side is to represent the number of scattered particles per unit volume of scatterer).

If more than two particles emerge from the reaction, then the differential cross section, normalized to unit incident flux of the particle of mass m_i, is given by

$$d\sigma = (2\pi)^4 \frac{m_i}{p_i} |T_{fi}|^2 \left[\prod_f d\mathbf{p}'_f \right] \delta(E' - E_i)\delta(\mathbf{P}' - \mathbf{P}_i) \tag{8.15}$$

if T_{fi} is calculated with the use of wave functions normalized as in (8.10) for each particle, E' is the total final energy, and $\mathbf{P}'$ is the total final momentum. The term

$$\left[\prod_f d\mathbf{p}'_f \right] \delta(E' - E)\delta(\mathbf{P}' - \mathbf{P}) \tag{8.16}$$

is usually referred to as the *phase space factor*.

The differential cross sections given by (8.9), (8.14), and (8.15) are those for a 100 percent polarized beam to a specific polarization direction. If the initial beam is unpolarized, one must *average* over the initial spin directions (and other initial quantum numbers); if the final spins (or other quantum numbers) are not observed, i.e., all are counted, then we must *sum* over final spins, etc. (See Sec. 8.3.3 for more details.)

Similarly, if some of the final particles are unobserved, (8.15) must be integrated over their momenta (as well as summed over their discrete quantum numbers). If one assumes that T_{fi} is independent of these variables, the resulting integral over a region in momentum space can be carried out and yields a function of the energies of the observed particles. This function contains the kinematical aspects of the reaction. The *dynamical* effects are visible in a comparison of the integral that contains $|T_{fi}|^2$ with the integral that does not contain $|T_{fi}|^2$. Since the kinematics produce an energy dependence of their own (or "shape" of a spectrum), resonances or other dynamical effects can be reliably observed only in such a comparison.[2]

8.2 Relativistic Generalization

The results of Section 8.1 are easily generalized to the relativistic case. If W_1 and W_2 are the relativistic energies of particles 1 and 2 (including their rest energies), then

$$W_i/c = (\mathbf{p}_i^2 + m_i^2 c^2)^{1/2},$$

2 An alternative method of deriving cross-section formulas such as (8.9), (8.14), and (8.15) uses the "golden rule" of time-dependent perturbation theory. See, e.g., L. I. Schiff (1968), pp. 314/315. This requires enclosing the physical system in a finite box of volume V in order to replace the continuous spectrum of the hamiltonian by a discrete one. One must then calculate the *density of final states* in the limit in which $V \to \infty$. This density factor is identical with the phase space factor, and the dependence upon V cancels out in the final result. For critical studies of the approach of the discrete spectrum to a continuous one as $V \to \infty$ we refer to Fukuda and Newton (1956), B. S. DeWitt (1956), and F. Iwamoto (1963).

The entire finite-volume approach must be regarded as unnatural to scattering theory. It is a remnant of the dominant place historically occupied in quantum mechanics by bound-state problems or by the discrete spectrum.

and the total energy in the center-of-mass frame is

$$W/c = (\mathbf{p}^2 + m_1^2c^2)^{1/2} + (\mathbf{p}^2 + m_2^2c^2)^{1/2},$$

where c is the velocity of light. It follows that for one particle

$$\delta(\mathbf{p} - \mathbf{p}') = \frac{c^2}{pW}\delta(W - W')\,\delta_\Omega(\hat{\mathbf{p}},\hat{\mathbf{p}}'), \tag{8.17}$$

and in the center-of-mass frame of two,

$$\delta(\mathbf{p} - \mathbf{p}') = \frac{Wc^2}{pW_1W_2}\delta(W - W')\,\delta_\Omega(\hat{\mathbf{p}},\hat{\mathbf{p}}'). \tag{8.18}$$

Therefore in (8.9) m is replaced by W/c^2 for a collision with a fixed target, and by W_1W_2/Wc^2 for a collision of two particles in their center-of-mass frame.

The phase-space factor (8.16) becomes relativistically

$$\delta^4(P' - P)\prod_f [d^4p'_f\delta(W_f'^2 + m_f^2c^4 - \mathbf{p}_f'^2c^2)2cW'_f]. \tag{8.19}$$

Let us next consider the question of Lorentz invariance.

If we bombard a stationary target of ρ_1 particles per unit volume with a beam of density ρ_2 and velocity $\mathbf{v}_2$, then the number of scattered particles counted in a given solid angle element $d\Omega$, per unit time and per unit volume (of target), is

$$dN = d\sigma\, F \tag{8.20}$$

where $d\sigma$ is the differential cross section and

$$F = \rho_1\rho_2v_2.$$

The beam density ρ_2 is expressed in terms of the density $\rho_2^{(0)}$ in its rest frame by

$$\rho_2 = \rho_2^{(0)}\left(1 - \frac{v_2^2}{c^2}\right)^{-1/2}$$

because of the Lorentz contraction; hence

$$F = \rho_1^{(0)}\rho_2^{(0)}v_2\left(1 - \frac{v_2^2}{c^2}\right)^{-1/2} = \frac{\rho_1^{(0)}\rho_2^{(0)}p_2}{m_2}. \tag{8.21}$$

As the scattering process is viewed in another reference frame which moves with a uniform velocity with respect to the first, dN is unchanged. It is the number of particles counted per unit four-dimensional volume, and the four-dimensional-volume element is invariant under Lorentz transformations. Suppose that the motion of the new frame is parallel to v_2, with the velocity v_1, so that in it the target has the velocity $-\mathrm{v}_1$. Then the cross section $d\sigma'$ in the new frame is defined by

$$dN = d\sigma'\, F'$$

with

$$F' = \rho_1'\rho_2'(v_1 + v_2')$$
$$= \rho_1^{(0)}\rho_2^{(0)}(v_1 + v_2')\left(1 - \frac{v_1^2}{c^2}\right)^{-1/2}\left(1 - \frac{v_2'^2}{c^2}\right)^{-1/2}.$$

But the beam velocity v_2' in the new frame is related to that in the old by

$$v_2' = \frac{v_2 - v_1}{1 - \left(\frac{v_1 v_2}{c^2}\right)}.$$

Inserting this value in the expression for F', we get

$$F' = \rho_1^{(0)}\rho_2^{(0)}v_2\left(1 - \frac{v_2^2}{c^2}\right)^{-1/2} = F.$$

Thus F is invariant under Lorentz transformations to all frames in which beam and target motions are collinear. Consequently so is the differential cross section.

When the scattering process is observed in a corrdinate frame in which the colliding particles are not initially moving collinearly, then we do not have any clear idea a priori of how to define the cross section. In view of its invariance under Lorentz transformations that keep the initial momenta collinear, it is customary to define it to be invariant under *all* Lorentz transformations. To achieve this, it is necessary only to make F invariant. But the invariant function whose value is (8.21) in the coordinate system in which $\mathbf{p}_1 = 0$ is (using the customary summation convention)

$$F = \frac{\rho_1^{(0)}\rho_2^{(0)}[(p_\mu^{(1)}p_\mu^{(2)})^2 - m_1^2 m_2^2 c^4]^{1/2}}{m_1 m_2 c}.$$

If this function is inserted in (8.20), then

$$d\sigma = \frac{d\sigma}{d\Omega}d\Omega = \frac{dN}{F}$$

is thereby *defined* to be relativistically invariant.

8.3 Scattering of Incoherent Beams

8.3.1 The Density Matrix. We have so far discussed only the scattering of particles described by a wave function, i.e., in a *pure state*. In particular, the state of the particles was to be characterized by specified quantum numbers, in addition to energy and momentum, such as spin, isotopic spin, their projections on given directions, etc. An experimentally prepared beam is not necessarily of this nature. It may, instead, contain these quantum numbers α with a certain probability distribution. One then has to distinguish between situations in which these quantum states are mixed coherently or incoherently. The proper tool for the description of both is the *density matrix*.

Let the description of a system be in terms of quantum numbers α; that is, we use an α representation. If the system is in a pure state, it is chartacterized by a vector

$$\Psi = \int d\alpha f(\alpha)\Psi_\alpha \qquad \|\Psi\|^2 = \int d\alpha |f(\alpha)|^2 = 1$$

on the assumption that

$$(\Psi_\alpha, \Psi_\beta) = \delta(\alpha - \beta).$$

The expectation value of any operator A is then obtained as

$$\langle A \rangle = (\Psi, A\Psi) = \int d\alpha \int d\alpha' f^*(\alpha) f(\alpha') A_{\alpha\alpha'}$$

where

$$A_{\alpha\alpha'} = (\Psi_\alpha, A\Psi_{\alpha'}).$$

The vector Ψ, of course, describes not a single system but an ensemble. The statement that the ensemble is in a pure state, or is characterized by a state vector, is equivalent to saying that all the systems in the ensemble are identically prepared with, literally, as much care as possible. In other words, our knowledge of the ensemble is *maximal*.[3]

If our knowledge of the ensemble is less than maximal, then we can say only that it consists of a mixture of subensembles, each characterized by a distinct state vector

$$\Psi_n = \int d\alpha f_n(\alpha)\Psi_\alpha \tag{8.22}$$

with

$$\|\Psi_n\|^2 = \int d\alpha |f_n(\alpha)|^2 = 1 \qquad |(\Psi_n, \Psi_m)| = 0 \qquad n \neq m \tag{8.23}$$

and present in proportion to P_n. That is, P_n is the probability of picking a member of the nth subensemble from among the whole. The numbers P_n must thus be such that

$$0 \leq P_n \leq 1 \qquad \sum_n P_n = 1.$$

3 I do not mean to imply here any element of subjectivity. The convenience of using the word *knowledge* in quantum mechanics has often been misinterpreted by the philosophically inclined. It would be much less ambiguous always to refer to the *preparation* of the state of a physical system. A pure state is prepared with maximal care in the sense that all values of a complete set of commuting observables have been fixed. Each system of the ensemble is then known to us as completely as quantum mechanics allows, and the ensemble is described by a *state vector*.

The expectation value of the variable A is now given by

$$\begin{aligned}\langle A\rangle &= \sum_n P_n(\Psi_n, A\Psi_n)\\ &= \sum_n P_n \int d\alpha \int d\alpha' f_n^*(\alpha) f_n(\alpha') A_{\alpha\alpha'}.\end{aligned}$$

This is conveniently expressed in terms of the *density operator*

$$\rho = \sum P_n Q_n \tag{8.24}$$

where

$$Q_n = \Psi_n \Psi_n^\dagger = \int d\alpha \int d\alpha' f_n^*(\alpha) f_n(\alpha') \Psi_{\alpha'} \Psi_\alpha^\dagger \tag{8.25}$$

is the orthogonal projection onto the state of the nth subensemble. Using ρ, we may write for the expectation value

$$\langle A\rangle = \operatorname{tr} \rho A. \tag{8.26}$$

Because

$$\operatorname{tr} Q_n = \int d\alpha |f_n(\alpha)|^2 = 1$$

we have

$$\operatorname{tr} \rho = 1. \tag{8.27}$$

The evaluation of the trace may be done in the α representation

$$\langle A\rangle = \sum_{\alpha,\alpha'} \rho_{\alpha\alpha'} A_{\alpha'\alpha} \tag{8.26a}$$

where

$$\begin{aligned}\rho_{\alpha\alpha'} &= (\Psi_\alpha, \rho\Psi_{\alpha'})\\ &= \sum_n P_n f_n(\alpha) f_n^*(\alpha')\end{aligned} \tag{8.28}$$

is the *density matrix*. The trace being invariant under a basis transformation, it may, of course, be calculated equally well in any other representation. The density matrix is hermitian and positive semidefinite and can therefore always be diagonalized by a unitary transformation; its eigenvalues are all nonnegative.

Similarly we may determine the probability of finding the value b upon measurement of the dynamical variable B by using the normalized eigenstate Φ_b of B with the eigenvalue b,

$$\begin{aligned}P_b &= \sum_n P_n |(\Phi_b, \Psi_n)|^2 = \sum_n P_n (\Phi_b, \Psi_n)(\Psi_n, \Phi_b)\\ &= (\Phi_b, \rho\Phi_b) = \langle\rho\rangle_b.\end{aligned} \tag{8.29}$$

This result implies (8.26), because it follows from (8.29) that the expectation value of B is

$$\langle B\rangle = \sum_b bP_b = \sum_b (\Phi_b, \rho B\Phi_b) = \operatorname{tr}\rho B.$$

In the Schrödinger picture the state vectors of the subensembles depend on the time; hence so does the density operator. Its equation of motion is obtained immediately from the Schrödinger equation for Ψ_n,

$$\frac{d}{dt}\rho = \frac{\partial}{\partial t}\rho + i[\rho, H] \tag{8.30}$$

where the partial derivative refers to the possibility that the probabilities P_n depend on the time,

$$\frac{\partial\rho}{\partial t} = \sum Q_n \frac{\partial P_n}{\partial t}.$$

Now by (8.23)

$$\operatorname{tr} Q_n Q_m = \delta_{nm} \tag{8.31}$$

and therefore

$$\operatorname{tr}\rho^2 = \sum_{n,m} P_n P_m \operatorname{tr} Q_n Q_m = \sum_n P_n^2 \leq 1$$

with

$$\operatorname{tr}\rho^2 = 1 \tag{8.32}$$

if and only if all but one P_n are zero. This is the case of the *pure state*. If the ensemble is in a *mixed state*, then

$$\operatorname{tr}\rho^2 < 1. \tag{8.32a}$$

Particle Beams. A beam of particles may be considered a statistical ensemble. It is therefore described most compactly by a density operator. On the assumption first that no quantum numbers other than energy are needed, the beam is represented at a fixed time $t = 0$ by

$$\rho = \sum_n P_n \int_0^\infty dE \int_0^\infty dE' f_n(E) f_n^*(E') \Psi(E) \Psi^\dagger(E')$$

and therefore at the time t

$$\rho(t) = \sum_n P_n \int_0^\infty dE \int_0^\infty dE' f_n(E) f_n^*(E') e^{-i(E-E')t} \Psi(E) \Psi^\dagger(E'). \tag{8.33}$$

The functions $f(E)$ may represent the fact that the particles were emitted by a system of finite lifetime and are therefore not monoenergetic. But this does not necessarily put them into a mixed state. The most common method by which a particle is produced in a mixed state is by being produced together with, or correlated with, another one. Suppose that a *pair* of particles is in a pure state of given total energy. Then the density operator of the pair is given by the expression

$$\rho_{12} = \int_0^E dE' \int_0^E dE'' f(E')f^*(E'')[\Psi_1(E')\Psi_1^\dagger(E'')] \times [\Psi_2(E - E')\Psi_2^\dagger(E - E'')].$$

If measurements are now performed on particle 1 alone, without regard to particle 2, which may in fact be far away, then all traces of operators on particle 1 and expectation values in states of particle 1 will collapse the part of ρ_{12} that refers to particle 2. The effective density operator of particle 1 is therefore the partial trace of ρ_{12} with respect to the states of particle 2,

$$\rho_1 = \mathrm{tr}_2\, \rho_{12} = \int_0^E dE' \, |f(E')|^2 \Psi_1(E')\Psi_1^\dagger(E')$$

because of the orthogonality of the Ψ_2 at different energies. The density operator of particle 1 is now that of a mixed state. This result is due to the loss of information inherent in disregarding the second particle.

A simple example is the emission of a particle in a decay process. If the source is stationary, then the emitted particle may be in a pure state, even though its energy is not sharp. But if the source *recoils*, then the emitted particle alone is necessarily in a mixed state, even if it and the source together are in a pure state. Electrons emitted in β decay are always in mixed states because of the associated neutrino, even if the emitting nucleus is considered infinitely massive.

For emitted particles to be in a pure state it is necessary, of course, for the source to have been in a pure state. This is generally not the case, for example, in ordinary emission of light from an atom or of α particles from a nucleus. Only if the atoms are in a crystal lattice and phase-correlated can they be described by a wave function. An important example of this is a laser.

A beam of particles in a pure state is also called *coherent*; in a mixed state, *incoherent*. There are clearly different degrees of incoherence. If P_n is close to unity for a single n and small for the others, then the beam is nearly coherent. If P_n is equally small for all n (and the Ψ_n are all mutually orthogonal), then it is as incoherent as possible. The value of $\mathrm{tr}\,\rho^2$ may be taken as a measure of the coherence, though, depending on the quantum numbers, it need not be zero when the incoherence is maximal.

If other quantum numbers in addition to the energy and the momentum are involved, then the density matrix may be discrete and may have to be used even for monoenergetic beams. A particularly important case is that of particles with *spin*.

8.3.2 Particles with Spin. Suppose that the particles under consideration have spin s, so that their spin functions are χ_ν, with

$$\mathbf{S}^2\chi_\nu = s(s+1)\chi_\nu \tag{8.34}$$

$$S_z\chi_\nu = \nu\chi_\nu, \tag{8.35}$$

S being the spin operator. If the particle at rest is described by χ, and there exists a coordinate system $\mathscr{S}_0$ such that χ is an eigenstate of the z component of the spin, with eigenvalue $\nu \geq 0$, then

$$\chi = \chi_\nu^{(0)}$$
$$S_z^{(0)}\chi_\nu^{(0)} = \nu\chi_\nu^{(0)}.$$

The direction of the z axis of this coordinate system is what we mean by the *direction of the spin* of the particle. In any other coordinate system $\mathscr{S}$, in which the particle is at rest (or moving slowly), the components of the spin operator are given by

$$S_i = \sum_j T_{ij} S_j^{(0)} \tag{8.36}$$

in terms of those in $\mathscr{S}_0$, where T_{ij} is the three-dimensional rotation matrix relating $\mathscr{S}$ to $\mathscr{S}_0$. The expectation value of S_i in the state χ is therefore

$$\begin{aligned} \langle S_i \rangle = (\chi, S_i \chi) &= \sum_j T_{ij}(\chi_\nu^{(0)}, S_j^{(0)}\chi_\nu^{(0)}) \\ &= \nu T_{i3} \end{aligned}$$

because

$$S_x = \frac{S_+ + S_-}{2} \qquad S_y = \frac{S_+ - S_-}{2i} \tag{8.37}$$

and

$$S_\pm \chi_\nu = [s(s+1) - \nu(\nu \pm 1)]^{1/2} \chi_{\nu \pm 1}. \tag{8.38}$$

As a result we find that, when the particle is in such a pure state, then

$$\mathbf{n} \equiv \frac{\langle \mathbf{S} \rangle}{\nu} = (\cos\varphi \sin\theta, \sin\varphi \sin\theta, \cos\theta), \tag{8.39}$$

φ and θ being the polar angles of the z axis of $\mathscr{S}_0$ in $\mathscr{S}$. The unit vector **n** is thus the particle's *spin direction*. But this is a special situation.

Now in an arbitrary rest-coordinate system let χ be described by the expansion on the basis of the eigenfunctions of $S_3 = S_z$

$$\chi = \sum_\mu a_\mu \chi_\mu \qquad \sum_\mu |a_\mu|^2 = 1. \tag{8.40}$$

In this system

$$\langle S_z \rangle = \sum_\mu \mu |a_\mu|^2 \tag{8.41}$$

$$\langle S_+ \rangle = \sum_\mu a_\mu a_{\mu+1}^* (s - \mu)^{1/2} (s + \mu + 1)^{1/2} = \langle S_- \rangle^*. \tag{8.42}$$

The spin direction is then calculated from

$$\cos\theta = \langle S_z \rangle [|\langle S_z \rangle|^2 + |\langle S_+ \rangle|^2]^{-1/2} \tag{8.43}$$

$$e^{2i\varphi} = \frac{\langle S_+ \rangle}{\langle S_- \rangle} \tag{8.44}$$

In the evaluation of the right-hand sides of (8.43) and (8.44), the a_μ need not be normalized to unity.

For spin $\frac{1}{2}$, we may write

$$\chi = \begin{pmatrix} a_+ \\ a_- \end{pmatrix} \qquad |a_+|^2 + |a_-|^2 = 1.$$

Then

$$\langle S_z \rangle = \tfrac{1}{2}(|a_+|^2 - |a_-|^2) \tag{8.45}$$

$$\langle S_+ \rangle = a_- a_+^* \tag{8.46}$$

and, of course,

$$|\langle S_z \rangle|^2 + |\langle S_+ \rangle|^2 = \tfrac{1}{4}.$$

Hence

$$\cos\theta = |a_+|^2 - |a_-|^2 \tag{8.47}$$

$$e^{2i\varphi} = \frac{a_-/a_-^*}{a_+/a_+^*}. \tag{8.48}$$

Thus φ is the phase of a_- relative to a_+.

The spin density operator for a particle of spin s may be written

$$\rho = \sum_n P_n \xi_n \xi_n^\dagger$$

if ξ_n, $n = 1, \ldots, 2s + 1$, are the eigenstates of ρ, and $0 \leq P_n \leq 1$ are the corresponding eigenvalues. We may expand ξ_n on the basis of the eigenstates χ_μ of S_z as in (8.40),

$$\xi_n = \sum_\mu a_\mu^{(n)} \chi_\mu.$$

Then

$$\rho = \sum_{\mu,\nu} \rho_{\mu\nu} \chi_\mu \chi_\nu^\dagger$$

and

$$\rho_{\mu\nu} = \sum_n P_n a_\mu^{(n)} a_\nu^{(n)*} = \langle \chi_\nu \chi_\mu^\dagger \rangle \tag{8.49}$$

is the density matrix on the basis of eigenfunction of S_z. The spin direction can now be calculated by means of (8.43) and (8.44), using

$$\langle S_z \rangle = \operatorname{tr} \rho\, S_z = \sum_\mu \mu \rho_{\mu\mu}, \tag{8.50}$$

$$\langle S_+ \rangle = \sum_\mu [(s - \mu)(s + \mu + 1)]^{1/2} \rho_{\mu\mu+1}. \tag{8.51}$$

In the special case in which for some choice of $\mathbf{n}$ ρ commutes with $\mathbf{S}\cdot\mathbf{n}$ we have on the basis of the eigenfunctions of this $\mathbf{S}\cdot\mathbf{n}$, $\rho_{\mu\nu} = \delta_{\mu\nu}P_\mu$ and

$$\begin{aligned} \rho &= \sum_{\mu=-s}^{s} P_\mu \chi_\mu \chi_\mu^\dagger \\ &= \sum_{\mu=-s}^{s} P_\mu \prod_{\nu\neq\mu} \frac{\mathbf{S}\cdot\mathbf{n} - \nu}{\mu - \nu}. \end{aligned} \tag{8.52}$$

The $(2s + 1)$ numbers P_μ may in this case be calculated by measuring the $2s$ quantities $s^{(n)} = \langle(\mathbf{S}\cdot\mathbf{n})^n\rangle$, $n = 1, \ldots, 2s$, and solving the $2s + 1$ equations

$$s^{(n)} = \sum_{\mu=-s}^{s} \mu^n P_\mu, \qquad n = 0, \ldots, 2s, \tag{8.53}$$

for P_μ.

In the general case it is also possible[4] to express the density matrix in terms of directly measurable quantities. These may be either expectation values of the components of multipole moments, i.e., of products of S_x, S_y, and S_z, of order less than or equal to $2s$, or alternatively, in terms of the relative weights of eigenvalues of dipole quantities. For $s = \frac{1}{2}$ (8.52) expresses the general case.

Spin $\frac{1}{2}$. The simplest nontrivial example is that of a spin $\frac{1}{2}$ particle. In this case we need only

$$s^{(1)} = \langle \mathbf{S}\cdot\mathbf{n}\rangle = \tfrac{1}{2}(P_{+1/2} - P_{-1/2})$$

and according to (8.52) the density operator is given by

$$\rho = P_{1/2}(\tfrac{1}{2} + \mathbf{S}\cdot\mathbf{n}) + P_{-1/2}(\tfrac{1}{2} - \mathbf{S}\cdot\mathbf{n})$$

or

$$\rho = \tfrac{1}{2}(\mathbb{1} + \mathbf{P}\cdot\boldsymbol{\sigma}). \tag{8.54}$$

In this expression

$$\boldsymbol{\sigma} = 2\mathbf{S}$$

is the set of Pauli spin matrices, and

$$\mathbf{P} = 2s^{(1)}\mathbf{n} = \langle\boldsymbol{\sigma}\rangle = \frac{P_{1/2} - P_{-1/2}}{P_{1/2} + P_{-1/2}}\,\mathbf{n} \tag{8.55}$$

is the *polarization vector*, whose length is the degree of polarization and whose direction is the polarization direction. Equation (8.54) implies that the degree of polarization $p = |\mathbf{P}|$ is directly related to the density matrix by

$$\rho^2 - \rho = \tfrac{1}{4}(P^2 - 1)\mathbb{1}. \tag{8.56}$$

4 Newton and Young (1968).

This implies that the eigenvalues of ρ are

$$\rho' = \tfrac{1}{2}(1 \pm P)$$

and consequently

$$\det \rho = \tfrac{1}{4}(1 - P^2).$$

The degree of polarization can thus be obtained from

$$P^2 = 1 - 4 \det \rho. \tag{8.57}$$

Equation (8.56) also shows explicitly the significance of $\operatorname{tr} \rho^2$ in terms of the polarization,

$$\operatorname{tr} \rho^2 = \tfrac{1}{2}(1 + P^2). \tag{8.58}$$

Example of Reduction of Pure State of Two Particles to Mixed State of One. As an example of the reduction of a pure state of two particles to a mixed state of one, we may consider that in which two spin $\frac{1}{2}$ particles are produced in a pure state of spin 0. The two-particle spin 0 (singlet) wave function is

$$\chi_0 = 2^{-1/2}(\chi_-^{(1)}\chi_+^{(2)} - \chi_+^{(1)}\chi_-^{(2)}).$$

Hence the spin-density operator for the pair is

$$\begin{aligned}\rho_{12} = \tfrac{1}{2}[(\chi_-^{(1)}\chi_-^{(1)\dagger})(\chi_+^{(2)}\chi_+^{(2)\dagger}) + (\chi_+^{(1)}\chi_+^{(1)\dagger})(\chi_-^{(2)}\chi_-^{(2)\dagger}) \\ - (\chi_-^{(1)}\chi_+^{(1)\dagger})(\chi_+^{(2)}\chi_-^{(2)\dagger}) - (\chi_+^{(1)}\chi_-^{(1)\dagger})(\chi_-^{(2)}\chi_+^{(2)\dagger})].\end{aligned}$$

If particle 2 is ignored, we need the density operator for particle 1 alone, which is obtained by taking the partial trace with respect to particle 2,

$$\rho_1 = \tfrac{1}{2}(\chi_-\chi_-^\dagger + \chi_+\chi_+^\dagger) = \tfrac{1}{2}\mathbb{1}.$$

The beam of particles 1 so obtained is therefore completely unpolarized or incoherent. The same is found to be true if the pair is produced in an incoherent triplet state, with equal probability for all three projections.

8.3.3 The Cross Section and the Density Matrix of the Scattered Wave. If the incident beam in a scattering experiment is in a mixed state described by the density matrix ρ_{inc}, then (8.29) shows that the probability (8.7) becomes

$$P = \cdots \int_C d\hat{\mathbf{p}} \, \langle T\rho_{\text{inc}} T^\dagger \rangle_f, \tag{8.59}$$

the expectation value being taken in the final state. This tells us at the same time that the density matrix of the scattered wave is given by

$$\rho_{\text{scatt}} = \frac{T\rho_{\text{inc}} T^\dagger}{\operatorname{tr}' \langle T\rho_{\text{inc}} T^\dagger \rangle_f}. \tag{8.60}$$

The trace in the denominator is only a partial one, in which the final directions of the particles are fixed. Only then can we speak of a beam scattered in a given direction whose density matrix is (8.60). This is indicated by the prime on the trace.

According to (8.14) the quantity

$$\sigma = \frac{(2\pi)^4}{p_i^2} \operatorname{tr}' \langle T\rho_{\text{inc}} T^\dagger \rangle_f \tag{8.61}$$

is the total differential cross section for scattering of the beam described by ρ_{inc} in the direction indicated by f; that is, all spin directions and other quantum numbers are summed over.

If the initial beam is coherent, so that all but one of the eigenvalues of ρ_{inc} are zero, then

$$\begin{aligned} \operatorname{tr}' \langle T\rho_{\text{inc}} T^\dagger T\rho_{\text{inc}} T^\dagger \rangle &= \sum_\alpha |T_{\alpha i}|^2 \sum_\beta |T_{\beta i}|^2 \\ &= (\operatorname{tr}' \langle T\rho_{\text{inc}} T^\dagger \rangle)^2. \end{aligned}$$

Consequently, by (8.60),

$$\operatorname{tr} \rho_{\text{scatt}}^2 = 1,$$

and the scattered wave is coherent too. The scattering cannot depolarize the particles. This is not to say, of course, that a beam of polarized missiles may not be depolarized by being scattered off an unpolarized target. Moreover, a mixed state of one (or more) of the final particles may be the result of ignoring others, as pointed out before.

The converse is not true. If for a given direction all but one of the **T**-matrix elements are zero then

$$\operatorname{tr} \rho_{\text{scatt}}^2 = 1$$

and the scattered wave is coherent, or 100 percent polarized, no matter how incoherent the incident beam is. A familiar example (discussed in Sec. 3.1) is the polarization of incoherent light scattered at right angles by small particles.

If the incident beam is unpolarized, or incoherent, then ρ_{inc} is a multiple of the unit matrix. Hence in that instance

$$\rho_{\text{scatt}} = \frac{TT^\dagger}{\operatorname{tr}' \langle TT^\dagger \rangle_f}. \tag{8.60a}$$

Notes and References

8.1 For relevant discussions in the literature, see C. Møller (1945); Jauch and Rohrlich (1950); Lippmann and Schwinger (1950); Gell-Mann and Goldberger (1953); Brenig and Haag (1959); S. S. Schweber (1961). The arguments of this section largely follow Newton and Shtokhamer (1974), where they are also generalized to N particles. For scattering from many centers, see also Agassi and Gal (1973a); S. Ström (1976).

For a discussion of phase-space integrals, see A. Krzywicki (1965).

8.2 The primary reference, on which all others are based, is C. Møller (1945).

For a general review of relativistic particle kinematics, see R. Hagedorn (1963).

8.3.1 The density matrix was conceived by J. von Neumann (1927).

For an extensive review and many references, see U. Fano (1957); R. McWeeny (1960); D. ter Haar (1961).

For discussions of coherence in quantum mechanics, especially with regard to laser beams, see R. J. Glauber (1963, 1965, and 1966); Mandel and Wolf (1963); E. C. G. Sudarshan (1963); Titulaer and Glauber (1965 and 1966).

8.3.2 For discussions of spin and polarization of electrons, see M. E. Rose (1961) and, in more detail, H. A. Tolhoek (1956). A more general treatment, applicable to other values of the spins, is W. H. McMaster (1961). Methods by which the spin density matrix of a beam of particles of arbitrary spin can be determined from measurements are discussed by Newton and Young (1968).

A covariant polarization matrix for spin 1 has been discussed by D. Zwanziger (1964). See also C. B. van Wyck (1958).

8.3.3 The density matrix of scattered particles is discussed by R. G. Newton (1979b).

For a somewhat different kind of analysis of collision experiments and their information content see Alhassid and Levine (1978); Levine and Alhassid (1979).

Problems

1 Prove (8.17) and (8.18).

2 Show that for two particles (8.15) reduces to (8.9). Calculate the relativistic phase-space factor for a particle of zero mass and another of mass m.

3 Compare the nonrelativistic phase-space factor for the emission of two particles with that for the emission of three, in which the third particle is unobserved. Assume that the **T**-matrix element is independent of the energies of the emitted particles.

4 (*a*) Suppose that an infinitely massive excited atom of angular momentum zero emits a photon. Assuming that the atom was originally in a pure state, is the photon in a pure state? Does the state of the photon depend on whether the angular momentum of the residual atom was measured or not? Does the state of the combined system atom + photon depend on it? (*b*) Suppose that the atom has one unit of angular momentum and it is 100 percent "polarized" by a magnetic field. In the course of a transition to the ground state of zero angular momentum it emits a photon. Is the photon in a pure state? Explain and discuss.

5 Suppose that a system has N possible states. What is the value of $\operatorname{tr} \rho^2$ when the system is in as incoherent a state as possible? Consequently define the degree of polarization of such a system in terms of the value of $\operatorname{tr} \rho^2$. Compare with (8.58).

6 Prove Eq. (8.38).

7 Suppose that a pair of particles of spin $\frac{1}{2}$ is produced in a pure triplet state. What is the spin-density matrix for one of the particles alone?

8 Particles of spin $\frac{1}{2}$ in a 100 percent polarized beam decay each into a pair of particles of spin 1 and spin $\frac{1}{2}$. What are the spin-density matrices for each of the decay products alone?

CHAPTER 9

Formal Methods of Solution and Approximations

We shall now discuss formally several standard methods of solving the scattering problem. These amount to procedures for constructing the complete Green's function, or the resolvent. Although they are in principle exact, they give rise in a natural manner to approximation methods.

9.1 Perturbation Theory

9.1.1 The Born Series. For the present purpose it is convenient to introduce a parameter γ which multiplies the interaction H'. Its physical significance is therefore that of the interaction strength, or of the *coupling constant*. The Lippmann–Schwinger equation then reads

$$\Psi = \Psi_0 + \gamma G H' \Psi \tag{9.1}$$

or

$$(\mathbb{1} - \gamma K)\Psi = \Psi_0$$

where the kernel is defined by

$$K = GH'. \tag{9.2}$$

Similarly for the T operator,

$$T(\mathbb{1} - \gamma K) = \gamma H'.$$

The solution of (9.1) is therefore equivalent to the construction of the operator

$$(\mathbb{1} - \gamma K)^{-1}.$$

The simplest method of constructing $(\mathbb{1} - \gamma K)^{-1}$ is that of expanding it in a *power series* in γ,

$$(\mathbb{1} - \gamma K)^{-1} = \sum_{n=0}^{\infty} \gamma^n K^n. \tag{9.3}$$

For $\gamma = 1$ this is the *Born series.*[1] The question is under what conditions such an expansion is legitimate.

We must immediately recognize that we cannot restrict ourselves to real values of γ when investigating the convergence properties of (9.3), even though physically only real γ can be meaningful. The convergence of the expansion of a function in a power series of some real variable depends in an essential way upon the analytic behavior of that function when the variable takes on *complex* values. Now notice that the replacement of γ by $1/\alpha$ converts the operator $(\mathbb{1} - \gamma K)^{-1}$ into the resolvent of K, whose behavior as a function of α was discussed in some detail in Sec. 7.3.

If K is *bounded*, then its spectrum is necessarily confined to a finite region in the complex α plane. Let us define

$$r_\sigma(K) = \sup |\alpha| \leq \|K\| \tag{9.4}$$

for all α in the spectrum of K. The positive number $r_\sigma(K)$ is called the *spectral radius* of K. It follows from the discussion in Sec. 7.3 that $(\mathbb{1} - \gamma K)^{-1}$ is an analytic function of γ for all $|\gamma| < 1/r_\sigma(K)$, and it can be represented there by the series[2] (9.3) which converges for all $|\gamma| < 1/r_\sigma(K)$. This statement, in the context of an expansion of an analytic operator-valued function, means that the remainder of the series tends to zero *in the norm*. It is, in other words, *uniform* with respect to the choice of different (normalized) vectors on which the operator acts. This implies that, if a normalized basis is used (in contrast to a continuous one), the convergence is uniform with respect to different matrix elements. For $|\gamma| > 1/r_\sigma(K)$ the series diverges.

The discussion is simplest if we assume that K is *compact*. Then the radius of convergence of the series (9.3) is determined by the eigenvalues of K with largest magnitude, i.e., by the characteristic values with the smallest magnitude. (In contradistinction to an eigenvalue, γ is called a *characteristic* value of the integral kernel K if there exists a normalizable Φ such that $\gamma K \Phi = \Phi$.) There are always a finite number of these (of equal magnitude, that is), and we may consider it an accident if there is more than one. Since r_σ is finite (K being compact, it is necessarily bounded) the radius of convergence of (9.3) is always nonzero, i.e., the series converges if only $|\gamma|$ is small enough. On the other hand, if (9.3) converges for all finite γ,

1 The *Born series* is also referred to as the *Neumann series*.

2 Taylor (1958), p. 262.

then it is an entire analytic function of γ and the spectrum of K (which cannot be empty if K is bounded and everywhere defined[3]) consists of the point $\alpha = 0$ only.

Modified Kernels. In specific cases to be considered in more detail later, the operator K defined in (9.2) is not compact, but we may get a compact kernel by investigating, instead of (9.1), the equation

$$\Phi = \Phi_0 + \gamma \mathfrak{K} \Phi \tag{9.5}$$

where

$$\begin{aligned} \Phi &= H'^{1/2} \Psi \\ \mathfrak{K} &= H'^{1/2} G(E) H'^{1/2} = H'^{1/2} K H'^{-1/2}. \end{aligned} \tag{9.6}$$

The square root of the interaction is not well defined, it is true, unless H' is positive definite. But this does not cause any serious difficulties. If H' is not positive definite, then an operator $H'^{1/2}$ such that

$$H'^{1/2} H'^{1/2} = H'$$

still *exists* and we simply pick one of them. (Of course, we assume that H' is hermitian; $H'^{1/2}$, however, is hermitian only if H' is positive definite.) In many important instances, $\mathfrak{K}$ is in the Hilbert–Schmidt class,[4] but K is not. If both H' and H'^{-1} are bounded operators, then, of course, the spectra of K and $\mathfrak{K}$ are the same. Henceforth we shall simply consider K, but the discussion may easily be shifted to $\mathfrak{K}$ when necessary.

A simple variant of the use of (9.5) and (9.6) is applicable if H' is a local potential,

$$H' = V(\mathbf{r}) = |V(\mathbf{r})| v(\mathbf{r}),$$

$v(\mathbf{r})$ being a sign factor equal to $+1$ where V is positive, -1 where V is negative. Then K is a "polar" kernel. We multiply Ψ by $|V(r)|^{1/2}$ and use the new kernel

$$\mathscr{K} = |V|^{1/2} G |V|^{1/2} v. \tag{9.6a}$$

This has all the advantages of $\mathfrak{K}$ and avoids the ambiguity of $H'^{1/2}$ if V changes sign. At the same time the mathematical theory of polar kernels is well established and can be used for the study of $\mathscr{K}$.[5]

Compact Kernels and Their Eigenvalues. On the assumption now that K is compact, the radius of convergence of (9.3) is simply the characteristic value of K with the smallest magnitude. Since the kernel $K = G(E)H'$ depends on the energy E, so, in general, will each of its characteristic values and each of its eigenvalues. Every eigenvalue $\alpha(E)$ of K signifies at the same time a pole of the resolvent $(E - H_0 - \gamma H')^{-1}$ of $H_0 + \gamma H'$ for $\gamma = 1/\alpha$. As E varies from $-\infty$ to $+\infty$, each eigenvalue describes a "trajectory" in

3 *Ibid.*, p. 261.
4 Defined in Sec. 7.3.
5 See K. Meetz (1961).

the complex α plane (and each characteristic value, in the complex γ plane.) Since

$$G(E) = (E - H_0)^{-1} \to 0 \quad \text{as } E \to -\infty \tag{9.7}$$

each eigenvalue of $K(E)$ tends to zero in that limit and they all do so uniformly. Furthermore, when E is negative, then $-G$ is positive definite and bounded and the eigenvalues of K are the same as those of the hermitian operator,

$$\begin{aligned} \mathsf{K} &= -(-G)^{1/2} H'(-G)^{1/2} \\ &= -(-G)^{-1/2} K(-G)^{1/2}. \end{aligned} \tag{9.8}$$

Thus all its eigenvalues are *real* when $E < 0$. Let us differentiate the equation

$$(\Phi(E), [(E - H_0)\alpha(E) - H']\Phi(E)) = 0 \tag{9.9}$$

with respect to E, $\Phi(E)$ being an eigenvector of K with the eigenvalue $\alpha(E)$. We then find that

$$\frac{1}{\alpha}\frac{d\alpha}{dE} = \frac{1}{\langle H_0 - E\rangle} > 0 \tag{9.10}$$

with the expectation value on the right evaluated in the state $\Phi(E)$. This shows that, when the energy increases from $-\infty$ to 0, *each positive eigenvalve $\alpha(E)$ moves toward the right and each negative one toward the left.* If H' is positive semidefinite (everywhere repulsive interaction), then so is $-\mathsf{K}$ when $E < 0$; thus every nonzero $\alpha(E)$ must be negative and move toward the left as E increases. Conversely, if $-H'$ is positive semidefinite (everywhere attractive), then so is K when $E < 0$; thus every nonzero $\alpha(E)$ must be positive and move toward the right. These statements, of course, merely reiterate the physically obvious fact that an everywhere repulsive interaction cannot produce bound states and, when the strength of an everywhere attractive one is increased, the binding energy increases (Remember that the strength parameter is $\gamma = 1/\alpha$.) One may therefore call the negative eigenvalues of K (for $E < 0$) repulsive and the positive ones attractive. If H' is neither positive nor negative semidefinite, then K must have both repulsive and attractive eigenvalues.

If one of the eigenvalues $\alpha(E)$ passes through unity, at some energy $-|E_0|$, then the hamiltonian $H = H_0 + H'$ has a bound state of binding energy $|E_0|$. As E increases, $\alpha(E)$ moves farther toward the right. Thus the number of eigenvalues $\alpha(0)$ of K which at $E = 0$ are greater than 1 is equal to the number of bound states produced by $H = H_0 + H'$; and the number of eigenvalues $\alpha(0)$ which at $E = 0$ are less than -1 is equal to the number of bound states produced by $H = H_0 - H'$.

When E passes through zero, the eigenvalues $\alpha(E)$ leave the real axis. Because $G(E)$ has a branch point at $E = 0$, so does $\alpha(E)$ and we must carefully go around it. When E moves off the negative real axis into the upper half plane, then (9.10) tells us that α moves into the upper half plane if on the right and into the lower if on the left. Once E is complex, (9.10) no longer holds, but then α cannot be real, because, for real α, $H' + \alpha H_0$ is hermitian and thus cannot have the complex eigenvalue αE. Consequently, as E circles the branch point $E = 0$ via the upper half of the complex plane, the attractive α's move in the upper half plane, the repulsive ones in the lower. That must then still be true when E reaches the positive real axis, so that we are on the upper rim of the cut of $G(E)$, where $K = G^+(E)H'$. It is possible for an α to move to the real axis at a positive energy. But so long as E remains real, it cannot traverse it, because any α in the wrong half plane would have to cross the real axis going back into its "natural" half plane as soon as E moves off the real axis to $E + i\varepsilon$.

Suppose that an eigenvalue α arrives just to the left of unity at $E = 0$ and that as E increases it still moves to the right. Then it will pass close to unity, slightly above it in the complex plane. In other words, when the real part of α equals 1, its imaginary part is small and positive. We then expect that, by giving E a suitable small imaginary part, we can maneuver $\alpha(E)$ to 1. The imaginary part of E necessary to accomplish this must be *negative*, for if it were positive, then we would have a complex eigenvalue of the hermitian operator H. If Im $E < 0$, on the other hand, then E must have moved through the cut of $(E - H_0 - \gamma H')^{-1}$ that runs along the continuous spectrum, and, on this second sheet of its Riemann surface, poles of the analytically continued resolvent, or eigenvalues $\alpha(E)$ of K, do not signify bound states of $H_0 + \gamma H'$. Instead, such poles on the second sheet, if sufficiently close to the real axis, cause *resonances* in the scattering, as will be shown in detail in Sec. 16.6. The physical significance, therefore, of the passing of an $\alpha(E)$ near unity, or of its crossing of the unit circle, with a small imaginary part, is the occurrence of a resonance. It will surely happen whenever $\alpha(0)$ is almost 1, and $\alpha(E)$ continues to move to the right with increasing energy.

Now in the case of scattering of two particles interacting via a "reasonable" spherically symmetric potential we shall see later (see Sec. 12.2) that $\alpha(E)$ continues to move to the right as E passes through zero if it is an eigenvalue of K of angular momentum larger than 0 but that it moves into the complex plane *at right angles* if the angular momentum is 0. Consequently we expect that, if there is an "almost bound" state of angular momentum 1 or higher, then there is a low-energy resonance; but if there is an "almost bound" s state, then no such low-energy resonance occurs. This well-known effect is of course easily explained on physical grounds. It is caused by the fact that the combination of an attractive potential with the centrifugal force produced by the angular momentum results in a barrier that can support an unstable state in which the particle can leak out only through the tunnel effect. In an s state the centrifugal barrier is absent and thus cannot produce an unstable state.

The question of where $\alpha(E)$ moves as the energy increases to $+\infty$ cannot be answered easily in general. If we allow the phase of E to be nonzero, arg $E \geq \varepsilon > 0$, then of course $\|K(E)\| \to 0$ as $|E| \to \infty$ and consequently all its eigenvalues $\alpha(E)$ must uniformly move to the origin. However, as $\varepsilon \to 0$, an α trajectory may, for example, osculate the imaginary axis more and more closely and, in the limit, "hang" at a finite imaginary value. At the present general level we cannot rule out such behavior. In the case of a well-behaved local spherically symmetric potential we shall see later that it cannot occur and that all $\alpha(E)$ tend to zero. The α trajectories are then closed as E changes from $-\infty$ to $+\infty$. What is more, they tend to zero *uniformly* and the spectral radius vanishes as $E \to +\infty$.

Convergence of the Born Series. The question of the convergence of the Born expansion of the Green's function $\mathscr{G}(E)$ at the fixed energy E is now answered simply by *whether $K(E)$ has any eigenvalues $\alpha(E)$ outside the unit circle.* If not, then the radius of convergence of (9.3) is larger than 1, and the Born series converges. If there are eigenvalues α outside the unit circle, then the radius of convergence of (9.3) is less than 1 and the Born series diverges. The number of such eigenvalues $|\alpha| > 1$ is always finite. When the energy is negative, the criterion is simply that of the existence of bound states. Any eigenvalue $\alpha(E)$ whose magnitude is greater than unity at an energy $E < 0$ must have moved through $+1$ or -1 at some lower energy. Therefore, at $E_0 < 0$ the Born series of $\mathscr{G}$ converges if and only if neither $H_0 + H'$ nor $H_0 - H'$ has bound states of energy $E < E_0$. If neither $H_0 + H'$ nor $H_0 - H'$ produces any bound states, then the Born series converges at zero energy and even somewhat above.

For positive energy the situation is more complicated because eigenvalues $\alpha(E)$ of K may move out of the unit circle and back in. Certainly, if either $H_0 + H'$ or $H_0 - H'$ produces bound states, then there must exist an energy range above (and including) zero in which the Born series diverges. Similarly, if either $H_0 + H'$ or $H_0 - H'$ produces a sharp resonance at $E_0 > 0$, then there must be an energy range near (usually above) E_0 in which the Born series diverges. The repulsive eigenvalues are as important as the attractive ones, and both $H_0 + H'$ and $H_0 - H'$ must be examined for "resonances" (including wide ones). It must be particularly emphasized that there is no "experimental" way of answering the convergence question, in the sense of looking merely for bound states and resonances of H. Even if H has neither, the Born series may diverge, either because α has crossed the unit circle too far from the real axis to cause an observable resonance or because the experimentally inaccessible hamiltonian $H_0 - H'$ produces bound states or resonances.

If the eigenvalues $\alpha(E)$ vanish *uniformly* as $E \to \infty$, then there must exist an energy above which the Born series converges uniformly. As mentioned before, this is the case for well-behaved local spherically symmetric potentials.

The question of whether the Born series converges uniformly for *all* energies is answered by determining whether or not any α trajectory leaves the interior of the unit circle. If and only if none leads outside is the convergence uniform in E, from $E = -\infty$ to $E = +\infty$. In Sec. 10.3 we shall derive a simple *sufficient* criterion for the uniform convergence of the Born series for a local spherically symmetric potential $V(r)$, namely, that $-|V|$ should produce no s-wave bound states (and no zero-energy s-wave resonance). This criterion is, of course, also *necessary* if the potential is either everywhere positive or everywhere negative. Its relation to the test in terms of the eigenvalues α is not clear, but it leads to the conjecture, at present unproved, that in the local spherically symmetric potential case no trajectory of angular momentum greater than 0 can cross the "leading" s-wave trajectory.

When $E < 0$, then the operator K defined in (9.8) is hermitian and we have the sum rules

$$\sum_n [\alpha_n(E)]^N = \operatorname{tr}[\mathsf{K}(E)]^N = \operatorname{tr}[K(E)]^N \tag{9.11}$$

for all integers $N \geq 2$. Since $\operatorname{tr} KK^\dagger$ was assumed to be finite, the righthand side of (9.11) is finite. It may now be analytically continued to complex E and then to positive E on the upper rim of the cut of G, and it continues to hold. These sum rules are sometimes useful for estimating the positions of some of the eigenvalues $\alpha(E)$.

Analyticity of $\alpha(E)$. In the foregoing discussion it was taken for granted that the eigenvalue $\alpha(E)$ of the analytic function $K(E)$ is an analytic function, regular wherever $K(E)$ is regular, i.e., everywhere except on the positive real axis. As a matter of fact, this need not be true. In order to test whether or not $\alpha(E)$ has a finite derivative at some complex value of E, we take the inner product of the equation

$$(\alpha - K)\Phi_\alpha = 0$$

with $\overline{\Phi}_{\alpha^*}$, which satisfies

$$(\alpha^* - K^\dagger)\overline{\Phi}_{\alpha^*} = 0.$$

This inner product is a function of E alone, and not of its real and imaginary parts separately. For it follows from the fact that for real negative E

$$K^\dagger = G^{-1}KG \tag{9.12}$$

and from Schwarz's reflection principle that on the first sheet of the Riemann surface

$$K^\dagger(E^*) = G^{-1}(E)K(E)G(E) \tag{9.13}$$

and that there exists a branch of $\alpha(E)$ for which

$$\alpha^*(E) = \alpha(E^*). \tag{9.14}$$

We also have according to (9.12)

$$G^\dagger(E)\overline{\Phi}_{\alpha^*}(E) = \Phi_\alpha(E^*) \tag{9.15}$$

if

$$K(E^*)\Phi_\alpha(E^*) = \alpha^*\Phi_\alpha(E^*).$$

Therefore

$$\overline{\Phi}^{\dagger}_{\alpha*}(E) = \Phi^{\dagger}_{\alpha}(E^*)G^{-1}(E).$$

Now differentiate

$$(\overline{\Phi}_{\alpha*}, (\alpha - K)\Phi_{\alpha}) = 0$$

with respect to E. This gives

$$\frac{d\alpha}{dE}(\overline{\Phi}_{\alpha*}, \Phi_{\alpha}) = -(\overline{\Phi}_{\alpha*}, (E - H_0)^{-2}H'\Phi_{\alpha}). \tag{9.16}$$

The right-hand side is always finite (for complex E), but there is no guaranty that the inner product on the left does not vanish. Wherever $\overline{\Phi}_{\alpha*}$ is orthogonal to Φ_{α}, there $\alpha(E)$ may fail to be analytic.

Now we have seen at the end of Sec. 7.3.3 that an orthogonality of Φ_{α} and $\overline{\Phi}_{\alpha*}$ occurs wherever $(K - \alpha)^{-1}$ has a *double* pole. Since $\mathsf{K}(E)$ is hermitian for negative E, the poles of $(K - \alpha)^{-1}$ must be simple there. We may therefore assume that the poles remain simple for other values of E, save accidentally. Such accidents are clearly coincidences of two poles at a given energy, so that a double pole occurs.[6] This can be stated simply by saying that $\alpha(E)$ is analytic as a function of E, except where two trajectories cross (at the same energy). For a discussion of this from the point of view of the Fredholm determinant, see Sec. 9.3.

The function $\alpha(E)$ has the following properties, as we have seen:

1. For negative E, it is real.
2. When $\operatorname{Im} E > 0$ (< 0), then $\operatorname{Im} \alpha > 0$ (< 0) if α is an "attractive" eigenvalue and $\operatorname{Im} \alpha < 0$ (> 0) if it is "repulsive."
3. As $|E| \to \infty$ with $\varepsilon \leq \arg E \leq 2\pi - \varepsilon$, $\varepsilon > 0$, we have

$$E\alpha(E) \to \text{const} = C.$$

Property 3 follows from the fact that in that limit

$$\|E(E - H_0)^{-1}\| \to 1$$

if we assume that H' is bounded. If we now add the further *assumption* that

4. $\alpha(E)$ is everywhere analytic in the finite E plane cut along the positive real axis,

then it admits the *spectral representation*, or *dispersion relation*,

$$\alpha(E) = \int_0^{\infty} \frac{d\eta(E')}{E' - E}, \tag{9.17}$$

6 More exactly, the situation is this: If two trajectories cross at $E = E_0$ then either the eigenvalue α_0 of $K(E_0)$ is nondegenerate and $(K - \alpha)^{-1}$ has a double pole at $\alpha = \alpha_0$ or else the eigenvalue α_0 of $K(E_0)$ is doubly degenerate and $(K - \alpha)^{-1}$ has a simple pole at $\alpha = \alpha_0$. In the first instance $\alpha(E)$ has a branch point at $E = E_0$; there is no unique way of following a trajectory across the intersection. In the second instance there is no branch point; each trajectory can be identified by its eigenvector, since these do not coincide at $E = E_0$.

where $\eta(E')$ is real, bounded, and nondecreasing. For real positive E we get from this, in the limit as $\operatorname{Im} E \to 0+$,

$$\begin{aligned} 2 \operatorname{Im} \alpha(E) &= \alpha(E + io) - \alpha(E - io) = A(E) \\ &= 2\pi \frac{d\eta(E)}{dE} \geq 0. \end{aligned}$$

We also know that:

5. $\alpha(E + io)$ is continuous.

Hence we can write

$$\alpha(E) = \frac{1}{2\pi} \int_0^\infty \frac{dE'\,A(E')}{E' - E}. \tag{9.17a}$$

When E is negative, then G is hermitian and so is the operator K of (9.8). The orthonormal eigenvectors Ψ_α of K,

$$(\Psi_\alpha, \Psi_\beta) = \delta_{\alpha\beta}$$

differ from those of K by

$$\Phi_\alpha = (-G)^{1/2} \Psi_\alpha.$$

Since Φ_α is an eigenvector of GH', we have

$$(E - H_0)\Phi_\alpha = \frac{1}{\alpha} H'\Phi_\alpha \tag{9.18}$$

and hence

$$(\Phi_\alpha, H'\Phi_\beta) = -\alpha\delta_{\alpha\beta}. \tag{9.19}$$

The Ψ_α form a complete set,[7]

$$\sum_\alpha \Psi_\alpha \Psi_\alpha^\dagger = \mathbb{1}$$

and consequently so do the Φ_α. From (9.18) and (9.19) we get

$$\sum_\alpha \Phi_\alpha \Phi_\alpha^\dagger = -G \tag{9.20}$$

and by (9.15)

$$\sum_\alpha \Phi_\alpha \overline{\Phi}_\alpha^\dagger = -\mathbb{1}. \tag{9.21}$$

The fact that

$$\mathscr{G} = (\mathbb{1} - K)^{-1} G$$

allows us immediately to obtain a representation also of the complete Green's function from (9.20),

7 We use a simplified notation. The sum here runs over all the α_n, from $n = 1$ to ∞.

$$\mathscr{G} = -\sum_\alpha \frac{\Phi_\alpha \Phi_\alpha^\dagger}{1-\alpha}$$
$$= G - \sum_\alpha \frac{\alpha}{1-\alpha} \Phi_\alpha \Phi_\alpha^\dagger. \tag{9.22}$$

Now we have seen that $\Phi_\alpha(E)$ and $\bar{\Phi}_\alpha(E)$ can both be analytically continued off the negative real E axis. The same is of course true of $\Phi_\alpha{}^\dagger(E)$. Suppose that H is invariant under time reversal, so that it commutes with the time-reversal operator ϑ,

$$[\vartheta, H] = 0$$

as does H_0. Then for negative E

$$(K - \alpha)\vartheta\Phi_\alpha = 0.$$

If there is no degeneracy, it follows that

$$\vartheta\Phi_\alpha = \varphi_\alpha \Phi_\alpha \tag{9.23}$$

where φ_α is a phase factor that may be chosen to be unity;[8] if there is degeneracy, then we may choose the basis in the degenerate subspace so that

$$\vartheta\Phi_{\alpha i} = \Phi_{\alpha j}. \tag{9.23a}$$

But because of the property (7.70) of ϑ, we have

$$\vartheta\Psi = U\Psi^\dagger \tag{9.24}$$

where U is unitary, and consequently

$$\Phi_\alpha^\dagger = U^\dagger \Phi_\alpha. \tag{9.25}$$

The operator U can be chosen to be independent of the energy.[9]

The analytic continuation of Φ_α thus leads to an analytic continuation of $\Phi_\alpha{}^\dagger$, and Eqs. (9.20) to (9.22) can be continued to complex E, around $E = 0$, and to positive real E above or below the cut. Equation (9.22) may then also be used for the **T** matrix and yields

$$T_{\gamma\beta} = (\Psi_{0\gamma}, H'\Psi_{0\beta}) - \sum_\alpha \frac{(\Psi_{0\gamma}, H'\Phi_\alpha)(\vartheta\Psi_{0\beta}, H'\Phi_\alpha)}{1-\alpha} \tag{9.26}$$

where all quantities are evaluated at $E + i\varepsilon$, $E > 0$. It must be remembered that the eigenvalues α, too, depend on E.

The analytic continuation of the completeness was predicated on the assumption that

$$\frac{1}{\alpha}(\Phi_\alpha(E^*), H'\Phi_\alpha(E)) = (\bar{\Phi}_{\alpha^*}, \Phi_\alpha) \neq 0. \tag{9.27}$$

8 If $\varphi_\alpha = e^{2i\varepsilon}$ we multiply both sides of (9.23) by $e^{-i\varepsilon}$ and use the fact that $e^{-i\varepsilon}\,\vartheta = \vartheta e^{i\varepsilon}$. Thus $e^{i\varepsilon}\Phi_\alpha = \Phi_\alpha{}'$ obeys

$$\vartheta\Phi_\alpha' = \Phi_\alpha'.$$

9 See E. P. Wigner (1959), pp. 329ff.

As we have seen before, the analyticity breaks down at an energy where (9.27) is violated. That this must be so can also be seen from the continuation of (9.21). Left multiplication of Φ_α by (9.21) would yield $\Phi_\alpha = 0$ if (9.27) were not true.

For further details in the local potential case we refer to Chap. 11.

9.1.2 The Born Approximation. The Born series gives rise in a natural way to an approximation. If only the first term in γ is kept, it is called the *first Born approximation* or, simply, the *Born approximation*. For the **T** matrix this is

$$T_{\beta\alpha}(E) \simeq (\Psi_0(E,\beta), H'\Psi_0(E,\alpha)) \tag{9.28}$$

once γ is set equal to 1.

The question of whether or not the Born term is a good approximation to the exact answer is of course related to the convergence question of the series, but it is not identical with it. Not only is it possible for the series to converge and the first term be far off the sum, but the first term (or terms) may even be a good approximation if the series diverges. In a general sense the relevant criterion is evidently one of *weakness* of the interaction. In the coordinate representation, what enters is always an integral over the interaction. Therefore it is the product of the *range* R and the (average) *strength* $\overline{H}'$ that counts. Moreover, if $\|K(E)\|$ tends to zero as $E \to +\infty$, as it does in many important cases, then, no matter what the *strength* of $\overline{H}'$, the Born approximation will be good provided that the energy is high enough. In this sense it is inherently a *high-energy* approximation.

On dimensional grounds we expect, as a result, that the general criterion for the validity of the Born approximation is

$$\frac{\overline{H}'R}{\hbar v} \simeq \frac{\overline{H}'Rp}{E\hbar} \ll 1 \tag{9.29}$$

if p is the particle momentum and v its velocity.[10] In simple cases this is borne out and will be corroborated later for local potentials. However, in general each instance has to be examined separately; there exists no rule applicable to all cases. A simple practical test is to examine the size of the second Born approximation relative to the first, but even this is not foolproof. The Coulomb potential is an important case in point. In this extraordinary instance the Born approximation to the scattering cross section is the exact answer. The second Born approximation to the amplitude actually modifies its *phase* only. But this is not apparent from the calculation alone, and if it is used straightforwardly for the cross section, it is seriously misleading. For further discussion of Coulomb scattering, see Sec. 14.6.

10 Another way of stating (9.29) is that the time R/v which the particle spends inside the potential range should be small compared with the characteristic time $\hbar/\overline{H}'$ required by the interaction to influence the particle appreciably.

9.1.3 The Distorted-Wave Born Approximation. The procedure of constructing the Green's function, the **T** matrix, or the state vector by means of a power-series expansion need not start from the solution in the total absence of interaction. Instead of (9.1) we may write

$$\Psi = \Psi_1 + \gamma \mathscr{G}_1 H_2' \Psi \tag{9.30}$$

or

$$\mathscr{G} = \mathscr{G}_1 + \gamma \mathscr{G}_1 H_2' \mathscr{G} \tag{9.31}$$

if the hamiltonian is split up as in Sec. 7.2.5,

$$H = H_0 + H_1' + \gamma H_2'$$

and Ψ_1 and $\mathscr{G}_1$ are the appropriate solution of

$$(H_0 + H_1')\Psi_1 = E\Psi_1$$

and

$$\mathscr{G}_1 = (E - H_0 - H_1')^{-1}.$$

We are then led to expand the operator

$$(\mathbb{1} - \gamma K)^{-1}$$

in a power series in γ, where the kernel is now

$$K = \mathscr{G}_1 H_2'.$$

On the assumption that the interaction H_1' is such that $\mathscr{G}_1$ can be explicitly constructed, the expansion in powers of H_2' is just as practical as the Born series. If, in addition, H_1' is "almost" the entire interaction, so that H_2' is weak, then this expansion of course converges much more rapidly than the Born series.

Suppose however that H_1' causes bound states. Then there must exist an energy range is which the new series diverges. This is because at a bound-state energy the Green's function $\mathscr{G}_1$ blows up, and so does K. In this region this series may then actually be worse than the Born series even for a weak H_2'.

If the power series in H_2' is broken off after one term, the result is known as the *distorted-wave Born approximation*. Equation (7.76b) shows that the corresponding **T** matrix is

$$T_{\beta\alpha} \simeq T_{1\beta\alpha} + (\Psi_{1\beta}^{(-)}, H_2' \Psi_{1\alpha}^{(+)}). \tag{9.32}$$

This is of particular usefulness in the scattering of slow charged particles. One may then take H_1' to be the Coulomb interaction and H_2' the additional forces. If the latter are weak, the resulting **T** matrix is well approximated by the Coulomb **T** matrix plus the matrix element of the additional interaction between Coulomb state vectors. The importance of this modification of the Born approximation lies in the fact that the Coulomb effect on the scattering is very large at small relative velocities. It should be particularly noticed that, for the initial state, an outgoing wave state is required, and for the final state, an incoming wave state.

9.1.4 Bound States from the Born Approximation. The fact that the Born series of the T operator begins to diverge when the energy passes the first bound state may be exploited for the calculation of the binding energy. Let us consider the **T** matrix in its Born expansion,

$$T_{\beta\alpha} = \sum_n (T_{\beta\alpha})_n \tag{9.33}$$

where

$$(T_{\beta\alpha})_n = (\Psi_0(\beta), H'K^n\Psi_0(\alpha)) \tag{9.34}$$

the Ψ_0 being free states of the same energy E as K. When the energy is less than that of the lowest bound state, then the series (9.33) converges. The ratio test therefore tells us that

$$\lim_{n\to\infty} \frac{|(T_{\beta\alpha})_{n+1}|}{|(T_{\beta\alpha})_n|} < 1. \tag{9.35}$$

When the energy passes the bound state, then this criterion must be violated and at the binding energy itself we must have

$$\lim_{n\to\infty} \frac{(T_{\beta\alpha})_{n+1}}{(T_{\beta\alpha})_n} = 1 \tag{9.36}$$

because there the series must in fact tend to infinity.

It is true that not all matrix elements of the T operator have a divergent Born expansion beyond the bound state. In fact, unless there is degeneracy, the subspace on which T has a pole, and on which the series therefore diverges, is one-dimensional. But only accidentally will one of the *free* states Ψ_0 used for the evaluation of the **T** matrix be orthogonal to the bound state. Hence these matrix elements can all be expected to have divergent Born expansions. Of course, if matrix elements of fixed angular momentum are taken, then only the bound state of the same angular momentum is relevant.

The lowest approximation to (9.36) is evidently

$$(T_{\beta\alpha})_1 = (T_{\beta\alpha})_0$$

or, in detail,

$$(\Psi_0(\beta, E), H'\Psi_0(\alpha, E)) = (\Psi_0(\beta, E), H'G(E)H'\Psi_0(\alpha, E)). \tag{9.37}$$

Since the solution of this equation occurs at a negative energy, we need not distinguish between G^+ and G^-. Furthermore, the nature of the states Ψ_0 is quite irrelevant, so long as they are not orthogonal to the bound state. Specifically, they need not be made to depend on the energy. Indeed, if they are chosen to resemble the bound-state wave function, the approximation can be expected to be improved.

Equation (9.37) is easily recognized as another form of the exact equation

$$(\Psi, (K - \mathbb{1})\Phi) = 0$$

for the bound-state vector Φ, which is stationary with respect to small variations in Φ if Ψ is chosen equal to $H'\Phi$,

$$\delta(\Phi, H'(K - \mathbb{1})\Phi) = 0 \qquad \text{if} \quad (K - \mathbb{1})\Phi = 0$$

because then

$$K^{\dagger} H'\Phi = H'\Phi.$$

Hence, after (9.37) is solved for E with a family of trial functions

$$\Psi_0(\alpha) = \Psi_0(\beta) = \Phi$$

the bound-state vector may be approximated by finding the trial function for which (9.37) is stationary. The lowest approximation to the exact statement (9.36) is therefore different from the customary variational method only in the replacement of $H - E$ by $H'(K - \mathbb{1})$. The detailed procedure of improving the approximation, of course, is different.

9.2 The Schmidt Process (Quasi Particles)

If there exists an eigenvalue α of the kernel K defined in (9.2) which is larger in magnitude than unity, then we know that the Born series diverges. We now discuss a method of remedying this.

The general basis of the procedure is property 6 of compact operators, listed in Sec. 7.3.2. Accordingly, if K is compact and ε any given positive number, there exists an operator K' of finite rank such that $K = K' + M$ and $\|M\| < \varepsilon$. The resolvent of K can therefore be constructed in terms of that of K' (which can be explicitly written down by algebraic methods) and a convergent power series expansion in M, by using the identity

$$(1 - \gamma K)^{-1} = (1 - \gamma K')^{-1}[1 - \gamma M(1 - \gamma K')^{-1}]^{-1}$$

as in the "distorted wave Born series." We shall do so in detail.

Separable Change of the Hamiltonian. Suppose that the interaction is altered in the following way,

$$H'' = H' - H'\Phi\Psi^{\dagger}H' \tag{9.38}$$

in the dyadic notation. Then

$$H''\Phi = \rho H'\Phi \qquad \Psi^{\dagger}H'' = \rho\Psi^{\dagger}H' \tag{9.39}$$

where

$$\rho = 1 - (\Psi, H'\Phi). \tag{9.40}$$

We wish to express the original T operator, which satisfies the equation

$$T = H' + TGH'$$

in terms of the new T operator, which obeys

$$T' = H'' + T'GH''. \tag{9.41}$$

Insertion of the definition (9.38) of H'' in T gives

$$T = H'' + TGH'' + T\Phi\Psi^{\dagger}H'.$$

Transfer the second term on the right to the left, and use the definition (9.41) of T' as well as (9.39) and you get

$$T = T' + \frac{1}{\rho}T\Phi\Psi^{\dagger}T'.$$

Apply this equation to the vector Φ,

$$T\Phi = \frac{\rho}{c}T'\Phi$$

where

$$c = \rho - (\Psi, T'\Phi). \tag{9.42}$$

Therefore

$$T = T' + \frac{1}{c}T'\Phi\Psi^{\dagger}T'. \tag{9.43}$$

It is worthwhile to rewrite this compact result in another way, even at the expense of greater length. This is because we are particularly interested, as we shall see, in the case in which ρ and c are small. Insertion of (9.41) for T' in (9.42) yields

$$c = \rho - \rho(1 - \rho) - \rho(\Psi, T'GH'\Phi) = \rho^2 c'$$

and similarly

$$T'\Phi\Psi^{\dagger}T' = \rho^2 B_2\Phi\Psi^{\dagger}B_1$$

where

$$c' = 1 - (\Psi, B_1GH'\Phi) = 1 - (\Psi, H'GB_2\Phi) \tag{9.44}$$

$$B_1 = H'(1 + GT') \qquad B_2 = (1 + T'G)H'. \tag{9.45}$$

The result

$$T = T' + \frac{1}{c'}B_2\Phi\Psi^{\dagger}B_1 \tag{9.46}$$

can be used even when $\rho = 0$.

In a similar way we get an expression for the Green's function $\mathscr{G}(E)$ in terms of the new one,

$$\mathscr{G}'(E) = (E - H_0 - H'')^{-1}$$

namely,

$$\mathscr{G} = \mathscr{G}' + \frac{1}{\Delta}\mathscr{G}'H'\Phi\Psi^{\dagger}H'\mathscr{G}' \tag{9.47}$$

with

$$\Delta = 1 - (\Psi, H'\mathscr{G}'H'\Phi) = c'. \tag{9.48}$$

Substitution of (9.47) gives the alternative expression

$$\frac{1}{\Delta} = 1 + (\Psi, H'\mathscr{G}H'\Phi). \tag{9.49}$$

Hamiltonian Change of Finite Rank. It is a simple matter to generalize these results to the case in which the difference between H'' and H' is of finite rank, i.e., a finite sum of separable operators,

$$H'' = H' - \sum_i H'\Phi_i\Psi_i^{\dagger}H' \tag{9.38a}$$

instead of (9.38). Then we have, in place of (9.39),

$$H''\Phi_j = \sum_i H'\Phi_i\rho_{ij} \qquad \Psi_j^{\dagger}H'' = \sum_i \rho_{ji}\Psi_i^{\dagger}H' \tag{9.39a}$$

with

$$(\Psi_i, H'\Phi_j) = \delta_{ij} - \rho_{ij}. \tag{9.40a}$$

Equation (9.43) is replaced by

$$T = T' + T'\left[\sum_{ij}\Phi_i(\mathbf{c}^{-1})_{ij}\Psi_j^{\dagger}\right]T' \tag{9.43a}$$

with

$$c_{ji} = \rho_{ji} - (\Psi_j, T'\Phi_i) \tag{9.42a}$$

and instead of (9.46) we have

$$T = T' + B_2\left[\sum_{ij}\Phi_i(\mathbf{c}'^{-1})_{ij}\Psi_j^{\dagger}\right]B_1 \tag{9.46a}$$

with

$$\begin{aligned} c'_{ji} &= \delta_{ij} - (\Psi_j, B_1GH'\Phi_i) \\ &= \delta_{ij} - (\Psi_j, H'GB_2\Phi_i) \end{aligned} \tag{9.44a}$$

$$c'_{ji} = \delta_{ij} - (\Psi_j, H'\mathscr{G}'H'\Phi_i) = \Delta_{ij}. \tag{9.48a}$$

Similarly,

$$\mathscr{G} = \mathscr{G}' + \mathscr{G}'H'\left[\sum_{ij}\Phi_i(\mathbf{\Delta}^{-1})_{ij}\Psi_j^{\dagger}\right]H'\mathscr{G}' \tag{9.47a}$$

and

$$(\mathbf{\Delta}^{-1})_{ij} = \delta_{ij} + (\Psi_i, H'\mathscr{G}H'\Phi_j). \tag{9.49a}$$

Shift of Single Eigenvalues. We must now choose the vectors Φ and Ψ

appropriately. Let α be an eigenvalue of $K = GH'$, and let α^* be an eigenvalue of $K^\dagger$. Take Φ and Ψ to be the vectors

$$\Phi = \Phi_\alpha(E) \qquad K(E)\Phi_\alpha(E) = \alpha(E)\Phi_\alpha(E) \tag{9.50}$$

$$\Psi = b\Phi_\alpha(E^*) \qquad K(E^*)\Phi_\alpha(E^*) = \alpha(E^*)\Phi_\alpha(E^*) \tag{9.51}$$

or, according to (9.15),

$$H'\Psi = \alpha^* b\overline{\Phi}_{\alpha^*}(E). \tag{9.52}$$

We then have the orthogonality relation

$$b^*\alpha(\overline{\Phi}_{\alpha^*}, \Phi_\beta) = (\Psi, H'\Phi_\beta) = 0 \qquad \alpha \neq \beta \tag{9.53}$$

and, according to (9.40),

$$\rho = 1 - b^*(\Phi_\alpha(E^*), H'\Phi_\alpha(E)) = 1 - b^*\alpha(\overline{\Phi}_{\alpha^*}, \Phi_\alpha). \tag{9.53a}$$

If K is compact and H' is bounded, then the new kernel

$$K' = K - K\Phi\Psi^\dagger H' = K[\mathbb{1} - \Phi\Psi^\dagger H'] \tag{9.54}$$

is also compact. It has the same eigenvalues as K, *except for* α. This can be seen very easily. Let K' act on an eigenvector Φ_β of K for $\beta \neq \alpha$. Then, according to (9.53),

$$K'\Phi_\beta = K\Phi_\beta = \beta\Phi_\beta$$

and β is an eigenvalue also of K'. Moreover, let β be an eigenvalue of K' but *not* of K, so that

$$K'\chi = \beta\chi.$$

Then, by (9.54),

$$(K - \beta)\chi = \alpha(\Psi, H'\chi)\Phi \tag{9.55}$$

and since β is not in the spectrum of K,

$$\chi = \alpha(\Psi, H'\chi)(K - \beta)^{-1}\Phi = \frac{\alpha}{\alpha - \beta}(\Psi, H'\chi)\Phi.$$

Let H' act on this, and take the inner product with Ψ,

$$(\Psi, H'\chi)(\alpha - \beta) = \alpha(1 - \rho)(\Psi, H'\chi).$$

Since we assumed that β is not in the spectrum of K, (9.55) shows that $(\Psi, H'\chi) \neq 0$. Consequently

$$\beta = \alpha\rho.$$

The only remaining possibility is that K' has the eigenvalue α too. Then take $\beta = \alpha$, let H' act on (9.55), and take the inner product with Ψ. According to (9.51) and (9.53)

$$0 = (\Psi, H'\chi)\alpha(1 - \rho).$$

We conclude that either $\rho = 1$, that is,

$$(\Psi, H'\Phi_\alpha) = 0$$

or else $(\Psi, H'\chi) = 0$. In the latter case, (9.55) says that χ is an eigenvector of K with eigenvalue α (χ may or may not be equal to Φ_α), and it is orthogonal to an eigenvector of $K^\dagger$ with eigenvalue α^*. According to Sec. 7.3.3 this is associated with a pole of the resolvent $(K - \alpha)^{-1}$ of order greater than unity.

To summarize, then: The operator K' has exactly the same spectrum as K, except for the eigenvalue α, which is shifted to $\alpha\rho$. The argument breaks down only if $(K - \alpha)^{-1}$ has a pole of order greater than 1. The last possibility may be regarded as accidental and we shall not pursue it further.

Making the Born Series Converge. Now, if $K = GH'$ has but one eigenvalue α of magnitude greater than 1, then we choose its eigenvectors to change H' according to (9.38), (9.50), and (9.51). The size of ρ is under our control by means of the number b [unless $\alpha(E)$ happens to be a crossing point of trajectories], and we can make $|\alpha\rho| < 1$. The Born series for T' or $\mathcal{G}'$ now converges, and (9.43) or (9.46) gives us T in terms of T' and (9.47) $\mathcal{G}$ in terms of $\mathcal{G}'$. It is clear that in some sense the "optimal" choice of ρ is $\rho = 0$. We then expect the Born series to converge fastest, for then the "size" of H'' is "smallest." In this case (9.46) must be used for T, and not (9.43).

The expression (9.48) for c' or Δ can be simplified by the use of (9.49), (9.50), and (9.53). We have

$$\frac{1}{\Delta} = 1 + (\Psi, H'(\mathbb{1} - K)^{-1}K\Phi) = \frac{1 - \alpha\rho}{1 - \alpha}. \tag{9.56}$$

The change (9.38) in the hamiltonian has removed exactly one pole of the Green's function $\mathcal{G}$. If we replace H' by $\gamma H'$ and H'' by $\gamma H''$, then $\mathcal{G}$ has a pole at $\gamma = 1/\alpha$ but $\mathcal{G}'$ does not. Equation (9.47) shows that therefore Δ must have a zero there. Now the introduction of the strength parameter γ changes (9.56) to

$$\frac{1}{\Delta} = \frac{1 - \alpha\gamma\rho}{1 - \alpha\gamma} \tag{9.57}$$

which shows explicitly that $\Delta = 0$ when $\gamma = 1/\alpha$. What is more, $\mathcal{G}'$ has a pole at $\gamma = 1/\alpha\rho$, but $\mathcal{G}$ does not. Equation (9.57) shows that at that point $1/\Delta = 0$. This changes the apparent double pole in (9.47) to a simple one, which, in turn, cancels against that in the first term.

If there are more than one α outside the unit circle, the procedure must be based on (9.38a) to (9.48a). Since we know that at each energy the number of α's outside the unit circle is finite, the Born series can always be made to converge in a finite number of steps. If $|K(E)| \to 0$ as $E \to +\infty$, in fact, this number is *uniformly* bounded. What is more, in this case at every E_0

the number of $\alpha(E)$ which *have been* outside the unit circle at $E < E_0$ is uniformly bounded. The Born series can therefore be made uniformly convergent for all energies in a finite number of steps.

Increasing the Rapidity of Convergence. The same procedure may be used to increase the rapidity of convergence of the Born series when it does converge. If there are only a small number of α's near the unit circle, then we move them to the origin and the remaining series can be expected to converge more rapidly. In fact, we may go to the limit of moving *all* the eigenvalues to zero. When $E < 0$, then, K of (9.8) is hermitian and the Φ_α form a complete set, as in (9.21), i.e.,

$$\sum_i \Phi_i \Psi_i H' = \mathbb{1} \tag{9.58}$$

provided that the Φ_α are normalized as in (9.19). We must choose each b so that $\rho = 0$. Then (9.38a) shows that $H'' = 0$ and $\mathscr{G}' = G$. Equation (9.48a) thus becomes simply

$$\begin{aligned} \mathscr{G} &= G + GH' \left[\sum_{ij} \Phi_i (\mathbf{\Delta}^{-1})_{ij} \Psi_j^\dagger \right] H'G \\ &= G - \sum_{ij} \alpha_i \Phi_i (\mathbf{\Delta}^{-1})_{ij} \Phi_j^\dagger. \end{aligned} \tag{9.59}$$

Of course, the matrix $\mathbf{\Delta}$ is now of infinite dimension. We may now analytically continue (9.58) to positive energies, as in Sec. 9.1.

Using (9.49a), (9.50), and (9.40a), we find the analog of (9.56),

$$(\mathbf{\Delta}^{-1})_{ij} = \frac{\delta_{ij} - \alpha_i \rho_{ij}}{1 - \alpha_j} \tag{9.60}$$

or, after introducing the strength parameter γ,

$$(\mathbf{\Delta}^{-1})_{ij} = \frac{\delta_{ij} - \gamma \alpha_j \rho_{ij}}{1 - \gamma \alpha_j}. \tag{9.60a}$$

But since $\rho_{ij} = 0$,

$$(\mathbf{\Delta}^{-1})_{ij} = \frac{\delta_{ij}}{1 - \gamma \alpha_i} \tag{9.61}$$

and therefore

$$\mathscr{G} = G - \sum_i \frac{\gamma \alpha_i \Phi_i \Phi_i^\dagger}{1 - \gamma \alpha_i}, \tag{9.62}$$

which is identical to the previous result (9.22).

Physical Interpretation. The physical interpretation of these results is based upon the significance of the Green's functions as *propagators* and of its poles, as bound states, or particles. We recall the interpretation of the iteration solution of Eq. (6.15a), for example. In (9.47) or (9.47a), the complete propagator for two particles, say, is exhibited in the form of a reduced propagator plus a term reasonably interpreted as showing the "reduced" propagation of the particles, followed by a "vertex" $\Psi^\dagger H'$ at which the two particles emerge and form a quasi-bound state or quasi particle. The latter then moves along according to the "propagator" $1/\Delta$. Its subsequent "decay" is described by the vertex $H'\Phi$, and the emitted pair is propagated via the Green's function $\mathscr{G}'$. On the basis of this physical picture the procedure is also referred to as the *quasi-particle method.*

Use as an Approximation. Again it is clear that the exact process gives rise to a natural approximation procedure. Once all but those eigenvalues α of K whose magnitude is small compared with unity have been removed, the remaining Born series for $\mathscr{G}'$ or T' may be expected to converge rapidly. The real problem, from a practical point of view, is how to construct the eigenvectors Φ_α and $\Phi_{\alpha*}$ of K. However, that problem need not be solved exactly either. All we need in (9.38) is *approximate* solutions of (9.50) and (9.51). If there is but one "large" α, then the reason why H'' in contrast to H' is weak is that, while GH', when acting on Φ_α is large, GH'' is not. But the same can be accomplished by choosing for Φ and $H'\Psi$ *approximate* eigenvectors of GH'. In this way the Schmidt process becomes a practical approximation procedure. The ease with which it can be applied does not depend on the *strength* of the interaction H' so much as on how many vectors in the Hilbert space feel the strength of H', so to speak. If H' is strong on only a single vector but weak on all others (in an inexact but obvious way of speaking), then the quasi-particle method will work very rapidly. But if H' is strong on a large subspace of the whole Hilbert space, then it is cumbersome and impractical.

Bound States. The Schmidt method can be used also for the calculation of bound states. If H' causes a single bound state, then $\mathscr{G}$ has one pole outside the unit circle for negative energies. Its position is determined by the equation

$$\Delta = 0,$$

Δ being given by (9.48). In the lowest approximation this becomes

$$(\Psi, H'G(E)H'\Phi) = 1. \tag{9.63}$$

The vectors Ψ and Φ are trial functions optimally normalized so that $\rho = 0$, or, by (9.40),

$$(\Psi, H'\Phi) = 1. \tag{9.64}$$

These two equations are seen to be identical to (9.37). In the lowest order the bound-state approximation based on the Schmidt process is the same as that based on the Born series.

9.3 The Fredholm Method

The Fredholm method is a straightforward generalization of the construction of the inverse of

$$M = \mathbb{1} - \gamma K = \mathbb{1} - \gamma GH'$$

when K is a finite-dimensional matrix. In this instance M^{-1} is a matrix N whose transpose consists of the cofactors of M, divided by the determinant,

$$\begin{aligned} NM &= MN = \mathbb{1}\Delta \\ \Delta &= \det M. \end{aligned} \tag{9.65}$$

Both Δ and N may be expanded in power series in γ (which in the finite dimensional case, of course, reduce to polynomials),

$$\begin{aligned} N &= \sum_n \gamma^n N_n \qquad N_0 = \mathbb{1} \\ \Delta &= \sum_n \gamma^n \Delta_n \qquad \Delta_0 = 1. \end{aligned} \tag{9.66}$$

Using the familiar differentiation rule for the determinant, we have

$$\frac{d\Delta}{d\gamma} = \operatorname{tr} N \frac{dM}{d\gamma} = -\operatorname{tr} NK. \tag{9.67}$$

The left-hand side is now evaluated by means of (9.65),

$$-\mathbb{1}\operatorname{tr} NK = -KN + M\frac{dN}{d\gamma} = -NK + \frac{dN}{d\gamma}M.$$

Insertion of the expansions (9.66) and term-by-term comparison yields the recursion relation

$$\begin{aligned} N_n &= KN_{n-1} - \frac{1}{n}\mathbb{1}\operatorname{tr} KN_{n-1} \\ &= N_{n-1}K - \frac{1}{n}\mathbb{1}\operatorname{tr} KN_{n-1} \qquad n \geq 1 \end{aligned} \tag{9.68}$$

whereas (9.66) substituted in (9.67) gives

$$\Delta = 1 - \sum_1^\infty \frac{\gamma^n}{n} \operatorname{tr} KN_{n-1}. \tag{9.69}$$

Equations (9.66), (9.68), and (9.69) can now be used to construct the inverse

$$M^{-1} = \frac{N}{\Delta} \tag{9.70}$$

or sometimes more conveniently

$$M^{-1} = \mathbb{1} + \frac{\gamma}{\Delta} Y \tag{9.71}$$

with

$$\begin{aligned} Y &= KN = \sum_0^\infty \gamma^n Y_n \\ Y_n &= KY_{n-1} + K\Delta_n \\ &= Y_{n-1}K + K\Delta_n \qquad n \geq 1 \\ Y_0 &= K \end{aligned} \tag{9.72}$$

$$\Delta = \sum_0^\infty \gamma^n \Delta_n \qquad \Delta_n = -\frac{1}{n} \operatorname{tr} Y_{n-1} \qquad \Delta_0 = 1. \tag{9.73}$$

Δ is called the *Fredholm determinant* and Y the *first Fredholm minor*.

When K is an operator, or an infinite-dimensional matrix, then it must be assumed that

$$\operatorname{tr} K^n \qquad n = 1, 2, \ldots$$

are all finite. What is more, the power series in γ have to be shown to converge.

Convergence Proof. To prove the convergence of (9.72) and (9.73), it is convenient to solve the recursion relations (9.72). Supposing that the Hilbert space is separable, we introduce an orthonormal basis. Then (9.72) is solved by

$$(Y_n)_{\alpha\beta} = \frac{(-)^n}{n!} (\varkappa_n)_{\alpha\beta}$$

$$(\varkappa_n)_{\alpha\beta} = \sum_{\alpha_1, \ldots, \alpha} \begin{vmatrix} K_{\alpha\beta} & K_{\alpha\alpha_1} & \cdots & K_{\alpha\alpha_n} \\ K_{\alpha_1\beta} & K_{\alpha_1\alpha_1} & \cdots & K_{\alpha_1\alpha_n} \\ \cdots & \cdots & \cdots & \cdots \\ K_{\alpha_n\beta_n} & K_{\alpha_n\alpha_1} & \cdots & K_{\alpha_n\alpha_n} \end{vmatrix}. \tag{9.74}$$

That this indeed solves (9.72) is shown by developing the determinant according to the first row,

$$\begin{aligned} &(\varkappa_n)_{\alpha\beta} \\ &= \sum_{\alpha_1, \ldots, \alpha_n} \left(K_{\alpha\beta} \begin{vmatrix} K_{\alpha_1\alpha_1} & \cdots & K_{\alpha_1\alpha_n} \\ \cdots & \cdots & \cdots \\ K_{\alpha_n\alpha_1} & \cdots & K_{\alpha_n\alpha_n} \end{vmatrix} - K_{\alpha\alpha_1} \begin{vmatrix} K_{\alpha_1\beta} & K_{\alpha_1\alpha_2} & \cdots \\ \cdots & \cdots & \cdots \\ K_{\alpha_n\beta} & \cdots & \cdots \end{vmatrix} + - \cdots \right) \\ &= K_{\alpha\beta} \operatorname{tr} \varkappa_{n-1} - n(K\varkappa_{n-1})_{\alpha\beta}. \end{aligned}$$

Therefore the Y_n of (9.74) satisfy the recursion relation (9.72).

For the convergence proof of the series in (9.72) and (9.73) we estimate (9.74) by Hadamard's inequality, which states that, if all elements a_{nm} of a p-dimensional determinant are less than unity in magnitude, then

$$|\det(a_{nm})| \le p^{1/2p}.$$

Applied to (9.74), this shows that if there are f_α and f_β such that

$$|K_{\alpha\beta}| \le f_\alpha f_\beta \tag{9.75}$$

then

$$\begin{aligned} |(Y_n)_{\alpha\beta}| &\le \frac{1}{n!} \sum_{\alpha_1, \ldots, \alpha_n} f_\alpha f_\beta f_{\alpha_1}^2 \cdots f_{\alpha_n}^2 (n+1)^{1/2(n+1)} \\ &\underset{n \to \infty}{\simeq} (2\pi)^{-1/2} f_\alpha f_\beta \left(\sum_\sigma f_\sigma^2 \right)^n e^{-1/2n \ln n + n + 1/2} \end{aligned}$$

by Stirling's formula. Consequently, provided that

$$\sum |K_{\alpha\alpha}| < \sum_\alpha f_\alpha^2 < \infty \tag{9.75a}$$

the series in (9.72) and (9.73) converge *no matter how large* γ is. Both series therefore represent *entire* analytic functions of γ.

Insertion of (9.74) in (9.73) gives the more explicit formula for the Fredholm determinant

$$\Delta = 1 + \sum_1^\infty \frac{(-\gamma)^n}{n!} \sum_{\alpha_1,\ldots,\alpha_n} \begin{vmatrix} K_{\alpha_1\alpha_1} & \cdots & K_{\alpha_1\alpha_n} \\ \cdots & \cdots & \cdots \\ K_{\alpha_n\alpha_1} & \cdots & K_{\alpha_n\alpha_n} \end{vmatrix}. \tag{9.76}$$

This procedure works just as well if the indices run over a continuous range rather than taking on discrete values, in other words, if we are solving an integral equation. In this case (9.75a) of course becomes

$$\int dx\,|K(x,x)| \leq \int dx f^2(x) < \infty. \tag{9.75b}$$

Furthermore, it is applicable to "indices" which are sets of variables, some of which are discrete and others continuous, i.e., the solution of sets of coupled integral equations. In such a case (9.74) is awkward to write down, but (9.72) and (9.73) are unaltered.

Other Expressions for the Fredholm Determinant. The formula (9.73) for the Fredholm determinant Δ can be rewritten by inserting (9.72) in it. Repeated use of (9.72) gives, for $n \geq 1$,

$$(n+1)\Delta_{n+1} + \sum_{m=0}^{n} \Delta_m \kappa_{n+1-m} = 0 \tag{9.77}$$

with

$$\kappa_p = \operatorname{tr} K^p.$$

If we now write

$$\Delta = e^{-g}$$

so that

$$\frac{d\Delta}{d\gamma} + \Delta \frac{dg}{d\gamma} = 0$$

then repeated differentiation with respect to γ shows that[11]

$$\sum_{m=0}^{n} \frac{n!}{m!(n-m)!} g^{(m+1)}\Delta^{(n-m)} + \Delta^{(n+1)} = 0. \tag{9.78}$$

We now expand

$$g = \sum_1^\infty \gamma^n C_n \qquad C_n = \frac{1}{n!} g^{(n)}\big|_{\gamma=0}$$

and set $\gamma = 0$ in (9.78),

11 The notation here means $\Delta^{(m)} = d^m\Delta/d\gamma^m$, etc.

$$\sum_{m=0}^{n} (m+1)C_{m+1}\Delta_{n-m} + (n+1)\Delta_{n+1} = 0.$$

Comparison with (9.77) shows that

$$C_m = \frac{\kappa_m}{m}$$

or

$$\Delta = \exp\left(-\sum_{n=1}^{\infty} \frac{\gamma^n}{n}\kappa_n\right). \tag{9.79}$$

This result also follows from (9.67), that is, from

$$\frac{d}{d\gamma}\ln\det M = \operatorname{tr} M^{-1}\frac{d}{d\gamma}M$$

by integration,

$$\Delta = \exp\left[-\int_0^\gamma d\gamma' \operatorname{tr} K(\mathbb{1} - \gamma' K)^{-1}\right] = \exp\operatorname{tr}\ln(\mathbb{1} - \gamma K) \tag{9.79a}$$

and subsequent expansion in γ.

In comparing (9.79) with (9.73) it must be remembered that, although (9.79) is more explicit and seems more useful, the series in it converges only under much more restrictive conditions than that in (9.73). In fact, (9.79) diverges as soon as γ is larger in magnitude than the smallest zero of $\Delta(\gamma)$, whereas the convergence of (9.73) is independent of the value of γ. In view of the remarks below concerning the significance of the zeros of Δ, this means that (9.79) converges if and only if the Born series of M converges.

The recursion relations (9.77) allow us to represent the Fredholm determinant in another useful form. They can be solved directly and yield

$$\Delta_n = \frac{(-)^n}{n!}\begin{vmatrix} \kappa_1 & 1 & 0 & 0 & \cdots \\ \kappa_2 & \kappa_1 & 2 & 0 & \cdots \\ \kappa_3 & \kappa_2 & \kappa_1 & 3 & \cdots \\ \cdots & \cdots & \cdots & \cdots & n-1 \\ \kappa_n & \kappa_{n-1} & \kappa_{n-2} & \kappa_{n-3} & \kappa_1 \end{vmatrix} \tag{9.80}$$

in terms of the traces of powers of K. The analogous expression for $\varkappa_n$ is

$$\varkappa_n = \begin{vmatrix} \kappa_1 & 1 & 0 & \cdots & 0 & 0 \\ \kappa_2 & \kappa_1 & 2 & \cdots & 0 & 0 \\ \cdots & \cdots & \cdots & \cdots & \cdots & \cdots \\ \kappa_n & \kappa_{n-1} & \kappa_{n-2} & \cdots & \kappa_1 & n \\ K^{n+1} & K^n & K^{n-1} & \cdots & K^2 & K \end{vmatrix}. \tag{9.80a}$$

Equations (9.68) to (9.70) or (9.71) to (9.73) represent the operator-valued function $(\mathbb{1} - \gamma K)^{-1}$ of γ in the form of an *entire* analytic operator-valued function, divided by an entire complex-number-valued one. The singularities of $(\mathbb{1} - \gamma K)^{-1}$ must therefore all arise from zeros of the denominator $\Delta(\gamma)$. Thus the zeros of $\Delta(\gamma)$ are the characteristic values of K. If (9.75a) is satisfied K is in the Hilbert-Schmidt class,

$$\operatorname{tr} K^\dagger K < \infty$$

and K is compact. We then know that its spectrum is a denumerable point spectrum and $(\mathbb{1} - \gamma K)^{-1}$ has poles only, with no point of accumulation except at infinity. The function $\Delta(\gamma)$ is an entire function whose zeros are these poles.

The Spectrum in Terms of Δ. The fact that the poles of the resolvent $(K - \alpha)^{-1}$ are identical with the zeros of $\Delta(\gamma)$ for $\gamma = 1/\alpha$ allows us to discuss the spectrum of K in terms of Δ. What is the significance of the occurrence of a *multiple* zero of Δ at $\gamma = \gamma_0$? It may indicate either that there is degeneracy or that the pole of $(K - \alpha)^{-1}$ is not simple. In the finite-demensional case the proof of the spectral theorem consists precisely in demonstrating that for hermitian operators the *algebraic multiplicity* (the multiplicity of the zero of Δ) is equal to the geometric one (the degeneracy). The same holds for the infinite-dimensional case. For hermitian operators a zero of Δ of order n indicates an n-fold degeneracy of the corresponding eigenvalue (or characteristic value). The numerator of the resolvent has a zero of order $n - 1$ at the same point, so that the resulting pole is simple. In a more general instance the algebraic and geometric multiplicities may differ. An n-fold zero of Δ may then indicate either degeneracy or an ascent of $K - \alpha$ greater than unity [i.e., a pole of $(K - \alpha)^{-1}$ of order greater than 1] or both.

We may also discuss the analyticity of the pole positions of $(K\gamma - \mathbb{1})^{-1}$ as functions of the energy E in terms of Δ. The Fredholm determinant Δ is regular analytic as a function both of γ and E for all γ and on the physical E sheet. Hence

$$\Delta(\gamma, E) = 0$$

defines $\gamma(E)$ as an analytic function, except where Δ has a double zero as a function of γ. But the zeros of Δ are almost everywhere simple because they are so on the negative real E axis. (In case of essential degeneracy, Δ can be factored for all E, and we consider one of these factors at a time.) Hence double zeros can be expected where zero trajectories cross one another at the same energy. Such crossings are necessary but not sufficient conditions for the analyticity of $\gamma(E)$ to break down, because we may have $\partial\Delta/\partial E = 0$ at the same point. (See also footnote 6.)

If K is an operator of finite rank D, then Y and Δ are polynomials in γ of degree $D - 1$ and D, respectively. Furthermore, it is evident that

$$\Delta(0) = 1.$$

Consequently the polynomial $\Delta(\gamma)$ must be the product

$$\Delta(\gamma) = \prod_{i=1}^{D} (1 - \gamma\alpha_i) \tag{9.81}$$

if α_i are the D eigenvalues of K. Comparison of (9.71) with (9.62) then shows that

$$Y = \sum_i \left[\prod_{j \neq i} (1 - \gamma\alpha_j)\right] |\alpha_i|^2 \Phi_i \Psi_i^\dagger, \tag{9.82}$$

where Φ_i and Ψ_i are given by (9.50) and (9.51) with the normalization

$$(\Psi_i, H'\Phi_i) = 1. \tag{9.83}$$

These results can be generalized to the case in which the rank of K is infinite, provided that the product in (9.81) converges. The necessary and sufficient condition for its absolute convergence is that

$$\sum_i |\alpha_i| < \infty.$$

If this criterion is satisfied, then K has a finite trace and so does $K^\dagger K$. Thus, not only is K in the Hilbert-Schmidt class, but the terms in the expansions (9.72) and (9.73) are all finite. If, on the other hand, the trace of K is infinite, then (9.73) does not exist either. We therefore expect that

$$\Delta(\gamma) = \prod_i (1 - \gamma\alpha_i) \tag{9.81a}$$

whenever the definition (9.73) is meaningful.

Pulling the Poison Tooth. A modification of the straightforward Fredholm formula (9.74) is necessary if the diagonal elements of K are undefined or infinite or if $\operatorname{tr} K$ does not exist. It consists in replacing the "dangerous" elements on the main diagonal of the determinant by zeros,

$$(\varkappa_n')_{\alpha\beta} = \sum_{\alpha_1,\ldots,\alpha_n} \begin{vmatrix} K_{\alpha\beta} & K_{\alpha\alpha_1} & \cdots & K_{\alpha\alpha_n} \\ K_{\alpha_1\beta} & 0 & \cdots & K_{\alpha_1\alpha_n} \\ \cdots & \cdots & \cdots & \cdots \\ K_{\alpha_n\beta} & K_{\alpha_n\alpha_1} & \cdots & 0 \end{vmatrix} \tag{9.84}$$

and similarly for the determinant,

$$\Delta' = 1 + \sum_1^\infty \frac{(-\gamma)^n}{n!} \sum_{\alpha_1,\ldots,\alpha_n} \begin{vmatrix} 0 & K_{\alpha_1\alpha_2} & \cdots & K_{\alpha_1\alpha_n} \\ K_{\alpha_2\alpha_1} & 0 & \cdots & K_{\alpha_2\alpha_n} \\ \cdots & \cdots & \cdots & \cdots \\ K_{\alpha_n\alpha_1} & K_{\alpha_n\alpha_2} & \cdots & 0 \end{vmatrix} \tag{9.85}$$

Expressions analogous to (9.80) and (9.80a) are obtained by replacing κ_1 by 0 there. The modified matrix

$$Y' = K + \sum_1^\infty \frac{(-\gamma)^n}{n!} \varkappa_n'$$

differs from the old by a numerical factor which is an entire function of γ with no zeros,

$$Y' = Y \exp(\gamma \operatorname{tr} K) \tag{9.86}$$

and so does the determinant,

$$\Delta' = \Delta \exp(\gamma \operatorname{tr} K). \tag{9.87}$$

Consequently Y' and Δ' serve the same purpose and can be used in place of Y and Δ in (9.71). Since they do not contain tr K, (9.84) and (9.85) are applicable even when tr K does not exist. The convergence proof, of course, has to be given a new form, but we shall not do so here.

The proof of (9.86) consists in a repeated application of the circumstance that

$$\varkappa_n^{(m+1)} = \varkappa_n^{(m)} - C\varkappa_{n-1}^{(m)} \tag{9.88}$$

if by $\varkappa_n{}^{(m)}$ is meant the sum of determinants defined in (9.74), except that m elements on the main diagonal other than the first are replaced by zero, and

$$C = \operatorname{tr} K.$$

Equation (9.88) is readily derived. It follows that

$$\varkappa_n^{(m)} = \sum_{k=0}^{m} \binom{m}{k} \varkappa_k (-C)^{n-k}$$

where $\binom{m}{k}$ is the binomial coefficient

$$\binom{m}{k} = \frac{m!}{k!(m-k)!}.$$

As a result, because $\varkappa'_n = \varkappa_n^{(n)}$, we find that

$$Y' = \sum_{n=0}^{\infty} \frac{(-\gamma)^n}{n!} \varkappa'_n = \sum_{n=0}^{\infty} \sum_{k=0}^{n} \frac{(-)^k \varkappa_k \gamma^n}{k!(n-k)!} C^{n-k}$$

$$= \sum_{k=0}^{\infty} \frac{(-\gamma)^k \varkappa_k}{k!} \sum_{n=k}^{\infty} \frac{(\gamma C)^{n-k}}{(n-k)!} = Y e^{\gamma C}.$$

Equation (9.87) is demonstrated similarly.

The modified Fredholm determinant Δ', also expressed as $\det_2(1 - \lambda K)$, does not satisfy the usual rule for the determinant of a product. Instead it follows from (9.87) that

$$\det_2[(\mathbb{1} - A)(\mathbb{1} - B)] = \det_2(\mathbb{1} - A)\det_2(\mathbb{1} - B)\exp[-\operatorname{tr} AB]. \tag{9.89}$$

By Schwarz's inequality the trace on the right exists if A and B are in the Hilbert–Schmidt class.

Iteration. Another way of dealing with the Fredholm method when tr K does not exist but tr $K^\dagger K$ does is to iterate the equation

$$(\mathbb{1} - K)^{-1} = \mathbb{1} + K(\mathbb{1} - K)^{-1} \tag{9.90}$$

once,

$$(\mathbb{1} - K)^{-1} = \mathbb{1} + K + K^2(\mathbb{1} - K)^{-1}.$$

This amounts to writing $(\mathbb{1} - K)^{-1}$ in the form

$$(\mathbb{1} - K)^{-1} = (\mathbb{1} + K)(\mathbb{1} - K^2)^{-1} \tag{9.90a}$$

and then constructing $(\mathbb{1} - K^2)^{-1}$ by the usual Fredholm procedure. The convergence proof is now directly applicable to Hilbert–Schmidt operators.

There is the slight difficulty that

$$\det(\mathbb{1} - \gamma^2 K^2) = 0$$

both when K and when $-K$ have the eigenvalue $1/\gamma$. However, it was shown by Poincaré[12] that the zeros of $\det(\mathbb{1} - \gamma^2 K^2)$ corresponding to the eigenvalues of $-K$ cancel against zeros in the numerator of the resolvent, so that the right-hand side of (9.89a) has just the poles it should have according to the left-hand side.

This can, of course, be generalized to

$$(\mathbb{1} - K)^{-1} = (\mathbb{1} + K + \cdots + K^{n-1})(\mathbb{1} - K^n)^{-1}$$

which is applicable when $\operatorname{tr} K^n$ exists but $\operatorname{tr} K^m$, $m < n$, do not.

9.4 Singularities of an Operator Inverse

Suppose that $F(z)$ is an analytic finite-dimensional matrix function of z. Then the matrix inverse $[F(z)]^{-1}$ is necessarily a meromorphic function of z. Its only singularities can be poles at those values of z where $\det F = 0$. The same is not true if $F(z)$ is operator-valued, in a Hilbert space such as L^2.

Suppose, for example, that $F(z)$ is an integral kernel on the interval $0 \leq x < \infty$, of the form

$$F(z;x,x') = g(z,x)\delta(x - x') + H(x,x') \tag{9.91}$$

where H is in the Hilbert–Schmidt class. As a function of z the inverse $G(z;x,x')$ of the operator F then has *branch points* at those values z_0 of z where

$$g(z_0,0) = 0 \qquad g(z_0,\infty) = 0. \tag{9.92}$$

The branch points are connected by cuts whose natural positions are given by the curves C defined by

$$g(z,x) = 0 \qquad 0 \leq x < \infty. \tag{9.93}$$

In order to show this, we compute the discontinuity of G across one of the cuts C. Let z lie on C, and assume that g has a simple zero as a function of x at $x_0(z)$. We then expand

$$\frac{1}{g(z,x)} = \frac{R(z)}{x - x_0(z)} + \cdots. \tag{9.94}$$

Now provided that

12 H. Poincaré (1910).

$$\frac{dx_0(z)}{dz} = -\left.\frac{\partial g/\partial z}{\partial g/\partial x}\right|_{x=x_0} \neq 0$$

x_0 changes by a purely imaginary amount when z is shifted at right angles to C,

$$x_0(z) \to x_0(z) \pm i\varepsilon,$$

The difference between the values of $1/g$ on the two sides of C is therefore

$$\Delta g^{-1} = 2\pi i\delta(x - x_0)R. \tag{9.95}$$

Let us now define the two inverse operators

$$N^{(L)} = (\mathbb{1} + g^{-1}H)^{-1} \qquad N^{(R)} = (\mathbb{1} + Hg^{-1})^{-1} \tag{9.96}$$

which satisfy the integral equations

$$N^{(L)} = \mathbb{1} - N^{(L)}g^{-1}H \qquad N^{(R)} = \mathbb{1} - Hg^{-1}N^{(R)}. \tag{9.96a}$$

These allow us to calculate the discontinuity of $G = N^{(L)}g^{-1}$ across C. After some simple algebra we obtain

$$\Delta G = N_+^{(L)}\Delta g^{-1}N_-^{(R)} = N_-^{(L)}\Delta g^{-1}N_+^{(R)} \tag{9.97}$$

where the subscripts + and − indicate the values above and below the cut, respectively. Written out explicitly this means that

$$\Delta G(z;x,x') = 2\pi i N_\pm^{(L)}(z;x,x_0)R(z)N_\mp^{(R)}(z;x_0,x'). \tag{9.97a}$$

The result (9.97a) not only shows that G generally has a branch cut along C, but it also shows that the discontinuity across the cut is *factorable* into two terms, one of which depends on x only and the other on x'. The precise nature of the branch points (9.92) at the ends of the cut depends, of course, on the behavior of g and of H at the end points of the integration interval. We also find that

$$\int_0^\infty dx'\, F(z;x,x')N^{(L)}(z;x',x_0) = 0 \tag{9.98}$$

and

$$\int_0^\infty dx'\, N^{(R)}(z;x_0,x')F(z;x',x) = 0 \tag{9.98a}$$

so that $N^{(L)}(z;x,x_0)$ and $N^{(R)}(z;x_0,x)$ [for fixed z and $x_0(z)$] are "vectors" which are annihilated by $F(z)$ on the left and on the right, respectively. These vectors are, however, usually not normalizable and hence not members of the Hilbert space.

An alternative version of (9.97) is

$$\Delta G = \Delta g^{-1} - G_\pm H\Delta g^{-1} - \Delta g^{-1}HG_\mp + G_\pm H\Delta g^{-1}HG_\mp . \tag{9.97b}$$

After insertion of (9.95) the first three terms are seen to contain δ functions of x or x'. Just as (9.95) is a manifestation of (9.94), so these indicate the presence of poles in $G(z;x,x')$ on the cuts, whose positions depend on x and x'. In addition, G may have fixed (that is, x and x' independent) poles where the Fredholm determinant of F vanishes.

Simple examples are furnished by resolvents of integral operators. For instance, let z be the energy E, and

$$F = E - \frac{p^2}{2m} - V(p,p')$$

in the momentum representation. Then

$$g = E - \frac{p^2}{2m}$$

and we conclude that the resolvent

$$\mathscr{G} = \left(E - \frac{p^2}{2m} - V\right)^{-1}$$

has a branch cut in the E plane from $E = 0$ to $E = \infty$. Equation (9.98) shows that $N^{(L)}(E;p,p_0)$ is a *continuum eigenvector*. So the result (9.97) is identical with that of the conventional Green's-function representation, and (9.96a) is identical with the Lippmann–Schwinger equation.

Notes and References

9.1.1 The original reference for the *Born series* and the *Born approximation* is M. Born (1926). Discussions of its convergence, or of the analyticity of the resolvent, are T. Kato (1949 and 1950a); W. Kohn (1952 and 1954); T. Kikuta (1953a); Zemach and Klein (1958); Aaron and Klein (1960); H. Davies (1960); W. Hunziker (1961); K. Meetz (1962); M. Rotenberg (1963); S. Weinberg (1963b and 1964a); F. Coester (1964a); C. Lovelace (1964b); Scadron, Weinberg, and Wright (1964); S. Tani (1965); J. V. Corbett (1968); W. G. Faris (1971); H. Rabitz (1972); P. J. Bushell (1972); Graves-Morris and Rennison (1974).

The trick of using the kernel defined in (9.6) instead of K was introduced by J. Schwinger (1961) and for the present purpose by M. Scadron et al. (1965). It is intimately related to the methods of K. Meetz (1962), who uses the theory of polar kernels, and of F. Coester (1964). For the original mathematical analysis of *polar* kernels, see D. Hilbert (1912); E. Garbe (1915).

The discussion of the behavior of the eigenvalues $\alpha(E)$ of $K(E)$ as functions of E is based on the work of S. Weinberg (1963 and 1964a). The present discussion of the analyticity of $\alpha(E)$ goes somewhat beyond his. The analytic continuation of the completeness relation is an extension of that of Meetz.

For a variational formulation of the eigenvalues $\alpha(E)$ see Wright and Scadron (1964), and Y. Hahn (1965), appendix C.

The following papers deal with rearrangements of the Born series for the purpose of more rapid convergence: M. Rotenberg (1963); M. Wellner (1963); S. Weinberg (1964b).

The authority sometimes adduced for Eq. (9.17) is a theorem by A. Herglotz (1911). Functions that can be written in that form are called *Herglotz functions*. They are intimately related to Wigner's R functions; see E. P. Wigner (1951 and 1952a and b); Wigner and Von Neumann (1954).

9.1.2 The paper by R. H. Dalitz (1951) discusses higher Born approximations. For the second Born approximation to relativistic Coulomb scattering, see McKinley and Feshbach (1948). For an examination of the Born approximation at low energies, see P. Swan (1963a). See also Calogero and Charap (1964); A. C. Yates (1979).

9.1.3 For discussions and applications of the distorted-wave Born approximation see, for example, H. S. W. Massey (1956a); Bassel and Gerjuoy (1960); L. Rosenberg (1964a). For other distorted-wave approximations, see P. Swan (1960, 1961, and 1963b); Austern and Blair (1965) and references therein.

9.1.4 The idea of this section is due to J. Mazo (1964), unpublished.

9.2 See, for example, Courant and Hilbert (1953), p. 155. The treatment here is based on S. Weinberg (1963a and b and 1964a and b), who resurrected the Schmidt method and, for the physical reason given, called it the *quasi-particle method.* See also Scadron and Weinberg (1964); T. Sasakawa (1963); Hahn and Luddy (1981).

9.3 For exhibitions of the Fredholm method, see any book on integral equations, for example, Whittaker and Watson (1948), chap. 11; Courant and Hilbert (1953), chap. 3; Morse and Feshbach (1953), chaps. 8 and 9; F. Smithies (1958).

The Fredholm method was first applied to scattering problems by Jost and Pais (1951).

The recursion relations for the terms in the Fredholm expansion in terms of traces were first given by J. Plemelj (1904). The present treatment follows R. G. Newton (1961a).

Hadamard's inequality can be found in most of the books already mentioned. It is needed in the *complex* form, as stated in F. Smithies (1958).

The modification of the Fredholm formulas consisting in replacing the elements on the main diagonal by zeros was introduced by O. Kellogg (1902). Equations (9.84) and (9.85) are there credited to a mathematician named Haskins. The entire procedure was referred to by Hilbert as "pulling the poison tooth" of the Fredholm determinant (private communication from R. Jost). Other relevant mathematical papers on Fredholm equations and their solutions are J. Tamarkin (1926); Hille and Tamarkin (1930, 1931, 1934); C. van Winter (1971).

For an application of Fredholm theory to the iterated Lippmann-Schwinger equation, see N. N. Khuri (1957). See also J. Schwinger (1954); M. Baker (1958); I. Manning (1965); W. P. Reinhardt (1970).

9.4 The content of this section is based on R. G. Newton (1963b). For other perturbation treatments see Gammel and McDonald (1966) (on Padé approximants); R. Blankenbecler (1966); B. Michalik (1968); I. Manning (1968); Saskawa and Sawada (1977a).

Problems

1 For what values of γ does the series $\sum_0^\infty \gamma^n K^n$ converge absolutely if K is the operator (*a*) a of Prob. 5, Chap. 7; (*b*) a of Prob. 6, Chap. 7; (*c*) $aa^\dagger$ of Prob. 6, Chap. 7; (*d*) $a^\dagger a$ of Prob. 6, Chap. 7?

2 Show that, if K is compact, hermitian, and defined everywhere in the Hilbert space, then $\sum \gamma^n K^n$ must have a finite radius of convergence; i.e., if $|\gamma|$ is chosen large enough, the series must diverge.

3 Show that unless H' is positive semidefinite, the hamiltonian $H = H_0 + \gamma H'$

must produce at least one bound state when γ is chosen sufficiently large (positive). If H' is negative semidefinite, does it follow that by increasing γ one must be able to introduce more than one bound state?

4 Calculate the first Born approximation to the differential scattering cross section in the center-of-mass and in the laboratory frame for the following potentials: (*a*) the Yukawa potential $V(r) = \gamma e^{-\mu r}/r$; (*b*) the exponential potential $V(r) = \gamma e^{-\mu r}$; (*c*) the square well (or wall) $V(r) = \gamma(r < R)$, $0(r > R)$; (*d*) the δ-function shell $V(r) = \gamma\delta(r - r_0)$; (*e*) $V(r) = \gamma e^{-\mu r} r^{-3/2}$; (*f*) $V(r) = \gamma e^{-\mu^2 r^2}/r$; (*g*) $V(r) = \gamma e^{-\mu^2 r^2}$ (gaussian); (*h*) $V(r) = \gamma r^{-1}$ (Coulomb). Draw polar diagrams. Calculate the first Born approximations to the total scattering cross sections for the same potentials. Estimate in each case for what energies you expect these approximations to be good.

5 One may be tempted to calculate approximate bound-state energies by calculating the Fredholm determinant to first order, $\Delta \simeq \gamma \operatorname{tr} K$, and then equating it to zero. But the result may be very bad because when $\gamma \operatorname{tr} K = 1$, the expansion can be expected to converge slowly. Nevertheless this procedure is simple as well as sometimes reliable. Give a simple condition under which it can be expected to work. [*Hint*: Consider Eq. (9.81a)].

Can you apply this method to the calculation of bound states of the potentials of Prob. 4, for example? If not, suggest a simple modification. Does this agree with (9.37)?

6 Suppose that the interaction between two particles contains a part $H_1(\mathbf{p})$ that depends only on the *relative* momentum $\mathbf{p}$. Is the continuous spectrum of H (in the center-of-mass system) still the same as that of H_0, or is it changed? If so, in what way?

7 Prove that if $\operatorname{tr}[G(E)H']^4 > \operatorname{tr}[G(E)H']^2$ for a given $E \leq 0$, then the hamiltonian $H = H_0 + H'$ has at least one bound state of energy less than E.

CHAPTER 10

Single-Channel Scattering (Three-Dimensional Analysis in Specific Representations)

10.1 The Scattering Equation in the One-Particle Case

10.1.1 Preliminaries. It is simplest to start the discussion of the integral equations for the scattering-state vectors from the time-dependent point of view again. We are interested primarily in the integral equations (6.15) and (6.17) or (6.21) and (6.22). These equations were seen in Sec. 6.3 to be equivalent to (6.33) and (6.34),

$$\Psi_{\substack{\text{in}\\\text{out}}}(t) = \Omega^{(\pm)\dagger}\Psi(t).$$

Furthermore it was shown there that $\Omega^\dagger$ annihilates the bound states of H [(6.40)]. If this statement is written with (6.24) or (6.29) for $\Omega^{(+)}$ or $\Omega^{(-)}$, it amounts to the integral equation

$$\Psi_{\text{bd}}(t) = \int_{-\infty}^{\infty} dt'\, G^{(\pm)}(t - t')H'\Psi_{\text{bd}}(t') \tag{10.1}$$

for bound states of H. Therefore, if H has bound states, the homogeneous analogs of Eqs. (6.15) and (6.17) have nontrivial solutions and consequently the solutions of the inhomogeneous equations, if they exist, certainly are not unique. How can they be made unique?

Since the bound states of H have a purely oscillatory time dependence,

$$\Psi_{\text{bd}}(t) = \Psi_{\text{bd}}(E)e^{-iEt}$$

the integral

$$\int_{-\infty}^{t} dt' \, e^{-iH_0(t-t')} H' \Psi_{bd}(t')$$

surely does not converge absolutely. The integral equation (10.1) can therefore be given at most a rather formal meaning. Since a "reasonable" H has only a finite number of bound states,[1] these cannot be used to form wave packets that tend to zero as $t \to \pm\infty$ (weakly).[2] In fact, it was argued in Sec. 6.8.3 that the differences between $\Psi^{(+)}$ and Ψ_{in} and between $\Psi^{(-)}$ and Ψ_{out} should vanish *strongly* as $t \to \pm\infty$, respectively. One device to ensure this is adiabatic switching. If H' is sufficiently localized (what this means exactly is to be discussed later) and Ψ_{in} and Ψ_{out} are *wave packets* then that device is not necessary. There exist, then, solutions of (6.15) and (6.17) for which

$$\Psi^{(+)}(t) - \Psi_{in}(t) \underset{t\to-\infty}{\Longrightarrow} 0$$

$$\Psi^{(-)}(t) - \Psi_{out}(t) \underset{t\to+\infty}{\Longrightarrow} 0.$$

Since Ψ_{in} and Ψ_{out} tend to zero weakly and the $\Psi_{bd}(t)$ oscillate, these demands serve to make the solutions of (6.15) and (6.17) unique. There are no other solutions of the homogeneous equations, except those made up of the bound states of H. Otherwise $\Omega^\dagger$ would annihilate other states and Ω could not be isometric.

We now take the Fourier transforms of the time-dependent equations and arrive at the Lippmann–Schwinger equations (7.15) and (7.15a). The situation is then somewhat different because the energy is fixed. The troublesome bound states of H may lie at energies other than those for which the equation is to be solved. Since we are at present concerned with the case in which a single particle is scattered by an external target or force, the scattering, or continuous, spectrum is well separated from the discrete, or bound-state, spectrum. We are interested in solving (7.15) for $E > 0$, whereas the bound states lie in the region $E < 0$. There is, then, no essential difficulty in the existence and uniqueness of solutions to (7.15), or to (7.11) for the Green's functions, or to (7.47) for the T operator. We shall examine this point more explicitly below.

10.1.2 The Coordinate Representation. Let us first consider the free states and the Green's functions in the coordinate representation. For a single free particle of mass m and zero spin, the simultaneous eigenstates of the kinetic energy H_0 and of the momentum $\mathbf{p}$ are

1 See Chap. 12 for a more exact criterion under which H has only a finite number of bound states. The Coulomb interaction is, of course, *not* "reasonable" in this sense. Its slowly decreasing tail leads to an infinite number of bound states.

2 See Sec. 6.8 for the definition of weak and strong convergence.

$$\psi_0(\mathbf{k},\mathbf{r}) = \frac{(mk)^{1/2}}{(2\pi)^{3/2}} \exp(i\mathbf{k}\cdot\mathbf{r}) \tag{10.2}$$

if $\mathbf{k}$ is the eigenvalue of $\mathbf{p}$. The energy, of course, is related to $\mathbf{k}$ by

$$E = \frac{k^2}{2m}.$$

If the particle has spin s and z component of spin v, then

$$\psi_0(\mathbf{k}sv,\mathbf{r}) = \frac{(mk)^{1/2}}{(2\pi)^{3/2}} \chi_v^s \exp(i\mathbf{k}\cdot\mathbf{r}) \tag{10.2a}$$

if χ is the relevant normalized spin function. These free wave functions are normalized as in (8.8),

$$\int d\mathbf{r}\, \psi_0^*(\mathbf{k}sv,\mathbf{r}) \cdot \psi_0(\mathbf{k}'s'v',\mathbf{r}) = \delta(E - E')\delta_\Omega(\hat{\mathbf{k}},\hat{\mathbf{k}}')\delta_{ss'}\delta_{vv'} \tag{10.3}$$

where δ_Ω is the solid-angle δ defined by (2.15) and the dot indicates the inner product in spin space. The variables $\mathbf{k}sv$ take the place of what previously was collectively called α. The set of free wave functions is complete,

$$\sum_{sv} \int_0^\infty dE \int d\Omega_k\, \psi_0(\mathbf{k}sv,\mathbf{r})\psi_0^*(\mathbf{k}sv,\mathbf{r}') = \delta(\mathbf{r} - \mathbf{r}'). \tag{10.4}$$

In practice it is usually not necessary to include the sum over s since one is mostly interested in expanding a given eigenfunction of the spin (magnitude).

The completeness (10.4) immediately yields coordinate representations of the time-dependent Green's functions (6.5) for a particle of spin s. For $t > 0$,

$$\begin{aligned} G_s^+(t;\mathbf{r},\mathbf{r}') &= -i\sum_v \int_0^\infty dE\, e^{-iEt} \int d\Omega_k\, \psi_0(\mathbf{k}sv,\mathbf{r})\psi_0^*(\mathbf{k}sv,\mathbf{r}') \\ &= -\frac{1}{4\pi^2}\frac{1}{|\mathbf{r}-\mathbf{r}'|}\sum_v \int_{-\infty}^\infty dk\, k \exp\left[-i\left(\frac{k^2 t}{2m} - k|\mathbf{r}-\mathbf{r}'|\right)\right]\chi_v^s\chi_v^{s*} \\ &= -\frac{m^{3/2}(1-i)}{4\pi^{3/2}t^{3/2}}\sum_v \exp\left(i\frac{m|\mathbf{r}-\mathbf{r}'|^2}{2t}\right)\chi_v^s\chi_v^{s*} \end{aligned} \tag{10.5}$$

and for $t < 0$

$$\begin{aligned} G_s^-(t;\mathbf{r},\mathbf{r}') &= i\sum_v \int_0^\infty dE\, e^{-iEt} \int d\Omega_k\, \psi_0(\mathbf{k}sv,\mathbf{r})\psi_0^*(\mathbf{k}sv,\mathbf{r}') \\ &= -\frac{m^{3/2}(1+i)}{4\pi^{3/2}|t|^{3/2}}\sum_v \exp\left(-i\frac{m|\mathbf{r}-\mathbf{r}'|^2}{2|t|}\right)\chi_v^s\chi_v^{s*}. \end{aligned} \tag{10.5a}$$

The time-independent Green's functions are obtained similarly,

$$\begin{aligned} G_s^{\pm}(E;\mathbf{r},\mathbf{r}') &= \sum_\nu \int_0^\infty dE' \int d\Omega_k \frac{\psi_0(\mathbf{k}s\nu,\mathbf{r})\psi_0^*(\mathbf{k}s\nu,\mathbf{r}')}{E - E' \pm i\varepsilon} \\ &= -\frac{m}{2\pi}\frac{\exp(\pm ik|\mathbf{r}-\mathbf{r}'|)}{|\mathbf{r}-\mathbf{r}'|}\sum_\nu \chi_\nu^s \chi_\nu^{s*}. \end{aligned} \tag{10.6}$$

They satisfy the equation

$$\left(\frac{1}{2m}\nabla^2 + E\right)G_s^{\pm}(E;\mathbf{r},\mathbf{r}') = \delta(\mathbf{r}-\mathbf{r}') \tag{10.7}$$

and boundary conditions that are given by their asymptotic behavior as $r \to \infty$, namely,

$$G_s^{+}(E;\mathbf{r},\mathbf{r}') \simeq -\frac{e^{ikr}}{r}\left(\frac{2\pi m}{k}\right)^{1/2} \sum_\nu \chi_\nu^s \psi_0^*(\mathbf{k}''s\nu,\mathbf{r}') \tag{10.8}$$

$$G_s^{-}(E;\mathbf{r},\mathbf{r}') \simeq -\frac{e^{-ikr}}{r}\left(\frac{2\pi m}{k}\right)^{1/2} \sum_\nu \psi_0(\mathbf{k}''s\nu,\mathbf{r}')\chi_\nu^{s*} \tag{10.8a}$$

where $\mathbf{k}'' = \hat{\mathbf{r}}k$. This identifies the physical significance of the superscripts $\pm$ from the time-independent point of view. The Green's function G^+ is that wave produced by a "unit source"[3] which at large distances contains *outgoing* spherical waves only, whereas G^- there contains only *incoming* spherical waves.

The derivation of the asymptotic values (10.8) and (10.8a) is based on the approximation

$$k|\mathbf{r}-\mathbf{r}'| \underset{r\to\infty}{\simeq} k(r - \mathbf{r}\cdot\mathbf{r}'r^{-1} + \cdots) = kr - \mathbf{k}''\cdot\mathbf{r}' + \cdots$$

in the exponential and $|\mathbf{r}-\mathbf{r}'| \simeq r$ in the denominator. To retain these values only is therefore a good approximation provided that

$$r \gg r' \tag{10.9}$$

and in addition

$$kr \gg (kr')^2. \tag{10.9a}$$

The first inequality is needed in the denominator, the second in the exponential.

We notice that the Green's functions in the coordinate representation depend on the coordinate *difference* $\mathbf{r} - \mathbf{r}'$ only. From an operator point of view this is because H_0, and therefore also $(E - H_0)^{-1}$, commutes with the momentum, which is the generator of displacements in space. From the point of view of the differential equation obeyed by G it is due to the fact that the differential operator on the left of (10.7) does not contain $\mathbf{r}$ explicitly and the inhomogeneity on the right contains $\mathbf{r} - \mathbf{r}'$ only, together with the imposition of the boundary condition at $r = \infty$.

3 We are talking here in electromagnetic language. There are no "sources" of the wave function.

The Scattering Amplitude. We may now look at the Lippmann–Schwinger integral equation (7.15) in the coordinate representation

$$\psi^{(+)}(\mathbf{k}sv;\mathbf{r}) = \psi_0(\mathbf{k}sv;\mathbf{r}) + \int d\mathbf{r}'\,d\mathbf{r}''\,G_s^+(E;\mathbf{r},\mathbf{r}'')(\mathbf{r}''|H|\mathbf{r}')\psi^{(+)}(\mathbf{k}sv;\mathbf{r}') \tag{10.10}$$

if $(\mathbf{r}''|H'|\mathbf{r}')$ is a (partial) matrix element of the (possibly spin-dependent) interaction in the coordinate representation. Deferring the question of the existence and uniqueness of solutions to (10.10) for the moment, we obtain the asymptotic behavior of $\psi^{(+)}$ as $r \to \infty$ from that of G^+. According to (10.8) and (10.2a)

$$\psi^{(+)}(\mathbf{k}sv;\mathbf{r}) \simeq \sum_{v'} \frac{(mk)^{1/2}}{(2\pi)^{3/2}} \chi_{v'}^s [\delta_{vv'} \exp(i\mathbf{k}\cdot\mathbf{r}) + r^{-1}\exp(ikr)A(\mathbf{k}'v',\mathbf{k}v)] \tag{10.11}$$

where $\mathbf{k}' = k\mathbf{r}/r$ and

$$\begin{aligned} A(\mathbf{k}'v',\mathbf{k}v) &= -\frac{(2\pi)^2}{k}\int d\mathbf{r}'\,d\mathbf{r}\,\psi_0^*(\mathbf{k}'sv';\mathbf{r}')(\mathbf{r}'|H'|\mathbf{r})\psi^{(+)}(\mathbf{k}sv;\mathbf{r}) \\ &= -\frac{(2\pi)^2}{k}T(\mathbf{k}'v',\mathbf{k}v). \end{aligned} \tag{10.12}$$

In view of (10.9) and (10.9a), this asymptotic value is a good approximation provided that r is large compared with the size D of the interaction region *and* large compared with D multiplied by the ratio of D to the wavelength,

$$r \gg D \qquad r \gg \frac{D^2}{\lambda}. \tag{10.13}$$

At low energies, when $kD \ll 1$, only the first inequality matters. At all other energies (10.13) implies

$$r \gg \lambda, \tag{10.13a}$$

and at high energies, when $kD \gg 1$, the second inequality in (10.13) is stronger than (10.13a).

Equation (10.11) shows immediately the physical significance of the superscript + on ψ, from a time-independent point of view. The function $\psi^{(+)}$ differs from ψ_0 at large distances by an *outgoing spherical wave* only. This is in accordance with its physical use and with the meaning of + from the time-dependent point of view. The free wave function ψ_0 describes the collimated beam sent in and controlled. The complete wave function differs from this at large distances by the outgoing scattered wave only. Furthermore, the spherical wave observed at the large distance r from the target can be considered "locally plane" and (10.11) shows that the scattered flux of particles with spin projection v', relative to the incident flux, is

$$\frac{d\sigma}{d\Omega} = |A(\mathbf{k}'v',\mathbf{k}v)|^2 \tag{10.14}$$

if the incident beam has the wave vector $\mathbf{k}$ and spin projection ν. The explicit formulas (10.12) and (10.14) argee with (8.14) and (7.44).

One may now make the connection with the time-dependent theory by forming wave packets out of (10.10) and considering $t \to \pm\infty$. The argument being identical to that of Sec. 1.3.1, we shall not go through it again.

It is worthwhile to write out the simple special case of spin 0 and local interaction explicitly. If we use a wave function that satisfies the integral equation

$$\psi^{(+)}(\mathbf{k},\mathbf{r}) = \exp(i\mathbf{k}\cdot\mathbf{r}) - \frac{1}{4\pi}\int d\mathbf{r}'\,\frac{\exp(ik|\mathbf{r}-\mathbf{r}'|)}{|\mathbf{r}-\mathbf{r}'|}\mathscr{V}(\mathbf{r}')\psi^{(+)}(\mathbf{k},\mathbf{r}') \tag{10.10a}$$

and hence behaves asymptotically for large r as

$$\psi^{(+)}(\mathbf{k},\mathbf{r}) \simeq \exp(i\mathbf{k}\cdot\mathbf{r}) + r^{-1}\exp(ikr)A(\mathbf{k}',\mathbf{k}) \tag{10.11a}$$

then

$$A(\mathbf{k}',\mathbf{k}) = -\frac{1}{4\pi}\int d\mathbf{r}\exp(-i\mathbf{k}'\cdot\mathbf{r})\mathscr{V}(\mathbf{r})\psi^{(+)}(\mathbf{k},\mathbf{r}). \tag{10.12a}$$

The function $\mathscr{V}(r)$ is connected to the interaction hamiltonian by

$$\mathscr{V}(\mathbf{r})\delta(\mathbf{r}-\mathbf{r}') = 2m(\mathbf{r}|H'|\mathbf{r}').$$

Note that the equality of the two versions of (7.40), together with the fact that, for spin 0, $\psi^{(-)*}(\mathbf{k},\mathbf{r}) = \psi^{(+)}(-\mathbf{k},\mathbf{r})$, implies the special form

$$A(\mathbf{k}',\mathbf{k}) = A(-\mathbf{k},-\mathbf{k}') \tag{10.15}$$

of the reciprocity theorem.

In the case of particles with spin, one may also write the amplitude in the form of a spin matrix. For this purpose it is simplest to use (7.41) and write the amplitude matrix

$$\begin{aligned} A(\mathbf{k}',\mathbf{k}) &= -\frac{m}{2\pi}\int d\mathbf{r}'\,d\mathbf{r}\exp(i\mathbf{k}\cdot\mathbf{r} - i\mathbf{k}'\cdot\mathbf{r}')(\mathbf{r}'|T|\mathbf{r}) \\ &= -(2\pi)^2 m(\mathbf{k}'|T|\mathbf{k}) \equiv -(2\pi)^2 m T_{\mathbf{k}'\mathbf{k}} \end{aligned} \tag{10.16}$$

where

$$T = H' + H'\mathscr{G}^+H' = H' + H'G^+H' + H'G^+H'G^+H' + \cdots.$$

The cross section for an unpolarized beam, with all final polarizations counted, is then given by

$$\begin{aligned} \frac{d\sigma}{d\Omega}(\mathbf{k}',\mathbf{k}) &= \frac{(2\pi)^4 m^2 \operatorname{tr}(\mathbf{k}'|T|\mathbf{k})(\mathbf{k}|\mathbf{T}^\dagger|\mathbf{k}')}{2s+1} \\ &= \frac{(2\pi)^4 m^2 \operatorname{tr} T_{\mathbf{k}'\mathbf{k}}T^*_{\mathbf{k}'\mathbf{k}}}{2s+1}, \end{aligned} \tag{10.17}$$

s being the initial spin and the trace being taken over the spin matrices only. It arises from a use of the completeness relation

$$\sum_{\nu} \chi_\nu \chi_\nu^\dagger = \mathbb{1}$$

of the spin functions. If the initial beam is fully or partially polarized, then we must insert its density matrix (8.49) in the trace:

$$\frac{d\sigma}{d\Omega} = (2\pi)^4 m^2 \operatorname{tr} T_{\mathbf{k}'\mathbf{k}} \rho T^*_{\mathbf{k}'\mathbf{k}}. \tag{10.18}$$

Finally, if the spins of the emerging particles are measured, the cross section for a particular spin direction, or for a particular spin-projection quantum number, is obtained by inserting the corresponding projection matrix P' in the trace,

$$\frac{d\sigma}{d\Omega} = (2\pi)^4 m^2 \operatorname{tr} P' T_{\mathbf{k}'\mathbf{k}} \rho T^*_{\mathbf{k}'\mathbf{k}}.$$

This means that the density matrix of the beam scattered in the direction $\mathbf{k}'$ is given by

$$\rho' = \frac{\sum_\nu P_\nu \operatorname{tr} P_\nu T_{k'k} \rho T^*_{k'k}}{\operatorname{tr} T_{k'k} \rho T^*_{k'k}} \tag{10.19}$$

if $P_\nu = \chi_\nu \chi_\nu{}^\dagger$ are the projections for the various spin-projection quantum numbers.

Double-scattering cross sections can be easily calculated directly from (10.16). If the two targets are sufficiently far apart [in the sense of (10.13)] and the vector $\mathbf{k}''$ leads from the first to the second, then the double-scattering cross section for an unpolarized beam is given by

$$\frac{d\sigma}{d\Omega}(\mathbf{k}', \mathbf{k}'', \mathbf{k}) = \frac{(2\pi)^8 m^4 \operatorname{tr} T^{(2)}_{\mathbf{k}'\mathbf{k}''} T^{(1)}_{\mathbf{k}''\mathbf{k}} T^{(2)*}_{\mathbf{k}'\mathbf{k}''} T^{(1)*}_{\mathbf{k}''\mathbf{k}}}{2s+1}. \tag{10.20}$$

The superscripts here refer to the two targets, which need not be identical. If the first scattering produces a transverse polarization, the second scattering serves to analyze it by showing an azimuthal asymmetry.[4]

10.1.3 The Momentum Representation. In the momentum representation the free state of a particle of spin (s, ν) and momentum $\mathbf{k}$ is given by

$$\psi_0(\mathbf{k}s\nu; \mathbf{p}) = (mk)^{1/2} \chi^s_\nu \delta(\mathbf{k} - \mathbf{p}). \tag{10.21}$$

This is normalized according to (8.8),

$$\int d\mathbf{p}\, \psi_0^*(\mathbf{k}s\nu; \mathbf{p}) \psi_0(\mathbf{k}'s'\nu'; \mathbf{p}) = \delta(E - E')\delta_\Omega(\hat{\mathbf{k}}, \hat{\mathbf{k}}')\delta_{ss'}\delta_{\nu\nu'} \tag{10.22}$$

4 A polarization perpendicular to the scattering plane may arise even if the target is spherically symmetric. A simple example is that of fast electrons scattered by a spherically symmetric electric field (say, a Coulomb field). The moving electrons see an effective magnetic field which interacts with their magnetic moment and gives rise to a term proportional to $\boldsymbol{\sigma} \cdot \mathbf{r} \times \mathbf{p}$ in the hamiltonian (*spin-orbit* interaction). The result is a polarization. See N. F. Mott (1929 and 1932).

which is checked by using (8.9). These wave functions form a complete set,

$$\sum_{sv}\int_0^\infty dE\int d\Omega_k\,\psi_0(\mathbf{k}sv;\mathbf{p})\psi_0^*(\mathbf{k}sv;\mathbf{p}') = \delta(\mathbf{p}-\mathbf{p}'). \tag{10.23}$$

The Green's functions are easily obtained from this by application of (6.5) to (10.23). For $t > 0$,

$$\begin{aligned} G_s^+(t;\mathbf{p},\mathbf{p}') &= -i\sum_v\int_0^\infty dE\,e^{-iEt}\int d\Omega_k\psi_0(\mathbf{k}sv;\mathbf{p})\psi_0^*(\mathbf{k}sv;\mathbf{p}') \\ &= -i\sum_v \chi_v^s\chi_v^{s*}e^{-iEt}\delta(\mathbf{p}-\mathbf{p}') \end{aligned} \tag{10.24}$$

where, of course, $E = \mathbf{p}^2/2m$. Similarly, for $t < 0$,

$$G_s^-(t;\mathbf{p},\mathbf{p}') = \sum_v \chi_v^s\chi_v^{s*}e^{-iEt}\delta(\mathbf{p}-\mathbf{p}'). \tag{10.24a}$$

In an analogous fashion we get the time-independent Green's functions,

$$\begin{aligned} G_s^\pm(E;\mathbf{p},\mathbf{p}') &= \sum_v\int_0^\infty dE'\int d\Omega_{k'}\frac{\psi_0(\mathbf{k}'sv;\mathbf{p})\psi_0^*(\mathbf{k}'sv;\mathbf{p}')}{E-E'\pm i\varepsilon} \\ &= \frac{\sum_v \chi_v^s\chi_v^{s*}\delta(\mathbf{p}-\mathbf{p}')}{E-(p^2/2m)\pm i\varepsilon}. \end{aligned} \tag{10.25}$$

As they must, the Green's functions are diagonal in $\mathbf{p}$, because the momentum operator commutes with H_0.

The Lippmann–Schwinger equation in the momentum representation reads

$$\begin{aligned} \psi^{(+)}(\mathbf{k}sv;\mathbf{p}) &= (mk)^{1/2}\chi_v^s\delta(\mathbf{k}-\mathbf{p}) \\ &\quad + \sum_{v'}\int d\mathbf{p}'\,\chi_{v'}^s\chi_v^{s*}2m\frac{(\mathbf{p}|H'|\mathbf{p}')}{k^2-p^2+i\varepsilon}\psi^{(+)}(\mathbf{k}sv;\mathbf{p}'). \end{aligned} \tag{10.26}$$

The momentum-space matrix elements of the interaction are connected with those in coordinate space by Fourier transformation,

$$(\mathbf{p}|H'|\mathbf{p}') = (2\pi)^{-3}\int d\mathbf{r}\,d\mathbf{r}'\exp[i(\mathbf{p}'\cdot\mathbf{r}'-\mathbf{p}\cdot\mathbf{r})](\mathbf{r}|H'|\mathbf{r}'). \tag{10.27}$$

If H' is a local potential,

$$(\mathbf{r}|H'|\mathbf{r}') = V(\mathbf{r})\delta(\mathbf{r}-\mathbf{r}')$$

then

$$(\mathbf{p}|H'|\mathbf{p}') = (2\pi)^{-3}\int d\mathbf{r}\exp[i(\mathbf{p}'-\mathbf{p})\cdot\mathbf{r}]V(\mathbf{r}) = \overline{V}(\mathbf{p}-\mathbf{p}') \tag{10.28}$$

which is a function only of $\mathbf{p}-\mathbf{p}'$. This is so simply because H' commutes with the position operator, which generates displacements in momentum space.

10.1.4 Separable Interactions. If the interaction is *separable*, i.e.,

$$(\mathbf{p}|H'|\mathbf{p}') = -f(\mathbf{p})f^*(\mathbf{p}'), \tag{10.29}$$

then (10.26) can be solved explicity. (This being negative semidefinite, H' is attractive; to make it repulsive, the minus sign must be replaced by a plus sign.) Assume for simplicity that the spin is 0 (or that H' is spin-independent). Then (10.26) is solved by a simple calculation to yield

$$\psi^{(+)}(\mathbf{k};\mathbf{p}) = (mk)^{1/2}\delta(\mathbf{k}-\mathbf{p}) - \frac{2m^{3/2}k^{1/2}}{k^2-p^2+i\varepsilon}\frac{f(\mathbf{p})f^*(\mathbf{k})}{1+\int d\mathbf{p}'\dfrac{2m|f(\mathbf{p}')|^2}{k^2-p'^2+i\varepsilon}}. \tag{10.30}$$

Similarly we can write down explicitly the complete Green's functions,

$$\mathscr{G}^{\pm}(E;\mathbf{p},\mathbf{p}') = \frac{\delta(\mathbf{p}-\mathbf{p}')}{E-(p^2/2m)\pm i\varepsilon} - \frac{f(\mathbf{p})f^*(\mathbf{p}')}{[(E-(p^2/2m)\pm i\varepsilon][E-(p'^2/2m)\pm i\varepsilon]\Delta^{\pm}(E)} \tag{10.31}$$

$$\Delta^{\pm}(E) = 1 + \int d\mathbf{p}\frac{|f(\mathbf{p})|^2}{E-(p^2/2m)\pm i\varepsilon} \tag{10.32}$$

and the **T** matrix according to (7.41)

$$T(\mathbf{p}',\mathbf{p}) = \frac{-mpf(\mathbf{p}')f^*(\mathbf{p})}{\Delta^+(E)}. \tag{10.33}$$

Notice that the **T** matrix, and consequently the differential cross section, differs from its Born-approximation value by an angle-independent factor. If $f(\mathbf{p})$ is invariant under rotation, then **T** is independent of the angle between $\mathbf{p}$ and $\mathbf{p}'$. The scattering cross section is isotropic.

The explicit formulas (10.30) and (10.33) may be used to illustrate a number of general results. The Born series, for instance, is obtained by the expansion of $1/\Delta^+(E)$ in powers of the integral in it. It is easy to check explicitly by means of (9.79a) that Δ^+ is the Fredholm determinant of the kernel of the Lippmann–Schwinger equation. Since $\Delta(E)$ monotonically decreases from 1 at $E = -\infty$ to

$$\Delta(0) = 1 - 2m\int d\mathbf{p}\,p^{-2}|f(\mathbf{p})|^2$$

at $E = 0$, it may have at most a single zero in that region. Thus an attractive separable interaction produces exactly one bound state if $\Delta(0) \le 0$, that is, if

$$2m\int d\mathbf{p}\,p^{-2}|f(\mathbf{p})|^2 \ge 1$$

and none otherwise.

Generalization. These simple results are easily generalized to an interaction which is a sum of separable terms, such as

$$(\mathbf{p}|H'|\mathbf{p}') = \sum_n \iota_n f_n(\mathbf{p}) f_n^*(\mathbf{p}') \tag{10.34}$$

where

$$\iota_n = \pm 1.$$

The wave function is then found to be

$$\psi^{(+)}(\mathbf{k};\mathbf{p}) = (mk)^{1/2}\delta(\mathbf{k}-\mathbf{p}) + 2m^{3/2}k^{1/2}\frac{\sum_{nm}\iota_n f_n(\mathbf{p})[(\mathbf{\Delta}^+)^{-1}]_{nm} f_m^*(\mathbf{k})}{k^2 - p^2 + i\varepsilon} \tag{10.35}$$

where $\mathbf{\Delta}^{\pm}$ is the matrix,

$$\Delta_{nm}^{\pm} = \delta_{nm} - \int d\mathbf{p}\frac{f_n^*(\mathbf{p}) f_m(\mathbf{p})}{E - (p^2/2m) \pm i\varepsilon}\iota_m. \tag{10.36}$$

The complete Green's functions are given by

$$\begin{aligned}\mathscr{G}^{\pm}(E;\mathbf{p},\mathbf{p}') = {} & \frac{\delta(\mathbf{p}-\mathbf{p}')}{E - (p^2/2m) \pm i\varepsilon} \\ & - \frac{\sum_{nm}\iota_n f_n(\mathbf{p})[(\mathbf{\Delta}^{\pm})^{-1}]_{nm} f_m^*(\mathbf{p}')}{[E - (p^2/2m) \pm i\varepsilon][E - (p'^2/2m) \pm i\varepsilon]}\end{aligned} \tag{10.37}$$

and the $\mathbf{T}$ matrix,

$$T(\mathbf{p}',\mathbf{p}) = mp\sum_{nm}\iota_n f_n(\mathbf{p}')[(\mathbf{\Delta}^+)^{-1}]_{nm} f_m^*(\mathbf{p}). \tag{10.38}$$

The Fredholm determinant of the Lippmann Schwinger kernel is now

$$\Delta^+(E) = \det \mathbf{\Delta}^+(E).$$

As a simple illustration we may take the interaction

$$(\mathbf{p}|H'|\mathbf{p}') = -f(p)f^*(p')\mathbf{p}\cdot\mathbf{p}' \tag{10.39}$$

in which $f(p)$ is a function of the magnitude of $\mathbf{p}$ only, so that the hamiltonian is rotationally invariant. In this case (10.36) shows that

$$\begin{aligned}\mathbf{\Delta}^{\pm} &= \mathbb{1}\Delta^{\pm} \\ \Delta^{\pm} &= 1 + 4\pi\int_0^{\infty} dp\frac{p^2|f(p)|^2}{E - (p^2/2m) \pm i\varepsilon},\end{aligned} \tag{10.40}$$

$\mathbb{1}$ being a 3×3 unit matrix, and

$$T(\mathbf{p}',\mathbf{p}) = \frac{-mp\,\mathbf{p}'\cdot\mathbf{p}f(p')f^*(p)}{\Delta^+(E)}. \tag{10.41}$$

The resulting cross section is proportional to the square of the cosine of the angle between $\mathbf{p}$ and $\mathbf{p}'$. Other angle dependences are easily constructed similarly.

10.2 The Scattering Equations in the Two-Particle Case (Elimination of Center-of-Mass Motion)

When we consider the physically realistic case of two particles in mutual interaction, without external forces, and we go through a discussion like that of Sec. 10.1.1 for single particles, we arrive at difficulties typical of all cases of more than one particle. If there is a bound state of energy E_{bd} in the center-of-mass coordinate system, it manifests itself in an arbitrary coordinate frame at all energies E greater than E_{bd}, because the difference $E - E_{bd}$ may simply be the kinetic energy of the bound system. In the time-independent theory this has the consequence that the bound state can no longer be made innocuous by fixing the total energy to be different from E_{bd}, as it could in the single-particle case. For all energies $E > E_{bd}$ the homogeneous version of the Lippmann–Schwinger equation now has a solution, and hence the solution of the inhomogeneous equation is ill defined.

Another, more general way of seeing the trouble is by realizing that, if there is an interparticle interaction only, then the *total* hamiltonian, as well as the kinetic energy alone, is invariant under displacement of the whole system. As a result, both H_0 and H must commute with the displacement generator, the total momentum $\mathbf{P}$,

$$[H_0, \mathbf{P}] = 0 \qquad [H, \mathbf{P}] = 0 \tag{10.42}$$

and, similarly, the kernel of the Lippmann–Schwinger equation,

$$[K(E), \mathbf{P}] = [(E - H_0 + i\varepsilon)^{-1} H', \mathbf{P}] = 0. \tag{10.43}$$

The lemma of Sec. 7.3.2 then immediately tells us that $K(E)$ cannot be a compact operator. Stated more concretely, (10.43) implies that matrix elements of K contain a δ function of the total momentum:

$$(\mathbf{P} \cdots |K| \mathbf{P}' \cdots) = \delta(\mathbf{P} - \mathbf{P}')(\cdots |K| \cdots). \tag{10.44}$$

This, of course, is nothing but the statement of the physical fact that the *total momentum is conserved* by such an interaction. A kernel that contains a δ function cannot be compact, and the Lippmann–Schwinger equation seems to be in trouble.

The remedy, however, is very simple just because of (10.44). Both terms on the right-hand side of the Green's-function equation,

$$\mathscr{G} = G + K\mathscr{G},$$

contain that momentum-conserving δ function, and so does $\mathscr{G}$. We may therefore factor it out,

$$\mathscr{G} = \delta(\mathbf{P} - \mathbf{P}')\mathscr{G}_R$$
$$G = \delta(\mathbf{P} - \mathbf{P}')G_R \qquad K = \delta(\mathbf{P} - \mathbf{P}')K_R$$

and get an equation for the reduced function,

$$\mathscr{G}_R = G_R + K_R\mathscr{G}_R$$

in which the total momentum is *fixed* and the troublesome δ function has disappeared. The same holds for the other scattering equations, for the state vector, and for the **T** matrix. A special case is that in which the fixed total momentum is set equal to zero. One then works in the *center-of-mass*, or *barycentric*, coordinate system.

The trouble inherent in the scattering equations for two particles is thus seen to be circumvented by rather trivial means. The reason for going through the argument as we did, rather than going directly to the center-of-mass system, is that the difficulty here is of the same nature as for more than two particles. In that case, though, it cannot be solved so simply, as we shall see in Sec. 17.4.

Center-of-Mass Coordinates. We proceed to the explicit introduction of the center of mass as a coordinate. If the two particles have masses m_1 and m_2 and coordinates $\mathbf{r}_1$ and $\mathbf{r}_2$, then one introduces the center-of-mass coordinate $\mathbf{R}$ and the relative coordinate $\mathbf{r}$,

$$\mathbf{R} = \frac{m_1\mathbf{r}_1 + m_2\mathbf{r}_2}{m_1 + m_2}, \qquad \mathbf{r} = \mathbf{r}_2 - \mathbf{r}_1, \tag{10.45a}$$

as a new set of coordinates. The Jacobian of this transformation is

$$\frac{\partial(\mathbf{R},\mathbf{r})}{\partial(\mathbf{r}_1,\mathbf{r}_2)} = 1,$$

and the canonically conjugate momenta are the total momentum $\mathbf{K}$ and the momentum of the relative motion $\mathbf{k}$, given by

$$\mathbf{K} = \mathbf{k}_1 + \mathbf{k}_2, \qquad \mathbf{k} = \frac{\mathbf{k}_2 m_1 - \mathbf{k}_1 m_2}{m_1 + m_2}, \tag{10.45b}$$

while the total energy is expressed as

$$E_{\text{tot}} = E_{CM} + E_{12} = \frac{K^2}{2M} + \frac{k^2}{2\mu}. \tag{10.46}$$

The kinetic-energy operator in the coordinate representation becomes

$$\begin{aligned} H_0 &= -\frac{1}{2m_1}\nabla_1^2 - \frac{1}{2m_2}\nabla_2^2 \\ &= -\frac{1}{2M}\nabla_R^2 - \frac{1}{2\mu}\nabla_r^2 \end{aligned} \tag{10.47}$$

with an obvious meaning for the various Laplace operators and

$$M = m_1 + m_2 \qquad \mu = \frac{m_1 m_2}{m_1 + m_2}. \tag{10.48}$$

The free wave function for a particle of spin 0 can now be written in the coordinate representation,

$$\psi_0(\mathbf{K},\mathbf{k};\mathbf{R},\mathbf{r}) = (2\pi)^{-3/2} \exp[i(\mathbf{K}\cdot\mathbf{R})]\psi_0(\mathbf{k};\mathbf{r}). \tag{10.49}$$

The normalization of (10.49) is such that

$$\int d\mathbf{R}\, d\mathbf{r}\, \psi_0^*(\mathbf{K},\mathbf{k};\mathbf{R},\mathbf{r})\psi_0(\mathbf{K}',\mathbf{k}';\mathbf{R},\mathbf{r}) = \delta(\mathbf{K}-\mathbf{K}')\delta(E_{12}-E'_{12})\delta_\Omega(\hat{\mathbf{k}},\hat{\mathbf{k}}') \tag{10.50}$$

and

$$\int d\mathbf{K} \int_0^\infty dE_{12} \int d\Omega_k \psi_0(\mathbf{K},\mathbf{k};\mathbf{R},\mathbf{r})\psi_0^*(\mathbf{K},\mathbf{k};\mathbf{R}',\mathbf{r}') = \delta(\mathbf{R}-\mathbf{R}')\delta(\mathbf{r}-\mathbf{r}') \tag{10.51}$$

since

$$\mathbf{r}_1\cdot\mathbf{k}_1 + \mathbf{r}_2\cdot\mathbf{k}_2 = \mathbf{R}\cdot\mathbf{K} + \mathbf{r}\cdot\mathbf{k}$$

and

$$d\mathbf{k}_1\, d\mathbf{k}_2 = d\mathbf{K}\, d\mathbf{k}$$
$$d\mathbf{r}_1\, d\mathbf{r}_2 = d\mathbf{R}\, d\mathbf{r}.$$

The factor $(2\pi)^{-3/2}e^{i\mathbf{K}\cdot\mathbf{R}}$ produces just the needed momentum-conserving δ function. As a result, $|\mathbf{K}|$ or E_{CM} can be fixed and conveniently set equal to zero. We are then in the barycentric coordinate system. The hamiltonian looks just like that of a single particle, except that the single-particle mass m is replaced everywhere by the reduced mass μ, the meaning of the coordinate $\mathbf{r}$ is that of the interparticle distance, and $\mathbf{k}$ is the momentum of one of the two particles.

If the particles have spins s_1 and s_2, (10.49) must be multiplied by the direct product of the two spin functions, $\chi_{\nu_1}{}^{s_1}\chi_{\nu_2}{}^{s_2}$. This is then written in terms of the eigenstates of the total spin by means of Clebsch-Gordan coefficients,

$$\chi_{\nu_1}^{s_1}\chi_{\nu_2}^{s_2} = \sum_{s\nu} C(s_1 s_2 s, \nu_1 \nu_2 \nu)\chi_\nu^{ss_1s_2} \tag{10.52}$$

where s runs from $|s_1 - s_2|$ to $s_1 + s_2$ and ν from $-s$ to s. The $\chi_\nu{}^{ss_1s_2}$ are eigenstates of the individual spin magnitudes,

$$\mathbf{S}_1^2\chi_\nu^{ss_1s_2} = s_1(s_1+1)\chi_\nu^{ss_1s_2} \qquad \mathbf{S}_2^2\chi_\nu^{ss_1s_2} = s_2(s_2+1)\chi_\nu^{ss_1s_2} \tag{10.53}$$

of the total spin magnitude

$$\mathbf{S}^2\chi_\nu^{ss_1s_2} = (\mathbf{S}_1+\mathbf{S}_2)^2\chi_\nu^{ss_1s_2} = s(s+1)\chi_\nu^{ss_1s_2} \tag{10.54}$$

and of the z component of the total spin,

$$S_z\chi_v^{ss_1s_2} = (S_{1z} + S_{2z})\chi_v^{ss_1s_2} = v\chi_v^{ss_1s_2}. \tag{10.55}$$

The free barycentric wave function now carries the labels

$$\psi_0(\mathbf{k}; s_1, s_2, s, v; \mathbf{r})$$

and so does the complete one. In the simple instance of elastic scattering of particles with intrinsic spin, the quantum numbers s_1 and s_2 are conserved and may be suppressed. If the interaction is spin-dependent, however, both s and v may be changed in the scattering. Two spin $\frac{1}{2}$ particles initially in a singlet state ($s = 0$) may emerge in the triplet state ($s = 1$).

Similar considerations apply to other internal degrees of freedom of the scattered particles, such as *isotopic spin*, *strangeness*, etc.

We have thus reduced the one- and two-particle scattering cases to similar formalisms in which only one coordinate appears. The only essential difference is that in the latter case the total spin may change, whereas in the former it usually cannot. The content of Secs. 10.1.2 to 10.1.4 is now transferable to the scattering of two particles without further delay.

Equations (10.11) and (10.12) must be replaced by

$$\psi^{(+)}(\mathbf{k}sv; \mathbf{r}) \simeq \sum_{s'v'} \frac{(\mu k)^{1/2}}{(2\pi)^{3/2}} \chi_{v'}^{s'} [\delta_{ss'}\delta_{vv'} \exp(i\mathbf{k}\cdot\mathbf{r}) + r^{-1}\exp(ikr)A(\mathbf{k}'s'v', \mathbf{k}sv)] \tag{10.56}$$

and

$$\begin{aligned} A(\mathbf{k}'s'v', \mathbf{k}sv) &= -\frac{(2\pi)^2}{k} \int d\mathbf{r}'\, d\mathbf{r}\, \psi_0^*(\mathbf{k}'s'v'; \mathbf{r}')(\mathbf{r}'|H'|\mathbf{r})\psi^{(+)}(\mathbf{k}sv; \mathbf{r}) \\ &= -\frac{(2\pi)^2}{k} T(\mathbf{k}'s'v', \mathbf{k}sv). \end{aligned} \tag{10.57}$$

10.3 Three-Dimensional Analysis of Potential Scattering

We return to the Lippmann–Schwinger equation (10.10) and assume that the interaction is described by a local spin-independent potential,

$$(\mathbf{r}|H'|r') = V(\mathbf{r})\delta(\mathbf{r} - \mathbf{r}').$$

We then need not carry along the spin labels on the wave function and write simply

$$\psi^{(+)}(\mathbf{k}; \mathbf{r}) = \psi_0(\mathbf{k}; \mathbf{r}) + \int d\mathbf{r}'\, K(E; \mathbf{r}, \mathbf{r}')\psi^{(+)}(\mathbf{k}; \mathbf{r}') \tag{10.58}$$

where according to (10.6)

$$K(E; \mathbf{r}, \mathbf{r}') = -\frac{\mu}{2\pi} \frac{\exp(ik|\mathbf{r} - \mathbf{r}'|)}{|\mathbf{r} - \mathbf{r}'|} V(\mathbf{r}').$$

The solution of this equation, of course, solves the three-dimensional

Schrödinger equation,

$$-\nabla^2\Psi + 2\mu V(\mathbf{r})\Psi = k^2\Psi \tag{10.59}$$

and the discussion of its properties could be based on the theory of partial differential equations. It is, however, much simpler to work directly with the integral equation (10.58), which contains the boundary condition also.

10.3.1 Born Series. For the solution of (10.58) we may bring the full weight of Chap. 9 to bear. Let us test whether or not the kernel K is in the Hilbert–Schmidt class,

$$\begin{aligned}\operatorname{tr} K^\dagger K &= \left(\frac{\mu}{2\pi}\right)^2 \int d\mathbf{r} \int d\mathbf{r}' \frac{\exp(-2|\mathbf{r}-\mathbf{r}'|\,\mathrm{Im}\,k)}{|\mathbf{r}-\mathbf{r}'|^2} |V(\mathbf{r})|^2 \\ &= \left(\frac{\mu}{2\pi}\right)^2 \int d\mathbf{r}\, |V(\mathbf{r})|^2 \int d\mathbf{r}'\, r'^{-2} \exp(-2r'\,\mathrm{Im}\,k)\end{aligned} \tag{10.60}$$

after a shift in the variable $\mathbf{r}'$ of integration. Provided, therefore, that

$$\int d\mathbf{r}\, |V(\mathbf{r})|^2 < \infty \tag{10.61}$$

the kernel is L^2 and all the previous theory is applicable *so long as the energy is complex*. Specifically we must choose $\mathrm{Im}\,k > 0$, which implies that $E = k^2/2\mu$ stays on the plane cut along the positive real axis, or on the first sheet of the Riemann surface defined by the mapping $E = k^2/2\mu$ which takes the upper half of the complex k plane onto the E plane cut from $E = 0$ to $E = +\infty$. This is the so-called *physical sheet*. As k approaches the real axis, or E approaches the cut, however, $\operatorname{tr} K^\dagger K$ tends to infinity, no matter how well behaved or weak the potential is. In accordance with the results of Chap. 9 there are a number of ways to circumvent this difficulty.

The first is to multiply (10.58) by $[V(\mathbf{r})]^{1/2}$ and write (in a simplified manner)

$$V^{1/2}\psi^{(+)} = V^{1/2}\psi_0 + \int d\mathbf{r}'\, \mathfrak{K}(\mathbf{r},\mathbf{r}')\, V^{1/2}(\mathbf{r}')\psi^{(+)}(\mathbf{r}')$$

with

$$\mathfrak{K}(\mathbf{r},\mathbf{r}') = -\frac{\mu}{2\pi} V^{1/2}(\mathbf{r}) \frac{\exp(ik|\mathbf{r}-\mathbf{r}'|)}{|\mathbf{r}-\mathbf{r}'|} V^{1/2}(\mathbf{r}').$$

This is slightly complicated, though, by the fact that, if $V(\mathbf{r})$ changes sign, the square root is not uniquely determined. As mentioned in Sec. 9.1, this is not a serious complication, but there is a simple alternative without this blemish. Let us write, as before (9.6a)

$$V(\mathbf{r}) = |V(\mathbf{r})|\,v(\mathbf{r}) \tag{10.62}$$

so that $v(\mathbf{r})$ is a sign factor, $+1$ where V is positive, -1 where V is negative.

We multiply (10.58) by $|V|^{1/2}$ and arrive at an integral equation with the kernel

$$\mathscr{K}(\mathbf{r},\mathbf{r}') = -\frac{\mu}{2\pi}|V(\mathbf{r})|^{1/2}\frac{\exp(ik|\mathbf{r}-\mathbf{r}'|)}{|\mathbf{r}-\mathbf{r}'|}|V(\mathbf{r}')|^{1/2}v(\mathbf{r}') = -\overline{\mathscr{K}}(\mathbf{r},\mathbf{r}')v(\mathbf{r}') \tag{10.63}$$

and for this, when $\operatorname{Im} k \geq 0$,

$$\begin{aligned}\operatorname{tr}\mathscr{K}^\dagger\mathscr{K} &= \left(\frac{\mu}{2\pi}\right)^2\int d\mathbf{r}\int d\mathbf{r}'\,|V(\mathbf{r})||V(\mathbf{r}')|\frac{\exp[-2|\mathbf{r}-\mathbf{r}'|\operatorname{Im}k]}{|\mathbf{r}-\mathbf{r}'|^2} \\ &\leq \left(\frac{\mu}{2\pi}\right)^2\int d\mathbf{r}\int d\mathbf{r}'\,|V(\mathbf{r})V(\mathbf{r}')||\mathbf{r}-\mathbf{r}'|^2.\end{aligned} \tag{10.64}$$

This is certainly finite even for $\operatorname{Im} k = 0$ if

$$\|V\|_R^2 \equiv \int d\mathbf{r}\int d\mathbf{r}'\,|V(\mathbf{r})V(\mathbf{r}')|\,|\mathbf{r}-\mathbf{r}'|^{-2} < \infty. \tag{10.65}$$

Potentials that satisfy this condition are said to be in the *Rollnik class*. For example, if for some $\varepsilon > 0$, $a > 0$, and C,

$$|V(\mathbf{r})| < C(a + r)^{-3-\varepsilon} \tag{10.65'}$$

for all $\mathbf{r}$, then (10.65) holds.

If the potential is weak enough so that $\mathscr{K}(E)$ has no eigenvalues $\alpha(E)$ outside the unit circle, the Born series

$$(\mathbb{1} - \mathscr{K})^{-1} = \mathbb{1} + \mathscr{K} + \mathscr{K}^2 + \cdots$$

converges. Let us now apply the considerations of Sec. 9.1.

Let $\alpha(E)$ and $\phi_\alpha(E,\mathbf{r})$ be an eigenvalue and eigenfunction, respectively, of $\mathscr{K}(E;\mathbf{r},\mathbf{r}')$,

$$\int d\mathbf{r}'\,\mathscr{K}(E;\mathbf{r},\mathbf{r}')\phi_\alpha(E,\mathbf{r}') = \alpha(E)\phi_\alpha(E,\mathbf{r}'),$$

so that (in a short-hand notation)

$$\iint \phi_\alpha^*(E)\mathscr{K}(E)\phi_\alpha(E) = \alpha(E)\int|\phi_\alpha(E)|^2.$$

But (10.63) shows that for all E

$$|\mathscr{K}(E;\mathbf{r},\mathbf{r}')| \leq \overline{\mathscr{K}}(0;\mathbf{r},\mathbf{r}),$$

so that

$$\left|\iint \phi_\alpha^*(E)\mathscr{K}(E)\phi_\alpha(E)\right| \leq \iint |\phi_\alpha(E)|\overline{\mathscr{K}}(0)|\phi_\alpha(E)| \leq \beta_{\max}\int|\phi_\alpha(E)|^2,$$

where $\beta_{\max}$ is the maximal eigenvalue of the self-adjoint compact operator $\overline{\mathscr{K}}(0)$. It follows that for all E

$$|\alpha(E)| < \beta_{\max},$$

that is, no eigenvalue of $K(E)$ for the potential V can surpass in magnitude the leading eigenvalue of $K'(0)$ for the potential $-|V|$. The immediate implication of this statement is that *for the uniform convergence of the Born series of the potential V for all energies it is sufficient that $-|V|$ cause no bound states.* If V never changes sign, *then this condition is evidently also necessary.*

High-Energy Limit. Next we examine the behavior of the kernel $\mathscr{K}$ in the high-energy limit. Consider

$$\mathscr{K}^2(\mathbf{s},\mathbf{t}) = \left(\frac{\mu}{2\pi}\right)^2 |V(\mathbf{s})V(\mathbf{t})|^{1/2} v(\mathbf{t}) \int d\mathbf{r}\, V(\mathbf{r}) \frac{\exp[ik(|\mathbf{r}-\mathbf{s}| + |\mathbf{r}-\mathbf{t}|)]}{|\mathbf{r}-\mathbf{s}|\,|\mathbf{r}-\mathbf{t}|}.$$

We use spheroidal coordinates in the integral,

$$\begin{aligned} \xi &= \tfrac{1}{2}(|\mathbf{r}-\mathbf{s}| + |\mathbf{r}-\mathbf{t}|) \\ a\eta &= \tfrac{1}{2}(|\mathbf{r}-\mathbf{s}| - |\mathbf{r}-\mathbf{t}|) \qquad a = |\mathbf{s}-\mathbf{t}| \\ \varphi &= \text{angle about } \mathbf{s}-\mathbf{t} \end{aligned}$$

for which

$$\frac{d\mathbf{r}}{|\mathbf{r}-\mathbf{s}|\,|\mathbf{r}-\mathbf{t}|} = d\xi\, d\eta\, d\varphi$$

and we get

$$\mathscr{K}^2(\mathbf{s},\mathbf{t}) = \left(\frac{\mu}{2\pi}\right)^2 |V(\mathbf{s})V(\mathbf{t})|^{1/2} v(\mathbf{t}) \int_a^\infty d\xi \int_{-1}^1 d\eta \int_0^{2\pi} d\varphi\, e^{2ik\xi} V.$$

The integral tends to zero as $k \to \infty$, by the Riemann–Lebesgue lemma. What is more, it does so uniformly in $\mathbf{s}$ and $\mathbf{t}$. Rather than proving this under the most general conditions, we may assume that V is differentiable and perform an integration by parts. (We leave the details to the reader.[5]) Consequently we find that

$$\begin{aligned} \operatorname{tr} \mathscr{K}^2 \mathscr{K}^{2\dagger} &= \left(\frac{\mu}{2\pi}\right)^4 \int d\mathbf{s} \int d\mathbf{t}\, |V(\mathbf{s})V(\mathbf{t})| \left| \int d\xi\, d\eta\, d\varphi\, e^{2ik\xi} V \right|^2 \qquad (10.66) \\ &\to 0 \qquad \text{as } k \to \infty. \end{aligned}$$

Now by Schwarz's inequality applied to the sixfold integration involved in the trace we have for every integral kernel A[6]

$$\|A\|^2 \leq \operatorname{tr} AA^\dagger. \qquad (10.67)$$

Setting $A = \mathscr{K}^2$, we conclude that

$$\|\mathscr{K}^2\| \to 0 \qquad \text{as } k \to \infty. \qquad (10.68)$$

5 See Zemach and Klein (1958), appendix and sec. 4; and N. N. Khuri (1957), appendix; see also B. Simon (1971), p. 23.

6 $\|A\|$ here means the operator norm defined in Sec. 6.8.2.

It follows that all the eigenvalues of $\mathscr{K}^2$, and hence all the eigenvalues $\alpha(E)$ of K, must *uniformly* tend to zero as $E \to +\infty$. Therefore there exists an energy E_0 above which *the Born series converges uniformly*, and, what is more, *if E is large enough, the first Born approximation must be good.*

In the complex k plane the situation is, of course, much simpler. When $\operatorname{Im} k > 0$, that is, when we are on the physical sheet of the E surface, K is in the Hilbert–Schmidt class. Equation (10.64) shows that

$$\operatorname{tr} K^\dagger K \to 0 \qquad \text{as } \operatorname{Im} k \to +\infty \tag{10.69}$$

and therefore, by (10.67),

$$\|K\| \to 0 \qquad \text{as } \operatorname{Im} k \to +\infty. \tag{10.69$'$}$$

It is clear that (10.68) holds also for $\operatorname{Im} k > 0$, as $\operatorname{Re} k \to \infty$. It is true in all directions in the upper half of the k plane. Consequently

$$\mathscr{G}(E) \to G(E) \qquad \text{as } |E| \to \infty \tag{10.70}$$

everywhere on the physical sheet, including the boundary, i.e., on the cut.

10.3.2 Fredholm Theory. We shall now study the properties of the wave function and of the bound states, by means of Fredholm theory.

The Wave Function. Multiplying (10.10a) by $|\mathscr{V}(\mathbf{r})|^{1/2}\exp(-i\mathbf{k}\cdot\mathbf{r})$ we obtain

$$\phi(\mathbf{k},\mathbf{r}) = |\mathscr{V}(\mathbf{r})|^{1/2} + \int d\mathbf{r}'\, \mathscr{K}'(\mathbf{k};\mathbf{r},\mathbf{r}')\phi(\mathbf{k},\mathbf{r}'), \tag{10.71}$$

where

$$\mathscr{K}'(\mathbf{k};\mathbf{r},\mathbf{r}') = -\frac{1}{4\pi}|\mathscr{V}(\mathbf{r})\mathscr{V}(\mathbf{r}')|^{1/2}v(\mathbf{r}')\frac{\exp\{i[k|\mathbf{r}-\mathbf{r}'| - \mathbf{k}\cdot(\mathbf{r}-\mathbf{r}')]\}}{|\mathbf{r}-\mathbf{r}'|}. \tag{10.72}$$

We will from now on write $\mathbf{k} = k\boldsymbol{\theta}$, where $\boldsymbol{\theta}$ is a unit vector in the direction of $\mathbf{k}$. It is easily seen as in (10.64) that if $\mathscr{V}$ is in the Rollnik class R then $\operatorname{tr} \mathscr{K}'^\dagger\mathscr{K}' < \infty$ for all k with $\operatorname{Im} k \geq 0$. Therefore, if $\mathscr{V} \in R \cap L^1$ then (10.71) is a Fredholm equation of the second kind.

The modified Fredholm determinant

$$\begin{aligned}\Delta(k) &= \det{}_2[\mathbb{1} - \mathscr{K}'(k)]\\ &= \det{}_2[\mathbb{1} - K(k)],\end{aligned} \tag{10.73}$$

defined as in (9.85) by "pulling the poison tooth" is well defined for all k with $\operatorname{Im} k \geq 0$, and unless $\Delta(k) = 0$, a square integrable solution of (10.71) exists

$$\int d\mathbf{r}\,|\phi(\mathbf{k};\mathbf{r})|^2 < \infty, \tag{10.74}$$

is unique, and constructable by the Fredholm method. What is more, since the kernel in (10.72) defines an analytic operator-valued function of k that is holomorphic in the open upper half-plane, and one easily shows[7] that the Hilbert–Schmidt norm of dK/dk is finite if $\mathscr{V} \in L^1$, the solution $\phi(k, \boldsymbol{\theta}; \mathbf{r})$ of (10.71) is, for almost all $\boldsymbol{\theta}$ and $\mathbf{r}$, the continuous boundary value of an analytic function of k that is meromorphic in the upper half-plane. Similarly, the Fredholm determinant $\det_2 [\mathbb{1} - \lambda K(k)]$, as a uniformly convergent power series in λ each term of which is analytic in k, is the boundary value of an analytic function holomorphic in the upper half-plane. For $\operatorname{Im} k > 0$ it follows from (10.66), (10.69), and the fact that also[7]

$$\lim_{|k| \to \infty} \operatorname{tr} \mathscr{K}^3 = 0,$$

that

$$\lim_{|k| \to \infty} \Delta(k) = 1. \tag{10.75}$$

Furthermore, since

$$G^*(-k^*; \mathbf{r}, \mathbf{r}') = G(k; \mathbf{r}, \mathbf{r}'), \tag{10.76}$$

$\Delta(k)$ has the symmetry

$$\Delta^*(-k^*) = \Delta(k). \tag{10.77}$$

The solution $\psi^{(+)}$ of (10.10a) is now obtained form

$$\psi^{(+)}(k, \boldsymbol{\theta}; \mathbf{r}) = e^{ik\boldsymbol{\theta}\cdot\mathbf{r}} + \int d\mathbf{r}' \mathscr{K}'(k, \boldsymbol{\theta}; \mathbf{r}, \mathbf{r}') \phi(k, \boldsymbol{\theta}; \mathbf{r}'). \tag{10.78}$$

Under somewhat stronger conditions on the potential (stated below) one can prove[8] that the function

$$\gamma(k, \boldsymbol{\theta}; \mathbf{r}) = \psi^{(+)}(k, \boldsymbol{\theta}; \mathbf{r}) \exp(-ik\boldsymbol{\theta} \cdot \mathbf{r}) \tag{10.79}$$

for all $\boldsymbol{\theta}$ and $\mathbf{r}$ is the boundary value of an analytic function of k meromorphic in the upper half-plane, with

$$\lim_{|k| \to \infty} \gamma(k, \boldsymbol{\theta}; \mathbf{r}) = 1. \tag{10.80}$$

On the real axis the approach of $\psi^{(+)}$ to its high-energy limit $\exp(i\mathbf{k} \cdot \mathbf{r})$ is such that for all $k_0 > 0$ there exists a constant C and

$$\int_{k_0}^{\infty} dk \, |\psi^{(+)}(\mathbf{k}, \mathbf{r}) - \exp(i\mathbf{k} \cdot \mathbf{r})|^2 = \int_{k_0}^{\infty} dk \, |\gamma(k, \boldsymbol{\theta}; \mathbf{r}) - 1|^2 < C \tag{10.81}$$

for all $\boldsymbol{\theta}$ and $\mathbf{r}$. Similarly,

$$\int_{-\infty}^{\infty} dk \, |\Delta(k) - 1|^2 < \infty. \tag{10.82}$$

The precise conditions under which Eqs. (10.80) to (10.82) have been proved are:

7 R. G. Newton (1977a).
8 R. G. Newton (1980b).

a) there exists positive constants a and C such that

$$\int d\mathbf{r}\,|\mathscr{V}(\mathbf{r})|^2 + \int d\mathbf{r}\,|\mathscr{V}(\mathbf{r})|\left(\frac{a + r + r'}{|\mathbf{r} - \mathbf{r}'|}\right)^2 < C \tag{10.83}$$

for all $\mathbf{r}'$;

b) there exist an $\mathbf{r}'$ and a monotonely decreasing function $M(t) \in L^1(0, \infty) \cap L^2(0, \infty)$ such that for all $\mathbf{r}$

$$|\mathscr{V}(\mathbf{r} + \mathbf{r}')| \leq M(r); \tag{10.84}$$

c) there exist positive constants l and C such that for all unit vectors and all $\mathbf{r}'$

$$\int d\mathbf{r}\,[\mathscr{V}(\mathbf{r} + \mathbf{r}')]^2/(1 + \boldsymbol{\theta} \cdot \mathbf{r}) < C. \tag{10.85}$$

Bound States and Exceptional Points. The points $k = i\kappa$ at which $\psi^{(+)}$ and γ have poles are the zeros of the Fredholm determinant

$$\Delta(i\kappa) = 0. \tag{10.86}$$

If $i\kappa$ is in the upper half-plane, $\operatorname{Re}\kappa > 0$, where the kernel K is in the Hilbert–Schmidt class, then the homogeneous form of (10.58) has a square integrable solution and hence $-\kappa^2$ is an eigenvalue of (10.59), or a *bound state*. It follows from the self-adjointness of (10.59) that therefore κ must be real. Conversely, if $-\kappa^2 < 0$ is a bound state, the homogeneous form of (10.58) must have a nontrivial square integrable solution and hence (10.86) must hold. The multiplicity of the zero of $\Delta(k)$ at $k = i\kappa$ equals the degeneracy of the corresponding bound state.[9]

Suppose that $-\kappa^2 < 0$ is a bound state eigenvalue of degeneracy N, and let $u_\kappa{}^b(\mathbf{r})$, $b = 1, \ldots, N$, be any orthonormal set of eigenfunctions with the eigenvalue $-\kappa^2$. The asymptotic form of $u_\kappa{}^b$ for large r can be obtained by assuming an asymptotic expansion of the form

$$u_\kappa^b(\mathbf{r})e^{\kappa r} = f(r)\,Y_\kappa^b(\hat{\mathbf{r}}) + \cdots,$$

where $Y_\kappa{}^b(\hat{\mathbf{r}}) \not\equiv 0$. The Schrödinger equation (10.59) leads to the result that $f = \text{const.}\, r^{-1}$, and we may normalize $Y_\kappa{}^b(\hat{\mathbf{r}})$ so that

$$u_\kappa^b(\mathbf{r}) = -(e^{-\kappa r}/4\pi r)\,Y_\kappa^b(\hat{\mathbf{r}}) + o(e^{-\kappa r}r^{-1}). \tag{10.87}$$

The homogeneous form of (10.58), which $u_\kappa{}^b$ must satisfy, shows that

$$Y_\kappa^b(\boldsymbol{\theta}) = \int d\mathbf{r}\,\mathscr{V}(\mathbf{r})e^{\kappa\boldsymbol{\theta}\cdot\mathbf{r}}\,u_\kappa^b(\mathbf{r}). \tag{10.88}$$

If $Y_\kappa{}^b$ and $Y_\kappa{}^{b'}$ are linearly independent then so are $u_\kappa{}^b$ and $u_\kappa{}^{b'}$. Since the asymptotic expansion of $u_\kappa{}^b$ implies that $Y_\kappa{}^b(\hat{\mathbf{r}})$ cannot vanish identically, linear dependence of $Y_\kappa{}^b$ and $Y_\kappa{}^{b'}$ implies linear dependence of the corresponding functions $u_\kappa{}^b$ and $u_\kappa{}^{b'}$. There is therefore a one-to-one correspondence between the bound-state eigenfunctions and the angle-functions

9 R. G. Newton (1974d).

$Y_\kappa{}^b$, $b = 1, \ldots, N$; the N functions $Y_\kappa{}^b$ uniquely characterize the bound-state eigenfunctions;[10] we call them their *characters*. Since $\mathscr{V}$ is real, $u_\kappa{}^b$ and $Y_\kappa{}^b$ can be chosen to be real.

Let us now look at (10.86) for real values of $i\kappa$. It has been proved[11] that for a class of potentials that includes those that satisfy (10.65′) the only point κ on the real axis for which (10.85) can hold is $\kappa = 0$. This implies that *there can be no positive-energy bound states* for such potentials, and the scattering wave function $\psi^{(+)}(\mathbf{k};\mathbf{r})$ exists for all positive energies.

If (10.86) holds for $\kappa = 0$, the origin is called an *exceptional point*. The homogeneous equation

$$\phi(\mathbf{r}) = -\frac{1}{4\pi}\int d\mathbf{r}'\frac{|\mathscr{V}(\mathbf{r})\mathscr{V}(\mathbf{r}')|^{1/2}v(\mathbf{r}')}{|\mathbf{r}-\mathbf{r}'|}\phi(\mathbf{r}') \tag{10.89}$$

then has nontrivial square integrable solutions. There are then three possibilities:[7]

a) If all n nontrivial linearly independent solutions $v\phi$ of the adjoint of (10.89) are orthogonal to the inhomogeneity in (10.71) for $k = 0$, i.e., if

$$\int d\mathbf{r}\, v(\mathbf{r})|\mathscr{V}(\mathbf{r})|^{1/2}\phi(\mathbf{r}) = 0, \tag{10.90}$$

then there still exists a well defined solution of (10.71). At the same time (10.90) implies that the r^{-1} term of each solution $\psi(\mathbf{r})$ of the homogeneous form of (10.58) vanishes and $\psi(\mathbf{r})$ is square integrable. Thus $E = 0$ is an n-fold degenerate *bound state*, and $k = 0$ is called an *exceptional point of the first kind*. The order of the zero of $\Delta(k)$ at $k = 0$ in this case in $2n$:

$$\Delta(k) = k^{2n}c + o(k^{2n}), \tag{10.91a}$$

where c is real and $c \neq 0$.

b) If (10.89) has exactly one (up to normalization) nontrivial solution and for it (10.90) fails then the corresponding solution ψ of the homogeneous form of (10.58) is of order r^{-1} as $r \to \infty$ and is not square integrable. Then $E = 0$ is *not* a bound state (it may be called a "half-bound" state) and $k = 0$ is called an *exceptional point of the second kind*. The order of the zero of $\Delta(k)$ at $k = 0$ is given by

$$\Delta(k) = ikc + o(k), \tag{10.91b}$$

where c is real and $c \neq 0$.

c) If (10.89) has n linearly independent nontrivial solutions for which (10.90) holds, and one for which it fails then $E = 0$ has an n-fold degenerate bound state, and also a half-bound state. [If there are $n' > 1$ linearly independent solutions for which (10.90) fails one can always form $n' - 1$

10 If $\mathscr{V}$ is spherically symmetric, the functions Y_κ^b become spherical harmonics and are independent of κ.

11 T. Kato (1959); R. Lavine (1973); Alsholm and Schmidt (1971); R. G. Newton (1977).

linear combinations of them for which (10.90) holds.] In that case $k = 0$ in an exceptional point of both the first and second kind, and near $k = 0$

$$\Delta(k) = ik^{2n+1}c + o(k^{2n+1}), \tag{10.91c}$$

where c is real and $c \neq 0$. The degeneracy n in cases a) and c) is necessarily finite.

It has been proved[12] under specific (reasonable) assumptions on the potential that if there are no bound states of negative energy and zero is an exceptional point then it must be of the *second kind.* This is the generalization of the s-wave property for spherically symmetric potentials discussed in Chapter 12, above Eq. (12.53).

We can conclude from the analyticity of $\Delta(k)$ together with (10.75) and its behavior (10.91) near $k = 0$ that $\Delta(k)$ cannot have an accummulation point of zeros for $\operatorname{Im} k > 0$. It follows that under these conditions on the potential the total number of bound states must be finite.[13]

The Green's Function. The complete Green's function $\mathscr{G}(k;\mathbf{r},\mathbf{r}')$ satisfies the resolvent equation (7.11). If we define

$$\check{\mathscr{K}}(k;\mathbf{r},\mathbf{r}') = |\mathscr{V}(\mathbf{r})|^{1/2}\mathscr{G}^+(k;\mathbf{r},\mathbf{r}')|\mathscr{V}(\mathbf{r}')|^{1/2}v(\mathbf{r}'), \tag{10.92}$$

then (7.11) becomes the analogue of (10.71),

$$\check{\mathscr{K}}(k;\mathbf{r},\mathbf{r}') = \mathscr{K}(k;\mathbf{r},\mathbf{r}') + \int d\mathbf{r}''\, \mathscr{K}(k;\mathbf{r},\mathbf{r}'')\check{\mathscr{K}}(k;\mathbf{r}'',\mathbf{r}'). \tag{10.93}$$

Hence it follows from (10.64) that if $\mathscr{V} \in R$ then for almost all $\mathbf{r}$ and $\mathbf{r}'$, $\check{\mathscr{K}}(k;\mathbf{r},\mathbf{r}')$ is the continuous (except at $k = 0$ if that is an exceptional point of the second kind) boundary value of an analytic function of k meromorphic in the upper half-plane. In fact, for real $k \neq 0$ it is in the Hilbert–Schmidt class,

$$\int d\mathbf{r}\, d\mathbf{r}' |\check{\mathscr{K}}(k;\mathbf{r},\mathbf{r}')|^2 < \infty. \tag{10.94}$$

It has poles at the points $k = i\kappa$ that correspond to the bound states $E = -\kappa^2 < 0$, and because $\mathscr{G}^+$ is the boundary value of the resolvent of a self-adjoint operator (see Sec. 7.3), these poles are necessarily *simple*, even when there is degeneracy. If $k = 0$ is an exceptional point of the second kind (and possibly also of the first kind) then $\mathscr{G}(k;\mathbf{r},\mathbf{r}')$ diverges at $k = 0$ like k^{-1}; in all other instances it remains finite at $k = 0^7$.

It is clear from the fact that $\mathscr{G}$ is the kernel of the resolvent $(E - H)^{-1}$, that the operator norm $\|\mathscr{G}\|$ tends to zero as $|E| \to \infty$ away from the positive real axis. For k on the real axis it follows from (10.74), (10.93), and (10.94) that

12 Klaus and Simon (1980).

13 For sufficient conditions for the existence of bound for purely attractive potentials, see Chadan and Martin (1980) and Chadan and De Mol (1980).

$$\lim_{k \to \pm\infty} \int d\mathbf{r}\, d\mathbf{r}' |\mathscr{K}(k;\mathbf{r},\mathbf{r}')|^2 = 0. \tag{10.95}$$

We shall not prove here the completeness of the spectrum of (10.58) and take its absolute continuity for granted. Since it follows from (10.10a) that $(2\pi)^{-3} \int d\mathbf{r}\, \psi^{(+)}(\mathbf{k},\mathbf{r})^* \psi^{(+)}(\mathbf{k}',\mathbf{r}) = \delta(\mathbf{k} - \mathbf{k}')$, the complete Green's function can then be expanded as

$$\mathscr{G}^{\pm}(k;\mathbf{r},\mathbf{r}') = (2\pi)^{-3} \int d\mathbf{k}' \frac{\psi^{(+)}(\mathbf{k}',\mathbf{r})\psi^{(+)}(\mathbf{k}',\mathbf{r}')^*}{k^2 - k'^2 \pm i\varepsilon} + \sum_{n,b} \frac{u^b_{\kappa_n}(\mathbf{r}) u^b_{\kappa_n}(\mathbf{r}')^*}{k^2 + \kappa_n^2}. \tag{10.96}$$

We may use $\mathscr{G}^+$ to express the solution of (10.58) in the form

$$\psi^{(+)}(\mathbf{k};\mathbf{r}) = \psi_0(\mathbf{k};\mathbf{r}) + \int d\mathbf{r}' \mathscr{G}^+(k;\mathbf{r},\mathbf{r}') \mathscr{V}(\mathbf{r}') \psi_0(\mathbf{k};\mathbf{r}'). \tag{10.97}$$

It follows that the pole of $\psi^{(+)}$ at $k = i\kappa$ if $E = -\kappa^2 < 0$ is the energy of a bound state, is *simple* and its principal part there is

$$\psi^{(+)}(k, \boldsymbol{\theta};\mathbf{r}) = \frac{1}{2i\kappa(k - i\kappa)} \sum_{b=1}^{N} Y^b_\kappa(-\boldsymbol{\theta})^* u^b_\kappa(\mathbf{r}) + \cdots, \tag{10.98}$$

where $Y_\kappa{}^b$ are the functions defined by (10.87) and given by (10.88).

If the potential is spherically symmetric then each angular-momentum state can be treated separately, as will be done in Chapter 12. The analog of $\Delta(k)$ for the angular momentum l is the Jost function f_l defined in (12.142). It has been proved[14] that $\Delta(k)$ may then be written as a convergent product in the form

$$\Delta = \prod_{l=0}^{\infty} (f_l e_l)^{2l+1},$$

where

$$e_l = \exp\left[(-1)^l k^{-1} \int_0^\infty dr\, \mathscr{V}(r) u_l(kr) w_l^{(+)}(kr)\right]$$

in terms of the functions u_l and w_l defined in (2.58) and (2.60).

10.3.3 Scattering Amplitude, Cross Section, and S Matrix. The scattering amplitude is defined by the asymptotic behavior of the wave function $\psi^{(+)}$ for large r, as in (10.11a) for spinless particles, and expressed as in (10.12a). In terms of the solution $\phi(k,r)$ of (10.71) it may be written as

$$A(\mathbf{k}',\mathbf{k}) = -\frac{1}{4\pi} \int d\mathbf{r} |\mathscr{V}(\mathbf{r})|^{1/2} v(\mathbf{r}) \phi(\mathbf{k},\mathbf{r}) \exp(-i\mathbf{k}' \cdot \mathbf{r}).$$

It follows that if $\mathscr{V} \in R \cap L^1$ then A exists and is finite for all real k and all

14 R. G. Newton (1972b).

directions $\hat{\mathbf{k}}$ and $\hat{\mathbf{k}}'$, except possibly at $k = 0$. Since A may also be written in the form

$$A(\mathbf{k}',\mathbf{k}) = -\frac{1}{4\pi}\int d\mathbf{r}\,\mathscr{V}(\mathbf{r})\exp\left[i(\mathbf{k} - \mathbf{k}')\cdot\mathbf{r}\right] + A'(k;\boldsymbol{\theta}',\boldsymbol{\theta}) \tag{10.99}$$

where

$$-4\pi A'(k;\boldsymbol{\theta}',\boldsymbol{\theta}) = \int d\mathbf{r}\,d\mathbf{r}'\,\mathscr{V}(\mathbf{r})\mathscr{V}(\mathbf{r}')\mathscr{G}^{+}(k;\mathbf{r},\mathbf{r}')\exp\left[ik(\boldsymbol{\theta}\cdot\mathbf{r}' - \boldsymbol{\theta}\cdot\mathbf{r})\right],$$

the symmetry (10.76) implies that A has the symmetry

$$A(-k;\boldsymbol{\theta}',\boldsymbol{\theta}) = A^{*}(k;\boldsymbol{\theta}',\boldsymbol{\theta}). \tag{10.100}$$

It follows from this and the properties of $\mathscr{G}^{+}$ that if $k = 0$ is an exceptional point of the second kind then near $k = 0$

$$A = ick^{-1} + o(k^{-1}) \tag{10.101}$$

where c is real, $c \neq 0$. The differential cross section being given by (10.14) as the squared modulus of A, it is well defined and finite for all energies, except at zero energy if $k = 0$ is an exceptional point of the second kind. In the exceptional case the differential cross section diverges at low energy like E^{-1}.

The high-energy behavior of the amplitude depends critically on the relative directions of $\mathbf{k}$ and $\mathbf{k}'$. Eq. (10.94) implies by Schwarz's inequality that A' in (10.99) vanishes in the limit as $k \to \pm\infty$, uniformly in $\boldsymbol{\theta}$ and $\boldsymbol{\theta}'$. Therefore for the forward scattering amplitude

$$\lim_{k\to\pm\infty} A(\mathbf{k},\mathbf{k}) = -\frac{1}{4\pi}\overline{\mathscr{V}}, \tag{10.102}$$

where

$$\overline{\mathscr{V}} = \int d\mathbf{r}\,\mathscr{V}(\mathbf{r}). \tag{10.103}$$

On the other hand, in all other directions the Born-approximation to A, i.e., the first term on the right-hand side of (10.99), oscillates to zero by the Riemann–Lebesgue lemma. The result is the strong forward diffraction peak of the differential cross section at high energies.

For the second term in (10.99) we obtain

$$\int d\boldsymbol{\theta}\,d\boldsymbol{\theta}'\,|A'(k;\boldsymbol{\theta},\boldsymbol{\theta}')|^{2} = \frac{1}{k^{2}}\int d\mathbf{r}\,d\mathbf{r}'\,d\bar{\mathbf{r}}\,d\bar{\mathbf{r}}'\,|\mathscr{V}(\mathbf{r})\mathscr{V}(\mathbf{r}')$$

$$\mathscr{V}(\bar{\mathbf{r}})\mathscr{V}(\bar{\mathbf{r}}')|^{1/2}\frac{\mathscr{K}(k;\mathbf{r},\mathbf{r}')\mathscr{K}^{*}(k;\bar{\mathbf{r}},\bar{\mathbf{r}}')}{|\mathbf{r} - \bar{\mathbf{r}}|\,|\mathbf{r}' - \bar{\mathbf{r}}'|}\sin(k|\mathbf{r} - \bar{\mathbf{r}}|)\sin(k|\mathbf{r}' - \bar{\mathbf{r}}'|)$$

$$\leq \frac{1}{k^{2}}\int d\mathbf{r}\,d\bar{\mathbf{r}}\frac{|\mathscr{V}(\mathbf{r})\mathscr{V}(\bar{\mathbf{r}})|}{|\mathbf{r} - \bar{\mathbf{r}}|^{2}}\int d\mathbf{r}'\,d\bar{\mathbf{r}}'\,|\mathscr{K}(k;\mathbf{r}',\bar{\mathbf{r}}')|^{2}$$

by Schwarz's inequality. Therefore it follows from (10.94) that

$$\lim_{k\to\pm\infty} k^2 \int d\boldsymbol{\theta} d\boldsymbol{\theta}' \, |A'(k;\boldsymbol{\theta},\boldsymbol{\theta}')|^2 = 0,$$

where the integrals extend over all solid angles. For the Born term we have

$$k^2 \int d\boldsymbol{\theta}\, d\boldsymbol{\theta}' \left| \frac{1}{4\pi} \int d\mathbf{r}\, \mathscr{V}(\mathbf{r}) \exp\left[ik(\boldsymbol{\theta} - \boldsymbol{\theta}')\cdot\mathbf{r}\right] \right|^2$$
$$= \int d\mathbf{r}\, d\mathbf{r}'\, \mathscr{V}(\mathbf{r})\mathscr{V}(\mathbf{r}') \frac{\sin^2(k|\mathbf{r} - \mathbf{r}'|)}{|\mathbf{r} - \mathbf{r}'|^2} = \tfrac{1}{2}\mathscr{V}_R^2 + o(1)$$

as $k \to \infty$, where

$$\mathscr{V}_R^2 = \int d\mathbf{r}\, d\mathbf{r}' \frac{\mathscr{V}(\mathbf{r})\mathscr{V}(\mathbf{r}')}{|\mathbf{r} - \mathbf{r}'|^2}.$$

It follows from these two results that

$$\lim_{k\to\infty} k^2 \int d\boldsymbol{\theta}\, d\boldsymbol{\theta}' \, |A(k;\boldsymbol{\theta},\boldsymbol{\theta}')|^2 = \tfrac{1}{2}\mathscr{V}^2.$$

The integral is 4π times the average total scattering cross section $\bar{\sigma}(E)$. We therefore have at high energy

$$\bar{\sigma}(E) = \mathscr{V}_R^2/8\pi E + o(E^{-1}). \tag{10.104}$$

In spite of the fact that the forward amplitude approaches a constant, and the "average amplitude" in some sense goes as k^{-1} for large k, in a weaker sense the amplitude goes to zero faster. Suppose that $f(\boldsymbol{\theta})$ is a square integrable function of $\boldsymbol{\theta}$, and we regard $A(k;\boldsymbol{\theta},\boldsymbol{\theta}')$ as the integral kernel of an operator $A(k)$ on such functions f. Assuming conditions (10.83) and (10.84), in addition that there exists a monotone function $M'(t)$ such that

$$\int_0^\infty dt\,(1 + t)M'(t) < \infty$$

and an $\mathbf{r}'$ such that for all $\mathbf{r}$

$$|\nabla\mathscr{V}(\mathbf{r} + \mathbf{r}')| \leq M'(r), \tag{10.105}$$

and that the Fourier transform $\tilde{V}(\mathbf{k})$ of $V(\mathbf{r})$ is $O(k^{-2})$ as $|k| \to \infty$, it has been proved [14a] that then

$$\int_{-\infty}^\infty dk\, k^2 \|A(k)f\|_p^2 < C\|f\|_p^2. \tag{10.106}$$

Let us next consider analyticity properties of $A(k;\boldsymbol{\theta},\boldsymbol{\theta}')$. In general we cannot expect $A(k;\boldsymbol{\theta},\boldsymbol{\theta}')$ to be the boundary value of an analytic function of k for fixed $\boldsymbol{\theta}$ and $\boldsymbol{\theta}'$. On the other hand, let us change variables and use, instead of the two vectors $\mathbf{k}$ and $\mathbf{k}'$, the *momentum transfer* vector

$$\boldsymbol{\tau} = \mathbf{k}' - \mathbf{k}, \tag{10.107}$$

14a R. G. Newton (1980b and 1982); $\|\cdot\|_p$ here is the L^p-norm.

whose magnitude is $\tau = 2k \sin \frac{1}{2}\theta$ if $\boldsymbol{\theta} \cdot \boldsymbol{\theta}' = \cos\theta$, and the vector

$$\boldsymbol{\pi} = \tfrac{1}{2}(\mathbf{k} + \mathbf{k}'), \tag{10.108}$$

whose magnitude is $\pi = (k^2 - \frac{1}{4}\tau^2)^{1/2}$ and which is orthogonal to $\boldsymbol{\tau}$, because $k = k'$. In place of the five variables $k, \boldsymbol{\theta}, \boldsymbol{\theta}'$, we shall use $k, \boldsymbol{\tau}$, and $\hat{\boldsymbol{\pi}}$ and write $A(k, \boldsymbol{\theta}, \boldsymbol{\theta}') = \mathscr{A}(k, \boldsymbol{\tau}, \hat{\boldsymbol{\pi}})$. Then (10.99) reads

$$\mathscr{A}(k, \boldsymbol{\tau}, \hat{\boldsymbol{\pi}}) = -\frac{1}{4\pi} \int d\mathbf{r}\, \mathscr{V}(\mathbf{r}) \exp(-i\boldsymbol{\tau} \cdot \mathbf{r}) + \mathscr{A}'(k, \boldsymbol{\tau}, \hat{\boldsymbol{\pi}}) \tag{10.109}$$

where

$$-4\pi \mathscr{A}'(k, \boldsymbol{\tau}, \hat{\boldsymbol{\pi}}) = \int d\mathbf{r}\, d\mathbf{r}'\, |\mathscr{V}(\mathbf{r})\mathscr{V}(\mathbf{r}')|^{1/2} v(\mathbf{r})$$

$$\exp[-\tfrac{1}{2} i \boldsymbol{\tau} \cdot (\mathbf{r} + \mathbf{r}')] \mathfrak{G}(k, \boldsymbol{\pi}; \mathbf{r}, \mathbf{r}'),$$

$$\mathfrak{G}(k, \boldsymbol{\pi}; \mathbf{r}, \mathbf{r}') = |\mathscr{V}(\mathbf{r})\mathscr{V}(\mathbf{r}')|^{1/2} \exp[i\boldsymbol{\pi} \cdot (\mathbf{r}' - \mathbf{r})] \mathscr{G}(k; \mathbf{r}, \mathbf{r}') v(\mathbf{r}').$$

The kernel $\mathfrak{G}$ satisfies an integral equation like (10.93) except that $\mathscr{K}$ is replaced by

$$\mathfrak{G}_0(k, \boldsymbol{\pi}; \mathbf{r}, \mathbf{r}') = -\frac{1}{4\pi} |\mathscr{V}(\mathbf{r})\mathscr{V}(\mathbf{r}')|^{1/2} v(\mathbf{r}') \frac{\exp[ik|\mathbf{r} - \mathbf{r}'| + i\boldsymbol{\pi} \cdot (\mathbf{r} - \mathbf{r}')]}{|\mathbf{r} - \mathbf{r}'|}.$$

If we write $\boldsymbol{\pi} \cdot (\mathbf{r} - \mathbf{r}') = \pi|\mathbf{r} - \mathbf{r}'|\xi$ then the function in the exponential is $i|\mathbf{r} - \mathbf{r}'|h(k)$, where

$$h(k) = k + \xi(k^2 - \tfrac{1}{4}\tau^2)^{1/2},$$

with $|\xi| \le 1$. The function $h(k)$ is analytic in the complex plane cut from $k = -\frac{1}{2}\tau$ to $k = \frac{1}{2}\tau$. For real τ and $\operatorname{Im} k \ge 0$

$$\operatorname{Im}(k^2 - \tfrac{1}{4}\tau^2)^{1/2} = 2^{-1/2}(|k^2 - \tfrac{1}{4}\tau^2| - \operatorname{Re} k^2 + \tfrac{1}{4}\tau^2)^{1/2}$$

$$\le 2^{-1/2}(|k|^2 - \operatorname{Re} k^2 + \tfrac{1}{2}\tau^2)^{1/2} \le \operatorname{Im} k + \tfrac{1}{2}$$

and hence $\operatorname{Re}[i|\mathbf{r} - \mathbf{r}'|h(k)] \le \frac{1}{2}\tau(r + r')$ for all $\operatorname{Im} k \ge 0$. Therefore, if we assume that $e^{\alpha r}\mathscr{V}(\mathbf{r}) \in R$ for some $\alpha > 0$ then $\mathfrak{G}_0$ is an analytic operator-valued function of k regular in the upper half-plane so long as $\tau < 2\alpha$. As a result, so is $\mathfrak{G}$, except for simple poles at the bound states. In addition we have the analogs of (10.76) and (10.78). We conclude that $\mathscr{A}'$, for fixed $\boldsymbol{\tau}$ and $\hat{\boldsymbol{\pi}}$, and $\tau < 2\alpha$, is the boundary value of an analytic function of k, meromorphic in the upper half-plane with simple poles at the bound states; as $|k| \to \infty$ with $\operatorname{Im} k \ge 0$, it tends to naught. The Born term in (10.109), of course, is independent of k. We note that $\mathscr{V} \in R$ is sufficient to conclude the analyticity of the *forward* amplitude $A(k; \boldsymbol{\theta}, \boldsymbol{\theta}) = \mathscr{A}(k, 0, \hat{\boldsymbol{\pi}})$. From (10.96), (10.99), and (10.88) we find that the principal part of $A(k; \boldsymbol{\theta}, \boldsymbol{\theta})$ at a bound-state pole $k = i\kappa$ is given by[8]

$$A(k; \boldsymbol{\theta}, \boldsymbol{\theta}) = \frac{i}{8\pi\kappa(k - i\kappa)} \sum_{b=1}^{N} Y_\kappa^b(\boldsymbol{\theta}) Y_\kappa^b(-\boldsymbol{\theta})^* + \cdots \tag{10.110}$$

in terms of the functions defined by (10.87).

It is of some interest (see Sec. 8.1) to consider the effect of a *shift* of the potential on the scattering amplitude. Eq (10.10a) shows that if the potential $V(\mathbf{r})$ is replaced by $V^s(\mathbf{r}) = V(\mathbf{r} - \mathbf{s})$ then the wave function $\psi^{s(+)}$ that corresponds to V^s is

$$\psi^{s(+)}(\mathbf{k},\mathbf{r}) = \psi^{(+)}(\mathbf{k},\mathbf{r} - \mathbf{s})e^{i\mathbf{k}\cdot\mathbf{s}}.$$

As a result the scattering amplitude becomes by (10.12a)

$$A^s(\mathbf{k}',\mathbf{k}) = A(\mathbf{k}',\mathbf{k})e^{i\mathbf{s}\cdot(\mathbf{k}-\mathbf{k}')}.$$

Neither the differential cross section nor the forward scattering amplitude is affected by such a translation.

The S Matrix. The difference between the outgoing-wave and the incoming-wave Green's functions can be written, according to (10.96)

$$\mathcal{G}^+(k;\mathbf{r},\mathbf{r}') - \mathcal{G}^-(k;\mathbf{r},\mathbf{r}') = -\frac{ik}{8\pi^2}\int d\boldsymbol{\theta}\psi^{(+)}(k,\boldsymbol{\theta};\mathbf{r})\psi^{(+)}(k;\boldsymbol{\theta};\mathbf{r}')^*, \qquad (10.111)$$

or both $\psi^{(+)}$ may be replaced by $\psi^{(-)}$, the incoming solution of (10.10) in which G^+ is replaced by G^-. Since $\psi^{(-)}$ can be expressed in terms of $\mathcal{G}^-$ as $\psi^{(+)}$ is in terms of $\mathcal{G}^+$ in (10.97) we find from (10.97) and (10.111)

$$\psi^{(+)}(k,\boldsymbol{\theta};\mathbf{r}) - \psi^{(-)}(k,\boldsymbol{\theta};\mathbf{r}) = \frac{ik}{2\pi}\int d\boldsymbol{\theta}'\,\psi^{(-)}(k,\boldsymbol{\theta}';\mathbf{r})A(k;\boldsymbol{\theta}',\boldsymbol{\theta}), \qquad (10.112)$$

because it follows from (10.99) and the fact that

$$\mathcal{G}^- = (\mathcal{G}^+)^\dagger \qquad (10.113)$$

that A can also be expressed as

$$A(\mathbf{k}',\mathbf{k}) = -\frac{1}{4\pi}\int d\mathbf{r}\,\psi^{(-)}(\mathbf{k}',\mathbf{r})^*\mathscr{V}(\mathbf{r})\exp(i\mathbf{k}\cdot\mathbf{r}). \qquad (10.114)$$

But (7.58) and (10.12) imply that the S matrix is related to the scattering amplitude as

$$S(k;\boldsymbol{\theta},\boldsymbol{\theta}') = \delta(\boldsymbol{\theta},\boldsymbol{\theta}') - \frac{k}{2\pi i}A(k;\boldsymbol{\theta},\boldsymbol{\theta}'). \qquad (10.115)$$

Therefore (10.112) can be written in the form

$$\psi^{(+)}(k,\boldsymbol{\theta};\mathbf{r}) = \int d\boldsymbol{\theta}'\,\psi^{(-)}(k,\boldsymbol{\theta}';\mathbf{r})S(k;\boldsymbol{\theta}',\boldsymbol{\theta}), \qquad (10.116)$$

which is a representation of (7.65). We note that it follows from the analog of (10.10a) for $\psi^{(-)}$ that

$$\psi^{(-)}(k,\boldsymbol{\theta};\mathbf{r}) = \psi^{(+)}(-k,-\theta;\mathbf{r}), \qquad (10.117)$$

so that (10.116) may also be written

$$\psi^{(+)}(k,\boldsymbol{\theta};\mathbf{r}) = \int d\boldsymbol{\theta}'\,\psi^{(+)}(-k,-\boldsymbol{\theta}';\mathbf{r})S(k;\boldsymbol{\theta}',\boldsymbol{\theta}). \qquad (10.116')$$

The operator whose integral kernel is $kA(k;\boldsymbol{\theta},\boldsymbol{\theta}')$ is Hilbert–Schmidt and has a finite trace. Therefore the Fredholm determinant $\det S(k)$ of (10.115) is well defined. It may be expressed in terms of $\Delta(k)$ as follows.

From (10.10a) and (10.73)

$$\begin{aligned}\Delta^* &= \det{}_2(\mathbb{1} - G^-\mathscr{V}) \\ &= \det{}_2\{(\mathbb{1} - G^+\mathscr{V})[\mathbb{1} - (\mathbb{1} - G^+\mathscr{V})^{-1}(G^- - G^+)\mathscr{V}]\}.\end{aligned}$$

But as in (10.111)

$$G^- - G^+ = \frac{ik}{8\pi^2}\int d\boldsymbol{\theta}\,\psi_0\psi_0^*$$

and therefore by (10.10a)

$$\begin{aligned}H(\mathbf{r},\mathbf{r}') &\equiv [(\mathbb{1} - G^+\mathscr{V})^{-1}(G^- - G^+)](\mathbf{r},\mathbf{r}') \\ &= \frac{ik}{8\pi^2}\int d\boldsymbol{\theta}\,\psi^{(+)}(k,\boldsymbol{\theta};\mathbf{r})\exp(-ik\boldsymbol{\theta}\cdot\mathbf{r}').\end{aligned}$$

Thus we easily find by (10.12a)

$$\operatorname{tr}(H\mathscr{V})^n = \left(\frac{k}{2\pi i}\right)^n \operatorname{tr}[A(k)]^n,$$

where the trace on the left means $\operatorname{tr} O = \int d\mathbf{r}\, O(\mathbf{r},\mathbf{r})$, and the trace on the right, $\operatorname{tr} P = \int d\boldsymbol{\theta}\, P(\boldsymbol{\theta},\boldsymbol{\theta})$. Therefore by (9.80a) and (10.115)

$$\det(\mathbb{1} - H\mathscr{V}) = \det S.$$

It follows from (9.87) that

$$\det{}_2[(\mathbb{1} - A)(1 - B)] = \det{}_2(\mathbb{1} - A)\det(\mathbb{1} - B)\exp\{\operatorname{tr}[(\mathbb{1} - A)B]\}.$$

For $A = G^+\mathscr{V}$ and $B = H\mathscr{V}$ we have

$$\operatorname{tr}[(\mathbb{1} - A)B] = \frac{ik}{8\pi^2}\int d\boldsymbol{\theta}\int d\mathbf{r}\,\mathscr{V}(\mathbf{r}) = \frac{ik}{2\pi}\overline{\mathscr{V}},$$

where $\overline{\mathscr{V}}$ is defined by (10.103), and therefore

$$\det S(k) = [\Delta(-k)/\Delta(k)]\exp\left(-\frac{ik}{2\pi}\overline{\mathscr{V}}\right). \tag{10.118}$$

The exponential term serves to reconcile (10.102) with (10.75).

The kernel $A(\boldsymbol{\theta},\boldsymbol{\theta}')$ defines A as a compact operator; hence it has a point spectrum only (see Sec. 7.3.2). Consequently the same is true for the unitary operator S, which has a complete set of orthonormal eigenfunctions $Y_n(\boldsymbol{\theta})$. If we write $\exp(2i\delta_n)$ for the eigenvalues of S then those of A, by (10.115) are the $4\pi k^{-1}\sin\delta_n \exp i\delta_n$, and A may be expanded in a strongly convergent series[15] (see Sec. 6.8.2),

15 Akhiezer and Glazman (1961), vol. 1, p. 131.

$$A(\boldsymbol{\theta},\boldsymbol{\theta}') = 4\pi k^{-1} \sum_{n=1}^{\infty} e^{i\delta_n} \sin \delta_n Y_n(\boldsymbol{\theta}) Y_n(\boldsymbol{\theta}')^*. \tag{10.119}$$

If the potential is spherically symmetric the functions $Y_n(\boldsymbol{\theta})$ are the spherical harmonics and they are independent of the energy. In general the Y_n depend on k and on the details of the dynamics, which makes the expansion (10.119) less useful than in the central case.

Since det S is of unit modules we may define a real function $\delta(k)$ by

$$\det S(k) = \exp [2i\delta(k)], \tag{10.120}$$

and we have

$$\delta = \sum_{n=1}^{\infty} \delta_n. \tag{10.121}$$

[If an eigenstate of S is N-fold degenerate the corresponding eigenvalue appears N times in (10.121).] $A(k;\boldsymbol{\theta},\boldsymbol{\theta}')$ being a continuous function of k, except at $k = 0$ if that is an exceptional point of the second kind, $\delta(k)$ can be defined continuously for $0 < k < \infty$. For negative k we use (10.100) to define δ as an odd function of k.

If $k = 0$ is an exceptional point of the second kind then (10.91b), (10.91c), and (10.118) show that $\det S(0) = -1$, whereas otherwise $\det S(0) = 1$. Therefore in the first case $\delta(0) = \frac{1}{2}\pi (\text{mod}\, \pi)$, while otherwise $\delta(0) = 0 (\text{mod}\, \pi)$.

Let us call $\eta(k) = \arg \Delta(-k)$. Then it follows from (10.118) that

$$\delta(k) = \eta(k) - \frac{k}{4\pi} \overline{\mathscr{V}}. \tag{10.122}$$

We may now use the analyticity of $\Delta(k)$ in the upper half-plane and (10.75) to connect its phase, $-\eta(k)$ with the number m of its zeros there by an argument that will be given in detail at the end of Sec. 12.1.3. The result is that

$$\eta(0) - \eta(\infty) = \pi(n' + m + \tfrac{1}{2}q), \tag{10.123}$$

where $q = 1$ if zero is an exceptional point of the second kind (i.e., if there is a "half-bound state" of zero energy), $q = 0$ otherwise, and n' is the number of linearly independent zero-energy bound states [from (10.91)]. But the number of zeros (counting their multiplicities) of Δ equals the number of negative-energy bound states (counting their degeneracy). Hence $n = n' + m$ is the *total number of bound states*. From (10.122) and (10.123) we therefore obtain the generalized *Levinson theorem*

$$\delta(0) - \lim_{k\to\infty} \left[\delta(k) + \frac{k}{4\pi} \overline{\mathscr{V}}\right] = \pi(n + \tfrac{1}{2}q). \tag{10.124}$$

We may also note that because of (10.75) $\eta(\infty)/\pi$ is necessarily an integer, say, p. Therefore

$$\lim_{k\to\infty}\left[\delta(k) + \frac{k}{4\pi}\overline{\mathscr{V}}\right] = p\pi, \tag{10.125}$$

and δ may be defined so that $p = 0$.

10.3.4 Dispersion Relations. Let us now assume that the potential is spherically symmetric, except for the case of forward scattering where we can be more general. The scattering amplitude is then a function of E and τ and does not depend on $\hat{\boldsymbol{\tau}}$ and $\hat{\boldsymbol{\pi}}$; we shall write it $A(E, \tau)$. We found in the last section that if $V \exp(\alpha r) \in R$ then the function

$$B(E, \tau) \equiv A(E, \tau) + \frac{\mu}{2\pi}\int d\mathbf{r} \exp(-i\boldsymbol{\tau}\cdot\mathbf{r})V(r)$$

for $\tau < 2\alpha$ is an analytic function of E, regular on the open physical sheet, $\operatorname{Im} k > 0$, continuous there and including the boundary $E > 0$, except for simple poles at the bound-state energies on the negative real axis. As $|E| \to \infty$ anywhere on the cut plane, B tends to zero.

These properties of the function $B(E, \tau)$ may be used to evaluate the integral

$$\frac{1}{2\pi i}\oint dE' \frac{B(E', \tau)}{E' - E}$$

extended in the counterclockwise direction over a contour around the cut from $E' = 0$ to $E' = \infty$ and closed at infinity. The energy E here is real. We get by Cauchy's theorem

$$\frac{1}{2\pi i}\oint dE' \frac{B(E', \tau)}{E' - E} = \sum_n \frac{R_n(\tau)}{E_n - E}$$

if E_n are the bound-state energies and R_n the corresponding residues of B. The part of the integral over the contour at infinity vanishes. For the other part we use the fact that, because of the reality and the rotational invariance of V,

$$\begin{aligned} A(E^*, \tau) &= A^*(E, \tau) \\ B(E^*, \tau) &= B^*(E, \tau) \end{aligned} \tag{10.126}$$

and hence

$$\int_0^\infty dE' \frac{B(E' + i\varepsilon, \tau)}{E' - E + i\varepsilon} - \int_0^\infty dE' \frac{B(E' - i\varepsilon, \tau)}{E' - E - i\varepsilon}$$
$$= -2\pi i \operatorname{Re} B(E, \tau) + 2i\mathscr{P}\int_0^\infty dE' \frac{\operatorname{Im} B(E', \tau)}{E' - E}$$

where $\mathscr{P}$ stands for Cauchy's principal value. Here we have used (7.7) and the fact that the physical amplitude is to be evaluated with the *outgoing-wave* Green's function, i.e., on the upper rim of the cut. We conclude that

$$\operatorname{Re} B(E,\tau) = \frac{1}{\pi}\mathscr{P}\int_0^\infty dE' \frac{\operatorname{Im} B(E',\tau)}{E'-E} + \sum_n \frac{R_n(\tau)}{E-E_n}$$

and since the Born approximation

$$\begin{aligned} A_{\mathbf{BA}}(\tau) &= -\frac{\mu}{2\pi}\int d\mathbf{r}\exp(-i\boldsymbol{\tau}\cdot\mathbf{r})V(r) \\ &= -\frac{2\mu}{\tau}\int_0^\infty dr\, r\sin r\tau\, V(r) \end{aligned} \tag{10.127}$$

is real,

$$\operatorname{Re} A(E,\tau) = A_{\mathbf{BA}}(\tau) + \sum_n \frac{R_n(\tau)}{E-E_n} + \frac{1}{\pi}\mathscr{P}\int_0^\infty dE' \frac{\operatorname{Im} A(E',\tau)}{E'-E}. \tag{10.128}$$

Because of its analogy with the electromagnetic case (see Sec. 4.2) this is usually called a *dispersion relation*. Just as for the electromagnetic amplitude, it has most physical content when the scattering angle is zero, i.e., for *forward scattering*. We may then avail ourselves of the optical theorem (7.55). Equation (10.12) shows that, in terms of the amplitude A, it reads,

$$\begin{aligned} \operatorname{Im} A(\mathbf{k},\mathbf{k}) &= \frac{k}{4\pi}\int d\Omega_{k'}\,|A(\mathbf{k}',\mathbf{k})|^2 \\ &= \frac{k}{4\pi}\sigma_{\text{total}}(\mathbf{k}) \end{aligned} \tag{10.129}$$

(If the potential is not spherically symmetrix σ_{total} generally depends on the direction $\mathbf{k}$ of incidence.) Thus the forward-dispersion relation reads

$$\operatorname{Re} A(E,0) = A_{\mathbf{BA}}(0) + \sum_n \frac{R_n(0)}{E-E_n} + \frac{1}{4\pi^2}\mathscr{P}\int_0^\infty dE' \frac{k'\sigma_{\text{total}}(E')}{E'-E} \tag{10.130}$$

where

$$A_{\mathbf{BA}}(0) = -2\mu\int_0^\infty dr\, r^2 V(r). \tag{10.127a}$$

A knowledge of the total cross section as a function of the energy thus allows us to calculate the forward-scattering amplitude. Equation (10.130) should be compared with (4.35).

The nonforward-dispersion relation (10.128) is in practice considerably less useful than (10.130), for two reasons. In the first place, the right-hand side is no better known, experimentally or otherwise, than the left. We do not have an optical theorem to simplify it. In the second place, and more important, the integral on the right includes an "unphysical region." A look at (10.107) shows that we must have

$$\tau \le 2k$$

for a real scattering angle to exist. Obviously, no elastically scattered particle can change its momentum by more than twice its initial momentum. Hence, for a given τ, the energy must exceed

$$E_{\min} = \frac{\tau^2}{8\mu}$$

for the amplitude to describe a physically realizable process.

It should also be stressed that, whereas (10.130) holds for potentials that satisfy the relatively weak condition $V(\mathbf{r}) \in R$, for (10.128) to be true we need the much stronger statement $V \exp(\alpha r) \in R$ and spherical symmetry. Unless the potential decreases asymptotically faster than *all* exponentials (for example, like a gaussian) there usually exists a limiting value of the momentum transfer beyond which no dispersion relation is true.

10.3.5 An Example (the Yukawa Potential). Apart from the Coulomb force, which will be treated in Sec. 14.6, no local potentials are known for which the full Schrödinger equation is exactly soluble in closed form. For illustrative purposes we may, however, examine some approximations for the special case of the Yukawa potential.

Suppose that the potential is of the Yukawa form

$$V(r) = \gamma r^{-1} e^{-r/R} \tag{10.131}$$

or, in the momentum representation,

$$(\mathbf{p}|V|\mathbf{p}') = (2\pi)^{-3} \int d\mathbf{r} \exp\left[i(\mathbf{p}' - \mathbf{p}) \cdot \mathbf{r}\right] V(r) = \frac{\gamma}{2\pi^2[R^{-2} + (\mathbf{p} - \mathbf{p}')^2]}. \tag{10.132}$$

According to (10.127) the Born approximation to the scattering amplitude is therefore

$$A_{\mathrm{BA}}(\tau) = -\frac{2\mu\gamma}{R^{-2} + \tau^2}. \tag{10.133}$$

The second Born approximation is somewhat more complicated to evaluate. Because of (10.25) and (10.132), it is given by the integral

$$A_2(\mathbf{p}', \mathbf{p}) = \frac{2\mu^2\gamma^2}{\pi^2} \int d\mathbf{p}'' \{[R^{-2} + (\mathbf{p}' - \mathbf{p}'')^2][R^{-2} + (\mathbf{p} - \mathbf{p}'')^2](p^2 + i\varepsilon - \mathbf{p}''^2)\}^{-1}.$$

Here $|\mathbf{p}| = |\mathbf{p}'| = k$. The solid-angle part of the integral is evaluated by using Feynman's trick, namely, the identity

$$\frac{1}{ab} = \frac{1}{2} \int_{-1}^{1} d\alpha \left(a \frac{1 + \alpha}{2} + b \frac{1 - \alpha}{2}\right)^{-2}.$$

Setting

$$a = R^{-2} + (\mathbf{p}' - \mathbf{p}'')^2 \qquad b = R^{-2} + (\mathbf{p} - \mathbf{p}'')^2$$

we get for the integral in A_2

$$\frac{1}{2}\int_{-1}^{1} d\alpha \int d\mathbf{p}'' \, (R^{-2} + k^2 + p''^2 - 2\mathbf{p}'' \cdot \mathbf{P})^{-2}(k^2 + i\varepsilon - p''^2)^{-1}$$

where

$$\mathbf{P} = \mathbf{p}\frac{1-\alpha}{2} + \mathbf{p}'\frac{1+\alpha}{2}.$$

The angle integrations $d\Omega_{p''}$ may now be easily done, and we get

$$\pi \int_{-1}^{1} d\alpha \int_{-\infty}^{\infty} dp'' \, p''^2$$
$$[(k^2 - p''^2 + i\varepsilon)(p''^2 + R^{-2} + k^2 - 2p''P)(p''^2 + R^{-2} + k^2 + 2p''P)]^{-1}.$$

At this point we may close the contour of integration in the upper half of the p'' plane by a large semicircle and perform the integration by the residue theorem:

$$-2i\pi^2 k \int_0^1 d\alpha (f^2 - k^2\tau^2\alpha^2)^{-1}$$
$$- 2\pi^2 R^{-2} \int_0^1 d\alpha (4R^{-2} + \tau^2 - \alpha^2\tau^2)^{-1/2}(f^2 - k^2\tau^2\alpha^2)^{-1}$$

where

$$f^2 = R^{-4} + 4R^{-2}k^2 + k^2\tau^2. \tag{10.134}$$

The α integrals are finally carried out easily, with the result

$$A_2(E, \tau) = \frac{2\mu^2\gamma^2}{\tau f}\left(2\tan^{-1}\frac{\tau}{2fR} + i\ln\frac{f + k\tau}{f - k\tau}\right). \tag{10.135}$$

The second approximation to the scattering cross section, of order γ^3, is obtained from the cross term between $A_{\mathbf{BA}}$ and A_2 in the evaluation of $|A|^2$. Since $A_{\mathbf{BA}}$ is real, only the real part of A_2 contributes to this,

$$\frac{d\sigma}{d\Omega} = \frac{4\mu^2\gamma^2}{(R^{-2} + \tau^2)^2} + \frac{16\mu^3\gamma^3}{(R^{-2} + \tau^2)\tau f}\tan^{-1}\frac{\tau}{2fR} + \cdots. \tag{10.136}$$

The function f is defined in (10.134). Suppose that the momentum transfer is fixed at a value that is not too small. Then, when the wavelength is small compared with the range of the potential, i.e.,

$$\frac{1}{k} \ll R \tag{10.137}$$

the arctangent may be replaced by its small argument, and the ratio of the second Born approximation to the first is

$$\frac{2\mu\gamma}{k^2 R}\frac{R^{-2}+\tau^2}{4R^{-2}+\tau^2} \approx \frac{\gamma}{RE}. \tag{10.138}$$

The quantity γ/R is, roughly, a measure of the strength of the Yukawa potential, in energy units. As we expect, the Born approximation is dominant at high energies, and the combination of (10.137) and (10.138) is the more exact formulation of (9.29) in this case. The assumption (10.137) is unnecessary in the (almost) forward direction, when the momentum transfer is small compared with the inverse of the range,

$$\tau \ll R^{-1}. \tag{10.137a}$$

In this cone of scattering angles, which becomes narrower and narrower the higher the energy, the cross section has the constant value

$$\frac{d\sigma}{d\Omega} \simeq 4\mu^2\gamma^2 R^4 + \frac{8\mu^3\gamma^3 R^5}{1+4k^2R^2} + \cdots. \tag{10.136a}$$

This is the forward *diffraction peak*, whose width is given by (10.137a).

Let us look at the behavior of A_2 as a function of the energy for complex values of E, with τ fixed. Equation (10.135) seems to indicate a singularity at $f = \pm k\tau$, which occurs at

$$k = \pm i\tfrac{1}{2}R^{-1}.$$

This, however, is partly spurious. One of the singularities present in the second term of (10.135) is canceled by one in the first. This can be seen by combining the two terms in the right-hand parentheses into the single expression

$$\ln\frac{(f+k\tau)(2f-i\tau R^{-1})}{(f-k\tau)(2f+i\tau R^{-1})}.$$

The function f, defined to be positive for positive k, is equal to $+i\tau/2R$ when k approaches $+i/2R$ along path (1) of Fig. 10.1 but is equal to $-i\tau/2R$ when k approaches $+i/2R$ along path (2). As a result the zeros in the numerator and denominator of the logarithm (both of which are simple) cancel, for both paths. As k tends to $-i/2R$ along path (3), however, f equals $-i\tau/2R$, and as k tends to $-i/2R$ along path (4), f equals $+i\tau/2R$. In these cases the zeros do not cancel, and the logarithm does have a singularity.

The only other singularity A_2 appears to have is at $f = 0$. But this, too, is spurious since the combined logarithm vanishes here. Thus, contrary to the first impression, A_2 is an analytic function of E, regular everywhere on the physical sheet ($\operatorname{Im} k > 0$), as required by our general arguments of Sec. 10.3.3. The bound-state poles, of course, do not appear in any order of the perturbation theory, but they make themselves felt by a lack of convergence of the Born expansion.

On the second sheet of the Riemann energy surface ($\operatorname{Im} k < 0$), A_2 is not everywhere regular. Our previous argument shows that here A_2 has a

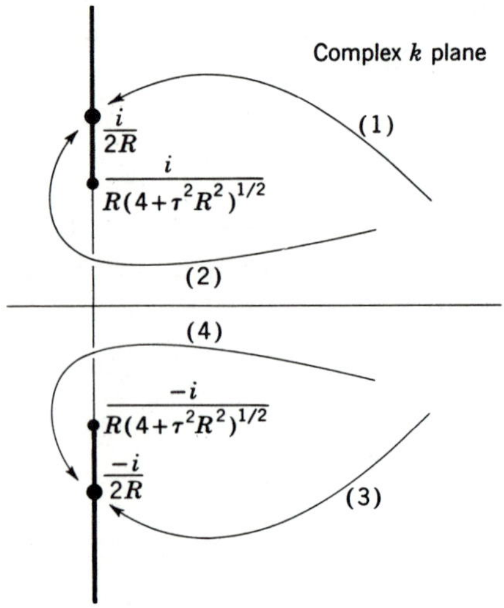

Figure 10.1. Singularities in the second Born approximation to the scattering amplitude for a Yukawa potential, Eq. (10.135).

logarithmic singularity at $k = -i/2R$. As a result there is a cut on the second sheet from $E = -(8\mu R^2)^{-1}$ to infinity, which is conveniently put along the negative real axis.

It is worthwhile at this point to anticipate some of our later results. We shall see in Chap. 12 that the partial-wave amplitudes for the Yukawa potential have a "left-hand cut" on the first sheet of the E surface. Here we see explicitly to the second order, and on general grounds saw in Sec. 10.3.3, that the full amplitude *has no such cut.* It will be useful to keep this fact in mind as a warning not to draw unwarranted conclusions from a dispersion relation.

We may also calculate the first terms of the Fredholm determinant Δ' for the Yukawa potential. According to (9.85),

$$\begin{aligned}\Delta' &= 1 - \operatorname{tr} GVGV + \cdots\\ &= 1 - \frac{\gamma^2\mu^2}{\pi^4}\int d\mathbf{p}\,d\mathbf{p}'\,(k^2 - p^2 + i\varepsilon)^{-1}(k^2 - p'^2 + i\varepsilon)^{-1}\\ &\qquad\qquad [R^{-2} + (\mathbf{p} - \mathbf{p}')^2]^{-2} + \cdots.\end{aligned}$$

The integrals are easily carried out, with the result that

$$\Delta' = 1 + \frac{2\mu^2\gamma^2R^2}{2ikR - 1} + \cdots. \tag{10.139}$$

As it should be, this is regular analytic for all k in the upper half of the k plane, i.e., in the physical sheet. On the second sheet, it has a *pole* at $k = -\frac{1}{2}iR^{-1}$.

Bound States. We may use (10.139) to calculate approximately the coupling strength γ_0 necessary to introduce a bound state. At the bound-state energy, the Fredholm determinant must vanish, and at the coupling strength γ_0, this must happen at zero energy. Equation (10.139) therefore tells us that

$$2\mu\gamma_0 R \simeq \sqrt{2} = 1.41 \cdots \tag{10.140}$$

whereas the "exact" value is known to be[16]

$$2\mu\gamma_0 R \simeq 1.68 \cdots.$$

This agreement is fortuitously good.

Once a bound state exists, we may calculate its binding energy E_B from (10.139):

$$|k_B|R \simeq (\mu\gamma R)^2 - \tfrac{1}{2} \qquad E_B = \frac{k_B^2}{2\mu}. \tag{10.141}$$

Obviously, this approximation becomes worse and worse the larger γ is, and it can be expected to be good only for very small binding energies. Already for $\mu\gamma_0 R = 1$ the approximate value $|k_B|R \simeq 0.5$ is far larger than the exact value[17] $|k_B|R = 0.2$.

If we let $R \to \infty$, the Yukawa potential goes over into the Coulomb potential. The first Born approximation (10.133) to the amplitude then becomes

$$A_{\mathbf{BA}} = -\frac{2\mu\gamma}{\tau^2} \tag{10.142}$$

and the corresponding cross section

$$\frac{d\sigma}{d\Omega} = \left[\frac{\gamma}{4E\sin^2(\theta/2)}\right]^2 \tag{10.143}$$

is the Rutherford cross section for $\gamma = eZ$. But look at the second Born approximation. The real part tends to zero as $R \to \infty$, but since then $f \to k\tau$, the imaginary part "blows up." As we shall see in Sec. 14.6, the explanation of this is that the slow decrease of the Coulomb potential at $r \to \infty$ prevents a conventional scattering amplitude from being definable. Its phase would be infinite. Its magnitude, however, is finite and, it so happens, equal to its first Born approximation. Now, in the Born series, both magnitude and phase are expanded in powers of γ, and it is the expansion of the phase, which does not contribute to the cross section, that causes the trouble. The Fredholm determinant (10.139) tends to infinity as $R \to \infty$.

Notes and References

10.1 For a partial wave expansion of the scattering amplitude for noncentral potentials, see A. D. Boardman et al. (1967). Complex potentials are discussed by A. Ramm (1970).

16 Blatt and Jackson (1949), eq. (4.5Y).

17 Sachs and Goeppert-Mayer (1938).

10.1.4 The following papers deal with separable potentials: Gourdin and Martin (1957); T. R. Morgan (1966); Cassola and Koshel (1968); Gutkowski and Scalia (1968); F. Tabakin (1968); Leung and Park (1969); R. C. Fuller (1969); Grubb and Pearson (1970); P. Beregi (1971); Arnold and MacKellar (1971); Ahmad and Pisent (1974). More general nonlocal interactions are treated by M. Bertero et al. (1967); G. Talenti (1967); Gutkowski and Scalia (1969); M. Coz et al. (1970); F. Capuzzi (1972); Saenz and Zachary (1976); B. Mulligan et al. (1976); B. Bagchi et al. (1977); R. G. Newton (1977b); Foldy and Lock (1979); Talukdar, Das, Mallick, and Sen Gupta (1979); Talukdar, Das, and Mukhopadhyaya (1979); B. Pätzel et al. (1980); C. Deskaloyannis (1981). Bound states in the continuum for separable potentials are discussed by Jain and Shastry (1975); and transparent nonlocal potentials, by K. Chadan (1967).

10.2 The specific difficulties with the Lippmann-Schwinger equation for more than a single particle referred to here are discussed in Foldy and Tobocman (1957). For Clebsch-Gordan coefficients, see the Notes and References to Sec. 2.1.4.

10.3 Most of the general references given in Chap. 9 are applicable to this section. See also H. P. Noyes (1965); K. L. Kowalski (1965); T. A. Osborn (1971); and H. E. Moses (1979a).

10.3.1 The method of using (10.63) for polar kernels is due to E. Garbe (1915). See specifically K. Meetz (1962) for the use of this method. The treatment here is based to some extent on that of Meetz. In particular the analytic continuation of the completeness relation is his. However, his generalization of (10.73) and his conclusion based on it are incorrect. The sufficient condition for the uniform convergence of the Born expansion stated below (10.73) is a slight generalization of that first proved by H. Davies (1960) in a different way. See also R. Huby (1963); Huby and Mines (1964). It is shown in the last-mentioned paper that the stated sufficient condition for the convergence of the Born series for *all* energies is, in general, not necessary.

A demonstration that the Born series always converges at sufficiently high energy can be found in Zemach and Klein (1958). The present argument leading to (10.66) is essentially theirs. M. Bertero et al. (1966) deal with nonlocal potentials, and Gesztesy and Thaller (1981) with Coulomb admixtures.

10.3.2 The fundamental paper on the use of Fredholm theory in scattering problems is that by Jost and Pais (1951). See also N. N. Khuri (1957); Grossmann and Wu (1961 and 1962); A. Grossman (1961); C. L. Dolph et al. (1966); B. Simon (1971); Holt and Santoso (1973). This section largely follows R. G. Newton (1977a and 1980b).

10.3.3 Estimates and bounds for cross sections can be found in D. Villarroel (1970); Cornille and Martin (1972a); A. Martin (1979 and 1981); Amrein and Pearson (1979a and b); A. Ramm (1980b); Enss and Simon (1980 and 1981). See also V. Enss (1980); Klaus and Simon (1980); S. Albeverio et al. (1981). For a general proof that the scattering amplitude defines a Hilbert-Schmidt operator, see T. Ikebe (1965).

10.3.4 The argument of this section is a somewhat simplified version of N. N. Khuri (1957), who first proved the dispersion relation (10.128) under the conditions of this section. Other papers on dispersion relations and analyticity for nonrelativistic potential scattering are L. D. Faddeev (1958b); Klein and Zemach (1959); Bowcock and Martin (1959); Bowcock and Walecka (1959); Blankenbecler, Goldberger, Khuri, and Treiman (1960); A. Klein (1960); W. Hunziker (1961); Grossmann and Wu (1961 and 1962); A. Grossmann (1961); D. I. Fivel (1962); W. W. Bell (1963); Belinfante and Unal (1963); A. R. Swift (1964); M. Jameel (1964); S. O. Aks (1965); I. Saavedra (1966); J. Nuttall (1967c); Lawrynowicz and Wojtcak (1967); and for nonlocal potentials, A. N. Mitra (1961); T. H. Yao (1973, 1974, 1975, 1976); see also G. R. Screaton

(1968 and 1969). Extensions to relativistic potential scattering are Gasiorowicz and Noyes (1958); Khuri and Treiman (1958). See also J. Hilgevoord (1960). For the connection with causality, see N. G. van Kampen (1953b and 1954); H. M. Nussenzveig (1960 and 1969a); I. Saavedra (1962); Aly and Okubo (1967). The last mentioned paper deals with complex potentials.

10.3.5 The Yukawa potential was first introduced by H. Yukawa (1935) as a model for the nucleon-nucleon force. It is obtained in the first approximation if the latter is thought of as produced by the exchange of a meson of mass $(cR)^{-1}$ (c being the velocity of light) in analogy with the electromagnetic force being due to the exchange of a photon.

For a calculation and discussion of higher-order approximations to scattering by Yukawa potentials, see R. Dalitz (1951). See also Jost and Pais (1951); S. Weinberg (1963b); Müller and Schilcher (1968); C. M. Rosenthal (1974).

In the literature of chemical physics, the Yukawa potential is known as the Debye-Hückel potential.

Different approaches to three-dimensional scattering are presented by J. V. Corbett (1966) and by Pattanayak and Wolf (1976); R. G. Newton (1980a). For scattering by δ-function potentials, see V. I. Karloukovski (1976).

Problems

1 Show that, if the interaction is a sum of N separable terms, the hamiltonian has at most N bound states.

2 An operator is separable in all representations if it is so in one. State the essential property of such an operator without reference to a representation.

3 Assume that the interaction is separable. Find the wave function in the coordinate representation and, from it, the scattering amplitude.

4 Calculate the scattering amplitude for the interaction $(\mathbf{r}|H'|\mathbf{r}') = \gamma f(r) f^*(r')\mathbf{r}\cdot\mathbf{r}'$.

5 Calculate the scattering amplitude for the interaction $(\mathbf{p}|H'|\mathbf{p}') = \gamma f(p) f^*(p') P_n(\hat{\mathbf{p}}\cdot\hat{\mathbf{p}}')$ (P_n being the Legendre polynomial of order n).

6 What does the interaction of Prob. 5 look like in the coordinate representation?

7 The flux of particles at the point r is given by $\mathbf{S} = -(ih/m)\,[\psi^*(\mathbf{r})\nabla\psi(\mathbf{r}) - \psi(\mathbf{r})\nabla\psi^*(\mathbf{r})]$. Use the asymptotic form (10.11a) of the wave function at large distances to express the scattered flux in terms of the scattering amplitude A.

8 Prove the expression for the volume element in spheroidal coordinates, given above (10.66).

9 Supply the missing details in the derivation of (10.66).

10 Supply the missing details in the argument for (10.67).

11 Consider the causality arguments given in Sec. 4.2 in the derivation of dispersion relations for electromagnetic radiation. Are they applicable to the case of quantum particle scattering? If not, why not?

12 Do the analyticity arguments for the scattering amplitude given in Sec. 10.3.3 (or in Sec. 10.3.5 in the special case of the Yukawa potential) imply that $\int_{-1}^{1} d\cos\theta\, A(E, \cos\theta)$ is an analytic function of E on the physical sheet? If not, why not?

CHAPTER 11

Single-Channel Scattering of Spin 0 Particles, I

11.1 Partial-Wave Expansion

11.1.1 The S Matrix and Traveling Waves. We now assume that the interaction is local and spherically symmetric, i.e., described by a potential function of $r = |\mathbf{r}|$ only,

$$V(\mathbf{r}) = V(r).$$

Furthermore the particle (or particles, if we are talking about two particles interacting via the potential V) has no intrinsic angular momentum. It is then very useful to expand the wave function and the scattering amplitude on the basis of the spherical harmonics.

We start with the expansion of the free Green's function,[1]

$$G^{\pm}(E;\mathbf{r},\mathbf{r}') = 2\mu \sum_{lm} Y_l^m(\hat{\mathbf{r}})\, Y_l^{m*}(\hat{\mathbf{r}}') r^{-1} r'^{-1} G_l^{\pm}(k;r,r') \tag{11.1}$$

with

$$G_l^{\pm}(k;r,r') = -k^{-1} u_l(kr_<) w_l^{(\pm)}(kr_>) \exp\left[-\tfrac{1}{2} i\pi l(1 \pm 1)\right]. \tag{11.2}$$

The Riccati–Bessel functions u_l and $w_l^{(\pm)}$ were defined in (2.58) and (2.60). The notation is such that $r_<$ indicates the smaller of r and r', and $r_>$ the larger. The fact that the $G_l^{\pm}$ are independent of m expresses the circumstance that $G^{\pm}$ depends on the direction of the vectors $\mathbf{r}$ and $\mathbf{r}'$ only via the angle

1 We recall that μ is the reduced mass of the particles.

between them. Because of (2.18), (11.1) may also be written

$$G^{\pm} = \frac{\mu}{2\pi}\sum_{l=0}^{\infty}(2l+1)P_l(\hat{\mathbf{r}}\cdot\hat{\mathbf{r}}')r^{-1}r'^{-1}G_l^{\pm}(k;r,r'). \tag{11.1a}$$

The partial-wave Green's function $G_l^{\pm}$ satisfies the ordinary differential equation

$$\left[-\frac{\partial^2}{\partial r^2} + \frac{l(l+1)}{r^2} - k^2\right] G_l^{\pm}(k;r,r') = -\delta(r-r'). \tag{11.3}$$

The boundary conditions specify that at $r = 0$ (for fixed r') it is zero and at $r \to \infty$ it contains *outgoing* waves only, i.e., no e^{-ikr} contributions. Its asymptotic value as $r \to \infty$ is, in more detail,

$$G_l^{\pm}(k;r,r') \underset{r\to\infty}{\simeq} -e^{\mp i1/2\pi l}k^{-1}u_l(kr')e^{\pm ikr}. \tag{11.4}$$

It is symmetric under the interchange of r and r'.

Similarly we expand the free wave function. Because of (2.57) it can be written as the series

$$\begin{aligned}\psi_0(\mathbf{k};\mathbf{r}) &= \left(\frac{2\mu k}{\pi}\right)^{1/2}(kr)^{-1}\sum_{lm} Y_l^m(\hat{\mathbf{r}})Y_l^{m*}(\hat{\mathbf{k}})i^l u_l(kr)\\ &= \left(\frac{2\mu k}{\pi}\right)^{1/2}(4\pi kr)^{-1}\sum_{l=0}^{\infty}(2l+1)i^l P_l(\hat{\mathbf{r}}\cdot\hat{\mathbf{k}})u_l(kr).\end{aligned} \tag{11.5}$$

The full wave function, too, can be expanded in this manner. Because of the rotational invariance of V (and of ∇^2) it, too, can depend on the direction of $\mathbf{r}$ only relative to that of the incident beam, i.e., relative to $\mathbf{k}$. Hence it must be writable in the form

$$\begin{aligned}\psi^{(\pm)}(\mathbf{k};\mathbf{r}) &= \left(\frac{2\mu k}{\pi}\right)^{1/2}(4\pi kr)^{-1}\sum_{l=0}^{\infty}(2l+1)i^l P_l(\hat{\mathbf{r}}\cdot\hat{\mathbf{k}})\psi_l^{(\pm)}(k,r)\\ &= \left(\frac{2\mu k}{\pi}\right)^{1/2}(kr)^{-1}\sum_{lm} Y_l^m(\hat{\mathbf{r}})Y_l^{m*}(\hat{\mathbf{k}})i^l\psi_l^{(\pm)}(k,r)\end{aligned} \tag{11.6}$$

with certain unknown coefficients $\psi_l^{(\pm)}(k,r)$.

We now insert all these expansions in the Lippmann-Schwinger integral equation (10.58) and use the orthonormality of the spherical harmonics as well as the rotational invariance of V. Comparing the series on both sides term by term, we get a linear integral equation for $\psi_l^{(\pm)}$,

$$\psi_l^{(\pm)}(k,r) = u_l(kr) + 2\mu\int_0^{\infty} dr'\, G_l^{\pm}(k;r,r')V(r')\psi_l^{(\pm)}(k,r'). \tag{11.7}$$

Use of (11.3) easily allows us to show that ψ_l satisfies the radial version of the Schrödinger equation,

$$\left[-\frac{d^2}{dr^2} + \frac{l(l+1)}{r^2} + 2\mu V(r)\right]\psi_l(k,r) = k^2\psi_l(k,r). \tag{11.8}$$

Furthermore, owing to the behavior of u_l and G_l, $\psi_l^{(+)}$ vanishes at $r = 0$, and at $r \to \infty$ its incoming wave contribution is that of u_l only. Under sufficiently strong conditions on the potential, to be specified in more detail later, this means that as $r \to \infty$

$$\psi_l^{(+)}(k,r) \simeq \tfrac{1}{2}e^{i1/2\pi(l+1)}(e^{-ikr} - e^{-i\pi l}S_l e^{ikr}) \tag{11.9}$$

where S_l is an unknown constant.[2]

Let us be sure to understand the physical picture described by the lth partial-wave functions $\psi_l^{(+)}(r)\, Y_l^m(\theta,\varphi)$. In the asymptotic region it contains radially traveling waves going inward and outward. The inward-traveling amplitude is fixed by the strength of the beam; the outward-going amplitude is determined by the scattering. In addition these waves have an angle dependence impressed upon them which varies with l and m. For $l = 0$ they are isotropic. For higher l values there are *standing* waves in the θ dependence whose l zero surfaces form fixed cones about **k**. The azimuthal variation finally is one of waves *traveling* circularly about **k**. The m nodal surfaces are planes through **k**, and they rotate about it.

The physical interpretation of l is based on the fact that the spherical harmonics are eigenfunctions of the orbital angular-momentum operator [(2.16a) and (2.17)]. The fixing of l therefore amounts to fixing the angular momentum of the particles, and the term $l(l+1)r^{-2}$ in (11.8) is the quantum-mechanical analog of the centrifugal-force term in the classical equation of motion when the angular momentum is given. Because of the physical demand that the wave function be well behaved for *all* directions of **r**, the number l must be restricted to be an *integer*.[3]

The Amplitude. The scattering amplitude $A(\mathbf{k}',\mathbf{k})$ may similarly be expanded on the basis of the spherical harmonics. Because the scatterer is rotationally invariant, its angle dependence must be limited to a dependence upon the angle between **k** and **k**′. It can therefore be written in the form of the series

$$\begin{aligned} A(\mathbf{k}',\mathbf{k}) &= \sum_{l=0}^{\infty} (2l+1)a_l(k)P_l(\hat{\mathbf{k}}\cdot\hat{\mathbf{k}}') \\ &= 4\pi \sum_{lm} Y_l^m(\hat{\mathbf{k}}')\, Y_l^{m*}(\hat{\mathbf{k}})a_l(k). \end{aligned} \tag{11.10}$$

The coefficients a_l may be obtained by expanding both the plane wave and the scattering amplitude in the asymptotic form (10.11) of the wave function according to (2.57) and (11.10) and then comparing the result term by term with the expansion (11.6) of the left-hand side, inserting there the asymptotic form (11.9). The result is

$$a_l = \frac{S_l - 1}{2ik}. \tag{11.11}$$

2 It depends on k, but not on r.

3 We recall the customary nomenclature whose origin goes back to the names of atomic spectral-line series: $l = 0$, s wave; $l = 1$, p wave; $l = 2$, d wave; etc.

The **S** *Matrix and Phase Shifts*. Insertion of the expansion (11.10) in the relation (10.12) between the scattering amplitude and the **T** matrix, and then in the expression (7.58) for the **S** matrix in terms of the **T** matrix, leads to the series

$$S(\mathbf{k}',\mathbf{k}) = \sum_{lm} Y_l^m(\hat{\mathbf{k}}')S_l(k)Y_l^{m*}(\hat{\mathbf{k}}) \tag{11.12}$$

after using (11.11). This shows the significance of the numbers S_l. Because of the rotational symmetry of the hamiltonian, the **S** matrix is diagonal in an angular-momentum representation. This expresses the conservation of angular momentum in the scattering. The eigenvalues of the **S** matrix are just the numbers S_l. Now we have seen in Sec. 7.2.4 that as a result of the conservation of flux, whose ultimate origin is the hermiticity of the hamiltonian, **S** is unitary. Consequently, its eigenvalues must be of magnitude unity,

$$|S_l| = 1. \tag{11.13}$$

We may therefore define a real phase angle δ_l to within an integral multiple of π by

$$S_l = e^{2i\delta_l}. \tag{11.14}$$

The significance of the δ_l is learned by inserting (11.14) in (11.9) and combining the exponentials. The asymptotic form of the wave function $\psi_l^{(+)}$ is thus found to be

$$\psi_l^{(+)} \simeq e^{i\delta_l}\sin(kr - \tfrac{1}{2}\pi l + \delta_l). \tag{11.15}$$

Compare this with the asymptotic form of the *free* wave function, in the absence of the scatterer,

$$u_l \simeq \sin(kr - \tfrac{1}{2}\pi l).$$

The only material difference between the radial wave function in the presence of the scatterer and in its absence is a *shift in phase* by δ_l. The superposition of infinitely many such spherical waves, each shifted in phase and added with the appropriate overall phase as shown in (11.15), makes up the complete solution, which differs at large distances from the incoming plane wave by outgoing spherical waves only.

The partial-wave analysis may now be divorced from the original three-dimensional physically realistic scattering picture. The scattering process may be thought of as proceeding with a fixed value of the angular momentum as though an inward-traveling spherical wave of multipolarity 2^l were sent toward the origin. According to (11.9), the **S**-matrix element S_l measures the response of the scattering target by determining the phase of the resulting outward-traveling spherical wave relative to what it would be if there were no target. The fact that S_l is of magnitude unity directly expresses the absence of absorption or emission. Corresponding to this self-contained physical picture (which, of course, lacks experimental reality) the calculation of S_l or of δ_l may be based entirely on the differential equation (11.8).

Expansion of Cross Sections. Once the phase shifts are calculated, the series (11.10), or, explicitly,

$$\begin{aligned} A(k,\cos\theta) &= (2ik)^{-1}\sum_{l=0}^{\infty}(2l+1)(e^{2i\delta_l}-1)P_l(\cos\theta) \\ &= k^{-1}\sum_{l=0}^{\infty}(2l+1)e^{i\delta_l}\sin\delta_l P_l(\cos\theta) \end{aligned} \tag{11.10a}$$

is used to calculate the scattering amplitude. The differential-scattering cross section is then obtained from (8.16) as the square of the magnitude of A. If this is to be expressed as a series in the Legendre polynomials $P_l(\cos\theta)$, the result is complicated.[4] Only for the total scattering cross section do we get another simple series. From the second version of (11.10) we readily find

$$\begin{aligned} \sigma_{\text{total}} &= \int d\Omega_{k'}\frac{d\sigma(\mathbf{k}',\mathbf{k})}{d\Omega} \\ &= 4\pi\sum_{l=0}^{\infty}(2l+1)|a_l|^2 = 4\pi k^{-2}\sum_{l=0}^{\infty}(2l+1)\sin^2\delta_l. \end{aligned} \tag{11.16}$$

The numbers

$$\sigma_l = 4\pi(2l+1)k^{-2}\sin^2\delta_l \tag{11.17}$$

are usually called *partial* cross sections, and the total scattering cross section is their sum. The unitarity of the **S** matrix, expressing itself in the reality of the phase shifts δ_l, entails an upper limit on the partial cross section of a given angular momentum and at a fixed energy. This is sometimes referred to as the *unitarity limit*, and its value is evidently reached when the phase shift has the value $\frac{1}{2}\pi$ (modulo π),

$$\sigma_l^{\max} = 4\pi(2l+1)k^{-2} = \frac{(2l+1)\lambda^2}{\pi} \tag{11.18}$$

in terms of the de Broglie wavelength of the particle. Because of the infinite summation in (11.16), this does not, however, imply any limitation whatever on the total scattering cross section.

The Optical Theorem. A comparison of the second version of (11.10a) with (11.16) immediately leads to a simple derivation of the optical theorem (10.99) in this special case,

$$4\pi k^{-1}\operatorname{Im}A(0) = \sigma_{\text{total}},$$

$A(0)$ being the forward-scattering amplitude.

An Integral Representation. Before proceeding we note that the expression (10.12) for the amplitude leads to an integral representation of the **S**-matrix element S_l. If we insert the expansions (11.5) and (11.6) in (10.12) and compare the result with (11.10a), we get

4 See Whittaker and Watson (1948), p. 331, prob. 11.

$$S_l = 1 - 4ik^{-1}\mu \int_0^\infty dr\, u_l(kr) V(r) \psi_l^{(+)}(k,r). \tag{11.19}$$

11.1.2 The K Matrix and Standing Waves. The **K** matrix, defined in Sec. 7.2.3, can also be readily expressed in terms of the phase shift. The **S** matrix being diagonal in an angular-momentum representation, so is **K**. Equation (7.59) allows us immediately to express its eigenvalues in terms of (11.14) as

$$K_l = \tan \delta_l. \tag{11.20}$$

The **K** matrix is hermitian; the eigenvalues K_l are therefore real.

As we have seen in (7.56), **K** can be calculated from a *principal-value* wave function $\psi^{(P)}$ which satisfies the integral equation (7.20). The principal-value Green's function can be expanded as G^+ was in (11.1),

$$G^P(E;\mathbf{r},\mathbf{r}') = 2\mu \sum_{lm} Y_l^m(\hat{\mathbf{r}}) Y_l^{m*}(\hat{\mathbf{r}}') r^{-1} r'^{-1} G_l^P(k;r,r') \tag{11.21}$$

and the radial principal-value Green's function is obtained from (7.8), (11.21), and (2.66),

$$G_l^P(k;r,r') = k^{-1} u_l(kr_<) v_l(kr_>) \tag{11.22}$$

where v_l is defined in (2.65). The function G_l^P is real and symmetric and satisfies (11.3). At infinity it goes as

$$G_l^P(k;r,r') \underset{r\to\infty}{\simeq} -k^{-1} u_l(kr') \cos(kr - \tfrac{1}{2}\pi l).$$

Similarly we expand the principal-value wave function,

$$\psi^{(P)}(\mathbf{k};\mathbf{r}) = \left(\frac{2\mu k}{\pi}\right)^{1/2} (kr)^{-1} \sum_{lm} Y_l^m(\hat{\mathbf{r}}) Y_l^{m*}(\hat{\mathbf{k}}) i^l \psi_l^{(P)}(k,r). \tag{11.23}$$

According to (7.20), the radial principal-value functions are then found to satisfy the integral equation

$$\psi_l^{(P)}(k,r) = u_l(kr) + 2\mu \int_0^\infty dr'\, G_l^P(k;r,r') V(r') \psi_l^{(P)}(k,r'). \tag{11.24}$$

They also obey the radial Schrödinger equation (11.8), and they vanish at $r = 0$. At large distances they behave as

$$\psi_l^{(P)}(k,r) \underset{r\to\infty}{\simeq} \sin(kr - \tfrac{1}{2}\pi l) + K_l \cos(kr - \tfrac{1}{2}\pi l)$$
$$= \frac{\sin(kr - \tfrac{1}{2}\pi l + \delta_l)}{\cos \delta_l}$$

where

$$K_l = -2\mu k^{-1} \int_0^\infty dr\, u_l(kr) V(r) \psi_l^{(P)}(k,r).$$

The principal-value function $\psi_l^{(P)}$ is *real*. Therefore, if it is multiplied by e^{-iEt}, it forms a wave whose nodes are *fixed in space*. Consequently, it represents a standing wave, in contrast to $\psi_l^{(\pm)}$, which (at large distances at any rate) is formed of inward- and outward-traveling waves $e^{\mp ikr-iEt}$. Although its reality has certain calculational advantages, the fact that it is a standing-wave solution of (11.8) makes it physically less suitable to scattering problems. Inherently these deal with inward- and outward-moving waves.

11.1.3 Time Delay. The time spent by a particle in the finite spatial region R may be expressed as the integral

$$t_R = \int_{-\infty}^{\infty} dt \int_R d\mathbf{r}\, |\psi(t, \mathbf{r})|^2.$$

If P_R denotes the projection operator onto R, so that in the coordinate representation $P_R\psi(\mathbf{r}) = \psi(\mathbf{r})$ if $\mathbf{r} \in R$ and $P_R\psi(\mathbf{r}) = 0$ otherwise, then

$$t_R = \int_{-\infty}^{\infty} dt\, (\Psi(t), P_R\Psi(t)).$$

If $\Psi(t)$ is a scattering state then this integral can be expected to converge by the mean ergodic theorem.[4a] If $\Psi(t)$ is taken to be $\Psi^{(+)} = \Omega^{(+)}\Psi_0(t)$ then

$$t_R = \int_{-\infty}^{\infty} dt\, (\Psi_0(t), \Omega^{(+)\dagger} P_R \Omega^{(+)} \Psi_0(t))$$

and it becomes relevant to compare t_R to the time spent by the free particle in R:

$$t_R^0 = \int_{-\infty}^{\infty} dt\, (\Psi_0(t), P_R\Psi_0(t))$$

so that

$$\begin{aligned} t_R - t_R^0 &= \int_{-\infty}^{\infty} dt\, (\Psi_0(t), (\Omega^{(+)\dagger} P_R \Omega^{(+)} - P_R)\Psi_0(t)) \\ &= \left(\Psi_0, \int_{-\infty}^{\infty} dt\, e^{iH_0t}(\Omega^{(+)\dagger} P_R \Omega^{(+)} - P_R)e^{-iH_0t}\Psi_0\right). \end{aligned}$$

The *time delay* is finally obtained by taking a limit, in some appropriate average sense, as $|R| \to \infty$ (by which we mean that the "radius" of R is allowed to grow to infinity and $P_R \to \mathbb{1}$). The time delay then becomes expressed as

$$t_D = \langle Q \rangle = (\Psi_0, Q\Psi_0),$$

where

$$Q = \lim_{|R|\to\infty} \int_{-\infty}^{\infty} dt\, e^{iH_0t}(\Omega^{(+)\dagger} P_R \Omega^{(+)} - P_R)e^{-iH_0t}. \tag{11.25}$$

4a J. von Neumann (1932).

In the momentum representation the time-delay operator Q becomes

$$Q(\mathbf{k}',\mathbf{k}) = \delta(E' - E)q(E,\mathbf{k}',\mathbf{k})/\mu k$$

where $q(E,\mathbf{k}',\mathbf{k})$ is the momentum-space representation of the operator

$$q(E) = 2\pi\mu k \lim_{|R|\to\infty} \{[\mathbb{1} + T^\dagger(E)G^-(E)]P_R[\mathbb{1} + G^+(E)T(E)] - P_R\}.$$

One then finds[4b] that $q(E)$ may be expressed directly in terms of the S matrix (7.58), namely

$$q(\mathbf{k}',\mathbf{k}) = -i\int d\hat{\mathbf{k}}''\, S(\mathbf{k}'',\mathbf{k}')^* \frac{d}{dE} S(\mathbf{k}'',\mathbf{k}). \tag{11.26}$$

(It is understood here that $k^2 = k'^2 = k''^2 = 2\mu E$.) In the case of a central potential the expansion (11.12) gives

$$q(\mathbf{k}',\mathbf{k}) = 2\sum_{l,m} Y_l^m(\hat{\mathbf{k}}') \frac{d\delta_l}{dE} Y_l^{m*}(\hat{\mathbf{k}}). \tag{11.27}$$

We will see in the next section that this expression for the time delay in scattering can be obtained, for resonances, in another manner. The result stated in (11.26) has very general validity and has been applied in a variety of contexts.

11.2 Heuristic Survey of Phase-Shift Behavior

11.2.1 General Properties. Before making a more rigorous investigation, we want to analyze some of the properties of the phase shifts in a heuristic manner.

First, let us write the radial Schrödinger equation in the form

$$\frac{\psi_l''}{\psi_l} = W(r) - k^2$$

$$W(r) = 2\mu V + l(l+1)r^{-2}.$$

At distances where k^2 is larger than $W(r)$, there the wave function is curving toward the abscissa; where k^2 is less than $W(r)$, it is curving away from it. Compared with the situation when $V = 0$, an attractive (negative) potential decreases W; thus, if in a given region k^2 is larger than $l(l+1)r^{-2}$, then the curvature of ψ_l is increased by V, i.e., the "local" wavelength is decreased. The effect is that the wave is "pulled in." A repulsive (positive) potential, on the contrary, "pushes" the wave out. Now, no matter how small k^2 is, it is larger than $l(l+1)r^{-2}$ almost everywhere. Hence we expect a "predominantly" attractive potential to cause *positive* phase shifts, and a "predominantly" repulsive potential to cause *negative* ones.

An increase in the value of the angular momentum will tend to make a

4b J. M. Jauch and J.–P. Marchand (1967); D. Bollé and T. A. Osborn (1975).

given potential relatively less important, and so will an increase in energy. Therefore we expect the phase shifts to approach zero as $l \to \infty$, for a given energy, and as $E \to \infty$, for a given l value. That $S_l \to 1$ as $E \to \infty$ is also to be expected after a glance at the expression (11.19). Since increases in energy and angular momentum to some extent compensate one another, owing to their opposite sign in ψ_l''/ψ_l, it is to be expected that the energy at which the phase shifts are "negligible" increases with the value of l.

To say that $\delta_l \to 0$ is, of course, partly a matter of convention, since they are defined by (11.14) only within an integral multiple of π. What we expect on the basis of the foregoing argument is that $S_l \to 1$ as either $l \to \infty$ or $E \to \infty$. It is customary to *define*

$$\lim_{E\to\infty} \delta_l = 0. \tag{11.28}$$

This makes $\delta_l(E)$ unique for every l value if we require it to be continuous as a function of E. (It remains to be established, of course, that S_l is continuous and hence that δ_l *can* be defined to be continuous.) It will have to be shown that the thus defined phase shifts tend to *zero* as $l \to \infty$, at a fixed value of E, rather than to an integral multiple of π.

11.2.2 Discussion of Low-Energy Phase-Shift Behavior. Now consider the low-energy region. According to (2.64), the function $u_l(kr)$ is proportional to k^{l+1} as $k \to 0$. Because of (2.66) and (2.67) the Green's functions have the finite real limit

$$G_l^{\pm}(0;r,r') = G_l^P(0;r,r') = \frac{-r_<^{l+1} r_>^{-l}}{(2l+1)}. \tag{11.29}$$

On the basis of the integral equation (11.7) we therefore expect $\psi_l^{(+)}$ also to be real and proportional to k^{l+1} as $k \to 0$. Similarly for $\psi_l^{(P)}$, on the basis of (11.25). The representation (11.19) for S_l therefore tells us that, unless the potential is such that the resulting factor c_l is infinite, the low-energy behavior of S_l is

$$S_l = e^{2i\delta_l} \simeq 1 - 2ic_l k^{2l+1} \tag{11.30}$$

or

$$a_l \simeq -c_l k^{2l}$$

or

$$K_l = \tan\delta_l \simeq -c_l k^{2l+1}$$

and the constant c_l is real. This implies, first of all, that the diagonal **S**-matrix elements tend to 1 at zero energy, and they do so at least linearly in k. Equation (11.10a) shows that this is a prerequisite for the scattering cross section to remain finite in the low-energy limit. What is more, (11.30) says that, as k increases from zero, the partial-wave amplitudes a_l vary more and more slowly as l increases. That is, at a given small energy, only a few amplitudes differ appreciably from zero, and the smaller the energy, the smaller that number. In the zero energy limit, in fact, all but the s-wave amplitude vanish. It is because of this circumstance that the partial-wave analysis is particularly useful for *low-energy* scattering. Obviously the series expansion (11.10) is unwieldy if many terms contribute to the sum. There always exists an energy region, though, in which a small number of terms in (11.10a) and (11.16) suffices to give a good approximation. As the energy increases, this number goes up. Thus the partial-wave analysis forms an important complement to the Born approximation, which is good at high energies.

The Physical Explanation. The physical reason for the dominance of waves of low angular momentum at low energies is easy to understand. The larger the angular momentum, the more formidable is the effective barrier produced by the centrifugal force, and the harder it is for the particles to penetrate to the central region where most of the potential energy is concentrated. Thus it takes a higher kinetic energy to "feel out" the potential at a large angular momentum than at a small one. It is clear from this picture that if the potential decreases very slowly at infinity, so that much of the scattering is caused by its long tail, then the centrifugal barrier is less effective and (11.30) may break down. We shall see in the subsequent more rigorous discussion that this is correct [see (12.160)].

Another physical way of understanding the dominance of the first terms in (11.10) at low energies is based on an analogy with electromagnetic scattering by small obstacles. If the wavelength is large compared with the size of the target, which in this instance is the "range" of the potential, then the wave will not "see" its detailed structure and the scattering is *as simple as possible*. In the electromagnetic case this is given by the Rayleigh formula (3.6), whose angle dependence is due to the spin of the photon; in the present case it is isotropic. The smaller the wavelength relative to the size of the target, the more structure appears in the scattering. We expect that the relevant criterion for the neglect of the p wave relative to the s wave is that

$$(kR)^2 \ll 1 \tag{11.31}$$

if R is in some sense the range of the potential.

The Scattering Length. If one inserts (11.30) in (11.10) at zero energy, he obtains the cross section

$$\sigma_{\text{total}}(0) = 4\pi c_0^2. \tag{11.32}$$

The number c_0 is called the (s-wave) *scattering length.*

For the purpose of comparison we may calculate the scattering amplitude for the simple case of a rigid sphere of radius R. Since the potential is $+\infty$ at $r = R$, the radial Schrödinger equation is solved with the boundary condition $\psi_l = 0$ at $r = R$; for $r > R$ we have $V = 0$. That yields immediately

$$\psi_l = c[u_l(kr)v_l(kR) - u_l(kR)v_l(kr)]$$

for $r > R$. Hence by comparison with (11.26) and the asymptotic values (2.59) and (2.68) of u_l and v_l, we find

$$\tan \delta_l = \frac{u_l(kR)}{v_l(kR)}. \tag{11.33}$$

In the zero energy limit we use (2.64) and (2.67) and get[5]

$$\tan \delta_l \simeq -\frac{(kR)^{2l+1}}{(2l+1)!!(2l-1)!!} \tag{11.34}$$

corroborating both (11.30) and the criterion (11.31). The (s-wave) scattering length is therefore equal to the radius, $c_0 = R$. Hence we conclude that in the zero energy limit all obstacles scatter as though they were rigid spheres of a radius equal to the scattering length.

It may at first seem surprising that a rigid sphere of radius R blocks out an area four times as large as its geometrical cross section. Remember, though, that the long-wavelength region is that of the extreme quantum-mechanical regime, far removed from the classical geometrical-optics limit. We shall return to the latter in Chap. 18 and note here only that a rigid sphere does not constitute a "well-behaved" scatterer in the present sense. The high-energy limit of (11.33) is

$$\tan \delta_l \simeq -\tan(kR - \tfrac{1}{2}\pi l) \tag{11.35}$$

so that the phase shifts tend to *infinity* as $k \to \infty$, rather than to zero.

The Effective-Range Approximation. A very useful analysis of the low-energy s-wave scattering is based on multiplying the radial Schrödinger equation for $\psi(0,r)$ by $\psi(0,r)$ and subtracting that for $\psi(0,r)$ multiplied by $\psi(k,r)$. (We drop the reference to l for simplicity; l equals zero for now.) We get

$$\frac{d}{dr} W[\psi(k,r), \psi(0,r)] = k^2 \psi(k,r)\psi(0,r)$$

where

$$W(F,G) = FG' - F'G \tag{11.36}$$

is the *wronskian*. Now the asymptotic form of $\psi(k,r)$ for large r is necessarily

$$\psi(k,r) \simeq C \sin(kr + \delta) \equiv \varphi(k,r), \tag{11.37}$$

and φ satisfies the radial Schrödinger equation for $V = 0$. Hence we also have

$$\frac{d}{dr} W[\varphi(k,r), \varphi(0,r)] = k^2 \varphi(k,r)\varphi(0,r).$$

If we choose $C = 1/\sin \delta$, then

5 $(2l+1)!! = (2l+1)(2l-1)\cdots 3 \cdot 1 = \Gamma(2l+2)/2^l\Gamma(l+1)$.

$$\varphi(0,r) = 1 - \frac{r}{c_0}. \tag{11.38}$$

Subtract the two wronskian equations, and integrate them from zero to infinity. Since

$$W[\varphi(k,r),\varphi(0,r)]_{r=0} = -\frac{1}{c_0} - k\cot\delta$$

we find that

$$k\cot\delta(k) = -\frac{1}{c_0} + k^2\int_0^\infty dr\,[\varphi(k,r)\varphi(0,r) - \psi(k,r)\psi(0,r)]. \tag{11.39}$$

To get a low-energy approximation, we now set $k = 0$ in the integral,

$$k\cot\delta(k) \simeq -\frac{1}{c_0} + \tfrac{1}{2}k^2 r_0 \tag{11.40}$$

where

$$r_0 = 2\int_0^\infty dr\,[\varphi^2(0,r) - \psi^2(0,r)]. \tag{11.41}$$

Equation (11.40) can always be written down and says almost nothing. That $k\cot\delta$ is an even function of k and hence that the linear term is absent follows from the fact that $\delta(-k) = -\delta(k)$, to be discussed further below. The physical significance of these formulas lies in the simple interpretation of (11.41). Suppose that the potential vanishes for $r > R$. Then $\psi = \varphi$ for all $r > R$ and the integral in r_0 extends only to the radius R. The function $\varphi(0,r)$ is a straight line from 1 at $r = 0$, crossing the abscissa at $r = c_0$; ψ is equal to it for $r > R$ but bends down to zero at $r = 0$. Hence we expect the integral in r_0 to have a value roughly equal to *half* the range of the potential. The quantity r_0 is therefore called the *effective range* of the potential and (11.40) the *effective-range approximation.*[6] It is particularly useful when the potential is *strong* because we expect the integral in (11.41) to be rather insensitive to k until the energy approaches a value comparable with the potential strength. Hence (11.40) can be expected to be a good approximation for energies small compared with the potential strength. For this reason the effective-range formula has been especially useful in neutron-proton scattering.

The effective-range approximation can be generalized to higher angular momenta, where we may expand

$$k^{2l+1}\cot\delta_l = -\frac{1}{c_l} + \tfrac{1}{2}r_l k^2 + \cdots. \tag{11.40a}$$

For two reasons it is, however, considerably less useful there. First of all, when higher-angular-momentum waves become appreciable, the energy

6 Since it depends on the range and depth of the potential only, it is also sometimes called the *shape-independent* approximation.

must be so high that (11.40) is no longer a good approximation to the s wave. Hence we must improve it, and the next terms will depend on the *shape* of the potential. Second, if φ is defined by the analog of (11.37) appropriate to other angular momenta, the analog of the integral in (11.39) does not vanish beyond the potential range on account of the centrifugal term. If, on the other hand, φ is taken to be the exact form of the wave function in the outside region, i.e., a combination of regular and irregular Riccati-Bessel functions, then it tends to infinity at $r \to 0$ for $l > 0$. The integral can then not be taken to $r = 0$. In other words, although (11.40a) is just as valid as (11.40), the parameter r_l for $l > 0$ does not admit of so simple an interpretation as r_0.

Detailed Phase-Shift Behavior at Low Energies. We now want to see in more detail how the phase shifts themselves behave in the low-energy region. Particularly, the question arises if the fact (11.30) that (usually) the **S** matrix approaches unity at zero energy implies that the phase shifts vanish there. Since they have been defined to be zero at infinite energy and to be continuous (if possible), we are not free to *make* them vanish.

To get general qualitative information on the shape of the phase shift in the low-energy domain, we resort to a simple model, namely, that of the *square-well* potential,

$$V(r) = \begin{cases} -V_0 & \text{for } r < R \\ 0 & \text{for } r > R. \end{cases} \tag{11.42}$$

The Schrödinger equation can now easily be solved. Inside the potential well we have

$$\psi_l = C_l u_l(Kr) \qquad K^2 = k^2 + K_0^2 \qquad K_0^2 = 2\mu V_0 \tag{11.43}$$

and outside

$$\psi_l = A_l u_l(kr) + B_l v_l(kr). \tag{11.43a}$$

Since both ψ_l and its derivative must be continuous at $r = R$, we obtain the coefficients by matching logarithmic derivatives there,

$$k\frac{A_l u_l'(kR) + B_l v_l'(kR)}{A_l u_l(kR) + B_l v_l(kR)} = \beta_l, \tag{11.44}$$

where the prime indicates differentiation with respect to the argument kR of the functions and β_l is the logarithmic derivative of the inside wave function at $r = R$,

$$\beta_l = \frac{Ku_l'(KR)}{u_l(KR)}. \tag{11.45}$$

According to (11.26) and the asymptotic values (2.59) and (2.68), we find

$$\tan\delta_l = -\frac{B_l}{A_l} = \frac{u_l}{v_l}\frac{\beta_l - k(u_l'/u_l)}{\beta_l - k(v_l'/v_l)}, \tag{11.46}$$

where all the Bessel functions are evaluated at kR.

We consider first the s wave. For $l = 0$ we get

$$\tan \delta_0 = \frac{k \tan KR - K \tan kR}{K + k \tan kR \tan KR} \tag{11.47}$$

or

$$\delta_0 = \tan^{-1}\left(\frac{k}{K} \tan KR\right) - kR. \tag{11.47a}$$

When $K/k \simeq 1$ in the sense that $|K - k|R \ll 1$, then the phase shift is small; and when $KR \ll 1$ as well as $kR \ll 1$, then $\tan \delta_0$ is small. For a weak potential, in the sense that $|K_0|R \ll 1$, these two regions overlap, and the phase shift itself must everywhere remain small. We are then certain that δ_0 approaches zero as $k \to 0$. Suppose now that the potential, which we assume for the moment to be attractive, is multiplied by γ and we *increase* that coupling strength. The phase shift will then increase. In the low-energy region, $kR \ll 1$, it is approximated by

$$\delta_0 \simeq kR\left(\frac{\tan KR}{KR} - 1\right). \tag{11.48}$$

When KR lies just below the value $\frac{1}{2}\pi$, the function in parentheses is large and positive until KR goes through $\frac{1}{2}\pi$. Then the approximation (11.48) breaks down; the phase shift quickly turns over and then decreases. When $K_0 R = \frac{1}{2}\pi$, the phase shift has the anomalous value $\frac{1}{2}\pi$ at zero energy. When $K_0 R$ is slightly above $\frac{1}{2}\pi$, then the phase shift goes through $\frac{1}{2}\pi$ and, as the energy decreases, keeps on increasing to π. Equation (11.48) must then be replaced by

$$\delta_0 \simeq \pi - kR\left(1 - \frac{\tan KR}{KR}\right). \tag{11.48a}$$

Figure 11.1 shows the phase-shift behavior in the low-energy region.

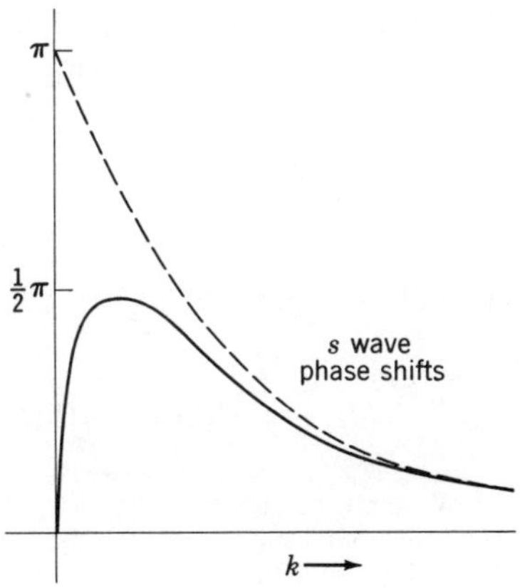

Figure 11.1. The solid line represents the low-energy s-wave phase shift for a potential not quite strong enough to bind; the dashed curve, for a potential that causes one loosely bound state.

Now, $K_0 R = \frac{1}{2}\pi$ is precisely the minimal value necessary for a *bound state* to exist. This is easily seen from the equation

$$K \cot KR = -|k|$$

for the bound-state energy $E = -|k|^2/2\mu$. We conclude therefore that the s-wave phase shift at zero energy has the value *zero* if there is no bound state and the value π if there is one. At the transitional value of the potential strength the value of the s-wave phase shift at $E = 0$ is $\pi/2$; $e^{2i\delta_0}$ is then -1, and the cross section is infinite.[7]

The discussion can be generalized, and one finds that every introduction of an additional bound state raises the phase shift at zero energy by an additional π. This observation says that

$$\delta_0(0) = m\pi \tag{11.49}$$

if m is the number of s-wave bound states, except at the "transitional" strengths when

$$\delta_0(0) = (m + \tfrac{1}{2})\pi. \tag{11.49a}$$

These equations are known as *Levinson's theorem.*

Notice that, if the potential does not quite bind,[8] the phase shift does not reach $\frac{1}{2}\pi$ and the partial cross section does not quite go up to its unitarity limit. If the potential barely binds, the phase shift goes through $\frac{1}{2}\pi$ in the low-energy region and the cross section is maximal. This is *not* to be confused with a resonance, though. We shall see in the discussion below that a resonance is associated with a cross-section peak at which the phase shift *increases* through $\frac{1}{2}\pi$.

The existence or nonexistence of an s-wave bound state is correlated with the scattering length. If there is no bound state, then δ_0 *increases* at $k = 0$ and the scattering length is *negative.* At the potential strength for which $\delta_0 = \frac{1}{2}\pi$, the scattering length is infinite. If there is a loosely bound state, then $c_0 > 0$. Now, before the potential reaches the strength to introduce a second bound state, the scattering length goes through zero and changes sign. Near each bound state it gets large, and it goes through infinity when a new bound state appears. When the potential is (predominantly) repulsive, the scattering length is positive.

Higher Angular Momenta. We now repeat these considerations for the higher angular momenta. Let us expand β_l of (11.45) near $k = 0$,

$$R\beta_l(k) = b_l - e_l(Rk)^2 + \cdots. \tag{11.50}$$

Then (11.46) gives us at low energies

$$\tan \delta_l \simeq \frac{(Rk)^{2l+1}}{(2l+1)!!(2l-1)!!} \frac{b_l - l - 1}{(b_l + l) - (Rk)^2[e_l + 1/(2l-1)]}. \tag{11.51}$$

7 We shall see in Sec. 12.1.2 that, at that transitional potential strength for $l = 0$, there is usually *no* bound state of zero binding energy. Without much physical justification the situation is sometimes called a *zero-energy resonance.*

8 For $l = 0$ it is customary to say then that there is a *virtual state.*

When $Rk \ll 1$, none of these phase shifts contributes much compared with that for $l = 0$, *except* when for some specific l value the denominator in (11.51) vanishes. Then the corresponding phase shift goes through $\frac{1}{2}\pi$ (modulo π), and the cross section is *maximal.* A necessary condition for this to happen (at low energies) is evidently that

$$|b_l + l| \ll 1,$$

i.e., according to (11.45) and the recursion relations (2.69),

$$\left| \frac{K_0 R u_{l-1}(K_0 R)}{u_l(K_0 R)} \right| \ll 1. \tag{11.52}$$

If the potential strength is such that $u_{l-1}(K_0 R)$ is in the vicinity of one of its zeros, then $\tan \delta_l$ moves through infinity, but not through zero at low energies. Hence, as E decreases, the phase shift rises through $\frac{1}{2}\pi$ (not necessarily at low energy), peaks below π, and then rapidly decreases through $\frac{1}{2}\pi$ to zero. As the potential strength increases, the peak shifts toward the left and rises. At the strength such that exactly

$$u_{l-1}(K_0 R) = 0 \tag{11.53}$$

the zero-energy phase shift "hangs" at π. Equation (11.53) is just the condition for a bound state of angular momentum l to be introduced. Thus the phase-shift behavior for $l > 0$ is as indicated in Fig. 11.2.

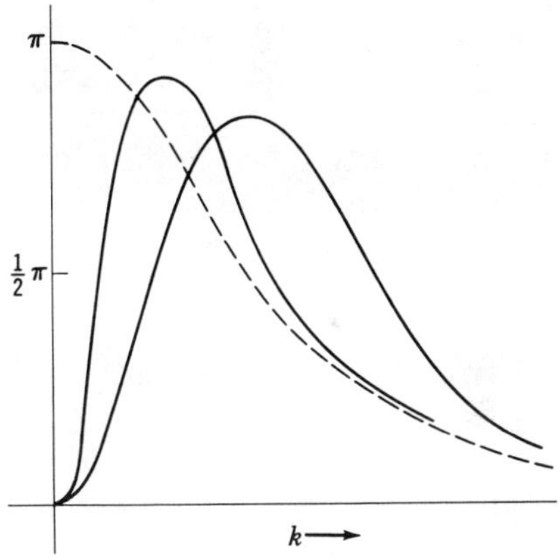

Figure 11.2. Low-energy phase-shift behavior for $l > 0$. The two solid curves are for "almost" bound states; the dashed, for a loosely bound state.

Just as in the case of the s wave, we conclude, from the fact that each newly introduced bound state raises the zero-energy phase shift by π, that Levinson's theorem (11.49) must hold. The particular deviation (11.49a) at the transitional strength,[9] though, is characteristic of $l = 0$ only. For reasonable potentials (to be specified more exactly later) the zero-energy phase shifts of the higher angular momenta are always integral multiples of π.

9 For $l > 0$, the "transitional" potential strength causes a true bound state of zero binding energy (see Sec. 12.2).

Resonances. In contrast to the s wave, the partial cross section of an angular momentum $l > 0$ attains its unitarity limit at low energies if the potential is almost strong enough to introduce a new bound state of angular momentum l. Since, then, all other partial cross sections are small, the phenomenon observed is striking. The scattering cross section in a narrow energy region about that at which the phase shift sharply rises through $\frac{1}{2}\pi$ is large and dominated by the angular distribution characteristic of l; according to (11.10a) and (11.51)

$$\frac{d\sigma}{d\Omega} \simeq \lambda\!\!\!^{-2}(2l+1)^2[P_l(\cos\theta)]^2 \frac{\frac{1}{4}\Gamma^2}{(E-E_0)^2+\frac{1}{4}\Gamma^2} \tag{11.54}$$

where

$$\begin{aligned} E_0 &= \frac{(b_l+l)(2l-1)}{2\mu R^2[e_l(2l-1)+1]} = \frac{k_0^2}{2\mu} \\ \Gamma &= \frac{2E_0(Rk_0)^{2l+1}}{(2l+1)!!(2l-1)!!}\left(\frac{2l+1}{b_l+l}-1\right). \end{aligned} \tag{11.55}$$

The shape of the cross section as a function of the energy is indicated in Fig. 3.2. It is sharply peaked at $E = E_0$, and the half width of the peak is $\frac{1}{2}\Gamma$. At this energy distance from the maximum the cross section has decreased to one-half of its maximal size. An energy dependence such as (11.54) is usually called a *Breit–Wigner formula*.

The phenomenon described is a *resonance*. Its characteristic is, not only the sharp peak in the cross section, but the fact that the phase shift *increases* with energy as it goes through $\frac{1}{2}\pi$ (modulo π). To understand the importance of this particular property, we must return for a moment to the time-dependent description of the scattering.

Time Delay. Let us refine the stationary-phase argument of Sec. 1.3.1. At large distance from the target a wave packet of central momentum k_0 is described by

$$\int d\mathbf{k}\, f(\mathbf{k}) \exp[i(\mathbf{k}\cdot\mathbf{r} - Et)] + \int d\mathbf{k}\, f(\mathbf{k}) r^{-1} \exp[i(kr - Et)] A(\mathbf{k}',\mathbf{k}),$$

and we assume that the function f is sharply peaked about $\mathbf{k} = \mathbf{k}_0$. It was argued in Sec. 1.3.1 that for large t the second term contributes only at a distance at which the phase of the exponential is stationary near $\mathbf{k}_0$. In this case that would mean

$$r \simeq \frac{k_0}{\mu} t = v_0 t$$

if v_0 is the central velocity of the particles. However, attention must also be paid to the phase φ of the scattering amplitude A. Then we find that the emerging packet is centered at the distance

$$r \simeq v_0 t - \left.\frac{d\varphi}{dk}\right|_{k=k_0}. \tag{11.56}$$

Therefore, if the phase of the amplitude sharply *increases* as a function of k, there is a significant *delay* in the emergence of the scattered particles. On the other hand, if φ *decreases* with increasing energy, there is a time *advance* in their emergence. We expect on the basis of causality that the possibility of such an advance is limited by the size of the scatterer.

If the target "size" (which in this case roughly means the range of the potential) is D, then an advance of

$$t_0 > \frac{D}{v_0} = \frac{D\mu}{k_0}$$

would mean that the scattered wave left the target before the incident wave reached it. This is causally impossible. Consequently we require that always

$$\frac{d\varphi}{dk} \gtrsim -D.$$

In the vicinity of a sharp peak in a single partial cross section this implies that the derivative of the corresponding phase shift is limited by the inequality

$$\frac{d\delta_l}{dk} \gtrsim -D. \tag{11.57}$$

A true physical resonance phenomenon is caused by a *capture* of the incident missiles in the scattering region, analogous to the classical occurrence of *orbiting* (see Sec. 5.4). Its quantum-mechanical mechanism is based on the fact that at specific wavelengths an almost standing wave can be set up in the target region. It is thus also intimately related to the electromagnetic discussion of Sec. 3.3.1. The fact that the missiles are almost caught shows that a *delay* of their emergence is an integral part of the phenomenon. In spite of the fact that experimentally usually only the peak is observed, the time delay is just as essential to it as the large size of the cross section. For practical reasons the relevant experiments usually do not measure emergence times, and it is often assumed that every sharp peak in a cross section is in fact evidence for a resonance. This assumption is justified only if the peak is narrower than would be reasonable on causality grounds if the phase were to decrease. The width Γ of the peak is approximately the energy range within which the phase shift changes by π; therefore

$$\Gamma \simeq \frac{\pi}{|d\delta_l/E|_{E=E_0}}.$$

Thus, if the peak is caused by a *decreasing* phase shift, according to (11.57) we expect its width to be at least

$$\Gamma_{\min} \simeq \frac{\pi k_0}{\mu D} = E_0 \frac{\lambda_0}{D}. \tag{11.58}$$

If the size D of the target is known and a resonance peak of much smaller width is experimentally observed, one may safely conclude without measurement of the time dependence of the flux that the phase shift must have been *increasing* and that the peak is a genuine resonance.

In any case, we see that only if the phase shift sharply *increases* through $\frac{1}{2}\pi$ (modulo π) may we speak of a resonance. Thus for angular momenta other than zero, an "almost" bound state causes a resonance, but for $l = 0$, it does not. The physical reason for this is easy to understand. Let us for the moment idealize the centrifugal barrier produced by the angular momentum as an infinitely high wall surrounding the target. At certain specific energies, this would allow standing waves inside, or bound states. Now the centrifugal barrier is not infinitely high. As a result the tunnel effect allows particles to get in and, once inside, eventually to leak out. As a result the *bound-state energies* are not sharp, and the corresponding states have a finite lifetime. It is these vestiges of bound states that allow the particles to be caught for a long time and then to emerge. As we deepen the potential well, the resonance energies move downward, where the centrifugal barrier is thicker. As a result their lifetimes increase, and thus their widths decrease, an effect that is clearly visible in (11.55). Finally, at a specific potential depth a resonance may move to zero and, thereafter, to negative energy. Then, of course, no "leakage" can occur, and we have a genuine bound state. It is for this reason that low-energy resonances are associated with the introduction of bound states. The absence of the centrifugal barrier explains the non-occurrence of low-energy resonances for s waves. We must not forget, though, that, if the potential contains its own barrier, it can cause low-energy resonances also in the s wave.

According to (11.56) the time delay caused by the sharply changing term $e^{2i\delta_l}$ in the amplitude is given by [see also (11.27)]

$$t_D = 2\frac{d\delta_l}{dk}\bigg|_{k=k_0} \frac{\mu}{k_0}. \tag{11.59}$$

The Heisenberg uncertainty principle tells us that the ensuing *line width* Γ is given by

$$\Gamma \simeq \frac{1}{t_D}.$$

The width thus calculated agrees with (11.55). Nevertheless, as we shall see in detail in Chap. 19, the delay time (11.59) must not be taken quantitatively seriously.

It is useful to describe the resonance phenomenon also in terms of the **S**-matrix element connected to the phase shift by (11.14). If δ_l sharply increases by about π in the vicinity of $k = k_0$, going through $\frac{1}{2}\pi$ (modulo π) at k_0, so that

$$\tan\delta_l \simeq \frac{k_1}{k_0 - k} \qquad k_1 > 0 \tag{11.60}$$

describes the essential energy dependence there, then S_l has the form

$$S_l = e^{2i\delta_l} \simeq \frac{k - k_0 - ik_1}{k - k_0 + ik_1}. \tag{11.61}$$

The S-matrix element thus has a *pole* at $k = k_0 - ik_1$ in the lower half of the complex k plane, close to the real axis. It then follows that near $k = k_0$

$$\sin^2 \delta_l = \tfrac{1}{4}|S_l - 1|^2 \simeq \frac{k_1^2}{(k - k_0)^2 + k_1^2} \simeq \frac{\frac{1}{4}\Gamma^2}{(E - E_0)^2 + \frac{1}{4}\Gamma^2} \tag{11.62}$$

as in (11.54), with

$$E_0 = \frac{k_0^2}{2\mu} \qquad \Gamma = \frac{4k_0^2 k_1^2}{\mu^2}.$$

The asymptotic behavior (11.9) of the wave function shows that when S_l has a pole at $k = k_0 - ik_1$ then we have outgoing spherical waves only, i.e., a "source" of flux. Of course the energy is complex, and so this is not physically realizable. Indeed, the source wave function blows up exponentially at infinity. The resonance phenomenon is due to the fact that at the (physically possible) energy ${k_0}^2/2\mu$ we are "near" such a source term and it is to be expected that the wave function is particularly sensitive there to small changes in the energy. Of course, (11.61) shows that we are also near a *zero* of S_l at $k = k_0 + ik_1$, where the wave function has incoming waves only. But this too blows up exponentially as $r \to \infty$ and thus adds to the sensitivity of the wave near k_0.

If k_1 were negative, then S_l would have a pole in the *upper* half plane and the wave function would *decrease* exponentially. Thus we would have a *bound state*. Since its energy is real, we must have $k_0 = 0$. So, if E_0 is the binding energy of a bound state, we expect S_l to have a pole at $k = i(2\mu E_0)^{1/2} = ik_B$. Now

$$e^{2i\delta_l} = \frac{\cot \delta_l + i}{\cot \delta_l - i},$$

and therefore we expect that at $k = ik_B$

$$\cot \delta_l = i. \tag{11.63}$$

Suppose now that the binding energy is small relative to the potential depth. Then we expect the effective-range approximation (11.40) to be valid, even for negative energies. Therefore, according to (11.63),

$$-k_B \simeq -\frac{1}{c_0} - \tfrac{1}{2}k_B^2 r_0$$

or

$$r_0 \simeq 2k_B^{-1} - 2k_B^{-2}c_0. \tag{11.64}$$

Thus the binding energy determines the effective range, if the scattering length is given.

An example in which these approximations are experimentally well verified is furnished by the deuteron. It must be remembered, though, that the approximations contain a number of assumptions not always valid. One of them in particular is that the **S**-matrix element possesses an analytic continuation into the complex k plane and that there are no other nearby singularities, except the bound-state pole. As we shall see in Chap. 12, this is certainly not always the case.

11.3 Variational Approaches

11.3.1 General Introduction. The purpose of variational formulas is to facilitate approximations. The state vector or the wave function of a physical system is almost never known exactly, even if we assume that the forces are. The scattering amplitude, **T** matrix, phase shifts, or whatever is to be calculated in order to predict the outcome of an experiment must therefore practically always be approximated. A number of systematic approximation procedures, based on truncating an infinite series, have already been discussed. A method that is often more practical and that may lead to fairly reliable results more simply is to assume an approximate wave function whose shape is considered physically reasonable. Its use in equations such as (7.40), for example, then leads to an approximation for the scattering amplitude. This method is, of course, quite familiar in the calculation of binding energies.

In such a procedure, as in the corresponding one for discrete energy levels, it is of great advantage to perform the actual calculation by means of a formula that is *stationary* with respect to infinitesimal variations in the form of the wave function in the neighborhood of the correct one. In this case the result can be expected to be relatively insensitive to the precise shape of the wave function, and the use of a reasonable trial function leads to a good approximation.

There are many ways of achieving the stationary character sought, and their relative advantages depend on the circumstances. We shall give only a small number and refer the reader to the literature for others.

There is another use of a stationary formula. One may use a *family* of trial functions which depends on one or more parameters. If the calculated quantity is known to be stationary near the correct result, the approximation is expected to be optimized by picking those values of the parameters in whose vicinity the calculated answer is in fact stationary with respect to small variations. This usually leads to a straightforward extremum problem. Again, this is a familiar procedure commonly used in bound-state calculations (e.g., the Rayleigh-Ritz method). But for this argument to be effective, we really have to know more than a *stationary* character of the equation used.

Suppose, for example, that a quantity A is a functional of the wave function

$$A = f[\psi]$$

which is stationary when ψ satisfies the correct Schrödinger equation. Let us now calculate a set of quantities A_α by using a family of wave functions ψ_α which does not include the correct one,

$$A_\alpha = f[\psi_\alpha].$$

Then the maximal A_α is the closest approximation to A, provided that we know, not only that $\delta A/\delta\psi = 0$ at the correct ψ, but that for all ψ_α the calculated A_α *is less than* A. Otherwise the largest A_α may in fact be the worst approximation to A among the family.

To cite an example familiar in bound-state calculations: If I want to calculate the energy of the first excited state, to pick an extremal value from a family of trial functions does me no good, unless I know that the energy calculated from any member of the family always lies *above* that sought. This is assured if they are all orthogonal to the ground-state wave function. This is why we need more than a local property of stationarity. We want an actual *minimum* or *maximum principle*.

In addition to the need for such principles for the foregoing reason, it is clear that they add important information to an approximate calculation. If a certain approximation formula is rigorously known to produce a result that always lies above (or below) the correct one, the reliability of the calculation is greatly increased, at least in one of the two possible directions of deviation. This, too, is familiar in the bound-state case, when the Rayleigh–Ritz theorem is used. We shall therefore also give one such minimum principle.

11.3.2 The T Matrix, K Matrix, and the Green's Function. The **T** matrix of a scattering process is defined by (7.40),

$$T_{\beta\alpha} = (\Psi_{0\beta}, H'\Psi_\alpha^{(+)}) = (\Psi_\beta^{(-)}, H'\Psi_{0\alpha}) \tag{11.65}$$

where H' is the interaction, $\Psi_{0\beta}$ the "unperturbed" state (usually a plane wave), and $\Psi_\alpha^{(+)}$ the complete state defined by

$$\Psi_\alpha^{(\pm)} = \Psi_{0\alpha} + G^\pm H'\Psi_\alpha^{(\pm)}. \tag{11.66}$$

The states $\Psi_{0\beta}$ or $\Psi_{0\alpha}$ and the interaction H' are assumed known. If $T_{\beta\alpha}$ is calculated by inserting an approximation for $\Psi_\alpha^{(+)}$ or $\Psi_\beta^{(-)}$ in (11.65), an error in the state produces a concomitant error of "comparable size"[10] in $T_{\beta\alpha}$. We are looking for an expression for $T_{\beta\alpha}$ that is stationary with respect to small variations of $\Psi_\alpha^{(+)}$ or $\Psi_\beta^{(-)}$ near the correct states.[11]

Let us insert (11.66) in (11.65), using (7.9),

$$T_{\beta\alpha} = (\Psi_\beta^{(-)}, H'\Psi_\alpha^{(+)}) - (\Psi_\beta^{(-)}, H'G^+H'\Psi_\alpha^{(+)}) \tag{11.67}$$

and subtract the result from

10 If the wave function deviates from the correct one by the infinitesimal amount $\delta\psi$, then the resulting error in $T_{\beta\alpha}$ is a linear functional of $\delta\psi$.

11 So that the error produced in $T_{\beta\alpha}$ is of the order of $(\delta\psi)^2$.

$$2T_{\beta\alpha} = (\Psi_{0\beta}, H'\Psi_\alpha^{(+)}) + (\Psi_\beta^{(-)}, H'\Psi_{0\alpha})$$

to get

$$T_{\beta\alpha} = (\Psi_{0\beta}, H'\Psi_\alpha^{(+)}) + (\Psi_\beta^{(-)}, H'\Psi_{0\alpha}) - (\Psi_\beta^{(-)}, H'\Psi_\alpha^{(+)}) + (\Psi_\beta^{(-)}, H'G^+H'\Psi_\alpha^{(+)}). \tag{11.68}$$

We now calculate the first variation of the right-hand side due to small changes in $\Psi_\alpha{}^{(+)}$ and $\Psi_\beta{}^{(-)}$,

$$\delta T_{\beta\alpha} = (\delta\Psi_\beta^{(-)}, H'(\Psi_{0\alpha} - \Psi_\alpha^{(+)} + G^+H'\Psi_\alpha^{(+)})) + ((\Psi_{0\beta} - \Psi_\beta^{(-)} + G^-H'\Psi_\beta^{(-)}), H'\delta\Psi_\alpha^{(+)}) = 0$$

because of (11.66). Equation (11.68) for $T_{\beta\alpha}$ therefore is *stationary* with respect to small variations of the exact state vector in the neighborhood of its correct value. It is thus more useful for an approximate calculation than (11.65).

Another way of proceeding with the same equations (11.65) and (11.67) is to *divide*

$$(T_{\beta\alpha})^2 = (\Psi_{0\beta}, H'\Psi_\alpha^{(+)})(\Psi_\beta^{(-)}, H'\Psi_{0\alpha})$$

by (11.67) and thus obtain

$$T_{\beta\alpha} = \frac{(\Psi_{0\beta}, H'\Psi_\alpha^{(+)})(\Psi_\beta^{(-)}, H'\Psi_{0\alpha})}{(\Psi_\beta^{(-)}, H'\Psi_\alpha^{(+)}) - (\Psi_\beta^{(-)}, H'G^+H'\Psi_\alpha^{(+)})}. \tag{11.69}$$

It is easily checked that this equation, too, is stationary. It has the advantage over (11.68) in being independent of the normalization of the state vectors $\Psi_\alpha{}^{(+)}$ and $\Psi_\beta{}^{(-)}$.

In the case of elastic scattering of spin 0 particles we use the momenta as labels, and then in the coordinate representation $\Psi^{(-)*}(\mathbf{k},\mathbf{r}) = \Psi^{(+)}(-\mathbf{k},\mathbf{r})$. Hence (11.69) becomes, by (10.12a),

$$A(\mathbf{k}',\mathbf{k}) = -\frac{\int d\mathbf{r}' \exp(-i\mathbf{k}'\cdot\mathbf{r}')\mathscr{V}(\mathbf{r}')\psi(\mathbf{k},\mathbf{r}') \int d\mathbf{r}\,\psi(-\mathbf{k}',\mathbf{r})\mathscr{V}(\mathbf{r})\exp(i\mathbf{k}\cdot\mathbf{r})}{4\pi\left[\int d\mathbf{r}\,\psi(-\mathbf{k}',\mathbf{r})\mathscr{V}(\mathbf{r})\psi(\mathbf{k},\mathbf{r}) + \int d\mathbf{r}\,d\mathbf{r}'\,\psi(-\mathbf{k}',\mathbf{r})\mathscr{V}(\mathbf{r})\frac{\exp(ik|\mathbf{r}-\mathbf{r}'|)}{|\mathbf{r}-\mathbf{r}'|}\mathscr{V}(\mathbf{r}')\psi(\mathbf{k},\mathbf{r}')\right]}, \tag{11.69a}$$

it being understood that $\psi = \psi^{(+)}$.[12]

Similar equations are easily obtained for the **K** matrix, defined by (7.56) and (7.56a) in terms of the principal-value state vectors (7.20).

$$K_{\beta\alpha} = -\pi\frac{(\Psi_{0\beta}, H'\Psi_\alpha^{(P)})(\Psi_\beta^{(P)}, H'\Psi_{0\alpha})}{(\Psi_\beta^{(P)}, H'\Psi_\alpha^{(P)}) - (\Psi_\beta^{(P)}H'G^{(P)}H'\Psi_\alpha^{(P)})}. \tag{11.70}$$

Notice that an approximate **K** matrix calculated by this stationary equation is automatically hermitian.

12 It should be noted that any **T** matrix or scattering amplitude calculated from (11.68), (11.69), or (11.69a) automatically satisfies the reciprocity theorem.

The Green's Function. The same approach can be used even for the calculation of a Green's function. Let us insert

$$\mathcal{G} - G = GH'\mathcal{G} \tag{11.71}$$

in

$$\mathcal{G} - G = \mathcal{G}H'G \tag{11.71a}$$

to obtain the nonlinear equation

$$\mathcal{G} - G = \mathcal{G}H'\mathcal{G} - \mathcal{G}H'GH'\mathcal{G} \tag{11.72}$$

which we then subtract from the sum of the first two,

$$\mathcal{G} = G + GH'\mathcal{G} + \mathcal{G}H'G - \mathcal{G}H'\mathcal{G} + \mathcal{G}H'GH'\mathcal{G}. \tag{11.73}$$

The first variation of the left-hand side produced by small changes in G on the right-hand side is given by

$$\boldsymbol{\delta}\mathcal{G} = (G - \mathcal{G} + \mathcal{G}H'G)H'\boldsymbol{\delta}\mathcal{G} + \boldsymbol{\delta}\mathcal{G}H'(G - \mathcal{G} + GH'\mathcal{G}) = 0$$

by virtue of (11.71) and (11.71a). Thus, if we start with a trial Green's function and insert it on the right of (11.73), the calculated $\mathcal{G}$ will be a considerable improvement. Again dependence on an overall factor can be removed by *dividing* the product of (11.71) and (11.71a) by (11.72). (This is meant, not in an operator sense, but as functions in a particular representation.)

The T Operator. The same approach leads to the replacement of the Lippmann-Schwinger equations for the T operator, (7.47),

$$T = H' + TG^+H' = H' + H'G^+T$$

by the stationary equations

$$T = H' + TG^+H' + H'G^+T - TG^+T + TG^+H'G^+T \tag{11.74}$$

or

$$T_{\beta\alpha} = H'_{\beta\alpha} + \frac{(TG^+H')_{\beta\alpha}(H'G^+T)_{\beta\alpha}}{[T(G^+ - G^+H'G^+)T]_{\beta\alpha}}. \tag{11.75}$$

The latter is independent of the normalization of the T operators inserted on the right.[13]

11.3.3 Variational Formulations of the Phase Shift. In a partial-wave analysis it is necessary to calculate phase shifts. If a regular radial wave function of angular momentum l has the asymptotic form

$$\varphi \simeq a\cos(kr - \tfrac{1}{2}\pi l) + k^{-1}\sin(kr - \tfrac{1}{2}\pi l) \tag{11.76}$$

then the phase shift is given by

13 The price one has to pay for a stationary equation is always nonlinearity. It is obvious that this is unavoidable.

$$\tan \delta_l = ak. \tag{11.77}$$

Consider now the integral

$$I = \int_0^\infty dr\, \varphi_1(\mathscr{H} - k^2)\varphi_2 \qquad \mathscr{H} = -\frac{d^2}{dr^2} + \frac{l(l+1)}{r^2} + \mathscr{V}(r) \tag{11.78}$$

whose value is zero if $\varphi_2(r)$ obeys the radial Schrödinger equation for the angular momentum l and for the potential $\mathscr{V}(r)$. Suppose, however, that φ_1 and φ_2 are general functions of r, subject only to the conditions of regularity at the origin,

$$\varphi_1(0) = \varphi_2(0) = 0$$

and the normalization (11.76) (with arbitrary values of a). Provided that φ_1 and φ_2 approach their asymptotic form (11.76) sufficiently rapidly and that $\mathscr{V}(r)$ is absolutely integrable, the integral I exists, and we may perform an integration by parts,

$$\int_0^\infty dr\, \varphi_1(\mathscr{H} - k^2)\varphi_2 - \int_0^\infty dr\, \varphi_2(\mathscr{H} - k^2)\varphi_1 = [\varphi_1'\varphi_2 - \varphi_2'\varphi_1]_0^\infty = a_2 - a_1. \tag{11.79}$$

If $\varphi_2 - \varphi_1 = \boldsymbol{\delta}\varphi$ is infinitesimal and $\varphi_1 = \varphi$ is an exact solution of the Schrödinger equation, then this equation implies that

$$\boldsymbol{\delta} \int_0^\infty dr\, \varphi(\mathscr{H} - k^2)\varphi = \boldsymbol{\delta} a.$$

In other words, the first variation of the quantity

$$a - \int_0^\infty dr\, \varphi(\mathscr{H} - k^2)\varphi$$

vanishes.

Consequently we choose a trial function, subject to the two conditions that it vanish at the origin and that it be normalized to behave asymptotically as (11.76), then calculate the phase shift from the equation

$$k^{-1} \tan \delta_l = a - \int_0^\infty dr\, \varphi(\mathscr{H} - k^2)\varphi. \tag{11.80}$$

If the function from which the integral and a are computed is an exact solution of the Schrödinger equation, then the integral vanishes and (11.80) gives us the correct phase shift. If the trial function is a good approximation, then (11.80) gives us a better approximation to the exact phase shift than (11.77), because the right-hand side is stationary near the correct value.

11.3.4 The *s*-Wave Scattering Length. The calculation may be carried out for $l = 0$ and at zero energy. In that event $-a$ is the $l = 0$ *scattering length c*.

Hence a stationary formula for the scattering length is

$$c = -a + \int_0^\infty dr\, \varphi \mathscr{H} \varphi \tag{11.81}$$

if the trial function has the asymptotic form

$$\varphi \simeq a + r \tag{11.82}$$

and vanishes at $r = 0$. Note that for the integral to exist we must now require the second absolute moment of $V(r)$ to be finite.

Equation (11.81) is easily replaced by another formula which not only is stationary but which leads to a *minimum* principle. Let φ be the exact wave function,

$$\mathscr{H}\varphi = 0, \qquad \varphi(0) = 0, \qquad \varphi \underset{r\to\infty}{\simeq} -c + r, \tag{11.83}$$

φ_t a trial function subject only to the boundary conditions

$$\varphi_t(0) = 0 \qquad \varphi_t \underset{r\to\infty}{\simeq} a + r, \tag{11.84}$$

and $u \equiv \varphi - \varphi_t$. Then (11.80) tells us that the exact scattering length is given by

$$c = -a + \int_0^\infty dr\, \varphi_t \mathscr{H} \varphi_t - \int_0^\infty dr\, u \mathscr{H} u. \tag{11.85}$$

In contrast to (11.81) this equation is correct for *any* φ_t that satisfies (11.84).

If $\mathscr{H}$ has no bound states, then it is positive semidefinite. If it were not for the fact that the function u has the asymptotic value

$$u \underset{r\to\infty}{\simeq} -(c + a)$$

and is therefore not square-integrable, we could immediately conclude that always

$$\int_0^\infty dr\, u \mathscr{H} u \geq 0. \tag{11.86}$$

That this must be true also for our u's is seen by approximating them more and more closely by a sequence of square-integrable functions. Consequently we learn from (11.85) that

$$c \leq -a + \int_0^\infty dr\, \varphi_t \mathscr{H} \varphi_t \tag{11.87}$$

if the potential supports no bound state. We are therefore certain that, among a family of trial functions subject to (11.84), the one that minimizes the right-hand side of (11.81) gives the closest approximation. Furthermore (11.87) tells us the direction in which the error lies.

The Case of Bound States. The argument may now be extended to the case in which the potential produces bound states.

Assume first that there is one bound state (of $l = 0$) of energy $E_0 < 0$. Then $\mathscr{H}$ is no longer positive semidefinite, but still (11.86) holds for all u orthogonal to the (normalized) bound-state wave function ψ_b; using the inner-product notation for simplicity,

$$u = \varphi - (\psi_b, \varphi)\psi_b.$$

Insertion of this form of u in (11.86) shows that

$$\begin{aligned}(\varphi, \mathscr{H}\varphi) &\geq (\psi_b, \varphi)(\varphi, \mathscr{H}\psi_b) \\ &= \frac{|(\psi_b, \mathscr{H}\varphi)|^2}{(\psi_b, \mathscr{H}\psi_b)}\end{aligned}$$

for arbitrary functions φ. The trouble with this inequality is that the exact bound-state wave function ψ_b, and hence the right-hand side, is usually unknown.

Now suppose that ψ is an arbitrary (normalized) "trial" bound-state function that sufficiently well approximates the exact bound state to give $\mathscr{H}$ a negative expectation value,

$$(\psi, \mathscr{H}\psi) < 0.$$

Then we claim that it is still true that

$$(\varphi, \mathscr{H}\varphi) \geq \frac{|(\psi, \mathscr{H}\varphi)|^2}{(\psi, \mathscr{H}\psi)} \tag{11.88}$$

for arbitrary φ. The proof of this is based on the following algebraic theorem (sometimes referred to as the *Hylleraas–Undheim theorem*)[14]:

Let M' be a hermitian matrix obtained from M by eliminating one row and one column. Then the eigenvalues of M and M' interlace one another; i.e., if $\{\lambda_n'\}$ are the eigenvalues of M' and $\{\lambda_n\}$ of M, they can be ordered so that

$$\lambda_1 \leq \lambda_1' \leq \lambda_2 \leq \lambda_2' \leq \cdots \leq \lambda_{N-1}' \leq \lambda_N. \tag{11.89}$$

This theorem implies that no (hermitian) submatrix of a given hermitian matrix M can have more negative (or more positive) eigenvalues than M. If we know that M has exactly *one* negative eigenvalue, then no 2×2 submatrix can have two. Suppose that M_{11} is a negative diagonal element (one-dimensional submatrix of eigenvalue $\lambda_1'' = M_{11} < 0$). Then enlarge it by forming a two-dimensional submatrix M'. The theorem says that the two eigenvalues of M' are such that

$$\lambda_1' \leq \lambda_1'' \leq \lambda_2'$$

so that $\lambda_1' \leq 0$, $\lambda_2' \geq 0$, and hence

$$M_{11}M_{22} - |M_{12}|^2 = \det M' = \lambda_1'\lambda_2' \leq 0$$

14 To be proved in the Appendix to this chapter.

or

$$M_{22} \geq \frac{|M_{12}|^2}{M_{11}}.$$

The basis being arbitrary, and not necessarily orthogonal, this inequality is just (11.88).

Insertion of (11.88) in (11.85) gives us the inequality

$$c \leq -a + \int_0^\infty dr\, \varphi_t \mathscr{H} \varphi_t - \frac{\left(\int_0^\infty dr\, \psi \mathscr{H} \varphi_t\right)^2}{\int_0^\infty dr\, \psi \mathscr{H} \psi} \tag{11.90}$$

because $\mathscr{H} u = -\mathscr{H} \varphi_t$. The only demand on ψ is that $\int_0^\infty dr\, \psi \mathscr{H} \psi < 0$. We therefore fix ψ and, among a family of trial functions φ_t, minimize the right-hand side of (11.90). This gives the best approximation to the exact c. If φ_t is exact, then the integrals containing φ_t in (11.90) vanish and the equality sign holds.

More than One Bound State. The generalization to more than one bound state is straightforward. Assume that there are N. Then we must pick a set of N orthogonal states $\{\psi_n\}$ so that the matrix $\mathfrak{H}$ given by the elements

$$\mathscr{H}_{nm} = \int_0^\infty dr\, \psi_n \mathscr{H} \psi_m$$

is diagonal and negative definite. We add another, linearly independent vector φ (not necessarily orthogonal to the set $\{\psi_n\}$) and enlarge the matrix $\mathfrak{H}$ to the $(N + 1) \times (N + 1)$ matrix $\mathfrak{H}'$. It must have N negative and one positive eigenvalue, so that

$$\frac{\det \mathfrak{H}'}{\det \mathfrak{H}} > 0.$$

But from the form of $\mathfrak{H}'$ we see that

$$\frac{\det \mathfrak{H}'}{\det \mathfrak{H}} = \mathscr{H}_{N+1\, N+1} - \sum_1^N \frac{|\mathscr{H}_{n\, N+1}|^2}{\mathscr{H}_{nn}}$$

and consequently

$$\mathscr{H}_{N+1\, N+1} \geq \sum_1^N \frac{|\mathscr{H}_{n\, N+1}|^2}{\mathscr{H}_{nn}}.$$

In other words, for arbitrary vectors φ,

$$(\varphi, \mathscr{H} \varphi) \geq \sum_1^N \frac{|(\psi_n, \mathscr{H} \varphi)|^2}{(\psi_n, \mathscr{H} \psi_n)} \tag{11.91}$$

provided that the ψ_n are such that

$$
\begin{aligned}
(\psi_n, \mathscr{H}\psi_m) &= 0 \qquad n \neq m \\
(\psi_n, \mathscr{H}\psi_n) &< 0 \qquad n = 1, \ldots, N.
\end{aligned}
\tag{11.92}
$$

If φ is not linearly independent of the $\{\psi_n\}$, it is easily seen that the equality sign holds in (11.91).

We therefore have the inequality

$$
c \leq -a + \int_0^\infty dr\, \varphi_t \mathscr{H} \varphi_t - \sum_1^N \frac{\left(\int_0^\infty dr\, \psi_n \mathscr{H} \varphi_t\right)^2}{\int_0^\infty dr\, \psi_n \mathscr{H} \psi_n}
\tag{11.93}
$$

if there are N bound states. The only demands on the functions ψ_n is that they are orthogonal to one another and that (11.92) holds.

It should be emphasized that the given bounds on the scattering length are independent of any assumption of a local nature of the interaction. This is clear from the fact that they are based on spectral arguments. Generalizations to energies other than zero and to many-channel processes have been given in the literature.[15] Applications to nuclear physics as well as to low-energy scattering of electrons by atoms can be found in many of the references given in the Notes and References.[16]

Appendix: Proof of the Hylleraas–Undheim Theorem

Let us introduce a basis (not necessarily orthogonal) in which M' is diagonal and situated on the main diagonal of M. This can be done without disturbing the additional row and column by which the $N \times N$ matrix M differs from M'. Without loss of generality we may envisage these to be the last. The secular determinant of M is then easily seen to be

$$
\begin{aligned}
f(\lambda) = \det(M - \lambda\mathbb{1}) &= \begin{vmatrix} \lambda_1' - \lambda & & \cdots & M_{1N} \\ \cdots & \lambda_2' - \lambda & \cdots & M_{2N} \\ \cdots & \cdots & \cdots & \cdots \\ M_{N1} & M_{N2} & \cdots & M_{NN} - \lambda \end{vmatrix} \\
&= (M_{NN} - \lambda) \prod_{n=1}^{N-1} (\lambda_n' - \lambda) - \sum_{m=1}^{N-1} |M_{mN}|^2 \prod_{n \neq m} (\lambda_n' - \lambda).
\end{aligned}
$$

Let us evaluate it at $\lambda = \lambda_k'$,

$$
\begin{aligned}
f(\lambda_k') &= \det(M - \lambda_k'\mathbb{1}) \\
&= -|M_{kN}|^2 \prod_{n \neq k} (\lambda_n' - \lambda_x') \begin{cases} \leq 0 & \text{if } k \text{ is odd} \\ \geq 0 & \text{if } k \text{ is even} \end{cases}
\end{aligned}
$$

if the basis has been so chosen that the eigenvalues of M' are ordered,

$$
\lambda_1' \leq \lambda_2' \leq \cdots \leq \lambda_{N-1}'.
$$

15 See particularly Hahn, O'Malley, and Spruch (1963 and 1964).

16 See also Rosenberg, Spruch, and O'Malley (1960b).

Hence $f(\lambda)$ approaches $+\infty$ as $\lambda \to -\infty$; it approaches $(-)^N\infty$ as $\lambda \to +\infty$, and it has an odd number of zeros between any two successive eigenvalues of M'. Since it must have *exactly* N (real) zeros λ_n, the inequalities (11.89) follow.

Notes and References

11.1 The partial-wave expansion of the scattering amplitude was introduced into quantum mechanics by Faxen and Holtsmark (1927). It is, of course, patterned after its previous use in the theory of sound and electromagnetic waves. A. D. Boardman et al. (1967) present a partial wave expansion for noncentral potentials. The use of *Padé approximants* to sum the partial wave series is dealt with by J. Fleischer (1972 and 1973); K. Alder et al. (1973); O. D. Corbella et al. (1976); Common and Stacey (1978); F. F. Grinstein (1980). For other partial wave expansions, see E. G. Kalnins et al. (1973); Daumens and Perroud (1979); and for relativistic generalizations, J. Bystricky et al. (1976); M. Daumens et al. (1979); Daumens and Winternitz (1980). The two-dimensional analog can be found in I. R. Lapidus (1982).

11.1.3 The following papers deal with the issue of time delay in quantum-mechanical two-particle scattering: B. A. Lippmann (1966); Jauch and Marchand (1967); J. L. Agudin (1968); H. M. Nussenzveig (1972); J. M. Jauch et al. (1972); Bollé and Osborn (1975); Ph.A. Martin (1975); Nowakowski and Osborn (1975); Tsang and Osborn (1975); Ph.A. Martin (1976); R. Lavine (1978); D. Bollé (1979a); J. O. Hirschfelder (1979); H. Narnhofer (1980); see also MacMillan and Osborn (1980). For time delay in n-particle scattering, see Osborn and Bollé (1974, 1975, 1976, 1978); Bollé and Osborn (1979); S. Bosanac (1981a); and in classical particle scattering, T. A. Osborn et al. (1980); Narnhofer and Thirring (1981).

11.2.2 The effective-range theory is due to J. Schwinger (1947) and his unpublished lecture notes [see Blatt and Jackson (1949)]. The method given here is due to H. Bethe (1949). See also Jackson and Blatt (1950).

The fact that the introduction of a new bound state causes the scattering length to be infinite has been exploited for the numerical calculation of those potential strengths at which new bound states of various angular momenta appear by Schey and Schwartz (1965) for Yukawa potentials as well as for Wood-Saxon potentials. For investigations of the low-energy behavior of the scattering amplitude in the case of a long-range potential, when the effective-range theory breaks down, see Spruch, O'Malley, and Rosenberg (1960 and 1962); Levy and Keller (1963); Hinkelman and Spruch (1971).

The most detailed study of the Schrödinger equation for the square-well potential is that of H. M. Nussenzveig (1959); see also S. Tani (1968).

Although the Levinson theorem was known to a number of persons before, it was first proved rigorously by N. Levinson (1949b).

The Breit-Wigner resonance formula is the quantum-mechanical analog of the electromagnetic Lorentz resonance formula. The two line shapes are identical. The original reference in the present context is Breit and Wigner (1936).

The lower limit on the phase-shift derivative, due to causality arguments, as expressed in (11.57), was first given by E. P. Wigner (1955).

A relativistically invariant partial-wave analysis was discussed by H. Munczek (1963).

11.3 The variational formulation of scattering theory was independently introduced by Hulthén and Schwinger: L. Hulthén (1944, 1947, and 1948); J. Schwinger (1947a, b, and 1950); Lippmann and Schwinger (1950). The general method based on the integral equation is Schwinger's, and so are specifically Eqs. (11.68) and (11.70). Eq. (11.69) is due to B. A. Lippmann (1956). Hulthén's procedure uses differential equations, as in Sec. 11.3.3 See also I. Tamm (1948 and 1949); J. M. Blatt (1948); W. Kohn (1948 and 1951); Blatt and Jackson (1949); S. Huang (1949); M. Verde (1949); Hulthén and Olsson (1950); T. Kato (1950b); Massey and Moiseiwitch (1951); J. L. Jackson (1951); Troesch and Verde (1951); H. E. Moses (1953b and 1957); Borowitz and Friedman (1953); Turner and Makinson (1953); Boyet and Borowitz (1954); Moe and Saxon (1958); C. Joachain (1965a and b); C. Schwartz (1966); Y. Hahn (1968b); H. Morawitz (1970); Sloan and Brady (1972); Rabitz and Conn (1973); Carew and Rosenberg (1973); Oberoi and Nesbeth (1973); L. Rosenberg (1973); D. G. Truhlar et al. (1974); Rosenberg and Spruch (1974); Singh and Stauffer (1975); Madan and Blankenbecler (1978); R. K. Nesbet (1978 and 1980); Darewych and Pooran (1978); M. Kawai et al. (1978); Spruch and Rosenberg (1978); Darewych and Sokoloff (1979); K. Takatsuka et al. (1981); Takatsuka and McKoy (1981); R. Goldflam et al. (1981).

For similar methods in problems of neutron diffusion, see R. E. Marshak (1947); B. Davison (1947); in electromagnetic problems, Levine and Schwinger (1948 and 1949); Kalikstein and Spruch (1964).

The development of *bounds*, or extremal principles, for scattering parameters was started by T. Kato (1951b) and considerably expanded in the following papers: T. Kikuta (1953b); V. Risberg (1956); I. C. Percival (1957 and 1960); Spruch and Kelly (1958); L. Spruch (1958 and 1962); Spruch and Rosenberg (1959 and 1960a and b); Rosenberg and Spruch (1960, 1961, and 1962); Rosenberg, Spruch, and O'Malley (1960a); Bartram and Spruch (1962); Hahn, O'Malley, and Spruch (1962, 1963, and 1964); Y. Hahn (1965); L. Rosenberg (1965 and 1970); Hahn and Spruch (1967).

11.3.3 The stationary formula (11.80) is due to W. Kohn (1948). See also F. E. Harris (1967); R. K. Nesbet (1968); Brownstein and McKinley (1968); Y. Tikochinsky (1970); Takatsuka and Kueno (1979). For variational *bounds*, see I. Aronson et al. (1967), J. N. Bardsley et al. (1972); K. J. Miller (1971); R. Blau et al. (1975); Darewych and Pooran (1980).

11.3.4 The arguments leading to the bounds (11.87), (11.90), and (11.93) follow Y. Hahn et al. (1963). See also A. M. Arthurs (1968 and 1975); D. Gelman (1974); Hahn and Spruch (1974); R. Blau et al. (1974 and 1977); Rosenberg and Spruch (1975); I. Aronson et al. (1979).

The Hylleraas–Undheim theorem is due to Hylleraas and Undheim (1930). The theorem is a variant of facts that are well known in algebra and referred to in this paper. The proof given here follows J. K. L. MacDonald (1933).

The following papers apply variational methods to resonances: A. Ronveau (1968b); Y. Hahn (1972).

Problems

1 Exactly what is wrong with the wave function $u_l(kr)r^{-1}Y_l^m(\theta,\varphi)$ if m is not an integer? If l is not an integer?

2 Explain why the scattering amplitude for a spherically symmetric target and spinless particles is axially symmetric about the incoming beam even though there are wave functions which are not.

3 Suppose that a given target has the property that at a specific energy all phase shifts vanish for $l > L$. What is the largest scattering cross section it can have?

4 Can the unitarity limit (11.18) be exceeded if the interaction is *absorptive*? Suppose that it is totally absorptive in the lth partial wave (that is, Im $\delta_l = +\infty$). What is the value of the lth partial cross section? Compare your result with (11.32) and (3.47). If it is totally absorptive in all partial waves, what is the differential cross section?

5 Equation (11.15) may be used as a definition of the phase shift δ_l. Note that this refers explicitly to the number $k = (2\mu E/\hbar^2)^{1/2}$ and thus to the energy. How do you reconcile this with the fact that physically the energy can be shifted by a constant without effect? Is the zero of the energy fixed in scattering theory? If so, by what?

6 Calculate the scattering length and the effective range for a square-well potential of range R and depth V_0. Calculate the next term in the expansion of the s-wave phase shift in powers of k^2, and then estimate for what values of the energy the effective-range approximation should be good. Use the effective-range approximation also to calculate the binding energy of a bound state. Is it a good approximation?

7 Consider a potential "shell" of value V_0 between $r = R_1$ and $r = R_2$ and zero otherwise. Calculate the s-wave phase shift, and show that for large V_0 (with respect to what?) resonances occur approximately at the energies which would be bound states if V_0 were infinite. Calculate the *leakage probability* (assuming that it is small) by a simple matching of wave functions across boundaries. Show that the resulting lifetime agrees with the result obtained from the resonance width.

8 Approximately calculate the scattering length for the exponential potential $V = \gamma e^{-\mu r}$, using as a trial function

$$\varphi_t = \begin{cases} r + a & r > -a \qquad a < 0 \\ -\dfrac{a}{\pi}\sin\dfrac{\pi r}{a} & r < -a. \end{cases}$$

Optimize the result by minimizing the calculated scattering length.

9 Do the same calculation as in Prob. 8 for the square well

$$V(r) = \begin{cases} \gamma & r < R \\ 0 & r > R. \end{cases}$$

10 Do the same calculation as in Prob. 8 for the harmonic oscillator well

$$V(r) = \begin{cases} \gamma(r^2 - R^2) & r < R \\ 0 & r > R. \end{cases}$$

11 Approximately calculate the $l = 0$ phase shift at the momentum k for the potential of Prob. 10 by using as a trial function

$$\varphi_t = \begin{cases} a\cos kr + k^{-1}\sin kr & r > r_0 \\ c\sin Kr & r < r_0. \end{cases}$$

How would you pick the parameters a, c, and K in order to optimize the result?

12 Do the same calculation as in Prob. 11 for the exponential potential $V = \gamma e^{-\mu r}$.

13 Two particles of mass 3×10^{-25} g interact with one another with a force of a range of 10^{-12} cm. A scattering experiment is carried out at 200 kev center-of-mass energy. If the differential cross section is measured with an accuracy of a few percent, what do you expect the general form of its angle dependence to be?

14 Suppose that a peak is found in the nucleon-nucleon scattering cross section at about 500 Mev center-of-mass energy, and its size and angle dependence are such as to "saturate" the unitarity limit in one partial wave. If the width is about 5 Mev, is it safe to assume that it is a resonance? What if it is 200 Mev wide?

15 Two particles of masses 1,500 Mev and 1,800 Mev interact with one another via an attractive square-well potential of 8.23 Mev depth and a range of 5.2×10^{-13} cm. At what center-of-mass energy do they have a *p*-wave resonance? What is its width? At what energy do they have a *d*-wave resonance, if any? What is its width?

CHAPTER 12

Single-Channel Scattering of Spin 0 Particles, II

12.1 Rigorous Discussion of s-Wave Scattering

12.1.1 The Regular and Irregular Solutions. We now return to the radial Schrödinger equation for a more rigorous discussion of the properties of the radial wave functions and of the S-matrix elements, as functions of the momentum k and of the potential strength γ. For simplicity we first take up the case of $l = 0$.

The differential equation to be solved is

$$-\psi'' + \gamma \mathscr{V} \psi = k^2 \psi \tag{12.1}$$

where

$$\mathscr{V} = 2\mu V$$

and eventually we set $\gamma = 1$. Supposing that the potential is less singular at $r = 0$ than r^{-2}, that is,

$$\lim_{r \to 0} r^2 \mathscr{V} = 0,$$

the point $r = 0$ is "regular" in the sense of the theory of ordinary differential equations of the second order, and we may specify arbitrary boundary conditions there. It is customary to call those solutions which vanish at $r = 0$ *regular* and those which do not *irregular*. Only regular solutions are acceptable for physical purposes. The reason for this is very simple when $l \geq 1$, but for $l = 0$ it must be sought in arguments other than the usual demand for square-integrability of the physical wave function.

Let us take $\mathscr{V} = 0$ for the moment. The reason why $\cos kr$ is not an "acceptable" solution of (12.1) is that, although considered as a three-dimensional wave function $\psi(\mathbf{r}) = r^{-1} \cos kr$ is square-integrable in any finite region of space, it does *not* satisfy the Schrödinger equation

$$-\nabla^2\psi = k^2\psi.$$

Instead it obeys the inhomogeneous equation

$$(\nabla^2 + k^2)\psi = -4\pi\delta(\mathbf{r}).$$

The error arises from the circumstance that the ordinary differential equation (12.1) has $r = 0$ as one of its boundary points, and it is understood that the values of the equation and of its solution at $r = 0$ are the corresponding limits as $r \to 0$. This does not allow a δ function there. From the point of view of the three-dimensional equation the point $\mathbf{r} = 0$ is no different from any other, and a term $\delta(\mathbf{r})$ can be easily handled. The same arguments, of course, are applicable when $\mathscr{V} \neq 0$. The essential fact is that, although $\nabla^2 r^{-1} = 0$ when $r > 0$, it has the value $-4\pi\delta(\mathbf{r})$ when $\mathbf{r} = 0$ is included.

The Regular Solution. Let us now define a regular solution $\varphi(k,r)$ by the boundary condition at $r = 0$

$$\begin{aligned} \varphi(k,0) &= 0 \\ \varphi'(k,0) &= 1, \end{aligned} \tag{12.2}$$

the prime indicating differentiation with respect to r, evaluated at $r = 0$. Note that φ is different from the "physical" wave function in (11.6) which is defined by mixed boundary conditions, one at $r = 0$ (regularity) and one at $r = \infty$ [(11.15)]. The function φ has simpler properties because of the simpler boundary condition.

We may now replace the differential equation (12.1) and the boundary condition (12.2) by an integral equation. Because (12.2) refers entirely to $r = 0$, the appropriate Green's function is *not* that given in (11.2) but differs from it by the addition of a solution of the homogeneous equation (12.1). We want

$$\mathfrak{G}(k;r,r') = 0 \qquad r' > r.$$

In the region $r' < r$ it must then be given by

$$\mathfrak{G}(k;r,r') = k^{-1} \sin k(r - r') \tag{12.3}$$

so that the integral equation reads

$$\varphi(k,r) = k^{-1} \sin kr + \gamma k^{-1} \int_0^r dr' \sin k(r - r')\mathscr{V}(r')\varphi(k,r'). \tag{12.4}$$

It is readily verified that a solution of this equation satisfies both (12.1) and (12.2), provided that differentiation under the integral sign is justified by uniform convergence.

The virtue of (12.4) is that it is an integral equation of the *Volterra* type. As a result, under extremely general conditions, which, specifically, are independent of the value of γ, it can be solved by iteration. From the Fredholm point of view this is so because the kernel is "triangular." [This means it is the generalization of a triangular matrix, $K(x,x') = 0$ when $x' > x$.] Hence the Fredholm determinant is identically equal to unity. The resolvent must consequently be an entire analytic function of γ. Notice that for

$$K = \mathfrak{G}\mathscr{V}$$
$$\operatorname{tr} K^2 = 0.$$

In order to see the precise conditions under which the series of successive approximations to (12.4) converges, we solve the integral equation by iteration,

$$\begin{aligned} \varphi &= \sum_{n=0}^{\infty} \gamma^n \varphi^{(n)} \\ \varphi^{(0)} &= k^{-1} \sin kr \\ \varphi^{(n)} &= k^{-1} \int_0^r dr' \sin k(r - r') \mathscr{V}(r') \varphi^{(n-1)}(r') \qquad n \geq 1. \end{aligned} \tag{12.5}$$

There exists a constant C such that for all values of k and all real $r \geq 0$

$$|\sin kr| \leq C e^{|v|r} \frac{|k|r}{1 + |k|r} \tag{12.6}$$

and for all k and all real $r \geq r' \geq 0$

$$|\sin k(r - r')| \leq C e^{|v|(r-r')} \frac{|k|r}{1 + |k|r} \tag{12.7}$$

if we write $\operatorname{Im} k = v$. These inequalities express the continuity of the left-hand sides and their behavior at infinity and at the origin. It follows that if we define

$$|\varphi^{(n)}(r)| \equiv e^{|v|r} \frac{r}{1 + |k|r} \psi^{(n)}(r)$$

then

$$\begin{aligned} \psi^{(n)}(r) &\leq C \int_0^r dr' |\mathscr{V}(r')| \frac{r'}{1 + |k|r'} \psi^{(n-1)}(r') \\ &\leq C^{n+1} \int_0^r dr_n \cdots \int_0^{r_2} dr_1 |\mathscr{V}(r_1) \cdots \mathscr{V}(r_n)| \frac{r_1}{1 + |k|r_1} \cdots \frac{r_n}{1 + |k|r_n} \\ &= C^{n+1} \frac{1}{n!} \left[\int_0^r dr' |\mathscr{V}(r')| \frac{r'}{1 + |k|r'} \right]^n \end{aligned}$$

and therefore

$$\begin{aligned} |\varphi(k,r)| &\leq \sum_{n=0}^{\infty} |\gamma|^n |\varphi^{(n)}| = \exp(|v|r) \frac{r}{1 + |k|r} \sum_{n=0}^{\infty} |\gamma|^n \psi^{(n)} \\ &\leq C \exp(|v|r) \frac{r}{1 + |k|r} \exp\left[|\gamma| C \int_0^r dr' |\mathscr{V}(r')| \frac{r'}{1 + |k|r'} \right]. \end{aligned} \tag{12.8}$$

Provided, therefore, that

$$\int_0^\infty dr\, r|\mathscr{V}(r)| < \infty \tag{12.9}$$

the power series in γ for $\varphi(k,r)$ converges absolutely for all values of γ and k and for all real $r \geq 0$. Moreover, in every closed region in the complex k plane and in every closed interval on the real r axis, the convergence is *uniform*. Each term in the series is an entire analytic function of k. Hence, for each fixed k and r, φ is an entire function of γ and, for each fixed γ and r, an entire function of k. The analytic nature of φ as a function or k is an illustration of a general theorem by Poincaré,[1] and it is due to the absence of k dependence of the boundary condition (12.2). Notice that for simplicity we have put the upper limit in the integral in (12.9) equal to infinity. It is clear from (12.4), though, that $\varphi(k,r)$ is quite ignorant of $\mathscr{V}(r')$ for $r' > r$ and depends on $\mathscr{V}(r)$ only in the region $0 \leq r' \leq r$. It should also be emphasized that the general properties of φ depend on "global" properties of the potential only, such as (12.9). They do not, specifically, depend on continuity, differentiability, or other such *local* properties of $\mathscr{V}$.

What is demanded by (12.9), in a vaguer but more intuitive way, is that the potential be less singular near the origin than r^{-2}. At infinity it should vanish faster than r^{-2}. But this requirement is not necessary for the existence of φ at any finite value of r. If it is violated, φ may behave badly at infinite distance. However, (12.8) shows that even this is true only for *small* k. So long as we restrict ourselves to $k \neq 0$, the demand

$$\int^\infty dr\,|\mathscr{V}(r)| < \infty \tag{12.10}$$

is quite sufficient to ensure that φ behaves no worse at $r \to \infty$ than does $\varphi^{(0)}$. The class (12.10) includes potentials which behave almost like r^{-1} at infinity. For the Coulomb potential, or even longer tails, there is trouble at $r \to \infty$ no matter what value k has.[2]

We may now insert our result (12.8) in the integral equation (12.4). This tells us, because of (12.7) and (12.9), that

$$|\varphi - k^{-1}\sin kr| \leq |\gamma| C' e^{|\nu| r}\frac{r}{1+|k|r}\int_0^r dr'\,|\mathscr{V}(r')|\frac{r'}{1+|k|r'}. \tag{12.11}$$

By dividing the integral into the ranges from 0 to a and from a to ∞, it is easily seen that

$$\lim_{|k|\to\infty}\int_0^\infty dr\frac{r}{1+|k|r}|\mathscr{V}(r)| = 0$$

if (12.9) holds. Hence the right-hand side of (12.11) is[3] $o(|k|^{-1}e^{|\nu|r})$ as

1 The theorem is stated by E. Hilb (1915), and the original reference is H. Poincaré (1884).

2 See Sec. 14.6.

3 We recall the definitions: $f(x) = o[g(x)]$ as $x \to 0(\infty)$ means that $f(x)/g(x) \to 0$ in that limit; $f(x) = O[g(x)]$ as $x \to 0(\infty)$ means that $|f(x)/g(x)|$ is bounded there.

$|k| \to \infty$, or

$$\varphi(k,r) = k^{-1} \sin kr + o(|k|^{-1} e^{|\nu| r}). \tag{12.12}$$

Thus φ approaches its "unperturbed" value $\varphi^{(0)}$ as $|k| \to \infty$ in all directions in the complex plane, with the error tending to zero *relative* to $\varphi^{(0)}$. It is clear from (12.11) that, if $\mathscr{V}(r)$ is absolutely integrable at $r = 0$, then the difference between φ and $\varphi^{(0)}$ is $O(|k|^{-2} e^{|\nu| r})$. Furthermore, (12.8) shows that the approach of φ to $\varphi^{(0)}$ is nonuniform in the strength γ. The larger $|\gamma|$, the larger C' in (12.11), and hence the larger must $|k|$ be chosen for $|\varphi - \varphi^{(0)}|$ to be smaller than a preassigned number. Notice also that the integral in (12.9) need not extend to infinity; (12.10) is sufficient for (12.12) to be true uniformly in r.

It follows from the fact that the differential equation (12.1) depends on k only via k^2, and that the boundary condition (12.2) is independent of k, that φ is an everywhere regular function of k^2,

$$\varphi(-k,r) = \varphi(k,r). \tag{12.13}$$

Furthermore, for real k^2 and real γ both (12.1) and (12.2) are real; hence so is φ. Schwarz's reflection principle then implies that the analytic continuation of φ to complex values of k satisfies the equation

$$\varphi^*(k^*,r) = \varphi(k,r) \tag{12.14}$$

provided that γ is real. For complex coupling constants γ it follows that

$$\varphi^*(\gamma^*,k^*;r) = \varphi(\gamma,k;r). \tag{12.14a}$$

If we finally make the much stronger assumption that the potential permits an analytic continuation to complex r and that there too, (12.9) holds, then (12.5) shows that φ can also be analytically continued to the same region in the complex r plane.

Irregular Solutions. In view of the fact that the **S** matrix, or the phase shift, is defined by the behavior of φ at large distances, it is convenient to introduce other solutions of (12.1) by boundary conditions at infinity. These are in general, of course, not regular at $r = 0$. But $r = \infty$ is an irregular singular point of the differential equation (12.1), and k^2 multiplies the term of highest singularity there. It is therefore unavoidable that the boundary condition at $r = \infty$ is a function of k. From our knowledge of the solutions of (12.1) when $\mathscr{V} = 0$ we gather that in general one cannot hope to do better than requiring that[4]

$$\lim_{r\to\infty} e^{\mp ikr} f_\pm(k,r) = 1. \tag{12.15}$$

To be sure, it remains to be shown whether or not solutions $f_\pm$ of (12.1) exist that satisfy the boundary conditions (12.15) and are uniquely determined by

4 With apologies to Res Jost, I am here changing his original definition, which has become customary. It is much more convenient to be able to refer to the *upper* half of the k plane both as the *physical sheet* and in discussing $f(k,r)$. I trust that this will not confuse the reader. In referring to previous literature, it should be kept in mind that what is here called $f_+(k,r) = f(k,r)$ is sometimes called $f(-k,r)$.

them. If f_+ and f_- exist, they are linearly independent, except when $k = 0$.

Again we convert (12.1) and (12.15) into integral equations. This time the appropriate Green's function vanishes for $r' < r$. In the region $r' > r$ it is therefore given by

$$\mathfrak{G}(k;r,r') = -k^{-1} \sin k(r - r').$$

The difference between this and (12.3) is a solution of the homogeneous equation (12.1) for the whole region $0 < r < \infty$. The integral equations for $f_\pm$ are therefore

$$f_\pm(k,r) = e^{\pm ikr} - \gamma k^{-1} \int_r^\infty dr' \sin k(r - r') \mathscr{V}(r') f_\pm(k,r'). \quad (12.16)$$

That a solution of this satisfies both (12.1) and (12.15) is easily verified, provided again that the integral converges absolutely.

Since (12.16), too, is a Volterra equation, we may solve it by iteration,

$$\begin{aligned} f_\pm &= \sum_{n=0}^{\infty} \gamma^n f_\pm^{(n)} \\ f_\pm^{(0)} &= e^{\pm ikr} \\ f_\pm^{(n)} &= -k^{-1} \int_r^\infty dr' \sin k(r - r') \mathscr{V}(r') f_\pm^{(n-1)}(r') \qquad n \geq 1. \end{aligned} \quad (12.17)$$

Defining

$$h_\pm^{(n)}(r) \equiv f_\pm^{(n)}(r) e^{\mp ikr}$$

we find, by the use of (12.7) as we obtained (12.8), that

$$\begin{aligned} &|f_\pm(k,r)| \\ &\leq \sum_{n=0}^{\infty} |\gamma|^n |f_\pm^{(n)}| \\ &\leq C \exp(\mp vr) \exp\left\{ |\gamma| C \int_r^\infty dr' |\mathscr{V}(r')| \frac{r'}{1 + |k| r'} \exp\left[(|v| \mp v)(r' - r)\right] \right\}. \end{aligned} \quad (12.18)$$

The solutions $f_\pm$ therefore exist for all k and r and are obtainable by absolutely convergent power series in γ, provided that

$$\alpha_\pm \equiv \int_0^\infty dr\, r |\mathscr{V}(r)| e^{(|v| \mp v)r} \quad (12.19)$$

is finite. For real values of k this demand is the same as (12.9). The quantity α_+ is then surely finite when $v = \operatorname{Im} k \geq 0$, and α_- when $v \leq 0$. Thus f_+ is well defined in the upper half of the k plane, and f_- in the lower. In the other half planes nothing can be said about the existence of f_+ or f_- unless stronger conditions are imposed on the potential. For example, if $\mathscr{V}$ decreases exponentially at infinity so that

$$\int_0^\infty dr\, r|\mathscr{V}(r)|e^{2ar} < \infty \tag{12.20}$$

for some $a > 0$, then α_- is finite for positive v up to $v = a$, and α_+ for negative v down to $v = -a$. In that case f_+ is well defined in the lower half of the k plane for $\operatorname{Im} k \geq -a$, and f_- in the upper for $\operatorname{Im} k \leq +a$. As in the case of φ, we notice that so long as $k \neq 0$ and k is real, (12.10) is sufficient at infinity. Only at $k = 0$ do we need the stronger demand (12.9) at $r \to \infty$. In contrast to φ, the functions $f_\pm(k,r)$ do not depend upon the potential $\mathscr{V}(r')$ in the region $r' < r$. Hence, so long as $r \neq 0$, the demand (12.9) at $r = 0$ is irrelevant. It enters only in relation to the behavior of $f_\pm$ at $r = 0$.

That under ordinary circumstances f_+ is well defined only when $\operatorname{Im} k \geq 0$ and f_- when $\operatorname{Im} k \leq 0$ should not come as a surprise. A look at the boundary conditions (12.15) shows that when $\operatorname{Im} k \geq 0$ then f_- is as big as or bigger than f_+ at $r \to \infty$ and thus any admixture of it is eliminated by the boundary condition for f_+. But for $\operatorname{Im} k < 0$, f_- is asymptotically negligible compared with f_+. The boundary condition (12.15) therefore does not serve to make f_+ unique. The same holds for f_- in $\operatorname{Im} k > 0$. Under strong assumptions on the potential, however, the integral equations enable us to go beyond where the boundary conditions serve their purpose.

We should like to conclude, as we did for φ, that $f_\pm$ are analytic functions of k. Because of the improper integrals in (12.17), however, we cannot do so as directly. Instead, we differentiate with respect to k and go through the iteration procedure again. Since the integral in (12.16) converges uniformly with respect to k in every closed region in the k plane inside the "allowed" half plane, the differentiation under the integral sign is justified. The argument then goes as for $f_\pm$, except that the differentiation with respect to k introduces additional factors of r' in the integrals. Consequently the continuity of the derivative of $f_\pm$ for real k requires not only (12.10) but (12.9), and if it is to include the point $k = 0$, the additional demand that

$$\int_0^\infty dr\, r^2|\mathscr{V}(r)| < \infty. \tag{12.21}$$

If (12.9) and (12.21) hold, therefore, $f_+(k,r)$ is for each r an analytic function of k regular for $\operatorname{Im} k > 0$ and continuous with a continuous k derivative in the region $\operatorname{Im} k \geq 0$. If the potential satisfies (12.20), then f_+ is analytic for $\operatorname{Im} k > -a$. A fortiori, if the potential has a finite range, in the sense that beyond some finite distance it vanishes identically, then f_+ is an *entire* analytic function of k. The analogous statements hold, *mutatis mutandis*, for f_-.

If the potential obeys the condition (12.20) for a_0, but not for $a > a_0$, then the foregoing statements are not meant to imply that $\operatorname{Im} k = a_0$ is necessarily a natural boundary of the region of analyticity of f_+ as a function of k. In fact, it is often quite easy to enlarge the region in which f_- is known not to have any singularities by use of the iteration procedure. For example, suppose that

$$\mathscr{V}(r) = e^{-ar} \qquad r > R$$

and is arbitrary [except for (12.9)] when $r < R$. Then, for $r > R$,

$$\begin{aligned} h_+^{(1)} &= -\tfrac{1}{2}ik^{-1} \int_r^\infty dr' \, e^{-ar'}(e^{2ik(r'-r)} - 1) \\ &= \frac{e^{-ar}}{a(a - 2ik)} \end{aligned}$$

and, for $r < R$,

$$h_+^{(1)} = -\tfrac{1}{2}ik^{-1} \int_r^R dr' \, \mathscr{V}(r')(e^{2ik(r'-r)} - 1) + \frac{e^{-aR}}{a(a - 2ik)}.$$

This function may now be analytically continued to $\operatorname{Im} k < -\frac{1}{2}a$, except for a simple pole at $k = -\frac{1}{2}ia$. Furthermore, it provides an additional convergence factor of e^{-ar}. We may therefore continue the iteration, using the fact that for each ι there exists a constant C such that

$$|h_+^{(1)}| \leq Ce^{-ar}$$

for all r and all $|k - \frac{1}{2}ia| > \iota$. This allows us to extend the analyticity of the remainder of the series to $\operatorname{Im} k > -a$, except for the pole at $k = -\frac{1}{2}ia$. This idea may be repeated for ${h_+}^{(2)}$, etc., and we find that h_+, and hence f_+, has no singularities in the lower half plane other than simple poles at $k = -\frac{1}{2}ian$, $n = 1, 2, \ldots$ (see also Sec. 14.3).

It is clear that the method is not restricted to the example given and is, in fact, applicable to a very large class of potentials whose asymptotic form is specified. Notice that the exact position of the singularities and their nature is extremely sensitive to the precise form of the potential tail. If we cut the exponentially decreasing potential off at any distance, no matter how large, all the poles of f_+ disappear.

An example in which the singularities of f_+ are not poles is furnished by assuming that $\mathscr{V}(r)$ is a continuous superposition of exponentials, i.e., a Laplace transform,

$$\mathscr{V}(r) = \int_{a_0}^\infty da \, \rho(a) e^{-ar}. \tag{12.22}$$

Then we get

$$h_+^{(1)} = \int_{a_0}^\infty \frac{da \, \rho(a) e^{-ar}}{a(a - 2ik)}.$$

Such an integral in general has a *branch point* at $k = -\frac{1}{2}ia_0$, and the "natural" position of the cut is along the negative imaginary axis, where the poles would lie if the integral in (12.22) were replaced by a sum. Provided only that $\rho(a)$ decreases sufficiently rapidly at $a \to \infty$, we may now repeat the previous arguments and find that $f_+(k, r)$ is regular in the whole k plane, cut from $k = -\frac{1}{2}ia_0$ to $-i\infty$.

A special case of the potential (12.22) is the *Yukawa* potential, obtained by setting $\rho(a) = \text{const}$,

$$\mathscr{V}(r) = \text{const}\,\frac{e^{-a_0 r}}{r}.$$

The function f_+ then has a logarithmic branch point at $k = -\frac{1}{2}ia_0$. Because of the physical significance of Yukawa potentials from the field theoretical point of view, potentials of the form (12.22) are also often written as superpositions of Yukawa potentials. If we assume that $\rho(a)$ is differentiable and $\rho(a_0) = 0$, we get, by integrating (12.22) by parts,

$$\mathscr{V}(r) = \int_{a_0}^{\infty} da\,\rho'(a)\frac{e^{-ar}}{r}. \tag{12.22a}$$

Analytic Potentials. The region of analyticity of $f(k,r)$ in the k plane may be extended quite generally if the potential is known to admit an analytic continuation into the complex r plane. Suppose that $\mathscr{V}(r)$ is regular in the region

$$-\tfrac{1}{2}\pi \leq -\sigma < \arg r < \sigma \leq \tfrac{1}{2}\pi$$

and such that for real x and $|\varphi| < \sigma$ the complex function

$$\mathscr{V}_\varphi(x) = e^{2i\varphi}\mathscr{V}(xe^{i\varphi})$$

has the properties (12.9) and (12.21). We shall refer to such potentials as *analytic* (in the σ class). Now write $r = xe^{i\varphi}$, $k = Ke^{-i\varphi}$, and

$$f_\varphi(K,x) = f(Ke^{-i\varphi}, xe^{i\varphi}).$$

Then f_φ satisfied the differential equation (12.1), with the complex potential $\mathscr{V}_\varphi$, and the boundary condition (12.15), both with x and K in place of r and k, respectively. We may now treat $f_\varphi(K,x)$ as we did $f(k,r)$ and conclude that, for real x, $f_\varphi(K,x)$ is an analytic function of K regular in the upper half of the K plane. But under the stated conditions imposed on the potential, the integral equation (12.16) shows that x may be taken into the complex plane, so long as $|\arg x| < \sigma$. The function $f_\varphi(K,x)$ is therefore regular in the upper half of the K plane even for $x = |x|e^{-i\varphi}$, $|\varphi| < \sigma$, and consequently, for real r,

$$f(k,r) = f_\varphi(ke^{i\varphi}, re^{-i\varphi})$$

is regular as a function of k so long as

$$-\sigma < \arg k < \sigma + \pi.$$

The singularities of $f(k,r)$ are thus confined to the wedge

$$\sigma - \pi < \arg k < -\sigma.$$

In particular, if $\mathscr{V}$ can be analytically continued all the way up to the imaginary axis and it obeys (12.9) and (12.21) along all rays whose angle with the positive real axis is less than $\frac{1}{2}\pi$, f can have singularities only along the negative imaginary axis.

Potentials of the class (12.22) or (12.22a) are "analytic" with $\sigma = \frac{1}{2}\pi$, provided that the weight function ρ is sufficiently well behaved at $a \to \infty$. Hence for such continuous superpositions of Yukawa potentials f is everywhere regular in the k plane, except along the negative imaginary axis. Furthermore, since (12.20) holds for $2a < a_0$, the singularities there cannot start until $\text{Im}\, k = -i\frac{1}{2}a_0$. In general there will be a branch cut from $k = -i\frac{1}{2}a_0$ to $-i\infty$, in agreement with our previous argument. This branch line is usually referred to as the *Yukawa cut.*

Connection Between f_+ *and* f_-. We may now connect the functions f_+ and f_- by analytic continuation. Starting with $f_+(k,r)$ for k on the positive real axis and going to the negative real axis via the upper half plane, we arrive at $f_+(-k,r)$ which satisfies the same boundary condition as $f_-(k,r)$. We conclude that for $k > 0$

$$f_-(k,r) = f_+(ke^{+i\pi},r). \tag{12.23}$$

Since f_+ and f_- are thus connected by analytic continuation, we shall sometimes write simply $f(k,r)$ for $f_+(k,r)$. If (12.20) holds for *some a*, then, of course, we need not be careful about the direction in which k circumscribes the origin in (12.23). Furthermore, (12.15) and the reality of (12.1) imply that for real k and γ

$$f_-(k,r) = f_+^*(k,r). \tag{12.24}$$

Therefore, if a part of the lower half plane is accessible to an analytic continuation of f_+ from the real axis, then we must have there

$$f_-(k,r) = f_+^*(k^*,r) \tag{12.24a}$$

for real γ and

$$f_-(\gamma,k;r) = f_+^*(\gamma^*,k^*;r) \tag{12.24b}$$

for complex γ.

Behavior of f for Large $|k|$. The behavior of f for large $|k|$ is obtained by inserting (12.18) back in the integral equation (12.16). This shows that

$$|f(k,r) - e^{ikr}| \leq |\gamma| Ce^{-\nu r} e^{|\gamma| C\alpha_+} \int_r^\infty dr' |\mathscr{V}(r')| e^{(|\nu|-\nu)(r'-r)} \frac{r'}{1 + |k|r'}. \tag{12.25}$$

As $|k| \to \infty$ in $\text{Im}\, k \geq 0$, the integral on the right approaches zero if (12.9) holds [or if (12.9) holds near the origin, and (12.10) at infinity], by the same argument as above (12.12). Therefore, as $|k| \to \infty$ for $\text{Im}\, k \geq 0$,

$$f(k,r) = e^{ikr} + o(e^{-\nu r}). \tag{12.26}$$

The approach is again nonuniform in γ. The stronger the potential, the larger $|k|$ has to be for $f(k,r)$ to be near e^{ikr} by a prescribed amount. Equation (12.26) is true *uniformly* in r for all $r \geq 0$. If the potential obeys (12.10) with the lower limit equal to zero, $o(e^{-\nu r})$ may be replaced by $O(|k|^{-1}e^{-\nu r})$. It should be noted also that (12.26) holds generally only for $\text{Im}\, k \geq 0$, even when the potential allows f to be analytically continued to $\text{Im}\, k < 0$.

12.1.2 The Jost Function and the Complete Green's Function. The two solutions f_+ and f_- of the second-order differential equation (12.1) being linearly independent, every solution may be written as a linear combination of them. In particular, the regular solution φ may be expressed as

$$\varphi = af_+ + bf_-.$$

The coefficients a and b are most simply obtained by evaluating the wronskians between φ and f_+ and between φ and f_-. We have evidently[5]

$$W(f_+,\varphi) = bW(f_+,f_-)$$

and

$$W(f_-,\varphi) = aW(f_-,f_+).$$

The wronskian of f_+ with f_- is easily calculated. Since it is independent of r, it may be evaluated in the limit as $r \to \infty$ by means of the boundary conditions (12.15). The result is

$$W(f_+,f_-) = -2ik. \tag{12.27}$$

Therefore, if we define

$$\mathscr{F}_\pm(k) = W(f_\pm,\varphi) \tag{12.28}$$

we get

$$\varphi(k,r) = \frac{1}{2ik}[\mathscr{F}_- f_+(k,r) - \mathscr{F}_+ f_-(k,r)]. \tag{12.29}$$

The function $\mathscr{F} = \mathscr{F}_+$ is called the *Jost function.*[6] Equation (12.28) tells us that

$$\mathscr{F}_-(k) = \mathscr{F}(ke^{i\pi}) \tag{12.30}$$

because of (12.23) and (12.13). If the wronskian in (12.28) is evaluated at $r = 0$, the boundary condition (12.2) shows that

$$\mathscr{F}(k) = f(k,0). \tag{12.31}$$

The expansion of f in powers of γ leads immediately to an expansion of $\mathscr{F}$, and the analyticity statements for f are applicable to $\mathscr{F}$ as well. For each fixed k it is an entire analytic function of γ, and for each fixed γ an analytic function of k regular in $\text{Im}\, k > 0$ and continuous, with a continuous derivative, in $\text{Im}\, k \geq 0$. The conditions on the potential for these statements to be true in general are (12.9) and (12.21). If (12.20) is true, then $\mathscr{F}$ is

5 We recall the definition of the wronskian: $W(f,g) = fg' - f'g$. If f and g are both solutions of the same second-order differential equation, then the wronskian is a constant, i.e., independent of r.

6 See footnote 4. It applies equally to the Jost function.

analytic also in $0 \geq \operatorname{Im} k > -a$. The extensions of the domain of analyticity of f discussed in Sec. 12.1.1 carry over to $\mathscr{F}$.

Since φ is an even function of k and real on the real axis [or because of (12.31)], the Jost function has the same symmetry properties as f. From (12.24) we get, for real k and γ,

$$\mathscr{F}_+^*(k) = \mathscr{F}_-(k) \tag{12.32}$$

for real γ and complex k,

$$\mathscr{F}_+^*(k^*) = \mathscr{F}_-(k) \tag{12.32a}$$

and, for complex γ and k,

$$\mathscr{F}_+^*(\gamma^*, k^*) = \mathscr{F}_-(\gamma, k). \tag{12.32b}$$

The high-energy behavior of $\mathscr{F}$ is immediately obtained from (12.26), which is uniform in r. Hence for $\operatorname{Im} k \geq 0$,

$$\lim_{|k| \to \infty} \mathscr{F}(k) = 1. \tag{12.33}$$

The approach is nonuniform in γ.

We may use the integral equation (12.16) to derive an integral representation for $\mathscr{F}$, namely,

$$\mathscr{F}(k) = 1 + \gamma k^{-1} \int_0^\infty dr \sin kr \mathscr{V}(r) f(k, r). \tag{12.34}$$

If the potential is analytic ($\sigma = \frac{1}{2}\pi$), we may use the arguments below (12.22a) to extend (12.34) into the upper half of the k plane. It then shows that, because of the analog of (12.26) for complex k and r, (12.33) holds also in the lower half plane, except possibly on the cut.

We also see from (12.26) and (12.34) that as $|k| \to \infty$ in the upper half of the k plane and on the real axis (by the Riemann-Lebesgue lemma)

$$\mathscr{F}(k) = 1 - \gamma(2ik)^{-1} \int_0^\infty dr \mathscr{V}(r) + o(|k|^{-1}) \tag{12.33a}$$

if that integral exists. If the potential vanishes for $r > R$, then we may use (12.25) to get

$$[\mathscr{F}(k) - 1]e^{-2ikR} = o(1) \quad \text{as } |k| \to \infty \tag{12.33b}$$

in the lower half of the k plane.

Another representation follows from the asymptotic behavior of φ as $r \to \infty$. According to (12.29) and (12.15)

$$\varphi(k, r) \underset{r \to \infty}{\simeq} \frac{1}{2ik} (\mathscr{F}_- e^{ikr} - \mathscr{F}_+ e^{-ikr}). \tag{12.35}$$

If we let $r \to \infty$ in the integral equation (12.4) for φ and separate the outgoing and incoming wave parts on the right-hand side, we get by comparison with (12.35)

$$\mathscr{F}(k) = 1 + \gamma \int_0^\infty dr\, e^{ikr} \mathscr{V}(r)\varphi(k,r). \tag{12.36}$$

The Physical Wave Function. The asymptotic form (12.35) of φ may be compared with that of the *physical* wave function $\psi^{(+)}$ [(11.9)] for $l = 0$. This shows that

$$\psi^{(+)}(k,r) = \frac{k\varphi(k,r)}{\mathscr{F}(k)}. \tag{12.37}$$

Thus the inverse of the square of the modulus of $\mathscr{F}$ measures the probability of finding the particle near $r = 0$, relative to what it would be if there were no forces.

Notice that, in contrast to φ, the physical wave function ψ is not everywhere an analytic function of k. Even in the upper half of the k plane it has poles where $\mathscr{F}$ vanishes.

The Principal-Value Wave Function. The principal-value wave function of (11.25) and (11.26) is related to φ by

$$\psi^{(P)}(k,r) = \frac{2k\varphi(k,r)}{\mathscr{F}_+(k) + \mathscr{F}_-(k)}. \tag{12.37a}$$

Because of (12.32) this is real for real k, as it should be.

Green's functions. The Jost function, together with the solutions f and φ, may be used to construct complete Green's functions that satisfy the inhomogeneous equation

$$\left[-\frac{\partial^2}{\partial r^2} + \mathscr{V}(r) - k^2\right]\mathscr{G} = -\delta(r - r') \tag{12.38}$$

and various boundary conditions. Equation (12.38) says that, for $r \neq r'$, $\mathscr{G}(r,r')$ obeys the homogeneous equation (12.1). At $r = r'$ it is continuous, but its derivative has a discontinuity of 1,

$$\frac{\partial}{\partial r}\mathscr{G}(r,r')\Bigg|_{r=r'-0}^{r=r'+0} = 1. \tag{12.39}$$

Since we want $\mathscr{G}$ to be regular at $r = 0$, it must be of the form

$$\mathscr{G}(r,r') = \varphi(k,r)h_1(r') \quad \text{for } r < r'.$$

Suppose $\mathscr{G}$ is to contain outgoing waves only for $r \to \infty$. We are then constructing $\mathscr{G}^+$, and it must read

$$\mathscr{G}^+(r,r') = f(k,r)h_2(r') \quad \text{for } r > r'.$$

The two unknown functions h_1 and h_2 are calculated by using the continuity and (12.39). This gives

$$h_1(r) = -\frac{f(k,r)}{\mathscr{F}(k)}$$

$$h_2(r) = -\frac{\varphi(k,r)}{\mathscr{F}(k)}$$

and therefore

$$\mathscr{G}^+(k;r,r') = -\frac{\varphi(k,r_<)f(k,r_>)}{\mathscr{F}(k)} \tag{12.40}$$

in the notation used in (11.2). For $\gamma = 0$ this reduces, of course, to (11.2). We may write it also in terms of the physical wave function. According to (12.37)

$$\mathscr{G}^+(k;r,r') = -k^{-1}\psi^{(+)}(k,r_<)f(k,r_>). \tag{12.40a}$$

This Green's function allows us to write the solution $\psi^{(+)}$ of the integral equation (11.7) in the form

$$\psi^{(+)}(k,r) = \sin kr + \int_0^\infty dr'\,\mathscr{G}^+(k;r,r')\mathscr{V}(r')\sin kr. \tag{12.41}$$

Other Green's functions are now easily constructed by adding a solution of (12.1) to $\mathscr{G}^+$. For example, the Green's function which vanishes for $r' > r$ must be, for $r' < r$,

$$\begin{aligned}\mathscr{G}(r,r') &= \frac{\varphi(k,r)f(k,r') - \varphi(k,r')f(k,r)}{\mathscr{F}(k)} \\ &= \frac{f_-(k,r')f_+(k,r) - f_-(k,r)f_+(k,r')}{2ik}.\end{aligned} \tag{12.42}$$

The Fredholm Determinant. We now want to show that the Jost function $\mathscr{F}$ is identical to the Fredholm determinant of the radial integral equation (11.7). In order to make the formal manipulations more transparent, we use a simple operator notation.

Differentiation of (12.4) with respect to γ gives an integral equation for $\varphi_\gamma \equiv d\varphi/d\gamma$,

$$\varphi_\gamma = G\mathscr{V}\varphi + \gamma G\mathscr{V}\varphi_\gamma$$

the solution of which is evidently

$$\varphi_\gamma = \mathscr{G}\mathscr{V}\varphi$$

since

$$\mathscr{G} = G + \gamma G\mathscr{V}\mathscr{G}.$$

Now differentiate the integral representation (12.36) with respect to γ,

$$\begin{aligned}\mathscr{F}_\gamma &= \int f^{(0)}\mathscr{V}\varphi + \int \gamma f^{(0)}\mathscr{V}\varphi_\gamma \\ &= \int (f^{(0)} + \gamma f^{(0)}\mathscr{V}\mathscr{G})\mathscr{V}\varphi.\end{aligned}$$

But according to (12.16) the function in parentheses is f. Therefore

$$\frac{d\mathscr{F}}{d\gamma} = \int f\mathscr{V}\varphi.$$

Comparison with the explicit form (12.40) of the complete Green's function now tells us that

$$\frac{d\mathscr{F}/d\gamma}{\mathscr{F}} = \frac{d\ln\mathscr{F}}{d\gamma} = -\operatorname{tr}\mathscr{G}^{+}\mathscr{V}.$$

In addition we know that $\mathscr{F} = 1$ for $\gamma = 0$. A look at (9.79a) therefore shows that

$$\Delta(k) = \mathscr{F}(k) \tag{12.43}$$

since

$$\mathscr{G}^{+} = (\mathbb{1} - \gamma G^{+}\mathscr{V})^{-1}G^{+}$$

and $G^{+}\mathscr{V}$ is the kernel of (11.7).

The identity (12.43) furnishes us with other representations of the Jost function based on those of Sec. 9.3 for Δ. The kernel of the Lippmann–Schwinger equation (11.7) reads, explicitly, for $l = 0$,

$$\begin{aligned} K(r,r') &= G^{+}(r,r')\mathscr{V}(r') \\ &= -k^{-1}\sin kr_{<}e^{ikr_{>}}\mathscr{V}(r') \end{aligned} \tag{12.44}$$

and the Fredholm determinant can be computed as in (9.73) by

$$\begin{aligned} \mathscr{F} &= \sum_{n=0}^{\infty}\gamma^{n}\Delta_{n} \\ \Delta_{n} &= -\frac{1}{n}\int_{0}^{\infty}dr\,Y_{n-1}(r,r) \qquad \Delta_{0} = 1 \\ Y_{n}(r,r') &= \int_{0}^{\infty}dr''\,Y_{n-1}(r,r'')K(r'',r') + \Delta_{n}K(r,r') \\ Y_{0}(r,r') &= K(r,r'). \end{aligned} \tag{12.45}$$

Alternatively we may use (9.76),

$$\mathscr{F} = 1 + \sum_{1}^{\infty}\frac{(-\gamma)^{n}}{n!}\int_{0}^{\infty}dr_{1}\cdots\int_{0}^{\infty}dr_{n}\begin{vmatrix} K(r_{1},r_{1}) & \cdots & K(r_{1},r_{n}) \\ \cdots & \cdots & \cdots \\ K(r_{n},r_{1}) & \cdots & K(r_{n},r_{n}) \end{vmatrix} \tag{12.46}$$

or, in fact, (9.80) may be used. Another representation is that of (9.79),

$$\ln\mathscr{F} = -\sum_{1}^{\infty}\frac{\gamma^{n}}{n}\int_{0}^{\infty}dr\,[K^{n}](r,r). \tag{12.47}$$

A further representation of $\mathscr{F}$ is based on (9.81a) in terms of the eigenvalues α_{n} of the kernel K defined in (12.44),

$$\mathscr{F} = \prod_{n}(1 - \alpha_{n}) \tag{12.48}$$

or, if the potential is multiplied by γ,

$$\mathscr{F} = \prod_n (1 - \gamma\alpha_n) \tag{12.48a}$$

The α_n, of course, depend on k.

The Zeros of the Jost Function. The identity of the Jost function and the Fredholm determinant tells us that the characteristic values of the kernel $G^+\mathscr{V}$, or, for $\gamma = 1$, the bound states produced by $\mathscr{V}$, are the *zeros* of $\mathscr{F}$ in the *upper half of the k plane*. Only there do they have that significance, because, from the energy point of view, that is the *physical sheet* of Δ. That the zeros of $\mathscr{F}$ have this meaning can be seen quite directly from (12.31), for example, or from (12.28). In the upper half of the k plane, $\operatorname{Im} k > 0, f(k,r)$ is exponentially decreasing at infinite r. If $\mathscr{F}(k_0) = 0$ there, then $f(k_0,r)$ is also regular at the origin. Hence we have a regular square-integrable wave function and hence an eigenvalue. The solutions f and φ must then be multiples of one another,

$$f(k_0,r) = c\varphi(k_0,r). \tag{12.49}$$

We therefore know that, for real γ, the zeros of $\mathscr{F}$ in the upper half plane must be confined to the imaginary axis so that the binding energy is real.[7] If $-k_0 = -i|k_0|$ lies in a region to which $f(k,r)$ can be analytically continued, then c is easily calculated from (12.29),

$$c = \frac{-2|k_0|}{\mathscr{F}(-i|k_0|)}. \tag{12.50}$$

Zeros of $\mathscr{F}$ in the lower half plane have no such significance because there f *increases* exponentially with r. Zeros of $\Delta(E)$ on the second sheet do not signify eigenvalues. There is then no general reason for them to be confined to the imaginary axis. If they do not lie there, though, the symmetry (12.32a) forces them (for real γ) to occur in pairs symmetric with respect to the imaginary axis.

The zeros of $\mathscr{F}(k)$ in the upper half plane must necessarily be simple. This can be seen as follows:

The wronskian of two solutions of the radial Schrödinger equation at different energies is easily found to obey the equation

$$\frac{d}{dr} W[f(k,r),\varphi(k',r)] = (k^2 - k'^2)f(k,r)\varphi(k'r) \tag{12.51}$$

and similarly for any other two solutions. Differentiate this with respect to k (indicated by a dot), subsequently see $k' = k = k_0$ with $\mathscr{F}(k_0) = 0$, and use (12.49) and (12.28),

$$\begin{aligned}\dot{\mathscr{F}}(k_0) &= c^{-1} W[\dot{f}(k_0,r), f(k_0,r)] + cW[\varphi(k_0,r), \dot{\varphi}(k_0,r)] \\ &= -2k_0\left[c\int_0^r dr'\,\varphi^2(k_0,r') + c^{-1}\int_r^\infty dr'\,f^2(k_0,r')\right] \\ &= -2k_0 c\int_0^\infty dr\,\varphi^2(k_0,r).\end{aligned} \tag{12.52}$$

7 Of course, this can also be easily proved in the present context.

We know that, for real γ, k_0 must be purely imaginary, and hence $\varphi\ (k_0,r)$ is real. Furthermore c cannot vanish because that would contradict the boundary condition (12.15). Hence the right-hand side of (12.52) cannot vanish for $\operatorname{Im} k_0 > 0$, and $\mathscr{F}(k_0) \neq 0$. The zero *must* be simple.

The simplicity of the zeros of $\Delta(k)$ at bound states implies that the eigenvalues of the *radial* Schrödinger equation for $l = 0$ *cannot* be degenerate. An accidental degeneracy is possible only between different l values.

The point $k = 0$ is the only one on the real axis at which $\mathscr{F}$ may vanish for real γ. This is because $\mathscr{F}(k) = 0$ for real k implies, according to (12.32), that $\mathscr{F}(-k) = 0$, also. But according to (12.29) the vanishing of both $\mathscr{F}_+$ and $\mathscr{F}_-$ at $k \neq 0$ leads to $\varphi \equiv 0$ for all r. This contradicts the boundary condition (12.2). In contrast to $\operatorname{Im} k > 0$, $\mathscr{F}(0) = 0$ does *not* signify a bound state.[8] The function $f(0,r)$ then vanishes at the origin, but at infinity it approaches unity. We shall see in Sec. 12.2 that for $l > 0$ this is not so.

If $\mathscr{F}(0) = 0$, then the multiplicity of the zero requires special consideration. It can be shown[9] that (for $l = 0$) if k approaches zero from the upper half plane then both

$$\mathscr{F}(k) = O(k)$$

and

$$\frac{1}{\mathscr{F}(k)} = O(k^{-1}). \tag{12.53}$$

In other words, $\mathscr{F}(k)$ tends to zero *exactly* as k. If $\mathscr{F}(k)$ is analytic at $k = 0$, then this means of course that if $\mathscr{F}(0) = 0$ the zero is *simple*. Furthermore, it follows from (12.30) and (12.32) that the factor multiplying k is *purely imaginary*.

Under conditions (12.9) and (12.21) on the potential, the number of zeros of $\mathscr{F}(k)$ in the upper half of the k plane must be *finite*. Otherwise they would have to have an accumulation point, at infinity, or at a finite value of $k \neq 0$, or at $k = 0$. The first would contradict (12.33); the second, the regularity of $\mathscr{F}$ for $\operatorname{Im} k > 0$; the third, (12.53). Hence the number of bound states of $l = 0$ must be finite.

Dispersion Relations for the Jost Function. The analyticity of $\mathscr{F}(k)$ together with (12.33) may be expressed in a *dispersion relation*. The argument proceeds as in Sec. 4.2.2. Since[10]

$$\frac{1}{2\pi i}\int_{-\infty}^{\infty} dk' \frac{\mathscr{F}(k') - 1}{k' - k} = \mathscr{F}(k) - 1 \tag{12.54}$$

when $\operatorname{Im} k > 0$, the real and imaginary parts of $\mathscr{F} - 1$ are one another's Hilbert transforms. For real k

8 If the potential is not "reasonable" in the sense of (12.9) and (12.21), $\mathscr{F}(0) = 0$ may indicate a bound state of $l = 0$. An example is given in Sec. 14.7.1.

9 See, for example, R. G. Newton (1960).

10 We close the contour by a large semicircle in the upper half plane and evaluate the integral by the residue theorem.

$$\operatorname{Re}\mathscr{F}(k) = 1 + \frac{1}{\pi}\mathscr{P}\int_{-\infty}^{\infty} \frac{dk' \operatorname{Im}\mathscr{F}(k')}{k' - k} \tag{12.55}$$

if $\mathscr{P}$ stands for the Cauchy principal value. This equation is not especially useful since neither the real nor the imaginary part of $\mathscr{F}$ is experimentally accessible.

If the potential is of the Yukawa type (12.22a), then $\mathscr{F}$ can be continued into the lower half plane, except for the Yukawa cut from $k = -ia_0/2$ to $-i\infty$. Furthermore, (12.33) holds there too. Hence the path of integration in (12.54) may be distorted to run up along the left of the cut and back down on the right. This gives

$$\mathscr{F}(k) = 1 + \frac{1}{2\pi i}\int_{-i\infty}^{-ia_0/2} dk' \frac{\Delta\mathscr{F}(k')}{k' - k} \tag{12.56}$$

if by $\Delta\mathscr{F}(k')$ we mean the discontinuity of $\mathscr{F}$ across the Yukawa cut,

$$\Delta\mathscr{F}(k') = \lim_{\varepsilon\to 0+} [\mathscr{F}(k' - \varepsilon) - \mathscr{F}(k' + \varepsilon)] \tag{12.57}$$

for $ik' = |k'| > \frac{1}{2}a_0$. According to (12.32a) the discontinuity $\Delta\mathscr{F}(k')$ is purely imaginary.

$$\Delta\mathscr{F}(k') = -2i \lim_{\varepsilon\to 0+} \operatorname{Im}\mathscr{F}(k' + \varepsilon) = -2ih(|k'|) \tag{12.58}$$

so that

$$\mathscr{F}(k) = 1 + \frac{1}{\pi}\int_{1/2a_0}^{\infty} dK \frac{h(K)}{K - ik}. \tag{12.59}$$

The variable k may now be taken anywhere in the cut k plane.

A more useful dispersion relation is the analog of (12.54) for $\ln\mathscr{F}(k)$. Suppose first that $\mathscr{F}$ has no zeros in the upper half plane; i.e., there are no bound states. Then $\ln\mathscr{F}$ is regular in the upper half plane and vanishes at $|k| \to \infty$. Consequently

$$\operatorname{Re}\ln\mathscr{F}(k) = \frac{1}{\pi}\mathscr{P}\int_{-\infty}^{\infty} dk' \frac{\operatorname{Im}\ln\mathscr{F}(k')}{k' - k}. \tag{12.60}$$

Now write

$$\mathscr{F}(k) = |\mathscr{F}(k)|e^{-i\delta(k)}.$$

The phase δ of $\mathscr{F}$, as we shall see in Sec. 12.1.3, is the (s-wave) *phase shift*. The dispersion relation (12.60) thus reads

$$|\mathscr{F}(k)| = \exp\left[\frac{1}{\pi}\mathscr{P}\int_{-\infty}^{\infty} dk' \frac{\delta(k')}{k - k'}\right] \tag{12.61}$$

or, in view of (12.32),

$$|\mathscr{F}(k)| = \exp\left[\frac{1}{\pi}\mathscr{P}\int_{0}^{\infty} dE' \frac{\delta(E')}{E - E'}\right]. \tag{12.61a}$$

We can also write this as

$$\begin{aligned}\mathscr{F}(k) &= \exp\left[\frac{1}{\pi}\int_{-\infty}^{\infty} dk' \frac{\delta(k')}{k - k'}\right] \\ &= \exp\left[\frac{1}{\pi}\int_{0}^{\infty} dE' \frac{\delta(E')}{E - E'}\right]\end{aligned} \tag{12.62}$$

for $\operatorname{Im} k \geq 0$.

These equations hold only if $\mathscr{F}$ has no zeros in the upper half plane. Suppose now that there are bound states at $E_n = -K_n^2/2\mu$. Then we know that $\mathscr{F}(k)$ has simple zeros at $k = ik_n$. Let us define

$$\mathscr{F}^{red}(k) = \prod_n \frac{k + iK_n}{k - iK_n} \mathscr{F}(k). \tag{12.63}$$

Then $\mathscr{F}^{red}(k)$ has all the properties of $\mathscr{F}(k)$ without bound-state zeros, and we may write

$$\mathscr{F}^{red}(k) = \exp\left[\frac{1}{\pi}\int_{-\infty}^{\infty} dk' \frac{\delta^{red}(k')}{k - k'}\right].$$

But since

$$\delta^{red}(k) = \delta(k) + 2\sum_n \cot^{-1}\frac{k}{K_n}$$

we find

$$\mathscr{F}(k) = \prod_n \left(1 - \frac{E_n}{E}\right) \exp\left[\frac{1}{\pi}\int_{-\infty}^{\infty} \frac{dk'\,\delta(k')}{k - k'}\right] \tag{12.64}$$

for $\operatorname{Im} k \geq 0$. This equation allows us to express $\mathscr{F}(k)$ explicitly in terms of its phase on the real axis and of the bound-state energies.

A Different Approach. The phase shift δ may also be obtained directly from a knowledge of the absolute magnitude $|f(k,r)|$ of the function $f(k,r)$. This is done by writing

$$f(k,r) = h(k,r)e^{-ik\delta(k,r)} \tag{12.65}$$

in which expression the functions h and δ are meant to be real. Insertion in the radial Schrödinger equation, and separation of real and imaginary parts, gives us the two equations

$$h'' - \mathscr{V}h + k^2(1 - \delta'^2)h = 0 \tag{12.66}$$

$$\delta''h + 2\delta'h' = 0. \tag{12.67}$$

Because of (12.15), the boundary conditions are

$$\delta \simeq -r \qquad h \simeq 1 \quad \text{as } r \to \infty.$$

The second equation is easily solved,

$$\delta' = -h^{-2}$$

because of the boundary conditions. Therefore

$$\delta(k,r) = \int_r^\infty dr'\,(h^{-2} - 1) - r. \tag{12.68}$$

Now (12.31) shows that $k\delta(k,0)$ is the phase of $f^*(k)$, and hence it is the s-wave phase shift. We have therefore found that the phase shift is given by

$$\delta(k) = k\int_0^\infty dr\,[|f(k,r)|^{-2} - 1] = -k\int_0^\infty dr\,r\frac{d}{dr}|f|^{-2} \tag{12.69}$$

while $h = |f|$ obeys the nonlinear equation

$$-h'' + \mathscr{V}h + k^2(h^{-4} - 1)h = 0. \tag{12.70}$$

No difficulty arises from the fact that h^{-2} and h^{-4} occur in these equations. For real values of $k \neq 0$, the function f cannot vanish. That follows from the expression of a solution ϕ of the s-wave Schrödinger equation with the boundary condition $\phi(r_0) = 0$, $\phi'(r_0) = 1$, in terms of f_+ and f_-. Equation (12.24) then rules out $f(k,r_0) = 0$ just as it does $\mathscr{F} = 0$.

The value of Eq. (12.70) combined with (12.69) lies mostly in its usefulness for numerical computation.

12.1.3 The S Matrix. The S-matrix element for $l = 0$ is readily expressed in terms of $\mathscr{F}$ by comparing (12.35) with (11.9),

$$S(k) = \frac{\mathscr{F}_-(k)}{\mathscr{F}_+(k)}. \tag{12.71}$$

Its unitarity for real γ and k follows from (12.32),

$$S^*(k) = \frac{1}{S(k)}. \tag{12.72}$$

We conclude that the phase shift δ is the phase of the complex function $\mathscr{F}_-$. For real k and γ

$$\mathscr{F}_- = |\mathscr{F}|e^{i\delta}. \tag{12.73}$$

Moreover, the connection (12.30) between the positive and negative real k axis via the upper half plane leads to

$$S(-k) = S^*(k) = \frac{1}{S(k)}. \tag{12.74}$$

This, however, is largely a matter of definition, because, under the weak conditions (12.9) and (12.21), $S(k)$ does not necessarily allow an analytic continuation off the real axis. There is then no analytic continuation from positive to negative k either. It takes a potential such as (12.20), no matter how small a, in order to be sure *in general* that $S(k)$ may be taken to complex k.

The unitarity (12.72), of course, assures the reality of the phase shift δ. Equation (12.74) says that $\delta(k)$ may be chosen an *odd* function of k,

$$\delta(-k) = -\delta(k). \tag{12.75}$$

Normally, when we speak of the **S** matrix for real positive energy, positive values of k are meant. That implies that E lies on the *upper* rim of the right-hand cut from $E = 0$ to $E = \infty$, in conformance with the prescription that it is to be calculated with the *outgoing*-wave Green's function G^+.

Representation of S. The integral representations (12.34) and (12.36) lead to an integral representation of S which is identical with (11.19), or, by using (12.41),

$$\begin{aligned} S = 1 &- 2ik^{-1}\int_0^\infty dr \sin^2 kr\, \mathscr{V}(r) \\ &- 2ik^{-1}\int_0^\infty dr \int_0^\infty dr' \sin kr\, \mathscr{V}(r)\mathscr{G}^+(k;r,r')\mathscr{V}(r')\sin kr'. \end{aligned} \tag{12.76}$$

This is the $l = 0$ analog of (7.46).

The power series in γ generated by the insertion of that for $\mathscr{G}^+$ in (12.76), i.e., the *Born series* for S, does not necessarily converge for all values of k at a given value of γ. Equation (12.71) shows why. For the expansion of $1/\mathscr{F}$ to converge at a given value of k and $\gamma = 1$, $\mathscr{F}$ must not have zeros for any $|\gamma| < 1$. For the Born expansion of the phase shift, the condition is even more stringent. Neither $\mathscr{F}_+$ nor $\mathscr{F}_-$ must have any zeros for $|\gamma| < 1$. If this is true, then (12.47) gives a simple series for the phase shift,[11]

$$\delta = \sum_1^\infty \frac{\gamma^n}{n}\int_0^\infty dr\, \mathrm{Im}\, K^n(r,r), \tag{12.77}$$

the kernel K being given by (12.44).

Another representation of the phase shift, in terms of the eigenvalues α_n of K, is based upon (12.48). If we define

$$\delta^{(n)} = \arg(1 - \alpha_n), \tag{12.78}$$

i.e.,

$$\tan\delta^{(n)} = \frac{\mathrm{Im}\,\alpha_n}{1 - \mathrm{Re}\,\alpha_n}, \tag{12.78a}$$

then the phase shift is evidently given by the sum of the $\delta^{(n)}$,

$$\delta = \sum_n \delta^{(n)}. \tag{12.79}$$

According to the discussion of Sec. 9.1, the α's are classified into attractive and repulsive eigenvalues depending on their algebraic sign when $E < 0$.

11 To be more explicit about it,

$$K^n(r,r) = \int_0^\infty \cdots \int_0^\infty dr_1 \cdots dr_n K(r,r_1)K(r_1,r_2)\cdots K(r_n,r).$$

For the former we *always* have

$$0 \leq \delta^{(n)} \leq \pi \tag{12.80}$$

and for the latter

$$-\pi \leq \delta^{(n)} \leq 0. \tag{12.80a}$$

This shows quite directly that an everywhere attractive potential or a negative definite interaction, even if nonlocal, always causes a positive phase shift, and an everywhere repulsive potential a negative one. It also shows the "cooperative" nature of any phase shift greater than π in magnitude. No single eigenvalue of K can contribute more than π. In fact, the eigenvalues near the origin contribute very little to the sum.

High-Energy Behavior. The high-energy behavior of S is obtained immediately from (12.33),

$$\lim_{k \to \pm\infty} S(k) = 1. \tag{12.81}$$

As was remarked there, the "stronger" the potential, the larger k must be in order for S to differ from unity by a given amount. Moreover, the error term is $O(|k|^{-1})$ only if the potential is integrable at infinity [(12.10)]. Equation (12.33a) shows that then

$$\delta = -\tfrac{1}{2}k^{-1} \int_0^\infty dr\, \mathscr{V}(r) + o(|k|^{-1}). \tag{12.82}$$

Notice that this confirms that at high energies the first Born approximation dominates and that an attractive potential causes a positive phase shift.

Analyticity. On the basis of the spectral theory we know that $\mathscr{G} = (E - H)^{-1}$ is an analytic operator function of E, regular everywhere on the (right-hand) cut plane, except at the bound states. Why, then, is S not necessarily regular there? The distinction between the behavior of $\mathscr{G}$ on the physical sheet and that of S is due to the fact that the matrix elements of S are evaluated by means of energy-dependent wave functions, which, for complex energies, are not normalizable. This is responsible for a possible lack of regularity of S where $\mathscr{G}$ is regular, as well as for the possible appearance of multiple poles where $\mathscr{G}$ must have simple ones. Furthermore, since the relevant matrix element of the residue of $\mathscr{G}$ may vanish, $S(k)$ need not have a pole where $\mathscr{G}$ does. The study of $\mathscr{G}$ as an operator-valued analytic function of E is therefore much simpler than that of the **S** matrix. For the former one may have recourse to general and powerful operator techniques. The latter is studied more conveniently by the methods used in this chapter.

If the potential is known to obey more stringent conditions than (12.9) and (12.21), the **S**-matrix elements may be analytically continued to complex energies. If $\mathscr{V}$ satisfies (12.20), then $S(k)$ has an analytic continuation to a strip in the k plane with $|\mathrm{Im}\, k| < a$, or, equivalently, to the inside of the parabola,

$$(\operatorname{Im} E)^2 = \frac{2a^2}{\mu} \operatorname{Re} E + \frac{a^4}{\mu^2} \tag{12.83}$$

on the physical sheet as well as on the second sheet of the E plane. In that region *on the physical sheet* S has simple poles exactly at the bound states, and nowhere else. The residue of S at a bound-state pole at $k = ik_n$ is easily calculated by means of (12.71), (12.52), (12.50), and (12.37) in terms of the analytic continuation of the physical wave function $\psi^{(+)}$,

$$\begin{aligned} \operatorname{res}_n = \operatorname{res} S &= \lim_{k \to iKn} (k - iK_n) S(k) \\ &= \frac{i}{4 \int_0^\infty dr \, [\psi^{(+)}(-iK_n, r)]^2}. \end{aligned} \tag{12.84}$$

Since φ is real and so is $\mathscr{F}$, according to (12.32a), $\psi^{(+)}(-iK_n,r)$ is purely imaginary and

$$i \operatorname{res}_n > 0. \tag{12.85}$$

On the second sheet, inside the parabola, the **S**-matrix element is regular, except for poles. Suppose that such a pole in the lower half of the k plane sits at $k = k_1 - ik_2$, close to the real axis. Then the **S**-matrix element near it must have the form

$$S(k) \simeq \frac{k - k_1 - ik_2}{k - k_1 + ik_2} \tag{12.86}$$

or

$$\delta \simeq \text{const} + \tan^{-1} \frac{k_2}{k_1 - k}. \tag{12.87}$$

If the pole is sufficiently isolated and close enough to the real axis, the phase shift will sharply *increase* by π in a momentum interval of approximate width k_2 about k_1. In so doing it must go through $\frac{1}{2}\pi$ (*modulo* π). According to the discussion in Sec. 11.2.2 the result has both characteristics of a *resonance*. The partial-wave amplitude attains its unitarity limit, and the efflux of particles is delayed. The precise criteria under which a pole of S on the second sheet (or in the lower half of the k plane) causes such a resonance depend on the remaining circumstances, such as the existence and nearness of other poles, for example. Nevertheless, one often refers to *all* the poles of S there as resonance poles. It must be kept in mind, though, that not all resonances in that sense are necessarily experimentally observable.

A phase shift such as (12.87), which, in a certain energy interval increases sharply by π, must, in so doing, traverse not only $\frac{1}{2}\pi$ but also π (modulo π). At that energy the partial-wave amplitude is *zero*. In other words, a sharp resonance peak is usually accompanied by a nearby zero of the partial amplitude (the Ramsauer–Townsend effect). That zero, however, is usually not so readily observable, for it is more easily obscured by the presence of other phase shifts. A look at Fig. 11.2 shows that the typical low-energy resonance for $l > 0$ before the introduction of a bound state lacks this zero satellite.

From the point of view of the eigenvalues α of the kernel K and (12.79), the resonance is caused by the fact that an individual α_n passes near unity. Then $\delta^{(n)}$ rapidly increases by π and thus causes the peak as well as the flux delay. If the phase shift passes through $\frac{1}{2}\pi$ (modulo π) by cooperation of several α's, it is unlikely to do so rapidly as a function of the energy. The partial cross section then has a peak, but the absence of a long delay prevents it from being properly called a resonance.

Note that each pole term such as (12.86) causes a rise as well as a fall of the phase shift. If δ increases through $\frac{1}{2}\pi$ (modulo π), it must somewhere decrease through $\frac{1}{2}\pi$. This, too, causes a large partial cross section. But since it is associated with a flux advance rather than a delay, it is not a resonance. On the basis of the causality argument given in Sec. 11.2.2 we expect that therefore the possible downward slope of the phase is limited if the potential has a finite range. That that is so we can now see as follows.

Differentiate (12.51) with respect to k', then set $k = k'$, and integrate from zero to r. Suppose that $\mathscr{V} = 0$ for $r > R$. Then, for $r > R$, according to (12.28) and (12.35)

$$\dot{\mathscr{F}}(k) = -2k \int_0^r dr' f(k,r')\varphi(k,r') + ir\mathscr{F}(k) - ik^{-1}e^{ikr}\sin(kr + \delta)\,|\mathscr{F}(k)|$$

if the dot indicates differentiation with respect to k. Therefore, by (12.29) and (12.37)

$$\begin{aligned} \frac{d}{dk}\ln S(k) &= 2i\dot{\delta}(k) \\ &= \frac{\dot{\mathscr{F}}_-}{\mathscr{F}_-} - \frac{\dot{\mathscr{F}}_+}{\mathscr{F}_+} \\ &= 4ik^2 \int_0^r \frac{dr'\,\varphi^2(k,r')}{|\mathscr{F}(k)|^2} - 2ir + ik^{-1}\sin 2(kr + \delta) \\ &= 4i \int_0^r dr'\,|\psi^{(+)}(k,r')|^2 - 2ir + ik^{-1}\sin 2(kr + \delta). \end{aligned} \tag{12.88}$$

This implies that for $r > R$

$$|\psi^{(+)}_{\text{outside}}(k,r)|^2 \equiv |\psi^{(+)}(k,r)|^2 = \tfrac{1}{2} - \tfrac{1}{2}k^{-1}\cos 2(kr + \delta)$$

and therefore

$$\frac{d}{dk}\delta(k) = 2\int_0^\infty dr\,[|\psi^{(+)}(k,r)|^2 - |\psi^{(+)}_{\text{outside}}(k,r)|^2] + \tfrac{1}{2}k^{-1}\sin 2\delta. \tag{12.89}$$

The function $\psi_{\text{outside}}{}^{(+)}$ is the exact form of $\psi^{(+)}$ in the outside region. $r > R$, continued to the inside.

Equation (12.89) directly indicates the physical significance of a rapidly upward-varying phase shift. The right-hand side shows that it means that there is a large probability for finding the particles inside the region of interaction. At the same time (12.88) tells us that

$$\begin{aligned} \frac{d\delta}{dk} &= 2\int_0^R dr\,|\psi^{(+)}(k,r)|^2 - R + \tfrac{1}{2}k^{-1}\sin 2(kR+\delta) \\ &> -R + \tfrac{1}{2}k^{-1}\sin 2(kR+\delta) \\ &> -(R + \tfrac{1}{2}k^{-1}). \end{aligned} \tag{12.90}$$

This is the exact statement of (11.57) in this case.

If the potential is "analytic" ($\sigma = \frac{1}{2}\pi$), such as a superposition of Yukawa potentials (12.22a), then S can be analytically continued to the whole first and second sheet of the E surface, except for the *Yakawa branch cuts* on both sheets which run along the real axes, from $E = -a_0{}^2/8\mu$ to $-\infty$. This branch line on the physical sheet is usually referred to as the *left-hand cut*. In addition, of course, S may (and generally does) have poles on the second sheet. Bound-state poles on the physical sheet necessarily have negative imaginary residues only if they occur before the onset of the left-hand cut, i.e., if $E_{\text{binding}} < a_0{}^2/8\mu$. Otherwise (12.32a) does not imply that $\mathscr{F}$ is real, and hence $\psi^{(+)}$ need not be purely imaginary.

Dispersion Relations. In general the lack of analyticity prevents $S(k)$ from satisfying a dispersion relation. This is true even if the potential is a superposition of Yukawa functions (12.22a) or exponentials (12.22). It is well to recall the results of Secs. 10.3.2 and 10.3.3 at this point. For such potentials the full amplitude generally does obey a dispersion relation. Nevertheless, the partial-wave amplitudes do not. First appearances notwithstanding, a dispersion relation for the full amplitude does not imply dispersion relations for each angular momentum. Let us also understand that if in a particular instance $S(E)$ is analytic everywhere on the first sheet, then (12.74) tells us that it must be meromorphic on the second sheet. The Yukawa potential shows that this is not true for the full amplitude.

If the potential is a superposition of exponentials (12.22) or Yukawa functions (12.22a), then one may get an analog of a dispersion relation by using (12.56). Since $\mathscr{F}$ does not have a cut in the upper half plane, we may write

$$\mathscr{F}(k) = 1 - \frac{1}{2\pi i}\int_{ia_0/2}^{i\infty} dk'\,\frac{\mathscr{F}(k')\Delta S(k')}{k'+k} \tag{12.91}$$

if by ΔS we mean the discontinuity of S across the Yukawa cut, for $-ik = |k| > \frac{1}{2}a_0$,

$$\Delta S(k) = \lim_{\varepsilon\to 0+}\left[S(k+\varepsilon) - S(k-\varepsilon)\right],$$

which, according to (12.74), is purely imaginary. Equation (12.91) is a linear inhomogeneous integral equation for $\mathscr{F}$ on the Yukawa cut, and it represents $\mathscr{F}$ explicitly everywhere else. If

$$\int_{ia_0/2}^{i\infty} dk\, k^{-1}|\Delta S(k)|^2 < \infty$$

then the kernel is of the Hilbert–Schmidt type and the Fredholm method may be employed to construct $\mathscr{F}$. The procedure of constructing the whole amplitude, or $S(k)$, from a knowledge of its discontinuity across the left-hand cut by means of (12.91) and (12.71) is usually referred to as the *N/D method.* The solution is, of course, not necessarily obtainable by interation.

If the potential vanishes for $r > R$, then it follows from (12.33) and (12.33b) that

$$\lim_{|E|\to\infty} (S - 1)e^{2ikR} = 0 \tag{12.92}$$

and furthermore S is regular everywhere on the physical sheet, except for simple poles at the bound states. Hence $(S - 1)e^{2ikR}$ is subject to a dispersion relation. We get

$$\begin{aligned}\operatorname{Re}\{[S(E) - 1]e^{2ikR}\} = \sum_n \frac{(E_n/K_n)\operatorname{res}_n e^{-2K_nR}}{E - E_n} \\ + \frac{1}{\pi}\mathscr{P}\int_0^\infty dE' \frac{\operatorname{Im}\{[S(E') - 1]e^{2ik'R}\}}{E' - E}\end{aligned} \tag{12.93}$$

if $E_n = -K_n{}^2/2\mu$ are the bound-state energies and res_n are the residues of S at $k = iK_n$.

The Levinson Theorem. We now want to prove the connection between the zero-energy value of the phase shift and the number of bound states.

Since $\mathscr{F}(k)$ is analytic in the upper half of the k plane, we have

$$\frac{1}{2\pi i}\int_C d\ln \mathscr{F}(k) = n \tag{12.94}$$

if n is the number of zeros of $\mathscr{F}$ enclosed in the contour C. For C we take a path along the real axis from $-\infty$ to $+\infty$, avoiding the origin by a small semicircle of radius ε, and closed by a large semicircle of radius R in the upper half plane. Each discrete eigenvalue produces a simple zero of $\mathscr{F}$ in the upper half plane; hence n is the number of bound states of angular momentum zero.

Because of (12.33) the large semicircle contributes nothing in the limit as $R \to \infty$. Near $k = 0$ we use (12.53) and get for the contribution from the small semicircle, as $\varepsilon \to 0$,

$$\int d\ln \mathscr{F} \to q\int d\ln k = -i\pi q$$

where $q = 0$ if $\mathscr{F}(0) \neq 0$ and $q = 1$ if $\mathscr{F}(0) = 0$. Consequently (12.94) becomes by (12.73)

$$\lim_{\varepsilon\to 0}\lim_{R\to\infty}\left[-\delta(R)+\delta(\varepsilon)-\delta(-\varepsilon)+\delta(-R)\right.$$
$$\left.+\, i\ln\left|\frac{\mathscr{F}(-R)}{\mathscr{F}(R)}\frac{\mathscr{F}(\varepsilon)}{\mathscr{F}(-\varepsilon)}\right|\right]=2\pi(n+\tfrac{1}{2}q).$$

The imaginary part vanishes because of (12.32); because of (12.75) and the fact that $\delta(\infty)$ is defined to be zero, we get *Levinson's theorem*,

$$\delta(0)=\begin{cases}\pi(n+\frac{1}{2}) & \text{if } \mathscr{F}(0)=0\\ \pi n & \text{if } \mathscr{F}(0)\neq 0.\end{cases} \tag{12.95}$$

This theorem is valid under the extremely general conditions (12.9) and (12.21).

12.1.4 The Poles of S. The poles of the **S** matrix, as we have seen, are of special significance. If strong enough assumptions concerning the potential are made so that all poles on the physical sheet are caused by zeros of $\mathscr{F}$, then these signify bound states. Those on the second sheet may cause observable resonances if they are close enough to the positive real axis. If on the negative real axis, the poles of S on the second sheet are sometimes called *virtual states* and also *antibound states*. According to (12.74) each pole of S on the second sheet corresponds to a *zero* of S on the first. We may therefore confine the study of S to the physical sheet, but we must then pay attention to both its poles and its zeros. If we allow the coupling strength γ to be complex, then the zeros of $\mathscr{F}$ are the characteristic values of the kernel of the radial Lippmann–Schwinger equation and thus they determine the convergence properties of the Born series for S.

It is therefore important to get us much general information as possible concerning the poles of S, or the zeros of $\mathscr{F}$, both in the k plane (or on the E surface) as a function of γ and in the γ plane as a function of k.

The Bound-State Poles. Let us first consider the *bound-state* poles of S, or the zeros of $\mathscr{F}$ in the upper half of the k plane. If we define

$$U=\begin{cases}-\mathscr{V} & \text{where } \mathscr{V}<0\\ 0 & \text{where } \mathscr{V}\geq 0\end{cases}$$

we certainly do not decrease their number by replacing $\mathscr{V}$ by $-U$. The number of bound states of $-U$ is equal to the number m of eigenvalues α of the kernel K of the radial Lippmann-Schwinger equation (with $-U$) which at $E=0$ lie outside the unit circle (see Sec. 9.1.1). This is because as the energy decreases from zero these must move along the real axis toward the left and at some energy must go through $\alpha=1$. Each thus causes one bound state.

Instead of $K=-GU$ we take the kernel of (9.6),

$$\mathfrak{K}=-U^{1/2}GU^{1/2}.$$

The Green's function for $E = 0$ is

$$G(r,r') = -r_<$$

and hence

$$\sum \alpha = \operatorname{tr} \mathfrak{K} = \int_0^\infty dr\, rU(r).$$

Now, $-U$ being everywhere attractive, each α is positive; therefore at $E = 0$

$$\sum \alpha \geq m.$$

Consequently the number n of ($l = 0$) bound states of $\mathscr{V}$ is

$$n \leq \int_0^\infty dr\, rU(r). \tag{12.96}$$

Suppose now that the potential is multiplied by the strength parameter γ and that, keeping γ positive, we decrease it. Then the zeros of $\mathscr{F}$ must move downward along the positive imaginary axis and, one after another, disappear from the upper half plane through the point $k = 0$. If $\mathscr{V}$ is such that $\mathscr{F}$ is analytic in a neighborhood of $k = 0$, then (12.53) shows that the zero is still simple as it moves through the origin. It must then remain on the negative imaginary axis until it can get off by colliding with another zero, so that both can "bounce off" symmetrically into the third and fourth quadrant in accordance with (12.32a). Thus, as the potential strength *increases*, an $l = 0$ bound state must be a *virtual* state before becoming a bound state (unless, of course, $\mathscr{F}$ is not analytic near $k = 0$). Since the zero does not get near the real axis at a positive energy, it does not cause a low-energy resonance. This agrees with the results of the heuristic physical discussion of Sec. 11.2.2. We shall see in Sec. 12.2 why the situation for $l \neq 0$ is different.

The simplicity of the zero of $\mathscr{F}$ at $k = 0$ also tells us the behavior of the eigenvalues α of the kernel K as functions of k. If we multiply the potential by γ and consider $\mathscr{F}$ as a function of both γ and k, then the equation

$$\mathscr{F}(\gamma, k) = 0$$

for a given value of k determines $\alpha(k) = 1/\gamma(k)$. Now, where $\mathscr{F}$ is an analytic function of both variables and the zero is simple, there $\alpha(k)$ is regular and defines a conformal mapping. Hence, as k moves from the positive imaginary axis to the positive real axis through the origin, α must move at right angles too. We conclude that the α trajectories leave the real axis at *right angles* as the energy becomes positive. From the point of view of these eigenvalues, this is the reason why an almost bound s state causes no low-energy resonance. An α that leaves the real axis just to the left of 1 generally does not intersect the unit circle. We shall see in Sec. 12.2 why for $l > 0$ this is not so.

Virtual States. About the distribution of zeros of $\mathscr{F}$ in the lower half of the k plane nothing can be said unless stronger assumptions than (12.9) and (12.21) are made concerning the potential. If it satisfies (12.20), then a strip of width a is accessible. Let us assume that the potential decreases in fact faster than *all* exponentials, so that $\mathscr{F}$ is regular in the whole k plane. The representation (9.22) of the complete Green's function in terms of the eigenvalues α_n of the kernel K of the radial Lippmann–Schwinger equation allows us immediately to get some information concerning the virtual states. Equation (9.22), together with (9.18), leads to the following solution of (11.7) (in a hybrid notation employing both abstract-state vectors and coordinate-space radial wave functions):

$$\psi^{(+)} = \sum_\alpha \frac{\Phi_\alpha(\Phi_\alpha, \mathscr{V}\psi_0)}{\alpha(\gamma\alpha - 1)}$$

where

$$\psi_0 = \sin kr = i \sinh Kr \qquad K > 0$$

for negative energies on the physical sheet. Insertion in (11.19) gives

$$S = 1 + \frac{2\gamma}{K}\sum_\alpha \frac{|(\Phi_\alpha, \mathscr{V}\psi_0)|^2}{\alpha(\gamma\alpha - 1)}. \tag{12.97}$$

Suppose first that $\mathscr{V}$ is everywhere *positive*. Then all the α's must be negative, and (12.97) shows that S can have no zeros for $\gamma > 0$. Thus an everywhere positive (repulsive) potential has, not only no bound states, but also *no virtual states.*

Consider now (12.97) as a function of $1/\gamma$, at a fixed (negative) energy. If the potential is everywhere negative, so that all the α are positive, then S is an everywhere increasing function of $1/\gamma$ with poles at $1/\gamma = \alpha_n$. Thus it must have exactly one zero between every two successive poles. If we call $\bar{\gamma}_n$ the strengths for which the potential has virtual states (at the given negative energy), then

$$\frac{1}{\alpha_{n-1}} < \bar{\gamma}_n \equiv \frac{1}{\bar{\alpha}_n} < \frac{1}{\alpha_n}.$$

Since $S \to 1$ when $1/\gamma \to \infty$ and $S \to -\infty$ when $1/\gamma$ approaches the largest α from the right, there must be exactly one zero to the right of all the poles. Thus, as γ increases from zero, it first produces a virtual state, then bound and virtual states in alternating succession. If $\mathscr{V}$ is not definite so that K has both attractive and repulsive eigenvalues, then (12.97) need not be monotone as a function of $1/\gamma$. Hence S may have any *odd* number of zeros between two successive poles, and an odd number of zeros must lead. This statement holds for both positive and negative γ.

Consider now the sequence of points α_n and $\bar{\alpha}_n$ as the energy increases to zero. The positive α_n move monotonely to the right and the negative to the left. Hence so must, *on the average*, the $\bar{\alpha}_n$. At an energy at which an α moves through $+1(-1)$, $+\mathscr{V}(-\mathscr{V})$ has a bound state; when an $\bar{\alpha}$ moves through $+1(-1)$, $+\mathscr{V}(-\mathscr{V})$ has a virtual state. This means that, between any two energies at which bound states of a given potential occur, there must be an *odd* number of virtual states. Since the motion of the virtual states, as functions of the energy, need not be monotone, there may be more than one such between successive bound states, even when $\mathscr{V}$ is everywhere negative. Furthermore, a priori there may be any number of virtual states between threshold and the first bound state. The leading state, i.e., that with the largest value of $-E$, must be virtual.

Before visualizing the motion of the bound and virtual states on the E surface we must add the realization that a bound and a virtual state must never coincide. This is because according to (12.29) the simultaneous vanishing of $\mathscr{F}(k)$ and of $\mathscr{F}(-k)$ would make $\varphi(k,r)$ identically zero, which would contradict its boundary condition. What is more, $\mathscr{F}$ being an analytic function of both k and γ, new zeros on the imaginary axis can be introduced only by a collision of two symmetric complex zeros there, or by moving in from $E = -\infty$. Since we know that, as γ increases, s-wave bound states are virtual prior to their introduction, we must picture the motion of the virtual and bound states for $l = 0$ as follows (we assume that the potential is somewhere negative and that γ is positive):

When γ is sufficiently small, there are no bound states and no virtual states. As γ increases, either one virtual state is introduced at minus infinity or a pair is produced by a collision of symmetric complex poles of S on the negative E axis (or a combination of any number of both). In the first instance a new virtual state must appear at $-\infty$ at the same value of γ at which the virtual state crosses the threshold and becomes bound. In the second instance one of the two virtual states becomes bound, and the other remains virtual. We must then have an odd number of virtual states to the left of the bound one. As γ increases further, a pair of complex poles of S collides to the right of the bound state, thus producing a pair of virtual states. One of them then becomes bound, and the other remains trapped between the two bound states. This process continues, with possible additional pair production and reannihilation of virtual states between successive bound states. For $l \geq 1$, as we shall see, "pairs" are produced right at $k = 0$, one becoming virtual, the other bound. Thus, for $l = 0$, the number of virtual states between the loosest bound state and threshold is even; for $l \geq 1$, odd.

It must be remembered that these conclusions generally hold only if the potential decreases faster than *all* exponentials. If $\mathscr{V}$ is a superposition of Yukawa potentials, for example, the left-hand cut may destroy our conclusions. For example, if a pair of symmetric complex poles on the second sheet move toward the left-hand cut, either they must move *through* the branch line, or else they move along it together, i.e., in parallel, rather than in opposite, directions. When they reach the end of the cut, one may move along the real axis on the first sheet and the other must go through the cut; or else both disappear in the swamp. The previous results are still true, though, for the bound and virtual states to the right of the left-hand cut.

Resonance Poles. For a simple understanding of an aspect of inelastic or absorptive processes to be discussed later in more detail, we may consider a pair of complex **S**-matrix poles which are not exactly symmetric with respect to the imaginary k axis, such as

$$S' = \frac{k + K - i\varepsilon}{k - K + i\varepsilon} \frac{k - K^* + i\varepsilon}{k + K^* - i\varepsilon} \cdots$$

where $k > 0$, $K_1 = \operatorname{Re} K > 0$, $K_2 = \operatorname{Im} K < 0$. For small ε we have

$$|S'|^2 = 1 + \frac{8\varepsilon K_1 K_2 k}{|k - K|^2 |k + K|^2}$$

so that S' is absorptive when $\varepsilon > 0$. If we let $K_1 \to 0$ in the unitary case of $\varepsilon = 0$, the two symmetric poles collide, and an identification of which subsequently moves up the imaginary axis and becomes bound, and which down, is impossible. But for $\varepsilon > 0$ the poles remain separate and never collide, provided that we let the one from the *left* move *up* and the one from the *right* move *down*. But this implies that the pole which eventually produces the bound state is *not* the one which was responsible for visible resonances. These, after all, are produced by poles in the fourth quadrant close to the positive real axis, i.e., just below the *upper* rim of the right-hand cut. In other words, if we consider the elastic **S** matrix the limiting case of an absorptive one and we decrease the potential strength, the bound-state poles move into the *upper* half plane of the second sheet of the E surface and are therefore not responsible for resonances. The latter are produced by those poles which for stronger forces are *virtual* states.

If the potential has a finite range R, that is, $\mathscr{V} = 0$ for $r > R$, then we can draw further general conclusions concerning the zeros of $\mathscr{F}$. Equations (12.36) and (12.12) shows that, as $\operatorname{Im} k \to -\infty$,

$$\mathscr{F}(k) \simeq \frac{1}{2ik} e^{2ikR} \int_0^R dr\, \mathscr{V}(r) e^{2ik(r-R)}.$$

Suppose that near $r = R$ the potential has an asymptotic expansion whose first term is

$$\mathscr{V}(r) = c(R - r)^\sigma + \cdots \qquad \sigma > 0. \tag{12.98}$$

Then we get, as $\operatorname{Im} k \to -\infty$,

$$\mathscr{F}(k) \simeq \text{const} \times |k|^{-\sigma-2} e^{2ikR}.$$

Consider now the function

$$g(k^2) = \mathscr{F}(k)\mathscr{F}(-k).$$

It is an entire analytic function of k^2 which at infinity goes as

$$g(k^2) \begin{cases} \simeq \text{const} \times |k|^{-\sigma-2} e^{2ikR} & \text{as } \operatorname{Im} k \to -\infty \\ \to 1 & \text{as } \operatorname{Re} k \to \pm\infty, \operatorname{Im} k \text{ fixed.} \end{cases} \tag{12.99}$$

Hence it is of *order*[12] $\rho = \frac{1}{2}$. An entire function of nonintegral order has necessarily an infinite number of zeros.[13] Because of (12.99), only a finite number of them can be on the imaginary axis. So g must have an infinite number of *complex* zeros, symmetric with respect to both the real and the imaginary axis. Those in the upper half plane must be zeros of $\mathscr{F}(-k)$ and those in the lower of $\mathscr{F}(k)$.

The same argument which excludes infinitely many zeros of $\mathscr{F}$ on the imaginary axis also excludes infinitely many zeros below any ray through the origin, since there also the right-hand side of (12.99) has no zeros. Furthermore the lower part of (12.99) implies that the number of zeros in any strip below the real axis is finite. In other words, although the total number of zeros is infinite, for any given positive μ and v there is only a finite number of them with imaginary parts above $-v$ or with a ratio of the magnitudes of imaginary to the real parts greater than μ. If $\{k_n\}$ are the zeros of $\mathscr{F}(k)$, then

$$\left.\begin{aligned} &\operatorname{Im} k_n \to -\infty \\ &\frac{\operatorname{Im} k_n}{\operatorname{Re} k_n} \to 0 \end{aligned}\right. \qquad \text{as}\, n \to \infty. \tag{12.100}$$

It follows also from the lower part of (12.99) that

$$\int_1^\infty dr\, k^{-2} \ln g(k^2) < \infty.$$

Therefore[14]

$$\sum_n |\operatorname{Im}(k_n^{-1})| < \infty$$

and, by (12.100),

$$\sum_n \frac{|\operatorname{Im} k_n|}{(\operatorname{Re} k_n)^2} < \infty. \tag{12.101}$$

The distribution of zeros k_n in the right half plane can be shown in more detail to be such that[15]

$$\begin{aligned} \operatorname{Re} k_n &= \frac{n\pi}{R} + O(1) \\ \operatorname{Im} k_n &= \frac{\sigma + 2}{2R} \ln n + O(1) \end{aligned} \tag{12.102}$$

where σ is defined by (12.98).

12 The definition of the order ρ of an entire function is

$$\rho = \limsup_{r\to\infty} \frac{\ln \ln M(r)}{\ln r}$$

if $M(r)$ is the maximal modulus of the function for $|z| = r$.

13 See R. P. Boas (1954), p. 24.

14 *Ibid.*, p. 134.

15 See J. Humblet (1952), p. 45; and T. Regge (1958a). See W. J. Romo (1978) for restrictions on the potential needed for this.

It should be remarked at this point that these general results for cutoff potentials permit no inferences whatever concerning the situation when the cutoff (i.e., range) is allowed to approach infinity. To be quite explicit on this: The singularity structure of the **S** matrix for any given potential of infinite range must generally be expected to differ from the limit of the singularity structure for the corresponding cutoff potential when the cutoff tends to infinity. A potential of finite range R produces no poles in the Jost function, no matter how large R. But the Jost function of the potential obtained by letting R approach infinity usually has poles in the lower half plane, in fact generally infinitely many. This situation not only applies to straight cutoffs but can be expected to be true in general. It applies equally well, for example, to the screened Coulomb field when the screening radius is allowed to increase beyond bounds [see Ferreira and Teixeira (1965)]. The mathematical reason is always that the limits $R \to \infty$ and $|k| \to \infty$ are approached nonuniformly and are not interchangeable.

Representations in Terms of Poles (for Potentials of Finite Range). The entire function $\mathscr{F}(k)$ may now be written in the form of an infinite product. According to Hadamard's form of the Weierstrass factorization theorem[16] we may write

$$\mathscr{F}(k) = \mathscr{F}(0)e^{ick}\prod_1^\infty \left(1 - \frac{k}{k_n}\right)e^{k/k_n} \tag{12.103}$$

if we assume that $\mathscr{F}(0) \neq 0$. The constant c is evaluated by using a theorem of Pfluger[17] which tells us that the asymptotic behavior of (12.103) for large $|k|$ along the imaginary axis, $k = \pm i|k|$, is

$$|k|^{-1}\ln\left|\frac{\mathscr{F}(k)}{\mathscr{F}(0)}\right| = A \mp \sum \operatorname{Im} k_n^{-1} \mp c + o(1).$$

Comparison with (12.99) and (12.33) shows that the left-hand side equals $2R$ for $k = i|k|$ and zero for $k = -i|k|$. Hence

$$c - i\sum k_n^{-1} = R$$

and consequently

$$\mathscr{F}(k) = \mathscr{F}(0)e^{ikR}\prod_1^\infty \left(1 - \frac{k}{k_n}\right). \tag{12.104}$$

If we write $-i\varkappa_n$ for the N' virtual states, iK_n for the N bound states, and k_n for the "resonance" states, $\operatorname{Re} k_n > 0$, then

$$\mathscr{F}(k) = \mathscr{F}(0)e^{ikR}\prod_1^{N'}\left(1 - i\frac{k}{\varkappa_n}\right)\prod_1^{N}\left(1 + i\frac{k}{K_n}\right)\prod_1^{\infty}\left(1 + \frac{2ik \operatorname{Im} k_n - k^2}{|k_n|^2}\right). \tag{12.104a}$$

16 See, for example, Boas, *op. cit.*, p. 22. We assume for simplicity that $\mathscr{F}(0) \neq 0$. Otherwise we must replace $\mathscr{F}(0)$ by const $\times\, k$.

17 A. Pfluger (1943), theorem 6*B*.

The phase shift therefore is given by the series

$$\delta(k) = -kR - \sum_{n=1}^{N} \tan^{-1}\frac{k}{K_n} + \sum_{n=1}^{N'} \tan^{-1}\frac{k}{\varkappa_n} + \sum_{n=1}^{\infty} \tan^{-1}\frac{2k\,\mathrm{Im}\,k_n}{|k_n|^2 - k^2} \tag{12.105}$$

and its derivative

$$\frac{d\delta}{dk} = -R - \sum_{n=1}^{N}\frac{K_n}{k^2 + K_n^2} + \sum_{n=1}^{N'}\frac{\varkappa_n}{k^2 + \varkappa_n^2} + \sum_{n=1}^{\infty}\frac{2\,\mathrm{Im}\,k_n(k^2 + |k_n|^2)}{(k^2 - |k_n|^2)^2 + 2k^2(\mathrm{Im}\,k_n)^2}. \tag{12.106}$$

This shows explicitly how each resonance pole contributes to the *positive* slope of the phase shift and that the eventual downward slope is determined primarily by the radius R. (See the heuristic discussion in Sec. 11.2.2.) As $k \to \infty$, the right-hand side of (12.106) tends to zero by virtue of a cancellation of the first term against the last series.

In spite of the appearance of infinitely many "resonance" terms in (12.105), there can be only a finite number of energies at which $\sin\delta = 1$. This follows from the analyticity of $\mathscr{F}(k)$ on the real axis, together with (12.33). *There always exists an energy beyond which no resonances occur.* Furthermore, (12.100) shows that for large n the maximal contributions of the series in (12.106) occur at $k \simeq \mathrm{Re}\,k_n$ and they decrease as $2/\mathrm{Im}\,k_n$. Consequently there exists an energy *beyond which the phase shift is monotone.*[18]

We may use (12.104) to represent the S-matrix element as an infinite product,

$$S(k) = e^{-2ikR}\prod_{1}^{\infty}\frac{k_n + k}{k_n - k} \tag{12.107}$$

or in more detail as in (12.104a)

$$S(k) = e^{-2ikR}\prod_{1}^{N'}\frac{\varkappa_n + ik}{\varkappa_n - ik}\prod_{1}^{N}\frac{K_n - ik}{K_n + ik}\prod_{1}^{\infty}\frac{|k_n|^2 - k^2 - 2ik\,\mathrm{Im}\,k_n}{|k_n|^2 - k^2 + 2ik\,\mathrm{Im}\,k_n}, \tag{12.107a}$$

it being understood that $\varkappa_n > 0$, $K_n > 0$, $\mathrm{Re}\,k_n > 0$. These representations explicitly incorporate the unitarity of S for real k.

Another way of representing $S(k)$ if $\mathscr{V} = 0$ for $r > R$ is a Mittag-Leffler expansion.[19] For this we need an estimate of the residues of S at $k = k_n$ if $\mathscr{F}(k_n) = 0$. Using (12.34) as we did to derive (12.33a) and because of (12.100) we get for the leading terms of $\mathscr{F}(k)$ in the vicinity of k_n, as $n \to \infty$,

18 The reader must be reminded at this point that we are still assuming that the potential has a finite range. For the number of points at which $\sin^2\delta = 1$ to be finite, it is sufficient to assume that the potential asymptotically decreases exponentially.

19 See, for example, C. Caratheodory (1950), pp. 215ff.

$$\mathscr{F}(k) = 1 + (2ik)^{-1} \int_0^R dr\, \mathscr{V}(r) e^{2ikr} + O(k_n^{-1})$$

and

$$\mathscr{F}(k) = k^{-1} \int_0^R dr\, r\mathscr{V}(r) e^{2ikr} + O(k_n^{-1}).$$

Assuming (12.98) again, this leads to

$$\dot{\mathscr{F}}(k) = 2iR[\mathscr{F}(k) - 1] + O(k_n^{-1})$$

since for large k only the vicinity of $r = R$ contributes to the integral. Hence,

$$\dot{\mathscr{F}}(k_n) = -2iR + O(k_n^{-1}). \tag{12.108}$$

Consequently the residue of $S(k)$ at $k = k_n$ is

$$\begin{aligned} R_n &= \frac{\mathscr{F}(-k_n)}{\dot{\mathscr{F}}(k_n)} \\ &= \frac{i}{2R} + O(k_n^{-1}). \end{aligned} \tag{12.109}$$

The S-matrix element may therefore be written in the form

$$\begin{aligned} S(k) = 1 + kP(k) + k\Bigg[\sum_1^{N'} \frac{iR_n'}{\varkappa_n(k + i\varkappa_n)} - \sum_1^{N} \frac{i\,\mathrm{res}_n}{K_n(k - iK_n)} \\ + \sum_1^{\infty}\left(\frac{R_n/k_n}{k - k_n} + \frac{R_n^*/k_n^*}{k + k_n^*}\right)\Bigg] \end{aligned} \tag{12.110}$$

with $\varkappa_n > 0$, $K_n > 0$, $\operatorname{Re} k_n > 0$, and $P(k)$ is an entire function of k. The R_n' are the residues at the virtual state poles. Since according to (12.109) and (12.100) the terms in (12.110) for large n go as

$$\frac{R_n}{k_n^2} - \frac{R_n^*}{k_n^{*2}} \simeq \frac{i}{2R}\left(\frac{1}{k_n^2} + \frac{1}{k_n^{*2}}\right) \simeq \frac{i}{R}\frac{1}{(\operatorname{Re} k_n)^2}$$

the inequality (12.101) shows that the Mittag–Leffler expansion (12.110) converges.

The residues in (12.110) may in principle be expressed in terms of the pole positions. According to (12.107a)

$$iR_n' = -2\varkappa_n e^{-2\varkappa_n R} \prod_{m \neq n} \frac{\varkappa_m + \varkappa_n}{\varkappa_m - \varkappa_n} \prod_p \frac{K_p - \varkappa_n}{K_p + \varkappa_n} \prod_s \frac{|k_s|^2 + \varkappa_n^2 + 2\varkappa_n \operatorname{Im} k_s}{|k_s|^2 + \varkappa_n^2 - 2\varkappa_n \operatorname{Im} k_s} \tag{12.111}$$

$$i\,\mathrm{res}_n = 2K_n e^{2K_n R} \prod_m \frac{\varkappa_m - K_n}{\varkappa_m + K_n} \prod_{p \neq n} \frac{K_p + K_n}{K_p - K_n} \prod_s \frac{|k_s|^2 + K_n^2 - 2K_n \operatorname{Im} k_s}{|k_s|^2 + K_n^2 + 2K_n \operatorname{Im} k_s}. \tag{12.111a}$$

The total number of virtual states and bound states to the right of any given bound state is even and that to the right of any given virtual state odd. Hence every $i\,\text{res}_n$ and iR_n' is *positive* [the first in agreement with (12.85)]. We also have

$$R_n = 2ik_n \frac{\operatorname{Im} k_n}{\operatorname{Re} k_n} e^{-2ik_nR} \prod_m \frac{\varkappa_m + ik_n}{\varkappa_m - ik_n} \prod_p \frac{K_p - ik_n}{K_p + ik_n} \prod_{s \neq n} \frac{|k_s|^2 - k_n^2 - 2ik_n \operatorname{Im} k_s}{|k_s|^2 - k_n^2 + 2ik_n \operatorname{Im} k_s}. \tag{12.111b}$$

The expansion (12.110) may be used for the sine of the phase shift. Using (12.85), we get

$$\begin{aligned} 2\sin^2\delta(k) &= \operatorname{Re}[1 - S(k)] \\ &= EQ(E) + E\sum_1^N \frac{A_n}{E + |E_n^{\text{bd}}|} - E\sum_1^{N'} \frac{B_n}{E + |E_n^{\text{virt}}|} \\ &\quad + E\sum_1^\infty \frac{C_n(E - E_n) + \frac{1}{2}D_n\Gamma_n}{(E - E_n)^2 + \frac{1}{4}\Gamma_n^2} \end{aligned} \tag{12.112}$$

where

$$\begin{aligned} A_n &= \frac{i\,\text{res}_n}{K_n} > 0 \\ B_n &= \frac{iR_n'}{\varkappa_n} > 0 \\ C_n - iD_n &= \frac{2R_n}{k_n} \\ E_n + \tfrac{1}{2}i\Gamma_n &= \frac{k_n^2}{2\mu} \end{aligned} \tag{12.113}$$

and

$$Q(E) = \frac{\frac{1}{2}[P(-k) - P(k)]k}{E}$$

is an entire function of E. The relation between $P(k)$ and the poles of S is hard to disentangle, and no simple expression is known. Equations (12.110) and (12.112) have the advantage over (12.107) of expressing the separate resonance contributions *additively*, but they do not incorporate the unitarity in any simple way.

We recall at this point that all relations from (12.98) on were derived and are generally vaid only on the assumption that the potential has the finite range R.

Analytic Potentials. Let us assume now that the potential is a superposition of exponentials, (12.22), or of Yukawas, (12.22a). We may then resort to the method outlined below (12.22a) of writing $r = xe^{i\varphi}$ and $k = Ke^{-i\varphi}$. From the radial Schrödinger equation we then get, by the standard method of deriving a wronskian,

$$\frac{1}{2i}\frac{d}{dx} W(f_\varphi^*, f_\varphi) = |f_\varphi|^2(\operatorname{Im} \gamma\mathscr{V}_\varphi - \operatorname{Im} K^2) \tag{12.114}$$

in the notation used below (12.22a). (We have multiplied $\mathscr{V}$ by the strength parameter γ.) Suppose now that $\mathscr{F} = 0$. We may then make r complex, $r = xe^{i\varphi}$ (x real), and the wronskian that defines $\mathscr{F}$ remains zero. Thus

$$f_\varphi(ke^{i\varphi}, x) = f(k, xe^{i\varphi})$$

decreases exponentially at infinity if $\operatorname{Im} ke^{i\varphi} > 0$, and it vanishes at $r = 0$. We may therefore integrate (12.114) and obtain

$$\int_0^\infty dx\,|f_\varphi|^2 A(x, \varphi, k) = 0 \tag{12.115}$$

where

$$A = \operatorname{Im} \gamma \mathscr{V}_\varphi - \operatorname{Im} k^2 e^{2i\varphi}.$$

Similarly, by *adding* the radial Schrödinger equation for f_φ multiplied by f_φ^* to its complex conjugate, we get

$$\frac{1}{2}\frac{d}{dx}\left[\frac{1}{x}|f_\varphi|^2 - \frac{d}{dx}|f_\varphi|^2\right] + |f_\varphi' - \frac{1}{2x}f_\varphi|^2 + |f_\varphi|^2 B(x, \varphi, k) = 0$$

where

$$B = \operatorname{Re} \gamma \mathscr{V}_\varphi - \operatorname{Re} k^2 e^{2i\varphi} + \tfrac{1}{4}x^{-2}.$$

Integration yields

$$\int_0^\infty dx\,|f_\varphi|^2 B = -\int_0^\infty dx\,|f_\varphi' - (2x)^{-1} f_\varphi|^2. \tag{12.116}$$

Now multiply (12.116) by $\cos\beta$, (12.115) by $\sin\beta$, with $0 \le \beta \le \frac{1}{2}\pi$, and add the results. The ensuing equation can be satisfied only if for some value of x

$$\tfrac{1}{4}x^{-2}\cos\beta + \cos\beta \operatorname{Re} \gamma \mathscr{V}_\varphi + \sin\beta \operatorname{Im} \gamma \mathscr{V}_\gamma - \operatorname{Re}\left[k^2 e^{i(2\varphi - \beta)}\right] < 0. \tag{12.117}$$

Suppose that the potential can be continued to the imaginary axis, $\varphi = \pm\frac{1}{2}\pi$, and that there exists a constant M such that

$$|\mathscr{V}_{1/2\pi}(x)| = |\mathscr{V}(ix)| \le Mx^{-1}. \tag{12.118}$$

Then (12.117) says that for $\varphi = \frac{1}{2}\pi$

$$\tfrac{1}{4}x^{-2}\cos\beta + \operatorname{Re} k^2 e^{-i\beta} < \sqrt{2}|\gamma| M x^{-1}$$

or since

$$AB \le \tfrac{1}{2}(A + B)^2$$

$$\cos\beta \operatorname{Re} k^2 e^{-i\beta} < 4|\gamma|^2 M^2.$$

In other words, every zero of $\mathscr{F}_-$ must occur at a value of the energy $E = E_1 + iE_2$ for which

$$E_1 \cos\beta + E_2 \sin\beta < \frac{2|\gamma|^2 M^2}{\mu \cos\beta}$$

for every choice of β between zero and $\frac{1}{2}\pi$. Hence E must lie inside the *envelope* of this set of curves, which is the parabola

$$E_2^2 = \frac{8|\gamma|^2 M^2}{\mu}\left(\frac{2|\gamma|^2 M^2}{\mu} - E_1\right). \tag{12.119}$$

All the poles of S on the second sheet are confined to this domain if the potential is "analytic" ($\sigma = \frac{1}{2}\pi$) and satisfied (12.118). (Yukawa potentials are in this class, but potentials that vanish beyond a finite distance are not.) A fortiori all resonances poles must occur at an energy for which

$$\operatorname{Re} E < \frac{2|\gamma|^2 M^2}{\mu}. \tag{12.120}$$

We may use the same argument to find a limitation on the eigenvalues $\alpha = 1/\gamma$ of the kernel $G\mathscr{V}$ of the radial Lippmann-Schwinger equation. For positive energy (12.120) tells us that

$$|\alpha| < 2Mk^{-1} \tag{12.121}$$

which shows quite directly that the α's uniformly tend to zero as $E \to +\infty$, and hence the Born series must converge when

$$k^2 > 4M^2. \tag{12.122}$$

12.1.5 Completeness. At this point we have all the tools necessary for a proof of the completeness of the bound and scattering-wave functions. Such a proof may be based entirely on abstract vector-space techniques. We have stated the spectral theorem in Sec. 7.3.3 in such a context without proof. Here we want to give a demonstration using the tools of complex analysis. It is instructive to see the analytic properties that have been discussed applied in this way. The only assumptions to be made on the potential are (12.9) and (12.21).

The basic idea of the proof is to evaluate the integral

$$\int dE\, \mathscr{G}(E; r, r')$$

over a closed contour running along the two rims of the right-hand cut in the (physical sheet of the) E surface and closed by a large circle at infinity. The integral is written down in terms of its various contributions, and it is evaluated by Cauchy's residue theorem in terms of the bound-state poles on the real axis. The whole procedure is a little simpler in the complex k plane.

Consider the integral

$$I(r) = \int_C dk\, k \int_0^\infty dr'\, h(r') \mathscr{G}(k; r, r')$$

where $h(r)$ is square-integrable and $\mathscr{G}$ is given by (12.40). The contour C runs along the real k axis from $-\infty$ to $+\infty$, avoiding the origin by a small semicircle of radius ε in the upper half plane, and closed by a large semicircle there of radius R. We write the integral in two pieces,

$$I = I_1 + I_2$$

$$I_1 = -\int_C dk\, k \int_0^r dr'\, h(r') \frac{\varphi(k,r')f(k,r)}{\mathscr{F}(k)}$$

$$I_2 = -\int_C dk\, k \int_r^\infty dr'\, h(r') \frac{f(k,r')\varphi(k,r)}{\mathscr{F}(k)}.$$

Take I_1 first.

Suppose that the discrete eigenvalues of H are $-K_n{}^2/2\mu$, $K_n > 0$. Then we write

$$\varphi^{(n)}(r) \equiv \varphi(iK_n, r)$$

$$f^{(n)}(r) \equiv f(iK_n, r)$$

$$C_n = \dot{\mathscr{F}}(iK_n).$$

The functions φ, f, and $\mathscr{F}$ being regular in the upper half of the k plane, and $\mathscr{F}$ having simple zeros at $k = iK_n$ only, the integral I_1 is evaluated immediately by the residue theorem,

$$\begin{aligned} I_1 &= -2\pi i \sum_n \int_0^r dr'\, h(r')\varphi^{(n)}(r')f^{(n)}(r)\frac{iK_n}{C_n} \\ &= i\pi \sum_n \int_0^r dr'\, h(r')\varphi^{(n)}(r')\varphi^{(n)}(r)N_n^{-2} \end{aligned} \tag{12.123}$$

if we write

$$N_n^2 \equiv \int_0^\infty dr\, [\varphi^{(n)}(r)]^2$$

and use (12.52) and (12.49).

On the other hand, we may evaluate I_1 directly. The contribution $I_{1\varepsilon}$ from the small semicircle vanishes in the limit as its radius tends to zero, except possibly when $\mathscr{F}(0) = 0$. Because of (12.53) it also vanishes when $\mathscr{F}(0) = 0$.

The contribution I_{1R} to I_1 from the large semicircle is evaluated by using the asymptotic form of the functions for large $|k|$, (12.12), (12.26), and (12.33). We get

$$\begin{aligned} I_{1R} &\underset{R\to\infty}{\simeq} -\tfrac{1}{2}i \int_0^r dr'\, h(r') \int_{\text{sc}} dk\, (e^{ik(r-r')} - e^{ik(r+r')}) \\ &\underset{R\to\infty}{\simeq} \tfrac{1}{2}h(r) \int_{\text{sc}} dk\, k^{-1} = \tfrac{1}{2}\pi i h(r). \end{aligned} \tag{12.124}$$

The remaining contribution I_{1E} to I_1 is the integral over the real axis. There we use the fact that φ is an even function of k and (12.29) and (12.32).

This gives

$$\begin{aligned} I_{1E} &= -\int_0^r dr'\, h(r') \left(\int_{-\infty}^{-\varepsilon} + \int_{+\varepsilon}^{\infty} \right) dk\, k \frac{\varphi(k,r') f(k,r)}{\mathscr{F}(k)} \\ &= -2i \int_0^r dr'\, h(r') \int_{+\varepsilon}^{\infty} dk\, k^2 \frac{\varphi(k,r')\varphi(k,r)}{|\mathscr{F}(k)|^2}. \end{aligned} \tag{12.125}$$

At this point we may let $\varepsilon \to 0$, add the contributions (12.124) and (12.125), and equate them to (12.123). The result is

$$h(r) = 2\int_0^r dr'\, h(r') \left[2\int_0^\infty dk\, k^2 \frac{\varphi(k,r')\varphi(k,r)}{\pi |\mathscr{F}(k)|^2} + \sum_n \frac{\varphi^{(n)}(r')\varphi^{(n)}(r)}{N_n^2} \right]. \tag{12.126}$$

We then go through the same arguments for I_2, in which we replace the upper limit of the r' integration by $r + \mu$, $\mu > 0$. The result is

$$h(r) = 2\int_r^{r+\mu} dr'\, h(r') [\quad], \tag{12.127}$$

the content of the bracket being the same as in (12.126). We now add (12.126) and (12.127) and let $\mu \to \infty$. If $h(r)$ is square-integrable, the resulting improper integral converges.

We may write the outcome in the customary δ-function notation

$$\frac{2}{\pi}\int_0^\infty dk\, k^2 \frac{\varphi(k,r)\varphi(k,r')}{|\mathscr{F}(k)|^2} + \sum_n \frac{\varphi^{(n)}(r)\varphi^{(n)}(r')}{N_n^2} = \delta(r - r'). \tag{12.128}$$

This shows that the set of continuum and discrete wave functions is complete and what the proper weight functions are. If we define the *spectral function* $\rho(E)$ by

$$\frac{d\rho}{dE} = \begin{cases} \dfrac{2\mu k}{\pi |\mathscr{F}(k)|^2} & E > 0 \\ \displaystyle\sum_n \frac{\delta(E - E_n)}{N_n^2} & E < 0 \end{cases} \tag{12.129}$$

together with the boundary condition $\rho(-\infty) = 0$, the completeness relation may be written as a Stieltjes integral,

$$\int d\rho\,(E)\varphi(k,r)\varphi(k,r') = \delta(r - r'). \tag{12.128a}$$

Alternatively, the completeness may be written in terms of the physical wave functions of (11.6), with (12.37) used for the scattering functions. For the bound-state wave functions we use

$$\psi^{(n)}(r) = \frac{\varphi^{(n)}}{N_n}$$

so that they are normalized to unity. Then

$$\frac{2}{\pi}\int_0^\infty dk\, \psi^{(+)}(k,r)\psi^{(+)*}(k,r') + \sum_n \psi^{(n)}(r)\psi^{(n)}(r') = \delta(r - r'). \tag{12.128b}$$

The Green's Function. Equations (12.128) to (12.128b) may immediately be used for a representation of the complete Green's function $\mathscr{G}$ or resolvent (12.40). Evidently,

$$\begin{aligned}\mathscr{G}(E;r,r') &= \frac{1}{2\mu}\int d\rho\,(E')\frac{\varphi(k',r)\varphi(k',r')}{E - E'}\\ &= \frac{2}{\pi}\int_0^\infty dk'\frac{\psi^{(+)}(k',r)\psi^{(+)*}(k',r')}{k^2 - k'^2} + \sum_n \frac{\psi^{(n)}(r)\psi^{(n)}(r')}{k^2 + k_n^2}\\ &= \frac{1}{\pi}\int_{-\infty}^{\infty}\frac{dk'}{k'}\frac{\psi^{(+)}(k',r)\psi^{(+)*}(k',r')}{k - k'} + \sum_n \frac{\psi^{(n)}(r)\psi^{(n)}(r')}{k^2 + k_n^2}.\end{aligned} \tag{12.130}$$

As k approaches the real positive k axis (from above), or as E approaches the right-hand cut from above, $\mathscr{G}$ becomes $\mathscr{G}^+$; as k approaches the negative real axis (from above), or E the right-hand cut from below, $\mathscr{G}$ becomes $\mathscr{G}^-$.

12.2 Higher Angular Momenta

In this section we want to generalize the results of Sec. 12.1 to angular momenta other than zero. Rather than going through all the steps again we shall simply point out which features are the same and which are not, indicating how the changed results are obtained. Since in the next chapter we shall have occasion to use angular momenta of nonintegral values, we shall here not restrict ourselves necessarily to integral l. The particular points which arise only when l is not an integer, however, will be discussed primarily in Sec. 12.3 and Chap. 13.

The Regular Solution. The methods of Secs. 12.1.1 and 12.1.2 are applicable to $l > 0$ as well. (They are, in fact, applicable to all values of l for which[20] $\operatorname{Re} l > -\frac{1}{2}$.) The regular solution of the radial Schrödinger equation

$$-\psi'' + l(l + 1)r^{-2}\psi + \gamma\mathscr{V}\psi = k^2\psi \tag{12.131}$$

is uniquely defined by the boundary condition

$$\lim_{r\to 0} r^{-l-1}\varphi_l(k,r) = 1 \tag{12.132}$$

instead of (12.2). The reason why (12.132) necessarily depends on l is quite analogous to the reason why (12.15) for $f(k,r)$ depends on k. When $l \neq 0$, the point $r = 0$ is a (regular) singular point of the second-order differential equation (12.131), and $l(l + 1)$ multiplies the term of the highest singularity in it. This will have a crucial bearing on the dependence of φ_l upon l, but this topic will be discussed in Chap. 13. So long as l is fixed (and $l > -\frac{1}{2}$), it has no influence on the general dependence of φ_l on k.

The combination (12.131) and (12.132) is again most simply replaced by an integral equation of the Volterra type,

$$\varphi_l(k,r) = \varphi_l^{(0)}(k,r) + \gamma\int_0^r dr'\, g_l(k;r,r')\mathscr{V}(r')\varphi_l(k,r'). \tag{12.133}$$

20 The reason for this restriction is discussed in Sec. 12.3.

The inhomogeneity, or zero-order function, is now

$$\begin{aligned}\varphi_l^{(0)}(k,r) &= (2l+1)!!k^{-l-1}u_l(kr)\\ &= r^{1/2}(\tfrac{1}{2}k)^{-l-1/2}\Gamma(l+\tfrac{3}{2})J_{l+1/2}(kr)\end{aligned} \tag{12.134}$$

and the Green's function for $r' < r$

$$\begin{aligned}g_l(k;r,r') &= k^{-1}[u_l(kr')v_l(kr) - u_l(kr)v_l(kr')]\\ &= ie^{-i\pi l}(2k)^{-1}[w_l^{(+)}(kr')w_l^{(-)}(kr) - w_l^{(+)}(kr)w_l^{(-)}(kr')]\\ &= \frac{\tfrac{1}{2}\pi(rr')^{1/2}[J_{l-1/2}(kr)J_{-l-1/2}(kr') - J_{l+1/2}(kr')J_{-l-1/2}(kr)]}{\cos\pi l}.\end{aligned} \tag{12.135}$$

Here J_λ are Bessel functions, and the functions u_l, v_l, and $w_l^{(\pm)}$ are defined in Sec. 2.2.1. The functions $\varphi_l^{(0)}$ and g_l are subject to the following inequalities, which are uniform in k and r for all (complex) k and all $r > 0$, $0 < r' \leq r$, for fixed values of l,

$$\begin{aligned}|\varphi_l^{(0)}(k,r)| &\leq Ce^{|\nu|r}|k|^{-l-1}[L(|k|r)]^{l+1}\\ |g_l(k;r,r')| &\leq Ce^{|\nu|(r-r')}|k|^{-1}[L(|k|r)]^{l+1}[L(|k|r')]^{-l}\end{aligned} \tag{12.136}$$

and similarly

$$|w_l^{(+)}(kr)| \leq Ce^{\nu r}[L(|k|r)]^{-l} \tag{12.136a}$$

where

$$L(x) = \frac{x}{1+x}$$

and $\nu = \operatorname{Im} k$. [For complex l we must replace l everywhere in (12.136) and (12.136a) by $\operatorname{Re} l$.] The series of successive approximations to (12.133) converges under the same conditions on $\mathscr{V}$ as for $l = 0$, and φ_l has all the analyticity and symmetry properties as a function of k and γ shown in Sec. 12.1.1. The behavior for large $|k|$ is now, of course, not (12.12), but

$$\varphi_l(k,r) = (2l+1)!!k^{-l-1}\sin(kr - \tfrac{1}{2}\pi l) + o(|k|^{-l-1}e^{|\nu|r}). \tag{12.137}$$

The Irregular Solutions. The irregular solutions of $f_{l\pm}(k,r)$ are still defined by the boundary conditions (12.15). Their integral equations are

$$f_{l\pm}(k,r) = f_{l\pm}^{(0)}(k,r) - \gamma\int_r^\infty dr'\, g_l(k;r,r')\mathscr{V}(r')f_{l\pm}(k,r') \tag{12.138}$$

where

$$f_{l\pm}^{(0)}(k,r) = w_l^{(\pm)}(kr)e^{-i1/2\pi l} \tag{12.139}$$

which are given by (2.60) in terms of Hankel functions. The Green's function g_l is the same as (12.135). The series of successive approximations to (12.138) converges under the same conditions as that to (12.16), and the analyticity and symmetry properties of $f_{l\pm}$ as functions of k are the same as those of $f_\pm$, except for the point $k = 0$, where it has a pole of order l [or if the potential is not sufficiently well behaved to make the functions $k^l f_{l\pm}(k,r)$ analytic in a neighborhood of $k = 0$, they are continuous there]. The arguments based on analyticity of the potential, and particularly on (12.22) or (12.22a), are quite the same as those following the latter equation. The behavior (12.26) for large $|k|$ when $\operatorname{Im} k \geq 0$ is unaltered.

The Jost Function. Jost functions $\mathfrak{F}_{l\pm}(k)$ are defined as in (12.28), so that (12.29) holds. They then have again the properties (12.30) and the symmetries (12.32) to (12.32b) (for real l), but, of course, (12.31) is now not true. Instead we have

$$\mathfrak{F}_l(k) = (2l + 1) \lim_{r\to 0} r^l f_l(k,r). \tag{12.140}$$

The limit for large k, (12.33), is also changed; because of (12.137) and (12.26) we have, for $\operatorname{Im} k \geq 0$,

$$\lim_{|k|\to\infty} k^l \mathfrak{F}_l(k) = (2l + 1)!!e^{-i1/2\pi l}. \tag{12.141}$$

It is therefore convenient to define another Jost function

$$\begin{aligned} \mathscr{F}_{l+}(k) = \mathscr{F}_l(k) &\equiv \frac{k^l e^{-i1/2\pi l}\mathfrak{F}_l(k)}{(2l + 1)!!} \\ &= \frac{k^l e^{-i1/2\pi l} W(f_l, \varphi_l)}{(2l + 1)!!} \end{aligned} \tag{12.142}$$

which approaches unity at large $|k|$ and is continuous at $k = 0$. If we also define

$$\begin{aligned} \mathscr{F}_{l-}(k) &\equiv \frac{k^l e^{i1/2\pi l}\mathfrak{F}_{l-}(k)}{(2l + 1)!!} \\ &= \frac{k^l e^{i1/2\pi l} W(f_{l-}, \varphi_l)}{(2l + 1)!!}, \end{aligned} \tag{12.142a}$$

then $\mathscr{F}_{l-}$ is related to $\mathscr{F}_{l+}$ as in (12.30). The function $\mathscr{F}_l$ satisfies symmetries (12.32) to (12.32b).[21]

The integral representations (12.34) and (12.36) have their analog for $l \neq 0$. Substitution of (12.133) and (12.138) in (12.142) and evaluation of the wronskian at $r = 0$ or $r = \infty$ yields

$$\mathscr{F}_l(k) = 1 + \gamma k^{-1} e^{-i1/2\pi l} \int_0^\infty dr\, u_l(kr) \mathscr{V}(r) f_l(k,r) \tag{12.143}$$

and

$$\mathscr{F}_l(k) = 1 + \frac{\gamma k^l e^{-i\pi l}}{(2l + 1)!!} \int_0^\infty dr\, w_l^{(+)}(kr) \mathscr{V}(r) \varphi_l(k,r). \tag{12.144}$$

These representations may be used to derive the more detailed high-energy behavior of $\mathscr{F}_l$. Equations (12.33a) and (12.33b) are then found still

21 For real l.

to be true, and so is therefore the high-energy phase-shift behavior (12.82). [See (12.154) below for the connection with the **S** matrix.] These formulas have yet to be shown to be uniform with respect to l, though. Although each phase shift eventually approaches zero in the same way, as in (12.82), the larger the l value, the higher the energy may have to be for this to be true to a given degree of approximation (see Sec. 12.3).

The Physical Wave Function. The relation of the physical wave functions $\psi_l^{(+)}$ to φ_l is obtained by comparison of the asymptotic form of φ_l, which is given by (12.35) in terms of $\mathfrak{F}_{l\pm}$, with (11.9). The result is that

$$\begin{aligned}\psi_l^{(+)}(k,r) &= \frac{k\varphi_l(k,r)e^{i1/2\pi l}}{\mathfrak{F}_l(k)}\\ &= \frac{k^{l+1}\varphi_l(k,r)}{\mathscr{F}_l(k)(2l+1)!!}.\end{aligned} \tag{12.145}$$

The Green's Function. The complete Green's function is constructed as in Sec. 12.1.2. We get the analog of (12.40), which is

$$\begin{aligned}\mathscr{G}_l^+(k;r,r') &= -\frac{\varphi_l(k,r_<)f_l(k,r_>)}{\mathfrak{F}_l(k)}\\ &= -k^{-1}\psi_l^{(+)}(k,r_<)f_l(k,r_>)e^{-1/2i\pi l}.\end{aligned} \tag{12.146}$$

It satisfies the equation

$$\left[-\frac{\partial^2}{\partial r^2} + \frac{l(l+1)}{r^2} + \mathscr{V}(r) - k^2\right]\mathscr{G}_l^+(k;r,r') = -\delta(r-r') \tag{12.147}$$

and the outgoing-wave boundary condition. Other Green's functions are easily written down in analogy with (12.42).

The Fredholm Determinant. The demonstration that the Jost function is identical with the Fredholm determinant of (11.7) goes through just as before. It must, of course, be $\mathscr{F}_l$ and not $\mathfrak{F}_l$,

$$\mathscr{F}_l(k) = \Delta_l(k) \tag{12.148}$$

because $\mathscr{F}_l = 1$ when $\gamma = 0$. The kernel of the radial Lippmann–Schwinger equation is now given by

$$\begin{aligned}K_l(r,r') &= G_l^+(r,r')\mathscr{V}(r')\\ &= -e^{-i\pi l}k^{-1}u_l(kr_<)w_l^{(+)}(kr_>)\mathscr{V}(r')\end{aligned} \tag{12.149}$$

in place of (12.44). Apart from that, the representations (12.45) to (12.48a) are still valid.

Bound States. The discussion of Sec. 12.1.2 concerning the bound states and zeros of $\mathscr{F}_l$ in the upper half plane holds for $l > 0$ as well. Instead of (12.50), though, we have

$$c_l = \frac{-2|k_0|}{\mathfrak{F}_l(-i|k_0|)} \tag{12.150}$$

if $\mathfrak{F}_l(i|k_0|) = 0$ and hence

$$f_l(k_0,r) = c_l\varphi_l(k_0,r). \tag{12.151}$$

It is still true that the bound-state zeros of $\mathfrak{F}_l$ must be *simple*. The demonstration of (12.52) holds again, except that the left-hand side is $\dot{\mathfrak{F}}_l(k_0)$. Similarly, $\mathfrak{F}_{l+}$ and $\mathfrak{F}_{l-}$ cannot both simultaneously vanish, except at $k = 0$. This again rules out zeros of $\mathscr{F}_l$ on the real axis, except at the origin.

The Point $k = 0$. The situation at $k = 0$ is now somewhat different. If $\mathscr{F}_l(0) = 0$, then (12.142) shows that $\varphi_l(0,r)$ is a multiple of $\lim k^l f_l(k,r)$ as $k \to 0$. Because of the factor k^l this tends, not to unity at infinite distance, but to *zero*. In fact, it is easily seen from the analog of (12.18) that, as $r \to \infty$,

$$\lim_{k\to 0} k^l f_l(k,r) = O(r^{-l}). \tag{12.152}$$

Hence $\mathscr{F}_l(0) = 0$ signifies a bound state of zero binding energy when $l > \frac{1}{2}$. The function $\varphi_l(0,r)$ is then square-integrable. In contrast to bound-state wave functions of positive binding energy, though, this function does not decrease exponentially.

If $\mathscr{F}_l(0) = 0$, the multiplicity of the zero also differs from that in the $l = 0$ case. If k approaches zero from the upper half plane, then it can be shown that for $l \geq 1$

$$\mathscr{F}_l(k) = O(k^2)$$

and

$$1/\mathscr{F}_l(k) = O(k^{-2}) \tag{12.153}$$

so that $\mathscr{F}_l$ goes to zero *exactly* as k^2. If $\mathscr{F}_l$ is analytic in the vicinity of $k = 0$, this means that there is a *double* zero.

Dispersion Relations. The dispersion relations (12.54) to (12.64) are valid also for $l > 0$. Since $\mathfrak{F}_l$ is not continuous at $k = 0$ and does not approach unity at $|k| \to \infty$, they must all refer to the function $\mathscr{F}_l$.

The **S** *Matrix*. The S-matrix element for $l \neq 0$ is obtained by comparing the analog of (12.35) with (11.9); thus

$$\begin{aligned} S_l(k) &= \frac{e^{i\pi l}\mathfrak{F}_{l-}(k)}{\mathfrak{F}_{l+}(k)} \\ &= \frac{\mathscr{F}_{l-}(k)}{\mathscr{F}_{l+}(k)} \end{aligned} \tag{12.154}$$

because of (12.142) and (12.142a). The symmetry properties of $S_l(k)$, (12.72) and (12.74), are unaltered, and the phase shift δ_l is the phase of $\mathscr{F}_{l-}$ or $-\delta_l$ that of $\mathscr{F}_l$. The integral representation (12.76), being nothing but a rewriting of (11.19) together with the solution of (11.7) by means of $\mathscr{G}_l^+$, we need not write down its generalization.

The residues of $S_l(k)$ at bound-state poles are expressible as in (12.84) in terms of $\psi_l^{(+)}$. Equation (12.145), however, shows that $\psi_l^{(+)}$ is not always purely imaginary at $k = iK_n$. In fact we get

$$ie^{i\pi l} \operatorname{res}_n > 0 \tag{12.155}$$

instead of (12.85).

The limitation (12.90) on the derivative of the phase shift with respect to k can be generalized, but for $l > 0$ it is not very transparent. Rather than writing it down, we refer to the literature.[22]

The dispersion relations and N/D-method (12.91) to (12.93) hold also for $l > 0$, provided we use $\mathscr{F}_l$ everywhere in place of $\mathscr{F}$.

The Levinson Theorem and Limits on the Number of Bound States. The proof of the Levinson theorem is the same as for $l = 0$, except when $\mathscr{F}_l(0) = 0$. In that case (12.153) must be used in place of (12.53). This makes $q = 2$ in the equation preceding (12.95). The result is that

$$\delta_l(0) = \pi n_l \tag{12.156}$$

if n_l is the number of bound states of angular momentum l, including those of zero binding energy. This corresponds to the fact that for $l \geq 1$ the latter are really normalizable eigenvalues, whereas for $l = 0$ they are not.

The number of bound states of angular momentum l is limited by an analog of (12.96). The argument is the same as there, except that the zero-energy Green's function is

$$G(r,r') = \frac{-r_<^{l+1} r_>^{-l}}{2l+1}$$

so that

$$\operatorname{tr} \mathfrak{K} = \frac{\int_0^\infty dr\, r U(r)}{2l+1}$$

and consequently

$$n_l \leq \frac{\int_0^\infty dr\, r U(r)}{2l+1}. \tag{12.157}$$

The function $U(r)$ is defined as for (12.96). The inequality (12.157) shows that if the integral in it is finite, then there always exists a maximum value of l for which the potential produces bound states,

$$2l_{\max} \leq \int_0^\infty dr\, r U(r) - 1. \tag{12.158}$$

The *total* number of bound states is therefore finite.

Low-Energy Behavior. The most important modification of the results of Sec. 12.1 for angular momenta other than zero are those concerning the low-energy behavior. In view of the discussion of Sec. 11.2.2 this is not surprising. It is primarily at low energies that the centrifugal barrier makes itself felt.

22 See E. P. Wigner (1955), G. Lüders (1955), and R. G. Newton (1964).

Let us insert (12.136a) and the analog of (12.8) for $l > 0$ in the representation (11.19) of $S_l(k)$, using (12.145). This yields

$$|S_l(k) - 1| \leq \frac{C_l |k|^{2l+1} \int_0^\infty dr\, |\mathscr{V}(r)| \left(\dfrac{r}{1 + |k|r}\right)^{2l+2}}{|\mathscr{F}_l(k)|} \tag{12.159}$$

for fixed l. If $\mathscr{F}_l(0) \neq 0$ *and* if

$$\int_0^\infty dr\, r^{2l+2} |\mathscr{V}(r)| < \infty \tag{12.160}$$

we may conclude that

$$S_l(k) - 1 = 2ie^{i\delta l} \sin \delta_l = O(k^{2l+1}) \tag{12.161}$$

as $k \to 0$, in agreement with (11.30). For l values for which (12.160) breaks down, (12.161) can in general be expected to break down too. When $\mathscr{F}_l(0) = 0$, then (12.153) together with (12.159) shows that for $l \geq 1$

$$S_l(k) - 1 = O(k^{2l-1}) \tag{12.161a}$$

as $k \to 0$.

Equation (12.161) implies that unless $\mathscr{F}_l(0) = 0$, we have

$$K_l(k) = O(k^{2l+1}) \tag{12.161b}$$

as $k \to 0$. It is therefore convenient to define a reduced **K** matrix by

$$\begin{aligned} \mathfrak{K}_l(k) &= (-)^{l+1} k^{-2l-1} K_l(k) \\ &= i(-)^l k^{-2l-1} \frac{S_l(k) - 1}{S_l(k) + 1} \end{aligned} \tag{12.162}$$

so that

$$S_l(k) = \frac{1 + (-)^{l+1} ik^{2l+1} \mathfrak{K}_l}{1 - (-)^{l+1} ik^{2l+1} \mathfrak{K}_l}. \tag{12.162a}$$

Because of (12.74), the function $\mathfrak{K}_l$ is an *even* function of k,

$$\mathfrak{K}_l(-k) = \mathfrak{K}_l(k). \tag{12.163}$$

Hence it has no right-hand cut in the E plane. Perhaps the simplest way of representing the unitary **S** matrix in terms of the function $\mathfrak{K}_l$ without right-hand cut is to define a function $\mathfrak{M}_l$ by

$$S_l = 1 - k^{2l+1} \mathfrak{M}_l. \tag{12.164}$$

Then

$$\mathfrak{M}_l^{-1} = \tfrac{1}{2} k^{2l+1} + \tfrac{1}{2} i (-)^{l+1} \mathfrak{K}_l^{-1}. \tag{12.165}$$

The Motion of Poles of S_l. The difference between the low-energy behavior of $\mathscr{F}_l$ when $\mathscr{F}_l(0) = 0$ for $l = 0$ and for $l \geq 1$ has important consequences on the motion of zeros of $\mathscr{F}_l$ as the potential strength γ is changed. For

$l = 0$ a zero of $\mathscr{F}_l$ remains simple as it passes through the origin. From this we have concluded that before a new bound state is introduced it is a *virtual* state. For $l \geq 1$, (12.153) says that, when a zero of $\mathscr{F}_l$ lands on $k = 0$, it must be *double*. This implies that two zeros of $\mathscr{F}_l$ coincide at that value of γ. The collision of two symmetric complex zeros of $\mathscr{F}_l$, which for $l = 0$ takes place on the negative imaginary axis, occurs right at the origin when $l \geq 1$. If the potential is not quite strong enough to introduce a new bound state, there are therefore *two* complex poles of S_l on the second sheet (provided that the potential allows an analytic continuation of S_l) very close to the origin, with a small imaginary contribution. These cause the low-energy resonance discussed in Sec. 11.2.2. From the present mathematical point of view we explain the fact that the resonance is the sharper the lower its energy as follows. The two zeros of $\mathfrak{F}_l$ move toward the origin *at right angles*, appropriate to the square-root branch point of the zero of $\mathfrak{F}_l$ as a function of γ. The trajectories of the zeros $k_l(\gamma)$ therefore *osculate* the real k axis at $k = 0$. Thus $\operatorname{Im} k_l$ decreases faster than $\operatorname{Re} k_l$ as γ decreases. These remarks can of course be explicitly verified for the square-well potential discussed in Sec. 11.2.2.

The foregoing considerations may be made equally well from the point of view of the eigenvalues α of the kernel of the radial Lippmann-Schwinger equation. The equation

$$\mathscr{F}_l(\gamma, k) = 0$$

determines $\alpha(k) = 1/\gamma(k)$ for a given value of k. Now, since for $l \geq 1$ at $k = 0$ the zero of $\mathscr{F}_l$ is *double* as a function of k, we have

$$\frac{d\gamma}{dk} = O(k)$$

there. If we assume that the potential obeys (12.160), then the imaginary part of $\mathscr{F}_l$ goes as

$$\operatorname{Im} \mathscr{F}_l(k) = O(k^{2l+1})$$

when $k \to 0+$. Since

$$\frac{d\alpha}{dk} = -\frac{\partial \mathscr{F}_l/\partial k}{\partial \mathscr{F}_l/\partial \alpha}$$

the α trajectory leaves the real axis at zero angle, according to

$$\frac{\operatorname{Im} \alpha(k)}{\operatorname{Re}\left[\alpha(k) - \alpha(0)\right]} = O(k^{2l-1}). \tag{12.166}$$

It does so in the *forward* directions, i.e., in the same direction in which $\alpha(E)$ was moving when $E < 0$. From this point of view, that is why an "almost bound" state of $l \geq 1$ causes a sharp low-energy resonance. An eigenvalue α *almost* reaches $+1$ or -1 at $E = 0$. As E increases, it crosses the unit circle at low energy, and with a small imaginary part. For $l = 0$ this is not so, for α leaves the real axis at right angles.

Pole Distribution. The discussion in Sec. 12.14 concerning the distribution of zeros of $\mathscr{F}(k)$ carries over to $\mathscr{F}_l(k)$ with no change. Consequently so do the representations (12.104) to (12.107) and the partial-fraction formulas (12.110) to (12.112).

The restrictions on the region of the k plane where $\mathscr{F}$ may have zeros if the potential is analytic carry over to $l \neq 0$. So long as l is real, A of (12.115) is unchanged, but B of (12.116) now reads

$$B = \operatorname{Re} \gamma \mathscr{V}_\varphi - \operatorname{Re} k^2 e^{2i\varphi} + x^{-2}(l + \tfrac{1}{2})^2.$$

The result is that the parabola (12.119) is replaced by

$$E_2^2 = 4\xi(\xi - E_1) \tag{12.167}$$

where $E = E_1 + iE_2$ and

$$\xi = \frac{2|\gamma|^2 M^2}{\mu(2l + 1)^2}. \tag{12.168}$$

If the potential is analytic ($\sigma = \frac{1}{2}\pi$) and satisfies (12.118) on the imaginary axis, the poles of S_l on the second sheet are confined to the inside of the parabola (12.167). Thus, the larger the angular momentum, the more the poles are confined to be near the real E axis. Furthermore, (12.167) implies that all resonance poles lie at an energy for which

$$\operatorname{Re} E < \frac{2|\gamma|^2 M^2}{(2l + 1)^2 \mu}. \tag{12.169}$$

The larger l, the smaller this limit.

Similarly, (12.121) is replaced by the fact that for $E > 0$ all eigenvalues α of the kernel of the Lippmann–Schwinger equation must obey the inequality

$$|\alpha| < \frac{2M}{(2l + 1)k}. \tag{12.170}$$

Thus, the larger l, the less large E need be chosen for all $|\alpha|$ to be smaller than a given number. The Born series must converge whenever all $|\alpha| < 1$, i.e., when

$$k^2 > \frac{4M^2}{(2l + 1)^2}. \tag{12.171}$$

The larger l, the less stringent is the requirement on k.

Completeness. The completeness proof also works as before, except that (12.146) must be used for the complete Green's function. There is a small change near $k = 0$ if $\mathscr{F}_l(0) = 0$, so that (12.153) must be used. The result is that (12.128a) still holds, but the spectral function has the derivative

$$\frac{d\rho}{dE} = \begin{cases} \dfrac{2\mu k}{\pi |\mathfrak{F}_l(k)|^2} & E > 0 \\ \displaystyle\sum_n \frac{\delta(E - E_n)}{N_n^2} & E \leq 0. \end{cases} \tag{12.172}$$

Equations (12.128b) and (12.130) are unchanged. They, as well as (12.172), hold in fact for all real $l > -\frac{1}{2}$. For complex l, $\text{Re}\, l > -\frac{1}{2}$, or complex γ, if the zeros of $\mathfrak{F}_l$ remain simple, the only change is that $|\mathfrak{F}_l(k)|^2$ in (12.172) must be replaced by $\mathfrak{F}_{l+}(k)\mathfrak{F}_{l-}(k)$ and φ_l, E_n, and N_n may all be complex.

12.3 Continuous Angular Momenta

In spite of the fact that the physical angular momentum of a particle, or of several particles, is quantum mechanically restricted to be an integral or half-integral multiple of $\hbar$, there is no reason why this restriction may not be relaxed for the purpose of a clearer mathematical understanding of the properties of the scattering amplitude. From a mathematical point of view, the quantization of the orbital angular momentum l originates in the partial-wave analysis. It is based on the properties of the spherical harmonics, which for nonintegral values of l are not well-behaved functions of $\cos\theta$ at $|\cos\theta| = 1$. The requirement that l be an integer is based on the demand that the three-dimensional wave function be "regular" in all directions.

Once the partial-wave analysis has been carried out we are left with the radial Schrödinger equation (11.8), in which l appears as a parameter multiplying the centrifugal term r^{-2}. If this equation is divorced from its origin in the full Schrödinger equation, there is no reason for l to be restricted to integral values. But since for the calculation of the scattering amplitude we need the **S**-matrix elements, or the phase shifts, of integral l only, there may appear to be no reason for us to consider other values of l. That it is nevertheless interesting to make l a continuous (and even complex) variable can be seen in a number of ways.

The crucial ingredient of the partial-wave analysis is the assumption of spherical symmetry of the scatterer (even though it can be usefully generalized to cases in which that spherical symmetry is somewhat relaxed, as in Chap. 15, this does not alter the argument). The calculation of the scattering amplitude is thereby reduced to that of the partial-wave amplitudes. When these are computed or their properties are discussed, a fixed l value at a time, one important fact is never used, that the potential which appears in the radial Schrödinger equation is the same for each angular momentum. For all the partial-wave analysis cares, we could adopt an arbitrarily different potential for each physical value of l. The circumstance that the potential in each radial Schrödinger equation is the same must produce powerful correlations between the **S**-matrix elements or phase shifts. These correlations are most easily investigated by allowing l to change *continuously* from one integer to another. As we shall see in Chap. 13 there are good reasons, even, to allow l to be *complex*.

The power of the correlations between the various partial-wave amplitudes can be appreciated by considering the dispersion relations satisfied by the full amplitude. Suppose that the potential decreases exponentially at infinity.

We have seen in Sec. 10.3.2 that then the forward-scattering amplitude is an analytic function of E regular on the physical sheet, except for simple poles at the bound-state energies. Consequently it satisfies a dispersion relation. But this same amplitude is a sum of partial-wave amplitudes each of which may, and in general do, have infinitely many singularities on the physical sheet. There must therefore be sufficiently strong correlations between the positions of the singularities of the various **S**-matrix elements, and between their residues, so that in the forward-scattering amplitude all these manage to cancel. What is more, since there is a dispersion relation even when the momentum transfer is not zero, the singularities must cancel also in nonforward amplitudes (up to a certain finite momentum transfer). Evidently this implies a strong dependence of these singularities upon each other.

In this section we want to study the dependence of the **S**-matrix elements $S_l(k)$ upon l as a continuous (real) parameter. As was pointed out there, the results of Sec. 12.2 hold even when l is not an integer. The only restriction to be imposed is that $l > -\frac{1}{2}$. This is because of the boundary condition (12.132) for the regular solution. The situation is analogous to that for $f(k,r)$ with respect to k. We know that the behavior of the irregular solution at $r = 0$ is as r^{-l}. When $l > -\frac{1}{2}$ this dominates r^{l+1}. Consequently (12.132) serves to define the regular solution uniquely. When $l < -\frac{1}{2}$, on the other hand, r^{l+1} dominates r^{-l} and (12.132) does not rule out an arbitrary admixture of the other solution.

Let us first consider the behavior of the **S**-matrix elements for large values of l. This situation is complicated by nonuniformity with respect to the energy. It is analogous to, and caused by, the nonuniform behavior of the Bessel functions with respect to their order and variable. The function $h(\lambda,z) = \lambda J_\lambda(z)$ tends to zero both as $\lambda \to \infty$ for fixed z and as $z \to \infty$ for fixed λ. But it has a maximum in the vicinity of $z \simeq \lambda$ which does not decrease. Thus h does *not* tend to zero uniformly in λ as $z \to \infty$.

In order to get an estimate for S_l valid for large l, we avail ourselves of the inequality[23]

$$|J_{l+1/2}(x_<)H^{(1)}_{l+1/2}(x_>)| < 2[\pi(2l + 1)]^{-1/2} \tag{12.173}$$

for real x, x', and l. This tells us that the Green's function (11.2) has the uniform bound

$$|G_l^+(k;r,r')| \leq \left(\frac{\pi r r'}{2l + 1}\right)^{1/2} \tag{12.174}$$

for real k and l. Substitution in the integral equation (11.7) yields

$$|\psi_l^{(+)}(k,r)| \leq |u_l(kr)| + \left(\frac{\pi r}{2l + 1}\right)^{1/2} \int_0^\infty dr'\, r'^{1/2}|\mathscr{V}(r')|\,|\psi_l^{(+)}(k,r')|$$

and hence

23 This inequality is due to A. Martin (1962 and 1964).

$$\int_0^\infty dr\, r^{1/2}|\mathscr{V}(r)|\,|\psi_l^{(+)}(k,r)| \leq \frac{\int_0^\infty dr\, r^{1/2}|\mathscr{V}(r)|\,|u_l(kr)|}{1-(2l+1)^{-1/2}A} \leq \frac{(2k)^{1/2}A}{1-(2l+1)^{-1/2}A} \tag{12.175}$$

with

$$A = \pi^{1/2}\int_0^\infty dr\, r|\mathscr{V}(r)|$$

since it follows from Bessel's integral[24] that $J_\lambda(x)$ is uniformly bounded by 2 and therefore

$$|u_l(kr)| \leq (2\pi kr)^{1/2} \tag{12.176}$$

for real k and l. Substitution of (12.174) and (12.175) in (11.7) then shows that

$$|\psi_l^{(+)}(k,r) - u_l(kr)| \leq \frac{(2\pi kr)^{1/2}A}{(2l+1)^{1/2}-A}. \tag{12.177}$$

This implies that

$$\left|(2ik)^{-1}\int_0^\infty dr\, u_l(kr)\mathscr{V}(r)[\psi_l^{(+)}(k,r) - u_l(kr)]\right| \leq \frac{\pi^{1/2}A^2}{(2l+1)^{1/2}-A}. \tag{12.178}$$

We now use (12.178) in the integral representation (11.19) for $S_l(k)$,

$$\left|S_l(k) - 1 + 2ik^{-1}\int_0^\infty dr\,\mathscr{V}(r)[u_l(kr)]^2\right| \leq \frac{\pi^{1/2}A^2}{(2l+1)^{1/2}-A}. \tag{12.179}$$

As $l\to\infty$, the right-hand side tends to zero uniformly in k. Consequently for $l \gg 1$ the function S_l is dominated by its Born approximation, and the error term can be made arbitrarily small independently of k,

$$S_l(k) - 1 = -i\pi\int_0^\infty dr\, r\mathscr{V}(r)[J_{l+1/2}(kr)]^2 + O(l^{-1/2}) \tag{12.180}$$

as $l\to\infty$. If we assume that the potential is a superposition of Yukawa functions, (12.22a), then the first Born approximation can be carried out[25] and we have

$$S_l - 1 = -ik^{-1}\int_{a_0}^\infty da\,\rho'(a)Q_l\left(1+\frac{a^2}{2k^2}\right) + O(l^{-1/2}) \tag{12.181}$$

where $Q_l(x)$ is the Legendre function of the second kind. Using the asymptotic form[26] of $Q_l(x)$ as $l\to\infty$, we find that the integral I in (12.181) is

24 See, for example, Whittaker and Watson (1948), p. 362.

25 G. N. Watson (1958), p. 389.

26 A. Erdelyi (1953), vol. 1, p. 136, eq. (44).

$$I = O(l^{-1/2}c^{-2l}) \qquad \text{as } l \to \infty \tag{12.182}$$

with

$$c = \frac{a_0}{2k} + \left(1 + \frac{a_0^2}{4k^2}\right)^{1/2}$$

provided that

$$\int_0^\infty dr\, r\mathscr{V}(r) = \int_{a_0}^\infty da\, \rho'(a)a^{-1}$$

is finite. We conclude that

$$S_l - 1 = O(l^{-1/2}) \tag{12.182a}$$

as $l \to \infty$ uniformly in k for $|k| \geq k_0 > 0$.

Equation (12.81), and its analog for $l > 0$ as discussed in Sec. 12.2, shows that, for each l, $S_l(k)$ approaches unity as $k \to \pm\infty$. Together with (12.182), this implies that

$$\lim_{k \to \pm\infty} S_l(k) = 1 \tag{12.183}$$

uniformly in l. To see this, we choose ε and then find L so that $|S_l - 1| < \varepsilon$ for all $l > L$. The number L is independent of k since (12.182) is uniform. We then find K so that $|S_l(k) - 1| < \varepsilon$ for all $k > K$ and all $l < L$. This is possible because the approach of S_l to 1 for large k must be uniform in $0 \leq l \leq L$ if it is true for each l.

Dependence of the Phase Shift on l. The uniformity of (12.183) implies that if we demand continuity of the phase $2\delta_l(k)$ of $S_l(k)$ as a function of both k and l, for $k > 0$ and $l > -\frac{1}{2}$, then defining it to be zero at $k \to \infty$ for one value of l uniquely determines it to be zero there for all $l > -\frac{1}{2}$. If we include also $k < 0$, then of course we have to relinquish continuity at $k = 0$, owing to the Levinson theorem. It might be thought that it would be simpler to define $\delta_l(0) = 0$ irrespective of the bound states; but this would be inconsistent with the continuity in l and the uniformity of (12.183).

The **S**-matrix element may be considered as a function of *three* variables, l, k, and the potential strength γ. The inequality (12.179), as well as (12.180), shows, as is to be expected, that the approach of $S_l(k)$ to unity is *not* uniform with respect to increasing γ. But in every interval $0 \leq \gamma \leq \gamma_0 < \infty$ it is uniform. Specifically, the phase shift defined by $\delta_0(\infty) = 0$ and continuity in k and l is continuous in γ and approaches zero when $\gamma \to 0$ for all $l > -\frac{1}{2}$ and all $k > 0$. At $k = 0$, of course, it is discontinuous, with jumps occurring every time a new bound state appears.

Let us consider now the phase shift $\delta_l(k)$ for real k and l. Differentiate the radial Schrödinger equation (11.8) for the regular solution φ_l with respect to l, multiply by φ_l, and subtract the equation for φ_l multiplied by $\partial\varphi_l/\partial l$. The result is

$$\frac{d}{dr} W\left(\varphi_l, \frac{\partial \varphi_l}{\partial l}\right) = (2l + 1)r^{-2}\varphi_l^2,$$

W being the wronskian. Integrate this from $r = 0$ to $r = \infty$, and use (12.29). This gives

$$-k^{-1}|\mathfrak{F}_l|^2 \frac{\partial \delta_l'}{\partial l} = (2l + 1) \int_0^\infty dr\, r^{-2}[\varphi_l(k,r)]^2 > 0 \qquad (12.184)$$

if by δ_l' we mean the phase of $\mathfrak{F}_{l-}$, which according to (12.142) and (12.73) is given by

$$\delta_l' = \delta_l - \tfrac{1}{2}\pi l$$

in terms of the phase shift δ_l. Therefore we find that

$$\frac{\partial \delta_l(k)}{\partial l} < \tfrac{1}{2}\pi \qquad (12.185)$$

and consequently

$$\delta_{l+1}(k) - \delta_l(k) < \tfrac{1}{2}\pi. \qquad (12.186)$$

The inequality (12.185) is not surprising from a classical point of view. As we shall see in Chap. 18, the semiclassical approximation of the phase shift is connected to the classical deflection function (5.4) by

$$\Theta = 2\frac{\partial \delta_l}{\partial l}.$$

In that limit (12.185) therefore merely states the fact that $\Theta < \pi$, evident from (5.4).

The phase shifts of a fixed l-value being defined only modulo π, the inequality (12.186) restricts successive δ_l at a fixed k not at all. Only when (12.186) is used together with the demand of continuity of each δ_l as a function of k does it say anything.

We assume now that $\delta_l(\infty) = 0$ for all l. Then (12.186) for $k \to 0$, together with the Levinson theorem (12.156), says that if n_l is the number of bound states of angular momentum l, then

$$n_{l+1} < n_l + \tfrac{1}{2}$$

which implies that

$$n_{l+1} \le n_l. \qquad (12.187)$$

That the number of bound states cannot be increased by raising the repulsive centrifugal potential is of course physically obvious as well as provable in a number of other ways. But here we can obtain a closely connected result in a similar manner by applying (12.186) at a *positive* energy.

Assume that the p wave has a sharp resonance at some energy E_0 in the vicinity of which the s and d waves are smoothly varying functions of E and below which the phase shift is positive. The p-wave phase shift then looks as indicated schematically in Fig. 12.1. Applying (12.186) to $l = 0$ just above E_0, we find that $\delta_0(E_0) > A > \tfrac{1}{2}\pi$. Consequently, either there is an s-wave bound state, or else the s wave must have a resonance at a lower energy. Notice that "resonance" here means the experimentally observable phenomenon of a maximal cross section and not just a (possibly unobservable) pole of S in the complex plane.

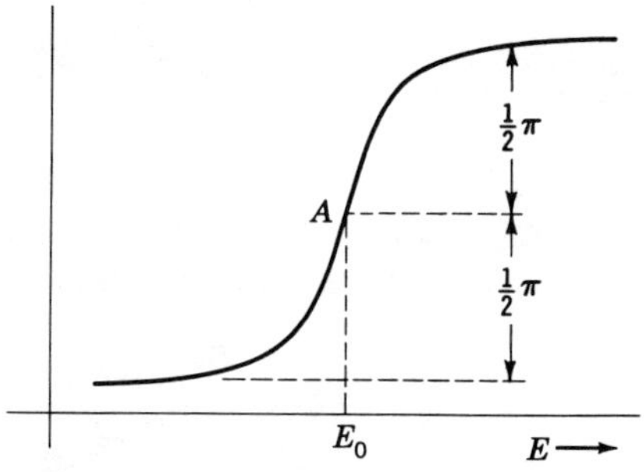

Figure 12.1. Schematic representation of p-wave phase shift as a function of the energy near a sharp resonance.

Motion of **S**-*Matrix Poles.* We may now study the motion of the zeros of $\mathscr{F}_l(k)$ as a function of the angular momentum l. For each fixed value of l the roots of

$$\mathscr{F}_l(k) = 0 \tag{12.188}$$

determine certain values of $k(l)$. If $\operatorname{Im} k > 0$, then we are concerned with the motion of the bound states as functions of l; if $\operatorname{Im} k < 0$, of the virtual states or resonances. [We make sufficiently strong assumptions on the potential so that $S_l(k)$ can be taken into the complex k plane.] Thus (12.188) determines zero trajectories of $\mathscr{F}_l$ or pole trajectories of S_l in the complex k plane or on the E surface.

Take $\operatorname{Im} k > 0$ first. Differentiation of the radial Schrödinger equation with respect to l yields in the standard manner for the wronskian

$$\frac{d}{dr} W\left(\frac{\partial f_l}{\partial l}, f_l\right) = \left(\frac{dk^2}{dl} - \frac{2l+1}{r^2}\right) f_l^2$$

if we "sit" on a zero (12.188) so that k^2 is a function of l. This may be integrated from zero to infinity. Because of (12.188) and the assumption that $\operatorname{Im} k > 0$, the left-hand side gives zero, and we find that

$$\frac{dk^2}{dl} = (2l+1)\frac{\displaystyle\int_0^\infty dr\, f_l^2 r^{-2}}{\displaystyle\int_0^\infty dr\, f_l^2} = (2l+1)\langle r^{-2}\rangle. \tag{12.189}$$

(The expectation value on the right must, of course, be evaluated at the same energy and angular momentum as the left.) Since f_l is real on the positive imaginary axis, the right-hand side of (12.189) is real and *positive*. Thus, as is physically obvious, the energy of a bound state increases, or the binding energy *decreases*, as the angular momentum rises. The stronger centrifugal force necessarily makes the state less tightly bound.

So if we start with a bound-state pole of S_l on the first sheet, it moves toward the right as l increases. Eventually it must pass the threshold $E = 0$, because we know that there exists a maximal value of l beyond which there are no more bound states. What happens to a zero of $\mathscr{F}_l(k)$ as it passes through $k = 0$ from the upper to the lower half plane depends upon the value l_0 of l at which it does so.

Consider the k dependence of the function $\mathscr{F}_l(k)$ for a moment. As is visible in (12.144), it is made up of k^l multiplying $w_l^{(+)}(kr)$, which in turn is multiplied by $\varphi_l(k,r)$. The function $w_l^{(+)}$, being $(kr)^{1/2}$ times a Hankel function, contains k^{-l} and k^{l+1}, both multiplied by analytic functions of k^2; φ_l is an analytic function of k^2. Hence, near $k = 0$, $\mathscr{F}_l$ can be expanded in the form

$$\mathscr{F}_l(k) = a_0 + a_1 k^2 + \cdots + b_1 k^{2l+1} + b_2 k^{2l+3} + \cdots. \tag{12.190}$$

For nonintegral values of l, $\mathscr{F}_l(k)$ always has a branch point at $k = 0$. We put the branch line along the negative imaginary axis. Now, if $\mathscr{F}_l(0) = 0$, we must have $a_0 = 0$. If $l < \frac{1}{2}$, the k^{2l+1} term dominates; if $l > \frac{1}{2}$, the k^2 term does. Hence

$$\mathscr{F}_l(k) - \mathscr{F}_l(0) = \begin{cases} O(k^{2l+1}) & \text{if } l < \frac{1}{2} \\ O(k^2) & \text{if } l > \frac{1}{2} \end{cases} \tag{12.190a}$$

in agreement with (12.53) and (12.153) for integral l.

Suppose that a zero of $\mathscr{F}_l(k)$ traverses $k = 0$ at $l = l_0$. Then it follows from (12.190) that as $l \to l_0-$,

$$k \simeq ai(l_0 - l)^{1/(1+2l_0)} \qquad a > 0 \tag{12.191}$$

if $-\frac{1}{2} < l_0 < \frac{1}{2}$. Unless $l_0 = 0$, this has a branch point at $l = l_0$; if l_0 is irrational, there are infinitely many branches. On the first sheet we have, for $l > l_0$,

$$k \simeq a(l - l_0)^{1/(1+2l_0)} \left(\pm \sin \frac{2\pi l_0}{1 + 2l_0} - i \cos \frac{2\pi l_0}{1 + 2l_0} \right), \tag{12.192}$$

the sign of the real part depending on the direction in which l circumscribed l_0. In addition there are infinitely many other zeros on the other sheets. These we ignore. As the zero of $\mathscr{F}_l$ moves down the imaginary axis and passes $k = 0$ at $l = l_0$, it collides with (infinitely many) other zeros coming from the other sheets. The visible effect is that at $l > l_0$ there are now *two* zeros on the first sheet, symmetric with respect to the imaginary axis. Their paths make the angle $2\pi l_0/(1 + 2l_0)$ with the negative imaginary axis. When $l_0 \to 0$, the cut "fuses" together and the zero moves straight down the imaginary axis. If $l_0 \to \frac{1}{2}$, the two paths become tangent to the real axis and they stay that way when $l_0 > \frac{1}{2}$. For $-\frac{1}{2} < l_0 < 0$, the bound-state zero disappears through the cut and does not appear on the first sheet at all when $l > l_0$.

As the bound-state zero of $\mathscr{F}_l$ moves into the lower half plane, it becomes a complex pair which may get close to the real axis when l is an integer. It then causes a visible resonance of that angular momentum. Because of (12.32) the coefficients a_n in (12.190) must be real. Hence for $l_0 > \frac{1}{2}$ the imaginary part of the zero in $\operatorname{Im} k < 0$ comes from the k^{2l+1} term. We therefore conclude that

$$\operatorname{Im} E_r = \text{const} \times (\operatorname{Re} E_r)^{l_0+1/2} \tag{12.193}$$

if E_r is a low-energy resonance pole of S_l. This is in agreement with (11.55). From the present point of view the reason for the sharpening of the low-energy resonance as l_0 increases is that the pole trajectories osculate the real k axis at $k = 0$ more and more closely the higher the l value at which they cross the threshold to become bound.

As l increases further, the zeros in the lower half plane move along trajectories about which not much can be said in detail. They are, however, subject to restrictions if the potential is analytic ($\sigma = \frac{1}{2}\pi$) (see Sec. 12.1.1) and obeys (12.118). Then the resonance poles of S_l must lie inside (12.165), or

$$|\operatorname{Re} k| < \frac{2|\gamma|M}{2l + 1}. \tag{12.194}$$

Thus each trajectory must approach the negative imaginary k axis as $l \to \infty$. What is more, they all do so uniformly. Of course, the approach is *not* uniform with respect to γ. The larger $|\gamma|$, the larger l must be chosen for all the resonance poles to lie within a given distance of the imaginary k axis.

The picture of the pole trajectories of S_l in the k plane, or on the E surface, as l changes continuously is physically very appealing. It allows us to assign various bound states and resonances to the same "cause" if they lie on the same trajectory. We may start with a deeply bound s state, if there is one, and let l increase. If the S-matrix pole is still on the first sheet when $l = 1$, it causes a p-wave bound state and we may say that these two bound states are really manifestations of the same phenomenon, namely, a single pole moving on the E surface. As l increases, the pole may cause more bound states of higher l values, but eventually it must move through the threshold. Unless it does so exactly at $l = 0$, it then manifests itself as a pair of symmetric complex poles on the second sheet of the E surface, the second member of the pair having come from another sheet through the "kinematic" left-hand cut that is there only for nonintegral values of l. We follow only the lower pole, which is "near" the upper rim of the right-hand cut. If the threshold value of l was greater than $\frac{1}{2}$, the trajectory now osculates the cut. If at the next integral value of l it is still close to the real E axis, it causes a visible resonance of the corresponding angular momentum. This resonance then has the same cause as the bound states, etc. Eventually, the pole must turn around and disappear toward the left, getting closer and closer to the negative real E axis on the second sheet. Each pole of S_l must describe such a trajectory, but of course not all of them start on the first sheet when l is zero, or even near the right-hand cut.

Circuit Relations. Because the Hankel function satisfies the circuit relation[27]

$$H_\lambda^{(1)}(ze^{2\pi i}) = -H_\lambda^{(1)}(z) + 2\cos\pi\lambda\, H_\lambda^{(1)}(ze^{i\pi}) \tag{12.195}$$

the function $f_l(k,r)$ satisfies a similar equation

$$f_l(ke^{2\pi i},r) = f_l(k,r) - 2i\sin\pi l f_l(ke^{-i\pi},r). \tag{12.196}$$

This follows from its integral equation. As a result $\mathfrak{F}_l$ satisfies the same circuit relation, and $\mathscr{F}_l$ obeys the equation

$$\mathscr{F}_{l+}(ke^{2\pi i}) = e^{2\pi il}\mathscr{F}_{l+}(k) + (1 - e^{2\pi il})\mathscr{F}_{l-}(k) \tag{12.197}$$

provided that there are no intervening cuts. For superpositions of Yukawa potentials, (12.22a), for example, the magnitude of k must be less than $\frac{1}{2}a_0$. The resulting circuit relation for S_l is

$$S_l(ke^{i\pi}) = e^{2\pi il}S_l^{-1}(k) + 1 - e^{2\pi il}. \tag{12.198}$$

When l is an integer, this goes over into (12.74).

Let us now define the **K**-matrix element for general values of l, instead of by (12.162), by

$$\mathfrak{K}_l(k) = e^{i\pi(l+1/2)}k^{-2l-1}\frac{S_l(k) - 1}{S_l(k) + e^{2\pi il}} \tag{12.199}$$

so that

$$S_l(k) = \frac{1 - i\mathfrak{K}_l k^{2l+1}e^{i\pi l}}{1 + i\mathfrak{K}_l k^{2l+1}e^{-i\pi l}}. \tag{12.199a}$$

It then follows from (12.199) that $\mathfrak{K}_l$ satisfies the simple equation

$$\mathfrak{K}_l(ke^{i\pi}) = \mathfrak{K}_l(k). \tag{12.200}$$

Thus, not only has it no right-hand cut as a function of E; it also has no left-hand kinematical cut for general values of l. The generalization of (12.164) to general values of l is

$$S_l = 1 - k^{2l+1}e^{2\pi il}\mathfrak{M}_l \tag{12.201}$$

so that

$$\mathfrak{M}_l^{-1} = \tfrac{1}{2}k^{2l+1} - \tfrac{1}{2}ie^{i\pi l}\mathfrak{K}_l^{-1}. \tag{12.202}$$

α Trajectories. In order to understand the convergence properties of the Born series for S_l at a given energy as a function of l, we may consider the eigenvalues α_l of the kernel (12.149) of the radial Lippmann–Schwinger equation as functions of l. They describe certain trajectories in the complex α plane. At an l value for which all α's are inside the unit circle, the Born series for S_l converges. If at the given energy no α trajectories lead outside the unit circle, then the Born series for each S_l converges. Now according

27 From A. Erdelyi (1953), vol. 2, p. 80.

to (12.170) there is always a finite angular momentum

$$L = Mk^{-1} - \tfrac{1}{2} \tag{12.203}$$

beyond which all α_l are inside the unit circle, provided that the potential is analytic ($\sigma = \frac{1}{2}\pi$) and obeys (12.118). If

$$k > 2M \tag{12.204}$$

then no α trajectories can lead outside the unit circle and hence the Born series for the full scattering amplitude converges. This is because the spectrum of the kernel of the full three-dimensional Lippmann–Schwinger equation consists of the union of the spectra of all the radial kernels.

12.4 Singular Potentials

12.4.1 The Difficulties. We have so far restricted the discussion to potentials which obey (12.9). In a vaguer manner of speaking this means that they are less singular at the origin than r^{-2}. From the point of view of the radial Schrödinger equation the r^{-2} behavior at $r = 0$ clearly divides two classes with quite different properties. So long as $r^2\mathscr{V}$ is bounded, $r = 0$ is a *regular singular* point of the second-order differential equation (for $l \neq 0$). When $r^2\mathscr{V}$ is not bounded as $r \to 0$ the origin is an *irregular singular* point. In the latter case the situation is quite analogous to what we already know to be true at $r \to \infty$. Let us summarize the situation there.

Replacement of r by $z = r^{-1}$ as a variable in the radial Schrödinger equation produces an equation of the same form but with k^2 multiplied by z^{-4}. Thus $r = \infty$ is an irregular singular point. As a consequence all solutions of the equation have an essential singularity there. What is more important for us is that the behavior of the solutions depends crucially on the algebraic sign of the factor k^2 of the singular term in the equation. If k^2 is negative, one solution strongly vanishes (it is regular) and the other blows up (it is irregular); if k^2 is positive, both solutions oscillate and no distinction between regular and irregular solutions is possible.

The same is true at $r = 0$ if the potential is more singular there than r^{-2}. That a behavior of the solution as a power of r is then impossible follows immediately upon insertion of the usual expansion

$$\psi = \sum_n a_n r^{\sigma+n}$$

in the equation. We search for the indicial equation. Since the second derivative lowers the first power by 2, there is no matching power for the potential term. The same can be seen perhaps more clearly in the equation obtained if the radial wave function is written in the form[28]

28 This equation defines G. It puts no restriction on ψ. It is the form we shall use again in the context of the WKB approximation [see (18.2) and (18.3)].

$$\psi = G^{-1/4} \exp\left(i \int^r dr' \, G^{1/2}\right) \tag{12.205}$$

which is

$$G + \frac{1}{4}\frac{G''}{G} - \frac{5}{16}\left(\frac{G'}{G}\right)^2 = F \tag{12.206}$$

with

$$F = k^2 - \mathscr{V} - l(l+1)r^{-2}.$$

If G goes as r^σ, then the second and third terms on the left of (12.206) are of order r^{-2}. Hence, when $\mathscr{V}$ is more singular than r^{-2}, it must be matched on the left by G. In other words, G must behave like F, that is, like

$$G \underset{r\to 0}{\simeq} -\mathscr{V}. \tag{12.207}$$

The same is true if $\mathscr{V}$ is more singular than r^{-2} but does not behave as a power.

As $r \to 0$ the integral in the exponential of (12.205) now diverges. It is real if $\mathscr{V}$ is negative and imaginary if $\mathscr{V}$ is positive. Hence, if the potential is attractive, ψ oscillates infinitely rapidly and the outside factor $G^{-1/4}$ makes it tend to zero. This must be true for *both* solutions of the equation; they are one another's complex conjugates. But if the potential is repulsive near the origin, then the integral in (12.205) tends to $\pm i\infty$ and one solution strongly tends to zero while the other rapidly blows up. The relatively weakly vanishing outside factor cannot alter this.

A strongly singular repulsive potential therefore produces no difficulties at all.[29] A strongly singular attractive potential, on the other hand, leads to two linearly independent solutions of the radial Schrödinger equation, each of which, from a physical point of view, is equally acceptable. Both are square-integrable near the origin, and hence there is no physical basis for preferring one to the other. We may combine these solutions into real functions. The arbitrariness then lies in the *phase* of the infinitely rapidly oscillating solution, or the choice of r_0, in

$$\psi = G^{-1/4} \cos \int_{r_0}^{r} dr' \, G^{1/2}(r'). \tag{12.205a}$$

The absence of a criterion of regularity at $r = 0$ makes it impossible to find a spectrum. We cannot find either bound states or phase shifts. Fixing the phase, or r_0, amounts to the assignment of a boundary condition at $r = 0$. A spectrum[30] (and a phase shift) is then determined, and so is a concomitant set of orthogonal wave functions.[31] But there is no *physical*

29 Other than the absence of a power behavior of the wave function. This is not really important.

30 The spectrum is not bounded below. This is an important difference, but not relevant from the point of view of scattering theory.

31 See K. M. Case (1950).

preference of one phase to another, and hence no physical way of selecting one spectrum (and phase shift).

The situation is quite different when discussed from the point of view of the full three-dimensional Schrödinger equation. It is then analogous to that of the second solution of the s-wave equation with a well-behaved potential, which was discussed at the beginning of Sec. 12.1.1. Remember the crux of the matter there: Although a (nonzero) constant obeys the radial equation

$$\nabla\psi'' = 0$$

the corresponding three-dimensional wave function cr^{-1} does not satisfy the three-dimensional equation

$$\nabla^2 cr^{-1} \neq 0.$$

Instead

$$\nabla^2 cr^{-1} = -4\pi c\delta(\mathbf{r}).$$

A similar result follows for any function of r^{-1},

$$\nabla^2 f(r^{-1}) = r^{-4}f''(r^{-1}) - 4\pi\delta(\mathbf{r})f'(r^{-1}). \tag{12.208}$$

The second term on the right is absent in the radial equation, for which $r = 0$ is an end point to be approached continuously. In other words, whenever r^{-1} times the solution of a *radial* Schrödinger equation is a function of r^{-1} whose derivative does not vanish at $r = 0$[32], it *does not* solve the corresponding three-dimensional Schrödinger equation. This fact alone is sufficient to make all the irregular solutions of the radial equations unacceptable, without recourse to the requirement of square-integrability. In the case of potentials that at $r = 0$ are attractive and as singular as r^{-4} or more so, it means that *no* solutions of the radial equations[33] solve the full three-dimensional Schrödinger equation. The latter, with such potentials in it, simply has no solution.[34]

It should be noted that these difficulties with a singular attractive potential have no classical analog. We saw in Sec. 5.5 that if the potential is more singular than r^{-2} at the origin, and attractive, then a classical particle falls through the origin with an infinite velocity. But that does not necessarily present it from emerging at a finite angle. Only at one particular angular momentum (or impact parameter) does orbiting occur. This *may* produce an infinite cross section, but if it does, it will do so even if the potential is

32 In other words, if $r^2 d(r^{-1}\psi)/dr \nrightarrow 0$ as $r \to 0$, for a radial function ψ.

33 After multiplication by the appropriate angle function.

34 Strictly speaking we may be jumping to a conclusion here. The radial equation may be appropriately altered by the inclusion of a δ-function term in such a way that a solution of this unorthodox radial equation would solve the correct three-dimensional equation. It does not look, however, as though the radial equation thus changed has a solution, except in the case of a well-behaved potential when the regular solution makes the δ-function term vanish. This point could bear some mathematical scrutiny. Also note that for potentials between r^{-2} and r^{-4}, whether the δ-function term is effective or not depends on our choice of test functions.

cut off near the origin. In the absence of a new physical idea it appears therefore that we have here a failure of quantum mechanics. It seems unable to deal with a well-defined situation which classical mechanics can handle. The fact that nonrelativistic quantum mechanics with local interactions loses its physical relevance in the inner region of an attractive singular force saves the situation from being *practically* bother-some. But that makes it no less unsatisfactory as a matter of principle.[34a]

The limiting case of a potential that goes exactly as r^{-2} near the origin requires special consideration. It then interferes with the centrifugal term. Suppose that

$$\lim_{r \to 0} r^2 \mathscr{V}(r) = a. \tag{12.209}$$

The ar^{-2} may be combined with the centrifugal $l(l+1)r^{-2}$ into a single term $(b^2 - \frac{1}{4})r^{-2}$, with

$$b^2 = a + (l + \tfrac{1}{2})^2.$$

We know that two linearly independent solutions exist which at the origin go as $r^{1/2 \pm b}$. If b^2 is positive, one of these is regular and the other irregular.[35] This will be true whenever $a > 0$, that is, when $\mathscr{V}$ is repulsive near the origin, or when $a < 0$ if nevertheless $-a < (l + \frac{1}{2})^2$, i.e., when $\mathscr{V}$ is sufficiently weakly attractive. However, if $\mathscr{V}$ is strongly attractive, in the sense that $-a > (l + \frac{1}{2})^2$, then b is imaginary and the two solutions behave as $r^{1/2} \exp(\pm i|b| \ln r)$. Again they both oscillate, and no physical criterion exists by which one is more acceptable than the other. What is more, both solutions of the radial equation actually solve the three-dimensional Schrödinger equation.

In this special instance the difficulty has a classical analog. We saw in Sec. 5.6[36] that if the potential has the property (12.209) and $-a > J^2$, then the classical particle spirals toward the origin and the deflection angle is infinite. A cross section could then be defined only for a "hollow" beam which had all angular momenta $J < \sqrt{-a}$ missing. The same is true in quantum mechanics if we replace J by $l + \frac{1}{2}$.

12.4.2 Singular Repulsive Potentials. If the potential is more singular near the origin than r^{-2}, but positive, then, we have seen, a regular solution of the radial Schrödinger equation exists and it solves the three-dimensional one. Nevertheless the usual techniques of obtaining it and many of its properties break down. The integral equation (12.4), for example, though still true, becomes useless. It cannot be solved by iteration, nor is it even a Fredholm equation. The properties of the solution, of the Jost function,

34a Both in classical and in quantum mechanics the region r^{-2} to r^{-4} is ambiguous.

35 If $b \leq \frac{1}{2}$, the second solution is irregular only in the sense of being dominant over the other solution. In this case the irregular solution does solve the three-dimensional equation, and it is square-integrable.

36 In the comparison it must be remembered that $\mathscr{V} = 2mV$.

and of the **S** matrix as functions of k are consequently quite changed. Furthermore, the boundary condition depends upon the potential. We saw in the last section that the behavior of the two solutions of the radial equation behave near the origin as

$$\psi \simeq \mathscr{V}^{-1/4} \exp\left(\pm \int^r dr' \, \mathscr{V}^{1/2}\right). \tag{12.210}$$

A relatively simple procedure is therefore the following:

We first solve the s-wave equation for zero energy,

$$\psi_0'' = \mathscr{V}\psi_0. \tag{12.211}$$

Actually, this need be done only for $r \leq r_0$, where r_0 is any positive number. Suppose that we know both solutions of (12.211). We then form the Green's function

$$G(r,r') = \begin{cases} \psi_{01}(r)\psi_{02}(r') - \psi_{01}(r')\psi_{02}(r) & r' < r \\ 0 & r' > r \end{cases} \tag{12.212}$$

if ψ_{01} and ψ_{02} are regular and irregular solutions of (12.211), defined so that their wronskian equals unity,

$$\psi_{01}'(r)\psi_{02}(r) - \psi_{01}(r)\psi_{02}'(r) = 1. \tag{12.213}$$

The regular solution of the lth angular-momentum radial equation for arbitrary energy can then be defined by the integral equation

$$\varphi_l(k,r) = \psi_{01}(r) + \int_0^r dr' \, G(r,r')[l(l+1)r'^{-2} - k^2]\varphi_l(k,r') \tag{12.214}$$

for all $r \leq r_0$. In contrast to the nonsingular case this can be done now because, according to (12.210), the product of a regular and an irregular solution goes as

$$\psi_{01}\psi_{02} \simeq \text{const} \times |\mathscr{V}|^{-1/2}$$

near $r = 0$. Therefore the integral in (12.214) converges if $\mathscr{V}$ is more singular than $r^{-2-\varepsilon}$, $\varepsilon > 0$, near $r = 0$.[37] The solution of (12.214) can be obtained by iteration, and it serves the same purpose as $\varphi_l(k,r)$ in the nonsingular case. It is clearly an entire function of k^2, as well as of $(l + \frac{1}{2})^2$. The only ambiguity in it is that of a constant factor in ψ_{01}. Since this is independent of both l and k, it is of no consequence. It will cancel in the **S** matrix.

The irregular solutions $f_{l\pm}(k,r)$ of the radial Schrödinger equation can be defined as in the nonsingular case. This is so because the integral equation (12.138) is independent of the behavior of $\mathscr{V}$ for small r. In fact we need to solve it only for $r \geq r_0$. Once $\varphi_l(k,r)$ and $f_{l\pm}(k,r)$ are known, the Jost functions $\mathfrak{F}_{l\pm}(k)$ and $\mathscr{F}_{l\pm}(k)$ are calculated as before, by the wronskian of φ_l and $f_{l\pm}$, (12.28). This may be done at r_0. The integral representations (12.143) and (12.144), of course, are no longer true, because the integral

37 This criterion can be strengthened. But it is not worth our while to do so.

equations for φ_l and f_l are now quite different from one another. The **S** matrix is expressed in terms of the Jost functions, as before.

It is clear from this construction that all previous statements concerning analyticity of the Jost function and of the **S** matrix in the finite k plane (or on the finite E surface) are still true in the singular case. What is altered is the behavior for large k, and the expansibility of the Jost function in a power series in the coupling constant. The first is changed because $\varphi_l(k,r)$ does not approach its zero-order value as $k \to \infty$ now. Equation (12.214) shows that its large energy behavior is hard to disentangle and depends on the potential. The phase shift does not tend to a multiple of π as the energy increases.[38] The dependence on the potential strength γ is changed because ψ_{01} cannot be expanded in a series in powers of γ without destroying the regularity of the wave function at the origin, which is essential to the convergence of the integral in (12.214). If we insist on a power series in γ, the individual terms in it must be infinite.

On the other hand, the strong coupling limit of the wave function will be seen in Sec. 18.2.1 to be the WKB form. The latter has the correct behavior near $r = 0$ even when the potential is singular. This follows from a comparison of (18.2) with (12.210). Hence the WKB phase shift remains valid as the strong coupling limit.

Another Representation. The approach of Eqs. (12.65) to (12.70) is particularly useful in case the potential is singular, because (12.69) expresses the phase shift directly in terms of $|f(k,r)|$, which may be obtained from (12.16) (by iteration) even in the singular case or from (12.70). A slight modification is obtained by defining

$$S(k,r) \equiv e^{2ik\delta(k,r)}$$

so that according to (12.65) and (12.68)

$$S' = -2ikf^{-2}.$$

This is solved by

$$S(k,r) = e^{-2ikr} + 2ik \int_r^\infty dr' \left[f(k,r')^{-2} - e^{-2ikr'} \right].$$

Therefore the s-wave **S**-matrix element is given by

$$S(k) = 1 + 2ik \int_0^\infty dr \left[f(k,r)^{-2} - e^{-2ikr} \right]. \tag{12.215}$$

12.4.3 An Example. As an example we consider the family of highly singular potentials

$$\mathscr{V}(r) = a^2(b^2 + c^2 e^{\alpha/r}) r^{-4} \qquad \alpha > 0.$$

38 As in the repulsive-core case, which will be discussed in Sec. 14.2, the phase shift will generally tend to $-\infty$. A repulsive singular potential will therefore cause an infinite number of peaks in the partial cross sections. But since the phase shift decreases, none of them are resonances.

The change of variables $r = 1/z$ puts (12.211) into the form

$$\frac{d^2}{dz^2}(z\psi_0) = a^2(e^{\alpha z}c^2 + b^2)(z\psi_0).$$

Now set

$$x = \frac{2iac}{\alpha}e^{\alpha z/2} = \frac{2iac}{\alpha}e^{\alpha/2r} \qquad z\psi = h.$$

Then we get Bessel's equation

$$\frac{d^2}{dx^2}h + x^{-1}\frac{d}{dx}h + \left[1 - \left(\frac{2ab}{\alpha}\right)^2 x^{-2}\right]h = 0.$$

The solutions are therefore

$$\psi_{01,2} = rH_\nu^{(1),(2)}(x) \qquad \nu = \frac{2ab}{\alpha}.$$

As $r \to 0$, $x \to +i\infty$; hence the Hankel function of the first kind is regular at $r = 0$ and that of the second kind irregular.

The asymptotic behavior of the regular solution as $r \to \infty$ is obtained by expansion,

$$H_\nu^{(1)}\left[\frac{2iac}{\alpha}\left(1 + \frac{\alpha}{2r} + \cdots\right)\right] = H_\nu^{(1)}\left(\frac{2iac}{\alpha}\right) + \frac{ia}{r}H_\nu^{(1)\prime}\left(\frac{2iac}{\alpha}\right) + \cdots$$

$$\psi_{01} \simeq rH_\nu^{(1)}\left(\frac{2iac}{\alpha}\right) + iaH_\nu^{(1)\prime}\left(\frac{2iac}{\alpha}\right) + \cdots$$

which means that the s-wave scattering length is given by[39]

$$\frac{-iaH_\nu^{(1)\prime}(2iac/\alpha)}{H^{(1)}(2iac/\alpha)}.$$

If a is imaginary, we are dealing with the case of an *attractive* potential. The variable x is then real (and may be chosen positive), and we get the highly oscillatory behavior of ψ_0 near $r = 0$ that we expect. Both solutions ψ_0 tend to zero equally rapidly and are therefore regular from the point of view of the radial equation. But the corresponding three-dimensional wave functions are $H_\nu^{(1),(2)}(x)$, and (12.208) shows that they do not solve the Schrödinger equation.

If $c = 0$, the potential is proportional to r^{-4}. The radial Schrödinger equation for arbitrary l and k may then be reduced to the Mathieu equation and solved in terms of Mathieu functions. For this we refer to the literature.[40] Both the classical and the quantum-mechanical cross sections are inversely proportional to the velocity of the particle.

39 See (11.40).

40 See Vogt and Wannier (1954) and R. M. Spector (1964).

Notes and References

12.1.1 For a discussion of the boundary condition for the radial wave function of $l = 0$, see Armstrong and Power (1963). The argument given in this section is essentially identical with that of P. A. M. Dirac (1947).

The procedure of defining regular and irregular functions by boundary conditions (12.2) and (12.15) and then proving their properties by solving the integral equations by iteration is due to R. Jost (1947) and N. Levinson (1949b). For a study of other boundary conditions, see H. E. Moses (1978). The asymptotic form of radial wave functions for cut-off potentials is examined in W. J. Romo (1979); L. Trlifaj (1980) studies Jost solutions for Yukawa potentials.

The argument below (12.21) that leads to an enlarged domain of analyticity of $f(k,r)$, is due to T. Regge (1958b). The method used for *analytic potentials* was invented by T. Regge (1959 and 1960); see also Bottino, Longoni, and Regge (1962). It has been very much generalized and is now known as *dilation analyticity*; see Balslev and Combes (1971); E. Balslev (1972); B. Simon (1972, 1973, 1978); I. W. Herbst (1979).

12.1.2 The original paper in which Jost first used the function $\mathscr{F}(k)$ is that of 1947. For generalizations, see H. E. Moses (1964); M. Petráš (1964a, b and 1965). See also De Alfaro, Regge, and Rossetti (1962). For a discussion of the Hilbert transform, see, for example, E. C. Titchmarsh (1939).

A paper relevant to the discussion of the Fredholm determinant is I. Manning (1964).

Proofs of the absence of positive-energy bound states under various assumptions were given by A. Ramm (1963) and 1965a); J. Weidmann (1964); B. Simon (1969b); F. Zirilli (1976); Ramm and Taylor (1980).

A dispersion relation for $\mathscr{F}$ such as (12.55) was formulated by Giambiagi and Kibble (1959).

The expression (12.64) of $\mathscr{F}$ in terms of δ is slightly more explicit than that given by Jost and Kohn (1952a). For a more general discussion of related procedures, see R. Omnes (1958).

The procedure following (12.65) is sometimes called the *phase-amplitude method.* It was first used by W. E. Milne (1930). See also H. A. Wilson (1930); L. A. Young (1931); J. A. Wheeler (1937b).

For another approach using nonlinear equations, see P. O. Olsson (1952); S. Franchetti (1957); F. Calogero (1963 and 1967); R. F. Dashen (1963); Levy and Keller (1963); Calogero and Ravenhall (1964); Chadan and Guennégués (1964); Matese and Henry (1972); U. Landman (1972); Tikochinsky and Koren (1973). For generalizations to spin and coupled channels, where the advantages for numerical calculations may be considerable, see G. J. Kynch (1952); C. Zemach (1964a); A. Degasperis (1964); Cox and Perlmutter (1965); J. R. Cox (1965); V. V. Babikov (1965); M. P. Faifman (1977).

The Jost function of the harmonic oscillator potential was discussed by Allessandrini and Giambiagi (1963); for momentum dependent potentials, by C. S. Warke (1972); for nonlocal potentials, by Warke and Bhaduri (1971); Mulligan and Qadri (1981). M. G. Fuda (1970) and Fuda and Whiting (1973) introduced the notion of an off-shell Jost function; see also Warke and Srivastava (1977 and 1978); B. Talukdar et al. (1978); H. van Haeringen (1978a).

The Jost function and Fredholm methods have been used extensively for numerical calculations by W. P. Reinhardt (1970); Reinhardt and Szabo (1970); Adelman and Reinhardt (1972); Heller and Reinhardt (1972); W. P. Reinhardt et al. (1972); Holt and Santoso (1972); E. J. Heller et al. (1973).

12.1.3 The result (12.85) for the residue of an S-matrix element at a bound-state pole was discovered by H. A. Kramers (1938), as quoted by Jost (1947), and independently by C. Møller (1946) and by W. Heisenberg (1946). The more detailed (12.84) was shown by C. Møller (1946). See also D. ter Haar (1946); R. Jost (1946).

E. M. Ferreira et al. (1968) studied S-matrix properties for velocity-dependent potentials; D. Masson (1967) examined its analyticity in the potential strength; and W. M. Frank (1971) and Frank and McLaughlin (1971) analysed it as a functional of the potential. For phase-shift calculation in momentum space, see M. Bolsterli (1974). See also K. W. McVoy (1968) for a discussion of phase shifts at resonances.

The inequality (12.90) was first given by E. P. Wigner (1955). See also G. Lüders (1955); A. Martin (1956); K. Chadan (1965).

The N/D method is due to Chew and Mandelstam (1960). A dispersion relation such as (12.93) was written down, for example, by E. Corinaldesi (1956).

The Levinson theorem was first proved by N. Levinson (1949b). The proof given here follows his. For a more general proof, see J. Jauch (1957). See also J. C. Polkinghorne (1958); A. Martin (1958); M. Ida (1959); E. Kazes (1959); Konisi and Ogimoto (1959); Vaughn, Aaron, and Amado (1961); J. R. Cox (1962); R. L. Warnock (1963); Ghirardi, Pauri, and Rimini (1963); M. Ciafaloni (1963); J. M. Charap (1965); P. Swan (1968); Lopez and Massidda (1970); P. Beregi et al. (1973); T. Dreyfus (1976); Bollé and Smeesters (1977); D. Bollé (1979b). It was generalized to noncentral potentials by T. Dreyfus (1976 and 1978); R. G. Newton (1977a); M. Wollenberg (1977); Osborn and Bollé (1977); L. Guillope (1981); and to nonlocal potentials by T. Dreyfuss (1976); R. G. Newton (1977b). G. -J. Ni (1979) proved it for the Dirac equation, and T. A. Osborn et al. (1980) derived its analog for classical particle scattering. For generalizations to more than two particles, see J. A. Wright (1965); P. Beregi et al. (1973); R. G. Newton (1975); Glöckle and LeTourneux (1976).

12.1.4 The limitation (12.96) on the number of bound states of a given angular momentum, and its generalization to other l values, (12.157), is due to V. Bargmann (1952). The proof given here follows J. Schwinger (1961). See also L. D. Faddeev (1957); M. Sh. Birman (1959, 1961a, b); J. H. E. Cohn (1965); Ghirardi and Rimini (1965a); F. Calogero (1965a, b, c, 1967); Calogero and Cozenza (1966); Fonda and Ghirardi (1966); A. Ronveaux (1968a); B. Simon (1969 and 1976); N. Seto (1974); K. Chadan (1976); E. Elizalde (1976); V. Glaser et al. (1976); H. Van Haeringen et al. (1977); H. Grosse (1980); Y. Colin de Verdiere (1981). Generalizations to three-particle systems were given by D. R. Yafaev (1972 and 1976); G. M. Zhislin (1972).

A detailed discussion of the trajectories of the eigenvalues α as functions of k, for a fixed l value, is given by Nataf and Cornille (1963).

The representation (12.97) and the arguments based upon it are due to Ciafaloni and Menotti (1965).

The remarkable fact that it is not the resonance pole which becomes responsible for the bound state, but the "other" one, was pointed out by a number of workers whose papers are listed in the Notes and References to Sec. 17.1.

That each S-matrix element for a potential of finite range has infinitely many poles was first shown by J. Humblet (1952) and independently by H. Rollnik (1956). The more general proof given here, as well as the subsequent arguments concerning the distribution of poles if the range of the potential is finite, follows T. Regge (1958a). For an extension to potentials which decrease faster than exponentials, but do not vanish identically, see L. Sartori (1963).

The factorization (12.104) was first proved by T. Regge (1958a).

The principal reference for properties of entire analytic functions is R. P. Boas (1954).

The product representation (12.107) of $S(k)$ was first formulated by N. Hu (1948) without proof. It was first proved under the conditions of this section by T. Regge (1958a).

The Mittag-Leffler expansion of $S(k)$ was also formulated first without proof by Hu (1948). It was proved first by Humblet (1952). The argument given is a somewhat simplified and less rigorous version of that of Humblet.

The arguments leading to the limitations (12.119) and (12.120) on the pole positions of $S(k)$ for analytic potentials are due to T. Regge (1960) and Bottino and Longoni (1962).

For further work on resonances see S. Tani (1968); C. A. Nelson et al. (1971); A. Jensen (1977).

12.1.5 The completeness proof given is the type presented by E. C. Titchmarsh (1946). In particular it follows Jost and Kohn (1952a). For completeness proofs for *complex* potentials, see M. A. Naimark (1954); V. E. Ljance (1964); M. G. Gasymov (1965); Bertero and Dillon (1971); Nikolayev and Olkhovsky (1973); see also Kok and van Haeringen (1981).

12.2 The inequalities (12.136) are due to N. Levinson (1949b). Inequality (12.136a) is derived in the same way, and was first given by R. G. Newton (1955).

For remarks concerning the inequality (12.155), see those about (12.85) in the Notes and References to Sec. 12.1.3. References for (12.156) and (12.157) are given in the Notes and References to Secs. 12.1.3 and 12.1.4.

The first rigorous demonstration of (12.161) under the assumption (12.160) was given by D. S. Carter (1952). The representation (12.165) was given by R. G. Newton (1959) and by Matthews and Salam (1959). For phase-shift behavior under weaker conditions, see P. S. Ganas (1972).

The references given for (12.119) and (12.120) apply equally to (12.167).

Formulas for the differences between phase shifts of different angular momenta can be found in T. Tietz (1963); Calogero and Fradkin (1966); Fradkin and Calogero (1966); see also Goodmanson and Taylor (1973).

12.3 The idea of considering the angular momentum as a continuous, even complex parameter is due to T. Regge (1959). For a discussion of trajectories of S-matrix poles in the complex k-plane, see P. Sanz et al. (1981).

The most useful general reference for the properties of many transcendental functions is A. Erdelyi (1953).

The inequalities (12.185) and (12.186) for the phase shifts as functions of l are due to T. Regge (1959).

12.4 The first general discussion of singular potentials in quantum mechanics was given by K. M. Case (1950). See also Vogt and Wannier (1954), where the radial equation for an r^{-4} potential is solved and both the classical and the quantum-mechanical case are discussed in detail. An independent later solution of the Schrödinger equation with a repulsive r^{-4} potential can be found in R. M. Spector (1964); see also N. A. W. Holzwarth (1973); W. Buhring (1974).

Singular potentials have received a considerable amount of attention, primarily because of their similarity to certain aspects of particle interactions of relativistic field theory. We list here the following references: F. L. Scarf (1958); T. Tietz (1959); Predazzi and Regge (1962); N. Limić (1962, 1963, and 1965); Bastai et al. (1963); Bertocchi, Fubini, and Furlan (1964 and 1965b); H. Cornille (1964 and 1965); Arbuzov, Filippov, and Krustalev (1964); Cornille and Predazzi (1964 and 1965a and b); Khuri and Pais (1964); Tiktopoulos and Treiman (1964); Pais and Wu (1964a and b); T. T. Wu (1964); Giffon and Predazzi (1964); F. Calogero (1964, 1965c and d); K. Meetz (1964);

Charap and Dombey (1964); Aly, Riazuddin, and Zimerman (1964 and 1965); Calogero and Cassandro (1964 and 1965); Del Giudice and Galzenati (1965); Fubini and Stroffolini (1965); N. Dombey (1965); Ahmed and Fairlie (1965); R. J. Jabbur (1965); Cornille, Burdet, and Giffon (1965); G. Tiktopolous (1965); Arbuzov and Filippov (1965); Jakšić and Limić (1966); Aly and Muller (1966); Calogero and DeStefano (1966); W. A. Gale (1966 and 1967); Kupsch and Sandhas (1966); R. M. Spector (1966); W. Hunziker (1967); Paliov and Rosendorff (1967); H. Behncke (1968); Dombey and Jones (1968); Frank and Land (1970); R. F. Bishop (1973); T. Dolizanszky (1974); Baeteman and Chadan (1976); F. F. Naundorf (1978); Behncke and Focke (1978); Fröman and Thylwe (1979).

For discussions of potentials that are singular in the sense of having a long range, see Ross and Taylor (1969); Freeman and Nuttall (1973); Prugovečki and Zorbas (1973b); S. S. Wald et al. (1975); Combescure and Ginibre (1976); Masson and Prugovečki (1976); M. Schechter (1977); Dollard and Friedman (1978); Ben-Artzi and Devinatz (1979); Chadan and Martin (1979); H. van Haeringen (1979a); M. Arai (1980); A. Devinatz (1980); Kitada and Yajima (1981); see also references for Sec. 14.6.

Equation (12.215) is due to Calogero (1965d), and so is the suggestion of its usefulness in the singular case.

General References

The following general surveys of the material in this chapter will be found useful: R. G. Newton (1960); H. M. Nussenzveig (1962); T. Regge (1963); A. Martin (1964).

For other approaches, see O. Hellmann (1960); F. Calogero (1963); R. F. Dashen (1963).

Discussions of the partial wave amplitude as a function of the potential strength parameter γ in the complex γ plane are given by Konisi and Ogimoto (1963) and by Lomsadze, Kimich, and Shuba (1965); and for the strong-coupling limit by W. M. Frank (1967 and 1968); see also H. Narnhofer (1975).

For studies of complex potentials, see A. Martin (1962); Kok and van Haeringen (1981a).

Problems

1 Assuming that for $r < r_0$ the potential is given by $\mathscr{V} = \gamma r^m$, $m \geq -2$, expand $\varphi_l(k,r)$ in a series about $r = 0$, and find the recursion relations for the coefficients. Do the nearest possible analog for an irregular solution.

2 Assuming that for $r > r_0$ the potential is given by $\mathscr{V} = \gamma r^{-m}$, $m > 2$, expand $f_l(k,r)e^{-ikr}$ in a series about $r = \infty$, and find the recursion relations for the coefficients.

3 Suppose that the potential is complex: $\mathscr{V} = U(r) + iW(r)$, where U and W both satisfy (12.9) and (12.21). How do the properties of φ, f, the Jost function, and the **S** matrix differ from those for a real potential? List as many general properties as you can which are unchanged and as many as you can which are altered.

4 Suppose that the kernel K of the Lippmann-Schwinger equation has a finite number of eigenvalues α_i other than zero. What is the nature of the interaction?

5 Suppose that the potential is such that

$$\int^{\infty} dr\, e^{\alpha r}|V(r)|$$

diverges for $\alpha > a$ and converges for $\alpha < a$, $a > 0$. What conclusions can you draw concerning the singularities of the **S** matrix (say, for $l = 0$) in the complex plane? Answer separately for $|\mathrm{Im}\, k| \leq \frac{1}{2}a$ and $|\mathrm{Im}\, k| \geq \frac{1}{2}a$, and distinguish between the kinds of singularities that *cannot* occur, *may* occur, and *must* occur. If possible, give examples.

6 Generalize the arguments of (12.88) to (12.90) to the p wave.

7 The regular solution $\varphi(k,r)$ of the s-wave radial equation satisfies the Volterra equation

$$\varphi(k,r) = r + \int_0^r dr'\,(r - r')[\mathscr{V}(r') - k^2]\varphi(k,r').$$

Discuss the properties of φ in terms of this equation, and give its advantages and disadvantages compared with (12.4). Write down its analog for $l \neq 0$.

8 The irregular solution $f_l(k,r)$ satisfies the equation

$$f_l(k,r) = e^{ikr} - k^{-1}\int_r^{\infty} dr' \sin k(r - r')[\mathscr{V}(r') + l(l+1)r'^{-2}]f_l(k,r').$$

Discuss the advantages and disadvantages of this equation compared with (12.138).

9 Construct the solutions of the s-wave Schrödinger equation, the Jost function, and the **S** matrix, for the nonlocal interaction

$$(\mathbf{r}|H'|\mathbf{r}') = \delta_\Omega(\hat{\mathbf{r}},\hat{\mathbf{r}}')\gamma A(r)A^*(r').$$

Discuss their properties.

10 Do the same for

$$(\mathbf{r}|H'|\mathbf{r}') = \delta_\Omega(\hat{\mathbf{r}},\hat{\mathbf{r}}')\sum_{i=1}^{N}\gamma_i A_i(r)A_i^*(r').$$

11 Show that (12.119) is the envelope of the set of curves defined by the previous inequality.

12 Suppose that two (spinless) particles of mass 10^{-27} g attract one another with a force which is such that $\int_0^{\infty} dr\, rV(r) = 6 \times 10^{-21}$ Mev cm^2. How many bound states can the two-particle system have, at most? Suppose that the two particles are identical; how many bound states can the system have?

13 Particles of mass 170 Mev are accelerated to a velocity of 10^9 cm/sec and scattered by particles of mass 285 Mev initially at rest. They interact via a Yukawa potential $V = \gamma r^{-1}e^{-r/r_0}$, where $\gamma = 10^{-13}$ Mev cm. Can the first Born approximation to the scattering amplitude be expected to be good? Does the Born series converge? What if the beam velocity were 10^8 cm/sec?

14 Calculate the s-wave scattering length for the potential

$$V = a^2(b^2 + c^2 e^{-\alpha/r})r^{-4} \qquad \alpha > 0.$$

15 Two particles of mass 2×10^{-24} g and 3×10^{-24} g interact with one another via the potential

$$V = -\gamma r^{-4}e^{-\alpha/r}$$

where $\alpha = 10^{-10}$ cm, $\gamma = 5 \times 10^{-44}$ Mev cm^4. How many s-wave bound states does the system have?

16 Solve the zero-energy radial Schrödinger equations of all angular momenta for the singular potential

$$V(r) = \gamma r^{-s} \qquad s > 2$$

by the technique of Prob. 18, Chap. 14. Discuss.

CHAPTER 13

The Watson–Regge Method (Complex Angular Momentum)

13.1 The Watson Transform

The Watson method of transforming the multipole expansion of the electromagnetic amplitude into a contour integral was discussed in Sec. 3.8. Here we want to deal with its analog in the quantum-mechanical case.

The starting point is the partial-wave expansion (11.10a), which may be written in the form

$$A = -ik^{-1} \sum_{l=0} (l + \tfrac{1}{2})(-)^l (S_l - 1) P_l(-z)$$

if $z = \cos\theta$. The series is now replaced by an integral in the λ plane, with $\lambda = l + \frac{1}{2}$, over a contour C such as indicated in Fig. 3.7 except that it crosses the real axis between 0 and 1. We then get

$$A = \frac{1}{2k} \int_C d\lambda\, \lambda [S(\lambda) - 1] P_{\lambda-1/2}(-z) \sec \pi\lambda \tag{13.1}$$

as in (3.99), provided that the infinite set of numbers S_l, $l = 0, 1, \ldots,$ may be interpolated by an analytic function of $\lambda = l + \frac{1}{2}$ which is regular in the vicinity of the real axis. The Legendre polynomials are interpolated in a standard fashion by (3.100).

Supposing that the functions in the integrand in (13.1) are sufficiently well behaved when $|\lambda| \to \infty$, we may add to C two quarter circles at large distances in the first and fourth quadrants and then distort the contour to lie along the imaginary axis. In addition to the integral we then obtain a sum of terms coming from the residues of the poles of the integrand in the first and fourth quadrants. If the integrand contains branch points, there will appear a sum of integrals over the discontinuities across the concomitant branch cuts. In the case of elastic scattering of two particles interacting via a "reasonable" potential, we shall see that there exists an interpolation of S_l which in the relevant region has poles only, say at $l = \alpha_n$, with residues β_n. Thus we obtain for the amplitude

$$A(k,z) = \frac{1}{2k}\int_{-\infty}^{\infty} d\lambda'\,\lambda'\,\frac{[S(i\lambda',k) - 1]P_{i\lambda'-1/2}(-z)}{\cosh \pi\lambda'} + i\pi k^{-1}\sum_n \frac{(\alpha_n + \frac{1}{2})\beta_n P_{\alpha n}(-z)}{\sin \pi\alpha_n}. \tag{13.2}$$

The poles of S in the complex l plane are usually referred to as *Regge poles.*

Interpolation of the Legendre Functions. The standard interpolation (3.100) of the Legendre polynomials $P_\nu(z)$, based on the simplest interpolation of Legendre's differential equations, is for each fixed z an entire analytic function of ν. It has the asymptotic behavior[1] (for fixed z)

$$P_\nu(z) \simeq (2\pi)^{-1/2}\nu^{-1/2}(z^2 - 1)^{-1/4}e^{\xi} \tag{13.3}$$

as $|\nu| \to \infty$ for $\operatorname{Re}\nu \geq 0$, where the function ξ is defined by

$$e^{\xi} = 2^{-1/2-\nu}\{[(z + 1)^{1/2} + (z - 1)^{1/2}]^{2\nu+1} + i[(z + 1)^{1/2} - (z - 1)^{1/2}]^{2\nu+1}\} \tag{13.4}$$

and therefore for real z, with $\nu = \nu_1 + i\nu_2$,

$$\xi = \begin{cases} (2\nu_1 + 1)\ln[2^{-1/2}(z + 1)^{1/2} + 2^{-1/2}(z - 1)^{1/2}] & \text{for } z > 1 \\ 2|\nu_2|\tan^{-1}\left(\dfrac{1 - z}{1 + z}\right)^{1/2} & \text{for } z^2 < 1. \end{cases} \tag{13.4a}$$

For fixed ν, $P_\nu(z)$ is an analytic function of z regular everywhere in the finite z plane cut from $z = -\infty$ to $z = -1$, with the asymptotic behavior[2]

$$P_{\lambda-1/2} \simeq \frac{\pi^{-1/2}(2z)^{\lambda-1/2}\Gamma(\lambda)}{\Gamma(\lambda + \frac{1}{2})} \tag{13.5}$$

as $|z| \to \infty$ for $\operatorname{Re}\lambda \geq 0$.

Interpolation of the Partial-Wave Amplitudes and the Lehmann Ellipse. The problem of justifying the Watson transform therefore reduces to finding a suitable interpolation of the **S**-matrix elements S_l with the required properties. In Sec. 12.3 we have already discussed an interpolation based on a simple extension of the radial Schrödinger equation. This is quite analogous to the usual extension of the Legendre polynomials, which is based on that of the Legendre differential equation. It is both mathematic-

1 A. Erdelyi (1953), vol. 1, p. 142, eq. (21).
2 *Ibid.*, p. 126, eq. (23).

ally the simplest and physically the most appealing procedure because it is based on the equation that incorporates the dynamics of the scattering process. Nevertheless, as was pointed out in Sec. 3.8, the interpolation is not unique and the choice of the "right" one must ultimately be justified by the results. The method used in Sec. 12.3 does have useful consequences; that is why we keep it. Of course, once it has been adopted for a continuous real interpolation such as in Sec. 12.3, the analytic continuation to *complex* values of l is no longer arbitrary.

The virtue of (13.2) as compared with the series (11.10a) is that it allows us to discuss much more directly the properties of the amplitude as a function of $z = \cos\theta$, or the momentum transfer τ, given by (10.84), i.e.,

$$\tau = 2^{1/2}k(1 - z)^{1/2}.$$

It is convenient to introduce the negative square of the momentum transfer, customarily called t,

$$t = -\tau^2 = 2k^2(z - 1). \tag{13.6}$$

The scattering amplitude may be considered as a function of E and t.

If the potential is a superposition of Yukawa functions (12.22a), then the asymptotic behavior of S_l as $\mathrm{Re}\, l \to \infty$ is given by (12.182a). Insertion of this together with (13.3) in (13.1) shows that the integral in (13.1) converges absolutely for values of z that obey the inequalities

$$\begin{gathered} |(z + 1)^{1/2} \pm (z - 1)^{1/2}| < 2^{1/2} C \\ C = \left| \frac{a_0}{2k} + \left(1 + \frac{a_0^2}{4k^2}\right)^{1/2} \right|. \end{gathered} \tag{13.7}$$

The boundary of this domain in the complex z plane is an ellipse with foci at $z = \pm 1$ and semimajor axis

$$z_0 = \tfrac{1}{2}(C^2 + C^{-2}). \tag{13.8}$$

It is called the *Lehmann ellipse*. For real values of k, that is, for positive energy, (13.8) simplifies to

$$z_0 = 1 + \frac{a_0^2}{2k^2}. \tag{13.8a}$$

The semiminor axis is then

$$z_1 = \frac{a_0}{k}\left(1 + \frac{a_0^2}{4k^2}\right)^{1/2}.$$

In other words, the partial-wave expansion of the amplitude converges absolutely, and A is an analytic function of t in the Lehmann ellipse in the complex t plane. This ellipse has foci at $t = 0$ and $t = -4k^2$; it is centered at $t = -2k^2$; its semimajor axis for real k is $a_0{}^2 + 2k^2$ and its semiminor axis $a_0(a_0{}^2 + 4k^2)^{1/2}$. The form (13.1) of the amplitude may make it appear to have a cut in the z plane, extending from 1 to ∞, because $P_{\lambda-1/2}(-z)$

has such a cut when $\lambda - \frac{1}{2}$ is not an integer. Nevertheless A has no such branch cut. This is clear from the original series (11.10a), which contained Legendre *polynomials* only, and it can be shown also directly.[3] For $z > z_0$, though, the original series diverges and cannot be used.

The transition from (13.1) to (13.2) is justified if the potential is a superposition of Yukawa functions such as (12.22a). In that case it can be demonstrated[4] that, as $\operatorname{Im} \lambda \to \pm\infty$,[5]

$$S - 1 = O(|\lambda|^{-1/2}). \tag{13.9}$$

This together with (12.182) and (13.3) shows that, when z is in the Lehmann ellipse, the contributions of the large quarter circles at infinity vanish, and (13.2) follows from (13.1). Since, according to (13.3), $P_\nu(z)$ increases at most as $\exp(|\operatorname{Im} \lambda| \arg z)$ when $\operatorname{Im} \lambda \to \pm\infty$, the "background" integral in (13.2) converges absolutely for all z not lying between $z = 1$ and $z = \infty$. For a large class of potentials we shall see that the sum in (13.2) contains a finite number of terms only. Thus (13.2) may be analytically continued and used even for values of z or t for which (11.10a) or (13.1) fails to converge.

Asymptotic Behavior for Large $|t|$. The form (13.2) of the amplitude A allows us to extract very simply its asymptotic behavior as a function of t for large $|t|$. According to (13.5) the background term in (13.2) is of order $z^{-1/2}$ as $|z| \to \infty$. Assuming that the sum of pole terms in (13.2) has only a finite number of contributions, we obtain the leading term as $|z| \to \infty$ from the pole with the largest real part. Suppose that the "leading" pole occurs at $l = \alpha$. Then (13.2) and (13.5) show that

$$A(k, z) \underset{|z| \to \infty}{\simeq} F(k) z^{\alpha(k)} \tag{13.10}$$

where $F(k)$ is a function of k only. The pole position $\alpha(k)$ in general depends on k, of course. In terms of the momentum transfer we may write, with $\alpha = \alpha_R + i\alpha_I$,

$$A \underset{|t| \to \infty}{\simeq} G(E) t^{\alpha_R(E)} e^{i\alpha_I(E) \ln t}. \tag{13.10a}$$

Momentum-Transfer Dispersion Relation. Now, according to (13.2) at a fixed value of E, the amplitude for a potential of Yukawa type (12.22a) is an analytic function of t regular everywhere in the finite t plane cut from $t = t_0 = a_0{}^2$ to infinity. According to (13.10a) it behaves asymptotically for large $|t|$ no worse than a power $n - \varepsilon$ of t, $n > \alpha_R$, $\varepsilon > 0$. It is therefore subject to a "subtracted" *momentum-transfer dispersion relation.* For an arbitrary choice of n points t_i not on the cut we have

$$\frac{1}{2\pi i} \int_C dt' \frac{A(t')}{t' - t} \prod_{i=1}^{n} \frac{t - t_i}{t' - t_i} = A(t) - \sum_{j=1}^{n} A(t_j) \prod_{i \neq j} \frac{t - t_i}{t_j - t_i}$$

3 See, for example, R. G. Newton (1964), p. 7.

4 See, for example, *ibid.*, chap. 6.

5 We recall the definition of O given in footnote 3, Sec. 12.1.

if the contour C runs clockwise around the cut and is closed by a large circle at infinity. Therefore, denoting the discontinuity of $A(t)$ across the cut by

$$\Delta A(t) = \lim_{\varepsilon \to 0+} [A(t + i\varepsilon) - A(t - i\varepsilon)]$$

we have

$$A(t) = \sum_{j=1}^{n} A(t_j) \prod_{i \neq j} \frac{t - t_i}{t_j - t_i} + \frac{1}{2\pi i} \int_{a_0{}^2}^{\infty} dt' \frac{\Delta A(t')}{t' - t} \prod_{i=1}^{n} \frac{t - t_i}{t' - t_i}. \quad (13.11)$$

The energy dependence has been suppressed here. Although the points t_i can be chosen to be independent of E, the number n may have to depend on it because α_R does. However, if the real part α_R of the leading pole in the complex λ plane is bounded as the energy increases to infinity, then there exists a finite integer n larger than the maximal value of α_R. If such as n is chosen in (13.11), then the equation holds for all energies. Of course it must be realized that the integral on the right of (13.11) extends entirely over nonphysical values of the scattering amplitude. According to (13.6) the physical region of t extends from $-4k^2$ to zero. Only in that interval does a physically possible scattering angle exist which produces the given value of the momentum transfer. Nevertheless, a representation such as (13.11) is not devoid of physical interest, especially if only a small number of subtractions are needed. (See Sec. 13.4 for the double dispersion relation.)

Resonances. The "physical" **S**-matrix element may be recovered from (13.2) by the integral

$$a_l = \frac{S_l - 1}{2ik} = \frac{1}{2} \int_{-1}^{1} dz\, A(k, z) P_l(z) \quad (13.12)$$

by using the fact[6] that for integral values of l

$$\int_{-1}^{1} dz\, P_l(z) P_\alpha(-z) = -\frac{2}{\pi} \frac{\sin \pi\alpha}{(l - \alpha)(l + \alpha + 1)}. \quad (13.13)$$

The nth pole term in (13.2) thus gives rise to

$$S_l^{(n)} - 1 = \frac{\beta_n(2\alpha_n + 1)}{(l - \alpha_n)(l + \alpha_n + 1)}. \quad (13.14)$$

When α_n gets close to the integer l, then we expect (13.14) to contribute a strong energy dependence to S_l. The result is a *resonance*. Expanding $\alpha_n(E)$ about the energy E_0 at which $\operatorname{Re} \alpha_n = l$, we obtain

$$[l - \alpha_n(E)]^{-1} \simeq (\alpha_n')^{-1}(E - E_0 + \Delta E + \tfrac{1}{2} i\Gamma)^{-1} \quad (13.15)$$

where

$$\Delta E = \frac{\operatorname{Im} \alpha_n \operatorname{Im} \alpha_n'}{(\operatorname{Re} \alpha_n')^2 + (\operatorname{Im} \alpha_n')^2}$$
$$\Gamma = \frac{2 \operatorname{Im} \alpha_n \operatorname{Re} \alpha_n'}{(\operatorname{Re} \alpha_n')^2 + (\operatorname{Im} \alpha_n')^2} \quad (13.16)$$

6 Erdelyi, *op. cit.*, p. 170, eq. (7).

the primes indicating derivatives with respect to E, evaluated at $E = E_0$. Insertion in (13.14) yields a standard Breit–Wigner formula such as (11.54) for the dominant energy dependence in the vicinity of E_0. The number ΔE is the shift from the energy E_0 at which the pole passes the angular momentum l to the real part of the corresponding pole in the E plane; and Γ is the resonance width. Only if Γ is small relative to the scale on which all other quantities in the amplitude change significantly is the "resonance" really visible as a peak in the cross section. In addition, ΔE must be small for the result to be meaningful. If ΔE is not small, then no resonance occurs near E_0 and near $E_0 - \Delta E$ the terms neglected in the expansion in (13.15) become important.

We shall see in Sec. 13.2 that $\operatorname{Im} \alpha_n > 0$. Therefore the width in (13.15) is positive if the pole moves toward the right as the energy increases and negative if it moves toward the left. Since Γ must be positive for the phase of S to *increase* near E_0, only a rightward-moving pole in the l plane causes a real resonance. A leftward motion of a pole past an integer l corresponds at most to a downward passage of the phase shift through $\frac{1}{2}\pi$.

Regge Trajectories. The curves described by the Regge poles in the complex angular-momentum plane as a function of the energy are called *Regge trajectories*. They allow us to construct a similar kind of connection between various bound states and resonances of different l values as do the trajectories of the poles of S in the k plane as l varies. For negative energies the Regge poles in the region $\operatorname{Re} l > -\frac{1}{2}$ must lie on the real l axis (as we shall show in the next section). When a pole passes through an integer, S is infinite and there is a *bound state*. (We assume that the potential is sufficiently well behaved so that poles of S must be zeros of $\mathscr{F}_l$.) Several bound states may lie on the same trajectory and are thus caused by the same moving pole. When the energy increases and takes on a positive value, the pole must move off the real l axis and become complex (as we shall see). At an energy at which it passes near an integer, close to the real axis, it is likely to produce a resonance. Several resonances and bound states may thus be caused by the same pole. They are then thought of as being manifestations of the same underlying cause.

The picture produced in this way has as much intuitive appeal as the one discussed in Sec. 12.3. There is, of course, no contradiction between them. The question whether a resonance pole is "really" a pole moving in the E plane as a function of l, or a pole in the l plane moving as a function of E, is meaningless. Both are different mathematical descriptions of the same thing. As a function of the two complex variables E and l, $S_l(E)$ has certain *surfaces of singularity*. The pole trajectories in the E plane (as a function of l) are their intersections with the hyperplane $\operatorname{Im} l = 0$, projected onto the complex E plane. The pole trajectories in the l plane are their intersections with the hyperplane $\operatorname{Im} E = 0$, projected onto the complex l plane.

As the energy increases, each pole in the l plane moves farther away from the real axis. If the potential is of the Yukawa type (12.22a), then each trajectory must eventually turn around and cross the positive imaginary λ

axis and disappear from view, so to speak. For potentials of finite range, such as the square well, this need not be the case, and in fact in this instance the Watson transform cannot be carried out. In the present context such potentials that are not "analytic" are not "reasonable."

13.2 Uniqueness of the Interpolation

Before proceding to the demonstrations of some of the essential properties of the Regge poles we may briefly discuss the uniqueness problem of the interpolation of the physical S-matrix elements S_l.

It was pointed out already in Sec. 3.8 that, if there is one way, then there are infinitely many ways of constructing an analytic function that takes on the prescribed values S_l when l assumes integral values. The interpolation of Sec. 12.3 via an "obvious" interpolation of the radical Schrödinger equation is one method of doing this. Is it the only one?

The answer is implied by a theorem due to *Carlson*:[7]

If $f(z)$ is regular in the right half plane $\mathrm{Re}\, z > 0$ and if it is $O(e^{k|z|})$ there as $|z| \to \infty$, with $k < \pi$, and if $f(z) = 0$ for $z = 0, 1, 2, \ldots$, then $f(z) \equiv 0$.

It follows that any two interpolations of S_l which at infinity behave sufficiently well to allow us to carry out the Watson transform and which in the region $\mathrm{Re}\, \lambda \geq 0$ have only a finite number of poles must be identical. In order to see this, we apply Carlson's theorem to their difference multiplied by a polynomial that vanishes at the poles of each interpolation. In other words, although there may be infinitely many possible interpolations of S_l, if there is one that permits the Watson transform, it is the only one that does.

The "correct" interpolation may in some interesting cases be obtained directly from the partial-wave amplitudes. The most tempting formula to use would be (13.12) with the standard interpolation for P_l on the right. But this would be incorrect. That it cannot be right can be seen simply by realizing that the right-hand integral always converges for all finite real values of k, since the integrand is bounded and the integration interval finite. Hence (13.12) would never lead to the Regge poles. There is, however, another way of obtaining the a_l from A. Using the Legendre functions of the second kind and a contour C which goes around the points -1 and $+1$ in the positive sense, we have[8]

$$a_l = \frac{1}{2\pi i} \int_C dz\, A(z) Q_l(z). \tag{13.17}$$

Assume now that the amplitude $A(z)$ is known to vanish asymptotically at least like $|z|^{-1/2}$ as $|z| \to \infty$ and that it has the other properties of an amplitude associated with a Yukawa-like potential (12.22a). It is then subject to an

7 See E. C. Titchmarch (1939), p. 186.

8 See Whittaker and Watson (1948), p. 322. The Q_l are Legendre functions of the second kind.

unsubtracted momentum-transfer dispersion relation such as (13.11), or

$$A(z) = \frac{1}{2\pi i}\int_{z_0}^{\infty} dz' \frac{\Delta A(z')}{z' - z}. \tag{13.18}$$

Insertion of this in (13.17) and inversion of the order of integration gives us

$$\begin{aligned} a_l &= \frac{1}{2\pi i}\int_{z_0}^{\infty} dz\, \Delta A(z) Q_l(z) \\ &= \frac{1}{4\pi i k^2}\int_{a_0{}^2}^{\infty} dt\, \Delta A(t) Q_l\left(1 + \frac{t}{2k^2}\right). \end{aligned} \tag{13.19}$$

In this equation we may now let l be nonintegral and use the standard interpolation of the Legendre functions.

That (13.19) defines the "right" interpolation follows from the asymptotic behavior of Q_l as $|z| \to \infty$, which is

$$Q_l(z) = O(z^{-l-1}). \tag{13.20}$$

The integral in (13.19) therefore converges absolutely for $\operatorname{Re} l > -\frac{1}{2}$, and a_l has no singularities for $l > -\frac{1}{2}$. What is more, the asymptotic behavior of Q_l for $|l| \to \infty$ in the right half plane is such that a_l vanishes in that limit. Hence the Watson transform can be carried out, and the Carlson theorem tells us that (13.19) is the only correct interpolation in this instance.

This argument may be extended to the case in which $A(z)$ in asymptotically $O(|z|^\delta)$ and it obeys a subtracted momentum-transfer dispersion relation. Again (13.19) defines the correct interpolation, but only for $\operatorname{Re} l > \delta$. At some value of l with $\operatorname{Re} l = \delta$, (13.19) then has a pole. In the region to the left of it, a_l can be obtained only by analytic continuation of (13.19). The Carlson theorem again tells us that this interpolation is the only one which allows the Watson transform. However, since it depends only upon the partial-wave amplitudes a_l for integers $l > \delta$, there is no guarantee that the a_l for integral $l < \delta$, obtained from analytic continuation of a_l for $l > \delta$, agree with the originally given a_l, $l < \delta$. If they do not, then there exists no interpolation of *all* the a_l that allows the Watson transform. But this does not necessarily rule out the existence of a momentum-transfer dispersion relation with a finite number of subtractions for the corresponding scattering amplitude. This may still exist but it is then not provable by means of the Watson transform.

For example, we may consider a set $\{a_l\}$ that belongs to a Yukawa potential. Then changing a single one of them, say a_L by Δa_L, alters the amplitude by $\Delta a_L P_L(z)$, which goes asymptotically as $|z|^L$ when $|z| \to \infty$. Hence the amplitude obeys a momentum-transfer dispersion relation with a finite number of subtractions. But the Watson transform no longer exists. We shall see in Sec. 20.3 that there still is a local potential that produces the new set $\{a_l'\}$. It is its Yukawa nature that has been destroyed. We conclude that *the existence of a Watson transform is more restrictive than that of a t dispersion relation.*

13.3 Regge Poles

Under the supposition that the potential is sufficiently well behaved, the poles of $S_l(k)$ in the complex l plane are due to zero of $\mathscr{F}_l(k)$, because of (12.154). The arguments of Secs. 12.2 and 12.3 are easily extended to complex values of l. They then show that, for fixed k and r, $f_l(k,r)$ and $\varphi_l(k,r)$ are analytic functions of l. Since the boundary condition (12.15) is independent of l, f_l is regular in the entire complex l plane. For the reason discussed below (12.132) and in Sec. 12.3 the boundary condition for φ_l necessarily depends upon l in such a way that φ_l need be regular only in the region $\operatorname{Re} l > -\frac{1}{2}$. Under very general conditions on the potential, there exists an analytic continuation of φ_l to $\operatorname{Re} l < -\frac{1}{2}$ which has simple poles there only at the negative integral and half integral values of l.[9] But since for the Watson transform we need only the region $\operatorname{Re} l > -\frac{1}{2}$, we shall not pursue this here.

The functions f_l and φ_l being analytic in l, so is the Jost function $\mathscr{F}_l(k)$ defined in (12.142). The equation

$$\mathscr{F}_l(k) = 0 \tag{13.21}$$

thus defines $l(k)$ as an analytic function of k so long as its values are finite and lie in the half plane $\operatorname{Re} l > -\frac{1}{2}$. In contrast to the trajectories discussed in Sec. 12.3, the Regge trajectories are obtained by prescribing a (varying) positive real (for positive energies) or positive imaginary (for negative energies) value of k and then calculating l from (13.21).

For real k^2 we easily calculate the wronskian in the standard manner from the radial Schrödinger equation,

$$\frac{d}{dr} W(\varphi_l^*, \varphi_l) = 2i \operatorname{Im} \lambda^2 r^{-2} |\varphi_l|^2$$

since for $\lambda = l + \frac{1}{2}$ we have $l(l+1) = \lambda^2 - \frac{1}{4}$. Integration gives

$$2i \operatorname{Im} \lambda^2 \int_0^\infty dr\, r^{-2} |\varphi_l|^2 = \lim_{r\to\infty} W(\varphi_l^*, \varphi_l). \tag{13.22}$$

Negative Energies. Assume first that k is positive imaginary and that l has a value for which (13.21) is true. Then φ_l vanishes exponentially at infinity; hence the right-hand side of (13.22) is zero. Since the integral on the left cannot vanish, we must conclude that $\operatorname{Im} \lambda^2 = 0$ and thus for $\operatorname{Re} \lambda > 0$ it follows that $\operatorname{Im} \lambda = 0$. For negative energies the Regge poles (in the half plane $\operatorname{Re} l > -\frac{1}{2}$) must lie on the real l axis. They must there be *simple* poles. This is shown similarly to the demonstration via (12.52) that the bound-state poles must be simple as functions of k. Furthermore, (12.189) shows that they *must* move toward the right as the energy increases.

Positive Energies. Now let us take $E > 0$, that is, $k > 0$, so that we are on the upper rim of the right-hand cut, where the physics lies. Then (13.22)

9 See R. G. Newton (1962b) and (1964, chap. 4).

together with (12.29) shows that

$$8k \operatorname{Re}\lambda \operatorname{Im}\lambda \int_0^\infty dr\, r^{-2}|\varphi_l|^2 = |\mathfrak{F}_{l-}|^2 - |\mathfrak{F}_{l+}|^2. \tag{13.23}$$

Since the integral is positive, $\mathfrak{F}_{l+}$ can vanish in $\operatorname{Re}\lambda > 0$ only for $\operatorname{Im}\lambda > 0$. At $E = 0$ the Regge trajectory leaves the real l axis and turns into the first quadrant of the complex λ plane. At positive energy it can never cross the positive real axis into the fourth quadrant.

Threshold Behavior. The manner in which the trajectory leaves the real axis at threshold is found similarly to (12.192)[10]. As $E \to 0+$, we find that, if the threshold value of l is l_0, then

$$\frac{d\operatorname{Re} l}{dE} = \begin{cases} a & \text{if } l_0 > \frac{1}{2} \\ aE^{l_0 - 1/2}\sin\pi l_0 & \text{if } -\frac{1}{2} < l_0 < \frac{1}{2} \end{cases} \tag{13.24}$$

with $a > 0$. The angle γ under which the trajectory leaves the real axis,

$$\cot\gamma = \frac{d\operatorname{Re} l/dE}{d\operatorname{Im} l/dE}$$

is given by

$$\cot\gamma = \begin{cases} O(E^{1/2 - l_0}) & \text{if } l_0 > \frac{1}{2} \\ O(1) & \text{if } -\frac{1}{2} < l_0 < \frac{1}{2} \end{cases} \tag{13.25}$$

as $E \to 0+$. Thus, when $-\frac{1}{2} < l_0 < 0$, the trajectory leaves the real axis at a finite angle toward the left; when $0 < l_0 < \frac{1}{2}$, toward the right; and when $\frac{1}{2} < l_0$, it osculates the real axis more and more closely toward the right the larger l_0. This explains in the present context why an almost bound s state causes no low-energy resonance, but an almost bound state of $l > 0$ does, and the higher its l value the sharper the resonance. The equation of the trajectory that $E = 0$ is

$$\operatorname{Im}(l - l_0) = c\operatorname{Re}(l - l_0)^{l_0 + 1/2}. \tag{13.26}$$

From this we conclude via (13.16) that the width of the low-energy resonance is

$$\Gamma = O(E^{l_0 + 1/2}) \tag{13.27}$$

if it occurs at E. This conclusion agrees with (12.193) and (11.55). It should be noted, though, that both in (13.27) and in (12.193) l_0 is the (nonintegral) threshold value of l and not that of the resonance. The result, however, can be expected to be applicable only when these two l values are almost the same.

Trajectory Turning. As the energy increases further, the trajectories may or may not turn back toward the left. In the case of the square well or other potentials of finite range they usually do not. But if the potential is analytic ($\sigma = \frac{1}{2}\pi$) as defined below (12.22a) and it obeys the inequality (12.118), then

10 See R. G. Newton (1964), chap. 9.

we may proceed as below (12.114). In the present instance, however, we assume that k is positive and that l is complex. Then the function A in (12.115) is given by

$$A = \operatorname{Im} \gamma \mathscr{V}_\varphi - k^2 \sin 2\varphi + x^{-2} \operatorname{Im} \lambda^2$$

and B in (12.116) by

$$B = \operatorname{Re} \gamma \mathscr{V}_\varphi - k^2 \cos 2\varphi + x^2 \operatorname{Re} \lambda^2.$$

The result, found by the same reasoning as (12.119), is that if $l = \lambda(k) - \frac{1}{2}$ is the position of a Regge pole at the momentum k, then

$$\operatorname{Re} \lambda(k) \le |\gamma| M k^{-1} \tag{13.28}$$

the constant M being defined by (12.118).

The inequality (13.28) clearly forces each trajectory to turn back. What is more, at each positive energy (13.28) provides a bound on the real parts of the positions of all Regge poles. As the energy increases, every trajectory must approach the positive imaginary λ axis. For superpositions of Yukawa potentials (12.22a), it can be shown[11] that each pole must in fact cross that axis and thus "disappear from view." It is clear from (13.28) that the turning and disappearance are nonuniform in γ. The stronger the potential, the later the poles disappear from the right half plane.

All the essential points needed in Sec. 13.1 for the Watson transform and the t dispersion relation have now been proved.

13.4 The Mandelstam Representation

The Watson–Regge method led us to the proof of a momentum-transfer dispersion relation (13.11). We may now combine this with the previously derived energy dispersion relation (10.98).

The Double Dispersion Relation. Let us assume that the potential is the superposition of Yukawas (12.22). Then (10.98) holds when the negative square t of the momentum transfer is fixed at a value $t > -4a_0^2$. Remember that

$$2i \operatorname{Im} A(E, t) = A(E + i\varepsilon, t) - A(E - i\varepsilon, t)$$

in the limit as $\varepsilon \to 0$, and imagine this resubstituted in (10.98). Now both $A(E + i\varepsilon, t)$ and $A(E - i\varepsilon, t)$ are analytic functions of t, regular everywhere in the complex t plane cut from $t = a_0^2$ to infinity. Furthermore according to (13.10a) they both behave asymptotically (for large $|t|$) no worse than a power $m - \iota$ of t, with $m > \alpha_R$, $\iota > 0$. We have seen in Sec. 13.2 that, under weak additional assumptions on the potential, each α_R is *uniformly* bounded as the energy increases from zero to infinity. Hence $A(E + i\varepsilon, t)$ and $A(E - i\varepsilon, t)$ are both uniformly bounded by $|t|^m$, and so is their differ-

11 See *ibid.*, p. 58.

ence. We may therefore write down a momentum-transfer dispersion relation for $\operatorname{Im} A(E, t)$ similar to (13.11). Let us for simplicity take all the $t_i = 0$. Then

$$\operatorname{Im} A(E', t) = \sum_{j=1}^{m} t^j g_j(E') + \frac{t^m}{\pi} \int_{a_0{}^2}^{\infty} dt' \frac{\rho(E', t')}{t'^m(t' - t)} \tag{13.29}$$

where

$$g_j(E') = \frac{1}{j!} \frac{d^j}{dt^j} \operatorname{Im} A(E', t) \bigg|_{t=0}.$$

Insertion in (10.98) yields

$$\begin{aligned} A(E, t) = A_{\mathrm{BA}}(t) + \sum_n \frac{R_n(t)}{E - E_n} + \sum_{j=1}^{m} t^j \frac{1}{\pi} \int_0^{\infty} dE' \frac{g_j(E')}{E' - E - i\varepsilon} \\ + \frac{1}{\pi^2} t^m \int_0^{\infty} dE' \int_{a_0{}^2}^{\infty} dt' \frac{\rho(E', t')}{t'^m(t' - t)(E' - E - i\varepsilon)} \end{aligned} \tag{13.30}$$

which can now, of course, be continued to the entire cut t plane. This is the *double dispersion relation*, or *Mandelstam representation*. Since we have assumed that the potential is of the form (12.22), the Born term is given by the simple expression

$$A_{\mathrm{BA}}(t) = \int_{a_0}^{\infty} da \frac{\rho(a)}{t - a^2}. \tag{13.31}$$

If there are no bound states and if no subtractions in the t dispersion relation are needed, then (13.30) reads simply

$$A(E, t) = \int_{a_0}^{\infty} da \frac{\rho(a)}{t - a^2} + \frac{1}{\pi^2} \int_0^{\infty} dE' \int_{a_0{}^2}^{\infty} dt' \frac{\rho(E', t')}{(t' - t)(E' - E - i\varepsilon)}. \tag{13.30a}$$

Unitarity. Let us now consider the double dispersion relation in conjunction with unitarity. The generalized optical theorem (7.54) reads

$$4\pi k^{-1} \operatorname{Im} A(\mathbf{k}', \mathbf{k}) = \int d\Omega_{k''} A^*(\mathbf{k}'', \mathbf{k}') A(\mathbf{k}'', \mathbf{k}) \tag{13.32}$$

in terms of the scattering amplitude. Insertion of (13.30a) on the right-hand side leads to integrals of the general form[12]

$$\begin{aligned} I &= \tfrac{1}{4} \int d\Omega_{k''} (c_1 - \mathbf{k}'' \cdot \mathbf{k})^{-1} (c_2 - \mathbf{k}'' \cdot \mathbf{k}')^{-1} \\ &= \tfrac{1}{2} \int_{-1}^{1} du \int d\Omega_{k''} \{c_1(1 - u) + c_2(1 + u) - \mathbf{k}'' \cdot \\ &\quad [\mathbf{k}(1 - u) + \mathbf{k}'(1 + u)]\}^{-2} \end{aligned}$$

12 The second line is a device for carrying out such integrations due to R. P. Feynman. See also Sec. 10.3.3.

$$= \pi s^{-1} \int_{-1}^{1} du(1 - u^2)^{-1}(v - \mathbf{k}\cdot\mathbf{k}')^{-1}$$

$$vs = \frac{1}{2}\left[c_1\left(\frac{1-u}{1+u}\right)^{1/2} + c_2\left(\frac{1+u}{1-u}\right)^{1/2}\right]^2 - s^2\frac{1+u^2}{1-u^2}$$

where $s = k^2 = 2\mu E$. Use of v as the new variable of integration leads to

$$\frac{du}{1-u^2} = \frac{1}{2}\frac{dv\, s}{[(vs - c_1c_2)^2 - (c_1^2 - s^2)(c_2^2 - s^2)]^{1/2}}$$

and hence

$$I = \pi \int_{v_0}^{\infty} \frac{dv}{v - \mathbf{k}\cdot\mathbf{k}'} \frac{1}{[(vs - c_1c_2)^2 - (c_1^2 - s^2)(c_2^2 - s^2)]^{1/2}}$$

$$v_0 s = c_1c_2 + [(c_1^2 - s^2)(c_2^2 - s^2)]^{1/2}.$$

We express this in terms of t, setting

$$t_1 = 2(c_1 - s) \qquad t_2 = 2(c_2 - s) \qquad t' = 2(v - s)$$

and get

$$\begin{aligned} I &= \int d\Omega_{k''}\,[t_1 + (\mathbf{k}'' - \mathbf{k})^2]^{-1}[t_2 + (\mathbf{k}'' - \mathbf{k}')^2]^{-1} \\ &= 4s^{-1/2}\int_0^{\infty} \frac{dt'}{t' - t} F(t', s; t_1, t_2) \end{aligned} \tag{13.33}$$

with

$$F(t, s; t_1, t_2) = \begin{cases} \frac{1}{2}\pi[s(t_1 + t_2 - t)^2 - t_1t_2(4s + t)]^{-1/2} & t > T \\ 0 & t < T \end{cases} \tag{13.34}$$

and

$$2sT(s; t_1, t_2) = 2s(t_1 + t_2) + t_1t_2 + [t_1t_2(4s + t_1)(4s + t_2)]^{1/2}.$$

This result is now used in the generalized optical theorem (13.32) together with the unsubtracted double dispersion relation (13.30a). It yields

$$\begin{aligned} \rho(s,t) &= \int da \int da'\, \rho(a)\rho(a')F(t, s; a^2, a'^2) \\ &\quad - \frac{2}{\pi^2}\mathcal{P}\int \frac{ds'}{s' - s}\int dt' \int da\, \rho(a)\rho(s', t')F(t, s; a^2, t') \\ &\quad + \frac{1}{\pi^4}\int \frac{ds'}{s' - s - i\varepsilon}\int \frac{ds''}{s'' - s + i\varepsilon}\int dt' \int dt''\, \rho(s', t')\rho(s'', t'')F(t, s; t', t'') \end{aligned} \tag{13.35}$$

after dropping the t integration.

Because $F(t, s; t_1, t_2) = 0$ for $t < T(s; t_1, t_2)$ and T is a monotonely increasing function of t_1 and t_2, this expression shows that $\rho(s, t) \equiv 0$ for

$t < T_0(s) = T(s;a_0{}^2,a_0{}^2) = 4a_0{}^2 + a_0{}^4/s$. If this fact is used in the remainder of (13.35), it shows that the second term contributes nothing so long as $t < T_1(s) = T(s;a_0{}^2,4a_0{}^2)$ and the third term vanishes for all $t < T(s;4a_0{}^2,4a_0{}^2)$. Hence the boundary curve below which $\rho(s,t)$ vanishes is given by $t = T_0(s)$. Between it and the curve $t = T_1(s)$ the function $\rho(s,t)$ exactly equals the first term in (13.35). At $t = T_1(s)$ the second term begins to come in. But between $t = T_1(s)$ and $t = T_2(s) = T(s;a_0{}^2,9a_0{}^2)$ its contribution is given exactly by the insertion of the first term for $\rho(s',t')$ in

$$-\frac{2}{\pi^2}\mathscr{P}\int\frac{ds'}{s'-s}\int_{T_0(s')}^{T_1(s')}dt'\int da\,\rho(a)\rho(s',t')F(t,s;a^2,t')$$

etc. Thus the entire (s,t) plane consists of pieces in each of which the function $\rho(s,t)$ can be expressed in closed form in terms of $\rho(a)$. The larger t (for given s), the more complicated these expressions become, i.e., the higher are the "powers" of $\rho(a)$ that occur. In other words, for each fixed value of s and t, $\rho(s,t)$ is given *exactly by a finite polynomial of the potential strength* γ [to which $\rho(a)$ is proportional]. This is the main utility of the Mandelstam representation for calculational purposes. If there are subtractions or bound states, the scheme works similarly but it is more complicated. We refer the interested reader to the literature.[13]

Notes and References

13.1 The introduction of the Watson method in quantum scattering theory is due to T. Regge (1959). However, see also Vogt and Wannier (1954). It was an extremely fashionable and useful tool for some years, and the literature about it is extensive. A complete bibliography as well as a much more detailed treatment than is possible in this book can be found in R. G. Newton (1964). Other detailed discussions are E. J. Squires (1963); Omnes and Froissart (1963); S. Frautschi (1963). For a discussion from the *R*-matrix point of view, see R. Roskies (1966).

The analyticity of the scattering amplitude as a function of the momentum transfer or as a function of the scattering angle was discussed by H. Lehmann (1959). The ellipse whose semimajor axis is (13.8) is the smaller of the two discovered there.

Another method by Gilbert and Shieh (1966) relates the behavior of the partial wave expansion directly to that of the corresponding power series.

The asymptotic behavior (13.10) or (13.10a) of the amplitude for large $\cos\theta$ or large t was first derived by T. Regge (1959).

For an investigation of S as a function of the *two* complex variables l and E, see Bottino and Longoni (1962).

13.2 The result (13.19), which may be used for the correct interpolation of the partial-wave amplitudes, is known as the *Froissart-Gribov* formula, after M. Froissart, unpublished; V. N. Gribov (1962). For a study of the *stability* of the angular-momentum interpolation, see Miller and Viano (1971).

13 See, for example, Blankenbecler, Goldberger, Khuri, and Treiman (1960).

13.3 The arguments in this section are mostly due to T. Regge (1959 and 1960); Bottino, Longoni, and Regge (1962).

The regular solution φ_l was analytically continued to the left half l plane and its properties there established by R. G. Newton (1962b). The threshold behavior of trajectories was first derived in the same paper and independently by Barut and Zwanziger (1962).

The arguments using analyticity of the potential are due to T. Regge (1960) and Bottino, Longoni, and Regge (1962).

Levinson's theorem was discussed from the point of view of the Watson-Regge method by M. Ciafaloni (1963).

Additional references are G. N. Minerbo (1965); M. Weigel (1965); H. Chen (1966); Lopez and Saavedra (1966); W. B. Kaufmann (1967); D. D. Brayshaw (1968a); C. A. Lopez (1968); Mukherjee and Shastry (1967 and 1968); P. J. Kelemen (1971); E. A. Remler (1971); G. R. Shaffer (1971); Tani and Inokuti (1974); Delos and Carlson (1975); M. Dillig (1975). For an application to atomic scattering, see J. N. L. Connor et al. (1981).

13.4 The double dispersion relation was first introduced in the relativistic context by S. Mandelstam (1958) and has found many useful applications there. It was first derived in the nonrelativistic domain by Blankenbecler, Goldberger, Khuri, and Treiman (1960); A. Klein (1960). See also T. Regge (1959); H. M. Nussenzveig (1963); O. Brander (1964); D. Bessis (1965); Bertero and Kreuzer (1967). The present treatment follows mostly the paper by Blankenbecler et al. For other references see the later papers listed in the Notes and References to Sec. 10.3.2.

For extensions to an off-the-energy-shell amplitude, see D. I. Fivel (1961) and to two-channel processes Fonda, Radicati, and Regge (1961); J. Underhill (1962 and 1963).

Problems

1 Show that the domain in the complex z plane which is defined by the inequality (13.7) is bounded by an ellipse with foci at $z = \pm 1$ and semimajor axis given by (13.8). What is the semiminor axis?

2 If the result (13.10a) of the Watson-Regge method is assumed to hold also in relativistic particle physics, then it becomes a statement that is subject to observation. This is because of the "crossing property" (or the *substitution law*) of relativistic field theory. According to this, if an amplitude $A(s,t)$ describes the scattering process $A + B \to A + B$, then, apart from some simple factors depending on spins, etc., $A(t,s)$ describes the reaction process $A + \bar{A} \to B + \bar{B}$. The variable s here is the total relativistic energy. Make the corresponding substitution in (13.10a), and discuss the behavior of the cross section at high energies and small momentum transfer as a function of both. Compare the result with ordinary diffraction scattering.

3 Making use of Sec. 14.6, discuss the Regge trajectories for a Coulomb potential. What is pathological about them, and why? What would you expect the corresponding trajectories for a Yukawa potential to look like?

4 Suppose that the **S** matrix $S(l,k)$ is assumed to have a finite number of Regge poles only, that number being fixed for all k. Would this be compatible with all the general properties the function $S(l,k)$ is supposed to have? If not, which property is violated?

CHAPTER 14

Examples

In this chapter we wish to consider a number of illustrative examples for which some of the general results discussed previously can be derived or studied directly. What we are particularly interested in is potentials for which the Schrödinger equation can be solved explicitly in terms of known functions. There is a large class of these for which the solution of (11.8) is known for a fixed value of l and all values of the energy. But there are essentially only four kinds of known potentials for which the radial Schrödinger equation can be solved explicitly for all l and all E; these are the square well, the harmonic oscillator well, the Coulomb potential, and the r^{-4} potential. It would be very desirable for many purposes to know more such potentials. Of course, the Schrödinger equation can always be solved numerically, if necessary on an electronic computer. But for the purpose of investigating the properties of the scattering amplitude it is often very useful to have explicitly solvable examples. What is more, these can be used as starting points for perturbation calculations.

Here we shall primarily pay attention to solving the radial Schrödinger equation in closed form for a fixed value of l. The square well has already been discussed in Chap. 11, and we shall not go into it in more detail. The first two examples are not really "reasonable" potentials at all, but they are sometimes useful for the construction of models.

14.1 The Zero-Range Potential

If the potential vanishes beyond $r = R$, then the wave function in the outside region is determined by assigning to it a logarithmic derivative c at $r = R$ which is equal to that of the inside wave function. The smaller R, the less energy-dependent c becomes. In the limit as $R \to 0$ we then obtain $\varphi(k,r)$ for all $r > 0$ by the boundary condition

$$\lim_{r\to 0} \frac{\varphi'(k,r)}{\varphi(k,r)} = c \tag{14.1}$$

where c is independent of k. This works only for $l = 0$, though. For $l > 0$, c tends to zero as $R \to 0$, and the higher angular-momentum wave functions are simply free ones.

The $l = 0$ wave function that satisfies the boundary condition (14.1) is

$$\varphi(k,r) = \text{const} \times \sin(kr + \delta)$$

with

$$\tan \delta = \frac{k}{c}.$$

Thus the phase shift δ does not approach zero at infinite energy. If we define $\delta(0) = 0$, then $\delta(\infty) = \pm\frac{1}{2}\pi$, depending on the sign of c. The number $-1/c$ is the scattering length (see Sec. 11.2.2).

The irregular solution is simply

$$f(k,r) = e^{ikr},$$

and the Jost function must now be obtained from (12.28),

$$f(k) = e^{-i\delta}$$

if we set the constant in φ equal to k^{-1}. It does not tend to unity as $k \to \pm\infty$. The **S** matrix is given by

$$S = e^{2i\delta} = \frac{c + ik}{c - ik}$$

so that

$$S(\infty) = -1.$$

If $c < 0$, then S has a pole in the upper half of the k plane at

$$k = i|c|,$$

and there is a bound state of binding energy $c^2/2\mu$. If $c > 0$, then the pole is in the lower half of the k plane, or on the second sheet of the E surface; there is a virtual state at $E = c^2/2\mu$.

The reason for the anomalous behavior of the amplitude is that the zero-range limit is not uniform in k. So long as the range R is finite, we have

$$\tan \delta = \frac{k \cot kR - c}{k + c \cot kR}.$$

If $k \to \infty$ at a fixed value of R, then c approaches its free value $k \cot kR$ and $\tan \delta \to 0$. Subsequently we may let $R \to 0$, and there is no change. But if we let $R \to 0$ *first*, with c fixed, then c becomes independent of k and $\tan \delta = k/c$. Letting $k \to \infty$ at this stage gives the anomalous result $\tan \delta = \infty$.

14.2 The Repulsive Core

An impenetrable sphere of radius R_c is described by a potential that is positive infinite for $r \leq R_c$. Such a condition of impenetrability manifests itself in the form of a boundary condition on the wave function,

$$\psi(\mathbf{k}, \mathbf{r}) = 0 \quad \text{at } |\mathbf{r}| = R_c.$$

This must then hold for each angular momentum. We may therefore define the solution φ_l by the boundary conditions

$$\begin{aligned} \varphi_l(k, R_c) &= 0 \\ \varphi_l'(k, R_c) &= 1. \end{aligned} \tag{14.2}$$

The solution f_l, of course, is unaware of the core when $r > R_c$. Its analyticity properties are therefore unchanged. The Jost function is still defined by (12.142), but instead of (12.140) we get

$$\mathscr{F}_l(k) = \frac{k^l e^{-i 1/2 \pi l} f_l(k, R_c)}{(2l + 1)!!} \tag{14.3}$$

and therefore as $|k| \to \infty$ in $\operatorname{Im} k \geq 0$

$$\mathscr{F}_l(k) \simeq \frac{k^l e^{-i 1/2 \pi l} e^{ikR_c}}{(2l + 1)!!}. \tag{14.4}$$

The $\mathbf{S}$ matrix elements are

$$S_l(k) = \frac{e^{i\pi l} f_{l-}(k, R_c)}{f_{l+}(k, R_c)}. \tag{14.5}$$

Thus, as $k \to \pm\infty$, the phase shift behaves as

$$\delta_l(k) \simeq -kR_c + \tfrac{1}{2}\pi l.$$

It no longer tends to a multiple of 2π, but to infinity. Note that the scattering cross section has infinitely many peaks as the energy increases. But since the phase shift is decreasing, they are *not* resonances.

For a *pure* repulsive core, i.e., no additional forces, f_l is replaced by $w_l^{(+)}$. We then have, for example,

$$S_0(k) = e^{-2ikR_c}$$

$$S_1(k) = -e^{-2ikR_c}\frac{k - iR_c^{-1}}{k + iR_c^{-1}} \tag{14.6}$$

$$S_2(k) = e^{-2ikR_c}\frac{(kR_c)^2 - 3ikR_c - 3}{(kR_c)^2 + 3ikR_c - 3}.$$

Thus for $l > 0$ the repulsive core produces virtual states (e.g., for $l = 1$) and resonance states (e.g., for $l = 2$). Of course, the latter are not sharp.

The Levinson theorem (12.156) breaks down for potentials with a hard core. However, one may prove a similar theorem for the difference between the actual phase shift and the pure-core phase shift for the same core radius.

14.3 The Exponential Potential

If the potential has the form

$$\mathscr{V}(r) = -\mathscr{V}_0 e^{-r/a} \tag{14.7}$$

then the radial Schrödinger equation for $l = 0$ is explicitly solvable by the substitution $x = e^{-r/2a}$. The result is

$$f(k,r) = \exp[iak\ln(a^2\mathscr{V}_0)]\Gamma(1 - 2iak)J_{-2iak}(2a\mathscr{V}_0^{1/2}e^{-r/2a}) \tag{14.8}$$

and hence the Jost function is given by

$$\mathscr{F}(k) = \exp[iak\,\mathrm{In}\,(a^2\mathscr{V}_0)\Gamma(1 - 2iak)J_{-2iak}(2a\mathscr{V}_0^{1/2}). \tag{14.9}$$

We see here explicitly that $\mathscr{F}$ has poles in the lower half plane, at $k = -in/2a$, $n = 1, 2, \ldots$. The **S** matrix element S_0 therefore has infinitely many poles *on the first sheet* of the E surface, at $E = -n^2/8\mu a^2$, $n = 1, 2, \ldots$, and these do not signify bound states. For a more general continuous superposition of exponential potentials such as (12.22) those poles become the left-hand cut.

The bound states, virtual states, and resonances are obtained from those values of k for which

$$J_{-2iak}(2a\mathscr{V}_0^{1/2}) = 0 \tag{14.10}$$

depending upon whether $\mathrm{Im}\,k > 0$ (bound states) or $\mathrm{Im}\,k < 0$. The values of a and $\mathscr{V}_0$ for which new bound states are introduced are obtained from

$$J_0(2a\mathscr{V}_0^{1/2}) = 0. \tag{14.11}$$

Generalization. The presence of infinitely many poles in $f(k,r)$ at $k = -in/2a$, $n = 1, 2, \ldots$, is a general feature of potentials[1] whose asymptotic tail is proportional to $e^{-r/a}$. If the potential is identical with (14.7) for $r > R$ but differs from it for $r < R$, this is easily seen. We then calculate the "inside" function $\varphi(k,r)$, and the "outside" function $f(k,r)$ is given by

1 That it need not *always* be so is shown by an example in Sec. 14.7.1.

(14.8). The Jost function is given by (12.28), which may be evaluated from φ and f at $r = R$. Since f contains $\Gamma(1 - 2iak)$ as a factor, so will generally $\mathscr{F}$. In exceptional circumstances, of course, some (or all) of these poles may be absent.

The same infinite sequence of poles can generally be expected to be present for higher angular momenta. To see this, we may calculate $f_l(k,r)$ from $f(k,r)$ of (14.8) by the integral equation

$$f_l(k,r) = f(k,r) + l(l+1)\int_r^\infty dr'\,\mathscr{G}(k;r,r')r'^{-2}f_l(k,r') \tag{14.12}$$

where the Green's function is calculated by (12.42) from (14.8). The first version of (12.42) shows that, because φ is everywhere regular, the Green's function has no singularities at the points $k = -in/2a$. The solution of (14.12) will contain the same Γ function as f and thus will contain the same poles.

These considerations are all generalizable to the case in which the potential is a finite sum of exponentials. For example, if $e^{-r/a}$ and $e^{-r/b}$ appear, poles generally occur at $k = -i[(n/2a) + (m/2b)]$. This includes the oscillatory case, when a or b is complex.

14.4 The Hulthén Potential

Suppose that the potential has the form

$$\mathscr{V}(r) = \frac{\mathscr{V}_0 e^{-r/a}}{1 - e^{-r/a}}. \tag{14.13}$$

At large distances this behaves like $\mathscr{V}_0 e^{-r/a}$ and at the origin like $a\mathscr{V}_0 r^{-1}$. It is a special instance of the class of *Eckart potentials*

$$\mathscr{V}(r) = \mathscr{V}_1 \frac{e^{-r/a}}{1 + ce^{-r/a}} + \mathscr{V}_2 \frac{e^{-r/a}}{(1 + ce^{-r/a})^2} \tag{14.14}$$

all of which allow the s-wave Schrödinger equation to be solved exactly. In the case of (14.13) it is solved by the substitutions $x = e^{-r/a}$, $f(k,r) = e^{ikr}g(k,r)$. It then goes over into the hypergeometric equation, and we find that[2]

$$f(k,r) = e^{ikr}F(A,B,C;e^{-r/a}) \tag{14.15}$$

with

$$A = -iak + ia(k^2 + \mathscr{V}_0)^{1/2}$$
$$B = -iak - ia(k^2 + \mathscr{V}_0)^{1/2}$$
$$C = 1 - 2iak.$$

2 A. Erdelyi (1953), vol. 1, p. 56.

Consequently[3]

$$\begin{aligned}\mathscr{F}(k) = F(A,B,C;1) &= \frac{\Gamma(1 - 2iak)\Gamma(1)}{\Gamma(1 + B)\Gamma(1 + A)} \\ &= \prod_{n=1}^{\infty}\left[1 + \frac{a^2\mathscr{V}_0}{n(n - 2iak)}\right].\end{aligned} \tag{14.16}$$

Thus $\mathscr{F}$ has poles in the lower half of the k plane, at

$$k = -\frac{in}{2a} \qquad n = 1, 2, \cdots.$$

There are infinitely many virtual states and their positions are independent of the potential strength. If the potential is attractive, $\mathscr{V}_0 < 0$, and $a^2|\mathscr{V}_0| > 1$, then there are bound states at

$$k = k_n = \frac{i(a^2|\mathscr{V}_0| - n^2)}{2an} \qquad n = 1, 2, \cdots < a|\mathscr{V}_0|^{1/2}.$$

The eigenvalues of the kernel of the s-wave Lippmann-Schwinger equation are

$$\alpha_n = \frac{a^2\mathscr{V}_0}{n(2iak - n)} \qquad n = 1, 2, \cdots .$$

If (for real k)

$$a^2|\mathscr{V}_0| < (1 + 4a^2k^2)^{1/2}$$

then they are all inside the unit circle and the Born series converges at the energy $k^2/2\mu$. Hence we see explicitly that in this case the necessary and sufficient conditions for the convergence of the s-wave Born series at all energies is the absence of s-wave bound states of the attractive version of (14.13).

14.5 Potentials of the Yukawa Type

If the potential is of the form (12.22),

$$\begin{aligned}\mathscr{V}(r) &= \int da\,\rho(a)e^{-ar} \\ \rho(a) &= 0 \quad \text{for } a < a_0\end{aligned} \tag{14.17}$$

or (12.22a), then the Schrödinger equation in general cannot be solved in closed form. Many of the important properties of the solutions, of the Jost function, and of the **S** matrix that follow from (14.17) have already been discussed. Here we want to outline a method of calculating the **S**

3 *Ibid.*, pp. 104 and 105.

matrix elements more explicitly under the assumption (14.17). We shall assume that $l = 0$. For $l > 0$ we refer to the literature.[4]

In accordance with (14.17) it is natural to try to write the solution $f(k,r)$ in the form

$$f(k,r) = [1 + \int da\, \varsigma(a,k)e^{-ar}]e^{ikr} \tag{14.18}$$

where we expect that $\varsigma(a,k) = 0$ for $a < a_0$. Insertion of (14.17) and (14.18) in the s-wave Schrödinger equation yields an equation for $\varsigma(a,k)$,

$$a(a - 2ik)\,\varsigma(a,k) = \rho(a) + \int da'\, \rho(a - a')\,\varsigma(a',k). \tag{14.19}$$

This shows, first of all, that it is consistent to assume that $\varsigma(a,k) \equiv 0$ for $a < a_0$. The integral in (14.19) then extends from $a' = a_0$ to $a' = a - a_0$, and it is absent for $a \leq 2a_0$. Equation (14.19) is therefore not really an integral equation, but it can be solved explicitly for every given value of a. For example, for $a_0 < a < 2a_0$

$$\varsigma(a,k) = \frac{\rho(a)}{a(a - 2ik)} \tag{14.20}$$

and for $2a_0 < a < 3a_0$

$$\varsigma(a,k) = \frac{\rho(a)}{a(a - 2ik)} + \int_{a_0}^{a-a_0} da' \frac{\rho(a - a')\rho(a')}{a(a - 2ik)a'(a' - 2ik)}. \tag{14.20a}$$

The situation here is quite analogous to that of the Mandelstam representation together with unitarity (see Sec. 13.4).

The Jost function is then given by

$$\mathscr{F}(k) = 1 + \int_{a_0}^{\infty} da\, \varsigma(a,k). \tag{14.21}$$

In order to calculate it we must, of course, know ς for *all* values of a, which involves infinitely many iterations of (14.19). Insertion of (14.20), (14.20a), etc., in (14.21) shows explicitly that $\mathscr{F}(k)$ has a branch point at $k = -ia_0/2$, when the pole of $\varsigma(a,k)$ as a function of a coincides with the end point of integration. This is the beginning of the lefthand cut. Only if $\rho(a)$ is a distribution (δ function) rather than an ordinary function, so that $\mathscr{V}$ is a single exponential or a sum of them, does this branch point become a pole.

There remains the question of the convergence of the integral in (14.21). It is not difficult to show[5] that, if $|\rho(a)|$ is uniformly bounded, then the integral converges absolutely and uniformly in k. What is more, $\mathscr{F}_-(k)$ approaches unity as $|k| \to \infty$ in the upper half plane too, even in the vicinity of the imaginary axis; more precisely,[5]

4 See A. Martin (1960) and Fivel and Klein (1960).

5 See A. Martin (1959).

$$\lim_{|k|\to\infty} \mathscr{F}(k) = 1$$

provided that

$$|\mathrm{Re}\, k| > c(\mathrm{Im}\, k)^{-n}.$$

where n is an arbitrary positive number. Hence $S(k)$ approaches unity at infinity on the entire physical sheet of the E plane.

14.6 The Coulomb Potential

14.6.1 The Pure Coulomb Field. The case of the Coulomb potential is anomalous. Because of its slow decrease as r^{-1} at $r \to \infty$, it does not belong to the class of well-behaved potentials in any of the senses used in this book. Nevertheless, since it plays an important role in nature and since, moreover, the solutions of the Schrödinger equation can be written down explicitly for all l and all k, we wish to treat it here.

Parabolic Coordinates. We first look at the full Schrödinger equation in parabolic coordinates defined by

$$\xi = r - z \qquad \zeta = r + z$$

and the azimuthal angle φ about the z axis. The equation reads

$$\left[-\frac{4}{\xi+\zeta}\left(\frac{\partial}{\partial\xi}\xi\frac{\partial}{\partial\xi} + \frac{\partial}{\partial\zeta}\zeta\frac{\partial}{\partial\zeta}\right) - \frac{1}{\xi\zeta}\frac{\partial^2}{\partial\varphi^2} + \frac{nk}{\xi+\zeta}\right]\psi = k^2\psi \tag{14.22}$$

where

$$n = \frac{Z_1 Z_2 e^2 \mu}{k} = \frac{Z_1 Z_2 e^2}{v}$$

if the two particles have charges $Z_1 e$ and $Z_2 e$ and v is their relative velocity. The z axis is defined by the direction of the incoming beam.

If there is no field, the solution we want is just e^{ikz}. We are now seeking a solution which differs from this by an *outgoing* spherical wave only. Therefore we are led to try a solution of the form

$$\psi = e^{ikz} f(\xi)$$

in the hope that we can select an f which asymptotically differs from unity by something like $e^{ik\xi} = e^{ik(r-z)}$. Insertion in (14.22) yields the confluent hypergeometric equation

$$\xi f'' + (1 - ik\xi)f' - nkf = 0. \tag{14.23}$$

Its regular solution is therefore[6]

$$f(\xi) = c\Phi(-in, 1; ik\xi).$$

6 Erdelyi, *op. cit.*, p. 252; we are using the notation of that reference.

Φ is the confluent hypergeometric function

$$\Phi(a,b;x) = \sum_{s=0}^{\infty} \frac{\Gamma(a+s)\Gamma(b)}{\Gamma(b+s)\Gamma(a)} \frac{x^s}{s!}, \tag{14.24}$$

and its asymptotic behavior for large x is[7]

$$\begin{aligned}\Phi(a,b;x) = {} & \frac{\Gamma(b)}{\Gamma(b-a)} e^{-a \ln(-x)} \left[1 + \frac{a(a+1-b)}{x} + \cdots\right] \\ & + \frac{\Gamma(b)}{\Gamma(a)} e^{x+(a-b)\ln x} \left[1 + \frac{(1-a)(b-a)}{x} + \cdots\right].\end{aligned} \tag{14.25}$$

The wave function ψ therefore goes asymptotically as

$$\begin{aligned}\psi^{(+)} \simeq \frac{ce^{1/2n\pi}}{\Gamma(1+in)} \Bigg\{ & e^{i[kz+n\ln k(r-z)]} \left[1 - \frac{n^2}{ik(r-z)} + \cdots\right] \\ & + A_c(\theta) r^{-1} e^{i[kr - n \ln 2kr]} \left[1 - \frac{(1+in)^2}{ik(r-z)} + \cdots\right]\Bigg\}\end{aligned} \tag{14.26}$$

where

$$A_c(\theta) = -\frac{n}{2k\sin^2 \frac{1}{2}\theta} e^{-ni \ln \sin^2 1/2\theta + 2i\eta_0} \tag{14.27}$$

$$\eta_0 = \arg\Gamma(1+in). \tag{14.28}$$

For the leading terms in both brackets in (14.26) to dominate, we must assume that

$$\left|\frac{n^2}{k(r-z)}\right| \ll 1. \tag{14.29}$$

This presupposes not only that r is large but also that we are not looking too closely in the forward direction. The larger the distance from the scattering center, though, the smaller may the scattering angle be chosen. Since in a practical experiment kr is always a very large number, the excluded angle is too small to be measurable.

The Cross Section. In contrast to the behavior (10.11) of the wave function for a well-behaved potential, that in the presence of the slowly decreasing Coulomb field contains logarithmically oscillating phase contributions. In other words, the Coulomb field is never completely "turned off." The scattering cross section must consequently be defined anew. It is clear from (14.26) that, if the proper precautions are taken to avoid the forward direction, the flux ratio of the outgoing spherical wave to the incoming plane wave is

$$\frac{d\sigma}{d\Omega} = |A_c(\theta)|^2 = \frac{n^2}{4k^2 \sin^4 \frac{1}{2}\theta} = \left(\frac{Z_1 Z_2 e^2}{4E\sin^2 \frac{1}{2}\theta}\right)^2. \tag{14.30}$$

7 *Ibid.*, p. 278

This is the Rutherford cross section. Comparison with (5.16) shows that it is identical with the corresponding classical formula. Notice that it is independent of whether the charges attract or repel one another. The phase of $A_c(\theta)$, however, depends on this, and especially strongly so in the forward direction and at low energy.

A difference between the classical and the quantum-mechanical result arises if the two colliding particles are identical. Whereas classically the resulting flux is the sum of the direct and recoil fluxes,

$$\begin{aligned}\frac{d\sigma}{d\Omega} &= \frac{d\sigma_0(\theta)}{d\Omega} + \frac{d\sigma_0(\pi - \theta)}{d\Omega} \\ &= \left(\frac{Z^2e^2}{4E}\right)^2 (\csc^4 \tfrac{1}{2}\theta + \sec^4 \tfrac{1}{2}\theta)\end{aligned} \tag{14.31}$$

quantum mechanically the *amplitudes* must be added,

$$\begin{aligned}\frac{d\sigma}{d\Omega} &= |A_c(\theta) + A_c(\pi - \theta)|^2 \\ &= \left(\frac{Z^2e^2}{4E}\right)^2 [\csc^4 \tfrac{1}{2}\theta + \sec^4 \tfrac{1}{2}\theta + 8 \csc^2 \theta \cos (n \ln \tan^2 \tfrac{1}{2}\theta)].\end{aligned} \tag{14.32}$$

The additional term is called an *exchange* contribution, because of the indistinguishability of the two particles. The resulting formula (14.32) is usually referred to as the *Mott* scattering cross section. If we reintroduce Planck's constant, we have

$$n = \frac{Z^2e^2}{\hbar v}.$$

When $\hbar \ll Z^2e^2$, then the exchange term oscillates extremely rapidly as a function of the scattering angle and thus in an experimental observation it averages to zero. Notice that in the Coulomb case the classical result is obtained in the limit of *small* velocities, i.e., *long* wavelengths, whereas usually it is the analog of the geometrical-optics limit of physical optics, namely, the *short*-wavelength limit (see Chap. 18).

A word should be said about the Born approximation. Equation (14.27) shows that the magnitude of A_c is proportional to e^2. Hence it is equal to its first Born approximation. However, the *phase* of A_c is not. Furthermore, the asymptotic form (14.26) of the wave function contains additional terms such as $\exp[-in \ln 2kr]$. Suppose, then, that we were to calculate the Born series to $\psi^{(+)}$ and extract its asymptotic form, assuming that it is of the usual form (10.11). Then the amplitude would have to contain the factor $\exp[-in \ln 2kr]$, which is infinite in the limit as $r \to \infty$. This is the reason why a conventionally calculated amplitude has an infinite phase factor. What is more, since the magnitude itself is the first Born approximation, the infinity will not appear until the *second* Born approximation, which contains $-in \ln 2kr$ because of the expansion of the phase factor.

The Gamow Factor. According to (14.26) the Coulomb wave function is normalized to unit incident flux by setting

$$c = e^{-1/2n\pi}\Gamma(1 + in)v^{1/2}. \tag{14.33}$$

It is then given by

$$\psi_c^{(+)}(\mathbf{k},\mathbf{r}) = v^{-1/2}e^{-1/2n\pi}\Gamma(1 + in)\exp(i\mathbf{k}\cdot\mathbf{r})\Phi[-in, 1; i(kr - \mathbf{k}\cdot\mathbf{r})]. \tag{14.34}$$

Its absolute value squared at $\mathbf{r} = \mathbf{0}$ is given by

$$|\psi_c^{(+)}(\mathbf{k},\mathbf{0})|^2 = v^{-1}\frac{2n\pi}{e^{2n\pi} - 1} \tag{14.35}$$

because

$$\Gamma(1 + in)\Gamma(1 - in) = \frac{in\pi}{\sin in\pi} = \frac{2n\pi}{e^{n\pi} - e^{-n\pi}}.$$

For very small velocities this becomes

$$\frac{2\pi Z_1 Z_2 e^2}{v^2}$$

if the force is *attractive*, that is, $n < 0$; and it becomes

$$2\pi Z_1 Z_2 e^2 v^{-2} e^{-2\pi Z_1 Z_2 e^2/v}$$

if the force is *repulsive*. In spite of the fact that the cross sections do not depend on whether the charges have equal or opposite sign, it is physically obvious that the detailed behavior of the particles does, and particularly the probability of finding the particles closely together. The factor by which the particle density at $r = 0$ differs in the two cases, namely,

$$\exp\left[-\frac{2\pi Z_1 Z_2 e^2}{v}\right]$$

is called the *Gamow factor*.

Partial Waves. We now perform a partial-wave analysis. The radial Schrödinger equation then contains the potential

$$\mathscr{V}(r) = 2nkr^{-1}.$$

The substitution

$$\varphi_l^{(c)}(k,r) = r^{l+1}e^{ikr}g_l$$

brings it into the form of the confluent hypergeometric equation

$$rg_l'' + 2(l + 1 + ikr)g_l' + 2[ik(l + 1) - nk]g_l = 0$$

from which we conclude that

$$\begin{aligned}\varphi_l^{(c)}(k,r) &= r^{l+1}e^{ikr}\Phi(l + 1 + in, 2l + 2; -2ikr)\\ &= r^{l+1}e^{-ikr}\Phi(l + 1 - in, 2l + 2; 2ikr)\end{aligned} \tag{14.36}$$

because of the boundary condition (12.132) and (14.23), and by Kummer's transformation.[8] Thus $\varphi_l^{(c)}(k,r)$ is the *Whittaker function*[9]

$$\varphi_l^{(c)}(k,r) = (2ik)^{-l-1} M_{in,l+1/2}(2ikr) = (-2ik)^{-l-1} M_{-in,l+1/2}(-2ikr), \tag{14.36a}$$

and it can also be expressed in terms of *Laguerre functions*[10]

$$\varphi_l^{(c)}(k,r) = r^{l+1} e^{ikr} \Gamma(-in - l) \mathsf{L}_{-in-l-1}^{(2l+1)}(-2ikr). \tag{14.36b}$$

Except for a factor, the Laguerre functions go over into the *Laguerre polynomials* $\mathsf{L}_m{}^\alpha$ when $in - l$ is a positive integer m,[11]

$$\mathsf{L}_m^{(\alpha)} = \frac{\Gamma(\alpha + 1)}{\Gamma(\alpha + 1 + m)} \mathsf{L}_m^\alpha.$$

Similarly, we may define an irregular solution $f_l(k,r)$, except that the boundary condition (12.15) no longer serves. It is easy to see from the radial Schrödinger equation that the closest we can come to it is to demand that

$$\lim_{r\to\infty} e^{-i(kr - n\ln 2kr)} f_l^{(c)}(k,r) = 1. \tag{14.37}$$

The logarithmic term is the spherical-wave analog of that present in (14.26). The solution $f_l^{(c)}$ is therefore

$$f_l^{(c)}(k,r) = -(2kr)^{l+1} i e^{i(kr - 1/2\pi l)} e^{1/2\pi n} \Psi(l + 1 + in, 2l + 2; -2ikr), \tag{14.38}$$

where Ψ is the irregular confluent hypergeometric function defined by[12]

$$\Psi(a,b;x) = \frac{\Gamma(1-b)}{\Gamma(a-b+1)} \Phi(a,b;x) + \frac{\Gamma(b-1)}{\Gamma(a)} x^{1-b} \Phi(a - b + 1, 2 - b; x). \tag{14.39}$$

8 *Ibid.*, p. 253.

9 *Ibid.*, p. 264.

10 *Ibid.*, p. 268.

11 We are using the notation of A. Erdelyi (1953), who defines the Laguerre polynomials $\mathsf{L}_m{}^\alpha$ on p. 188 of vol. 2. Other sources use Laguerre polynomials $\mathfrak{L}_m$ defined by the generating function

$$(1 - s)^{-1} \exp[xs(s-1)^{-1}] = \sum \frac{s^m \mathfrak{L}_m(x)}{m!}$$

and then associated Laguerre polynomials defined by

$$\mathfrak{L}_m^n = \left(\frac{d}{dx}\right)^n \mathfrak{L}_m.$$

These are related to those of Erdelyi's definition by

$$\mathfrak{L}_m^n = m!(-)^n \mathsf{L}_{m-n}^n.$$

12 Erdelyi, *op. cit.*, p. 257.

Since its asymptotic value is[13]

$$\Psi(a,b;x) \simeq x^{-a},$$

(14.38) satisfies (14.37). Thus $e^{-1/2\pi n} f_l^{(c)}$ is the other Whittaker function,[14]

$$f_l^{(c)}(k,r) = e^{1/2\pi n} W_{-in,l+1/2}(-2ikr). \tag{14.38a}$$

The function $\varphi_l^{(c)}$ may now be expressed in terms of $f_{l+}^{(c)}$ and $f_{l-}^{(c)}$. We have, for $\operatorname{Re} k > 0$,[15]

$$\begin{aligned}\varphi_l^{(c)}(k,r) = e^{1/2n\pi}(2k)^{-l-1}\Bigg[&\frac{\Gamma(2l+2)}{\Gamma(l+1-in)} e^{-i\pi 1/2(l+1)} f_{l+}^{(c)}(k,r) \\ &+ \frac{\Gamma(2l+2)}{\Gamma(l+1+in)} e^{i\pi 1/2(l+1)} f_{l-}^{(c)}(k,r)\Bigg]\end{aligned} \tag{14.40}$$

from which we conclude that the nearest analog of the Jost function is given by

$$\mathfrak{F}_{l\pm}^{(c)}(k) = \frac{(2k)^{-l} e^{1/2n\pi \pm i1/2\pi l} \Gamma(2l+2)}{\Gamma(l+1 \pm in)} \tag{14.41}$$

and $\mathscr{F}_{l\pm}^{(c)}$, defined by (12.142) and (12.142a),

$$\mathscr{F}_{l\pm}^{(c)}(k) = \frac{e^{1/2n\pi}\Gamma(l+1)}{\Gamma(l+1 \pm in)}. \tag{14.41a}$$

The Coulomb **S** matrix is then given by (12.154) as

$$S_l^{(c)} = \frac{\Gamma(l+1+in)}{\Gamma(l+1-in)}. \tag{14.42}$$

Notice that $\mathscr{F}_l^{(c)}$ approaches unity as $k \to \pm\infty$. However, at $k = 0$ it is highly singular, both on account of the Γ function and on account of the factor[16] $e^{1/2n\pi}$.

The asymptotic behavior of $\varphi_l^{(c)}$ as $r \to \infty$ is evidently

$$\varphi_l^{(c)}(k,r) \simeq 2(2k)^{-l-1} e^{1/2n\pi} \frac{\Gamma(2l+2)}{|\Gamma(l+1+in)|} \sin\left(kr - n\ln 2kr - \tfrac{1}{2}\pi l + \eta_l\right) \tag{14.43}$$

where

$$\eta_l = \arg\Gamma(l+1+in). \tag{14.44}$$

Again there appears the ubiquitous logarithmic phase, so that the phase shift must be defined differently from (11.15). Equations (14.42) and (14.44) show that

13 *Ibid.*, p. 278.
14 *Ibid.*, p. 264.
15 *Ibid.*, p. 259.
16 Remember that n contains k^{-1}.

$$S_l^{(c)} = e^{2i\eta_l}. \tag{14.45}$$

The physical wave function which is related to the three-dimensional $\psi^{(+)}(\mathbf{k},\mathbf{r})$ by (11.6) is obtained by comparing (14.43) with (11.9),[17]

$$\begin{aligned}\psi_l^{(c)(+)}(k,r) &= \frac{1}{2}(2k)^{l+1}e^{-1/2n\pi}\frac{\Gamma(l+1+in)}{(2l+1)!}\varphi_l^{(c)}(k,r) \\ &= \frac{1}{2}(2kr)^{l+1}e^{-1/2n\pi}\frac{\Gamma(l+1+in)}{(2l+1)!}e^{ikr}\Phi(l+1+in,\, 2l+2;\, -2ikr).\end{aligned} \tag{14.46}$$

The following integral representation is sometimes useful for calculations with Coulomb wave functions[18]

$$\psi_l^{(c)(+)}(k,r) = \frac{(2kr)^{l+1}e^{-1/2n\pi}}{2\Gamma(l+1-in)}\int_0^1 du\, e^{ikr(1-2u)}u^{l+in}(1-u)^{l-in}. \tag{14.47}$$

Bound States. According to (14.42) the poles of $S_l^{(c)}$ are the poles of the Γ function,

$$\frac{1}{\Gamma(l+1+in)} = 0. \tag{14.48}$$

If $Z_1Z_2 < 0$, the bound states are therefore given by

$$k_m = i\frac{Z_1Z_2e^2\mu}{m+l+1} \qquad m = 0, 1, 2, \ldots. \tag{14.49}$$

According to (14.36b) the bound-state wave functions are expressible in terms of generalized Laguerre polynomials L_m^{2l+1}.

Owing to the long tail of the Coulomb field, there are infinitely many bound states for each angular momentum, the ground state having the binding energy

$$E_0 = \tfrac{1}{2}(Z_1Z_2e^2)^2\mu = \tfrac{1}{2}(Z_1Z_2)^2\alpha^2\mu c^2$$

in terms of the fine structure constant

$$\alpha = \frac{e^2}{\hbar c} = \text{“}1/137\text{”}.$$

They accumulate at $k = 0$, which is possible now because $\mathscr{F}_l^{(c)}$ has an essential singularity there. Furthermore, since the energy depends only on the integer $N = m + l + 1$, there is an N-fold degeneracy (in addition to the degeneracy due to rotational invariance).

Pole Trajectories. The Regge trajectories are given by

$$l = -iZ_1Z_2e^2\mu k^{-1} - m - 1 \qquad n = 0, 1, \ldots.$$

17 This $\psi_l^{(+)}$ is normalized as in (10.11) and *not* to unit incident flux.

18 Erdelyi, *op. cit.*, p. 255.

In the attractive case they start, for $E = -\infty$, at $l = -m - 1$, move to $l = +\infty$ as $E \to 0-$, and as E increases from 0 they come down parallel to the imaginary axis, ending at $-m - 1$ as $E \to +\infty$. In the repulsive case they move toward the left, to $l = -\infty$, and move *up* to $-m - 1$ as $E \to +\infty$.

In the k plane, as a function of l, each bound-state pole moves from its $l = 0$ position to the origin as $l \to \infty$. There are no other poles. In the repulsive case there are only virtual states, and they move up the imaginary axis to $k = 0$.

The α trajectories of the eigenvalues of the Lippmann–Schwinger kernel are described by

$$\alpha = \frac{-iZ_1Z_2e^2\mu}{(m + l + 1)k}.$$

In the attractive case they start at the origin for $E = -\infty$; as E increases, each α moves to $+\infty$ as $E \to 0-$, and as E increases from zero, it comes down the positive imaginary axis and ends at the origin as $E \to +\infty$. When

$$\frac{k}{\mu} > Z_1Z_2e^2$$

then all α's are inside the unit circle. However, in this case this does not imply that the Born series converges. The kernel GV is now not in the Hilbert–Schmidt class, and, for reasons already discussed, the Born series never converges.

The Coulomb Green's Function. We may now explicitly construct the complete Coulomb Green's function. According to (12.146), (14.36), (14.38), and (14.41) we have

$$\mathscr{G}_l^{(c)+}(r,r') = i(-)^l(4k^2rr')^{l+1}e^{ik(r+r')}\frac{\Gamma(l + 1 + in)}{2k(2l + 1)!}$$
$$\Phi(l + 1 + in, 2l + 2; -2ikr_<)\Psi(l + 1 + in, 2l + 2; -2ikr_>). \tag{14.50}$$

For calculational purposes we may avail ourselves of the integral representation (14.47) and the corresponding one for[19] Ψ. This yields

$$\mathscr{G}_l^{(c)+}(r,r') = (2k)^{-1}i(-)^l(4k^2rr')^{l+1}|\Gamma(l + 1 + in)|^{-2}\int_0^1 du\int_0^\infty dv$$
$$\exp\left[ikr_<(1 - 2u) + ikr_>(1 + 2v)\right](uv)^{l+in}[(1 - u)(1 + v)]^{l-in}. \tag{14.51}$$

14.6.2 Coulomb Admixtures. Since the addition of the Coulomb potential to any reasonable one still contains the characteristic slow decrease of the Coulomb field, the scattering of charged particles subject to other forces cannot be described by the same formalism used for neutral ones. One must then split the Coulomb term off from the rest of the interaction and write

19 *Ibid.*, p. 255.

$$H' = H_1' + H_2'$$

where H' is the electrostatic interaction alone. From a formal point of view the appropriate treatment is that given in (7.76) or (7.76a) which expresses the **T** matrix as a sum of the pure Coulomb **T** matrix and a matrix just like **T**, except that the plane wave is replaced by a Coulomb wave function. Both the first term and the wave function $\Psi_1^{(-)}$ are thus explicitly known, and we may want to calculate the remainder by any of the standard methods. If $\Psi^{(+)}$ is replaced by the Coulomb wave function, the result is known as the *distorted-wave Born approximation.*

If H_2' is spherically symmetric, we may expand the second term in (7.76) in partial waves. The first is explicitly known and need not be expanded. In other words, we may write the amplitude

$$A(k,\theta) = A_c(k,\theta) + A'(k,\theta) \tag{14.52}$$

where A_c is (14.27), and we expand A' as in (11.10) or (11.10a).

The cross section now contains an interference term between A_c and A', and since the phase of A_c strongly depends on whether the charges are equal or opposite, so will the interference term. This is true particularly near the forward direction and at low energy. That this should be so is clear on physical grounds. If the particles are slow and repel one another, they cannot penetrate to the region where the other force is strong; if they are oppositely charged, though, they can.

In the calculation of A', all previously "free" wave functions, as in (11.5), now become Coulomb wave functions; $u_l(kr)$ is replaced by $\varphi_l^{(c)}$ of (14.36), and $w_l^{(+)}(kr)$ by $f_l^{(c)}$ of (14.38). The total wave function is obtained from the integral equation (11.7), except for the replacement of u_l by $\varphi_l^{(c)}$ and of $G_l^{\pm}$ by $\mathcal{G}_l^{(c)\pm}$ given by (14.50); V is now the *additional* potential. The asymptotic behavior of the lth radial wave function is proportional to

$$\sin(kr - \tfrac{1}{2}\pi l - n \ln 2kr + \eta_l + \delta_l)$$

the Coulomb phase shifts η_l being given by (14.44). This defines the *additional* phase shift δ_l. The additional amplitude A' can then be expanded in the form

$$A'(k,\theta) = k^{-1} \sum_{l=0} (2l + 1)e^{i(2\eta_l+\delta_l)} \sin \delta_l \, P_l(\cos\theta). \tag{14.53}$$

Functions φ_l and f_l are defined as in Chap. 12, with the appropriate replacements of u_l and $w_l^{(\pm)}$ by Coulomb functions and of the boundary condition (12.15) by (14.37). Since (12.27) is still true, the Jost function $\mathfrak{F}_l$ is defined as before, by (12.28), and φ_l is expressed in terms of f_l by (12.29). Hence the **S**-matrix element is still given by (12.154) in terms of $\mathfrak{F}_l$. In our new notation we have

$$S_l = \exp 2i(\eta_l + \delta_l).$$

The integral representations of the Jost function are no longer (12.143) and (12.144), but use of the integral equations for f_l and φ_l and of the Green's-function representation (12.146) gives

$$\begin{aligned}\mathfrak{F}_l(k) &= \mathfrak{F}_l^{(c)}(k) + \int_0^\infty dr\, \varphi_l^{(c)}(k,r)\mathscr{V}(r)f_l(k,r) \\ &= \mathfrak{F}_l^{(c)}(k) + \int_0^\infty dr\, f_l^{(c)}(k,r)\mathscr{V}(r)\varphi_l(k,r)\end{aligned} \tag{14.54}$$

if $\mathscr{V}$ is the potential *in addition* to the Coulomb field.

14.7 Bargmann Potentials and Generalizations

14.7.1 General Procedure. We now wish to discuss a very large class of potentials for each of which the radial Schrödinger equation of one specified l value can be solved explicitly (for all energies).

Let us suppose that for a given value of l the solutions $\varphi_l^{(0)}$ and $f_l^{(0)}$ of the radial Schrödinger equation with the potential $\mathscr{V}^{(0)}$ are known, and so is the corresponding **S**-matrix element $S_l^{(0)}$. For example, $\mathscr{V}^{(0)}$ may be the Coulomb potential, or it may be zero. We wish to find the potential $\Delta\mathscr{V}$ which, if added to $\mathscr{V}^{(0)}$, produces an **S**-matrix element S_l which differs from $S_l^{(0)}$ by a factor of the form

$$S_l(k) = S_l^{(0)}(k)\frac{R(-k)}{R(k)}. \tag{14.55}$$

Here $R(k)$ is a *rational* function which has N simple poles at $k = \beta_n$, $\operatorname{Im}\beta_n < 0$, and N simple zeros at $k = \alpha_n$ and which tends to unity at infinity,[20]

$$R(k) = \prod \frac{k - \alpha}{k - \beta}. \tag{14.56}$$

Among the α's, those in the lower half plane are called γ, $\operatorname{Im}\gamma < 0$, and those in the upper $\varkappa$, $\operatorname{Im}\varkappa > 0$. In other words,

$$S_l(k) = S_l^{(0)}(k)\prod \frac{k + \gamma}{k - \gamma}\frac{k + \varkappa}{k - \varkappa}\frac{k - \beta}{k + \beta} \tag{14.55a}$$

with as many β's as there are γ's and $\varkappa$'s together. Now, the phase difference of the function $R(-k)/R(k)$ between $k = \infty$ and $k = -\infty$ is equal to $-2m\pi$, if m is the number of its poles minus the number of its zeros in the upper half plane. This number equals twice the number of $\varkappa$'s. Hence by Levinson's theorem, the number of bound states of angular momentum l in the new system described by S_l is equal to that of the system described by $S_l^{(0)}$, plus the number of $\varkappa$'s (if the new potential is reasonable). We are looking for a potential that produces bound states at $E = \varkappa^2/2\mu$, in addition to those of $\mathscr{V}^{(0)}$.

Let us form the wronskian functions

20 We use a simplified notation in which sums and products run over all α's, β's, etc., without employing subscripts that enumerate them.

$$x_\beta(k,r) \equiv \frac{W[\varphi_l^{(0)}(\beta,r),\ f_l^{(0)}(k,r)]}{\beta^2 - k^2}$$
$$= \int_0^r dr'\, \varphi_l^{(0)}(\beta,r') f_l^{(0)}(k,r') + \mathfrak{F}_l^{(0)}(k)(k^2 - \beta^2)^{-1} \tag{14.57}$$

$$y_\beta(k,r) \equiv \frac{W[\varphi_l^{(0)}(\beta,r),\ \varphi_l^{(0)}(k,r)]}{\beta^2 - k^2}$$
$$= \int_0^r dr'\, \varphi_l^{(0)}(\beta,r')\varphi_l^{(0)}(k,r'), \tag{14.58}$$

where the second lines follow from the Schrödinger equation, the boundary conditions, and (12.28). We also define

$$x_{\gamma\beta}(r) \equiv x_\beta(-\gamma,r)$$
$$x_{\varkappa\beta}(r) \equiv x_\beta(-\varkappa,r) - C_\varkappa y_\beta(\varkappa,r), \tag{14.59}$$

the $C_\varkappa$ being a set of arbitrary real constants with the same algebraic sign as $\mathfrak{F}_l^{(0)}(\varkappa)$.

Now calculate N functions $K_\beta(r)$ from the N linear equations,

$$\sum_\beta x_{\gamma\beta}(r)K_\beta(r) = -f_l^{(0)}(-\gamma,r)$$
$$\sum_\beta x_{\varkappa\beta}(r)K_\beta(r) = -f_l^{(0)}(-\varkappa,r) + C_\varkappa\varphi_l^{(0)}(\varkappa,r). \tag{14.60}$$

In view of the arbitrary nature of the numbers γ, $\varkappa$, and $C_\varkappa$, these equations are in general linearly independent, i.e.,

$$\det\,[x_{\alpha\beta}(r)] \not\equiv 0 \tag{14.61}$$

and can therefore be solved for the K_β.

We claim that if

$$\Delta\mathscr{V}(r) \equiv 2\frac{d}{dr}\sum_\beta K_\beta(r)\varphi_l^{(0)}(\beta,r) \tag{14.62}$$

is added to $\mathscr{V}^{(0)}$, it produces the **S**-matrix element $S_l(k)$ given by (14.55). Furthermore, there are, in addition to those of $\mathscr{V}^{(0)}$, bound states of energies $\varkappa^2/2\mu$.

A few steps of simple algebra show that the two functions

$$h(k,r) \equiv f_l^{(0)}(k,r) + \sum_\beta K_\beta(r)x_\beta(k,r)$$
$$g(k,r) \equiv \varphi_l^{(0)}(k,r) + \sum_\beta K_\beta(r)y_\beta(k,r) \tag{14.63}$$

satisfy the differential equations

$$-h'' + [l(l+1)r^{-2} + \mathscr{V}^{(0)} + \Delta\mathscr{V} - k^2]h = \sum_\beta \rho_\beta(r)x_\beta(k,r) \tag{14.64}$$

$$-g'' + [l(l+1)r^{-2} + \mathscr{V}^{(0)} + \Delta\mathscr{V} - k^2]g = \sum_\beta \rho_\beta(r)y_\beta(k,r) \tag{14.65}$$

where

$$\rho_\beta(r) \equiv -K''_\beta + [l(l+1)r^{-2} + \mathscr{V}^{(0)} + \Delta\mathscr{V} - \beta^2]K_\beta. \qquad (14.66)$$

Now, according to (14.63) and (14.60),

$$h(-\gamma,r) = 0 \qquad h(-\varkappa,r) - C_\varkappa g(\varkappa,r) = 0. \qquad (14.67)$$

We insert these equations in (14.64) and (14.65) and conclude that

$$\sum_\beta x_{\alpha\beta}(r)\rho_\beta(r) = 0$$

for all α. Because of (14.61) it follows that $\rho_\beta(r) = 0$ for all β. The functions h and g therefore satisfy the radial Schrödinger equation with the potential $\mathscr{V} = \mathscr{V}^{(0)} + \Delta\mathscr{V}$.

Next we look at the boundary values. As $r \to \infty$ we find that

$$x_\beta(k,r) \simeq -\frac{\mathfrak{F}_{l-}^{(0)}(\beta)}{2\beta(k+\beta)}e^{i(k+\beta)r}$$

$$y_\beta(\varkappa,r) \simeq \frac{i\mathfrak{F}_{l-}^{(0)}(\beta)\mathfrak{F}_{l+}^{(0)}(\varkappa)}{4\beta\varkappa(\varkappa-\beta)}e^{i(\beta-\varkappa)r}.$$

Thus for large values of r Eqs. (14.60) for $K_\beta(r)$ become

$$\sum_\beta \frac{K_\beta(r)e^{ir\beta}\mathfrak{F}_{l-}^{(0)}(\beta)}{2\beta(\alpha-\beta)} = -1,$$

from which it follows that there exists a set of N constants a_β such that

$$\lim_{r\to\infty} \frac{K_\beta(r)e^{ir\beta}\mathfrak{F}_{l-}^{(0)}(\beta)}{2\beta} = -a_\beta \qquad (14.68)$$

and

$$1 - \sum_\beta \frac{a_\beta}{\alpha-\beta} = 0 \qquad (14.69)$$

for all α. We may infer from this that

$$R(k) = 1 - \sum_\beta \frac{a_\beta}{k-\beta}, \qquad (14.70)$$

both sides being rational functions of k with the same zeros and poles and unity at infinity.

The asymptotic behavior of $h(k,r)$ is now found to be, as $r \to \infty$,

$$h(k,r) \simeq e^{ikr}R(-k).$$

We conclude that for the potential $\mathscr{V}(r)$,

$$f_l(k,r) = \frac{h(k,r)}{R(-k)}. \qquad (14.71)$$

The solution $g(k,r)$, on the other hand, is regular, and (14.58), (14.60), and

(14.63) show that at $r \simeq 0$ it satisfies the same boundary condition (12.132) as $\varphi_l^{(0)}$. Hence

$$\varphi_l(k,r) = g(k,r). \tag{14.72}$$

Furthermore, use of (12.35) in (14.58), together with (14.68) and (14.69), shows that as $r \to \infty$

$$g(k,r) \simeq \frac{1}{2ik}\left[\mathfrak{F}_{l-}^{(0)} R(-k)e^{ikr} - \mathfrak{F}_{l+}^{(0)} R(k)e^{-ikr}\right]. \tag{14.73}$$

This confirms that (14.55) is the S-matrix element for the potential $\mathscr{V}$. What is more, comparison with (12.35) shows that

$$\mathfrak{F}_l(k) = \mathfrak{F}_l^{(0)}(k)R(k). \tag{14.74}$$

To see whether or not $\varkappa^2/2\mu$ is indeed a bound-state energy, we take (14.71) at $k = \varkappa$. According to (14.67) and (14.72),

$$f_l(\varkappa,r) = C_\varkappa \prod \frac{\varkappa + \beta}{\varkappa + \alpha} \varphi_l(\varkappa,r) \tag{14.75}$$

so that $f_l(\varkappa,r)$ is both regular and exponentially decreasing at infinity. We can even evaluate the normalization integral. Using (12.49) and (12.52) we get

$$\int_0^\infty dr\,|\varphi_l(\varkappa,r)|^2 = \frac{\prod_{\alpha \neq \varkappa} (\alpha^2 + |\varkappa|^2)}{\prod_\beta (\beta^2 + |\varkappa|^2)} \frac{\mathfrak{F}_l^{(0)}(\varkappa)}{C_\varkappa}. \tag{14.76}$$

Notice that if $\mathscr{V}^{(0)}$ had bound states then $\mathscr{V}$ also has bound states at the same energy, because $\mathfrak{F}_l$ vanishes in the upper half plane where $\mathfrak{F}_l^{(0)}$ does. (If we want to remove a bound state of $\mathscr{V}^{(0)}$, we must choose a β in the *upper* half plane so that $\beta^2/2\mu$ is the energy of the old bound state.)

The expression (14.62) for the addition to the potential can be simplified. If we write

$$F_\gamma(r) \equiv f_l^{(0)}(-\gamma,r)$$
$$F_\varkappa(r) \equiv f_l^{(0)}(-\varkappa,r) - C_\varkappa \varphi_l^{(0)}(\varkappa,r)$$

then (14.60) is solved by

$$K_\beta(r) = -\sum_\alpha [x^{-1}(r)]_{\beta\alpha} F_\alpha(r) \tag{14.77}$$

if by $(x^{-1})_{\beta\alpha}$ is meant the inverse of the matrix $x_{\alpha\beta}$. Now it follows from the second lines in (14.57) and (14.58) that

$$F_\alpha(r)\varphi_l^{(0)}(\beta,r) = \frac{d}{dr} x_{\alpha\beta}(r).$$

Hence (14.77) inserted in (14.62) gives[21]

21 We use the differentiation rule for determinants, $(\det M)' = \operatorname{tr} M'M^{-1} \det M$.

$$\Delta\mathscr{V}(r) = -2\frac{d^2}{dr^2}\ln\det\left[x_{\alpha\beta}(r)\right]. \tag{14.78}$$

Summary of Procedure. We recapitulate the procedure. The potential $\mathscr{V} = \mathscr{V}^{(0)} + \Delta\mathscr{V}$, with $\Delta\mathscr{V}$ given by (14.78), allows us to solve the radial Schrödinger equation of angular momentum l explicitly. The regular and irregular solutions are (14.72) and (14.71); the bound-state energies are $\varkappa^2/2\mu$, the bound-state wave functions $\varphi_l(\varkappa,r)$, and their normalization (14.76). The Jost function is (14.74), and the **S**-matrix element (14.55). The produre works even if $\Delta\mathscr{V}$ is complex and S_l not unitary. For S_l to be unitary, we must choose all the $\varkappa$'s purely imaginary and the γ's and β's either purely imaginary or else in pairs symmetric with respect to the imaginary axis. The potential is then real. There is nothing to prevent us from choosing a γ equal to a $-\varkappa$. If we do, then S_l has no pole at the bound state but $\mathfrak{F}_l$ still has a zero there.

Bargmann Potentials. If $S_l^{(0)}$ is taken to be unity, $\mathscr{V}^{(0)} = 0$, then the resulting potentials are those of Bargmann. They are characterized by the fact that, for *one* value l_0 of l, their **S**-matrix element is a rational function of k. For the other angular momenta, of course, the **S**-matrix elements are not rational and the Schrödinger equation cannot be explicitly solved. If $l_0 = 0$ the functions that enter in $\mathscr{V}$ are exponentials (multiplied by sines and cosines if we choose complex β's and γ's). The Bargmann potentials for the s wave therefore usually have exponential tails. (As we shall see in the examples below, there are cases in which they do not.) This shows that an exponential asymptotic form of the potential does not *necessarily* lead to infinitely many poles of each S_l in the upper half plane, although, as we have seen in Sec. 14.3, it generally does.

For $l_0 \neq 0$ the functions from which the Bargmann potentials are constructed are spherical Bessel functions, and thus they contain inverse powers of r. As a result they generally have asymptotic tails of the forms r^{-n}, $n \geq 3$. It has been shown[22] that a sufficient condition for a Bargmann potential to have an expotential tail is that $\mathscr{F}_l(k) = \mathscr{F}_l^{(0)} + O(k^{2l})$ as $k \to 0$.

14.7.2 Special Cases *Two Poles, no Bound States.* The simplest special case is one in which $l_0 = 0$ and

$$\mathfrak{F}_0(k) = R(k) = \frac{k + ia}{k + ib}, \tag{14.79}$$

i.e., one γ, with $\gamma = ia$, and one $\beta = ib$, with $a \geq 0$, $b \geq 0$. The s-wave phase shift is given by

$$k\cot\delta_0 = \frac{ab}{b - a} + \frac{k^2}{b - a} \tag{14.80}$$

so that the effective-range approximation is exact. The potential that produces this phase shift is a special case of those of Eckart, (14.14),

22 By T. Fulton (unpublished) and R. G. Newton (1956).

$$\mathscr{V}(r) = -8b^2\beta\frac{e^{-2br}}{(1 + \beta e^{-2br})^2} \tag{14.81}$$

with

$$\beta = \frac{b - a}{b + a}.$$

Its asymptotic behavior as $r \to \infty$ is

$$\mathscr{V}(r) \simeq -8b^2\beta e^{-2br}.$$

The regular s-wave function is given by

$$\varphi_0(k,r) = \frac{\sin kr}{k} + \frac{b^2 - a^2}{k(k^2 + b^2)}\frac{k\sinh br\cos kr - b\cosh br\sin kr}{b\cosh br + a\sinh br}. \tag{14.82}$$

Degenerate Cases. If we set $a = 0$, there is a zero energy resonance and

$$\tan\delta_0 = \frac{b}{k}.$$

The potential that produces this phase shift is

$$\mathscr{V}(r) = -2b^2\operatorname{sech}^2 br. \tag{14.83}$$

On the other hand, we may let $b = 0$. The potential then becomes

$$\mathscr{V}(r) = 2a^2(1 + ar)^{-2}, \tag{14.84}$$

which possesses neither a second nor a first absolute moment at infinity. As a result the Jost function has a *pole* at $k = 0$, and the phase shift

$$\tan\delta_0 = -\frac{a}{k}$$

violates Levinson's theorem, since $\delta_0(0) = -\frac{1}{2}\pi$.

One Bound State. If in (14.79) we replace a by $-\varkappa$, $\varkappa > 0$, then the effective-range approximation is still exact, but we have a bound state with binding energy $\varkappa^2/2\mu$. There is now a one-parameter family of potentials that produces the same phase shift and the same bound state. They are

$$\mathscr{V}_c(r) = -4\varkappa\frac{d}{dr}\left[\sinh br\frac{g_c(\varkappa,r)}{g_c(\varkappa + b,r) - g_c(\varkappa - b,r)}\right], \tag{14.85}$$

where

$$g_c(k,r) = k^{-1}(e^{-kr} + c\sinh kr).$$

The normalized bound-state wave function is

$$\phi_c(r) = 2\left(\frac{c\varkappa}{b^2 - \varkappa^2}\right)^{1/2}\frac{\sinh br}{g_c(\varkappa + b,r) - g_c(\varkappa - b,r)}. \tag{14.86}$$

The larger the positive number c, the more the bound-state wave function is concentrated near the origin. In fact it is not hard to see that

$$\lim_{c\to\infty} \langle r^2 \rangle_c = 0$$

$$\lim_{c\to 0} \langle r^2 \rangle_c = \infty.$$

The potentials in (14.85) all have asymptotic tails proportional to $e^{-2\varkappa r}$, except when $c = 2$, in which unique case it decreases more rapidly. It is a general property of the potentials producing a given phase shift and given bound states of smallest binding energy $\varkappa_1{}^2/2\mu$ and largest binding energy $\varkappa_2{}^2/2\mu$ that, if one of them decreases asymptotically more rapidly than $\exp(-2\varkappa_2 r)$, then it is the only one with that property and, if one of them decreases asymptotically less rapidly than $\exp(-2\varkappa_1 r)$, then they all do (see Sec. 20.3.2).

The potential family (14.85) may be used, for example, in a neutron-proton model to reproduce the effective range and the deuteron binding energy. In this instance one must choose

$$|\varkappa|^{-1} = 4.31 \times 10^{-13}\ \text{cm}$$
$$b = 0.944 \times 10^{13}\ \text{cm}^{-1}.$$

The experimental value of the photodisintegration cross section of the deuteron then indicates that c should have the unique value 2 which produces the potential with the shortest tail.[23]

Degenerate Cases. An amusing special case is obtained from (14.85) by letting $b \to 0$ and $\varkappa \to 0$ while $\varkappa^2 c = -3N^2$ is fixed. We then get the potentials

$$\mathscr{V}(r) = -6r\frac{2N^2 - r^3}{(N^2 + r^3)^2}. \tag{14.87}$$

They cause no s-wave scattering whatever, at any energy, i.e.,

$$S_0(k) = 1$$

but they have a zero-energy bound state whose normalized wave function is

$$\phi(r) = \frac{\sqrt{3}Nr}{N^2 + r^3}. \tag{14.88}$$

The asymptotic tail of (14.87) is proportional to r^{-2}, so the potential is not reasonable. In contrast to the situation for potentials with a finite first and second moment, the s-wave bound state at zero energy is normalizable. The Levinson theorem, of course, is violated. The regular s-wave function at positive energy is

$$\varphi_0(k,r) = k^{-1}\sin kr - \frac{3rk^{-3}}{N^2 + r^3}(\sin kr - kr\cos kr) \tag{14.89}$$

which shows directly that $\delta_0(k) \equiv 0$.

23 See K. Chadan (1955 and 1956); R. G. Newton (1957), Levinger and Rustgi (1957).

Notes and References

14.1 The zero-range potential was first used by Bethe and Peierls (1935).

14.3 The exponential potential was first employed and the Schrödinger equation solved for it by Bethe and Bacher (1936), particularly p. 110. See also R. Jost (1947). The general demonstration that potentials with an exponential tail $e^{-r/a}$ usually lead to poles at $k = in/2a$, $n = 1, 2, \ldots$, was given in R. E. Peierls (1959). For a discussion of potentials which are finite sums of exponentials, see A. Martin (1959); see also M. G. Fuda (1971a). For a generalization to the Klein-Gordon equation, see Bawin and Lavine (1974).

For some numerical values of low-energy phase shifts and a comparison of various approximations for exponential, Yukawa, and gaussian potentials, see L. Wojtczak (1963).

14.4 The potentials (14.14) were first used and the Schrödinger equation solved with them by C. Eckart (1930). The special case of the potential (14.13) was introduced by L. Hulthén (1942). See also Jost and Pais (1951); M. G. Fuda (1971b).

14.5 The method of this section is due to A. Martin (1959). Its extension to $l > 0$ is given in A. Martin (1960) and by a different technique in Fivel and Klein (1960).

14.6 The Rutherford scattering formula was first derived quantum mechanically in the Born approximation by taking the limit of a screened Coulomb field by G. Wentzel (1926b) and subsequently without screening by J. R. Oppenheimer (1927). The first *exact* solution was given by N. F. Mott (1928), and both in the partial-wave expansion and with parabolic coordinates, by W. Gordon (1928).

The symmetrized scattering formula (14.32) is due to N. F: Mott (1930). For discussions of higher Born approximations in the Coulomb case, see R. H. Dalitz (1951); and Gesztesy and Thaller (1981). The Gamow factor was introduced by G. Gamow (1928); see also Gurney and Condon (1928). For very detailed discussions of Coulomb wave functions see Bethe and Salpeter (1957); Hull and Breit (1959); Curtis (1964); Luk'yanov, Teplov, and Akimova (1965).

For treatments of the Coulomb Green's function, see Wichmann and Woo (1961); Hostler and Pratt (1963); L. Hostler (1964); J. Schwinger (1964); and, in the relativistic case, M. E. Rose (1961); Swam and Biedenharn (1963); V. G. Gorshkov (1964); Fradkin, Weber, and Hammer (1964). See also J. H. Hetherington (1963).

An interesting discussion of the pole distribution of the **S** matrix for a screened Coulomb field, in the limit of large screening radius, was given by Ferreira and Teixeira (1966). They found that this limiting pole distribution does not coincide with that of the Coulomb **S** matrix given by (14.42). For other discussions of the screened Coulomb potential, see W. F. Ford (1964); Holdeman and Thaler (1965); W. F. Ford (1966); J. D. Dollard (1968); Prugovečki and Zorbas (1973a); J. R. Taylor (1974); J. A. Zorbas (1974a); E. A. Bartnik et al. (1975); Goodmanson and Taylor (1980).

The time-dependent scattering theory was extended to the Coulomb case by J. D. Dollard (1964).

Relativistic Coulomb scattering was discussed, for example, by V. G. Gorshkov (1961); Gluckstern and Lin (1964); Dollard and Velo (1966).

Detailed studies of analyticity properties of partial-wave scattering amplitudes for charged particles are to be found in Cornille and Martin (1962); Yu. L. Mentkovsky (1962, 1963, 1964, and 1965).

For further work on scattering by pure and admixed Coulomb potentials, see L. A. Sakhnovich (1965 and 1972); P. Swan (1967a); G. B. West (1967); D. Zwanziger (1967); Mulherin and Zinnes (1970); A. M. Veselova (1970a); J. D. Dollard (1971 and 1972);

Nuttall and Stagat (1971); W. W. Zachary (1971); S. Ali et al. (1972); Marchesin and O'Carroll (1972); L. Marquez (1972); I. H. Sloan (1973); Chandler and Gibson (1974); I. Herbst (1974b); M. C. Li (1974); P. Hillion (1975); Semon and Taylor (1975); van Haeringen and van Wageningen (1975); H. van Haeringen (1976, 1977, 1978b, 1979a and b); Prugovečki and Zorbas (1978); M. J. Seaton (1978); Banerjee and Chakravorty (1978); J. Zorbas (1977 and 1978); Y. Saito (1979 and 1980a); C. Chandler (1980); F. W. Gesztesy et al. (1980); Gesztesy and Lang (1981); van Haeringen and Kok (1981a and b); Kok and van Haeringen (1981b); S. H. Patil (1981); Coulomb interference is discussed by P. R. Auvil (1968); and V. Franco (1973). See also references for Sec. 12.4. For other extensions, see Luming and Predazzi (1966).

1.4.7 The treatment of this section is a generalization of that of W. R. Theis (1956). The Bargmann potentials were first described (in a more restricted form) in V. Bargmann (1949b). The special cases described in 14.7.2 were also discussed there. The potential (14.85) was discussed by K. Chadan (1955 and 1956); and by R. G. Newton (1957), as a simple model for the deuteron. In the same context, see M. Blazek (1962a). For a discussion of pathological cases such as (14.87) and others, see Moses and Tuan (1959). For generalizations to coupled systems (spin $\frac{1}{2}$ with spin $\frac{1}{2}$), see Fulton and Newton (1956); and to coupled channel equations, J. R. Cox (1964); Cox and Garcia (1975); E. B. Plekhanov et al. (1981); B. N. Zakhariev et al. (1981).

General: For other potentials for which the Schrödinger equation is explicitly solvable, see Vogt and Wannier (1954); R. M. Spector (1964). They discuss the r^{-4} potential, which leads to a solvable equation for all k and all l. Much larger classes of potentials are treated by Bhattacharjie and Sudarshan (1962); A. K. Bose (1964); Aly and Spector (1965).

Problems

1 Suppose that two particles interact with one another via a potential $V(r)$ which contains a repulsive core. Set up the integral equations for the regular and irregular solutions, and discuss the properties of the Jost function, of the **S** matrix, and of the phase shift. Prove the Levinson theorem for the difference between the phase shift and the corresponding pure-core phase shift.

2 Calculate the Green's function that appears in Eq. (14.12) for the exponential potential.

3 Calculate the Jost function for $l = 0$ and the potential

$$\mathscr{V} = \begin{cases} -\mathscr{V}_0 & r \le R \\ -\mathscr{V}_0 e^{(R-r)/a} & r \ge R, \mathscr{V}_0 > 0. \end{cases}$$

Discuss its properties.

4 Calculate the s-wave Jost function for the potential

$$\mathscr{V} = \begin{cases} -\mathscr{V}_0 e^{-r/a} & r \le R \\ 0 & r \ge R. \end{cases}$$

Discuss its properties, especially in the limit as $R \to \infty$.

5 Discuss the Jost function of the Hulthén potential in the limit when $a \to \infty$ and $\mathscr{V}_0 \to 0$ such that $\mathscr{V}_0 a = c$ is fixed. Discuss the way the singularity structure approaches this limit.

6 Solve the s-wave Schrödinger equation for the Eckart potential by making the same substitutions as in the Hulthén case. Go to the limit $c = -1$, $a \to \infty$, $a\mathscr{V}_1 = b$, $a^2\mathscr{V}_2 = \alpha$, and discuss. Take the special case $c = 0$, and discuss.

7 Express the Eckart potentials as superpositions of Yukawas, and calculate the weight functions.

8 Solve the Schrödinger equation for $k = 0$ and arbitrary angular momentum when the potential is given by

$$\mathscr{V}(r) = \frac{\gamma r^{b-2}}{(C + r^b)^2}.$$

Find the regular solution and calculate the scattering length. (*Hint*: Change variables to $z = \ln r$.)[24]

9 Discuss the behavior of the Coulomb phase shift η_l of (14.44) in the limit of small energies. What is the relevant criterion for smallness of the energy?

10 Prove the analog of Levinson's theorem for the phase shift δ_l of (14.53).

11 Calculate the potential whose s-wave **S** matrix is ($\operatorname{Im}\beta < 0$, $\operatorname{Im}\gamma < 0$)

$$S(k) = \frac{k + \gamma}{k + \gamma^*}\,\frac{k - \gamma^*}{k - \gamma}\,\frac{k - \beta}{k - \beta^*}\,\frac{k + \beta^*}{k + \beta}.$$

Consider the special cases in which $\operatorname{Re}\gamma \to 0$ or $\operatorname{Re}\beta \to 0$ or $\operatorname{Im}\gamma \to 0$ or $\operatorname{Im}\beta \to 0$. Discuss.

12 Study the behavior of the potential as you change the position of γ in Prob. 11. Let it become purely imaginary and then a virtual and a bound state. Compare with the discussion in Sec. 12.1.4. What happens if the virtual and the bound state coincide?

13 Two spinless particles of mass 1.4 Gev and 2 Gev interact in such a way that their scattering cross section exhibits an s-wave resonance of 1 Mev width at a center-of-mass energy of 100 Mev. Construct a family of possible potentials for the interaction, assuming that there are no bound states. Plot at least one of them, and discuss.

14 Using the potential (14.85), show that for the bound state $\langle r^2 \rangle$ approaches zero as $c \to \infty$ and infinity as $c \to 0$.

15 Suppose that two particles interact with one another via the potential $\mathscr{V}_0$ and that they have no bound states. Construct a family of potentials which cause the same s-wave phase shift as $\mathscr{V}_0$ but which also cause a bound state of energy $-\varkappa^2/2\mu$. Are these potentials well behaved? Discuss.

16 Suppose that the s-wave phase shift vanishes for all energies and that there is one s-wave bound state of binding energy E_0. Construct the Jost function and a family of potentials that causes it, together with the bound-state and scattering wave functions. Discuss.

17 By making the substitutions

$$r = x^2 \qquad \psi(r) = x^{1/2}\varphi(x)$$

convert the radial Schrödinger equation with the Coulomb potential into that of another familiar and explicitly solvable potential.[25] Discuss.

18 By making the substitutions

$$r = x^n \qquad \psi(r) = x^m\varphi(x) \qquad n = \frac{2}{2 - s} \qquad m = \frac{s}{2(2 - s)}$$

24 See H. Cheng (1966).

25 This technique is due to D. I. Fivel (1966).

convert the radial Schrödinger equation for $E = 0$ and the potential

$$V(r) = \gamma r^{-s}$$

into another familiar equation. Thereby calculate the scattering lengths of this class of potentials for all angular momenta, and use them to find the strengths at which new bound states are introduced. Are there any limitations on s? If so, what are they? Discuss.

CHAPTER 15

Elastic Scattering of Particles with Spin

15.1 Partial-Wave Analysis

15.1.1 Expansion in j and s. In Sec. 10.1.2 we wrote down the form of the free wave function and of the Green's functions for particles with arbitrary spin. We now wish to subject them to a partial wave analysis similar to that used in Sec. 11.1.1 for particles of spin 0. For this purpose it is useful to introduce generalizations of the spherical harmonics analogous to the vector spherical harmonics of electromagnetic theory.

The Basis Functions. Let $\chi_\nu{}^s$ be the normalized eigenvectors of the spin, as discussed in Sec. 8.3.2, and let $C(lsJ,m\nu M)$ be the Clebsch–Gordan coefficients, as used in Sec. 2.1.4. We define the functions

$$\mathscr{Y}^M_{jls}(\hat{\mathbf{r}}) = \sum_{m\nu} C(lsj,m\nu M)\, Y_l^m(\hat{\mathbf{r}})\chi_\nu^s. \tag{15.1}$$

They are eigenfunctions of the magnitude of the total angular-momentum operator,

$$\mathbf{J}^2\mathscr{Y}^M_{jls} = j(j+1)\mathscr{Y}^M_{jls}, \tag{15.2}$$

of its z component (with respect to a given z axis),

$$J_z\mathscr{Y}^M_{jls} = M\mathscr{Y}^M_{jls}, \tag{15.3}$$

of the magnitude of the orbital angular-momentum operator (2.16),

$$\mathbf{L}^2\mathscr{Y}^M_{jls} = l(l+1)\mathscr{Y}^M_{jls}, \tag{15.4}$$

and of the magnitude of the spin operator,

$$\mathbf{S}^2 \mathscr{Y}_{jls}^M = s(s+1)\mathscr{Y}_{jls}^M. \tag{15.5}$$

In addition, let us define auxiliary functions

$$\begin{aligned} \mathfrak{Y}_j^M(lsv;\hat{\mathbf{k}}) &= \sum_{m=-l}^{l} C(lsj, mvM) i^{-l} Y_l^m(\hat{\mathbf{k}}) \\ &= i^{-1} \chi_v^{s\dagger} \cdot \mathscr{Y}_{jls}^M(\hat{\mathbf{k}}). \end{aligned} \tag{15.6}$$

The dot here indicates an inner product in spin space.

Expansion of States and Green's Function. A plane wave such as (10.2a) can be expanded by using (2.57) and the completeness of the Clebsch–Gordan coefficients,[1]

$$\sum_{mv} C(lsj, mvM) C(lsj', mvM') = \delta_{jj'} \delta_{MM'}. \tag{15.7}$$

We get

$$\psi_0(\mathbf{k}sv;\mathbf{r}) = \left(\frac{2\mu k}{\pi}\right)^{1/2} (kr)^{-1} \sum_{jMl} u_l(kr) \mathscr{Y}_{jls}^M(\hat{\mathbf{r}}) \mathfrak{Y}_j^{M*}(lsv;\hat{\mathbf{k}}) \tag{15.8}$$

in place of (11.5); μ is the mass of the particle.

In the case of two particles of spins s_1 and s_2 in the center-of-mass systems, we first form total spin functions as in (10.52). The part of the free-wave function that refers to the relative coordinates then carries the labels s_1 and s_2 in addition to s, as below (10.55). Since these are conserved in the scattering, we may suppress them. In other words, the function $\chi_v{}^s$ in (15.1) is then actually $\chi_v{}^{ss_1s_2}$ of (10.52). The only other change is that μ is now the *reduced* mass.

In a similar manner we expand the Green's function. Instead of (11.1) we write

$$G^{\pm}(E;\mathbf{r},\mathbf{r}') = 2\mu \sum_{jMls} \mathscr{Y}_{jls}^M(\hat{\mathbf{r}}) \mathscr{Y}_{jls}^{M\dagger}(\hat{\mathbf{r}}') r^{-1} r'^{-1} G_l^{\pm}(k;r,r') \tag{15.9}$$

and the $G_l^{\pm}$ are given by (11.2).

We assume that the hamiltonian is invariant under rotations. This implies that it commutes with the system's generators of rotations, which are the components of the total angular momentum,

$$[J_x, H] = [J_y, H] = [J_z, H] = 0. \tag{15.10}$$

Consequently it commutes also with $\mathbf{J}^2$,

$$[\mathbf{J}^2, H] = 0. \tag{15.11}$$

As a result, j and M are conserved, or good quantum numbers; l and s need not be. We can consequently expand the full-wave function in the form

1 See, for example, M. E. Rose (1957).

$$\psi^{(\pm)}(\mathbf{k}sv;\mathbf{r}) = \left(\frac{2\mu k}{\pi}\right)^{1/2}(kr)^{-1}\sum_{jMll's'}\psi^{j(\pm)}_{l's',ls}(k,r)\mathscr{Y}^M_{jl's'}(\hat{\mathbf{r}})\mathfrak{Y}^{M*}_j(lsv;\hat{\mathbf{k}}). \tag{15.12}$$

The fact that the coefficients ${\psi_{l's',ls}}^{j(\pm)}$ do not require $l = l'$, $s = s'$ expresses the nonconservation of the spin and orbital angular momenta. That they are independent of M, and that $\mathscr{Y}$ and $\mathfrak{Y}$ in (15.12) carry the same j, is due to rotational invariance (Wigner-Eckart theorem[2]).

The Radial Equations. Insertion of all these expansion in the Lippmann–Schwinger equation (10.58) gives us a set of coupled integral equations for the radial functions,

$$\psi^{j(\pm)}_{l's',ls}(k,r) = u_l(kr)\delta_{ll'}\delta_{ss'} + \sum_{l''s''}\int_0^\infty dr'\, G^\pm_{l'}(k;r,r')\,\mathscr{V}^j_{l's',l''s''}(r')\psi^{j(\pm)}_{l''s'',ls}(k,r') \tag{15.13}$$

where

$$\mathscr{V}^j_{l's',l''s''}(r) = 2\mu\int d\Omega\,\mathscr{Y}^{M\dagger}_{jl's'}(\hat{\mathbf{r}})H'(\hat{\mathbf{r}})\mathscr{Y}^M_{jl''s''}(\hat{\mathbf{r}}). \tag{15.14}$$

The interaction H' in general depends on the spin. The fact that $\mathscr{V}$ is independent of M and that the numbers j are equal for both $\mathscr{Y}$'s expresses the rotational invariance of H'. Since as a result M does not appear in (15.13), ψ does not depend on it either, which confirms our previous assertion.

Equations (15.12) and (15.13) show the meaning of the two sets of subscripts on ψ. The first set, l', s' indicates the component of ψ belonging to specific orbital and spin angular momenta, i.e., the relative probability amplitude of finding the particles with these angular momenta. The second set, l, s refers to the angular momenta of the incident beam, i.e., to the boundary condition. That is, they indicate the angular momenta at the time $t \to -\infty$.

The solution of the set of integral equations (15.13) satisfies the set of coupled Schrödinger equations

$$-\frac{d^2}{dr^2}\psi^j_{l's',ls} + \sum_{l''s''}\mathscr{V}^j_{l's',l''s''}\psi^j_{l''s'',ls} + \frac{l'(l'+1)}{r^2}\psi^j_{l's',ls} = k^2\psi^j_{l's',ls}. \tag{15.15}$$

For a given value of j, the values of l and l' are restricted by the properties of the Clebsch–Gordan coefficients, which enter Eqs. (15.13) and (15.15) via ${\mathscr{V}_{l's',ls}}^j$, and by the possible values of s and s'. The latter in turn are restricted by the spins s_1 and s_2 of the individual particles. There are additional restrictions if H' is invariant under reflection and hence conserves parity.

For example, let the two particles each have spin $\frac{1}{2}$. Then s and s' can be only 0 or 1. For $s = 0$, we must have $l = j$; for $s = 1$, $l = j$, and $l = j \pm 1$ are possible. If H' conserves parity, then $\mathscr{V}$ has matrix elements only between states of equal parity. Hence the states of parity $(-)^j$ are uncoupled

2 See, for example, *ibid.*, p. 85.

from those of parity $(-)^{j+1}$. The former have $l = l' = j$ and s and s' either 0 or 1; the latter, $l = j \pm 1$, $l' = j \pm 1$, $s = s' = 1$. All other combinations are uncoupled from these and of no interest; they may be set equal to zero.

The Scattering Amplitudes. The asymptotic behavior of the components of the wave function is found by means of (15.13) and (11.4),

$$\psi^{j(+)}_{l's',ls}(k,r) \underset{r\to\infty}{\simeq} \tfrac{1}{2}e^{i1/2\pi(l'+1)}(e^{-ikr}\delta_{ll'}\delta_{ss'} - S^{j}_{l's',ls}e^{-i\pi l'}e^{ikr}) \qquad (15.16)$$

with

$$S^{j}_{l's',ls} = \delta_{ll'}\delta_{ss'} - 2ik^{-1}\sum_{l''s''}\int_0^\infty dr\, u_{l'}(kr)\mathscr{V}^{j}_{l's',l''s''}(r)\psi^{j(+)}_{l''s'',ls}(k,r). \qquad (15.17)$$

The large-distance behavior of the full-wave function $\psi^{(+)}(\mathbf{k}sv,\mathbf{r})$, on the other hand, gives us the scattering amplitude (10.56). If we expand (10.57), we get

$$A(\mathbf{k}'s'v',\mathbf{k}sv) = 4\pi\sum_{jMll'}\mathfrak{Y}^{M}_{j}(l's'v',\hat{\mathbf{k}}')a^{j}_{l's',ls}(k)\mathfrak{Y}^{M*}_{j}(lsv,\hat{\mathbf{k}}) \qquad (15.18)$$

with the partial-wave amplitudes

$$a^{j}_{l's',ls} = \frac{S^{j}_{l's',ls} - \delta_{ll'}\delta_{s's}}{2ik}. \qquad (15.19)$$

According to (15.6) we may also write

$$A(\mathbf{k}'s'v',\mathbf{k}sv) = \chi^{s'\dagger}_{v'}\cdot\mathscr{A}(\mathbf{k}'s',\mathbf{k}s)\cdot\chi^{s}_{v} \qquad (15.20)$$

$$\mathscr{A}(\mathbf{k}'s',\mathbf{k}s) = 4\pi\sum_{jMll'} i^{l-l'}\mathscr{Y}^{M}_{jl's'}(\hat{\mathbf{k}}')a^{j}_{l's',ls}\mathscr{Y}^{M\dagger}_{jls}(\hat{\mathbf{k}}). \qquad (15.21)$$

The functions $\mathscr{Y}_{jls}{}^{M}$ are vectors in the spin space, and their product in (15.21) is a *direct* one; $\mathscr{A}$ is a spin operator or matrix. We then have the analog of (11.12),

$$S(\mathbf{k}'s'v',\mathbf{k}sv) = \sum_{jMll'}\mathfrak{Y}^{M}_{j}(l's'v',\hat{\mathbf{k}}')S^{j}_{l's',ls}\mathfrak{Y}^{M*}_{j}(lsv,\hat{\mathbf{k}}) \qquad (15.22)$$

and the spin matrix

$$\mathscr{S}(\mathbf{k}'s',\mathbf{k}s) = \sum_{jMll'}\mathscr{Y}^{M}_{jl's'}(\hat{\mathbf{k}}')S^{j}_{l's',ls}\mathscr{Y}^{M\dagger}_{jls}(\hat{\mathbf{k}}). \qquad (15.22a)$$

Helicity Amplitudes. A more convenient expression than (15.21) is obtained as follows. The functions $\mathscr{Y}_{jl's'}{}^{M}(\hat{\mathbf{k}}')$ we have used are expressed relative to a given coordinate system $\mathfrak{S}$, the same in which the $\mathscr{Y}_{jls}{}^{M}(\hat{\mathbf{k}})$ are expressed. They are related to the $[\mathscr{Y}_{jls}{}^{M}]'$ in a new coordinate system $\mathfrak{S}'$ by

$$\mathscr{Y}^{M}_{jl's'}(\hat{\mathbf{k}}') = \sum_{M'}[\mathscr{Y}^{M'}_{jl's'}(\hat{\mathbf{k}}')]'D^{j}_{M'M}(\mathscr{R}) \qquad (15.23)$$

where $\mathscr{R}$ is the rotation that converts $\mathfrak{S}'$ into $\mathfrak{S}$ and $D_{M'M}{}^{j}$ is the corresponding irreducible rotation matrix of dimensionality $2j + 1$. $\mathscr{R}$ is most conveniently specified in terms of Euler angles, which we use with the following convention.

In order to convert an old system into a new one whose z axis has the polar angles θ and φ in the old system, and in which the old z axis has the polar angles θ and φ', we rotate (1) the old system by φ about its z axis, (2) the system thus obtained by θ about its y axis, and (3) the system thus obtained by $\varphi' - \pi$ about its (the new) z axis. All rotations are meant in the positive screw sense.

Let us take $\mathbf{k}$ to lie along the z axis in $\mathfrak{S}$ and $\mathbf{k}'$ along the z' axis in $\mathfrak{S}'$. The simplest way of rotating $\mathbf{k}$ into $\mathbf{k}'$ is to rotate $\mathfrak{S}$ by the scattering angle θ about the normal $\mathbf{n} = \mathbf{k}' \times \mathbf{k}$ to the scattering plane. This implies that the intersection of the xy and $x'y'$ planes is the invariant line, and since the first rotation takes the y axis into it, the first and third Euler angles must be equal and opposite (see Fig. 15.1). Thus, if the direction of $\mathbf{k}'$ in $\mathfrak{S}$ is

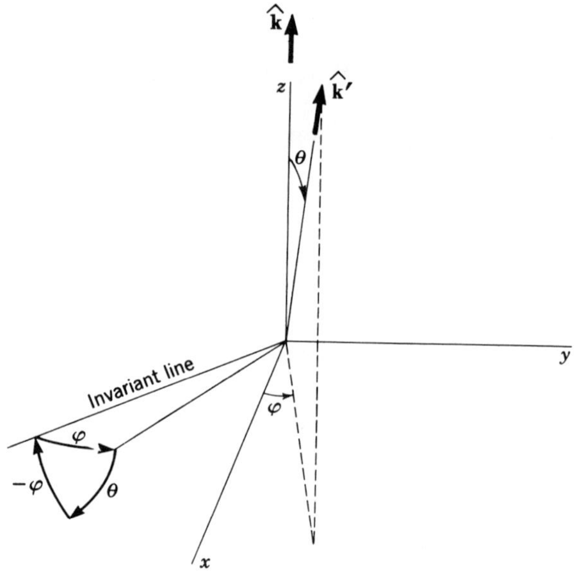

Figure 15.1. The rotation of the coordinate system associated with the incident momentum $\mathbf{k}$ into that associated with the final momentum $\mathbf{k}'$.

described by the polar angles θ and φ, then $\mathfrak{S}'$ must be such that $\mathbf{k}$ is described in it by the angles θ and $\varphi' = \pi - \varphi$. We may express this by saying that this particular rotation from $\mathbf{k}'$ to $\mathbf{k}$ is given by

$$\mathscr{R}(\mathbf{k}',\mathbf{k}) = (\varphi,\theta,-\varphi). \tag{15.24}$$

The corresponding rotation that takes $\mathbf{k}$ into $\mathbf{k}'$ must then be

$$\mathscr{R}(\mathbf{k},\mathbf{k}') = (\varphi',\theta,-\varphi') = (\pi - \varphi,\theta,\varphi - \pi). \tag{15.24a}$$

Now let us write down the $\mathscr{Y}$ functions in $\mathfrak{S}$ and $\mathfrak{S}'$,

$$\mathscr{Y}^M_{jls}(\hat{\mathbf{k}}) = \left(\frac{2l+1}{4\pi}\right)^{1/2} i^l C(lsj,0MM)\chi^s_M \tag{15.25}$$

$$\mathscr{Y}^M_{jl's'}(\hat{\mathbf{k}}') = \sum_{M'} D^j_{M'M}(\mathscr{R})\left(\frac{2l'+1}{4\pi}\right)^{1/2} i^{l'} C(l's'j,0M'M')[\chi^{s'}_{M'}]'. \tag{15.25a}$$

Equation (15.21) therefore becomes

$$\mathscr{A}(\mathbf{k}'s',\mathbf{k}s) = \sum_{jMM'll'} D^j_{M'M}(\mathscr{R})[(2l'+1)(2l+1)]^{1/2} a^j_{l's',ls} \\ C(l's'j,0M'M')C(lsj,0MM)[\chi^{s'}_{M'}]'\chi^{s\dagger}_M \tag{15.26}$$

or

$$\begin{aligned} A^h(\mathbf{k}'s'v',\mathbf{k}sv) &= A^h(\theta,\varphi;s'v',sv) \\ &= \sum_j (2j+1)a^j(s'v',sv)D^j_{v'v}(\varphi,\theta,-\varphi) \\ &= e^{i(v-v')\varphi}\sum_j (2j+1)a^j(s'v',sv)d^j_{v'v}(\theta) \end{aligned} \tag{15.27}$$

with the new partial wave amplitudes

$$\begin{aligned} a^j(s'v',sv) &= \sum_{ll'} \frac{[(2l+1)(2l'+1)]^{1/2}}{2j+1} C(l's'j,0v'v')a^j_{l's',ls}C(lsj,0vv) \\ &= (-)^{s+s'+v+v'}\sum_{ll'} C(js'l',-v'v'0)a^j_{l's',ls}C(jsl,-vv0). \end{aligned} \tag{15.28}$$

The spin components here are taken relative to the forward direction, both for the incoming and for the outgoing beam. They are *helicities*. This is the significance of the h on A. The x axis, with respect to which φ is measured, has an arbitrary direction perpendicular to $\mathbf{k}$, but it is fixed once and for all. The direction of the x axis relative to which the spin phases of the outgoing beam are measured has been fixed in the discussion preceding (15.24).

The functions $d_{v'v}{}^j(\theta)$ are Wigner's simplified rotation matrices. They are given by

$$\begin{aligned} d^j_{vv'}(\theta) &= \left[\frac{(j+v')!(j-v')!}{(j+v)!(j-v)!}\right]^{1/2} (\cos\tfrac{1}{2}\theta)^{v'+v}(\sin\tfrac{1}{2}\theta)^{v'-v}P^{(v'-v,v'+v)}_{j-v'}(\cos\theta) \\ &= (-)^{v'-v}d^j_{v'v}(\theta) \end{aligned} \tag{15.29}$$

when $v' \geq v$. Here $P_n^{(\alpha,\beta)}$ are Jacobi polynomials,[3] which are special cases of the hypergeometric function. When $\alpha = \beta = 0$, they go over into Legendre polynomials,

$$P_n = P_n^{(0,0)}.$$

We list here a number of these rotation matrices in terms of Legendre polynomials,[4]

$$\begin{aligned} d^j_{00} &= P_j \\ d^j_{1/2\,1/2} &= d^j_{-1/2-1/2} = \tfrac{1}{2}\sec\tfrac{1}{2}\theta(P_{j-1/2}+P_{j+1/2}) \\ d^j_{-1/2\,1/2} &= -d^j_{1/2-1/2} = \tfrac{1}{2}\csc\tfrac{1}{2}\theta(P_{j-1/2}-P_{j+1/2}) \\ d^j_{01} &= -d^j_{10} = d^j_{-10} = -d^j_{0-1} = [j(j+1)]^{1/2}\sin\theta\,\pi_j \\ d^j_{11} &= d^j_{-1-1} = \pi_j + \tau_j \\ d^j_{-1+1} &= d^j_{+1-1} = \pi_j - \tau_j, \end{aligned} \tag{15.30}$$

3 A. Erdelyi (1953), vol. 2, p. 168.

4 From *ibid.*, p. 173, and the recursion relations of the Legendre polynomials.

the functions π_j and τ_j having been defined in (2.56). Furthermore, for integral values of j,[5]

$$d^j_{\nu 0} = (-)^\nu \left[\frac{(j-\nu)!}{(j+\nu)!}\right]^{1/2} P_j^\nu \tag{15.31}$$

where P_j^ν is the associated Legendre function defined in (2.6).

The expansion (15.27) is evidently the simplest generalization of the standard series in the spin 0 case, the first line of (11.10). It differs from it merely by the replacement of the Legendre polynomials by the rotation functions $D_{\nu'\nu}{}^j(\varphi, \theta, -\varphi)$ whose θ dependence is expressed in terms of Jacobi polynomials. It adds to our physical insight to realize that the functions $D_{\nu'\nu}{}^j$ are the eigenfunctions of the Schrödinger equation for a *symmetrical top*. The significance of j in this context is the quantum number of the magnitude of the angular momentum and that of ν and ν', the quantum numbers of its projection onto space and body fixed axis, respectively.

If the initial beam has a spin direction other than that of the beam, and a given spin-component eigenvalue, it is described by a spin function as in (8.40). The scattering amplitude is then given by

$$a'_{\nu'} = A^h(\mathbf{k}'s'\nu') = \sum_\nu A^h(\mathbf{k}'s'\nu', \mathbf{k}s\nu)a_\nu. \tag{15.32}$$

The spin direction of the beam scattered in the direction $\mathbf{k}'$ is now calculated from (8.43) and (8.44) by using a_ν' in (8.41) and (8.42). It is not necessary to normalize the a's.

15.1.2 Amplitudes for Individual Spins. The scattering amplitude need not be labeled as in (15.18). Instead it may refer to the z components of the spins of the individual particles. Let us first disentangle (15.18) by substituting the explicit formula (15.6) and inverting the order of the summations:

$$A(\mathbf{k}'s'\nu', \mathbf{k}s\nu) = 4\pi \sum_{ll'mm'} i^{l-l'} Y_{l'}^{m'}(\hat{\mathbf{k}}')a(l'm's'\nu', lms\nu) Y_l^{m*}(\hat{\mathbf{k}}) \tag{15.33}$$

where the new partial wave amplitudes are given by

$$a(l'm's'\nu', lms\nu) = \sum_{jM} C(l's'j, m'\nu'M)a^j_{l's',ls}C(lsj, m\nu M). \tag{15.34}$$

These are evidently the (relative) probability amplitudes for finding the particles with the angular momenta $(l'm's'\nu')$ if they were initially $(lms\nu)$. The amplitudes for finding angular momenta $(l'm'\nu_1'\nu_2')$ if they were initially $(lm\nu_1\nu_2)$ are therefore

$$\begin{aligned} a(l'm'\nu_1'\nu_2', lm\nu_1\nu_2) = \sum_{jM\nu\nu'ss'} & C(s_1s_2s', \nu_1'\nu_2'\nu')C(l's'j, m'\nu'M) \\ & a^j_{l's',ls}C(lsj, m\nu M)C(s_1s_2s, \nu_1\nu_2\nu), \end{aligned} \tag{15.35}$$

and they vanish unless $m + \nu_1 + \nu_2 = m' + \nu_1' + \nu_2'$. The summations

5 From (2.7) and Erdelyi, *op. cit.*, p. 169, eq. (10).

over M, ν, and ν' can, of course, be carried out instantly, since $\nu' = \nu_1' + \nu_2'$, $\nu = \nu_1 + \nu_2$, $M = \nu_1 - \nu_2 + m = \nu_1' + \nu_2' + m'$. We may then write

$$a^M(l'\nu_1'\nu_2', l\nu_1\nu_2) = a(l'm'\nu_1'\nu_2', lm\nu_1\nu_2). \tag{15.36}$$

The scattering amplitude is now given by the expression

$$A(\mathbf{k}'\nu_1'\nu_2', \mathbf{k}\nu_1\nu_2) = 4\pi \sum_{ll'M} Y_{l'}^{M-\nu_1'-\nu_2'}(\hat{\mathbf{k}}')i^{l-l'}a^M(l'\nu_1'\nu_2', l\nu_1\nu_2)Y_l^{M-\nu_1-\nu_2*}(\hat{\mathbf{k}}). \tag{15.37}$$

Helicity Amplitudes. In order to express the spin components with respect to the direction of motion we first reinsert (15.35) in (15.37) and get

$$A(\mathbf{k}'\nu_1\nu_2', \mathbf{k}\nu_1\nu_2) = \chi_{\nu_1'}^{s_1\dagger}\chi_{\nu_1'}^{s_2\dagger} \cdot \mathscr{A}(\mathbf{k}', \mathbf{k}) \cdot \chi_{\nu_1}^{s_1}\chi_{\nu_2}^{s_2} \tag{15.38}$$

where

$$\mathscr{A}(\mathbf{k}', \mathbf{k}) = 4\pi \sum_{jMll'ss'} i^{l-l'}\mathscr{Y}_{jl's'}^{M}(\hat{\mathbf{k}}')a_{l's',ls}^{j}\mathscr{Y}_{jls}^{M\dagger}(\hat{\mathbf{k}}). \tag{15.39}$$

This differs from (15.21) only by the additional summations over s and s'. We may now use (15.25) and (15.25a) and subsequently eliminate the spin functions as in the step from (15.26) to (15.27). The result is a scattering amplitude in which both the initial and the final spins refer to the direction of relative motion. In order to make them into helicity amplitudes, we merely change the signs of ν_2 and ν_2'. In the original amplitude both ν_1 and ν_2 (as well as ν_1' and ν_2') refer to the same axis. In the helicity amplitude in the center-of-mass coordinate system, ν_1 and ν_2 (as well as ν_1' and ν_2') refer to axes of *opposite* direction, because the directions of motion of the two particles are opposite. We get

$$\begin{aligned} A^h(\mathbf{k}'\nu_1'\nu_2', \mathbf{k}\nu_1\nu_2) &= A^h(\theta, \varphi; \nu_1'\nu_2', \nu_1\nu_2) \\ &= e^{i(\nu-\nu')\varphi}\sum_j (2j+1)d_{\nu'\nu}^j(\theta)a^j(\nu_1'\nu_2', \nu_1\nu_2) \end{aligned} \tag{15.40}$$

where the partial-wave amplitudes are given by

$$\begin{aligned} a^j(\nu_1'\nu_2', \nu_1\nu_2) &= \sum_{ll'} \frac{[(2l+1)(2l'+1)]^{1/2}}{2j+1} \sum_{ss'} C(l's'j, 0\nu'\nu') \\ &\qquad C(s_1s_2s', \nu_1' - \nu_2'\nu')a_{l's',ls}^j C(lsj, 0\nu\nu)C(s_1s_2s, \nu_1 - \nu_2\nu) \\ &= (-)^{\nu_1+\nu_2-\nu_1'-\nu_2'} \sum_{ll'ss'} C(js'l', -\nu'\nu'0)C(s_2s_1s', -\nu_2'\nu_1'\nu') \\ &\qquad a_{l's',ls}^j C(s_2s_2s, -\nu_2\nu_1\nu)C(jsl, -\nu\nu 0) \\ &= \sum_{ss'} C(s_1s_2s', \nu_1' - \nu_2'\nu')a^j(s'\nu', s\nu)C(s_1s_2s, \nu_1 - \nu_2\nu) \end{aligned} \tag{15.41}$$

and

$$\nu = \nu_1 - \nu_2 \qquad \nu' = \nu_1' - \nu_2'.$$

The result (15.40) can also be derived relativistically.[6] In that case, however, the split-up of j into l and s is impossible, and (15.41) is no longer valid.

Since the most important special cases of interest concern particles of spin $\frac{1}{2}$, the Clebsch–Gordan coefficients in which $s_2 = \frac{1}{2}$ are listed in Table 15.1.

Table 15.1 Clebsch–Gordan Coefficients $C(lsj, m\nu M)$ for $s = \frac{1}{2}$

	$\nu = \frac{1}{2}$	$\nu = -\frac{1}{2}$
$j = l + \frac{1}{2}$	$\left(\frac{j+M}{2j}\right)^{1/2}$	$\left(\frac{j-M}{2j}\right)^{1/2}$
$j = l - \frac{1}{2}$	$-\left(\frac{j+1-M}{2j+2}\right)^{1/2}$	$\left(\frac{j+1+M}{2j+2}\right)^{1/2}$

Specifically,

$$\begin{aligned} C(\tfrac{1}{2}\tfrac{1}{2}1, \tfrac{1}{2}\,-\tfrac{1}{2}0) &= C(\tfrac{1}{2}\tfrac{1}{2}1, -\tfrac{1}{2}\tfrac{1}{2}0) = C(\tfrac{1}{2}\tfrac{1}{2}0, \tfrac{1}{2}\,-\tfrac{1}{2}0) \\ &= -C(\tfrac{1}{2}\tfrac{1}{2}0, -\tfrac{1}{2}\tfrac{1}{2}0) \\ &= 2^{-1/2} \\ C(\tfrac{1}{2}\tfrac{1}{2}1, \tfrac{1}{2}\tfrac{1}{2}1) &= C(\tfrac{1}{2}\tfrac{1}{2}1, -\tfrac{1}{2}-\tfrac{1}{2}-1) = 1. \end{aligned} \tag{15.42}$$

15.1.3 Unitarity, Reciprocity, Time-Reversal Invariance, and Parity Conservation. As we have seen in Sec. 7.2.4, the conservation of flux which results from a hermitian hamiltonian entails a unitary **S** matrix. It follows from the orthogonality and completeness of the $\mathscr{Y}$'s that consequently the matrix $S_{l's',ls}{}^j$ must be unitary,

$$\sum_{l''s''} S^j_{l's',l''s''} S^{j*}_{ls,l''s''} = \delta_{ll'}\delta_{ss'} \tag{15.43}$$

for each value of j.

Let us assume now that H' is invariant under *time reversal*. It must then commute with the antiunitary time-reversal operator ϑ discussed in Sec. 7.2.4:

$$[\vartheta, H'] = 0. \tag{15.44}$$

The Phase Conventions. It is convenient to use the following explicit convention for the phases of the wave functions: For spin functions of spin $\frac{1}{2}$, use

$$\chi^{1/2}_{+1/2} = \begin{pmatrix}1\\0\end{pmatrix} \qquad \chi^{1/2}_{-1/2} = \begin{pmatrix}0\\1\end{pmatrix} \tag{15.45}$$

and the time-reversal operator

6 See Jacob and Wick (1959).

$$\vartheta = i\sigma_y \mathsf{K} = \begin{pmatrix} 0 & 1 \\ -1 & 0 \end{pmatrix} \mathsf{K} \tag{15.46}$$

where K is the (antiunitary) complex conjugation operator. Then

$$\vartheta \chi_\nu^{1/2} = (-)^{1/2+\nu} \chi_{-\nu}^{1/2}.$$

For spin 0 we use $\chi_0{}^0 = 1$, and for spin 1 the functions (2.20). The time-reversal operator in these cases is simply

$$\vartheta = \mathsf{K}. \tag{15.46a}$$

According to (2.20), then, in all three instances,

$$\vartheta \chi_\nu^s = (-)^{s+\nu} \chi_{-\nu}^s. \tag{15.47}$$

In the cases of higher spins we proceed similarly, so that (15.47) is always true.

If we now combine two spin functions as in (10.52) in order to form χ's for two particles and if we use Clebsch–Gordan coefficients with the phase convention as in Rose,[7] for example, so that they obey (2.45), then the two-particle spin functions also obey (15.47), provided, of course, that we use the proper time-reversal operator. For example, for two particles of spin $\frac{1}{2}$,

$$\vartheta = (i\sigma_y^{(1)})(i\sigma_y^{(2)})\mathsf{K}$$

etc.

Next, the spin functions are combined according to (15.1) with the spherical harmonics, which satisfy (2.12). It then follows that

$$\vartheta \mathscr{Y}_{jls}^M = (-)^{j+M} \mathscr{Y}_{jls}^{-M}. \tag{15.48}$$

What is more, we have

$$\mathscr{Y}_j^{M*}(ls\nu, \hat{\mathbf{k}}) = (-)^{s+\nu-j-M} \mathscr{Y}_j^{-M}(ls\,-\nu, -\hat{\mathbf{k}}). \tag{15.49}$$

Therefore, according to (15.8),

$$\vartheta \psi_0(\mathbf{k}s\nu; \mathbf{r}) = (-)^{s+\nu} \psi_0(-\mathbf{k}s\,-\nu; \mathbf{r}) \tag{15.50}$$

and hence by (10.10) and (15.44),

$$\vartheta \psi^{(+)}(\mathbf{k}s\nu; \mathbf{r}) = (-)^{s+\nu} \psi^{(-)}(-\mathbf{k}s\,-\nu; \mathbf{r}) \tag{15.51}$$

because

$$G^{+*} = G^{-}. \tag{15.52}$$

Reciprocity. We now use the property (7.70) of the time-reversal operator in (10.57), together with (15.50), (15.51), and (7.44). This tells us that, as a consequence of time-reversal invariance [(15.44)], the scattering amplitude obeys the symmetry relation

7 Rose, *op. cit.*

$$A(\mathbf{k}'s'\nu', \mathbf{k}s\nu) = (-)^{s+s'+\nu+\nu'} A(-\mathbf{k}s -\nu, -\mathbf{k}'s' -\nu') \qquad (15.53)$$

which is usually referred to as the *reciprocity theorem* (see also Sec. 7.2.4).

The application of (7.70) and (15.48) in (15.14) shows that time reversal invariance, together with the hermiticity and rotational invariance of H', implies that the potential matrix (15.14) is *symmetric*,

$$\mathscr{V}^j_{l's',ls}(r) = \mathscr{V}^j_{ls,l's'}(r). \qquad (15.54)$$

Since it is hermitian, it must be *real.* According to (15.18) and (15.49), the reciprocity (15.53) is also equivalent to the *symmetry* of the partial wave amplitudes and of the **S** matrix,

$$S^j_{l's',ls} = S^j_{ls,l's,}. \qquad (15.55)$$

For the helicity amplitudes the situation is somewhat different from (15.53). If the hamiltonian is invariant under time reversal, then we may use the resulting symmetry of the partial wave amplitudes $a_{ls,l's'}{}^j$ together with the properties of the rotation matrices[8]

$$d^j_{\nu\nu'}(\theta) = d^j_{\nu'\nu}(-\theta) = (-)^{\nu'-\nu} d^j_{\nu'\nu}(\theta) \qquad (15.56)$$

to derive the reciprocity relation

$$A^h(\theta,\varphi; s'\nu', s\nu) = A^h(\theta, \pi - \varphi; s\nu, s'\nu'). \qquad (15.57)$$

But according to (15.24a) this means that

$$A^h(\mathbf{k}'s'\nu', \mathbf{k}s\nu) = A^h(\mathbf{k}s\nu, \mathbf{k}'s'\nu'). \qquad (15.57a)$$

Thus for the helicity amplitudes (with the coordinate definitions we have adopted) the reciprocity is of a much simpler form than (15.53).

Similarly for (15.41) and (15.40). Time-reversal invariance implies that

$$a^j(\nu_1'\nu_2', \nu_1\nu_2) = a^j(\nu_1\nu_2, \nu_1'\nu_2') \qquad (15.58)$$

and

$$A^h(\theta,\varphi; \nu_1'\nu_2', \nu_1\nu_2) = A^h(\theta, \pi - \varphi; \nu_1\nu_2, \nu_1'\nu_2') \qquad (15.59)$$

or, by (15.24a),

$$A^h(\mathbf{k}'\nu_1'\nu_2', \mathbf{k}\nu_1\nu_2) = A^h(\mathbf{k}\nu_1\nu_2, \mathbf{k}'\nu_1'\nu_2'). \qquad (15.59a)$$

Parity Conservation. In most physical problems of interest the hamiltonian is invariant under space reflection, and, as a result, *parity* is conserved. The potential matrix $\mathscr{V}_{ls,l's'}{}^j$ defined in (15.14) then couples only l values which differ by an even integer, and as a result so will the **S** matrix. Since according to (2.13) and (15.6)

$$\mathfrak{Y}^M_j(ls\nu; -\hat{\mathbf{k}}) = (-)^l \mathfrak{Y}^M_j(ls\nu; \hat{\mathbf{k}}) \qquad (15.60)$$

we get for the scattering amplitude (15.18)

$$A(-\mathbf{k}'s'\nu', -\mathbf{k}s\nu) = A(\mathbf{k}'s'\nu', \mathbf{k}s\nu). \qquad (15.61)$$

8 *Ibid.*, p. 53.

The reciprocity theorem can then be somewhat simplified to read

$$A(\mathbf{k}'s'\nu',\mathbf{k}s\nu) = (-)^{s+s'+\nu+\nu'}A(\mathbf{k}s\ -\nu,\mathbf{k}'s'\ -\nu'). \tag{15.62}$$

In the helicity amplitude (15.27) we use the facts that[9]

$$d^j_{-\nu-\nu'} = (-)^{\nu'-\nu}d^j_{\nu\nu'} \tag{15.63}$$

and

$$C(lsj,-m\ -\nu\ -M) = (-)^{l+s-j}C(lsj,m\nu M). \tag{15.64}$$

Then parity conservation leads to

$$A^h(\theta,\varphi;s'\ -\nu',s\ -\nu) = (-)^{s-s'}A^h(\theta,\pi - \varphi;s'\nu',s\nu) \tag{15.65}$$

or

$$A^h(\mathbf{k}'s'\ -\nu',\mathbf{k}s\ -\nu) = (-)^{s-s'}A^h(\mathbf{k}s'\nu',\mathbf{k}'s\nu). \tag{15.65a}$$

It should be noted that $s - s'$ is always an integer.

Similarly, (15.41) and (15.40) show that parity conservation entails the symmetries

$$a^j(-\nu_1'\ -\nu_2',-\nu_1\ -\nu_2) = a^j(\nu_1'\nu_2',\nu_1\nu_2) \tag{15.66}$$

$$A^h(\theta,\varphi;-\nu_1'\ -\nu_2',-\nu_1\ -\nu_2) = A^h(\theta,\pi - \varphi;\nu_1'\nu_2',\nu_1\nu_2) \tag{15.67}$$

$$A^h(\mathbf{k}'\ -\nu_1'\ -\nu_2',\mathbf{k}\ -\nu_1\ -\nu_2) = A^h(\mathbf{k}\nu_1'\nu_2',\mathbf{k}'\nu_1\nu_2). \tag{15.67a}$$

Together with time-reversal invariance, (15.57a), (15.65), (15.59a), and (15.67a) show that

$$A^h(\mathbf{k}'s'\ -\nu',\mathbf{k}s\ -\nu) = (-)^{s-s'}A^h(\mathbf{k}'s\nu,\mathbf{k}s'\nu') \tag{15.68}$$

$$A^h(\mathbf{k}'\ -\nu_1'\ -\nu_2',\mathbf{k}\ -\nu_1\ -\nu_2) = A^h(\mathbf{k}'\nu_1\nu_2,\mathbf{k}\nu_1'\nu_2'). \tag{15.69}$$

The Generalized Optical Theorem. The flux-conserving unitarity was expressed in (7.67) in terms of the **T** matrix. Using (10.57) it reads, in terms of the scattering amplitude,

$$\begin{aligned} -2\pi ik^{-1}[A(\mathbf{k}'s'\nu',\mathbf{k}s\nu) - A^*(\mathbf{k}s\nu,\mathbf{k}'s'\nu')] &\\ = \sum_{s''\nu''}\int d\Omega_k''\, A(\mathbf{k}'s'\nu',\mathbf{k}''s''\nu'')A^*(\mathbf{k}s\nu,\mathbf{k}''s''\nu'') &\qquad (15.70)\\ = \sum_{s''\nu''}\int d\Omega_k''\, A^*(\mathbf{k}''s''\nu'',\mathbf{k}'s'\nu')A(\mathbf{k}''s''\nu'',\mathbf{k}s\nu). & \end{aligned}$$

For the helicity amplitudes we find that the unitarity (15.43) expresses itself first of all via (15.41) in the form[10]

$$\sum_{\nu_1''\nu_2''} a^{j*}(\nu_1''\nu_2'',\nu_1'\nu_2')a^j(\nu_1''\nu_2'',\nu_1\nu_2) = \frac{1}{ik}[a^j(\nu_1'\nu_2',\nu_1\nu_2) - a^{j*}(\nu_1\nu_2,\nu_1'\nu_2')]. \tag{15.71}$$

9 *Ibid.*, pp. 38 and 54.

10 Using Blatt and Weisskopf (1952), p. 791, eq. (5.10).

We then use the fact that the rotation matrices form a representation of the rotation group, so that if $\mathscr{R}_{ij}$ denotes the rotation from coordinate system $\mathfrak{S}_j$ to system $\mathfrak{S}_i$,

$$D^j_{\nu\nu'}(\mathscr{R}_{20}) = D^j_{\nu\nu'}(\mathscr{R}_{21}\mathscr{R}_{10}) = \sum_\mu D^j_{\nu\mu}(\mathscr{R}_{21})D^j_{\mu\nu'}(\mathscr{R}_{10})$$

and since they are unitary,

$$D^j_{\nu\nu'}(\mathscr{R}) = D^{j*}_{\nu'\nu}(\mathscr{R}^{-1}). \tag{15.72}$$

Furthermore, using the Clebsch–Gordan series[11]

$$d^j_{\nu\nu'}d^{j'}_{\mu\mu'} = \sum_{J\lambda\lambda'} (-)^{\mu-\mu'}C(jj'J,\nu\,-\mu\lambda)C(jj'J,\nu'\,-\mu'\lambda')d^J_{\lambda\lambda'} \tag{15.73}$$

and the integral[12]

$$\int_{-1}^{1} d\cos\theta\, d^j_{\nu\nu}(\theta)d^{j'}_{\mu\mu}(\theta) = \frac{2}{2j+1}\delta_{jj'} \tag{15.74}$$

we find that

$$\int_0^{2\pi} d\varphi \int_{-1}^{1} d\cos\theta\, D^j_{\nu\nu'}(\varphi,\theta,-\varphi)D^{j*}_{\mu\mu'}(\varphi,\theta,-\varphi) = \frac{2\pi}{2j+1}\delta_{jj'}\delta_{\nu\mu}\delta_{\nu'\mu'}. \tag{15.75}$$

As a result we get the unitarity expression for helicity amplitudes,

$$-2\pi ik^{-1}[A^h(\mathbf{k}'\nu_1'\nu_2',\mathbf{k}\nu_1\nu_2) - A^{h*}(\mathbf{k}\nu_1\nu_2,\mathbf{k}'\nu_1'\nu_2')]$$
$$= \sum_{\nu_1''\nu_2''} \int d\Omega_k''\, A^{h*}(\mathbf{k}''\nu_1''\nu_2'',\mathbf{k}'\nu_1'\nu_2')A^h(\mathbf{k}''\nu_1''\nu_2'',\mathbf{k}\nu_1\nu_2) \tag{15.70a}$$

If the hamiltonian is invariant under time reversal, then (15.59a) shows that this can be written more simply in the form

$$4\pi k^{-1}\,\mathrm{Im}\, A^h(\mathbf{k}'\nu_1'\nu_2',\mathbf{k}\nu_1\nu_2)$$
$$= \sum_{\nu_1''\nu_2''} \int d\Omega_k'' A^{h*}(\mathbf{k}''\nu_1''\nu_2'',\mathbf{k}'\nu_1'\nu_2')A^h(\mathbf{k}''\nu_1''\nu_2'',\mathbf{k}\nu_1\nu_2). \tag{15.70b}$$

If we chose $\mathbf{k}' = \mathbf{k}$ and $\nu_1' = \nu_1$, $\nu_2' = \nu_2$ or $s = s'$, $\nu' = \nu$, then both (15.70) and (15.70a) go over into the optical theorem. Time reversal invariance is not required for this.

The Eigenphase Shifts. The unitarity (15.43) together with the symmetry (15.55) of **S** implies that it can be diagonalized by an *orthogonal real* matrix B,

$$S^j_{ls,l's'} = \sum_\alpha B^j_{ls,\alpha}\exp(2i\delta^j_\alpha)B^j_{l's',\alpha} \tag{15.76}$$

where the *eigenphase shifts* $\delta_\alpha{}^j$ are real. The angles in B are called the *mixture parameters*.

11 Rose, *op. cit.*, p. 58, together with (4.19) on p. 54.

12 Rose *op. cit.*, p. 74.

The physical meaning of the eigenphase shifts is seen by forming a new set of wave functions

$$\psi_{ls,\alpha}^{j(+)}(k,r) = \sum_{l's'} \psi_{ls,l's'}^{j(+)}(k,r) B_{l's',\alpha}^{j} \tag{15.77}$$

which also satisty (15.15) and an outgoing-wave boundary condition. Whereas $\psi_{ls,l's'}{}^{j}$ has an incoming wave of angular momenta l', s' only, $\psi_{ls,\alpha}{}^{j}$ has incoming waves of all angular momenta (compatible with j). However, its asymptotic behavior, according to (15.16) and (15.76) is given by

$$\psi_{ls,\alpha}^{j(+)} \underset{r\to\infty}{\simeq} B_{ls,\alpha}^{j} e^{i\delta_\alpha{}^j} \sin(kr - \tfrac{1}{2}\pi l + \delta_\alpha^j). \tag{15.78}$$

Its characteristic property is therefore that all its components experience the same phase shifts $\delta_\alpha{}^j$. Both $\delta_\alpha{}^j$ and $B_{ls,\alpha}{}^j$ generally depend upon the energy.

15.1.4 Special Cases. *Spins 0 and $\frac{1}{2}$.* We write down explicitly the most important special amplitudes. First, take the case of two particles, one of spin $\frac{1}{2}$ and the other of spin 0. In this case the total spin s must be $\frac{1}{2}$, and the only possible values of l for a given j are $j \pm \frac{1}{2}$. If the hamiltonian conserves parity, then there can be no amplitudes connecting $l = j + \frac{1}{2}$ with $l = j - \frac{1}{2}$; in other words, we must have $l = l'$. The amplitudes are then found to be

$$\begin{aligned} A_{++}^{h} = A_{--}^{h} &= \tfrac{1}{4}\sec\tfrac{1}{2}\theta \sum_{j=1/2}^{\infty} (2j+1)(a_-^j + a_+^j)(P_{j-1/2} + P_{j+1/2}) \\ A_{\substack{-+\\+-}}^{h} &= \pm\tfrac{1}{4}e^{\pm i\varphi}\csc\tfrac{1}{2}\theta \sum_{j=1/2}^{\infty} (2j+1)(a_-^j - a_+^j)(P_{j-1/2} - P_{j+1/2}). \end{aligned} \tag{15.79}$$

The labels on A^h refer to forward (+) or backward (−) spin and the labels on a^j to $l = j + \frac{1}{2}$, (+), or $l = j - \frac{1}{2}$, (−). If the interaction is spin-independent, then a^j depends on l only, and $a_-{}^j = a_+{}^{j-1}$. In this case (15.79) reduces to (11.10), multiplied by the amplitude for finding the spin in the final forward (backward) direction if it was originally in the initial forward (or backward) direction.

Two Particles of Spin $\frac{1}{2}$. In the physically important instance of two spin $\frac{1}{2}$ particles complications begin to occur. There are, however, the following simplifying restrictions:

The two spins may combine in a singlet state ($s = 0$) or a triplet state ($s = 1$). The parity of the singlet state of angular momentum j is necessarily $(-)^j$, since $l = j$. For the triplet state $l = j$ or $l = j \pm 1$; hence there are a state of parity $(-)^j$ and two of parity $(-)^{j+1}$. If parity is conserved, then the matrix $a_{l's',ls}{}^j$ falls apart into two uncoupled pieces: a 2×2 matrix for the triplet states of parity $(-)^{j+1}$, and a 2×2 matrix for the singlet state, the triplet state of parity $(-)^j$, and singlet-triplet transitions. If we further assume that the two particles are *identical*, then the singlet-triplet transition terms must vanish. This is because the singlet spin function is antisymmetric and the triplet symmetric under exchange [see Eq. (15.42)]. Since parity conservation forces the final orbital angular momentum to be equal to the initial, if the initial state is a singlet, the only triplet state that can be reached from a singlet without parity violation has the opposite symmetry under exchange. [Remember that the "space exchange" carries a factor $(-)^l$.]

If furthermore the hamiltonian is invariant under time reversal, then the 2×2 matrix for the triplet state of parity $(-)^{j+1}$ is symmetric.

For identical particles in parity-conserving interaction, then, the partial-wave amplitudes (15.41) fall into five classes: those in which no helicity flips occur, those in which one particle flips its helicity, and those in which both do; among the first and the last we must distinguish between equal and opposite helicities of the two particles. We label these partial-wave amplitudes (and similarly the corresponding full amplitudes) by $a_{o0}{}^j$, $a_{e0}{}^j$, $a_1{}^j$, $a_{o2}{}^j$, and $a_{e2}{}^j$. We find that

$$\begin{aligned} a_{o0}^j &= \tfrac{1}{2}(\bar{a}_{11}^j + b^j) \\ a_{e0}^j &= \tfrac{1}{2}(\bar{a}_{22}^j + a_s^j) \\ a_{o2}^j &= \tfrac{1}{2}(\bar{a}_{11}^j - b^j) \\ a_{e2}^j &= \tfrac{1}{2}(\bar{a}_{22}^j - a_s^j) \\ a_1^j &= \tfrac{1}{2}[j(j+1)]^{1/2}\bar{a}_{12}^j \end{aligned} \tag{15.80}$$

where $a_s{}^j$ is the singlet amplitude, b^j the triplet amplitude for parity $(-)^j$, and the $\bar{a}_{mn}{}^j$ are transformed amplitudes for the triplet states of parity $(-)^{j+1}$. Let $\mathfrak{a}^j$ be the matrix made up of $a_{j-1\,j-1}{}^j = a_{--}{}^j$, etc.,

$$\mathfrak{a}^j = \begin{pmatrix} a_{--}^j & a_{-+}^j \\ a_{-+}^j & a_{++}^j \end{pmatrix}.$$

Then

$$\bar{\mathfrak{a}}^j = U_j \mathfrak{a}^j U_j^{-1} \tag{15.81}$$

with

$$U_j = U_j^{-1} = (2j+1)^{-1/2}\begin{pmatrix} (j+1)^{1/2} & j^{1/2} \\ j^{1/2} & -(j+1)^{1/2} \end{pmatrix}. \tag{15.82}$$

Explicitly,

$$\begin{aligned} \bar{a}_{11}^j &= \frac{(j+1)a_{--}^j + ja_{++}^j + 2[j(j+1)]^{1/2}a_{-+}^j}{2j+1} \\ \bar{a}_{22}^j &= \frac{ja_{--}^j + (j+1)a_{++}^j - 2[j(j+1)]^{1/2}a_{-+}^j}{2j+1} \\ \bar{a}_{12}^j = \bar{a}_{21}^j &= [j(j+1)]^{1/2}\frac{a_{--}^j - a_{++}^j - [j(j+1)]^{1/2}a_{-+}^j}{2j+1}. \end{aligned} \tag{15.82a}$$

According to (15.40) and (15.30) the full amplitudes are given by

$$
\begin{aligned}
A^h_{oo} &= \sum_j (2j+1) a^j_{oo}(\pi_j + \tau_j) \\
A^h_{eo} &= \sum_j (2j+1) a^j_{eo} P_j \\
A^{h\prime}_{o2} &= \sum_j (2j+1) a^j_{o2}(\pi_j - \tau_j) \\
A^h_{e2} &= \sum_j (2j+1) a^j_{e2} P_j \\
A^{h\prime}_1 &= \sin\theta \sum_j (2j+1)[j(j+1)]^{1/2} a^j_1 \pi_j.
\end{aligned}
\tag{15.83}
$$

The "one-flip" and the "opposite two-flip" amplitudes are primed, because the various members of these two classes have different phases, namely

$$
\begin{aligned}
A^h(+-,-+) &= e^{-2i\varphi} A^{h\prime}_{o2} \\
A^h(-+,+-) &= e^{2i\varphi} A^{h\prime}_{o2} \\
A^h(+-,++) &= A^h(+-,--) = A^h(++,-+) \\
&= A^h(--,-+) = e^{-i\varphi} A^{h\prime}_1 \\
A^h(-+,++) &= A^h(++,+-) = A^h(--,+-) \\
&= A^h(-+,--) = -e^{i\varphi} A^{h\prime}_1.
\end{aligned}
\tag{15.84}
$$

These phase factors, of course, are irrelevant for the cross sections, but they matter when spin directions are calculated.

15.1.5 Cross Sections. In order to calculate differential cross sections, the magnitude of the corresponding amplitude is squared. The resulting dependence upon the scattering angle can again be expanded on the basis of the d^j functions by use of the Clebsch–Gordan series (15.73). The angles of all three d^j functions there are the same; the λ and λ' sums can be carried out immediately. Since we need only $\mu = \nu$, $\mu' = \nu'$, the relevant formula is

$$
d^j_{\nu\nu'} d^{j'}_{\nu\nu'} = (-)^{\nu-\nu'} \sum_J C(jj'J, \nu\, -\nu 0) C(jj'J, \nu'\, -\nu' 0) P_J \tag{15.85}
$$

so that the expansion is directly in terms of Legendre polynomials.

The density matrix of the scattered beam can now be calculated by the method of Sec. 8.3.3 and polarizations as shown in Sec. 8.3.2.

Total scattering cross sections are obtained by integrating and using the fact that[13]

$$
\int_{-1}^{1} d\cos\theta\, d^j_{\nu\nu'}(\theta) d^{j'}_{\nu\nu'}(\theta) = \frac{2}{2j+1}\delta_{jj'}. \tag{15.86}
$$

We then obtain the total scattering cross sections for specific helicities from (15.40),

$$
\sigma(\nu_1'\nu_2', \nu_1\nu_2) = \pi \sum_j (2j+1) |a^j(\nu_1'\nu_2', \nu_1\nu_2)|^2. \tag{15.87}
$$

13 This follows from (15.85) and *ibid.*, p. 38.

If the initial beam is unpolarized, then the total cross section is given by the expression

$$\sigma = (2s_1 + 1)^{-1}(2s_2 + 1)^{-1}\pi \sum_{jv_1v_2v_1'v_2'} (2j + 1)|a^j(v_1'v_2', v_1v_2)|^2. \quad (15.88)$$

The Zero-Energy Limit. In the low-energy limit the centrifugal barrier permits $l = l' = 0$ only. Equation (15.41) then shows that

$$a^j(v_1'v_2', v_1v_2) = \frac{2}{2j+1} C(s_1s_2j, v_1' - v_2'v')a^j_{0j,0j}C(s_1s_2j, v_1 - v_2v), \quad (15.89)$$

and the total cross section at zero energy is given by

$$\sigma = \frac{4\pi}{(2s_1 + 1)(2s_2 + 1)} \sum_s \frac{|a^s_{0s,0s}|^2}{2s + 1} \quad (15.90)$$

where s runs over all values of the total spin that can be formed from s_1 and s_2. In the case of $s_1 = s_2 = \frac{1}{2}$, for example, the total cross section at zero energy is

$$\sigma = \pi(|a^0_{00,00}|^2 + \tfrac{1}{3}|a^1_{01,01}|^2).$$

But the zero-energy cross section is now not necessarily isotropic. The fact that j need not be zero when $l = 0$ makes it possible for Legendre polynomials of higher order to appear.

15.1.6 Double Scattering. If two successive scatterings occur on centers whose mutual distance is large compared with the wavelength k^{-1}, the double-scattering amplitude is given by

$$A(\mathbf{k}'v_1'v_2', \mathbf{k}v_1v_2) = \sum_{v_1''v_2''} A_2(\mathbf{k}'v_1'v_2', \mathbf{k}''v_1''v_2'')A_1(\mathbf{k}''v_1''v_2'', \mathbf{k}v_1v_2) \quad (15.91)$$

where $\mathbf{k}''$ is the vector from the first to the second scatterer. A similar formula holds for amplitudes labeled by s and v instead of v_1 and v_2.

Notice that, since φ appears in (15.40) only in an overall phase factor, an ordinary scattering cross section for unpolarized particles (or for longitudinally polarized ones) cannot have any azimuthal dependence. In the double-scattering cross sections obtained by squaring the magnitude of (15.91), however, the azimuthal angle φ does *not* disappear, and the cross section usually depends upon it. The physical explanation, of course, is simple. In the single scattering no transverse direction is preferred to any other unless there is transverse polarization in the incoming beam. Hence the scattering cross section must be rotationally invariant about the direction of incidence. In the double scattering, the first scattering plane defines a preferred direction other than that of the beam falling on the second target (unless the two targets are collinear with the incident beam). Hence, in the second scattering, there may be an azimuthal dependence about the line joining the targets. However, this is possible only if the particles carry spin and if the first scattering produces a transverse polarization.

Otherwise there is no agent to carry a "memory" of the first scattering plane to the second target. The double-scattering experiment thus serves to analyze the transverse polarization produced in the first scattering.

15.2 Solution of the Coupled Schrödinger Equations

15.2.1 The Matrix Equation. We now consider the calculation of the partial wave-scattering amplitudes by solving the coupled radial Schrödinger equations (15.15). It is convenient to use a matrix notation in which $\psi_{l's',ls}{}^j$ is an element of the square matrix Ψ_j and $\mathscr{V}_{l's',l''s''}{}^j$ [(15.14)] of the square matrix V_j. The dimensionality of the matrices depends on the possible values of l and s for given j and s_1, s_2. Conservation laws may reduce these dimensionalities. For example, if H' conserves parity, then V_j couples l values of the same parity only and the equations fall apart into two uncoupled sets which may be treated separately, one set in which $j - l$ and $j - l'$ are both even, and one set in which they are both odd. Additional restrictions may arise for identical particles, as discussed in Sec. 15.1.4.

The set of coupled Schrödinger equations now is a matrix equation

$$-\Psi_j'' + L(L+1)r^{-2}\Psi_j + V_j\Psi_j = k^2\Psi_j \tag{15.92}$$

where L is a diagonal matrix, with the various values of l on the main diagonal. Each column of the square matrix Ψ_j solves (15.92), but the various columns differ by their boundary condition. According to (15.16) the meaning of the second indices is the angular-momentum state of the incoming wave at large distances.

As an example, take the case of two identical particles of spin $\frac{1}{2}$ in parity-conserving time-reversal invariant interaction. (We shall refer to this simply as the *neutron-proton case*.) In order to make the interaction as local as possible and still have it spin-dependent, we assume that

$$2\mu H' = \mathscr{V}_c(r) + \mathscr{V}_\sigma(r)\boldsymbol{\sigma}_1 \cdot \boldsymbol{\sigma}_2 + \mathscr{V}_t(r)S_{12} \tag{15.93}$$

where $\frac{1}{2}\boldsymbol{\sigma}_1$ and $\frac{1}{2}\boldsymbol{\sigma}_2$ are the spin operators for the two particles, and S_{12} is the *tensor operator*,

$$S_{12} = 3\boldsymbol{\sigma}_1 \cdot \hat{\mathbf{r}}\boldsymbol{\sigma}_2 \cdot \hat{\mathbf{r}} - \boldsymbol{\sigma}_1 \cdot \boldsymbol{\sigma}_2. \tag{15.94}$$

As discussed in Sec. 15.1.4, the interaction matrix V_j then decomposes itself into three "blocks": two diagonal elements, for the singlet state and for the triplet state of parity $(-)^j$; and a 2×2 block for the coupled triplet states of parity $(-)^{j+1}$. The first two equations being uncoupled, they are solved by the methods of Chap. 12. We are now interested in the coupled equations for the triplet state of parity $(-)^{j+1}$. The orbital angular momenta being $j-1$ and $j+1$, we have

$$L(j) = \begin{pmatrix} j-1 & 0 \\ 0 & j+1 \end{pmatrix}. \tag{15.95}$$

The potential matrix is calculated by using (15.14). We get

$$V_j = \frac{1}{2j+1}\begin{pmatrix} (2j+1)\mathscr{V}_d - 2(j-1)\mathscr{V}_t & 6[j(j+1)]^{1/2}\mathscr{V}_t \\ 6[j(j+1)]^{1/2}\mathscr{V}_t & (2j+1)\mathscr{V}_d - 2(j+2)\mathscr{V}_t \end{pmatrix} \tag{15.96}$$

with

$$\mathscr{V}_d = \mathscr{V}_c + \mathscr{V}_\sigma.$$

An additional interaction of importance in nuclear applications is the spin-orbit force,

$$V_{LS} = \mathbf{L}\cdot\mathbf{S}V_0(r) \tag{15.97}$$

where $\mathbf{L}$ and $\mathbf{S}$ are the orbital and total spin angular momenta. Inserted in (15.14), it gives rise to the additional contribution

$$V_j = \begin{pmatrix} j-1 & 0 \\ 0 & -j-2 \end{pmatrix}\mathscr{V}_0(r), \tag{15.98}$$

$\mathscr{V}_0 = 2\mu V_0$. Note that it is diagonal but that it contributes forces of opposite sign in the two orbital angular-momentum states.

The state $j = 0$ must be treated separately. For it, the $l = j - 1$ component is a "nonsense" state of no physical interest. Since for $j = 0$ the off-diagonal terms in (15.95) vanish, the equations are uncoupled and we need to pay attention to the $l = j + 1$ state only. It is treated by the methods of Chap. 12.

15.2.2 Solutions. *The Regular Solution.* The matrix differential equation (15.92) is solved by the methods employed in Chap. 12. The only difficulty, and it is a minor one, arises from the fact that the equations coupled together have different centrifugal terms and hence the solutions have a different behavior at $r = 0$. The first effect of this is that there is no simple way of defining a regular solution by a boundary condition at the origin. For the triplet state of two spin $\frac{1}{2}$ particles, with parity $(-)^{j+1}$, the dominant term in one regular column solution is the first, and it is of order r^j. That of the other column is the second and of order r^{j+2}. The other elements are of an intermediate order, and they depend on the potential. Thus there is no way of specifying the first column by a boundary condition in a way that rules out an addition of an arbitrary multiple of the second.

The altered situation at $r = 0$ can be seen also by an expansion of the solution in a power series in r if the potential is a square well. One then finds that all but one of the regular column solutions contain the logarithmic terms of the Fuchs theory.

The absence of a simple boundary condition for the regular solution is not an insurmountable obstacle, though. We simply write down immediately the integral equation that is the obvious analog of (12.133). In the neutron-proton case the zero-order solution is taken to be

$$\Phi_j^{(0)}(k,r) = \begin{pmatrix} (2j-1)!!k^{-j}u_{j-1}(kr) & 0 \\ 0 & (2j+3)!!k^{-j-2}u_{j+1}(kr) \end{pmatrix} \tag{15.99}$$

and the Green's function, for $r' < r$,

$$\mathfrak{G}_j(k;r,r') = \begin{pmatrix} g_{j-1}(k;r,r') & 0 \\ 0 & g_{j+1}(k;r,r') \end{pmatrix} \tag{15.100}$$

if g_l is defined by (12.135). It is clear what these functions should be in more general instances. The diagonal matrix L introduced in (15.92) may be used to write, in a symbolism that should cause no difficulty,

$$\Phi_j^{(0)}(k,r) = (2L+1)!!k^{-L-1}u_L(kr) \tag{15.99a}$$

$$\mathfrak{G}_j(k;r,r') = g_L(k;r,r'). \tag{15.100a}$$

We shall make use of the matrix L in a similar manner further on.

One would now be tempted to write down the integral equation for a regular solution Φ_j that is the analog of (12.133), simply as

$$\Phi_j(k,r) = \Phi_j^{(0)}(k,r) + \int_0^r dr'\, \mathfrak{G}_j(k;r,r')V_j(r')\Phi_j(k,r').$$

This, however, works only if the off-diagonal elements of V_j strongly tend to zero at the origin. The difficulty is that the matrix product in the integrand contains products of a regular solution of one orbital angular momentum and an irregular solution of a higher one (from the Green's function). As a result the integral diverges at $r = 0$ unless strong assumptions are made concerning the potential elements.

A method to aviod this difficulty is to modify the inhomogeneity in the integral equation and to replace $\Phi_j^{(0)}$ by a judiciously chosen (constant) matrix multiple of $\Phi_j^{(0)}$. In the neutron-proton case this is

$$\Phi_j^{(0)}\left[\mathbb{1} - (2j+3)^{-1}\int_0^{R_0} dr'\, r'^{-1}V_j^{\mathrm{OFF}}(r')\right]$$

where

$$V_j^{\mathrm{OFF}} = V_{j-1,j+1}^j \begin{pmatrix} 0 & 0 \\ 1 & 0 \end{pmatrix}$$

and R_0 is an arbitrary fixed number. It may, for example, be taken to be infinite. The integral equation is now written by splitting the integral

$$\int_0^{R_0} = \int_0^r + \int_r^{R_0}$$

that is,

$$\Phi_j(k,r) = \Phi_j^{(0)}(k,r)\left[\mathbb{1} + (2j+3)^{-1}\int_{R_0}^r dr'\, r'^{-1}V_j^{\mathrm{OFF}}(r')\right]$$
$$+ \int_0^r dr'\left[\mathfrak{G}_j(k;r,r')V_j(r')\Phi_j(k,r') - (2j+3)^{-1}\Phi_j^{(0)}(k,r)V_j^{\mathrm{OFF}}(r')r'^{-1}\right]. \tag{15.101}$$

In this form it is no longer necessary to assume that

$$\int_0 dr\, r^{-1} V_j^{\mathrm{OFF}}$$

exists. The troublesome term cancels out, and the assumptions on the potential matrix V_j need be no stronger than in Chap. 12. The series of successive approximations to (15.101) converges absolutely under the same conditions as that for φ_l, and Φ_j has the same analyticity properties.

In the more general case, a similar trick of introducing counterterms to ensure convergence works equally well. They have to be more complicated, of course, if the difference between the largest and smallest l value coupled to the same j is larger.

Irregular Solutions. Irregular solutions $F_{j\pm}$ that are the analog of $f_{l\pm}$ are easily defined by the boundary condition

$$\lim_{r\to\infty} e^{\mp ikr} F_{j\pm}(k,r) = \mathbb{1} \tag{15.102}$$

or by the integral equation

$$F_{j\pm}(k,r) = F_{j\pm}^{(0)}(k,r) - \int_r^\infty dr'\, \mathfrak{G}_j(k;r,r')V_j(r')F_{j\pm}(k,r') \tag{15.103}$$

where

$$F_{j\pm}^{(0)}(k,r) = e^{-i1/2\pi L} w_L^{(\pm)}(kr). \tag{15.104}$$

Equation (15.103) can be solved by successive approximations under the same conditions as (12.138), and F_j has the same regularity properties as f_l.

15.2.3 Jost Matrix and S Matrix. The next step is to define Jost-matrix functions $\mathfrak{F}_{j\pm}$ by the analog of (12.28). The wronskian must now be defined, however, as[14]

$$W(F,\Phi) \equiv \tilde{F}\Phi' - \tilde{F}'\Phi. \tag{15.105}$$

The transpose is necessary in order that the wronskian of two matrix solutions of the same differential equation (with a symmetric potential matrix) be a constant. With this definition we set

$$\mathfrak{F}_{j\pm}(k) = W(F_{j\pm}, \Phi_j). \tag{15.106}$$

Then Φ_j is expressible in the form

$$\Phi_j(k,r) = \frac{1}{2ik}(F_{j+}\mathfrak{F}_{j-} - F_{j-}\mathfrak{F}_{j+}) \tag{15.107}$$

and the **S** matrix is obtained by comparison with (15.16) and (15.102),

$$S^j = e^{i\pi L}\mathfrak{F}_{j-}\mathfrak{F}_{j+}^{-1}. \tag{15.108}$$

The diagonal matrix $e^{i\pi L}$ commutes with S^j if parity is conserved, because only l values that differ by an even integer are coupled. In the neutron-proton case of interest we may write $e^{i\pi L} = -e^{i\pi j}$.

14 The tilde denotes the transpose of a matrix.

The symmetry of S^j is demonstrated by using the equation

$$W(\Phi_j, \Phi_j) = 0 \tag{15.109}$$

which follows from the integral equation (15.101). Insertion of (15.101) yields

$$\tilde{\mathfrak{F}}_{j-}\mathfrak{F}_{j+} = \tilde{\mathfrak{F}}_{j+}\mathfrak{F}_{j-}$$

and therefore

$$\tilde{S}^j = S^j. \tag{15.110}$$

This confirms that, because of time-reversal invariance and our choice of phases, the **S** matrix is symmetric. Time-reversal invariance manifests itself in (15.92) in the symmetry, and therefore also reality, of the potential matrix. Without that (15.110) would not hold.

Of course, the **S** matrix also satisfies

$$S^j(-k) = S^{j*}(k) = S^j(k)^{-1} \tag{15.111}$$

for real k, just like (12.74). Hence it is unitary, as it should be.

The physical wave function of (15.13) asymptotically behaves like (15.16). Comparison with (15.107) shows that

$$\Psi^{j(+)} = k\Phi_j \mathfrak{F}_{j+}^{-1} e^{i1/2\pi L}. \tag{15.112}$$

Jost-matrix functions analogous to (12.142) and (12.142a) are defined by

$$\mathscr{F}_{j\pm}(k) = \mathfrak{F}_{j\pm}(k)[(2L+1)!!]^{-1}k^L e^{\mp i1/2\pi L}. \tag{15.113}$$

The matrix function $\mathscr{F}_j$ has an integral representation similar to (12.144), but it is somewhat complicated because of the counterterms in (15.101). In the neutron-proton case,

$$\mathscr{F}_j(k) = \mathbb{1} + \int_0^\infty dr\{e^{-i1/2\pi L}w_L^{(+)}(kr)V_j(r)\Phi_j(k,r)e^{-i1/2\pi L}[(2L+1)!!]^{-1}k^L + (2j+1)V_j^{\mathrm{OFF}}(r)r^{-1}k^{-2}\} \tag{15.114}$$

provided that R_0 in (15.101) is chosen to be infinite. It may be checked that the counterterm in (15.114) is just right to cancel an otherwise present divergence at $r = 0$. Equation (15.114) is easily seen to lead to (15.17). An analog of (12.143) is more complicated, and we shall not write it down.

High- and Low-Energy Limits. The matrix $\mathscr{F}_j$ has the high-energy limit

$$\mathscr{F}_j(k) \xrightarrow[|k|\to\infty]{} \mathbb{1} \tag{15.115}$$

for $\operatorname{Im} k \geq 0$ and since by (15.108) and (15.113)

$$S^j = \mathscr{F}_{j-}\mathscr{F}_{j+}^{-1} \tag{15.116}$$

then also

$$S^j \xrightarrow[|k|\to\infty]{} \mathbb{1} \tag{15.117}$$

for real values of k. Hence we can define the eigenphase shifts $\delta_\alpha{}^j$ of (15.76) so that they tend to zero at infinite energy.

The low-energy behavior of the **S** matrix is obtained as was (12.161). We now find that as $k \to 0$

$$S^j_{ll'}(k) - \delta_{ll'} = O(k^{l+l'+1}) \tag{15.118}$$

provided that

$$\int_0^\infty dr\, r^{2l+2} |V^j_{l'l''}| < \infty \tag{15.119}$$

for all l, l', and l'' coupled to j, and provided that det $\mathfrak{F}_j(0) \neq 0$. The situation is complicated if det $\mathfrak{F}_j(0) = 0$, and we refer to the literature for that special case.[15]

The K Matrix. Equation (15.118) implies that near zero energy the **K** matrix goes as the right-hand side. We therefore define a reduced **K** matrix that is the analog of (12.162),

$$\begin{aligned}\mathscr{K}^j &= i(-ik)^{-L-1/2} K^j(-ik)^{-L-1/2} \\ &= (-ik)^{-L-1/2}(S^j - \mathbb{1})(S^j + \mathbb{1})^{-1}(-ik)^{-L-1/2}\end{aligned} \tag{15.120}$$

so that

$$\begin{aligned}S^j &= [\mathbb{1} + (-ik)^{L+1/2}\mathscr{K}^j(-ik)^{L+1/2}][\mathbb{1} - (-ik)^{L+1/2}\mathscr{K}^j(-ik)^{L+1/2}]^{-1} \\ &= (-ik)^{L+1/2}[(-ik)^{2L+1} + \mathscr{K}^j][(-ik)^{2L+1} - \mathscr{K}^j]^{-1}(-ik)^{-L-1/2}.\end{aligned} \tag{15.120a}$$

The matrix $\mathscr{K}^j$ is real (for real k) symmetric and, because of (15.111), an even function of k. In analogy with (12.164) we define a symmetric matrix $\mathfrak{M}^j$ by

$$S^j = \mathbb{1} - k^{L+1/2}\mathfrak{M}^j k^{L+1/2} \tag{15.121}$$

so that

$$(\mathfrak{M}^j)^{-1} = \tfrac{1}{2}k^{2L+1} - \tfrac{1}{2}i^{L+1/2}(\mathscr{K}^j)^{-1}i^{L+1/2} \tag{15.122}$$

Eigenphase Shifts and Fredholm Determinant. For some purposes it is convenient to consider directly the determinant of the **S** matrix. According to (15.76) it can be written in terms of the sum of the eigenphase shifts,

$$\det S^j = e^{2i\delta^j} \tag{15.123}$$

where

$$\delta^j = \sum_\alpha \delta^j_\alpha. \tag{15.124}$$

According to (15.116) it can also be written as

$$\det S^j = \frac{\Delta^j_-}{\Delta^j_+} \tag{15.125}$$

15 See R. G. Newton (1955).

where

$$\Delta_+^j = \Delta^j = \det \mathscr{F}_j \tag{15.126}$$

is the Fredholm determinant of the set of coupled integral equations (15.13). [This is shown exactly as (12.43).]

An expression for the complete Green's function is obtained in the same way as (12.40) and (12.146). Let us first define an auxiliary matrix solution $\Lambda(k,r)$ of (15.92) by the boundary conditions

$$\Lambda_j(k,r_0) = 0 \qquad \Lambda_j'(k,r_0) = \mathbb{1} \tag{15.127}$$

at an arbitrary point $r_0 \neq 0$. It can be expressed in terms of $F_{j\pm}$ by

$$\Lambda_j = \frac{1}{2ik}(F_{j+}a_{j-} - F_{j-}a_{j+}) \tag{15.128}$$

where

$$a_{j\pm} = W(F_{j\pm}, \Lambda_j).$$

This wronskian can be calculated at $r = r_0$, which gives, according to (15.127),

$$a_{j\pm} = \tilde{F}_{j\pm}(k,r_0). \tag{15.129}$$

Substitution in (15.128) and use of (15.127) then tells us that

$$F_{j+}\tilde{F}_{j-} - F_{j-}\tilde{F}_{j+} = 0 \tag{15.130}$$

$$F_{j+}'\tilde{F}_{j-} - F_{j-}'\tilde{F}_{j+} = 2ik\mathbb{1}. \tag{15.130a}$$

Since r_0 could have any positive value, these equations must hold for all r.

Equations (15.130) and (15.130a) are now used to verify that

$$\mathscr{G}_j^+(k;r,r') = \begin{cases} -\Phi_j(k,r)\mathfrak{F}_{j+}^{-1}(k)\tilde{F}_{j+}(k,r') & r < r' \\ -F_{j+}(k,r)\mathfrak{F}_{j+}^{-1}(k)\tilde{\Phi}_j(k,r') & r > r' \end{cases} \tag{15.131}$$

has the properties required of a Green's function. Because of (15.112) this may also be written

$$\mathscr{G}_j^+(k;r,r') = \begin{cases} -k^{-1}\Psi^{j(+)}(k,r)e^{-i1/2\pi L}\tilde{F}_{j+}(k,r') & r < r' \\ -k^{-1}F_{j+}(k,r)e^{-i1/2\pi L}\tilde{\Psi}^{j(+)}(k,r') & r > r'. \end{cases} \tag{15.131a}$$

Diagonalization of the Potential. Before considering bound states we want to note here that in some important cases it is possible to diagonalize the potential matrix V_j by an r-independent transformation. In the neutron-proton case, for example, the matrix V_j of (15.96), in the absence of a spin-orbit potential (15.97), is diagonalized by the matrix U_j given by (15.82),

$$W(r) = U_j V_j U_j^{-1} = \begin{pmatrix} \mathscr{V}_d + 2\mathscr{V}_t & 0 \\ 0 & \mathscr{V}_d - 4\mathscr{V}_t \end{pmatrix}. \tag{15.132}$$

Notice that W is independent of j. We may subject the entire equation (15.92), or the integral equations, to the same transformation. The coupling then comes from the centrifugal term only, since

$$U_j L(L+1) U_j^{-1} = \begin{pmatrix} \beta^2 & -2\beta \\ -2\beta & \beta^2+2 \end{pmatrix} \tag{15.133}$$

where $\beta^2 = j(j+1)$. New solutions are defined by

$$\begin{aligned} \bar{F}_{j\pm} &= U_j F_{j\pm} U_j^{-1} \\ \bar{\Phi}_j &= U_j \Phi_j U_j^{-1} \end{aligned} \tag{15.134}$$

and Jost functions

$$\bar{\mathfrak{F}}_{j\pm} = U_j \mathfrak{F}_{j\pm} U_j^{-1} = W(\bar{F}_{j\pm}, \bar{\Phi}_j) \tag{15.135}$$

because $\tilde{U}_j = U_j = U_j^{-1}$. It is a remarkable fact that the transformed **S** matrix

$$\begin{aligned} \bar{S}_j &= U_j S_j U_j^{-1} \\ &= e^{i\pi L} \bar{\mathfrak{F}}_{j-} \bar{\mathfrak{F}}_{j+}^{-1} \end{aligned} \tag{15.136}$$

is exactly the one needed directly for the calculation of the helicity amplitudes, as seen in (15.80), (15.81), and (15.83). If a spin-orbit force is present, the same transformation U_j may still be used but it no longer diagonalizes the potential matrix. What is more, the transformed potential matrix now depends on j.

15.2.4 Bound States. Suppose that at some value k_0 of k, with Im $k_0 > 0$,

$$\det \mathfrak{F}_j(k_0) = 0. \tag{15.137}$$

Then there exists a vector (or column matrix) a such that

$$\tilde{a}\mathfrak{F}_j(k_0) = 0. \tag{15.138}$$

According to (15.106) this says that the wronskian of the vector $F_j(k_0, r)a$ with each column of Φ_j vanishes. Thus $F_j(k_0, r)a$ is a linear combination of those columns, or

$$F_j(k_0, r)a = \Phi_j(k_0, r)b. \tag{15.139}$$

The left-hand side decreases exponentially at $r \to \infty$, and the right-hand side is regular at the origin. Hence $F_j(k_0, r)a$ is normalizable, and $k_0{}^2/2\mu$ is the energy of a bound state. The vector $F_j(k_0, r)a$ is the (unnormalized) bound-state wave function. The significance of the vector a is seen from the asymptotic value for large r,

$$F_j(k_0, r)a \simeq e^{-|k_0|r} \begin{pmatrix} a_1 \\ a_2 \\ \cdot \\ \cdot \\ \cdot \end{pmatrix}. \tag{15.140}$$

In that sense, the ratio of the components of a determines the asymptotic mixture of angular momenta that forms a bound state.

Simplicity of the Pole of $\mathfrak{F}_j^{-1}$. It follows from (15.139) that the vector b is such that

$$\mathfrak{F}_j(k_0)b = 0. \tag{15.138a}$$

Now a straightforward analog of the derivation of (12.52) shows that

$$\tilde{a}\dot{\mathfrak{F}}_j(k_0)b = -2k_0 \int_0^\infty dr\, \tilde{b}\tilde{\Phi}_j\Phi_j b \tag{15.141}$$

the dot indicating the derivative with respect to k and Φ_j on the right being evaluated at k_0. Since k_0 must be purely imaginary and hence Φ_j real, the right-hand side cannot vanish. Consequently,

$$\tilde{a}\dot{\mathfrak{F}}_j(k_0)b \neq 0 \tag{15.142}$$

if a and b are such that (15.138) and (15.138a) are true.

We conclude from (15.142) that *the pole of* $\mathfrak{F}_j^{-1}$ *at* $k = k_0$ *is simple.* To see this, we expand

$$\mathfrak{F}_j(k) = A_0 + (k - k_0)A_1 + \cdots$$
$$\mathfrak{F}_j^{-1}(k) = B_{-n}(k - k_0)^{-n} + \cdots + B_{-1}(k - k_0)^{-1} + B_0 + \cdots$$

with

$$\begin{aligned} A_0 B_{-n} &= 0 \\ A_0 B_{-n+1} + A_1 B_{-n} &= 0 \\ \cdots\cdots\cdots\cdots\cdots\cdots\cdots \\ A_0 B_0 + A_1 B_{-1} + \cdots + A_n B_{-n} &= \mathbb{1}. \end{aligned} \tag{15.143}$$

Multiply the second line by $\tilde{a}$ on the left, where $\tilde{a}A_0 = 0$. It then reads $\tilde{a}A_1 B_{-n} = 0$. According to the first line of (15.143), B_{-n} is made of columns which annihilate A_0 from the right. But according to (15.142) it then follows that $B_{-n} = 0$. In the same manner it follows also that $B_{-n+1} = 0$, etc., except for B_{-1}, for which

$$\tilde{a}A_1 B_{-1} = \tilde{a}$$

because of the last line in (15.143).

Let us consider now the function $\mathfrak{F}_j^{-1}(k)$. The simplicity of its pole at k_0 should not be interpreted as a necessary absence of degeneracy. If there is no degeneracy and $\mathfrak{F}_j^{-1}c$ has a pole at k_0, then there is no other vector d orthogonal to c such that $\mathfrak{F}_j^{-1}d$ has a pole, too. That implies that the residue B_{-1} annihilates all vectors orthogonal to c, and therefore it can be written as the direct product,

$$B_{-1} = c'\tilde{c}.$$

According to the last line of (15.143),

$$\tilde{a} = \tilde{a}A_1B_{-1} = (\tilde{a}A_1c')\tilde{c}$$

so that

$$\lim_{k\to k_0}(k - k_0)\mathfrak{F}_j^{-1}(k) = a'\tilde{a}.$$

As a result, the residue of S^j at the bound-state pole (if $\mathfrak{F}_{j-}$ can be analytically continued there) is factorable in the sense that

$$\lim_{k\to k_0}(k - k_0)S^j(k) = a\tilde{a} \tag{15.144}$$

where a is defined by (15.138) and has the significance (15.140). We have here made use of the symmetry of S^j.

If there is degeneracy, then the pole of S^j is still simple but there exist several orthogonal vectors a_n that satisfy (15.138). The residue of S^j is then of the form

$$\sum_\alpha a_n\tilde{a}_n.$$

Equation (15.131) shows that the simplicity of the poles of $\mathfrak{F}_j^{-1}$ at bound states implies that the poles of the resolvent $(E - H)^{-1}$ are simple, too. This, of course, is as it should be, as we have seen in Chap. 7. The hamiltonian being hermitian, the resolvent necessarily has *simple* poles at the eigenvalues.

Levinson's Theorem. We may now derive Levinson's theorem. The determinant $\Delta^j(k)$ of (15.126) has the same properties $\mathscr{F}_l$ has in the spin 0 case, and it is used in the same manner as in the derivation of (12.95). Since at a bound state the pole of $\mathscr{F}_j^{-1}$ is simple, the multiplicity of the zero of $\Delta^j(k)$ there is equal to the degeneracy of the bound states. We may therefore simply say that the number n_j of zeros of $\Delta^j(k)$ in the upper half of the k plane equals the number of bound states, counting each zero as many times as its multiplicity and each bound state as many times as its degeneracy. The only complication arises at $k = 0$. For that point we refer to the literature.[16]

Assuming, then, that $\Delta^j(0) \neq 0$, we have the analog of (12.156),

$$\delta^j(0) = \pi n_j. \tag{15.145}$$

It should be recalled from (15.124) that δ^j is the *sum of the eigenphase shifts* of angular momentum j.

At this point we may ask if there is any general condition on the *individual* eigenphase shifts in addition to (15.145). We know from (15.117) that each eigenphase shift may be defined to be zero at infinite energy. Then (15.118) tells us that at zero energy each must be an integral multiple of π. No general condition is known concerning what these multiples should be, other than the restriction (15.145) on their sum. Some of them may even be negative. It should be noted that in general the eigenvalues of S^j need not be eigenvalues of $\mathscr{F}_j$. The only conclusion we can draw from (15.116), (15.110), and

16 See *ibid.*

(15.111) is that $\mathscr{F}_j$ can be written in the form

$$\mathscr{F}_j = B^j e^{-i\Lambda^j} N^j \tag{15.146}$$

where B^j is the real orthogonal matrix that diagonalizes S^j according to (15.76), Λ^j is the diagonal matrix of the eigenphase shifts, and N^j is another real matrix (for real k).

Bound States of $j = 1$ without Bound States of $j = 0$. A further remark should be made concerning the bound states. For spin 0 particles the existence of a bound state of angular momentum 1 implies the existence of at least one bound state of angular momentum 0. In the presence of spin this is not so. This is most easily understood in terms of continuously varying angular momenta. For example, in the neutron-proton case, as j decreases from unity to zero, the bound state must become either purely $l = j + 1$ or $l = j - 1$. If it is the former, it is physical; if the latter, it is "nonsense" and not physically real. Hence there may be a bound state of $j = 1$ without the necessary existence of one of $j = 0$.

15.2.5 Miscellaneous Remarks. *Decomposability of the S matrix.* Unitarity. symmetry, (reasonably fast) high-energy approach to unity, and the Levinson theorem (15.145) do not exhaust the restrictions on the **S** matrix imposed by its being correlated with a "reasonable" potential matrix. In the spin 0 case they did. Every function of magnitude unity on the real axis, with the correct high- and low-energy behavior, can be decomposed into $\mathscr{F}$'s according to (12.71), with the appropriate regularity. The unique $\mathscr{F}_+$ is given by (12.64). In the matrix case the problem of finding $\mathscr{F}_j$ from (15.116) is considerably less trival. For its solution we refer to the literature.[17] It turns out in its course that *not every* matrix function subject to the restrictions mentioned can be so decomposed. However, no general method is known which allows one to recognize, by inspection, whether it can or not.

The N/D Method. The N/D method (12.91) is readily generalized to the present case. Of course, it now is a matrix equation and reads

$$\mathscr{F}_j(k) = \mathbb{1} \frac{1}{2\pi i} \int_{ia_0/2}^{i\infty} dk' \frac{\Delta S^j(k')\mathscr{F}_j(k')}{k' + k} \tag{15.147}$$

if the potential matrix is of the Yukawa type (12.22a) and where ΔS^j is the discontinuity across the left-hand cut, as defined below (12.91).

Closed-Form Solutions. We may finally ask for potential matrices that permit a solution of (15.92) in closed form (for all energies). The only such potential matrices known are generalizations of the Bargmann potentials. We refer the interested reader to the literature.[18] It should be noted that (15.92) *cannot* be solved in closed form if V_j is a square-well matrix (unless it contains no coupling between different orbital angular momenta).

17 See Newton and Jost (1955).

18 See Fulton and Newton (1956).

Notes and References

15.1 For properties of the Clebsch-Gordan coefficients, and of the rotation matrices, see J. Schwinger (1952); Blatt and Weisskopf (1952), appendix A; M. E. Rose (1957); A. R. Edmonds (1957); E. P. Wigner (1959). Our notation and phase convention are those of Rose. For the Jacobi polynomials, see also Morse and Feshbach (1953), pp. 1730ff and 1754ff.

Following is a list of papers concerning scattering (and reaction) amplitudes for particles with spin, including polarizations. It is not exhaustive but should suffice in guiding the student to the literature: J. Schwinger (1948a); C. N. Yang (1948); L. Wolfenstein (1949 and 1956); R. J. Blin-Stoyle (1951); Blatt and Biedenharn (1952); Simon and Welton (1953); A. Simon (1953); M. I. Shirokov (1957 and 1959); Chou Kuang-Chao and Shirokov (1958); Yu. M. Shirokov (1958); Chou Kuang-Chao (1958 and 1959); Jacob and Wick (1959).

Dirac potential scattering was discussed by Fradkin, Hammer, and Weber (1964). A Lorentz-group expansion of the scattering amplitude was given by Shukre and Winternitz (1972a); and spin-dependent spheroidal potentials are discussed by M. C. Li (1973).

15.2 For the first detailed discussion of the *tensor force*, see Rarita and Schwinger (1941). M. G. Fuda (1972) discusses a boundary-condition model for the **T** matrix in the presence of a tensor force. An exposition of Fuchs theory can be found in E. L. Ince (1927), p. 356. The treatment of the coupled equations, their Jost matrix, and their **S** matrix is due to Newton and Jost (1955) and specifically for coupled orbital angular momenta, with the introduction of the counterterm, R. G. Newton (1955). See also R. G. Newton (1960).

The Bargmann potentials were generalized to the case of coupled angular momenta by Fulton and Newton (1956) and applied to the neutron-proton case by K. Chadan (1955 and 1956); and by Newton and Fulton (1957).

For the extension of the Regge method to the case of two particles with spin $\frac{1}{2}$, see Desai and Newton (1963) and also R. G. Newton (1964), chap. 16.

The following annotated bibliography on positron scattering should be mentioned: H. F. Kaiser (1964).

Problems

1 Calculate the spin-density matrix of the scattered beam in terms of that of the incident beam (in the helicity language) for the scattering of spin $\frac{1}{2}$ particles by spin 0 particles. Assuming that the incident beam is unpolarized, give the polarization as a function of the angle if the only contributions come from $l = 0$ and $l = 1$.

2 Particles A of spin $\frac{1}{2}$ are scattered by particles B of spin $\frac{1}{2}$. Assume that they are initially unpolarized and that the energy is so low that only s, p, and d waves contribute. Calculate the spin-density matrix for particles A as a function of the angle (using helicities).

3 Under conditions similar to those in Prob. 2, suppose that coincidence measurements of the scattered particles A and B are made and only particles B of positive helicity are accepted. What is the spin-density matrix for the scattered particles A? What is their degree and direction of polarization as a function of the scattering angle?

4 Unpolarized particles A are scattered off unpolarized particles B initially at rest.

(*a*) The recoil particles B are unobserved. What is the direction of polarization, if any, of the scattered particles? (*b*) Suppose that coincidence measurements are made and thus the momenta of both the scattered and recoil particles are observed. What is the direction of polarization of those scattered particles whose recoil partner goes in a given direction? (*c*) Suppose that the spins of the recoil particles are also measured. What is the direction of polarization of those scattered particles whose recoil partners have positive helicity?

5 What is the possible angle dependence of the differential scattering cross section for spin s_1 particles on spin s_2 particles in the limit of low energy? What if the incident particles are unpolarized? What if the target particles are unpolarized but the beam particles are polarized?

6 Suppose that particles of spin $\frac{1}{2}$ are scattered by a target with which they interact via $H' = V_1(r) + \boldsymbol{\sigma}\cdot\mathbf{r}\times\mathbf{p}V_2(r) + V_2(r)\boldsymbol{\sigma}\cdot\mathbf{r}\times\mathbf{p}$. In the Born approximation calculate the polarization of the scattered beam as a function of the scattering angle if the incident beam (*a*) is unpolarized; (*b*) has positive helicity.

7 For the particles of Prob. 6 calculate the double-scattering cross section (in Born approximation) for two targets whose connecting line makes a given angle with the incident beam. Assume that the incident particles are unpolarized and that no final spins are measured.

8 (*a*) Express the elements of a 2×2 **S** matrix in terms of the eigenphase shifts and the mixture parameter. (*b*) Do the same for a 3×3 **S** matrix. How many mixture parameters are there?

9 Prove (15.96) and (15.98).

10 Assuming that the potentials $\mathscr{V}_d$ and $\mathscr{V}_t$ in (15.96) are constant for $r < r_0$, calculate the regular square-matrix solution Φ_1 for $j = 1$ by expansion about $r = 0$. Similarly calculate a linearly independent irregular solution.

11 Check that Eq. (15.101) works whenever all elements of the potential matrix possess absolute first and second moments.

12 Prove Eq. (15.118) under the condition (15.119).

13 Show that $\det \mathscr{F}_j$ is the Fredholm determinant of the set of coupled integral equations (15.13).

14 Show that (15.131) is a Green's function.

15 Prove from (15.116) that in the neutron-proton case of parity $(-)^{j+1}$ the low-energy behavior of the eigenphase shifts must be

$$e^{2i\delta_1^j} - 1 = O(k^{2j-1})$$
$$e^{2i\delta_2^j} - 1 = O(k^{2j+3})$$

and that of the mixture angle,

$$\sin 2\theta = O(k^2).$$

16 Write down an explicit example of a 2×2 **S** matrix for the neutron-proton case of $j = 1$ which satisfies unitarity, symmetry, (15.117), (15.118), and the Levinson theorem for no bound states and each element of which is a rational function of k.

CHAPTER 16

Inelastic Scattering and Reactions (Multichannel Theory), I

16.1 Descriptive Introduction

For easier understanding and general orientation let us first discuss the case of three particles in mutual interaction and with bound states between them. Schematically this may be thought of as a hydrogen atom in interaction with an electron. However, we really have in mind interactions of shorter range than the Coulomb force, and we may want the two "electrons" to attract one another so that they, too, can form bound states. Furthermore the two electrons may be distinguishable. Let us call the electrons particles 1 and 2 and the "nucleus" particle 3.

If particle 1 impinges upon a bound state (2,3), a number of processes are possible. Suppose that initially the (2,3) system is in its ground state, and assume that its binding energy is larger than those of the bound systems (1,3) and (1,2). Then there is an energy region in which only elastic scattering of particle 1 on (2,3) can take place. As the energy of the incident beam is raised, we may reach a point above which it is energetically possible to raise the (2,3) system to an excited state while particle 1 loses a corresponding amount of kinetic energy. We are at the *threshold* of an inelastic process. One says that a new *channel* has opened up in the scattering. From now on, elastic and inelastic scattering compete for their share of incoming flux.

If the (2,3) system has other bound states that can be excited, other inelastic processes of a similar kind may become possible as the energy is raised further. On the other hand, if the (1,2) system has bound states, too, additional inelastic processes of another kind may become feasible. Particle

1 may pick up particle 2, and the final state may consist of particle 3 and the (1,2) system flying away from one another. This is a *rearrangement collision.* Other channels may consist of rearrangements in which particle 2 and the (1,3) system emerge, and, of course, neither of the final bound configurations need be ground states. All these various channels are distinguished by different internal energies of the two fragments in the final state (except in cases where some of these accidentally coincide). The rearrangement collisions are in addition distinguished by the different nature of the two emerging fragments.

Finally, if the incident kinetic energy is large enough, the atom may be "ionized." That is, the (2, 3) system may be raised to its own energy continuum, and the final state consists of three particles flying off in various directions. At lower energies the loss in kinetic energy of each finally emerging fragment (compared with the incident kinetic energy, say, in the barycentric system) is a fixed, discrete amount that depends upon the binding energies of the bound systems. Above the "ionization" threshold at each fixed total energy a continuity of energy distributions is possible among the three particles. This complicates matters even more than at lower energies. It is therefore customary and for many purposes good enough to truncate such problems by ignoring the possibility of ionization. In other words, one acts as though the bound systems had several excited states, but could not be dissociated (even though this, of course, contradicts the possibility of rearrangements). Thus truncated, it is usually called the *many-channel problem.* The procedure is not unreasonable when the energy is far below the ionization threshold, so that even virtual transitions to the region above it can be ignored. But as that threshold is approached, it breaks down.

Resonances and Bound States Embedded in the Continuum. Let us consider now the bound states of the system (1, 2, 3). Suppose for a moment that there is no interaction between particles 1 and 2. Then particles 1 and 2 are bound with particle 3 at the same energy levels each would separately have with particle 3. For example, in the absence of μ-electron interaction the energy levels of the μ-mesic helium atom are simply combinations of those of the two ions. This includes levels in which both particles 1 and 2 are so highly excited that their combined energy is sufficient to enable particle 1 to be ionized at the expense of the energy of particle 2, which simultaneously drops to a lower level. This combined excited state is thus degenerate with a continuum state in which particle 1 is free but particle 2 is more tightly bound. One says that the bound state is *embedded in the continuum.* Although its dissociation is energetically possible, it is perfectly stable because of the assumed absence of interaction between the two particles 1 and 2. The two channels, which differ by the binding energy of particle 2, are uncoupled.

Suppose now that a weak interaction between particles 1 and 2 is "turned on." Then we expect that the transition from the excited combined bound state to the state in which particle 1 is free and particle 2 more tightly bound really "goes." As a result the previously stable state will have acquired a *finite lifetime*. It is thus no longer a bound state. The weaker the interaction, or the coupling between the channels, the longer we expect the lifetime to be. In atomic physics this is called the *Auger effect*; in molecular physics, *predissociation*; and in a combination of nuclear and atomic physics, *internal conversion*.

Let us look at the same effect in the opposite order in time. Instead of having particle 1 initially bound and emerging freely at a specific energy E_0 depending on that of the level to which particle 2 drops, we bombard the resulting (2, 3) state with missiles 1 at the energy E_0. In that case the missile may be caught in the almost bound state while particle 2 is excited. Of course, eventually particle 1 must emerge again and fly off. But the fact that it spent a long time bound produces a large scattering cross section, as well as a long delay between incidence and emergence. Both are characteristic of a *resonance* (see Sec. 11.2.2). The uncertainty principle tells us that, since the almost bound state has a finite lifetime τ, its energy is not sharp but uncertain by an amount $\Delta E \simeq 1/\tau$. Hence the resonance effect must be visible within an energy interval of width $\Gamma \simeq 1/\tau$. The weaker the coupling between the channels, the sharper we expect the resonance to be.

At this point we may consider what happens as we turn on the interaction between particles 1 and 2, or between the channels, ever more strongly. We expect that when this coupling is strong (whatever this should mean more precisely) the originally narrow resonances have become so broad as to be unobservable. This need not necessarily be so, however. In fact it is entirely possible for a real, stable bound state embedded in the continuum to exist even when the coupling is strong. The objection that in the presence of coupling there must be a finite probability for the bound particle to "leak out" is not necessarily valid. The matrix element whose squared magnitude measures that escape probability is simply zero *at the energy* of the bound state. In another sense, however, such a bound state embedded in the continuum (in the presence of coupling) is not stable. If it is subjected to small perturbations (for example, from other particles passing at a distance), these will generally destroy the delicate balance that keeps it stable and the state will acquire a finite lifetime. But it is characteristic of such states, in contrast to other almost stable states, that their lifetime depends entirely upon the environment.

What interests us most about these bound states embedded in the continuum is, not so much the possibility of their existence, as the fact that, if *they* can be formed in the presence of strong coupling, then a fortiori it must be possible to produce sharp resonances. In other words, we wish to emphasize that the presence of sharp resonances is not necessarily restricted to a regime of weak coupling. All that is required is that there exist a small perturbation of the hamiltonian which converts it into one with a bound state embedded in the continuum. If the channel coupling is weak, this perturbation may consist simply of turning off the coupling. If it is strong, it must be something more complicated.

16.2 Time-Dependent Theory

16.2.1 The Schrödinger Picture. Attempting to describe a collision between a particle and a bound state, or between two bound states, let us try to think through a time-dependent theory such as that of Chap. 6. It is clear that in the split-up (6.2) of the hamiltonian, to take H_0 as the kinetic energy alone is not useful. The state in the infinite past has several particles bound. Only the "missile" is free. What we want then is a split-up of the hamiltonian into a part H_a that is left when the two initial fragments are taken far apart and the remainder, H_a'. Let $\mathbf{R}_1$ be the center of mass of the particles in fragment 1 and $\mathbf{R}_2$ that of those in fragment 2. Then we define in the coordinate representation

$$H_a \equiv \lim_{|\mathbf{R}_1 - \mathbf{R}_2| \to \infty} H \tag{16.1}$$

and H_a' by

$$H = H_a + H_a'. \tag{16.2}$$

It is understood that the limit in (16.1) is to be taken in such a way that all coordinates of the particles in fragment 1 and all coordinates of the particles in fragment 2 are kept fixed *relative* to $\mathbf{R}_1$ and $\mathbf{R}_2$, respectively.

We may now go through the arguments of Chap. 6 concerning the development of $\Psi(t)$ from $\Psi_{\text{in}}(t)$. When we come to the development in the distant future, there is another difficulty.

If rearrangements are possible, or break-ups, then the "channel hamiltonian" that governs the state of interest as $t \to \infty$ may not be the same as the one that governs the initial state as $t \to -\infty$. We handle the various possibilities for an n-particle system by defining first a *partition* of them into k clusters or groups; we denote it by a_k. If a_k can be obtained from a_l, $l \geq k$, by combining none, some, or all of the clusters of a_l into larger or equal ones (i.e., if a_l can be obtained from a_k by subdividing its clusters or leaving them unchanged) then we say a_l is included in a_k, $a_l \subset a_k$. If a is a given partition, then we define H_a as a generalization of (16.1); it is obtained by allowing all distances between clusters independently to tend to infinity. Thus H_a contains all the interactions internal to clusters in a but none between them. The interaction H_a' is then defined by (16.2), and we have for any two partitions a and b,

$$H = H_a + H_a' = H_b + H_b'. \tag{16.3}$$

To each partition correspond Green's functions defined by

$$\left(i\frac{\partial}{\partial t} - H_a\right) G_a^{\pm}(t) = \delta(t)\mathbb{1} \tag{16.4}$$

and the boundary conditions (6.4).

If a given partition a is such that the hamiltonian H_a, after removal of the kinetic energy of the center-of-mass motion and of the relative motion of the centers of mass of its clusters, has at least one bound state, then we call a an *arrangement channel*. We will also call the n cluster partition an arrangement channel and denote it by "o" since in that case $H_a = H_0$, the kinetic energy. A partition that is not an arrangement channel is of no interest as an initial or final scattering state. If H_a, in the sense mentioned above, has more than one bound state then each of them defines a *channel* in the ordinary sense. Thus in each channel the clusters are in a specific bound state but freely moving relative to one another.

Let $\mathscr{H}_a$ be the space of arrangement channel a. If a has m fragments, then each state in $\mathscr{H}_a$ has m groups of particles bound. Only if a is the n-fragment arrangement channel is $\mathscr{H}_a$ the whole Hilbert space. Otherwise the ionized eigenstates of H_a are missing. Since each $\mathscr{H}_a$ is defined by a different channel hamiltonian H_a, they are generally not orthogonal to one another.

It is convenient to introduce orthogonal projections P_a onto the channel spaces $\mathscr{H}_a$. That is,

$$P_a^2 = P_a \qquad P_a^\dagger = P_a \qquad P_a\mathscr{H}_a = \mathscr{H}_a \tag{16.5}$$

and the null space of P_a is the space spanned by the ionized eigenstates of H_a. For the n-cluster arrangement channel we have $P_0 = \mathbb{1}$.

In and Out States. We start with the definition of an *a-state*, which is a state in $\mathscr{H}_a$ that develops according to H_a,

$$\left(i\frac{\partial}{\partial t} - H_a\right)\Psi_a(\alpha, t) = 0. \tag{16.6}$$

The label α contains all other necessary information in addition to the arrangment channel. [It is notationally convenient to include in α also the arrangement channel, even though in $\Psi_a(\alpha, t)$ this is redundant.] The state $\Psi^{(+)}(\alpha, t)$ is then defined as a state in $\mathscr{H}$ that develops according to

$$\left(i\frac{\partial}{\partial t} - H\right)\Psi^{(+)}(\alpha, t) = 0 \tag{16.6a}$$

and for which there exists an a-state such that

$$\lim_{t\to-\infty} (\Psi_a(\alpha, t), \Psi^{(+)}(\alpha, t)) = 1. \tag{16.7}$$

This defines the meaning of α and of $\Psi_a(\alpha, t)$ as the in-state $\Psi_{\text{in}}(\alpha, t)$ in relation to $\Psi^{(+)}$. The demand (16.7), that the probability for finding the system in the state $\Psi_a(\alpha, t)$ in the remote past approach 1, is equivalent to[1]

$$\Psi^{(+)}(\alpha, t) \underset{t\to-\infty}{\Longrightarrow} \Psi_a(\alpha, t) \tag{16.7$'$}$$

and to

$$iG_a^+(t - t')\Psi^{(+)}(\alpha, t') \underset{t'\to-\infty}{\Longrightarrow} \Psi_{\text{in}}(\alpha, t). \tag{16.7$''$}$$

1 The double arrow denotes the strong limit; see Sec. 6.8.

As in Sec. 6.2, it follows that

$$\begin{aligned}\Psi^{(+)}(\alpha, t) &= \Psi_{\text{in}}(\alpha, t) + \int_{-\infty}^{\infty} dt'\, G_a^+(t - t')H_a'\Psi^{(+)}(\alpha, t') \\ &= \Psi_{\text{in}}(\alpha, t) + \int_{-\infty}^{\infty} dt'\, \mathscr{G}^+(t - t')H_a'\Psi_{\text{in}}(\alpha, t').\end{aligned} \tag{16.8}$$

Similarly,

$$-iG_a^-(t - t')\Psi^{(-)}(\alpha, t') \underset{t\to+\infty}{\Longrightarrow} \Psi_{\text{out}}(\alpha, t), \tag{16.7''a}$$

where ψ_{out} is an a-state, and hence

$$\begin{aligned}\Psi^{(-)}(\alpha, t) &= \Psi_{\text{out}}(\alpha, t) + \int_{-\infty}^{\infty} dt'\, G_a^-(t - t')H_a'\Psi^{(-)}(\alpha, t') \\ &= \Psi_{\text{out}}(\alpha, t) + \int_{-\infty}^{\infty} dt'\, \mathscr{G}^-(t - t')H_a'\Psi_{\text{out}}(\alpha, t').\end{aligned} \tag{16.8a}$$

Wave Operators and S Operators. The wave operators are next defined by the mappings

$$\Psi^{(\pm)}(\alpha, t) = \Omega_a^{(\pm)}\Psi_a(\alpha, t) \tag{16.9}$$

which relate $\Psi^{(+)}$ to its in state $\Psi_{\text{in}} = \Psi_a$ and $\Psi^{(-)}$ to its out state $\Psi_{\text{out}} = \Psi_a$. However only those states Ψ_a will be admitted in (16.9) which are members of the channel space $\mathscr{H}_a$. On its orthogonal complement (i.e., on the ionized eigenstates of H_a) we define $\Omega_a{}^{(\pm)}$ to be zero:

$$\Omega_a^{(\pm)}P_a = \Omega_a^{(\pm)}.$$

From (16.8) we find, on the space $\mathscr{H}_a$,

$$\begin{aligned}\Omega_a^{(+)} &= P_a + K_a^{(+)} \\ K_a^{(+)} &= -i\int_{-\infty}^{\infty} dt\, \mathscr{G}^+(-t)H_a'G_a^-(t)P_a \\ &= -i\int_{-\infty}^{0} dt\, e^{iHt}H_a'e^{-iH_at}P_a\end{aligned}$$

or

$$\Omega_a^{(+)} = \lim_{t\to-\infty} e^{iHt}e^{-iH_at}P_a. \tag{16.10}$$

Similarly, on $\mathscr{H}_a$,

$$\begin{aligned}\Omega_a^{(-)} &= P_a + K_a^{(-)} \\ K_a^{(-)} &= i\int_{-\infty}^{\infty} dt\, \mathscr{G}^-(-t)H_a'G_a^+(t)P_a \\ &= i\int_{0}^{\infty} dt\, e^{iHt}H_a'e^{-iH_at}P_a\end{aligned}$$

or

$$\Omega_a^{(-)} = \lim_{t\to\infty} e^{iHt} e^{-iH_a t} P_a. \tag{16.10a}$$

The *range* of $\Omega_a^{(+)}(\Omega_a^{(-)})$ is the space of all full states that develop from (into) arrangement channel a. Let us call it $\mathscr{R}_a^{(+)}(\mathscr{R}_a^{(-)})$ and the orthogonal projection on it $Q_a^{(+)}(Q_a^{(-)})$. The wave operators $\Omega_a^{(\pm)}$ map $\mathscr{H}_a$ onto $\mathscr{R}_a^{(\pm)}$, and it follows from (16.8) and (16.8a) that on $\mathscr{R}_a^{(+)}$ (or $\mathscr{R}_a^{(-)}$, respectively)

$$\begin{aligned}\Psi_a(\alpha, t) &= \Psi_{\text{in}}(\alpha, t) = \Omega_a^{(+)\dagger}\Psi^{(+)}(\alpha, t)\\ &= \Psi_{\text{out}}(\alpha, t) = \Omega_a^{(-)\dagger}\Psi^{(-)}(\alpha, t).\end{aligned}$$

Consequently,

$$\Omega_a^{(\pm)\dagger}\Omega_a^{(\pm)} = P_a \tag{16.11}$$

since the $\Psi_a(\alpha, t)$ span the space $\mathscr{H}_a$. Equation (16.11) is expressed by saying that the $\Omega_a^{(\pm)}$ are *partially isometric* (from the space $\mathscr{H}_a$). The operators $\Omega_a^{(\pm)\dagger}$, on the other hand, are partially isometric from the ranges $\mathscr{R}_a^{(\pm)}$ of the $\Omega_a^{(\pm)}$,

$$\Omega_a^{(\pm)}\Omega_a^{(\pm)\dagger} = Q_a^{(\pm)}. \tag{16.12}$$

Now the full states that develop from (into) one arrangement channel are orthogonal to those which develop from (into) another. This can be seen by evaluating $(\Psi^{(+)}(\beta, t), \Psi^{(+)}(\alpha, t)) = (\Psi^{(+)}(\beta, t'), \Psi^{(+)}(\alpha, t'))$ in the limit as $t' \to -\infty$. If the two arrangement channels are different, then there must be at least one particle for which the "overlap" of the two states was negligible in the remote past because it belonged to a different fragment. Hence that inner product must vanish for all times. This shows that the spaces $\mathscr{R}_a^{(+)}(\mathscr{R}_a^{(-)})$ are mutually orthogonal,[2]

$$Q_a^{(\pm)}Q_b^{(\pm)} = 0 \qquad a \neq b \tag{16.13}$$

and hence by (16.11)

$$\Omega_a^{(\pm)\dagger}\Omega_b^{(\pm)} = P_a\delta_{ab}. \tag{16.11a}$$

Note that the same argument shows that the inner product $(\Psi_b(\beta, t), \Psi_a(\alpha, t))$ approaches zero as $t \to \pm\infty$ (for $a \neq b$). But since $H_a \neq H_b$, it is not independent of t and hence it does not generally vanish for finite times. Thus in general $P_a P_b \neq 0$.

From the Schrödinger equation one concludes that for each arrangement channel

$$H\Omega_a^{(\pm)} = \Omega_a^{(\pm)}H_a, \tag{16.14}$$

i.e., $\Omega_a^{(\pm)}$ is an intertwining operator.

We may include among our channel definitions the single-cluster arrangment channel, which consists of all the (n-particle) bound states of H. If

2 For a more rigorous proof, see the appendix to this chapter.

Λ is the orthogonal projection onto that subspace then for all a

$$Q_a^{(\pm)}\Lambda = 0.$$

Since every nonbound state must be decomposable into states that come from (go into) one of the other arrangments we expect that

$$\Lambda + \sum_a Q_a^{(\pm)} = \mathbb{1}. \tag{16.15}$$

Eq. (16.15) is known as *asymptotic completeness*. Its rigorous proof under specific assumptions on the hamiltonian is one of the most difficult problems of n-particle theory and we shall take it for granted in this book.

The state $\Psi^{(+)}(\alpha, t)$ approaches an a state $\Psi_a(\alpha, t)$ as $t \to -\infty$, but as $t \to +\infty$ it is complicated. Using (16.10a), (16.14), and (16.15) one easily sees that

$$\Psi^{(+)}(\alpha, t) \underset{t\to+\infty}{\Longrightarrow} \Psi_{\text{out}}(t) = \sum_b S_{ba}\Psi_a(\alpha, t), \tag{16.16}$$

where

$$S_{ba} = \Omega_b^{(-)\dagger}\Omega_a^{(+)}. \tag{16.17}$$

It immediately follows from (16.14) that

$$S_{ba}H_a = H_b S_{ba}. \tag{16.18}$$

We remark here that it is also possible[3], and convenient for some purposes, to define wave operators without projections P_a in them:

$$\hat{\Omega}_a^{(\pm)} = \lim_{t\to\mp\infty} e^{iHt}e^{-iH_a t}. \tag{16.19}$$

Then $\hat{\Omega}_a{}^{(\pm)}$ maps the orthogonal complement of $\mathscr{H}_a$ on a proper *subspace* of the orthogonal complement of the range of $\Omega_a{}^{(\pm)}$, and $\hat{\Omega}_a{}^{(\pm)\dagger}$ inverts this mapping. If $b \not\subset a$ then the range of $\Omega_b{}^{(\pm)}$ is orthogonal to that of $\hat{\Omega}_a{}^{(\pm)}$, so that

$$\hat{\Omega}_a^{(\pm)\dagger}Q_b^{(\pm)} = 0 \quad \text{if } b \not\subset a. \tag{16.20}$$

The reason is essentially the same as for (16.13). As $t \to \mp\infty$ there is at least one particle for which the "overlap" of a state $\Psi^{(\pm)}(\beta, t)$ and a state $\Psi^{(\pm)}(\alpha, t)$ vanishes. Eq. (16.20) may be regarded[3] as the origin of the nonuniqueness of the solution of the Lippmann–Schwinger equation discussed in Sec. 17.4.1.

We now ask the scattering question. Just as in (6.45) we get

$$\begin{aligned}(\Psi^{(-)}(\beta, t), \Psi^{(+)}(\alpha, t)) &= (\Psi_{\text{out}}(\beta, t), \Omega_b^{(-)\dagger}\Omega_a^{(+)}\Psi_{\text{in}}(\alpha, t)) \\ &= (\Psi_b(\beta, t), S_{ba}\Psi_a(\alpha, t)) \\ &= (\Psi_b(\beta, 0), S_{ba}\Psi_a(\alpha, 0))\end{aligned} \tag{16.21}$$

from (16.17). The last line in (16.21) follows from (16.18).

3 W. Sandhas (1978).

We may check the unitarity. Using (16.17), (16.15), and (16.11a), we get

$$\sum_b S_{ab} S_{cb}^\dagger = \Omega_a^{(-)\dagger}(\mathbb{1} - \Lambda)\Omega_c^{(-)} = \Omega_a^{(-)\dagger}\Omega_c^{(-)} = P_a \delta_{ac}$$

$$\sum_b S_{ba}^\dagger S_{bc} = \Omega_a^{(+)\dagger}(\mathbb{1} - \Lambda)\Omega_c^{(+)} = \Omega_a^{(+)\dagger}\Omega_c^{(+)} = P_a \delta_{ac}.$$

These equations imply the unitarity of the **S** matrix.

The situation we have now is that **S** is the matrix, not of a single S operator, but of a set of operators S_{ba}. For each transition from one arrangement channel to another a different scattering operator must be chosen, and the states on the basis of which the **S** matrix is evaluated are just the ones most convenient for this purpose: bound-state eigenfunctions of the hamiltonians that bind each fragment, multiplied by plane waves for the fragment motion.

16.2.2 The Heisenberg Picture. To each arrangement channel there corresponds a different Heisenberg picture. First we define operators that vanish on the orthogonal complement of the channel space $\mathscr{H}_a$, whose range is $\mathscr{H}_a$,

$$\mathsf{A}_{0,a} = P_a \mathsf{A}_{0,a} P_a$$

and which develop according to the channel hamiltonian H_a,

$$i\frac{d}{dt}\mathsf{A}_{0,a}(t) = [\mathsf{A}_{0,a}(t), H_a]. \tag{16.22}$$

We shall call such operators *a operators.*

Now the full Heisenberg operator which at the time $t = t_0$ equals $\mathsf{A}_{0,a}(t_0)$ is given by

$$\mathsf{A}_a(t) = e^{iH(t-t_0)}e^{-iH_a(t-t_0)}\mathsf{A}_{0,a}(t)e^{iH_a(t-t_0)}e^{-iH(t-t_0)}$$

in terms of the corresponding a operator at the same time. For large negative times this becomes

$$\mathsf{A}_a(t) \xrightarrow[t\to-\infty]{} \Omega_a^{(+)}\mathsf{A}_{0,a}(t)\Omega_a^{(+)\dagger}$$

and for large positive times

$$\mathsf{A}_a(t) \xrightarrow[t\to+\infty]{} \Omega_a^{(-)}\mathsf{A}_{0,a}(t)\Omega_a^{(-)\dagger}.$$

In an idealized scattering problem the operators used for the preparation of the initial and for the detection of the final states commute with the respective channel hamiltonians, so that $\mathsf{A}_{0,a}$ is independent of the time,

$$\mathsf{A}_a(\mp\infty) = \Omega_a^{(\pm)}\mathsf{A}_{0,a}\Omega_a^{(\pm)\dagger}. \tag{16.23}$$

The scattering question now is: The system was prepared in the remote past as an eigenstate $\Psi(\alpha)$ of some operator $\mathsf{A}_a(-\infty)$; hence it is in the state

$$\Psi(\alpha) = \Omega_a^{(+)}\Psi_a(\alpha)$$

if $\Psi_a(\alpha)$ is an eigenstate of $\mathsf{A}_{0,a}$. What is the probability amplitude for finding it, in the far future, to be in an eigenstate $\Psi(\beta)$ of an operator $\mathsf{B}_b(+\infty)$, that is, in the state

$$\Psi(\beta) = \Omega_b^{(-)}\Psi_b(\beta)$$

where $\Psi_b(\beta)$ is an eigenstate of $\mathsf{B}_{0,b}$? The answer is evidently

$$(\Psi(\beta), \Psi(\alpha)) = (\Psi_b(\beta), \Omega_b^{(-)\dagger}\Omega_a^{(+)}\Psi_a(\alpha))$$

as in (16.21).

16.2.3 Two-Hilbert-Space Formulation. The lack of orthogonality of the space $\mathscr{H}_a$ and the resulting absence of a single S operator on the Hilbert space $\mathscr{H}$ may be remedied by introducing a second Hilbert space $\hat{\mathscr{H}}$, called the *asymptotic space*, or the *super Hilbert space*, which is the direct sum of the spaces $\mathscr{H}_a$,

$$\hat{\mathscr{H}} = \oplus_a \mathscr{H}_a.$$

This means that to every vector $\Phi \in (\mathbb{1} - \Lambda)\mathscr{H}$ there corresponds a vector $\hat{\Phi} \in \hat{\mathscr{H}}$ given by

$$\hat{\Phi} = \oplus_a P_a \Phi \tag{16.24}$$

with the inner product

$$(\hat{\Phi}, \hat{\Psi}) = \sum_a (\Phi_a, \Psi_a) = \sum_a (\Phi, P_a \Psi).$$

The projections $\hat{P}_a$ on $\hat{\mathscr{H}}$ defined by

$$\hat{P}_a \hat{\Phi} = 0 \oplus \cdots \oplus \Phi_a \oplus 0 \cdots \oplus 0$$

are mutual orthogonal,

$$\hat{P}_a \hat{P}_b = \hat{P}_a \delta_{ab},$$

and complete on $\hat{\mathscr{H}}$,

$$\sum_a \hat{P}_a = \hat{\mathbb{1}}.$$

Thus to every vector $\hat{\Phi} \in \hat{\mathscr{H}}$ there corresponds the vector $\Phi_a \in \mathscr{H}$ which is such that $P_a \Phi_a$ is defined by $\hat{P}_a \hat{\Phi}$. This mapping from $\hat{\mathscr{H}}$ to $\mathscr{H}$ is accomplished by the *injection operator* J defined by

$$J\hat{\Phi} = J \oplus_a \Phi_a = \sum_a P_a \Phi_a.$$

The adjoint operator from $\mathscr{H}$ to a proper subset of $\hat{\mathscr{H}}$ is given by

$$J^\dagger \Phi = \oplus_a P_a \Phi$$

so that

$$(\Psi, JJ^\dagger \Phi) = (J^\dagger \Psi, J^\dagger \Phi) = \sum_a (\Psi, P_a \Phi)$$

and hence

$$JJ^{\dagger} = \sum_a P_a. \tag{16.25}$$

The full hamiltonian $\hat{H}$ on $\hat{\mathscr{H}}$ is defined by

$$\hat{H}\hat{\Phi} = \oplus_a H\Phi_a. \tag{16.26}$$

We also define an asymptotic hamiltonian on $\mathscr{H}$

$$\hat{H}_0\hat{\Phi} = \oplus_a H_a\Phi_a, \tag{16.27}$$

so that the interaction operator on $\hat{\mathscr{H}}$ is

$$\hat{H}'\Phi = \oplus_a H_a'\Phi_a. \tag{16.28}$$

If $\hat{\Phi} = J^+\Phi$ so that $\Phi_a = P_a\Phi$ then (12.26) implies that

$$\hat{H}J^+\Phi = \oplus_a HP_a\Phi,$$

$$J\hat{H}J^+ = \sum_a P_a HP_a.$$

Similarly, from (16.27) and (16.28)

$$J\hat{H}_0J^+ = \sum_a H_a P_{a'}$$

$$J\hat{H}'J^+ = \sum_a P_a H_a' P_a.$$

In accordance with (16.6) a free state on $\hat{\mathscr{H}}$ develops in time by the equation

$$\left(i\frac{\partial}{\partial t} - \hat{H}_0\right)\hat{\Psi}_0 = 0, \tag{16.29}$$

while a fully developing state goes as

$$\left(i\frac{\partial}{\partial t} - \hat{H}\right)\hat{\Psi} = 0. \tag{16.29$'$}$$

Scattering theory may now be done on $\hat{\mathscr{H}}$ just as in Chapter 6. We define wave operators on $\hat{\mathscr{H}}$ by

$$e^{i\hat{H}t}e^{-i\hat{H}_0t} \xrightarrow[t\to\mp\infty]{} \hat{\Omega}^{(\pm)}. \tag{16.30}$$

Then $\hat{\Omega}^{(\pm)}$ takes $\hat{\Psi}_0(\alpha, t)$ into the full state $\hat{\Psi}^{(\pm)}(\alpha, t)$ from (into) which it develops,

$$\Psi^{(\pm)}(\alpha, t) = \hat{\Omega}^{(\pm)}\hat{\Psi}_0(\alpha, t), \tag{16.31}$$

$$\lim_{t\to\mp\infty} (\hat{\Psi}_0(\alpha, t), \hat{\Psi}^{(\pm)}(\alpha, t)) = 1. \tag{16.32}$$

Finally, the S operator is defined on $\hat{\mathscr{H}}$ by

$$\hat{S} = \hat{\Omega}^{(-)\dagger}\hat{\Omega}^{(+)}. \tag{16.33}$$

It answers the scattering question

$$
\begin{aligned}
(\hat{\Psi}^{(-)}(\beta,t),\hat{\Psi}^{(+)}(\alpha,t)) &= (\hat{\Psi}_0(\beta,t),\hat{S}\hat{\Psi}_0(\alpha,t)) \\
&= (\hat{\Psi}_0(\beta,0),\hat{S}\hat{\Psi}_0(\alpha,0)).
\end{aligned}
\tag{16.34}
$$

The last line follows from the fact that

$$\hat{\Omega}^{(\pm)}\hat{H}_0 = \hat{H}\hat{\Omega}^{(\pm)}$$

and hence

$$[\hat{S},\hat{H}_0] = 0.$$

Since (16.30) implies that the action of $\hat{\Omega}^{(\pm)}$ on $\mathscr{H}$ is such that

$$\hat{\Omega}^{(\pm)}\hat{\Phi} = \oplus_a \Omega_a^{(\pm)}\Phi_a,$$

and we are interested only in states $\Psi^{(+)}$ that develop from a single channel (and states $\Psi^{(-)}$, into a single channel) we have

$$\hat{\Psi}^{(\pm)}(\alpha,t) = 0 \oplus \cdots \oplus \Omega_a^{(\pm)}\Psi_a(\alpha,t) \oplus 0 \oplus \cdots \oplus 0.$$

Thus the left-hand side of (16.34) equals $(\Psi^{(-)}(\beta,t),\ \Psi^{(+)}(\alpha,t))$, and the right-hand side equals $(\Psi_b(\beta,0),\ S_{ba}\Psi_a(\alpha,0))$. So (16.34) agrees with (16.19).

16.3 Time-Independent Theory

16.3.1 Formal Theory. We may proceed from the time-dependent description to the time-independent one, as in Chap. 7. The fact that the simple hamiltonian H_a in each arrangement channel includes some particle interactions does not add anything essential.

Equations (7.15) and (7.15a) become

$$
\begin{aligned}
\Psi^{(\pm)}(E,\alpha) &= \Psi_a(E,\alpha) + G_a^{\pm}(E)H_a'\Psi^{(\pm)}(E,\alpha) \\
&= \Psi_a(E,\alpha) + \mathscr{G}^{\pm}(E)H_a'\Psi_a(E,\alpha).
\end{aligned}
\tag{16.35}
$$

We may directly go to (7.39). The starting point is now (16.19), or

$$(\Psi^{(-)}(E_\beta,\beta),\Psi^{(+)}(E_\alpha,\alpha)) = (\Psi_b(E_\beta,\beta),S_{ba}\Psi_a(E_\alpha,\alpha)). \tag{16.36}$$

From the second line of (16.35) we get

$$\Psi^{(+)}(E,\alpha) - \Psi^{(-)}(E,\alpha) = -2\pi i\delta(E - H)H_a'\Psi_a(E,\alpha) \tag{16.37}$$

so that

$$
\begin{aligned}
(\Psi_b(E_\beta,\beta),S_{ba}\Psi_a(E_\alpha,\alpha)) &= (\Psi^{(+)}(E_\beta,\beta),\Psi^{(+)}(E_\alpha,\alpha)) \\
&\quad -2\pi i\delta(E_\alpha - E_\beta)(\Psi_b(E_\alpha,\beta),H_b'\Psi^{(+)}(E_\alpha,\alpha)) \\
&= (\Psi^{(-)}(E_\beta,\beta),\Psi^{(-)}(E_\alpha,\alpha)) \\
&\quad -2\pi i\delta(E_\alpha - E_\beta)(\Psi^{(-)}(E_\alpha,\beta),H_a'\Psi_a(E_\alpha,\alpha)).
\end{aligned}
\tag{16.38}
$$

Now according to (16.13) and the normalization of the a states,

$$(\Psi^{(\pm)}(E_\beta, \beta), \Psi^{(\pm)}(E_\alpha, \alpha)) = \delta_{ab}\delta_{\alpha\beta}\delta(E_\alpha - E_\beta). \tag{16.39}$$

The Kronecker $\delta_{\alpha\beta}$ becomes a Dirac δ function for continuous variables, and δ_{ab} is meant symbolically.

We now define the **T** matrix as in Sec. 7.2.2, by

$$(\Psi_b(E_\beta, \beta), S_{ba}\Psi_a(E_\alpha, \alpha)) = \delta_{ab}\delta_{\alpha\beta}\delta(E_\alpha - E_\beta) - 2\pi i\delta(E_\alpha - E_\beta)\mathfrak{T}_{\beta\alpha}(E) \tag{16.40}$$

with

$$\begin{aligned}\mathfrak{T}_{\beta\alpha}(E) &= (\Psi_b(E, \beta), H_b'\Psi^{(+)}(E, \alpha)) \\ &= (\Psi^{(-)}(E, \beta), H_a'\Psi_a(E, \alpha)).\end{aligned} \tag{16.41}$$

Finally, the δ function that conserves the total momentum is removed as in (7.43). The result is simply the same kind of change as in (7.44). That is, $T_{\beta\alpha}(E)$ is given by the same expressions as $\mathfrak{T}_{\beta\alpha}(E)$ in (16.41), but the states are taken at a fixed total momentum (the same in the initial and in the final state). In the barycentric system this total momentum is zero. In the future we simply write $T_{\beta\alpha}$.

The T Operators. Equation (16.41) may also be written in the form

$$T_{\beta\alpha}(E) = (\Psi_b(E, \beta), [H_b' + H_b'\mathcal{G}^+(E)H_a']\Psi_a(E, \beta))$$

so that $T_{\beta\alpha}$ is a matrix element between eigenstates of the simple hamiltonians H_a and H_b of the operator

$$T_{\underline{b}a}(E) = H_b' + H_b'\mathcal{G}^+(E)H_a' \tag{16.42}$$

or of the operator

$$\begin{aligned}T_{b\underline{a}}(E) &= H_a' + H_b'\mathcal{G}^+(E)H_a' \\ &= T_{\underline{b}a} + H_a' - H_b'.\end{aligned} \tag{16.42a}$$

The two operators $T_{b\underline{a}}$ and $T_{\underline{b}a}$ have the same "simple" matrix elements between states of equal energy.

Insertion of the various integral equations for $\mathcal{G}^+$ in (16.42) and (16.42a) gives us not only "integral equations" for $T_{\underline{b}a}$ and $T_{b\underline{a}}$,

$$\begin{aligned}T_{\underline{b}a}(E) &= H_b' + T_{\underline{b}a}(E)G_a^+(E)H_a' \\ T_{b\underline{a}}(E) &= H_a' + H_b'G_b^+(E)T_{b\underline{a}}(E)\end{aligned} \tag{16.43}$$

but also equations that determine them in terms of the ordinary T operators T_{aa} and T_{bb},

$$\begin{aligned}T_{\underline{b}a}(E) &= H_b' + H_b'G_a^+(E)T_{aa}(E) \\ &= H_b' + T_{bb}(E)G_b^+(E)H_a'\end{aligned} \tag{16.44}$$

and similar equations for $T_{b\underline{a}}$.

The nonexistence of a single scattering operator can now be understood from the point of view of the T operator. It is, of course, always possible to define $T \equiv H' + H'\mathcal{G}^+H'$ in terms of the entire interaction H'. One may then be tempted to express the **T** matrix as a set of elements of T on the basis of the eigenstates of H_0, even for rearrangements, instead of using elements of the operator T_{ba} on the bases Ψ_a and Ψ_b. This leads to an expression of T_{ba} in terms of T. However, the requisite formal manipulations use Lippmann-Schwinger equations to express Ψ_a and Ψ_b in terms of Ψ_0, or vice versa. And just for the states of interest (i.e., "partial" bound states) there are no such Lippmann–Schwinger equations! There is no state Ψ_0 of (approximately) fixed energy from (or into) which Ψ_a or Ψ_b develops.

16.3.2 Distorted-Wave Rearrangement Theory. We may now duplicate the distorted-wave theory of Sec. 7.2.5 for rearrangement collisions. In such a calculation it is assumed that H_a and H_b contain an interaction between the fragments. On can still define arrangement channels, but in these the fragments do not move quite independently. Each fragment may, for example, be charged, as in ion-ion scattering. We should then put the effective Coulomb interaction between the fragments into H_a and H_b.

In such a situation we proceed as in Sec. 7.2.5, and in Eq. (16.35) Ψ_a must be replaced by $\Psi_a^{(+)}$, respectively. With fragment interaction present, the $\Psi_a^{(+)}(E,\alpha)$ are related to $\Psi_0(E,\alpha)$ by a Lippmann-Schwinger equation in terms of $H_a'' \equiv H_a - H_0$:

$$\Psi_a^{(\pm)}(E,\alpha) = \Psi_0(E,\alpha) + G^{\pm} H_a'' \Psi_a^{(\pm)}(E,\alpha). \tag{16.45}$$

One then finds the analog of (7.76) and (7.76a),

$$\begin{aligned} T_{\beta\alpha}(E) &= (\Psi_0(E,\beta), T(E)\Psi_0(E,\alpha)) \\ &= (\Psi_b^{(-)}(E,\beta), T_{b\underline{a}}(E)\Psi_a^{(+)}(E,\alpha)) + (\Psi_b^{(-)}(E,\beta), H_b''\Psi_0(E,\alpha)) \\ &= (\Psi_b^{(-)}(E,\beta), T_{\underline{b}a}(E)\Psi_a^{(+)}(E,\alpha)) + (\Psi_0(E,\beta), H_a''\Psi_a^{(+)}(E,\alpha)) \end{aligned} \tag{16.46}$$

with $T_{\underline{b}a}$ and $T_{b\underline{a}}$ as given by (16.42) and (16.42a). Whether H_a and H_b are equal does not matter now.

It must be remarked that with the inclusion of a fragment interaction the definition of arrangement channels must be handled with care. For example, in the three-body problem, a clear definition of arrangement channels is based on setting

$$\begin{aligned} H_a &= H_0 + V_{13} \qquad & H_a' &= V_{23} + V_{12} \\ H_b &= H_0 + V_{23} \qquad & H_b' &= V_{12} + V_{13}. \end{aligned}$$

In this way one may describe for instance, particle 1 impinging on a bound state of (2, 3), with particle 2 and the bound state (1, 3) emerging. The fragments 1 and (2, 3) then do not interact at all in the initial state, and 2 and (1, 3) do not interact in the final state. In a distorted-wave description, on the other hand, one may set

$$H_a = H_0 + V_{13} + V_{23} \qquad H_a' = V_{12}.$$

In this case both the initial state of 1 and (2,3) and the final state of 2 and (1,3) are eigenstates of the same hamiltonian H_a, with "fragment interaction" V_{13} in the first and V_{23} in the second. In this case *the two channel hamiltonians are the same*, and (7.76) and (16.46) are applicable. Note that now the two channel spaces are orthogonal.

Alternatively, one may want to include in H_a only part of the interaction of 2 with 3, such as an electrostatic one,

$$H_a = H_0 + V_{13}^{(1)} + V_{13}^{(2)} + V_{23}^{(1)} \qquad H_a' = V_{12} + V_{23}^{(2)}$$
$$H_b = H_0 + V_{23}^{(1)} + V_{23}^{(2]} + V_{13}^{(1)} \qquad H_b' = V_{12} + V_{13}^{(2)}.$$

In this event the two channel hamiltonians for a and b are *not* the same, but (16.46) is still applicable.

16.3.3 Identical Particles. If any of the particles involved in the collision are identical, the initial- and final-state vectors have to be made symmetric or antisymmetric with respect to their interchange, depending upon whether they are bosons or fermions. Every arrangement is then best treated as belonging to the same channel hamiltonian as the one that differs from it by an exchange of identical particles. For example, in the three-body problem in which particles 1 and 2 are identical, we handle things most symmetrically by setting

$$H_a = H_0 + V_{23} + V_{13} \qquad H_a' = V_{12}$$

so that 1 and (2,3) have the same channel hamiltonian as 2 and (1,3). Since H_a and H_a' are symmetric for particles 1 and 2, it is not necessary to symmetrize or antisymmetrize the initial *and* the final states explicitly. If one of them is made symmetric or antisymmetric, the matrix element of S_{aa} automatically picks out only the symmetric or antisymmetric part of the other. The only point that has to be remembered is the proper normalization factor.

However, this is not the only way of proceeding. It would be no help at all, for example, in a three-body problem in which all three particles are identical. We may envisage solving equations such as (16.35) without attention to the fact that particles are indistinguishable. The symmetrization or antisymmetrization is performed at the end. Let us write $\overline{\Psi}$ for the symmetrized or antisymmetrized state. For example,

$$\overline{\varphi}(1,2) = 2^{-1/2}[\varphi(1,2) \pm \varphi(2,1)],$$

the sign depending upon the statistics of the particles. Then (16.36) must be replaced by

$$(\overline{\Psi}^{(-)}(E_\beta,\beta), \overline{\Psi}^{(+)}(E_\alpha,\alpha)) = \frac{1}{n!}\sum_{ij}(\Psi_{b_i}(E_\beta,\beta), S_{b_ia_j}\Psi_{a_j}(E_\alpha,\alpha))\zeta_i\zeta_j \qquad (16.47)$$

where the sums run over all permutations of the identical particles and ζ_i is the corresponding sign factor. For the **T** matrix this means that (16.41) must be replaced by

$$\begin{aligned} T_{\beta\alpha} &= (n!)^{-1/2} \sum_i \zeta_i(\Psi_{b_i}(\beta), H'_{b_i}\overline{\Psi}^{(+)}(\alpha)) \\ &= (n!)^{-1/2} \sum_i \zeta_i(\overline{\Psi}^{(-)}(\beta), H'_{a_i}\Psi_{a_i}(\alpha)) \end{aligned} \tag{16.48}$$

in which expression the energy dependence has not been indicated explicitly. Now, since

$$\sum_i \zeta_i H'_{a_i}\Psi_{a_i} \quad \text{and} \quad \sum_i \zeta_i H'_{b_i}\Psi_{b_i}$$

are symmetric (or antisymmetric) under the interchange of the identical particles, the vectors $\Psi^{(+)}$ and $\Psi^{(-)}$ in (16.48) need not be explicitly symmetrized (or antisymmetrized). Each of the $n!$ terms gives the same result. Hence

$$\begin{aligned} T_{\beta\alpha} &= \sum_i \zeta_i(\Psi_{b_i}(\beta), H'_{b_i}\Psi^{(+)}(\alpha)) \\ &= \sum_i \zeta_i(\Psi^{(-)}(\beta), H'_{a_i}\Psi_{a_i}(\alpha)). \end{aligned} \tag{16.49}$$

If there are several groups of identical particles, then the normalization factor $(n!)^{-1/2}$, of course, is different. But the result (16.49), which does not contain it, is still true.

16.3.4 Large-Distance Behavior of the Two-Cluster Wave Function.[3a] Let a be a fixed arrangement channel in which there exist two fragments whose internal hamiltonians are $H_1{}^{\text{int}}$ and $H_2{}^{\text{int}}$. The channel hamiltonian is then

$$H_a = H_1^{\text{int}} + H_2^{\text{int}} + H_{0a},$$

H_{0a} being the kinetic energy of the centers of mass of the two fragments. We assume we are in the total center-of-mass system so that H_{0a} refers to the relative coordinates $\mathbf{r}$ of the two centers of mass.

Let $\Phi_1(\varepsilon_1, s_1, v_1)$ and $\Phi_2(\varepsilon_2, s_2, v_2)$ be two eigenstates of the internal hamiltonians. We are interested only in the case in which Φ_1 and Φ_2 are *bound* states, and we assume they are normalized. The labels s and v are the spins of the fragments and their z components. They are their total angular momenta, made up partially of orbital momenta and partially of particle spins, as the case may be. The number ε is the internal energy. Out of these individual fragment state vectors we form a total fragment state,

$$\Phi_a(\alpha) = \Phi_a(\varepsilon_\alpha, s_1 s_2 s v) = \sum_{v_1 v_2} C(s_1 s_2 s, v_1 v_2 v)\Phi_1(\varepsilon_1 s_1 v_1)\Phi_2(\varepsilon_2 s_2 v_2), \tag{16.50}$$

the product being meant as a *direct* one, and $\varepsilon_\alpha = \varepsilon_1 + \varepsilon_2$. These states are now used in place of $\chi_v{}^s$ in Sec. 15.1.1. The states $\Phi_a(\alpha)$ form a complete set on the space of internal coordinates of the fragments in arrangement channel a, provided that we include among them all the continuum states, i.e., even those in which the fragments are dissociated. The sum in

3a See also Sec. 17.4.5.

$$\sum_{\alpha} \Phi_a(\alpha)\Phi_a^\dagger(\alpha) = \mathbb{1}_0 \tag{16.51}$$

is therefore partly an integral. Here $\mathbb{1}_0$ is the projection on the subspace in which both fragments are at rest.

The Green's functions $G_a^{\pm}$ are given by expressions quite analogous to (10.6),

$$G_a^{\pm}(E;\mathbf{r},\mathbf{r}') = \frac{\mu_a}{2\pi}\sum_{\alpha}\frac{\exp(\pm ik_a|\mathbf{r}-\mathbf{r}'|)}{|\mathbf{r}-\mathbf{r}'|}\Phi_a(\alpha)\Phi_a^\dagger(\alpha) \tag{16.52}$$

where

$$k_\alpha = [2\mu_a(E-\varepsilon_\alpha)]^{1/2} \tag{16.53}$$

is the channel wave number and μ_a is the reduced mass of the fragments. The asymptotic behavior of G_a^+ for large r is the analog of (10.8),

$$G_a^+(E;\mathbf{r},\mathbf{r}') \simeq -\sum_{\alpha}\frac{e^{ik_\alpha r}}{r}\left(\frac{2\pi\mu_a}{k_\alpha}\right)^{1/2}\Phi_a(\alpha)\Psi_a^\dagger(\mathbf{k}''\alpha,\mathbf{r}') \tag{16.54}$$

where Ψ_a is the partial coordinate representation of a plane-wave state (with respect to the relative fragment center-of-mass coordinates) that satisfies the Schrödinger equation

$$(H_a - E)\Psi_a = 0.$$

The vector $\mathbf{k}''$ points in the direction $\mathbf{r}$, and its length is k_α. Explicitly,

$$\Psi_a(\mathbf{k}\alpha,\mathbf{r}) = \frac{(\mu_a k_\alpha)^{1/2}}{(2\pi)^{3/2}}\exp(i\mathbf{k}_\alpha\cdot\mathbf{r})\Phi_a(\alpha) \tag{16.55}$$

which is normalized as (8.13), i.e., the analog of (10.4).

The Reaction Amplitudes. We now write down the Lippmann–Schwinger equation for $\Psi^{(+)}(\alpha)$ in a partial coordinate representation,

$$\Psi^{(+)}(\mathbf{k}\alpha,\mathbf{r}) = \Psi_a(\mathbf{k}\alpha,\mathbf{r}) + \int d\mathbf{r}'\,d\mathbf{r}''\,G_a^+(E;\mathbf{r},\mathbf{r}')H_a'(\mathbf{r}',\mathbf{r}'')\Psi^{(+)}(\mathbf{k}\alpha,\mathbf{r}''). \tag{16.56}$$

For large values of r (see Sec. 10.1.2 for a more detailed discussion) we obtain from (16.54)

$$\Psi^{(+)}(\mathbf{k}\alpha,\mathbf{r}) \simeq \sum_{\alpha'}\frac{(\mu_a k_{\alpha'})^{1/2}}{(2\pi)^{3/2}}\Phi_a(\alpha')\,[\delta_{\alpha\alpha'}\exp(i\mathbf{k}_\alpha\cdot\mathbf{r}) + r^{-1}\exp(ik_\alpha' r)\bar{A}(\mathbf{k}'\alpha',\mathbf{k}\alpha)] \tag{16.56a}$$

with $\hat{\mathbf{k}}' = \hat{\mathbf{r}}$, $k_\alpha' = k_{\alpha'}$, and

$$\begin{aligned}\bar{A}(\mathbf{k}'\alpha',\mathbf{k}\alpha) &= -(2\pi)^2(k_\alpha k_\alpha')^{-1/2}\int d\mathbf{r}\,d\mathbf{r}'\,\Psi_a^\dagger(\mathbf{k}'\alpha',\mathbf{r}')(\mathbf{r}'|H_a'|\mathbf{r})\Psi^{(+)}(\mathbf{k}\alpha,\mathbf{r})\\ &= -(2\pi)^2(k_\alpha k_\alpha')^{-1/2}\,T(\mathbf{k}'\alpha',\mathbf{k}\alpha).\end{aligned} \tag{16.57}$$

The differential cross section is given by (8.14)

$$\frac{d\sigma}{d\Omega} = \frac{k'_\alpha}{k_\alpha} |\bar{A}(\mathbf{k}'\alpha', \mathbf{k}\alpha)|^2. \tag{16.58}$$

Rearrangements. In the case of a rearrangement collision we cannot so simply divide the coordinates into internal fragment coordinates and the relative distance. The meaning of the relative distance $\mathbf{r}$ is different in the initial and final configurations, being the distance between the centers of mass of different groups of particles. The wave function of a state that initially was entirely in arrangement channel a must satisfy (16.56) if $\mathbf{r}$ is the fragment distance in a. We may now express that same wave function as a function of the coordinates which in arrangement channel b are internal, and the fragment distance $\mathbf{r}_b$ there. Let us write this in the form

$$\Psi^{(+)}(\mathbf{k}\alpha, \mathbf{r}) = \boldsymbol{\Psi}^{(+)}(\mathbf{k}\alpha, \mathbf{r}_b).$$

As a function of $\mathbf{r}_b$ it does not satisfy a simple integral equation such as (16.56). However, because one readily finds from the Lippmann–Schwinger equations that

$$\Psi^{(+)}(\alpha) = [1 - G_b^+(E - H_b)]\Psi_a(\alpha) + G_b^+ H'_b \Psi^{(+)}(\alpha) \tag{16.59}$$

its asymptotic form for large $|\mathbf{r}_b|$ is given by

$$\boldsymbol{\Psi}^{(+)}(\mathbf{k}\alpha, \mathbf{r}_b) \simeq \zeta + \sum_\beta \frac{(\mu_a k_\alpha)^{1/2}}{(2\pi)^{3/2}} \Phi_b(\beta) r^{-1} e^{ik_\beta r_b} \bar{A}(\mathbf{k}'\beta, \mathbf{k}\alpha)$$

where $\hat{\mathbf{k}}' = \hat{\mathbf{r}}_b$ and ζ is the first term in (16.59). We shall see in section 17.4.1 that $\zeta = 0$. The expression we get for $\bar{A}$ is

$$\begin{aligned} \bar{A}(\mathbf{k}'\beta, \mathbf{k}\alpha) &= -c \int d\mathbf{r}\, d\mathbf{r}_b \Psi_b^\dagger(\mathbf{k}'\beta, \mathbf{r}_b)(\mathbf{r}_b|H'_b|\mathbf{r})\Psi^{(+)}(\mathbf{k}\alpha, \mathbf{r}) \\ &= -cT(\mathbf{k}'\beta, \mathbf{k}\alpha) \\ c &= (2\pi)^2 \left(\frac{\mu_b}{\mu_a k_\alpha k_\beta}\right)^{1/2} \end{aligned} \tag{16.57a}$$

and the differential cross section is given by

$$\frac{d\sigma}{d\Omega} = \frac{\mu_a k_\beta}{\mu_b k_\alpha} |\bar{A}(\mathbf{k}'\beta, \mathbf{k}\alpha)|^2. \tag{16.60}$$

Redefined Amplitudes. In view of (16.58) and (16.57a) it is useful to define the scattering amplitude, both in the rearrangement case and in the non-rearrangement case, by

$$\begin{aligned} A(\mathbf{k}'\beta, \mathbf{k}\alpha) &= \left(\frac{\mu_a k_\beta}{\mu_b k_\alpha}\right)^{1/2} \bar{A}(\mathbf{k}'\beta, \mathbf{k}\alpha) \\ &= -(2\pi)^2 k_\alpha^{-1} T(\mathbf{k}'\beta, \mathbf{k}\alpha) \\ &= -(2\pi)^2 \left(\frac{\mu_a \mu_b k_\beta}{k_\alpha}\right)^{1/2} T'(\mathbf{k}'\beta, \mathbf{k}\alpha) \end{aligned} \tag{16.61}$$

if T' is the **T**-matrix element evaluated by using state vectors normalized as in (8.10). We then have

$$\frac{d\sigma}{d\Omega} = |A(\mathbf{k}'\beta, \mathbf{k}\alpha)|^2. \tag{16.62}$$

The factor $\mu_a k_\beta/\mu_b k_\alpha$ is easily recognized as the ratio of the *relative velocities* in the incident and outgoing beams. Its origin, of course, is the flux ratio in the cross section.

Let us put our expressions together. According to (16.55), (16.57), and (16.61), the scattering amplitude is given by the integral

$$A(\mathbf{k}'\beta, \mathbf{k}\alpha) = -\frac{\mu_a}{2\pi}\left(\frac{k_\beta}{k_\alpha}\right)^{1/2} \int d\mathbf{r}\, d\mathbf{r}_b \Phi_b^\dagger(\beta) \exp(-i\mathbf{k}_\beta \cdot \mathbf{r}_b)(\mathbf{r}_b|H_b'|\mathbf{r})\Psi^{(+)}(\mathbf{k}\alpha, \mathbf{r}) \tag{16.57b}$$

if the exact solution $\Psi^{(+)}$ is normalized so that its plane-wave part is given by $\Phi_a(\alpha)\exp(i\mathbf{k}_\alpha \cdot \mathbf{r})$. The notation is hybrid in that the inner product with respect to the internal coordinates is not explicitly indicated.

16.4 Partial-Wave Analysis

16.4.1 The Coupled Equations. The appropriate eigenfunctions of the total angular momentum and of the internal hamiltonian in arrangement channel a are

$$\mathscr{Y}_{jls}^M(a, \varepsilon_\alpha; \hat{\mathbf{r}}) = \sum_{mv} C(lsj, mvM)\, Y_l^m(\hat{\mathbf{r}})\Phi_a(\varepsilon_\alpha, s_1 s_2 sv) \tag{16.63}$$

and the functions $\mathfrak{Y}$ are defined as in (15.6). In place of the second line of (15.6) we now have

$$\mathfrak{Y}_j^M(lsv; \hat{\mathbf{k}}) = i^{-l}\Phi_a^\dagger(\varepsilon, s_1 s_2 sv) \cdot \mathscr{Y}_{jls}^M(a, \varepsilon; \hat{\mathbf{k}}). \tag{16.63a}$$

The dot is meant to indicate an inner product with respect to the internal coordinates.

The plane-wave eigenfunction (16.55) of H_a, the analog of (15.8), is now given by

$$\psi_a(\mathbf{k}\alpha, \mathbf{r}) = \left(\frac{2\mu_a k_\alpha}{\pi}\right)^{1/2} (k_\alpha r)^{-1} \sum_{jMl} u_l(k_\alpha r)\mathscr{Y}_{jls}^M(a, \varepsilon_\alpha; \hat{\mathbf{r}})\mathfrak{Y}_j^{M*}(lsv; \hat{\mathbf{k}}). \tag{16.64}$$

The Green's functions can be written in the form

$$G_a^\pm(E; \mathbf{r}, \mathbf{r}') = 2\mu_a \sum_{jMls\varepsilon\alpha} \mathscr{Y}_{jls}^M(a, \varepsilon_\alpha; \hat{\mathbf{r}})\mathscr{Y}_{jls}^{M\dagger}(a, \varepsilon_\alpha; \hat{\mathbf{r}}')r^{-1}r'^{-1}G_l^\pm(k_a; r, r'), \tag{16.65}$$

where $G_l^\pm$ is given by (11.2). We have used here the completeness relation (16.51) for the space of the internal coordinates of the two fragments. The summation in (16.65) should really include an integration over the continuum states in which the fragments are dissociated. We are not primarily interested in that part of G_a, though, and hence we do not explicitly indicate it.

Assuming again that the entire hamiltonian is invariant under rotation, we can write the analog of (15.12),

$$\Psi^{(\pm)}(\mathbf{k}\alpha;\mathbf{r}) = \left(\frac{2\mu_a k_\alpha}{\pi}\right)^{1/2}(k_\alpha r)^{-1}\sum_{jMll's'\varepsilon'_\alpha}\psi^{j(\pm)}_{l's'\varepsilon'_\alpha,ls\varepsilon_\alpha}(E,r)\mathscr{Y}^M_{jl's'}(a,\varepsilon'_\alpha;\hat{\mathbf{r}})\mathfrak{Y}^{M*}_j(lsv;\hat{\mathbf{k}}). \tag{16.64a}$$

The letter α combines all the necessary information concerning the initial state: the arrangement channel a, the internal energy ε_α of the fragments, their spins, the toal spin, and its z component.

Insertion of (16.64a) in the Lippmann–Schwinger equation gives us a set of coupled integral equations for the radial functions,

$$\begin{aligned}\psi^{j(\pm)}_{l's'\varepsilon'_\alpha,ls\varepsilon_\alpha}(E,r) &= u_l(k_\alpha r)\delta_{ll'}\delta_{ss'}\delta_{\varepsilon_\alpha\varepsilon'_\alpha}\\ &\quad + \sum_{l''s''\varepsilon''}\int_0^\infty dr'\,G^\pm_{l'}(k'_\alpha;r,r')\mathfrak{V}^j_{l's'\varepsilon'_\alpha,l''s''\varepsilon''_\alpha}(r')\psi^{j(\pm)}_{l''s''\varepsilon''_\alpha,ls\varepsilon_\alpha}(k,r')\end{aligned} \tag{16.66}$$

where

$$\mathfrak{V}^j_{l's'\varepsilon'_\alpha,ls\varepsilon_\alpha}(r) = 2\mu_a\int d\Omega\,\mathscr{Y}^{M\dagger}_{jl's'}(a,\varepsilon'_\alpha;\hat{\mathbf{r}})H'_a\mathscr{Y}^M_{jls}(a,\varepsilon_\alpha;\hat{\mathbf{r}}).$$

These are the exact analogs of (15.13) and (15.14). The only differences are that the indices now include a reference to the excitation energy of the fragments and that the individual spins of the fragments are *not* fixed. The collision may raise one (or both) of the fragments to an excited state of a different angular momentum.

It is easy to see that a solution of (16.66) satisfies the set of coupled radial Schrödinger equations,

$$\begin{aligned}-\frac{d^2}{dr^2}\psi^j_{l's'\varepsilon'_\alpha,ls\varepsilon_\alpha} + \frac{l'(l'+1)}{r^2}\psi^j_{l's'\varepsilon'_\alpha,ls\varepsilon_\alpha} + \sum_{l''s''\varepsilon''_\alpha}\mathfrak{V}^j_{l's'\varepsilon'_\alpha,l''s''\varepsilon''_\alpha}\psi^j_{l''s''\varepsilon''_\alpha,ls\varepsilon_\alpha}\\ = k'^2_\alpha\psi^j_{l's'\varepsilon'_\alpha,ls\varepsilon_\alpha},\end{aligned} \tag{16.67}$$

where it must be recalled that k_α is given by (16.53).

The Conventional Many-Channel Problem.[4] In (16.66) and (16.67) we are confronted by infinite sets of coupled integral and differential equations, respectively. In fact they form a *continuity* of such equations, since ε_α must run at least partially over a continuous range. This aspect is usually not taken into account. As mentioned before, the conventional many-channel problem consists of a finite set of equations (16.66) or (16.67) obtained by truncation. Their handling can be simplified by using a matrix notation analogous to that of Sec. 15.2.1. We combine $\psi_{l's'\varepsilon'_\alpha,ls\varepsilon_\alpha}{}^j$ into the square matrix Ψ_j, $\mathfrak{V}_{l's'\varepsilon'_\alpha,ls\varepsilon_\alpha}{}^j$ into the square matrix V_j; L is the diagonal matrix defined below (15.92), and K is the diagonal matrix of the channel wave numbers (16.53). The notation for them is of the same kind as for L in Sec. 15.2.1. Then (16.66) reads more simply

$$\Psi^{(\pm)}_j(E,r) = u_L(Kr) + \int_0^\infty dr'\,G^\pm_L(K;r,r')V_j(r')\Psi^{(\pm)}_j(E,r') \tag{16.66a}$$

4 Among chemists this is known as the *close-coupling* method.

and (16.67) reads

$$-\Psi_j'' + L(L+1)r^{-2}\Psi_j + V_j\Psi_j = K^2\Psi_j. \tag{16.67a}$$

16.4.2 The S Matrix. The asymptotic behavior of the radial function for large r is found from (16.66a) and (11.4),

$$\Psi_j^{(+)}(K,r) \simeq \tfrac{1}{2}e^{i1/2\pi(L+1)}[e^{-iKr} - e^{iKr}e^{-i\pi L}\bar{S}^j]$$

where

$$\bar{S}^j = \mathbb{1} - 2iK^{-1}\int_0^\infty dr\, u_L(Kr)V_j(r)\Psi_j^{(+)}(K,r). \tag{16.68}$$

In contrast to the situation in (15.16) the position of e^{iKr} now matters; it must stand on the left. On the other hand we expand (16.56a) and find the analog of (15.18),

$$\bar{A}(\mathbf{k}'\alpha',\mathbf{k}\alpha) = 4\pi\sum_{jMll'}\mathfrak{Y}_j^M(l's'v';\hat{\mathbf{k}}')\bar{a}^j_{l's'\varepsilon_\alpha',ls\varepsilon_\alpha}(E)\mathfrak{Y}_j^{M*}(lsv;\hat{\mathbf{k}}) \tag{16.69}$$

in which the matrix $\bar{a}^j$ is given by

$$\bar{a}^j = (\bar{S}^j - \mathbb{1})(2iK)^{-1}$$

in terms of the matrix $\bar{S}$.

Because of (16.60), however, it is more convenient to expand the amplitude A of (16.61) directly. We call the matrix of its coefficients in an expansion like (16.69), a^j. Then

$$a^j = K^{1/2}\bar{a}^jK^{-1/2} = (S^j - \mathbb{1})(2iK)^{-1} \tag{16.70}$$

with

$$S^j = K^{1/2}\bar{S}^jK^{-1/2}. \tag{16.71}$$

Since we are now not dealing with rearrangements, the initial and final channel masses are equal.

Reciprocity and Unitarity. The arguments of Sec. 15.1.1 to 15.1.3 can now be carried through for the amplitude $\bar{A}$ as before, with the appropriate replacements of the functions $\mathscr{Y}$ and summations over internal energies. Specifically, time-reversal invariance implies the *symmetry* of the matrix $\bar{a}^j$ and hence of S^j defined in (16.71) (*not* of $\bar{S}^j$). Similarly, it implies the symmetry of the potential matrix V_j. The only difference arises in the symmetries due to parity conservation. Since the initial and final individual fragment spins need not be the same, the right-hand sides of (15.66) to (15.67a) acquire a factor of

$$(-)^{s_1+s_2-s_1'-s_2'}.$$

The flux-conserving unitarity (15.70) must be handled with care because of the difference between initial and final momenta. If time-reversal invariance holds, then it reads simplest for the helicity amplitudes for individual spins. We find the analog of (15.70b),

$$4\pi k_{\alpha'}^{-1}\,\mathrm{Im}\,A^h(\mathbf{k}'\alpha',\mathbf{k}\alpha) = \sum_{\nu_1''\nu_2''\varepsilon''} \int d\Omega_k''\, A^{h*}(\mathbf{k}''\alpha'',\mathbf{k}'\alpha')A^h(\mathbf{k}''\alpha'',\mathbf{k}\alpha). \tag{16.72}$$

These are the unbarred amplitudes defined by (16.61), and α must refer (among other things) to the individual fragment helicities. If we insert in this the partial wave expansions, we find that it is equivalent to the unitarity of the matrix S^j (*not* of $\bar{S}^j$),

$$S^jS^{j*} = S^{j*}S^j = \mathbb{1}. \tag{16.73}$$

It must be remembered, though, that these equations include *integrations* over the "continuous" indices, i.e., over the range of partial energies in which there are more than two final (or initial) fragments.

16.4.3 Rearrangements. In order to describe rearrangement collisions, $\mathbf{\Psi}^{(+)}(\alpha,\mathbf{r})$ is expanded on the basis of the eigenfunctions of the internal hamiltonian in arrangement channel b. Instead of (16.64a) we write

$$\mathbf{\Psi}^{(+)}(\mathbf{k}\alpha,\mathbf{r}_b) = \left(\frac{2\mu_a k_\alpha}{\pi}\right)^{1/2}(k_\alpha r_b)^{-1}\sum_{jMll's'\varepsilon_\beta}\psi^{j(+)}_{l's'\varepsilon_\beta,ls\varepsilon_\alpha}(E,r_b)$$
$$\mathcal{Y}^M_{jl's'}(b,\varepsilon_\beta;\hat{\mathbf{r}}_b)\mathfrak{Y}^{M*}_j(ls\nu;\hat{\mathbf{k}}). \tag{16.74}$$

The radial functions $\psi^{j(+)}(E,r_b)$ do not satisfy a simple integral equation. They can be computed only by inverting (16.74). Use of (16.64a) gives an expression that is too complicated to be of much use.

The only feasible way is to expand (16.41) directly, using (16.64) and (16.64a). The result is

$$T_{\beta\alpha} = \sum_{jll'}\mathfrak{Y}^M_j(l's'\nu';\hat{\mathbf{k}}')T^j_{l's'\varepsilon_\beta,ls\varepsilon_\alpha}\mathfrak{Y}^{M*}_j(ls\nu;\hat{\mathbf{k}}) \tag{16.75}$$

with

$$T_{l's'\varepsilon_\beta,ls\varepsilon_\alpha} = \frac{2}{\pi}\left(\frac{\mu_a\mu_b}{k_\alpha k_\beta}\right)^{1/2}\sum_{l''s''\varepsilon_\alpha'}\left[r_b^{-1}u_{l'}(k_\beta r_b)\mathcal{Y}^M_{jl's'}(b,\varepsilon_\beta;\hat{\mathbf{r}}_b),\,H_b'\psi^{j(+)}_{l''s''\varepsilon_\alpha'}\mathcal{Y}^M_{jls}(a,\varepsilon_\alpha;\hat{\mathbf{r}}_a)\right]. \tag{16.76}$$

The bracket here indicates an inner product in a hybrid notation. Notice that in the left member $\mathbf{r}_b$ is explicit and all other coordinates are suppressed; in the right member $\mathbf{r}_a$ is explicit and the others suppressed. For instance, in a three-body collision that leads from 1 and (2,3) to 2 and (1,3), $\mathbf{r}_b$ would be the distance from particle 2 to the center of mass of (2,3); and $H_b' = V_{12} + V_{32}$.

16.5 General Scattering Rates

In most practical situations the initial number of clusters in a collision is two. If the final number is also two, so that the collision is elastic, inelastic (in the narrow sense), or a rearrangement, then the considerations of Sec. 8.1 are applicable and the differential cross section is given by (8.9) or (8.14) or their barycentric analog (16.62). However, for n-particle systems there are more possibilities. The number of final particles and even the number of initial particles, may be larger than two. While from a practical point of view it may seem unrealistic to envisage scattering experiments in which three (or more) particle-beams are directed at each other (or two at a fixed target), such multi-particle collision probabilities are needed, for instance, for the calculation of certain chemical reaction rates. We shall therefore derive here the scattering rates for the collision of arbitrary numbers of clusters of particles.

The ideas of the derivation are analogous to those of Sec. 8.1. If the initial wave packet is in channel α (arrangement channel i) its momentum space wave function is given by

$$\hat{\psi}_{\text{in}}(\alpha',k,t) = \delta_{\alpha\alpha'}g_b(k)\exp\{-i[E(k) + \varepsilon_\alpha]t\},$$

where k is a momentum vector in $3m$-dimensional Euclidean space if there are m fragments in arrangement channel i, g is normalized to unity and peaked at k^0 with a width $\Delta \ll |k^0|$, ε_α is the total bound-state energy of the clusters in channel α, and $E(k)$ is the kinetic energy of the fragments. The subscript b on g refers to the generalized impact parameter to be discussed below. It follows from (16.16) that the projection of the out-state momentum-space wave function on the arrangement channel f, channel β, is given by

$$\hat{\psi}_{\text{out},f}(\beta,k',t) = \int d^{3m}k S_{fi}(\beta,k';\alpha,k)g_b(k)\exp\{-i[E(k) + \varepsilon_\alpha]t\}.$$

According to (6.82) and (6.83) the probability that the n final fragments will eventually cross a distant surface that subtends the cone C (in configuration space) is equal to

$$P_b = \int_C d^{3m}k\left|\int d^{3m}k' S_{fi}(\beta,k;\alpha,k')g_b(k')\right|^2.$$

If the cone C does not include the initial direction of k' this may be written

$$P_b = (2\pi)^2 \int d^{3m}k \int d^{3m}k'\, d^{3m}k''\; T_{fi}(E;\beta,k,\alpha,k')T^*_{fi}(E;\beta,k,\alpha,k'')$$
$$g_b(k')g_b^*(k'')\delta(E + \varepsilon_f - E' - \varepsilon_i)\delta(E + \varepsilon_f - E'' - \varepsilon_i)\delta^3(\mathbf{K}' - \mathbf{K})\delta^3(\mathbf{K}'' - \mathbf{K}), \tag{16.77}$$

where $\mathbf{K}$ is the total momentum,

$$\mathbf{K} = \sum_{j=1}^{n} \mathbf{k}_j, \qquad \mathbf{K}' = \sum_{j=1}^{m} \mathbf{k}_j', \qquad \mathbf{K}'' = \sum_{j=1}^{m} \mathbf{k}_j''.$$

We must now be more specific about the impact parameter b and present arguments that are analogous to those leading from (8.4) to (8.7).

The impact parameter of a single particle is a vector in three-dimensional Euclidean space that is orthogonal to the particle's momentum, because a shift along its trajectory will merely displace its time of collision. For m colliding clusters a generalized impact parameter is defined so that its change will not shift their time of collision nor their center of mass. Any displacement of the fragments along their tracks in proportion to their (central) velocities will merely affect the collision time. Therefore the impact-parameter space is the orthogonal complement (in $3m$-dimensional space with a conveniently chosen metric) of the four-dimensional space spanned by the 3-momentum of the center of mass and a projection of the (central) cluster velocities relative to the center of mass. We implement this as follows.

Let $\mathbf{k}_i^0$, $i = 1, \ldots, m$, be the momenta of the m initial clusters at the center of the wave packet; let $\mathbf{K}^0$ be their total momentum,

$$\mathbf{K}^0 = \sum_{j=1}^{m} \mathbf{k}_j^0;$$

let $\mathbf{V}_i^0$ be the velocities relative to the center of mass,

$$\mathbf{V}_i^0 = \frac{\mathbf{k}_i^0}{m_i} - \frac{\mathbf{K}^0}{M}, \qquad i = 1, \ldots, m,$$

where $M = \Sigma m_i$; and let E^0 be the energy in the barycentric system,

$$E^0 = \sum_{i=1}^{m} \frac{|\mathbf{k}_i^0|^2}{2m_i} - \frac{|\mathbf{K}^0|^2}{2M} = \sum_{i=1}^{m} \tfrac{1}{2} m_i |\mathbf{V}_i^0|^2.$$

Let ξ be any vector in $3m$-dimensional Euclidean space made up of the m 3-vectors $\boldsymbol{\xi}_i$, $i = 1, \ldots, m$. Then form

$$\boldsymbol{\xi}_i = M^{-1/2} m_i^{1/2} \boldsymbol{\xi}_i, \qquad i = 1, \ldots, m,$$

and perform a rotation so that the new vector $\bar{\xi}$ has the first four components

$$\bar{\xi}_i = M^{-1/2} \sum_{j=1}^{m} m_j^{1/2} \xi_j^{(i)}, \qquad i = 1, 2, 3,$$

$$\bar{\xi}_4 = (2E^0)^{-1/2} \sum_{1}^{m} m_j^{1/2} \mathbf{V}_j^0 \cdot \boldsymbol{\xi}_j.$$

We shall call the first three components of $\bar{\xi}$ simply $\bar{\boldsymbol{\xi}}$. The remaining components $\bar{\xi}_i$, $i = 5, \ldots, m$, are arbitrarily related to the $\tilde{\xi}_i$, except that the transformation from ξ to $\bar{\xi}$ is to be *orthogonal*. Then

$$\sum_{i=1}^{3m} \bar{\xi}_i^2 = |\xi|^2 = M^{-1} \sum_{i=1}^{m} m_i |\boldsymbol{\xi}_i|^2$$

and the Jacobian is given by

$$\partial(\bar{\xi}_1, \ldots, \bar{\xi}_{3m})/\partial(\xi_1, \ldots, \xi_{3m}) = M^{-3m/2} \prod_i m_i^{3/2}.$$

We define

$$\xi = (\bar{\xi}_1, \ldots, \bar{\xi}_4, 0, \ldots, 0), \qquad \mathring{\xi} = (0, \ldots, 0, \xi_5, \ldots, \xi_{3m}).$$

Then the volumn element is

$$d^{2m}\xi = M^{3m/2}\left(\prod_i m_i^{-3/2}\right) d^4\breve{\xi}\, d^{3m-4}\mathring{\xi} \tag{16.78}$$

and the δ-function

$$\delta^{3m}(\xi) = M^{-3m/2}\left(\prod_i m_i^{3/2}\right)\delta^4(\breve{\xi})\delta^{3m-4}(\mathring{\xi}). \tag{16.79}$$

We also have

$$\delta^3(\bar{\boldsymbol{\xi}}) = M^3\delta^3\left(\sum_{i=1}^m m_i\boldsymbol{\xi}_i\right)$$

$$\delta(\bar{\xi}_4) = (2ME^0)^{1/2}\,\delta\left(\sum_{i=1}^m m_i\mathbf{V}_i^0\cdot\xi_i\right).$$

In the special case when $\boldsymbol{\xi}_i = \mathbf{v}_i - \mathbf{v}_i'$, the particle velocities, this means

$$\delta^3(\bar{\mathbf{v}} - \bar{\mathbf{v}}') = M^3\delta^3(\mathbf{K} - \mathbf{K}')$$

and

$$\delta(\bar{v}_4 - \bar{v}_4') = (2ME^0)^{1/2}\delta[\Sigma V_i^0\cdot(\mathbf{k}_i - \mathbf{k}_i')].$$

If both $\mathbf{k}_i$ and $\mathbf{k}_i'$ are restricted to be near $\mathbf{k}_i^0$ then we get

$$\delta^4(\breve{v} - \breve{v}') = M^{7/2}(2E^0)^{1/2}\delta^3(\mathbf{K} - \mathbf{K}')\delta(E - E'). \tag{16.80}$$

We may now write (16.77), using $g_b(k) = g(k)\exp(-i\Sigma\mathbf{b}_i\cdot\mathbf{k}_i)$, in the form

$$\begin{aligned} P_b = (2\pi)^2 M^{-7/2}(2E^0)^{-1/2}\int_C d^{3m}k \int d^{3m}k'\, d^{3m}k''\, T_{fi}(E;\beta,k,\alpha,k') \\ T_{fi}^*(E;\beta,k,\alpha,k'')g(k')g^*(k'')\delta(E - E' + \varepsilon_f - \varepsilon_i) \\ \delta^4(\breve{v} - \breve{v}')\delta^3(\mathbf{K} - \mathbf{K}'')\exp[iM\mathring{b}\cdot(\mathring{v}'' - \mathring{v}')]. \end{aligned} \tag{16.81}$$

As in Sec. 8.1 we take the initial state to be *mixed*; the probability for the impact parameter b to lie in the element $d^{3m}b$ is $\mu(b)d^{3m}b$. Then

$$P = \int d^{3m}b\mu(b)P_b.$$

Assume that the beam of particles i has a density $\rho_i(\mathbf{b}_i)$. We decompose $\mathbf{b}_i$ into its component along $\mathbf{v}_i^0$, $\mathbf{b}_i\cdot\mathbf{v}_i^0 = c_i$, and $\mathbf{b}_i^\perp$, orthogonal to $\mathbf{v}_i^0$, $\mathbf{b}_i^\perp = \mathbf{b}_i - c_i\mathbf{v}_i^0$. Then ρ_i is taken to be different from zero only when

$$\pi|\mathbf{b}_i^\perp|^2 \le a_i \quad \text{and } t_i \le c_i|\mathbf{v}_i^0| \le t_i + T_i,$$

where a_i is the cross-sectional area and T_i the duration of the (incoherent) beam-pulse of particles i. Inside this region of volume $V_i = a_i|\mathbf{v}_i^0|T_i$, ρ_i has the constant value $\rho_i^0 = 1/V_i$. Then

$$\mu = \prod_{i=1}^m \rho_i. \tag{16.82}$$

At this point we must make two assumptions that can be expected to be very well satisfied in any physical situation: (A) The radii $(a_i/\pi)^{1/2}$ of the beams are very large compared to the maximal impact parameters $|\mathbf{b}_i^{\perp}|$ for which the integral in (16.81) significantly differs from zero. This is analogous to the assumption made below (8.5) and amounts to the recognition that the particle forces of interest are microscopic and beam sizes, macroscopic. There is, however, the following caveat.

One of the difficulties in multi-particle scattering arises from the possibility of on-energy-shell multiple scattering.[4] Such processes, in which in a three-particle case, for example, particle 1 is first scattered by particle 2, conserving energy and momentum, and subsequently by particle 3, are effectively of infinite range. Therefore, in order for assumption (A) to be satisfied the cone C must not include those momenta that meet the multiple-scattering conditions. Since the T matrix becomes infinite[4a] at those momenta (in a nonintegrable way) there is another reason to exclude them from C anyway. This will therefore be part of assumption (A).

Assumption (B) is that the beam-pulse durations T_i are so large that the "edge effects" that occur at the beginning and end of the pulse-intersections can be neglected.

Let us now calculate the $3m$-dimensional integral $\int d^{3m}b\mu(b)P_b$. Because of assumption (A) the only values of $\mathring{b}$ that significantly contribute to the integral are much smaller than those allowed by μ. Hence we may, with little error, set $\mathring{b} = 0$ in the evaluation of $\int d^4\check{b}$, and allow the integral over $\mathring{b}$ to range to infinity.

When $\mathring{b} = 0$ then

$$\mathbf{b}_i = \mathbf{b} + t\mathbf{v}_i^0,$$

where

$$\mathbf{b} = \bar{\mathbf{b}} - t\mathbf{K}^0/M, \quad t = (M/2E^0)^{1/2}\bar{b}_4.$$

Thus $\mathbf{b}$ and t are the point and instant of collision, respectively. Therefore, by assumption (B)

$$\int d^4\check{b}\prod_i \rho_i = \int d^3\mathbf{b}\,dt\,(2E^0/M)^{1/2}\prod_i \rho_i \cong V_{\text{int}}T(2E^0/M)^{1/2}\prod_i \rho_i^0,$$

where V_{int} is the volume of the spatial intersection of the beams and T is the length of time during which all beam pulses intersect. Again using assumption (A), then by (16.78), (16.81), and (16.82),

$$\begin{aligned} P &= \int d^{3m}b_\mu(b)P_b = M^{3m/2}\left(\prod_i m_i^{-3/2}\right)\int d^4\check{b}\prod_i \rho_i \int d^{3m-4}\mathring{b}P_b \\ &= (2\pi)^{3m-2}V_{\text{int}}T\left(\prod_i \rho_i^0\right)\int_C d^{3n}k|T_{fi}(E^0;\beta,k,\alpha,k^0)|^2 \\ &\qquad \delta^3(\mathbf{K}-\mathbf{K}^0)\delta(E-E^0+\varepsilon_f-\varepsilon_i), \end{aligned} \tag{16.83}$$

4a E. Gerjuoy (1970 and 1971); J. Nuttall (1971); S. P. Merkuriev (1971); R. G. Newton (1972 and 1974). See also discussion in Sec. 17.4.5.

where we have also used the sharp peaking of $g(k)$ at k^0 and the fact that g is normalized to one.

If N_i fragments of kind i are sent in then the number of m-tuples of particles that can be expected to be found in the cone C is $(\Pi N_i)P$, and since $\Pi \rho_i^0 N_i = \Pi (N_i/V_i) = \Pi d_i$, where d_i is the fragment-density of the ith beam, the rate at which clusters of type f will be found crossing a distant surface intersecting C is given by

$$R = V_{\text{int}}\left(\prod_i d_i\right)\sigma, \tag{16.84}$$

where

$$\sigma = (2\pi)^{3m-2}\int_C d^{3n}k|T_{fi}(E^0;\beta,k,\alpha,k^0)|^2\delta^3(\mathbf{K}-\mathbf{K}^0)\delta(E-E^0+\varepsilon_f-\varepsilon_i) \tag{16.85}$$

is the *specific scattering rate*.

The quantity σ has the dimensions (distance)$^{3m-3}$ (time)$^{-1}$. In the case of two incident clusters this becomes (distance)^3 (time)$^{-1}$, which, when divided by the incident relative velocity, gives an area, the scattering cross section.

In order to relate (16.85) to physical measurements the $3n$-dimensional cone C in configuration space has to be made up of n three-dimensional cones C_i and the cone $\bar{C}$ formed by simultaneous arrival times of all the fragments at their counters, so that $r_i = v_i t$. The corresponding cone in momentum space is "the same" provided that a metric is used in which $|r|^2 = \Sigma m_i \mathbf{r}_i^2$ and $|k|^2 = \Sigma k_i^2/m_i$. Thus in momentum space $\bar{C}$ is formed by

$$\int_{\bar{C}} dk_1^2 \cdots dk_m^2 \cdots = \int_{\bar{C}} d\xi_1 \cdots d\xi_n(\Pi\, 2m_i)\int_0^\infty dEE^n\,\delta(E-\Sigma E_i)$$

where $\xi_i = E_i/E$ is the energy fraction of the ith fragment. Therefore

$$\begin{aligned}\sigma = (2\pi)^{3m-2}\int_{C_1} d^2\hat{\mathbf{k}}_1 \cdots \int_{C_n} d^2\hat{\mathbf{k}}_n \int_{\bar{C}} d^n\xi \int_0^\infty dEE^{n-1}\left(\prod_i m_i k_i\right)\delta(1-\Sigma\xi_i)\\ \delta(E-E^0+\varepsilon_f-\varepsilon_i)\delta^3(\mathbf{K}-\mathbf{K}^0)|T_{fi}(E^0;\beta,k,\alpha,k^0)|^2.\end{aligned} \tag{16.86}$$

Because of the multiple-scattering difficulties mentioned earlier, these cones cannot be opened up to include all space to define a "total scattering rate". The resulting integral would in general diverge. (We shall return to this point in the case of three particles in Sec. 17.4.5.)

16.6 Formal Resonance Theory

Let us start with a reminder of the definition of a *channel* as used in this book. First, *arrangement channels* are defined by allowing one or more particle coordinates to approach infinity. As the distance between two groups of particles (called fragments) tends to infinity, the hamiltonian approaches a "simple" channel hamiltonian H_a. The part that tends to zero is H_a',

$$H = H_a + H'_a.$$

We similarly defined arrangement channels of more than two fragments by "ripping" the hamiltonian apart into more than two pieces. A somewhat extended notion of the same idea allows one or more particles to join either fragment; their interaction with them is then included in H_a and the arrangement channel thus defined combines two (or more) of the previous ones. In the three-body problem, for example, three arrangement channels are defined by

$$H_a = H_0 + V_{12} \qquad H_b = H_0 + V_{23} \qquad H_c = H_0 + V_{13}.$$

However, we may combine the first two by setting instead

$$H_a = H_0 + V_{12} + V_{23}.$$

Within each arrangement channel the various channels proper are defined by the internal energies of the fragments. If, in the three-body problem, the interaction V_{12} leads to two bound states of particles 1 and 2, then arrangement channel a contains two discrete channels. The states in which particles 1 and 2 are not bound form a continuum. Usually we are not explicitly interested in these but rather consider them in the context of a new arrangement channel of three fragments. In the three-body problem this would be defined by

$$H_d = H_0.$$

For *virtual* or *intermediate states*, however, the inclusion of continuous channels within an arrangement cannot always be avoided.

At a given total energy it may be possible to excite some internal energy levels and still have some (positive) kinetic energy of the fragments left over; other levels may not be so accessible. The channels corresponding to the former are said to be *open*, and those corresponding to the latter are said to be *closed*. The energy at which a channel opens up is its *threshold*.

Effective Hamiltonians. We now focus our attention on a fixed two-fragment arrangement defined by

$$H = H_a + H'_a.$$

We assume that there are several channels within this arrangement defined by fragment excitations. Let P_α be an orthogonal projection operator on a *group* α of channels (which may consist of a single channel, or more than one). Such a projection can be constructed, for example, from the internal state vectors of (16.50). All other channels are combined into a group β with projection P_β so that

$$P_\alpha + P_\beta = \mathbb{1}. \tag{16.87}$$

These projections are used in the definition of operators such as

$$\begin{aligned} V_\alpha &\equiv P_\alpha H'_a P_\alpha \qquad & V_{\alpha\beta} &\equiv P_\alpha H'_a P_\beta \\ H_\alpha &\equiv P_\alpha H P_\alpha \qquad & H_{a\alpha} &\equiv P_\alpha H_a P_\alpha, \text{ etc.} \end{aligned} \tag{16.88}$$

Next we define Green's functions on the spaces of these channel groups,

$$\begin{aligned}(\mathscr{E}_\alpha - H_\alpha)G_\alpha(\mathscr{E}_\alpha) &= P_\alpha \\ (\mathscr{E}_\beta - H_\beta)G_\beta(\mathscr{E}_\beta) &= P_\beta\end{aligned} \tag{16.89}$$

and the outgoing-wave boundary condition. Furthermore, define

$$G(E) \equiv G_\alpha(E) + G_\beta(E). \tag{16.90}$$

Neither G_α nor G_β depends upon the interaction $V_{\alpha\beta}$ between the two groups of channels, and hence neither does G. Since the two spaces on which P_α and P_β project are orthogonal, G is simply

$$G(E) = (E - H_\alpha - H_\beta)^{-1}. \tag{16.91}$$

The complete Green's function

$$\mathscr{G} = (E - H)^{-1} \tag{16.92}$$

(and outgoing-wave boundary condition) can be expressed in terms of G by the equation

$$\mathscr{G} = G + G(V_{\alpha\beta} + V_{\beta\alpha})\mathscr{G} \tag{16.93}$$

so that

$$\begin{aligned}P_\alpha\mathscr{G} &= G_\alpha + G_\alpha V_{\alpha\beta}P_\beta\mathscr{G} \\ P_\beta\mathscr{G} &= G_\beta + G_\beta V_{\beta\alpha}P_\alpha\mathscr{G}.\end{aligned} \tag{16.94}$$

Substitution of the second equation in the first and subsequent left multiplication by G_α^{-1} yields

$$(\mathscr{E}_\alpha - H_\alpha - V_{\alpha\beta}G_\beta V_{\beta\alpha})P_\alpha\mathscr{G} = P_\alpha + V_{\alpha\beta}G_\beta. \tag{16.95}$$

Now define a pseudo hamiltonian on the channel group α,

$$\mathscr{H}_\alpha(\mathscr{E}_\beta) \equiv H_\alpha + V_{\alpha\beta}G_\beta(\mathscr{E}_\beta)V_{\beta\alpha} \tag{16.96}$$

and a corresponding Green's function by

$$[\mathscr{E}_\alpha - \mathscr{H}_\alpha(\mathscr{E}_\beta)]\mathfrak{G}_\alpha = P_\alpha \tag{16.97}$$

and the outgoing-wave boundary condition. Then (16.95) becomes

$$P_\alpha\mathscr{G} = \mathfrak{G}_\alpha + \mathfrak{G}_\alpha V_{\alpha\beta}G_\beta \tag{16.98}$$

which, in turn, is substituted in the second line of (16.94),

$$P_\beta\mathscr{G} = G_\beta + G_\beta V_{\beta\alpha}\mathfrak{G}_\alpha + G_\beta V_{\beta\alpha}\mathfrak{G}_\alpha V_{\alpha\beta}G_\beta. \tag{16.99}$$

Finally we add (16.98) and (16.99),

$$\mathscr{G} = G_\beta + (\mathbb{1} + G_\beta V_{\beta\alpha})\mathfrak{G}_\alpha(V_{\alpha\beta}G_\beta + \mathbb{1}). \tag{16.100}$$

This [or (16.98)] shows that the complete Green's function $\mathscr{G}$ restricted to channel group α is exactly given by $\mathfrak{G}_\alpha$,

$$\mathscr{G}_\alpha \equiv P_\alpha\mathscr{G}P_\alpha = \mathfrak{G}_\alpha. \tag{16.101}$$

The result (16.101) means that, as far as channel group α is concerned, the entire effect of the coupling to other channels is the replacement of the original hamiltonian by the *energy-dependent* pseudo hamiltonian $\mathscr{H}_\alpha$. Of course, both the numbers $\mathscr{E}_\alpha$ and $\mathscr{E}_\beta$ in (16.97) must be set equal to the total energy E in the evaluation of $\mathscr{G}(E)$. The physical interpretation of $\mathscr{H}_\alpha$ is simple enough. In addition to the original H_α there is a term that evidently describes a transition from channel group α to β, propagation there according to the total hamiltonian H_β with energy $\mathscr{E}_b(=E)$, and return to group α.

Notice the fact, visible in the pseudo hamiltonian $\mathscr{H}_\alpha$, that coupling to closed channels is *always* equivalent to an *attractive* contribution to the hamiltonian. (When $\mathscr{E}_\beta$ is below the thresholds of the β channels, G_β is negative definite; furthermore, $V_{\beta\alpha} = V_{\alpha\beta}{}^\dagger$.) Therefore, bound states in existence when such coupling is absent can only be *deepened* when it is taken into account. They cannot disappear. What is more, it is quite possible for bound states to exist in the presence of such coupling which disappear when it is removed.

If any channels in group β are open, then G_β is not a hermitian operator, and hence neither is $\mathscr{H}_\alpha$. Its nonhermitian character may be explicitly exhibited by writing (for real values of E)

$$\mathscr{H}_\alpha(E) = H_\alpha + V_{\alpha\beta}\mathscr{P}(E - H_\beta)^{-1}V_{\beta\alpha} - i\pi V_{\alpha\beta}\delta(E - H_\beta)V_{\beta\alpha} \qquad (16.102)$$

where $\mathscr{P}$ stands for the *principal value*. This shows that the skew-hermitian part of $\mathscr{H}_\alpha(E)$ is always negative semidefinite, a circumstance which clearly expresses the effective *absorption* felt in channels α owing to the loss of flux to channels β. If E lies below the continuous spectrum of H_β, the δ function in (16.102) vanishes. All channels in group β are then closed, and $\mathscr{H}_\alpha$ is hermitian. Furthermore, whenever $\operatorname{Im} E > 0$ ($\operatorname{Im} E < 0$) the skew-hermitian part of $\mathscr{H}_\alpha$ is negative (positive) semidefinite; hence $\mathfrak{G}_\alpha(E)$ cannot have any complex poles, in agreement with the fact that, H being hermitian, $\mathscr{G}(E)$ has none.

Resonances and Bound States Embedded in the Continuum. Suppose now that the hamiltonian is such that, in spite of the nonhermitian nature of $\mathscr{H}_\alpha$ for real E above the threshold of some channels in β, $\mathscr{H}_\alpha(E_0)$ has E_0 as a discrete eigenvalue. The operator function $\mathfrak{G}_\alpha(E)$ then has a *pole* at E_0, and according to (16.101), so does $\mathscr{G}(E)$. This means that E_0 is a bound state of H *embedded in the continuum.* Such a state cannot show up in the scattering; it is normalizable and hence asymptotically vanishes.

Watch now what happens if the hamiltonian is changed continuously from the "accidental" situation just envisaged. For example, we may suppose that H_a' is multiplied by a coupling parameter ξ, and we alter its value. If $\mathfrak{G}_\alpha(E)$ is assumed to have an analytic continuation to a neighborhood of E_0, its pole must move as a function of ξ and only under very special conditions will it remain on the real axis. But as we have seen (for real ξ) it cannot move into the upper half plane, because, for $\operatorname{Im} E > 0$, the skew-hermitian part of $\mathscr{H}_\alpha(E)$ is negative semidefinite. Hence it must move *through the cut* that $\mathfrak{G}_\alpha(E)$ has along the continuous spectrum of H

and through the cut of G_β along the continuous spectrum of H_β. The outgoing-wave boundary condition for $\mathfrak{G}_\alpha$ and G_β placed E on the upper rims of both cuts. Consequently, if a small alteration of the hamiltonian is able to produce a real bound state, and if $\mathfrak{G}_\alpha$ permits an analytic continuation through its cut, then we expect it to have a *pole* there close to the real axis.

We can look at the same problem from the point of view of the eigenvalue of $\mathscr{H}_\alpha$. Assuming that for some value of the parameter ξ, $\mathscr{H}_\alpha(E_0, \xi)$ has the eigenvalue $A = E_0$,

$$[\mathscr{H}_\alpha(E_0, \xi) - A]u(\xi) = 0 \tag{16.103}$$

we search for an analytic continuation of this equation as ξ varies. If $G_\beta(E)$ is analytic in a neighborhood of E_0, then so is $\mathscr{H}_\alpha(E, \xi)$ and we expect to be able to continue (16.103) at least a short distance. The value of E_0 will of course change, and in general it will not remain real. But it cannot go into the upper half plane, the "physical sheet" of G_β, because then $\mathscr{G}$ would have a pole on the physical sheet and H would have a complex eigenvalue. So E_0 must move into the lower half plane, which takes it through the cut of G_β into the "second sheet." There G_β no longer is a resolvent as on the first sheet, and the fact that the $\mathscr{H}_\alpha(E)$ so continued has E as an eigenvalue does not imply that E is also an eigenvalue of H.

We have therefore found (provided that all the analyticity assumptions are justified, which depends on the hamiltonian) that, because the pseudo-hamiltonian $\mathscr{H}_\alpha(E)$ has a right-hand branch cut itself, a pole of $\mathfrak{G}_\alpha$ at E_0 *on the second sheet* which by a small change in H can be maneuvered onto the real axis (upper rim of the right-hand cut) signifies an *eigenvalue* A of $\mathscr{H}_\alpha$ which is such that $A(E_0) = E_0$. The residue of $\mathfrak{G}_\alpha$ at E_0 is consequently of the form

$$\frac{uv^\dagger}{(v, u)} \tag{16.104}$$

if

$$[\mathscr{H}_\alpha^\dagger - A^*]v = 0 \tag{16.103a}$$

and we use the dyadic notation. In other words, near $E = E_0$, according to (16.100),

$$\mathscr{G} = G_\beta + \frac{(\mathbb{1} + G_\beta V_{\beta\alpha})uv^\dagger(V_{\alpha\beta}G_\beta + \mathbb{1})}{(v, u)[E - A(E)]} + \cdots \tag{16.105}$$

where $A(E)$ is such that

$$A(E_0) = E_0.$$

If E_0 is not far from the real axis, it can be expected that there exists a *real* value E_1 of the energy at which

$$R \equiv \operatorname{Re} A(E_1) = E_1$$

and at which $I \equiv -\operatorname{Im} A(E_1)$ is small (compared with the energy region over which the other quantities in $\mathscr{G}$ or T vary by substantial fractions). If such an E_1 exists, we expand $A(E)$ about it,

$$A(E) = R + iI + (E - E_1)(R' + iI') + \cdots$$

and obtain

$$E - A(E) = (E - E_r + i\tfrac{1}{2}\Gamma)[1 - A'(E_1)] + \cdots \tag{16.106}$$

with

$$\begin{aligned} E_r &= E_1 - \frac{II'}{(1 - R')^2 + I'^2} \\ \Gamma &= \frac{2(1 - R')I}{(1 - R')^2 + I'^2}. \end{aligned} \tag{16.107}$$

When $I = 0$, we must have $I' = 0$ because the eigenvalue cannot cross the real axis. Hence we expect I' to be small when I is small and, unless R' happens to be close to 1,

$$\begin{aligned} E_r &\simeq E_1 - \frac{II'}{(1 - R')^2} \\ \Gamma &\simeq \frac{2I}{1 - R'} \end{aligned} \tag{16.107a}$$

$$E - A(E) \simeq (1 - R')(E - E_r + i\tfrac{1}{2}\Gamma). \tag{16.106a}$$

Breit–Wigner Formula. Let us use these results in an evaluation of the **T** matrix. Write $\Psi_\beta{}^{(f)(\pm)}$ and $\Psi_\beta{}^{(i)(\pm)}$ for the final and initial scattering states of the channel hamiltonian H_a,

$$\Psi_\beta^{(f)(\pm)} = \Psi_0^{(f)} + G_\beta^\pm V_\beta \Psi_0^{(f)} \tag{16.108}$$

if $\Psi_\beta{}^{(f)(\pm)}$ is in channel group β, etc. After some calculation we get, near $E = E_1$,

$$T_{fi} = (\Psi^{(f)}, [H_a' + H_a' G_\beta H_a'] \Psi^{(i)}) + \frac{B_{fi}}{(v, u)(E - E_r + i\tfrac{1}{2}\Gamma)} + \cdots. \tag{16.109}$$

The coefficient B_{fi} is found to be given by

$$B_{fi} = \frac{\bar{\gamma}^f \bar{\gamma}^i}{1 - R'} \tag{16.110}$$

provided that both the initial and final states are contained in channel group β, and

$$\bar{\gamma}^i = (\Psi_\beta^{(i)(-)}, V_{\beta\alpha} u). \tag{16.111}$$

If the hamiltonian is time-reversal invariant, the phases can be chosen so that

$$\bar{\gamma}^i = (v, V_{\alpha\beta}\Psi_\beta^{(i)(+)}) \tag{16.111a}$$

also. Similar expressions hold for $\bar{\gamma}^f$. If either the initial or the final state is not in group β, then B_{fi} contains a factor of $E - A$, and the pole of T_{fi} is not explicitly visible. But we are free, in any given energy region, to include in β *all* open channels. Then each matrix element of **T** of physical interest for scattering behaves as though it had a pole at the complex energy

$$E = E_r - i\tfrac{1}{2}\Gamma.$$

Equation (16.102) allows us to evaluate I. Assuming that u is normalized to unity, and using the completeness of the outgoing- and the incoming-wave eigenstates of H_β, we get

$$I = \pi \sum_f |\bar{\gamma}^f|^2. \tag{16.112}$$

The sum, of course, may mean, at least in part, an integration over directions. The energy is fixed and not summed over.

Finally we realize that always

$$|(v,u)|^2 \leq 1$$

if both u and v are normalized to unity. When the eigenvalue A is real, v is a multiple of u and

$$|(v,u)|^2 = 1$$

(because of the semidefinite nature of the skew-hermitian part of $\mathscr{H}_\alpha$). Hence $|(u,v)|^2$ is stationary with respect to small variations near the particular hamiltonian which leads to a real A. We therefore expect that

$$|(v,u)| \simeq 1$$

to a good approximation when $\operatorname{Im} A$ is small. As a result the main energy dependence of T_{fi} near $E = E_r$ is given by

$$T_{fi} \simeq \frac{1}{2\pi} \frac{\gamma^f \gamma^i}{E - E_r + i\frac{1}{2}\Gamma} + \cdots \tag{16.113}$$

with

$$\gamma = \bar{\gamma}\left(\frac{2\pi}{1 - R'}\right)^{1/2}$$
$$\Gamma = \sum_f |\gamma^f|^2. \tag{16.114}$$

Equation (16.113) describes the rapid energy dependence of the amplitude near a *resonance*. It is usually referred to as a *Breit–Wigner formula*. Near the resonance energy the total cross section, averaged over all incident polarizations, etc., is given by

$$\sigma \propto \frac{\frac{1}{4}\Gamma^2}{(E - E_r)^2 + \frac{1}{4}\Gamma^2} + \cdots \tag{16.115}$$

according to (16.113) and (16.114). The shape of this curve (as a function of the energy) is the same as the lorentzian in electromagnetic theory (see Fig. 3.2). It is peaked at $E = E_r$ and symmetric about it, with a width (at half maximum) Γ.

Other Shapes. It must be borne in mind that the derivation of (16.115) rests on a number of specific assumptions that are not always fulfilled. The most important of them is that the width Γ and the "shift" from E_1 to E_r in (16.107) are both small relative to the energy scale on which the other quantities in the amplitude vary appreciably. If the shift is not small, the expansion (16.106) is no longer useful near E_r. If the width Γ is not small, the resonance peak is superimposed upon a background that varies as rapidly as the peak itself. Its shape may then be quite distorted; in fact, it may effectively become invisible.

A special case in which the assumption of a smoothly varying background breaks down is that of two resonances whose distance is comparable with or smaller than their widths. This may produce important changes in their individual shapes. Another case is that in which the resonance appears at low energies or near a threshold, where the amplitude has an additional "natural" rapid energy dependence (see Sec. 17.2).

A second assumption that went into the derivation of (16.115) is that the pole of the Green's function, in (16.105), is *simple*. If the hamiltonian that produces the "nearby" bound state embedded in the continuum is hermitian, then the pole of *its* resolvent must be simple, of course. But even then it may happen that two such poles coincide when they shift into the complex plane. A *double* pole produces a resonance-line shape entirely different from (16.115). It may be much flatter on top, or it may even produce a double maximum. However, such a case may be considered as an extreme instance of two "overlapping" resonances. (We shall return to this point in Sec. 19.3.)

Factorization of the Residue. Turning from the total cross section to that of a specific reaction, we notice that (16.113) shows that *every* collision amplitude that is not zero (or "forbidden") at that energy exhibits the resonance at the same energy and with the same width. If near E_r some γ's are more rapidly energy-dependent than others, this fact may of course be obscured in a practical observation. Notice also the characteristic factorization of the residue in (16.113). Each individual cross section at the resonance energy is factorable into two terms, one of which depends upon the initial state only and the other upon the final state only. Furthermore, if the hamiltonian is rotationally invariant then the bound-state wave function u has a single value of the total angular momentum (except in the case of an accidental degeneracy). Equation (16.111) shows that as a result only that same single angular-momentum state in both the initial and the final state exhibits the resonance behavior. The angular distribution of the fragments emerging at the resonance energy is consequently characteristic of that particular angular momentum. The same holds for other quantum numbers that are "good" in the reaction. Whenever the hamiltonian is such that a mixture of some quantum numbers would be considered accidental in a bound state, their mixture in a resonance is to be considered equally accidential.

In the light of the mechanism that produces the resonance, the form of (16.113) together with (16.114) has a simple physical interpretation. Equation (16.111) shows that the numbers $|\gamma|^2$ have the significance of a leakage probability, per unit time, $|\gamma^i|^2$ from the beam *into* the bound state and $|\gamma^f|^2$ from the bound state back out. [The factor $(1 - R')^{1/2}$ between γ and $\bar{\gamma}$ merely takes care of the energy dependence of u and v and the shift from E_0 to E_1.] Equation (16.114) shows that Γ is the sum of all these transition probabilities (which are referred to as *partial widths*) out of the bound state. It is the total *leakage probability per unit time*, and thus its inverse is the average lifetime of the bound state. Therefore, (16.113) gives us just the result expected on the basis of Heisenberg's uncertainty relation.

Time Delay. Finally, we must ask whether or not (16.113) really justifies calling the observable peak in the cross section a resonance. If so, it should be associated with a time *delay* caused by the particles being caught in the bound state. As we mentioned before, such a delay is visible in the time-independent description in the form of a sharp rise in the phase of the amplitude when the energy increases. Since (16.113) implies that near $E = E_r$ the phase of T_{fi} behaves as

$$\delta \simeq \tan^{-1} \frac{\frac{1}{2}\Gamma}{E_r - E}$$

this phase sharply increases as E increases through E_r provided that Γ is *positive*. The number I is necessarily positive. So according to (16.107a) the sign of Γ depends upon that of $1 - R'$.

Suppose that the hamiltonian is such that at $E = E_0$ the pseudo hamiltonian $\mathscr{H}_\alpha(E)$ has a *real* eigenvalue A. Then (16.102) shows that the derivative of $A = R$ with respect to E at that point $E = E_0$ is given by

$$R' = \frac{1}{\pi} \frac{d}{dE} \int \frac{dE' |h(E')|^2}{E - E'} \bigg|_{E=E_0}$$

where $|h(E_0)|^2 = I(E_0) = 0$. This derivative is clearly nonpositive. Hence $R' \leq 0$ and $1 - R' \geq 1$ when there is a bound state embedded in the continuum. Now, when there is a narrow resonance, we are supposedly "near" such a bound state, and hence $1 - R'$ is surely not negative.

Appendix

Proof of Eq. (16.13). Let us assume that $H_a \neq H_b$ in the sense that the only vector Φ for which

$$e^{iH_a t}\Phi = e^{iH_b t}\Phi \tag{16.116}$$

is true for all t is the null vector. Roughly speaking this means that H_a and H_b should have no common eigenvectors with the same eigenvalue. If this requirement should be "accidentally" violated, then for that specific state the special considerations of the rearrangement theory are unnecessary anyway, and one may act as though H_a and H_b were equal.

Consider the operator

$$W_t \equiv \Omega_t^{a\dagger}\Omega_t^b = P_a e^{iH_a t} e^{-iH_b t} P_b$$

with

$$\Omega_t^a \equiv e^{iHt} e^{-iH_a t} P_a \qquad \Omega_t^b \equiv e^{iHt} e^{-iH_b t} P_b$$

so that (16.10) and (16.10a) together with (6.84) read[5]

$$\Omega_t^a \Rightarrow \Omega_a^{(\pm)} \qquad \Omega_t^b \Rightarrow \Omega_b^{(\pm)} \tag{16.117}$$

as $t \to \mp\infty$. Now

$$W_t \to \Omega_a^{(\pm)\dagger}\Omega_b^{(\pm)} \tag{16.118}$$

in the same limits. That follows from the identity

$$(\Psi, (W_t - \Omega_a^{(\pm)\dagger}\Omega_b^{(\pm)})\Phi) = (\Omega_t^a\Psi, (\Omega_t^b - \Omega_b^{(\pm)})\Phi) + ((\Omega_a^{(\pm)} - \Omega_t^a)\Psi, \Omega_b^{(\pm)}\Phi).$$

As $t \to \mp\infty$ the second term vanishes according to (16.117), and the first term does by Schwarz's inequality and (16.117).

The next step is to show that as $t \to \pm\infty$

$$W_t \to 0. \tag{16.119}$$

For that we must make use of the special nature of the channel spaces $\mathscr{H}_a$ and $\mathscr{H}_b$. Let us write H_a in two pieces:

$$H_a = H_a^{\text{kin}} + H_a^{\text{int}}.$$

H_a^{int} is the part that refers to the interval coordinates of those fragments which in the arrangement channel a are bound; H_a^{kin} refers to the kinetic energy of the bound fragments. Similarly for H_b. For example, in the three-body problem let particles 2 and 3 be bound and 1, free in a. Then in the center-of-mass system

$$H_a^{\text{int}} = -\frac{1}{2\mu_{23}}\nabla^2 + V_{23}(\mathbf{r})$$

$$H_a^{\text{kin}} = -\frac{1}{2\bar{\mu}_1}\nabla'^2$$

$$\mu_{23} = \frac{m_2 m_3}{m_2 + m_3} \qquad \bar{\mu} = \frac{m_1(m_2 + m_3)}{m_1 + m_2 + m_3}$$

if $\mathbf{r}$ is the distance between particles 2 and 3, and $\mathbf{r}'$ that between 1 and the center of mass of 2 and 3. We define the space $\mathscr{H}_a^{(n)}$ as the direct product of the nth *bound state* (i.e., discrete eigenvector) of H_a^{int} and the space spanned by the normalizable wave packets made up of the "continuum eigenstates"[6] of H_a^{kin}. The operator $P_a^{(n)}$ is the orthogonal projection on $\mathscr{H}_a^{(n)}$, and we have

5 A reminder: The double arrow denotes a strong limit, the single arrow, a weak limit; see Sec. 6.8.

6 H_a^{kin} of course has no bound states. What we mean is the Hilbert space of square integrable functions of the relative coordinates of the fragment centers of mass.

$$H_a P_a^{(n)} = (E_{an}^{\text{int}} + H_a^{\text{kin}}) P_a^{(n)}, \tag{16.120}$$

$E_{an}{}^{\text{int}}$ being the nth (discrete) eigenvalue of $H_a{}^{\text{int}}$. If $H_a{}^{\text{int}}$ is "reasonable," then the number N of its bound states may be assumed to be finite, and

$$P_a = \sum_{n=1}^{N} P_a^{(n)}. \tag{16.121}$$

Since the various *kinetic parts* surely commute,

$$[H_a^{\text{kin}}, H_b^{\text{kin}}] = 0 \tag{16.122}$$

it follows from (16.120) that

$$P_a^{(n)} W_t P_b^{(m)} = P_a^{(n)} W_t^{nm} P_b^{(m)}$$

where

$$W_t^{nm} = \exp\left[i(E_{an}^{\text{int}} + H_a^{\text{kin}} - E_{bm}^{\text{int}} - H_b^{\text{kin}})t\right]$$

so that

$$W_t^{nm} W_s^{nm} = W_{t+s}^{nm}.$$

We may therefore apply the mean ergodic theorem[7]:

$$\lim_{T\to\pm\infty} \frac{1}{T} \int_0^T dt\, (\Psi, P_a^{(n)} W_t P_b^{(m)} \Phi) = (P_a^{(n)} \Psi, P P_b^{(m)} \Phi) \tag{16.123}$$

where P is the projection onto the eigenspace of W_t with eigenvalue 1,

$$W_t P = P$$

or

$$e^{iH_a t} P = e^{iH_b t} P.$$

Since (16.116) was assumed to imply $\Phi = 0$, the right-hand side of (16.123) vanishes. But because of (16.118) the left-hand side is clearly equal to

$$(\Psi, P_a^{(n)} \Omega_a^{(\pm)\dagger} \Omega_b^{(\pm)} P_b^{(m)} \Phi).$$

The result (16.119) now follows from (16.121). The only assumptions made are (16.122) and the finiteness of the number of bound states of $H_a{}^{\text{int}}$ and $H_b{}^{\text{int}}$. The former is always true, and the latter could be relaxed if necessary.

Notes and References

16.1 For an n-particle system in its CM frame the onset of the continuous spectrum, i.e., the break-up threshold, is at the lowest energy of any of its two-fragment subsystems. This is variously known as *Hunziker's theorem* or as the *HVZ theorem*; see G. M. Zhislin (1960); C. van Winter (1964); W. Hunziker (1966); V. Enss (1977).

16.2 For the treatment of inelastic processes, and especially of rearrangement collisions, from a time-dependent point of view the following papers should be particularly mentioned: H. Eckstein (1956a); J. M. Jauch (1958b); Coester and Kummel (1958); A. Galindo Tixaire (1959); S. Sunakawa (1960); Grawert and Petzold (1960); T. F. Jordan (1962); Jauch and Marchand (1966). See also Brenig and Haag (1959); A. Ramakrishnan et al. (1961). The procedure in this section, although independently developed, bears a resemblance to that of Sunakawa (1960).

7 J. von Neumann (1932).

For discussions of Coulomb potentials, see L. A. Sakhnovich (1972); E. Prugovečki (1973 and 1974); Chandler and Gibson (1974); J. Zorbas (1974b, 1976b, 1977); W. W. Zachary (1976); as well as the references on this topic listed in the Notes to Sec. 17.4.

16.2.3 For the two-Hilbert space formulation of scattering theory, see T. Kato (1967a), Chandler and Gibson (1973, 1977, 1978a, b, 1980); E. Prugovečki (1973); I. M. Sigal (1978c, 1979, 1980); D. R. Yafaev (1978); V. D. Koshmanenko (1978). For generalization to relativistic scattering, see Coester and Schlessinger (1973); F. Coester (1980).

16.3 In addition to the papers treating formal scattering theory in general, listed in the Notes and References of Chap. 9, and to those already given in the Notes and References to Sec. 16.2, the following should be particularly mentioned. They deal specifically with inelastic processes and rearrangement collisions, primarily from a time-independent point of view: J. R. Oppenheimer (1928); Morse and Allis (1933); Bates, Fundaminsky, and Massey (1951); D. Layzer (1951); Ta-You Wu (1952 and 1956); Corinaldesi, Trainor, and Wu (1952); H. E. Moses (1953a); S. Altshuler (1953); P. A. M. Dirac (1955); B. A. Lippmann (1956); Foldi and Tobocman (1957); S. T. Epstein (1957); R. G. Newton (1958); E. Gerjuoy (1958a and b); G. Mohan (1959); Y. Yamaguchi (1959); Day, Rodberg, Snow, and Sucher (1961); M. Bolsterli (1963); T. Sasakawa (1963 and 1967); T. F. Jordan (1964); M. Mittleman (1964); Poluéktov, Presnyakov, and Sobelman (1964); C. Joachain (1965b); M. Coz (1965); F. S. Levin (1966); J. C. Y. Chen (1966); Grassberger and Sandhas (1968); D. G. Truhlar et al. (1968); N. Austern (1969); Y. Hahn (1968a, 1969b, 1971b, 1974, 1976a); D. J. Kouri (1969); Chen and Chung (1970a); Kazaks and Koshel (1970); G. Payne (1970); F. Coester (1971); D. G. Truhlar (1971); K. L. Kowalski (1972b); Baer and Kouri (1973); Blasczak and Fuda (1973); P. J. R. Soper (1974); L. Rosenberg (1976); Rule and Hahn (1976); Bolsterli and Rinat (1977); Levin and Greben (1978); S. K. Adhikari (1979b and 1981a); W. H. Klink (1979); Austern and Vincent (1981); A. M. Badalyan et al. (1981); J. W. Evans (1981); W. J. G. Thijssen et al. (1981); U. Fano (1981).

16.3.3 Several of the above-mentioned articles, including that by Oppenheimer, contain treatments of symmetrization or antisymmetrization or take the Pauli's principle into account in the collisions. Additional papers particularly concerned with this point are Trainor and Wu (1953); Takeda and Watson (1955). See also E. Corinaldesi (1962); F. S. Levin (1963, 1965, 1966b); S. Saito (1968 and 1969); Kharchenko and Kuzmichev (1972, 1973, 1974); V. F. Kharchenko et al. (1974); Kouri and Levin (1974a, c, and 1976); Bencze and Redish (1975 and 1978); W. Tobocman (1975b); R. B. Raphael et al. (1976); Gy. Bencze (1978); Goldflam and Tobocman (1978); Kukulin and Pomerantsev (1978a, b, and c); Levin and Kruger (1978); V. I. Kukulin et al. (1978 and 1979); Goldflam and Kowalski (1980b); Bencze and Chandler (1982); Kowalski and Picklesimer (1981); Leeb and Schmid (1981); E. W. Schmid (1981). One procedure is known as the *resonating group method*; see J. A. Wheeler (1937c); Hill and Wheeler (1953); Griffin and Wheeler (1957); Wildermuth and McClure (1966); Y. C. Tang (1969); Heiss and Hackenbroich (1970 and 1973); H. Friedrich (1974); E. W. Schmid (1980); S. K. Adhikari (1981b).

16.4 For partial-wave expansion of two-fragment to two-fragment scattering, see A. Lindner (1970).

Variational methods for multichannel scattering are discussed by L. Rosenberg (1968); Carew and Rosenberg (1969 and 1972); W. H. Miller (1969a); R. K. Nesbet (1969); G. Payne (1969); Dirks and Hahn (1970); Y. Hahn (1970a, 1971a, 1972a, b, 1975, 1976b); Pieper, Schlessinger, and Wright (1970); Pieper, Wright, and Schlessinger (1970); Chung and Chen (1972); M. Lieber et al. (1972); Nesbet and Oberoi (1972); Oberoi and Nesbet (1973 and 1974); L. Rosenberg (1973); Rosenberg and Tolchin

(1973); S. P. Merkuriev (1974); Pollak and Levine (1974); L. Rosenberg (1974) W. Tobocman (1974a); Kouri and Levin (1975a); A. L. Zubarev (1976); M. Kamimura (1977); R. Goldflam et al. (1981). For the use of Fredholm methods, see W. P. Reinhardt (1970). High-energy approximations are given by R. T. Ling et al. (1972); Shakeshaft and Spruch (1973); Chen and Hahn (1975); Hahn and Rule (1975); Laing and Freed (1975). Other relevant papers are W. H. Klink (1971 and 1973); H. Rabitz (1971); Conn and Rabitz (1973).

16.5 Relevant work can be found in J. D. Dollard (1969, 1973); J. M. Jauch et al. (1972); Newton and Shtokhamer (1974); J. H. Hendrikson (1976). For bounds on total cross sections, see W. O. Amrein et al. (1979); Osborn and Bollé (1979).

16.6 The literature on the theory of resonance reactions is very large. Furthermore, there are many different approaches to the problem, some of which offer advantages in some contexts, some in others. The approach of this section follows for the most part Fonda and Newton (1960c).

The first paper using the modern approach of analytic continuation of the partial wave amplitude off the real energy axis was by A. J. F. Siegert (1939).

The following list of papers constitutes an outline of the literature. These papers with their bibliographies should make it possible for the interested reader to explore even the most special topics in this area. Breit and Yost (1935); Breit and Wigner (1936); Bethe and Placzek (1937); Kalckar, Oppenheimer, and Serber (1937); Bohr and Kalckar (1937); H. A. Bethe (1937); Kapur and Peierls (1938); G. Breit (1940, 1946, 1959); E. P. Wigner (1946); Wigner and Eisenbud (1947); S. Flügge (1948); Teichman and Wigner (1952); Blatt and Weisskopf (1952), chaps. VIII–X; Feshbach, Porter, and Weisskopf (1954); B. Zumino (1956); H. Rollnik (1956); C. Bloch (1957); G. E. Brown (1957); Sano, Yoshida, and Teresawa (1958); Lane and Thomas (1958); R. G. Newton (1958); H. Feshbach (1958 and 1962); Lee and Klein (1959); W. Brenig (1959); R. E. Peierls (1959); Agodi and Eberle (1960); L. Fonda (1961b and 1964); Humblet and Rosenfeld (1961); L. Rosenfeld (1961); L. Rodberg (1961); R. Lipperheide (1962); Davies and Baranger (1962); J. Humblet (1962 and 1964); Herzenberg and Mandl (1963); A. Herzenberg et al. (1964); H. A. Weidenmüller (1964); C. van Winter (1964); W. M. McDonald (1964); J. P. Jeukenne (1964); C. Mahaux (1965); M. Coz (1965a and b); Mahaux and Weidenmüller (1965); Nath and Srivastava (1965); Fulco, Shaw, and Wong (1965); Rajamaran and Susskind (1965); Y. Hahn (1966, 1967a, b, 1968a, 1972a, b, c); A. Ramm (1966, 1972); Goebel and McVoy (1967); L. Fonda (1967); I. G. Haliday (1967b); D. A. Micha (1967); A. Ronveau (1968b); M. H. Mittleman (1969); Takeuchi and Moldauer (1970); W. H. Klink (1974); D. L. Huestis (1975); L. Durand (1976); Garcia–Calderon and Peierls (1976); G. Garcia–Calderon (1977); R. Lavine (1978); Horn and Novoseller (1978); B. J. Hoenders (1979); A. Jensen (1980); Fortes and Garcia–Calderon (1980); W. J. Romo (1980); A. M. Badalyan et al. (1981); S. Bosanac (1981b); B. I. Schneider (1981); Albeverio and Høegh–Krohn (1981).

Bound states embedded in the continuum were further discussed by Fonda and Ghiradi (1964b); M. Bolsterli (1969); B. Simon (1973 and 1974); Stillinger and Herrick (1975); W. P. Reinhardt (1977).

Appendix. The proof given here is an adaptation of that given by J. M. Jauch (1958b) to our present procedure.

General Note. There is a large amount of literature concerned particularly with electron-atom, electron-molecule, and ion-atom scattering. The following provides a partial list from which more can be traced: H. S. W. Massey (1930 and 1956b); Massey and Mohr (1932); R. Roscoe (1938 and 1941); Mott and Massey (1949); E. E. Salpeter (1950); Massey and Burhop (1952); E. H. Kerner (1953); Baranger and Gerjuoy (1957);

A. Temkin (1957, 1959, and 1962); E. V. Ivash (1958); Craggs and Massey (1959); C. Schwartz (1961); B. L. Moiseiwitch (1962); M. J. Seaton (1962); Burke and Smith (1962); Burke and Schey (1962); K. Takayanagi (1963); V. I. Ochkur (1963); B. M. Smirnov (1963); R. G. Breene (1963); M. R. C. McDowell (1963 and 1964); Glassgold and Lebedeff (1964); J. M. Peek (1964); Damburg and Peterkop (1964); G. L. Nutt (1964); Motz, Olsen, and Koch (1964); E. Gerjuoy (1965b); Fulton and Mittleman (1965); J. M. Lévy-Leblond (1967); Lévy-Leblond and Provost (1967); J. S. Ball et al. (1968); J. C. Y. Chen et al. (1969); Chen and Chung (1970b); B. Schneider et al. (1970); Chen and Kramer (1972); V. S. Potapov (1972); J. C. Y. Chen (1973); L. R. Dodd (1974); Y. Hahn et al. (1975); K. Matsuda et al. (1975); J. –M. Combes (1978); R. K. Nesbet (1978); J. Nuttall (1978); C. Bottcher (1981); B. H. Choi et al. (1981); Suck and Emmons (1981); U. Fano (1981). See also the paper on neutron scattering by molecules, Zemach and Glauber (1956).

For the scattering of electrons by atoms or molecules in a laser field, see Hertel and Stoll (1974); Macek and Hertel (1974); N. K. Rahman (1974); L. Rosenberg (1977, 1979, 1981, 1982); Kelsey and Rosenberg (1979); G. Ferrante (1979); G. Ferrante et al. (1980).

Problems

1 Prove Eqs. (16.52) and (16.54).

2 Prove Eqs. (16.73) and (16.75).

3 Prove that, if the initial and final fragment spins are s_1, s_2, s_1', and s_2', then the right-hand sides of (15.66) and (15.67a) acquire factors of $(-)^{s_1+s_2-s_1'-s_2'}$.

4 Prove (16.72) and (16.73).

5 Suppose that $L = 0$ in (16.67a), that V_j is a constant real symmetric matrix for $r < R$, and that it vanishes for $r > R$. Solve the matrix equations, and calculate the S-matrix elements for $l = l' = 0$.

6 Suppose that the $N \times N$ potential matrix V_j does not couple different orbital angular momenta and that it is of the special form

$$V_j(r) = \sum_{n=1}^{N} v_n(r) M_n$$

where the $v_n(r)$ are a set of numerical functions of r and the M_n are N constant, mutually commuting matrices. Assume further that the energy is so large that the difference between the various channel momenta can be neglected. Calculate the eigenphase shifts and the mixture parameters.

7 Show that in the two-channel case the necessary and sufficient condition for the potential matrix to be of the form assumed in Prob. 6 is that its off-diagonal element be a constant multiple of the difference between its diagonal elements.

8 Assume that the potential matrix V_j is constant for $r < R$ and that it vanishes for $r > R$. Calculate a regular solution of (16.67a) by expanding about $r = 0$.

9 Consider a two-channel problem, and assume that V_j is a positive semidefinite matrix for all r. Does it follow that the phases $\delta_{11}{}^j$ and $\delta_{22}{}^j$ of the two diagonal S-matrix elements of a given j value are negative? As an example, take the model of Prob. 6, and express these phases in terms of the mixture parameters and the eigenphase shifts. Discuss.

10 Prove Eqs. (16.110), (16.111), (16.111a), and (16.112).

11 Show that a sharp resonance line may contain two peaks if the **S** matrix has a double pole.

CHAPTER 17

Inelastic Scattering and Reactions (Multichannel Theory), II

17.1 Analyticity in Many-Channel Problems

17.1.1 The Coupled Equations. We now want to approach the manychannel problem from the point of view of the coupled radial Schrödinger equations (16.67) or (16.67a). Their structure resembling that of (15.92), the procedure of solving them is very much like that of Sec. 15.2.2. The main complication there arose from the coupling between different orbital angular momenta. Because it is generally possible to excite internal energy levels of different angular momenta (i.e., different fragment spins), that complication is present here also. In many practical cases, however, when the fragment spin is really an internal *oribital* angular momentum, it is somewhat less serious.

For example, consider the scattering of a particle 1 by the bound state of two others, 2 and 3. (All three have spin 0.) Suppose that the potential between particles 1 and 2 is a superposition of Yukawas as in (12.22a) and that, for simplicity, particle 3 is infinitely massive, at rest at the origin. Then we may write

$$\begin{aligned} 2\mu V_{12}(|\mathbf{r}_1 - \mathbf{r}_2|) &= 2\mu \int da\, \rho'(a) \frac{\exp(-a|\mathbf{r}_1 - \mathbf{r}_2|)}{|\mathbf{r}_1 - \mathbf{r}_2|} \\ &= -4\pi \int da\, \rho'(a) G(ia; \mathbf{r}_1, \mathbf{r}_2), \end{aligned} \tag{17.1}$$

where G is the free two-particle Green's function (10.6) evaluated at $k = ia$. It may be expanded according to (11.1),

$$V_{12} = 4\pi \int da\, \rho'(a) \sum_{LM} (-)^L Y_L^M(\hat{\mathbf{r}}_1)\, Y_L^{M*}(\hat{\mathbf{r}}_2) \frac{u_L(iar_<) w_L^{(+)}(iar_>)}{iar_1 r_2}. \tag{17.2}$$

Inserting this in (16.66) we get, according to (16.63),

$$\begin{aligned} \mathscr{V}^j_{l's'\beta,ls\alpha}(r_1) = &-8\pi i\mu \sum_{LMmm'vv'} (-)^L C(l's'j, m'v'M) \\ &C(lsj, mvM) A^{m'mM}_{l'lL} A^{v'vm*}_{s'sL} \int da\, a^{-1} \rho'(a) B(a,r_1) \end{aligned} \tag{17.3}$$

where

$$B(a,r_1) = \int_0^\infty dr_2\, r_1^{-1} r_2^{-1} u_L(iar_<) w_L^{(+)}(iar_>) \psi_s(r_2) \psi_{s'}^*(r_2),$$

ψ_s and $\psi_{s'}$ are the bound-state radial wave functions and[1]

$$\begin{aligned} A^{m'mM}_{l'lL} &= \int d\Omega\, Y_{l'}^{m'*}(\hat{\mathbf{r}})\, Y_l^m(\hat{\mathbf{r}})\, Y_L^M(\hat{\mathbf{r}}) \\ &= \left[\frac{(2l+1)(2L+1)}{4\pi(2l'+1)}\right]^{1/2} C(lLl', mMm') C(lLl', 000) \\ &= 0 \quad \text{unless } |l - l'| \le L \le l + l'. \end{aligned} \tag{17.4}$$

The integral B falls apart into two ranges,

$$\begin{aligned} B = r_1^{-1} u_L(iar_1) &\int_{r_1}^\infty dr_2\, r_2^{-1} w_L^{(+)}(iar_2) \psi_s \psi_{s'}^* \\ &+ r_1^{-1} w_L^{(+)}(iar_1) \int_0^{r_1} dr_2\, r_2^{-1} u_L(iar_2) \psi_s \psi_{s'}^*. \end{aligned}$$

The first integrand is $O({r_2}^{s+s'+1-L})$ as $r_2 \to 0$. Because of the second factor A in (17.3) only $L \le s + s'$ contributes; hence the first integral converges as $r_1 \to 0$, and the first term in B is $O({r_1}^L)$. According to (17.4) it is therefore $O({r_1}^{|l-l'|})$. The second integrand in B is $O({r_2}^{L+s+s'+2})$ as $r_2 \to 0$, and hence the second integral is $O({r_1}^{L+s+s'+3})$ as $r_1 \to 0$ and the second term is $O({r_1}^{s+s'+2})$. But unless $|l_1 - l_2| \le s + s'$ at least one of the two A factors in (17.3) always vanishes. Consequently the first term in B dominates as $r_1 \to 0$, and we find that, as $r \to 0$,

$$\mathscr{V}^j_{l's'\beta,ls\alpha}(r) = O(r^{|l-l'|}). \tag{17.5}$$

In this example therefore the coupling terms in the potential matrix in (16.67) vanish sufficiently rapidly as $r \to 0$ so that the counter terms used in Sec. 15.2.2 are unnecessary. This can be expected to be true under more general conditions. We shall therefore simply assume that the potential matrix V_j is sufficiently well behaved at the origin and shall not use these terms.

The set (16.67) consists of infinitely many equations, including in fact a

1 See M. E. Rose (1957), p. 62, eq. (4.34).

continuous range due to the internal energy region in which the fragments are dissociated (either really or virtually). For the present we shall ignore this and shall truncate the set so that it consists of a finite number of equations only.

For rearrangement collisions the original equations cannot be reduced to simple radial equations like (16.67). There is no single distance r that serves as a fragment separation in the initial and in the final state. Nevertheless it is sometimes useful to simulate effects that arise in rearrangements by a model described by equations like (16.67a), except that the reduced masses in the various channels are not the same. The equations then read

$$-\tfrac{1}{2}M^{-1}\overline{\Psi}_j'' + M^{-1}L(L+1)r^{-2}\overline{\Psi}_j + \overline{V}_j\overline{\Psi}_j = \mathscr{E}\overline{\Psi}_j \tag{17.6}$$

instead of (16.67a). Here M is the diagonal matrix of the channel reduced masses and $\mathscr{E}$ that of the kinetic energies. This matrix equation can be reduced to the form (16.67a) by defining

$$\begin{aligned} \Psi_j &\equiv M^{-1/2}\overline{\Psi}_j \\ V_j &\equiv 2M^{1/2}\,\overline{V}_j M^{1/2}. \end{aligned} \tag{17.7}$$

We may therefore confine ourselves to the considerations of (16.67a).

The procedures of Secs. 15.2.2 and 15.2.3 may now be taken over directly, except for the fact that K is a (diagonal) matrix and that its position must be handled with care.

Using a symbolism similar to that of Sec. 15.2.2, but including the wave numbers in it, we write down integral equations for regular and irregular solutions. Since our primary attention is now not on the angular momentum, we drop the index j. The equations refer to a fixed j value.

We have for the regular solution

$$\Phi(K,r) = \Phi^{(0)}(K,r) + \int_0^r dr'\,\mathfrak{G}(K;r,r')V(r')\Phi(K,r') \tag{17.8}$$

and for the irregular one

$$F_\pm(K,r) = F_\pm^{(0)}(K,r) - \int_r^\infty dr'\,\mathfrak{G}(K;r,r')V(r')F_\pm(K,r') \tag{17.9}$$

with

$$\begin{aligned} \Phi^{(0)}(K,r) &= (2L+1)!!K^{-L-1}u_L(Kr) \\ F_\pm^{(0)}(K,r) &= e^{-i1/2\pi L}w_L^{(\pm)}(Kr) \\ \mathfrak{G}(K;r,r') &= g_L(K;r,r'), \end{aligned} \tag{17.10}$$

the functions g_l being defined in (12.135). The irregular solution $F = F_+$ obeys the boundary condition

$$F_{\alpha\beta}(K,r) \underset{r\to\infty}{\simeq} \delta_{\alpha\beta}e^{ik_\beta r} \tag{17.11}$$

which says that the β column has no particles at infinity except in the β channel, and there outgoing waves only.

The integral equations (17.8) and (17.9) can be solved by iteration under conditions that are identical to those on the single equations in Chap. 12. We shall pay no further attention to them. As in Sec. 15.2.3, Φ is expressed in terms of F_+ and F_- by the wronskian matrix [defined in (15.105)]

$$\mathfrak{F}_\pm = W(F_\pm, \Phi). \tag{17.12}$$

The result is

$$\Phi = \tfrac{1}{2}i[F_- K^{-1}\mathfrak{F}_+ - F_+ K^{-1}\mathfrak{F}_-]. \tag{17.13}$$

Comparison with (17.11) and the equation before (16.68) shows that

$$\overline{S} = e^{i\pi L} K^{-1}\mathfrak{F}_- \mathfrak{F}_+^{-1} K \tag{17.14}$$

and hence, according to (16.71),[2]

$$S = e^{i\pi L} K^{-1/2}\mathfrak{F}_- \mathfrak{F}_+^{-1} K^{1/2}. \tag{17.15}$$

The fact that $W(\Phi, \Phi) = 0$ leads to the equation

$$\tilde{\mathfrak{F}}_+ K^{-1}\mathfrak{F}_- = \tilde{\mathfrak{F}}_- K^{-1}\mathfrak{F}_+ \tag{17.16}$$

which implies the symmetry of S. [The assumed time-reversal invariance enters into this argument in the form of the symmetry of the potential matrix, without which (17.12) would not be independent of r.]

Another Jost matrix is introduced in analogy with (15.113),

$$\mathscr{F}_\pm = \mathfrak{F}_\pm [(2L + 1)!!]^{-1} K^L e^{\mp i1/2\pi L}. \tag{17.17}$$

In terms of this S can be written more simply as

$$S = K^{-1/2}\mathscr{F}_- \mathscr{F}_+^{-1} K^{1/2}. \tag{17.18}$$

(because of parity conservation[3]), and it has the integral representations

$$\mathscr{F}_+ = \mathbb{1} + \int_0^\infty dr\, e^{-i1/2\pi L} w_L^{(+)}(Kr) V(r) \Phi(K, r) e^{-i1/2\pi L} K^L [(2L + 1)!!]^{-1} \tag{17.19}$$

$$\mathscr{F}_+ = \mathbb{1} + \int_0^\infty dr\, \tilde{F}(K, r) V(r) u_L(Kr) K^{-1} e^{-i1/2\pi L}. \tag{17.19a}$$

The boundary condition (17.11) implies that the matrix F has the symmetry

$$F(-K, r) = F^*(K, r) \tag{17.20}$$

when all k's are real, and consequently in general

$$F(-K^*, r) = F^*(K, r). \tag{17.20a}$$

2 If the original equations are of the form (17.6), then (17.7) shows that $\overline{S} = e^{i\pi L} M^{1/2} K^{-1}\mathfrak{F}_- \mathfrak{F}_+^{-1} K M^{-1/2}$. But according to (16.71) the S matrix is then still given by (17.15); the mass factors drop out. Equation (17.15) therefore holds whether the channel masses are equal or not.

3 Parity conservation implies that only l values of equal parity are coupled. Hence $e^{i\pi L}$ commutes with all matrices that enter the discussion.

Therefore similarly

$$\mathfrak{F}(-K^*) = \mathfrak{F}^*(K) \tag{17.21}$$

and

$$\mathscr{F}(-K^*) = \mathscr{F}^*(K). \tag{17.21a}$$

Furthermore,

$$\mathscr{F}_-(K) = \mathscr{F}_+(-K). \tag{17.22}$$

It should be noted that at a given energy not all the N channels of the problem need be open. Nevertheless, the matrix defined in (17.15) is an $N \times N$ matrix. Its submatrix S_+, which refers to open channels only, is customarily called the **S** matrix. The matrix S_+ ought to be unitary, but S of course generally is not.

Equations (17.15) and (17.18) represent S and S_+ in terms of $N \times N$ matrices $\mathfrak{F}$ and $\mathscr{F}$. But S_+ can also be decomposed into matrices which refer to open channels only. This can be done as follows.

Let us multiply the integral equation (17.8) on the right by a matrix I which is chosen so that the "reduced" solution $\Phi^{(R)} \equiv \Phi I$ contains no terms which for large r exponentially increase on account of the imaginary channel momenta k_α. According to (17.13) the condition on I is evidently

$$P\mathfrak{F}I = 0 \tag{17.23}$$

if by P is meant the projection onto the closed channels,

$$P = \tfrac{1}{2}\left(\mathbb{1} - \frac{K^2}{|K^2|}\right).$$

One may choose

$$\begin{aligned} I &= \mathscr{I}(\mathbb{1} - P) \\ \mathscr{I} &= [\mathbb{1} + P(\mathfrak{F} - \mathbb{1})]^{-1}. \end{aligned} \tag{17.24}$$

Then the asymptotic behavior of $\Phi^{(R)}$ is

$$\Phi^{(R)} \simeq \tfrac{1}{2}i(\mathbb{1} - P)[e^{-iKr}K^{-1}\mathfrak{F}\mathscr{I} - e^{iKr}K^{-1}\mathfrak{F}_-\mathscr{I}](\mathbb{1} - P) \tag{17.25}$$

and therefore the open-channel part of S,

$$S_+ = e^{i\pi L}K^{-1/2}(\mathfrak{F}_-\mathscr{I})_+(\mathfrak{F}\mathscr{I})_+^{-1}K^{1/2}. \tag{17.26}$$

The subscript + everywhere indicates the open-channel submatrix.[4]

Now notice that the boundary condition (17.11) implies that $F_{\alpha\beta}$ is an even function of all the k's except k_β. Because of the transposition in the wronskian, therefore, $\mathfrak{F}_{\alpha\beta}$ is an even function of all k's except k_α. (The function Φ is even in all k's.) Equation (17.24) shows that consequently the matrix $\mathscr{I}$ is an even function of all the open-channel momenta. Because

4 This + is not to be confused with the subscript + on $\mathfrak{F}_+$, which has been left off in Eqs. (17.25) and (17.26) in order to avoid confusion.

of (17.21) this implies that, for real energies (all k's either real or positive imaginary), $\mathscr{I}$ is *real*. Equation (17.21) together with the symmetry of S therefore shows that for real energies S_+ is unitary,

$$S_+^\dagger = S_+^{-1}. \tag{17.27}$$

17.1.2 An Alternative Procedure. Instead of solving the Eqs. (16.67a) as in (17.8) and (17.9), we may start by first solving the uncoupled equations. This is particularly useful if the uncoupled equations can be solved exactly and the coupling is to be treated as a perturbation. An important case in point is that of charged particles when the diagonal part of the potential includes the Coulomb field.

Let us write, accordingly,

$$V = V_D + \bar{V} \tag{17.28}$$

where V_D contains the diagonal elements of V only and $\bar{V}$ only the off-diagonal ones. (It is not essential that V_D contain *all* the diagonal elements of V. $\bar{V}$ may still have diagonal elements, but V_D must be diagonal.) Let $\Phi^{(1)}(K,r)$ and $F_\pm{}^{(1)}(K,r)$ be the corresponding solutions as in (17.8) and (17.9), but with V_D only. (Of course, each element of these diagonal matrices depends on *its* channel wave number only.) Out of these solutions we form Jost functions $\mathfrak{F}^{(1)}$ and $\mathscr{F}^{(1)}$ as in (17.12) and (17.17) and a Green's function

$$\mathfrak{G}^{(1)}(K;r,r') = [\Phi^{(1)}(K,r)F^{(1)}(K,r') - \Phi^{(1)}(K,r')F^{(1)}(K,r)]\mathfrak{F}^{(1)-1}(K) \tag{17.29}$$

as in (12.42).

Once the uncoupled equations are solved, we proceed to set up the solutions of the coupled equations in terms of the uncoupled ones. The place of (17.8) and (17.9) is taken by

$$\Phi(K,r) = \Phi^{(1)}(K,r) + \int_0^r dr'\, \mathfrak{G}^{(1)}(K;r,r')\bar{V}(r')\Phi(K,r') \tag{17.30}$$

$$F(K,r) = F^{(1)}(K,r) - \int_r^\infty dr'\, \mathfrak{G}^{(1)}(K;r,r')\bar{V}(r')F(K,r'). \tag{17.31}$$

The Jost matrix $\mathfrak{F}$ is found by (17.12), into which we substitute (17.30) and (17.31). This gives us the integral representations

$$\begin{aligned}\mathfrak{F}(K) &= \mathfrak{F}^{(1)}(K) + \int_0^\infty dr\, F^{(1)}(K,r)\bar{V}(r)\Phi(K,r)\\ &= \mathfrak{F}^{(1)}(K) + \int_0^\infty dr\, \tilde{F}(K,r)\bar{V}(r)\Phi^{(1)}(K,r)\end{aligned} \tag{17.32}$$

in place of (17.19) and (17.19a). Let us multiply the first equation by $\mathfrak{F}^{(1)-1}$ on the left and subtract the analogous equation for $\mathfrak{F}$. Using (17.13) and (17.15), we obtain

$$S = S^{(1)} - 2ie^{i\pi L}K^{1/2}\mathfrak{F}^{(1)-1}\int_0^\infty dr\,\Phi^{(1)}\overline{V}\Phi\mathfrak{F}^{-1}K^{1/2}. \tag{17.33}$$

Final-State Interaction. In a practical case we may want to treat the off-diagonal potential $\overline{V}$ by perturbation theory. Then to first order

$$S = S^{(1)} - 2ie^{i\pi L}K^{1/2}\mathfrak{F}^{(1)-1}\int_0^\infty dr\,\Phi^{(1)}\overline{V}\Phi^{(1)}\mathfrak{F}^{(1)-1}K^{1/2} + \cdots.$$

The off-diagonal elements of S are therefore approximated by

$$S_{\beta\alpha} \simeq -2ie^{i\pi l_\beta}(k_\alpha k_\beta)^{1/2}\int_0^\infty dr\frac{\Phi_\beta^{(1)}(k_\beta,r)\,V_{\beta\alpha}\Phi_\alpha^{(1)}(k_\alpha,r)}{\mathfrak{F}_\alpha^{(1)}(k_\alpha)\mathfrak{F}_\beta^{(1)}(k_\beta)}. \tag{17.34}$$

The important aspect to be noticed is the presence of the Jost functions for the initial and final channels in the denominator.

Suppose that the final fragments of the reaction have an interaction (such as an electrostatic one, for example) that produces *resonances*. Then the Jost function $\mathfrak{F}_\beta{}^{(1)}$ has a zero in the lower half of the complex k_β plane, close to the real axis. At an energy at which the final kinetic energy of the fragments is near the resonance, the denominator in the approximation (17.34) will then be small. The **S**-matrix element is consequently much larger than it would otherwise be, and the cross section is *enhanced*. This effect is called *final-state interaction*. Of course, a look at (17.34) shows that the same argument is applicable to the incident energy. There is then, an analogous *initial-state interaction*.

In order to be able to draw the conclusion that the size of $S_{\beta\alpha}$ is enhanced by a zero of the Jost function near the real axis, we must be sure that the numerator in (17.34) is not small, too. Now if $|\mathfrak{F}_+{}^{(1)}|$ is small (at a real energy), then so is $|\mathfrak{F}_-{}^{(1)}| = |\mathfrak{F}_+{}^{(1)}|$, and hence so is Φ for most values of r. Only in the region near the origin (small r) is it unaffected by the smallness of $\mathfrak{F}$, because there Φ is fixed by the boundary conditions. This, of course, expresses the fact that the resonance signifies a "pulling in" of the wave function; the particles stay close to one another. Hence the integral in the numerator of (17.34) will not change much in size when $\mathfrak{F}$ gets small, *provided* that the range of $V_{\beta\alpha}$ is small (compared with the region within which V_D influences the wave function appreciably).

If the potential $V_{\beta\alpha}$ is sufficiently short-ranged so that the numerator in (17.34) is unchanged by the smallness of $\mathfrak{F}^{(1)}$, then it is also independent of the energy. This is because the boundary condition on $\Phi^{(1)}$ does not depend on k. The energy dependence of $S_{\beta\alpha}$ is then essentially contained in the outside factors,

$$S_{\beta\alpha} \simeq \text{const} \times \frac{(k_\alpha k_\beta)^{1/2}}{\mathfrak{F}_\alpha^{(1)}(k_\alpha)\mathfrak{F}_\beta^{(1)}(k_\beta)}.$$

What is more, the integral is well approximated by substituting simply the boundary value r^{l+1} for $\Phi^{(1)}$. It is therefore insensitive to the diagonal potential. The influence of the latter is felt only via the Jost functions in

(17.34). They represent "sticking factors" whose origin is the probability of finding the fragments in each other's vicinity. (We recall that this is the physical significance of the inverse of the square of the magnitude of $\mathfrak{F}$.)

Physically it is easy to understand the origin of these effects. The potential element $V_{\beta\alpha}$ is responsible for the inelastic reaction, and we assume it is weak. But before and after the reaction the initial and final fragments interact via a force that is not weak. If this causes a resonance, then the fragments are almost caught in a bound state with one another and they will spend a great deal of time close together. It is clear that this must have an appreciable effect upon the reaction cross section, whether it occurs before or after the collision.

The final- (or initial-) state interaction effect is most important at low relative final (or initial) fragment velocities. This is mostly because resonances *in a one-channel problem* are usually produced by centrifugal barrier effects and hence are concentrated at low energies. Therefore, $\mathfrak{F}_\alpha{}^{(1)}$ and $\mathfrak{F}_\beta{}^{(1)}$ are expected to have zeros near the real axis only near the origin.[5] Nevertheless it must be recognized that this need not be so, and it is not impossible for the effect to occur in other energy regions.

17.1.3 Analyticity Properties. Consider now the analytic properties of Φ, F, and $\mathscr{F}$. It is useful to start by assuming that the channel wave numbers k_α are all independent complex variables. The regular solution Φ is evidently an even function of all the k's, and it is an entire analytic function of each of them. If we relate each k_α to the total energy by

$$k_\alpha = [2\mu_\alpha(E - \mathbf{\Delta}_\alpha)]^{1/2}, \tag{17.35}$$

$\mathbf{\Delta}_\alpha$ being a fixed number (equal to the excitation energy of the internal state of the fragments in channel α), then Φ is consequently an entire analytic function of E.

The irregular solution F has no simple general analyticity properties. The trouble is easily visible in the version of (17.9) for $l = 0$ in all channels. Then

$$F(K,r) = e^{iKr} - \int_r^\infty dr'\, K^{-1} \sin K(r - r') V(r') F(K,r') \tag{17.36}$$

from which it is clear that $F_{\alpha\beta}$ is an analytic function of k_β, regular in the upper half of the k_β plane (always under similarly weak assumptions as in Chap. 12). But we cannot in general make any of the other k's complex without producing a divergence because of the exponential increase in $\sin K(r - r')$. This means that after insertion of (17.35) for each k_α the energy cannot be taken below the highest threshold on the real axis. At least one of the k_α would then be imaginary, and all elements of F would generally cease to exist.

5 Furthermore, the higher the relative velocity, the shorter the range of the offdiagonal potential has to be for the integral in (17.34) to be both constant and independent of $V^{(1)}$.

Similarly it is evident from the $l = 0$ version of (17.19)

$$\mathscr{F} = \mathbb{1} + \int_0^\infty dr\, e^{iKr} V(r)\Phi(K,r) \tag{17.37}$$

that $\mathscr{F}_{\alpha\beta}$ is an analytic function of k_α, regular in the upper half plane, but that none of the other k's may be taken into the complex plane with impunity.

One way of circumventing this difficulty is to use the method of Sec. 16.6. If, in a certain energy region, the channel group α comprises all open channels and group β all the closed ones, then we may restrict all considerations for the present to group α, and the effective hamiltonian is $\mathscr{H}_\alpha$, given by (16.96). That this is nonlocal has no important consequences. Furthermore, since $\mathscr{E}_\beta$ is below threshold in the β group, G_β is analytic in a neighborhood of the real axis. We may then introduce Φ, F, and $\mathfrak{F}$ as before, but the dimensionality of all the matrices is that of the number of *open* channels only. In this manner we obtain a Jost matrix that is well defined in the physical region. The only drawback, which is serious for analytic continuation, is that this procedure has to be repeated and new matrix functions, of different dimensionalities, defined every time the energy crosses a threshold. We shall therefore not pursue this approach in more detail.

The **S** *Matrix in Terms of the Fredholm Determinant.* Instead we wish to look at the *determinant* of the Jost matrix (17.19). Just as in the single-channel case (Sec. 12.1.2) this determinant (which there, of course, is just the Jost function itself) is equal to the Fredholm determinant Δ of the radial integral equation (16.73a). The argument is the same as the one that leads to (12.43), except that the equation preceding (12.43) must now read

$$\frac{d \ln \det \mathscr{F}}{d\gamma} = \operatorname{tr} \mathscr{F}^{-1} \frac{d\mathscr{F}}{d\gamma} = -\operatorname{tr} \mathscr{G}^+ \mathscr{V}.$$

As a result (9.79a), which is applicable to a set of coupled integral equations as well as to a single one, shows that

$$\det \mathscr{F}(k) = \Delta(k). \tag{17.38}$$

Now look back to Sec. 9.3. The expansion of the Fredholm determinant shows that it is an analytic function of all the channel momenta k_α regular in the intersection of all the upper halves of the complex planes. Equation (17.38) therefore shows that, although the matrix $\mathscr{F}$ generally has no simple regularity properties for complex values of the channel momenta, its determinant does. Specifically there is no difficulty in crossing any thresholds, so long as the channel momentum that vanishes there is continued along the *positive* imaginary axis below.

The problem now is to construct the **S** matrix. Since the matrix $\mathscr{F}$ is not necessarily well defined below any thresholds, in the energy region where some channels are closed (17.18) gives us a useful definition of **S** only under very restrictive assumptions on the potential matrix. It is therefore important to show the remarkable fact that all open-channel elements of S can be calculated directly from the Fredholm determinant Δ.

Let us first write (17.18) in the form

$$S = K^{-1/2-L} S' K^{1/2+L}$$
$$S' = \frac{\mathscr{F}'_- X}{\Delta} \tag{17.39}$$

where

$$\mathscr{F}' = K^L \mathscr{F} K^{-L} \tag{17.40}$$

and

$$X = \mathscr{F}'^{-1}\Delta \tag{17.41}$$

is the transpose of the matrix made up of the cofactors of $\mathscr{F}$. Now Eq. (17.19) shows that, Φ being an even function of all k's, $\mathscr{F}_{\alpha\beta}'$ is an even function of all k's except k_α,

$$\mathscr{F}'_{\alpha\beta}(\ldots, -k_\gamma, \ldots) = \mathscr{F}'_{\alpha\beta}(\ldots, k_\gamma, \ldots)$$

for all $\gamma \neq \alpha$. Hence $X_{\alpha\beta}$ is an *even function of* k_β.

Let us write the diagonal elements of S explicitly according to (17.39) and develop the determinant,

$$\begin{aligned} S_{\alpha\alpha} &= \frac{\sum_\gamma (\mathscr{F}'_-)_{\alpha\gamma} X_{\gamma\alpha}}{\sum_\gamma \mathscr{F}'_{\alpha\gamma} X_{\gamma\alpha}} \\ &= \frac{\sum_\gamma \mathscr{F}'_{\alpha\gamma}(\ldots, -k_\alpha, \ldots) X_{\gamma\alpha}(\ldots, -k_\alpha, \ldots)}{\sum_\gamma \mathscr{F}'_{\alpha\gamma} X_{\gamma\alpha}} \\ &= \frac{\Delta(\ldots, -k_\alpha, \ldots)}{\Delta} \\ &= \frac{1}{S_{\alpha\alpha}(\ldots, -k_\alpha, \ldots)}. \end{aligned} \tag{17.42}$$

Thus one can obtain all the diagonal elements of the **S** matrix from the Fredholm determinant Δ, provided that it is given as a function of the channel momenta independently.

Before we can prove a similar result for the off-diagonal elements, a number of symmetry properties must be derived. Let us define, for $\alpha \neq \beta$,

$$h_{\alpha\beta} \equiv \sum_\gamma (\mathscr{F}'_-)_{\alpha\gamma} X_{\gamma\beta} k_\alpha^{-1-2l_\alpha} \tag{17.43}$$

so that,

$$S'_{\alpha\beta} = \frac{k_\alpha^{1+2l_\alpha} h_{\alpha\beta}}{\Delta} \tag{17.44}$$

and

$$h_{\alpha\beta} = h_{\beta\alpha} \tag{17.45}$$

expresses the symmetry of the **S** matrix. The element $X_{\gamma\beta}$ being even in k_β and $\mathscr{F}_{\alpha\gamma}{}'$ being even in all k's but k_α, $h_{\alpha\beta}$ is an even function of k_β. Because it is equal to $h_{\beta\alpha}$ it is an even function also of k_α. Consequently, using (17.42), we get for $\alpha \neq \beta$

$$\begin{aligned} S_{\alpha\alpha}S'_{\beta\alpha}(\ldots,-k_\alpha,\ldots) &= S'_{\beta\alpha} \\ S_{\alpha\alpha}S'_{\alpha\beta}(\ldots,-k_\alpha,\ldots) &= -S'_{\alpha\beta}. \end{aligned} \tag{17.46}$$

Furthermore, for $\alpha \neq \beta$, $\alpha \neq \gamma$, we evaluate

$$[\mathscr{F}'_- X\Delta(\ldots,-k_\alpha,\ldots) - \mathscr{F}'(\ldots,-k_\alpha,\ldots)X(\ldots,-k_\alpha,\ldots)\Delta]_{\beta\gamma}$$

once by inserting

$$\mathbb{1}\Delta = X\mathscr{F}' = \mathscr{F}'X$$

in the middle and once on the right,

$$\begin{aligned} [\quad]_{\beta\gamma} &= [\mathscr{F}'_- X(\ldots,-k_\alpha,\ldots)\mathscr{F}'(\ldots,-k_\alpha,\ldots)X \\ &\qquad - \mathscr{F}'(\ldots,-k_\alpha,\ldots)X(\ldots,-k_\alpha,\ldots)\mathscr{F}'X]_{\beta\gamma} \\ &= (\mathscr{F}'_- X)_{\beta\alpha}(\mathscr{F}'_- X)_{\alpha\gamma} = h_{\beta\alpha}h_{\alpha\gamma}k_\beta^{1+2l_\beta}k_\alpha^{1+2l_\alpha} \end{aligned}$$

because of the evenness properties of $\mathscr{F}$ and X. According to (17.39) this means that

$$S'_{\beta\alpha}(\ldots,-k_\alpha,\ldots)S'_{\alpha\gamma} = S'_{\beta\gamma} - S'_{\beta\gamma}(\ldots,-k_\alpha,\ldots). \tag{17.47}$$

We may combine Eqs. (17.42), (17.46), and (17.47) in the form

$$S'P^{(\alpha)}S'(\ldots,-k_\alpha,\ldots) = P^{(\alpha)} - S'Q^{(\alpha)} + Q^{(\alpha)}S'(\ldots,-k_\alpha,\ldots) \tag{17.48}$$

if $P^{(\alpha)}$ is the projection onto the α channel and $Q^{(\alpha)} = \mathbb{1} - P^{(\alpha)}$. This equation contains its own generalization. Let us write $P^{(\alpha\beta\cdots)}$ for the projection onto the $\alpha, \beta, \ldots$ channels. Then repeated application of (17.48) yields

$$\begin{aligned} S'P^{(\alpha\beta\cdots)}S'(\ldots,-k_\alpha,-k_\beta,\ldots) &= P^{(\alpha\beta\cdots)} \\ &\quad - S'Q^{(\alpha\beta\cdots)} + Q^{(\alpha\beta\cdots)}S'(\ldots,-k_\alpha,-k_\beta,\ldots). \end{aligned} \tag{17.49}$$

The most important special case of (17.49) is obtained by multiplying the equations from both sides by $P^{(\alpha\beta\cdots)}$,

$$P^{(\alpha\beta\cdots)}S'P^{(\alpha\beta\cdots)}S'(\ldots,-k_\alpha,-k_\beta\ldots)P^{(\alpha\beta\cdots)} = P^{(\alpha\beta\cdots)} \tag{17.50}$$

which says that the $(\alpha, \beta \ldots)$ submatrix of S' is the inverse of the $(\alpha, \beta \ldots)$ submatrix of S' $(\ldots, -k_\alpha, -k_\beta \ldots)$.[6] If the channels $(\alpha, \beta \ldots)$ are all open and no others are open, then it follows from (17.18), (17.21a), (17.39), and the symmetry of S that this is equivalent to the *unitarity* of the open-channel part of the **S** matrix.

We now insert (17.42) and (17.44) in (17.47), for $\beta = \gamma$. Using the evenness of $h_{\alpha\beta}$ as a funtion of k_α and k_β, we find

6 This is Peierls's version of the unitarity condition; see R. E. Peierls (1959).

$$(h_{\alpha\beta})^2 k_\alpha^{1+2l_\alpha} k_\beta^{1+2l_\beta} = \Delta_\alpha \Delta_\beta - \Delta\Delta_{\alpha\beta} \tag{17.51}$$

or, equivalently,

$$(S_{\alpha\beta})^2 = S_{\alpha\alpha} S_{\beta\beta} - \frac{\Delta_{\alpha\beta}}{\Delta} \tag{17.51a}$$

if we use the abbreviations

$$\begin{aligned} \Delta_\alpha &\equiv \Delta(\ldots, -k_\alpha, \ldots) \\ \Delta_{\alpha\beta} &\equiv \Delta(\ldots, -k_\alpha, -k_\beta, \ldots) \end{aligned} \tag{17.52}$$

etc. Equations (17.42) and (17.51a) completely determine all elements of the **S** matrix from Δ as a function of the independent channel momenta.

Restrictions on the Fredholm Determinant. The question now arises whether the function Δ can be arbitrarily prescribed or whether there are restrictions on it. We find in fact that they are not quite arbitrary. For $\beta \neq \gamma$, (17.47) shows that

$$h_{\beta\gamma}(\ldots, -k_\alpha, \ldots)\Delta = h_{\beta\gamma}\Delta(\ldots, -k_\alpha, \ldots) - k_\alpha^{1+2l_\alpha} h_{\alpha\beta} h_{\alpha\gamma}.$$

Squaring and inserting (17.51) yields an expression for Δ with three momenta reversed in terms of Δ with one or two reversed,

$$\begin{aligned} \Delta^2 \Delta_{\alpha\beta\gamma} = {} & \Delta(\Delta_\alpha \Delta_{\beta\gamma} + \Delta_\beta \Delta_{\alpha\gamma} + \Delta_\gamma \Delta_{\alpha\beta}) - 2\Delta_\alpha \Delta_\beta \Delta_\gamma \\ & + 2[(\Delta_\alpha \Delta_\beta - \Delta\Delta_{\alpha\beta})(\Delta_\alpha \Delta_\gamma - \Delta\Delta_{\alpha\gamma})(\Delta_\beta \Delta_\gamma - \Delta\Delta_{\beta\gamma})]^{1/2}. \end{aligned} \tag{17.53}$$

This equation comes into play only for three or more channels. Together with (17.42) and (17.51) it is equivalent to (17.49) and hence implies (17.50).

There remains the question of the unitarity of the **S** matrix constructed via (17.42) and (17.51a). If the open-channel **S**′ matrix so constructed has the property

$$S'(-K^*) = S'^*(K) \tag{17.54}$$

then it is unitary, because it satisfies (17.50). Now it follows from (17.21a) that the determinant must have the same property,

$$\Delta(-K^*) = \Delta^*(K). \tag{17.55}$$

If this is used in (17.51), it follows that

$$[h_{\alpha\beta}(-K^*)]^{2*} = [h_{\alpha\beta}(K)]^2. \tag{17.56}$$

What we need for (17.54) to be true is the more stringent condition

$$h_{\alpha\beta}^*(-K^*) = -h_{\alpha\beta}(K). \tag{17.57}$$

Now let k_α be positive and all other k's either positive or positive imaginery. Then (17.42) implies that Δ must be such that

$$|\Delta_\alpha| = |\Delta(\ldots, -k_\alpha, \ldots)| \leq |\Delta| \tag{17.58}$$

for S to be unitary. Suppose that k_α and k_β are positive and all other k's positive imaginary. Then (17.51) together with (17.55) shows that

$$(h_{\alpha\beta})^2 k_\alpha^{1+2l_\alpha} k_\beta^{1+2l_\beta} = |\Delta_\alpha|^2 - |\Delta|^2. \tag{17.59}$$

According to (17.58) the right-hand side is *negative*. Therefore $h_{\alpha\beta}$ is purely *imaginary* and obeys (17.57). So (17.58) implies (17.57) for positive k_α and k_β and all other k's positive imaginary. Equation (17.57) can now be analytically continued to other values of the remaining k's, specifically to real ones. As a result (17.57) will hold, and so will the unitarity of the open-channel part of S.

The only remaining property of the function Δ is that if all $\operatorname{Im} k_n \geq 0$ then

$$\lim_{|k_1|,\ldots,|k_n| \to \infty} \Delta = 1. \tag{17.60}$$

This is easily seen to be true from the definition of the Fredholm determinant.

These are all the restrictions the function $\Delta(k_1, \ldots, k_n)$ has to satisfy. Equations (17.53), (17.55), and (17.58), together with the regularity of Δ in the upper half planes, are necessary and sufficient conditions for (17.42) and (17.51a) to lead to a unitary and symmetric **S** matrix. Equation (17.60) assures that S approaches unity at infinite energy.

Equations (17.42) and (17.51a) in general allow us to construct only the open-channel part of S. That is to say, the elements $S_{\alpha\alpha}$, $S_{\alpha\beta}$, and $S_{\beta\beta}$ are in general well defined only if k_α and k_β are real. In order to define S in a larger domain, Δ would have to be continued to a region where some of the k's are in the *lower* half plane. There we know nothing about its behavior. Of course, if the potential matrix is known to obey more stringent conditions than assumed so far, then Δ may have a larger region of regularity and then (17.42) and (17.51a) define closed-channel elements of S, too. The situation in this respect is quite similar to that in the simpler one-channel problem. If all elements of V vanish beyond a finite distance, for example, then Δ is regular for all finite values of the k's.

17.1.4 Bound States. Let us consider an irregular matrix solution of (16.67a) that obeys the integral equation

$$H(K,r) = F^{(0)}(K,r)\,\Delta(K) + \int_0^\infty dr'\, G^+(K;r,r')V(r')H(K,r') \tag{17.61}$$

where $F^{(0)}$ is given by (17.10) and G^+ is the outgoing-wave Green's function used in (16.66a),

$$G^+(K;r,r') = -e^{-i\pi L} K^{-1} u_L(Kr_<) w_L^{(+)}(Kr_>)$$

according to (11.2). Since at infinity H contains outgoing waves only, it must be a (right) multiple of F,

$$H = FA.$$

Its asymptotic behavior for large r is given by

$$H \simeq e^{iKr}\left[\Delta\mathbb{1} - e^{-i1/2\pi L}K^{-1}\int_0^\infty dr'\, u_L(Kr')V(r')F(K,r')A\right] = e^{iKr}A.$$

This is solved for the matrix A. By consulting (17.19) one finds that

$$A = \Delta\tilde{\mathscr{F}}^{-1} = K^L\tilde{X}K^{-L}$$

i.e.,

$$H = FK^L\tilde{X}K^{-L}. \tag{17.62}$$

Now suppose that all k's are either real or in the upper half plane; $\operatorname{Im} k \geq 0$. Let k_α be among the former, $\operatorname{Im} k_\alpha > 0$. Then (17.62) shows that $H_{\alpha\gamma}(K,r)$ vanishes exponentially at large distance. On the other hand, let k_β be real. Then we get from (17.61) and (17.62), for large r,

$$H_{\beta\gamma}(K,r) \simeq e^{ik_\beta r}X_{\gamma\beta}k_\beta^{l_\beta}k_\gamma^{-l_\gamma} - \left[\int_r^\infty dr'\, g_L(K;r,r')V(r')H(K,r')\right]_{\beta\gamma} \tag{17.63}$$

where g_L is defined in (12.135).

Suppose now that for some set of values $\mathscr{K} = \mathscr{K}_1, \ldots, \mathscr{K}_n$ of the channel wave numbers k_α, all in the upper right-hand quandrant or positive, the Fredholm determinant Δ vanishes,

$$\Delta(\mathscr{K}) = 0 \qquad \operatorname{Re}\mathscr{K}_\alpha \geq 0 \qquad \operatorname{Im}\mathscr{K}_\alpha \geq 0 \qquad \alpha = 1, \ldots, n. \tag{17.64}$$

Then it may be concluded from (17.61) that $H(\mathscr{K},r)$ is *regular* at the origin. Furthermore, we easily calculate the wronskian of H and H^* in the standard way from the Schrödinger equation,

$$W[H(\mathscr{K},r), H^*(\mathscr{K},r)] = 2i\int_0^r dr'\, \tilde{H}(\mathscr{K},r')\operatorname{Im}\mathscr{K}^2\, H^*(\mathscr{K},r'). \tag{17.65}$$

But the wronskian of F and F^*, in the limit of large values of r, is easily calculated from the boundary condition. If $\operatorname{Im} k_\alpha \geq 0$ for all α, then

$$\lim_{r\to\infty} W(F,F^*) = -2iK\mathscr{R}, \tag{17.66}$$

$\mathscr{R}$ being the projection matrix onto all those channels for which $\operatorname{Im} k_\alpha = 0$. Insertion of (17.62) and (17.66) in (17.65) then gives us in the limit as $r \to \infty$

$$X\mathscr{R}\mathscr{K}^{2L+1}\tilde{X}^* = -\int_0^\infty dr\, \mathscr{K}^L\tilde{H}\operatorname{Im}\mathscr{K}^2\, H^*\mathscr{K}^{*L}.$$

The left-hand side is positive semidefinite and the right-hand side negative semidefinite. Hence both must vanish. From the vanishing of the right-hand side we conclude that for each β either $\operatorname{Im}\mathscr{K}_\beta{}^2 = 0$ or else $H_{\beta\alpha}(\mathscr{K},r) = 0$ for all α and all r. According to (17.62) this means that, if $\operatorname{Im}\mathscr{K}_\beta{}^2 \neq 0$, we must have $X_{\alpha\beta} = 0$ for all α. The vanishing of the left-hand side implies that if $\operatorname{Im}\mathscr{K}_\beta = 0$, that is, if $\mathscr{K}_\beta$ is positive, then $X_{\alpha\beta} = 0$ for all α. In other words, it follows from (17.63) that

$$X_{\alpha\beta} = 0 \tag{17.67}$$

for all α and all β, except for those values of β for which $\mathscr{K}_\beta$ is positive imaginary.

We now look at (17.63) and discover that (17.67) ensures that all those terms in H which would fail to vanish at large distance if some $\mathscr{K}_\beta$ in (17.64) were real drop out. Hence (17.64) implies not only that, at $K = \mathscr{K}$, H is regular at the origin but that it vanishes at infinity, and sufficiently rapidly so as to be square-integrable. We thus have a *bound state*. If none of the $\mathscr{K}_\beta$ is real, it is a conventional bound state and the asymptotic decrease of the bound-state wave function is exponential. If any of them are real, then it is a *bound state embedded in the continuum*. Equation (17.63) shows that in this instance the wave function does not generally fall off exponentially. It may do so like an inverse power of r, depending upon the asymptotic form of the potential.

The Fredholm determinant Δ cannot vanish at an energy where all channels are open. This is easily seen as follows: When all k's are real, then the whole matrix S of (17.15) is unitary. Hence

$$e^{i\pi L}K^{-1/2}\mathfrak{F}_- = SK^{-1/2}\mathfrak{F}_+$$

shows that, if $\mathfrak{F}_+ a = 0$, then $\mathfrak{F}_- a = 0$. Consequently, if $\det \mathfrak{F} = 0$ for all k's real, and thus there exists a vector a such that $\mathfrak{F}_+ a = 0$, then (17.13) shows that $\Phi a \equiv 0$ for all r. But this contradicts the integral equation (17.8).

It may appear puzzling that in contrast to the single-channel case we could not conclude that at the bound state all $\mathscr{K}^2$'s must be real. That this cannot be a general requirement follows immediately from a consideration of the special case in which a channel γ is entirely uncoupled. At a bound state of the remaining channels we may then give k_γ any value we please, and the bound state remains. However, at least *one* of the $\mathscr{K}$'s must be purely positive imaginary for (17.64) to signify a bound state, because otherwise *all* elements of the matrix X vanish and hence so does H. If the $\mathscr{K}$'s are related to one another by energy conservation (17.35) then this is enough to ensure that all the other $\mathscr{K}$'s are also either real or imaginary and the energy, real.

The generalization of *Levinson's theorem* to many channels is straightforward. It generally holds for the sum δ of the eigenphase shifts only, as in (15.145). If δ is to be continuous, then the bound states must include those embedded in the continuum.

17.1.5 The Riemann Surface of the Many-Channel S Matrix. In the foregoing discussion the channel momenta were considered as independent variables. Let us assume now that the interparticle forces are as well behaved as possible, so that the solutions of (17.6) have as large a regularity domain as can ever be expected. In other words, all elements of the auxiliary **S** matrix (17.14) are meromorphic functions of all the channel momenta.

The functions we are really interested in are obtained by expressing each channel momentum in terms of the energy as in (17.35). Each **S**-matrix element then becomes a function of E, but unless it happens to be an even function of k_α, it has a *branch point* of the square-root type at the threshold energy $E = E_\alpha$ of the α channel. If none of the channel couplings is zero, each element of S is generally a function of *all* k's and hence each element generally has a branch point at all thresholds. Needless to say, the resulting Riemann surface for many channels is quite complicated.

"Physics" lies on the upper rim of all the cuts along the *real axis, running from each threshold branch point to* $+\infty$. As we move from a real energy E toward the left, around the α threshold, then to the right back to E, without circumscribing any other threshold branch point in so doing (see Fig. 17.1),

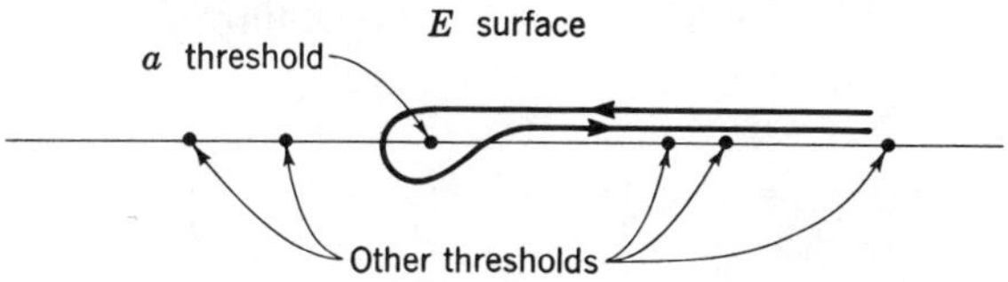

Figure 17.1. The path from the upper to the lower rim of the α cut.

we are on the lower rim of the cut starting at the α threshold, but on the upper rim of all others. In order to get to the lower rims of several branch cuts, all the corresponding branch points must be included in the contour; their momenta are now negative. As we traverse the threshold of the α channel (downward), k_α must be made *positive* imaginary, so that the same boundary condition which above threshold eliminates *incoming* spherical waves (except in the channel of incidence) assures the absence of exponentially *increasing* terms below. The "physical sheet" of the Riemann surface is therefore defined by the demand that each channel momentum have a positive imaginary part, and its physical boundary by the requirement that the real channel momenta be positive and the imaginary ones positive imaginary.

As an example of the relevance of the nonphysical boundary of the physical sheet we look back at the construction of the elements of the open-channel **S** matrix from the Fredhold determinant Δ. Equations (17.42) and (17.51a), once (17.35) is inserted in them, refer to the values of Δ on the lower rims of the various cuts. Equation (17.53) says that the values of Δ on the lower rims of *three* cuts are not independent of those on the lower rims of one and two.

As we have seen, the nonphysical sheets of the Riemann surface, reached by analytic continuation through the various cuts, become relevant for the description of resonances. Of course, unless the forces are rather more well behaved than one has a right to expect physically, there may be all kinds of singularities on these sheets. But for purposes of orientation we may neglect such complications. It should be added that, so long as one is restricting himself to a truncated problem of a *finite* number of channels, the behavior of the **S** matrix on the other sheets of its Riemann surface cannot really be much worse than on the physical sheet. This follows from unitarity. In fact we can see it directly by inspecting (17.42). If the diagonal elements of S are to be regular on the physical sheet (except for bound-state poles), then Δ must have no singularities on any of the other sheets. As a result (17.42) shows that the diagonal elements of S must be meromorphic on the other sheets. The only singularities can be poles at the zeros of Δ. Equation (17.51a) shows that the off-diagonal elements may, in addition, have branch points.

Uniformization. For the purpose of parametrization, or of the construction of phenomenological models, it is very convenient to introduce another variable, in place of the energy, in terms of which the **S** matrix is *single-valued.* Such a procedure is called *uniformization.* It allows us, so to speak, to "open up" the Riemann surface by mapping it on the complex plane. In the single-channel case this procedure is trivial. The simplest such variable is the momentum k. The function

$$E = \frac{k^2}{2\mu}$$

maps both sheets of the E surface onto the complex k plane.

In the case of two channels the uniformization is still relatively simple. It can be done as follows.

Let us first consider the **S**-matrix elements as functions of k_1, writing

$$k_2 = \left(\frac{\mu_2}{\mu_1}\right)^{1/2} (k_1^2 - \boldsymbol{\Delta}^2)^{1/2} \qquad \boldsymbol{\Delta}^2 \equiv 2\mu_1(\boldsymbol{\Delta}_2 - \boldsymbol{\Delta}_1). \tag{17.68}$$

After k_2 has been replaced by (17.68), an analytic function of k_1 and k_2 then has two branch points as a function of k_1: at $k_1 = \pm\boldsymbol{\Delta}$. The transformation

$$u = \frac{k_1 - \boldsymbol{\Delta}}{k_1 + \boldsymbol{\Delta}}$$

moves one of these branch points to the origin and the other to infinity. We now have

$$k_1 = \boldsymbol{\Delta}\frac{1 + u}{1 - u}$$

$$k_2 = 2\boldsymbol{\Delta}\left(\frac{\mu_2}{\mu_1}\right)^{1/2} \frac{u^{1/2}}{1 - u}.$$

The new Riemann surface is opened up by setting $u = t^2$, so that

$$\begin{aligned} k_1 &= \boldsymbol{\Delta}\frac{1 + t^2}{1 - t^2} \\ k_2 &= 2\boldsymbol{\Delta}\left(\frac{\mu_2}{\mu_1}\right)^{1/2} \frac{t}{1 - t^2}. \end{aligned} \tag{17.69}$$

A more symmetric form is obtained by setting

$$z = \frac{1 - t}{1 + t}$$

so that

$$\mu_1^{1/2}k_1 + \mu_2^{1/2}k_2 = \mathbf{\Delta}\mu_2^{1/2}z \qquad \mu_1^{1/2}k_1 - \mu_2^{1/2}k_2 = \mathbf{\Delta}\mu_2^{1/2}z^{-1}. \tag{17.69a}$$

Equation (17.69) constitutes a mapping of the Riemann energy surface to the complex t plane in such a way that an everywhere analytic function of the two variables k_1 and k_2 is an everywhere analytic function of t. The mapping is shown in Fig. 17.2.

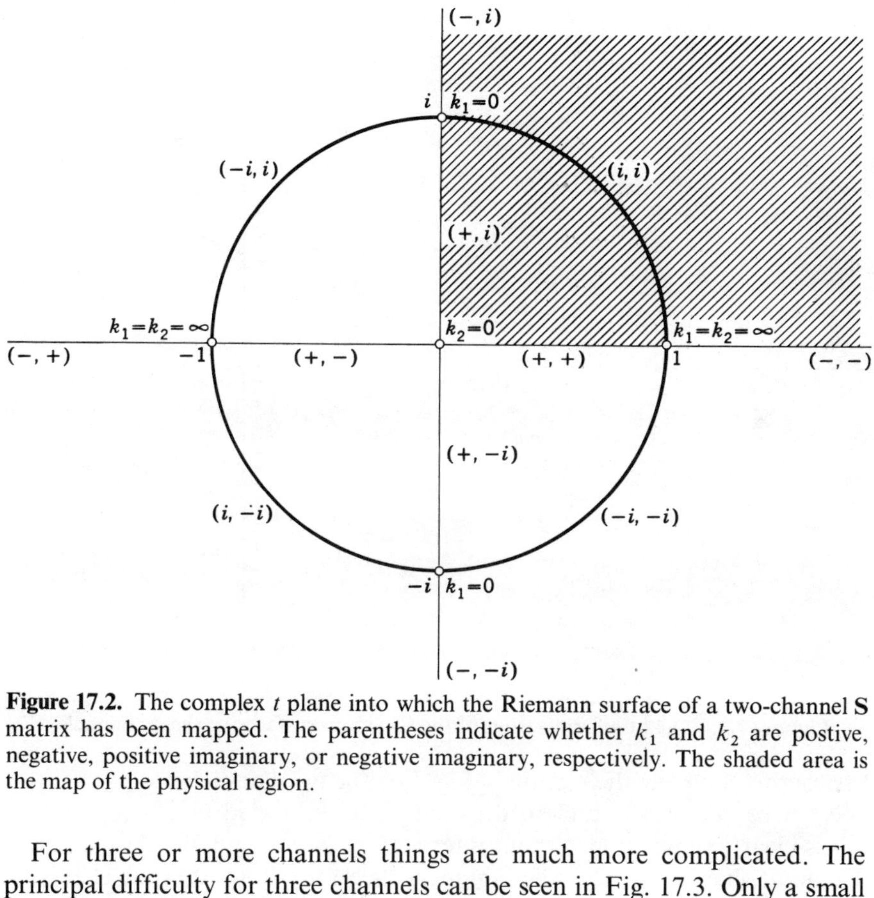

Figure 17.2. The complex t plane into which the Riemann surface of a two-channel **S** matrix has been mapped. The parentheses indicate whether k_1 and k_2 are postive, negative, positive imaginary, or negative imaginary, respectively. The shaded area is the map of the physical region.

For three or more channels things are much more complicated. The principal difficulty for three channels can be seen in Fig. 17.3. Only a small part of an analog of Fig. 17.2 is pictured there schematically, with a few of the branch points and the sign and reality of the channel wave numbers. The latter are understood by consulting the relations between them. If we write $\mathbf{\Delta}_1$, $\mathbf{\Delta}_2$, and $\mathbf{\Delta}_3$ for the threshold energies, assuming that

$$\mathbf{\Delta}_1 < \mathbf{\Delta}_2 < \mathbf{\Delta}_3$$

and

$$z_\alpha = k_\alpha(2\mu_\alpha)^{-1/2}$$

then

$$z_1 = [z_2^2 + (\boldsymbol{\Delta}_2 - \boldsymbol{\Delta}_1)]^{1/2} = [z_3^2 + (\boldsymbol{\Delta}_3 - \boldsymbol{\Delta}_1)]^{1/2}$$
$$z_2 = [z_1^2 - (\boldsymbol{\Delta}_2 - \boldsymbol{\Delta}_1)]^{1/2} = [z_3^2 + (\boldsymbol{\Delta}_3 - \boldsymbol{\Delta}_2)]^{1/2}$$
$$z_3 = [z_1^2 - (\boldsymbol{\Delta}_3 - \boldsymbol{\Delta}_1)]^{1/2} = [z_2^2 - (\boldsymbol{\Delta}_3 - \boldsymbol{\Delta}_2)]^{1/2}.$$

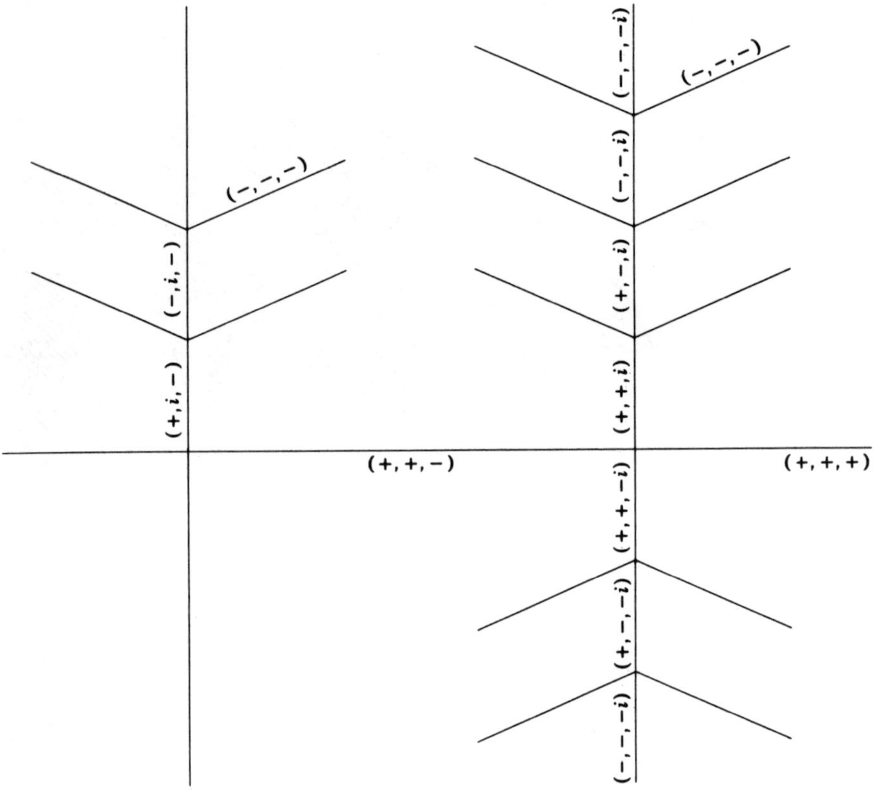

Figure 17.3. Local uniformization of the Riemann surface for three channels.

In order to complete the picture in Fig. 17.3 the two lines labeled $(-, -, -)$ have to be connected, and so do the two lines labeled $(-, -, -i)$. But this is clearly impossible without intersection. It can be done, though, on a doughnut-shaped surface. The argument therefore shows that the Riemann surface for a three-channel **S** matrix cannot be topologically mapped onto a plane (or a sphere), but it can be mapped on a torus. This is expressed by saying that its *genus* is equal to 1. The genus of the Riemann surface of an n-channel **S** matrix is $(n/2) - 1$ or $[(n + 1)/2] - 1$ depending on whether n is even or odd.[7]

Motion of Resonance Poles. Let us now examine the motion of a resonance pole of the **S** matrix as the hamiltonian is changed continuously. Before proceeding we realize that the symmetry (17.21) obliges the **S** matrix

7 See G. Springer (1957).

to have several poles symmetric with respect to imaginary k axes if it has one, unless the latter is situated on the imaginary axis of all k's.

Consider a three-channel system (all angular momenta zero) with channel 1 open, channel 3 closed, and channel 2 open but near threshold. Suppose that channel 3 is uncoupled and has a bound state there. The **S** matrix then has a pole on the physical sheet of channel 3, i.e., on the positive imaginary k_3 axis, and this pole appears both on the upper and on the lower rims of the cuts starting at the thresholds of channels 1 and 2. As $-V_{33}$ is strengthened, the two poles on the upper and lower rim of the channel 2 cut collide at $k_2 = 0$ and bounce off, one into the upper, the other into the lower half plane (see Sec. 12.1.4), i.e., into the physical and unphysical sheets.

Now let there be a small amount of coupling to channel 3. Then for *positive* k_1 (that is, on the upper rim of the channel 1 cut) the pole in the upper half of the k_3 plane must be pushed to the *left*, because if

$$\frac{k_3}{(2\mu_3)^{1/2}} = ia - b \qquad a > 0 \qquad b > 0 \qquad b \ll a$$

then

$$\frac{k_2}{(2\mu_2)^{1/2}} = \pm(\mathbf{\Delta}_{23} - a^2 - 2iab)^{1/2}. \tag{17.70}$$

The pole near the upper rim of the channel 2 cut is thus in the lower half plane (i.e., on the second sheet), as it should be. We now allow $-V_{33}$ to increase, so that the poles move below the channel 2 threshold. Then

$$\frac{k_2}{(2\mu_2)^{1/2}} = \pm i(a^2 - \mathbf{\Delta}_{23} + 2iab)^{1/2} \tag{17.71}$$

and the path from (17.70) to (17.71) is as indicated in Fig. 17.4. This is so because if we write k_2 in the form

$$\frac{k_2^2}{2\mu_2} = \left|\frac{k_2^2}{2\mu_2}\right| e^{i\varphi}$$

then φ changes from $-\varepsilon$ to $-\pi + \varepsilon$ as $\mathbf{\Delta}_{23} - a^2$ changes sign, and hence $\frac{1}{2}\varphi$ changes from $-\frac{1}{2}\varepsilon$ to $-\frac{1}{2}\pi + \frac{1}{2}\varepsilon$.

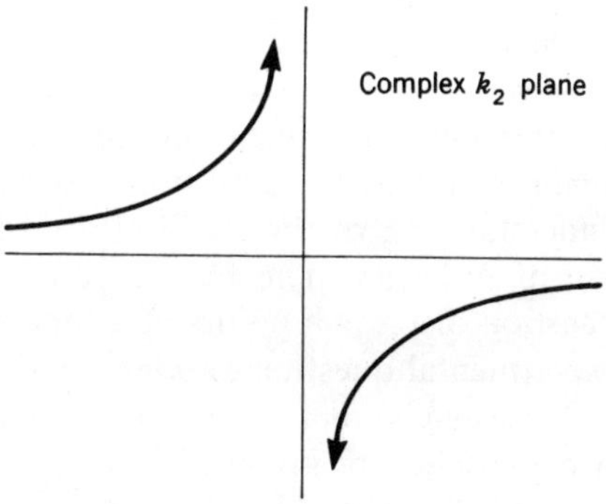

Figure 17.4. Pole motion near threshold.

We therefore find that the pole which is "responsible" for the resonance between the channel 2 and 3 thresholds, i.e., the one near the upper rims of both cuts, remains on the unphysical sheet of the E surface. At the same time there is another pole near the upper rim of the channel 1 cut (on the second sheet of channel 1) which lies near the *lower* rim of the channel 2 cut on the *first sheet* of channel 2. As the forces are strengthened and it moves past the channel 2 threshold, it remains there and becomes responsible for the resonance between the thresholds of channels 1 and 2. It is clear that this argument can easily be generalized. It shows that, contrary to first expectations, as the forces change continuously and a resonance shifts its position, it is not always the same pole of S that "causes" the resonance. Every time a threshold is passed, a new pole takes over and the old one retires to a less exposed position far from the physical sheet. Figure 17.5 shows the situation in a local uniformization.

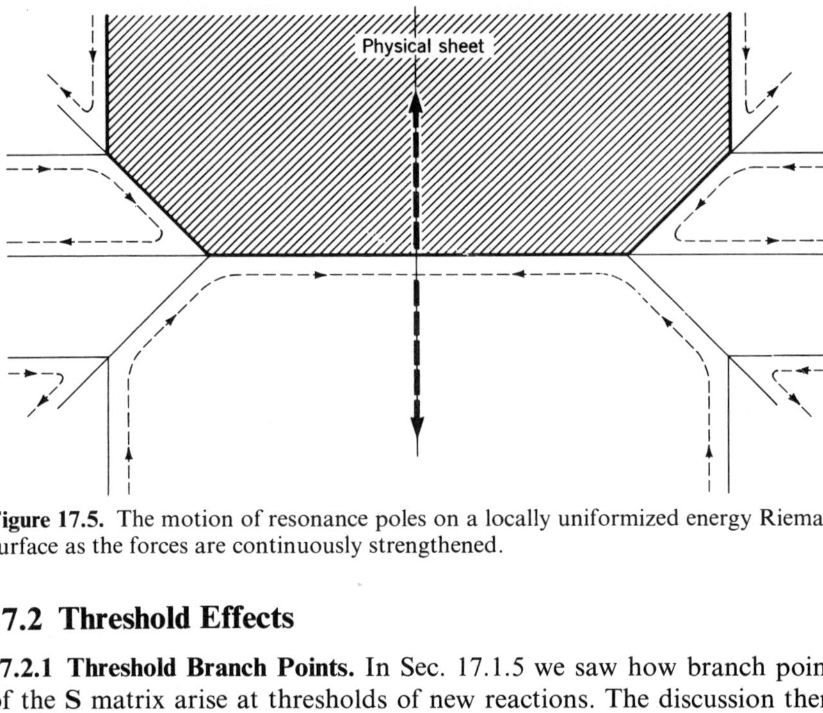

Figure 17.5. The motion of resonance poles on a locally uniformized energy Riemann surface as the forces are continuously strengthened.

17.2 Threshold Effects

17.2.1 Threshold Branch Points. In Sec. 17.1.5 we saw how branch points of the **S** matrix arise at thresholds of new reactions. The discussion there, however, was tied to a specific truncated many-channel formulation treated in more detail in the earlier sections. Here we wish to look at the situation again from two different points of view and add another dimension to our understanding of threshold effects. Both these approaches make essential use of unitarity. The first is concerned primarily with the mathematical question of branch points of a function of the energy; the second, with the experimental question of *observable* effects at channel thresholds.

A many-channel description of scattering processes in which more than two particles are involved is complete only if it is not truncated. In other words, in order to be sure that no aspects of the process are lost, we have to include among the channels, not only those defined by means of the discrete internal excitations of the fragments, but also those in which the fragments are *dissociated.* There being then more than two final particles, the energy distribution is continuous, and we are confronted with a *continuity of channels* so to speak. If continuous channels are included in a description using a single scattering operator, it is complete and includes rearrangements. Of course, the latter are hidden in such a formulation and cannot be extracted easily to be treated in a convenient manner. Nevertheless, it is possible to disentangle some of their effects.

We may remark at this point that the matrix formulation of Sec. 17.1.1 is not restricted to discrete channels. We may formally incorporate in it continuous ones by including among the matrix elements of the interaction (16.66) those between scattering states. Equation (16.67) then contains an *integral* over the energies ε_α. We may still use a matrix notation as in (16.67a), but it must be remembered that the matrices are in part integral operators (i.e., "continuous" matrices). The price for this, however, is high. We shall see in Sec. 17.4.1 that the inclusion of the full range of all intermediate energies makes the kernel of the integral equation that replaces (16.67) fall outside the Hilbert–Schmidt class. Fortunately, this need not concern us for the moment. The argument to be given presently uses the open "continuous" channels only. The energy integral involved is therefore always finite.

Suppose now that the **S** matrix has been calculated. Its unitarity is expressed most easily in terms of the **K** matrix, as in (7.59).

$$\mathbf{S} = (\mathbb{1} + i\mathbf{K})(\mathbb{1} - i\mathbf{K})^{-1}.$$

We must therefore calculate the inverse of the *integral operator* $1 - i\mathbf{K}$. Now the **K** matrix, just like the **S** matrix, contains a diagonal part. That part of **K** simply describes a process in which some particle (or particles) fails to interact. For a clearer understanding of this, let us for the moment consider the three-body problem as an example.

A general element of the **S** matrix describes a process which may be schematically represented by a graph as in Fig. 17.6. Three particles come in, interact, and go out. Among these processes there are those in which one of the particles, say particle 1, does not interact at all. It simply moves along freely, while particles 2 and 3 interact by themselves. If the entire interaction among the three particles consisted of three-body forces, this would be a negligible part of the **S** matrix. As it is, however, at least some

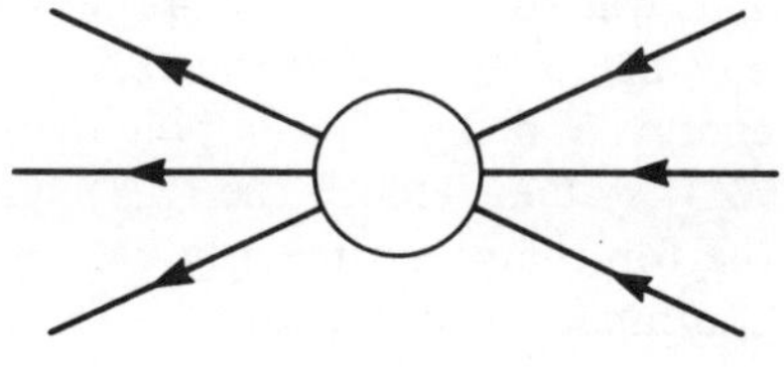

Figure 17.6. A general three-body scattering process.

of the interparticle forces (possibly all) are *binary*. The hamiltonian thus contains terms which commute with the single-particle operators of all but two particles. As a result, the "disconnected" diagrams such as are shown in Fig. 17.7 give rise in the **S** matrix to δ-function contributions with respect to the partial energies of the disconnected fragments.

Let us assume then that a particular parametrization of the **S** matrix has been adopted. That is to say, the variables in the "free" states have been chosen, and corresponding matrix elements of the S operator taken which allow us to speak of the **S** matrix. This can be done in a variety of ways, but it *must* include a continuous variable such as the internal energy distribution among the three particles. Suppose that this choice has been made to determine the internal energy by fixing that of particle 1. Given the total energy, this also fixes the energy of the (2, 3) system. Then the disconnected diagram of Fig. 17.7 produces a contribution to **K** which is a multiple of $\delta(E_1 - E_1')$. Clearly the constant of proportionally must be the **K** matrix for the (2, 3) system alone, since the disconnected part describes just that.

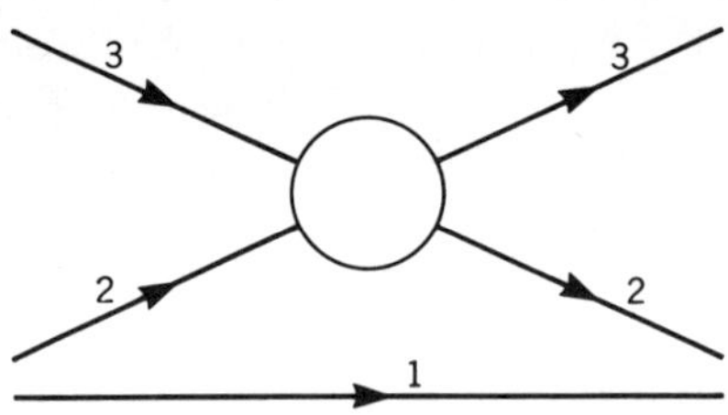

Figure 17.7. A disconnected diagram.

The structure of the integral operator $\mathbb{1} - i\mathbf{K}$ is therefore

$$\mathbb{1} - i\mathbf{K} = \delta(x - x')g(E,x) - M(E,x,x') \equiv F(E;x,x') \quad (17.72)$$

where x is the partial energy of particle 1 (defined so that the continuous range starts at $x = 0$), g is the **K** matrix of the (2, 3) system alone, E is the total energy, and M contains no δ-function contribution in x. (In fact, we may assume that it is continuous as a function of x.) The functions g and M are generally still matrices with other indices.

The inverse of an integral operator of a structure such as $F(E;x,x')$, considered as a function of E, has generally *branch points* at those values of E where $g(E,0)$ vanishes, or, more generally, where $g^{-1}(E,0)$ has poles. The "natural" positions of the branch cuts are described by $g^{-1}(E,x) = \infty$, or

$$\det g(E,x) = 0 \qquad 0 \leq x. \quad (17.73)$$

This was shown in Sec. 9.4. Let us see what it means.

When $x = 0$, then the (2, 3) system has all the energy; that is, E is *its* energy. When g^{-1} has a pole there, then so has the **S** matrix of the isolated (2, 3) system. Hence the (2, 3) system has a *bound state*. (We assume that the forces are sufficiently well behaved so that all the necessary analytic continuations are justified.) Thus in the present parametrization the three-body **S** matrix has branch points at the bound-state energies of the (2, 3) system. These, of course, are just the thresholds of the new channels defined by the excitations of the (2, 3) system. In other words, these are the minimal energies for the (2, 3) system to be raised to a specific excited level and still have some kinetic energy left over for the relative motion of 1 and (2, 3). It is the disconnected diagram 17.7 that is responsible for branch points at the bound states of (2, 3).

Now there are two other disconnected graphs: that in which particle 2 is free, and that in which particle 3 is free. However, *in the parametrization adopted*, these do not cause branch points. The reason is that, when **S** is considered as a function of E and E_1, these diagrams do not correspond to diagonal terms. To fix the energy of particle 2 does not determine that of particle 1. We conclude that whether or not a specific disconnected diagram gives rise to a branch point in **S** depends upon the variables used in the description. It depends, that is to say, on what other variables *are kept fixed* while the total energy is allowed to change. Figure 17.7 is responsible for a branch point in **S** as a function of E *if* E_1 *is held constant*. But if, as E changes, E_2 is held constant and E_1 varies, then it causes no such singularity.

It is clear that this argument lends itself easily to generalizations beyond the three-body problem.

It remains for us to look at the details of the branch points in the inverse of F defined by (17.72). We refer back to Sec. 9.4.

The function g really depends only on the internal energy E_{23} of the (2, 3) system,[8] which is linearly related to the total energy E and the energy x of particle 1 (in the center-of-mass system),

$$g = g(E_{23})$$

$$E_{23} = E - \frac{x}{\alpha} \qquad \alpha = \frac{m_2 + m_3}{m_1 + m_2 + m_3}.$$

Let g^{-1} have a simple pole at $E_{23} = E_B$,

$$g^{-1} = \frac{R(E)}{x - (E - E_B)\alpha}. \tag{17.74}$$

If we pay attention to the other variables (angular momenta, etc.), then $R(E)$ is generally factorable as a matrix [see (15.144)],

$$R_{ij} = R_i^{(L)} R_j^{(R)}. \tag{17.75}$$

The result (9.97a) for the discontinuity of $G \equiv F^{-1}$ then becomes in more detail

$$\Delta G(E; x, x')_{ij} = 2\pi i \left[\sum_k N_\pm^{(L)}(E; x, x_0)_{ik} R_k^{(L)}(E)\right]\left[\sum_l R_l^{(R)}(E) N_\pm^{(R)}(E; x_0, x')_{lj}\right]$$

$$x_0 = \alpha(E - E_B). \tag{17.76}$$

8 This is not essential to the argument, but it simplifies it a little.

This shows that (17.73) indeed defines a branch cut. In addition, (17.76) generalizes the factorization of the residue at a pole.

It must be realized that F^{-1}, as a function of E, has not only cuts but also poles. Some of these manifest themselves as δ functions in the discontinuity and are explicitly visible in the first three terms of the alternative form (9.97b). The poles there visible depend on x and x'. In addition, G has x- and x'-independent poles at those values of E where the Fredholm determinant vanishes. They are the bound states of the three-body system.

These considerations can obviously be generalized to more than three particles. The important conclusion to be drawn is that unitarity generally forces elements of the (open-channel) **S** matrix for n particles to have branch points at the bound states of less than n particles. Which of these bound states actually do produce branch points depends (among other things) on the parametrization adopted. Only if, as the total energy changes, the energy of a potentially isolated subsystem remains fixed do the bound states of that subsystem cause branch cuts in E. For *all* bound states of proper subsystems to cause branch points (and cuts), it is necessary to adopt a parametrization of the **S** matrix in which all partial energies, in the subsystems' own center-of-mass frame, are used as variables.

17.2.2 Physical Threshold Phenomena; General Arguments. Consider an inelastic scattering event, or a reaction, taking place at low kinetic energy of the incoming fragments but at high internal energy so that an exoergic[9] reaction is possible in which the new fragments have a higher kinetic energy. According to (16.61) the corresponding cross section (if calculated with wave functions which contain no troublesome factor) contains a factor of $k_\alpha{}^{-1}$, if k_α is the relative momentum of the two initial fragments. Because of the slowness of the incident flux, the cross section for an exoergic reaction thus becomes infinite at its own threshold.

Now consider an endoergic reaction. Owing to the flux factor, its cross section is linear in the final momentum k_β. If the internal excitation energy is $\boldsymbol{\Delta}_\beta$, then this is related to the incident energy E by (17.35),

$$k_\beta = [2\mu_\beta(E - \boldsymbol{\Delta}_\beta)]^{1/2}.$$

Its derivative with respect to E is therefore *infinite* at the threshold energy $E_T = \boldsymbol{\Delta}_\beta$. For this reason endoergic cross sections, considered as functions of the incident energy, start with an *infinite slope* at their threshold.

The infinite slope of the new cross section that opens up at E_T must necessarily have an influence upon the energy dependence of the other cross sections for the "old" reactions and elastic scattering. All of them, after all, "feed" off the same incident flux. If at the threshold E_T the new cross section starts to diminish the flux available to the old channels at a rate whose energy derivative is infinite, we expect that the old cross section *decrease*,

9 The terms *exoergic* and *endoergic* are analogous to *exothermic* and *endothermic* in chemistry. In the first instance the new fragments have more kinetic energy than the old; in the second, less.

with a correspondingly infinite slope. If only one channel is open below E_T, this argument, which leads us to expect a *negatively infinite* slope of the old cross section at E_T (as approached from above), is unassailable. If several channels are open, however, we cannot rule cut the possibility that some cross sections have positively infinite slopes and others negatively.

This argument, based as it is on flux conservation alone, is essentially classical. Quantum mechanically we expect to find that the vicinity of a threshold for a new process makes itself felt already at an energy *below* that at which transitions into the new channel are energetically possible. That is because "virtual" transitions can take place, and the less they violate the conservation of energy, the stronger their influence. It should therefore come as no surprise to find that at a threshold the slopes of the old cross sections as functions of the energy are generally infinite, when approached not only from above but also from below. The cross sections may then exhibit a behavior such as (*a*), (*b*), (*c*), or (*d*) of Fig. 17.8. The first two are *cusps* and the second *rounded steps.*

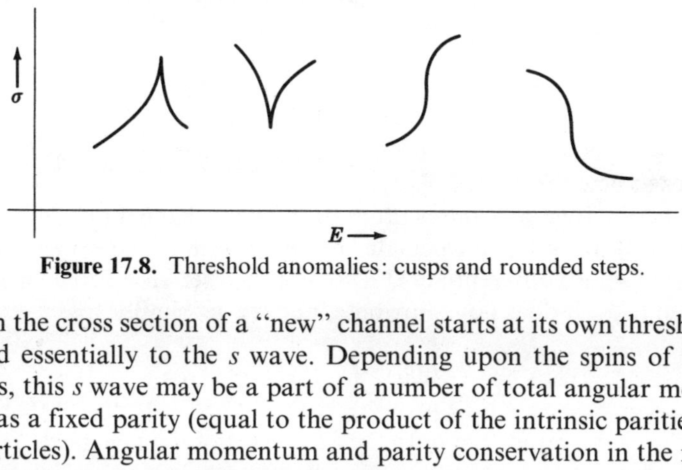

Figure 17.8. Threshold anomalies: cusps and rounded steps.

When the cross section of a "new" channel starts at its own threshold it is confined essentially to the *s* wave. Depending upon the spins of the new particles, this *s* wave may be a part of a number of total angular momenta, but it has a fixed parity (equal to the product of the intrinsic parities of the new particles). Angular momentum and parity conservation in the reaction (if applicable) then allow this new flux to feed off only quite particular states in the incident beam, which in turn feed only some special final states in the old channels. In other words, the threshold anomaly (cusp or rounded step) will be visible only in those partial cross sections which are allowed to be coupled to the new *s* wave by angular momentum and parity conservation. If the intrinsic parities of the old fragments are known, experimental determination of the states which show the anomaly at a new threshold may therefore permit an inference upon the parities of the new fragments. This is one of the practical uses to which these cusps may be put.

17.2.3 Details of the Anomaly. From a mathematical point of view, the threshold anomalies are caused by the branch points of the elements of the S matrix at a threshold. (These have been discussed in Sec. 17.2.1.) However, we wish to base our mathematical discussion here more directly upon the physical arguments of the last section.

The mathematical expression of the conservation of flux in a reaction being the *unitarity* of the **S** matrix, the qualitative arguments of the last section concerning threshold anomalies should be translatable into quantitative statements using that unitarity as the essential tool. The unitarity, together with the behavior of each **S**-matrix element at its own threshold, is conveniently expressed by introducing the matrix $\mathfrak{M}^j$ as in (15.121),

$$S_+^j = \mathbb{1} - K^{L+1/2}\mathfrak{M}^j K^{L+1/2}. \tag{17.77}$$

(Note that $S_+{}^j$ refers to *open* channels only.) It is clear that in this case the number k must be replaced by the (diagonal) matrix K of the channel wave numbers. The unitarity statement then reads as (15.122),

$$(\mathfrak{M}^j)^{-1} = \tfrac{1}{2}K^{2L+1} - \tfrac{1}{2}i^{L+1/2}(\mathscr{K}^j)^{-1}i^{L+1/2} \tag{17.78}$$

where $\mathscr{K}^j$ is a *real symmetric* matrix. (We assume time-reversal invariance.) As in (15.122), $\mathscr{K}^j$ is furthermore *an even function of all open-channel momenta*. This is seen by inserting (17.26) in (17.78),

$$\mathscr{K}^j = -i^{L+1/2}K^{-L-1}[(e^{i\pi L}\mathfrak{F}_- - \mathfrak{F})\mathscr{I}]_+[(e^{i\pi L}\mathfrak{F}_- + \mathfrak{F})\mathscr{I}]_+^{-1}i^{L+1/2}K^{-L} \tag{17.79}$$

and using parity conservation, the evenness of $\mathscr{I}$ as a function of the open-channel momenta, and (17.22). Hence all old elements of the matrix $\mathscr{K}^j$ can be expected to be well-behaved functions of the energy (for real E) without anomalies at new thresholds.

Let k_t be the fragment momentum in the new channel (whose threshold lies at $k_t = 0$). In order to calculate the linear term in the expansion of $\mathfrak{M}^j$ in powers of k_t, we differentiate (17.78) and then set $k_t = 0$. The matrix $\mathscr{K}^j$ being even in k_t, it does not contribute, and we get simply

$$\frac{d}{dk_t}(\mathfrak{M}^j)^{-1}\big|_{k=0} = \tfrac{1}{2}P^{(t,0)}$$

where $P^{(t,0)}$ is the projection onto the $l = 0$ part of the t channel. Therefore

$$\lim_{k_t\to 0}\frac{\partial}{\partial k_t}\mathfrak{M}^j = -\tfrac{1}{2}\mathfrak{M}^j P^{(t,0)}\mathfrak{M}^j, \tag{17.80}$$

the right-hand side being evaluated at threshold. This implies that, just above the new threshold, the leading term $\Delta\mathfrak{M}^j$ in the difference between $\mathfrak{M}^j$ and its value at threshold is

$$\Delta\mathfrak{M}^j = -\tfrac{1}{2}\mathfrak{M}^j P^{(t,0)}\mathfrak{M}^j k_t. \tag{17.81}$$

Consider now $\mathfrak{M}^j$ as a function of the incident energy E. It is related to k_t by

$$k_t = [2\mu_t(E - E_t)]^{1/2}$$

so that

$$\frac{dk_t}{dE} = \frac{\mu_t}{k_t} \tag{17.82}$$

which tends to infinity as $E \to E_t$. Hence it is precisely the linear term (17.81) which gives rise to the infinite derivative of $\mathfrak{M}^j$ at the threshold.

In order to find the behavior of the old cross sections as the energy approaches the threshold from below, we must analytically continue (17.81) to that region. Since there $E < E_t$, k_t becomes purely imaginary. We must take it to be *positive* imaginary to be on the physical sheet, as argued in Sec. 17.1.5. Equation (17.81) then becomes

$$\Delta\mathfrak{M}^j = -\tfrac{1}{2}i\mathfrak{M}^j P^{(t,0)}\mathfrak{M}^j |k_t|. \tag{17.83}$$

Equations (17.81) and (17.83) may be combined in a single equation by the symbolism $\binom{1}{i}$ to indicate "above and below threshold,"

$$\Delta\mathfrak{M}^j = -\tfrac{1}{2}\begin{pmatrix}1\\ i\end{pmatrix}|k_t|\mathfrak{M}^j P^{(t,0)}\mathfrak{M}^j$$

or for the partial-wave amplitudes (16.70)

$$\Delta a^j = \begin{pmatrix}i\\ -1\end{pmatrix}|k_t| a^j P^{(t,0)} a^j. \tag{17.84}$$

Now look at (16.69) evaluated at threshold. For an endoergic reaction this means that only $l' = 0$ contributes, and hence j must be equal to s',

$$\begin{aligned} A(\mathbf{k}'\alpha', \mathbf{k}\alpha) &= (4\pi)^{1/2} \sum_l a^{s'}_{0s'\varepsilon_t, ls\varepsilon_\alpha} \mathfrak{Y}^{v'*}_{s'}(lsv, \hat{\mathbf{k}}) \\ &\equiv A_t(s'v', \mathbf{k}\alpha). \end{aligned} \tag{17.85}$$

For an exoergic reaction instead $l = 0$, and therefore $j = s$,

$$\begin{aligned} A(\mathbf{k}'\alpha', \mathbf{k}\alpha) &= (4\pi)^{1/2} \sum_{l'} \mathfrak{Y}^{v}_{s}(l's'v', \hat{\mathbf{k}}') a^{s}_{l's'\varepsilon'_\alpha, 0s\varepsilon_t} \\ &\equiv A_t(\mathbf{k}'\alpha', sv). \end{aligned} \tag{17.86}$$

In (17.85) the partial wave amplitude contains a factor of $k_t^{1/2}$, and in (17.86) it contains a factor of $k_t^{-1/2}$. So (17.85) and (17.86) cannot be evaluated separately at the threshold; only their product can. If we now insert (17.84) in (16.69) and use (17.85) and (17.86), we get the linear term in k_t of A near the new threshold,

$$\Delta A(\mathbf{k}'\alpha', \mathbf{k}\alpha) = \begin{pmatrix}i\\ -1\end{pmatrix}|k_t| \sum_{s''v''} A_t(\mathbf{k}'\alpha', s''v'')A_t(s''v'', \mathbf{k}\alpha). \tag{17.87}$$

(The notation here is mixed. The exoergic amplitude at threshold is independent of the direction of incidence, but the spin of the incident fragments has been explicity indicated. The endoergic amplitude at threshold is independent of the angle of emergence, but the spin of the outgoing fragments is explicitly shown.)

The Cross Sections. Equation (17.87) is now used to obtain the linear term in k_t in the cross section near the threshold. Because

$$\Delta\sigma = 2\,\mathrm{Re}\,(A^*\Delta A)$$

we find that

$$\Delta \frac{d\sigma}{d\Omega}(\mathbf{k}'\alpha', \mathbf{k}\alpha) = -2|k_t| \begin{pmatrix} \text{Im} \\ \text{Re} \end{pmatrix} A^*(\mathbf{k}'\alpha', \mathbf{k}\alpha) \sum_{s''\nu''} A_t(\mathbf{k}'\alpha', s''\nu'') A_t(s''\nu'', \mathbf{k}\alpha). \tag{17.88}$$

The linear term in the total cross section is most easily found from the optical theorem,

$$\Delta\sigma^{\text{total}}(\mathbf{k}\alpha) = 4\pi|k_t|k_\alpha^{-1} \begin{pmatrix} \text{Re} \\ -\text{Im} \end{pmatrix} \sum_{s''\nu''} A_t(\mathbf{k}\alpha, s''\nu'') A_t(s''\nu'', \mathbf{k}\alpha).$$

If the hamiltonian is time-reversal invariant, then S^j is symmetric and, according to (16.70), a^j has the property

$$\tilde{a}^j = K^{-1} a^j K.$$

Assume furthermore that parity is conserved. Then l in (17.85) is restricted to even or odd values depending on whether the intrinsic parities of the two fragments in channel α are equal or opposite to those in the threshold channel. If we choose the z axis to lie along the direction of $\mathbf{k}$, then

$$\mathfrak{Y}_j^{M*}(ls\nu, \hat{\mathbf{k}}) = (-)^l \mathfrak{Y}_j^M(ls\nu, \hat{\mathbf{k}})$$

and both sides vanish unless $M = \nu$. Hence

$$A_t(s'\nu', \mathbf{k}\alpha) = \pm A_t(\mathbf{k}\alpha, s'\nu') k_t k_\alpha^{-1}$$

with the upper sign if the intrinsic parity products of the two fragments in the initial and final channels are equal and the lower if they are opposite. Therefore, we get for the linear term in k_t of the total cross section, for a beam in which the projection of the total spin of the incident fragments along the direction of incidence is ν,

$$\Delta\sigma^{\text{total}}(\alpha\nu) = \pm 4\pi \begin{pmatrix} \text{Re} \\ -\text{Im} \end{pmatrix} \sum_{s''} [A_t(s''\nu, \nu)]^2. \tag{17.89}$$

(The channel α of incidence has not been explicitly indicated in the amplitude.) If the incident beam is unpolarized, of course,

$$\Delta\sigma^{\text{total}}(\alpha) = \pm \frac{4\pi}{2s+1} \begin{pmatrix} \text{Re} \\ -\text{Im} \end{pmatrix} \sum_{s''\nu} [A_t(s''\nu, \nu)]^2. \tag{17.89a}$$

The sign in (17.89) and (17.89a) depends upon whether the intrinsic parity products in the channel of incidence and in the threshold channel are equal or opposite. Now

$$\begin{aligned} \sum_\nu [A_t(s''\nu, \nu)]^2 &= (2s''+1) \sum_l (-)^l (a_{0s'', ls}^{s''})^2 \\ &= \pm(2s''+1) \sum_l (a_{0s'', ls}^{s''})^2 \end{aligned}$$

and the cross section for an unpolarized beam in channel α to total spin s'' in the threshold channel is

$$\sigma(ts'',\alpha) = 4\pi \frac{2s''+1}{2s+1} \sum_l |a^{s''}_{0s'',ls}|^2. \tag{17.90}$$

We can therefore write (17.89) in the form

$$\Delta\sigma^{\text{total}}(\alpha) = \sum_{s''} \sigma(ts'',\alpha) \begin{pmatrix} \text{Re} \\ -\text{Im} \end{pmatrix} \frac{\sum_l (a^{s''}_{0s'',ls})^2}{\sum_l |a^{s''}_{0s'',ls}|^2}. \tag{17.91}$$

The usefulness of the expressions (17.88) to (17.91) lies in the fact that they allow us experimentally to determine separately the real and imaginary parts of threshold amplitudes. What is required is separate measurements of the shape of the anomaly above and below a threshold.

We may also use (17.84) for an inelastic cross section leading to the threshold channel. The result is that near threshold the total inelastic cross section leading from the α channel and total fragment spin s to the threshold channel and fragment spin s' is given by

$$\sigma(ts',\alpha s) = c(k_t - c'k_t^2 + \cdots) \tag{17.92}$$

where

$$c = \lim_{k_t \to 0} k_t^{-1} \sigma(ts',\alpha s)$$

$$c' = \lim_{k_t \to 0} \frac{k_t \sigma^{\text{total}}(ts')}{2\pi}.$$

Detailed measurement of the energy dependence of a cross section leading to a newly opened channel therefore allows the determination of the *total* cross section from the new channel at threshold.

Examples. As an example we consider the simple case in which both fragments in the old channel and both fragments in the threshold channel have spin 0. If the product of the intrinsic fragment parities in the threshold channel is unequal to that in the incident one, there is no anomaly. If they are equal, then

$$A_t(t,\alpha) = \tfrac{1}{2} i k_t^{1/2} k_\alpha^{-1/2} \mathfrak{M}^0_{t\alpha}$$

$$A_t(\alpha,t) = \tfrac{1}{2} i k_\alpha^{1/2} k_t^{-1/2} \mathfrak{M}^0_{t\alpha}$$

and

$$\Delta \frac{d\sigma}{d\Omega}(\mathbf{k}'\alpha,\mathbf{k}\alpha) = \tfrac{1}{2}|k_t| \begin{pmatrix} \text{Im} \\ \text{Re} \end{pmatrix} [e^{2i\varphi}|\mathfrak{M}^0_{t\alpha}|^2 A^*(\mathbf{k}'\alpha,\mathbf{k}\alpha)], \tag{17.93}$$

the angle φ being the phase of $\mathfrak{M}_{t\alpha}{}^0$,

$$\varphi = \arg \mathfrak{M}^0_{t\alpha}.$$

Measurement of the detailed energy dependence of the elastic scattering cross section in the α channel on both sides of the new threshold thus leads to a determination of the inelastic cross section in the new channel at threshold, as well as of the elastic *amplitude* to within a constant (i.e., angle-independent) phase φ. The latter can in principle be determined from (17.91) by an additional measurement of the anomaly in the total cross section in the α channel,

$$\Delta\sigma^{\text{total}}(\alpha) = \sigma(t,\alpha)\begin{pmatrix} -\cos 2\varphi \\ \sin 2\varphi \end{pmatrix}. \tag{17.94}$$

For more complicated examples with nonzero spins, we refer to the literature[10] and to the problems.

Experimentally the threshold cusp effects are usually hard to observe. This is either because the "bump" is small or because it occurs over a narrow energy interval and hence requires very good resolution. Nevertheless, *in principle* the effect need not be small. Figure 17.9 shows the threshold cusp calculated in a simple three-channel model of particles of the reduced masses of π meson and proton, Σ and K, and Λ and K, interacting via square-well potential elements of a range of 0.7×10^{-13} cm and the following depths: $V(\pi p \to \pi p) = -200$ Mev, $V(\pi p \to \Lambda K) = 40$ Mev, $V(\pi p \to \Sigma K) = 50$ Mev, $V(\Lambda K \to \Lambda K) = 100$ Mev, $V(\Lambda K \to \Sigma K) = -500$ Mev, $V(\Sigma K \to \Sigma K) = 50$ Mev. It is assumed that the relative $\Sigma\Lambda$ parity and the relative $\Lambda K p$ parity are both odd. (The model is not meant to have any significance about the real world but is intended only as an illustration.) The figure shows the cusp in the P_1 wave of the ΛK system at the ΣK threshold.

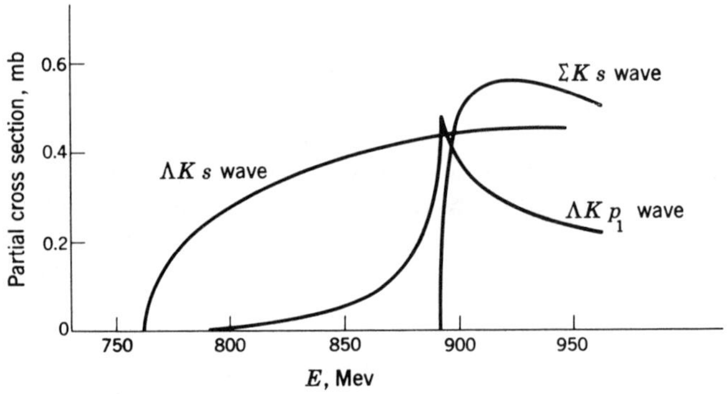

Figure 17.9. Example of a calculated threshold effect. [*From Fonda and Newton* (1959a). *This reference contains a number of other numerically calculated examples.*]

17.2.4 The Threshold Anomaly for Charged Particles. In the foregoing discussion it was assumed that the interparticle forces are well behaved. This excluded the presence of the Coulomb force. It is easy to understand why the attractive Coulomb potential especially causes difficulties at thresholds. After all, for each angular momentum it produces infinitely many bound states whose energies accummulate at $E = 0$. This implies that each **S**-matrix element has an accumulation point of poles there. If we consider **S** as a function both of E and of the product of the charges (coupling constant),

10 See R. G. Newton (1959) and also Baz, Puzikov, and Smorodinskii (1962).

then changing the latter through complex values from negative to positive does not eliminate that essential singularity, even though there are no bound states now.

In order to find the threshold dependence of the **S** matrix in the presence of a Coulomb field, we use the method of Sec. 17.1.2, together with (14.41), which can be written in matrix form,

$$\mathfrak{F}_\pm^{(c)} = (2K)^{-L} e^{1/2N\pi \pm i1/2\pi L} \frac{\Gamma(2L+2)}{\Gamma(L+1 \pm iN)} \tag{17.95}$$

with

$$N = AK^{-1}$$

and the matrix A given by

$$A_{\alpha\beta} = (e^2 Z_{1\alpha} Z_{2\alpha} \mu_\alpha)\delta_{\alpha\beta}.$$

The functions labeled (1) there are now all Coulomb functions, and we label them by (c). According to (17.32) we write

$$\mathfrak{F}' \equiv \mathfrak{F}^{(c)-1}\mathfrak{F} = \mathbb{1} + \int_0^\infty dr\, \mathfrak{F}^{(c)-1} F^{(c)} \overline{V}\Phi \tag{17.96}$$

and (17.33) becomes

$$\begin{aligned} e^{-i\eta_0}(S - S^{(c)})e^{-i\eta_0} = &-ie^{i\pi L/2}(2K)^{1/2+L} \frac{\Gamma(L+1+iN)e^{-i\eta_0}}{\Gamma(2L+2)} e^{-\pi N/2} \\ &\mathfrak{M}(2K)^{1/2+L} e^{-i\pi L/2} \frac{\Gamma(L+1+iN)e^{-i\eta_0}}{\Gamma(2L+2)} e^{-\pi N/2} \end{aligned} \tag{17.97}$$

where

$$\mathfrak{M} = \int_0^\infty dr\, \Phi^{(c)} \overline{V} \Phi \mathfrak{F}'^{-1}. \tag{17.98}$$

The phase factors $e^{-i\eta_0}$ are introduced for convenience [η_0 is defined in (14.28)]. In the cross section they are unobservable.

The matrix $\mathfrak{M}$ is finite in the limit as $k \to 0+$. This is because the regular Coulomb functions $\varphi_l^{(c)}$ of (14.36) are finite there, as can be seen from the integral representation (14.47) [together with (14.46)]. Similarly, the irregular function $f_l^{(c)}/\mathfrak{F}_l^{(c)}$ is finite, which can be seen by using the integral representation[11]

$$\begin{aligned} &f_l^{(c)}(k,r)/\mathfrak{F}_l^{(c)} \\ &= -ir^{l+1}(2k)^{2l+1} e^{ikr - i\pi l} \Psi(l+1+in, 2l+2; -2ikr) \frac{\Gamma(l+1-in)}{\Gamma(2l+2)} \\ &= \frac{-ir^{l+1}(2k)^{2l+1} e^{ikr - i\pi l}}{\Gamma(2l+2)} \int_0^\infty dt\, e^{2ikrt} t^l (1+t)^l \exp[-in\ln(1+t^{-1})]. \end{aligned} \tag{17.99}$$

11 See A. Erdelyi (1953), vol. 1, p. 255.

As $k \to 0+$, the integral is $O(k^{-2l-1})$. Hence the result is finite. Therefore, the Coulomb Green's function also is well behaved as $k \to 0+$. Consequently, so is Φ, and if $\bar{V}$ decreases sufficiently rapidly, so are $\mathfrak{F}'$ and $\mathfrak{M}$. The same is true as k tends to zero along the positive imaginary axis.

Therefore all the threshold effects, both from below and from above, are contained in the outside factors in (17.97). Consider an individual element of this; aside from a constant factor, it is

$$\frac{e^{-i\eta_0}k^{l+1/2}e^{-\pi n/2}\Gamma(l+1+in)}{l!}$$
$$= e^{i(\eta_l-\eta_0)}\left(\frac{2\pi a}{e^{2\pi n}-1}\right)^{1/2}\left\{\left[k^2+\left(\frac{a}{1}\right)^2\right]\cdots\left[k^2+\left(\frac{a}{l}\right)^2\right]\right\}^{1/2}$$
$$\xrightarrow[k\to 0]{} \begin{cases} \dfrac{\sqrt{2\pi}\, e^{-\pi n} a^{l+1/2} i^l}{l!} & \text{if } a > 0 \\[2ex] \dfrac{\sqrt{2\pi}\,|a|^{l+1/2}(-i)^l}{l\;!} & \text{if } a < 0 \\[2ex] k^{l+1/2} & \text{if } a = 0 \end{cases} \tag{17.100}$$

a being an element of the diagonal matrix A. These limits contain all the low-energy statements about elastic, exoergic, and endoergic cross sections near their threshold, for equally charged, oppositely charged, and neutral channels.

Notice that in the Coulomb repulsive case ($a > 0$) the threshold factor approaches zero extremely rapidly[12] and that it is not equal to zero at all in the attractive one. In the latter instance, the partial wave amplitudes of all angular momenta start with nonzero values at threshold! Since in partical case a is small, however, $l = 0$ usually dominates.

Effects on Old Channels. We now wish to examine the behavior of an element of **S** near the threshold of *another* channel. Since the outside factors in (17.97) depend upon the momentum of the same channel only, the effect must come from $\mathfrak{M}$.

Let us orient ourselves again, recalling the arguments of Sec. 17.2.2. In the Coulomb repulsive case the particles are kept so far apart at low energy that the new cross section comes in extremely slowly. In fact, all its energy derivatives vanish at threshold. Hence flux is withdrawn from other channels only very gradually, and *no threshold anomaly should be visible there.* In the attractive case, however, the new cross section comes in *discontinuously*, and a finite amount of flux is suddenly withdrawn from other channels. We therefore expect to find, not cusps (or rounded steps) in the old cross section, but *jumps.* This expectation is essentially borne out by the mathematics. But it is considerably complicated by the presence of infinitely many resonances owing to the Coulomb bound states which accumulate at threshold. For a detailed mathematical treatment of the

12 Remember that $n = a/k$.

phenomenon we refer to the literature.[13] Here it will be discussed only in a qualitative manner.

It is clear that, in an experimental measurement of a total cross section, it is always only an *average* over the infinitely many Coulomb resonances that is accessible. One finds that the averages of the old amplitudes are continuous across the new threshold. The optical theorem therefore tells us that the total cross sections (averaged over the Coulomb resonances) are also continuous there. But if the total cross sections are continuous, then the old cross sections must have jumps in order to compensate for the discontinuous onset of the new, inelastic ones.

X-Ray Scattering. An experimentally observable example[14] of these statements is that of X-ray scattering by an atom in the vicinity of the threshold for the photoelectric effect (i.e., ionization). The total cross section, and hence the mass-absorption coefficient, is continuous at the photothreshold.[15] The so-called *X-ray absorption edge* occurs at the frequency corresponding to raising the atom to its first unoccupied excited state. It is then followed by a series of closely spaced resonances at the higher Coulomb bound states, the "secondary structure" of the absorption edge on its short-wavelength side.[16]

The elastic-scattering cross section for X rays on an atom, on the other hand, does have a discontinuity at the photothreshold. In a reasonable approximation one finds that the sudden onset of the K-shell photo cross section is compensated for almost entirely by an equal and opposite discontinuity in the (averaged) elastic X-ray cross section. The situation is schematically represented in Fig. 17.10.

From an experimental point of view all these threshold statements for charged particles must be considered with some caution. In the repulsive case it is mathematically true that there is no cusp, i.e., the slope of an old cross section is finite at the new threshold. And yet it is clear that if we envisage slowly "turning on" a Coulomb repulsion where there was originally a large cusp, it will not suddenly disappear. All that will happen is a rounding of the initially sharp peak and, as the Coulomb potential gets stronger, the cusp will gradually be washed out. The same will happen for an attraction. In other words, whether an observable cusp-like (or rounded-step-like) threshold phenomenon exists or not depends on the strength of the Coulomb field relative to the "strong forces" responsible for the threshold. Exactly what that means more quantitatively is not clear, especially since the very

13 See Fonda and Newton (1959b and 1960a).

14 See Fonda and Newton (1960a).

15 As far as the index of refraction is concerned, one must distinguish between its direct measurement via the beam deflection by a prism and its indirect inference from the angle of total reflection. In the former one gets directly the real part of the forward-scattering amplitude, while in the latter one observes its square root (if $n \simeq 1$). Both measurements represent different averages over the Coulomb resonances, and the results are not the same. The deflection angle is continuous at the threshold, while the angle of total reflection has a discontinuity there.

16 It should be stressed that the occasionally advanced explanation of the absorption edge as due to the onset of the photoeffect is incorrect.

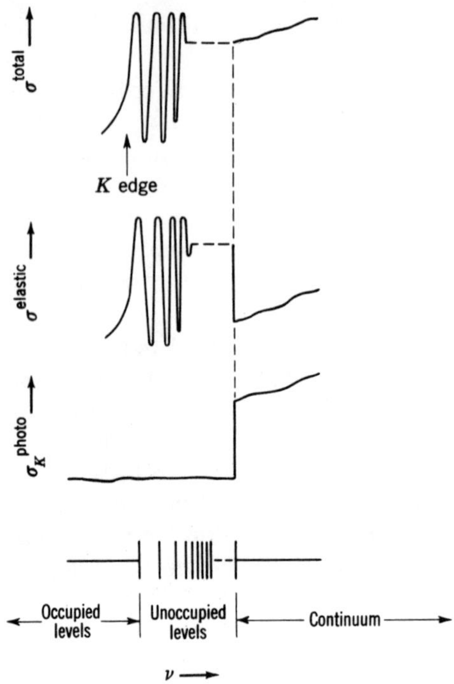

Figure 17.10. Schematic represenation of X-ray scattering cross sections by atoms. [*From Fonda and Newton* (1960a).]

meaning of the "size" of a cusp is vague. However, we may assume that the order of magnitude of the energy ΔE around a threshold within which Coulomb effects should be important is equal to the depth of the ground state produced by its attractive version. This energy is about 27 ev multiplied by the channel's reduced mass in Mev and by the square of the product of the two charges (in units of electronic charge). If the experimental energy resolution is wide compared to ΔE and an observable threshold phenomenon is expected when Coulomb forces are neglected, then it should be observable also with Coulomb forces present.

17.3 Examples

17.3.1 The Square Well. Let us consider the simple case of any number of $l = 0$ channels coupled by a square-well-potential matrix V of range[17] r_0. Then (16.67a) reads

$$\psi'' + A\psi = 0 \tag{17.101}$$

$$A = \begin{cases} K^2 - V & r < r_0 \\ K^2 & r > r_0. \end{cases} \tag{17.102}$$

17 If the ranges of the elements of V are unequal, the problem can be solved similarly but the algebra is considerably more complicated.

Let us diagonalize the real symmetric matrix A for $r < r_0$ by a real orthogonal matrix T,

$$A = Ta^2T^{-1} \tag{17.103}$$

so that a^2 is the diagonal matrix of the eigenvalues of A. The solution $F(K,r)$ must be of the form

$$F(K,r) = \begin{cases} T(e^{-iar}B + e^{iar}C) & r < r_0 \\ e^{iKr} & r > r_0. \end{cases} \tag{17.104}$$

The constant matrices B and C are determined by the requirement of continuity of F and of its first derivative. One finds that

$$F = [T\cos a(r - r_0)T^{-1} + iTa^{-1}\sin a(r - r_0)T^{-1}K]e^{iKr_0} \tag{17.105}$$

and therefore the Jost matrix

$$\begin{aligned} \mathfrak{F} &= e^{iKr_0}[T\cos ar_0\tilde{T} - iKTa^{-1}\sin ar_0\tilde{T}] \\ &= e^{iKr_0}[Ta\cot ar_0\tilde{T} - iK]Ta^{-1}\sin ar_0\tilde{T}. \end{aligned} \tag{17.106}$$

The matrices T and a being even functions of the k's, we get the **S** matrix from (17.15) as

$$S = e^{-iKr_0}K^{-1/2}[Ta\cot ar_0\tilde{T} + iK][Ta\cot ar_0\tilde{T} - iK]^{-1}K^{1/2}e^{-iKr_0}. \tag{17.107}$$

The Fredholm determinant is obtained by calculating the determinant of $\mathfrak{F}$,

$$\Delta = e^{ikr_0}\prod_n\left(\frac{\sin a_nr_0}{a_n}\right)\det[Ta\cot ar_0\tilde{T} - iK] \tag{17.108}$$

where $k = \sum_\alpha k_\alpha$.

Two Channels. Let us now take the case of two channels and set

$$e_n = a_nr_0 \qquad \rho_\alpha = k_\alpha r_0 \qquad R_\alpha = A_{\alpha\alpha}^{1/2}r_0 = (\rho_\alpha^2 - V_{\alpha\alpha}r_0^2)^{1/2}.$$

Then

$$e_{1,2}^2 = \tfrac{1}{2}(R_1^2 + R_2^2) \pm \tfrac{1}{2}[(R_1^2 - R_2^2)^2 + 4V_{12}^2r_0^4]^{1/2}$$

and

$$\Delta(k_1,k_2) = \frac{e^{i(\rho_1+\rho_2)}}{\alpha_2 - \alpha_1}\frac{\sin e_1}{e_1}\frac{\sin e_2}{e_2}g(\rho_1,\rho_2) \tag{17.109}$$

where

$$\begin{aligned} g(\rho_1,\rho_2) = i\rho_1(\mathscr{C}_1\alpha_1 - \mathscr{C}_2\alpha_2) + i\rho_2(\mathscr{C}_2\alpha_1 - \mathscr{C}_1\alpha_2) \\ + (\alpha_1 - \alpha_2)(\rho_1\rho_2 - \mathscr{C}_1\mathscr{C}_2) \end{aligned} \tag{17.110}$$

with the abbreviations

$$\alpha_1 = e_1^2 - R_1^2 = R_2^2 - e_2^2 \qquad \alpha_2 = e_2^2 - R_1^2 = R_2^2 - e_1^2$$

$$\mathscr{C}_1 = e_1 \cot e_1 \qquad \mathscr{C}_2 = e_2 \cot e_2.$$

The elements of the **S** matrix are thus calculated from (17.42) and (17.51a),

$$\begin{aligned} S_{11} &= \frac{g(-\rho_1, \rho_2)}{g(\rho_1, \rho_2)} e^{-2i\rho_1} \qquad S_{22} = \frac{g(\rho_1, -\rho_2)}{g(\rho_1, \rho_2)} e^{-2i\rho_2} \\ S_{12} = S_{21} &= \frac{2(\mathscr{C}_1 - \mathscr{C}_2)(\alpha_1 \alpha_2 \rho_1 \rho_2)^{1/2}}{g(\rho_1, \rho_2)} e^{-i(\rho_1 + \rho_2)}. \end{aligned} \tag{17.111}$$

Suppose that channel 2 has a higher threshold than channel 1. Under what conditions is there a bound state embedded in the continuum? We must look for a zero of the Fredholm determinant Δ at a positive value of ρ_1 and at a positive imaginary value of ρ_2. The numbers $R_1{}^2$ and $R_2{}^2$ are then real, and so are e_1, e_2, α_1, α_2, $\mathscr{C}_1$, and $\mathscr{C}_2$. Let us rewrite (17.110) in the form

$$\begin{aligned} i\rho_1 g = (i\rho_1 - i\rho_2)[(\mathscr{C}_1 - i\rho_2)\alpha_1 \mathscr{C}_2 - (\mathscr{C}_2 - i\rho_2)\alpha_2 \mathscr{C}_1] \\ - i\rho_1(\alpha_1 - \alpha_2)(\mathscr{C}_1 - i\rho_2)(\mathscr{C}_2 - i\rho_2). \end{aligned} \tag{17.110a}$$

Since the real and imaginary parts of this must vanish separately, and for $V_{12} \neq 0$ always $\alpha_1 \neq \alpha_2$, $\alpha_1 \neq 0$, $\alpha_2 \neq 0$, the following two equations must be simultaneously satisfied,

$$\mathscr{C}_1 = i\rho_2 \qquad \mathscr{C}_2 = i\rho_2.$$

In other words,

$$e_1 \cot e_1 = -|\rho_2| \qquad e_2 \cot e_2 = -|\rho_2|. \tag{17.112}$$

A necessary condition for these equations to hold is that $e_2 > \frac{1}{2}\pi$.[18]

In order to show that there exist hamiltonians for which the simultaneous equations (17.112) have solutions, let us arbitrarily choose positive numbers ρ_1 and $-i\rho_2$ and a real V_{12}. We then solve the two equations (17.112) for two numbers e_1 and e_2 such that $e_1{}^2 - e_2{}^2 > 2|V_{12}|r_0{}^2$. Not only do such solutions always exist, but there are infinitely many of them. We then find $R_1{}^2$ and $R_2{}^2$ from

$$R_{1,2}^2 = \tfrac{1}{2}(e_1^2 + e_2^2) \pm \tfrac{1}{2}[(e_1^2 - e_2^2)^2 - 4V_{12}^2 r_0^4]^{1/2}$$

and, finally, V_{11} and V_{22} from

$$V_{\alpha\alpha} r_0^2 = \rho_\alpha^2 - R_\alpha^2.$$

For every given positive value of k_1 and $-ik_2$, and for every given V_{12}, there are therefore infinitely many diagonal potential terms V_{11} and V_{22} which, together with V_{12}, produce a bound state embedded in the continuum.

Bound states of the conventional kind are found similarly by searching for zeros of (17.110), except that now $i\rho_1$ and $i\rho_2$ are both negative, so that g is real,

18 This implies in this instance that V_{22} alone causes a bound state. But such an implication is accidental and does not hold in general.

$$g = |\rho_1|(\alpha_2\mathscr{C}_2 - \alpha_1\mathscr{C}_1) + |\rho_2|(\alpha_2\mathscr{C}_1 - \alpha_1\mathscr{C}_2) + (\alpha_2 - \alpha_1)(|\rho_1|\,|\rho_2| + \mathscr{C}_1\mathscr{C}_2)$$
$$= 0.$$

It is easy to see that, no matter what the diagonal potentials are, if the off-diagonal one is large enough in magnitude, it will bind. (Note that only the square of V_{12} enters in g.) Let us fix ρ_1, ρ_2, V_{11}, and V_{22}, and let V_{12} increase. Then $e_2{}^2$ eventually becomes negative and $\mathscr{C}_2 \to |V_{12}|^{1/2}$, while $\alpha_1 \to -\alpha_2 \to |V_{12}|$. Meanwhile $\mathscr{C}_1$ keeps going from $+\infty$ to $-\infty$, and on every trip g must go through zero. This corroborates the fact that the coupling term of the hamiltonian always is *effectively* attractive. No matter what the diagonal terms are, if only the magnitude of the off-diagonal term is big enough, the compound system of the two channels always has a bound state. There is, of course, no way of identifying the channel in which the bound state occurs. In a most essential manner it belongs to both.

17.3.2 Potentials of Yukawa Type. Suppose that the matrix potential is a continuous superposition of Yukawa functions or exponentials,

$$V(r) = \int da\, \rho(a) e^{-ar} \qquad (17.113)$$

where $\rho(a)$ is a matrix with

$$\rho_{\alpha\beta}(a) = 0 \quad \text{for } a \le a_{0\alpha\beta}.$$

The Schrödinger equations can now, of course, not be solved in closed form. But we may calculate the Fredholm determinant according to Sec. 9.3. Let us label the kernel we there called K by the letter R in order to avoid confusion with the wave-number matrix K. Then

$$R(K;r,r') = -K^{-1} \sin Kr < e^{iKr_<} \int da\, \rho(a) e^{-ar'}. \qquad (17.114)$$

It is convenient to do the calculation in the Fourier-transform language,

$$R(K;k,k') \equiv \frac{1}{2\pi}\int_0^\infty dr \int_0^\infty dr'\, e^{ikr - ik'r'} R(K;r,r')$$
$$= \frac{1}{2\pi i}\int da\, (K+k)^{-1}(k + ia - k')^{-1}(k' - K - ia)^{-1}\rho(a).$$

Then

$$\operatorname{tr} R(K) = -\frac{1}{2\pi}\int da\, a^{-1}\int_{-\infty}^{\infty} dk\, \operatorname{Tr}(k+K)^{-1}(k - K - ia)^{-1}\rho(a)$$
$$= -i\int da\, a^{-1}\operatorname{Tr}(2K + ia)^{-1}\rho(a) \qquad (17.115)$$
$$= -i\int da\, a^{-1}\sum_\alpha (2k_\alpha + ia)^{-1}\rho_{\alpha\alpha}(a)$$

if Tr stands for the trace over the discrete matrix indices only. Thus as a function of k_α there is a branch line along the negative imaginary axis, starting at $k_\alpha = -\frac{1}{2}ia_{0\alpha\alpha}$.

Next we look at R^2,

$$R^2(k,k') = \frac{1}{(2\pi i)^2}\int da \int da' \int_{-\infty}^{\infty} \frac{dk''}{(k''-k-ia)(k'-k''-ia)}$$
$$(k''-K-ia)^{-1}(k+K)^{-1}\rho(a)(k'-K-ia')^{-1}(k''+K)^{-1}\rho(a').$$

The integrations in the trace are readily carried out, and one obtains, after some algebra,

$$\operatorname{tr} R^2(K) = -2i\int da \int da' \sum_{\alpha\beta} \rho_{\alpha\beta}(a)\rho_{\beta\alpha}(a')$$
$$\frac{k_\alpha + k_\beta + ia + ia'}{(a+a')(2k_\alpha + ia + ia')(2k_\beta + ia + ia')(k_\alpha + k_\beta + ia)(k_\alpha + k_\beta + ia')}. \tag{17.116}$$

The only new branch line comes from

$$k_\alpha + k_\beta + ia = 0 \tag{17.117}$$

or the same with $a \to a'$. It starts at

$$k_\beta^{(0)} = -i(a_{0\alpha\beta}^2 + 2\mu_\alpha\Delta_{\alpha\beta})\left\{a_{0\alpha\beta} + \left(\frac{\mu_\alpha}{\mu_\beta}\right)^{1/2}[a_{0\alpha\beta}^2 - 2\Delta_{\alpha\beta}(\mu_\beta - \mu_\alpha)]^{1/2}\right\}^{-1} \tag{17.118}$$

where

$$\Delta_{\alpha\beta} = \frac{k_\alpha^2}{2\mu_\alpha} - \frac{k_\beta^2}{2\mu_\beta}.$$

If

$$a_{0\alpha\beta} > [2\Delta_{\alpha\beta}(\mu_\beta - \mu_\alpha)]^{1/2} \tag{17.119}$$

then the branch point is negative imaginary and the branch line may be taken to run from there to $-i\infty$. However, if the inequality (17.119) is violated, then the branch point lies on the *real* axis.

The branch lines arise simply from the coincidence of singularities in the original integrants of R and R^2. It is therefore clear that $\operatorname{tr} R^n$ will contain no singularities other than $\operatorname{tr} R^2$. Hence neither will the nth term of the expansion of the Fredholm determinant Δ. We may conclude that Δ, considered as an overall function of k_β, that is, with all k's expressed as functions of k_β, has a branch line along the negative imaginary axis from $k_\beta{}^{(0)}$ on, provided that (17.119) holds for all $a_{0\alpha\beta}$. In addition, of course, there are the usual threshold branch points on the real and imaginary axes.[19]

19 We usually want to express all the k's in terms of the wave number or the energy of the channel that opens up *first*. Then the threshold branch points all lie on the real axis.

Finally we form the **S**-matrix elements from Δ by inverting the sign of one or two k's. As a result there are branch lines also along the positive imaginary axis. There criterion for branch lines on the imaginary axis only is still (17.119).

We find therefore that (17.119) is a necessary condition for all the cuts produced by the Yukawa potential to be left-hand cuts. If it is violated, then there are right-hand cuts in addition to the usual threshold branch cuts.

17.3.3 The Wigner-Weisskopf Model. A simple soluble model for inelastic particle processes was first considered by Wigner and Weisskopf. It consists of a motionless particle whose wave function is the constant χ and another particle of mass m and wave function $\psi(\mathbf{r})$. The two particles are one another's sources with a *form factor* $\Lambda(r)$. The postulated Schrödinger equations are

$$\left(E + \frac{1}{2m}\nabla^2\right)\psi(\mathbf{r},E) = \Lambda(r)\chi(E)$$

$$(E - E_0)\chi(E) = \int d\mathbf{r}\,\Lambda(r)\psi(\mathbf{r},E).$$

Because Λ is assumed to depend on the magnitude of the radius vector r only, a partial wave analysis shows immediately that the angular momenta $l > 0$ are uncoupled and unaffected by the source. Only the equations for $l = 0$ are of interest:

$$\left(k^2 + \frac{d^2}{dr^2}\right)\psi_0(k,r) = \lambda(r)\chi$$

$$(k^2 - k_0^2)\chi = \int_0^\infty dr\,\lambda(r)\psi_0(k,r)$$

$$\lambda(r) = 2mr\Lambda(r).$$

(From now on we drop the subscript 0, and $l = 0$ is understood.) The irregular solution is given by

$$f(k,r) = e^{ikr} - \int_r^\infty dr\,k^{-1}\sin k(r - r')\lambda(r')\chi$$

and therefore

$$(k^2 - k_0^2)\chi = \int_0^\infty dr\,\lambda(r)e^{ikr} - \int_0^\infty dr\int_r^\infty dr'\,k^{-1}\sin k(r - r')\lambda(r)\lambda(r')\chi$$

or

$$\chi = \frac{\int_0^\infty dr\,\lambda(r)e^{ikr}}{k^2 - k_0^2 + \int_0^\infty dr\int_r^\infty dr'\,k^{-1}\sin k(r - r')\lambda(r)\lambda(r')}.$$

The Jost function is given by

$$\mathscr{F}(k) = 1 + \int_0^\infty dr\, k^{-1} \sin kr\, \lambda(r)\chi,$$

which a little calculation shows is equal to

$$\mathscr{F}(k) = \frac{\overline{\mathscr{F}}(k)}{k^2 - k_0^2 + \int_0^\infty dr \int_r^\infty dr'\, \lambda(r)\lambda(r')\, k^{-1} \sin k(r - r')}$$

$$\overline{\mathscr{F}}(k) = k^2 - k_0^2 + \int_0^\infty dr \int_0^\infty dr'\, \lambda(r)\lambda(r') k^{-1} e^{ikr_>} \sin kr_<.$$

The denominator in $\mathscr{F}$ being an even function of k, the $l = 0$ eigenvalue of the **S** matrix is

$$S(k) = \frac{\overline{\mathscr{F}}(-k)}{\overline{\mathscr{F}}(k)} = \frac{k^2 - k_0^2 + \int_0^\infty dr \int_0^\infty dr'\, \lambda(r)\lambda(r') e^{-ikr_>} \sin kr_</k}{k^2 - k_0^2 + \int_0^\infty dr \int_0^\infty dr'\, \lambda(r)\lambda(r') e^{ikr_>} \sin kr_</k}.$$

For $k = i\varkappa$, $\varkappa > 0$,

$$\overline{\mathscr{F}}(i\varkappa) = -\varkappa^2 - k_0^2 + \int_0^\infty dr \int_0^\infty dr'\, \lambda(r)\lambda(r') e^{-\varkappa r_>} \sinh \varkappa r_</\varkappa.$$

The $\varkappa$ derivative of the integral kernel $e^{-\varkappa r_>} \sinh \varkappa r_</\varkappa$ is negative definite. Consequently, $\overline{\mathscr{F}}(i\varkappa)$ is a monotonely decreasing function of $\varkappa$. It therefore has at most one zero. If

$$\overline{\mathscr{F}}(0) = -k_0^2 + 2\int_0^\infty dr \int_0^r dr'\, \lambda(r)\lambda(r') r' > 0,$$

i.e., if

$$E_0 < 4m \int_0^\infty dr\, r \int_0^r dr'\, r'^2\, \Lambda(r)\, \Lambda(r')$$

then there is one zero and hence a bound state. Otherwise there is no bound state.

In the "local" case, in which the form factor $\Lambda(r)$ is concentrated near the origin, we may assume that

$$\lambda(r) = \begin{cases} \lambda & r < R \\ 0 & r > R \end{cases}$$

and get for $kR \ll 1$

$$\overline{\mathscr{F}}(k) = k^2 - k_0^2 + \tfrac{1}{3}\lambda^2 R^3 + \tfrac{1}{4} i\lambda^2 k R^4 + \cdots.$$

We now let R approach zero and λ infinity, in such a way that $\lambda R^2 = C =$ const. Then

$$\overline{\mathscr{F}}(k) = k^2 - k_0'^2 + i\tfrac{1}{4}kC^2$$

where

$$k_0'^2 = k_0^2 - \frac{\frac{1}{3}C^2}{R}.$$

In the strictly local limit, of course, $k_0'^2 = -\infty$.

In the absence of coupling $\overline{\mathscr{F}}$ has a zero at $k = k_0$, which represents the fact that the "stationary" particle χ has the energy E_0. The coupling to the ψ particle, which can be emitted and reabsorbed, produces a *shift* $-C^2/6mR$ in the energy of the χ to the value E_0'. Its energy has been "renormalized." If

$$E_0' < 0$$

then there is a bound state, i.e., the "clothed" χ particle is stable. But if

$$E_0' > \frac{C^4}{128m}$$

then the χ particle is unstable and there is a *resonance* pole of the **S** matrix at

$$E = E_0' - \frac{C^4}{64m} - i\frac{C^2}{8m}\left(2mE_0' - \frac{C^4}{64}\right)^{1/2}.$$

If $0 < E_0' < C^4/128m$ then there are two virtual states.

17.4 The Three-Body Problem

17.4.1 Failure of the Multichannel Method and of the Lippmann–Schwinger Equation. It is, in principle, possible to use a canonical parametrization in terms of angular momenta (see Sec. 17.4.6) to subject the three-body Schrödinger equation

$$(H_0 + V_{12} + V_{13} + V_{23})\psi = E\psi \tag{17.120}$$

to a multichannel treatment as in Secs. 16.4.1 and 17.1, particularly subsection 17.1.1. The only difference between the procedure there and a complete description is that the equations ought not to be truncated. They then include a continuous range of internal energies of the (2,3) pair, as discussed in Sec. 17.2.1. We are consequently dealing with infinitely many coupled integrodifferential equations or, after the introduction of the Green's functions, with infinitely many coupled integral equations in the two variables, distance and internal energy. They are written out in (17.8) and (17.9) without explicit indication of the energy "matrix" variable.

The important question to be asked now is whether the kernel of these equations is in the Hilbert–Schmidt class. Before looking at this, let us return for a moment and examine the original kernel of the full Lippmann–

Schwinger equation, without partial wave analysis. I invite you to recall the discussion of Sec. 10.2. The fact that the binary potentials of interest conserve the total momentum of two interacting particles leads to the appearance of a δ function in the kernel of the Lippmann–Schwinger equation for two particles.[20] This difficulty was seen to be easily removable by factoring out an over-all momentum-conserving δ function. The remaining equation then looked like that for a single particle scattered by a fixed target. In the three- (or more) body problem there is an analogous difficulty owing to the fact that each binary potential in (17.120) conserves separately the momenta of one particle and of the other two combined. For example, V_{23} conserves both $\mathbf{q}_1$ and $\mathbf{k}_1$ [see (17.145)] separately and, in the corresponding momentum representation, contains a product of the two δ functions which express this fact. The other potentials do not contain these δ functions; hence they cannot be factored out and removed as can the *total* momentum-conserving one. As a result, the trace of the kernel multiplied by its conjugate is infinite, and it is not in the Hilbert–Schmidt class.[21] This time the trouble cannot be removed by simple means.

We return now to the kernel of Eqs. (17.8) and (17.9), which we write out more explicitly as

$$\mathfrak{K}_{\lambda' l' E'_{23}, \lambda l E_{23}}(E;r,r') = \begin{cases} g_{\lambda'}(E - E'_{23};r,r')\,\overline{V}_{\lambda' l' E'_{23}, \lambda l E_{23}}(r') & r' < r \\ 0 & r' > r. \end{cases}$$

The notation is adapted to the present case, but apart from that, $\overline{V}$ is given by (16.66) with the potential $V_{12} + V_{13}$ in place of H_a'. The total angular momentum is fixed and not especially indicated. Let us then compute

$$\begin{aligned}
\sum_{\lambda l} \int dE_{23} \int_0^\infty dr\,(\mathfrak{K}\mathfrak{K}^\dagger)_{\lambda l E_{23}, \lambda l E_{23}}(E;r,r) &= \sum_{\lambda l} \int dE_{23} \int_0^\infty dr \int_0^r dr'\, |g_\lambda(E - E_{23};r,r')|^2 \\
&\quad \sum_{\lambda' l'} \int dE'_{23}\, \overline{V}_{\lambda l E_{23}, \lambda' l' E'_{23}}(r')\, \overline{V}_{\lambda' l' E'_{23}, \lambda l E_{23}}(r') \\
&= \sum_{\lambda l} \int dE_{23} \int_0^\infty dr \int_0^r dr'\, |g_\lambda(E - E_{23},r,r')|^2\, [\overline{V}^2]_{\lambda l E_{23}, \lambda l E_{23}}(r').
\end{aligned}$$

(Even though the sum over λ' and l' and the integral over E_{23}' do not include a sum over j', this additional summation can be added without change because of angular-momentum conservation. The "intermediate" set is then complete.) The trace here differs from the corresponding one of the

20 See also the lemma near the end of Sec. 7.3.2.

21 If the system is described in the center-of-mass frame and the total momentum no longer appears as a dynamical variable, V_{23} still commutes with $\mathbf{p}_1$, etc. Hence, according to the lemma of Sec. 7.3.2 it is not compact. The total kernel is therefore a sum of noncompact operators, and thus not only is it generally not in the Hilbert–Schmidt class, but it is not compact.

Lippmann–Schwinger kernel in two respects: one is that j and its z component are fixed, and the other that the integral over r' runs up to r only. The last aspect is the one that essentially distinguishes the Volterra equations from those of Lippmann–Schwinger in the two-particle case. In the present instance it fails to save us. If the factor $|g_\lambda|^2$ were absent, the E_{23} integral would produce an infinity because

$$\int dE_{23}\,|\psi_l(E_{23},r)|^2 = \infty.$$

The presence of $|g_\lambda|^2$ cannot mitigate this because that factor increases exponentially for large E_{23} ($E - E_{23}$ is then negative).

We conclude that the integration over all intermediate energies makes the kernels of the nontruncated equations (17.8) and (17.9) fall outside the Hilbert–Schmidt class, even though in r they are of the Volterra type. Although this does not necessarily destroy their usefulness altogether, it makes conclusions based on them suspect and numerical calculations by their means unreliable. The specific unreliability may be seen directly by examining the Lippmann–Schwinger equation.

Let us write the hamiltonian of (17.120)

$$H = H_1 + H_1', \quad H_1 = H_0 + V_{23}, \quad H_1' = V_{12} + V_{13}, \tag{17.121}$$

and let

$$\mathscr{G}_1^+(E) = (E + i\varepsilon - H_1)^{-1}. \tag{17.122}$$

If we apply the equation

$$\hat{\Omega}_1^{(+)\dagger} = \mathbb{1} + i\int_{-\infty}^{0} dt e^{\varepsilon t} e^{iH_1 t} H_1' e^{-iHt}$$

with $\varepsilon > 0$, to a (nonnormalizable) state $\Psi(\alpha, E)$ we obtain the Lippmann–Schwinger equation

$$\Psi_1(\alpha, E) = \Psi(\alpha, E) - \mathscr{G}_1^+(E) H_1' \Psi(\alpha, E).$$

On the other hand, if we apply it to $\Psi(\beta, E)$, where β is in arrangement channel 2, then by (16.20),

$$0 = \Psi(\beta, E) - \mathscr{G}_1^+ H_1' \Psi(\beta, E).$$

Therefore a scattering state that corresponds to an initially bound pair 2 solves the *homogeneous* form of the Lippmann–Schwinger equation for a state that corresponds to an initially bound pair 1. Thus the solution of this equation is *not unique*. The effect of the disconnected-diagram difficulties, via the noncompactness of the kernel, is that, whereas the Lippmann–Schwinger equation is correct, it does not uniquely define its solution.

Another way of understanding the origin of these problems is to think of the three-particle Schrödinger equation in the barycentric system as an equation like the one-particle Schrödinger equation but in six dimensions, with a potential that is the sum of the pair-potentials. This effective potential then has the characteristic property of not tending to zero in all directions. If $|\mathbf{r}_a - \mathbf{r}_b|$ stays fixed, for any a, b, $V_c(\mathbf{r}_a - \mathbf{r}_b)$ remains unchanged. As a consequence the three-particle Hamiltonian is a much "worse" operator than that for two.

17.4.2 The Faddeev Method. Let us adapt the method given in Eqs. (7.80) and (7.87) to the present case.

In the three-body problem we have to deal necessarily with rearrangements. Accordingly it is convenient to use the T operators discussed in Sec. 16.3.1 and whose equations are (16.43) and (16.44). We now adopt the following notation and conventions: Let the initial state consist of particle 1 impinging on a bound state of 2 and 3. The split-up of the hamiltonian corresponding to this arrangement is that given by (17.121). The Green's function of the channel hamiltonian H_1 is given by (17.122). The T operator appropriate to the description of collisions in which the final arrangement is again 1(2, 3) is called T_{11}. The **T** matrix is obtained from it by taking matrix elements between eigenstates of H_1 as in (7.76b). Its equation is

$$T_{11} = H_1' + H_1'\mathscr{G}_1^+ T_{11} = V_{12} + V_{13} + (V_{12} + V_{13})\mathscr{G}_1^+ T_{11}. \quad (17.123)$$

T operators appropriate to rearrangements in which the final state is 2(1, 3) or 3(1, 2) are called T_{21} and T_{31}, respectively. [We omit the underlining of an index used in (16.42) to (16.44). A specific off-the-energy-shell extension is adopted here. It is clear that the choice is not unique.] According to (16.44) they can be obtained from T_{11} by the equations

$$\begin{aligned} T_{21} &= H_2' + H_2'\mathscr{G}_1^+ T_{11} = V_{23} + V_{21} + (V_{23} + V_{21})\mathscr{G}_1^+ T_{11} \\ T_{31} &= H_3' + H_3'\mathscr{G}_1^+ T_{11} = V_{31} + V_{32} + (V_{31} + V_{32})\mathscr{G}_1^+ T_{11}. \end{aligned} \quad (17.124)$$

We now define three auxiliary T operators,

$$\mathfrak{T}_{11} \equiv V_{23}\mathscr{G}_1^+ T_{11} \qquad \mathfrak{T}_{21} \equiv V_{13} + V_{13}\mathscr{G}_1^+ T_{11} \qquad \mathfrak{T}_{31} \equiv V_{12} + V_{12}\mathscr{G}_1^+ T_{11} \quad (17.125)$$

in terms of which the original T's are expressed as

$$T_{11} = \mathfrak{T}_{21} + \mathfrak{T}_{31} \qquad T_{21} = V_{23} + \mathfrak{T}_{11} + \mathfrak{T}_{31} \qquad T_{31} = V_{23} + \mathfrak{T}_{11} + \mathfrak{T}_{21}. \quad (17.126)$$

Then there is the T operator needed for the calculation of amplitudes for processes in which three free particles emerge. For it we take

$$H_0' = V_{12} + V_{13} + V_{23}$$

and the operator T_{01} is given by

$$T_{01} = H_0' + H_0'\mathscr{G}_1^+ T_{11} = V_{23} + \mathfrak{T}_{11} + \mathfrak{T}_{21} + \mathfrak{T}_{31}. \quad (17.127)$$

Remember that the **T** matrix for a process $1(2,3) \to 2(1,3)$ is obtained by taking matrix elements of T_{21} between eigenstates of H_1 and H_2,

$$H_2 = H_0 + V_{13}$$

for $1(2,3) \to 123$, of T_{01} between eigenstates of H_1 and H_0, etc. All four of the needed operators, T_{11}, T_{21}, T_{31}, and T_{01}, are obtained from $\mathfrak{T}_{11}$, $\mathfrak{T}_{21}$, and $\mathfrak{T}_{31}$.

The derivation of equations for $\mathfrak{T}_{i1}$, $i = 1, 2, 3$, proceeds just as in (7.80) to (7.87), and the result is almost the same. If we call the individual two-body T operators T_i, that is,

$$T_1 = V_{23} + V_{23}\mathscr{G}_1^+ V_{23}$$

etc., then

$$\begin{aligned} \mathfrak{T}_{11} &= T_1 G^+(\mathfrak{T}_{21} + \mathfrak{T}_{31}) \\ \mathfrak{T}_{21} &= T_2 + T_2 G^+(\mathfrak{T}_{11} + \mathfrak{T}_{31}) \\ \mathfrak{T}_{31} &= T_3 + T_3 G^+(\mathfrak{T}_{11} + \mathfrak{T}_{21}) \end{aligned} \tag{17.128}$$

or in matrix form,

$$\begin{pmatrix} \mathfrak{T}_{11} \\ \mathfrak{T}_{21} \\ \mathfrak{T}_{31} \end{pmatrix} = \begin{pmatrix} 0 \\ T_2 \\ T_3 \end{pmatrix} + \begin{pmatrix} 0 & T_1 & T_1 \\ T_2 & 0 & T_2 \\ T_3 & T_3 & 0 \end{pmatrix} G^+ \begin{pmatrix} \mathfrak{T}_{11} \\ \mathfrak{T}_{21} \\ \mathfrak{T}_{31} \end{pmatrix}. \tag{17.128a}$$

G^+ is the free Green's function, $G^+(E) = (E + i\varepsilon - H_0)^{-1}$. These equations differ from (7.87) only in the zero in the inhomogeneity. It should be noted, however, that, in contrast to the T of (7.83), the present operators are directly useful for the calculation of rearrangement cross sections.

In order to describe *all* possible such rearrangements, we need also the T operators for 2(13) and 3(12) incidence. They are obtained just as (17.128a). All of them can then be arranged in square matrix form, with $\mathfrak{T}$ standing for the square matrix $\mathfrak{T}_{ij}$. If T_{ij} describes the process $i(\cdot\cdot) \to j(\cdot\cdot)$, then, for $i \neq j$, $T_{ij} = \mathfrak{T}_{lj} + \mathfrak{T}_{mj} + V_{kn}$, with $l \neq m \neq i$, $k \neq n \neq j$, and $T_{ii} = \mathfrak{T}_{li} + \mathfrak{T}_{mi}$. The equation for the matrix $\mathfrak{T}$ reads

$$\mathfrak{T} = \mathfrak{B} + \mathfrak{B}G^+\mathfrak{T} \tag{17.129}$$

with

$$\mathfrak{B} = \begin{pmatrix} 0 & T_1 & T_1 \\ T_2 & 0 & T_2 \\ T_3 & T_3 & 0 \end{pmatrix}.$$

The great virtue of Eq. (17.128a) or (17.129) lies in the presence of the zeros on the main diagonal of the kernel matrix $\mathfrak{B}$. As a result, iterations of the equation expressed in diagrammatic form contain no disconnected graphs (except in the first term). This is visible in a single iteration, which yields

$$\mathfrak{K}^2 = \mathfrak{B}G^+\mathfrak{B}G^+ = \begin{pmatrix} T_1 G^+(T_2 + T_3) & T_1 G^+ T_3 & T_1 G^+ T_2 \\ T_2 G^+ T_3 & T_2 G^+(T_1 + T_3) & T_2 G^+ T_1 \\ T_3 G^+ T_2 & T_3 G^+ T_1 & T_3 G^+(T_1 + T_2) \end{pmatrix} G^+.$$

The only δ function contained in $\mathfrak{K}^2$ is that of overall momentum conservation. There are no dangerous partial momentum-conserving terms. As a consequence (17.128a) has very much better properties than the Lippmann–Schwinger equations or those of Sec. 17.4.1.

The Faddeev method works equally well in the presence of an additional intrinsic three-particle term in the hamiltonian, call it V_{123}. We assume, of course, that V_{123} contains no δ functions, i.e., that it is a well-behaved operator on the space of three-particle states. As a consequence we may define a three-body T operator called T_{123} by the Lippmann–Schwinger equation

$$T_{123} = V_{123} + V_{123}G^+T_{123}$$

and encounter no trouble in its solution. In the presence of V_{123} *only*, there are no disconnected diagrams.

The term V_{123} will occur in Eqs. (17.123) to (17.125) in addition to the other potentials. We must then define a fourth auxiliary T operator,

$$\mathfrak{T}_{01} \equiv V_{123} + V_{123}\mathscr{G}_1^+T_{11} \tag{17.125a}$$

which appears on the right-hand sides of all four equations (17.126) and (17.127) as an additive term. As a result one obtains *four* coupled equations in place of the three in (17.128). Equation (17.129) becomes a 4×4 matrix equation, with $\mathfrak{T}$ defined by $\mathfrak{T}_{ij}$, i, $j = 0, 1, 2, 3$, and

$$\mathfrak{B} \equiv \begin{pmatrix} 0 & T_{123} & T_{123} & T_{123} \\ T_1 & 0 & T_1 & T_1 \\ T_2 & T_2 & 0 & T_2 \\ T_3 & T_3 & T_3 & 0 \end{pmatrix}.$$

Otherwise (17.129) is unchanged, and so are its connectedness properties. For simplicity we shall assume from now on that $V_{123} \equiv 0$.

Actual calculations of three-body amplitudes, of course, have to be performed by replacing the operator equation (17.129) by an integral equation. The operator T must be replaced by an (off-the-energy-shell) **T** matrix.

It should be noted that the procedure of Faddeev is independent of whether any or all pairs of the three particles have bound states. In other words the labelling used in it does not refer to arrangement channels as defined in Sec. 16.2, but to a *partition* of the three particles into clusters, irrespective of the existence of bound states. The same will be true of other methods to be discussed.

17.4.3 Other Methods. The Faddeev method is not the only procedure for arriving at equations with connected kernels. In fact it is only one of a large class of procedures with this property.

The Channel Coupling Array Method. Let us label the three two-cluster partitions by latin indices, with the hamiltonian as in (17.121), and let W be any 3×3 matrix such that

$$\sum_b W_{ab} = 1. \tag{17.130}$$

Then we have

$$\begin{aligned}\mathcal{G}^+ &= \mathcal{G}_c^+ + \mathcal{G}_c^+ H_c' \mathcal{G}^+ \\ &= \sum_d W_{cd} \mathcal{G}_d^+ (\mathbb{1} + H_d' \mathcal{G}^+),\end{aligned}$$

and hence by (16.42a)

$$\mathcal{G}^+ H_a' = \sum_d W_{cd} \mathcal{G}_d^+ T_{d\underline{a}}, \tag{17.131}$$

where c is quite arbitrary. Inserting this in (16.42a) we get

$$T_{b\underline{a}} = H_a' + H_b' \sum_d W_{cd} \mathcal{G}_d^+ T_{d\underline{a}}. \tag{17.132}$$

With the notation

$$H_1' = V_2 + V_3 = \sum_{e \neq 1} V_e \tag{17.133}$$

this may also be written

$$T_{b\underline{a}} = H_a' + \sum_{e \neq b} \sum_d V_e W_{cd} \mathcal{G}_d^+ T_{d\underline{a}}. \tag{17.132a}$$

Since the array W_{cd} is arbitrary, subject only to (17.130), this represents a large number of equations for the operators $T_{b\underline{a}}$. Some of them have connected kernels, after a number of iterations, and some do not.

We may, for example, choose $c = e$ and let W be the unit matrix, $W_{ed} = \delta_{ed}$. The result is

$$\begin{aligned}T_{b\underline{a}} &= H_a' + \sum_{c \neq b} V_c \mathcal{G}_c^+ T_{c\underline{a}} \\ &= H_a' + \sum_{c \neq b} T_c G^+ T_{c\underline{a}}.\end{aligned} \tag{17.134}$$

Comparison with (17.128) shows that, apart from the inhomogeneity, these equations have a structure somewhat similar to the Faddeev equations. However, they are not equations for auxiliary T operators, but for the operators $T_{b\underline{a}}$ themselves. One iteration leads to

$$T_{b\underline{a}} = H_a' + \sum_{c \neq b} T_c G^+ H_a' + \sum_{\substack{c \neq b \\ d \neq c}} T_c G^+ T_d G^+ T_{d\underline{a}}, \tag{17.134a}$$

whose kernel is connected.

Another fruitful choice of W is to set $W_{12} = W_{23} = W_{31} = 1$ and all other elements equal to zero. This is called a *channel permuting array*. With the further choice of $c = b$, this leads to

$$\begin{aligned}T_{1\underline{a}} &= H_a' + H_1' \mathcal{G}_2^+ T_{2\underline{a}}, \\ T_{2\underline{a}} &= H_a' + H_2' \mathcal{G}_3^+ T_{3\underline{a}}, \\ T_{3\underline{a}} &= H_a' + H_3' \mathcal{G}_1^+ T_{1\underline{a}}.\end{aligned} \tag{17.135}$$

One straight-forward iteration does not suffice to arrive at a connected kernel in this case. However, two iterations not only lead to a connected kernel but to a complete decoupling of the three equations:

$$T_{1\underline{a}} = H_a' + H_1'\mathscr{G}_2^+ H_a' + H_1'\mathscr{G}_2^+ H_2'\mathscr{G}_3^+ H_a' + H_1'\mathscr{G}_2^+ H_2'\mathscr{G}_3^+ H_3'\mathscr{G}_1^+ T_{1\underline{a}}, \tag{17.135a}$$

and similarly for the other two.

On the other hand, if we write explicitly $H_1' = V_2 + V_3$ and substitute the second of Eqs. (17.135) in the first term and the third in the second, we obtain equations that are connected but not uncoupled:

$$T_{1\underline{a}} = H_a' + H_1'\mathscr{G}_2^+ H_a' + V_2\mathscr{G}_2^+ H_2'\mathscr{G}_3^+ T_{3\underline{a}} + V_3\mathscr{G}_2^+ H_3'\mathscr{G}_1^+ T_{1\underline{a}}, \tag{17.135b}$$

and analogously for the other two.

The channel permuting array is also easily used for the construction of an equation for the state vector. Setting $a = 1$ and using (16.35) we obtain

$$\Psi^{(+)} = \Psi_1 + \sum_b W_{cb}\mathscr{G}_b^+(H_1' - H_b')\Psi_1 + \sum_b W_{cb}\mathscr{G}_b^+ H_b'\Psi^{(+)}.$$

For $c = 1, 2$, this leads to two equations,

$$\begin{aligned}\Psi^{(+)} &= \Psi_1 + \mathscr{G}_2^+(V_2 - V_1)\Psi_1 + \mathscr{G}_2^+(V_1 + V_3)\Psi^{(+)}\\ &= \Psi_1 + \mathscr{G}_2^+(V_3 - V_1)\Psi_1 + \mathscr{G}_3^+(V_1 + V_2)\Psi^{(+)}.\end{aligned}$$

We insert the second into the first where V_3 multiplies $\Psi^{(+)}$, and the first of (16.35) in the other, and obtain

$$\begin{aligned}\Psi^{(+)} = \Psi_1 &+ \mathscr{G}_2^+(V_2 + V_3)\Psi_1 + \mathscr{G}_2^+ V_1\mathscr{G}_1^+(V_2 + V_3)\Psi_1\\ &+ \mathscr{G}_2^+[V_1\mathscr{G}_1^+(V_2 + V_3) + V_3\mathscr{G}_3^+(V_1 + V_2)]\Psi^{(+)},\end{aligned} \tag{17.136}$$

which is an equation with a connected kernel.

There are, of course, many possible variations on the two examples given, and similar techniques may be employed for $T_{\underline{b}a}$.

The Weinberg–van Winter Method. In this procedure the full Green's function is written in the form

$$\mathscr{G}^+ = G^+ + \sum_{a=1}^{3} L_a + C,$$

where

$$L_a = \mathscr{G}_a^+ - G^+ = G^+ T_a G^+.$$

One finds by straight-forward operator manipulations that

$$C = I\mathscr{G}^+,$$

where

$$I = \sum_a L_a H_a' = \sum_a G^+ T_a G^+ H_a', \tag{17.137}$$

which is connected. Therefore the equation

$$\mathscr{G}^+ = G^+ + G^+ \sum_a T_a G^+ + I\mathscr{G}^+ \tag{17.138}$$

has a connected kernel. The construction of $\mathscr{G}^+$ then leads both to the wave functions, by (16.35), and the T operators, by (16.42) or (16.42a).

Rosenberg's Method. Another "algebraic" procedure is the following. Let us write

$$U_a = G^+ V_a,$$

so that

$$\mathscr{G}^+ = \left(\mathbb{1} - \sum_{a=1}^{3} U_a\right)^{-1} G^+.$$

Then one finds that

$$\mathbb{1} - \sum_{a=1}^{3} U_a = (\mathbb{1} - U_1)(\mathbb{1} - W_{123})(\mathbb{1} - U_2)(\mathbb{1} - U_3), \tag{17.139}$$

where

$$\begin{aligned} W_{123} &\\ &= (\mathbb{1} - U_1)^{-1}(U_1 U_2 + U_2 U_3 + U_1 U_3 - U_1 U_2 U_3)(\mathbb{1} - U_3)^{-1}(\mathbb{1} - U_2)^{-1} \\ &= \mathscr{G}_1^+ V_1 \mathscr{G}_2^+ V_2 + \mathscr{G}_1^+(V_1 + V_2)\mathscr{G}_3^+ V_3 \mathscr{G}_2^+ G^{-1}. \end{aligned} \tag{17.140}$$

The operators $(\mathbb{1} - U_a)^{-1}$ are essentially two-particle Green's function, since $\mathscr{G}_a{}^+ = (\mathbb{1} - U_a)^{-1} G^+$. Thus the "heart" of the three-particle Green's function $\mathscr{G}^+$ is $(\mathbb{1} - W_{123})^{-1}$, and W_{123} is connected.

17.4.4 Fredholm Properties and Spurious Solutions. In Sections 17.4.2 and 17.4.3 we have constructed a variety of equations that can be expected to have better mathematical properties than the Lippmann–Schwinger equation because their kernels are connected. The question now is whether, in fact, they do have better properties. What is really wanted is integral equations whose kernels are square integrable or at least compact, so that the Fredholm alternative applies to their solution and matrix methods can be used in numerical procedures to solve them.

So long as the energy is taken to be a complex number on the physical sheet it is a relatively straight forward matter to prove such properties. However, physical energies are real, and for real energies, right in the continuous spectrum of the hamiltonian, such proofs are very much more difficult. Faddeev,[22] working in momentum space, proved that under certain specified conditions on the potentials the fifth iterate of the kernel of his equations is compact (for real E) as an operator on a suitable Banach space, provided that there are no two-body bound or half-bound states

22 L. D. Faddeev (1960–1963), also C. Lovelace (1964a and b).

of zero energy. If the potentials are superpositions of Yukawa potentials then Fredholm methods have been proved[23] to be applicable to the solution of the Faddeev equations.

The construction of the Green's function via (17.141) has been studied in more detail. The method employed is a generalization of that of Meetz. No matter how rapidly V_1 and V_2 decrease at infinity, an operator such as $G^+ V_1 G^+ V_2$ is not in the Hilbert–Schmidt class. Instead one has to consider[24] $|V_2|^{1/2} G^+ V_1 G^+ |V_2|^{1/2} V_2$, and similarly for the more general operators that occur in (17.140), in (17.137), and in the Faddeev equations. By detailed estimates of the integrals involved it has been proved[25] that if all three of the pair-wise potentials are such that for some C and $\varepsilon > 0$, and all $\mathbf{r}$,

$$|V_a(\mathbf{r})| \leq Mr^{-3-\varepsilon} \tag{17.141}$$

then the Fredholm theory for Hilbert–Schmidt operators is applicable to the construction of $(\mathbb{1} - W_{123})^{-1}$.

In particular, the modified Fredholm determinant

$$d_{123} = \det_2(\mathbb{1} - W_{123}) \tag{17.142}$$

exists as an absolutely convergent power series in λ of $\det_2(\mathbb{1} - \lambda W_{123})$ for all λ. Although d_{123} is not symmetric in its three indices, the function

$$\Delta = d_{abc} \exp \operatorname{tr} \{[\mathscr{G}_a^+ V_a, \mathscr{G}_b^+ V_b] G^+ V_b \mathscr{G}_c^+ V_c\} \tag{17.143}$$

is symmetric,[26] which implies that the zeroes of the six functions obtained from d_{123} by permutations of its indices are the same. It has also been proved[27] that the multiplicity of any zero of $\Delta(E)$ on the physical sheet is equal to the degeneracy of the corresponding three-particle bound state.

The function d_{123} is also related to the Fredholm determinant of the Faddeev equations. The Fredholm determinant $\det(\mathbb{1} - \mathfrak{B} G^+)$ of (17.129) has been shown[27] to differs from Δ by a factor that is an analytic function of E holomorphic on the physical sheet and without zeroes there.

Spurious Solutions. Once the relevant integral equations have been demonstrated to have compact kernels the Fredholm alternative applies to them. The crucial question then is whether the homogeneous form of the equation has a nontrivial solution. If so, the further question is whether the existence of such a solution has physical significance or is an artifact of the formalism.

In the Rosenberg method the answer to this question is easily given. Equations (17.138) to (17.140) show that if for some value of E $W_{123}(E)$ has the eigenvalue 1 then $\mathscr{G}^+$ has a pole, i.e., the homogeneous equation

$$\xi = \mathscr{G}\xi$$

23 Rubin, Sugar, and Tiktopoulos (1967).

24 The notation is the same as in (10.62).

25 R. G. Newton (1971a).

26 R. G. Newton (1971b). The bracket [,] in (17.143) denotes the commutator.

27 R. G. Newton (1974).

has a nontrivial solution, which means E is a bound-state eigenvalue. No pole of $(\mathbb{1} - W_{123})^{-1}$ could be removed by the factors in (17.139) since the resolvents $\mathcal{G}_a(E)$ or their boundary values $\mathcal{G}_a^+$ cannot have nontrivial null-spaces. Conversely, a pole of $\mathcal{G}(E)$ on the physical sheet must be a pole of $(\mathbb{1} - W_{123})^{-1}$ since there can be no bound states of the three-particle system in which only one pair interacts. [So there can be no poles of $(\mathbb{1} - U_a)^{-1}$.] Thus there is a one-to-one correspondence between the poles of $\mathcal{G}$ and the poles of $(\mathbb{1} - W_{123})^{-1}$.

Similarly for the Faddeev equations. Nontrivial solutions of (17.129) exist if and only if $\det_2(\mathbb{1} - \mathfrak{B}) = 0$, which is equivalent to $\det_2(\mathbb{1} - W_{123}) = 0$ and hence a bound state.

The same is not true for the Weinberg-van Winter equations. One easily verifies that the kernel I of (17.137) is such that

$$\mathbb{1} - I = \left(\mathbb{1} + G^+ \sum_{a=1}^{3} T_a\right)\left(\mathbb{1} - \sum_{a=1}^{3} U_a\right).$$

Therefore $(\mathbb{1} - I)^{-1}$ can be expected to have singularities not only when $\left(\mathbb{1} - \sum_a U_a\right)^{-1}$ has singularities, and hence $\mathcal{G}$, but also when $\left(\mathbb{1} + \sum_a \mathcal{G}_a V_a\right)^{-1}$ has such. In other words the homogeneous form of (17.138) may have nontrivial "spurious" solutions that do not signify bound states and have no physical meaning. This is refered to as the "Federbush disease" of these equations. Similar flaws are attached to some of the equations obtained by the channel-coupling array method.[28] If these spurious solutions occur at unphysical energies they are relatively harmless, but if they occur in the continuous spectrum they may be serious obstacles to numerical solution of the scattering equations.

17.4.5 The Asymptotic Form of Three-Particle Wave Functions. We now want to examine the behavior of three-particle wave functions in the coordinate representation. To this end we introduce the so-called *Jacobi coordinates* in the barycentric system. As the space is six-dimensional (nine for the three particles, minus three for the fixed center of mass) we use the relative coordinates of one pair, and the vector leading from their center of mass to the position of the third particle, as the two three-vectors to describe the position of the system. Let the three particles have masses m_1, m_2, m_3, let their positions be $\mathbf{R}_1, \mathbf{R}_2, \mathbf{R}_3$, and their momenta $\mathbf{p}_1, \mathbf{p}_2, \mathbf{p}_3$, with $\Sigma\mathbf{p}_i = \Sigma m_i\mathbf{R}_i = 0$. We define

$$\mu_1 = m_2 m_3/(m_2 + m_3), \qquad \bar{\mu}_1 = m_1(m_2 + m_3)/M, \qquad M = m_1 + m_2 + m_3 \tag{17.144}$$

$$\mathbf{r}_1 = (2\mu_1)^{1/2}(\mathbf{R}_2 - \mathbf{R}_3), \quad \rho_1 = (2/\bar{\mu}_1)^{1/2}(m_2\mathbf{R}_2 + m_3\mathbf{R}_3) = m_1(2/\bar{\mu}_1)^{1/2}\mathbf{R}_1$$
$$\mathbf{k}_1 = (2\bar{\mu}_1)^{-1/2}\mathbf{p}_1, \qquad \mathbf{q}_1 = (\mu_1/2)^{1/2}\left(\frac{\mathbf{p}_2}{m_2} - \frac{\mathbf{p}_3}{m_3}\right), \tag{17.145}$$

28 C. Chandler (1978), C. Chandler and I. H. Sloan (1980).

and their cyclic permutations. In these variables the kinetic energy operator is given by

$$H_0 = -\Delta_{\mathbf{r}_l} - \Delta_{\boldsymbol{\rho}_l}, \qquad l = 1, 2, 3. \tag{17.146}$$

A change from a description in terms of $\mathbf{r}_i$, $\boldsymbol{\rho}_i$ to one in terms of $\mathbf{r}_j$, $\boldsymbol{\rho}_j$ is accomplished by the rotation

$$\mathbf{r}_j = a_{ji}\mathbf{r}_i + b_{ji}\boldsymbol{\rho}_i, \qquad \boldsymbol{\rho}_j = -b_{ji}\mathbf{r}_i + a_{ji}\boldsymbol{\rho}_i \tag{17.147}$$

where

$$\begin{aligned} a_{ji} = a_{ij} &= -(\mu_i\mu_j)^{1/2}/m_k, \qquad k \neq i \neq j, \\ -b_{ij} = b_{ji} &= (\mu_j/\bar{\mu}_i)^{1/2}, \qquad i, j = 1, 2; 2, 3; 3, 1, \end{aligned} \tag{17.148}$$

so that

$$a_{ij}^2 + b_{ij}^2 = 1.$$

Similarly for the momenta

$$\mathbf{q}_j = a_{ji}\mathbf{q}_i + b_{ji}\mathbf{k}_i, \qquad \mathbf{k}_j = -b_{ji}\mathbf{q}_i + a_{ji}\mathbf{k}_i. \tag{17.149}$$

The six-dimensional coordinate vector will be denoted by $\mathbf{R}$ or $(\boldsymbol{\rho}_i, \mathbf{r}_i)$, and the six-dimensional momentum vector by $\mathbf{K}$ or $(\mathbf{k}_i, \mathbf{q}_i)$. Then

$$\begin{aligned} d^6\mathbf{R} &= d^3\boldsymbol{\rho}_i d^3\mathbf{r}_i, \qquad i = 1, 2, 3, \\ R^2 &= |\mathbf{R}|^2 = r_i^2 + \rho_i^2 = 2\sum_{l=1}^{3} m_l \mathbf{R}_l^2 \\ E = K^2 &= |\mathbf{K}|^2 = k_i^2 + q_i^2, \qquad i = 1, 2, 3 \\ \mathbf{K}\cdot\mathbf{R} &= \mathbf{k}_i\cdot\boldsymbol{\rho}_i + \mathbf{q}_i\cdot\mathbf{r}_i = \sum_{l=1}^{3} \mathbf{p}_l\cdot\mathbf{R}_l. \end{aligned}$$

The Green's Function. There are two different asymptotic limits of the Green's function $\mathscr{G}_a^+$ that are of interest. In the first R is allowed to grow to infinity by letting both r_a and ρ_a grow in such a way that their ratio is kept fixed. If we define $\mathbf{k}_a'$ and $\mathbf{q}_a'$ by

$$\mathbf{k}_a' = K\boldsymbol{\rho}_a/R, \qquad \mathbf{q}_a' = K\mathbf{r}_a/R$$

while $R \to \infty$ then $\mathbf{k}_a'$ and $\mathbf{q}_a'$ are the asymptotic momenta appropriate to observations at (large) $\boldsymbol{\rho}_a$ and $\mathbf{r}_a$. One then finds[29] that

$$\mathscr{G}_a^+(E;\mathbf{R},\mathbf{R}') = \frac{(1+i)K^{3/2}}{16\pi^{5/2}R^{5/2}} e^{iKR}\psi_a^{(-)*}(\mathbf{K}',\mathbf{R}') + o(R^{-5/2}), \tag{17.150}$$

where $K' = K$,

$$\psi_a^{(-)}(\mathbf{K},\mathbf{R}) = \exp(i\boldsymbol{\rho}_a\cdot\mathbf{k}_a)\varphi_a^{(-)}(\mathbf{q}_a,\mathbf{r}_a),$$

and $\varphi_a^{(-)}$ is the scattering wave function in channel a (in which only the pair a interacts).

29 R. G. Newton (1972a).

The second kind of limit of interest is one in which $R \to \infty$ while $\mathbf{r}_a$ remains fixed. This means that the pair a remains together while the third particle separates from it. As might be expected under these circumstances the leading contributions do not come from the scattering states in channel a but from the bound states. One finds[29] that

$$\mathscr{G}_a^+(E;\mathbf{R},\mathbf{R}') = \frac{1}{4\pi(2\mu_a)^{3/2}\rho_a}\sum_n e^{ik'_{an}\rho_a}\varphi_{an}(\mathbf{r}_a)\psi_a^*(\mathbf{k}'_{an},E_n;\mathbf{R}') + 0(\rho_a^{-2}), \tag{17.151}$$

where $\mathbf{k}_{an}' = \hat{\boldsymbol{\rho}}_a k_{an}$ and

$$\psi_a(\mathbf{k}'_a,E_n;\mathbf{R}) = e^{i\mathbf{k}'_a\cdot\boldsymbol{\rho}_a}\varphi_{an}(\mathbf{r}_a),$$

φ_{an} being the nth bound-state wave function of pair a normalized as

$$(2\mu_a)^{-3/2}\int d^3\mathbf{r}_a|\varphi_{an}(\mathbf{r}_a)|^2 = 1,$$

with $E = E_n + k_{an}'^2$.

The Wave Function for Bound-State Scattering. These asymptotic forms of the Green's functions are used for the calculation of the asymptotic form of the wave function for the scattering of particle a on the pair $a = (b,c)$ $a \neq b \neq c$. From the Lippmann–Schwinger equation

$$\psi^{(+)} = \psi_a + \mathscr{G}_a^+ H'_a\psi^{(+)}$$

and the asymptotic form (17.151) one finds that as $R_a \to \infty$

$$\begin{aligned}\psi^{(+)}(\mathbf{k}_a,E_a;\mathbf{R}) &= e^{i\mathbf{k}_a\cdot\boldsymbol{\rho}_a}\varphi_a(E_a,\mathbf{r}_a)\\ &\quad + R_a^{-1}\sum_n e^{ik'_{an}\rho_a}\varphi_{an}(\mathbf{r}_a)A(k'_{an}\hat{\mathbf{R}}_a,E_{an};\mathbf{k}_a,E_a) + 0(R_a^{-2}),\end{aligned} \tag{17.152}$$

where the amplitude for elastic or inelastic scattering (non-rearrangement) is given by

$$A(\mathbf{k}'_{an},E_{an};\mathbf{k}_a,E_a) = A_{aa} = -\frac{\bar{\mu}_a^{1/2}}{16\pi m_a\mu_a^{3/2}}(\psi_a,T_{aa}\psi_a). \tag{17.153}$$

We have here not written out all the variables, T_{aa} is the T-operator (16.42), and $E_a + k_a{}^2 = E_{an} + k_{an}{}^2$.

Rearrangements are described by utilizing the fact noted in Sec. 17.4.1 that $\psi^{(+)}(\mathbf{k}_a,E_a;\mathbf{R})$ also solves the homogeneous equation

$$\psi^{(+)} = \mathscr{G}_b^+ H'_b\psi^{(+)}$$

and then letting $\rho_b \to \infty$. Using (17.151) again, one obtains

$$\psi^{(+)}(\mathbf{k}_a,E_a;\mathbf{R}) = R_b^{-1}\sum_n e^{ik_{an}\rho_b}\varphi_{bn}(\mathbf{r}_b)A_{ba} + 0(R_b^{-2}), \tag{17.152a}$$

where

$$A_{ba} = -\frac{\bar{\mu}_b^{1/2}}{16\pi m_b\mu_b^{3/2}}(\psi_b,T_{ba}\psi_a). \tag{17.153a}$$

The scattering cross sections are related to these amplitudes by[29]

$$\frac{d\sigma}{d\Omega} = \frac{k'_{bn}}{k_a} \cdot \frac{m_a \bar{\mu}_b^{1/2}}{m_b \bar{\mu}_a^{1/2}} |A_{ba}|^2, \tag{17.154}$$

which for $a = b$ gives the elastic or inelastic (nonrearrangement) cross section.

On the other hand, if we let $R \to \infty$, with $r_a/\rho_a =$ const., then (17.150) must be used and we obtain the asymptotic form

$$\psi^{(+)}(\mathbf{k}_a, E_a; \mathbf{R}) = R^{-5/2} e^{iKR} A_{oa}(\mathbf{K}'; \mathbf{k}_a, E_a) + o(R^{-5/2}) \tag{17.155}$$

where

$$A_{oa} = \frac{(1+i)K^{3/2}}{16\pi^{5/2}} (\psi_o, T_{oa}\psi_a). \tag{17.156}$$

If the break-up cross section is defined as the number of particles a emerging per unit time per unit solid angle in the direction $\hat{\mathbf{k}}_a'$, accompanied by particles b and c, $a \neq b \neq c$, for which $\mathbf{q}_a$ lies in the element $(d^3\mathbf{q}_a')$ around $\mathbf{q}_a'$, relative to the number of incident particles a per unit area and unit time, it is given by

$$d\sigma/d\Omega = \frac{\bar{\mu}_a k'_a}{2^{5/2} m_a^2 \mu_a^{3/2} k_a K^3} |A_{oa}|^2 (d^3\mathbf{q}'_a). \tag{17.157}$$

The Wave Function for Three Free Incident Particles. The asymptotic form of the wave function for three free incident particles, which solves the Lippmann–Schwinger equation

$$\psi^{(+)} = \psi_o + G^+ \sum_a V_a \psi^{(+)}$$

with $\psi_o = \exp(i\mathbf{K} \cdot \mathbf{R})$, is complicated. It consists of several different pieces that decrease at different rates and describe different physical processes, and its fall-off depends critically on the direction of $\mathbf{R}$ in the six-dimensional space.

Let us isolate certain specific parts of $\psi^{(+)}$ and write it in the form

$$\psi^{(+)} = \psi_o + \psi_P + \psi_C + \psi', \tag{17.158}$$

where

$$\begin{aligned} \psi_P &= \sum_a G^+ T_a \psi_o \\ \psi_C &= \sum_a \mathscr{G}_a^B V_a H'_a \psi^{(+)} \end{aligned} \tag{17.159}$$

and $\mathscr{G}_a{}^B$ is the contribution to $\mathscr{G}_a{}^+$ that comes from the bound states of the pair a,

$$\mathscr{G}_a^B(E; \mathbf{R}, \mathbf{R}') = -\frac{1}{4\pi(2\mu_a)^{3/2}} \sum_n \varphi_{an}(\mathbf{r}_a)\varphi_{an}^*(\mathbf{r}'_a) \frac{e^{ik_{an}|\boldsymbol{\rho}_a - \boldsymbol{\rho}'_a|}}{|\boldsymbol{\rho}_a - \boldsymbol{\rho}'_a|}.$$

A dominant term in $\psi^{(+)}$ is, of course, the plane wave ψ_o that describes the incident free particles, unscattered; it does not decrease as R increases in any direction. The term ψ_P describes the scattering of the pair a by itself without participation by particle a. If $R \to \infty$ while $\mathbf{r}_a$ is fixed and $\rho_a \to \infty$, the dominant behavior of ψ_P is $\exp(ik_a \cdot \boldsymbol{\rho}_a)\varphi_a^{(+)}(\mathbf{q}_a, \mathbf{r}_a) + 0(\rho_a^{-1})$. This single pair-scattering can be physically distinguished by the fact that the momentum of one of the particles is unchanged.

After ψ_P is subtracted the dominant term as $\rho_a \to \infty$, with $\mathbf{r}_a$ fixed, comes from ψ_C and one finds[29] that its leading behavior is

$$\psi_C = R_a^{-1} \sum_m e^{ik'_{an}\rho_a} \varphi_{an}(\mathbf{r}_a) A_{ao}(\mathbf{k}_{an}, E_{an}; \mathbf{K}) + 0(R_a^{-2}), \qquad (17.160)$$

where

$$A_{ao} = -\frac{\bar{\mu}_a^{1/2}}{16\pi m_a \mu_a^{3/2}} (\psi_a, T_{ao}\psi_o) \qquad (17.161)$$

is the *capture amplitude*. If the cross section for the formation of the nth bound state of the pair a is defined as the number of particles a emerging per unit solid angle per unit time, accompanied by the pairs a in a bound state of energy E_n, relative to the number of incident particles a per unit time per unit area, it is given by

$$\frac{d\sigma}{d\Omega} = \frac{k'_{an}}{k_a} |A_{ao}|^2. \qquad (17.162)$$

If $R \to \infty$ in any direction that allows all three r_a's to increase to infinity then ψ_C decreases exponentially, because of the bound-state wave functions.

The remaining portion ψ' of $\psi^{(+)}$ decreases asymptotically as R^{-2}, but it is not only the leading term that is of physical interest.

Consider two successive on-energy-shell collisions, first by the pair a and then by the pair b. If $\mathbf{K}''$ describes the momenta after the first collision then the equations for conservation of energy and momentum in both collisions are

$$\begin{aligned} \mathbf{k}''_a &= \mathbf{k}_a, & q''_a &= q_a, \\ \mathbf{k}'_b &= \mathbf{k}''_b, & q'_b &= q''_b. \end{aligned} \qquad (17.163)$$

If the initial momenta $\mathbf{K}$ are given then these equations determine the magnitudes of the final momenta $k_b' = \bar{k}_b$ and $q_b' = \bar{q}_b$. Suppose now that $R \to \infty$ in such a way that $\rho_b/r_b = \bar{k}_b/\bar{q}_b$. One then finds that[29]

$$\psi' = R_b^{-1} R_{ac}^{-1} \exp[i(\bar{q}_b r_b + \bar{k}_b \rho_b)] A^{(ba)} + 0(R^{-5/2}), \qquad (17.164)$$

where $a \neq b \neq c$, $R_{ac} = |\mathbf{r}_a - \mathbf{r}_c|$, and

$$A^{(ba)}(\mathbf{K}', \mathbf{K}) = \frac{\bar{\mu}_b^{3/2}}{2m_b \mu_a^{1/2}} \frac{\bar{k}_b^2}{|\mathbf{k}'_b \cdot \mathbf{q}''_a|} A_b(\mathbf{q}'_b, \mathbf{q}''_b) A_a(\mathbf{q}''_a, \mathbf{q}_a). \qquad (17.165)$$

It is evident that this term describes the *double scattering* in which the on-energy-shell collision of pair a is followed by an on-energy-shell collision of pair b. There are six possibilities of such double collisions, and they define six cones in $\mathbb{R}^6$ along which ψ' goes as R^{-2} as indicated by (17.164), with particle momenta appropriate to the double scattering.

If R is allowed to increase to infinity in other directions, i.e., so that ρ_b/r_b is not equal to any double scattering ratios $\bar{k}_b/\bar{q}_b$ then

$$\psi' = 2R_b^{-1}R_{ac}^{-1} \exp\left[i(\bar{q}_b\rho_b + \bar{k}_b r_b)\right]A^{(ba)} + 0(R^{-5/2}) \quad (17.164a)$$

so that, in fact the dominant behavior is as R^{-2} in all directions; but away from the double-scattering cones the momenta of the particles do not correspond to their arrival times from the center. This is caused by the fact that there may be a large distance between the first and the second collision, so that the particles that arrive at the counter with double-scattering momenta may have had their second collision anywhere. In that sense the effective range of the double-scattering is infinite, which is the reason for some of the anomalies it produces.

What is left after the subtraction of the double-scattering terms from ψ' goes like $R^{-5/2}$. There are two kinds of contributions to this, and they are experimentally indistinguishable. The first kind comes from off-energy-shell double scattering; each of its terms is proportional to the product of two off-shell amplitudes. These terms contain *poles* at those final values of q_a that correspond to double scattering. The second kind comes from what might be called "true" three-particle scattering. The sum then has the following characteristics.

Away from the double-scattering cones, and after subtraction of the double-scattering terms, the wave function goes as $R^{-5/2}$. However, near the double scattering cones the asymptotic behavior is *nonuniform* with respect to the distance from these surfaces. The detailed behavior of the wave function there has been shown[30] to be described by Fresnel functions. On the cones they go like R^{-2}, but if the $R^{-5/2}$-behavior is isolated then the nonuniformity expresses itself as an infinity as the cones are approached.

The effect of this anomaly is that, while it is possible[29] to derive an expression for the differential three-three scattering rate from the $R^{-5/2}$-term in the wave function, this rate cannot be integrated to obtain a total three-three rate; such an integral would diverge. Now it is physically clear, and it also follows from the fact that the corresponding term in the wave function goes as R^{-2}, that among all the emerging triples of particles that have interacted the vast majority has undergone double scattering rather than real three-three scattering. This double scattering rate is, in fact, of interest in the calculation of some chemical reaction rates. But so long as the total three-three rate cannot be calculated without getting the nonsensical result that it is infinite, no double-scattering rate can be reliably calculated either. Methods have, in fact, been found to calculate the total rates unambiguously, and finite results have been obtained by these more refined methods.[31]

30 S. P. Merkuriev (1971).

31 Newton and Shtokhamer (1976); Potapov and Taylor (1977).

17.4.6 Angular Momentum Couplings. If the interactions are invariant under rotations so that the total angular momentum of the three-particle system is conserved, then it is useful to introduce angular momenta explicitly into the equations.

Let us first combine the spins of the particles in a two-particle channel into a total spin s, those of particles 2 and 3 in a three-body channel into σ_1, and σ_1 with s_1 into a total spin s. Similarly we combine the orbital angular momenta λ_1 (of ρ_1) and l_1 (of $\mathbf{r}_1$) into a total L. Let us use the following notation:

	For a two-body channel	For a three-body channel
$\mathbf{P} =$	$\mathbf{p}$	$\mathbf{k}_1, \mathbf{q}_1$
$\Lambda =$	l	λ_1, l_1, L
$\Sigma =$	s_1, s_2, s	$s_1, s_2, s_3, \sigma_1, s$
$\mathbf{v} =$	v_1, v_2	v_1, v_2, v_3

To save ourselves some subscripts we shall fix our attention on a given 1(2,3) division of the three particles and then not write the subscript 1 on **k**, **q**, λ, l, and σ. The eigenvalue of the z component of the total angular momentum j will be called $\mathfrak{m}$, those of λ, l, and L will be called μ, m, and M, respectively, that of the z component of σ will be called ι, and that of the z component of the total spin s, v.

The angle functions for two-body channels are defined in (15.1) and (15.6), except that we now call them $\mathfrak{Y}_j^{\mathfrak{m}}(\Lambda s v;\mathbf{P})$ and $\mathscr{Y}_{j\Lambda\Sigma}{}^{\mathfrak{m}}(\mathbf{P})$. For three-body states they are given by

$$\mathfrak{Y}_j^{\mathfrak{m}}(\Lambda s v;\mathbf{P}) = \sum_M C(Lsj, Mv\mathfrak{m})\, Y_\Lambda^M(\mathbf{P}) i^{-\lambda-l} \tag{17.166}$$

$$\mathscr{Y}_{j\Lambda\Sigma}^{\mathfrak{m}}(\mathbf{P}) = \sum_v i^{l+\lambda} \mathfrak{Y}_j^M(\Lambda s v;\mathbf{P}) \chi_v^\Sigma \tag{17.167}$$

where

$$Y_\Lambda^M(\mathbf{P}) = \sum_{\mu m} C(\lambda l L, \mu m M)\, Y_\lambda^\mu(\hat{\mathbf{k}})\, Y_l^m(\hat{\mathbf{q}}) \tag{17.168}$$

$$\chi_v^\Sigma = \sum_{v_1 v_2 v_3 \iota} C(s_2 s_3 \sigma, v_2 v_3 \iota) C(s_1 \sigma s, v_1 \iota v) \chi_{v_1}^{s_1} \chi_{v_2}^{s_2} \chi_{v_3}^{s_3}. \tag{17.169}$$

In the initial state the spin of what we there regard as particle 2 is of course made up of spin and orbital angular momenta of the bound (2,3) system. This simply means that we use the function $\Phi_a(\varepsilon_\alpha, s_2 s_3 s v)$ of (16.50) for the spin function of that "particle," as in (16.63).

The angular-momentum expansion of the break-up amplitude now becomes

$$A(\mathbf{P}'\Sigma' v', \mathbf{P}\Sigma v) = \frac{2\pi}{ip} \sum_{j\mathfrak{m}\lambda' l' L' l} \mathfrak{Y}_j^{\mathfrak{m}}(\Lambda' s' v', \mathbf{P}') S^j_{\Lambda'\Sigma', \Lambda\Sigma}(E_{23}, E) \mathfrak{Y}_j^{\mathfrak{m}*}(\Lambda s v, \mathbf{P}) \tag{17.170}$$

which is the analog of (15.18). We have written here directly S^j instead of $S^j - \mathbb{1}$ because these are amplitudes for two initial and three final fragments.

It is more useful to label the cross sections by the individual spin components. We write

$$\mathfrak{Y}_j^{\mathfrak{m}}(\Lambda\Sigma\mathbf{v},\mathbf{P}) = \sum_{M\nu} i^{-\lambda-l} C(s_2 s_3 \sigma, \nu_2\nu_3\iota) C(s_1\sigma s, \nu_1\iota\nu) C(Lsj, M\nu\mathfrak{m}) Y_\Lambda^M(\mathbf{P})$$
$$= i^{-\lambda-l}(\chi_{\nu_1}^{s_1\dagger}\chi_{\nu_2}^{s_2\dagger}\chi_{\nu_3}^{s_3\dagger}) \cdot \mathscr{Y}_{j\Lambda\Sigma}^{\mathfrak{m}}(\mathbf{P}). \tag{17.171}$$

Then the alternative (15.38), (15.39) has the analog

$$A(\mathbf{P}'\mathbf{v}',\mathbf{P}\mathbf{v}) = \frac{2\pi}{ip}\sum \mathfrak{Y}_j^{\mathfrak{m}}(\Lambda'\Sigma'\mathbf{v}',\mathbf{P}') S^j_{\Lambda'\Sigma',\Lambda\Sigma}\,\mathfrak{Y}_j^{\mathfrak{m}*}(\Lambda\Sigma\mathbf{v},\mathbf{P})$$
$$= (\chi_{\nu_1}^{s_1}\chi_{\nu_2}^{s_2}\chi_{\nu_3}^{s_3})^\dagger \cdot \mathscr{A}(\mathbf{P}',\mathbf{P}) \cdot (\chi_{\nu_1}^{s_1}\chi_{\nu_2}^{s_2})$$

with

$$\mathscr{A}(\mathbf{P}',\mathbf{P}) = \frac{2\pi}{ip}\sum \mathscr{Y}_{j\Lambda'\Sigma'}^{\mathfrak{m}}(\mathbf{P}') i^{l-\lambda'-l'} S^j_{\Lambda'\Sigma',\Lambda\Sigma}\mathscr{Y}_{j\Lambda\Sigma}^{\mathfrak{m}\dagger}(\mathbf{P}). \tag{17.172}$$

In both equations the sums run over j, $\mathfrak{m}$, λ', l', L', l, σ', s', and s.

Helicity Amplitudes. We now wish to use helicities. For the initial state we therefore go back to (15.25). For the final state the functions referring to particle 1 are expressed in a coordinate system whose z axis lies along $\mathbf{k}$ and the functions referring to particles 2 and 3 in a system whose z axis lies along $\mathbf{q}$. The direction of $\mathbf{k}$ is that of the momentum of particle 1, and that of $\mathbf{q}$ is the direction of the momentum of particle 2 as seen *in the center-of-mass system of particles* 2 *and* 3. The helicities are consequently not all going to be expressed with respect to the same coordinate system. The spin z axis for particle 1 is its momentum in the overall center-of-mass coordinate frame; those for particles 2 and 3 are their momenta as seen in an inertial system that moves along with the (2, 3) center of mass.

We express $\mathscr{Y}$ accordingly. Let us call f the angular momentum made up of l and σ; $\bar{f}$, that of λ and s_1; and their z components, o and $\bar{o}$, respectively. Then

$$\mathscr{Y}_{j\Lambda\Sigma}^{\mathfrak{m}}(\mathbf{P}) = \sum_{f\bar{f}o\bar{o}} \Gamma_{j\Lambda\Sigma}^{\mathfrak{m}}(f\bar{f}o\bar{o})\mathscr{Y}_{\bar{f}\lambda s_1}^{\bar{o}}(\hat{\mathbf{k}})\mathscr{Y}_{fl\sigma}^{o}(\hat{\mathbf{q}}) \tag{17.173}$$

with the recoupling coefficients

$$\Gamma_{j\Lambda\Sigma}^{\mathfrak{m}}(f\bar{f}o\bar{o}) = \sum_{\mu m\nu_1\iota M\nu} C(\lambda lL, \mu mM) C(s_1\sigma s, \nu_1\iota\nu)$$
$$C(Lsj, M\nu\mathfrak{m}) C(\lambda s_1 \bar{f}, \mu\nu_1\bar{o}) C(l\sigma f, m\iota o) \tag{17.174}$$

and refer the two $\mathscr{Y}$'s to coordinate systems in which $\mathbf{k}$ and $\mathbf{q}$ lie along the z axis,

$$\mathscr{Y}_{j\Lambda\Sigma}^{\mathfrak{m}}(\mathbf{P}) = \frac{1}{4\pi}\sum B_{j\Lambda\Sigma}^{\mathfrak{m}}(f\bar{f}\nu_1'\nu_2'\nu_3'o\bar{o}) i^{\lambda+l}[(2\lambda+1)$$
$$(2l+1)]^{1/2}[\chi_{\nu_1'}^{s_1}]'[\chi_{\nu_2'}^{s_2}\chi_{\nu_3'}^{s_3}]'' D_{\nu_1'\bar{o}}^{\bar{f}}(\mathscr{R}_k) D_{\iota' o}^{f}(\mathscr{R}_q) \tag{17.175}$$

where

$$B_{j\Lambda\Sigma}^{\mathfrak{m}}(f\bar{f}\nu_1'\nu_2'\nu_3'o\bar{o}) = \Gamma_{j\Lambda\Sigma}^{\mathfrak{m}}(f\bar{f}o\bar{o}) C(\lambda s_1\bar{f}, 0\nu_1'\nu_1') C(l\sigma f, 0\iota'\iota') C(s_2 s_3\sigma, \nu_2'\nu_3'\iota').$$

The summation in (17.175) runs over $f\bar{f}\nu_1'\nu_2'\nu_3'o\bar{o}$, and $\iota' = \nu_2' + \nu_3'$. In order to get (17.175) we used (15.25).

Insertion in (17.172) gives us the helicity amplitudes. Because the spin projections of the second initial and the third final particle now are taken along momenta $-\mathbf{p}_2$ and $-\mathbf{p}_3'$, we must change the signs of ν_2 and of ν_3', though. One obtains

$$A^h(\mathbf{P}'\boldsymbol{\nu}', \mathbf{P}\boldsymbol{\nu}) = \sum_{jf\bar{f}o\bar{o}} (2j+1) D^{\bar{f}}_{\nu_1'\bar{o}}(\mathscr{R}_{k'}) D^{f}_{\iota' o}(\mathscr{R}_{q'}) a^j(f\bar{f}; \boldsymbol{\nu}'\boldsymbol{\nu}, o\bar{o}) \tag{17.176}$$

with the partial wave amplitudes

$$a^j(f\bar{f}; \boldsymbol{\nu}'\boldsymbol{\nu}, o\bar{o}) = \sum \frac{[(2j+1)(2\lambda+1)(2L+1)]^{1/2}}{2ip\sqrt{4\pi}(2j+1)} S^j_{\Lambda'\Sigma',\Lambda\Sigma} \\ B^{\nu}_{j\Lambda\Sigma}(f\bar{f}\nu_1'\nu_2' - \nu_3'o\bar{o}) C(lsj, 0\nu\nu) C(s_1 s_2 s, \nu_1 - \nu_2 \nu). \tag{17.177}$$

Here $\nu = \nu_1 - \nu_2$, $\iota' = \nu_2' - \nu_3'$, and the summation in (17.140) runs over $\lambda' l' L' l s' s \sigma'$. The symbols $\mathscr{R}_k'$ and $\mathscr{R}_q'$ stand for the Euler angles of the rotation from $\mathbf{k}'$ and $\mathbf{q}'$, respectively, to $\mathbf{p}$ [about the axes $\mathbf{k}' \times \mathbf{p}$ and $\mathbf{q}' \times \mathbf{p}$; for discussion see just before Eq. (15.24)]. We repeat that the helicities of particles 2 and 3 are measured in the (2, 3) center-of-mass frame and that of 1, in the overall barycentric system.

The foregoing parametrization can be used for the amplitude of any process that leads from two fragments (a particle and a bound state) to three. For calculations of the partial wave amplitudes, however, the asymmetry in the treatment of the three particles is a serious difficulty. The parameters adopted are well suited to the description of processes in which initially and finally particles 2 and 3 are bound, as well as those in which finally all three are free. But for the description of rearrangement collisions in which initially 2 and 3 are bound and finally, say, 1 and 2, they are not. And the calculations of these amplitudes are quite inseparable. One is free to write down amplitudes in which the meaning of the variables in the initial and final states is different, but, except perhaps in the Born approximation, it is difficult to perform calculations in this way.

Symmetric Parametrization. In the barycentric system the three momenta $\mathbf{p}_1$, $\mathbf{p}_2$, and $\mathbf{p}_3$ form a triangle whose shape is entirely determined by the three magnitudes p_1, p_2, and p_3. Its orientation in space can be fixed by attaching a coordinate system to it and giving the Euler angles ψ, θ, and φ which rotate this "body-fixed"[32] system into a prescribed "space-fixed" one. The body-fixed system may be chosen to be right-handed and to have its y axis perpendicular to the plane of the triangle and its z axis along one of the momenta. Instead of the three momentum magnitudes we may specify the kinetic energies

$$E_1 = \frac{p_1^2}{2m_1} \qquad E_2 = \frac{p_2^2}{2m_2} \qquad E_3 = \frac{p_3^2}{2m_3}$$

32 This terminology has to be used with some caution. The "body" defined by the three momenta is not rigid.

to which we refer collectively as $\mathfrak{E}$. The six parameters ψ, θ, φ, E_1, E_2, E_3 take the place of the six components of, say, $\mathbf{k}_1$ and $\mathbf{q}_1$ of Sec. 17.4.5.

The first thing to remember is that the Green's functions, and hence the two-particle $\mathbf{T}$ matrices, contain the *entire* kinetic energy of all three particles. Hence in the over-all center-of-mass system, for example,

$$T_1(E) = V_{23} + V_{23}\left(E - \frac{p_1^2}{2m_1} - \frac{p_2^2}{2m_2} - \frac{p_3^2}{2m_3} - V_{23}\right)^{-1} V_{23}$$
$$= V_{23} + V_{23}\left(E - \frac{p_1^2}{2m_1} - \frac{(\mathbf{p}_2 + \mathbf{p}_3)^2}{2(m_1 + m_2)} - \frac{q_1^2}{2\mu_{23}} - V_{23}\right)^{-1} V_{23}.$$

When we take matrix elements between eigenstates of the momenta, this becomes

$$V_{23} + V_{23}\left(E - E_1\frac{m_1}{\bar{\mu}_1} - \frac{q_1^2}{2\mu_{23}} - V_{23}\right)^{-1} V_{23} = \hat{T}_1\left(E - E_1\frac{m_1}{\bar{\mu}_1}\right) \tag{17.178}$$

if by $\hat{T}_1$ we mean the two-body T operator in its own center-of-mass system.

Recalling our notation convention for two kinds of off-the-energy-shell extensions of $\mathbf{T}$, we realize that the matrix of $\mathfrak{T}$ is of the "plus" kind, whereas the two-body $\mathbf{T}$ matrices that enter in the kernel of (17.129) must be taken to be of the "completely off" variety defined in (7.48a). The number E_1 in (17.178) refers to the matrix element to be taken; $\hat{T}_1$ being diagonal in $\mathbf{p}_1$, this is the same on both sides.

We must next consider the angles. $\hat{T}_1$ is a function of the angle between $\mathbf{q}_1$ and $\mathbf{q}_1'$. If we choose p_1 as the z axis, then we find

$$\cos\theta_1 = \frac{\mathbf{p}_1 \cdot \mathbf{q}_1}{p_1 q_1} = \frac{(m_2 + m_3)(p_3^2 - p_2^2) + (m_2 - m_3)p_1^2}{2\{m_1 m_2 m_3 E_1[(m_2 + m_3)(E_2 + E_3) - m_1 E_1]\}^{1/2}} \tag{17.179}$$

which gives us the angle between $\mathbf{q}_1$ and $\mathbf{p}_1$ in terms of the initial energies E_1, E_2, E_3. A similar expression, of course, connects the final angle θ_1' between $\mathbf{q}_1'$ and $\mathbf{p}_1'$ to the final energies E_1', E_2', E_3'. The scattering angle $\bar{\theta}_1$ upon which $\hat{T}_1$ depends is given by

$$\cos\bar{\theta}_1 = \cos\theta_1\cos\theta_1' + \sin\theta_1\sin\theta_1'\cos(\varphi - \varphi'), \tag{17.180}$$

φ and φ' being the third Euler angles of the rotation of the space-fixed system into the initial and final body-fixed ones.

Angular-Momentum Analysis. We now perform an angular-momentum analysis, following Omnes.[33] For this purpose we require the transformation function

$$(\mathbf{p}_1'\mathbf{p}_2'\mathbf{p}_3'|0\mathfrak{E}jm\mathfrak{m}) = (\Psi(\mathbf{p}_1'\mathbf{p}_2'\mathbf{p}_3'), \Psi(0\mathfrak{E}jm\mathfrak{m}))$$

where the total momentum vanishes, $\mathbf{P} = 0$, and m and $\mathfrak{m}$ are the eigen-

33 R. L. Omnes (1964).

values of the projections of the angular momentum onto the space- and body-fixed z axes, respectively. (We assume for simplicity that the particles are spinless.) It must be of the form

$$(\mathbf{p}_1'\mathbf{p}_2'\mathbf{p}_3'|0\mathfrak{E}jm\mathfrak{m}) = C_j\delta(\mathbf{P}')\delta(E_1 - E_1')\delta(E_2 - E_2')\delta(E_3 - E_3')D^{j*}_{m\mathfrak{m}}(\mathscr{R}) \tag{17.181}$$

if $D_{m\mathfrak{m}}{}^j$ is the rotation matrix of dimensionality $2j + 1$ and $\mathscr{R}$ the rotation whose Euler angles are ψ, θ, and φ. The constant C_j depends on the normalizations, which we assume to be

$$\begin{aligned}(\Psi(\mathbf{p}_1'\mathbf{p}_2'\mathbf{p}_3'), \Psi(\mathbf{p}_1\mathbf{p}_2\mathbf{p}_3)) &= \delta(\mathbf{p}_1 - \mathbf{p}_1')\delta(\mathbf{p}_2 - \mathbf{p}_2')\delta(\mathbf{p}_3 - \mathbf{p}_3')\\ (\Psi(\mathbf{P}'\mathfrak{E}'j'm'\mathfrak{m}'), \Psi(\mathbf{P}\mathfrak{E}jm\mathfrak{m})) &= \delta(\mathbf{P} - \mathbf{P}')\delta(E_1 - E_1')\delta(E_2 - E_2')\\ &\qquad \delta(E_3 - E_3')\delta_{jj'}\delta_{mm'}\delta_{\mathfrak{m}\mathfrak{m}'}.\end{aligned} \tag{17.182}$$

We compute it by calculating

$$\begin{aligned}\int (0\mathfrak{E}jm\mathfrak{m}|\mathbf{p}_1''\mathbf{p}_2''\mathbf{p}_3'')\, d\mathbf{p}_1''\, d\mathbf{p}_2''\, d\mathbf{p}_3''\, (\mathbf{p}_1''\mathbf{p}_2''\mathbf{p}_3''|\mathbf{P}'\mathfrak{E}'j'm'\mathfrak{m}')\\ = \int d\mathbf{p}_1''\, d\mathbf{p}_2''\, d\mathbf{p}_3''|C_j|^2\delta(\mathbf{P}'')\delta(\mathbf{P}')\delta(E_1 - E_1'')\cdots\\ \delta(E_3' - E_3'')D^j_{m\mathfrak{m}}(\mathscr{R})D^{j'*}_{m'\mathfrak{m}'}(\mathscr{R}).\end{aligned}$$

Let $\mathbf{p}_1''$ be taken as the z axis of the body-fixed system. To indicate this specific assumption we write $\mathfrak{m}_1$ for $\mathfrak{m}$. Then the Euler angles θ and ψ are its polar angles in the space-fixed coordinate frame; thus

$$d\mathbf{p}_1'' = p_1''^2\, dp_1''\, d\theta \sin\theta\, d\psi.$$

The third Euler angle φ is the azimuthal angle of $\mathbf{p}_2''$, since the x axis of the body-fixed system lies in the plane of $\mathbf{p}_1''$ and $\mathbf{p}_2''$. The angle between $\mathbf{p}_1''$ and $\mathbf{p}_2''$ is given by

$$\cos\theta_{12} = \frac{p_1''^2 + p_2''^2 - p_3''^2}{2p_1''p_2''} \tag{17.183}$$

so that

$$d\cos\theta_{12} = \frac{p_3''\, dp_3''}{p_1''p_2''}.$$

We assume that the original integration over $\mathbf{p}_3''$ is carried out first, by using the δ function of $\mathbf{P}''$. Then the new p_3'' integral takes the place of the θ_{12} integration. So we get[34]

$$\begin{aligned}d\mathbf{p}_1''\, d\mathbf{p}_2'' &= p_1''p_2''p_3''\, dp_1''\, dp_2''\, dp_3''\, d\mathscr{R}\\ &= m_1m_2m_3\, dE_1''\, dE_2''\, dE_3''\, d\mathscr{R}\\ d\mathscr{R} &= \sin\theta\, d\psi\, d\theta\, d\varphi.\end{aligned} \tag{17.184}$$

34 The integration over $\mathbf{p}_2''$ is done in the twice rotated system, in which the z-axis is $\mathbf{p}_1''$ and the x-axis is in the $(\hat{\mathbf{z}}\mathbf{p}_1'')$ plane.

Finally we use the orthonormality of the rotation matrices,[35]

$$\int d\mathscr{R}\, D^j_{m\mathfrak{m}}(\mathscr{R}) D^{j'*}_{m'\mathfrak{m}'}(\mathscr{R}) = \delta_{jj'}\delta_{mm'}\delta_{\mathfrak{m}\mathfrak{m}'}\frac{8\pi^2}{2j+1} \tag{17.185}$$

and compare the result with (17.182). The conclusion is that (to within an arbitrary phase factor)

$$C_j = \frac{(2j+1)^{1/2}}{2\pi(2m_1m_2m_3)^{1/2}}. \tag{17.186}$$

An instructive check on this value of C_j is obtained by calculating (for $\mathbf{p}_1 + \mathbf{p}_2 + \mathbf{p}_3 = 0$)

$$\begin{aligned}\int d\mathbf{P}\int dE_1\,dE_2\,dE_3 \sum_{jm\mathfrak{m}} (\mathbf{p}_1'\mathbf{p}_2'\mathbf{p}_3'|\mathbf{P}\mathfrak{E}jm\mathfrak{m})(\mathbf{P}\mathfrak{E}jm\mathfrak{m}|\mathbf{p}_1''\mathbf{p}_2''\mathbf{p}_3'') \\ = \delta(\mathbf{P}'')\frac{\delta(E_1' - E_1'')\delta(E_2' - E_2'')\delta(E_3' - E_3'')}{8\pi^2 m_1m_2m_3} \\ \sum_{jm\mathfrak{m}} D^{j*}_{m\mathfrak{m}}(\mathscr{R}')D^j_{m\mathfrak{m}}(\mathscr{R}'')(2j+1).\end{aligned} \tag{17.187}$$

But because of the unitarity of the D's,

$$\begin{aligned}\sum_{jm\mathfrak{m}} (2j+1)D^{j*}_{m\mathfrak{m}}(\mathscr{R}')D^j_{m\mathfrak{m}}(\mathscr{R}'') &= \sum_{jm\mathfrak{m}} (2j+1)D^j_{\mathfrak{m}m}(\mathscr{R}'^{-1})D^j_{m\mathfrak{m}}(\mathscr{R}'') \\ &= \sum_j (2j+1)\chi^j(\mathscr{R}'^{-1}\mathscr{R}'') \\ &= \sum_j \chi^{j*}(\mathbb{1})\chi^j(\mathscr{R}'^{-1}\mathscr{R}''),\end{aligned}$$

χ^j being the trace of $D_{\mathfrak{m}\mathfrak{m}}{}^j$,

$$\chi^j = \sum_{\mathfrak{m}} D^j_{\mathfrak{m}\mathfrak{m}}.$$

Now the *group-character theorem*[36] says that

$$\sum_j \chi^j(\mathscr{R}_1)\chi^{j*}(\mathscr{R}_2) = 8\pi^2\delta(\psi_1 - \psi_2)\delta(\varphi_1 - \varphi_2)\delta(\cos\theta_1 - \cos\theta_2) \tag{17.188}$$

and therefore

$$\sum_{jm\mathfrak{m}} (2j+1)D^{j*}_{m\mathfrak{m}}(\mathscr{R}')D^j_{m\mathfrak{m}}(\mathscr{R}'') = 8\pi^2\delta(\mathscr{R}' - \mathscr{R}'') \tag{17.189}$$

the right-hand side being shorthand for the product of δ functions in (17.188). Equation (17.189) is the generalization of (2.14).

The right-hand side of (17.187) now becomes

35 See Rose, *op. cit.*, p. 75.

36 Equation (17.188) is the rotation-group analog of eq. (3.15) of M. Tinkham (1964), p. 26, in the same way in which eq. (10.13) of E. P. Wigner (1959), p. 101, is the analog of eq. (9.33), p. 83, there.

$$\delta(\mathbf{P}'')(m_1m_2m_3)^{-1}\delta(E_1' - E_1'')\delta(E_2' - E_2'')\delta(E_3' - E_3'')\delta(\psi' - \psi'')\delta(\varphi' - \varphi'')$$
$$\delta(\cos\theta' - \cos\theta'') = \delta(\mathbf{p}_2' - \mathbf{p}_2'')\delta(\mathbf{p}_3' - \mathbf{p}_3'')$$

by (17.184).

We can now calculate the matrix elements of the two-body **T** matrix on the angular-momentum basis,

$$(0\mathfrak{E}'j'm'\mathfrak{m}_1'|T_1(E)|0\mathfrak{E}jm\mathfrak{m}_1) = C_jC_j'\int d\mathbf{p}_1'' \cdots d\mathbf{p}_3''' \,\delta(\mathbf{P}'')$$
$$\delta(\mathbf{P}''')\delta(E_1' - E_1'') \cdots \delta(E_3 - E_3''')D^{j'}_{m'\mathfrak{m}_1'}(\mathscr{R}'')D^{j*}_{m\mathfrak{m}_1}(\mathscr{R}''')$$
$$(\mathbf{p}_1''\mathbf{p}_2''\mathbf{p}_3''|T_1(E)|\mathbf{p}_1'''\mathbf{p}_2'''\mathbf{p}_3''').$$

The integrals are done just as before and result in

$$\frac{[(2j+1)(2j'+1)]^{1/2}}{8\pi^2}m_1m_2m_3\int d\mathscr{R}''\,d\mathscr{R}'''\,D^{j'}_{m'\mathfrak{m}_1'}(\mathscr{R}'')D^{j*}_{m\mathfrak{m}_1}(\mathscr{R}''')(|T_1|).$$

But (in the barycentric system), according to (17.178),

$$(\mathbf{p}_1''\mathbf{p}_2''\mathbf{p}_3''|T_1(E)|\mathbf{p}_1'''\mathbf{p}_2'''\mathbf{p}_3''') = \delta(\mathbf{p}_1'' - \mathbf{p}_1''')\hat{T}_1\left(\mathbf{p}_2'',\mathbf{p}_2''';E - E_1\frac{m_1}{\bar{\mu}_1}\right)$$

and

$$\delta(\mathbf{p}_1'' - \mathbf{p}_1''') = \frac{\delta(E_1'' - E_1''')\delta(\psi'' - \psi''')\delta(\cos\theta'' - \cos\theta''')}{m_1p_1}.$$

We now perform the ψ''' and θ''' integrals and then use the fact that[37]

$$D^{j*}_{mm}(\psi''\theta''\varphi''') = e^{im(\varphi'''-\varphi'')}D^{j*}_{mm}(\psi''\theta''\varphi'').$$

The **T** matrix $\hat{T}_1$ depends on the difference $\varphi''' - \varphi'' = \alpha$ only [see (17.180)]; so the integrals over ψ'', θ'', and φ'' may be carried out by using (17.185). The result is

$$(0\mathfrak{E}'j'm'\mathfrak{m}_1'|T_1(E)|0\mathfrak{E}jm\mathfrak{m}_1) = \delta_{jj'}\delta_{mm'}\delta_{\mathfrak{m}_1\mathfrak{m}_1'}\delta(E_1 - E_1')$$
$$m_2m_3p_1^{-1}\mathfrak{I}_1^{\mathfrak{m}_1}(\mathfrak{E}',\mathfrak{E};E) \qquad (17.190)$$

with

$$\mathfrak{I}_1^{\mathfrak{m}_1}(\mathfrak{E}',\mathfrak{E};E) = \int_0^{2\pi} d\alpha\, e^{i\mathfrak{m}_1\alpha}\hat{T}_1\left(E_2'E_3', E_2E_3;\cos\bar{\theta}_1;E - E_1\frac{m_1}{\bar{\mu}_1}\right). \qquad (17.191)$$

The function $\hat{T}_1$ in the integrand is the ordinary (completely) off-the-energy-shell **T** matrix for the scattering of particles 2 and 3. The scattering angle $\bar{\theta}_1$ is expressed via (17.179) and (17.180) in terms of the energies E_1, E_2, E_2', E_3, E_3' and the angle $\varphi - \varphi' = \alpha$.

The result (17.190) and (17.191) was derived under the specific assumption

37 See Rose, *op. cit.*, p. 54.

that the body-fixed z axis lies along $\mathbf{p}_1$. This restriction must now be removed. Let $\mathbf{p}_1$ make the angles β_1 and β_1' with the initial and final body-fixed z axes, both of which lie in the common plane of the triangles $\mathbf{p}_1$, $\mathbf{p}_2$, $\mathbf{p}_3$ and $\mathbf{p}_1'$, $\mathbf{p}_2'$, $\mathbf{p}_3'$. Then

$$(j\mathfrak{m}_1|j\mathfrak{m}) = d^j_{\mathfrak{m}_1\mathfrak{m}}(\beta_1) \qquad (j\mathfrak{m}'|j\mathfrak{m}_1) = d^{j*}_{\mathfrak{m}_1\mathfrak{m}'}(\beta_1')$$

and (17.190) becomes

$$T^{j(1)}_{\mathfrak{E}'\mathfrak{m}',\mathfrak{E}\mathfrak{m}}(E) = (0\mathfrak{E}'jm\mathfrak{m}'|T_1(E)|0\mathfrak{E}jm\mathfrak{m}) = m_2 m_3 p_1^{-1}\delta(E_1 - E_1') \sum_{\mathfrak{m}''} d^{j*}_{\mathfrak{m}''\mathfrak{m}'}(\beta_1')d^j_{\mathfrak{m}''\mathfrak{m}}(\beta_1)\,\mathfrak{I}_1^{\mathfrak{m}''}(\mathfrak{E}',\mathfrak{E};E). \tag{17.192}$$

We get a simpler expression by using the fact that, owing to rotational invariance, (17.190) is diagonal in $\mathfrak{m}$, m, and j and independent of m. Hence, if we average over m right from the start, we can combine the two rotation matrices and get

$$T^{j(1)}_{\mathfrak{E}'\mathfrak{m}',\mathfrak{E}\mathfrak{m}}(E) = m_1 m_2 m_3 \int d\mathscr{R}\, D^{j*}_{\mathfrak{m}'\mathfrak{m}}(\mathscr{R})(\mathbf{p}_1'\mathbf{p}_2'\mathbf{p}_3'|T_1(E)|\mathbf{p}_1\mathbf{p}_2\mathbf{p}_3) \tag{17.193}$$

where $\mathscr{R}$ is the rotation that converts the coordinate system fixed in $\mathbf{p}_1'\mathbf{p}_2'\mathbf{p}_3'$ into that fixed in $\mathbf{p}_1\mathbf{p}_2\mathbf{p}_3$ and $d\mathscr{R}$ is given in (17.184). The inverse of (17.193) is obtained by using (17.189),

$$(\mathbf{p}_1'\mathbf{p}_2'\mathbf{p}_3'|T_1|\mathbf{p}_1\mathbf{p}_2\mathbf{p}_3) = (8\pi^2 m_1 m_2 m_3)^{-1} \sum_{jm\mathfrak{m}'} D^j_{\mathfrak{m}'\mathfrak{m}}(\mathscr{R})T^{j(1)}_{\mathfrak{E}'\mathfrak{m}',\mathfrak{E}\mathfrak{m}}(E) \tag{17.194}$$

These equations are clearly the analogs of (11.10) and (13.12). For calculations, (17.192) is somewhat more useful than (17.193).

It is clear that the matrix elements of T_2 and T_3 are analogous to (17.192),

$$T^{j(2)}_{\mathfrak{E}'\mathfrak{m}',\mathfrak{E}\mathfrak{m}}(E) \equiv (0\mathfrak{E}'jm\mathfrak{m}'|T_2(E)|0\mathfrak{E}jm\mathfrak{m}) = m_1 m_3 p_2^{-1}\delta(E_2 - E_2') \sum_{\mathfrak{m}''} d^{j*}_{\mathfrak{m}''\mathfrak{m}'}(\beta_2')d^j_{\mathfrak{m}''\mathfrak{m}}(\beta_2)\,\mathfrak{I}_2^{\mathfrak{m}''}(\mathfrak{E}',\mathfrak{E};E)$$

$$T^{j(3)}_{\mathfrak{E}'\mathfrak{m}',\mathfrak{E}\mathfrak{m}}(E) \equiv (0\mathfrak{E}'jm\mathfrak{m}'|T_3(E)|0\mathfrak{E}jm\mathfrak{m}) = m_1 m_2 p_3^{-1}\delta(E_3 - E_3') \sum_{\mathfrak{m}''} d^{j*}_{\mathfrak{m}''\mathfrak{m}'}(\beta_3')d^j_{\mathfrak{m}''\mathfrak{m}}(\beta_3)\,\mathfrak{I}_3^{\mathfrak{m}''}(\mathfrak{E}',\mathfrak{E};E). \tag{17.192a}$$

Here β_2, β_2', β_3, and β_3' are the angles between the body-fixed z axis and $\mathbf{p}_2$ and $\mathbf{p}_3$, respectively. We must now fix this z axis in relation to the $\mathbf{p}$'s. For example, it may be chosen always to lie along $\mathbf{p}_1$. Then $\beta_1 = \beta_1' = 0$ and β_2, β_2', β_3, and β_3' are given by (17.183) and its obvious analogs.

Equations (17.192) and (17.192a), together with (17.179), allow us to express both the inhomogeneity and the kernel of the Faddeev equation (17.129) in terms of the energy and angular-momentum variables. We then define similar matrix elements of the operators $\mathfrak{T}_{kl}$; call them $\mathfrak{T}_{\mathfrak{E}'\mathfrak{m}',\mathfrak{E}\mathfrak{m}}{}^{j(kl)}(E)$. Equations (17.192) and (17.193), of course, hold for these, too. Equation (17.128) now reads

$$\mathfrak{T}^j_{\mathfrak{E}'\mathfrak{m}',\mathfrak{E}\mathfrak{m}} = \mathfrak{B}^j_{\mathfrak{E}'\mathfrak{m}',\mathfrak{E}\mathfrak{m}} + \sum_{\mathfrak{m}''=-j}^{j} \int dE_1'' \, dE_2'' \, dE_3'' \, \mathfrak{B}^j_{\mathfrak{E}'\mathfrak{m}',\mathfrak{E}''\mathfrak{m}''} \frac{1}{E + i\varepsilon - E_1'' - E_2'' - E_3''} \mathfrak{T}^j_{\mathfrak{E}''\mathfrak{m}'',\mathfrak{E}\mathfrak{m}} \tag{17.195}$$

after the $\mathfrak{T}^{(kl)}$ and $T^{(k)}$ have been put again into 3×3 matrices. Note that in reality the integral in (17.195) extends over two variables only, but they are different variables for the various elements.

Suppose that Eq. (17.195) has been solved for each j. Use of (17.181) then gives us the various matrices $\mathfrak{T}_{kl}$ in the momentum representation. Equations (17.126) and their analogs give us the T_{ij}. We finally need matrix elements of the latter between states that represent one free particle and the other two bound. The latter have a given angular momentum, which must be combined with the orbital angular momentum of the free particle to give a total j. Since the free plane wave contains infinitely many orbital angular momenta, the calculation of the amplitude for a rearrangement collision from a given bound-state spin to another requires in principle the calculation of the **T** matrices of infinitely many values of j.

17.4.7 The S Matrix. The double-scattering singularities discussed in Sec. 17.4.5 raise the question whether, for three particles, the T matrix is a noncompact integral operator. The answer to this question has important consequencies for the nature of the spectrum of the S-matrix. In the two-particle case with central forces there is a phase shift for each angular momentum, which means that the S-matrix has a discrete spectrum of eigenvalues $\exp(2i\delta_l)$ that is complete. In the three-particle case this cannot be taken for granted.

If the T-matrix is defined by

$$\hat{S} = \mathbb{1} - 2\pi i \hat{T}$$

on the super Hilbert space defined in Sec. 16.2.3 then the first obstacle to compactness of $\hat{T}$ is the existence of "disconnected pieces" that describe the scattering of a pair of particles without participation of the third. Since these pieces conserve the momentum of a particle they cannot be compact (see the lemma of Sec. 7.3.2). It is therefore necessary to remove them. We accomplish this by first defining the three pairwise S matrices $\hat{S}_a$ such that

$$(\hat{S}_a)_{bc} = S_a \delta_{ob} \delta_{oc} + \delta_{bc}(1 - \delta_{ob}) \tag{17.196}$$

where S_a is the S matrix for the pair a alone. [In (17.196) b and c take on the values of the four arrangement channels.] One may then define a reduced S matrix and a correspondingly reduced T matrix by

$$\hat{S}' = \hat{S}_3^{-1} \hat{S}_2^{-1} \hat{S}_1^{-1} \hat{S} = \mathbb{1} - 2\pi i \hat{T}'. \tag{17.197}$$

Note that the same goal would be accomplished by factoring out the pair S matrices in a different order or on the other side, and $\hat{S}'$ depends on this order; the essential result, however, is independent of it.

It has been proved[38] that if the potential satisfies (17.141) and if there are no zero-energy bound or half-bound states in any of the pairs, then the reduced T matrix defined by (17.197) is *compact*. From this result if follows that $\hat{T}'$ has a pure point spectrum, which furthermore is complete, because $\hat{S}'$ is unitary. One may define (mod π) a denumerable set of eigenphase shifts by setting the eigenvalues of $\hat{S}'$ equal to $\exp(2i\delta_n)$. In fact, it has furthermore been shown[38] that $\hat{T}'\hat{T}'^{\dagger}$ is Hilbert-Schmidt, which implies that if d_n is the degeneracy of δ_n then

$$\sum_n d_n \sin^4 \delta_n < \infty. \tag{17.198}$$

As a consequence of these results there exist strongly convergent expansions of the various 3-particle T matrices on the basis of their eigenfunctions. For two reasons these expansions cannot be expected to be as useful as the partial-wave expansions in the central two-particle case: In the first place, the eigenfunctions depend on the dynamics; they are not "universal" functions like the spherical harmonics, and no feasible technique for calculating them is known. In the second place, the series cannot be expected to be well approximated by a few terms even near the three-particle threshold. That is because the double-scattering singularities discussed in Sec. 17.4.5 can easily be seen[38] to be (pointwise) divergencies of the expansion, no matter how close to the threshold.

Once compactness and a convergence such as (17.198) have been proved it is possible to prove a three-particle version of the Levinson theorem. It has to be formulated in terms of a series of powers of the eigen-phase shifts of $\hat{S}'$, or in terms of the phase of a modified Fredholm determinant of $\hat{S}'$. This has been done and we refer to the literature.[39]

17.4.8 The Efimov Effect. We have seen in Sec. 12.2 that reasonable central potentials cannot produce more than a finite number of bound states. For non-central potentials the same conclusion is drawn in Sec. 10.3.2. Furthermore, the number of bound states for a two-particle system is a monotonely increasing function of the potential strength (i.e., of λ if we replace V by λV). Both of these statements are generally false for three-particle systems. One of the surprising results of the analysis of three-particle dynamics is the so-called *Efimov effect*: If all three pairs have no bound states but zero-energy reasonances (half-bound states) then the three-particle system has infinitely many bound states which accumulate at the threshold. The same is true if none of the pairs has bound states, two of the pairs have half-bound states, and certain inequalities on the masses are satisfied.

Since the strengthening of the pairwise forces will transform the half-bound states into bound states of nonzero binding energy, and the number

38 R. G. Newton (1974c).

39 J. A. Wright (1965). This paper takes the discreteness of the spectrum and the necessary convergences for granted.

of three-particle bound states then is finite, the number of bound states is not a monotone function of a parameter multiplying all three potentials.

The physical reason for this effect is the fact that the wave function of a zero-energy resonance in the s-wave, or generally, of a half-bound state (see Sec. 10.3.2) decreases as r^{-1} at infinity. As a result the third particle finds itself in an effective potential due to the other two particles that decreases at infinity as r^{-2}, and if such a potential is sufficiently strong it produces infinitely many bound states whose energies accumulate at zero. [The hamiltonian $-\Delta - \lambda\theta(r - r_0)r^{-2}$, where $\theta(x)$ is the Heaviside function, has infinitely many bound states if $\lambda > \frac{1}{4}$.]

We note that under the stated conditions the three-particle threshold is the lower end of the continuous spectrum. However, as the potentials are strengthened a new two-fragment threshold is produced below the three-particle threshold and moves downward, passing the infinity of bound states that previously accumulated at the three-particle threshold. Their "ghosts", so to speak, are still there, but they are now in the continuous spectrum and not real bound states. This is the mechanism by which the number of bound states is reduced as the forces are strengthened.[40]

In more detail, if particles 1 and 2 have equal masses M, particle 3 has mass m, if $M/m > 0.9$, and if V_1 and V_2 are such that the pairs 1 and 2 have half-bound states, then the three-particle system has infinitely many bound states.[41] If the pairs 1 and 2 are close to half-bound states then the number N of bound states is such that[42]

$$N \geq \frac{0.0107}{\pi}\left(\frac{M}{m} - 0.9\right)^{3/2} |\ln \rho|$$

where $\rho = \min |c_a|$, and

$$1/c_a = -\lim_{k \to 0} \{[1/A_a(k)] + ik\},$$

$A_a(k)$ being the scattering amplitude for the pair a in its center-of-mass system. For central potentials $-1/c_a = \lim_{k\to 0} k \cot \delta_a$, where δ_a is the s-wave phase shift. Thus c_a is the s-wave scattering length of the pair a.

Notes and References

17.1.1 The argument leading to (17.5) was first given by R. G. Newton (1963a).

A coupled-Schrödinger-equation approach to reactions was employed first by G. Breit (1946).

The treatment here follows mostly R. G. Newton (1958). See also T. Sasakawa (1964) and Sasakawa and Tsukamoto (1965).

40 See B. Simon (1970).

41 Ovchinnikov and Sigal (1979).

42 This is a generalization of the result of Ref. 41 to noncentral potentials, which is easy to derive from their proof.

17.1.2 The final-state interaction theory is due to K. M. Watson (1952). See also sec. 9.3 of Goldberger and Watson (1964b). For a generalization to three-particle final states, see R. D. Amado (1967 and 1975), Adhikari and Amado (1974), and to four-body final states, Adhikari and Amado (1977). See also Aitchison and Kacser (1966); R. T. Cahill (1974).

17.1.3 The existence of a function from which all S-matrix elements can be obtained as in (17.42) and (17.51) was first demonstrated in a more special case by K. J. LeCouteur (1960). That the function is the Fredholm determinant of the scattering integral equation was shown by R. G. Newton (1961a). Section 17.1.3 essentially follows this reference. See also Chan Hong-Mo (1961, 1963); R. Blankenbecler (1964); Sugar and Blankenbecler (1965); W. Glöckle (1966); Dahmen and Reiner (1967); R. G. Newton (1967c); Benayoun and Leruste (1969 and 1970); W. P. Reinhardt (1973).

17.1.5 For mathematical treatments of uniformization, see, for example, G. Springer (1957), pp. 7–11. The specific uniformization for the two-channel problem given in (17.69) is due to J. R. Cox (1962). That in (17.69) was given by M. Kato (1965).

The argument connected with Fig. 17.3, concerning the impossibility of mapping the three-channel Riemann surface onto the plane (without use of doubly periodic functions), is due to C. Goebel (private communication). See also H. A. Weidenmüller (1964).

The fact that, as the forces change, the poles which are near the upper rim of the cut and thus cause the resonances change places as they pass a threshold was independently pointed out by M. Ross (1963); Eden and Taylor (1963); D. Amati (1963); Dalitz and Rajasekaran (1963); Nauenberg and Nearing (1964). See also R. L. Warnock (1964).

For generalizations of the double dispersion relation to many-channel processes, see Fonda, Radicati, and Regge (1961); J. Underhill (1962 and 1963). Generalizations of the N/D method to many-channel problems and discussions of them can be found in D. J. Bjorken (1960); Chew and Mandelstam (1960); M. Froissart (1961); Frye and Warnock (1963); E. J. Squires (1964); M. Luming (1964); H. Munczek (1964); Bander, Coulter, and Shaw (1965); Nath and Shaw (1965b); Bali, Munczek, and Pignotti (1965); Munczek and Pignotti (1965); K. Kikawa (1965); R. L. Warnok (1967). See also C. van Winter (1966) and F. Riahi (1969).

17.2 The arguments of Sec. 17.2.1 are due to R. G. Newton (1963b).

Among papers concerned with the analytic behavior of the S matrix near a threshold we may mention R. J. Eden (1949 and 1952). The threshold anomaly as an observable phenomenon was first pointed out by E. P. Wigner (1948) and is therefore sometimes referred to as a *Wigner cusp*; however, see also E. Fermi (1936). Applications can be found in V. Mamasakhlisov (1953); Capps and Holladay (1955), appendix B.

General treatments of the anomaly are to be found in Guier and Hart (1957); G. Breit (1957 and 1959), p. 274; A. I. Baz (1957 and 1959); R. G. Newton (1958 and 1959); Baz and Okun (1958); L. M. Delves (1958–1959); L. Fonda (1959, 1961a and b); Jackson and Wyld (1959); Y. Yamaguchi (1959); Nauenberg and Pais (1961); W. E. Meyerhof (1962 and 1963); T. F. Tuan (1963); S. K. Adhikari (1973); Bardsley and Nesbet (1973); I. Aronson et al. (1975). The usefulness of the effect for parity information was emphasized by R. K. Adair (1958). The influence of the Coulomb potential was investigated by Fonda and Newton (1959) and applied to X-ray scattering in Newton and Fonda (1960); see also F. H. M. Faisal (1968); Shakeshaft and Spruch (1979).

A similar threshold effect in three-body channels was treated in Fonda and Newton (1960b); I. G. Halliday (1966 and 1967a). See also Dragt and Karplus (1962); N. S. Kronfli (1963); R. Roskies (1966); Amado and Rubin (1970).

If the particles in the threshold channel are unstable, then the cusp is of course "washed out." This effect was emphasized in A. I. Baz (1961); Nauenberg and Pais (1962); Fonda and Ghirardi (1964a). In the graphic terminology of the second reference it is called a "woolly" cusp.

For references concerning Eq. (17.78) see the Notes and References to Sec. 12.2 referring to Eq. (12.165).

For threshold influences on resonances, see J. B. Ehrman (1951); R. G. Thomas (1952); A. I. Baz (1959); D. R. Inglis (1962); Barker and Traecy (1962); F. C. Barker (1964); H. A. Weidenmüller (1965).

17.3 The content of Sec. 17.3.1 follows mostly Fonda and Newton (1960c); R. G. Newton (1961); that of Sec. 17.3.2 the last reference. See also Fonda, Radicati, and Regge (1961).

The Wigner-Weisskopf model was introduced by Wigner and Weisskopf (1930) and studied also by M. Moshinsky (1951); M. Wellner (1960).

The *Bargmann potentials* have been generalized to the coupled-channel case by J. R. Cox (1964); Cox and Garcia (1975); E. B. Plekhanov et al. (1981).

The *effective-range theory* has been generalized to multichannel problems in the following articles: L. M. Delves (1958); Ross and Shaw (1960 and 1961); Shaw and Ross (1962); Nath and Shaw (1965a).

Other solvable models are discussed by J. B. McGuire (1964); Brezin and Zinn-Justin (1966); C. N. Yang (1968); F. Calogero (1969 and 1971a); J. R. Dodd (1970); C. Marchioro (1970); Kamal and Kreutzer (1970); C. K. Majumdar (1972); McGuire and Hurst (1972); M. G. Fuda (1973); Calogero and Marchioro (1974); Gerber and Rosenbach (1974); E. A. Rohlfing et al. (1980). See also R. F. Dashen et al. (1976).

17.4.2 The Faddeev equations are due to L. D. Faddeev (1960, 1961, 1962 and 1963). The slight modification of the Faddeev equations given in this section corresponds to that of C. Lovelace (1964a and b). The procedure is related to the older method of K. M. Watson (1953 and 1957). The modification necessary in the presence of three-body forces was first given by R. G. Newton (1966).

17.4.3 The Weinberg-van Winter equations are due to S. Weinberg (1964a), C. van Winter (1964, 1965, and 1970). The Rosenberg method was invented by L. Rosenberg (1964b); see also R. G. Newton (1967a). The coupled-channel array (CCA) method originated with Baer and Kouri (1973); Kouri and Levin (1974d); W. Tobocman (1974).

17.4.4 The use of Fredholm methods in the three-particle problem began with M. Rubin et al. (1966 and 1967). The discussion in this section is based on R. G. Newton (1971 and 1974d). The issue of *spurious solutions* was first raised by P. Federbush (1966). For its further discussion see J. V. Noble (1966); R. G. Newton (1967d); Y. Hahn et al. (1974b); Adhikari and Glöckle (1978, 1979, 1980b); C. Chandler (1978); K. L. Kowalski (1978b); V. Vanzani (1978b and c); S. K. Adhikari (1979c and 1982); Chandler and Sloan (1980).

Analyticity properties are discussed by M. M. Islam (1965); I. Sh. Vashakidze et al. (1965); M. Rubin et al. (1966 and 1967); D. D. Brayshaw (1968b); Hartle and Sugar (1968); R. D. Amado et al. (1971); Y. Avishai (1971b).

17.4.5 The method and results of this section are based on R. G. Newton (1972a) and S. P. Merkuriev (1971). See also F. A. Berezin (1965); J. Nuttall (1967a and 1971); Nuttall and Webb (1969); Doolen and Nuttall (1971); S. P. Merkuriev (1973); Osborn and Bollé (1973); C. Gignoux et al. (1974); W. Glöckle (1974); V. Vanzani (1978); Cattapan and Vanzani (1979a); S. Servadio (1981). The recognition of the "double scattering difficulties" is originally due to E. Gerjuoy (1970 and 1971) and J. Nuttall

(1971). That the resulting three-particle cross section is nevertheless finite was shown by Newton and Shtokhamer (1976) and Potapov and Taylor (1977). For an application, see R. G. Newton (1976b).

17.4.6 The introduction of the variables $\psi, \theta, \varphi, E_1, E_2, E_3$ into the Faddeev equations is due to Omnes, and so is the subsequent angular-momentum analysis. The present procedure essentially follows Omnes in this, except for the correction of some minor errors.

The following papers deal with angle-functions and angular-momentum analyses of 3-particle or n-particle systems: E. Fabri (1954); R. H. Dalitz (1954); V. I. Ritus (1961); A. J. Macfarlane (1962); W. Zickendraht (1964 and 1965); Badalyan and Simonov (1966); Lévy-Leblond and Lévy-Nahas (1965); Pustovalov and Simonov (1966 and 1968); Y. A. Simonov (1966); E. L. Surkov (1967); Raynal and Revai (1970); M. Fabre de la Ripelle (1971); Shukre and Winternitz (1972b); A. Lindner (1973); Bollé and Osborn (1974); A. Kupperman (1975); K. K. Fang (1977 and 1979); D. Eppel (1978); B. R. Johnson (1980); J. L. Friar et al. (1981). For relativistic generalizations see M. I. Shirokov (1961); G. C. Wick (1962); J. A. Lock (1975).

17.4.7 The results of this section are mostly due to R. G. Newton (1974c and 1975). See also J. A. Wright (1965), D. B. Pearson (1966); E. A. Remler (1966), D. R. Yafaev (1978).

17.4.8 The discovery of the Efimov effect in three-particle systems is due to V. Efimov (1970, 1973, 1976). A more rigorous proof of its existence was given by Amado and Noble (1971 and 1972). Amada and Greenwood (1973) showed that there is no analogous effect in n-particle systems ($n > 3$) at "zero-energy bound states" of its m-fragment subsystems for $m > 2$. There is, nevertheless, such an effect in an n-particle system if among one of its three-fragment states at least two have simultaneous zero-energy half-bound states; I. M. Sigal (unpublished); Kröger and Perne (1980). A rigorous proof of the Efimov effect was given by Ovchinnikov and Sigal (1979). See also B. Simon (1970); D. R. Yafaev (1974); A. C. Fonseca et al. (1979); Fonseca and Shanley (1979); Adhikari and Fonseca (1981). Numerical calculations were done by Stelbovics and Dodd (1972); A. G. Antunes et al. (1976).

17.4 General. Further references on three-particle scattering are: Feshbach and Rarita (1949); R. Clapp (1949 and 1961); M. Verde (1949); Troesch and Verde (1951); Massey and Moiseiwitch (1951); Borowitz and Friedman (1953); Boyet and Borowitz (1954); Skorniakov and Ter-Martirosian (1956); G. V. Skorniakov (1956); Derrick and Blatt (1958 and 1960); L. Eyges (1959); G. H. Derrick (1960 and 1962); Fonda and Newton (1960b); Cohen, Judd, and Riddell (1960); L. M. Delves (1960); G. S. Danilov (1961); Gallina, Nata, Bianchi, and Viano (1962); Barsella and Fabri (1962); F. T. Smith (1962); R. D. Amado (1963); Mitra and Bhasin (1963); J. R. Higgins (1963); A. I. Baz (1964); A. Chakrabarti (1964); Aaron, Amado, and Yam (1964 and 1965); Bhatia and Temkin (1964); M. McMillan (1964); M. H. Choudhury (1964); C. Zemach (1964b); R. L. Omnes (1964); Omnes and Alessandrini (1964); Bianchi and Favella (1964a); W. Hunziker (1964); Ghirardi and Rimini (1965b); J. T. Cushing (1965); Amadzadeh and Tjon (1965); L. Basdevant (1965); N. Mishima (1965); Mishima and Yamasaki (1965); Block and Gillet (1965); M. Bander (1965); Levy-Leblond and Levy-Nahas (1965); Tadic and Tuan (1965); G. Immirzi (1965); S. Mandelstam (1965); Berman and Jacob (1965); Alessandrini and Omnes (1965); C. Zupančič (1965); Hetherington and Schick (1965); A. Tucciarone (1966); Aaron and Shanley (1966); I. G. Halliday (1966); R. Roskies (1966); E. O. Alt et al. (1967); MacDonald and Mekjian (1967); Malfliet and Ruijgrok (1967); N. Mishina (1967); Tobocman and

Nagarajan (1967); van Nieuwenhuizen and Ruijgrok (1967); W. Zickendraht (1967); Ball and Wong (1968); G. Doolen (1968); J. R. Jasperse (1968); D. D. Brayshaw (1969); M. G. Fuda (1969); Grassberger and Sandhas (1969); V. Vanzani (1969, 1973, 1974); R. J. Yaes (1969); W. W. Zachary (1969); Kazaks and Greider (1970); McKee and Rolph (1970); J. C. Y. Chen et al. (1971); Chen and Hambro (1971); Chen and Joachain (1971); W. Glöckle (1971); Osborn and Kowalski (1971); S. C. Pieper et al. (1971); T. Sasakawa (1971, 1972, 1973, 1977a); E. O. Alt (1972); Baer and Kouri (1972); Bahethi and Fuda (1972); K. L. Kowalski (1972a and 1973); W. Sandhas (1972); K. Takeuchi (1972 and 1973); Kloet and Tjon (1973); R. D. Amado (1974); M. Fuda (1974); Hirooka and Sunakawa (1974); Klink and Johnson (1974); Kouri and Levin (1974a and 1975c); Larson and Hetherington (1974); Schmid and Ziegelman (1974); Adhikari and Slaon (1975); Karlsson and Zeiger (1975); Kim and Tubis (1975); D. Bollé (1976); H. Hannover (1976); D. J. Kouri et al. (1976); S. P. Merkuriev et al. (1976); J. Ginibre (1977); V. B. Mandelzweig (1977); Osborn and Wilk (1977); Sasakawa and Sawada (1977b); O. A. Yakubovskii (1977); V. I. Kukulin et al. (1978); Kukulin and Pomerantsev (1978a, b, c); V. E. Kuzmichev (1978); T. Preist (1978); A. G. Sitenko (1978); K. Yajima (1978); Adhikari and Glöckle (1980a); N. Austern (1980); Dodd and Nieukerke (1980); V. V. Komarov et al. (1980); M. Kawai et al. (1981); G. Payne (1981); I. H. Sloan (1981); Kröger and Fenske (1981); Fonseca and Shanley (1981); Faddeev and Yakubovskii (1981).

B. C. Eu (1971a) treated the classical three-body problem in analogy with Faddeev.

The following papers deal with electrically charged three or n-particle systems: J. V. Noble (1967); L. Schulman (1967); van Winter and Brascamp (1968); Hamza and Edwards (1969); A. M. Brodskii et al. (1970); A. M. Veselova (1970b, 1972, 1978); Gy. Bencze (1972); L. Rosenberg (1973); Gibson and Chandler (1974); E. O. Alt (1976 and 1978); E. O. Alt et al. (1976 and 1978); S. P. Merkuriev (1976, 1977, 1978a, b, 1979a, b, 1980); Alt and Sandhas (1978 and 1980); Z. Bajzer (1978); Chandler and Gibson (1978c); Kharchenko and Shadchin (1978); P. V. Sauer (1978); W. Thirring (1978); Bencze and Zankel (1979); Kuzmichev and Kharchenko (1979); V. Enss (1979); Sasakawa and Sawada (1979); M. L. Zepalova (1979); J. Zorbas (1979); C. Chandler (1980); L. P. Kok (1980); L. P. Kok et al. (1980); Kok and van Haeringen (1980); Evans and Hoffman (1981).

For bounds on the number of bound states see D. R. Yafaev (1972 and 1976); G. M. Zhislin (1972); see also Buslaev and Merkuriev (1969); Bollé and Chadan (1982).

Relativistic treatments of three-particle systems are given by J. Nuttall (1967b); Bhasin and Mitra (1970); P. du T. van der Merwe (1972).

For approximation methods, see R. Aaron et al. (1961); Mitra and Bhasin (1963); Sitenko and Karchenko (1963); Dodd and Greider (1966); Greider and Dodd (1966); Wong and Zambotti (1967); M. G. Fuda (1968); I. H. Sloan (1968); K. L. Kowalski (1969); Amado and Rubin (1970); Brodskii and Potapov (1970); J. A. Tjon (1970); Becker and Harrington (1971); R. L. Kelly (1971); I. Manning (1972); J. M. Namyslowski (1972); Sinfailam and Chen (1972); H. Ziegelmann (1972); S. K. Adhikari (1973); Bakker and Sandhas (1975); T. Sasakawa (1977b); Cattapan and Vanzani (1978); J. A. Lock (1978); V. S. Potapov (1980); Tomio and Adhikari (1980 and 1981); D. Eyre et al. (1981); G. A. Hagedorn (1981).

Treatments of four-particle equation are to be found in the following papers: A. N. Mitra et al. (1965); Takahashi and Mishima (1965); V. A. Allessandrini (1966); Mishima and Takahashi (1966); Popova and Komarov (1967); Yakubovskii (1967); E. O. Alt et al. (1970); I. Sloan (1972); D. Bessis (1974); W. Sandhas (1975); Fonseca and Shanley

(1976); T. Sasakawa (1976 and 1978); J. A. Tjon (1976 and 1980); Adhikari and Amado (1977); Kharchenko and Levachev (1977); Kholyavin and Yakubovskii (1977); S. K. Adhikari (1979a); Haberzettl and Sandhas (1981).

For reviews of n-particle scattering theory see L. D. Faddeev (1970); V. Vanzani (1976, 1978a, b); K. L. Kowalski (1978a); M. J. Moravcsik (1978); E. F. Redish (1978); W. Sandhas (1978); Gy. Bencze (1980); F. S. Levin (1981b). See also the conference proceedings B. Pisent et al. (1978); H. Zingl et al. (1978); and F. S. Levin (1981a). The following papers treat general n-particle scattering equations in detail: W. Hunziker (1965, 1968, 1977); J. Weyers (1966); Grassberger and Sandhas (1967); T. Sasakawa (1967, 1977, 1978); R. Omnes (1968); W. Bierter (1969); J.-M. Combes (1969 and 1974); K. Hepp (1969); Komarov and Popova (1969); W. Glöckle (1970); Y. Avishai (1971a); Narodetskii and Yakubovskii (1971); L. Rosenberg (1971); V. Vanzani (1971, 1976, 1978b); Y. Hahn (1972d, 1976a, 1977); Hahn and Watson (1972); Gy. Bencze (1973, 1976, 1977); Y. Hahn et al. (1974a); Karlson and Zeiger (1974); Kouri and Levin (1974d, 1975b, 1977); E. F. Redish (1974); W. Sandhas (1974, 1976); W. Tobocman (1974b, 1975b); P. Benoist-Gueutal (1975); Cotanch and Vincent (1976); E. F. Redish et al. (1976); J. Schwager (1976); Gy. Bencze et al. (1977); Bencze and Tandy (1977); Cattapan and Vanzani (1977, 1979b, 1981); M. L'Huillier et al. (1977); D. J. Kouri et al. (1977); K. L. Kowalski (1977); P. Benoist-Gueutal et al. (1978); K. L. Kowalski et al. (1978); Polyzou and Redish (1978 and 1979); I. M. Sigal (1978); Greben and Levin (1979); Hahn and Retter (1979); Lovitch and Vanzani (1979); Ludeking and Vary (1979); Goldflam and Kowalski (1980a); B. E. Grinyuk (1980); F. S. Levin (1980 and 1981c); W. N. Polyzou (1980 and 1981); G. Cattapan et al. (1981); K. L. Kowalski et al. (1981); Levin and Li (1981); S. F. J. Wilk et al. (1981); see also B. R. Karlsson (1982).

A relativistic version of n-particle scattering theory is discussed by L. Rosenberg (1966). N-particle scattering with *dilation analytic* potentials is treated by Balslev and Combes (1971); I. M. Sigal (1978a); C. van Winter (1978). *Optical-model* potentials are derived from n-particle scattering theory by K. L. Kowalski (1979) and by Goldflam and Kowalski (1980a). *Numerical tests* of n-particle scattering formalisms were performed by Lewanski and Tobocman (1978) and by W. Tobocman (1981a, b).

Problems

1 Prove Eqs. (17.32) and (17.33).

2 Suppose that particle A has two bound states of zero angular momentum and binding energies 10 and 11 Mev, with an infinitely heavy nucleus B and no other bound states with it. Particle C of 818 Mev mass has an attractive square-well interaction of 8.23 Mev strength and 5.2×10^{-13} cm range with the nucleus B, and it has an exponential interaction of 0.5×10^{-13} cm range and 100 kev strength with particle A. In an inelastic collision experiment particles C of an energy between 1 and 2 Mev are shot at the (AB) system in its ground state, leaving it finally in its excited state. At what energies do you expect the inelastic cross section to have peaks? Calculate the cross section in the relevant energy range, making appropriate approximations. Plot it as a function of energy.

3 Supply the missing details in the proof of (17.38).

4 Calculate the threshold anomalies in the differential and total cross sections (for neutral particles) if both particles in the incident channel have spin 0 and in the threshold channel one has spin J and the other either spin 0 or spin $\frac{1}{2}$.

5 Do the same as in Prob. 4 for the case in which both in the incident and in the threshold channel one particle has spin 0 and the other spin $\frac{1}{2}$.

6 Do the same as in Prob. 4 for incident spins 0 and $\frac{1}{2}$ and in the threshold channel spins J and 0.

7 Do the same as in Prob. 4 for spins 0 and 1 in the incident channel and both spins 0 in the threshold channel.

8 Verify Eq. (17.116).

9 Derive Eq. (17.128).

10 Verify Eq. (17.179).

11 Suppose that in the $N \times N$ square-well example of Sec. 17.3.1 the last threshold occurs at a very high energy E_N. Show that as $E_N \to \infty$ the scattering in every fixed energy region depends only on the $(N - 1) \times (N - 1)$ submatrix of V for the first $N - 1$ channels. In other words, the coupling to channels with faraway thresholds can be neglected.

CHAPTER 18

Short-Wavelength Approximations

18.1 Introduction

In this chapter we want to discuss approximation methods for scattering that are applicable when the de Broglie wavelength of the scattered particle (or of the relative motion of two) is small. In Secs. 18.2 and 18.3 this means small compared with the characteristic distance within which the forces change by an appreciable fraction of themselves. In Sec. 18.4, where we consider the scattering of a particle by a bound state, the energy scale is set by the binding energy E_B of the latter. If we call $(2mE_B)^{-1/2}$ the *size* of the bound state, in the sense in which one speaks of the size of the deuteron, we may say in all these cases that the wavelength of the missile is small compared with the size of the target. For given interparticle potentials we are therefore in the high-frequency, or *high-energy*, region.

For a given energy, the methods are applicable when the distance scale of the forces, i.e., the size of the scatterer, is large. The situation is therefore quite analogous to that of the geometrical-optics limit of electromagnetic scattering, as discussed in Sec. 3.5.

A slow change of the kinetic energy relative to the wavelength is of course a necessary condition for the concept of a *local wavelength* to be meaningful. The demand now is that *everywhere* in the spatial region of interest the local wavelength change slowly. Just as physical optics then allows us to speak of *rays* and goes over into geometrical optics, so quantum physics allows us to speak of *trajectories* and goes over into classical mechanics. The mathematical reason is the same; it was explained at the beginning of Sec. 3.5.

One of the characteristics of high-energy scattering is that it tends to be concentrated in the forward direction. This is physically obvious. It can be seen in a more quantitative manner by realizing that the small deflection angle of a classical particle of momentum p to which the target has transferred the momentum Δp is well approximated by

$$\theta \simeq \frac{\Delta p}{p}.$$

This "kick" in turn equals the exerted force, integrated over the transit time across the region of interaction of radius R,

$$\Delta p \simeq \frac{\bar{V}}{R}\frac{R}{v} = \frac{\bar{V}}{v}$$

so that

$$\theta \simeq \frac{\bar{V}}{pv} \simeq \frac{\bar{V}}{2E}$$

provided that this ratio is small compared with unity.

Quantum-mechanical effects make this angle uncertain by an amount $\delta\theta$, because the definition of the *transverse position* to within R by passage through the scatterer renders the transverse momentum uncertain by

$$\delta p \simeq \frac{\hbar}{R}.$$

Consequently

$$\delta\theta \simeq \frac{\hbar}{pR} = \frac{\lambdabar}{R}.$$

For trajectories to be well defined, this must be small compared with θ if θ itself is small and small compared with unity otherwise. We therefore require that

$$\lambdabar \ll R \quad \text{and} \quad \lambdabar \ll R\frac{\bar{V}}{E}$$

which latter inequality can also be written

$$\frac{\hbar v}{\bar{V}R} \ll 1.$$

Since the wavelength is connected with the momentum of a particle by

$$\lambda = \frac{2\pi\hbar}{p},$$

the short-wavelength regime (for given forces) may also formally be thought of as the limit in which $\hbar \to 0$.[1] This is mathematically very convenient and,

1 The considerations of Sec. 18.4 cannot be handled in this way, unless one introduces two different Planck's constants, one for the target bound state and another for the missile. This must not be done, of course.

of course, reiterates the fact that we are going into the classical region where the size of Planck's constant can be considered negligible. (In Secs. 18.2 and 18.3 we shall therefore not choose units in which $\hbar = 1$.) We expect to find that in this limit the equations of classical mechanics emerge from those of quantum mechanics and to learn in exactly what sense this is true. More importantly from a practical point of view, we want to stop short of the extreme classical limit to obtain quantum-mechanical high-energy approximations analogous to those described in Secs. 3.2, 3.4, and 3.5. These are the WKB and the eikonal approximations.

18.1.1 Diffraction from the Optical Theorem. The major qualitative features of the forward diffraction peak at high energies may be derived directly from the optical theorem (see Secs. 1.3.9, 7.2.2, 11.1.1). Since both are results of an interference between the unscattered beam and the forward-scattered particles this should not be surprising.

For spin-0 particles the optical theorem reads

$$\operatorname{Im} A(0) = (k/4\pi)\sigma_{\text{total}},$$

where $A(0)$ is the elastic scattering amplitude in the forward direction. Therefore the forward elastic cross section satisfies the inequality

$$\sigma_{el}(0) \geq (k/4\pi)^2 \sigma_{\text{total}}^2.$$

If $\bar{\sigma}_{el} = \sigma_{el}/4\pi$ is the average elastic cross section then $\sigma_{\text{total}} \geq \sigma_{el}$ implies therefore that

$$\sigma_{el}(0)/\bar{\sigma}_{el} \geq (k^2/4\pi)\sigma_{\text{total}}, \tag{18.1}$$

which shows that unless the total cross section falls off faster than k^{-2} as the energy increases the elastic cross section must become more and more sharply peaked in the forward direction.

Let us call Θ the angular width of the forward peak. Then (assuming for simplicity no φ-dependence), for large k,

$$\begin{aligned}\sigma_{el} &= 2\pi \int_{-1}^{1} d\cos\theta\, \sigma_{el}(\theta) \gtrsim 2\pi\sigma_{el}(0)(1 - \cos\Theta) \\ &= 4\pi\sigma_{el}(0)\sin^2 \tfrac{1}{2}\Theta.\end{aligned}$$

Consequently,

$$\sin^2 \tfrac{1}{2}\Theta \lesssim \bar{\sigma}_{el}/\sigma_{el}(0) \leq 4\pi/(k^2\sigma_{\text{total}}).$$

If $k^2\sigma_{\text{total}} \gg 1$ this means that the width of the forward peak is approximately bounded by

$$\Theta \lesssim 2\pi^{-1/2}\lambda(\sigma_{\text{total}})^{-1/2}. \tag{18.2}$$

Comparison with the earlier argument based on the uncertainty principle shows that this implies at the same time that at high energies the total cross section can be expected to be of the order of magnitude of the geometric cross section of the scattering region.

18.2 The WKB Method

The procedure consists of a number of stages, the first one of which is an approximate calculation of the phase shifts. The remaining steps are treatments of the partial wave expansion of the scattering amplitude to facilitate its summation.

18.2.1 The WKB Phase Shifts. Assuming that the interparticle potential is spherically symmetric (and the particles are spinless), the lth partial wave satisfies the radial wave equation

$$-\frac{\psi''}{\psi} = [2m(E - V) - \hbar^2 l(l+1)r^{-2}]\hbar^{-2} \equiv F\hbar^{-2} \tag{18.3}$$

so that $F^{1/2}\hbar^{-1}$ is the *local wave number*. Let us make the *ansatz*

$$\psi = \text{const}\, G(r)^{-1/4} \exp\left[i\hbar^{-1}\int^r dr'\, G^{1/2}(r')\right]$$

which by itself says nothing (so long as G may be complex); it defines G. Since l is fixed in the following, we have not indicated explicitly that ψ, F, and G depend upon it.

Two differentiations show that G must obey the nonlinear inhomogeneous differential equation

$$G + \tfrac{1}{4}\hbar^2 \frac{G''}{G} - \tfrac{5}{16}\hbar^2 \left(\frac{G'}{G}\right)^2 = F$$

which is equivalent to (18.3). One may attempt to solve it in the form of an expansion in powers of $\hbar^2$,

$$G = G_0 + G_1\hbar^2 + G_2\hbar^4 + \cdots \tag{18.4}$$

and one finds that

$$\begin{aligned} G_0 &= F \\ G_1 &= \frac{5}{16}\left(\frac{F'}{F}\right)^2 - \frac{1}{4}\frac{F''}{F}, \end{aligned} \tag{18.4a}$$

etc. In the first approximation, therefore, when $\hbar$ can be considered very small, the wave function is equal to

$$\psi \simeq \psi_{\text{WKB}} \equiv \text{const}\, F^{-1/4} \exp\left[i\hbar^{-1}\int^r dr'\, F^{1/2}(r')\right]. \tag{18.5}$$

Now (18.4a) shows that the first approximation will not be reliable if F is small in a region where F' or F'' is not. Since F approaches $-\infty$ as $r \to 0$ and $2mE$ as $r \to \infty$, there *always* is a point where $F = 0$, and hence the approximation (18.5) and the expansion (18.4) break down. This is the classical *turning point* of the motion; the radial kinetic energy there vanishes. But the fact that (18.5) is not a good approximation to the wave function near the turning point has no influence on the scattering. In order to calculate the phase shift, ψ has to be constructed as "smoothly" as possible. That is, it must be a continuous function, with a continuous derivative, from $r = 0$ to $r = \infty$, in both of which regions (18.5) approximates it well. Near the turning point the higher-order terms in powers of $\hbar$ become important, and the expansion (18.4) diverges. But in any case *each term* must there be made separately as smooth as possible. So our task is simply to "continue" (18.5).[2] This is done as follows.

Near $r = 0$ the centrifugal term dominates F, and it is *negative*. (We assume that V is less singular than r^{-2}.) Since ψ is to be regular there, we must choose

$$\psi = (-F)^{-1/4} \exp\left[\hbar^{-1} \int_{r_0}^{r} dr' \, (-F)^{1/2}\right] \tag{18.6}$$

near $r = 0$. Let us take r_0 to be the classical turning point, i.e., the value of r where $F = 0$. As we cross it, $(-F)^{1/2}$ must be replaced by $\pm iF^{1/2}$. The "inside" wave function (18.6) being real, so must the "outside" function be. Hence it does not matter which sign of i is chosen, but of course it must be the same in both places where F appears. Let us take $(-F)^{1/2} \to iF^{1/2}$ so that $(-F)^{-1/4} \to F^{-1/4} e^{-i\pi/4}$. Then, for $r > r_0$,[3]

$$\begin{aligned}
\psi &= F^{-1/4} \,\mathrm{Re} \exp\left[i\left(\hbar^{-1} \int_{r_0}^{r} dr' \, F^{1/2} - \tfrac{1}{4}\pi\right)\right] \\
&= F^{-1/4} \sin\left(\hbar^{-1} \int_{r_0}^{r} dr' \, F^{1/2} + \tfrac{1}{4}\pi\right) \\
&\underset{r\to\infty}{\simeq} p^{-1/2} \sin\left[kr + \hbar^{-1} \int_{r_0}^{\infty} dr' \, (F^{1/2} - p) - kr_0 + \tfrac{1}{4}\pi\right]
\end{aligned}$$

because F approaches p^2 as $r \to \infty$. Consequently, the phase shift in the WKB approximation is given by

$$\delta_l^{\mathrm{WKB}} = \tfrac{1}{2}\pi(l + \tfrac{1}{2}) + \hbar^{-1} \int_{r_0}^{\infty} dr \, (F^{1/2} - p) - kr_0 \tag{18.7}$$

or, after integrating by parts,

$$\delta_l^{\mathrm{WKB}} = \tfrac{1}{2}\pi(l + \tfrac{1}{2}) - \hbar^{-1} \int_{r_0}^{\infty} dr \, r \frac{d}{dr} F^{1/2}. \tag{18.7a}$$

A slight improvement in this expression is obtained by looking at the

2 Where the potential, or its first derivative, is discontinuous, there (18.4) shows G_0 also to be a poor approximation. This was to be anticipated because it is the slowness of the *change* in the potential energy that counts.

3 A word should be said about the absence of "connecting" formulas in the present derivation of the form of the outside wave function. The arguments given are intuitive and simple but, of course, they cannot be considered a rigorous justification. For this the customary derivation of the connecting formulas is indispensable. The standard approach being contained in any textbook of quantum mechanics [e.g., L. I. Schiff (1968b), pp. 272ff.], we dispense with it here and give only this heuristic derivation.

behavior of (18.6) near $r = 0$. Because there the centrifugal term dominates, we get

$$\psi \simeq \text{const}\, r^{1/2} \exp\left[l(l+1)\right]^{1/2} \ln r$$

whereas we know that it should be

$$\psi \simeq \text{const}\, r^{1+l}.$$

We get the correct behavior by replacing $l(l+1)$ in F by $(l+\frac{1}{2})^2$,

$$F = 2m(E - V) - \hbar^2(l+\tfrac{1}{2})^2 r^{-2}. \tag{18.8}$$

This expression for F should therefore be used in (18.7) and (18.7a). If we set

$$J = \hbar(l + \tfrac{1}{2}) \tag{18.9}$$

this means that

$$\begin{aligned} \delta_l^{\mathrm{WKB}} &= \hbar^{-1}\left(\tfrac{1}{2}\pi J - \int_{r_0}^{\infty} dr\, r \frac{d}{dr} F^{1/2}\right) \\ F &= 2m(E - V) - J^2 r^{-2}. \end{aligned} \tag{18.10}$$

Note that if J is interpreted as the magnitude of the angular momentum then the parentheses in (18.10) contain only quantities amenable to classical interpretation. The sole remnant of quantum mechanics is the $\hbar^{-1}$ in front. The additive $\frac{1}{2}$ in (18.9) is of very little practical consequence because in the energy region of interest small values of l play a negligible role. For $l = 0$, the whole argument in fact is likely to break down. It depends, after all, on the existence of a classical turning point r_0. Unless the potential tends to $+\infty$ at $r = 0$, there is always an energy above which for $l = 0$ no r_0 exists. The s wave has no real classical analog.

Let us finally differentiate (18.10) or (18.7) with respect to l,

$$\frac{d\delta_l^{\mathrm{WKB}}}{dl} = \tfrac{1}{2}\pi - J \int_{r_0}^{\infty} dr\, r^{-2} F^{-1/2}$$

which should be compared with (5.4). It shows that the WKB phase shift is connected with the classical deflection function of the same angular momentum by

$$\Theta = 2\frac{d\delta_l^{\mathrm{WKB}}}{dl}. \tag{18.11}$$

For a supplementary remark, we look back at Eq. (18.3). At fixed values of E and l, $|F|$ increases with the magnitude of the potential or with the coupling constant. This increases the first term on the left of (18.3) but not the other two. As a result $G \simeq F$, or (18.5), is also a *strong coupling* approximation. The WKB phase shift, given by (18.7a), is at the same time the limiting value of δ_l as the coupling constant or the potential strength becomes large. A look at F shows that for an attractive potential this statement is uniform in E, but not necessarily in l. For a repulsive potential, it is uniform in l, but not necessarily in E. In general it need not be uniform in either variable.

Multiplication of E and V by γ and of l by $\gamma^{1/2}$ amounts to multiplying F by γ. As $\gamma \to +\infty$, the WKB phase shift becomes exact, and according to (18.7a) it is *proportional* to γ. The **S** matrix then goes as $\exp(ic\gamma)$. This is the nature of the essential singularity of the **S** matrix for strong coupling, combined with high energy and large angular momentum.

18.2.2 The Scattering Amplitude. At high energies (in the sense explained in Sec. 18.1) large l values dominate the scattering amplitude for a given direction. Quantum mechanically this is due to the analog of the optical *localization principle*, or the property of the Bessel functions discussed at the beginning of Sec. 3.5. If the size of the interaction region is R, then the dominant l value is of the order of magnitude

$$l \simeq kR.$$

What is more, the phase shift is a relatively slowly varying function of l in the relevant region. As (18.11) shows, there is not simply a single phase shift (or a small number of them) that dominates the scattering. As a result the partial wave expansion of the amplitude consists of many relatively slowly varying terms, with the burden carried by terms of high order.

For these reasons we treat the partial wave expansion (11.10a) of the amplitude as in Sec. 3.5. The 1 term in (11.10a), first line, gives rise to the diffraction scattering confined to a small angular region in the forward direction, whose size depends on the range R of the interaction,

$$k\theta \lesssim R^{-1} \tag{18.12}$$

as in Sec. 3.5.1. For larger angles we use only the $e^{2i\delta}$ term and there replace the Legendre polynomials by their asymptotic value (3.73). Furthermore, as in Sec. 3.5, the relatively slow variation of the phase shift with l justifies replacing the sum by an integral, after insertion of a term $e^{2\pi inl}$ (n an integer). This gives us an expression analogous to (3.75),

$$\begin{aligned} A(\theta) &\simeq (ip)^{-1}(2\pi\hbar\sin\theta)^{-1/2}\int_0^\infty dJ\,J^{1/2}(e^{i\zeta_+} + e^{i\zeta_-}) \\ \zeta_\pm &= \hbar^{-1}[2\hbar\delta_l^{\mathrm{WKB}} \pm J\theta\hbar \mp \tfrac{1}{4}\pi + 2\pi n(J - \tfrac{1}{2}\hbar)]. \end{aligned} \tag{18.13}$$

Because of the small value of $\hbar$ (as discussed in Sec. 18.1) these integrals may be evaluated by the method of stationary phase; the result becomes exact as $\hbar \to 0$. Accordingly, for a given angle θ the major contribution to the $\pm$ integrals comes from those values of J where

$$2\frac{d\delta_l^{\mathrm{WKB}}}{dl} \pm \theta + 2\pi n = 0.$$

In view of (18.11), this means that the scattering angle θ is connected to the J value by

$$\theta = \varepsilon[\Theta(J) + 2\pi n] \qquad \varepsilon = \pm 1. \tag{18.14}$$

This is the classical result, according to which a given impact parameter, or angular momentum, leads to a specific deflection angle Θ, which may have any value less than π. The sign in (18.14) and the value of the integer n have to be chosen so that θ lies in the range $0 \leq \theta \leq \pi$.

For a given θ, let us call J_0 the value of J where $\zeta_\pm$ is stationary. We expand $\zeta_\pm$ about it,

$$\begin{aligned} \zeta_\pm &= \zeta_{\pm 0} + \tfrac{1}{2}(J - J_0)^2 \zeta_0'' + \cdots \\ \zeta_{\pm 0} &= 2\delta_{l_0}^{\text{WKB}} - \varepsilon l_0 \theta + \tfrac{1}{4}\pi\varepsilon + 2\pi n l_0 \\ \zeta_0'' &= 2\hbar^{-2} \left.\frac{d^2\delta_l}{dl^2}\right|_{l=l_0} = \hbar^{-1} \left.\frac{d\Theta}{dJ}\right|_{J=J_0}. \end{aligned} \tag{18.15}$$

The extremely rapid oscillations of the integrand everywhere, except near $J = J_0$, allow us to extend the integral from $-\infty$ to $+\infty$ with negligible error. The resulting gaussian integral is readily evaluated and yields

$$A(\theta)\hbar \simeq (ip)^{-1} e^{i\zeta_{\pm 0}} \left(\frac{J_0}{-i\hbar\zeta_0'' \sin\theta} \right)^{1/2}.$$

Whether ζ_{+0} or ζ_{-0} is to be used depends on which ζ is stationary, i.e., which sign must be used in (18.14). Hence we finally obtain

$$A(\theta) = |A(\theta)| e^{i\alpha}$$

where

$$\begin{aligned} |A(\theta)| &= k^{-1} \left[\frac{l + \frac{1}{2}}{2|d^2\delta_l^{\text{WKB}}/dl^2| \sin\theta} \right]^{1/2} \\ &= p^{-1} \left[\frac{J}{|d\Theta/dJ| \sin\theta} \right]^{1/2} \end{aligned} \tag{18.16}$$

$$\alpha = 2\delta_l^{\text{WKB}} - (2l + 1)\frac{d\delta_l}{dl} - \tfrac{1}{2}\pi(2n + s) + \tfrac{1}{4}\pi(\varepsilon - 1) \tag{18.17}$$

and s is the algebraic sign of the slope of the deflection function,

$$s = \operatorname{sgn} \frac{d^2\delta_l}{dl^2} = \operatorname{sgn} \frac{d\Theta}{db}.$$

It is understood that the value of l or J in (18.16) and (18.17) is l_0 or J_0, respectively. The differential cross section obtained from (18.16) is the *classical* one,

$$\frac{d\sigma}{d\Omega} = \frac{b \csc\theta}{|d\Theta/db|} \tag{18.18}$$

where $b = J/p = (l + \frac{1}{2})/k$ is the impact parameter [see Eq. (5.12)].

Interferences. The phase α of the scattering amplitude is of no consequence, *provided* that for the given value of θ there is but one J value at which ζ_+

or ζ_- is stationary. If more than one such value exists, then classically unobtainable interference effects arise. Suppose that we label the various different stationary points by subscripts. Then

$$\frac{d\sigma}{d\Omega} = 2\sum_{i<j}\left[\frac{d\sigma_i^{cl}}{d\Omega}\frac{d\sigma_j^{cl}}{d\Omega}\right]^{1/2}\cos(\alpha_i - \alpha_j) + \sum_i \frac{d\sigma_i^{cl}}{d\Omega}, \tag{18.19}$$

the $d\sigma_i^{cl}/d\Omega$ being the classical cross sections originating from the different impact parameters which give rise to the same deflection angle. Since an over-all phase is irrelevant, we may write the phase angles (8.17) in the form

$$\alpha_i = k\left[\int_c^b db\,\Theta(b) - b_i\varepsilon_i\theta + 2\pi n_i b_i\right] - \pi n_i - \tfrac{1}{2}\pi s_i - (\varepsilon_i - 1)\tfrac{1}{4}\pi. \tag{18.20}$$

We find therefore that unless the deflection function Θ is a monotone function of the impact parameter b and is confined to values between $-\pi$ and π, so that for a given θ there is but one value of b, we do *not* get the classical result for the cross section, no matter how short the wavelength. Instead there are additional interference terms. However, in truly macroscopic scattering the energies of the scattered particles are not all *exactly* the same. That is to say, they are sufficiently smeared out to make the relative uncertainty of the wavelength large compared with the ratio of wavelength to impact parameter,

$$\frac{\delta\lambda}{\lambda} \gg \frac{\lambda}{b}.$$

This leads to an averaging of the phase angles α_i over a range large compared with 2π, and the interference terms average to zero. In contrast to the magnitude of the scattering amplitude, which contains classical parameters only, its phase contains the microscopic variable k, or the wavelength. It is the averaging over this, due to *macroscopic* errors in the energy, which finally gives us the purely classical result.

We may also estimate the accuracy with which *angles* have to be measured in order to observe the interference effects. Let Θ have one maximum, so that the same scattering angle is obtained from two impact parameters b_1 and b_2 or two l values l_1 and l_2. Then the cross section lies between

$$[(\sigma_1^{cl})^{1/2} + (\sigma_2^{cl})^{1/2}]^2 \quad \text{and} \quad [(\sigma_1^{cl})^{1/2} - (\sigma_2^{cl})^{1/2}]^2.$$

The difference between the two corresponding phases is

$$\Delta\alpha = \int_{l_1}^{l_2} dl\,(\Theta - \theta) + \pi.$$

We wish to know by how much the observation angle θ has to be changed in order to shift the cross section from a maximum to a minimum, assuming that this change in angle is so small that the magnitudes of the individual classical cross sections remain fixed. The phase change from constructive to destructive interference being π, the necessary shift in angle is given by

$$\pi = \Delta\alpha - \Delta\alpha' = \int_{l_1}^{l_2} dl(\Theta - \theta) - \int_{l_1'}^{l_2'} dl(\Theta - \theta')$$
$$\simeq (l_2 - l_1)(\theta' - \theta)$$

so that

$$\Delta\theta \simeq \frac{\pi}{l_2 - l_1} = \frac{\frac{1}{2}\lambda}{b_2 - b_1}. \tag{18.21}$$

This shows the narrowness of the angular oscillations produced by the interference. Only if the difference between the two impact parameters that lead to the same deflection is microscopic, i.e., comparable with the wavelength, are these oscillations observable. If this difference is of macroscopic size, the size of the counter that defines the acceptance angle in a practical scattering experiment leads to an averaging over an angular region that is large compared with (18.21). The interference effect is then unobservable even if the energy of the particles is very well defined.

The arguments show that there are two good reasons why a scattering experiment carried out with, say, pebbles shot at a rock is well described by classical mechanics. But it also tells us that there is a large region of energies in which the scattering of microscopic particles is well described by the semiclassical approximations without being simply given by the classical cross section. This will be true whenever the classical deflection angle fails to be a monotone function of the impact parameter, or to remain between $-\pi$ and π, so that the interference effect can be observed. As was pointed out in Sec. 5.5, such a situation always exists, for instance, if the interparticle force is everywhere attractive. Contrary to statements frequently found in the literature, it does not require an "unusual" potential for the interferences to occur. An example in which this approximation has seen much practical application is that of the scattering of α particles by nuclei. For details and other examples we refer to the literature.[4]

18.2.3 The Rainbow. We saw in Sec. 5.5 that classical particle scattering exhibits the analog of a rainbow at an angle at which the deflection function Θ is stationary with respect to variations in the impact parameter, i.e., where

$$\frac{d\Theta}{db} = 0.$$

This happens where the deflection angle has a relative maximum or minimum. If Θ has such an extremum, then the cross section exhibits the interferences described in the last section. At the *rainbow angle*, however, the discussion breaks down, because there the phase ζ_0'' of (18.15) vanishes. We must therefore proceed as in Sec. 3.6.

If $\Theta(J)$ is expanded about the rainbow value J_r of J at which $\Theta' = 0$,

4 See Ford and Wheeler (1959b).

$$\Theta(J) = \Theta_r + C(J - J_r)^2\hbar^{-2} + \cdots$$

$$C = \tfrac{1}{2}\hbar^2 \frac{d^2\Theta}{dJ^2}\bigg|_{J=J_r} = \tfrac{1}{2}k^{-2}\frac{d^2\Theta}{db^2},$$

then the phase shift is expanded as

$$\hbar\delta = \hbar\delta_r + \tfrac{1}{2}\Theta_r(J - J_r) + \tfrac{1}{6}C(J - J_r)^3\hbar^{-2} + \cdots$$

and the phases $\zeta_\pm$ of (18.13) become

$$\zeta_\pm = \hbar^{-1}[2\hbar\delta_r + \Theta_r(J - J_r) + \tfrac{1}{3}\hbar^{-2}C(J - J_r)^3 \pm J\theta \mp \tfrac{1}{4}\pi + 2\pi n(J - \tfrac{1}{2}\hbar)].$$

The resulting integral is of the same form as that in Sec. 3.6. By the same method we find that

$$A(\theta) \simeq |C|^{-1/3}(ik)^{-1}\left[\frac{2\pi(l_r + \frac{1}{2})}{\sin\theta}\right]^{1/2} e^{i\alpha + i\varepsilon(\theta_r - \theta)(l_r + 1/2)}\mathrm{Ai}(z) \tag{18.22}$$

where again ε is defined by (18.14) and α is given by

$$\alpha = 2\delta_r - \Theta_r(l_r + \tfrac{1}{2}) - \pi n + \tfrac{1}{4}\pi\varepsilon.$$

The definition of the Airy function Ai is explicitly shown in (3.88), and

$$z = s\varepsilon(\theta_r - \theta)|C|^{-1/3} \qquad s = \operatorname{sgn} C. \tag{18.23}$$

The result (18.22) shows that, just as in the electromagnetic case, the angular dependence of the cross section near the rainbow angle is determined by the square of the Airy integral. For a discussion see Sec. 3.6. Which is the dark and which the light side of the rainbow depends upon the algebraic signs of C and ε. The phase of (18.22) becomes important only if for the same angle θ other parts of the deflection function also contribute to the cross section.

The value of the differential cross section near the rainbow angle can be written in the form

$$\frac{d\sigma}{d\Omega} \simeq k^{1/3}\frac{2^{5/3}\pi b_r}{\sin\theta}\left|\frac{d^2\Theta}{db^2}\right|^{-2/3}[\mathrm{Ai}(z)]^2. \tag{18.24}$$

This shows that it tends to infinity as $\lambda\!\!\!^{-}{}^{-1/3}$ in the extreme classical limit. Note also that

$$z = \frac{2^{1/3}k^{2/3}s\varepsilon(\theta_r - \theta)}{|d^2\Theta/db^2|} \tag{18.23a}$$

so that the spacing between successive maxima goes to zero as $\lambda\!\!\!^{-}{}^{2/3}$.

18.2.4 The Glory. It was pointed out in Sec. 5.5 that, when back scattering is possible from impact parameters other than zero, then the classical cross section is infinite in the backward direction.[5] The effect is the analog of

5 Also in the forward direction, but there the effect is masked by diffraction.

the glory in the scattering of light, which was discussed in Sec. 3.7. It occurs whenever Θ goes smoothly through zero or through an integral multiple of π.

Since we are now interested in small values of $\sin\theta$, the evaluation of the partial wave expansion must use the asymptotic form (3.59) of the Legendre polynomials, rather than (3.73). In all other respects the procedure is the same as in Sec. 18.2.2,

$$A(\theta) \simeq (2ik)^{-1} \sum_l (2l+1) e^{2i\delta_l} (-)^l J_0[(l+\tfrac{1}{2})\sin\theta]$$
$$\simeq (ik)^{-1} \int dl\,(l+\tfrac{1}{2}) e^{i\pi l + 2\pi i n l + 2i\delta_l} J_0[(l+\tfrac{1}{2})\sin\theta].$$

We expand about the *glory value*, indicated by the subscript g,

$$\Theta(l) = \Theta_g + a(l - l_g) + \cdots \qquad \Theta_g = m\pi$$
$$a = \left.\frac{d\Theta}{dl}\right|_{l=l_g} = \left.\frac{1}{k}\frac{d\Theta}{db}\right|_{l_g}$$
$$\delta_l = \delta_g + \tfrac{1}{2}\Theta_g(l - l_g) + \tfrac{1}{4}a(l - l_g)^2 + \cdots$$

and obtain

$$A(\theta) \simeq (ik)^{-1} \exp[2i\delta_g + i\pi(1+2n)l_g] \int dl\,(l+\tfrac{1}{2})$$
$$\exp[i(\Theta_g + \pi + 2\pi n)(l - l_g) + i\tfrac{1}{4}a(l - l_g)^2] J_0[(l+\tfrac{1}{2})\sin\theta]$$
$$\simeq \frac{1}{k}\frac{1 + 2l_g}{(2\pi|a|)^{1/2}} \exp(i\alpha) J_0[(l_g + \tfrac{1}{2})\sin\theta] \tag{18.25}$$
$$\alpha = 2\delta_g - \pi m l_g - \tfrac{1}{4}\pi - \tfrac{1}{2}\pi\,\mathrm{sgn}\,a.$$

Thus the differential cross section for angles close to the backward direction is given by

$$\frac{d\sigma}{d\Omega} = \frac{b_g^2}{\lambda|d\Theta/db|}[J_0(kb_g\sin\theta)]^2. \tag{18.26}$$

The number b_g is the impact parameter that leads to exact backward deflection, and the derivative in the denominator is evaluated at that point.

Equation (18.26) shows the angular oscillations near the backward direction, caused by interference between the two rays going that way if Θ goes through $m\pi$. The peak is reached in the backward direction,

$$\left.\frac{d\sigma}{d\Omega}\right|_{\sin\theta=0} \simeq \frac{b_g^2}{\lambda|d\Theta/db|} \tag{18.27}$$

which goes to infinity in the extreme classical limit. Again, the larger k, the more rapid the angular oscillations.

The same effect exists in the forward direction. But there is also Fraunhofer diffraction, which was calculated in Sec. 3.5.1. According to (3.61) the diffraction contribution to the differential cross section for $\theta = 0$ is

$$\left.\frac{d\sigma}{d\Omega}\right|_{\text{diffr}} = \tfrac{1}{4}R^4k^2 = \frac{\pi^2 R^4}{\lambda^2} \tag{18.28}$$

if the range of the potential is R. Since (18.28) goes as λ^{-2}, it is usually larger than the glory when the effects are visible.

18.2.5 Orbiting (Spiral Scattering). The discussion of Sec. 5.4 showed that, if the potential has a relative maximum V_0 at some nonzero value of the radial distance, then for every energy $E > V_0$ the phenomenon of *orbiting* takes place at some value b_0 of the impact parameter. At b_0 the energy is just equal to the top of the peak of the combination of external potential and centrifugal barrier, and the deflection angle is infinite (see Fig. 5.3). As a result the differential cross section is made up of infinitely many contributions, one from each circuit of the scattered particle about the target. The same effect occurs if the potential is monotonely increasing with distance (as well as for many other shapes) and E is larger than the critical energy. It is clear that in such a situation interference effects should be particularly pronounced, and in the semiclassical region one may expect large deviations from the classical result (5.18). This will, of course, depend upon the value of the constant c in (5.19). As (5.20) shows, if c is small compared with unity, then the first one or two terms of the series (5.18) approximate the result well and interferences cannot play an important role.

There is, however, another point to be considered. When Θ approaches infinity, then, according to (18.11), the WKB phase shift is a rapidly varying function of l. This destroys the basis of the method used in Sec. 18.2.2 to sum the partial wave series, namely, the assumption that the phase shifts vary relatively slowly with l. The angular-momentum region near which orbiting occurs must therefore be treated separately.

Since spiral scattering is caused by the fact that the energy is just at a maximum of the effective potential, it is clear that quantum-mechanical effects of *barrier penetration* and related interference phenomena just above such potential barriers must play an important role. The result of this complicated combination will not be described here in detail, but we refer to the literature.[6]

18.3 The Eikonal Approximation

We now wish to consider a semiclassical approximation scheme which is the analog of the Van de Hulst extension of Rayleigh–Gans scattering of light. Let us take particles with no spin and write out the Lippmann–Schwinger equation (10.11), explicitly using (10.6) and defining

$$\varphi(\mathbf{k},\mathbf{r}) = \psi(\mathbf{k},\mathbf{r})\exp(-i\mathbf{k}\cdot\mathbf{r}).$$

6 See Ford, Hill, Wakano, and Wheeler (1959) and Ford and Wheeler (1959a).

Then the integral equation for φ reads

$$\varphi(\mathbf{k},\mathbf{r}) = 1 - \frac{m}{2\pi\hbar^2}\int d\mathbf{r}' \frac{\exp\{i[k|\mathbf{r}-\mathbf{r}'| - \mathbf{k}\cdot(\mathbf{r}-\mathbf{r}')]\}}{|\mathbf{r}-\mathbf{r}'|} V(\mathbf{r}')\varphi(\mathbf{k},\mathbf{r}')$$

$$= 1 - \frac{m}{2\pi\hbar^2}\int d\mathbf{r}''\, r''^{-1} \exp[i(kr'' - \mathbf{k}\cdot\mathbf{r}'')]\, V(\mathbf{r}-\mathbf{r}'')\varphi(\mathbf{k},\mathbf{r}-\mathbf{r}'')$$

after the substitution $\mathbf{r}'' = \mathbf{r} - \mathbf{r}'$. For large values of $k(\lambda \ll R)$ the exponential in the integral oscillates rapidly, except where the phase is stationary. That happens in the *forward direction*, i.e., where $\mathbf{r}''$ is parallel to $\mathbf{k}$. We therefore get

$$\varphi(\mathbf{k},\mathbf{r}) \simeq 1 - \frac{mi}{\hbar^2 k}\int_0^\infty dr'\, V(\mathbf{r}-\mathbf{r}')\varphi(\mathbf{k},\mathbf{r}-\mathbf{r}')(1 - e^{2ikr'})$$

$$\simeq 1 - \frac{ik}{2E}\int_0^\infty dr'\, V(\mathbf{r}-\mathbf{r}')\varphi(\mathbf{k},\mathbf{r}-\mathbf{r}')$$

with the understanding that $\mathbf{r}' = \mathbf{k}r'/k$. Let us decompose the radial vector into two parts,

$$\mathbf{r} = \mathbf{b} + \hat{\mathbf{k}}z$$

where the (impact parameter) vector $\mathbf{b}$ is perpendicular to $\mathbf{k}$. Then

$$\varphi(\mathbf{k};\mathbf{b}+\hat{\mathbf{k}}z) \simeq 1 - \frac{ik}{2E}\int_{-\infty}^{z} dz''\, V(\mathbf{b}+\hat{\mathbf{k}}z'')\varphi(\mathbf{k};\mathbf{b}+\hat{\mathbf{k}}z'')$$

after the replacement $z'' = z - z'$. This equation is immediately solved by

$$\varphi(\mathbf{k};\mathbf{b}+\hat{\mathbf{k}}z) \simeq \exp\left[-\frac{ik}{2E}\int_{-\infty}^{z} dz'\, V(\mathbf{b}+\mathbf{k}z')\right]$$

so that the wave function is approximated by the simple expression

$$\psi(\mathbf{k},\mathbf{b}+\hat{\mathbf{k}}z) \simeq \exp\left\{ik\left[z - \int_{-\infty}^{z} dz' \frac{V(\mathbf{b}+\hat{\mathbf{k}}z')}{2E}\right]\right\}. \tag{18.29}$$

This has a very simple interpretation. The integral represents the shift in the phase of ψ along the straight path from $-\infty$ to r. As we have seen in Sec. 18.1, $V/E \ll 1$ is the condition for little deflection. Hence the local wave number is

$$k(\mathbf{r}) = \frac{[2m(E-V)]^{1/2}}{\hbar} \simeq k\left(1 - \frac{V}{2E}\right)$$

so that the phase has been shifted by the total amount shown in (18.29). This realization suggests that the approximation may be improved in two ways: (1) by integrating, not along a straight path, but along the classical particle trajectory that passes through the given point; (2) by replacing $-kV/2E$ by $\{[2m(E-V)]^{1/2} - p\}/\hbar$. Since such a procedure, however, would considerably complicate a calculation, we shall not pursue it.

A slight improvement, however, can be made cheaply. If we are interested in scattering in the direction $\mathbf{k}'$, we may choose the path of integration along a straight line lying halfway between $\mathbf{k}$ and $\mathbf{k}'$, parallel to the vector

$$\hat{\mathbf{K}} = \frac{\mathbf{k} + \mathbf{k}'}{|\mathbf{k} + \mathbf{k}'|}$$

and $\mathbf{b}$ perpendicular to it,

$$\psi(\mathbf{k}, \mathbf{b} + \hat{\mathbf{K}}z) \simeq \exp\left\{i\left[\mathbf{k}\cdot\mathbf{b} + \mathbf{k}\cdot\hat{\mathbf{K}}z - \frac{k}{2E}\int_{-\infty}^{z} dz'\, V(\mathbf{b} + \hat{\mathbf{K}}z')\right]\right\}. \tag{18.30}$$

We must now calculate the scattering amplitude

$$\begin{aligned} A(\mathbf{k}', \mathbf{k}) &= -\frac{m}{2\pi\hbar^2}\int d\mathbf{r}\exp(-i\mathbf{k}'\cdot\mathbf{r})\,V(\mathbf{r})\psi(\mathbf{k}, \mathbf{r}) \\ &\simeq -\left(\frac{m}{2\pi\hbar^2}\right)\int d\mathbf{b}\, dz\, V(\mathbf{b} + \hat{\mathbf{K}}z) \\ &\qquad \exp\left\{i\left[(\mathbf{k} - \mathbf{k}')\cdot(\mathbf{b} + \hat{\mathbf{K}}z) - \frac{k}{2E}\int_{-\infty}^{z} dz'\, V(\mathbf{b} + \hat{\mathbf{K}}z')\right]\right\} \end{aligned}$$

in which the $\mathbf{b}$ integral is two-dimensional. But because $k = k'$,

$$(\mathbf{k} - \mathbf{k}')\cdot\hat{\mathbf{K}} = 0.$$

Had we chosen the original approximation (18.29), we should now argue that $(\mathbf{k} - \mathbf{k}')\cdot\mathbf{k}$ is negligible because the scattering angle is small, and we would throw the term away. Hence using (18.30) or (18.29) leads to the same result from now on. But (18.30), in addition to being reasonably expected to be more accurate, also eliminates the necessity for a further approximation. In any case, we can now carry out the z integral,

$$A(\mathbf{k}', \mathbf{k}) \simeq \frac{-ik}{2\pi}\int d\mathbf{b}\exp(-i\boldsymbol{\tau}\cdot\mathbf{b})\left\{\exp\left[-\frac{ik}{2E}\int_{-\infty}^{\infty} dz\, V(\mathbf{b} + \hat{\mathbf{K}}z)\right] - 1\right\} \tag{18.31}$$

where $\boldsymbol{\tau}$ is the momentum transfer

$$\boldsymbol{\tau} = \mathbf{k}' - \mathbf{k} \qquad \tau = \sqrt{|\boldsymbol{\tau}|^2} = 2k\sin\tfrac{1}{2}\theta.$$

It lies in the plane over which the vector $\mathbf{b}$ ranges.

If the potential is spherically symmetric, the angle integration may be carried out, using Bessel's integral,[7]

$$\frac{1}{2\pi}\int_0^{2\pi} d\varphi\, e^{-i\tau b\cos\varphi} = J_0(\tau b) = J_0(2kb\sin\tfrac{1}{2}\theta).$$

The result is therefore

7 See Whittaker and Watson (1948), p. 362.

$$A(\mathbf{k}',\mathbf{k}) \simeq -ik \int_0^\infty db\, b J_0(2kb \sin \tfrac{1}{2}\theta)(e^{2i\chi(b)} - 1), \tag{18.32}$$

$$\begin{aligned} \chi(b) &= -\frac{k}{2E} \int_0^\infty dz\, V((b^2 + z^2)^{1/2}) \\ &= -\frac{k}{2E} \int_b^\infty \frac{dr\, r}{(r^2 - b^2)^{1/2}} V(r). \end{aligned} \tag{18.33}$$

The approximation (18.32) may be considered also from the point of view of the partial wave analysis. Since $kb = l + \frac{1}{2}$ in the semiclassical correspondence and $J_0[(l + \frac{1}{2})\theta]$ is the large-l small-angle approximation (3.59) of the Legendre polynomial of order l, (18.32) is simply the replacement of

$$A(\mathbf{k}',\mathbf{k}) \simeq \left(\frac{1}{2ik}\right) \sum_l (2l + 1)(e^{2i\chi_l} - 1) P_l(\cos\theta) \tag{18.34}$$

by an integral. This replacement is justified by the arguments used in Sec. 18.2. The function $\chi(b)$ therefore is an approximation to the lth angular-momentum phase shift. Its relation to the WKB phase shift is obtained from (18.7). If in that expression F is expanded in a power series in V/E and the limit r_0 replaced by b, the result is (18.33). The approximation (18.32) is therefore equivalent to the WKB approximation for small angles of scattering, but taken at a stage before the integral over impact parameters is carried out by the method of stationary phase. Its advantages are the simplicity of (18.33) as compared with (18.7), the somewhat extended range of energies at which it can be expected to be valid because (18.32) is not executed by the stationary-phase method, and its applicability, in the form (18.31), to noncentral forces. Its chief drawback seems to be the limited angular region in which it is applicable. However, even that is not a real limitation, as we shall see below. In fact, the total cross section

$$\sigma_{\text{total}} \simeq 8\pi \int_0^\infty db\, b \sin^2 \chi(b) \tag{18.35}$$

calculated from (18.32) by the optical theorem, is a good approximation to the correct result. If there is absorption, of course, (18.35) must be modified to read

$$\sigma_{\text{total}} \simeq 4\pi \int_0^\infty db\, b(1 - \operatorname{Re} e^{2i\chi(b)}). \tag{18.35a}$$

Generalization. The simple form of (18.31) or (18.32) and (18.33) makes the method easily applicable and extendable. For example, it can be used amost as simply in many-channel problems. In that case χ is a hermitian matrix, and the only difficulty is the calculation of $e^{2i\chi}$. In the two-channel instance that is easily accomplished by expanding it on the basis of the Pauli spin matrices,

$$\chi = c\mathbb{1} + \mathbf{a} \cdot \boldsymbol{\sigma} \qquad \mathbf{a} = \operatorname{tr} \boldsymbol{\sigma}\chi.$$

Then

$$e^{2i\chi} = e^{2ic}(\cos 2a + i\mathbf{a}\cdot\boldsymbol{\sigma}a^{-1}\sin 2a)$$

with

$$a^2 = a_1^2 + a_2^2 + a_2^2.$$

Hermiticity implies that $\mathbf{a}$ is real and symmetry that $a_2 = 0$. For more than two channels one may use similar tricks.

Relation to Born Approximation. In spite of the fact that χ is linear in the potential, it is not simply either the Born approximation or the high-energy approximation to the phase shift. However, if, for a given l, k tends to infinity, then $b \to 0$ and (18.33) shows that

$$\chi(0) = -m\hbar^{-1}p^{-1}\int_0^\infty dr\, V(r).$$

Comparison with (12.33a) and with the remarks below (12.144) shows that this agrees with the high-energy limit of the lth phase shift.[8]

The approximation method of this section can be extended to time-dependent interactions and to the scattering of a particle by a bound state. For this, as well as for the calculation of examples, we refer to the literature.[9]

Extension to All Energies and Angles. An *impact parameter representation* of the scattering amplitude,

$$A(\mathbf{k}',\mathbf{k}) = A(k,\cos\theta) = 2k^2\int_0^\infty db\, bJ_0(2bk\sin\tfrac{1}{2}\theta)a(k,b) \tag{18.36}$$

analogous to (18.32) may be used for all values of the energy and angle. Insertion of the expansion [10]

$$\sum_l (2l+1)J_{2l+1}(x)P_l(\cos\theta) = \tfrac{1}{2}xJ_0(x\sin\tfrac{1}{2}\theta) \tag{18.37}$$

inversion of the sum and the integral, and comparison with the partial wave series leads to

$$a_l(k) = \frac{e^{2i\delta_l}-1}{2ik} = 2k\int_0^\infty db\, a(k,b)J_{2l+1}(2bk). \tag{18.38}$$

This expression in turn is inverted by using the orthogonality relation[11]

$$\int_0^\infty dx\, x^{-1}J_{2l+1}(x)J_{2l'+1}(x) = \frac{\delta_{ll'}}{2(2l+1)} \tag{18.39}$$

which yields

8 Notice that $2m\hbar^{-2}V$ is there called $\mathscr{V}$.

9 See particularly the lectures by R. J. Glauber (1959). For extensions to the Dirac equation and to spin-orbit forces, see the papers by L. I. Schiff (1956a and b).

10 This follows from A. Erdelyi (1954), vol. 2, p. 13, eq. (1).

11 See A. Erdelyi (1953), vol. 2, p. 92, eq. (32).

$$a(k,b) = \frac{1}{bk}\sum_l (2l + 1)J_{2l+1}(2bk)a_l(k). \tag{18.40}$$

Finally, $a(k,b)$ may be expressed directly in terms of the scattering amplitude by use of the inversion (13.12) of the partial wave series in (18.40); reversing the order of the sum and the integral, and again using (18.37),

$$a(k,b) = \tfrac{1}{2}\int_0^1 d\theta \sin\theta\, A(k,\cos\theta)J_0(2bk\sin\tfrac{1}{2}\theta). \tag{18.41}$$

These representations are valid for all energies and angles. At high energy, however, we may use the "optical localization principle" discussed at the beginning of Sec. 3.5. When $2bk \gg 1$ then the only term in the series (18.40) that contributes appreciably is the one for which $2l + 1 \simeq 2bk$. Hence

$$a(k,b) \simeq \frac{a_l}{bk}\sum_l (2l + 1)J_{2l+1}(2bk) = a_l(k) \qquad l = bk - \tfrac{1}{2}.$$

On account of the high energy behavior of the phase shifts, small values of b make a negligible contribution to (18.36). Thus it becomes in the high energy limit

$$A(\mathbf{k}',\mathbf{k}) \simeq -ik\int_0^\infty db\, bJ_0(2bk\sin\tfrac{1}{2}\theta)(e^{2i\delta_l} - 1). \tag{18.42}$$

If the WKB phase shift is inserted here we get an improved version of (18.32) which goes over into (18.32) on approximating δ^{WKB} by $\chi(b)$. This derivation shows that (18.32), or (18.42), should be applicable for *all angles*.

The main disadvantage of using the representation (18.36) for all energies, as compared to the partial wave expansion, is that the relation between the scattering amplitude $A(k,\cos\theta)$ and the impact parameter amplitude $a(k,b)$ is not one to one. Because the variable $z = \cos\theta$ is restricted to be greater than minus one, the function $a(k,b)$ given by (18.40) or (18.41) is not the only one that leads to a prescribed A when inserted in (18.36). [In fact the amplitude defined by (18.40) or (18.41) must satisfy a Kapteyn integral equation, and if it is inserted in (18.36) the integral vanishes for $z < -1$.] What is more, the unitarity relation expressed in terms of $a(k,b)$ takes a much less transparent form than for the partial waves. For further details and discussion we refer to the literature.[12]

18.4 The Impulse Approximation

In the description of the scattering of a particle by a bound system there is an energy scale that is absent in the considerations of the other sections of this chapter: the binding energy of the target. The impulse approximation

12 See especially Adachi and Kotani (1965 and 1966), T. Adachi (1966), and E. Predazzi (1966). In comparing the two sets of papers the lack of uniqueness of $a(k,b)$ in (18.36) should be kept in mind. Predazzi uses a Hankel transform instead of (18.41).

is based on the assumption that the energy of the missile is high compared with this binding energy. In that event the particles bound in the target may be viewed as essentially free. This, of course, is in practice a perfectly familiar approach. In many experimental situations the target particles are not really free. But they are so loosely bound that for all but very low bombarding energies a description by means of ordinary scattering theory for free particles is always assumed to introduce a negligible error. High-energy electron-electron scattering is a case in point.

There is another way of saying the same thing. If during the interaction time the wavelength or momentum of the target particle has been little changed by its binding force, it may be treated as a free particle of fixed momentum. The wave function of the missile and target particle together is then obtained as an appropriate superposition of individual scattering wave functions, weighted according to the probability amplitude of finding individual momenta of the target particle in its bound state. Let us formulate this mathematically.

The wave function describing scattering of particle 1 by particle 2 in an arbitrary coordinate system, is given by

$$\psi(\mathbf{k}_1,\mathbf{k}_2;\mathbf{r}_1,\mathbf{r}_2) = \psi(\mathbf{k},\mathbf{r})\exp(i\mathbf{K}\cdot\mathbf{R}) \tag{18.43}$$

where

$$\begin{aligned} \mathbf{R} &= \frac{m_1\mathbf{r}_1 + m_2\mathbf{r}_2}{m_1 + m_2} & \mathbf{r} &= \mathbf{r}_1 - \mathbf{r}_2 \\ \mathbf{K} &= \mathbf{k}_1 + \mathbf{k}_2 & \mathbf{k} &= \frac{m_2\mathbf{k}_1 - m_1\mathbf{k}_2}{m_1 + m_2} \end{aligned} \tag{18.44}$$

and $\psi(\mathbf{k},\mathbf{r})$ is the scattering wave function in the center-of-mass coordinate system. It is the solution of the integral equation

$$\psi(\mathbf{k},\mathbf{r}) = \exp(i\mathbf{k}\cdot\mathbf{r}) - \frac{1}{4\pi}\int d\mathbf{r}'\frac{\exp(ik|\mathbf{r}-\mathbf{r}'|)}{|\mathbf{r}-\mathbf{r}'|}\mathscr{V}(\mathbf{r}')\psi(\mathbf{k},\mathbf{r}')$$

if $\mathscr{V}$ is the potential between particles 1 and 2. We now assume that the momentum of particle 2 is "smeared out," with an amplitude $f(\mathbf{k}_2)$ for momentum $\mathbf{k}_2$. Then the corresponding two-particle wave function is approximated by

$$\begin{aligned} \phi(k_1;\mathbf{r}_1,\mathbf{r}_2) &= (2\pi)^{-3/2}\int d\mathbf{k}_2 f(\mathbf{k}_2)\psi(\mathbf{k}_1,\mathbf{k}_2;\mathbf{r}_1,\mathbf{r}_2) \\ &= (2\pi)^{-3/2}\int d\mathbf{k}_2 f(\mathbf{k}_2)\exp(i\mathbf{K}\cdot\mathbf{R})\psi(\mathbf{k},\mathbf{r}). \end{aligned} \tag{18.45}$$

Specifically, we determine $f(\mathbf{k}_2)$ from the bound-state wave function $\phi(\mathbf{r}_2)$,

$$\phi(\mathbf{r}_2) = 2\pi)^{-3/2}\int d\mathbf{k}_2 f(\mathbf{k}_2)\exp(i\mathbf{k}_2\cdot\mathbf{r}_2)$$

so that

$$f(\mathbf{k}_2) = (2\pi)^{-3/2} \int d\mathbf{r}_2\, \phi(\mathbf{r}_2) \exp(-i\mathbf{k}_2 \cdot \mathbf{r}_2). \tag{18.46}$$

Thus $f(\mathbf{k}_2)$ is the bound-state wave function in momentum space.

The function (18.45) is, of course, not an exact solution of the two-particle Schrödinger equation. This is because the approach of the missile changes the momentum distribution of particle 2. If it is $f(\mathbf{k}_2)$ when the particles are far apart, it will not be so when they interact. The approximation (18.45) keeps the momentum distribution frozen during the interaction time.

In reality (18.45) satisfies the equation

$$\begin{aligned} H\phi(\mathbf{k}_1;\mathbf{r}_1,\mathbf{r}_2) &= (2\pi)^{-3/2} \int d\mathbf{k}_2\, f(\mathbf{k}_2) \left[\frac{K^2}{2M} + \frac{k^2}{2\mu} + V_2(\mathbf{r}_2)\right] \exp(i\mathbf{K}\cdot\mathbf{R})\psi(\mathbf{k},\mathbf{r}) \\ &= (2\pi)^{-3/2} \int d\mathbf{k}_2\, f(\mathbf{k}_2) \left[\frac{k_1^2}{2m_1} + \frac{k_2^2}{2m_2} + V_2(\mathbf{r}_2)\right] \exp(i\mathbf{K}\cdot\mathbf{R})\psi(\mathbf{k},\mathbf{r}) \end{aligned}$$

where

$$M = m_1 + m_2 \qquad \mu = \frac{m_1 m_2}{M}.$$

But if $k_1 \gg k_2$ for all $\mathbf{k}_2$ that appreciably contribute to the integral, i.e., for which $f(\mathbf{k}_2)$ is not negligibly small, then $\psi(\mathbf{k},\mathbf{r})$ varies little over the essential region of integration and $k_2{}^2/2m_2 + V_2(\mathbf{r}_2)$ may be replaced by E_2, the negative binding energy of particle 2. We then have

$$H\phi \simeq (E_1 + E_2)\phi$$

to a good approximation.

The function defined by (18.45) of course obeys the correct boundary conditions. As the interparticle distance $\mathbf{r}$ tends to infinity (or the time to $-\infty$), it approaches

$$\begin{aligned} \phi(\mathbf{k}_1;\mathbf{r}_1,\mathbf{r}_2) &\simeq (2\pi)^{-3/2} \int d\mathbf{k}_2 f(\mathbf{k}_2) \exp[i(\mathbf{K}\cdot\mathbf{R} + \mathbf{k}\cdot\mathbf{r})] \\ &= (2\pi)^{-3/2} \int d\mathbf{k}_2 f(\mathbf{k}_2) \exp[i(\mathbf{k}_1\cdot\mathbf{r}_1 + \mathbf{k}_2\cdot\mathbf{r}_2)] \\ &= \phi(\mathbf{r}_2)\exp(i\mathbf{k}_1\cdot\mathbf{r}_1). \end{aligned}$$

We use the approximate wave function (18.45) to calculate the scattering amplitude, elastic or inelastic. If initially particle 2 is in a bound state ϕ_1 and finally it is in the bound state (or continuum state, for that matter) ϕ_2, then according to (16.57b) the exact scattering or reaction amplitude is

$$A(\mathbf{k}_1,\mathbf{k}_1) = -\frac{m_1}{2\pi}\left(\frac{k_1'}{k_1}\right)^{1/2} \int d\mathbf{r}_1\, d\mathbf{r}_2 \exp(-i\mathbf{k}_1'\cdot\mathbf{r}_1)\phi_2^*(\mathbf{r}_2) V(\mathbf{r})\psi(\mathbf{k}_1;\mathbf{r}_1,\mathbf{r}_2)$$

if $\psi(\mathbf{k}_1;\mathbf{r}_1,\mathbf{r}_2)$ is the exact solution of the two-particle Schrödinger equation. Replacing ψ by the approximation (18.45), we get

$$A(\mathbf{k}_1',\mathbf{k}_1) \simeq -m_1(2\pi)^{-5/2}\left(\frac{k_1'}{k_1}\right)^{1/2}\int d\mathbf{k}_2\, d\mathbf{r}_1\, d\mathbf{r}_2$$

$$\exp(-i\mathbf{k}_1'\cdot\mathbf{r}_1)\phi_2^*(\mathbf{r}_2)V(\mathbf{r})f_1(\mathbf{k}_2)\exp(i\mathbf{K}\cdot\mathbf{R})\psi(\mathbf{k},\mathbf{r})$$

$$= -m_1(2\pi)^{-4}\left(\frac{k_1'}{k_1}\right)^{1/2}\int d\mathbf{k}_2\, d\mathbf{k}_2' f_2^*(\mathbf{k}_2')f_1(\mathbf{k}_2)$$

$$\int d\mathbf{R}\, d\mathbf{r} \exp\{i[(\mathbf{K}-\mathbf{K}')\cdot\mathbf{R} - \mathbf{k}'\cdot\mathbf{r}]\}\, V(r)\psi(\mathbf{k},\mathbf{r})$$

$$= \left(\frac{k_1'}{k_1}\right)^{1/2}\int d\mathbf{K}\, d\mathbf{K}'\, \delta(\mathbf{K}-\mathbf{K}')f_2^*(\mathbf{K}'-\mathbf{k}_1')f_1(\mathbf{K}-\mathbf{k}_1)t(\mathbf{k}',\mathbf{k})$$

or finally

$$A(\mathbf{k}_1',\mathbf{k}_1) \simeq \left(\frac{k_1'}{k_1}\right)^{1/2}\int d\mathbf{K}\, f_2^*(\mathbf{K}-\mathbf{k}_1')f_1(\mathbf{K}-\mathbf{k}_1)t(\mathbf{k}',\mathbf{k}) \tag{18.47}$$

where $t(\mathbf{k}',\mathbf{k})$ is the ordinary two-particle amplitude

$$t(\mathbf{k}',\mathbf{k}) = -\frac{m_1}{2\pi}\int d\mathbf{r} \exp(i\mathbf{k}'\cdot\mathbf{r})V(\mathbf{r})\psi(\mathbf{k},\mathbf{r})$$

and

$$\mathbf{k} = \mathbf{k}_1 - \left(\frac{m_1}{M}\right)\mathbf{K} \qquad \mathbf{k}' = \mathbf{k}_1' - \left(\frac{m_1}{M}\right)\mathbf{K}. \tag{18.48}$$

Note, however, that $t(\mathbf{k}',\mathbf{k})$ is taken *off the energy shell*. That is, even for elastic scattering ($k_1 = k_1'$) the magnitudes of $\mathbf{k}$ and $\mathbf{k}'$ are equal over only a negligible part of the range of integration of $\mathbf{K}$.

In some cases the approximation (18.47) may be further simplified by taking the function $t(\mathbf{k}',\mathbf{k})$ outside the integral. This can be done if approximately $t(\mathbf{k}',\mathbf{k})$ depends only upon the momentum transfer $\mathbf{k}' - \mathbf{k} = \mathbf{k}_1' - \mathbf{k}_1$. If V is weak enough, for example, so that $t(\mathbf{k}',\mathbf{k})$ is well approximated by its first Born approximation, this is always so. If the process described is a *dissociation* of the initial bound state, then ϕ_2 is a continuum wave function and it must refer to momentum $\mathbf{k}_2'$ of the ejected particle 2. In that case $f_2(\mathbf{k}_2)$ is the solution of the integral equation

$$f(\mathbf{k}_2) = \delta(\mathbf{k}_2 - \mathbf{k}_2') + \left(E_2 - \frac{k_2^2}{2m_2}\right)^{-1}\int d\mathbf{k}_2''(\mathbf{k}_2|U|\mathbf{k}_2'')f(\mathbf{k}_2'')$$

if U is the potential that binds particle 2 in the initial state, and $E_2 = k_2'^2/2m_2$. Thus it may be justifiable to evaluate $t(\mathbf{k}',\mathbf{k})$ in (18.47) at $\mathbf{K} = \mathbf{k}_1' + \mathbf{k}_2'$ and take it out of the integral. In either of these two instances $A(\mathbf{k}_1',\mathbf{k}_1)$ becomes a *product* of one term that depends only on the interaction between particles 1 and 2 and another that depends only on the binding force acting on particle 2.

The Cross Section. The squared magnitude of the amplitude (18.47) approximates the cross section for scattering of particle 1 in the direction $\mathbf{k}_1'$, accompanied by an excitation of particle 2 to a specific level. If the residual bound state is not observed, the cross section is obtained by summing over the final levels. We may then use *closure*, i.e., the fact that

$$\sum_n f_n(\mathbf{k}_2) f_n^*(\mathbf{k}_2') = \delta(\mathbf{k}_2 - \mathbf{k}_2')$$

and get

$$\begin{aligned} \frac{d\sigma}{d\Omega}(\bar{\mathbf{k}}_1', \mathbf{k}_1) &= \left(\frac{\bar{k}_1'}{k_1}\right) \int d\mathbf{k}_2 \, |f_1(\mathbf{k}_2)|^2 |t(\mathbf{k}', \mathbf{k})|^2 \\ &= \int d\mathbf{k}_2 \, |f_1(\mathbf{k}_2)|^2 \frac{d\sigma_0}{d\Omega}(\mathbf{k}', \mathbf{k}) \frac{\bar{k}_1' k}{k_1 k'} \end{aligned} \tag{18.49}$$

$$\mathbf{k} = \frac{m_2 \mathbf{k}_1 - m_1 \mathbf{k}_2}{M} \qquad \mathbf{k}' = \bar{\mathbf{k}}_1' - \mathbf{k}_1 + \mathbf{k}.$$

Since k_1' is unobserved we have replaced it by an appropriate average $\bar{k}_1'$ (so that $\bar{\mathbf{k}}_1' = \hat{\mathbf{k}}_1' \bar{k}_1'$). This had to be done before the use of closure and it constitutes a somewhat dubious approximation. As a consequence of this procedure all interferences have vanished, and the result is simply the sum of the various cross sections for scattering by target particles of different momenta, weighted according to their probabilities in the initial bound state.

For further details, extensions, other formulations, and applications, particularly to the scattering of fast neutrons by deuterons, we refer to the literature.[13]

Notes and References

18.2 The original references for the WKB method are H. Jeffreys (1923); G. Wentzel (1926a); H. A. Kramers (1926); L. Brillouin (1926). The name "JWKB method," which is occasionally used, is therefore more appropriate. See also R. E. Langer (1932 and 1937); W. H. Furry (1947).

In scattering theory this method is, of course, combined with the older procedures used in electromagnetic theory (see Notes and References to Sec. 3.5). For a detailed treatment of the semiclassical theory of scattering see particularly Ford and Wheeler (1959a), on which most of this section is based. For some modifications and extensions, see Hecht and Mayer (1957); A. Erdelyi (1960); S. A. Lebedeff (1964 and 1970); Rosen and Yennie (1964); L. Bertocchi et al. (1965a); Fröman and Fröman (1965); P. C. Sabatier (1965); D. R. Yennie et al. (1965); Bassichis and Dar (1966); R. J. Cross (1967, 1968, 1969); van Horn and Salpeter (1967); P. Pechukas (1969); B. C. Eu (1970, 1971b, 1972, 1973); P. Lu (1970); Berry and Mount (1972); Y. Hahn (1972e); Lu

13 G. F. Chew (1948 and 1950); Fernbach, Green, and Watson (1951); B. Segall (1951); Fujimoto and Yamaguchi (1951); Chew and Lewis (1951); W. Cheston (1952); see also the other references in the Notes and References.

and Measure (1972); Eu and Tsien (1973); Lu and Wald (1973); R. K. Nesbet (1973); H. G. Guerin et al. (1974); Wald and Lu (1974); A. R. P. Rau (1976); Fröman and Yngve (1980); S. F. J. Wilk et al. (1981).

Applications are given by Ford and Wheeler (1959b); Rabitz and Gordon (1970); Mizuno and Chen (1971); Yu. S. Sayasov (1975); Spruch and Shakeshaft (1979); L. D. Thomas (1980).

For semiclassical treatments of barrier phenomena, any textbook of quantum mechanics may be consulted. Particularly relevant details can be found in Goldman and Migdal (1954); Ford, Hill, Wakano, and Wheeler (1959). These methods are used in the first-mentioned paper by Ford and Wheeler for a study of the interferences encountered in orbiting.

For treatments of the *diffraction* part of the scattering amplitude, see Bethe and Placzek (1940); Greider and Glassgold (1960).

A discussion of the summation of partial waves at high energy was given by H. Brysk (1965) and of the polarization in the classical approximation by Lu, Fradkin, and Good (1964).

An application of the WKB method to the three-body problem was given by Vezzetti and Rubinow (1965); see also A. M. Brodskii et al. (1970); Brodskii and Potapov (1970).

18.3 The eikonal approximation and its extension have been developed in the following articles: G. Molière (1947 and 1948); Fernbach, Serber, and Taylor (1949); G. Parzen (1951); Montroll and Greenberg (1954); Goldman and Migdal (1954); L. I. Schiff (1956b); Saxon and Schiff (1957); Nigam, Sundaresan, and Wu (1959); R. J. Glauber (1959). The last-mentioned article is particularly recommended; our approach largely follows one part of it.

See also Blankenbecler and Goldberger (1962); Lynch and Carlson (1962); M. Ida (1962); R. Serber (1963 and 1964); Cottingham and Peierls (1965); Adachi and Kotani (1965, 1966a and b); (the last part of Sec. 18.3 follows these authors;) T. Adachi (1966); T. Adachi et al. (1966); R. Heinzi (1966); E. Predazzi (1966); F. T. Smith et al. (1966); R. J. Glauber (1967); M. M. Islam (1967 and 1976); Y. Hahn (1968b, 1969a, 1970b, 1973); D. R. Harrington (1969, 1970, 1971); Sugar and Blankenbecler (1969); H. Chew (1970); T. A. Osborn (1970 and 1972); R. A. Rudin (1970); B. Corrigall et al. (1971); Joachain and Mittleman (1971); R. L. Kelly (1971); E. Kujawski (1971); S. J. Wallace (1971, 1973b, 1974); A. Baker (1972 and 1973); R. T. Ling et al. (1972); I. Manning (1972); J. M. Namyslowki (1972); L. van Hove (1972); Agassi and Gal (1973b); Berriman and Castillejo (1973); Byron and Joachain (1973); F. W. Byron et al. (1973 and 1975); M. E. Riley (1973); Fried and Hahn (1974); A. R. Swift (1974); Storm and Rapp (1976); Semon and Taylor (1977); Darewych and Pooran (1978); Vall and Makeev (1978); F. T. Chan et al. (1979); K. Yajima (1979a, 1981a, b, c); R. E. Turner (1980a and b); A. C. Williams (1980); H. Cheng et al. (1981); Halpern and Franco (1981).

18.4 The impulse approximation was invented by E. Fermi (1936) and G. F. Chew (1950). See also Chew and Wick (1952); Ashkin and Wick (1952); Chew and Goldberger (1952); Ivanov and Sayasov (1963); L. Rosenberg (1963 and 1964); Goldberger and Watson (1964b), pp. 683ff; R. M. Hayborn (1971); R. E. Turner (1980a). The papers by Chew and Wick and by Ashkin and Wick especially should be consulted for a formulation of the impulse approximation in the language of formal scattering theory.

For other approaches to high-energy approximation, see G. Parzen (1950); E. Corinaldesi (1954b); M. Verde (1955, 1957, 1958, 1963); W. Hunziker (1963); K. Chadan (1968); P. Johnson (1968); L. I. Schiff (1968a); R. J. Moore (1970); Burt and Woods (1971); Y. Hahn (1968c); S. J. Wallace (1973a); Tobocman and Pauli (1972);

Reading and Bassichis (1972); A. C. T. Wu (1974); E. A. Bartnik et al. (1975); Chen and Hook (1975); M. M. Skriganov (1978).

Problems

1 Particles of mass 500 and 600 Mev interact with one another via an attractive square-well potential of depth 100 Mev and range 10^{-11} cm. Calculate and plot their total collision cross section at center-of-mass energies of 1 to 50 Mev in the eikonal approximation. Do you expect this approximation to be good in this energy range? At what energy will the Born approximation be good?

2 Suppose that the potential of Prob. 1 is complex,

$$V = -(80 \text{ Mev}) - i(50 \text{ Mev})$$

when $r < 10^{-11}$ cm, and zero otherwise. Calculate the total cross section in the eikonal approximation.

3 Suppose that the particles of Prob. 1 interact via a gaussian potential

$$V = -\gamma e^{-r^2/a^2}$$

where $\gamma = 120$ Mev, $a = 10^{-11}$ cm. Calculate and plot their total cross section in the eikonal approximation for the same energy range as in Prob. 1. Compare the two curves.

4 Let the particles of Prob. 1 interact via the potential

$$V = -\gamma(c^2 + r^2)^{-3/2}$$

where $\gamma = 130$ Mev, $c = 10^{-11}$ cm. Calculate and plot their total cross section in the eikonal approximation for the same energy range as in Prob. 1. Compare the curve with those of Probs. 1 and 3.

5 Calculate the differential scattering cross section in the eikonal approximation for a screened Coulomb potential

$$V = \begin{cases} \dfrac{Ze^2}{r} & r < R \\ 0 & r > R \end{cases}$$

in the limit when R is large. (What is the relevant criterion?) Discuss.

6 Suppose that in the scattering of particle A by a bound state of particle B with a "nucleus" the energy is high enough so that both the impulse approximation and the Born approximation for the interaction of A with B can be used. (What are the relevant criteria for this?) Calculate the inelastic amplitude for scattering accompanied by a specific excitation of the B-nucleus system. Is it a function only of the momentum transfer? If not, can it be made so by a simple transformation? What is the value of the forward-scattering cross section? What is the value of the total cross section?

CHAPTER 19

The Decay of Unstable States

19.1 Qualitative Introduction

Suppose that a scattering experiment is carried out at an energy at which the interacting fragments have a sharp resonance. We have seen in Sec. 11.2.2 that the elastically or inelastically scattered particles then are considerably delayed relative to those coming in. We now wish to examine in detail the time dependence of the emergent flux.

The subject is of physical importance, not so much because scattering or reaction experiments are usually carried out in a way which allows the measurement of such a delay—they rarely are—but every observation of radioactivity, or of the decay of any other unstable quantum-mechanical system of long lifetime, is an instance in point. If the delay between incoming and outgoing flux is long enough, the target may be removed from the apparatus and the efflux of the decay products observed in circumstances that allow us to forget all about the initial excitation of the resonance. This is why the problem is usually treated from a bound-state point of view. But it fits into a scattering framework much more naturally. The fact is that only a truly bound state, of infinite lifetime, is really independent of how it was formed. Decaying states are not bound states which after their formation were disturbed and therefore are unstable. They are always sharp resonances, and as such they are most reliably and physically most meaningfully treated in the context of the theory of scattering or reactions.

The fiction of supposing the decaying state to be bound immediately yields a simple answer for the time dependence of the emitted-particle flux. Since the properties of the emitting system are constant, and the rate of emission is proportional to the number of "compound" systems extant at the time, the decay law is necessarily one of *exponential* dependence on the time. Now the premise is incorrect, and so is the result based on it. Nevertheless, a compound state of long lifetime physically very much resembles a stable state and may in some situations be almost indistinguishable from it. We think of a neutron as very much like a stable particle such as a proton. The fact that it is subject to beta decay does not essentially alter this. The exponential law should therefore not be altogether wrong, and of course it is not. On the contrary, it is experimentally verified in some cases to an astonishing degree, much more so than one has a right to expect considering the rather flimsy basis on which it is usually derived. The question we should therefore answer is: If the exponential law is not exact, by how much does it differ from the correct time dependence, and on what do the deviations depend?

One of the reasons for the exponential time dependence not to be exact can be understood classically. Consider a system that at the time $t = 0$ starts to emit particles along a line. Suppose that the missile flux is not monoenergetic but has a velocity distribution $\rho(v)$[1] and that it decreases exponentially with time, as $e^{-t/\tau}$. Then the number of particles that cross a point at a distance x from the orifice, per unit time at the time t, is

$$N(x,t) = \int_{x/t}^{\infty} dv\, \rho(v) \exp\left(-\frac{t - x/v}{\tau}\right).$$

For a fixed value of x, this is not an exponential function of the time. For very large values of t, the highly singular nature of $\exp(x/v\tau)$ in fact produces large deviations from an exponential. By setting $y = t - x/v$ we readily find the asymptotic value as $t \to \infty$,

$$N(x,t) \simeq \rho(0)\frac{x}{\tau}\left(\frac{\tau}{t}\right)^2.$$

For a distribution $\rho(v)$ that is strongly peaked at $v = v_0$, the decay curve observed at x is very nearly exponential starting at the time t_F that it takes for the distribution center to arrive at x. So long as

$$R = \rho(0)\frac{x}{\tau}\left(\frac{\tau}{t}\right)^2 \exp\left(\frac{t - t_F}{\tau}\right)$$

remains very small, the curve retains its nearly exponential shape. Note that the length of the exponential part of the decay curve depends on the distance x from the decay center. If the observation takes place too far away, an exponential law is never seen.

The physical reason for the nonexponential tail is of course the late arrival of slow particles. Classically the slowly decreasing asymptotic tail can be easily removed by cutting the velocity distribution $\rho(v)$ off at some minimal value $v_{\min} > 0$. Quantum mechanically this would not help, because it would

1 We assume that $\rho(v)$ is normalized to unity.

necessarily add an infinitely long tail in configuration space. Indeed, the mere fact that quantum mechanically we do not have the full range of frequencies from minus infinity to plus infinity at our disposal makes it impossible to cut the signal off cleanly. This adds an unavoidable ambient excitation whose experimental consequences are small but which contributes to the impossibility of an exponential asymptotic decrease.

19.2 Exponential Decay and Its Limitations

Turning to the quantitative discussion, we are trying to describe a situation in which a missile is shot at a target, forms an unstable state with it, and subsequently decays into products that may or may not be identical with the initial particles. The efflux is observed as a function of the time at some distance r from the scattering center. If the entire event is described by means of time-dependent scattering theory, the question of precise initial conditions never arises.

The first assumptions necessary are those which enable us to use scattering theory. According to (10.13), they are that the distance r of the counter from the decay center be large compared with the size D if the interaction region

$$r \gg D \tag{19.1}$$

and that furthermore r be large compared with D multiplied by the ratio of D to all the wavelengths that appreciably contribute to the event,

$$r \gg \frac{D^2}{\lambda} \quad \text{that is } kr \gg (kD)^2. \tag{19.2}$$

In practice both these inequalities are usually extremely well satisfied.

With the assumptions (19.1) and (19.2) we may write the two-particle wave function of missile and target in the center-of-mass system in the asymptotic form

$$\psi(\mathbf{k},\mathbf{r}) \simeq \exp(i\mathbf{k}\cdot\mathbf{r}) + r^{-1}\exp(ikr)A(\mathbf{k}',\mathbf{k}).$$

For simplicity we shall restrict ourselves to two-particle decays. But the scattering need not be elastic. For inelastic instances, the first term on the right is missing, and $k \neq k'$. In any case, $\hat{\mathbf{k}} = \hat{\mathbf{r}}$.

We now suppose that the scattering amplitude has a sharp *resonance*. This, it will be remembered, means two things: Its magnitude has a sharp peak *and* its phase turns sharply upwards by π near the same energy. The first causes the large cross section, which is usually the only effect observed. As we have seen in the discussion of Eq. (11.56), the second is responsible for the time delay that is an integral part of the mechanism to which the resonance peak in the cross section is physically attributed. In principle these two effects are quite independent, and only when they both occur together is it legitimate to speak of a resonance. As a consequence the true energy dependence of the amplitude A in the vicinity of a Breit–Wigner resonance in the cross section, given by

$$\sigma \propto \frac{\frac{1}{4}\Gamma^2}{(E - E_0)^2 + \frac{1}{4}\Gamma^2} \tag{19.3}$$

is of necessity

$$A \propto \frac{\frac{1}{2}\Gamma}{E - E_0 + \frac{1}{2}i\Gamma} \tag{19.4}$$

with $\Gamma > 0$. The particular shape (19.3) is of course an additional assumption whose modification we shall discuss in the next section.

Let us assume that A contains a well-isolated sharp resonance (19.4). Since we are interested in the detailed time development of the delayed signal, it is clear what the relevant criterion of sharpness ought to be: The delay time at the resonance peak must be large relative to those at neighboring energies. In other words, the slope of the phase φ of the amplitude at E_0 must be much larger than at other energies nearby. Since for the Breit–Wigner shape Eq. (19.4) together with (11.56) or (11.59) gives us the delay time

$$t_D = \left(\frac{d\varphi}{dE}\right)_{E=E_0} = \frac{2}{\Gamma}$$

this means that $2/\Gamma$ must be large relative to the slope of φ elsewhere in this region. The relevant criterion for calling the resonance isolated must be that we can form a wave packet inside whose energy range only one resonance line appears. It is to be expected that this means that the next line is at least several widths away. A more precise conclusion is to be drawn from the analysis below.

The following word of caution about the stationary-phase argument given in Sec. 11.2.2: As far as the first term r/v_0 in (11.56) of the outgoing wave's peak time is concerned, the phase argument is exact in the limit as $r \to \infty$; it is excellent for practical values of r. The delay term $d\varphi/dk$, however, must not be taken quantitatively seriously. It is of qualitative significance only. The reason is that when $d\varphi/dk$ is large in a resonance the actual increase in φ is only about π so that the oscillation does not really have the effect of reducing the integrand appreciably. Furthermore the resonance *magnitude* of the integrand is sharply peaked at the same point. One must therefore *not* conclude from the phase argument that the observed activity should be maximal two lifetimes after the time of flight. The actual maximum will, of course, occur near the time of flight. The real meaning of the delay is contained in the exponential-decay law. For $\Gamma < 0$ there would be no such decay.

It is also incumbent upon us to say something about the shape of the exciting wave packet. It is physically clear that it should do two mutually contradictory things as best it can: In order not to interfere more than necessary with the time curve of the decaying system, it should be of short duration and should be cut off sharply. In order to isolate the resonance products from other delayed signals (such as from other resonances), it should be of narrow energy range and should be cut off sharply there. In the extreme case of a monochromatic beam there is obviously no decay but a steady state. The opposite extreme of all positive energies represented almost equally in the beam will be possible only in exceptional circumstances of extreme isolation of the line. One may expect that the energy width of the packet should be larger than the line width so as to "fill out" its shape well. In order to realize a sharp cutoff both in time and energy, we choose a gaussian shape (restricted to physically accessible energies). It is certainly possible to alter details of the ensuing results by minor alterations in the packet shape. Specifically, the asymptotic behavior for infinite time is sensitive to such changes. This is why we shall avoid basing any arguments on the exact shape of the asymptotic tail. But, as in many other contexts, the essential results, if gleaned carefully without recourse to accidental features, can be expected to be independent of the precise shape of the packet.

In order to be able to apply the results as generally as possible, we shall use relativistic kinematics in the following: If W is the total energy and m_1 and m_2 are the rest masses of the two emerging particles, the momentum in the barycentric system is expressed by

$$k^2 = \frac{(W^2 - m_1^2c^4 - m_2^2c^4)^2 - 4m_1^2m_2^2c^8}{4W^2c^2}. \tag{19.5}$$

We may also express k in terms of the final *kinetic* energy,

$$E = W - (m_1 + m_2)c^2$$

or of its relative deviation from the center E_0 of the packet. Set $k = k_0\zeta$ and $s = (E - E_0)/E_0$; then

$$\zeta = \left[(1+s)\left(1 + \frac{sE_0}{E_0 + 2(m_1+m_2)c^2}\right)\left(1 + \frac{sE_0}{E_0 + 2m_1c^2}\right)\left(1 + \frac{sE_0}{E_0 + 2m_2c^2}\right)\right]^{1/2}\left(1 + \frac{sE_0}{E_0 + (m_1+m_2)c^2}\right)^{-1}. \tag{19.6}$$

It will be useful to expand this equation in a series in powers of s. This gives

$$\zeta = 1 + \rho_1 s - \rho_2 s^2 + O(s^3) \tag{19.7}$$

with the coefficients

$$\begin{aligned} \rho_1 &= [(1+\alpha_1)(1+\alpha_2) + \beta_1\beta_2]^{-1} \\ \rho_2 &= \frac{1 + \beta_1\beta_2 - \beta_1^2 - \beta_2^2}{(1+\alpha_1)(1+\alpha_2)(1+\alpha_1\alpha_2+\beta_1\beta_2)} \end{aligned} \tag{19.8}$$

$$\beta_1 = \frac{v_1}{c} \qquad \beta_2 = \frac{v_2}{c} \qquad \alpha_1 = (1-\beta_1^2)^{1/2} \qquad \alpha_2 = (1-\beta_2^2)^{1/2},$$

v_1 and v_2 being the velocities of the final particles at the center of the distribution.

The Incoming Wave. Let us analyze first the time development of the gaussian wave packet as it would be seen in the absence of a scatterer. We do this in a way that can then be carried over to the case of real interest.

The time dependence of a gaussian wave packet of width Δ is expressed by the integral

$$I = \Delta^{-1} \int_A^\infty dW \exp\left[-\frac{(W - W_0)^2}{2\Delta^2} + i(kr - Wt)\right] \tag{19.9}$$

whose lower limit A is either $(m_1 + m_2)c^2$ or the incident particles' rest energy $(m_1 + m_2)_{\text{inc}}c^2$, whichever is larger. Let us rewrite (19.9) in the form

$$I = \exp\left[i(k_0 r - W_0 t) - \frac{(1 - i\gamma)(t - t_F)^2}{2T^2}\right](1 + i\gamma)^{-1/2} J \tag{19.9a}$$

with the following rotation,

$$J = (1 + i\gamma)^{1/2}\Delta^{-1} \int_A^\infty dW\, e^{i(z^2 + \xi)}$$

$$T = \frac{(1 + \gamma^2)^{1/2}}{\Delta}$$

$$2z = \varepsilon\left(\frac{sE_0}{\Delta} - \frac{t - t_F}{T}\right) + i\varepsilon^{-1}\left(\frac{sE_0}{\Delta} + \frac{t - t_F}{T}\right)$$

$$\gamma = 2\rho_2\left(\frac{\Delta}{E_0}\right)^2 k_0 r$$

$$\varepsilon = [(1 + \gamma^2)^{1/2} - \gamma]^{1/2}$$

$$\xi = k_0 r(\zeta - 1 - \rho_1 s + \rho_2 s^2)$$

$$t_F = \frac{r}{v_1 + v_2} = \frac{\rho_1 k_0 r}{E_0}$$

so that t_F is the time of flight of the decay products from the center of mass to the separation r. Using z as the new variable of integration we obtain

$$J = (1 - i)\int_B dz\, e^{i[z^2 + \xi(z)]}. \tag{19.10}$$

The lower limit of integration is now given by

$$B = -\tfrac{1}{2}\left[\left(\frac{E_0'}{\Delta} + \frac{t - t_F}{T}\right)\varepsilon + i\left(\frac{E_0'}{\Delta} - \frac{t - t_F}{T}\right)\varepsilon^{-1}\right]$$

with

$$E_0' = \begin{cases} E_0 = W_0 - (m_1 + m_2)c^2 & \text{if } m_1 + m_2 > (m_1 + m_2)_{\text{inc}} \\ E_{0\text{inc}} = W_0 - (m_1 + m_2)_{\text{inc}}c^2 & \text{if } m_1 + m_2 < (m_1 + m_2)_{\text{inc}} \end{cases}$$

and the integral runs over the "old contour" of Fig. 19.1, a straight line upward at an angle

$$\delta = \tfrac{1}{4}\pi + \tfrac{1}{2}\tan^{-1}\gamma.$$

Next we change the path of integration so as to run along the "new contour" of Fig. 19.1, on which always $\operatorname{Im} z^2 \geq 0$. Let us look at $\xi(z)$ on the distorted piece of the contour.

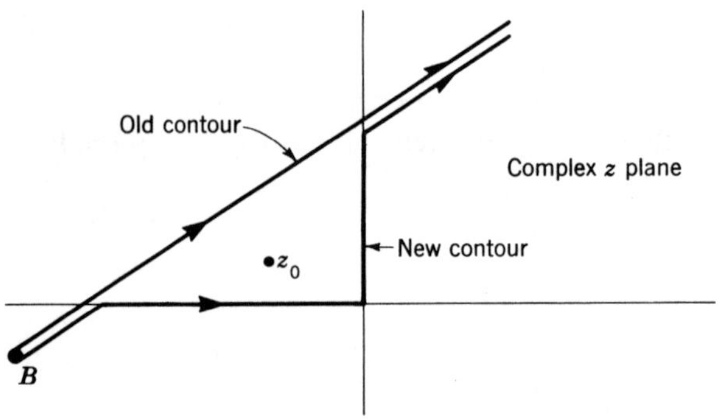

Figure 19.1. Showing the change of contour in the evaluation of the integrals I and I_R; z_o is the position of the pole in J_R when $(t - t_F)/\tau > T_1$.

Since ξ was chosen so that it is of order s^3 in an expansion in powers of s, we expect that $|\xi| \leq 1$ so long as

$$k_0 r |s|^3 \lesssim 1.$$

Now on the distorted portion of the path the magnitude of s is at most

$$|s| \simeq \frac{\Delta}{E_0} \frac{t - t_F}{T}$$

and its phase ranges between 0 and $-\pi$. For $i\xi$ not to become large anywhere on it, we must therefore demand that

$$k_0 r \left(\frac{\Delta}{E_0}\right)^3 \left(\frac{t - t_F}{T}\right)^3 \lesssim 1$$

or, in other words,

$$\frac{t - t_F}{T} \lesssim \frac{(E_0/\Delta)}{(k_0 r)^{1/3}}. \tag{19.11}$$

When this inequality is seriously violated, then the integral J increases rapidly. So long as it holds, J remains of order of magnitude 1, provided that $\operatorname{Im} B^2 \geqq 0$. The latter condition is fulfilled so long as

$$\frac{t - t_F}{T} \lesssim \frac{E_0'}{\Delta}, \tag{19.12}$$

which is usually far less stringent than (19.11)

We have thus found that, so long as (19.11) and (19.12) hold, the temporal development of $|I|^2$ is determined by the gaussian factor

$$\exp\left[-\frac{(t-t_F)^2}{T^2}\right]$$

which is explicitly visible in (19.9a). This is the analog of the familiar fact that the *Fourier transform* of a gaussian is again a gaussian. Notice that for $\gamma \ll 1$ the mean life of the packet is $T \simeq 1/\Delta$, but for $\gamma \gg 1$ it is

$$T \simeq 2\rho_2 k_0 r \frac{\Delta}{E_0}\frac{1}{E_0}$$

which *increases* with Δ.

After (19.11) is violated, the packet no longer falls off as rapidly as a gaussian. By that time it has dropped by a factor of

$$\exp\left[-\frac{(E_0/\Delta)^2}{(k_0 r)^{2/3}}\right]$$

[provided that $E_0' \geqq E_0/(k_0 r)^{1/3}$]. The shape is therefore never really gaussian in time unless

$$\frac{E_0}{\Delta} \gg (k_0 r)^{1/3}. \tag{19.13}$$

From then on the details of the tail depend on the experimental circumstances. The exact asymptotic behavior as $t \to \infty$, particularly, depends on how it is cut off at $E = 0$.

The Scattered Wave. We now turn to the scattered wave. The time dependence of a resonance term (19.4) is determined by the integral

$$I_R = \int_A^\infty dW \frac{\frac{1}{2}\Gamma/\Delta}{W - W_0 + \frac{1}{2}i\Gamma}\exp\left[-\frac{(W-W_0)^2}{2\Delta^2} + i(kr - Wt)\right]. \tag{19.14}$$

It is handled exactly like I. The only difference comes from the contour change. There is now a pole at

$$z = z_0 = \tfrac{1}{2}\varepsilon\left[\left(\frac{1}{2}\frac{\Gamma}{\Delta}\varepsilon^{-2} - \frac{t-t_F}{T}\right) - i\left(\frac{1}{2}\frac{\Gamma}{\Delta} - \varepsilon^{-2}\frac{t-t_F}{T}\right)\right]$$

which is such that z_0 lies in the second quadrant when

$$\frac{t-t_F}{\tau} > T_1 \equiv \frac{1}{2}\left(\frac{\Gamma}{\Delta}\right)^2 \varepsilon^{-2}(1+\gamma^2)^{1/2} \tag{19.15}$$

($\tau \equiv 1/\Gamma$). Consequently, as soon as the time satisfies (19.15), the pole lies between the old and the new contour and there is a residue contribution to I_R,

$$I_R = I_{\text{res}} + \frac{1}{2}\frac{\Gamma}{\Delta}\exp\left[i(k_0 r - W_0 t) - \frac{(1-i\gamma)(t-t_F)^2}{2T^2}\right] J_R$$

where

$$I_{\text{res}} = -i\pi\frac{\Gamma}{\Delta}\exp\left[i(k_0 r - W_0 t) + \frac{1}{8}\left(\frac{\Gamma}{\Delta}\right)^2(1+i\gamma) - \frac{t-t_F}{2\tau}\right]$$

$$J_R = \int_B dz\,(z - z_0)^{-1} e^{i(z^2+\xi)}.$$

Again J_R becomes larger than an order of magnitude 1 only when either (19.11) or (19.12) is violated. In other words, J_R is of magnitude 1 provided that

$$\frac{t - t_F}{\tau} \lesssim \frac{\Gamma}{\Delta}\frac{E_0}{\Delta}(1 + \gamma^2)^{1/2}(k_0 r)^{-1/2} \tag{19.11a}$$

and

$$\frac{t - t_F}{\tau} < \frac{\Gamma}{\Delta}\frac{E_0'}{\Delta}(1 + \gamma^2)^{1/2}. \tag{19.12a}$$

So long as the inequalities (19.11a) and (19.12a) are satisfied, I_R is therefore of the form

$$I_R = -i\pi\frac{\Gamma}{\Delta}\exp\left[i(k_0 r - W_0 t) + \frac{1}{8}\left(\frac{\Gamma}{\Delta}\right)^2(1 + i\gamma) - \frac{t - t_F}{2\tau}\right](1 + R), \tag{19.16}$$

where

$$|R| \leq C\exp(-\tfrac{1}{2}Y)$$

$$Y = (1 + \gamma^2)^{-1}\left(\frac{\Delta}{\Gamma}\right)^2\left(\frac{t - t_F}{\tau}\right)^2 - \frac{t - t_F}{\tau} - \frac{1}{4}\left(\frac{\Gamma}{\Delta}\right)^2,$$

C being a constant of order of magnitude 1. The decay curve is approximately exponential to the extent to which $|R|$ is small compared with 1. We have $Y = 16$ for

$$\frac{t - t_F}{\tau} = T_2 \equiv \frac{1}{2}\left(\frac{\Gamma}{\Delta}\right)^2(1 + \gamma^2)^{1/2}\left\{(1 + \gamma^2)^{1/2} + \left[\gamma^2 + \left(\frac{8\Delta}{\Gamma}\right)^2\right]^{1/2}\right\}$$
$$\geq 4\frac{\Gamma}{\Delta} + \frac{1}{2}\left(\frac{\Gamma}{\Delta}\right)^2. \tag{19.17}$$

By that time $|R|$ is quite small ($e^{-8} \simeq 1/3{,}000$). It decreases further until either (19.11a) or (19.12a) breaks down. From then on J_R decreases very slowly, say, as $(t - t_F)^{-n}$. Then

$$Y \simeq \left(\frac{E_0}{\Delta}\right)^2(k_0 r)^{-2/3} + 2n\ln\frac{t - t_F}{\tau} - \frac{t - t_F}{\tau}$$

which becomes negative and R dominates at the time when

$$\frac{t - t_F}{\tau} \gtrsim T_3 \equiv \left(\frac{E_0}{\Delta}\right)^2(k_0 r)^{-2/3} + 4n\ln\left[\frac{E_0}{\Delta}(k_0 r)^{-1/3}\right] \tag{19.18}$$

[assuming (19.13)]. The number n depends on the experimental circumstances, but T_3 is relatively insensitive to it.

We can now discuss the chape of the decay curve as seen at the separation distance r, provided that (19.13) holds. Since it will be seen below that we must have $\Delta \gtrsim \Gamma$, this means that we must assume that

$$k_0 r \ll \left(\frac{E_0}{\Gamma}\right)^3. \tag{19.13a}$$

For $(t - t_F)/\tau < T_1$ the residue term is absent, and the shape of the curve is roughly as it would be without the resonance, i.e., gaussian. Approximately from T_1 on, the overall features of the decay law are beginning to be dominated by the exponential; i.e., the decrease of activity is exponential in its gross features without necessarily being very nearly so in detail. By the time T_2 the decay is almost exactly exponential, to within about 0.1 percent. It then remains indistinguishable from a pure exponential until T_3, when it can be expected to flatten out.

As expected on the grounds of general arguments using the principle of uncertainty, the exponential

$$e^{-(t-t_F)/\tau}$$

by which the flux, or $|I|^2$, decreases in time has a mean life[2] τ that is the inverse of the resonance width. In general units,

$$\tau = \frac{\hbar}{\Gamma}.$$

Suppose first that $\Delta \lesssim \Gamma$; that is, the width of the incident packet (or the acceptance of the counter window) is smaller than the resonance width. Then T_2 tends to be large, and the exponential does not become pure until most of the activity has died out. Table 19.1 shows the rapid unobservability of the exponential under these circumstances. When γ is not small, then it is even less observable. It is clear that the excitation width Δ must be at least about equal to the line width Γ of the decaying state for a clean exponential curve to appear at a time of observable activity. This, of course, is only as one would expect. If the packet does not "fill out" the line shape, the time dependence cannot be determined by the resonance but must depend largely on the exciting wave.

2 One sometimes introduces the more exact *half-life* equal to $\tau \ln 2$ at which the activity has dropped by a factor of 2 instead of e. We shall not distinguish between these.

Table 19.1

Δ/Γ‡	T_1§	T_2§	$\lvert I_R\rvert^2/\lvert I_R\rvert^2_{max}$¶
1	$\frac{1}{2}$	4.5	$\sim\frac{1}{50}$
$\frac{2}{3}$	1	7	$\sim\frac{1}{500}$
$\frac{1}{2}$	2	10	$\sim\frac{1}{10,000}$
$\frac{1}{4}$	8	24	$\sim 10^{-8}$

‡Δ/Γ is the excitation width relative to the resonance width.

§T_1 and T_2 are the initial times of the roughly and almost exactly exponential curves, respectively, both measured in mean lives from the time of flight t_F.

¶$|I_R|^2/|I_R|^2_{max}$ is the relative value to which the observed activity has dropped by the time T_2. It was assumed that $\gamma \ll 1$.

If $\Delta > \Gamma$ and $\gamma \ll 1$, then the onset of the roughly exponential region is less than half a lifetime from the peak and the exponential is almost exact about $4\Gamma/\Delta$ mean lives after the time of flight. The decay curve then remains exponential for about $(E_0'/\Delta)^2$ mean lives (assuming $E_0' \gg \Gamma$). In order to maximize the time after which the decay curve levels off, it is therefore advantageous to make Δ as small as possible.

We see, therefore, that the wider the excitation is (in momentum space), the earlier the exponential starts; the narrower, the later it ends. The physical reason is clear: for an early start of the decay law, the excitation should be short and should be cut off sharply in time. For a long duration, however, the late straggling in of the slow particles must be minimized; this requires as monoenergetic as possible an excitation. We then expect to optimize things by making Δ larger than Γ by a factor between 1 and 10. But in many cases the observable results may be quite insensitive to upward changes in Δ. In other words, one may see an exponential curve for as long as the remaining activity permits observation, even though Δ is very much larger than Γ.

It is interesting to notice the dependence of the characteristics of the decay curve on the observation distance from the decay center. When the distance is made larger, then γ increases and the onset of the exponential is retarded. The smaller ρ_2 is, the larger r has to be for appreciable effects to occur. Since

$$\rho_2 = \tfrac{1}{8}$$

if both final particles are slow, since

$$\rho_2 = \tfrac{1}{2}(1 - \beta_1) + \tfrac{1}{2}(1 - \beta_2)$$

if both are highly relativistic, and since

$$\rho_2 = \frac{\beta(1 - \beta)}{(1 + \beta)(1 + \alpha)} \simeq \beta \quad \text{for } \beta \ll 1$$

if one is a photon, the decay law is least sensitive to the observation distance when one of the final particles is a photon and the other either suffers very little recoil or else recoils extremely rapidly. The parameter γ becomes comparable with or larger than unity when the observation distance is such that

$$k_0 r \gtrsim \frac{(E_0/\Delta)^2}{2\rho_0}.$$

Beyond such a distance the time T_2 becomes sensitive to it and increases, because then

$$T_2 \gtrsim 4\rho_2^2\left(\frac{\Gamma}{\Delta}\right)^2\left(\frac{\Delta}{E_0}\right)^4 (k_0 r)^2.$$

When the distance is so large that T_2 and T_3 are comparable, then no exponential-decay law is observed at all. This happens when

$$k_0 r \gtrsim \left(\frac{E_0}{\Gamma}\right)^{9/4} \left(\frac{\Gamma}{\Delta}\right)^{3/2} (2\rho_2)^{-3/4}. \tag{19.19}$$

It therefore depends primarily on the relative sharpness of the resonance line.

The Excitation Width. What, in practice, is the excitation width Δ? In most practical cases the unstable state is excited from another unstable state. In that event the shape of the packet is of course not gaussian but more nearly exponential in time, i.e., Breit–Wigner or lorentzian in energy. We do not expect this to have any serious consequences so far as the orders of magnitude of the times T_1, T_2, and T_3 are concerned. It is clear that, if anything, T_3 will decrease, because the exponential tail of the excitation decreases much less rapidly than a gaussian. The meaning of Δ then is the *width of the parent level.*

Mixed States. Another point to be cleared up is that in actuality most unstable systems undergoing decay are not in a pure state, but mixed. That is to say, they are not *coherently* excited and describable by a wave function. They should more appropriately be described by a density matrix. Consider a case in which the unstable state is formed as a decay product of an unstable parent level. Since the parent state must have emitted at least one other particle in addition to the daughter fragment, and these other particles fly off and are ignored from then on, the daughter may be in a mixed state even if the parent state was pure. Let us look at this a little more closely.

Supposing for a moment that the parent is in a pure state of sharp energy; let it decay into two fragments (as in α or γ decay) one of which is the daughter of interest. In such an event the daughter state is pure[3] and must also have a sharp energy, because the products of a two-particle decay have a uniquely determined energy. On the other hand, if the parent undergoes a three-particle (or more) decay (as in β decay), then each of the emitted particles, considered by itself, is in a mixed state, with components of sharp energy mixed with weights according to the *spectrum* of the decay.

But the initial supposition is of course false. The unstable parent cannot have a sharp energy, or it would not change in time. If it decays approximately exponentially, it must have its energy "smeared out" in Breit–Wigner fashion, of the appropriate width. Still, if it is in a pure state and undergoes a two-fragment decay, its daughter is in a pure state and the parent state's width takes the place of Δ in the foregoing discussion. If it decays by three-particle emission, the daughter is in a mixed state, each component of which has a smeared-out energy. The previous discussion then applies to each such component of the mixed state. In other words, Δ is again the width of the parent level and *not* the width of the energy spectrum of the decay. The latter contributes only to the further *incoherent* smearing out of the energy in the mixture but has no bearing on the coherent effects.

The same argument is applicable to the situation in which the parent, in

3 We assume for simplicity that there are no other degrees of freedom, such as spin. They would complicate nothing but the language used in the explanation.

addition to having a coherently uncertain energy as before, is in a mixed state of an additional energy spectrum. Such incoherent effects can be ignored in the discussion.

Another important method of producing systems in unstable states is that of *collisional excitation*. Examples of this are atomic excitations in a gas and production of unstable particles in nucleon-nucleon collisions. Let us concentrate on the latter case. For schematic simplicity we may imagine an experiment carried out with colliding proton beams in the center-of-mass frame, and we ignore the spin degree of freedom. If the protons were produced under identical circumstances and had a sharp energy, then the unstable fragments produced, when considered as single particles rather than as pairs, triplets, etc., would be in mixed states made up of states of sharp energy and weighted according to the spectrum. An exponential decay of the strange particles would then not be observed. In fact, the only particles that could be seen would have to be stable ones. Any unstable state must be a *coherent* superposition of different energies. Only if the original beam has a nonzero coherent energy width can unstable particles be produced. Of course, every particle beam made in an accelerator has such a width. The mere pulsing of the beam implies it. But the discussion shows that unstable states whose width is larger than the coherent width of the original beam cannot be produced, or if they are, a clean exponential decay law will not be observed on them.

In view of the frequency of occurrence of mixed states, it is important to ask whether for them the results found for pure states have to be modified. The answer is: Not in any essential way. If the mixture is made up of states whose energy is too sharp to produce an exponential decay by themselves, the mixing cannot produce it. In the same way, mixing cannot postpone the onset of a nonexponential tail produced by the pure states. Once the slowly decreasing terms dominate over the exponential ones in each of the constituents, mixing cannot produce any cancellation of that dominance. Similarly, during the time when each state in the mixture decays almost exactly exponentially, the mixture must, too. Therefore one can always discuss the question of the exponential decay of a mixture by looking at each constituent separately.

Examples. Let us now take some simple examples. Consider the case of a nuclear deexcitation by γ-ray emission. In a typical instance the energy may be $E_0 \simeq 200$ kev and the lifetime $\tau \simeq 10^{-8}$ sec so that $\Gamma/E_0 \simeq 3 \times 10^{-13}$. Then, for an observation distance $r = 10$ cm, $k_0 r \simeq 10^{11}$. Suppose that the level is excited from a parent whose width is twice that of the daughter level. Then (19.13) is very well fulfilled, and $\gamma \ll 1$. The decay curve should be roughly exponential after $\frac{1}{8}$ of a mean life from the peak and excellently exponential after 2. It then remains exponential for 10^{17} lifetimes! In order to destroy most of the exponential-decay curve, one would have to move the detector away to a distance of about 10^{22} miles. Note that things would not be observably different even if Δ were, say, 2,000 times as large as Γ. The onset of the exponential would be much faster, indeed only a small fraction of a lifetime, and it would last for 10^{11} lifetimes.

Consider a particle such as the K_1, with a mass of about 500 Mev and a lifetime $\tau \simeq 10^{-10}$ sec. If we consider only the decay into two pions, then $E_0 \simeq 220$ Mev, $E_0/\Gamma \simeq 4 \times 10^{13}$, and $k_0 \simeq 10^{13}$ cm^{-1}. If the excitation width is $\Delta = 10^4\Gamma$ and the observation is at a distance of 100 cm, then the exponential is almost exact after less than 1/1,000 of a lifetime and it remains so for about 10^9 lifetimes.

19.3 Multiple Poles of the S Matrix

The discussion in the last section was based on the assumption that the shape of the resonance line is of the Breit–Wigner type. That is to say, the amplitude has a simple isolated pole near the real axis, or at least it looks that way when viewed from the only physically accessible region, the real energy axis. In some cases two or more such poles are close together. It is obvious that in general this will, on the one hand, distort the shape of the lines and, on the other, produce decay curves that differ from exponentials.[4]

The most extreme instance of closely spaced resonances is that of *multiple* poles. Whereas bound-state poles of the **S** matrix must be simple, there is no general prohibition of multiple resonance poles in the complex plane. The first visible effect would be to alter the shape of the corresponding spectral line to be other than lorentzian; the second, to modify the decay curve in its central part to be nonexponential.

To see what kind of line shapes can be produced, let us look at the example of a *double* pole. Let the amplitude (or the **S** matrix) have the expansion

$$A = \frac{-\frac{1}{4}\Gamma^2 a}{(E - E_0 + \frac{1}{2}i\Gamma)^2} + \frac{i\frac{1}{2}\Gamma b}{E - E_0 + \frac{1}{2}i\Gamma} + \cdots. \tag{19.20}$$

The energy dependence of the cross section near E_0 is then given by

$$|A|^2 = [(E - E_0)^2 + \tfrac{1}{4}\Gamma^2]^{-2}\{\tfrac{1}{16}\Gamma^4|a|^4 + \tfrac{1}{4}\Gamma^2|b|^2[(E - E_0)^2 + \tfrac{1}{4}\Gamma^2] + \tfrac{1}{4}\Gamma^3 \operatorname{Re}[ab^* i(E - E_0 - \tfrac{1}{2}i\Gamma)]\}. \tag{19.21}$$

If $b = 0$, the entire energy dependence lies in the outside factor, the square of a Breit–Wigner term. Its shape resembles that of the usual line, except that it is flatter on top. But if $b \neq 0$, the energy dependence of the brace may produce quite different shapes. For example, unless the relative phase of a and b is 0 or π, the line is not symmetric. What is more, it may have two distinct peaks and thus resemble two closely spaced resonances of smaller width.

4 For a detailed example of the effect of two closely spaced resonances, see Treiman and Sachs (1956).

Now let us see what the concomitant decay curves are. The arguments of the last section show that the mathematical origin of the exponential time dependence is the residue of the integrand of an integral such as (19.14) at the resonance pole. If the amplitude has a double pole as in (19.20), then there will be two contributions, because both terms in (19.20) yield a residue when multiplied by the exponential in (19.14). Apart from an outside factor, we get the residue contribution to the amplitude

$$\left\{b + \frac{1}{2}a\left[(t - t_F)\tau^{-1} - \frac{1}{2}\left(\frac{\Gamma}{\Delta}\right)^2\right]\right\} e^{-1/2(t-t_F)/\tau} \tag{19.22}$$

This shows that the exponential law is replaced by a linear function of the time (or, in the cross section, a quadratic) multiplied by the exponential. Similarly, higher-order poles produce polynomials of higher order in t multiplied by the same exponential. These modifications will be important for large times in particular. At what time they begin to produce observable deviations from a pure exponential depends of course on the relative size of a and b. Although eventually the linear term in (19.22) will always dominate, the smaller a is relative to b, the later it will do so.

Decay curves such as (19.22) have so far not been observed experimentally. However, although in some instances the exponential law has been experimentally well verified over many lifetimes, it has certainly not been checked in many cases in which it is nevertheless assumed to be valid. Not that one should expect it to be frequently violated. Unless a fundamental reason can be found for some particular resonance poles to be other than simple, multiple poles must be regarded as extremely unlikely accidents.

Notes and References

19.2 The first pertinent investigations of the decay problem were by Weisskopf and Wigner (1930).

The following studies used the approach of *radiation damping*: Heitler and Ma (1949); Arnous and Zienau (1951); Arnous and Bleuler (1952); E. Arnous (1952); Arnous and Heitler (1953).

Other references are M. Moshinsky (1951); L. A. Khalfin (1957 and 1960); G. Höhler (1958); Matthews and Salam (1959a); J. Petzold (1959); Ya. B. Zel'dovich (1960); M. Lévy (1960); J. Schwinger (1960); Lozano and Moshinsky (1960); F. T. Smith (1960); Krzywicki and Szymanski (1960); Beck and Nussenzveig (1960); H. M. Nussenzveig (1961); Jacob and Sachs (1961); R. G. Newton (1961); R. G. Winter (1961); Goldberger and Watson (1962 and 1964b, chapt. 8); T. Ohmura (1964); Kerler and Petzold (1965); L. Rosenfeld (1965); M. N. Hack (1967); J. P. Jeukenne (1967); W. Kerler (1970); Fonda and Ghirardi (1970 and 1971); L. P. Horwitz et al. (1971); M. V. Terent'ev (1972); G. N. Fleming (1973); Fonda, Ghirardi and Rimini (1973); Fonda, Ghirardi, Rimini, and Weber (1973); A. Degasperis et al. (1974); R. Lavine (1974); S. T. Ali et al. (1975); P. Exner (1976); Misra and Sinha (1977); E. C. G. Sudarshan et al. (1978); A. Peres (1980); A. Bohm (1981); M. Hillery (1981).

Sequential decay processes were examined by L. Fonda et al. (1978). The ambiguities involved in any analytic continuation of the amplitude to a resonance pole were analyzed by G. Calucci et al. (1968); Calucci and Ghirardi (1968 and 1972). I. W. Herbst (1980) studied the exponential decay in the Stark effect.

The approach of this chapter follows the paper by the author.

19.3 For discussions of other line shapes and their effects on the decay curve, see Fonda and Newton (1960c) and particularly Goldberger and Watson (1964a); Bell and Goebel (1965).

Problems

1 A gaussian wave packet of 1 ev width, centered at an energy of 1 Mev, consisting of particles of mass 500 Mev, is sent toward an infinitely massive scattering center. It is observed at a distance of 1 cm from the center. What is the approximate time dependence of the measured signal while it is reasonably strong? If observations are made at a distance of 10 km, is the time dependence similar?

Suppose that another packet is twice as wide as the first. What is its duration, compared with the first, when observed at a distance of 10^{-2} cm from the center? What if the distance is 1 m?

2 A resonance of 1 kev width produces an unstable particle of mass 500 Mev and a lifetime of 10^{-12} sec which decays into two particles of 100 and 200 Mev mass. The decay products are observed at a distance of 1 m. Approximately when does the decay curve become exponential? For how many lifetimes does it stay exponential? What if the observation distance were 10 m?

3 Suppose that the **S** matrix has a triple pole near the real axis. What is the shape of the resonance line, and what is the decay law?

CHAPTER 20

The Inverse Scattering Problem

20.1 Introduction

In the usual scattering-theory problem the hamiltonian of the system or the interparticle forces are known and a cross section, polarization, etc., are to be calculated and subsequently confronted with experimental results. The "inverse" problem is posed in the opposite direction: given certain kinds of information obtained more or less directly from scattering experiments, we are to determine the interparticle forces. Or before this problem can be solved: Is the given amount of information sufficient to determine these forces uniquely? If not, what kinds of ambiquity are there?

Of course, the ultimate aim of a scattering experiment is *always* to determine the forces. In the more customary procedure a hamiltonian is chosen on general grounds of simplicity or from among a class thought to have particular virtues, being derived from a more fundamental theory or by any number of other criteria. The cross sections are thus calculated, and if they disagree with experiments, the hamiltonian is discarded or modified. Needless to say, it is very useful in such an approach to have a good feeling, acquired by experience or physical insight, for the effects produced in scattering experiments by certain features of interparticle forces. This is where such simple approximations as the effective-range theory, Born approximation, and others are of particular help. With such a physical feeling it is possible to make reasonably reliable inferences from the experimental results back to the potential. But clearly the most direct route to the desired result is to find a mathematical method of constructing the hamiltonian from the scattering information or, if this cannot be uniquely determined, to find the *class* of hamiltonians that would yield the same experimental result.

The first question that arises is: What kind of scattering information should be used as input? This must be determined both by its experimental accessibility and by the invention of a mathematical method of solving the problem, or by a compromise between both. No matter how much weight one would like to assign to the first, it must in reality always yield to the second. The most easily accessible information helps us not at all if we are not astute enough to find a procedure for obtaining the hamiltonian from it. On the other hand, if the only known procedure requires a rather inaccessible input, the inverse method loses most of its appeal.

The kinds of scattering information for which the inverse problem has been solved at least in part are, for central potentials, knowledge of the phase shift of a single angular momentum for all energies, or of all phase shifts (i.e., of the scattering amplitude) at one energy; and for noncentral potentials, knowledge of the scattering amplitude for all energies and all incoming and outgoing directions, or of the backward-scattering amplitude for all energies and all directions of incidence. It would be very desirable to have solutions for other kinds of input but they have not yet been found.

The starting points of the inverse problems that have been solved in principle are all one step removed from directly experimentally obtainable information. They make use of the scattering amplitude or the phase shifts rather than the cross section. There is always a phase ambiguity in going from the latter to the former. The phase of the scattering amplitude, as a function of the scattering angle and energy, can be practically determined in three ways. The first is applicable if the colliding particles are charged. In that case the interference effects with the known Rutherford amplitude serve to give information on the phase of the additional term.

The second method is to start at low enough energies, where not only is the cross section isotropic, and hence only the s wave contributes, but where furthermore the $l = 0$ phase shift is still a linear function of the momentum. Since, then, according to (11.11), the s-wave amplitude is connected to the phase shift by

$$a_0(k) = \frac{\delta_0(k) - \delta_0(0)}{k}$$

and $|a_0(k)|^2$ is observed, the only ambiguity left is a sign and $\delta_0(0)$. This is usually removed by recourse to the effective-range theory together with additional knowledge either of the sign of the potential, or of the existence of bound states, or both, etc. Once a choice of sign has been made, the s-phase shift can be "continuously" followed by measurements at somewhat higher energies. As the p wave begins to enter, it first produces an interference with the s wave. In principle, it can therefore be unambiguously determined and can be followed from then on. As higher partial waves successively enter, they, too, can be found from the measurements. It is clear that, although the phase shifts or the scattering amplitude can thus be determined *in principle* to within an overall sign, in practice this is not simple. This is particularly so because the procedure depends upon arguments of continuity, and rarely are measurements made at sufficiently small energy intervals.

The lack of uniqueness increases when the colliding particles possess spin. In such cases polarizations, measured possibly by double or triple scatterings, must serve to decrease the ambiguities. A most important case in point is that of nucleon-nucleon scattering over a wide range of energies. The history of its phase-shift determination can be followed in the literature[1] and should serve as a useful example.

The third method of determining the phase of the scattering amplitude utilizes the generalized optical theorem as a (nonlinear) integral equation for it. We shall describe this procedure in Sec. 20.1.2. In most practical cases, unfortunately, it is not applicable as the cross sections are too large. If one assumes that the potential is in a more restricted class this may be combined with dispersion relations to reduce residual ambiguities.[1a]

Another method available *in principle* for the determination of the phase of the amplitude (up to an angle-independent constant) is the use of the Hanbury Brown and Twiss effect (described in Sec. 4.3). This involves coincidence measurements at two different scattering angles. But this, too, is experimentally very difficult.

Despite these practical difficulties we must assume from now on that the scattering amplitude or the phase shifts are experimentally known.

There are other kinds of inverse problems, for which solutions are known, that start with entirely inaccessible input information, such as the off-the-energy-shell **T** matrix[2] or the spectra in the complex angular-momentum plane.[3] These are of mathematical interest, but their physical relevance is questionable. We shall not go into them.

Having fixed our attention on a particular kind of starting point from which to construct the interparticle hamiltonian, we must come to grips with the questions of existence and uniqueness. It is clear to start with that we cannot expect to be able to construct uniquely a general nonlocal interaction even from a knowledge of the full scattering amplitude as a function of the energy and all angles. The reason is simply that such an amplitude is a function of five parameters and the potential contains six. In fact, the inverse problem in such a context is essentially trivial. One merely chooses a family of twice differentiable functions $\psi(\mathbf{k},\mathbf{r})$ with the prescribed asymptotic form and lets the mapping from the functions $\psi(\mathbf{k},\mathbf{r})$ to the functions $(\nabla^2 + k^2)\psi(\mathbf{k},\mathbf{r})$ define the potential operator. If the class

1 See, for example, G. Breit (1962), and a list of references to be found there, especially on p. 776. Other references are MacGregor and Arndt (1965) and Noyes, Bailey, Arndt, and MacGregor (1965). See also N. P. Klepikov (1962).

1a See Itzykson and Martin (1973).

2 See H. Eckstein (1960) and F. Coester (1964b).

3 See Burdet, Giffon, and Predazzi (1964).

of potentials is restricted it becomes less trivial; separable, or sums of separable potentials, are one possibility, and the inverse problem has been solved there[4]. We shall confine our attention to *local potentials*.

For central potentials, with a single phase shift prescribed as a function of the energy, we shall see that there is essentially no existence problem. Any "reasonable" phase shift of a given angular momentum is associated with at least one (energy-independent) local potential. Similarly for all phase shifts at one energy. No matter how they are chosen (subject only to rather weak conditions on their decrease for large l) there always exist at least one (angular-momentum independent) local potential that can be considered as "causing" them. In both cases the demand that the potential be physically reasonable produce certain minor restrictions. Apart from that, the local nature of the potential produces no restrictions on a single phaseshift as a function of the energy or on the set of all phaseshifts at one energy.

The situation is quite different if we do not assume a central potential and the scattering data consist of the whole scattering amplitude for all energies and angles. Since the scattering amplitude is a function of five real parameters, and a local potential contains only three, it is clear that the existence of a local potential must strongly restrict the class of "permissible" amplitudes. If the only aim is to reconstruct a potential that is known to exist for a given amplitude then this is, of course, no problem. However, if one starts with an amplitude for which the existence of an underlying potential is not known then it constitutes a major stumbling block. Part of the solution of the inverse problem for noncentral potentials then is to solve the "characterization problem", that is, to characterize the class of amplitudes that are associated with local potentials in a certain set. At this writing the characterization problem has been solved only in part. What is known about it indicates that there appears to be no simple way of testing or ascertaining by inspection whether a local potential exists. The only known tests can be applied only after solving an integral equation. If one is looking to the inverse problem as a method for generating potentials for which the full three-dimensional Schrödinger equation is explicitly solvable (analogously to the generation of the Bargmann potentials in the radial case) then the lack of a simple characterization is particularly inconvenient.

The existence problem has a chance of having a much simpler solution if the data are taken to be only a part of the full amplitude function. In particular, it is known that the backward scattering amplitude $A(-\mathbf{k},\mathbf{k})$, given as a function of $\mathbf{k}$, determines the potential if it is, in some sense, weak enough. The equation that determines it is nonlinear and we small discuss it in Section 20.5.4.

The question of uniqueness is more complicated. We shall see that each phase shift of a given angular momentum for which there is no bound state is associated with a *unique* local potential. If there are N bound states of

4 See the Notes for references.

the same angular momentum, then we are free to assign their energies, and there exists an N-parameter family of "equivalent" potentials, each of which is associated with the given phase shift for all energies and with the given bound states. In other words, apart from the connection between the phase shift and the *number* of bound states of the same angular momentum, which is contained in the Levinson theorem, the bound states are entirely independent of the scattering. It is, in principle, impossible to obtain information about one from the other.[5] Similarly we shall see that each set of phase shifts of a given energy is associated with a one-parameter family of potentials.

These results raise the question: If the energy dependence of a single phase shift determines the potential (to within the stated ambiguity), then the energy dependence of two or more phase shifts of different angular momenta are not necessarily compatible with one another and with the assumption of a local l-independent potential. Similarly, the scattering amplitudes at two different energies are not necessarily compatible with one another and with the existence of a local energy-independent potential. It would be of great interest to have simple *direct* compatibility criteria. This compability problem, of course, is only a special case of the existence problem alluded to earlier when the full amplitude is given for all energies and angles.

The *uniqueness problem* if the entire amplitude is given, on the other hand, will be found to be much simpler. In that case there is no potential ambiguity analogous to that for a single phaseshift in the central case. In particular, the bound states are not independent of the amplitude. Their energies are determined from the forward amplitude as shown by (10.110), and their "characters" (10.87) have been shown[6] to be obtainable as well. The essential tool for this determination is the presence of the continuous rotation group in three dimensions, even though it is not an invariance group. Of course, no such *a priori* guaranty of uniqueness holds for the backward amplitude.

20.2 The Phase of the Amplitude

We shall outline here a method for determining the phase of the complex scattering amplitude from a knowledge of the angular dependence of the differential scattering cross section. The only assumption we shall make is that at the energy under consideration (which remains fixed) no inelastic processes are possible and that the underlying forces are such that the S matrix is unitary, i.e., that the scattering amplitude satisfies the generalized optical theorem. Specifically, no locality of the potential will be assumed, and no analyticity requirements are made.

5 This does not contradict the results of the effective-range approach. The approximation simply breaks down.

6 R. G. Newton (1980b).

Since the energy will remain fixed, we shall write the amplitude (for one spinless particle by a potential or for two spinless particles by one another in the c.m. system) for scattering from the direction $\boldsymbol{\theta}$ to the direction $\boldsymbol{\theta}'$

$$A(k,\boldsymbol{\theta}',\boldsymbol{\theta}) = k^{-1}F(\boldsymbol{\theta}',\boldsymbol{\theta}),$$

so that F is dimensionless. Then the generalized optical theorem (7.54), together with (10.16), reads

$$\operatorname{Im} F(\boldsymbol{\theta}',\boldsymbol{\theta}) = (1/4\pi)\int d\boldsymbol{\theta}''\, F^*(\boldsymbol{\theta}'',\boldsymbol{\theta}')F(\boldsymbol{\theta}'',\boldsymbol{\theta}), \tag{20.1}$$

in which the integration runs over all solid angles. Writing

$$F(\boldsymbol{\theta}',\boldsymbol{\theta}) = G(\boldsymbol{\theta}',\boldsymbol{\theta}) \exp\left[i\phi(\boldsymbol{\theta}',\boldsymbol{\theta})\right],$$

where $G > 0$ and ϕ is real, we assume that $G(\boldsymbol{\theta}',\boldsymbol{\theta})$ is a known function of $\boldsymbol{\theta}$ and $\boldsymbol{\theta}'$; its square is the differential cross section in units of the square of the reduced wavelength. Use of (10.15) then leads to the equation

$$\sin\phi(\boldsymbol{\theta}',\boldsymbol{\theta}) = \int d\boldsymbol{\theta}''\, H(\boldsymbol{\theta}',\boldsymbol{\theta},\boldsymbol{\theta}'') \cos\left[\phi(\boldsymbol{\theta}'',\boldsymbol{\theta}') - \phi(\boldsymbol{\theta}'',\boldsymbol{\theta})\right], \tag{20.2}$$

where

$$H(\boldsymbol{\theta}',\boldsymbol{\theta},\boldsymbol{\theta}'') = G(\boldsymbol{\theta}'',\boldsymbol{\theta}')G(\boldsymbol{\theta}'',\boldsymbol{\theta})/4\pi G(\boldsymbol{\theta}',\boldsymbol{\theta}).$$

This equation may be regarded as a nonlinear integral equation for the unknown phase function ϕ.

If the potential has spherical symmetry then F can depend only on the scattering angle θ, with $\boldsymbol{\theta}' \cdot \boldsymbol{\theta} = \cos\theta$. Writing $x = \cos\theta$, $F(\boldsymbol{\theta}',\boldsymbol{\theta}) = F(x)$, Eq. (20.2) then simplifies to the equation

$$\sin\phi(x) = \int dy\, dz\, H(x,y,z) \cos\left[\phi(y) - \phi(z)\right], \tag{20.2$'$}$$

where

$$H(x,y,z) = \frac{G(y)G(z)}{2\pi G(x)(1 - x^2 - y^2 - z^2 + 2xyz)^{1/2}}$$

and the region of integration is the interior of the ellipse $1 - x^2 - y^2 - z^2 + 2xyz = 0$.

The question to be answered is: *under what conditions does* (20.2) *or* (20.2′) *have a solution, and is it unique*? Both parts of this question are of importance. If there are functions G for which no solution exists then unitarity by itself imposes restrictions on the possible angle-dependence of the differential cross section. If the solution is unique then unitarity alone suffices to determine the phase of the amplitude.

One lack of uniqueness is immediately obvious. If F satisfies (20.1) then so does $-F^*$; the sign of its real part cannot be determined. In other words, if ϕ satisfies (20.2) then so does $\pi - \phi$. We shall say that the solution ϕ of

(20.2) or (20.2′) is *essentially* unique if $\sin\phi$ is unique. Explicit examples of lack of essential uniqueness are known.[6a]

As for existence of a solution, the optical theorem implies the obvious restriction

$$G(\boldsymbol{\theta},\boldsymbol{\theta}) \geq (1/4\pi) \int d\boldsymbol{\theta}'\, G^2(\boldsymbol{\theta}',\boldsymbol{\theta}) \tag{20.3}$$

on the forward scattering amplitude. It is not hard to see that this is not the only restriction on the cross section imposed by unitarity.[6b]

Let us write Eq. (20.2′), in the form

$$\phi(x) = \sin^{-1} \int dy\, dz\, H(x,y,z) \cos[\phi(y) - \phi(z)] \tag{20.4}$$

and regard the right-hand side as a non-linear mapping

$$\psi = \mathscr{M}(\phi),$$

$$\psi(x) = \sin^{-1} \int dy\, dz\, H(x,y,z) \cos[\phi(y) - \phi(z)].$$

If

$$Q(x) = \int dy\, dz\, H(x,y,z)$$

is such that $Q \leq M < 1$ for all $-1 \leq x \leq 1$ and we set $M = \sin\alpha$, $0 \leq \alpha < \frac{1}{2}\pi$, then we may restrict the domain and the range of the map to the interval $[0,\alpha]$. In this interval one readily sees that

$$\left|\frac{\sin\psi_1 - \sin\psi_2}{\psi_1 - \psi_2}\right| \geq \cos\alpha,$$

$$\left|\frac{\cos\psi_1 - \cos\psi_2}{\psi_1 - \psi_2}\right| \leq \sin\alpha.$$

Therefore

$$\begin{aligned}
&|\psi_1(x) - \psi_2(x)| \\
&\quad \leq \sec\alpha\, |\sin\psi_1(x) - \sin\psi_2(x)| \\
&\quad \leq \sec\alpha \int dy\, dz\, H(x,y,z)\, |\cos[\phi_1(y) - \phi_1(z)] - \cos[\phi_2(y) - \phi_2(z)]| \\
&\quad \leq \tan\alpha \int dy\, dz\, H(x,y,z)\, |[\phi_1(y) - \phi_2(y)] - [\phi_1(z) - \phi_2(z)]| \\
&\quad \leq 2\tan\alpha \int dy\, dz\, H(x,y,z) |\phi_1(y) - \phi_1(z)|.
\end{aligned}$$

6a The first were given by J. H. Crichton (1966).
6b See R. G. Newton (1968).

In terms of the norm

$$\|\phi\| = \sup_{-1 \leq x \leq 1} |\phi(x)|$$

it follows that

$$\|\psi_1 - \psi_2\| \leq \varepsilon \|\phi_1 - \phi_2\|,$$

where $\varepsilon = 2\sin^2\alpha \sec\alpha$. If $\varepsilon < 1$ then $\mathcal{M}$ is a contraction mapping on a Banach space and as such it is know[6c] to have a unique fixed point, i.e., $\phi = \mathcal{M}(\phi)$, which therefore solves (20.4). Moreover, this fixed point may be constructed by iteration[6d]. The condition $\varepsilon < 1$ is satisfied if $M < 0.62$. Therefore if $M < 0.62$ then (20.2′) [and similarly, (20.2)] has a unique solution and this solution can be obtained by successive approximations iterating the equation (20.2′) or (20.2).

For larger values of $M < 1$ more sophisticated estimates and fixed-point theorems may be employed to assure the existence, but not necessarily the uniqueness and constructibility, of a solution of (20.2). A variety of results have been obtained (and are referred to in the Notes), but there remains a small range of $M < 1$ for which existence has been proved, but not uniqueness. For $M > 1$ very little is known, except that many examples of nonuniqueness have been explicitly constructed.

The existence and uniqueness results that have been proved for $M < 1$ are of interest as a matter of principle if not so much for direct experimental application. Suppose that $G(\boldsymbol{\theta}, \boldsymbol{\theta}')$ is given (say, without zeros); then $\gamma G(\boldsymbol{\theta}, \boldsymbol{\theta}')$, for sufficiently small γ, will be such that $M < 1$. It follows that, if only the "average" cross section (in some sense) is small enough then unitarity implies no restriction at all on the angular distribution. From a phase-shift point of view (for central potentials) this is somewhat surprising, as it shows that the following naive but plausible argument is fallacious.

Assume that a given differential cross section is expressible as a finite linear combination of Legendre polynomials up to order $2L$, which requires $2L + 1$ real coefficients. It is then plausible to conclude that the scattering amplitude is expressible in terms of Legendre polynomials up to order L. This requires $L + 1$ coefficients, and unitarity implies that they can be chosen *real* (the phase shifts). Thus it appears as though unitarity implies strong restrictions on the possible angular distributions. The existence of a solution of (20.2′) for all reasonable cross sections that are small enough shows that in fact almost all (in a strict sense) cross sections expressible in terms of a finite number of Legendre polynomials require an *infinite* number of them for their amplitudes.

These arguments are, in principle, not hard to generalize to particles with spin. They have also been combined with the dependence of the amplitude on the energy, assuming analyticity. For such results we refer to the literature cited in the Notes.

6c See for example, P. M. Anselone (1964), p. 13.

6d For a simple proof, see for example, R. G. Newton (1968).

20.3 The Central Potential Obtained from a Phase Shift

20.3.1 The Gel'fand–Levitan Equations. Consider elastic scattering of spinless particles by a spherically symmetric local potential. The problem is to find the potential in the radial Schrödinger equation from a knowledge of the **S** matrix or of the asymptotic form of the regular solution for all positive energies. It is most elegantly solved by means of an integral equation. We shall derive it in a manner that is particularly amenable to being transferred to the context of Sec. 20.4.

Let us start with solutions $\varphi_1(E,r)$ of the radial Schrödinger equation of angular momentum l and for an arbitrary potential (well behaved in the sense of Chap. 12) $\mathscr{V}_1$,

$$D_1(r)\varphi_1(E,r) = -k^2\varphi_1(E,r) \tag{20.5}$$

in which the differential operator $D_1(r)$ is defined by

$$D_1(r) = \frac{d^2}{dr^2} - \frac{l(l+1)}{r^2} - \mathscr{V}_1(r)$$

and φ_1 satisfies the boundary condition (12.132). We define a function $g(r,r')$ by the Stieltjes integral

$$g(r,r') = \int_{-\infty}^{\infty} dh(E)\varphi_1(E,r)\varphi_1(E,r'). \tag{20.6}$$

The function $h(E)$ is for the moment quite arbitrary and is to be fixed later. Clearly g satisfies the partial differential equation

$$D_1(r)g(r,r') = D_1(r')g(r,r') \tag{20.7}$$

and the boundary conditions

$$g(r,0) = g(0,r') = 0. \tag{20.8}$$

Now suppose that the function $K(r,r')$ is the *unique* solution of the linear integral equation of Gel'fand and Levitan,

$$K(r,r') = g(r,r') - \int_0^r dr''\, K(r,r'')g(r'',r'). \tag{20.9}$$

That is to say, subject to later proof, we assume for the moment that (20.9) possesses a solution and that it is unique. Then introduce the function

$$\xi(r,r') \equiv D(r)K(r,r') - D_1(r')K(r,r')$$

with the differential operator $D(r)$ defined by

$$\begin{aligned} D(r) &= D_1(r) - \Delta\mathscr{V}(r) \\ \Delta\mathscr{V}(r) &= -2\frac{d}{dr}K(r,r). \end{aligned} \tag{20.10}$$

It is a matter of straightforward differentiation, integration by parts, and use of the differential equation (20.7), as well as of the boundary conditions

(20.8), to verify that $\xi(r,r')$ satisfies the homogeneous version of the integral equation (20.9). Since we assumed that the solution of (20.9) is unique, the homogeneous form can have only the trivial solution. We conclude that $\xi(r,r') \equiv 0$. In other words, $K(r,r')$ obeys the partial differential equation

$$D(r)K(r,r') = D_1(r')K(r,r'). \tag{20.11}$$

It follows from (20.8) that it satisfies the boundary conditions

$$K(r,0) = K(0,r') = 0. \tag{20.12}$$

We now use the solution $K(r,r')$ of the integral equation (20.9) to define a function

$$\varphi(E,r) \equiv \varphi_1(E,r) - \int_0^r dr'\, K(r,r')\varphi_1(E,r'). \tag{20.13}$$

Application of the differential operator $D(r)$ to (20.13), together with two integrations by parts and use of (20.11) and (20.5), shows that $\varphi(k,r)$ is a solution of the differential equation

$$D(r)\varphi(E,r) = -k^2\varphi(E,r).$$

Furthermore it follows from (20.12) that φ is a *regular* solution,

$$\varphi(E,0) = 0.$$

Notice that the integral equation (20.9) is an equation for $K(r,r')$ only in the region $r' \leq r$. But that is all we need for (20.10) and (20.13). Equation (20.9) could also serve to define $K(r,r')$ explicitly in the region $r' > r$ in terms of the same function in $r' \leq r$, once it is known there. But this extension is never required.

At this point we have constructed, via the explicit equation (20.13), a regular solution of the radial Schrödinger equation with the potential

$$\mathscr{V}(r) = \mathscr{V}_1(r) + \Delta\mathscr{V}(r). \tag{20.14}$$

We started with (20.6), and it is now incumbent upon us to relate the input $h(E)$ to the function $\varphi(E,r)$. Before we do so, we realize that the procedure could have been inverted. If the potential of (20.14) is well behaved, we could envisage having started with a function such as (20.6), but in terms of $\varphi(E,r)$ instead of $\varphi_1(E,r)$, and possibly with a different $h(E)$. It must be possible to choose such a new function $h(E)$ that we would then end up with $\varphi_1(E,r)$. In other words, since (20.13) is true in general, there must exist another kernel function $\overline{K}(r,r')$ such that

$$\varphi_1(E,r) = \varphi(E,r) + \int_0^r dr'\, \overline{K}(r,r')\varphi(E,r'). \tag{20.13a}$$

Note the remarkable fact that the kernel functions in (20.13) and (20.13a) *do not depend on the energy*.

Now let us insert (20.6) in (20.9). Then it follows that $K(r,r')$ must be expressible in the form

$$K(r,r') = \int dh(E)\,\mathscr{K}(E,r)\varphi_1(E,r')$$

where furthermore

$$\mathscr{K}(E,r) = \varphi_1(E,r) - \int_0^r dr'\,K(r,r')\varphi_1(E,r').$$

Comparison with (20.13) shows that $\mathscr{K}(E,r) = \varphi(E,r)$, so that

$$K(r,r') = \int dh(E)\,\varphi(E,r)\varphi_1(E,r'). \tag{20.15}$$

Now we know that the functions $\varphi_1(E,r)$ form a complete set. The completeness relation (12.128a) contains the weight function $\rho_1(E)$ which is defined in (12.172) in terms of the bound states and the Jost function for the potential $\mathscr{V}_1$. If $\mathscr{V}(r)$ of (20.14) possesses finite first and second absolute moments, then $\varphi(E,r)$ similarly forms a complete set, with the spectral function $\rho(E)$,

$$\begin{aligned} &\int d\rho_1(E)\,\varphi_1(E,r)\varphi_1(E,r') = \delta(r - r') \\ &\int d\rho(E)\varphi(E,r)\varphi(E,r') = \delta(r - r'). \end{aligned} \tag{20.16}$$

Let us multiply (20.13) by $\varphi_1(E,r')$ integrate with the weight $d\rho_1$, and subtract (20.13a) for $\varphi_1(E,r')$ multiplied by $\varphi(E,r)$, integrated with the weight $d\rho$. The result of using (20.16) is

$$\int d[\rho(E) - \rho_1(E)]\varphi(E,r)\varphi_1(E,r') = \begin{cases} K(r,r') & r' \le r \\ \bar{K}(r',r) & r' \ge r. \end{cases} \tag{20.17}$$

Comparison with (20.15) shows that the input function $h(E)$ is directly related to the new spectral function,

$$h(E) = \rho(E) - \rho_1(E). \tag{20.18}$$

We have therefore succeeded in constructing the potential $\mathscr{V}$ which is associated with a given spectral function $\rho(E)$. The procedure is to insert (20.18) in (20.6) and to solve the integral equation (20.9). Then (20.10) and (20.14) give the potential.

Existence of the Povsner–Levitan Representation. Starting with the ansatz (20.6) we have constructed the Povsner–Levitan representation (20.13) for a regular solution of the radial Schrödinger equation with the potential (20.14). In order to ascertain that we are able to reconstruct the potential that underlies a given spectral function ρ we must show that the regular solution φ in fact has such a representation.

Let $\varphi^{(0)}(k,r)$ be the solution of (12.1) with $V = 0$ that satisfies the same boundary conditions as $\varphi(k,r)$, and define $\varphi_S = \varphi - \varphi^{(0)}$. The results of Chapter 12 show that if $\mathscr{V}$ satisfies (12.9) then for fixed r $k\varphi_S$ is an entire analytic function of exponential type r. Furthermore, for $l = 0$ one easily proves from (12.11) that if

$$\int_0^r dr'\, r'^{1/2} |\mathscr{V}| < \infty$$

then for each $r \geq 0$, $k\varphi_S$ is square intergrable as a function of k on the real axis. For $l \geq 1$ (12.137) shows that the stronger condition on V is not needed. Therefore by the Paley–Wiener theorem[6e] the Fourier transform of φ_S is an L^2-function of support in $(-r,r)$:

$$k\varphi_S(k,r) = \int_{-r}^{r} dr'\, \tilde{\varphi}_S(r,r')e^{ikr'}$$

and since φ_S is an even function of k,

$$\varphi(k,r) = \varphi^{(0)}(k,r) - \int_0^r dr'\, K_0(r,r')\varphi^{(0)}(k,r'), \tag{20.13'}$$

where $K_0(r,r') = i[\tilde{\varphi}_S(r,r') - \tilde{\varphi}_S(r,-r')]$. For $l \geq 1$ the same representation follows from a Fourier-Bessel integral. So we have a Povsner–Levitan representation with $K_0(r,r')$, as a function of r', in L^2. For $l \geq 1$, in fact, (12.137) shows that φ_S is in L^1 as a function of k and therefore K is continuous. For $l = 0$ we insert (20.13′) in (12.4) and, after some algebra, obtain the integral equation for K_0,

$$K_0(r,r') = -\tfrac{1}{2}\int_u^v dt\, \mathscr{V}(t) + \int_u^v dt \int_0^u ds\, \mathscr{V}(t+s)\, K_0(t+s, t-s) \tag{20.19}$$

for $r' \leq r$, where $u = \frac{1}{2}(r - r')$, $v = \frac{1}{2}(r + r')$. This equation, which can be solved by successive approximations, can be used to show that for $r' < rK_0$ is continuous and differentiable if $\mathscr{V} \in L^1\,(a,b)$ for all $b > a > 0$.

Equation (20.13′) may be written in operator form

$$\varphi(k) = \omega_0 \varphi^{(0)}(k)$$

where ω_0 is the linear operator whose integral kernel is given by

$$\omega_0(r,r') = \delta(r - r') - K_0(r,r')\Theta(r - r')$$

and $\Theta(x)$ is the Heaviside function. Because of the "triangular" nature of $K_0(r,r')\Theta(r - r')$ the unique inverse of ω can be constructed as the solution of a Volterra equation,

$$\omega_0^{-1}(r,r') = \delta(r - r') + \bar{K}_0(r,r')\Theta(r - r'),$$

$$\bar{K}_0(r,r') = K_0(r,r') + \int_{r'}^{r} dr''\, K_0(r,r'')\,\bar{K}_0(r'',r').$$

It then follows that the kernel $K(r,r')$ of (20.13) is given by

6e See, for example, Chadan and Sabatier (1977), p. 36.

$$K(r,r') = K_0(r,r') - \bar{K}_1(r,r') + \int_{r'}^{r} dr''\, K_0(r,r'')\bar{K}_1(r'',r'),$$

where

$$\varphi^{(0)}(k) = \omega_1^{-1}\varphi_1(k),$$
$$\omega_1^{-1}(r,r') = \delta(r - r') + \bar{K}_1(r,r')\Theta(r - r').$$

Uniqueness of the Solution to (20.9). There is still one loose end: the integral equation (20.9) must be shown to have a unique solution. Suppose that there were in fact two solutions. Then the homogeneous equation would have to possess a nontrivial solution; call it $\chi_r(r')$,

$$\chi_r(r') + \int_0^r dr''\, g(r'',r')\chi_r(r'') = 0$$

or, according to (20.6) and (20.18),

$$\chi_r(r') + \int d[\rho(E) - \rho_1(E)]\varphi_1(E,r') \int_0^r dr''\, \varphi_1(E,r'')\chi_r(r'') = 0.$$

Multiply this by $\chi_r(r')$, and integrate,

$$\int_0^r dr'\, [\chi_r(r')]^2 + \int d\rho(E)\left[\int_0^r dr'\, \varphi_1(E,r')\chi_r(r')\right]^2 = \int d\rho_1(E)\left[\int_0^r dr'\, \varphi_1(E,r')\chi_r(r')\right]^2. \tag{20.20}$$

But the completeness relation

$$\int_0^\infty dr[f(r)]^2 = \int d\rho_1(E)\left[\int_0^\infty dr f(r)\, \varphi_1(E,r)\right]^2$$

applied to

$$f(r') \equiv \begin{cases} \chi_r(r') & r' < r \\ 0 & r' > r \end{cases}$$

says that

$$\int_0^r dr'\, [\chi_r(r')]^2 = \int d\rho_1(E)\left[\int_0^r dr'\, \varphi_1(E,r')\chi_r(r')\right]^2$$

and therefore by (20.20)

$$\int d\rho(E)\left[\int_0^r dr'\, \varphi_1(E,r')\chi_r(r')\right]^2 = 0.$$

Since $\rho(E)$ is monotone, this implies that $\chi_r(r')$ is orthogonal to all $\varphi_1(E,r')$; hence it must vanish. This proves that, if the integral equation (20.9) has a solution which leads to a well-behaved potential (so that its wave functions form a complete set) it must be unique.

Existence of a Solution. The character of Eq. (20.9) depends upon the rapidity with which $h(E)$ approaches zero for large energies. The kernel is given by

$$G_r(r',r'') = \begin{cases} g(r',r'') & r'' < r \\ 0 & r'' > r \end{cases}$$

and we may calculate the trace of $GG^\dagger$. Since it follows from (20.16) that

$$\int_0^\infty dr\, \varphi_1(E,r)\varphi_1(E',r) = \frac{\delta(E-E')}{d\rho_1/dE}$$

we find

$$\begin{aligned} tr\, G_r G_r^\dagger &= \int_0^\infty dr' \int_0^r dr''\, g^2(r',r'') \\ &= \int dE \frac{(d\rho/dE - d\rho_1/dE)^2}{d\rho/dE} \int_0^r dr' \,[\varphi_1(E,r')]^2. \end{aligned}$$

For reasonable spectral functions ρ and ρ_1 this integral converges and the kernel of (20.9) is in the Hilbert-Schmidt class. In other words, (20.9) is a *Fredholm equation*. The demonstration that the homogeneous version has no nontrivial solution therefore proves at the same time that a solution of the inhomogeneous equation (20.9) exists.

Let us write Eq. (20.9) in the form

$$K(r,r') = g(r,r') - \int_0^\infty dr''\, K(r,r'')G_r(r'',r') \tag{20.9'}$$

and express the solution in the form

$$K(r,r') = \int_0^\infty dr''\, g(r,r'')N_r(r'',r'),$$

where

$$N_r = (\mathbb{1} + G_r)^{-1}.$$

Now we have by (9.67)

$$\begin{aligned} \frac{d}{dr} \ln \det (\mathbb{1} + G_r) &= \operatorname{tr} (\mathbb{1} + G_r)^{-1} \frac{d}{dr} G_r \\ &= \int_0^\infty dr' \int_0^\infty dr''\, N_r(r',r'')\delta(r''-r)g(r'',r') \\ &= \int_0^\infty dr'\, N_r(r',r)g(r,r') = K(r,r). \end{aligned}$$

Therefore we can express the potential difference (20.10) directly in terms of the Fredholm determinant of (20.9):

$$\Delta \mathscr{V}(r) = -2 \frac{d^2}{dr^2} \ln \det (\mathbb{1} + G_r). \tag{20.21}$$

The Nonlinear Gel'fand–Levitan Equation. If we write (20.13) in operator form, $\varphi(k) = \omega\varphi_1(k)$, operate on the second Eq. (20.16) with ω^{-1} on the left and with $\omega_1{}^{\dagger-1}$ on the right, and subtract the first, we obtain

$$\omega^{-1}\omega^{\dagger-1} - \mathbb{1} = g,$$

where g has the kernel (20.6). This equation, written out in detail, reads for $r' < r$,

$$\bar{K}(r,r') = g(r,r') - \int_0^r dr''\, \bar{K}(r,r'')\bar{K}(r',r''), \tag{20.22}$$

a nonlinear integral equation for $\bar{K}$. According to (20.17) we have $K(r,r) = \bar{K}(r,r)$ and hence $\bar{K}$ serves to determine the potential by (20.10).

The Spectral Function from the **S** *matrix.* There remains only the question of obtaining the spectral function from the **S** matrix or from the phase shift. But this is accomplished by the dispersion relation (12.64), which allows us to calculate the magnitude of the Jost function from the phase shift and from the energies of the bound states of the same angular momentum. The expression (12.172) for the spectral function shows that, in addition, it contains one positive parameter for each bound state. Thus, for a given phase shift and a given set of N bound states, there is an N-parameter family of associated spectral functions and, thus, of potentials.

The significance of the parameters is the value of the normalization integral of the bound-state wave functions, if these are defined by the fixed boundary condition (12.132). For normalized bound-state wave functions, they fix the value of the first nonvanishing derivative at the coordinate origin. Therefore, they determine the amount of "spread" of the wave function or the "size" of the bound state, i.e., the value of the root-mean-square (rms) interparticle distance.[7]

Changes in the Normalization Parameters. Suppose in fact that the spectral function ρ differs from the comparison function ρ_1 only by one of these normalization parameters,

$$\frac{d\rho}{dE} - \frac{d\rho_1}{dE} = \delta(E - E_0)C$$

where

$$C = N^{-2} - N_1^{-2} \tag{20.23}$$

is the difference in the reciprocals of the two normalization integrals, as below (12.123). We are then going to get a potential $\mathscr{V}$ that produces the same phase shift and the same bound states (with the same binding energies) as $\mathscr{V}_1$.

With (20.23) the kernel g is separable, and so is $K(r,r')$,

$$g(r,r') = C\varphi_0(r)\varphi_0(r') \qquad K(r,r') = C\varphi(r)\varphi_0(r')$$

7 For a specific example, see Sec. 14.7.1.

if $\varphi_0(r)$ is the wave function of the bound state of $\mathscr{V}_1$ whose normalization is to be changed and $\varphi(r)$ is the corresponding new wave function. Equation (20.13) shows that

$$\varphi(r) = \frac{\varphi_0(r)}{1 + C\int_0^r dr'\, \varphi_0^2(r')} \tag{20.24}$$

and hence

$$K(r,r') = \frac{C\varphi_0(r)\varphi_0(r')}{1 + C\int_0^r dr''\, \varphi_0^2(r'')}$$

$$\Delta\mathscr{V}(r) = -2\frac{d^2}{dr^2}\ln\left[1 + C\int_0^r dr'\, \varphi_0^2(r')\right]. \tag{20.25}$$

The new scattering wave functions are given by

$$\varphi(E,r) = \varphi_1(E,r) - C\varphi_0(r)\frac{\int_0^r dr'\, \varphi_0(r')\varphi_1(E,r')}{1 + C\int_0^r dr'\, \varphi_0^2(r')}. \tag{20.26}$$

The last term tends to zero as $r \to \infty$, because $\varphi_0(r)$ is a bound-state function. This is why there is no change in the phase shift. It can be explicitly verified that

$$\frac{1}{\int_0^\infty dr\, \varphi^2(r)} = C + \frac{1}{\int_0^\infty dr\, \varphi_0^2(r)}.$$

Equation (20.25) shows that the asymptotic value of the potential change for large r is

$$\Delta\mathscr{V}(r) \simeq \text{const} \times e^{-2Kr}$$

if the energy of the bound state whose normalization was altered is $E = -K^2/2\mu$. It follows that, if there is one potential with given phase shift and given binding energies E_n, and with the property that

$$\lim_{r\to\infty} \mathscr{V}(r)e^{2Kr} = 0 \qquad K^2 = \max(2E_n\mu) \tag{20.27}$$

then this potential must be the only one with that property. Furthermore, it follows that if there exists one potential which is such that

$$\lim_{r\to\infty} \mathscr{V}(r)e^{2Kr} = \infty \qquad K^2 = \min(2E_n\mu)$$

then every potential with the same phase shift and bound states has the same asymptotic decrease. In the first instance there is a natural possibility of selecting a *unique* potential out of the family of those equivalent with respect to both phase shift and binding energies.

Now we have seen in Chap. 12 that Eq. (20.27) implies an **S** matrix which is analytic in a strip of width K above the real k axis, except for the bound-state poles on the imaginary axis. Therefore, a *necessary* condition for the existence of a *unique* potential, in the sense of (20.27), is that the **S** matrix is in fact analytic in a strip above the real axis which contains all the bound states, except for the points corresponding to the latter. All bound states are thus recognizable as poles of **S**, and no poles are "redundant." *Sufficient* criteria for the existence of such a unique potential are not known.[8]

In this context one may ask: Does knowledge of a phase shift over any finite range of energy tell us if the potential decreases asymptotically as an exponential or less rapidly. The answer must clearly be "no." Since analyticity of **S** in a strip of finite width along the real k axis is a necessary condition for exponential decrease of the force, no finite piece of δ_l can reveal it. A small "kink" in the phase shift at high energies is sufficient to alter radically the asymptotic behavior of the potential.[9]

20.3.2 Infinitesimal Variations. The method of the last section may be employed for the calculation of the infinitesimal changes that arise in a potential if the phase shift is changed infinitesimally (as a function of the energy). The result will be a variational derivative of the potential with respect to the phase shift. It effectively inverts the more usual perturbation theory.

If the spectral function is changed infinitesimally,

$$\rho - \rho_1 = \boldsymbol{\delta}\rho(E)$$

then (20.18) and (20.6) say that

$$g(r,r') = \int d\boldsymbol{\delta}\rho(E)\varphi_l(E,r)\varphi_l(E,r').$$

(We now omit the subscript 1 on the initial functions but indicate the angular momentum.) Equation (20.9) then shows that

$$K(r,r') = g(r,r')$$

and hence by (20.10)

$$\boldsymbol{\delta}\mathscr{V}(r) = -2\frac{d}{dr}\int d\boldsymbol{\delta}\rho(E)\,[\varphi_l(E,r)]^2. \tag{20.28}$$

Now the infinitesimal change in the spectral function $\rho(E)$ is to be due to an infinitesimal change in the phase shift. What is the connection?

The dispersion relation (12.64) shows that, if the phase shift is infinitesi-

8 Among the large class of Bargmann potentials a sufficient condition for exponential asymptotic decrease is known to be that the Jost function $\mathscr{F}_l(k)$ approach its limit as $k \to 0$ at least as fast as $O(k^{2l})$. See R. G. Newton (1956), appendix. Because of (12.153) this cannot be a necessary condition.

9 For an investigation of the asymptotic potential information obtainable from the phase shift, see R. Jost (1956).

mally altered by $\boldsymbol{\delta}\delta_l(k)$, then the concomitant change in the magnitude of the Jost function is

$$\frac{\boldsymbol{\delta}|\mathscr{F}_l(k)|}{|\mathscr{F}_l(k)|} = -\frac{1}{\pi}\mathscr{P}\int_{-\infty}^{\infty}\frac{dk'\,\boldsymbol{\delta}\delta_l(k')}{k'-k}$$

$$= -\frac{1}{\pi}\mathscr{P}\int_{0}^{\infty}\frac{dE'\,\boldsymbol{\delta}\delta_l(E')}{E'-E}.$$

Therefore, according to (12.172) the variation in the spectral function is given by

$$\boldsymbol{\delta}\frac{d\rho}{dE} = \frac{d\rho}{dE}\frac{2}{\pi}\mathscr{P}\int_0^{\infty}\frac{dE'\,\boldsymbol{\delta}\delta_l(E')}{E'-E}$$

for $E > 0$. For $E < 0$, of course, it vanishes. Thus we get, after an interchange in orders of integration,

$$K(r,r') = \frac{2}{\pi}\int_0^{\infty} dE'\,\boldsymbol{\delta}\delta_l(E')\mathscr{P}\int_0^{\infty}\frac{d\rho(E)}{E'-E}\varphi_l(E,r)\varphi_l(E,r').$$

But the second integral is part of a Green's function. According to (12.130), and (11.22)

$$2\mu\mathscr{G}_l^{P}(E';r,r') = \mathscr{P}\int_0^{\infty}\frac{d\rho(E)}{E'-E}\varphi_l(E,r)\varphi_l(E,r') + \sum_n\frac{\phi_n(r)\phi_n(r')}{E'+E_n}$$

if ϕ_n are the normalized bound-state wave functions and E_n the (positive) binding energies. That same Green's function is expressible as in (12.40),

$$\mathscr{G}_l^{P}(E;r,r') = \varphi_l(E,r_<)I_l(E,r_>)$$

in terms of an irregular solution $I_l(E,r)$ of the radial Schrödinger equation whose wronskian with $\varphi(E,r)$ is unity,

$$2I(E,r) = -\frac{f_{l+}(k,r)}{\mathscr{F}_{l+}(k)} - \frac{f_{l-}(k,r)}{\mathscr{F}_{l-}(k)}.$$

We therefore find for the variational derivative of the potential with respect to the phase shift, defined by

$$\boldsymbol{\delta}V(r) = \int_0^{\infty} dE\frac{\boldsymbol{\delta}V(r)}{\boldsymbol{\delta}\delta_l(E)}\boldsymbol{\delta}\delta_l(E)$$

$$\frac{\boldsymbol{\delta}V(r)}{\boldsymbol{\delta}\delta_l(E)} = -\frac{4}{\pi}\frac{d}{dr}\left[\varphi_l(E,r)I_l(E,r) - \frac{1}{2\mu}\sum_n\frac{\phi_n^2(r)}{E+E_n}\right]. \tag{20.29}$$

Similarly, according to (20.13), the variational derivative of the wave function with respect to the phase shift,

$$\frac{\boldsymbol{\delta}\varphi_l(E',r)}{\boldsymbol{\delta}\delta_l(E)} = -\frac{2}{\pi}\int_0^{r} dr'\,\varphi_l(E',r')\left[2\mu\varphi_l(E,r')I_l(E,r) - \frac{\phi_n(r')\phi_n(r)}{E+E_n}\right].$$

Because

$$\varphi_l(E,r)I_l(E,r) = -\frac{1}{2k}\operatorname{Im}[S_l(k)f_l^2(k,r)]$$

and

$$2\mu\int_0^r dr'\,\varphi(E,r')\varphi(E',r') = \frac{W[\varphi(E,r),\varphi(E',r)]}{E-E'}$$

we can simplify these results to

$$\frac{\boldsymbol{\delta} V(r)}{\boldsymbol{\delta}\delta_l(E)} = \frac{4}{\pi}\frac{d}{dr}\left\{k^{-1}\operatorname{Im}[S_l(k)f_l^2(k,r)] + \frac{1}{\mu}\sum_n\frac{\phi_n^2(r)}{E+E_n}\right\} \tag{20.30}$$

$$\frac{\boldsymbol{\delta}\varphi_l(E,r)}{\boldsymbol{\delta}\delta_l(E')} = \frac{2}{\pi}\left\{\frac{W[\varphi_l(E,r),\varphi_l(E',r)]}{E'-E}I(E,r) + \frac{1}{2\mu}\sum_n\frac{W[\varphi_l(E,r),\phi_n(r)]}{(E+E_n)^2}\phi_n(r)\right\}. \tag{20.31}$$

W here is the wronskian, defined in Sec. 12.1.2.

Application to $r = 0$. A very simple consequence of (20.29) is obtained by letting r approach zero. Then

$$\varphi_l(E,r)I_l(E,r) \to -\frac{r}{2l+1}$$

follows from

$$W[\varphi_l(E,r),I_l(E,r)] = 1.$$

Hence according to (20.29),

$$\frac{\boldsymbol{\delta} V(0)}{\boldsymbol{\delta}\delta_l(E)} = \frac{4}{\pi}\frac{1}{2l+1}.$$

This equation can be integrated to the exact result

$$V(0) = \frac{4}{\pi(2l+1)}\int_0^\infty dE\,\delta_l(E) \tag{20.32}$$

if both sides are finite and there are no bound states. Otherwise we can merely conclude that

$$V(0) - V_0(0) = \frac{4}{\pi(2l+1)}\int_0^\infty dE[\delta_l(E) - \delta_l^{(0)}(E)]$$

if $V_0(r)$ is another potential with the same bound states as $V(r)$ and whose phase shift has the same asymptotic value as δ_l. According to (12.82),[10]

$$\lim_{k\to\infty} k\delta_l(k) = -\mu\int_0^\infty dr\,V(r)$$

if both sides exist.

10 Recall that $\mathscr{V}(r) = 2\mu V(r)$.

At this point we simply use one of the Bargmann potentials of Sec. 14.7 for V_0. The phase shift

$$\delta_l^{(0)}(E) = \tan^{-1}\frac{\beta}{k} - \tan^{-1}\frac{\alpha}{k} + 2\sum_n \tan^{-1}\frac{K_n}{k}$$

belongs to a potential whose value at the origin is

$$V_0(0) = \frac{\alpha^2 - \beta^2}{l + \frac{1}{2}}$$

and whose bound-state energies are $E_n^{(l)} = K_n{}^2/2\mu$; α and β are positive constants. Moreover,

$$4\pi^{-1}\int_0^\infty dk\left(k\tan^{-1}\frac{\beta}{k} - \beta\right) = -\beta^2.$$

Therefore, upon choosing α and β so that

$$\beta - \alpha + 2\sum_n K_n = -\mu\int_0^\infty dr\, V(r)$$

we find the exact result

$$V(0) = \frac{4}{\pi(2l+1)}\int_0^\infty dE\left[\delta_l(E) + k^{-1}\mu\int_0^\infty dr\, V(r)\right] + \frac{4}{2l+1}\sum_n E_n^{(l)}. \tag{20.33}$$

The left-hand side of (20.33) is independent of l. Hence the equation shows (if there are no bound states) that, as l increases, the energy integral of the phase shifts, minus their asymptotic value, *increases* linearly—this in spite of the fact that they become smaller and smaller for small energies. The reason is that the phase-shift maxima occur at larger and larger energies as l increases.

20.3.3 The Marchenko Equation. For simplicity we shall work with the s-wave only. It follows from (12.25) and the analyticity of $f_+(k,r)$ in the upper half-plane that the Fourier transform of $f_+(k,r) - e^{ikr}$,

$$A(r,r') = (1/2\pi)\int_{-\infty}^{\infty} dk\, e^{-ikr'}\left[f_+(k,r) - e^{ikr}\right] \tag{20.34}$$

exists as an L^2-function of r' and vanishes for $r' < r$:

$$f_+(k,r) = e^{ikr} + \int_r^\infty dr'\, A(r,r')e^{ikr'}. \tag{20.35}$$

Changing k into $-k$ in (20.34) and using (12.23), (12.29), and (12.71), namely,

$$f_+(-k,r) = S(k)f_+(k,r) - 2ik\varphi(k,r)/\mathscr{F}_+(k)$$

we obtain for $r' > r$:

$$A(r,r') = (1/2\pi)\int_{-\infty}^{\infty} dk\, e^{ikr'}\{[S(k) - 1][f_+(k,r) - e^{ikr}] + e^{ikr}[S(k) - 1] + [f_+(k,r) - e^{ikr}] - 2ik[\varphi(k,r)/\mathscr{F}_+(k) - k^{-1}\sin kr]\}.$$

Define $\bar{h}(r)$ as the Fourier transform of $S(k) - 1$. Then the first term is the convolution of $\bar{h}$ with $A(r, -r')$; the second term is $g(r + r')$; the third term vanishes by (12.25) and the analyticity of f_+; and the fourth term, by (12.11), (12.33) and the analyticity of φ and $\mathscr{F}_+$, gives a sum of residue contributions from the zeros of $\mathscr{F}_+$ at $k = i\kappa_n$, i.e., from the bound states. One then uses (12.52) and the analytic continuation of (20.35) to $k = i\kappa_n$, and one arrives at the Marchenko equation for $r' > r$

$$A(r,r') = h(r + r') + \int_r^{\infty} dr''\, h(r' + r'')A(r,r''), \tag{20.36}$$

where

$$h(r) = (1/2\pi)\int_{-\infty}^{\infty} dk\, e^{ikr}[S(k) - 1] - \sum_n M_n e^{-\kappa_n r} \tag{20.37}$$

$$M_n = 1/\int_0^{\infty} dr[f_+(i\kappa_n, r)]^2.$$

Insertion of (20.35) in the radial Schrödinger equation shows that for $r' > r$ $A(r,r')$ must satisfy the partial differential equation

$$\left(\frac{\partial^2}{\partial r^2} - \frac{\partial^2}{\partial r'^2}\right)A(r,r') = \mathscr{V}(r)A(r,r') \tag{20.38}$$

and the boundary condition

$$-2\frac{d}{dr}A(r,r) = \mathscr{V}(r). \tag{20.39}$$

If h_r is the operator whose integral kernel is given by

$$h_r(r',r'') = h(r' + r'')\Theta(r'' - r)$$

where Θ is the Heaviside function, then one finds, in the same manner as (20.21),

$$\mathscr{V}(r) = -2\frac{d^2}{dr^2}\ln\det(\mathbb{1} - h_r). \tag{20.40}$$

Here $\det(\mathbb{1} - h_r)$ is the Fredholm determinant of (20.36).

We refer to the literature[10a] for detailed proofs and discussions of the properties of the Marchenko equation (20.36). As compared to the Gel'fand–Levitan equation (20.9) it has the virtue that it does not require the prior construction of the spectral function from the S-matrix. Its disadvantage is that the region of integration in it is (r, ∞), whereas in (20.9) it is $(0, r)$.

10a L. D. Faddeev (1963); Agranovich and Marchenko (1963), Chadan and Sabatier (1977).

20.4 The Central Potential Obtained from All Phase Shifts at One Energy

20.4.1 The Construction Procedure. Again we take elastic scattering of spinless particles by a spherically symmetric local potential. The problem is to find the potential which, inserted in each radial Schrödinger equation, produces the prescribed set of phase shifts for all angular momenta at the given energy. Before going into the actual method a word should be said about the choice of energy at which the potential determination is made.

It is physically clear that at very low energies the particles do not approach one another closely enough to be able to see the inner region of the potential very distinctly. Mathematically this expresses itself in a high degree of sensitivity to information that is experimentally inaccessible. In other words, *in principle*, even low-energy phase shifts serve to determine a potential. But then the s wave dominates to such an extent that the others cannot be measured, and the potential is sensitive to them even though they are small. In practice, therefore, one ought to work at an energy where all phase shifts that are ever going to be of significant size come into play.

Up to a certain point the procedure to be followed is quite analogous to that of Sec. 20.3.1. In order to simplify the equations, we measure distances in units of the reduced de Broglie wavelength. In other words, we set $k = 1$. In contrast to the situation in Sec. 20.3.1, the energy is now going to be fixed.

The differential operator to be used is defined by

$$D_1(r) = r^2\left[\frac{d^2}{dr^2} + 1 - \mathscr{V}_1(r)\right], \tag{20.41}$$

$\mathscr{V}_1$ again being a well-behaved "comparison" potential. The regular solutions $\varphi_l{}^{(1)}(r)$ of the radial Schrödinger equations

$$D_1(r)\varphi_l^{(1)}(r) = l(l+1)\varphi_l^{(1)}(r) \tag{20.42}$$

subject to the boundary conditions,

$$\lim_{r\to 0} \varphi_l^{(1)}(r)r^{-l-1}(2l+1)!! = 1$$

are assumed known. [These boundary conditions are more convenient for the present than (12.132).] In terms of them we define a function

$$g(r,r') = \sum_{l=0}^{\infty} c_l\varphi_l^{(1)}(r)\varphi_l^{(1)}(r') \tag{20.43}$$

with real coefficients c_l which for the moment are arbitrary and are to be fixed later. Evidently g satisfies the partial differential equation

$$D_1(r)g(r,r') = D_1(r')g(r,r') \tag{20.44}$$

and the boundary conditions

$$g(0,r') = g(r,0) = 0. \tag{20.45}$$

Now let the function $K(r,r')$ be the *unique* solution of the linear integral equation

$$K(r,r') = g(r,r') - \int_0^r dr''\, r''^{-2} K(r,r'') g(r'',r'), \tag{20.46}$$

which is the analog of the Gel'fand-Levitan equation. Again the assumption (to be proved below) is crucial that the solution of (20.46) is unique. Next we define the function

$$\xi(r,r') \equiv D(r)K(r,r') - D_1(r')K(r,r') \tag{20.47}$$

in which the differential operator $D(r)$ is defined by

$$D(r) = D_1(r) - \Delta\mathscr{V}(r)$$

$$\Delta\mathscr{V}(r) = -2r^{-1}\frac{d}{dr}[r^{-1}K(r,r)]. \tag{20.48}$$

By straightforward differentiation, integrations by parts, and use of the differential equation (20.44) and of the boundary condition (20.45), one verifies that $\xi(r,r')$ satisfies the homogeneous version of (20.46). Since that equation is assumed to have a unique solution, we conclude that $\xi(r,r') \equiv 0$. In other words, $K(r,r')$ satisfies the partial differential equation

$$D(r)K(r,r') = D_1(r')K(r,r'). \tag{20.49}$$

In addition, it follows from (20.46) that

$$K(0,r') = K(r,0) = 0. \tag{20.50}$$

Functions $\varphi_l(r)$ are now defined by means of the solution $K(r,r')$ of (20.46)[11]

$$\varphi_l(r) \equiv \varphi_l^{(1)}(r) - \int_0^r dr'\, r'^{-2} K(r,r')\varphi_l^{(1)}(r'). \tag{20.51}$$

If we apply $D(r)$ to (20.51), integrate twice by parts, and use (20.49) and (20.42), we find that $\varphi_l(r)$ obeys the differential equation

$$D(r)\varphi_l(r) = l(l+1)\varphi_l(r). \tag{20.52}$$

In addition it follows from (20.50) that

$$\varphi_l(0) = 0. \tag{20.53}$$

Thus the solution $K(r,r')$ of (20.46) has allowed us to construct, via (20.51), regular solutions of all the radial Schrödinger equations (20.52) with the potential

$$\mathscr{V}(r) = \mathscr{V}_1(r) + \Delta\mathscr{V}(r). \tag{20.54}$$

11 The temptation to conclude from the l independence of $K(r,r')$ that (20.51) can be summed and must hold for the three-dimensional wave functions must be firmly resisted! The functions $\varphi_l(r)$ and $\varphi_l^{(1)}(r)$ differ from the "physical" wave functions needed for this by different factors [see (12.145)].

Again we may express $K(r,r')$ in terms of the φ_l. It follows from (20.46) and (20.43) that it must be expressible in the form

$$K(r,r') = \sum_{l=0}^{\infty} c_l K_l(r)\varphi_l^{(1)}(r')$$

where

$$K_l(r) = \varphi_l^{(1)}(r) - \int_0^r dr'\, r'^{-2} K(r,r')\varphi_l^{(1)}(r').$$

Comparison with (20.51) therefore shows that $K_l(r) = \varphi_l(r)$, so that

$$K(r,r') = \sum_{l=0}^{\infty} c_l \varphi_l(r)\varphi_l^{(1)}(r'). \tag{20.55}$$

It is now necessary to relate the input, namely, the set of constants c_l, to the asymptotic properties of the wave function $\varphi_l(r)$. In the procedure of Sec. 20.3.1 this was done by means of the orthogonality and completeness of the set of radial functions of all energies. In the present context this cannot be done. The $\varphi_l{}^{(1)}$ are not orthogonal. We must therefore proceed differently.

Let us substitute (20.55) in (20.51) and invert the order of the summation and integration. This yields

$$\varphi_l(r) = \varphi_l^{(1)}(r) - \sum_{l'} L_{ll'}(r) c_{l'} \varphi_{l'}(r) \tag{20.56}$$

where

$$L_{ll'}(r) = \int_0^r dr'\, r'^{-2} \varphi_l^{(1)}(r')\varphi_{l'}^{(1)}(r'). \tag{20.57}$$

Multiplication of (20.42) by $\varphi_{l'}$, subtraction of (20.42) for $\varphi_{l'}$, multiplied by φ_l, and subsequent integration and use of the boundary condition yields an alternative expression for $L_{ll'}$,

$$L_{ll'} = \frac{\varphi_l^{(1)}\varphi_{l'}^{(1)\prime} - \varphi_{l'}^{(1)}\varphi_l^{(1)\prime}}{(l'-l)(l'+l+1)}. \tag{20.57a}$$

We now let r tend to infinity, setting

$$\begin{aligned} \varphi_l(r) &\simeq A_l \sin(r - \tfrac{1}{2}\pi l + \delta_l) \equiv \varphi_l^{\infty}(r) \\ \varphi_l^{(1)}(r) &\simeq A_l^{(1)} \sin(r - \tfrac{1}{2}\pi l + \delta_l^{(1)}) \equiv \varphi_l^{(1)\infty}(r) \\ L_{ll'}(r) &\simeq L_{ll'}^{\infty}. \end{aligned} \tag{20.58}$$

Insertion of the asymptotic form of $\varphi_l{}^{(1)}$ into (20.57) gives for $l \neq l'$,

$$L_{ll'}^{\infty} = A_l^{(1)} A_{l'}^{(1)} \frac{\sin[\frac{1}{2}\pi(l-l') + \delta_l^{(1)} - \delta_{l'}^{(1)}]}{(l'-l)(l'+l+1)}. \tag{20.59}$$

For $l = l'$ we can replace

$$L_{ll}^{\infty} = \int_0^{\infty} dr\, r^{-2} [\varphi_l^{(1)}(r)]^2$$

only by the somewhat useless expression

$$L_{ll}^{\infty} = A_l^{(1)2} \frac{\frac{1}{2}\pi + d\delta_l^{(1)}/dl}{2l+1}.$$

The subsequent discussion will be greatly simplified if we set the comparison potential $\mathscr{V}_1$ equal to zero from now on. In that case

$$\varphi_l^{(1)}(r) = u_l(r) \qquad A_l^{(1)} = 1 \qquad \delta_l^{(1)} = 0$$

and

$$L_{ll'}^{\infty} = \begin{cases} i^{l'-l-1} M_{ll'} & l \neq l' \\ \dfrac{\frac{1}{2}\pi}{2l+1} & l = l' \end{cases} \tag{20.60}$$

with the matrix

$$M_{ll'} = \begin{cases} [(l'-l)(l'+l+1)]^{-1} & l'-l \text{ odd} \\ 0 & l'-l \text{ even.} \end{cases} \tag{20.61}$$

Insertion of (20.58) in (20.56) gives us a set of equations for the asymptotic forms,

$$\varphi_l^{\infty}(r) = \varphi_l^{(1)\infty}(r) - \sum_{l'} L_{ll'}^{\infty} c_{l'} \varphi_{l'}^{\infty}(r).$$

We write the sine functions in their exponential form, equate the coefficients of e^{ir} and e^{-ir} separately, and use (20.61). This yields the equations

$$e^{i\delta_l} A_l = 1 - \tfrac{1}{2}\pi \frac{c_l A_l e^{i\delta_l}}{2l+1} - \sum_{l' \neq l} L_{ll'}^{\infty} i^{l-l'} c_{l'} e^{i\delta_{l'}} A_{l'} \tag{20.62}$$

and their complex conjugates. Now let us multiply this equation by $e^{-i\delta_l}$ and set

$$b_l \equiv c_l A_l. \tag{20.63}$$

Then we get

$$A_l = e^{-i\delta_l} - \frac{\frac{1}{2}\pi b_l}{2l+1} + i \sum_{l'} M_{ll'} b_{l'} e^{i(\delta_{l'} - \delta_l)}, \tag{20.64}$$

the real and imaginary parts of which are

$$\sin \delta_l = \sum_{l'} M_{ll'} b_{l'} \cos(\delta_{l'} - \delta_l) \tag{20.65}$$

$$A_l = \cos \delta_l - \frac{\frac{1}{2}\pi b_l}{2l+1} - \sum_{l'} M_{ll'} b_{l'} \sin(\delta_{l'} - \delta_l), \tag{20.66}$$

since the numbers A_l are all real. Equation (20.65) allows us to rewrite (20.66) in the somewhat simpler form

$$A_l = \sum_{l'} \csc \delta_l \, M_{ll'} \cos \delta_{l'} \, b_{l'} - \frac{\frac{1}{2}\pi b_l}{2l+1}. \tag{20.66a}$$

Now the assumed input information consists of the phase shifts δ_l; wanted are the coefficients c_l. The infinite set of algebraic equations (20.65) must therefore be solved for the numbers b_l. Then (20.66) expresses the A_l explicitly in terms of the b_l, and (20.63) gives us the c_l.

Comparison of (20.58) with (12.154) and (12.35) shows that for $k \neq 1$

$$A_l e^{-i\delta_l} = k^{-l-1} \mathscr{F}_l(k)$$

(because of the different boundary conditions on the φ's). Equation (20.62) therefore shows that the Jost functions of different l values and the same energy are connected by the equations

$$\left[1 + \tfrac{1}{2}\pi \frac{c_l(k)}{2l+1}\right] \mathscr{F}_l(k) = k^{l+1} - i \sum_{l'} M_{ll'} c_{l'}(k) k^{l-l'} \mathscr{F}_{l'}(k). \tag{20.62a}$$

These imply restrictions because the c's are real and the $\mathscr{F}$'s complex. In fact, (20.66) shows the restrictions. These equations, together with (20.65), express the magnitudes of all the Jost functions in terms of their phases. They perform the same function for the angular momentum as the dispersion relation (12.61) for the energy.

There remain two important points to be cleared up: one is to show that (20.46) has a unique solution, and the other to see whether or not (20.65) can be solved uniquely for the coefficients b_l.

Uniqueness of the Solution to (20.46). Let us look first at the uniqueness of the solution of (20.46). Suppose that there were two solutions. Then there would have to exist a nontrivial solution $\chi(r,r')$ of the homogeneous equation,

$$\chi(r,r') = -\int_0^r dr'' \, \chi(r,r'') g(r'',r').$$

It then follows from (20.43) that $\chi(r,r')$ can be expressed in the form of the series

$$\chi(r,r') = \sum_l c_l \chi_l(r) u_l(r')$$

(on the assumption that $\mathscr{V}_1 = 0$), with

$$\begin{aligned}\chi_l(r) &= -\int_0^r dr' \, r'^{-2} \chi(r,r') u_l(r') \\ &= -\sum_{l'} L_{ll'}(r) c_{l'} \chi_{l'}(r).\end{aligned}$$

Thus $\chi_l(r)$ is a solution of the homogeneous form of (20.56). In other words, the infinite matrix $L_{ll'} c_{l'}$ has the eigenvalue -1. But surely this cannot be true for all values of r. The coefficients c_l are constant, and the matrix $L_{ll'}$ tends to zero as r approaches the origin. More convincingly, we may form the square root of the symmetric positive definite matrix $L_{ll'}$ and then must have

$$\operatorname{tr} (L^{1/2} c L^{1/2})^2 \geq 1$$

if one of the eigenvalues of $L^{1/2}cL^{1/2}$ is -1. But

$$\operatorname{tr}(L^{1/2}cL^{1/2})^2 = \operatorname{tr}(Lc)^2 = \sum_{ll'} L_{ll'}^2 c_l c_{l'} \to 0$$

as $r \to 0$. We must conclude that $\chi_l(r) \equiv 0$, and therefore $\chi(r,r') \equiv 0$. The only solution of the homogeneous version of (20.46) is the trivial one.

Existence of a Solution. If the kernel $r''^{-2}g(r'',r')$ of the integral equation (20.46) is in the Fredholm class, then it follows from the absence of a solution of the homogeneous equation that (20.46) *has* a solution. A sufficient condition for this is that the kernel is in the Hilbert–Schmidt class,

$$\int_0^R dr\, r^{-2} \int_0^R dr'\, r'^{-2}[g(r,r')]^2 = \sum_{ll'} c_l c_{l'} [L_{ll'}(R)]^2 < \infty. \tag{20.67}$$

A similar condition, of course, is sufficient for the validity of the argument proving the absence of a solution of the homogeneous equation, namely, that the matrix $L_{ll'}(r)c_{l'}$ be in the Hilbert–Schmidt class,

$$\sum_{ll'} c_{l'}^2 [L_{ll'}(r)]^2 < \infty. \tag{20.67a}$$

Because the Riccati–Bessel functions have the asymptotic behavior[11a]

$$u_l(r) \simeq C\exp\left[l + l\ln\tfrac{1}{2}r - (l+1)\ln l\right]$$

for fixed r and large l, the inequalities (20.67) and (20.67a) hold if the coefficients c_l are bounded by a fixed power of l,

$$|c_l| \leq Cl^p \tag{20.68}$$

which evidently is a rather weak condition. They cannot in general be assumed to decrease rapidly as $l \to \infty$. One can show,[11b] for example, that unless the first moment of V vanishes, $\int d\imath\, V(\imath) = 0$, the sequence $\{c_l\}$ cannot be absolutely summable. It has also been shown[12] that if the phase shifts are such that for all l

$$|\tan \delta_l| < Cl^{-3-\varepsilon} \tag{20.69}$$

for some C and $\varepsilon > 0$ then the c_l are uniformly bounded and as $l \to \infty$

$$c_l = 4/\pi + 0(l^{-2}).$$

Thus they satisfy (20.68).

Existence and Uniqueness of a Solution of (20.65). The construction of the potential and of the wave functions for all angular momenta from the infinite set of numbers $\{c_l\}$ is therefore reduced to the solution of a Fredholm equation [Eq. (20.46)], if (20.69) is true. There remains the question of whether a set $\{c_l\}$ exists for every given set of phase shifts and whether it is unique, i.e., whether or not Eqs. (20.65) have a unique solution.

11a Whittaker and Watson (1948), p. 225.

11b R. G. Newton (1967).

12 Chadan and Sabatier (1977), p. 193.

Let us write (20.65) in matrix notation as follows: Let $\tan\Delta$ be the diagonal square matrix whose diagonal elements are $\tan\delta_l$, M the square matrix whose elements are $M_{ll'}$, $\mathbf{a}$ the column matrix whose elements are $b_l\cos\delta_l$, and $\mathbf{e}$ the column matrix whose elements are all equal to unity. Then (20.65) reads

$$(M + \tan\Delta\, M\tan\Delta)\mathbf{a} = \tan\Delta\,\mathbf{e} \tag{20.65a}$$

or

$$\begin{aligned}(\mathbb{1} + R)\mathbf{a} &= M^{-1}\tan\Delta\,\mathbf{e}\\ \mathbf{a} &= (\mathbb{1} + R)^{-1}M^{-1}\tan\Delta\,\mathbf{e}\end{aligned} \tag{20.70}$$

with

$$R = M^{-1}\tan\Delta\, M\tan\Delta. \tag{20.71}$$

According to (20.63) and (20.66a) the coefficients c_l are given by

$$\frac{1}{c_l} = \frac{\sum_{l'} M_{ll'}a_{l'}}{a_l\tan\delta_l} - \frac{\frac{1}{2}\pi}{2l+1}. \tag{20.72}$$

Once the matrix M^{-1} (which does not depend on the phase shifts) is constructed, then the inverse $(\mathbb{1} + R)^{-1}$ can be calculated by means of the Fredholm method, for example, or by a Neumann series expansion. This depends only on the rapidity with which the phase shifts tend to zero as $l \to \infty$. If only a finite number p of them differs from zero (as in any practical calculation will surely be the case), then (20.70) reduces to a finite set of p coupled equations and the remaining components of $\mathbf{a}$ are explicitly expressed in terms of the first p. Fredholm theory has been shown to be applicable to the construction of $(\mathbb{1} + R)^{-1}$ if the phase shifts obey (20.69), unless R "accidentally" has the eigenvalue -1; furthermore, if $\Sigma\tan^2\delta_l < 1$ then the Neumann series converges.[12a]

An explicit two-sided inverse of M is given by[13]

$$M^{-1} = -\mu M\mu, \tag{20.73}$$

where μ is a diagonal matrix with the elements

$$\begin{aligned}\mu_{2n} &= \frac{2}{\pi}(4n+1)\left[\frac{\Gamma(n+\frac{1}{2})}{\Gamma(n+1)}\right]^2\\ \mu_{2n+1} &= \frac{2}{\pi}(4n+3)\left[\frac{\Gamma(n+\frac{3}{2})}{\Gamma(n+1)}\right]^2\end{aligned} \tag{20.74}$$

on the main diagonal. However, it is not unique. There exists a nonnormalizable vector (column matrix) that is annihilated by it; its elements are

$$v_{2n} = \mu_{2n} \qquad v_{2n+1} = 0. \tag{20.75}$$

12a Chadan and Sabatier (1977), p. 191.
13 P. C. Sabatier (1966a).

As a result we draw the conclusion that *solutions of the inhomogeneous equations* (20.65) *exist, but they are not unique*. In fact, there exists a nonzero solution for the equations when all phase shifts vanish! What is more, the coefficients c_l behave sufficiently well for large values of l to make the inversion procedure work. In other words, there exists, at every energy, at least a one-parameter family of potentials[14] which produce no scattering whatever. These "transparent" potentials[13] asymptotically oscillate and decrease in magnitude as $r^{-3/2}$. If we restrict ourselves to potentials that decrease faster than $r^{-3/2}$ then the inversion procedure gives a unique answer.

For the demonstration of these existence and nonuniqueness statements, we refer to the literature.[15]

Consistency of the Method. The potentials that are constructed on the basis of the ansatz (20.43) are meromorphic functions that decrease asymptotically as $r^{-3/2}$. In order to be sure of the consistency of the construction procedure it has to be ascertained that the δ_l's used for the calculation of the c_l's and then for a potential are, in fact, the phase shifts of that potential. If the δ_l's satisfy (20.69) then it has been proved[16] that this is the case, unless the set of phase shifts is "exceptional" in the sense that R of (20.71) has the eigenvalue -1. The circle is thus closed and the method works. One, and only one, of the so constructed potentials decreases faster than $r^{-3/2}$. For numerical applications and tests of its sensitivity we refer to the literature.[17]

Generalizations. The class of potentials constructed, or reconstructed, by the method described is rather restricted. For example, none of them is an even function of r. The ansatz (20.43) has been generalized to include indices other than integers,[18] as well as to the inclusion of irregular solutions.[19] The method has been further generalized by Sabatier.[20]

20.4.2 Examples. Simple examples can be generated by starting with a finite number of coefficients c_l. One can then construct the potential and the wave functions on the one hand and the phase shifts on the other. Of course, the vanishing of all but a finite number of c's does not imply that of all but a finite number of δ's.

The simplest case is that in which a single c_L is given as a real number and all other $c_l = 0$, $L \neq l$. In that instance (20.56) shows that

$$\varphi_L(r) = \frac{u_L(r)}{1 + c_L L_{LL}(r)} \tag{20.76}$$

14 If $\{b_l\}$ solves (20.65) for $\{\gamma_l\} = 0$, then so does $\alpha\{b_l\}$. Equation (20.66) shows that this generates a one-parameter family of c_l's.

15 R. G. Newton (1962); P. J. Redmond (1964); P. C. Sabatier (1966a).

16 P. C. Sabatier (1966a and b); see also Chadan and Sabatier (1977).

17 Sabatier and van Phu (1971); C. Coudray (1977); Chadan and Sabatier (1977).

18 P. C. Sabatier (1967a).

19 Cox and Thompson (1969, 1970a and b).

20 P. C. Sabatier (1971, 1972a).

and for $l \neq L$,

$$\varphi_l(r) = u_l(r) - \frac{c_L L_{lL}(r) u_L(r)}{1 + c_L L_{LL}(r)}. \qquad (20.77)$$

Accoring to (20.48) and (20.55), these wave functions solve the radial Schrödinger equations with the potential,

$$\mathscr{V}(r) = -2r^{-1} c_L \frac{d}{dr}\left\{r^{-1} \frac{[u_L(r)]^2}{1 + c_L L_{LL}(r)}\right\}. \qquad (20.78)$$

This, of course, is not a physically particularly useful potential, because it oscillates as a function of r. The only way to avoid such oscillations is to have an infinite number of nonzero coefficients c_l.

The phase shifts are obtained from Eqs. (20.65) and (20.66), together with (20.61). Equation (20.65) tells us that $\sin \delta_l = 0$ for all even $l - L$, including $l = L$. Then (20.66) shows that

$$A_L = \frac{1}{1 + [\frac{1}{2}\pi c_L/(2L + 1)]}$$

and therefore by (20.63) and (20.65), for odd $l - L$,

$$\tan \delta_l = \frac{c_L}{1 + \frac{1}{2}\pi c_L/(2L + 1)} \frac{1}{(l - L)(l + L + 1)}. \qquad (20.79)$$

These phase shifts can equally well be obtained directly from (20.77).

Remarkably enough, the scattering amplitude produced by these phase shifts can be written down in closed form. The result is[13]

$$A(\theta) = (-)^{L+1} \tfrac{1}{4}\pi a k^{-1}[P_b(\cos\theta) - (-)^L P_b(-\cos\theta)] \qquad (20.80)$$

where

$$a = A_L c_L$$

$$b = [(L + \tfrac{1}{2})^2 + ia]^{1/2} - \tfrac{1}{2} \qquad \operatorname{Re} b > 0.$$

Similar methods can be used to construct examples for any number of coefficients. If we call C the diagonal matrix made up of the c_l on the diagonal, C' similarly made up of $\pi c_l/2(2l + 1)$, $L(r)$ the matrix of the $L_{ll'}(r)$, $\boldsymbol{\phi}(r)$ and $\mathbf{u}(r)$ the column matrixes made up of the $\varphi_l(r)$ and $u_l(r)$, respectively, we get from (20.56)

$$\boldsymbol{\phi}(r) = [\mathbb{1} + L(r)C]^{-1}\mathbf{u}(r) \qquad (20.81)$$

and from (20.48) and (20.55)

$$\mathscr{V}(r) = -2r^{-1}\frac{d}{dr}\{r^{-1}\tilde{\mathbf{u}}(r)C[\mathbb{1} + L(r)C]^{-1}\mathbf{u}(r)\}. \qquad (20.82)$$

From (20.64) we get for the phase shifts

$$\tan \delta_l = \frac{[M'(\mathbb{1} + M'^2)^{-1}\mathbf{e}]_l}{[(\mathbb{1} + M'^2)^{-1}\mathbf{e}]_l}, \qquad (20.83)$$

where

$$M' = MC(\mathbb{1} + C')^{-1}.$$

20.5 The Inverse Scattering Problem for Noncentral Potentials

20.5.1 Introduction. In Sections 20.3 and 20.4 we dealt with central potentials. We shall now consider the inverse scattering problem for local potentials without spherical symmetry. The first question that arises then is what would constitute suitable "data," i.e., what information should be considered as given. For the methods to be described in Sections 20.5.2 and 20.5.3 we will regard the scattering amplitude $A(k;\boldsymbol{\theta},\boldsymbol{\theta}')$ given as a function of the energy and of the directions of the incoming and outgoing particles. For another method briefly described in Sec. 20.5.4, we shall consider $A(k;-\boldsymbol{\theta},\boldsymbol{\theta})$ given as a function of k and $\boldsymbol{\theta}$.

The Uniqueness. We saw in Chapter 10, in the discussion following Eq. (10.109), that if $V \in R$, the Rollnik class, then at fixed momentum transfer $\boldsymbol{\tau}$ the scattering amplitude approaches the Fourier transform of the potential (the Born term) with $\boldsymbol{\tau}$ as the variable, when the energy tends to infinity. Furthermore, as $k \to \infty$, the limitation $|\boldsymbol{\tau}| \leq 2k$ becomes less and less restrictive, so that the high-energy limit of the amplitude determines the entire Fourier transform of V. It may therefore be concluded that if $A(k;\boldsymbol{\theta},\boldsymbol{\theta}')$ (on the energy shell) is known for all $k,\boldsymbol{\theta}$, and $\boldsymbol{\theta}'$, then the potential is uniquely determined. This important fact has to be kept in mind in the following.

20.5.2 The Generalized Marchenko Equation. Let us assume for simplicity at first that there are no bound states and that $k = 0$ is not an exceptional point.

The Reconstruction. The function $\gamma(k,\boldsymbol{\theta};\mathbf{r})$ defined by (10.79) has been shown to be the boundary value of an analytic function of k holomorphic (since we assumed no bound states) in the upper half plane with the asymptotic limit (10.80). It is convenient to think of $\gamma(k,\boldsymbol{\theta};\mathbf{r})$ as a function $\gamma(k)$ with values that are vectors in $L^2(S^2)$ (square-integrable functions of $\boldsymbol{\theta}$, a point on the unit sphere); it also depends parametrically on $\mathbf{r}$, which for the time being can be ignored. In that notation we shall write (10.80)

$$\lim_{|k|\to\infty} \gamma(k) = \hat{1},$$

where $\hat{1}$ is the function of $\boldsymbol{\theta}$ that equals 1 for all $\boldsymbol{\theta}$. In view of (10.81) (where we can replace k_0 by zero if that is not an exceptional point) the Fourier transform of $\gamma(k) - \hat{1}$ exists as an L^2–function,

$$\begin{aligned} \eta(\alpha) &= (1/2\pi)\int_{-\infty}^{\infty} dk\, e^{-ik\alpha}[\gamma(k) - \hat{1}] \\ &= (1/2\pi)\int_{-\infty}^{\infty} dk\, e^{ik\alpha}[\gamma(-k) - \hat{1}], \end{aligned} \tag{20.84}$$

and it vanishes for $\alpha < 0$, so that

$$\gamma(k) = \hat{1} + \int_0^\infty d\alpha\, e^{ik\alpha}\eta(\alpha). \tag{20.85}$$

Now let us write (10.116′) in the form

$$\gamma(-k) = \mathscr{S}(k)\gamma(k), \tag{20.86}$$

where $\mathscr{S}$ is the operator whose kernel is

$$\mathscr{S}(k;\mathbf{r};\boldsymbol{\theta},\boldsymbol{\theta}') = S^*(k;-\boldsymbol{\theta},\boldsymbol{\theta}')\exp[ik\mathbf{r}(\boldsymbol{\theta}' + \boldsymbol{\theta})]. \tag{20.87}$$

Therefore

$$\gamma_- - \hat{1} = (\mathscr{S} - \mathbb{1}')\hat{1} + (\mathscr{S} - \mathbb{1}')(\gamma - \hat{1}) + (\gamma' - \hat{1}),$$

where $\gamma_- = \gamma(-k,\boldsymbol{\theta},\mathbf{r})$, $\gamma' = \gamma(k,-\boldsymbol{\theta},\mathbf{r})$, and $\mathbb{1}' = \delta(-\boldsymbol{\theta},\boldsymbol{\theta}')$, which we insert in the second line of (20.84), defining

$$G(\alpha) = (1/2\pi)\int_{-\infty}^\infty dk\,[\mathscr{S}(k) - \mathbb{1}']e^{ik\alpha}. \tag{20.88}$$

Then for $\alpha > 0$ the first integral is $G(\alpha)\hat{1}$; the second integral is the convolution of $G(\alpha)$ with $\eta(-\alpha)$, and the third integral vanishes. Thus for $\alpha > 0$ Eq. (20.84) becomes

$$\eta(\alpha) = G(\alpha)\hat{1} + \int_0^\infty d\beta\, G(\alpha + \beta)\eta(\beta), \tag{20.89}$$

which is a three-dimensional generalization of the Marchenko equation.

More explicitly we define the real function

$$g(\alpha;\boldsymbol{\theta},\boldsymbol{\theta}') = -(1/2\pi)^2 i\int_{-\infty}^\infty dk\, kA^*(k;-\boldsymbol{\theta},\boldsymbol{\theta}')e^{ik\alpha}, \tag{20.88′}$$

so that by (10.115),

$$G(\alpha;r;\boldsymbol{\theta},\boldsymbol{\theta}') = g[\alpha + \mathbf{r}\cdot(\boldsymbol{\theta} + \boldsymbol{\theta}');\boldsymbol{\theta},\boldsymbol{\theta}']$$

and (20.89) reads for $\alpha > \mathbf{r}\cdot\boldsymbol{\theta}$

$$\check{\eta}(\alpha,\mathbf{r},\boldsymbol{\theta}) = \int d\boldsymbol{\theta}'\, g(\alpha + \mathbf{r}\cdot\boldsymbol{\theta}';\boldsymbol{\theta},\boldsymbol{\theta}') + \int d\boldsymbol{\theta}'\int_{\mathbf{r}\cdot\boldsymbol{\theta}'}^\infty d\beta\, g(\alpha + \beta;\boldsymbol{\theta},\boldsymbol{\theta}')\check{\eta}(\beta;\mathbf{r},\boldsymbol{\theta}') \tag{20.89′}$$

where

$$\check{\eta}(\alpha,\mathbf{r},\boldsymbol{\theta}) = \eta(\alpha - \mathbf{r}\cdot\boldsymbol{\theta},\mathbf{r},\boldsymbol{\theta})$$

and the integral over $\boldsymbol{\theta}'$ extends over all solid angles. Eq. (20.85) becomes

$$\psi(k;\boldsymbol{\theta},\mathbf{r}) = e^{ik\boldsymbol{\theta}\cdot\mathbf{r}} + \int_{\boldsymbol{\theta}\cdot\mathbf{r}}^\infty d\alpha\, e^{ik\alpha}\check{\eta}(\alpha,\mathbf{r},\boldsymbol{\theta}). \tag{20.85′}$$

If we insert this representation in the Schrödinger equation we find that $\check{\eta}$ must satisfy the partial differential equation

$$\left(\nabla^2 - \frac{\partial^2}{\partial\alpha^2}\right)\check{\eta}(\alpha,\mathbf{r},\boldsymbol{\theta}) = \mathscr{V}(\mathbf{r})\check{\eta}(\alpha,\mathbf{r},\boldsymbol{\theta}) \tag{20.90}$$

and the boundary condition

$$-2\boldsymbol{\theta}\cdot\nabla\check{\eta}(\boldsymbol{\theta}\cdot\mathbf{r},\mathbf{r},\boldsymbol{\theta}) = \mathscr{V}(\mathbf{r}). \tag{20.91}$$

The reconstruction of the potential that underlies a given scattering amplitude therefore proceeds by calculating g by means of (20.88′), solving the linear integral equation (20.89′), and then obtaining $\mathscr{V}$ from (20.91). It is remarkable that in spite of its appearance the lefthand side of (20.91) must be independent of $\boldsymbol{\theta}$, as the righthand side is. We refer to this as the *miracle.* Because of the uniqueness discussed at the end of Sec. 20.4.1 we are assured that the result obtained in this fashion is unique. However, this alone does not guarantee that (20.89′) or (20.89) has a unique solution; conceivably there might be others that are not *miraculous*, i.e., for which the lefthand side of (20.91) varies with $\boldsymbol{\theta}$.

Properties of (20.89). It has been proved[21] that if for a given potential there is a point $\mathbf{r}_0$, a C and an $\varepsilon > 0$ such that

$$|\nabla V(\mathbf{r} + \mathbf{r}_0)| < Cr^{\varepsilon-2}(1 + r)^{-2\varepsilon-2}$$

then the hermitean operator $\mathscr{G}$ on $L^2(R_+ \times S^2)$ whose integral kernel is given by

$$\mathscr{G}(\alpha,\beta;\boldsymbol{\theta},\boldsymbol{\theta}') = G(\alpha + \beta;\boldsymbol{\theta},\boldsymbol{\theta}')$$

as defined in (20.88), is compact and its square is in the Hilbert–Schmidt class,

$$\operatorname{tr}\mathscr{G}^4 < \infty. \tag{20.92}$$

Consequently, unless it happens that (for some $\mathbf{r}$)$\mathscr{G}$ is *exceptional*, i.e., $\mathscr{G}$ has the eigenvalue 1, the inverse $(\mathbb{1} - \mathscr{G})^{-1}$ exists, is a bounded operator, and can be constructed by Fredholm methods; if $\mathscr{G}$ does not have the eigenvalue -1 either it has even been shown to be expandable in a Neumann series. If we write (20.89) in the form

$$\eta = \eta_1 + \mathscr{G}\eta$$

with $\eta_1(\alpha) = G(\alpha)\hat{1}$, then its solution is given by

$$\eta = (\mathbb{1} - \mathscr{G})^{-1}\eta_1.$$

If[21]

$$|\nabla V(\mathbf{r} + \mathbf{r}_0)| < Cr^{\varepsilon-1}(1 + r)^{-2\varepsilon-3}$$

21 R. G. Newton (1981b). The result has been proved under somewhat weaker conditions that are more complicated to state.

then $\eta_1 \in L^2(\mathbb{R}_+ \times S^2)$. It follows that unless A is such that (for some $\mathbf{r}$) $\mathcal{G}$ is exceptional (20.89) has a unique solution, and this solution leads to the reconstruction of V by (20.91).

Construction. Suppose, on the other hand, we are given an arbitrary scattering amplitude $A(k;\boldsymbol{\theta},\boldsymbol{\theta}')$ (without bound states) subject to (10.15) and (10.100) that is such that $\mathcal{G}$ satisfies (20.92) and does not have the eigenvalue 1, and such that $\eta_1 \in L^2$. Furthermore, assume that the operator $\mathcal{G}'$ whose kernel is given by $\mathcal{G}'(\alpha,\beta;\boldsymbol{\theta},\boldsymbol{\theta}') = -G(-\alpha-\beta;\boldsymbol{\theta},\boldsymbol{\theta}')$, also does not have the eigenvalue 1. We then solve (20.89) or (20.89′) whose unique solution $\eta(\alpha) \in L^2$. The Fourier transform

$$\beta(k) = \hat{1} + \int_0^\infty d\alpha\, \eta(\alpha) e^{ik\alpha} \tag{20.93}$$

is such that for $\alpha > 0$

$$\eta(\alpha) = (1/2\pi) \int_{-\infty}^{\infty} dk\, [\beta(-k) - \hat{1}] e^{ik\alpha}$$

and according to (20.89)

$$\eta(\alpha) = (1/2\pi) \int_{-\infty}^{\infty} dk\, [\mathcal{S}(k) - \mathbb{1}] \beta(k) e^{ik\alpha}.$$

Therefore $\beta(k)$ is the boundary value of an analytic function that is holomorphic in the upper half-plane and tends to $\hat{1}$ there for large $|k|$, while

$$\beta_0(k) \equiv \beta(-k) - \mathcal{S}(k)\beta(k) \tag{20.94}$$

is the boundary value of an analytic function that is holomorphic in the upper half-plane and vanishes there as $|k| \to \infty$. The unitarity of the S matrix together with (10.15) leads from (20.94) to the equation

$$\beta_0(-k) = -\mathcal{S}'(k)\beta_0(k), \tag{20.95}$$

where the kernel of $\mathcal{S}'$ is given by

$$\mathcal{S}'(k;\mathbf{r};\boldsymbol{\theta},\boldsymbol{\theta}') = \mathcal{S}^*(k;\mathbf{r},\boldsymbol{\theta},\boldsymbol{\theta}').$$

Next define

$$\eta_0(\alpha) = (1/2\pi) \int_{-\infty}^{\infty} dk\, e^{-ik\alpha} \beta_0(k).$$

Then the properties of $\beta_0(k)$, together with (20.95), are used in the same manner that led from those $\gamma(k)$ and (20.86) to (20.89), to obtain the equation for $\alpha > 0$

$$\eta_0(\alpha) = -\int_0^\infty d\beta\, G(-\alpha-\beta)\eta_0(\beta).$$

Thus η_0 must satisfy the homogeneous version of (20.89), with $G(\alpha)$ replaced by $-G(-\alpha)$. Since we assumed that neither $\mathcal{G}$ nor $\mathcal{G}'$ has the eigenvalue 1, it follows that $\eta_0 \equiv 0$, and hence $\beta_0 \equiv 0$. Eq. (20.94) therefore shows that $\beta(k)$ solves (20.86). In other words, the Fourier transform (20.93) of the solution $\eta(\alpha)$ of (20.89) has the right analyticity and asymptotic properties, and it satisfies (20.89) with the S matrix that went into (20.88).

In order to arrive at the Schrödinger equation we apply the operator $\nabla^2 - 2\boldsymbol{\theta}\cdot\nabla\partial/\partial\alpha$ to (20.89). After an integration by parts and some algebra one finds that

$$\sigma(\alpha) = \sigma_0(\alpha) + \int_0^\infty d\beta\, G(\alpha+\beta)\sigma(\beta), \tag{20.96}$$

where

$$\begin{aligned} \sigma(\alpha,\mathbf{r},\boldsymbol{\theta}) &= (\nabla^2 - 2\boldsymbol{\theta}\cdot\nabla\partial/\partial\alpha)\eta(\alpha,\mathbf{r},\boldsymbol{\theta}) \\ \sigma_0(\alpha,\mathbf{r},\boldsymbol{\theta}) &= -2\int d\boldsymbol{\theta}'\, G(\alpha;\boldsymbol{\theta},\boldsymbol{\theta}')\boldsymbol{\theta}'\cdot\nabla\eta(0,\mathbf{r},\boldsymbol{\theta}'). \end{aligned} \tag{20.97}$$

Therefore, if η is *miraculous*, i.e., $\boldsymbol{\theta}\cdot\nabla\eta(0,\mathbf{r},\boldsymbol{\theta})$ is independent of $\boldsymbol{\theta}$, then we can define $\mathscr{V}(\mathbf{r})$ by

$$\mathscr{V}(\mathbf{r}) = -2\boldsymbol{\theta}\cdot\nabla\eta(0,\mathbf{r},\boldsymbol{\theta}) \tag{20.98}$$

and have

$$\sigma_0(\alpha) = \mathscr{V}G(\alpha)\hat{1}.$$

Since (20.89) was assumed to have a unique solution it follows that

$$\sigma(\alpha) = \mathscr{V}\eta(\alpha).$$

One then readily verifies by means of (20.93) and (20.97) that the function

$$\psi(k;\boldsymbol{\theta},\mathbf{r}) = \beta(k,\boldsymbol{\theta},\mathbf{r})\exp(ik\boldsymbol{\theta}\cdot\mathbf{r}) \tag{20.99}$$

satisfies the Schrödinger equation with the potential $\mathscr{V}$ given by (20.98). Because β obeys (20.86), ψ satisfies (10.116′) with the S matrix that had been initially assumed.

Furthermore it follows from (20.89′) that if $\check{\eta}$ satisfies that equation then

$$\lim_{\imath\to\infty} \imath\check{\eta}(\alpha+\imath,\mathbf{r},\boldsymbol{\theta}) = 2\pi\int_\alpha^\infty d\beta\, g(\beta;-\hat{\mathbf{r}},\boldsymbol{\theta}) = (1/2\pi)\int_{-\infty}^\infty dk\, e^{-ik\alpha}A(k;\hat{\mathbf{r}},\boldsymbol{\theta})$$

and therefore from (20.99) and (20.93),

$$\lim_{r\to\infty} r[\psi(k,\boldsymbol{\theta},\mathbf{r}) - \exp(ik\boldsymbol{\theta}\cdot\mathbf{r})]e^{-ikr} = A(k;\hat{\mathbf{r}},\boldsymbol{\theta}).$$

Thus the wave function ψ also satisfies the correct boundary conditions for large r, and the construction has succeeded.

Characterization. It follows from the construction and reconstruction that in some sense the *miracle*, together with (20.92), is a necessary and sufficient condition for the existence of an associated local potential if there are no bound states. It is not known at this writing what conditions on a given amplitude ensure that the potential is in the Rollnik class or in any reasonable class for which the procedure has been proved to work.

Bound States. If there are bound states [which is detectable from the S matrix by means of Levinson's theorem (10.124)] then they have to be "removed" before the reconstruction can work. An outline of how this can be done is as follows.[22]

The fact that the forward amplitude has an analytic continuation into the upper half-plane (see Sec. 10.3.2), vanishes as $|k| \to \infty$, and has simple poles at $k = i\kappa_n$ if there are bound states at $E = -\kappa_n^2$ allows us to find the bound-state energies by Fourier transformation of the forward amplitude. In addition, the use of infinitesimal rotations in the scattering amplitude near the forward direction enables us also to construct the character-functions $Y_{\kappa_n}{}^b(\boldsymbol{\theta})$ defined by (10.87).[23] Let $B_m^{\mathfrak{r}}$ be the (real, orthogonal) projection onto $(C_n^{\mathfrak{r}})^{-1}\mathscr{H}_n^{\mathfrak{r}}$, where $\mathscr{H}_n^{\mathfrak{r}}$ is the space spanned by the functions $Y_{\kappa_n}{}^b(-\boldsymbol{\theta})$ $\exp(\kappa_n \boldsymbol{\theta} \cdot \mathbf{r})$, $b = 1, \ldots, N_n$ (N_n being the degeneracy of the bound state at $-\kappa_n^2$), and

$$C_n^{\mathfrak{r}} = \left(\mathbb{1} + B_1^{\mathfrak{r}} \frac{2\kappa_1}{\kappa_n - \kappa_1}\right) \cdots \left(\mathbb{1} + B_{n-1}^{\mathfrak{r}} \frac{2\kappa_{n-1}}{\kappa_n - \kappa_{n-1}}\right). \tag{20.100}$$

Then define

$$\Pi^{\mathfrak{r}}(k) = \left(\mathbb{1} + B_1^{\mathfrak{r}} \frac{2i\kappa_1}{k - i\kappa_1}\right) \cdots \left(\mathbb{1} + B_l^{\mathfrak{r}} \frac{2i\kappa_l}{k - i\kappa_l}\right) \tag{20.101}$$

if there are l bound states, and a reduced function

$$\gamma'(k) = [\Pi^{\mathfrak{r}}(k)]^{-1}\gamma(k) \tag{20.102}$$

in terms of (10.79). The operators $\Pi^{\mathfrak{r}}(k)$ have been so defined that $\gamma'(k)$ is holomorphic in the upper half-plane, and tends to $\hat{1}$ there on the real axis. It satisfies the equation

$$\gamma'(-k) = \mathscr{S}'(k)\gamma'(k), \tag{20.103}$$

where

$$\mathscr{S}' = [\Pi^{\mathfrak{r}}(-k)]^{-1}\mathscr{S}\Pi^{\mathfrak{r}}(k) \tag{20.104}$$

and $\mathscr{S}$ is the operator whose kernel is given by (20.87).

The reconstruction now proceeds as before, using $\mathscr{S}'$ in place of $\mathscr{S}$ in (20.88), and solving Eq. (20.89). One then shows that the potential is given by

$$\mathscr{V}(x) = -2\boldsymbol{\theta} \cdot \nabla \eta(0, \mathbf{r}, \boldsymbol{\theta}) - 2\sum_n \int d\boldsymbol{\theta}' \, \boldsymbol{\theta} \cdot \nabla \, E_n^{\mathfrak{r}}(\boldsymbol{\theta}, \boldsymbol{\theta}') \tag{20.105}$$

where

$$E_n^{\mathfrak{r}} = 2\kappa_n C_n^{\mathfrak{r}} B_n^{\mathfrak{r}} D_n^{\mathfrak{r}},$$

$$D_n^{\mathfrak{r}} = \left(\mathbb{1} + B_{n+1}^{\mathfrak{r}} \frac{2\kappa_{n+1}}{\kappa_n - \kappa_{n+1}}\right) \cdots \left(\mathbb{1} + B_l^{\mathfrak{r}} \frac{2\kappa_l}{\kappa_n - \kappa_l}\right).$$

22 R. G. Newton (1980b and 1981b).

23 R. G. Newton (1980b).

The reconstruction thus works even if there are bound states. The construction from an arbitrarily given amplitude with bound states has also recently been accomplished.[24]

20.5.3 A Generalized Gel'fand–Levitan Equation. We ahall outline here what is known about a generalization of the Gel'fand–Levitan equation and refer to the literature[23] for details.

The first step is to define a generalized Jost function. The functions $J'(k)$ and $[J'(k)]^{-1}$ are operator-valued boundary values of analytic functions of k, holormorphic in the upper half-plane and approaching 1 there as $|k| \to \infty$, such that

$$J'(-k) = \bar{J}'(k)S'(k), \tag{20.106}$$

where $\bar{J}$ is the operator whose kernel is $J(-\boldsymbol{\theta}, -\boldsymbol{\theta}')$ if that of J is $J(\boldsymbol{\theta}, \boldsymbol{\theta}')$, and

$$S'(k) = [\bar{\Pi}(k)]^{-1}S(k)\Pi(-k). \tag{20.107}$$

[Here Π is the operator defined by (20.101) with $\mathbf{r} = 0$.] If the thereby posed generalized Hilbert problem has a solution it is unique and constructed by the same method used to solve (20.86). This leads to another generalized Marchenko equation quite similar to (20.89), except that G does not depend on $\mathbf{r}$. The Jost function $J(k)$ is given by $J = J'\Pi^{-1}$, so that it satisfies the equation

$$J(-k) = \bar{J}(k)S(k). \tag{20.108}$$

One then defines a solution $\phi(k, \boldsymbol{\theta}, \mathbf{r})$ of the Schrödinger equation by $\phi = J\psi^{(+)}$, or more explicitly,

$$\phi(k; \boldsymbol{\theta}, \mathbf{r}) = \int d\boldsymbol{\theta}'\, J(k; \boldsymbol{\theta}, \boldsymbol{\theta}')\psi^{(+)}(k; \boldsymbol{\theta}', \mathbf{r})$$

where $\psi^{(+)}$ is the solution defined by (10.78). It follows from (20.108) and (10.116) that

$$\phi(-k; \boldsymbol{\theta}, \mathbf{r}) = \phi(k; -\boldsymbol{\theta}, \mathbf{r}) \tag{20.109}$$

and ϕ is an entire analytic function of k. Its spectral function is given by[25]

$$\frac{d\rho_E}{dE} = \begin{cases} \frac{1}{2}(2\pi)^{-3}k[J(k)^*\, \tilde{J}(k)]^{-1}, & E > 0, \\ \sum_{n=1}^{l} M_{\kappa_n}\delta(E + \kappa_n^2), & E < 0, \end{cases} \tag{20.110}$$

where M_{κ_n} is constructed from the functions $Y_{\kappa_n}{}^b$ and J. We have the completeness relation

24 R. G. Newton (1982).

25 $\tilde{J}$ has the kernel $J(\boldsymbol{\theta}', \boldsymbol{\theta})$ if the kernel of J is $J(\boldsymbol{\theta}, \boldsymbol{\theta}')$, and J^* has the kernel $J^*(\theta, \theta')$.

$$\iint d\boldsymbol{\theta}\, d\boldsymbol{\theta}'\, \phi(k;\boldsymbol{\theta},\mathbf{r})\, d\rho_E(\boldsymbol{\theta},\boldsymbol{\theta}')\phi(k;\boldsymbol{\theta}',\mathbf{r}) = \delta^3(\mathbf{r}-\mathbf{r}'), \qquad (20.111)$$

and ϕ has the Povsner–Levitan representation

$$\phi(k;\boldsymbol{\theta},\mathbf{r}) = \exp(ik\boldsymbol{\theta}\cdot\mathbf{r}) - \int_{-r}^{r} d\alpha\, h(\mathbf{r},\boldsymbol{\theta},\alpha)e^{ik\alpha}, \qquad (20.112)$$

as well as

$$\phi(k;\boldsymbol{\theta},\mathbf{r}) = \exp(ik\boldsymbol{\theta}\cdot\mathbf{r}) - \int d\mathbf{r}'\,\check{h}(\mathbf{r},\mathbf{r}')e^{ik\boldsymbol{\theta}\cdot\mathbf{r}'}. \qquad (20.112')$$

From the fact that h vanishes for $|\alpha| > r$, one infers[25] that $\check{h}$ (of which h is the Radon transform) vanishes for $r' > r$. This gives h the required "triangularity" that leads to the generalized Gel'fand–Levitan equation for $r' < r$

$$\check{h}(\mathbf{r},\mathbf{r}') = \check{h}_0(\mathbf{r},\mathbf{r}') - \int_{r''<r} d\mathbf{r}''\,\check{h}(\mathbf{r},\mathbf{r}'')\check{h}_0(\mathbf{r}'',\mathbf{r}'), \qquad (20.113)$$

where

$$\check{h}_0(\mathbf{r},\mathbf{r}') = \int d\boldsymbol{\theta}\, d\boldsymbol{\theta}'\, d[\rho(E,\boldsymbol{\theta},\boldsymbol{\theta}') - \rho^0(E,\boldsymbol{\theta},\boldsymbol{\theta}')]\exp[ik(\boldsymbol{\theta}\cdot\mathbf{r} - \boldsymbol{\theta}'\cdot\mathbf{r}')],$$

and the potential is given by

$$V(r) = -2\hat{\mathbf{r}}\cdot\nabla[h(\hat{\mathbf{r}}r,\hat{\mathbf{r}}'r)r^2]. \qquad (20.114)$$

The advantage of (20.113) over (20.89′) is that the region of integration in it is finite. The price is that in order to construct J an equation like (20.89) has to be solved (once only, not for each $\mathbf{r}$) in addition.

This procedure has been generalized to other comparison potentials and a generalization of (20.22) has been constructed as well.[24] It has also been shown[24] that when V is central then the three- dimensional procedure reduces to that of Sec. 20.3.1.

20.5.4 Potential Obtained from Backscattering. Let us write Eq. (7.46) explicitly in the momentum representation

$$T(\mathbf{k}',\mathbf{k}) = \bar{V}(\mathbf{k}'-\mathbf{k}) + \int d\mathbf{s}\, d\mathbf{t}\, \bar{V}(\mathbf{k}'-\mathbf{s})\mathscr{G}^+(k;\mathbf{s},\mathbf{t})\bar{V}(\mathbf{t}-\mathbf{k})$$

where $k = |\mathbf{k}| = |\mathbf{k}'|$, $\mathbf{s}$ and $\mathbf{t}$ are momentum vectors,

$$\mathscr{G}^+(k;\mathbf{s},\mathbf{t}) = (\mathbf{s}|(k^2 + i\varepsilon - H_0 - V)^{-1}|\mathbf{t})$$

and $\bar{V}$ is defined by (10.28). Therefore we have

$$\bar{V}(\mathbf{k}) = T(\tfrac{1}{2}\mathbf{k}, -\tfrac{1}{2}\mathbf{k}) - \int d\mathbf{k}'\, d\mathbf{k}''\, G^+(\tfrac{1}{2}k;\tfrac{1}{2}\mathbf{k} - \mathbf{k}',\mathbf{k}'' - \tfrac{1}{2}\mathbf{k})\,\bar{V}(\mathbf{k}')\,\bar{V}(\mathbf{k}''). \qquad (20.115)$$

This may be regarded as a nonlinear equation for the Fourier transform of $V(\mathbf{r})$ if the backward scattering amplitude is given. The righthand side may be looked upon as mapping $\mathscr{M}(V)$, so that (20.115) reads

$$\mathscr{M}(V) = V. \tag{20.115'}$$

One is thus looking for a *fixed point* of this map. Note that the locality of the potential has been explicitly taken into account in Eq. (20.115). If the equation has a unique solution then the backscattering data are sufficient to determine V. Furthermore, since both $\bar{V}$ and $T(\frac{1}{2}\mathbf{k}, -\frac{1}{2}\mathbf{k})$ are functions of three real continuous parameters, the local nature of the potential is not expected to restrict the permissible backscattering data significantly.

If a suitable norm is introduced and V is restricted to a class of functions for which $\mathscr{M}(V)$ is a contraction mapping on the Banach space with that norm then a unique fixed point exists and (20.115) has a unique solution, which can be constructed by iteration. However, this can work at best only for "weak" potentials, and it cannot be expected to apply to potentials that cause bound states. We leave further details to the literature.[27] For "strong" potentials it is not known whether the backscattering data are sufficient for a unique inversion.

Notes and References

20.1 For a discussion of the ambiguities of inferring the scattering amplitudes from cross section measurements of particles with spin, see R. van Wageningen (1965); M. J. Moravcsik (1968); Moravcsik and Yu (1969).

The inverse scattering problem is treated in the WKB approximation in the following works: F. C. Hoyt (1939); O. B. Firsov (1953); J. W. Brackett et al. (1963); Sanders and Mueller (1963); P. C. Sabatier (1965 and 1973); Bernstein and O'Brien (1967); M. V. Berry (1969); Buck and Pauly (1969); W. H. Miller (1969 and 1971); G. Vollmer (1969); J. F. Boyle (1971); U. Buck (1971, 1972, 1974); Yu. N. Demkov et al. (1971); S. V. Khudyakov (1971); V. G. Rich et al. (1971); R. Klingbeil (1972); D. E. Pritchard (1972); Bernstein and LaBudde (1973); E. Kujawski (1973); M. Cuer (1977); Vasilevskii and Zhirnov (1978); R. B. Gerber et al. (1978); Chan and Lu (1980); L. Hüwel et al. (1981).

The inverse problem for separable potentials (or sums of such) was solved by Gourdin and Martin (1957 and 1958); K. Chadan (1958 and 1967); Bolsterli and McKenzie (1965); Mills and Reading (1969); H. Fiedeldey (1969); F. Tabakin (1969); M. G. Fuda (1981). See also Garcilazo and Wilde (1980).

An excellent general reference on the inverse scattering problem is Chadan and Sabatier (1977).

20.2 The first explicit example of a phase-shift ambiguity for spinless particles was given by J. H. Crichton (1966). The idea of using the generalized optical theorem as a means for determining the phase of the amplitude goes back to N. P. Klepikov (1962 and 1964) and, independently, to R. G. Newton (1968) and A. Martin (1969). For further analyses and applications, see L. M. Lazarev (1967); A. Gersten (1969); Gerber

27 R. T. Prosser (1980).

and Karplus (1970); H. Goldberg (1970); R. F. Alvarez-Estrada (1971); C. Eftimiu (1971, 1972, 1973); D. Atkinson et al. (1972); H. Burkhardt (1972); M. Tortorella (1972, 1974, 1975); Cornille and Martin (1972b); R. B. Gerber (1972); Alvarez-Estrada and Carreras (1973); Atkinson, Johnson, Mehta, and De Roo (1973); Atkinson, Johnson, and Warnock (1973); Atkinson, Mahoux, and Yndurain (1973 and 1975); G. R. Bart et al. (1973); Behrends and Ruijsenaars (1973); D. Atkinson et al. (1974); Boyce and Roberts (1974); Cornille and Drouffe (1974); M. DeRoo (1974); Atkinson, Kaebebeke, and DeRoo (1975); Tortorella and Leise (1976); D. Atkinson et al. (1977); I. A. Sakmar (1978, 1979, 1980); R. B. Gerber et al. (1980). For additional use of analyticity to resolve ambiguities, see Itzykson and Martin (1973); G. R. Bart et al. (1974); Burkhardt and Martin (1975). See also Bessis and Martin (1967).

20.3 The first attempts at a solution of the inversion problem of this section were made by C. E. Fröberg (1947, 1948, 1949, and 1951); E. A. Hylleraas (1948). It was then shown by V. Bargmann (1949a and b) that the phase shift alone does not necessarily uniquely determine the potential. The uniqueness question was resolved by N. Levinson (1949a and b) and independently by V. A. Marchenko (1950) and G. Borg (1949) [see also (1946)]. The previous work of Fröberg and Hylleraas was amended in the light of the new knowledge by B. Holmberg (1952). New construction procedures were then invented by Jost and Kohn (1952a and b).

Meanwhile the Gel'fand-Levitan equation had been introduced into the inverse Sturm-Liouville problem by Gel'fand and Levitan (1951a and b). It was applied to the inverse scattering problem by Jost and Kohn (1953) and by N. Levinson (1953). Since that time the problem has given rise to a rather extensive Russian mathematical literature, the main items in which are the following: L. A. Chudov (1949 and 1956); V. A. Marchenko (1952, 1953, 1955, 1960); Ju. M. Berezanskii (1953, 1955, 1958); A. Sh. Block (1953); M. G. Krein (1953a, b, c, 1954, 1955, 1956, 1957, 1958); M. G. Neuhaus (1955); B. Ya. Levin (1956); B. M. Levitan (1956); L. D. Faddeev (1956 and 1958); Agranovich and Marchenko (1957a and b, 1958); Gohberg and Krein (1958). See also P. Jauho (1950); K. Chadan (1955 and 1956); M. M. Crum (1955); T. Ohmura (1956); R. Jost (1956); R. G. Newton (1956); B. Friedman (1957); M. Petrás (1962); M. Blažek (1962, 1963, 1964); E. A. Hylleraas (1964); W. A. Pearce (1964); J. Lanik (1964); A. G. Ramm (1965b); Swan and Pearce (1966); Benn and Scharf (1969); J. Weiss (1969); H. Cornille (1970a); J. Underhill (1970); Gugushvili and Mentkovsky (1972); Coz and Coudray (1973); F. Lambert et al. (1975); V. V. Malyarov et al. (1975a); H. Cornille (1976); R. J. W. Hodgson (1978); J. Wiesner et al. (1978); H. Moses (1979b); Brander and Chadan (1981); M. Coz (1981). An attempt at summing the contributions from all l was made by M. Blazek (1966). P. B. Abraham et al. (1981) showed explicitly that in a wider class of potentials the inversion, even in the absence of bound states, is not unique. See also P. B. Abraham et al. (1982) for ambiguity if nonlocal potentials are admitted.

Generalizations of the Jost-Kohn (or Gel'fand-Levitan and Marchenko) procedures to coupled equations were given by Newton and Jost (1955); M. G. Krein (1956); to the scattering of spin $\frac{1}{2}$ particles with the tensor force present by R. G. Newton (1955); Agranovich and Marchenko (1958); [see also sec. 9 of the review by R. G. Newton (1960); V. E. Troitskii (1978)]; to the Klein-Gordon equation by E. Corinaldesi (1954a); A. Degasperis (1970); H. Cornille (1970b); J. Weiss (1971); Weiss and Scharf (1971); J. Jp. Leon (1980); to the Dirac Equation by Prats and Toll (1959); M. G. Gasimov (1966 and 1968); O. D. Corbella (1970); R. Weiss et al. (1972); to both relativistic equations by M. Verde (1958–1959); [see also S. N. Sokolov (1979); J. Jp. Leon (1981)]; to coupled channels by J. R. Cox (1962, 1966, 1967, 1975); M. Coz (1966); Cox and

Garcia (1975); Coz and Rochus (1977); P. Rochus (1979); to energy dependent potentials by V. I. Mal'cenko (1966); M. Jaulent (1972 and 1975); Jaulent and Jean (1972, 1975, 1982); M. Tsutsumi (1981); to complex potentials by M. Jaulent (1976); De Facio and Moses (1980); to include Coulomb potentials by P. Swan (1967b); to singular potentials by Baeteman and Chadan (1975); K. Chadan (1978).

For treatments of the inverse scattering problem in the momentum representation, see M. I. Sobel (1968); B. R. Karlsson (1972, 1974, 1978); for a generalization to other boundary conditions, see H. E. Moses (1978).

Eq. (20.21) was first proved, in the context of the Marchenko equation, by H. Cornille (1967).

For the solution of the inverse scattering problems in one dimension (from $-\infty$ to $+\infty$) see A. Sh. Bloch (1953); I. Kay (1955); Kay and Moses (1956, 1957); H. E. Moses (1956); L. D. Faddeev (1958 and 1964); J. C. Portinari (1965); Deift and Trubowitz (1979); R. G. Newton (1980c, 1981a, appendix); see also M. G. Krein (1953a).

For discussions of the inverse scattering problem in electromagnetic theory, see R. Mireles (1966); Wolf and Shewell (1967); Shewell and Wolf (1968); E. Wolf (1969); W. H. Carter (20.3).

Other useful general references are the review by L. D. Faddeev (1959) and the monographs Agranovitch and Marchenko (1963), and Chadan and Sabatier (1977).

20.3.2 The content of this section follows R. G. Newton (1956). See also N. Levinson (1949). Equation (20.40) is a special case of the information contained in the *moments* of the phase shift. For more general formulas of this nature see Buslaev and Faddeev (1960); V. S. Buslaev (1962, 1967, 1971); I. C. Percival (1962 and 1963); Percival and Roberts (1963); M. J. Roberts (1963, 1964, 1965); K. Chadan (1965 and 1966); Calogero and Degasperis (1968); Chadan and Montes (1968); F. Calogero (1971b); O. D. Corbella (1971); Kirhznitz and Takibaev (1972); R. D. Puff (1975); N. Zh. Takibaev (1977); D. Bollé (1979a); Bollé and Osborn (1980); see also F. Calogero et al. (1968).

20.4 The problem of obtaining the potential from a knowledge of all phase shifts at one energy was first approached by J. A. Wheeler (1955) from the point of view of the WKB approximation. Then T. Regge (1959) dealt with it by extending the angular momentum into the complex plane. The approach of Martin and Targonski (1961) is applicable to superpositions of Yukawa potentials only. The first general solution of the problem was given by R. G. Newton (1962a). The present section follows that paper for the most part. A mathematically amusing adjunct is provided by P. J. Redmond (1964). This work was carried further by P. C. Sabatier (1966, 1967, 1972b) [see also corrections on p. 337 of Chadan and Sabatier (1977)]; R. G. Newton (1967b); Cox and Thompson (1969 and 1970); G. A. Viano (1969); Coudray and Coz (1970); Jean and Sabatier (1973); I. Miodek (1976); C. Coudray (1978 and 1980); Lipperheide and Fiedeldey (1978 and 1981); Munchow and Scheidt (1980). A more general method was given by P. C. Sabatier (1971 and 1972a). For numerical calculations and applications, see F. Q. van Phu (1970); Sabatier and van Phu (1971); V. V. Malyarov et al. (1975b and 1977); C. Coudray (1977); V. V. Malyarov et al. (1978). The following papers generalized the method to include spin-orbit and tensor forces: P. C. Sabatier (1968); M. A. Hooshyar (1971, 1972, 1975, 1978, 1980); and Coudray and Coz (1972) gave a relativistic generalizations.

The Regge approach, working in the complex λ-plane, was carried further by G. Burdet et al. (1966); J.-J. Loeffel (1968); Malyarov and Poplavskii (1972); Heniger and Loeffel (1973); V. V. Malyarov et al. (1975c).

20.5 The first attempt at solving the full three-dimensional inverse scattering problem for noncentral potentials was made by Kay and Moses (1961a and b) based on their general method of (1955), but with incomplete success. A different approach was taken by L. D. Faddeev (1974) and R. G. Newton (1974a, b) [but see corrections mentioned on p. 334 of Chadan and Sabatier (1977)] on the basis of a new Green's function introduced by L. D. Faddeev (1965 and 1966). A new and more natural solution was given by R. G. Newton (1979a, 1980b, c, 1981b, 1982). Secs. 20.5.2 and 20.5.3 are based on these papers. See also A. J. Devaney (1978), and H. E. Moses (1980); Y. Saito (1980b and 1981).

20.5.4 The idea of reconstructing the potential from the back-scattering amplitude $A(-\mathbf{k},\mathbf{k})$ goes back to H. E. Moses (1956). His procedure was further developed by R. T. Prosser (1969, 1976, 1980). This Section is based on their work.

General. From a practical point of view the question of stability of inversion procedures and the handling of errors in the data, is of great importance. The following papers deal with this problem: V. A. Marchenko (1968); Lundina and Marchenko (1969); Di Salvo and Viano (1976); V. P. Zhigunov (1979); W. W. Symes (1980).

Other relevant references are Bernstein and Muckerman (1967); Calogero and Cox (1968); F. J. Dyson (1976); Zakhariev (1976); J. T. Cushing (1977); P. C. Sabatier (1978 and 1982); Gerber and Yinnon (1980).

Problems

1 Examine the differential cross section for unpolarized spin $\frac{1}{2}$ particles on unpolarized spin $\frac{1}{2}$ particles in the low-energy region. Discuss the ambiguities in any inference from the cross section to the amplitudes if no polarizations are measured. How much of this ambiguity can be removed by polarization measurements, and how?

2 Suppose that one of the constants N_n^2 in (12.129) is *negative* and you use this spectral function in (20.18). Does the Gel'fand–Levitan equation have a unique solution? Can you find the corresponding potential? What are its properties? Discuss.

3 Suppose that the comparison potential V_1 has no bound states and that its s-wave phase shift is δ_1. Let $\delta \equiv \delta_1$, but let V have a bound state at $E = E_0$. Construct V. Is it well behaved? Discuss.

4 Construct the Bargmann potential needed in the derivation of (20.33).

5 Starting with the phase shift (14.80), construct the Bargmann potential (14.81) and the wave function (14.82) by solving the Gel'fand–Levitan equation.

6 Do the same as in Prob. 5 for the case of a bound state; i.e., derive (14.85) and (14.86).

7 Derive the general class of Bargmann potentials with no bound states from (14.55a) with $S_l^{(0)} \equiv 1$ by solving the Gel'fand–Levitan equation. (*Hint*: Replace the integral equation by a differential equation).

8 Do the same as in Prob. 7, but with bound states present.

9 Calculate the potentials which cause an s-wave bound state of binding energy E_0 and for which $|\mathscr{F}_0(k)| \equiv 1$. Find the bound state and scattering wave functions and the phase shift.

10 Supposing that $c_l = 0$, $l \neq L$, calculate the Jost functions for all l values. What must be assumed to be the k dependence of c_L for $\mathscr{F}_l$ to satisfy (12.32) with (12.30)? Can these Jost functions have all the analyticity properties that go with a well-behaved potential?

11 Do the same as in Prob. 10, assuming $c_l = 0$ for $l > L$.

Bibliography

The numbers in parentheses at the end of each reference correspond to the sections or chapters in which the reference is cited.

Aaron, R., and A. Klein (1960): *J. Math. Phys.*, **1**:131. (9.1.1)
Aaron, R., R. D. Amado, and B. W. Lee (1961): *Phys. Rev.*, **121**:319. (17.4)
—, —, and Y. Y. Yam (1964): *Phys. Rev. Letters*, **13**:574, 701; *Phys. Rev.*, **136**:B650. (17.4.3)
—, —, and — (1965): *Revs. Modern Phys.*, **37**:516; *Phys. Rev.*, **140**:B1291. (17.4.3)
— and P. E. Shanley (1966): *Phys. Rev.*, **142**:608 (17.4.3)
Abraham, P. B., B. DeFacio, and H. E. Moses (1981): *Phys. Rev. Lett.*, **46**:1657. (20.3)
—, —, and — (1982): *Studies in Appl. Math.*, **66**:45. (20.3)
Adachi, T. (1966): *Progr. Theor. Phys.*, **35**:463. (18.3)
— and T. Kotani (1965): *Progr. Theor. Phys. Suppl., Extra No.*:316. (18.3)
— and — (1966a): *Progr. Theor. Phys.*, **35**:485. (18.3)
Adachi, T., and T. Kotani (1966b): *Progr. Theor. Phys. Suppl.*, **37** & **38**:297. (18.3)
—, —, and M. Takeda (1966): *Progr. Theor. Phys.*, **36**:745. (18.3)
Adair, R. K. (1958): *Phys. Rev.*, **111**:632. (17.2)
Adelman, S. A., and W. P. Reinhardt (1972): *Phys. Rev.*, **A6**:255. (12.1.2)
Adhikari, S. K. (1973): *Phys. Rev.*, **D8**:1195. (17.2, 17.4)
— (1979a): *Phys. Rev.*, **C19**:26. (17.4)
— (1979b): *Phys. Rev.*, **C19**:1729. (7, 16.3)
— (1979c): *Phys. Rev.*, **C19**:2121. (17.4.4)
— (1981a): *Phys. Rev.*, **C24**:379. (16.3)
— (1981b): *Phys. Rev.*, **C** (16.3.3)
— (1982): *Phys. Rev.*, **C25**:128. (17.4.4)
— and R. D. Amado (1974): *Phys. Rev.*, **D9**:1467. (17.1.2)

— and — (1977): *Phys. Rev.*, **C15:**498. (17.1.2, 17)
— and A. C. Fonseca (1981): *Phys. Rev.*, **D24:**416. (17.4.8)
— and W. Glöckle (1978): in H. Zingl et al. (1978), I, p. 365. (17.4.4)
— and — (1980a): *Phys. Rev.*, **C21:**54. (17.4)
— and — (1980b): *Phys. Rev.*, **C22:**309. (17.4.4)
— and I. H. Sloan (1975): *Phys. Rev.*, **C12:**1152. (17.4)
Agarwal, G. S. (1976): *Phys. Rev.*, **D14:**1168. (4.1)
Agassi, D., and A. Gal (1973a): *Ann. Phys. (N.Y.)*, **75:**56. (8.1)
— and — (1973b): *Phys. Rev. Lett.*, **31:**961. (18.3)
Agmon, S. (1975): *Ann. Scuol. Norm. Sup. Pisa*, **2:**151. (7.3)
Agodi, A., and E. Eberle (1960): *Nuovo Cimento*, **18:**718. (16.5)
Agranovich, Z. S., and V. A. Marchenko (1957a): *Dokl. Akad. Nauk SSSR*, **113:**951 (*Math. Rev.*, **19:**746). (20.3)
— and — (1957b): *Usp. Mat. Nauk (N.S.)*, **12**(1): 73, 143 (*Am. Math. Soc. Transl.*, **16:**355). (20.3)
— and — (1958): *Dokl. Akad. Nauk SSSR*, **118:**1055 (*Math. Rev.*, **21:**741). (20.3)
— and — (1963): "The Inverse Problem in Scattering Theory," Gordon and Breach, New York. (20.3)
Agudin, J. L. (1968): *Phys. Rev.*, **171:**1385. (11.1.3)
Ahmad, S. S., and G. Pisent (1974): *Nuovo Cim.*, **A22:**79. (10.14)
Ahmadzadeh, A., and J. A. Tjon (1965): *Phys. Rev.*, **139:**1085. (17.4.3)
Ahmed, M. A., and D. B. Fairlie (1965): *Nuovo Cimento*, **38:**547 (12.4)
Aitchison, I. J. R., and C. Kacser (1966): *Phys. Rev.*, **142:**1104 and **152:**158. (17.1.2)
Akhiezer, N. I., and I. M. Glazman (1961): "Theory of Linear Operators in Hilbert Space," Ungar, New York. (7.3)
Akimova, M. K.: See Luk'yanov, A. V., I. B. Teplov, and M. K. Akimova.
Aks, S. O. (1965): *Nuovo Cimento*, **38:**1794. (10.3.2)
Albeverio, S., F. Gesztesy, and R. Hoegh–Krohn (1981): preprint (10.3.3)
— and R. Hoegh–Krohn (1981): preprint. (16.6)
Alder, K., D. Trautmann, and R. D. Violliet (1973): *Z. Naturwiss.*, **28a:**321. (11.1)
Alessandrini, V. A. (1966): *J. Math. Phys.*, **7:**215. (17)
Alessandrini, V. A., and J. J. Giambiagi (1963): *Nuovo Cimento*, **29:**1353. (12.1.2)
— and R. L. Omnes (1965): *Phys. Rev.*, **139:**B167. (17.4)
Alexopoulos, N. G., and N. K. Uzunoglu (1978): *Appl. Optics*, **17:**235. (2.4)
Alhassid, Y., and R. D. Levine (1978): *Phys. Rev.*, **C18:**89. (8)
Ali, S. T., L. Fonda, and G. C. Ghirardi (1975): *Nuovo Cim.*, **25A:**134. (19.2)
—, M. Rahman, and D. Husain (1972): *Phys. Rev.*, **D6:**1178. (14.6.2)
Alt, E. O. (1972): *Acta Phys. Austr. Suppl.*, **9:**139. (17.4)
— (1976): in A. N. Mitra et al., *Few-Body Dynamics*, North-Holland Publ. Co., Amsterdam. (17.4)
— (1978): in H. Zingl et al. (1978), I, p. 181; and in G. Pisent et al. (1978), p. 271. (17.4)
—, P. Grassberger, and W. Sandhas (1967): *Nucl. Phys.*, **B2:**167. (17.4)
—, —, and — (1970): *Phys. Rev.*, **C1:**85. (17.4)
—, and W. Sandhas (1978): in H. Zingl et al. (1978), I, pp. 373 and 375. (17.4)
— and — (1980): *Phys. Rev.*, **C21:**1733. (17.4)
—, W. Sandhas, H. Zankel, and H. Ziegelmann (1976): *Phys. Rev. Lett.*, **37:**1537. (17.4)
—, W. Sandhas, and H. Ziegelmann (1978): *Phys. Rev.*, **C17:**1981. (17.4)
Altshuler, S. (1953): *Phys. Rev.*, **91:**1167; **92:**1157. (16.3)
— (1956): *Nuovo Cimento*, **3:**246. (7.22.)

Alvarez-Estrada, R. F. (1971): *Ann. Phys. (N.Y.)*, **68**:222. (20.2)
— and B. Carreras (1973): *Nucl. Phys.*, **B54**:237. (20.2)
Aly, H. H., and H. J. W. Miller (1966): *J. Math. Phys.*, **7**:1. (12.4)
— and S. Okubo (1967): *Z. Physik*, **200**:20. (10.3.4)
—, Riazuddin, and A. H. Zimerman (1964): *Phys. Rev.*, **136**:B1174. (12.4)
—, —, and — (1965): *Nuovo Cimento*, **35**:324; *J. Math. Phys.*, **6**:1115. (12.4).
— and R. M. Spector (1965): *Nuovo Cimento*, **38**:149. (14)
Amado, R. D. (1963): *Phys. Rev.*, **132**:485. (17.4.3)
— (1967): *Phys. Rev.*, **158**:1414. (17.1.2)
— (1970): *Phys. Rev.*, **C2**:2439. (7.2.2)
— (1974): *Phys. Rev. Lett.*, **33**:333. (17.4)
— (1975): *Phys. Rev.*, **C11**:719. (17.1.2)
—, D. F. Freeman, and M. H. Rubin (1971): *Phys. Rev.*, **D4**:1032. (17.4.4)
— and F. C. Greenwood (1973): *Phys. Rev.*, **D7**:2517. (17.4.8)
— and J. V. Noble (1971): *Phys. Lett.*, **B35**:25. (17.4.8)
— and — (1972): *Phys. Rev.*, **D5**:1992. (17.4.8)
— and M. H. Rubin (1970): *Phys. Rev. Lett.*, **25**:194. (17.2, 17.4)
Amati, D. (1963): *Phys. Letters*, **7**:290. (17.1)
Amrein, W. O., J. M. Jauch, and K. B. Sinha (1977): *Scattering Theory in Quantum Mechanics*, W. A. Benjamin, Reading, Mass. (6.8,7)
— and D. B. Pearson (1979a): *J. Phys. A*, **12**:1469. (10.3.3)
— (1979b): *Ann. Inst. H. Poincaré*, **XXXA**:89. (10.3.3)
—, D. B. Pearson, and K. B. Sinha (1979): *Nuovo Cim.*, **52A**:115. (16.5)
Andreo, R. H., and J. A. Krill (1980): in Varadan and Varadan (1980), p. 565. (1.5)
Anselone, P. M. (1964): *Nonlinear Integral Equations*, Univ. of Wisc. Press, Madison. (20.1.2)
Arbuzov, B. A., and A. T. Filippov (1965): *Phys. Letters*, **13**:95. (12.4)
—, —, and O. A. Krustalev (1964): *Phys. Letters*, **8**:205. (12.4)
Armstrong, B. H., and E. A. Power (1963): *Am. J. Phys.*, **31**:261. (12.1.1)
Arnous, E. (1952): *Helv. Phys. Acta*, **25**:631. (19.2)
— and K. Bleuler (1952): *Helv. Phys. Acta*, **25**:581. (19.2)
— and W. Heitler (1953): *Proc. Roy. Soc. (London)*, A220:290. (19.2)
— and S. Zienau (1951): *Helv. Phys. Acta*, **24**:279. (19.2)
Aronszajn, N. (1957): *Am. J. Math.*, **79**:597. (7.3)
Artunes, A. G., F. L. Balta, and E. M. Ferreira (1976): *Nucl. Phys.*, **A265**:365. (17.4.8)
Arai, M. (1980): *Publ. Res. Inst. Math. Sci. Kyoto*, **16**:35. (12.4)
Arnold, L. G., and A. D. MacKellar (1971): *Phys. Rev.*, **C3**:1095. (10.1.4)
Aronson, I., R. Blau, C. J. Kleinman, L. Rosenberg, and L. Spruch. (1979): *Phys. Rev.*, **A19**:1568. (11.3.4)
—, Y. Hahn, P. H. Henry, C. J. Kleinman, and L. Spruch (1967): *Phys. Rev.*, **153**:73. (11.3.3)
—, C. J. Kleinman, and L. Spruch (1975): *Phys. Rev.*, **A12**:349. (17.2)
Arthurs, A. M. (1968): *Phys. Rev.*, **176**:1730. (11.3.4)
— (1975): *Phys. Rev.*, **D11**:2316. (11.3.4)
Asano, S. (1979): *Appl. Optics*, **18**:712. (2)
— and M. Sato (1980): *Appl. Optics*, **19**:962. (2)
Ashkin, J., and G. C. Wick (1952): *Phys. Rev.*, **85**:686. (18.4)
Atkinson, D., P. W. Johnson, L. P. Kok, and M. DeRoo (1974): *Nucl. Phys.*, **B77**:109. (20.2)
—, N. Mehta, and M. DeRoo (1973): *Nucl. Phys.*, **B55**:125. (20.2)

—, —, and R. L. Warnock (1972): *Commun. Math. Phys.*, **28**:133. (20.2)
—, —, and — (1973): *Commun. Math. Phys.*, **33**:221. (20.2)
—, M. Kaebebeke, and M. DeRoo (1975): *J. Math. Phys.*, **16**:685. (20.2)
—, L. P. Kok, and M. DeRoo (1977): *Phys. Rev.*, **D17**:2492. (20.2)
—, G. Mahoux, and F. J. Yndurain (1973): *Nucl. Phys.*, **B54**:263 and **B66**:429. (20.2)
—, —, and — (1975): *Nucl. Phys.*, **B98**:521. (20.2)
Austern, N. (1969): *Phys. Rev.*, **188**:1595. (16.3)
— (1980): *Phys. Lett.*, **90B**:33. (17.4)
— and J. S. Blair (1965): *Ann. Phys. (N.Y.)*, **33**:15. (9.1.3)
— and C. M. Vincent (1981): *Phys. Rev.*, **C23**:1847. (16.3)
Auvil, P. R. (1968): *Phys. Rev.*, **168**:1568. (14.6)
Avishai, Y. (1971a): *Nucl. Phys.*, **A161**:621. (17)
— (1971b): *Phys. Rev.*, **D4**:400. (17.4.4)
Avron, J. E., and I. W. Herbst (1977): *Commun. Math. Phys.*, **52**:239. (6.3)
— and — (1978): *Duke Math. J.*, **45**:847. (6.3)
Babikov, V. V. (1965): *Yad. Fiz.*, **1**:369 (transl. *Soviet J. Nuclear Phys.*, **1**:261). (12.1.2)
Badalyan, A. M., L. P. Kok, M. I. Polikarpov, and Yu. A. Simonov (1981): preprint. (16.6)
—, —, and — (1981): preprints. (16.3)
— and Yu. A. Simonov (1966): *Yad. Fiz.*, **3**:1032 (transl. *Soviet J. Nucl. Phys.*, **3**:755). (17.4.6)
Baeteman, M.–L., and K. Chadan (1975): *Nucl. Phys.*, **A255**:35. (20.3)
— and — (1976): *Ann. Inst. H. Poincaré*, **24**:1. (12.4)
Baer, M., and D. J. Kouri (1972): *J. Chem. Phys.*, **56**:4840. (17.4)
— and — (1973): *J. Math. Phys.*, **14**:1637. (16.3, 17.4.3)
Bagchi, B., T. O. Krause, and B. Mulligan (1977) *Phys. Rev.*, **C15**:1623. (10.1.4)
Bahethi, O. P., and M. G. Fuda (1972): *Phys. Rev.*, **C6**:1956. (17.4)
Bajzer, Z. (1978): in G. Pisent et al. (1978), p. 365. (17.4)
Baker, A. (1972): *Phys. Rev.*, **D6**:3462. (18.3)
— (1973): *Phys. Rev.*, **D8**:1937. (18.3)
Baker, M. (1958): *Ann. Phys. (N.Y.)*, **4**:271. (9.3)
Bakker, B. L. G., and W. Sandhas (1975): *Czech. J. of Phys.*, **B25**:259. (17.4)
Bali, N., H. Munczek, and A. Pignotti (1965): *Nuovo Cimento*, **38**:374. (17.1)
Ball, J. S., J. C. Y. Chen, and D. Y. Wong (1968): *Phys. Rev.*, **173**:202. (16)
— and D. Y. Wong (1968): *Phys. Rev.*, **169**:1362. (17.4)
Balslev, E. (1972): *Ann. Phys. (N.Y.)*, **73**:49. (12.1.1)
— and J.–M. Combes (1971): *Commun. Math. Phys.*, **22**:280. (12.1.1, 17)
Bander, M. (1965): *Phys. Rev.*, **138**:B322. (17.4)
—, P. W. Coulter, and G. L. Shaw (1965): *Phys. Rev. Letters*, **14**:270. (17.1)
Banerjee, S. N., and S. N. Chakravorty (1978): *J. Math. Phys.*, **19**:2481. (14.6.2)
Baranger, E., and E. Gerjuoy (1957): *Phys. Rev.*, **106**:1182. (16)
Barber, P. W. (1980): in Varadan and Varadan (1980), p. 191. (4.1)
— and D.–S. Wang (1979): *Appl. Optics*, **17**:797. (3.2)
— and C. Yeh (1975): *Appl. Optics*, **14**:2864. (1.3)
Bardsley, J. N., and R. K. Nesbet (1973): *Phys. Rev.*, **A8**:203. (17.2)
—, E. Gerjuoy, and C. V. Sukumar (1972): *Phys. Rev.*, **A6**:1813. (11.3.3)
Bargmann, V. (1949a): *Phys. Rev.*, **75**:301. (20.3)
— (1949b): *Revs. Modern Phys.*, **21**:488. (14.7, 20.3)
— (1952): *Proc. Natl. Acad. Sci. U.S.*, **38**:961. (12.1.4)

Barker, F. C. (1964): *Proc. Phys. Soc. (London)*, **84:**681. (17.2)
— and P. B. Treacy (1962): *Nuclear Phys.*, **38:**33. (17.2)
Barsella, B., and E. Fabri (1962): *Phys. Rev.*, **126:**1561. (17.4)
Bart, G. R., P. W. Johnson, and R. L. Warnock (1973): *J. Math. Phys.*, **14:**1558. (20.2)
—, —, and — (1974): *Nucl. Phys.*, **B72:**329. (20.2)
Bartnik, E. A., Z. R. Iwinski, and J. M. Namislowski (1975): *Phys. Rev.*, **A12:**1785. (14.6, 18)
Bartram, R., and L. Spruch (1962): *J. Math. Phys.*, **36:**287. (11.3)
Barut, A. O., and D. Zwanziger (1962): *Phys. Rev.*, **127:**974. (13.3)
Basdevant, L. (1965): *Phys. Rev.*, **138:**B1198. (17.4.3)
Bassel, R. H., and E. Gerjuoy (1960): *Phys. Rev.*, **117:**749. (9.1.3)
Bassichis, W. H., and A. Dar (1966): *Ann. Phys. (N.Y,)*, **36:**130. (18.2)
Bastai, A., L. Bertocchi, G. Furlan, S. Fubini, and M. Tonin (1963): *Nuovo Cimento*, **30:**1512. (12.4)
Bateman Manuscript: See Erdelyi, A.
Bates, D. R., A. Fundaminsky, and H. S. W. Massey (1951): *Trans. Roy. Soc. (London)*, **243:**93. (16.3)
Bates, R. H. T. (1980): in Varadan and Varadan (1980), p. 21. (4.1)
Bawin, M., and J. P. Lavine (1974): *Nuovo Cim.*, **23A:**311. (14.3)
Baz, A. I. (1957): *Z. Eksp. i Teor. Fiz.*, **33:**923 (transl. *Soviet Phys. JETP* **6:**709). (17.2)
— (1959): *Advances in Phys. (Phil. Mag. Suppl.)*, **8:**349. (17.2)
— (1961): *Z. Eksp. i Teor. Fiz.*, **40:**1511 (transl. *Soviet Phys. JETP*, **13:**1058). (17.2)
— (1964): *Nuclear Phys.*, **51:**145. (17.4.3)
— and L. B. Okun (1958): *Z. Eksp. i Teor. Fiz.*, **35:**757 [transl. *Soviet Phys. JETP*, **35**(8):526]. (17.2)
—, L. D. Puzikov, and Ya. A. Smorodinskii (1962): *Z. Eksp. i Teor. Fiz.*, **42:**1249 (transl. *Soviet Phys. JETP*, **15:**865). (17.2.3)
Beck, G., and H. M. Nussenzveig (1960): *Nuovo Cimento*, **16:**416 and **20:**694. (19.2)
Becker, S. F., and D. R. Harrington (1971): *Phys. Rev.*, **D4:**411. (17.4)
Behncke, H. (1968): *Nuovo Cim.*, **55A:**780. (12.4)
— and H. Focke (1978): *Math. Z.*, **158:**87. (12.4)
Behrends, F. A., and S. N. M. Ruijsenaars (1973): *Nucl. Phys.*, **B56:**507 and 525. (20.2)
Belinfante, F. J., and C. Møller (1953): *Kgl. Danske Videnskab. Selskab, Mat.-fys. Medd.*, **28**(6). (6,7)
Belinfante, J. G. (1964): *J. Math. Phys.*, **5:**1070. (7.3)
— and B. C. Unal (1963): *J. Math. Phys.*, **4:**372. (10.3.4)
Bell, J. S., and C. J. Goebel (1965): *Phys. Rev.*, **138:**B1198. (19.3)
Bell, W. W. (1963): *Nuovo Cimento*, **29:**644. (10.3.4)
Ben-Artzi, M., and A. Devinatz (1979): *J. Math. Phys.*, **20:**594 and **21:**2472. (12.4)
Benayoun, M., and P. Leruste (1969): *Lett. Nuovo Cim.*, **1:**548 and *Nuovo Cim.*, **63A:**969. (17.1.3)
— and — (1970): *Nuovo Cim.*, **65A:**99. (17.1.3)
Bencze, Gy. (1972): *Nucl. Phys.*, **A196:**135. (17.4)
— (1973): *Nucl. Phys.*, **A210:**568. (17.4)
— (1976): *Phys. Lett.*, **61B:**139 and *Lett. Nuovo Cim.* **17:**91. (17.4)
— (1977): *Phys. Lett.*, **72B:**155. (17.4)

— (1978): in G. Pisent et al. (1978), p. 113. (16.3.3)
— (1980): *European Symp. on Few-Body Problems in Nucl. and Particle Phys.*, Sesimbra, Portugal. (17.4)
—, G. Cattapan, and V. Vanzani (1977): *Lett. Nuovo Cim.*, **20:**248. (17.4)
— and C. Chandler (1982): *Phys. Rev.*, **C25:**136. (16.3.3)
— and E. F. Redish (1975): *Nucl. Phys.*, **A238:**240. (16.3.3)
— and — (1978): *J. Math. Phys.*, **19:**1909. (16.3.3)
— and P. C. Tandy (1977): *Phys. Rev.*, **C16:**564. (17.4)
— and H. Zankel (1979): *Phys. Lett.*, **82B:**316. (17.4)
Benn, J., and G. Scharf (1969): *Nucl. Phys.*, **A134:**481. (20.3)
Benoist-Gueutal, P. (1975): *Phys. Lett.*, **56B:**413. (17.4)
—, M. L'Huillier, E. F. Redish, and P. C. Tandy (1978): *Phys. Rev.*, **C17:**1924. (17.4)
Beregi, P. (1971): *Lett. Nuovo Cim.*, **2:**233. (10.1.4)
—, B. N. Zakhar'ev, and S. A. Niyazgulov (1973): *Fiz. El. Chast. Atom. Yad.*, **4:**512 (transl. *Soviet J. Part. and Nucl.*, **4:**217). (12.1.3)
Berezanskii, Ju. M. (1953): *Dokl. Akad. Nauk SSSR*, **93:**591 (*Math. Rev.*, **15:**797). (20.3)
— (1955): *Dokl. Akad. Nauk SSSR*, **105:**197 (*Math. Rev.*, **17:**1210). (20.3)
— (1958): *Trudy Moskov. Mat. Obsc.*, **7:**2 (*Am. Math. Soc. Transl.*, **35:**167). (20.3)
Berezin, F. A. (1965): *Dokl. Akad. Nauk. SSSR*, **163:**795 (transl. *Soviet Math.*, **6:**997). (17.4.5)
Berman, S. M., and M. Jacob (1965): *Phys. Rev.*, **139:**B1023. (17.4)
Bernstein, R. B., and R. A. LaBudde (1973): *J. Chem. Phys.*, **58:**1109. (20.1)
— and J. T. Muckerman (1967): *Adv. Chem. Phys.*, **12:**389. (20)
— and J. T. P. O'Brien (1967): *J. Chem. Phys.*, **46:**1208. (20.1)
Berriman, B. J., and L. Castillejo (1973): *Phys. Rev.*, **D8:**4647. (18.3)
Berry, M. V. (1969): *J. Phys. B*, **2:**381. (20.1)
— and K. E. Mount (1972): *Rep. Progr. Phys.*, **35:**315. (18.2)
Bertero, M., and G. Dillon (1971): *Nuovo Cim.*, **2A:**1024. (12.1.5)
— and H. J. Kreuzer (1967): *Nucl. Phys.*, **B3:**195. (13.4)
—, G. Talenti, and G. A. Viano (1966): *Nuovo Cim.*, **46A:**337. (10.3.1)
—, —, and — (1967): *Commun. Math. Phys.*, **6:**128. (10.1.4)
Bertocchi, L., S. Fubini, and G. Furlan (1964): *Nuovo Cimento*, **32:**745. (12.4)
—, —, and — (1965a): *Nuovo Cimento*, **35:**599. (10.2)
—, —, and — (1965b): *Nuovo Cimento*, **35:**633. (12.4)
Bessis, D. (1965): *J. Math. Phys.*, **6:**637 and 820. (13.4)
— and A. Martin (1967): *Nuovo Cim.*, **52A:**719. (20.2)
—, P. Mery, and G. Turchetti (1974): *Phys. Rev.*, **D10:**199L. (17.4)
Bethe, H. A. (1937): *Revs. Modern Phys.*, **9:**69. (16.5)
— (1949): *Phys. Rev.*, **76:**38. (11.2.2)
— and R. Bacher (1936): *Revs. Modern Phys.*, **8:**82. (14.3)
— and R. E. Peierls (1935): *Proc. Roy. Soc.* (*London*), **148A:**146. (14.1)
— and G. Placzek (1937): *Phys. Rev.*, **51:**450. (16.5)
— and — (1940): *Phys. Rev.*, **57:**1075. (18.2)
— and E. E. Salpeter (1957): "Quantum Mechanics of One and Two Electron Atoms," Academic, New York [and in S. Flügge (ed.), "Handbuch der Physik," vol. 35, Springer, Berlin]. (14.6)
Bhasin, V. S., and A. N. Mitra (1970): *Phys. Rev.*, **D1:**3496. (17.4)
Bhatia, A. K., and A. Temkin (1964): *Revs. Modern Phys.*, **36:**1050. (17.4.3)
Bhattacharjie, A., and E. C. G. Sudarshan (1962): *Nuovo Cimento*, **25:**864. (14)

Bianchi, L., and L. Favella (1964): *Nuovo Cimento*, **34:**1823. (17.4.3)
Bierter, W. (1969): *Nucl. Phys.*, **A126:**675. (17.4)
Birman, M. Sh. (1959): *Dokl. Akad. Nauk. SSSR*, **129:**239 (*Math. Revs.* **22**, 817). (12.1.4)
— (1961a): *Vestnik Leningr. Univ.*, **16:**163 (*Math. Rev.*, **25:**541). (12.1.4)
— (1961b): *Mat. Sb.*, **55**(97):125 (*Am. Math. Soc. Transl.* Ser. 2, **53:**23). (12.1.4)
Bishop, R. F. (1973): *Phys. Rev.*, **C7:**479. (12.4)
Bjorken, D. J. (1960): *Phys. Rev. Letters*, **4:**473. (17.1)
Blais, N. C., and D. L. Bunker (1964): *J. Chem. Phys.*, **41:**2377. (5)
Blankenbecler, R. (1964): In R. G. Moorhouse (ed.), "Strong Interactions and High Energy Physics," p. 411, Oliver & Boyd, Edinburgh. (17.1)
— (1966): *Phys. Rev.*, **152:**1471. (9)
— and M. L. Goldberger (1962): *Phys. Rev.*, **126:**766. (18.3)
—, —, N. N. Khuri, and S. B. Treiman (1960): *Ann. Phys.* (*N.Y.*), **10:**62. (10.3.4, 13.4)
Blasczak, D., and M. G. Fuda (1973): *Phys. Rev.*, **C8:**1665. (16.3)
Blatt, J. M. (1948): *Phys. Rev.*, **74:**92. (11.3)
— and L. C. Biedenharn (1952): *Revs. Modern Phys.*, **24:**258. (15.1)
— and J. D. Jackson (1949): *Phys. Rev.*, **76:**18. (10.3.3, 11.2.2, 11.3)
— and V. Weisskopf (1952): "Theoretical Nuclear Physics," Wiley, New York. (2.1.3, 2.1.5, 7.2.4, 15.1, 16.5)
Blau. R., L. Rosenberg, and L. Spruch (1974): *Phys. Rev.*, **A10:**2246 and **A11:**200. (11.3.4)
—, —, and — (1975): *Phys. Rev.*, **A12:**1859. (11.3.3)
—, —, and — (1977): *Phys. Rev.*, **A15:**1475. (11.3.4)
Blažek, M. (1962a): *Czechoslov. J. Phys.*, **B12:**249. (14.7)
— (1962b): *Czechoslov. J. Phys.*, **B12:**258 and 497. (20.3)
— (1963): *Mateni.–Fys. Casopis SAV*, **13:**147. (20.3)
— (1964): *C. R. Acad. Bulgar Sci.*, **17:**1005. (20.3)
— (1966): *Commun. Math. Phys.*, **3:**282. (20.2)
Blin-Stoyle, R. J. (1951): *Proc. Phys. Soc.* (*London*), **64A:**700. (15.1)
Bloch, A. Sh. (1953): *Dokl. Akad. Nauk SSSR*, **92:**209 (*Math. Rev.*, **15:**708). (20.3)
Bloch, C. (1957): *Nuclear Phys.*, **4:**503. (16.5)
Block, C., and V. Gillet (1965): *Phys. Lett.*, **16:**62. (17.4)
Boardman, A. D., A. D. Hill, and S. Samathar (1967): *Phys. Rev.*, **160:**472. (10.1, 11.1)
Boas, R. P. (1954): "Entire Functions," Academic, New York. (12.1.4)
Boerner, W.–M. (1979): *Commun. Lab. Rep.*, **78-3**, Univ. of Illinois, Chicago Circle. (4)
Böhm, A. (1981): *J. Math. Phys.*, **22:**2813. (19.2)
Bohr, N., and F. Kalckar (1937): *Kgl. Danske Videnskab. Selskab, Mat.–fys. Medd.*, **14**(10). (16.5)
—, R. E. Peierls, and G. Placzek (1939): *Nature*, **144:**200. (7.2.2)
Bollé, D. (1976): *Phys. Rev.*, **D13:**1809. (17.4)
— (1979a): *Phys. Lett.*, **74A:**26. (11.1.3, 20.3.2)
— (1979b): *Ann. Phys.* (*N.Y.*), **121:**131. (12.1.3)
— and K. Chadan (1982): preprint. (17.4)
— and T. A. Osborn (1974): *Phys. Rev.*, **C9:**441. (17.4.6)
— and — (1975): *Phys. Rev.*, **D11:**3417. (11.1.3)
— and — (1976): *Phys. Rev.*, **D13:**299. (11.1.3)
— and — (1979): *J. Math. Phys.*, **20:**1121. (11.1.3)
— and — (1981): *J. Math. Phys.*, **22:**883. (5, 20.3.2)

— and H. Smeesters (1977): *Phys. Lett.*, **62A:**290. (12.1.3)
Bolsterli, M. (1963): *Phys. Rev.*, **131:**883. (16.3)
— (1969): *Phys. Rev.*, **182:**1095. (16.6)
— (1974): *Phys. Rev. Lett.*, **32:**436. (12.1.3)
— and J. MacKenzie (1965): *Physics*, **2:**141. (20.3)
— and A. S. Rinat (1977): *Phys. Rev.*, **C15:**40. (16.3)
Borg, G. (1946): *Acta Math.*, **78:**1. (20.3)
— (1949): Eleventh Congress of Scandinavian Mathematicians, Trondheim, p. 276 [as quoted by L. D. Faddeev (1959)]. (20.3)
Born, M. (1926): *Z. Phys.*, **38:**803. (9.1)
— and E. Wolf (1959): "Principles of Optics," Pergamon, New York. (1.2, 3, 4.3)
Borowitz, S., and B. Friedman (1953): *Phys. Rev.*, **89:**441. (11.3, 17.4)
Bosanac, S. (1981a): *Phys. Rev.*, **A24:**777. (11.1.3)
— (1981b): *J. Math. Phys.*, . (16.6)
Bose, A. K. (1964): *Nuovo Cimento*, **32:**679. (14)
Bottcher, C. (1981): *J. Phys. B*, **14:**L349. (16)
Bottino, A., and A. M. Longoni (1962): *Nuovo Cimento*, **24:**353. (12.1.4, 13.1)
—, —, and T. Regge (1962): *Nuovo Cimento*, **23:**954. (12.1.1, 13.3)
Bowcock, J., and A. Martin (1959): *Nuovo Cimento*, **14:**516. (10.3.4)
— and D. Walecka (1959): *Nuclear Phys.*, **12:**371. (10.3.4)
Bowman, J. J., T. B. A. Senior, and P. L. E. Uslenghi (1969): Eds., *Electromagnetic and Acoustic Scattering by Simple Shapes*, North-Holland Publ. Co., Amsterdam. (1.3)
Boyce, J. F., and S. A. Roberts (1974): *Acta Cryst.*, **A30:**543. (20.2)
Boyet, H., and S. Borowitz (1954): *Phys. Rev.*, **93:**1225. (11.3, 17.4)
Boyle, J. F. (1971): *Mol. Phys.*, **22:**993. (20.1)
Brackett, J. W., C. R. Mueller, and W. A. Sanders (1963): *J. Chem. Phys.*, **39:**2564. (20.1)
Brander, O. (1964): *Nuovo Cimento*, **32:**1059. (13.4)
— and K. Chadan (1981): preprints. (20.3)
Braun, G. (1956): *Acta Phys. Austriaca*, **10:**8. (1.3, 6.7.3)
Brayshaw, D. D. (1968a): *Phys. Rev.*, **167:**1505. (13.3)
— (1968b): *Phys. Rev.*, **176:**1855. (17.4.4)
— (1969): *Phys. Rev.*, **182:**1658. (17.4)
Breene, R. G. (1963): "Proton and Electron Scattering by Diatomic Molecules," Physical Studies, Centerville, Ohio. (17)
Breit, G. (1940): *Phys. Rev.*, **58:**506 and 1069. (16.5)
— (1946): *Phys. Rev.*, **69:**472. (16.5, 17.1)
— (1957): *Phys. Rev.*, **107:**1612. (17.2)
— (1959): In S. Flügge (ed.), "Handbuch der Physik," vol. 41, pt. 1, p. 1, Springer, Berlin. (16.5, 17.2)
— (1962): *Revs. Modern Phys.*, **34:**766. (20.1)
— and H. A. Bethe (1954): *Phys. Rev.*, **93:**888. (7.2.2)
— and E. P. Wigner (1936): *Phys. Rev.*, **49:**519. (11.2.2, 16.5)
— and F. L. Yost (1935): *Phys. Rev.*, **48:**203. (16.5)
Brenig, W. (1959): *Nuclear Phys.*, **13:**333. (16.5)
— and R. Haag (1959): *Fortschr. Phys.* **7:**183 [transl. in M. H. Ross (1963a)]. (6, 7, 8.1, 16.2)
Brezin, E., and J. Zinn-Justin (1966): *C. R. Acad. Sci. Paris*, **B263:**670. (17.3)
Brillouin, L. (1926): *Comptes Rend.*, **183:**24. (18.2)

Brodskii, A. M., and V. S. Potapov (1970): *Yad. Fiz.*, **12:**1163 (transl. *Soviet J. Nucl. Phys.*, **12:**639). (17.4, 18.2)
—, —, and V. V. Tolmachev (1970): *Zh. Eksp. Teor. Fiz.*, **58:**264 (transl. *Soviet Phys. JETP*, **31:**144). (17.4, 18.2)
Brown, G. E. (1957): *Proc. Phys. Soc.* (*London*), **A70:**681. (16.5)
Brown, R. Hanbury, and R. Q. Twiss (1954): *Phil. Mag.*, **45:**663. (4.3)
— and — (1956): *Nature*, **177:**27, 178, 1447. (4.3)
— and — (1957): *Proc. Roy. Soc.* (*London*), **242A:**300, **243A:**291. (4.3)
Brownell, F. H. (1962): *Pacific J. Math.*, **12:**47. (6.7)
Brownstein, K. R., and W. A. McKinley (1968): *Phys. Rev.*, **170:**1255. (11.3.3)
Brysk, H. (1965): *J. Math. Phys.*, **6:**51. (18.2)
Buck, V. (1971): *J. Chem. Phys.*, **54:**1923. (20.1)
— (1972): *J. Chem. Phys.*, **57:**578. (20.1)
— (1974): *Rev. Mod. Phys.*, **46:**369. (20.1)
— and H. Pauly (1969): *J. Chem. Phys.*, **51:**1662. (20.1)
Buhring, W. (1974): *J. Math. Phys.*, **15:**1451. (12.4)
Burdet, G., M. Giffon, and J. Goldberg (1966): *Nuovo Cim.*, **44:**138. (20.4)
—, —, and E. Predazzi (1965): *Nuovo Cimento*, **36:**1337. (20.1)
Burke, P. G., and H. M. Schey (1962): *Phys. Rev.*, **126:**147. (16)
— and K. Smith (1962): *Revs. Modern Phys.*, **34:**458. (16)
Burkhardt, H. (1972): *Nuovo Cim.*, **10A:**379. (20.2)
— and A. Martin (1975): *Nuovo Cim.*, **29A:**141. (20.2)
Burt, P. B., and F. J. Woods (1971): *J. Math. Phys.*, **12:**2275. (18)
Bushell, P. J. (1972): *J. Math. Phys.*, **13:**1540. (9.1.1)
Buslaev, V. S. (1962): *Dokl. Akad. Nauk SSSR*, **143:**1067 (transl. *Soviet Phys. Dokl.*, **7:**295). (20.3.2)
— and L. D. Faddeev (1960): *Dokl. Akad. Nauk SSSR*, **132:**13 (transl. *Soviet Math. Dokl.*, **1:**451). (20.3.2)
— (1967): in M. Sh. Birman, ed., *Topics in Math. Phys.*, vol. 1, Consultants Bureau, N.Y., p. 69. (20.3.2)
— (1971): in M. Sh. Birman, ed., *Topics in Math. Phys.*, vol. 4, Consultants Bureau, N.Y., p. 43. (20.3.2)
— and S. P. Merkuriev (1969): *Dokl. Akad. Nauk SSSR*, **189:**269 (transl. *Soviet Phys. Dokl.*, **14:**1055). (17.4)
Byron, F. W., and C. J. Joachain (1973): *Physica*, **66:**33. (18.3)
—, —, and E. H. Mund (1973): *Phys. Rev.*, **D8:**2622. (18.3)
—, —, and — (1975): *Phys. Rev.*, **D11:**1662. (18.3)
Bystricky, J., F. Lehar, J. Patera, and P. Winternitz (1976): *Phys. Rev.*, **D13:**1276. (11.1)
Cahill, R. T. (1974): *Phys. Rev.*, **C9:**473. (17.1.2)
Calogero, F. (1963): *Nuovo Cimento*, **27:**261, 1007, **28:**66, 320, 761. (12.1.2)
— (1964): *Phys. Rev.*, **135:**B693. (12.4)
— (1965a): *J. Math. Phys.*, **6:**161, 1105. (12.1.4)
— (1965b): *Nuovo Cimento*, **36:**199. (12.1.4)
— (1965c): *Phys. Rev.*, **139:**602. (12.4)
— (1965d): *Nuovo Cimento*, **37:**756. (12.4)
— (1965e): *Communs. in Math. Phys.*, **1:**80. (12.1.4)
— (1967): *Variable Phase Approach to Potential Scattering*, Academic Press, N.Y. (12.1.2, 12.1.4)
Calogero, F. (1969): *J. Math. Phys.*, **10:**2191. (17.3)

— (1971a): *J. Math. Phys.*, **12:**419. (17.3)
— (1971b): in L. Colin, ed., *Mathematics of Profile Inversion*, NASA TMX-62, 150, p. 5–33. (20.3.2)
— and M. Cassandro (1964): *Nuovo Cimento*, **34:**1712. (12.4)
— and — (1965): *Nuovo Cimento*, **37:**760. (12.4)
— and J. M. Charap (1964): *Ann. Phys.* (*N.Y.*), **26:**55. (9.1.2)
—, O. D. Corbella, A. Degasperis, and M. B. DeStefano (1968): *J. Math. Phys.*, **9:**1002. (20.3.2)
— and J. R. Cox (1968): *Nuovo Cim.*, **55A:**786. (20)
— and G. Cozenza (1966): *Nuovo Cim.*, **45A:**867. (12.1.4)
— and M. B. DeStefano (1966): *Phys. Rev.*, **146:**1195. (12.4)
— and A. Degasperis (1968): *J. Math. Phys.*, **9:**90. (20.3.2)
— and D. M. Fradkin (1966): *Nuclear Phys.*, **75:**470. (12.2)
— and C. Marchioro (1974): *J. Math. Phys.*, **15:**1425. (17.3)
— and D. G. Ravenhall (1964): *Nuovo Cimento*, **32:**1755. (12.1.2)
Calucci, G., L. Fonda, and G. C. Ghirardi (1968): *Phys. Rev.*, **166:**1719. (19.2)
— and G. C. Ghirardi (1968): *Phys. Rev.*, **169:**1339. (19.2)
— and — (1972): *Phys. Rev.*, **D6:**3346. (19.2)
Campesino–Romeo, E., and J. R. Taylor (1975): *J. Math. Phys.*, **16:**1227. (8.1)
Capps, R. H., and W. G. Holladay (1955): *Phys. Rev.*, **99:**931. (17.2)
Capuzzi, F. (1972): *Nuovo Cim.*, **A11:**801. (10.1.4)
Caratheodory, C. (1950): "Funktionentheorie I," Birkhäuser, Basel. (12.1)
Carew, J., and L. Rosenberg (1969): *Phys. Rev.*, **177:**2599. (16.4)
— and — (1972): *Phys. Rev.*, **C5:**658. (16.4)
— and — (1973): *Phys. Rev.*, **D7:**1113. (11.3)
Carter, D. S. (1952): Ph.D. thesis, Princeton University (unpublished). (12.2)
Carter, W. H. (1970): *J. Opt. Soc. Am.*, **60:**306. (20.3)
Case, K. M. (1950): *Phys. Rev.*, **80:**797. (12.4)
Cassola, R. L., and R. D. Koshel (1968): *J. Phys.*, **A1:**224. (10.1.4)
Cattapan, G., L. Lovitch and V. Vanzani (1981): *Nuovo Cim.*, **58A:**275. (17.4)
— and V. Vanzani (1977): *Nuovo Cim.*, **37A:**32. (17.4)
— and — (1978): in H. Zingl et al. (1978), I, p. 371. (17.4)
— and — (1979a): *Nuovo Cim.*, **51A:**509. (17.4.5)
— and — (1979b): *Phys. Rev.*, **C19:**1168; *Nuovo Cim.* **50A:**97. (17.4)
— and — (1981): *Nuovo Cim.*, **58A:**85. (17.4)
Chadan, K. (1955): *J. Phys. et le Radium*, **16:**843. (14.7.2, 15.2, 20.3)
— (1956): *C. R. Acad. Sci. Paris*, **242:**1964. (14.7.2, 15.2, 20.3)
— (1958): *Nuovo Cim.*, **10:**892. (20.1)
— (1965): *Nuovo Cim.*, **40A:**1194. (12.1.3, 20.3.2)
— (1966): *Nuovo Cim.*, **41A:**115 and **44A:**838. (12.1.3, 20.3.2)
— (1967): *Nuovo Cim.*, **47A:**510. (10.1.4, 20.1)
— (1968): *Nuovo Cim.*, **57A:**436. (18)
— (1976): *Lett. Math. Phys.*, **1:**281. (12.1.4)
— (1978): in P. C. Sabatier, ed., *Applied Inverse Problems*, Springer Verlag, Berlin, p. 404. (20.3)
— and Ch. De Mol (1980): *Ann. Phys.* (*N.Y.*), **129:**466. (10.3.2, 12.1.4)
— and J. Y. Guennéguès (1964): *Nuovo Cimento*, **34:**665. (12.1.2)
— and A. Martin (1979): *Commun. Math. Phys.*, **70:**1. (12.4)
— and — (1980): *J. Physique–Lett.*, **41:**L-205. (10.3.2, 12.1.4).
— and A. Montes (1968): *J. Math. Phys.*, **9:**1898. (20.3.2)

— and P. C. Sabatier (1977): *Inverse Problems in Quantum Scattering Theory*, Springer Verlag, N.Y. (20.1, 20.3)
Chakrabarti, A. (1964): *Ann. Inst. Henri Poincaré*, **1A:**301. (17.4)
Chan, C. K., and P. Lu (1980): *Phys. Rev.*, **D22:**1490. (20.1)
Chan, F. T., M. Lieber, G. Foster, and W. Williamson, Jr. (1979): *Adv. in Electr. and Electron Phys.*, **49:**133. (18.3)
Chan Hong-Mo (1961): *Proc. Roy. Soc.* (*London*), **A261:**329. (17.1)
— (1963): *J. Math. Phys.*, **4:**1042. (17.1)
Chandler, C. (1978): *Nucl. Phys.*, **A301:**1. (17.4.4)
— (1980): in F. S. Levin (1980), p. 129c. (14.6, 17.4)
— and A. G. Gibson (1973): *J. Math. Phys.*, **14:**1328. (16.2.3)
— and — (1974): *J. Math. Phys.*, **15:**291. (14.6, 16.2)
— and — (1977): *J. Math. Phys.*, **18:**2336. (16.2.3)
— and — (1978a): *J. Math. Phys.*, **19:**1610. (16.2.3)
— and — (1978b): in H. Zingl et al. (1978), I, P. 356. (16.2.3)
— and — (1978c): in J. Nuttall (1978), p. 189. (17.4)
— and — (1980): in J. A. DeSanto et al. (1980), p. 134. (16.2.3)
— and I. H. Sloan (1980): *Nucl. Phys.*, . (17.4.4)
Chandrasekhar, S. (1950): "Radiative Transfer," Oxford, Fair Lawn, N.J. (1.4)
Charap, J. M. (1965): *Nuovo Cimento*, **36:**419. (12.1.3)
— and N. Dombey (1964): *Phys. Letters*, **9:**210. (12.4)
Chen, A. C., and Y. Hahn (1975): *Phys. Rev.*, **A12:**823. (16.4)
Chen, H. (1966): *Phys. Rev.*, **144:**1237. (13.3)
Chen, J. C. Y. (1966): *Phys. Rev.*, **152:**1454. (16.3)
— (1973): *Case Studies in Atomic Physics*, **3:**305. (16)
—, A. C. Chen, and P. J. Kramer (1971): *Phys. Rev.*, **A4:**1982. (17.4)
— and K. T. Chung (1970a): *Phys. Rev.*, **A2:**1892. (16.3)
— and — (1970b): *Phys. Rev.*, **A2:**1449. (16)
—, —, and P. J. Kramer (1969): *Phys. Rev.*, **184:**64. (16)
— and L. Hambro (1971): *J. Phys. B*, **4:**191. (17.4)
— and C. J. Joachain (1971): *Physica*, **53:**333. (17.4)
— and P. J. Kramer (1972): *Phys. Rev.*, **A5:**1207. (16)
Chen, T. W., and D. W. Hook (1975): *Phys. Rev.*, **D12:**1765. (18)
Cheng, H. (1966): *Nuovo Cim.*, **45:**487. (14.7.1)
—, J. A. Dickenson, P. S. Yeung and K. Olaussen (1981): *Phys. Rev.*, **D23:**1411. (18.3)
Cheston, W. (1952): *Phys. Rev.*, **85:**952. (18.4)
Chew, G. F. (1948): *Phys. Rev.*, **74:**809. (18.4)
— (1950): *Phys. Rev.*, **80:**196. (18.4)
— and M. L. Goldberger (1952): *Phys. Rev.*, **87:**778. (18.4)
— and H. W. Lewis (1951): *Phys. Rev.*, **84:**779. (18.4)
— and S. Mandelstam (1960): *Phys. Rev.*, **119:**467. (12.1.3, 17.1)
— and G. C. Wick (1952): *Phys. Rev.*, **85:**636. (18.4)
Chew, H. (1970): *Phys. Rev.*, **D1:**719. (18.3)
Choi, B. H., R. T. Poe, J. C. Sun, K. T. Tang, and Y. Y. Yung (1981): *J. Chem. Phys.*, **74:**5686. (16)
Choudhury, M. H. (1964): *Nuovo Cimento*, **34:**956, **35:**339. (17.4)
Chou Kuang-Chao (1958): *Z. Eksp. i Teor. Fiz.*, **35:**783 (transl. *Soviet Phys. JETP* **8:**543). (15.1)
— (1959): *Z. Eksh. i Teor. Fiz.*, **36:**909 (transl. *Soviet Phys. JETP*, **9:**642). (15.1)

— and M. I. Shirokov (1958): *Z. Eksp. i Teor. Fiz.*, **34:**1230 (transl. *Soviet Phys. JETP*, **7:**851). (15.1)
Chudov, L. A. (1949): *Mat. Sbornik*, **25**(67):451. (20.3)
— (1956): *Dokl. Akad. Nauk SSSR*, **109:**40. (20.3)
Chung, K. T., and J. C. Y. Chen (1972): *Phys. Rev.*, **A6:**686. (16.4)
Ciafaloni, M. (1963): *Nuovo Cimento*, **29:**420. (12.1.3, 13.3)
— and P. Menotti (1965): *Nuovo Cimento*, **35:**160. (12.1.4)
Clapp, R. (1949): *Phys. Rev.*, **76:**873. (17.4.3)
— (1961): *Ann. Phys. (N.Y.)*, **13:**187. (17.4.3)
Clebsch, A. (1863): *J. f. reine u. andgew. Math.*, **61:**195. (2.4)
Coester, F. (1953): *Phys. Rev.*, **89:**619. (7.2.4)
— (1964a): *Phys. Rev.*, **133:**B1516. (9.1.1)
— (1964b): "Lectures on the Quantum Theory of Scattering," Argonne National Laboratory (unpublished). (20.1)
— (1965): *Helv. Phys. Acta*, **38:**7. (6)
— (1971): *Phys. Rev.*, **C3:**525. (16.3)
— (1980): in De Santo et al. (1980), p. 190. (6, 16.2.3)
—, M. Hamermesh, and K. Tanaka (1954): *Phys. Rev.*, **96:**1142. (6)
— and H. Kümmel (1958): *Nuclear Phys.*, **9:**225. (16.2)
— and L. Schlessinger (1973): *Ann. Phys. (N.Y.)*, **78:**90. (16.2.3)
Cohen, S., D. L. Judd, and R. J. Riddell (1960): *Phys. Rev.*, **119:**384. (17.4)
Cohn, J. H. E. (1965): *J. London Math. Soc.*, **40:**523. (12.1.4)
Colin de Verdiere, Y. (1981): preprint. (12.1.4)
Combes, J.–M. (1969): *Commun. Math. Phys.*, **12:**283 and *Nuovo Cim.*, **64A:**111. (17.4)
— (1974): in Lavita and Marchand (1974), p. 243. (17.4)
— (1978): in J. Nuttall (1978), p. 25. (16)
— (1980a): in K. Osterwalder, ed., *Mathematical Problems in Theoretical Physics*, Springer Verlag, Berlin, p. 1. (6.8)
— (1980b): in G. Velo and A. S. Wightman, eds., *Rigorous Atomic and Molecular Physics*, Plenum Press, N.Y. (7.3)
—, R. G. Newton, and R. Shtokhamer (1975): *Phys. Rev.*, **D11:**366. (6.7)
Combescure, M., and J. Ginibre (1976): *Ann. Inst. H. Poincaré*, **24:**17. (12.4)
Common, A. K., and T. Stacey (1978): *J. Phys. A*, **11:**259 and 275. (11.1)
Conn, R., and H. Rabitz (1973): *Phys. Rev.*, **A7:**658. (16.4)
Connor, J. N. L., D. Farrelly, and D. C. Mackay (1981): *J. Chem. Phys.*, **74:**3278. (13.3)
Conway, J. M. (1967): *Phys. Rev.*, **156:**1365. (4.3)
Cook, J. (1957): *J. Math. and Phys.*, **36:**82. (6.7)
Corbella, O. D. (1970): *J. Math. Phys.*, **11:**1695. (20.3)
— (1971): *J. Math. Phys.*, **12:**1873. (20.3.1)
—, C. R. Garibotti, and F. F. Grinstein (1976): *Z. Physik*, **A277:**1. (11.1)
Corbett, J. V. (1966): *Nucl. Phys.*, **85:**401. (10.3)
— (1968): *J. Math. Phys.*, **9:**891. (9.1.1)
Corinaldesi, E. (1954a): *Nuovo Cimento*, **11:**468, **12:**469. (20.3)
— (1954b): *Nuovo Cimento*, **11:**200, **12:**438. (18)
— (1956): *Nuclear Phys.*, **2:**420. (12.1.3)
— (1962): *Nuovo Cimento*, **29:**92. (16.3.3)
—, L. Trainor, and Ta-You Wu (1952): *Nuovo Cimento*, **9:**436. (16.3)
Cornille, H. (1964): *Nuovo Cimento*, **33:**434. (12.4)

— (1965): *Nuovo Cimento*, **36:**1316, **38:**1243. (12.4)
Cornille, H. (1967): *J. Math. Phys.*, **8:**2268. (20.3)
— (1970a): *J. Math. Phys.*, **11:**61. (20.3)
— (1970b): *J. Math. Phys.*, **11:**79. (20.3)
— (1976): *J. Math. Phys.*, **17:**2143. (20.3)
—, G. Burdet, and M. Giffon (1965): *Nuovo Cimento*, **38:**647. (12.4)
— and J. M. Drouffe (1974): *Nuovo Cim.*, **20A:**401. (20.2)
— and A. Martin (1962): *Nuovo Cimento*, **26:**298. (14.6)
— and — (1972a): *Nucl. Phys.*, **B48:**104. (10.3.3)
— and — (1972b): *Nucl. Phys.*, **B49:**413. (20.2)
— and E. Predazzi (1964): *Phys. Letters*, **10:**149. (12.4)
— and — (1965a): *Nuovo Cimento*, **35:**879. (12.4)
— and — (1965b): *J. Math. Phys.*, **6:**1730. (12.4)
Corrigall, B., B. Kuppers, R. Wallace (1971): *Phys. Rev.*, **A4:**977. (18.3)
Cotanch, S. R., and C. M. Vincent (1976): *Phys. Rev.*, **C14:**1739. (17)
Cottingham, W. N., and R. F. Peierls (1965): *Phys. Rev.*, **137:**B147. (18.3)
Coudray, C. (1977): *Nuovo Cim. Lett.*, **19:**319. (20.4)
— (1978): in P. C. Sabatier, ed., *Applied Inverse Problems*, Springer Verlag, Berlin, p. 351. (20.4)
— (1980): *J. Phys. A*, **13:**2937. (20.4)
— and M. Coz (1970): *Ann. Phys. (N.Y.)*, **61:**488. (20.4)
— and — (1972): *J. Math. Phys.*, **12:**1166. (20.4)
Courant, R., and D. Hilbert (1953): "Methods of Mathematical Physics," Interscience, New York. (9.2, 9.3)
Cox, J. R. (1962): "Construction of Potentials from the Many-channel S-matrix," Ph. D. thesis, Indiana University (unpublished). (12.1.3, 17, 20.3)
— (1964): *J. Math. Phys.*, **5:**1065. (14.7, 17)
— (1965): *Nuovo Cimento*, **37:**474. (12.1.2)
— (1966): *Ann. Phys. (N.Y.)*, **39:**216. (20.3)
— (1967): *J. Math. Phys.*, **8:**2327. (20.3)
— (1975): *J. Math. Phys.*, **16:**1410. (20.3)
— and H. R. Garcia (1975): *J. Math. Phys.*, **16:**1402. (14.7, 17.3, 20.3)
— and A. Perlmutter (1965): *Nuovo Cimento*, **37:**76. (12.1.2)
— and K. W. Thompson (1969): *Bull. Am. Phys. Soc.*, **14:**579. (20.4)
— and — (1970): *J. Math. Phys.*, **11:**805 and 815. (20.4)
Coz, M. (1965a): *Nuovo Cimento*, **35:**492. (16.3, 16.5)
— (1965b): *Ann. Phys. (N.Y.)*, **35:**53. (16.5)
— (1966): *Ann. Phys. (N.Y.)*, **39:**237. (20.3)
— (1981): *J. Math. Phys.* **22:**1596. (20.3)
—, L. G. Arnold, and A. D. MacKellar (1970): *Ann. Phys. (N.Y.)*, **59:**219. (10.1.4)
— and C. Coudray (1973): *J. Math. Phys.*, **14:**1574. (20.3)
— and P. Rochus (1977): *J. Math. Phys.*, **18:**2223 and 2232. (20.3)
Craggs, J. D., and H. S. W. Massey (1959): In S. Flügge (ed.), "Handbuch der Physik," vol. 37/1, p. 314, Springer, Berlin. (16)
Crichton, J. H. (1966): *Nuovo Cim.*, **45:**256. (20.2)
Cross, R. J. (1967): *J. Chem. Phys.*, **47:**3724. (18.2)
— (1968): *J. Chem. Phys.*, **48:**4838 and **49:**1753. (18.2)
— (1969): *J. Chem. Phys.*, **51:**5163. (18.2)
Crum, M. M. (1955): *Quart. J. Math.*, **6:**121. (20.3)
Cruzan, O. R. (1962): *Q. Appl. Math.*, **20:**33. (1.3)

Cuer, M. (1977): *J. Math. Phys.*, **18:**658. (20.1)

Curtis, A. R. (1964): "Coulomb Wave Functions," Royal Society Mathematical Tables, vol. 11, Cambridge, New York. (14.6)

Cushing, J. T. (1965): *Nuovo Cimento*, **36:**586 and 905; **38:**463. (17.4.3)

— (1977): *Phys. Rev.*, **D15:**1790. (20)

Dahmen, H. D., and A. S. Reiner (1967): *Ann. Phys.* (*N.Y.*), **44:**74. (17.1.3)

Dalitz, R. H. (1951): *Proc. Roy. Soc.* (*London*), **A206:**509. (9.1.2, 10.3.5, 14.6)

— (1954): *Phys. Rev.*, **94:**1046. (17.4.6)

— and G. Rajasekaran (1963): *Phys. Letters*, **7:**373. (17.1)

Damburg, R. Ya., and R. K. Peterkop (1964): *Z. Eksp. i Teor. Fiz.*, **47:**1602 (transl. *Soviet Phys. JETP*, **20:**1076). (16)

Danilov, G. S. (1961): *Z. Eksp. i Teor. Fiz.*, **40:**498 (transl. *Soviet Phys. JETP*, **13:**349) (17.4)

Darewych, J. W., and R. Pooran (1978): *Can. J. Phys.*, **56:**1358. (11.3, 18.3)

— and — (1980): *J. Math. Phys.*, **21:**2398. (11.3.3)

— and J. Sokoloff (1979): *J. Math. Phys.*, **20:**736. (11.3)

Dashen, R. F. (1963): *J. Math. Phys.*, **4:**388. (12.1.2)

—, J. B. Healey, and I. J. Muzinich (1976): *Phys. Rev.*, **D14:**2773. (17.3)

Daumens, M., and M. Perroud (1979): *J. Math. Phys.*, **20:**1077. (11.1)

—, —, and P. Winternitz (1979): *Phys. Rev.*, **D19:**3413. (11.1)

— and P. Winternitz (1980): *Phys. Rev.*, **D21:**1919. (11.1)

Davies, H. (1960): *Nuclear Phys.*, **14:**465. (9.1.1, 10.3.1)

Davies, K. T. D., and M. Baranger (1962): *Ann. Phys.* (*N.Y.*), **19:**383. (16.5)

Davison, B. (1947): *Phys. Rev.*, **71:**694. (11.3)

Day, T. B., L. S. Rodberg, G. A. Snow, and J. Sucher (1961): *Phys. Rev.*, **123:**1051. (7.2.5, 16.3)

DeAlfaro, V., T. Regge, and C. Rossetti (1962): *Nuovo Cimento*, **25:**701, **26:**1029. (12.1.2)

DeFacio, B., and H. E. Moses (1980): *J. Math. Phys.*, **21:**1716. (20.3)

DeRoo, M. (1974): Ph. D. thesis, Groningen, unpublished. (20.2)

DeSanto, J. A., A. W. Saenz, and W. W. Zachary (1980): eds., *Mathematical Methods and Applications of Scattering Theory*, Springer Verlag, N.Y. (6)

DeWitt, B. S. (1955a): "The Operator Formalism in Quantum Scattering Theory," *Univ. Calif.* (*Berkeley*) *Radiation Lab.* UCRL-2884. (6, 7)

— (1955b): *Phys. Rev.*, **100:**905. (7)

— (1956): *Phys. Rev.*, **103:**1565. (8.1.3)

Debye, P. (1909a): *Ann Phys.*, **30:**57 (3.3, 3.5, 4.1.1)

— (1909b): *Math. Ann.*, **67:**535. (3 App.)

— (1910): *Abhandl. Akad. Wiss., München, Math.-phys. Kl.* **5** [(3 (App.)]

Degasperis, A. (1964): *Nuovo Cimento*, **34:**1667. (12.1.2)

— (1970): *J. Math. Phys.*, **11:**551. (20.3)

—, L. Fonda, and G. C. Ghirardi (1974): *Nuovo Cim.*, **21A:**471. (19.2)

Deift, P., and E. Trubowitz (1979): *Commun. Pure & Appl. Math.*, **32:**121. (20.3)

Del Giudice, E., and E. Galzenati (1964): *Nuovo Cimento*, **38:**443, **40:**739. (12.4)

Delos, J. B., and C. E. Carlson (1975): *Phys. Rev.*, **A11:**210. (13.3)

Delves, L. M. (1958): *Nuclear Phys.*, **8:**358. (17.3)

— (1958–1959): *Nuclear Phys.*, **9:**391. (17.2)

— (1960): *Nuclear Phys.*, **20:**275. (17.3.4)

Demkov, Yu. N., V. N. Ostrovskii, and N. B. Berezina (1971): *Zh. Eksp. Teor. Fiz.* **60:**1604 (transl. *Soviet Phys. JETP*, **33:**867). (20.1)

Derrick, G. H. (1960): *Nuclear Phys.*, **16:**405, **18:**303. (17.4.3)
— (1962): *Nuclear Phys.*, **34:**543. (17.4.3)
— and J. M. Blatt (1958): *Nuclear Phys.*, **8:**310. (17.4.3)
— and — (1960): *Nuclear Phys.*, **17:**67. (17.4.3)
Desai, B. R., and R. G. Newton (1963): *Phys. Rev.*, **129:**1437. (15.2)
Devaney, A. J. (1978): *J. Math. Phys.*, **19:**1526. (20.5)
— (1980): *J. Math. Phys.*, **21:**2603. (1.4)
Devinatz, A. (1980): *J. Math. Phys.*, **21:**2406. (12.4)
Di Salvo, E., and G. A. Viano (1976): *Nuovo Cim.*, **33B:**547. (20)
Dialetis, D., and E. Wolf (1967): *Nuovo Cim.*, **47:**113. (4.3)
Dillig, M. (1975): *Nuovo Cim.*, **26A:**281. (13.3)
Dirac, P. A. M. (1926): *Proc. Roy. Soc.* (*London*), **A112:**661. (6.5)
— (1927): *Proc. Roy. Soc.* (*London*), **A114:**243. (6.5)
— (1947): "The Principles of Quantum Machanics," 3d ed., Oxford, Fair Lawn, N.J. (12.1.1)
— (1955): *Can. J. Phys.*, **33:**709. (16.3)
Dirks, J. F., and Y. Hahn (1970): *Phys. Rev.*, **A2:**1861 and **A3:**310. (16.4)
Dodd, L. R. (1970): *J. Math. Phys.*, **11:**207. (17.3)
— (1974): *Phys. Rev.*, **A9:**637. (16)
— and K. R. Greider (1966): *Phys. Rev.*, **146:**675. (17.4)
— and I. E. McCarthy (1964): *Phys. Rev.*, **134:**A1136. (6)
— and K. J. Nieukerke (1980): *Phys. Rev.*, **C21:**1724. (17.4)
Doetsch, G. (1956): "Handbuch der Laplace-Transformation," Birkäuser, Basel. (5.9)
Dolizanszky, T. (1974): *Nuovo Cim.*, **22A:**578. (12.4)
Dollard, J. D. (1964): *J. Math. Phys.*, **5:**729. (6, 14.6)
— (1968): *J. Math. Phys.*, **9:**620. (14.6)
— (1969): *Commun. Math. Phys.*, **12:**193. (6.7, 16.5)
— (1971): *Rocky Mtn. J. Math.*, **1:**5. (14.6)
— (1972): *Rocky Mtn. J. Math.*, **2:**317. (14.6)
— (1973): *J. Math. Phys.*, **14:**708. (6.7, 16.5)
— and C. Friedman (1978): *Ann. Phys.* (*N.Y.*), **111:**251. (12.4)
— and G. Velo (1966): *Nuovo Cim.*, **A45:**801. (14.6)
Dolph, C. L., J. B. McLeod and D. Thoe (1966): *J. Math. Anal. & Appl.*, **16:**311. (10.3.2)
Dombey, N. (1965): *Nuovo Cimento*, **37:**1743. (12.4)
— and R. H. Jones (1968): *J. Math. Phys.*, **9:**986. (12.4)
Doolen, G. (1968): *Phys. Rev.*, **166:**1651. (17.4)
— and J. Nuttall (1971): *J. Math. Phys.*, **12:**2198. (17.4.5)
Dragt, A. J., and R. Karplus (1962): *Nuovo Cimento*, **26:**168. (17.2)
Dreyfuss, T. (1976): Ph.D. thesis, Univ. of Geneva, unpubl., *J. Phys.*, **A9:**L187. (12.1.3)
— (1978): *J. Math. Anal. & Appl.* **64:**114; *Helv. Phys. Acta*, **51:**321. (12.1.3)
Dunford, N., and J. T. Schwartz (1958 and 1963): "Linear Operators," vols. 1 and 2, Interscience, New York. (7.3)
Durand, L. (1976): *Phys. Rev.*, **D14:**3174. (16.6)
Dyson, F. J. (1949): *Phys. Rev.*, **75:**486 and 1736. (6.4, 6.5)
— (1976): *Commun. Math. Phys.*, **47:**171. (20)
Eckart, C. (1930): *Phys. Rev.*, **35:**1303. (14.4)
Eckstein, H. (1954): *Phys. Rev.*, **94:**1063. (6.3)
— (1956a): *Phys. Rev.*, **101:**880. (6, 16.2)

— (1956b): *Nuovo Cimento*, **4:**1017. (6)
— (1960): *Phys. Rev.*, **117:**1590. (20.1)
Eden, R. J. (1949): *Proc. Roy. Soc.* (*London*), **A199:**256. (17.2)
— (1952): *Proc. Roy. Soc.* (*London*), **A210:**388. (17.2)
—, and J. R. Taylor (1963): *Phys. Rev. Letters*, **11:**516. (17.1)
Edmonds, A. R. (1957): "Angular Momentum in Quantum Mechanics," Princeton, Princeton, N. J. (2.1.4, 15.1)
Efimov, V. (1970): *Phys. Lett.*, **B33:**563; *Yad. Fiz.*, **12:**1080 (transl. *Soviet J. Nucl. Phys.*, **12:**589). (17.4.8)
— (1973): *Nucl. Phys.*, **A210:**157. (17.4.8)
— (1976): *Proc. 7th Intern. Conf. on Few Body Problems in Nuclear Physics*, North-Holland Publ., Amsterdam, p. 126. (17.4.8)
Eftimiu, C. (1971): *J. Math. Phys.*, **12:**2047. (20.2)
— (1972): *J. Math. Phys.*, **13:**458. (20.2)
— (1973): *Phys. Rev.*, **D7:**267. (20.2)
Ehrenfest, P. (1927): *Z. Phys.*, **45:**455. (6)
Ehrman, J. B. (1951): *Phys., Rev.*, **81:**412. (17.2)
Elizalde, E. (1976): *Nuovo Cim.*, **31A:**151. (12.1.4)
Enss, V. (1977): *Commun. Math. Phys.*, **52:**233. (16.1)
— (1978): *Commun. Math. Phys.*, **61:**285. (6.3)
— (1979): *Commun. Math. Phys.*, **65:**151. (17.4)
— (1980): in G. Velo and A. S. Wrightman, eds., *Rigorous Atomic and Molecular Physics*, Plenum, N.Y. (10.3.3)
— and B. Simon (1980): *Commun. Math. Phys.*, **76:**177. (10.3.3)
— and — (1981): in K. Gustafson and W. P. Reinhardt, eds., *Classical, Semiclassical, and Quantum Mechanical Problems in Mathematics, Chemistry, and Physics*, Plenum, N.Y. (10.3.3)
Eppel, D. (1978): in G. Pisent et al. (1978), p. 193. (17.4.6)
Epstein, S. T. (1955): *Phys. Rev.*, **98:**196. (7)
— (1957): *Phys. Rev.*, **106:**598. (16.3)
Erdelyi, A. (ed.) (1953): "Higher Transcendental Functions," McGraw-Hill, New York. (2.1.1, 2.2.1, 12.3, 13.1, 15.1.1, 17.2.4, 18.3)
— (1954): "Tables of Integral Transforms," McGraw-Hill, New York. (5.9, 18.3)
— (1955): "Asymptotic Expansions," Dept. of Math., Cal. Inst. of Tech. (1.3.1, 3 App., 6.7.3)
— (1960): *J. Math. Phys.*, **1:**16. (18.2)
Eu, B. C. (1970): *J. Chem. Phys.*, **52:**1882 and 3903. (18.2)
— (1971a): *J. Chem. Phys.*, **54:**559. (5, 17.4)
— (1971b): *J. Chem. Phys.*, **55:**5600. (18.2)
— (1972): *J. Chem. Phys.*, **56:**2507 and 5202. (18.2)
— (1973): *J. Chem. Phys.*, **58:**472. (18.2)
— and T. P. Tsien (1973): *Phys. Rev.*, **A7:**648. (18.2)
Evans, J. W. (1981): *J. Math. Phys.*, **22:**1672. (16.3)
— and D. K. Hoffman (1981): *J. Math. Phys.* **22:**2858. (17.4)
Exner, P. (1976): *Commun. Math. Phys.*, **50:**1. (19.2)
Eyges, L. (1959): *Phys. Rev.*, **115:**1643. (17.4)
Eyre, D., T. A. Osborn, and J. P. Svenne (1981): *Phys. Rev.* **C24:**2409. (17.4)
Fabre de la Ripelle, M. (1971): *C. R. Acad. Sci. Paris*, **273:**1007. (17.4.6)
Fabri, E. (1954): *Nuovo Cimento*, **11:**479. (17.4.6)
Faddeev, L. D. (1956): *Vestnik Leningrad Univ.*, **11:**126 (*Math. Rev.*, **18:**259). (20.3)

— (1957): *Vestnik Leningrad Univ.*, **12:**164 (*Math. Rev.*, **19:**661). (12.1.4)
— (1958a): *Dokl, Akad. Nauk SSSR*, **121:**63 (*Math, Rev.*, **20:**733). (20.3)
— (1958b): *Z. Eksp. i Teor. Fiz.*, **35:**433 (transl. *Soviet Phys. JETP*, **8:**299). (10.3.4)
— (1959): *Usp. Mat. Nauk*, **14:**57 [transl. *J. Math. Phys.*, **4:**72 (1963)]. (20.3)
— (1960): *Z. Eksp. i Teor. Fiz.*, **39:**1459 (transl. *Soviet Phys. JETP*, **12:**1014). (7.2.5, 17.4.2)
— (1961): *Dokl. Akad. Nauk SSSR*, **138:**565 (transl. *Soviet Phys. Dokl.*, **6:**384). (7.2.5, 17.4.2)
— (1962): *Dokl. Akad. Nauk SSSR*, **145:**301 (transl. *Soviet Phys. Dokl.*, **7:**600). (7.2.5, 17.4.2)
— (1963): Mathematical Aspects of the Three-body Problem in Quantum Scattering Theory, *Steklov Math. Inst., Leningrad Publ.* 69 (transl. D. Davey and Co., New York, 1965). (7.3, 17.4.3)
— (1964): *Trudy Mat. Inst. Steklov*, **73:**314 (*Am. Math. Soc. Transl.*, Ser. 2, **65:**139). (20.3)
— (1965): *Dokl. Akad. Nauk SSSR*, **165:**514 (transl. *Soviet Phys. Dokl.*, **10:**1033). (20.5)
— (1966): *Dokl. Akad. Nauk SSSR*, **167:**69 (transl. *Soviet Phys. Dokl.*, **11:**209). (20.5)
— (1970): in McKee and Rolph (1970), p. 154. (17)
— (1974): *Itogi Nauki i Tek. Sov. Pr. Mat.*, **3:**93 (transl. *J. of Soviet Math.*, **5:**334). (20.5)
— and D. A. Yakubovskii (1981): *Yad. Fiz.*, **33:**634 (trans). *Soviet J. Nucl. Phys.*, **33:**331. (17.4)
Faifman, M. P. (1977): *Yad. Fiz.*, **26:**433 (transl. *Soviet J. of Nucl. Phys.*, **26:**227). (12.1.2)
Faisal, F. H. M. (1968): *J. Phys. B*, **1:**181 and 187. (17.2)
Fang, K. K. (1977): *Phys. Rev.*, **C15:**1204. (17.4.6)
— (1979): *Phys. Rev.*, **C19:**1637. (17.4.6)
Fano, U. (1957): *Revs. Modern Phys.*, **29:**74. (8.3.1)
— (1981): *Phys. Rev.* **A24:**2402. (16, 16.3)
Faris, W. G. (1971): *Rocky Mtn. J. Math.*, **1:**637. (9.1.1)
Faxen, H., and J. Holtsmark (1927): *Z. Phys.*, **45:**307. (2.2.5, 11.1)
Federbush, P. (1966): *Phys. Rev.*, **148:**1551. (17.4.4)
Feenberg, E. (1932): *Phys. Rev.*, **40:**40. (1.3, 7.2.2)
— (1948): *Phys. Rev.*, **74:**664. (7)
Fermi, E. (1936): *Ric. Scientifica*, **7**(2):13 (transl. in E. Fermi, *Collected Papers*, Univ. of Chicago Press, 1962, p. 980). (17.2, 18.4)
Fernbach, S., T. A. Green, and K. M. Watson (1951): *Phys. Rev.*, **82:**980. (18.4)
—, R. Serber, and T. Taylor (1949): *Phys. Rev.*, **75:**1352. (18.3)
Ferrante, G. (1979): *Nuovo Cim.*, **52B:**229. (16)
—, E. Fiordilino, and C. Leone (1980): *Nuovo Cim.*, **56B:**237. (16)
Ferreira, E. M., N. Guillen, and J. Sasma (1968): *J. Math. Phys.*, **9:**1210. (12.1.3)
— and A. F. F. Teixeira (1966): *J. Math. Phys.*, **7:**1207. (14.6)
Ferretti, B. (1951): *Nuovo Cimento*, **8:**108. (6.7.3)
Feshbach, H. (1958): *Ann. Phys. (N.Y.)*, **5:**357. (16.5)
— (1962): *Ann Phys. (N.Y.)*, **19:**287. (16.5)
—, C. E. Porter, and V. F. Weisskopf (1954): *Phys. Rev.*, **96:**448. (16.5)
— and W. Rarita (1949): *Phys. Rev.*, **75:**1384. (17.4.3)
Feynman, R. P. (1949): *Phys. Rev.*, **76:**749 and 769. (6.4)
— (1951): *Phys. Rev.*, **84:**108. (6.5)

Fiedeldey, H. (1969): *Nucl. Phys.*, **A135:**353. (20.1)
Firsov, O. B. (1953): *Zh. Eksp. Teor. Fiz.*, **24:**279. (20.1)
Fivel, D. I. (1961): *Nuovo Cimento*, **22:**326. (13.4)
— (1962): *Phys. Rev.*, **125:**1085. (10.3.4)
— (1966): *Phys. Rev.*, **142:**1219. (14.6)
— and A. Klein (1960): *J. Math. Phys.*, **1:**274. (14.5)
Fleischer, J. (1972): *Nucl. Phys.*, **B347:**59 and **B44:**640. (11.1)
— (1973): *J. Math. Phys.*, **14:**276. (11.1)
Fleming, G. N. (1973): *Nuovo Cim.*, **16A:**232. (19.2)
Flügge, S. (1948): *Z. Naturforsch.*, **3a:**97. (16.5)
Foldy, L. L., and J. A. Lock (1979): *Phys. Rev.*, **C20:**418. (10.1.4)
— and W. Tobocman (1957): *Phys. Rev.*, **105:**1099. (7, 10.2, 16.3)
Fonda, L. (1959): *Nuovo Cimento*, **13:**956. (17.2)
— (1961a): *Nuovo Cimento*, **20:**116. (17.2)
— (1961b): *Ann. Phys.* (*N.Y.*), **12:**476. (16.5, 17.2)
— (1964): *Ann. Phys.* (*N.Y.*), **29:**401. (16.5)
— (1967): in *Fundamentals in Nuclear Theory*, I.A.E.A. Vienna, p. 793. (16.6)
— and G. C. Ghirardi (1964a): *Nuclear Phys.*, **58:**374. (17.2)
— and — (1964b): *Ann. Phys.* (*N.Y.*), **26:**240. (16.5)
— and — (1966): *Nuovo Cim.*, **46A:**47. (12.1.4)
— and — (1970): *Nuovo Cim.*, **67A:**257. (19.2)
— and — (1971): *Nuovo Cim.*, **6A:**553. (19.2)
—, —, C. Omero, A. Rimini, and T. Weber (1978): *Phys. Rev.*, **D18:**4757. (19.2)
—, — and A. Rimini (1973): *Nuovo Cim.*, **18B:**1. (19.2)
—, —, —, and T. Weber (1973): *Nuovo Cim.*, **15A:**689 and **18A:**805. (19.2)
—, —, T. Weber, and A. Rimini (1966): *J. Math. Phys.* **7:**1643. (7.3)
— and R. G. Newton (1959a); *Nuovo Cimento*, **14:**1027. (17.3.2)
— and — (1959b): *Ann. Phys.* (*N.Y.*), **7:**133. (17.2)
— and — (1960a): *Ann. Phys.* (*N.Y.*), **9:**416. (17.2.4)
— and — (1960b): *Phys. Rev.*, **119:**1394. (17.2, 17.4.3)
— and — (1960c): *Ann. Phys.* (*N.Y.*), **10:**490. (16.5, 17.3, 19.3)
—, L. A. Radicati, and T. Regge (1961): *Ann. Phys.* (*N.Y.*), **12:**68. (13.4, 17.1, 17.3)
Fong, R., and J. Sucher (1964): *J. Math. Phys.*, **5:**456. (6)
Fonseca, A. C., E. F. Redish, and P. E. Shanley (1979): *Nucl. Phys.*, **A320:**273. (17.4.8)
— and P. E. Shanley (1976): *Phys. Rev.*, **D13:**2255. (17.4)
— and — (1979): *Ann. Phys.* (*N.Y.*), **117:**268. (17.4.8)
— and — (1981): preprint. (17.4)
Ford, K. W., D. L. Hill, M. Wakano, and J. A. Wheeler (1959): *Ann. Phys.* (*N.Y.*), **7:**239. (18.2.5)
— and J. A. Wheeler (1959a): *Ann. Phys.* (*N.Y.*), **7:**259. (5.4, 18.2.5)
— and J. A. Wheeler (1959b): *Ann. Phys.* (*N.Y.*), **7:**287. (18.2)
Ford, W. F. (1964): *Phys. Rev.*, **133:**B1616. (14.6)
— (1966): *J. Math. Phys.*, **7:**626. (14.6)
Fortes, M., and G. Garcia–Calderon (1980): *Lett. Nuovo Cim.*, **28:**390. (16.6)
Fradkin, D. M., and F. Calogero (1966): *Nuclear Phys.*, **75:**475. (12.2)
—, C. L. Hammer, and T. A. Weber (1964): *J. Math. Phys.*, **5:**1645. (15.1)
—, T. A. Weber, and C. L. Hammer (1964): *Ann. Phys.* (*N.Y.*), **24:**338. (14.6)
Franchetti, S. (1957): *Nuovo Cimento*, **6:**601. (12.1.2)
Franco, V. (1973): *Phys. Rev.*, **D7:**215. (14.6)

Frank, W. M. (1967): *J. Math. Phys.*, **8:**466. (12)
— (1968): *J. Math. Phys.*, **9:**1890 and **10:**2128. (12)
— (1971): *J. Math. Phys.*, **12:**2348 and 2529. (12.1.3)
— and D. J. Land (1970): *J. Math. Phys.*, **11:**2041 and 2058. (12.4)
— and D. W. McLaughlin (1971): *J. Math. Phys.*, **12:**2525. (12.1.3)
Franz, W. (1954): *Z. Naturforsch*, **9a:**705. (3.8)
Frautschi, S. (1963): "Regge Poles," Benjamin, New York. (13.1)
Freeman, D. F. and J. Nuttall (1973): *J. Math. Phys.*, **14:**1883. (12.4)
Friar, J. L., E. L. Tomusiak, B. F. Gibson, and G. L. Payne (1981): *Phys. Rev.*, **C24:**677. (17.4.6)
Fried, H. M., and Y. Hahn (1974): *Phys. Rev.*, **D10:**1908. (18.3)
Friedman, B. (1957a) "Principles and Techniques of Applied Mathematics," Wiley, New York. (7.3)
— (1957b): *Mich. Math. J.*, **4:**137. (20.3)
Friedrich, H. (1974): *Nucl. Phys.*, **A224:**537. (16.3.3)
Friedrichs, K. O. (1948): *Communs. Pure Appl. Math.*, **1:**361. (7.3)
— (1952): *Nachr. Ges. Wiss. Göttingen, Math.-phys. Kl.*, **43**. (6.7)
— (1960): "Spectral Theory of Operators in Hilbert Space," Institute of Mathematical Science, New York University. (7.3)
Fröberg, C. E. (1947): *Phys. Rev.*, **72:**519. (20.3)
— (1948): *Arkiv Mat. Astron. Fys.*, **34A** (28). (20.3)
— (1949): *Arkiv Mat. Astron. Fys.*, **36A** (11). (20.3)
— (1951): *Arkiv Fys.*, **3**(1). (20.3)
Froissart, M. (1961): *Nuovo Cimento*, **22:**191. (17.1)
Fröman, N., and P. O. Fröman (1965): "JWKB Approximation," North-Holland, Amsterdam. (18.2)
— and K.-E. Thylwe (1979): *J. Math. Phys.*, **20:**1716. (12.4)
— and S. Yngve (1980): *Phys. Rev.*, **D22:**1375. (18.2)
Frye, G., and R. Warnock (1963): *Phys. Rev.*, **130:**478. (17.1)
Fubini, S. (1952a): *Nuovo Cimento*, **9:**846. (6.3)
— (1952b): *Atti Accad. Naz. Lincei*, **12:**298. (6.4)
— and R. Stroffolini (1965): *Nuovo Cimento*, **37:**1812. (12.4)
Fuda, M. G. (1968): *Phys. Rev.*, **166:**1064. (17.4)
— (1969): *Phys. Rev.*, **186:**1078. (17.4)
— (1970): *Phys. Rev.*, **C1:**1910. (12.1.2)
— (1971a): *J. Math. Phys.*, **12:**1163. (14.3)
— (1971b): *J. Math. Phys.*, **12:**2076. (14.4)
— (1971c): *Phys. Rev.*, **C3:**55. (7.2.2)
— (1972): *Phys. Rev.*, **C5:**275. (15.2)
— (1973): *Phys. Rev.*, **C7:**1365. (17.3)
— (1974): *Phys. Rev. Lett.*, **32:**620. (17.4)
— (1981): *Phys. Rev.*, **C24:**614. (20.1)
— and J. S. Whiting (1973): *Phys. Rev.*, **C8:**1255. (12.1.2)
Fujimoto, Y., and Y. Yamaguchi (1951): *Progr. Theor. Phys.*, **6:**166. (18.4)
Fujiwara, I. (1952): *Progr. Theor. Phys.*, **7:**433. (6.5)
Fukuda, N., and R. G. Newton (1956): *Phys. Rev.*, **103:**1558. (8.1.3)
Fulco, J. R., G. L. Shaw, and D. Y. Wong (1965): *Phys. Rev.*, **137:**B1242. (16.5)
Fuller, R. C. (1969): *Phys. Rev.*, **188:**1649. (10.1.4)
Fulton, M. J., and M. H. Mittleman (1965): *Ann. Phys. (N.Y.)*, **33:**65. (16)

Fulton, T., and R. G. Newton (1956): *Nuovo Cimento*, **3:**677. (14.6, 15.2)
Furry, W. H. (1947): *Phys. Rev.*, **71:**360. (18.2)
Gale, W. A. (1966): *J. Math. Phys.*, **7:**2171. (12.4)
— (1967): *J. Math. Phys.*, **8:**1050. (12.4)
Galindo Tixaire, A. (1959): *Helv. Phys. Acta*, **32:**412. (6.7, 16.2)
Gallina, V., P. Nata, L. Bianchi, and G. Viano (1962): *Nuovo Cim.*, **24:**835. (17.4)
Gammel, J. L. and F. A. McDonald (1966): *Phys. Rev.*, **142:**1245. (9)
Gamow, G. (1928): *Z. Phys.*, **52:**510. (14.6)
Ganas, P. S. (1972): *Phys. Rev.*, **A6:**1493. (12.2)
Gans, R. (1925): *Ann. Phys.*, **76:**29. (3.1.2)
Garbe, E. (1915): *Math. Ann.* **76:**527. (9.1.1, 10.3.1)
Garcia–Calderon, G. (1977): *Nucl. Phys.*, **A284:**209. (16.6)
— and R. Peierls (1976): *Nucl. Phys.*, **A265:**443. (16.6)
Garcilazo, H., and R. R. Wilde (1980): *Lett. Nuovo Cim.*, **29:**335. (20.1)
Gasiorowicz, S., and H. P. Noyes (1958): *Nuovo Cimento*, **10:**78. (10.3.4)
Gasymov, M. G. (1965): *Dokl. Akad. Nauk SSSR*, **165:**261 (transl. *Soviet Math. Dokl.*, **11:**676). (20.3)
— (1966): *Dokl. Akad. Nauk SSSR*, **169:**1037 (transl. *Soviet Phys. Dokl.*, **11:**676). (20.3)
— (1968): *Trudy Moscow Mat. Obs.*, **19:**41 (transl. *Trans. Moscow Math. Soc.*, **19:**41). (20.3)
— and B. M. Levitan (1966): *Dokl. Akad. Nauk SSSR*, **167:**967 and 1219 (transl. *Soviet Math. Dokl.*, **7:**495 and 543).
Gel'fand, I. M., and B. M. Levitan (1951a): *Dokl. Akad. Nauk SSSR*, **77:**557 (*Math. Rev.*, **13:**240). (20.2)
— (1951b): *Isvest. Akad. Nauk SSSR*, **15:**309 (*Am. Math. Soc. Transl.*, **1:**253). (20.2)
Gell-Mann, M., and M. L. Goldberger (1953): *Phys. Rev.*, **91:**398 [reprinted in M. H. Ross (1963a)]. (6, 8.1)
Gelman, D. (1974): *Phys. Rev.*, **D9:**931. (11.3.4)
Gerber, R. B. (1972): *Phys. Rev.*, **A5:**2151. (20.2)
—, V. Buch, and U. Buck (1980): *J. Chem. Phys.*, **72:**3596. (20.2)
— and M. Karplus (1970): *Phys. Rev.*, **D1:**998. (20.2)
— and N. C. Rosenbach (1974): *Phys. Rev.*, **A9:**301. (17.3)
—, M. Sapiro, U. Buck, and J. Schleusener (1978): *Phys. Rev. Lett.*, **41:**236. (20.1)
— and A. T. Yinnon (1980): *J. Chem. Phys.*, **73:**3232. (20)
Gerjuoy, E. (1958a): *Ann. Phys.* (*N.Y.*), **5:**58. (7, 16.3)
— (1958b): *Phys. Rev.*, **109:**1806. (7, 16.3)
— (1965a): *Ann. Phys.* (*N.Y.*), **32:**1. (4.2.1)
— (1965b): *Phys. Today*, **18**(5)**:**24. (16)
— (1965c): *J. Math. Phys.*, **6:**993. (6)
— (1970): *J. Phys. B*, **3:**L92. (17.4.5)
— (1971): *Phil. Trans. Roy. Soc. London*, **A270:**197. (17.4.5)
Gersten A. (1969): *Nucl. Phys.*, **B12:**537. (20.2)
Gesztesy, F., and C. B. Lang (1981): *J. Math. Phys.*, **22:**312. (14.6)
—, W. Plessas, and B. Thaller (1980): *J. Phys. A*, **13:**2659. (14.6)
— and B. Thaller (1981): *J. Phys. A*, **14:**639. (10.3.1, 14.6.2)
Ghirardi, G. C., M. Pauri, and A. Rimini (1963): *Ann. Phys.* (*N.Y.*), **21:**401. (12.1.3)
— and A. Rimini (1965a): *J. Math. Phys.*, **6:**40. (12.1.4)
— and — (1965b): *Nuovo Cimento*, **37:**450. (17.4.3)
Giambiagi, J. J., and T. W. B. Kibble (1959): *Ann. Phys.* (*N.Y.*), **7:**39. (12.1.2)

Gibbs, W. R. (1974): *Phys. Rev.*, **C10:**2166. (7.2.5)
Gibson, A. G., and C. Chandler (1974): *J. Math. Phys.*, **15:**1366. (17.4)
Giffon, M., and E. Predazzi (1964): *Nuovo Cimento*, **33:**1374. (12.4)
Gignoux, C., A. Laverne, and S. P. Merkuriev (1974): *Phys. Rev. Lett.*, **33:**1350. (17.4.5)
Gilbert, R. P., and S. Y. Shieh (1966): *J. Math. Phys.*, **7:**431. (13.1)
Ginibre, J. (1977): in W. Thirring and P. Urban, eds., *The Schrödinger Equation*, Springer Verlag, Vienna, p. 95. (17.4)
Glaser, V., A. Martin, H. Grosse, and W. Thirring (1976): in E. H. Lieb, B. Simon, and A. S. Wightman, eds., *Studies in Mathematical Physics*, Princeton Univ. Press, p. 169. (12.1.4)
Glassgold, A. E., and S. A. Lebedeff (1964): *Ann. Phys. (N.Y.)*, **28:**181. (16)
Glauber, R. J. (1959): In W. E. Brittin and L. G. Dunham (eds.), "Lectures in Theoretical Physics," vol. 1, p. 315, Interscience, New York. (18.3)
— (1963): *Phys. Rev. Letters*, **10:**84; *Phys. Rev.*, **130:**2529, **131:**2766. (4.3, 8.3.1)
— (1965): In C. de Witt et. al. (eds.), "Quantum Optics and Electronics," (Les Houches Lectures, 1964), Gordon and Breach, New York. (4.3, 8.3.1)
— (1966): In P. Kelley et al. (eds.), "Proc. of the Physics of Quantum Electronics Conference, San Juan, Puerto Rico, 1965," McGraw-Hill, New York. (4.3, 8.3.1)
— (1967): in *High Energy Physics and Nuclear Structure*, North-Holland Publ., Amsterdam, p. 311. (18.3)
Glöckle, W. (1966): *Z. Phys.*, **190:**391. (17.1)
— (1970): *Nucl. Phys.*, **A141:**620 and **A158:**257. (17)
— (1971): *Nucl. Phys.*, **A175:**337. (17.4)
— (1974): *Z. Physik*, **271:**31. (17.4.5)
— and J. LeTourneux (1976): *Nucl. Phys.*, **A269:**16. (12.1.3)
Gluckstern, R. L., and S. R. Lin (1964): *J. Math. Phys.*, **5:**1594. (14.6)
Goebel, C. J., and K. W. McVoy (1967): *Phys. Rev.*, **164:**1932. (16.6)
Gohberg, I. C., and M. G. Krein (1958): *Usp. Mat. Nauk (N.S.)*, **13:**(2):80 (*Math. Rev.*, **21:**286). (20.3)
Goldberg, H. (1970): *Phys. Rev.*, **D1:**1242. (20.2)
Goldberger, M. L. (1951a): *Phys. Rev.*, **82:**757. (7)
— (1951b): *Phys. Rev.*, **84:**929. (7)
—, H. W. Lewis, and K. M. Watson (1963): *Phys. Rev.*, **132:**2764. (4.3)
— and K. M. Watson (1962): *Phys. Rev.*, **127:**2284. (19.2)
— and — (1964a): *Phys. Rev.*, **136:**B1472. (19.3)
— and — (1964b): "Collision Theory," Wiley, New York. (17.1, 18.4, 19.2)
— and — (1964c), *Phys. Rev.*, **134:**B919. (4.3)
— and — (1965), *Phys. Rev.*, **137:**B1396, **140:**B500. (4.3)
Goldflam, R., and K. L. Kowalski (1980a): *Phys. Rev.*, **C21:**483 and **C22:**2341; *Phys. Rev. Lett.*, **44:**1044. (17.4)
— and — (1980b): *Phys. Rev.*, **C22:**949. (16.3.3)
— and W. Tobocman (1978): *Phys. Rev.*, **C17:**1914. (16.3.3)
— and — (1979): *Phys. Rev.*, **C20:**904. (17.4)
—, R. Thaler, and W. Tobocman (1981): *Nucl. Phys.* **A359:**122. (11.3, 16.4)
Gol'dman, I. I., and A. B. Migdal (1954): *Z. Eksp. i Teor. Fiz.*, **28:**394 (transl. *Soviet Phys. JETP*, **1:**304). (18.2, 18.3)
Goldstein, H. (1980): "Classical Mechanics," 2nd ed., Addison-Wesley, Reading, Mass. (5.1)
Goodmanson, D., and J. R. Taylor (1973): *Nucl. Phys.*, **A210:**211. (12.2)

— and — (1980): *J. Math. Phys.*, **21:**2202. (14.6)
Goodrich, R. F., B. A. Harrison, R. E. Kleinman, and T. B. A. Senior (1961): *Univ. Mich. Coll. Eng. Rept.* 3648-1-T. (2.4)
— and N. D. Kazarinoff (1963): *Mich. Math. J.*, **10:**105; *Proc. Cambridge Phil. Soc.*, **59:**167. (3.8)
Gordon, W. (1928): *Z. Phys.*, **48:**180. (14.6)
Gorshkov, V. G. (1961): *Zh. Eksp. Teor. Fiz.*, **40:**1481 (transl. *Soviet Phys. JETP* **13:**1037). (14.6)
— (1964): *Z. Eksp. i Teor. Fiz.*, **47:**352 and 1984 (transl. *Soviet Phys. JETP*, **20:**234 and 1331). (14.6)
Gourdin, M., and A. Martin (1957): *Nuovo Cim.*, **6:**757; *C. R. Acad. Sci. Paris*, **244:**1329. (10.1.4, 20.1)
— and — (1958): *Nuovo Cim.*, **8:**699. (20.1)
Grassberger, P., and W. Sandhas (1967): *Nucl. Phys.*, **B2:**181. (17)
— and — (1968): *Z. Physik*, **217:**9. (16.3)
— and — (1969): *Z. Physik*, **220:**29. (17.4)
Graves-Morris, P. R., and J. F. Rennison (1974): *J. Math. Phys.*, **15:**230. (9.1.1)
Grawert, G., and J. Petzold (1960): *Z. Naturforsch.*, **15a:**311. (16.2)
Greben, J. M., and F. S. Levin (1979): *Nucl. Phys.*, **A325:**145.
Green, T. A., and O. E. Lanford (1960): *J. Math. Phys.*, **1:**139. (6.7)
Greider, K. R., and L. R. Dodd (1966): *Phys. Rev.*, **146:**671. (17.4)
— and A. E. Glassgold (1960): *Ann. Phys. (N.Y.)*, **10:**100. (18.2)
Gribov, V. N. (1962): *Z. Eksp. i Teor. Fiz.*, **42:**1260 (transl. *Soviet Phys. JETP*, **15:**873). (13.3)
Griffin, J., and J. A. Wheeler (1957): *Phys. Rev.*, **108:**311. (16.3.3)
Grinstein, F. F. (1980): *J. Math. Phys.*, **21:**112. (11.1)
Grinyuk, B. E. (1980): *Teor. i Mat. Fiz.*, **43:**386 (transl. *Theor. Math. Phys.*, **43:**532). (17.4)
Grosse, H. (1980): *Acta Phys. Austr.* **52:**89. (12.1.4)
Grossmann, A. (1961): *J. Math. Phys.*, **2:**714. (10.3.2, 10.3.4)
— and T. T. Wu (1961): *J. Math. Phys.*, **2:**710. (10.3.2, 10.3.4)
— and — (1962): *J. Math. Phys.*, **3:**684. (10.3.2, 10.3.4)
Grubb, A. M., and D. B. Pearson (1970): *J. Math. Phys.*, **11:**2415. (10.1.4)
Gryzinski, M. (1965a): *Phys. Rev.*, **138:**A305 and A322. (5.7)
— (1965b): *Phys. Rev.*, **138:**A336. (5)
Guérin, H. G., T. P. Tsien, B. C. Eu, and R. E. Olsen (1974): *Phys. Rev.*, **A9:**995. (18.2)
Gugushvili, E. I., and Yu. L. Mentkovsky (1972): *Nuovo Cim.*, **10A:**277 and 292. (20.3)
Guier, W. H., and R. W. Hart (1957): *Phys. Rev.*, **106:**296. (17.2)
Guillope, L. (1981): These de doctorat, Grenoble. (12.1.3)
Gurney, R. W., and E. U. Condon (1928): *Nature*, **122:**439. (14.6)
Gutkowski, D., and A. Scalia (1968): *J. Math. Phys.*, **9:**588. (10.1.4)
— and — (1969): *J. Math. Phys.*, **10:**2306. (10.1.4)
Haag, R. (1961): In W. E. Brittin et al. (eds.), "Lectures in Theoretical Physics," pp. 326–352, Interscience, New York. (6, 7)
Haberzettl, H., and W. Sandhas (1981): *Phys. Rev.*, **C24:**659. (17.4)
Hack, M. N. (1954): *Phys. Rev.*, **96:**196. (7)
— (1958): *Nuovo Cimento*, **9:**731. (6.3)

— (1967): *Nuovo Cim.*, **47:**301. (19.2)
Hagedorn, G. A. (1981): preprint . (17.4)
Hagedorn, R. (1963): "Relativistic Kinematics," Benjamin, New York. (8.2)
Hahn, Y. (1965): *Phys. Rev.*, **139:**B212. (9.1.1, 11.3)
— (1966): *Phys. Rev.*, **142:**603. (16.6)
— (1967a): *Phys. Rev.*, **154:**981. (16.6)
— (1967b): *Phys. Rev.*, **159:**91. (16.6)
— (1968a): *Phys. Rev.*, **169:**794. (16.3)
— (1968b): *Phys. Rev.*, **174:**1135. (11.3, 18.3)
— (1968c): *Phys. Rev.*, **175:**197. (18)
— (1969a): *Phys. Rev.*, **184:**1022 and **187:**1365. (18.3)
— (1969b): *Nucl. Phys.*, **A132:**353. (16.3)
— (1970a): *Phys. Rev.*, **C1:**12, **A2:**1861, **A3:**310. (16.4)
— (1970b): *Phys. Rev.*, **C2:**775. (18.3)
— (1971a): *Phys. Rev.*, **A4:**1881. (16.4)
— (1971b): *Phys. Rev.*, **C3:**1432. (16.3)
— (1972a): *Phys. Rev.*, **C5:**1788. (11.3, 16.4, 16.6)
— (1972b): *Phys. Rev.*, **A5:**1607. (16.4, 16.6)
— (1972c): *Phys. Rev.*, **A5:**2452. (16.6)
— (1972d): *Phys. Rev.*, **A5:**309. (17.4)
— (1972e): *Phys. Rev.*, **A6:**548. (18.2)
— (1973): *Phys. Rev.*, **C8:**1672. (18.3)
— (1974): *Phys. Rev.*, **A9:**2014 and 2024. (16.3)
— (1975): *Phys. Rev.*, **A12:**2402. (16.4)
— (1976a): *Phys. Rev.*, **C14:**446. (16.3, 17.4)
— (1976b): *Phys. Rev.*, **C14:**1709. (16.4)
— (1977): *Phys. Rev.*, **C16:**2449. (17.4)
—, J. F. Dirks, and A. C. Chen (1975): *Phys. Rev.*, **A12:**816. (16)
—, D. J. Kouri, and F. S. Levin (1974a): *Phys. Rev.*, **C10:**1615. (17.4)
—, —, and — (1974b): *Phys. Rev.*, **C10:**1620. (17.4.4)
— and R. Luddy (1981): *Phys. Rev.*, **C24:**1. (9.2)
—, T. F. O'Malley, and L. Spruch (1962): *Phys. Rev.*, **128:**932. (11.3)
—, —, and — (1963): *Phys. Rev.*, **130:**381. (11.3)
—, —, and — (1964): *Phys. Rev.*, **134:**B397 and B911. (11.3)
— and J. A. Retter (1979): *Phys. Rev.*, **C19:**1174. (17.4)
— and D. W. Rule (1975): *Phys. Rev.*, **C12:**1607. and 1616. (16.4)
— and L. Spruch (1967): *Phys. Rev.*, **153:**1159. (11.3)
— and — (1974): *Phys. Rev.*, **A9:**229. (11.3.4)
— and K. M. Watson (1972): *Phys. Rev.*, **A5:**1718. (17.4)
Halliday, I. G. (1966): *Nuovo Cimento*, **42B:**38. (17.2, 17.4)
— (1967a): *Nuovo Cim.*, **43A:**982. (17.2)
— (1967b): *Nuovo Cim.*, **47A:**57. (16.6)
Halmos, P. R. (1942): "Finite Dimensional Vector Spaces," Princeton, Princeton, N. J. (7.3)
Halpern, A. M., and V. Franco (1981): *Phys. Rev. Lett.*, **46:**714. (18.3)
Hamza, K. A. A., and S. Edwards (1969): *Phys. Rev.*, **181:**1494. (17.4)
Hannover, H. (1976): *Z. Physik*, **A276:**127. (17.4)
Harrington, D. R. (1969): *Phys. Rev.*, **184:**1745. (18.3)
— (1970): *Phys. Rev.*, **D2:**2512. (18.3)

— (1971): *Phys. Rev.*, **D4:**840. (18.3)
Harris, F. E. (1967): *Phys. Rev. Lett.*, **19:**173. (11.3.3)
Hartle, J. B., and R. L. Sugar (1968): *Phys. Rev.*, **169:**1104. (17.4.4)
Hayborn, R. M. (1971): *Phys. Rev.*, **C4:**338. (18.4)
Hecht, C. E., and J. E. Mayer (1957): *Phys. Rev.*, **106:**1156. (18.2)
Heinzi, R. (1966): *Nuovo Cim.*, **46:**370. (18.3)
Heisenberg, W., (1943): *Z. Phys.*, **120:**513 and 673. (2.2.2, 6.4)
— (1944): *Z. Phys.*, **123:**93. (6.4)
— (1946): *Z. Naturforsch.*, **1:**608. (12.1.3)
Heiss, P., and H. H. Hackenbroich (1970): *Z. Physik*, **235:**422. (16.3.3)
— and — (1973): *Nucl. Phys.*, **A202:**353. (16.3.3)
Heitler, W. (1941): *Proc. Cambridge Phil. Soc.*, **37:**291. (6.4)
— (1954): "The Quantum Theory of Radiation," 3d ed., Oxford, Fair Lawn, N.J. (7)
— and S. T. Ma (1949): *Proc. Roy. Irish Acad.*, **52:**109. (19.2)
Heller, E. J., and W. P. Reinhardt (1972): *Phys. Rev.*, **A5:**757. (12.1.2)
—, T. N. Rescigno, and W. P. Reinhardt (1973): *Phys. Rev.*, **A8:**2946. (12.1.2)
Hellmann, O. (1960): *Kgl. Danske Videnskab. Selskab, Mat.-fys. Medd.*, **32**(4) (12)
Hendrickson, J. H. (1976): *J. Math. Phys.*, **17:**729. (16.5)
Heniger, A., and J.-J. Loeffel (1973): *Helv. Phys. Acta*, **46:**462. (20.4)
Hepp, K. (1969): *Helv. Phys. Acta*, **42:**425. (17.4)
Herbst, I. (1974a): *Commun. Math. Phys.*, **35:**193. (5.3)
— (1974b): *Commun. Math. Phys.*, **35:**181. (14.6)
— (1979): *Commun. Math. Phys.*, **64:**279. (12.1.1)
— (1980): *Commun. Math. Phys.*, **75:**179. (19.2)
Herglotz, A. (1911): *Ber. Verhandl. K. Sächs. Ges. Wiss., Leipzig, Math.-phys. Kl.*, **63:**501. (9.1.1)
Hertel, I. V., and W. Stoll (1974): *J. Phys. B*, **7:**570. (16)
Herzenberg, A., K. L. Kwok, and F. Mandl (1964): *Proc. Phys. Soc. London*, **84:**345 and 477. (16.6)
— and F. Mandl (1963): *Proc. Phys. Soc. London*, **A274:**253; *Phys. Lett.*, **6:**288. (16.6)
Hetherington, J. H. (1963): *J. Math. Phys.*, **4:**357. (14.6)
— and L. H. Schick (1965): *Phys. Rev.*,**137:**B935. (17.4.3)
Higgins, J. R. (1963) "Kinematic Analysis of Production Reactions," Ph.D. dissertation, Indiana University (unpublished). (17.4)
Hilb, E. (1915): In "Encyklopädie der Mathematischen Wissenschaften," vol. 2, pt. 2, p. 501, B. G. Teubner, Leipzig. (12.1)
Hilbert, D. (1912): "Grundzüge einer allgemeinen Theorie der linearen Integralgleichungen," B. G. Teubner, Leipzig. (9.1.1)
Hilgevoord, J. (1960): "Dispersion Relations and Causal Description," North-Holland, Amsterdam. (4.2.1, 10.3.4)
Hill, D. L., and J. A. Wheeler (1953): *Phys. Rev.*, **89:**1102. (16.3.3)
Hille, E., and R. S. Phillips (1957): "Functional Analysis and Semi-groups," American Mathematical Society, New York. (7.3)
— and J. Tamarkin (1930): *Ann. Math.*, **31:**479. (9.3)
— and — (1931): *Acta Math.*, **57:**1. (9.3)
— and — (1934): *Ann. Math.*, **35:**445. (9.3)
Hillery, M. (1981): *Phys. Rev.*, **A24:**933. (19.2)
Hillion, P. (1975): *J. Math. Phys.*, **16:**1920. (14.6)
Hinkelmann, O., and L. Spruch (1971): *Phys. Rev.*, **A3:**642. (11.2.2)
Hirooka, M., and S. Sunakawa (1974): *Progr. Theor. Phys.*, **52:**131. (17.4)

Hirschfelder, J. O. (1979): *Phys. Rev.*, **A19:**2463. (11.1.3)
—., C. F. Curtiss, and R. B. Bird (1954): "Molecular Theory of Gases and Liquids," Wiley, New York. (5.4)
Hizal, A. (1980): in Varadan and Varadan (1980), p. 169. (4.1)
— and A. Marincic (1979): *Proc. IEE*, **117:**1639. (1.3)
— and Z. Yasa (1973): *Proc. IEE*, **120:**181. (1.3)
Hodgson, R. J. W. (1978): *Phys. Rev.*, **C18:**1133. (20.3)
Hoenders, B. J. (1979): *J. Math. Phys.*, **20:**329. (16.6)
Höhler, G. (1958): *Z. Phys.*, **152:**542. (19.2)
Holdeman, J. T., and R. M. Thaler (1965): *Phys. Rev.*, **139:**B1186. (14.6)
Holmberg, B. (1952): *Nuovo Cimento*, **9:**597. (20.3)
Holt, A. R. (1980): in Varadan and Varadan (1980), p. 255. (4.1)
— and B. Santoso (1972): *J. Phys. B*, **5:**497. (12.1.2)
— and — (1973): *J. Phys. B*, **6:**2010. (10.3.2)
— and J. W. Shepherd (1979): *J. Phys.*, **A12:**159. (2)
—, N. K. Uzunoglu, and B. G. Evans (1978): *IEEE Trans. Antennas and Prop.*, **AP-26:** 706. (2)
Holzwarth, N. A. W. (1973): *J. Math. Phys.*, **14:**191. (12.4)
Hooshyar, M. A. (1971): *J. Math. Phys.*, **12:**2243. (20.4)
— (1972): *J. Math. Phys.*, **13:**1931. (20.4)
— (1975): *J. Math. Phys.*, **16:**257. (20.4)
— (1978): *J. Math. Phys.*, **16:**257. (20.4)
— (1980): *J. Math. Phys.*, **21:**1695. (20.4)
Horn, D., and D. E. Novoseller (1978): *Phys. Rev.*, **D17:**1763. (16.6)
Horwitz, L. P., J. A. LaVita, and J.–P. Marchand (1971): *J. Math. Phys.*, **12:**2537. (19.2)
Hostler, L. (1964): *J. Math. Phys.*, **5:**591 and 1235. (14.6)
— and R. H. Pratt (1963): *Phys. Rev. Letters*, **10:**469. (14.6)
Hoyt, F. C. (1939): *Phys. Rev.*, **55:**664. (5.9, 20.1)
Hu, N. (1948): *Phys. Rev.*, **74:**131. (12.1.4)
Huang, S. (1949): *Phys. Rev.*, **75:**980, **76:**477, 866, and 1878. (11.3)
Huby, R. (1963): *Nuclear Phys.*, **45:**473. (10.3.1)
— and J. R. Mines (1964): *Nuclear Phys.*, **54:**28. (10.3.1)
Huestis, D. L. (1975): *J. Math. Phys.*, **16:**2148. (16.6)
Hull, M. H., and G. Breit (1959): In S. Flügge (ed.), "Handbuch der Physik," vol. 41/1, pp. 408–465, Springer, Berlin. (14.6)
Hulthén, L. (1942): *Arkiv Mat. Astron. Fys.*, **28A**(5), **29B**(1). (14.4)
— (1944): *Kgl. Fysiograf. Sällskap Lund Förh.*, **14**(21) (*Math. Rev.*, **6:**111). (11.3)
— (1947): *Comptes Rend. Dixième Congr.-Mat. Scandinaves* 1946, Copenhagen, p. 201. (11.3)
— (1948): *Arkiv Mat. Astron. Fys.*, **35A**(25). (11.3)
— and P. O. Olsson (1950): *Phys. Rev.*, **79:**532. (11.3)
Humblet, J. (1952): *Mém. Soc. Roy. Sci. Liège*, **4:**12. (12.1.4)
— (1962): *Nuclear Phys.*, **31:**544. (16.5)
— (1964): *Nuclear Phys.*, **50:**1, **57:**386. (16.5)
— and L. Rosenfeld (1961): *Nuclear Phys.*, **26:**529. (16.5)
Hunziker, W. (1961): *Helv. Phys. Acta*, **34:**593. (9.1.1, 10.3.4)
— (1963): *Helv. Phys. Acta*, **36:**838. (18.3)
— (1964): *Phys. Rev.*, **135:**B800. (7.2.5, 17.4.3)
— (1965): *J. Math. Phys.*, **6:**6. (17.4)

— (1966): *Helv. Phys. Acta*, **39:**451. (16.1)
— (1967): *Helv. Phys. Acta*, **40:**1052. (12.4)
— (1968): in A. O. Barut and W. E. Brittin, eds., *Lectures in Theoretical Physics*, Gordon and Breach, N.Y., vol. X–A, p. 1. (17.4)
— (1974): in LaVita and Marchand (1974), p. 79. (5)
— (1977): in W. Thirring and P. Urban, eds., *The Schrödinger Equation*, Springer Verlag, Vienna, p. 43. (17.4)
Hüwel, L., J. Maier, and H. Pauly (1981): *J. Chem. Phys.*, **74:**5613. (20.1)
Hylleraas, E. A. (1948): *Phys. Rev.*, **74:**48. (20.3)
— (1964): *Nuclear Phys.*, **57:**208. (20.3)
— and B. Undheim (1930): *Z. Phys.*, **65:**759. (11.3)
Ida, M. (1959): *Progr. Theor. Phys.*, **21:**625. (12.1.3)
— (1962): *Progr. Theor. Phys.*, **28:**943 and 945. (18.3)
Ikebe, T. (1960): *Arch. Rat. Mech. Anal.*, **5:**1. (7.3)
— (1965): *Pac. J. Math.*, **15:**511. (10.3.3)
Immirzi, G. (1965): *Nuovo Cimento*, **34:**1361. (17.4)
Ince, E. L. (1927): "Ordinary Differential Equations," Longmans, New York. (15.2)
Inglis, D. R. (1962): *Nuclear Phys.*, **30:**1. (17.2)
Integrals of Airy Functions (1958), *Nat. Bur. Standards Appl. Math. Ser.* **52**. (3.6)
Ishimaru, A. (1978): Wave Propagation and Scattering in Random Media, Academic Press, N.Y. (1.5)
Islam, M. M. (1965): *Phys. Rev.*, **138:**B1222. (17.4.4)
— (1967): in A. O. Barut and W. E. Brittin, eds., *Lectures in Theoretical Physics*, vol. XB, Gordon and Breach Publ., N.Y., p. 97. (18.3)
— (1976): *Nucl. Phys.*, **B104:**511. (18.3)
Itzykson, C., and A. Martin (1973): *Nuovo Cim.*, **17A:**245. (20.2)
Ivanov, G. K., and Yu. S. Sayasov (1963): *Z. Eksp. i Teor. Fiz.*, **45:**1456 (transl. *Soviet Phys. JETP*, **18:**1006). (18.4)
Ivash, E. V. (1958): *Phys. Rev.*, **112:**155. (16)
Iwamoto, F. (1963): *J. Math. Phys.*, **4:**809. (8.1.3)
Jabbur, R. J. (1965): *Phys. Rev.*, **138:**B1525. (12.4)
Jackson, J. D. (1962): "Classical Electrodynamics," Wiley, New York. (2.1.3, 2.1.5, 4.1.1, 4.1.2)
— and J. M. Blatt (1950): *Revs. Modern Phys.*, **22:**77. (11.2.2)
— and H. W. Wyld (1959): *Nuovo Cimento*, **13:**85. (17.2)
Jackson, J. L. (1951): *Phys. Rev.*, **83:**301. (11.3)
Jacob, M., and G. C. Wick (1959): *Ann. Phys. (N.Y.)*, **7:**404. (15.1)
Jacob, R., and R. G. Sachs (1961): *Phys. Rev.*, **121:**350. (19.2)
Jain, K., and C. S. Shastry (1975): *Phys. Rev.*, **A12:**2237. (10.1.4)
Jakšić, B., and N. Limić (1966): *J. Math. Phys.*, **7:**88. (12.4)
Jameel, M. (1964): *Nuclear Phys.*, **56:**353. (10.3.4)
Jasperse, J. R. (1968): *J. Math. Phys.*, **9:**1931. (17.4)
Jauch, J. M. (1957): *Helv. Phys. Acta*, **30:**143. (12.1.3)
— (1958a): *Helv. Phys. Acta*, **31:**127. (6.7)
— (1958b): *Helv. Phys. Acta*, **31:**661. (16.2)
—, R. Lavine, and R. G. Newton (1972): *Helv. Phys. Acta*, **45:**325. (6.7, 16.5)
— and J.–M. March, and (1966): *Helv. Phys. Acta*, **39:**325. (16.2)
— and — (1967): *Helv. Phys. Acta*, **40:**217. (11.1.3)
— and F. Rohrlich (1955): "The Theory of Photons and Electrons," Addison-Wesley Reading, Mass. (6, 7, 8.1)

—, K. B. Sinha, and B. N. Misra (1972): *Helv. Phys. Acta*, **45:**398. (11.1.3)
— and I. Zinnes (1959): *Nuovo Cimento*, **11:**553. (6)
Jauho, P. (1950): *Ann. Acad. Sci. Fennicae*, ser. A, no. 80. (20.3)
Jaulent, M. (1972): *Ann. Inst. H. Poincaré*, **17:**363. (20.3)
— (1975): *C. R. Acad. Sci. Paris*, **280:**1467. (20.3)
— (1976): *J. Math. Phys.*, **17:**1351. (20.3)
— and C. Jean (1972): *Commun. Math. Phys.*, **28:**177. (20.3)
— and — (1975): *Ann. Inst. H. Poincaré*, **25:**105 and 119. (20.3)
— and — (1982): *J. Math. Phys.*, **23:**258. (20.3)
Jean, C., and P. C. Sabatier (1973): *Nuovo Cim.*, **18A:**105. (20.4)
Jeffreys, H. (1923): *Proc. London Math. Soc.*, **23**(2):428. (18.2)
Jensen, A. (1977): *J. Math. Anal. & Appl.*, **59:**505. (12.1.4)
— (1980): *Ann. Inst. H. Poincaré*, **A33:**209. (16.6)
Jeukenne, J. P. (1964): *Nuclear Phys.*, **58:**1. (16.5)
— (1967): *Nucl. Phys.*, **A101:**229. (19.2)
Joachain, C. (1965a): *Nuclear Phys.*, **64:**529. (11.3)
— (1965b), *Nuclear Phys.*, **64:**548. (11.3, 16.3)
— and M. H. Mittleman (1971): *Phys. Rev.*, **A4:**1492. (18.3)
Johnson, B. R. (1980): *J. Chem. Phys.*, **73:**5051. (17.4.6)
Johnson, P. (1968): *J. Math. Phys.*, **9:**1633. (18)
Jordan, T. F. (1962): *J. Math. Phys.*, **3:**414 and 429. (6.7, 7.3, 16.2)
— (1964): *J. Math. Phys.*, **5:**1345. (16.2)
—, A. J. Macfarlane, and E. C. G. Sudarshan (1964): *Phys. Rev.*, **133:**B487. (6)
Jost, R. (1946): *Physica*, **12:**509. (12.1.3)
— (1947): *Helv. Phys. Acta*, **20:**256. (12.1.1, 14.3)
— (1956): *Helv. Phys. Acta*, **29:**410. (20.3)
— and W. Kohn (1952a): *Phys. Rev.*, **87:**977. (12.1.2, 12.1.5, 20.3)
— and — (1952b): *Phys. Rev.*, **88:**382. (20.3)
— and — (1953): *Kgl. Danske Videnskab. Selskab, Mat.-fys. Medd.*, **27**(9). (20.2)
— and A. Pais (1951): *Phys. Rev.*, **82:**840. (9.3, 10.3.2, 10.3.5, 14.4)
Kaiser, H. F. (1964): "Annotated Bibliography on Positron Scattering," U.S. Naval Research Laboratory, Washington, D.C. (15)
Kalckar, F., J. R. Oppenheimer, and R. Serber (1937): *Phys. Rev.*, **52:**273 and 279. (16.5)
Kalikstein, K., and L. Spruch (1964): *J. Math. Phys.*, **5:**1261. (11.3)
Kalnins, E. G., J. Patera, R. T. Sharp, and P. Winternitz (1973): *Phys. Rev.*, **D8:**2552 and 3527. (11.1)
Kamal, A. N., and H. J. Kreuzer (1970): *Phys. Rev.*, **D2:**2033. (17.3)
Kamimura, M. (1977): *Progr. Theor. Phys. Suppl.* **62:**236. (16.4)
Kano, Y., and E. Wolf (1962): *Proc. Phys. Soc. London*, **80:**1273. (4.3)
Kapur, P. L., and R. Peierls (1938): *Proc. Roy. Soc. (London)*, **A166:**277. (16.5)
Karal, F. C., Jr., and J. B. Keller (1964): *J. Math. Phys.*, **5:**537. (1.5)
Karloukovski, V. I. (1976): *Rep. Math. Phys.*, **10:**87. (10.3)
Karlsson, B. R. (1972): *Phys. Rev.*, **D6:**1662. (7.2.2, 20.3)
— (1974): *Phys. Rev.*, **D10:**1985. (20.3)
— (1978): in P. C. Sabatier, ed., Applied *Inverse Problems*, Springer Verlag, Berlin, p. 226. (20.3)
— (1982): *J. Math. Phys.*, . (17.4)
— and E. M. Zeiger (1974): *Phys. Rev.*, **D9:**1761 and **D10:**1291. (17)
— and — (1975): *Phys. Rev.*, **D11:**939. (17.4)

Karplus, M., R. N. Porter, and R. D. Sharma (1964): *J. Chem. Phys.*, **40:**2033. (5)
—, —, and — (1965): *J. Chem. Phys.*, **43:**3259. (5)
— and L. M. Raff (1964): *J. Chem. Phys.*, **41:**1267. (5)
Kato, M. (1965): *Ann. Phys.* (*N.Y.*), **31:**130. (17.1)
Kato, T. (1949): *Progr. Theor. Phys.*, **4:**514. (9.1.1)
— (1950a): *Progr. Theor. Phys.*, **5:**207. (9.1.1)
— (1950b): *Phys. Rev.*, **80:**475. (11.3)
— (1951a): *Trans. Am. Math. Soc.*, **70:**195. (7.3)
— (1951b): *Progr. Theor. Phys.*, **6:**295 and 394. (11.3)
— (1957a): *J. Math. Soc. Japan*, **9:**239. (7.3)
— (1957b): *Proc. Japan. Acad.*, **33:**260. (7.3)
— (1967a): *J. Funct. Anal.*, **1:**342. (16.2.3)
— (1967b): *Progr. Theor. Phys. Suppl.*, **40:**3. (6.3)
— and S. T. Kuroda (1959): *Nuovo Cimento*, **14:**1102. (6.4, 6.7)
Kaufmann, W. B. (1967): *Phys. Rev.*, **164:**1991. (13.3)
Kawai, M., M. Ichimura, and N. Austern (1981): preprint. (17.4)
—, M. Kamimura, Y. Mito, and K. Takesako (1978): *Progr. Theor. Phys.*, **59:**674 and 676. (11.3)
Kay, I., and H. E. Moses (1955): *Nuovo Cimento*, **2:**917. (20.3)
— and — (1956): *Nuovo Cimento*, **3:**66 and 276. (20.3)
— and — (1957): *Nuovo Cimento Suppl.*, **5:**230. (20.3)
— and — (1961a): *Nuovo Cimento*, **22:**689. (20.5)
— and — (1961b): *Communs. Pure Appl. Math.*, **14:**435. (20.5)
Kazaks, P. A., and K. R. Greider (1970): *Phys. Rev.*, **C1:**856. (17.4)
— and R. D. Koshel (1970): *Phys. Rev.*, **D1:**1906. (16.3)
Kazarinoff, N. D., and R. K. Ritt (1959): *Ann. Phys.* (*N.Y.*), **6:**277. (3.8)
Kazes, E. (1959): *Nuovo Cimento*, **13:**983. (12.1.3)
Kelemen, P. J. (1971): *J. Math. Phys.*, **12:**1142. (13.3)
Keller, J. B., I. Kay, and J. Shmoys (1956): *Phys. Rev.*, **102:**557. (5.9)
—, S. I. Rubinow, and M. Goldstein (1963): *J. Math. Phys.*, **4:**829. (3.8)
Kellogg, O. (1902): *Nachr. kgl. Ges. Wiss. Göttingen, Math.-physik. Kl.*, 165. (9.3)
Kelly, R. L. (1971): *Phys. Rev.*, **D4:**1755. (17.4, 18.3)
Kelsey, E. J., and L. Rosenberg (1979): *Phys. Rev.*, **A19:**756. (16)
Kenner, R. D., E. A. Ogryzlo, and P. T. Wassell (1981): *Nature*, **291:**398. (17.4.5)
Kerker, M. (1963): (ed.) "Electromagnetic Scattering," Macmillan, New York. (1.4)
— (1969): *The Scattering of Light and Other Electromagnetic Radiation*, Academic Press, N.Y. (Pt. I)
— (1977): *J. of Colloidal and Interf. Sci.*, **58:**100. (3)
— (1978): *Appl. Optics*, **17:**3337. (1.3)
—, W. A. Farone, and E. Matijevic (1963): *J. Opt. Soc. Am.*, **53:**758. (3.2)
Kerler, W. (1970): *Z. Physik*, **234:**330. (19.2)
— and J. Petzold (1965): *J. Physik*, **186:**168. (19.2)
Kerner, E. H. (1953): *Phys. Rev.*, **91:**1174, **92:**1441. (16)
Khalfin, L. A. (1957): *Z. Eksp. i Teor. Fiz.*, **33:**1371 (transl. *Soviet Phys. JETP*, **6:**1053). (19.2)
— (1960): *Dokl. Akad. Nauk SSSR*, **132:**1051 (transl. *Soviet Phys. Dokl.*, **5:**515). (19.2)
Kharchenko, V. E., and V. E. Kuzmichev (1972): *Nucl. Phys.*, **A183:**606 and **A196:**636. (16.3.3)
— and — (1973): *Yad. Fiz.*, **17:**975 (transl. *Soviet J. Nucl. Phys.*, **17:**511). (16.3.3)

— and — (1974): *Czech. J. Phys.*, **B24:**1071; *Yad. Fiz.*, **20:**334 (transl. *Soviet J. Nucl. Phys.*, **20:**178). (16.3.3)
—, —, and S. A. Shadchin (1974): *Nucl. Phys.*, **A226:**71. (16.3.3)
— and V. P. Levachev (1977): *Yad. Fiz.*, **26:**703 (transl. *Soviet J. Nucl. Phys.*, **26:**371). (17.4)
— and S. A. Shadchin (1978): *Ukr. Fiz. Zh.*, **23:**1651. (17.4)
Khare, V., and H. M. Nussenzveig (1974): *Phys. Rev. Lett.*, **33:**976. (3.6)
Kholyavin, I. I., and O. A. Yakubovskii (1977): *Yad. Fiz.*, **26:**181 (transl. *Soviet J. Nucl. Phys.*, **26:**94). (17.4)
Khudyakov, S. V. (1971): *Zh. Eksp. Teor. Fiz.*, **60:**2040 (transl. *Soviet Phys. JETP*, **33:**1099). (20.1)
Khuri, N. N. (1957): *Phys. Rev.*, **107:**1148. (9.3, 10.3.2, 10.3.4)
— and A. Pais (1964): *Revs. Modern Phys.*, **36:**590. (12.4)
— and S. B. Treiman (1958): *Phys. Rev.*, **109:**198. (10.3.2)
Kikkawa, K. (1965): *Progr. Theor. Phys.*, **33:**1107. (17.1)
Kikuta, T. (1953a): *Progr. Theor. Phys.*, **10:**653. (9.1.1)
— (1953b): *Progr. Theor. Phys.*, **12:**225. (11.3)
Kim, Y. E., and A. Tubis (1975): *Phys. Rev.*, **D11:**947. (17.4)
King, R. W. P., and T. T. Wu (1959): "The Scattering and Diffraction of Waves," pt. I, Harvard, Cambridge, Mass. (Pt. I)
Kirhznitz, D. A., and N. Zh. Takibaev (1972): *Yad. Fiz.*, **16:**435 (transl. *Soviet J. Nucl. Phys.*, **16:**242). (20.3.2)
Kitada, H., and K. Yajima (1981): Princeton preprint. (12.4)
Klaus, M., and B. Simon (1980): *Ann. Phys. (N.Y.)*, **130:**251. (10.3.3)
Klein, A. (1960): *J. Math. Phys.*, **1:**41. (10.3.4, 13.4)
— and C. Zemach (1959): *Ann. Phys. (N.Y.)*, **7:**440. (10.3.4)
Kleinman, R. E., and T. B. A. Senior (1963): *Univ. Mich. Coll. Eng. Rept.* 3648-2-T. (2.4)
Klepikov, N. P. (1962): *Z. Eksp. i Teor. Fiz.*, **41:**1187 (transl. *Soviet Phys. JETP*, **14:**846). (20.1)
— (1964): *Zh. Eksp. Teor. Fiz.*, **47:**757 (transl. *Soviet Phys. JETP*, **20:**505). (20.2)
Klingbeil, R. (1972): *J. Chem. Phys.*, **56:**132. (20.1)
Klink, W. H. (1971): *Phys. Rev.*, **D4:**2260. (16.4)
— (1973): *Phys. Rev.*, **D7:**2980 and 2989. (16.4)
— (1974): *J. Math. Phys.*, **15:**565. (16.6)
— (1979): *Phys. Rev.*, **A20:**1864. (16.3)
— and L. E. Johnson (1974): *Phys. Rev.*, **D9:**1057. (17.4)
Kloet, W. M., and J. A. Tjon (1973): *Ann. Phys. (N.Y.)*, **75:**407. (17.4)
Kodeira, K. (1949): *Am. J. Math.*, **71:**921. (7.3)
— (1950): *Am. J. Math.*, **72:**502. (7.3)
Kohler, D., and L. Mandel (1970): *J. Opt. Soc. Am.*, **60:**280. (4.3)
Kohn, W. (1948): *Phys. Rev.*, **74:**1763. (11.3)
— (1951): *Phys. Rev.*, **84:**495. (11.3)
— (1952): *Phys. Rev.*, **87:**539. (9.1.1)
— (1954): *Revs. Modern Phys.*, **26:**292. (9.1.1)
Kok, L. P. (1980): in F. S. Levin (1980), p. 171c. (17.4)
—, D. J. Struik and H. van Haeringen (1980): preprint. (17.4)
— and H. van Haeringen (1980): *Phys. Rev.*, **C21:**512. (17.4)
— and — (1981a): *Ann. Phys. (N.Y.)*, **131:**426. (12)
— and — (1981b): *Phys. Rev. Lett.*, **46:**1257. (14.6)

Komarov, V. V. and A. M. Popova (1969): *Yad. Fiz.*, **10:**1089 (transl. *Soviet J. Nucl. Phys.* **10:**621); *Phys Lett.*, **28B:**476. (17.4)
—, V. L. Shablov, A. M. Popova, and T. A. Osborn (1980): *Phys. Rev.*, **C22:**976. (17.4)
Konisi, G., and T. Ogimoto (1959): *Progr. Theor. Phys.*, **22:**807. (12.1.3)
— and — (1963): *Progr. Theor. Phys.*, **29:**908. (12)
Koshmanenko, V. D. (1978): *Rep. Math. Phys.*, **14:**185. (16.2.3)
Kouri, D. J. (1969): *J. Chem. Phys.*, **51:**5204. (16.3)
—, H. Krüger, and F. S. Levin (1977): *Phys. Rev.*, **D15:**1156. (17.4)
— and F. S. Levin (1974a): *Phys. Rev.*, **A10:**1616. (16.3.3, 17.4)
— and — (1974b): *Ann. Phys.* (*N.Y.*), **83:**316. (7.2.3)
— and — (1974c): *Ann. Phys.* (*N.Y.*), **83:**326. (16.3.3)
— and — (1974d): *Phys. Lett.*, **50B:**421 and **53B:**27. (17.4, 17.4.3)
— and — (1975a): *Phys. Rev.*, **C11:**352. (16.4)
— and — (1975b): *Nucl. Phys.*, **A253:**395. (17.4)
— and — (1975c): *Nucl. Phys.*, **A250:**127. (17.4)
— and — (1976): *Phys. Rev.*, **C13:**14. (16.3.3)
— and — (1977): *Phys. Rev.*, **C16:**556. (17.4)
—, —, and W. Sandhas (1976): *Phys. Rev.*, **C13:**1825. (17.4)
Kowalski, K. L. (1965): *Phys. Rev. Lett.*, **15:**789. (10.3)
— (1969): *Phys. Rev.*, **188:**2235 and **D2:**812. (17.4)
— (1972a): *Phys. Rev.*, **D5:**395 and **D6:**3705. (17.4)
— (1972b): *Nucl. Phys.*, **A190:**645. (7, 16.3)
— (1973): *Phys. Rev.*, **D7:**1806 and 2957. (17.4)
— (1977): *Phys. Rev.*, **C16:**7 and 2073. (17.4)
— (1978a): in H. Zingl et al. (1978)II, p. 393. (17.4)
— (1978b): *Lett. Nuovo Cim.*, **22:**531. (17.4.4)
— (1979): *Ann. Phys.* (*N.Y.*), **120:**328. (17.4)
— and A. Picklesimer (1981): *Phys. Rev. Lett.*, **46:**228. (16.3.3)
—, W. N. Polyzou, and E. F. Redish (1981): *J. Math. Phys.* **22:**1965. (17.4)
—, E. R. Siciliano, and R. M. Thaler (1978): *Phys. Rev.*, **C18:**2017. (17.4)
Kramers, H. A. (1926): *Z. Phys.*, **39:**828. (18.2)
— (1927): *Atti. Congr. Intern. Fisici, Como*, **2:**545 (reprinted in H. A. Kramers, "Collected Scientific Papers," North-Holland, Amsterdam, 1956, p. 333). (1.3, 4.2.1)
— (1938): *Hand-u. Jahrb. chem. Phys.*, **1:**312. (12.1.3)
Krein, M. G. (1953a): *Dokl. Akad. Nauk SSSR*, **88:**405 (*Math. Rev.*, **15:**316). (20.3)
— (1953b): *Dokl. Akad. Nauk SSSR*, **93:**617 (*Math. Rev.*, **15:**796, **17:**740). (20.2)
— (1953c): *Dokl. Akad. Nauk SSSR*, **94:**987 (*Math. Rev.*, **16:**38). (20.3)
— (1954): *Dokl. Akad. Nauk SSSR*, **97:**21 (*Math. Rev.*, **16:**372). (20.3)
— (1955): *Dokl. Akad. Nauk SSSR*, **105:**433 (*Math. Rev.*, **17:**1210). (20.3)
— (1956): *Dokl. Akad. Nauk SSSR*, **111:**1167 (*Math. Rev.*, **19:**227). (20.3)
— (1957): *Dokl. Akad. Nauk SSSR*, **113:**970 (*Math. Rev.*, **19:**961). (20.3)
— (1958): *Usp. Math. Nauk*, **13**(5):83, (*Math. Rev.*, **21:**287). (20.3)
Kröger, H., and C. C. Fenske (1981): *Reps. on. Math. Phys.*, **17:**459. (17.4)
— and R. Perne (1980): *Phys. Rev.*, **C22:**21. (17.4.8)
Kronfli, N. S. (1963): *Nuovo Cimento*, **30:**1465. (17.2)
Kronig, R. (1926): *J. Optical Soc. Am.*, **12:**547. (4.2.1)
Krzywicki, A. (1965): *J. Math. Phys.*, **6:**485. (8.1.3)
—, and J. Szymanski (1960): *Progr. Theor. Phys.*, **23:**376. (19.2)
Kujawski, E. (1971): *Phys. Rev.*, **D4:**2573. (18.3)

— (1973): *Phys. Rev.*, **C8:**100. (20.1)
Kukulin, V. I., V. G. Neudatchin, and V. V. Pomerantsev (1978): *J. Phys. G*, **4:**1409. (16.3.3, 17.4)
—, —, and Yu. F. Smirnov (1979): *Fiz. Elem. Chastits At. Yadra*, **10:**1236 (transl. *Soviet J. Part. Nucl.*, **10:**492). (16.3.3)
— and V. V. Pomerantsev (1978a): *Ann. Phys.* (*N.Y.*), **111:**330. (16.3.3, 17.4)
— and — (1978b): in Zingl et al. (1978)I, p. 385. (16.3.3, 17.4)
— and — (1978c): *Yad. Fiz.*, **27:**1668 (transl. *Soviet J. Nucl. Phys.*, **27:**875). (16.3.3, 17.4)
Kupperman, A. (1975): *Chem. Phys. Lett.*, **32:**374. (17.4.6)
Kupsch, J., and W. Sandhas (1966): *Commun. Math. Phys.*, **2:**147. (12.4)
Kuroda, S. T. (1959a): *Nuovo Cimento*, **12:**431. (6.7)
— (1959b): *J. Math. Soc. Japan*, **11:**247. (7.3)
— (1960): *J. Math. Soc. Japan*, **12:**243. (7.3)
— (1962): *J. Math. Phys.*, **3:**933. (6.7)
Kuzmichev, V. E. (1978): in G. Pisent et al. (1978), p. 161. (17.4)
— and V. F. Kharchenko (1979): preprint . (17.4)
— and — (1981): *Teor. i Mat. Fiz.*, **47:**73 (transl. *Theor. and Math. Phys.*, **47:**324). (17.4)
Kynch, G. J. (1952): *Proc. Phys. Soc.* (*London*), **A65:**83. (12.1.2)
LaVita, J. A., and J.–P. Marchand (1974): eds., *Scattering Theory in Mathematical Physics*, Reidel Publ. Co., Dordrecht. (6)
Ladyzhenskaya, O. A., and L. D. Faddeev (1958): *Dokl. Akad. Nauk SSSR*, **120:**1187 (*Math. Rev.*, **21:**293). (7.3)
Laing, J. R., and K. F. Freed (1975): *Phys. Rev. Lett.*, **34:**849. (16.4)
Lambert, F., O. D. Corbella, and Z. D. Thome (1975): *Nucl. Phys.*, **B90:**267. (20.3)
Landman, U. (1972): *Phys. Rev.*, **A5:**1. (12.1.2)
Lane, A. N., and R. G. Thomas (1958): *Revs. Modern Phys.*, **30:**257. (16.5)
Langer, R. E. (1932): *Transact. Am. Math. Soc.*, **34:**447. (18.2)
— (1937): *Phys. Rev.*, **51:**669. (18.2)
Lanik, J. (1964): *Czech. J. Phys.*, **B14:**667. (20.3)
Lapidus, I. R. (1982): *Am. J. Phys.*, **50:**45. (11.1)
Laporte, O. (1923): *Ann. Phys.*, **70:**595. (3.8)
Larson, N. M., and J. H. Hetherington (1974): *Phys. Rev.*, **C9:**699. (17.4)
Lavine, R. (1974): in LaVita and Marchand (1974), p. 141. (19.2)
— (1978): in J. Nuttall (1978), p. 45. (11.1.3, 16.6)
Lawrynowicz, J., and L. Wojtcak (1967): *Ann. Physik*, **19:**76. (10.3.4)
Lawson, J. L., and G. E. Uhlenbeck (1950): "Threshold Signals," McGraw-Hill, New York. (4.3)
Lax, P. D., and R. S. Phillips (1964): *Bull. Am. Math. Soc.*, **70:**130. (Pt. I)
— and — (1967): *Scattering Theory*, Academic Press, N.Y. (Pt. I)
Layzer, D. (1951): *Phys. Rev.*, **84:**1221. (16.3)
Lazarev, L. M. (1967): *Yad Fiz.*, **5:**101 (transl. *Soviet J. Nucl. Phys.*, **5:**69). (20.2)
LeCouteur, K. J. (1960): *Proc. Roy. Soc.* (*London*), **A256:**115. (17.1)
Lebedeff, S. A. (1964): *Phys. Rev.*, **165:**1399. (18.2)
— (1970): *Phys. Rev.*, **D1:**1583. (18.2)
Lee, B. W., and A. Klein (1959): *Nuovo Cimento*, **13:**891. (16.5)
Leeb, H., and E. W. Schmid (1981): preprint . (16.3.3)
Lehmann, H. (1959): *Nuovo Cimento*, **10:**579. (13.1)
Leon, J. Jp. (1980): *Lett. Nuovo Cim.*, **29:**45. (20.3)

— (1981): *J. Math. Phys.*, **22:**965. (20.3)
Leung, C. C. H., and S. C. Park (1969): *Phys. Rev.*, **186:**1297. (10.1.4)
Levin, B. Ya. (1956): *Dokl. Akad. Nauk SSSR*, **106:**187 (*Math. Rev.*, **18:**35). (20.3)
Levin, F. S. (1963): *Nuclear Phys.*, **46:**275. (16.3.3)
— (1965): *Phys. Rev.*, **140:**B1099. (16.3.3)
— (1966a): *Phys. Rev.*, **141:**858. (16.3)
— (1966b): *Phys. Rev.*, **149:**1212. (16.3.3)
— (1980): *Ann. Phys. (N.Y.)*, **130:**139; *Phys. Rev.*, **C21:**2199. (17.4)
— (1981a): ed., *The Few Body Problem*, North-Holland Publ., Amsterdam; *Nucl. Phys.*, **A353**, Nos. 1 & 2. (17.4)
— (1981b): in F. S. Levin (1981a), p. 143c. (17.4)
— (1981c): *Ann. Phys. (N.Y.)*, **130:**139. (17.4)
— and J. M. Greben (1978): *Phys. Rev. Lett.*, **41:**1447. (16.3)
— and H. Krüger (1978): in H. Zingl et al. (1978) II, p. 185. (16.3.3)
— and C. T. Li (1981): preprint. (17.4)
Levine, H., and J. Schwinger (1948): *Phys. Rev.*, **74:**958. (11.3)
— and — (1949): *Phys. Rev.*, **75:**1423. (11.3)
— and — (1951): In "The Theory of Electromagnetic Waves," Interscience, New York. (4.1.2)
Levine, R. D., and Y. Alhassid (1979): *Phys. Rev.*, **C20:**1775. (8)
Levinger, J. S. (1973): *J. Math. Phys.*, **14:**1314. (7.2.2)
— and M. L. Rustgi (1957): *Phys. Rev.*, **106:**607. (14.7)
Levinson, N. (1949a): *Phys. Rev.*, **75:**1445. (20.3, 20.3.2)
— (1949b): *Kgl. Danske Videnskab. Selskab, Mat.-fys. Medd.*, **25**(9). (11.2.2, 12.1.1, 12.1.3, 12.2, 20.3)
— (1953): *Phys. Rev.*, **89:**755. (20.3)
Levitan, B. M. (1956): *Usp. Mat. Nauk (N.S.)*, **11**(6):72, 117 (*Am. Math. Soc. Transl.*, **18:**49). (20.2)
Levy, B. R., and J. B. Keller (1963): *J. Math. Phys.*, **4:**54. (11.2.2, 12.1.2)
Lévy, M. (1960): *Nuovo Cimento*, **14:**612. (19.2)
Lévy-Leblond, J.–M. (1967): *Phys. Rev.*, **153:**1. (16)
— (1977): *Phys. Lett.*, **64A:**159. (4.3)
— and M. Lévy–Nahas (1965): *J. Math. Phys.*, **6:**1372 and 1571. (17.4.6)
— and J. P. Provost (1967): *Phys. Lett.*, **26B:**104. (16)
Lewanski, A. J., and W. Tobocman (1978): *Phys. Rev.*, **C17:**423. (17.4)
L'Huillier, M., E. F. Redish, and P. C. Tandy (1977): *J. Math. Phys.*, **19:**1276. (17.4)
Li, M. C. (1973): *J. Math. Phys.*, **14:**1358. (15.1)
— (1974): *J. Math. Phys.*, **15:**570. (14.6)
Lieber, M., L. Rosenberg, and L. Spruch (1972): *Phys. Rev.*, **D5:**1330 and 1347. (16.4)
Lighthill, M. J. (1958): "Introduction to Fourier Analysis and Generalized Functions," Cambridge, New York. (7)
Limić, N. (1962): *Nuovo Cimento*, **26:**581. (12.4)
— (1963): *Nuovo Cimento*, **28:**1066. (12.4)
— (1965): *Nuovo Cimento*, **36:**100. (12.4)
— (1966): *Commun. Math. Phys.*, **1:**321. (7.3)
Lindner, A. (1970): *Nucl. Phys.*, **A155:**145. (16.4)
— (1973): *Nucl. Phys.*, **A199:**110. (17.4.6)
Ling, R. T., J. C. Y. Chen, J. L. Peacher, and K. M. Watson (1972): *Phys. Rev.*, **A6:**650. (16.4, 18.3)
Lipperheide, R. (1962): *Ann. Phys. (N.Y.)*, **17:**114. (16.5)

— and H. Fiedeldey (1978): *Z. Physik*, **A286:**45. (20.4)
— and — (1981): *J. Physik* **A301:**81. (20.4)
Lippmann, B. A. (1956): *Phys. Rev.*, **102:**264. (11.3, 16.3)
— (1965): *Phys. Rev., Letters*, **15:**11. (6)
— (1966): *Phys. Rev.*, **151:**1023. (11.1.3)
— and J. Schwinger (1950): *Phys. Rev.*, **79:**469 [reprinted in M. H. Ross (1963a)]. (6, 7.1, 8.1, 11.3)
Ljance, V. E. (1964): *Dokl, Akad. Nauk SSSR*, **154:**1030 (*Math. Revs.*, **28:**449). (12.1.5)
Ljunggren, T. (1948): *Arkiv Mat. Astron. Fysik*, **36A**(14). (3.6)
Lock, J. A. (1975): *Phys. Rev.*, **D12:**3319. (17.4.6)
— (1978): *Phys. Rev.*, **C18:**2465. (17.4)
Loeffel, J.–J. (1968): *Ann. Inst. H. Poincare*, **8A:**339. (20.4)
Logan, N. A. (1962): *J. Opt. Soc. Am.*, **52:**342. (2.4)
— (1965): *Proc. IEEE*, **53:**773. (2, 3)
Lomsadze, Yu. M., I. V. Khimich, and J. M. Shuba (1965): *Nuclear Phys.*, **67:**631 (12)
Lopez, C. A. (1968): *Nuovo Cim.*, **58A:**160. (13.3)
— and V. Massidda (1970): *Nucl. Phys.*, **A149:**33. (12.1.3)
— and I. Saavedra (1966): *Nucl. Phys.*, **87:**65. (13.3)
Lorenz, L. (1890): *Vidanskab, Selsk. Skrifter*, **6:**1 (French transl. *Oeuvres Scientifiques de L. Lorenz*, vol. 1, Libr. Lehmann, Copenhagen, 1896; reprinted, Johnson, N.Y. 1964, p. 405). (2.4)
Lovelace, C. (1964a): In R. G. Moorhouse (ed.), "Strong Interactions and High Energy Physics," p. 437, Oliver & Boyd, London. (17.4.2)
— (1964b): *Phys. Rev.*, **135:**B1225. (7.2.4, 9.1.1, 17.4.2)
Lovitch, L., and V. Vanzani (1979): *Lett. Nuovo Cim.*, **26:**65. (17.4)
Low, F. E. (1955): *Phys. Rev.*, **97:**1392. (7.2.2)
— (1959): in "Brandeis University Summer Institute in Theoretical Physics, 1959, Lecture Notes," pp. 3–79. (6, 7)
Lozano, J. M., and M. Moshinsky (1960): *Proc. Intern. Conf. on High Energy Phys., 1960, Rochester*, Interscience, New York, p. 271. (19.2)
Lu, P. (1970): *Phys. Rev.*, **C1:**468. (18.2)
—., D. M. Fradkin, and R. H. Good (1964): *Nuclear Phys.*, **34:**581. (18.2)
— and P. M. Measure (1972): *Phys. Rev.*, **D5:**2514. (18.2)
— and S. S. Wald (1973): *Phys. Rev.*, **D8:**4371. (18.2)
Ludeking, L., and J. P. Vary (1979): *Phys. Rev.*, **D20:**2188. (17.4)
Lüders, G. (1955): *Z. Naturforsch.*, **10a:**581. (12.1.3, 12.2)
Luk'yanov, A. V., I. B. Teplov, and M. K. Akimova (1965): "Tables of Coulomb Wave Functions," Pergamon, New York. (14.6)
Luming, M. (1964): *Phys, Rev.*, **136:**B1120. (17.1)
— and E. Predazzi (1966): *Nuovo Cim.*, **B44:**210; *Ann. Phys.* (*N.Y.*) **40:**221. (14.6)
Lundina, D. S., and V. A. Marchenko (1969): *Mat. Sb.*, **78:**475 (transl. *Math. USSR–Sbornik*, **7:**467). (20)
Lynch, P. J., and B. C. Carlson (1962): *J. Math. Phys.*, **3:**440. (18.3)
Ma, S. R. (1953): *Phys. Rev.*, **91:**392. (6)
MacDonald, J. K. L. (1933): *Phys. Rev.*, **43:**830. (11.3)
MacDonald, W., and A. Mekjian (1967): *Phys. Rev.*, **160:**730. (17.4)
McDonald, W. M. (1964): *Nuclear Phys.*, **54:**393, **56:**636 and 647. (16.5)
McDowell, M. R. C. (1963): Lectures on Ion-Atom Collisions, *Natl. Bur. Standards Tech. Note* 185. (16)

— (1964): "Atomic Collision Processes," North-Holland, Amsterdam. (16)
Macfarlane, A. J. (1962): *Nuclear Phys.*, **38:**504. (17.4.6)
MacGregor, M. H., and R. A. Arndt (1965): *Phys. Rev.*, **139:**B362. (20.1)
McGuire, J. B. (1964): *J. Math. Phys.*, **13:**1595. (17.3)
— and C. A. Hurst (1972): *J. Math. Phys.*, **13:**1595. (17.3)
McKee, J. S. C., and P. M. Rolph (1970); eds., *Three-Body Problem*, North-Holland Publ. Co., Amsterdam. (17.4)
McKinley, W. A., and H. Feshbach (1948): *Phys. Rev.*, **74:**1759. (9.1.2)
McMaster, W. H. (1961): *Revs. Modern Phys.*, **33:**8. (8.3.2)
McMillan, M. (1964): *Nuovo Cimento*, **32:**919. (17.4.3)
MacMillan, L. W., and T. A. Osborn (1980): *Ann. Phys.* (*N.Y.*), **126:**1. (11.1.3)
McVoy, K. W. (1968): *Nucl. Phys.*, **A115:**481. (12.1.3)
McWeeny, R. (1960): *Revs. Modern Phys.*, **32:**335. (8.3.1)
Macek, J., and I. V. Hertel (1974): *J. Phys. B*, **7:**2173. (16)
Madan, R. N., and R. Blankenbecler (1978): *Phys. Rev.*, **D17:**1155. (11.3)
Mahaux, C. (1965): *Nuclear Phys.*, **68:**481, **71:**241. (16.5)
— and H. A. Weidenmüller (1965): *Ann. Phys.* (*N.Y.*), **32:**259. (16.5)
Majumdar, C. K. (1972): *J. Math. Phys.*, **13:**705. (17.3)
Mal'cenko, V. I. (1966): *Ukr. Mat. Z.*, **18, 2:**66 (transl. *Am. Math. Soc. Transl.*, **75:**227). (20.3)
Malfliet, R., and Th. Ruijgrok (1967): *Physica*, **33:**607. (17.4)
Malyarov, V. V., V. N. Pivovarckik, I. V. Poplovskii, and M. N. Popushoi (1978): *Yad. Fiz.* **27:**1128 (transl. *Soviet J. Nucl. Phys.*, **27:**599). (20.4)
—, and I. V. Poplavskii (1972): *Yad. Fiz.* **16:**873 (transl. *Soviet J. Nucl. Phys.*, **16:**483). (20.4)
—, —, and M. N. Popushoi (1975a): *Yad. Fiz.*, **21:**987 (transl. *Soviet J. Nucl. Phys.*, **21:**506). (20.3)
—, —, and — (1975b): *Yad. Fiz.*, **22:**860 (transl. *Soviet J. Nucl. Phys.* **22:**445). (20.4)
—, —, and — (1975c): *Zh. Eksp. Teor. Fiz.*, **68:**432 (transl. *Soviet Phys. JETP*, **41:**210). (20.4)
—, —, and — (1977): *Yad. Fiz.* **25:**72 (transl. *Soviet J. Nucl. Phys.*, **25:**38). (20.4)
Mamasakhlisov, V. I. (1953): *Z. Eksp. i Teor. Fiz.*, **25:**36 (*Sci. Abstr.*, **A58:**63). (17.2)
Mandel, L., and E. Wolf (1963): *Phys. Rev. Letters*, **10:**276. (4.3, 8.3.1)
— and — (1965): *Revs. Modern Phys.*, **37:**231. (4.3)
Mandelstam, S. (1958): *Phys. Rev.*, **112:**1344. (13.4)
— (1965): *Phys. Rev.*, **140:**375. (17.4.3)
Mandelzweig, V. B. (1977): *Ann. Phys.* (*N.Y.*), **104:**1. (17.4)
Manning, I. (1964): *J. Math. Phys.*, **5:**1223. (9.3, 12.1.2)
— (1965): *Phys. Rev.*, **139:**495, **142:**1262. (9.1)
— (1968): *Phys. Rev.*, **168:**1875. (9)
— (1972): *Phys. Rev.*, **D5:**1472. (17.4, 18.3)
Mapleton, R. (1954): *Phys. Rev.*, **96:**415. (7.2.2)
Marchenko, V. A. (1950): *Dokl. Akad. Nauk SSSR*, **72:**457 (*Math. Rev.*, **12:**183). (20.3)
— (1952): *Trudy Moskov. Mat. Obsc.*, **1:**327 (*Math. Rev.*, **15:**315). (20.3)
— (1953): *Trudy Moskov. Mat. Obsc.*, **2:**3 (*Math. Rev.*, **15:**315). (20.3)
— (1955): *Dokl. Akad. Nauk SSSR*, **104:**695 (*Math. Rev.*, **17:**740). (20.3)
— (1960): *Mat. Sbornik* (*N.S.*), **52:**739 (*Am. Math. Soc. Transl.*, **25:**77). (20.3)
— (1968): *Mat. Sbornik*, **77:**139 (transl. *Math. USSR–Sbornik*, **6:**125). (20)
Marshak, R. E. (1947): *Phys. Rev.*, **71:**688. (11.3)

Marchesin, D., and M. L. O'Carroll (1972): *J. Math. Phys.*, **13:**982. (14.6)
Marchioro, C. (1970): *J. Math. Phys.*, **11:**2193. (17.3)
Marquez, L. (1972): *Am. J. Phys.*, **40:**1420. (14.6)
Martin, A. (1956): *Comtes Rend. Ac. Sci. Paris*, **243:**22. (12.1.3)
— (1958): *Nuovo Cimento*, **7:**607. (12.1.3)
— (1959): *Nuovo Cimento*, **14:**403. (14.3)
— (1960): *Nuovo Cimento*, **15:**99. (14.5)
— (1962): *Nuovo Cimento*, **23:**641. (12.3)
— (1965): In J. G. Wilson and S. A. Wouthuysen (eds.), "Progress in Elementary Particle and Cosmic Ray Physics," vol. 8, pp. 1–66, Wiley, New York. (12.3)
— (1969): *Nuovo Cim.*, **A59:**131. (20.2)
— (1979): *Commun. Math. Phys.*, **69:**89. (10.3.3)
— (1981): *Acta Phys. Austr. Suppl.* (10.3.3)
— and G. Targonski (1961): *Nuovo Cimento*, **20:**1182. (20.4)
Martin, Ph.A. (1975): *Nuovo Cim.*, **30B:**217. (11.1.3)
— (1976): *Commun. Math. Phys.*, **47:**221. (11.1.3)
Massey, H. S. W. (1930): *Proc. Roy. Soc.* (*London*), **A129:**616. (16)
— (1956a): *Revs. Modern Phys.*, **28:**199 [in S. Flügge (ed.), "Handbuch der Physik," vol. 36, p. 285, Springer, Berlin]. (9.1.3)
— (1956b): In S. Flügge (ed.), "Handbuch der Physik," vol. 36, p. 307, Springer, Berlin. (16)
— and E. H. S. Burhop (1952): "Electronic and Ionic Impact Phenomena," Oxford, Fair Lawn, N.J. (16)
— and C. B. S. Mohr (1932): *Proc. Roy. Soc.* (*London*), **A135:**258. (16)
— and B. L. Moiseiwitch (1951): *Proc. Roy. Soc.* (*London*), **A205:**483. (11.3, 17.4)
Masson, D. (1967): *J. Math. Phys.*, **8:**2308. (12.1.3)
— and E. Prugovečki (1976): *J. Math. Phys.*, **17:**297. (12.4)
Matese, J. J., and R. J. W. Henry (1972): *Phys. Rev.*, **A5:**222. (12.1.2)
Matsuda, K., M. Hirooka, and S. Sunakawa (1975): *Progr. Theor. Phys.*, **54:**79. (16)
Matthews, P. T., and A. Salam (1959a): *Phys. Rev.*, **115:**1079. (19.2)
— and — (1959b): *Nuovo Cimento*, **13:**381. (12.2)
Mavroyannis, C. (1967): *J. Math. Phys.*, **8:**1522. (1.5)
Medgyest-Mitschang, L. N., and C. Eftimiu (1979): *Appl. Physik*, **19:**275. (2.4)
Meetz, K. (1962): *J. Math. Phys.*, **3:**690. (7.3, 9.1.1, 10.3.1)
— (1964): *Nuovo Cimento*, **34:**690. (12.4)
Mehta, C. L. (1968): *J. Opt. Soc. Am.*, **58:**1233. (4.3)
Meixner, J. (1933): *Math. Z.*, **36:**677. (7.1.1)
— (1934): *Z. Phys.*, **90:**312. (7.1.1)
— (1937): *Ann. Phys.*, **29:**97. (7.1.1)
Mentkovsky, Yu. L. (1962): *Ukrayin. fiz. z.* (*SSSR*), **7:**593 and 966 (*Phys. Abstr.*, **65A:** No. 22452, **66A:** No. 19587). (14.6)
— (1963): *Ukrayin. fiz. z.* (*SSSR*), **8:**17 and 144 (*Phys. Abstr.*, **67A:** Nos. 5967 and 5968). (14.6)
— (1964): *Ukrayin fiz. z.* (*SSSR*), **9:**1169 (*Phys. Abstr.*, **68A:** No. 10861). (14.6)
— (1965): *Nuclear Phys.*, **65:**673. (14.6)
Merkuriev, S. P. (1971): *Teor. i Mat. Fiz.*, **8:**235 (transl. *Theor. and Math. Phys.*, **8:**798). (17.4.5)
— (1973): *Teor. i Mat. Fiz.*, **17:**221 (transl. *Theor. and Math. Phys.*, **17:**1105). (17.4.5)
— (1974): *Nucl. Phys.*, **A233:**395. (16.4)
— (1976): *Yad. Fiz.*, **24:**289 (transl. *Soviet J. Nucl. Phys.*, **24:**150). (17.4)

— (1977): *Teor. i Mat. Fiz.*, **32:**187 (transl. *Theor. and Math. Phys.*, **32:**680). (17.4)
— (1978a): *Zapisky nauch. sem. LOMI, Leningrad*, **77:**148. (17.4)
— (1978b): *Dokl. Akad. Nauk SSSR*, **241:**68 (transl. *Soviet Phys. Dokl.*, **23:**467). (17.4)
— ((1979a): *Teor. i Mat. Fiz.*, **38:**201 (transl. *Theor. and Math. Phys.*, **38:**134); *Lett. Math. Phys.*, **3:**141. (17.4)
— (1979b): *Yad. Fiz.*, **30:**941 (transl. *Soviet J. Nucl. Phys.*, **30:**489). (17.4)
— (1980): *Ann. Phys. (N.Y.)*, **130:**395. (17.4)
—, C. Gignoux, and A. Laverne (1976): *Ann. Phys. (N.Y.)*, **99:**30. (17.4)
Meyerhof, W. E. (1962): *Phys. Rev.*, **128:**2312. (17.2)
— (1963): *Phys. Rev.*, **129:**692. (17.2)
Micha, D. A. (1967): *J. Math. Phys.*, **8:**1716. (16.6)
Michalik, B. (1968): *Ann. Phys. (N.Y.)*, **48:**197 and **50:**106. (9)
Mie, G. (1908): *Ann. Phys.*, **25:**377. (2.4)
Miles, J. R. N., and J. S. Dahler (1970): *J. Chem. Phys.*, **52:**616. (5)
Miller, J. C. P. (1946): "The Airy Integral," British Association for the Advancement of Science, Mathematical Tables, pt.-vol. B, Cambridge, New York. (3.6)
Miller, K., and G. A. Viano (1971): *Nucl. Phys.*, **B25:**460. (13.2)
Miller, K. J. (1971): *Phys. Rev.*, **A3:**607. (11.3.3)
Miller, W. H. (1969a): *J. Chem. Phys.*, **50:**407. (16.4)
— (1969b): *J. Chem. Phys.*, **51:**3631. (5.9, 20.1)
— (1971): *J. Chem. Phys.*, **54:**4174. (20.1)
Mills, R. L., and J. F. Reading (1969): *J. Math. Phys.*, **10:**321. (20.1)
Milne, W. E. (1930): *Phys. Rev.*, **35:**864. (12.1.2)
Minerbo, G. N. (1965): *Nuovo Cim.*, **40A:**1055. (13.3)
— and M. E. Levy (1969): *SIAM J. Num. Anal.*, **6:**598. (5.9)
Miodek, I. (1976): *J. Math. Phys.*, **17:**168. (20.4)
Mireles, R. (1966): *J. Math. and Phys.*, **45:**179. (20.3)
Mishima, N. (1965): *Progr. Theor. Phys.*, **34:**74. (17.4.3)
— (1967): *Progr. Theor. Phys.*, **37:**745. (17.4)
— and Y. Takahashi (1966): *Progr. Theor. Phys.*, **35:**440. (17.4)
— and M. Yamazaki (1965): *Progr. Theor. Phys.*, **34:**284. (17.4.3)
Misra, B., and K. B. Sinha (1977): *Helv. Phys. Acta*, **50:**99. (19.2)
—, D. Speiser, and G. Targonski (1963): *Helv. Phys. Acta*, **36:**963. (7.3)
Mitra, A. N. (1961): *Phys. Rev.*, **123:**1892. (10.3.2)
— and V. S. Bhasin (1963): *Phys. Rev.*, **131:**1265. (17.4)
—, J. Gillespie, R. Sugar, and N. Panchapakesan (1965): *Phys. Rev.*, **140B:**1336. (17.4)
Mittleman, M. H. (1964): *Ann. Phys. (N.Y.)*, **28:**430. (16.3)
— (1969): *Phys. Rev.*, **182:**128. (16.6)
Mizuno, J., and J. C. Y. Chen (1971): *Phys. Rev.*, **A4:**1500. (18.2)
Moe, M., and D. S. Saxon (1958): *Phys. Rev.*, **111:**950. (11.3)
Mohan, G. (1959): *Nuovo Cimento*, **13:**1065. (16.3)
Moiseiwitch, B. L. (1962): In D. R. Bates (ed.), "Atomic and Molecular Processes," Academic, New York. (16)
Møller, C. (1945): *Kgl. Danske Videnskab. Selskab, Mat.-fys. Medd.*, **23**(1) [reprinted in M. H. Ross (1963a)]. (6.3, 6.4, 7, 8.1, 8.2)
— (1946): *Kgl. Danske Videnskab. Selskab, Mat.-fys. Medd.*, **22**(19). (6.3, 6.4, 7, 12.1.3)
Molière, G. (1947): *Z. Naturforsch.*, **2a:**133. (18.3)

— (1948): *Z. Naturforsch.*, **3a:**78. (18.3)
Mongan, T. R. (1966): *Phys. Rev.*, **147:**1117. (10.1.4)
Montroll, E. W., and J. M. Greenberg (1954): *Proc. Symposia Appl. Math.*, **5:**103. (18.3)
Moore, R. J. (1970): *Phys. Rev.*, **D2:**313. (18)
Moravcsik, M. J. (1968): *Phys. Rev.*, **170:**1440. (20.1)
— (1978): in H. Zingl et al. (1978) II, p. 452. (17.4)
— and W.-Y. Wu (1969): *J. Math. Phys.*, **10:**925. (20.1)
Morawitz, H. (1970): *J. Math. Phys.*, **11:**649. (11.3)
Morita, N. (1979): *Trans. IEEE.*, **AP–27:**97. (4.1)
Morrison, J. A., and M. J. Cross (1974): *Bell Syst. Tech. J.*, **53:**955. (2)
Morse, P. M., and W. P. Allis (1933): *Phys. Rev.*, **44:**269. (16.3)
— and H. Feshbach (1953): "Methods of Theoretical Physics," McGraw-Hill, New York. (3 App., 4.1.2, 15.1)
Moses, H. E. (1953a): *Phys. Rev.*, **91:**185. (16.3)
— (1953b): *Phys. Rev.*, **92:**817. (11.3)
— (1955): *Nuovo Cimento*, **1:**103. (6.7.3)
— (1956): *Phys. Rev.*, **102:**559. (20.3)
— (1957): *Nuovo Cimento Suppl.*, **5:**120 and 144. (11.3)
— (1964): *J. Math. Phys.*, **5:**833. (12.1.2)
— (1978): *Stud. Appl. Math.*, **58:**187. (20.3)
— (1979a): *J. Math. Phys.*, **20:**1151. (10.3)
— (1979b): *J. Math. Phys.*, **20:**2047. (20.3)
— (1980): *J. Math. Phys.*, **21:**83 and **22:**3010. (20.5)
— and S. F. Tuan (1959): *Nuovo Cimento*, **13:**197. (14.7)
Moshinsky, M. (1951a): *Phys. Rev.*, **81:**347. (17.3)
— (1951b): *Phys. Rev.*, **84:**525. (19.2)
Mott, N. F. (1928): *Proc. Roy. Soc. London*, **118:**542. (14.6)
— (1929): *Proc. Roy. Soc.* (*London*), **A124:**425. (10.1.2)
— (1930): *Proc. Roy. Soc.* (*London*), **A126:**259. (14.6)
— (1932): *Proc. Roy. Soc.* (*London*), **A135:**429. (10.1.2)
— and H. S. W. Massey (1949): "The Theory of Atomic Collisions," Oxford, Fair Lawn, N.J. (7, 16)
Motz, J. W., H. Olsen, and H. W. Koch (1964): *Revs. Modern Phys.*, **36:**88. (16)
Mukherjee, S., and C. S. Shastry (1967): *Nucl. Phys.*, **B3:**1. (13.3)
— and — (1968): *Phys. Rev.*, **169:**1234. (13.3)
Muldowney, J. S. (1965): *Nuovo Cimento*, **35:**1138. (13.1)
Mulherin, D., and I. I. Zinnes (1970): *J. Math. Phys.*, **11:**1402. (14.6)
Müller, H. J. W., and K. Schilcher (1968): *J. Math. Phys.*, **9:**255. (10.3.5)
Mulligan, B., L. G. Arnold, B. Bagchi, and T. O. Krause (1976): *Phys. Rev.*, **C13:**2131. (10.1.4)
— and S. B. Qadri (1981): *Phys. Rev.*, **C24:**874. (12.1.2)
Munchow, M., and W. Scheidt (1980): *Phys. Rev. Lett.*, **20:**1299. (20.4)
Munczek, H. (1963): *Nuovo Cimento*, **29:**1175. (11.1)
— (1964): *Phys. Letters*, **13:**92. (17.1)
— and A. Pignotti (1965): *Phys. Letters*, **16:**198. (17.1)
Naimark, M. A. (1954): *Trudy Mosc. Mat. Obs.*, **3:**181 (*Math. Rev.*, **15:**959). (12.1.5)
Namislowki, J. M. (1972): *Acta Phys. Austr. Suppl.*, **9:**184. (17.4, 18.3)
Narnhofer, H. (1975): *Nuovo Cim.*, **30B:**254. (12)
— (1980): *Phys, Rev.*, **D22:**2387. (11.1.3)

— and W. Thirring (1981): *Phys. Rev.*, **A23:**1688. (5, 11.1.3)
Narodetskii, I. M., and O. A. Yakubovskii (1971): *Yad. Fiz.*, **14:**315 (transl. *Soviet J. Nucl. Phys.*, **14:**178). (17.4)
Nataf, R., and H. Cornille (1963): *J. Phys. Radium*, **24:**591. (12.1.4)
Nath, P., and G. L. Shaw (1965a): *Phys. Rev.*, **138:**B702. (17)
— and — (1965b): *Phys. Rev.*, **137:**B711. (17.1)
— and Y. N. Srivastava (1965): *Phys. Rev.*, **138:**B404. (16.5)
Nauenberg, M., and J. C. Nearing (1964): *Phys. Rev. Letters*, **12:**63. (17.1)
— and A. Pais (1961): *Phys. Rev.*, **123:**1058. (17.2)
— and — (1962): *Phys. Rev.*, **126:**360. (17.2)
Naundorf, F. F. (1978): *J. Math. Phys.*, **19:**1426. (12.4)
Nelson, C. A., A. K. Rajagopal, and C. S. Shastry (1971): *J. Math. Phys.*, **12:**737. (12.1.4)
Nesbet, R. K. (1968): *Phys. Rev.*, **175:**134. (11.3.3)
— (1969): *Phys. Rev.*, **179:**60. (16.4)
— (1973): *J. Math. Phys.*, **14:**1522. (18.2)
— (1978): in J. Nuttall (1978), p. 137. (11.3, 16)
— (1980); *Variational Methods in Electron-Atom Scattering Theory*, Plenum Press. (11.3)
— and R. S. Oberoi (1972): *Phys, Rev.*, **A6:**1855. (16.4)
Neuhaus, M. G. (1955): *Dokl. Akad. Nauk SSSR*, **102:**25. (20.3)
Newton, R. G. (1955): *Phys. Rev.*, **100:**412. (12.2, 15.2, 20.3)
— (1956): *Phys. Rev.*, **101:**1588. (14.7.1, 20.2, 20.3.2)
— (1957): *Phys. Rev.*, **105:**763, **107:**1025. (14.7)
— (1958): *Ann. Phys. (N.Y.)*, **4:**29. (16.3, 16.5, 17.1, 17.2)
— (1959): *Phys. Rev.*, **114:**1611. (12.2, 17.2.3)
— (1960): *J. Math. Phys.*, **1:**319 [reprinted in M. H. Ross (1963a)]. (12.1, 12.2, 15.2, 20.2)
— (1961a): *J. Math. Phys.*, **2:**188. (9.3, 17.1, 17.3)
— (1961b): *Ann. Phys. (N.Y.)*, **14:**333. (19.2)
— (1962a): *J. Math. Phys.*, **3:**75. (20.3)
— (1962b): *J. Math. Phys.*, **3:**867, **4:**1342. (13.3)
— (1963a): *Nuovo Cimento*, **29:**400. (17.1)
— (1963b): *Phys. Letters*, **4:**11. (9.4, 17.2)
— (1964): "The Complex *j*-plane," Benjamin, New York. (12.2, 13.1, 15.2)
— (1966): In "1965 Brandeis University Summer Institute in Theoretical Physics," Gordon and Breach, New York. (17.4.2)
— (1967a): *J. Math. Phys.*, **8:**851. (17.4.3)
— (1967b): *J. Math. Phys.*, **8:**1566. (20.4)
— (1967c): *J. Math. Phys.*, **8:**2347. (17.1.3)
— (1967d): *Phys. Rev.*, **153:**1502. (17.4.4)
— (1968): *J. Math. Phys.*, **9:**2050 and **16:**1358. (20.2)
— (1971): in F. M. Renard and P. C. Sabatier, eds., *Interactions Fortes*, Colloque Nat. du CNRS, Montpellier, France; *J. Math. Phys.*, **12:**1552. (17.4.4)
— (1972a): *Ann. Phys. (N.Y.)*, **74:**324 and **78:**561. (17.4.5)
— (1972b): *J. Math. Phys.*, **13:**880. (10.3.2)
— (1974a): Inv. lect. at 1974 *Summer Seminar on Inverse Problems*, Am. Math. Soc., UCLA (unpublished). (20.5)
— (1974b): in LaVita and Marchand (1974), p. 193. (20.5)
— (1974c): *J. Math. Phys.*, **15:**338. (17.4.7)

— (1974d): *Czech. J. of Phys.*, **B24:**1195. (10.3.2, 17.4.4)
— (1975): *Trudy Mat. Inst. Steklov*, **136:**224. (12.1.3, 17.4.7)
— (1976a): *Am. J. Phys.*, **44:**639. (1.3.9, 7.2.2)
— (1976b): *J. Chem. Phys.*, **65:**3237. (17.4.5)
— (1977a): *J. Math. Phys.*, **18:**1348. (10.3.2, 12.1.3)
— (1977b): *J. Math. Phys.*, **18:**1582. (10.1.4, 12.1.3)
— (1979a): *Phys. Rev. Lett.*, **43:**541. (20.5)
— (1979b): *Found. Phys.*, **9:**929. (8.3.3)
— (1980a): in J. A. DeSanto et al. (1980), p. 123. (10)
— (1980b): *J. Math. Phys.*, **21:**1698 and **22:**631. (10.3.2, 20.5)
— (1980c): *J. Math. Phys.*, **21:**493. (20.3)
— (1981a): *Geophys. J. Roy. Astr. Soc.*, **65:**191. (20.3)
— (1981b): *J. Math. Phys.*, **22:**2191. (20.5)
— (1982): *J. Math. Phys.*, **23**. (10.3, 20.5)
— and L. Fonda (1960): *Ann. Phys.* (*N.Y.*), **9:**416. (17.2)
— and T. Fulton (1957): *Phys. Rev.*, **107:**1103. (15.2)
— and R. Jost (1955): *Nuovo Cimento*, **1:**590. (15.2, 20.3)
— and R. Shtokhamer (1974): in C. P. Enz and J. Mehra, eds., *Physical Reality and Mathematical Description*, Reidel Publ. Co., Dordrecht, p. 286. (8.1., 16.5)
— and — (1976): *Phys. Rev.*, **A14:**642. (17.4.5)
— and B.–L. Young (1968): *Ann. Phys.* (*N.Y.*), **49:**393. (8.3.2)
Ni, G.–J. (1979): *Phys. Energ. Fort. Nucl.* (*Beijing*), **3:**432. (12.1.3)
Nigam, B. P., M. K. Sundaresan, and Ta-You Wu (1959): *Phys. Rev.*, **115:**491. (18.3)
Nikolayev, M. V., and V. S. Olkhovsky (1973): *Lett. Nuov. Cim.*, **8:**703. (12.1.5)
Noble, J. V. (1966): *Phys. Rev* , **148:**1553. (17.4.4)
— (1967): *Phys. Rev.*, **161:**945. (17.4)
Nowakowski, J., and T. A. Osborn (1975): *Nucl. Phys.*, **A249:**301. (11.1.3)
Noyes, H. P. (1967): *Phys. Rev. Lett.*, **15:**538. (10.3)
—, D. S. Bailey, R. A. Arndt, and M. H. MacGregor (1965): *Phys. Rev.*, **139:**B380. (20.1)
Nussenzveig, H. M. (1959): *Nuclear Phys.*, **11:**499. (11.2.2)
— (1960): *Physica*, **26:**209. (10.3.4)
— (1961): *Nuovo Cimento*, **20:**694. (19.2)
— (1962): "Analytic Properties of Nonrelativistic Scattering Amplitudes," Escuela Latino Americana de Fisica, Universidad de Mexico, Mexico, D.F. (12)
— (1963): *Ann. Phys.* (*N.Y.*), **21:**344. (13.4)
— (1965): *Ann. Phys.* (*N.Y.*), **34:**23. (3.8)
— (1969a): *Phys. Rev.*, **177:**1848. (10.3.4)
— (1969b): *J. Math. Phys.*, **10:**82 and 125. (3.6)
— (1970): in J. E. Bowcock, ed., *Methods and Problems in Theoretical Physics*, North-Holland Publ. Co., Amsterdam, p. 203. (3.8)
— (1972): *Phys. Rev.*, **D6:**1534. (11.1.3)
Nutt, G. L. (1964): *Phys. Rev.*, **135:**A345. (16)
Nuttall, J. (1967a): *Phys. Rev. Lett.*, **19:**473. (17.4.5)
— (1967b): *Phys. Rev.*, **160:**1469. (17.4)
— (1967c): *J. Math. Phys.*, **8:**873. (7.2.2, 10.3.4)
— (1971): *J. Math. Phys.*, **12:**1896. (17.4.5)
— (1978): ed., *Atomic Scattering Theory*, Univ. of Western Ontario, London, Ont., Canada. (16)
— and R. W. Stagat (1971): *Phys. Rev.*, **A3:**1355. (14.6)

— and J. G. Webb (1969): *Phys. Rev.*, **178:**2226. (17.4.5)
Oberoi, R. S., and R. K. Nesbet (1973): *Phys. Rev.*, **A8:**215. (11.3, 16.4)
— and — (1974): *Phys. Rev.*, **A9:**2804. (16.4)
Ochkur, V. I. (1963): *Z. Eksp. i Teor. Fiz.*, **45:**734 (transl. *Soviet Phys. JETP*, **18:**503). (16)
Ohmura, T. (1956): *Progr. Theor. Phys.*, **16:**231. (20.3)
— (1964): *Progr. Theor. Phys., Suppl.*, **29:**108. (19.2)
Olsson, P. O. (1952): *Arkiv Fys.*, **4:**217. (12.1.2)
Omnes, R. (1958): *Nuovo Cimento*, **8:**316. (12.1.2)
— (1964): *Phys. Rev.*, **134:**B1358. (17.4.3)
— (1968): *Phys. Rev.*, **165:**1265. (17.4)
— and V. A. Alessandrini (1964): *Phys. Rev.*, **136:**B1137. (17.4.3)
— and M. Froissart (1963): "Mandelstam Theory and Regge Poles," Benjamin, New York. (13.1)
Oppenheimer, J. R. (1927): *Z. Phys.*, **43:**413. (14.6)
— (1928): *Phys. Rev.*, **32:**361. (16.3)
Osborn, T. A. (1969): *Nucl. Phys.*, **A138:**305. (7.2.2)
— (1970): *Ann. Phys.* (*N.Y.*), **58:**417. (18.3)
— (1971): *Phys. Rev.*, **D3:**395. (10.3)
— (1972): *Nucl. Phys.*, **B46:**538. (18.3)
— (1973): *J. Math. Phys.*, **14:**373 and 1485. (7.2.2)
— and D. Bollé (1973): *Phys. Rev.*, **C8:**1198. (17.4.5)
— and — (1974): *Phys. Lett.*, **52B:**13. (11.1.3)
— and — (1975): *J. Math. Phys.*, **16:**1533. (11.1.3)
— and — (1977): *J. Math. Phys.*, **18:**432. (12.1.3)
— and — (1978): in H. Zingl et al. (1978) I, p. 367. (11.1.3)
— and — (1979): *J. Math. Phys.*, **20:**1059. (16.5)
—, R. G. Froese, and S. F. Howes (1980): *Phys. Rev.*, **A22:**101. (5, 11.1.3, 12.1.3)
— and K. L. Kowalski (1971): *Ann. Phys.* (*N.Y.*), **68:**361. (17.4)
— and S. F. J. Wilk (1977): *Phys. Rev.*, **D16:**2545. (17.4)
Ovchinnikov, Yu. N., and I. M. Sigal (1979): *Ann. Phys.* (*N.Y.*), **123:**274. (17.4.8)
Pais, A., and T. T. Wu (1964a): *Phys. Rev.*, **134:**B1303. (12.4)
— and — (1964b): *J. Math. Phys.*, **5:**799. (12.4)
Paliov, A., and S. Rosendorff (1967): *J. Math. Phys.*, **8:**1829. (12.4)
Parzen, G. (1950): *Phys. Rev.*, **80:**261 and 355. (18)
— (1951): *Phys. Rev.*, **81:**808. (18.3)
Patil, S. H. (1981): *Phys. Rev.*, **A24:**2913. (14.6)
Pattanayak, D. N., and E. Wolf (1976): *Phys. Rev.*, **D13:**913 and 2287. (10.3)
Pätzel, B., M. Stingl, and A. Weiguny (1980): *Nucl. Phys.*, **A339:**175. (10.1.4)
Payne, G. (1969): *Phys. Rev.*, **179:**85. (16.4)
— (1970): *Phys. Rev.*, **A2:**775. (16.3)
— (1981): in F. S. Levin (1981), p. 61c. (17.4)
Pearce, W. A. (1964): *Progr. Theor. Phys.*, **32:**180. (20.3)
Pearson, D. B. (1966): *Nuovo Cim.*, **45:**607. (17.4.7)
Pechukas, P. (1969): *Phys. Rev.*, **181:**166. (18.2)
Peek, J. M. (1964): *Phys. Rev.*, **134:**A877. (16)
Peierls, R. E. (1959): *Proc. Roy. Soc.* (*London*), **A253:**16. (14.3, 16.5, 17.1.3)
Percival, I. C. (1957): *Proc. Phys. Soc.* (*London*), **A70:**494. (11.3)
— (1960): *Phys. Rev.*, **119:**159. (11.3)
— (1962): *Proc. Phys. Soc.* (*London*), **80:**1290. (20.3.2)

— (1963): *Proc. Phys. Soc.* (*London*), **82**: 528. (20.3.2)
— and M. Roberts (1963): *Proc. Phys. Soc. London*, **82**: 519. (20.3.2)
Peres, A. (1980): *Ann. Phys.* (*N.Y.*), **129**: 33. (19.2)
Peterson, B., and S. Ström (1973): *Phys. Rev.*, **D8**: 3661. (1.4, 4.1)
— and — (1974): *Phys. Rev.*, **D10**: 2670. (1.5, 4.1)
Petráš, M. (1962): *Czechoslov. J. Phys.*, **B12**: 87. (20.3)
— (1964a): *Nuovo Cimento*, **31**: 247. (12.1.2)
— (1964b): *Czechoslov. J. Phys.*, **B14**: 895. (12.1.2)
— (1965): *Czechoslov. J. Phys.*, **B15**: 627. (12.1.2)
Petzold, J. (1959): *Z. Phys.*, **155**: 422, **157**: 122. (19.2)
Pfluger, A. (1943): *Communs. Math. Helv.*, **16**: 1. (12.1)
Pieper, S. C., L. Schlessinger, and J. Wright (1970): *Phys. Rev.*, **D1**: 1674. (16.4)
—, J. Wright, and L. Schlessinger (1970): *Phys. Rev.*, **D2**: 1561. (16.4)
—, —, and — (1971): *Phys. Rev.*, **D3**: 2419. (17.4)
Pisent, G., V. Vanzani, and L. Fonda (1978): eds., *Few-Body Nuclear Physics*, IAEA, Vienna. (17.4)
Plekhanov, E. B., A. S. Suzko, and B. N. Zakhariev (1981): *J. Phys.*, **A**. (14.7, 17.3)
Plemelj, J. (1904): *Monatsh. Math. Phys.*, **15**: 93. (9.3)
Poincaré, H. (1884): *Acta Math*, **4**: 201. (12.1)
— (1910): *Acta Math.*, **33**: 57 [reprinted in "Oeuvres," vol. 3, p. 556, Gauthier-Villars, Paris (1934) (but dated erroneously)]. (9.3)
Polkinghorne, J. C. (1958): *Proc. Cambridge Phil. Soc.*, **54**: 560. (12.1.3)
Pollak, E., and R. D. Levine (1974): *Phys. Rev.*, **A9**: 2398. (16.4)
Poluéktov, I., L. Presnyakov, and I. Sobelman (1964): *Z. Eksp. i. Teor. Fiz.*, **47**: 181 (transl. *Soviet Phys. JETP*, **20**: 122). (16.3)
Polyzou, W. N. (1980): *J. Math. Phys.*, **21**: 506. (17.4)
— (1981): *Nucl. Phys.*, **A**; *J. Math. Phys.*, **22**: 798. (17.4)
— and E. F. Redish (1978): in H. Zingl et al. (1978) I, p. 379. (17.4)
— and — (1979): *Ann. Phys.* (*N.Y.*), **119**: 1. (17.4)
Popova, A. M., and V. V. Komarov (1967): *Nucl. Phys.*, **A90**: 634. (17.4)
Portinari, J. C. (1967): *Ann. Phys.* (*N.Y.*), **45**: 445. (20.3)
Potapov, V. S. (1972): *Zh. Eksp. Teor. Fiz.*, **63**: 431 and 2094 (transl. *Soviet Phys. JETP*, **36**: 228 and 1105). (16)
— (1980): *Teor. i Mat. Fiz.*, **43**: 65 (transl. *Theor. and Math. Phys.*, **43**: 323). (17.4)
— and J. R. Taylor (1977): *Phys. Rev.*, **A16**: 2264 and 2276. (17.4.5)
Povsner, A. Ya. (1953): *Mat. Sbornik* (*N.S.*), **32(74)**: 109 (*Math. Rev.*, **14**: 755). (7.3)
— (1955): *Dokl. Akad. Nauk SSSR*, **104**: 360 (*Math. Rev.*, **17**: 1205). (7.3)
Prats, F., and J. S. Toll (1959): *Phys. Rev.*, **113**: 363. (20.3)
Predazzi, E. (1966): *Ann. Phys.* (*N.Y.*), **36**: 228 and 250. (18.3)
— and T. Regge (1962): *Nuovo Cimento*, **24**: 518. (12.4)
Preist, T. (1978): *J. Chem. Phys.*, **69**: 1227. (17.4)
Prigogine, I., and F. Henin (1957): *Bull. Classe Sci. Acad. Roy. Belg.*, **43**: 814. (5)
— and — (1958): *Physica*, **24**: 214. (5)
Pritchard, D. E. (1972): *J. Chem. Phys.*, **54**: 4206. (20.1)
Prosser, R. T. (1964): *J. Math. Phys.*, **5**: 708. (6.7.3)
Prosser, R. T. (1969): *J. Math. Phys.*, **10**: 1819. (20.5.4)
— (1976): *J. Math. Phys.*, **17**: 1775. (20.5.4)
— (1980): *J. Math. Phys.*, **21**: 2648. (20.5.4)
Prugovečki, E. (1973): *J. Math. Phys.*, **14**: 957. (16.2, 16.2.3)
— (1974): *Phys. Lett.*, **49B**: 305. (16.2)

— and J. Zorbas (1973a): *Nucl. Phys.*, **A213:**541. (14.6)
— and — (1973b): *J. Math. Phys.*, **14:**1398 and **15:**268. (12.4)
— and — (1978): in J. Nuttall (1978), p. 63. (14.6)
Puff, R. D. (1975): *Phys. Rev.*, **A11:**154. (20.3.2)
Purcell, E. M. (1956): *Nature*, **178:**1449. (4.3)
Pustovalov, V. V., and Yu. A. Simonov (1966): *Zh. Eksp. Teor. Fiz.*, **51:**345 (transl. *Soviet Phys. JETP*, **24:**230). (17.4.6)
Rabitz, H. A. (1971): *J. Chem. Phys.*, **55:**407. (16.4)
— (1972): *Phys. Rev.*, **A5:**620. (9.1.1)
— and R. Conn (1973): *Phys. Rev.*, **A7:**577. (11.3)
— and R. Gordon (1970): *J. Chem. Phys.*, **53:**1815 and 1831; *J. Math. Phys.*, **11:**3339. (8.2)
Rahman, N. K. (1974): *Phys. Rev.*, **A10:**440. (16)
Rajaraman, R., and L. Susskind (1965): *Nuovo Cimento*, **38:**1201. (16.5)
Ramakrishnan, A., G. Ramachandran, and V. Devanathan (1961): *Nuovo Cim.*, **21:**145. (16.2)
Ramm, A. G. (1963): *Dokl. Akad. Nauk SSSR*, **152:**282 (transl. *Soviet Math. Dokl.*, **4:**1303). (12.1.2)
— (1965a): *Mat. Sb.* (*N.S.*), **66:**321 (*Math. Rev.*, **30**, #3297) and **72:**638 (*Math. Rev.*, **34**, #7994). (12.1.2)
— (1965b): *Dokl Akad. Nauk SSSR*, **157:**1073 (transl. *Soviet Phys. Dokl.*, **9:**645). (20.3)
— (1966): *Dokl. Akad. Nauk SSSR*, **166:**1319 (transl. *Soviet Phys. Dokl.*, **11:**114). (16.6)
— (1970): *Dokl. Akad. Nauk SSSR*, **191:**50 (transl. *Soviet Phys. Dokl.*, **15:**231). (10.1)
— (1972): *Dokl. Akad. Nauk SSSR*, **204:**1071 (transl. *Soviet Phys. Dokl.*, **17:**522). (16.6)
— (1980a): in Varadan and Varadan (1980), p. 537. (3.1)
— (1980b): *J. Math. Phys.*, **21:**308. (10.3.3)
— and B. A. Taylor (1980): *J. Math. Phys.*, **21:**2395. (12.1.2)
Raphael, R. B., P. C. Tandy, and W. Tobocman (1976): *Phys. Rev.*, **C14:**1355 and **C15:**463. (16.3.3)
Rarita, W., and J. Schwinger (1941): *Phys. Rev.*, **59:**436. (15.2)
Rau, A. R. P. (1976): *J. Math. Phys.*, **17:**1338. (18.2)
Rayleigh, Lord (J. W. Strutt) (1871): *Phil. Mag.*, **41:**107, 274, and 447. (3.1)
— (1881): *Phil. Mag.*, **12:**81. (3.1.2)
— (1899): *Phil. Mag.*, **47:**375. (3.1)
Raynal, J., and J. Revai (1970): *Nuovo Cim.*, **68:**612. (17.4.6)
Reading, J. F., and W. H. Bassichis (1972): *Phys. Rev.*, **D5:**2031. (18)
Rédei, L. B. (1965): *J. Math. Phys.*, **6:**487; *Arkiv. fys.*, **29:**1 and 11. (6)
Redish, E. F. (1974): *Nucl. Phys.* **A225:**16 and **A235:**82; *Phys. Rev.*, **C10:**67. (17.4)
— (1978): in H. Zingl et al. (1978) II, p. 427. (17.4)
—, P. C. Tandy, and M. L'Huillier (1976): *Phys. Lett.*, **61B:**413. (17.4)
Redmond, P. J. (1964): *J. Math. Phys.*, **5:**1547. (20.7)
Reed, M., and B. Simon (1972): *Methods of Modern Mathematical Physics I*: Functional Analysis, Academic Press, N.Y. (7.3.3)
Regge, T. (1958a): *Nuovo Cimento*, **8:**671. (12.1.4)
— (1958b): *Nuovo Cimento*, **9:**295. (12.1.1)
— (1959): *Nuovo Cimento*, **14:**951. (12.1.1, 12.3, 13.3, 13.4, 20.4)
— (1960): *Nuovo Cimento*, **18:**947. (12.1.1, 12.1.4, 13.3)

— (1963): In "Theoretical Physics," International Atomic Energy Agency, Vienna. (12)
Reinhardt, W. P. (1970): *Phys. Rev.*, **A2:**1767 and **A4:**429. (9.3, 12.1.2, 16.4)
— (1973): *Phys. Rev.*, **A8:**754. (17.1.3)
— (1977): *Phys. Rev.*, **A15:**802. (16.6)
—, D. W. Oxtoby, and T. N. Rescigno (1972): *Phys. Rev. Lett.*, **28:**401. (12.1.2)
— and A. Szabo (1970): *Phys. Rev.*, **A1:**1162. (12.1.2)
Remler, E. A. (1966): *J. Math. Phys.*, **7:**871. (17.4.1)
— (1971): *Phys. Rev.*, **A3:**1949. (13.3)
— (1975): *Ann. Phys.* (*N.Y.*), **95:**455. (6)
Résibois, P. (1959): *Physica*, **25:**725. (5)
Riahi, F. (1969): *Helv. Phys. Acta*, **42:**299. (17.1)
Rich, V. G., S. M. Bobbio, R. L. Champion, and L. Doversfihe (1971): *Phys. Rev.*, **A4:**2253 (20.1)
Riesz, F., and B. Sz-Nagy (1955): "Functional Analysis," Ungar, New York. (7.3)
Riley, M. E. (1973): *Phys. Rev.*, **A7:**626. (18.3)
Risberg, V. (1956): *Arch. Mat. Naturvidenskab*, **53:**1. (11.3)
Ritus, V. I. (1961): *Z. Eksp. i Theor. Fiz.*, **40:**352 (transl. *Soviet Phys. JETP*, **13:**240). (17.4.6)
Roberts, M. J. (1963): *Proc. Phys. Soc.* (*London*), **82:**594. (20.3.2)
— (1964): *Proc. Phys. Soc. London*, **13:**503 and **84:**825. (20.32)
— (1965): *Proc. Phys. Soc. London*, **86:**683. (20.3.2)
Rochus, P. (1979): *J. Math. Phys.*, **20:**781. (20.3)
Rodberg, L. S. (1961): *Phys. Rev.*, **124:**210. (16.5)
Rohlfing, E. A., J. G. Loeser, A. E. DePristo, and H. Rabitz (1980): *J. Chem. Phys.*, **72:**464. (17.3)
Rollnik, H. (1956): *Z. Phys.*, **145:**639 and 654. (7, 12.1.4, 16.5)
Romo, W. J. (1975): *Nucl. Phys.*, **A237:**275. (12.1.4)
— (1978): *Nucl. Phys.*, **A302:**61. (12.1.4)
— (1979): *J. Math. Phys.*, **20:**1210 and 2704. (12.1.1)
— (1980): *J. Math. Phys.*, **21:**311. (12.1.4, 16.6)
Ronveaux, A. (1968a): *C. R. Acad. Sci. Paris*, **266A:**306. (12.1.4)
— (1968b): *C. R. Acad. Sci. Paris*, **266A:**478. (11.3, 16.6)
Roscoe, R. (1938): *Phil. Mag.*, **26:**32. (16)
— (1941): *Phil. Mag.*, **31:**349. (16)
Rose, M. E. (1955): "Multipole Fields," Wiley, New York. (2.1.3, 2.1.4)
— (1957): "Elementary Theory of Angular Momentum," Wiley, New York. (2.1.4, 15.1, 17.4.3)
— (1961): "Relativistic Electron Theory," Wiley, New York. (8.3.2, 14.6)
Rosen, M., and D. R. Yennie (1964): *J. Math. Phys.*, **5:**1505. (18.2)
Rosenberg, L. (1963): *Phys. Rev.*, **131:**874. (17.4.3, 18.4)
— (1964a): *Phys. Rev.*, **134:**B937. (9.1.3)
— (1964b): *Phys. Rev.*, **135:**B715. (17.4.3, 18.4)
— (1965): *Phys. Rev.*, **138:**B1343. (11.3)
— (1966): *Phys. Rev.*, **147:**1016. (17.4)
— (1968): *Phys. Rev.*, **168:**1756. (16.4)
— (1970): *Phys. Rev.*, **D1:**1019. (11.3)
— (1971): *Phys. Rev.*, **A4:**734. (17.4)
— (1973): *Phys. Rev.*, **D8:**1833. (10.3, 16.4, 17.4)
— (1974): *Phys. Rev.*, **D9:**1789. (16.4)

— (1976): *Phys. Rev.*, **C13:**1406. (16.3)
— (1977): *Phys. Rev.*, **A16:**1941. (16)
— (1979): *Phys. Rev.*, **A19:**1568 and **A20:**457 and 1352. (16)
— (1981): *Phys. Rev.*, **A23:**2283. (16)
— (1982): *J. Phys. A*, . (16)
— and L. Spruch (1960): *Phys. Rev.*, **120:**474. (11.3)
— and — (1961): *Phys. Rev.*, **121:**1720. (11.3)
— and — (1962): *Phys. Rev.*, **125:**1407. (11.3)
— and — (1974): *Phys. Rev.*, **A10:**2002. (11.3)
— and — (1975): *Phys. Rev.*, **A12:**1297. (11.3.4)
—, —, and T. F. O'Malley (1960a): *Phys. Rev.*, **118:**184. (11.3)
—, —, and — (1960b): *Phys. Rev.*, **119:**164. (11.3)
— and S. Tolchin (1973): *Phys. Rev.*, **A8:**1702. (16.4)
Rosenblum, M. (1957): *Pacific J. Math.*, **7:**997. (7.3)
Rosenfeld, L. (1961): *Nuclear Phys.*, **26:**594. (16.5)
—. (1965): *Nucl. Phys.*, **70:**1. (19.2)
Rosenthal, C. M. (1974): *Phys. Rev.*, **A9:**273. (10.3.5)
Roskies, R. (1966): Ph.D. thesis, Princeton, Oak Ridge National Laboratory Report 3914. (13.1, 17.2, 17.4.3)
Ross. M. H. (1963a) (ed.): "Quantum Scattering Theory," Indiana University Press, Bloomington, Ind. (6)
— (1963b): *Phys. Rev. Letters.*, **11:**450 and 567. (17.1)
— and G. L. Shaw (1960): *Ann. Phys. (N.Y.)*, **9:**391. (17.3)
— and — (1961): *Ann. Phys. (N.Y.)*, **13:**147. (17.3)
Ross, W. R., and J. R. Taylor (1969): *J. Math. Phys.*, **10:**654. (12.4)
Rotenberg, M. (1963): *Ann. Phys. (N.Y.)*, **21:**579. (9.1.1)
Rubin, M., R. Sugar, and G. Tiktopoulos (1966): *Phys. Rev.*, **146:**1130. (17.4.4)
—, —, and — (1967): *Phys. Rev.*, **159:**1348 and **162:**1555. (17.4.4)
Rudin, R. A. (1970): *Phys. Rev.*, **D1:**2557 and **D4:**272. (18.3)
Rule, D. W., and Y. Hahn (1976): *Phys. Rev.*, **C14:**1102. (16.3)
Rutherford, E. (1911): *Phil. Mag.*, **21:**669. (5.3)
Rys, F. (1965): *Helv. Phys. Acta*, **38:**457. (6.7)
Saavedra, I. (1962): *Nucl. Phys.*, **29:**137. (10.3.4)
— (1966): *Nucl. Phys.*, **87:**267. (10.3.4)
Sabatier, P. C. (1965): *Nuovo Cimento*, **37:**1180. (18.2, 20.1)
— (1966): Ph.D. thesis, U. of Paris, *C. R. Acad. Sci. Paris*, **263:**788; *J. Math. Phys.*, **7:**1515 and 2079. (20.4)
— (1967): *C. R. Acad. Sci. Paris*, **265:**5; *J. Math. Phys.*, **8:**905 and 1957. (20.4)
— (1968): *J. Math. Phys.*, **9:**1241. (20.4)
— (1971): *J. Math. Phys.*, **12:**1303. (20.4)
— (1972a): *J. Math. Phys.*, **13:**675. (20.4)
— (1972b): in L. Colin, ed., *Mathematics of Profile Inversion*, **NASA TMX-62**, 150, p. 9.2. (20.4)
— (1973): *Phys. Rev.*, **A8:**589; *Internat. J. Quantum Chem. Symp.*, **7:**421. (20.1)
— (1978): *J. Math. Phys.*, **19:**2410. (20)
— (1982): *Radio Sci.*, . (20)
— and F. Q. Van Phu (1971): *Phys. Rev.*, **D4:**127. (20.4)
Sachs, R. G., and M. Goeppert-Mayer (1938): *Phys. Rev.*, **53:**991. (10.43)
Saenz, A. W., and W. W. Zachary (1976): *J. Math. Phys.*, **17:**409. (10.1.4)
Saito, S. (1968): *Progr. Theor. Phys.*, **40:**893. (16.3.3)

— (1969): *Progr. Theor. Phys.*, **41**:705. (16.3.3)
Saito, Y. (1979): *Spectral Representations of Schrödinger Operators with Long-Range Potentials*, Lecture Notes in Math. vol. 727, Springer-Verlag, Berlin. (14.6)
— (1980a): *J. Reine Angew. Math.*, **314**:99. (14.6)
— (1980b): *Osaka J. of Math.* (20.5)
— (1981): *Osaka J. of Math.* (20.5)
Sakhnovich, L. A. (1965): *Mat. Sb.* (*N.S.*), **67**:221 (*Math. Rev.*, **33**, 882). (14.6)
— (1972): *Teor. i Mat. Fiz.*, **13**:421 (transl. *Theor. Math. Phys.*, **13**:1239). (14.6, 16.2)
Sakmar, I. A. (1978): *J. Math. Phys.*, **19**:2124. (20.2)
— (1979): *J. Math. Phys.*, **20**:11. (20.2)
— (1980): *Lett. Nuovo Cim.*, **28**:118. (20.2)
Salpeter, E. E. (1950): *Proc. Phys. Soc.* (*London*), **63A**:1295. (16)
Sanders, W. A., and C. R. Muller (1963): *J. Chem. Phys.*, **39**:2572. (20.1)
Sandhas, W. (1972): *Acta. Phys. Austr. Suppl.*, **9**:57. (17.4)
— (1974): *Acta. Phys. Austr. Suppl.*, **13**:679. (17.4)
— (1975): *Czech. J. Phys.*, **B25**:251. (17.4)
— (1976): in A. N. Mitra et al., eds., *Few-Body Dynamics*, North-Holland Publ. Co., Amsterdam, p. 540. (17.4)
— (1978): in G. Pisent et al. (1978), p. 3. (16.2.1, 17.4)
Sano, M., S. Yoshida, and T. Teresawa (1958): *Nucl. Phys.*, **6**:20. (16.5)
Sanz, P., J. Sanudo, and J. Sesma (1981): *J. Math. Phys.*, **22**:2594. (12.3)
Sartori, L. (1963): *J. Math. Phys.*, **4**:1408. (12.1.4)
Sasakawa, T. (1963): *Suppl. Progr. Theor. Phys.*, **27**:1. (7, 16.3, 9.2)
— (1964): *Progr. Theor. Phys.*, **31**:787, **32**:565. (17.1)
— (1967): *Phys. Rev.*, **158**:1249. (16.3, 17.4)
— (1971): *Nucl. Phys.*, **A160**:321. (17.4)
— (1972): *Nucl. Phys.*, **A186**:417. (17.4)
— (1973): *Nucl. Phys.*, **A203**:496. (17.4)
— (1976): *Phys. Rev.*, **C13**:1801. (17.4)
— (1977a): *Sci. Rep.* (*Tohoku*), **60**:103. (17.4)
— (1977b): *Progr. Theor. Phys. Suppl.*, **61**:149. (17.4)
— (1978): *Phys. Rev.*, **C17**:2015 and 2056. (17.4)
— and T. Sawada (1977a): *Progr. Theor. Phys. Suppl.*, **61**:1. (9)
— and — (1977b): *Progr. Theor. Phys. Suppl.*, **61**:61. (17.4)
— and — (1979): *Phys. Rev.*, **C20**:1954. (17.4)
— and T. Tsukamoto (1965): *Progr. Theor. Phys.*, **34**:442. (17.1)
Sauer, P. V. (1978): in G. Pisent et al. (1978), p. 313. (17.4)
Saxon, D. S., and L. I. Schiff (1957): *Nuovo Cimento*, **6**:614. (18.3)
Sayasov, Yu. S. (1975): *Z. Physik*, **A275**:103. (18.2)
Scadron, M., and S. Weinberg (1964): *Phys. Rev.*, **133**:B1589. (9.2)
—, S. Weinberg, and J. A. Wright (1964): *Phys. Rev.*, **135**:B202. (9.1.1)
Scarf, F. L. (1958): *Phys. Rev.*, **109**:2170. (12.4)
Schechter, M. (1977): *Lett. Math. Phys.*, **2**:127. (12.4)
Schey, H. M., and J. L. Schwartz (1965): *Phys. Rev.*, **139**:B1428. (11.2)
Schiff, L. I. (1956a): *Phys. Rev.*, **103**:443. (18.3)
— (1956b): *Phys. Rev.*, **104**:1481. (3.4, 18.3)
— (1968a): *Phys. Rev.*, **176**:1390. (18)
— (1968b): "Quantum Mechanics," 3d ed., McGraw-Hill, New York. (6, 8.1.3, 18.2.1)

Schmid, E. W. (1980): *Phys. Rev.*, **C21:**691. (16.3.3)
— (1981): preprint. (16.3.3)
— and H. Ziegelmann (1974): *The Quantum Mechanical Three Body Problem*, Pergamon Press, Oxford. (17.4)
Schneider, B. I. (1981): *Phys. Rev.*, **A24:**1. (16.6)
Schneider, B., H. S. Taylor, and R. Yaris (1970): *Phys. Rev.*, **A1:**855. (16)
Schönberg, M. (1951): *Nuovo Cimento*, **8:**651. (7.1.1)
Schulman, L. (1967): *Phys. Rev.*, **156:**1129. (17.4)
Schwager, J. (1976): *Ann. Phys.* (*N.Y.*) **98:**14. (17.4)
Schwartz, C. (1961): *Phys. Rev.*, **124:**1468. (16)
— (1966): *Phys. Rev.*, **141:**1457. (11.3)
Schweber, S. S. (1961): "An Introduction to Quantum Field Theory," Harper & Row, New York. (6, 8.1)
Schwinger, J. (1947a): Harvard University lectures (unpublished). (11.3)
— (1947b): *Phys. Rev.*, **72:**742. (11.2.2, 11.3)
— (1948a): *Phys. Rev.*, **73:**407. (15.1)
— (1948b): *Phys. Rev.*, **74:**1439. (6.4, 6.5)
— (1950): *Phys. Rev.*, **78:**135. (11.3)
— (1952): *U.S. Atomic Energy Commission Rept.* NYO-3071, reprinted in L. C. Biedenharn and H. van Dam (eds.), "Quantum Theory of Angular Momentum," Academic, New York, 1965. (15.1)
— (1954): *Phys. Rev.*, **94:**1362. (9.2)
— (1960): *Ann. Phys.* (*N.Y.*), **9:**169. (19.2)
— (1961): *Proc. Natl. Acad. Sci. U.S.*, **47:**122. (9.1.1, 12.1.4)
— (1964): *J. Math. Phys.*, **5:**1606. (14.6)
Screaton, G. R. (1968): *Phys. Rev.*, **165:**1610. (10.3.4)
— (1969): *Phys. Rev.*, **182:**1415. (10.3.4)
Seaton, M. J. (1962): In D. R. Bates (ed.), "Atomic and Molecular Processes," Academic, New York. (16)
— (1978): in J. Nuttall (1978), p. 83. (14.6)
Segall, B. (1951): *Phys. Rev.*, **83:**1247. (18.4)
Semon, M. D., and J. R. Taylor (1975): *Nuovo Cim.*, **26A:**48. (14.6)
— and — (1977): *Phys. Rev.*, **A16:**33. (18.3)
Senior, T. B. A., and R. F. Goodrich (1964): *Proc. Inst. Electr. Engin.*, **111:**907. (3.8)
Serber, R. (1963): *Phys. Rev. Letters*, **10:**357. (18.3)
— (1964): *Revs. Modern Phys.*, **36:**649. (18.3)
Servadio, S. (1981): *Nuovo Cim.*, **65B:**57. (17.4.5)
Seto, N. (1974): *Publ. of RIMS* (*Kyoto*), **9:**429. (12.1.4)
Shaffer, G. R. (1971): *J. Math. Phys.*, **12:**2302. (13.3)
Shakeshaft, R., and L. Spruch (1973): *Phys. Rev.*, **A8:**206. (16.4)
— and — (1979): *Phys. Rev.*, **A20:**376. (17.2)
Shaw, G. L., and M. H. Ross (1962): *Phys. Rev.*, **126:**806 and 814. (17.3)
Shewell, J. R., and E. Wolf (1968): *J. Opt. Soc. Am.*, **58:**1596. (20.3)
Shirokov, M. I. (1957): *Z. Eksp. i Teor. Fiz.*, **32:**1022 (transl. *Soviet Phys. JETP*, **5:**835). (15.1)
— (1959): *Z. Eksp. i Teor. Fiz.*, **36:**1524 (transl. *Soviet Phys. JETP*, **9:**1081). (15.1)
— (1961): *Z. Eksp. i Teor, Fiz.*, **40:**1387 (transl. *Soviet Phys. JETP*, **13:**975). (17.4.6)
Shirokov. Yu. M. (1958): *Z. Eksp. i Teor. Fiz.*, **35:**1005 (transl. *Soviet Phys. JETP*, **8:**703). (15.1)
Shukre, C., and P. Winternitz (1972a): *Phys. Rev.*, **D6:**3592. (15.1)

— and — (1972b): *Phys. Rev.*, **D6:**3607. (17.4.6)
Siegert, A. J. F. (1939): *Phys. Rev.*, **56:**750. (16.5)
Sigal, I. M. (1978a): *Bull. Am. Math. Soc.*, **84:**152. (17.4)
— (1978b): *Mem. Am. Math. Soc.*, **16**, #209. (17.4)
— (1978c): *Lecture Notes in Mathematics*, Springer–Verlag, to be published. (16.2.3)
— (1979): preprint, Princeton. (16.2.3)
— (1980): in J. A. De Santo et al. (1980), p. 149. (16.2.3)
Simon, A. (1953): *Phys. Rev.*, **92:**1050, **93:**1435. (15.1)
— and T. A. Welton (1953): *Phys. Rev.*, **90:**1036, **93:**1435. (15.1)
Simon, B. (1969a): *J. Math. Phys.*, **10:**1123. (12.1.4)
— (1969b): *Commun. Pure Appl. Math.*, **22:**531. (12.1.2)
— (1970): *Helv. Phys. Acta*, **43:**607. (17.4.8)
— (1971): *Quantum Mechanics for Hamiltonians Defined as Quadratic Forms*, Princeton Univ. Press. (6.8, 7.3, 10.3.2)
— (1972): *Commun. Math. Phys.*, **27:**1. (12.1.1)
— (1973): *Ann. Math.* **97:**247. (12.1.1, 16.6)
— (1974): *Math. Ann.*, **207:**133. (16.6)
— (1976): in E. Lieb, B. Simon, and A. Wightman, eds., *Studies in Mathematical Physics*, Princeton Univ. Press. p. 305. (12.1.4)
— (1978): *Internat. J. Quantum Chem.*, **14:**529. (12.1.1)
Simonov, Yu. A. (1966): *Yad. Fiz.*, **3:**630 (transl. *Soviet J. Nucl. Phys.*, **3:**461). (17.4.6)
— (1968): *Yad. Fiz.*, **7:**1210 (transl. *Soviet J. Nucl. Phys.*, **7:**722). (17.4.6)
Sinfailam; A.–L., and J. C. Y. Chen (1972): *Phys. Rev.*, **A5:**1218. (17.4)
Singh, S. R., and A. D. Stauffer (1975): *J. Phys. A*, **8:**1379. (11.3)
Sitenko, A. G. (1978): in H. Zingl et al. (1978) II, p. 210. (17.4)
— and V. F. Karchenko (1963): *Nucl. Phys.*, **49:**15. (17.4)
Skorniakov, G. V. (1956): *Z. Eksp. i Teor. Fiz.*, **31:**1046 (transl. *Soviet Phys. JETP*, **4:**910). (17.4.3)
— and K. A. Ter-Martirosian (1956): *Z. Eksp. i Teor. Fiz.*, **31:**775 (transl. *Soviet Phys. JETP*, **4:**648). (17.4.3)
Skriganov, M. M. (1978): *Dokl. Akad. Nauk SSSR*, **241:**326 (transl. *Soviet Phys. Dokl.*, **23:**472). (18)
Sloan, I. H. (1968): *Phys. Rev.*, **165:**1587. (17.4)
— (1972): *Phys. Rev.*, **C6:**1945. (17.4)
— (1973): *Phys. Rev.*, **A7:**1016. (14.6)
— (1981): *Phys. Rev.*, **C23:**1289. (17.4)
— and T. J. Brady (1972): *Phys. Rev.*, **C6:**701. (11.3)
Smirnov, B. M. (1963): *Z. Eksp. i Teor, Fiz.*, **45:**155 (transl. *Soviet Phys. JETP*, **18:**111). (16)
Smith, F. T. (1960): *Phys. Rev.*, **118:**349, **119:**2098. (19.2)
— (1962): *J. Math. Phys.*, **3:**735. (17.4.3)
Smith, F. T., R. P. Marchi, and K. G. Dedrick (1966): *Phys. Rev.*, **150:**79. (18.3)
Smithies, F. (1958): "Integral Equations," Cambridge, New York. (9.3)
Snyder, H. S. (1951): *Phys. Rev.*, **83:**1154. (6.7.3)
Sobel, M. I. (1968): *J. Phys. A*, **1:**610. (20.3)
Sokolov, S. N. (1979): preprint, Serpukhov. (20.3)
Sommerfeld, A. (1910): *Physik. Z.*, **11:**1057. (4.1.2, 7.1.1)
— (1912): *Jahresber. deut. Math. Ver.*, **21:**309. (4.1.2, 7.1.1)
— (1931): *Ann. Phys.*, **11:**257. (7.1.1)
— (1933): "Atomic Structure and Spectral Lines," Dutton, New York. (5.3)

— (1949): "Partial Differential Equations of Physics," Academic, New York. (3.8, 4.1.2)
Soper, P. J. R. (1974): *Phys. Rev.*, **C10:** 1282. (16.3)
Spector, R. M. (1964): *J. Math. Phys.*, **5:** 1185. (12.4, 14)
— (1966): *J. Math. Phys.*, **7:** 2103. (12.4)
Springer, G. (1957): "Introduction to Riemann Surfaces," Addison–Wesley, Reading, Mass. (17.1)
Spruch, L. (1958): *Phys. Rev.*, **109:** 2149. (11.3)
— (1962): In Brittin, Downs, and Downs (eds.), "Lectures in Theoretical Physics," vol. 4, Interscience, New York. (11.3)
— and M. Kelly (1958): *Phys. Rev.*, **109:** 2144. (11.3)
—, T. F. O'Malley and L. Rosenberg (1960): *Phys. Rev. Letters*, **5:** 375. (11.2.2)
—, —, and — (1962): *Phys. Rev.*, **125:** 1300. (11.2.2)
— and L. Rosenberg (1959): *Phys. Rev.*, **116:** 1034. (11.3)
— and — (1960a): *Phys. Rev.*, **117:** 143 and 1095. (11.3)
— and — (1960b): *J. Appl. Phys.*, **31:** 2104. (11.3)
— and — (1978): in J. Nuttall (1978), p. 95. (11.3)
— and R. Shakeshaft (1979): *Phys. Rev.*, **A19:** 1023. (18.2)
Squires, E. J. (1963): "Complex Angular Momenta and Particle Physics," Benjamin, New York. (13.1)
— (1964): *Nuovo Cimento*, **34:** 1751. (17.1)
Stankevich, I. V. (1962): *Dokl. Akad. Nauk SSSR*, **144:** 279 (transl. *Soviet Math. Dokl.*, **3:** 719). (6.7)
Stelbovics, A. T., and L. R. Dodd (1972): *Phys. Lett.*, **39B:** 450. (17.4.8)
Stillinger, F. H., and D. R. Herrick (1975): *Phys. Rev.*, **A11:** 446. (16.6)
Stokes, G. C. (1852): *Cambridge Phil. Soc. Trans.*, **9:** 399. (1.2)
Stone, J. M. (1963): "Radiation and Optics," McGraw-Hill, New York. (Pt. I)
Storm, D., and D. Rapp (1976): *Phys. Rev.*, **A14:** 193. (18.3)
Stratton, J. A. (1941): "Electromagnetic Theory," McGraw-Hill, New York. (4.1.1)
Ström, S. (1974): *Phys. Rev.*, **D10:** 2685. (1.5, 4.1)
— (1976): *Phys. Rev.*, **D13:** 3485. (8.1)
Strutt, J. W.: See Rayleigh, Lord.
Stueckelberg, E. C. G. (1943): *Helv. Phys. Acta*, **17:** 3. (6.4)
— (1945): *Helv. Phys. Acta*, **18:** 195. (6.4)
— (1946): *Helv. Phys. Acta*, **19:** 242. (6.4)
Suck, S. H., and R. W. Emmons (1981): *Phys. Rev.*, **A24:** 129. (16)
Sudarshan, E. C. G. (1963): *Phys. Rev. Letters*, **10:** 277. (4.3, 8.3.1)
—, C. B. Chiu, and V. Gorini (1978): *Phys. Rev.*, **D18:** 2914. (19.2)
Sugar, R., and R. Blankenbecler (1965): *Phys. Rev.*, **136:** B472. (17.1)
— and — (1969): *Phys. Rev.*, **183:** 1387. (18.3)
Sunakawa, S. (1955): *Progr. Theor. Phys.*, **14:** 175. (6)
— (1960): *Progr. Theor. Phys.*, **24:** 963. (16.2)
Surkov, E. L. (1967): *Yad. Fiz.*, **5:** 908 (transl. *Soviet J. Nucl. Phys.*, **5:** 644). (17.4.6)
Suura, H., Y. Mimura, and T. Kimura (1952): *Progr. Theor. Phys.*, **7:** 171. (6.7.3)
Swam, N. V. V. J., and L. C. Biedenharn (1963): *Phys. Letters*, **6:** 315. (14.6)
Swan, P. (1960): *Nature*, **187:** 585; *Nuclear Phys.*, **18:** 245, **21:** 233. (9.1.3)
— (1961): *Nuclear Phys.*, **27:** 620. (9.1.3)
— (1963a): *Nuclear Phys.*, **42:** 134. (9.1.2)
— (1963b): *Australian J. Phys.*, **16:** 177. (9.1.3)
— (1967a): *Nucl. Phys.*, **A90:** 412 and 423. (14.6)

— (1967b): *Nucl. Phys.*, **A90:**436. (20.3)
— (1968): *Nucl. Phys.*, **A119:**40. (12.1.3)
— and W. A. Pearce (1966): *Nuclear Phys.*, **79:**77. (20.2)
Swift, A. R. (1964): *Nuovo Cimento*, **33:**1119. (10.3.4)
— (1974): *Phys. Rev.*, **D9:**1740. (18.3)
Symes, W. W. (1980): *MRC Tech. Rep.* #**1990** (Univ. of Wisc.). (20)
Tabakin, F. (1968): *Phys. Rev.*, **174:**1208. (10.1.4)
— (1969): *Phys. Rev.*, **177:**1443. (20.1)
Tadic, D., and T. F. Tuan (1965): *Nuovo Cimento*, **36:**463. (17.4)
Takahashi, Y., and N. Mishima (1965): *Progr. Theor. Phys.*, **34:**498. (17.4)
Takatsuka, K., and T. Kueno (1979): *Phys. Rev.*, **A19:**1011. (11.3.3)
— and V. McKoy (1981): *Phys. Rev.*, **A23:**2352, 2359, and **24:**2473. (11.3)
—, R. R. Lucchese, and V. McKoy (1981): *Phys. Rev.*, **A24:**1812. (11.3)
Takayanagi, K. (1963): *Progr. Theor. Phys., Suppl.*, **25:**1. (16)
Takeda, G., and K. M. Watson (1955): *Phys. Rev.*, **97:**1336. (16.3.3)
Takeuchi, K. (1972): *Phys. Lett.*, **40B:**73. (17.4)
— (1973): *Nucl. Phys.*, **A208:**1, 21, 46. (17.4)
— and P. A. Moldauer (1970): *Phys. Rev.*, **C2:**920. (16.6)
Takibaev, N. Zh. (1977): *Yad. Fiz.*, **26:**81 (transl. *Soviet J. Nucl. Phys.*, **26:**42). (20.3.2)
Talenti, G. (1967): *Rend. Sem. Mat. Univ. Padova*, **38:**305 (*Math. Rev.*, **36:**951). (10.1.4)
Talukdar, B., U. Das, and D. Chatterji (1978): *Phys. Rev.*, **C18:**1917. (12.1.2)
—, —, N. Mallick, and M. Sen Gupta (1979): *J. Math. Phys.*, **20:**378. (10.1.4)
—, —, and S. Mukhopadhyaya (1979): *J. Math. Phys.*, **20:**372. (10.1.4)
Tamarkin, J. (1926): *Ann. Math.*, **28:**127. (9.3)
Tamm, I. (1948): *Z. Eksp. i Teor. Fiz.*, **18:**337 (*Math. Rev.*, **9:**558). (11.3)
— (1949): *Z. Eksp. i Teor. Fiz.*, **19:**74 (*Math. Rev.*, **10:**665). (11.3)
Tang, Y. C. (1969): in *Proc. Internatl. Conf. on Clustering Phenomena in Nuclei*, IAEA, Vienna, p. 109. (16.3.3)
Tani, S. (1966): *Ann. Phys. (N.Y.)*, **37:**411 and 451. (9.1.1)
— (1968): *Phys. Rev.*, **174:**2054. (11.2.2, 12.1.4)
— and M. Inokuti (1974): *J. Chem. Phys.*, **61:**4422. (13.3)
Taylor, A. E. (1958): "Introduction to Functional Analysis," Wiley, New York. (6.7.1, 6.7.2, 7.3)
Taylor, J. R. (1974): *Nuovo Cim.*, **B23:**313. (14.6)
Teichman, T., and E. P. Wigner (1952): *Phys. Rev.*, **87:**123. (16.5)
Temkin, A. (1957): *Phys. Rev.*, **107:**1004. (16)
— (1959): *Phys. Rev.*, **116:**358. (16)
— (1962): *Phys. Rev.*, **126:**130. (16)
Ter Haar, D. (1946): *Physica*, **12:**500. (12.1.3)
— (1961): In A. C. Strickland (ed.), "Reports on Progress in Physics," The Physical Society, London. (8.3.1)
Terent'ev, M. V. (1972): *Ann. Phys. (N.Y.)*, **74:**1. (19.2)
Theis, W. R. (1956): *Z. Naturforsch.*, **11a:**889. (14.7)
Thijssen, W. J. G., B. J. Verhaar, and A. M. Schulte (1981): *Phys. Rev.*, **C23:**1984. (16.3)
Thirring, W. (1978): in H. Zingl et al. (1978) II, p. 353. (17.4)
Thomas, L. (1972): *Helv. Phys. Acta*, **45:**1057. (7.3)
Thomas, L. D. (1980): *J. Chem. Phys.*, **73:**5905. (18.2)
Thomas, R. G. (1952): *Phys. Rev.*, **88:**1109. (17.2)

Tietz, T. (1959): *Nuovo Cimento*, **11:**126. (12.4)
— (1963): *Nuclear Phys.*, **44:**633. (12.2)
Tikochinsky, Y. (1970): *J. Math. Phys.*, **11:**3019. (11.3.3)
— and U. Koren (1973): *Phys. Rev.*, **A7:**824. (12.1.2)
Tiktopolous, G. (1965): *Phys. Rev.*, **138:**B1550. (12.4)
— and S. B. Treiman (1964): *Phys. Rev.*, **134:**B844. (12.4)
Tinkham, M. (1964): "Group Theory and Quantum Mechanics," McGraw-Hill, New York. (17.4.3)
Titchmarsh, E. C. (1937): "Theory of Fourier Integrals," Oxford, Fair Lawn, N.J. (4.2.1)
— (1939): "The Theory of Functions," 2d ed., Oxford, Fair Lawn, N.J. (12.1.2, 13.2)
— (1946): "Eigenfunction Expansions I," Oxford, Fair Lawn, N.J. (12.1.5)
Titulaer, U. M., and R. J. Glauber (1965): *Phys. Rev.*, **140:**B676. (4.3, 8.3.1)
— and — (1966): *Phys. Rev.*, **145:**1041. (4.3, 8.31)
Tjon, J. A. (1970): *Phys. Rev.*, **D1:**2109. (17.4)
— (1976): in A. N. Mitra et al, eds., *Few Body Dynamics*, North-Holland Publ. Co., Amsterdam, p. 567. (17.4)
— (1980): in F. S. Levin (1980), p. 47c. (17.4)
Tobocman, W. (1965): *Phys. Rev.*, **136:**B1825. (9.2)
— (1974a): *Phys. Rev.*, **C10:**60. (16.4)
— (1974b): *Phys. Rev.*, **C9:**2466. (17.4, 17.4.3)
— (1975a): *Phys. Rev.*, **C11:**43. (17.4)
— (1975b): *Phys. Rev.*, **C12:**1146. (16.3.3)
— (1981a): preprint. (17.4)
— (1981b): *Phys. Rev.*, **C24:**2743. (17.4)
— and M. A. Nagarajan (1967): *Phys. Rev.*, **163:**1011. (17.4)
— and M. Pauli (1972): *Phys. Rev.*, **D5:**2088 and **D10:**2714. (18)
Tolhoek, H. A. (1956): *Revs. Modern Phys.*, **28:**277. (8.3.2)
Toll, J. S. (1956): *Phys. Rev.*, **104:**1760. (4.2.1)
Tomio, L., and S. K. Adhikari (1980): *Phys. Rev.*, **C22:**28 and 2359. (17.4)
— and — (1981): *Phys. Rev.*, **C24:**43. (17.4)
Tomonaga, S. (1946): *Progr. Theor. Phys.*, **1:**27. (6.5)
Tortorella, M. (1972): *J. Math. Phys.*, **13:**1764. (20.2)
— (1974): *J. Math. Phys.*, **15:**745 and 1153. (20.2)
— (1975): *J. Math. Phys.*, **16:**823. (20.2)
— and J. A. Leise (1976): *SIAM J. Appl. Math.*, **30:**407. (20.2)
Trainor, L. E. H., and Ta-You Wu (1953): *Phys. Rev.*, **89:**273. (16.3.3)
Treiman, S. B., and R. G. Sachs (1956): *Phys. Rev.*, **103:**1545. (19.3)
Trlifaj, L. (1980): *J. Math. Phys.*, **21:**1724. (12.1.1)
Troesch, A., and M. Verde (1951): *Helv. Phys. Acta*, **24:**39. (11.3, 17.4)
Troitskii, V. E. (1978): *Teor. i Mat. Fiz.*, **37:**243 (transl. *Theor. Math. Phys.* **37:**996). (20.3)
Truhlar, D. G. (1971): *Phys. Rev.*, **A4:**1886. (16.3)
—, J. Abdallah, Jr., and R. L. Smith (1974): *Adv. in Chem. Phys.* **25**, Wiley Publ. Co., N.Y. (11.3)
—, D. C. Cartwright, and A. Kuppermann (1968): *Phys. Rev.*, **175:**113. (16.3)
Tsang, T. Y., and T. A. Osborn (1975): *Nucl. Phys.*, **A247:**43. (11.1.3)
Tsutsumi, M. (1981): *J. Math. Anal. Appl.*, **83:**316. (20.3)
Tuan, T. F. (1963): *Proc. Roy. Soc. (London)*, **A276:**492. (17.2)

Tucciarone, A. (1966): *Nuovo Cimento*, **41**:204. (17.4.3)
Turner, J. S., and R. E. R. Makinson (1953): *Proc. Phys. Soc.* (*London*), **A66**:857 and 866. (11.3)
Turner, R. E. (1980a): *J. Phys. A*, **13**:2079. (18.3, 18.4)
— (1980b): *J. Phys. A*, **13**:3447. (18.3)
Twersky, V. (1978): *J. Math. Phys.*, **19**:215. (1.5)
Underhill, J. (1962): *Proc. Cambridge Phil. Soc.*, **58**:363. (13.4, 17.1)
— (1963): *Proc. Cambridge Phil. Soc.*, **59**:161. (13.4, 17.1)
— (1970): *J. Math. Phys.*, **11**:1409. (20.3)
Urban, P., and K. Wildermuth (1953): *Z. Naturforsch.*, **8a**:594. (9.1)
Uzunoglu, N. K., N. G. Alexopoulos, and J. G. Fikioris (1978): *J. Opt. Soc. Am.*, **68**:194. (1.3)
—, B. G. Evans, and A. R. Holt (1976): *Electronic Lett.*, **12**:312. (4.1)
— and A. R. Holt (1977): *J. Phy. A*, **10**:413. (1.3)
Vall, A. N., and N. A. Makeev (1978): *Yad. Fiz.*, **27**:558 (transl. *Soviet J. Nucl. Phys.*, **27**:298). (18.3)
van de Hulst, H. C. (1949): *Physica*, **15**:740. (1.3)
— (1957): "Light Scattering by Small Particles," Wiley, New York. (1.2, 2.1.5, 3.5, 3.6, 4.1.1)
van der Corput, J. G. (1934): *Comp. Math.*, **1**:15. (1.3, 6.7.3)
— (1936): *Comp. Math.*, **3**:328. (1.3, 6.7.3)
van der Merwe, P. du T. (1972): *Phys. Rev.*, **D5**:353. (17.4)
van der Pol, B., and H. Bremmer (1937): *Phil. Mag.*, **24**:141. (3.6)
van der Waerden, B. L. (1951): *Appl. Sci. Research*, **B2**:33. (3 App.)
van Haeringen, H. (1976): *J. Math. Phys.*, **17**:995. (14.6)
— (1977): *J. Math. Phys.*, **18**:927. (14.6)
— (1978a): *Phys. Rev.*, **A18**:56. (12.1.2)
— (1978b): *J. Math. Phys.*, **19**:1379. (14.6)
— (1979a): Ph.D. thesis, Groningen, unpubl. (12.4, 14.6)
— (1979b): *J. Math. Phys.*, **20**:861. (14.6)
— and L. P. Kok (1981a): *Phys. Lett.*, **82A**:317. (14.6)
— and — (1981b): *Phys. Rev.*, **C24**:1827. (14.6)
—, C. V. M. van der Mee, and R. van Wageningen (1977): *J. Math. Phys.*, **18**:941. (12.1.4)
— and R. van Wageningen (1975): *J. Math. Phys.*, **16**:1441. (14.6)
van Horn, H. M., and E. E. Salpeter (1967): *Phys. Rev.*, **157**:751. (18.2)
van Hove, L. (1955): *Physica*, **21**:**901**. (6)
— (1956): *Physica*, **22**:343. (6)
— (1972): *Nucl. Phys.*, **B46**:75. (18.3)
van Kampen, N. G. (1953a): *Phys. Rev.*, **89**:1072. (4.2.1)
— (1953b): *Phys. Rev.*, **91**:1267. (10.3.4)
— (1954): *Physica*, **20**:115. (10.3.4)
van Nieuwenhuizen, P., and Th. Ruijgrok (1967): *Physica*, **33**:595. (17.4)
van Phu, F. Q. (1970): Ph.D. thesis, Montpellier, unpubl. (20.4)
van Wageningen, R. (1965): *Ann. Phys.* (*N.Y.*), **31**:148. (20.1)
van Winter, C. (1964): *Nuclear Phys.*, **57**:134. (16.5)
— (1964): *Mat. Fys. Skr. Dan. Vid. Selsk.* **2**, No. 8. (16.1, 17.4.3)
— (1965): *Mat. Fys. Skr. Dan. Vid. Selsk.* **2**, No. 10. (17.4.3)
— (1966): *Nucl. Phys.*, **77**:99. (17.1)
— (1970): in R. P. Gilbert and R. G. Newton, eds., *Analytic Methods in Mathematical*

Physics, Gordon and Breach Publ. Co., N.Y., p. 569. (17.4.3)
— (1971): *Trans. Am. Math. Soc.*, **162:**103. (9.3)
— (1978): in J. Nuttall (1978), p. 181. (17.4)
— and H. J. Brascamp (1968): *Commun. Math. Phys.*, **11:**19. (17.4)
van Wyck, C. B. (1958): *Nuovo Cimento*, **9:**270. (8.3.2)
Vanzani, V. (1969): *Lett. Nuovo Cim.*, **2:**706. (17.4)
— (1971): *Nuovo Cim.*, **2A:**525. (17.4)
— (1973): *Nuovo Cim.*, **16A:**449. (17.4)
— (1974): *Lett. Nuovo Cim.*, **10:**610. (17.4)
— (1976): in A. N. Mitra et al., eds., *Few-Body Dynamics*, North-Holland Publ. Co., Amsterdam, p. 394; *Lett. Nuovo Cim.*, **16:**1. (17.4)
— (1978a): in G. Pisent et al. (1978), p. 57. (17.4, 17.4.5)
— (1978b): in H. Zingl et al. (1978), I, p. 360. (17.4, 17.4.4)
— (1978c): in *Lett. Nuovo Cim.*, **23:**586. (17.4.4)
Varadan, V. K. (1980): in Varadan and Varadan (1980), p. 103. (1.5)
—, V. N. Bringi, and V. V. Varadan (1979): *Phys. Rev.*, **D19:**2480. (1.5)
— and V. V. Varadan (1980): eds., *Acoustic Electromagnetic, and Elastic Wave Scattering-Focus on the T-Matrix Approach*, Pergamon Press, N.Y. (1.5)
Vashakidze, I. Sh., R. M. Muradyan, A. N. Takhelidze, G. A. Chilashvili, and V. P. Shelest (1965): *Dokl. Akad. Nauk SSSR*, **158:**1302 (transl. *Soviet Phys. Dokl.*, **9:**910). (17.4.4)
Vasilevskii, L. S., N. I. Zhirnov (1978): *Yad. Fiz.*, **27:**1122 (transl. *Soviet J. Nucl. Phys.*, **27:**595. (20.1)
Vaughn, M. T., R. Aaron, and R. D. Amado (1961): *Phys. Rev.*, **124:**1258. (12.1.3)
Verde, M. (1949): *Helv. Phys. Acta*, **22:**339. (11.3, 17.4)
— (1955): *Nuovo Cimento*, **2:**1001. (18)
— (1957): *Nuovo Cimento*, **6:**340. (18)
— (1958): *Nuovo Cimento*, **8:**560. (18)
— (1958–1959): *Nuclear Phys.*, **9:**255. (20.3)
— (1963): *Nuovo Cimento*, **28:**547. (18)
Veselova, A. M. (1970a): *Teor. i Mat. Fiz.*, **3:**326. (14.6)
— (1970b): *Teor. i Mat. Fiz.*, **3:**542. (17.4)
— (1972): *Teor. i Mat. Fiz.*, **13:**368. (transl. *Theor. Math. Phys.* **13:**1200). (17.4)
— (1978): *Teor. i Mat. Fiz.*, **35:**180 (transl. *Theor. Math. Phys.*, **35:**395). (17.4)
Vezzetti, D. J., and J. B. Keller (1967): *J. Math. Phys.*, **8:**1861. (1.3)
— and S. I. Rubinow (1965): *Ann. Phys.* (*N.Y.*), **35:**373. (18.2)
Viano, G. A. (1969): *Nuovo Cim.*, **63A:**581. (20.4)
Villaroel, D. (1970): *Nuovo Cim.*, **70A:**175. (10.3.3)
Vogt, E., and G. H. Wannier (1954): *Phys. Rev.*, **95:**1190. (5.6, 12.4, 13.1, 14.7)
Vollmer, G. (1969): *Z. Physik*, **226:**423. (20.1)
von Neumann, J. (1927): *Nachr. deut. Akad. Wiss. Göttingen, Math.-phys. Kl.*, 245 and 273. (8.3.1)
— (1932): *Proc. Natl. Acad. Sci. U.S.*, **18:**70. (16 App.)
— (1955): "Mathematical Foundations of Quantum Mechanics," Princeton, Princeton, N.J. (7.3)
Wald, S. S., and P. Lu (1974): *Phys. Rev.*, **D10:**3434. (18.2)
—, M. Schardt, and P. Lu (1975): *Phys. Rev.*, **D12:**224. (12.4)
Wallace, S. J. (1971): *Phys. Rev. Lett.*, **27:**622. (18.3)
— (1973a): *Phys. Rev.*, **D8:**1846. (18)
— (1973b): *Phys. Rev.*, **D8:**1934. (18.3)

— (1974): *Phys. Rev.*, **D9:** 406. (18.3)
Wang, D.–S., and P. W. Barber (1979): *Appl. Optics*, **18:** 1190. (1.3)
Warke, C. S. (1972): *J. Math. Phys.*, **13:** 1. (12.1.2)
— and R. K. Bhaduri (1971): *Nucl. Phys.*, **162A:** 289. (12.1.2)
— and M. K. Srivastava (1977): *Phys. Rev.*, **C15:** 561. (12.1.2)
— and — (1978): *Phys. Rev.*, **C18:** 1919. (12.1.2)
Warner, C., and A. Hizal (1976): *Radio Sci.*, **11:** 921. (2)
Warnock, R. L. (1963): *Phys. Rev.*, **131:** 1320. (12.1.3)
— (1964): *Nuovo Cimento*, **32:** 255. (17.1)
— (1967): *Nuovo Cim.*, **50A:** 894. (17.1)
— (1968): *Phys. Rev.*, **170:** 1323. (7.2.2)
Watanabe, S. (1955): *Revs. Modern Phys.*, **27:** 26, 40, and 179. (7.2.4)
Waterman, P. C. (1965): *Proc. IEEE*, **53:** 805. (4.1)
— (1969): *Alta Frequenza*, **38 (Spec.):** 348. (1.3)
— (1971): *Phys. Rev.*, **D3:** 825. (4.1)
— (1979): *J. Appl. Phys.*, **50:** 4550. (4.1)
— (1980): in Varadan and Varadan (1980), p. 61. (4.1)
Watson, G. N. (1918): *Proc. Roy. Soc. (London)*, **A95:** 83 and 546. (3.8)
— (1958): "Bessel Functions," Cambridge, New York. (2.2.1, 12.3)
Watson, K. M. (1952): *Phys. Rev.*, **88:** 1163. (17.1)
— (1953): *Phys. Rev.*, **89:** 575. (17.4.2)
— (1957): *Phys. Rev.*, **105:** 1388. (17.4.2)
— (1969): *J. Math. Phys.*, **10:** 688. (1.5)
Weidenmüller, H. A. (1964): *Ann. Phys. (N.Y.)*, **28:** 60, **29:** 378. (16.5, 17.1)
— (1965): *Nuclear Phys.*, **69:** 113. (17.2)
Weidmann, J. (1964): *Math. Z.*, **98:** 268. (12.1.2)
Weigel, M. (1965): *Z. Physik*, **185:** 186. (13.3)
Weinberg, S. (1963a): *Phys. Rev.*, **130:** 776. (9.1.1, 9.2)
— (1963b): *Phys. Rev.*, **131:** 440. (9.1.1, 9.2, 10.3.5)
— (1964a): *Phys. Rev.*, **133: B**232. (7.2.5, 7.3, 9.1.1, 9.2, 17.4.3)
— (1964b): *J. Math. Phys.*, **5:** 743. (9.1.1, 9.2)
Weiss, J. (1969): *Czech. J. Phys.*, **B19:** 1578. (20.3)
— (1971): *Czech J. Phys.*, **B21:** 1019 and 1032. (20.3)
Weiss, R., and G. Scharf (1971): *Helv. Phys. Acta*, **44:** 910. (20.3)
—, W. Stahel, and G. Scharf (1972): *Nucl. Phys.*, **A183:** 337. (20.3)
Weisskopf, V. F., and E. P. Wigner (1930): *Z. Phys.*, **63:** 54, **65:** 18. (19.2)
Wellner, M. (1960): *Phys. Rev.*, **118:** 875. (17.3)
— (1963): *Phys. Rev.*, **132:** 1848. (9.1.1)
Wentzel, G. (1926a): *Z. Phys.*, **38:** 518. (18.2)
— (1926b): *Z. Phys.*, **40:** 590. (14.6)
West, G. B. (1967): *J. Math. Phys.*, **8:** 942. (14.6)
Weyers, J. (1966): *Phys. Rev.*, **145:** 1236 and **151:** 1159. (17.4)
Wheeler, J. A. (1937a): *Phys. Rev.*, **52:** 1107. (2.2.2, 6.4)
— (1937b): *Phys. Rev.*, **52:** 1123. (12.1.2)
— (1937c): *Phys. Rev.*, **52:** 1083. (16.3.3)
— (1955): *Phys. Rev.*, **99:** 630. (20.3)
White, F. P. (1922): *Proc. Roy. Soc. (London)*, **100:** 505. (3.8)
Whittaker, E. T., and G. N. Watson (1948): "A Course of Modern Analysis," Cambridge, New York. (2.1.1, 11.1.1, 12.3, 13.2, 18.3, 20.3.1)
Wichmann, E. H., and C. H. Woo (1961): *J. Math. Phys.*, **2:** 178. (14.6)

Wick, G. C. (1962): *Ann. Phys. (N.Y.)*, **18:**65. (17.4.6)
Wiesner, J., E. P. Zhidkov, R. V. Malyshev, B. N. Khoromskii, E. Kh. Khristov, V. Lelek, and I. Ulehla (1978): *Fiz. Elem. Chast. At. Yadra (SSSR)*, **9:**710 (transl. *Soviet J. Part. Nucl.*, **9:**291). (20.3)
Wigner, E. P. (1946): *Phys. Rev.*, **70:**15 and 606. (16.5)
— (1948): *Phys. Rev.*, **73:**1002. (17.2)
— (1951): *Ann. Math.*, **53:**36. (9.1.1)
— (1952a): *Ann. Math.*, **55:**7. (9.1.1)
— (1952b): *Am. Math. Monthly*, **59:**669. (9.1.1)
— (1955): *Phys. Rev.*, **98:**145. (11.2.2, 12.1.3, 12.2)
— (1959): "Group Theory," Academic, New York. (2.1.4, 7.1.4, 7.2.4, 9.1.1, 15.1, 17.4.3)
— (1960): *J. Math. Phys.*, **1:**409 and 414. (7.2)
— and L. Eisenbud (1947): *Phys. Rev.*, **72:**29. (7.2.4, 16.5)
— and J. von Neumann (1954): *Ann. Math.*, **59:**418. (9.1.1)
— and V. F. Weisskopf (1930): *Z. Phys.*, **63:**62. (17.3)
Wilcox, C. H. (1965): *J. Math. Phys.*, **6:**611. (6.7)
Wildermuth, K., and W. McClure (1966): *Cluster Representations of Nuclei*, Springer Verlag, Berlin. (16.3.3)
Wilk, S. F. J., Y. Fujiwara, and T. A. Osborn (1981): *Phys. Rev.*, **A24:**2187. (17.4, 18.2)
Williams, A. C. (1980): *Ann. Phys. (N.Y.)*, **129:**22. (18.3)
Wilson, H. A. (1930): *Phys. Rev.*, **35:**948. (12.1.2)
Winter, R. G. (1961): *Phys. Rev.*, **123:**1503. (19.2)
Wojtczak, L. (1963): *Nuclear Phys.*, **48:**325. (14.3)
Wolf, E. (1962): *Proc. Phys. Soc. London*, **80:**1269. (4.3)
— (1969): *Optics Commun.*, **1:**153. (20.3)
— (1970): *J. Opt. Soc. Am.*, **60:**18. (4.3)
— (1973): in L. Mandel and E. Wolf, eds., *Coherence and Quantum Optics*, Plenum Press, N.Y. (1.3.9)
— and J. R. Shewell (1967): *Phys. Lett.*, **25A:**417. (20.3)
Wolfenstein, L. (1949): *Phys. Rev.*, **75:**1664. (15.1)
— (1956): In J. G. Beckerley (ed.), "Annual Review of Nuclear Science," vol. 6, p. 43, Annual Reviews, Palo Alto, Calif. (15.1)
Wollenberg, M. (1977): *Math. Nachr.*, **78:**223 and 369. (12.1.3)
Wong, D. Y., and G. Zambotti (1967): *Phys. Rev.*, **154:**1540. (17.4)
Wright, J. A. (1965): *Phys. Rev.*, **139:**B137. (12.1.3)
—and M. Scadron (1964): *Nuovo Cimento*, **34:**1571. (9.1.1)
Wu, A. C. T. (1974): *Phys. Rev.*, **D10:**2248 and 3871. (18)
Wu, T. T. (1964): *Phys. Rev.*, **136:**B1176. (12.4)
Wu, T. Y. (1952): *Phys. Rev.*, **87:**1012. (16.3)
— (1956): *Can. J. Phys.*, **34:**179. (16.3)
— and T. Ohmura (1962): "Quantum Theory of Scattering," Prentice-Hall, Englewood Cliffs, N.J. (6, 7)
Yaes, R. J. (1969): *Nucl. Phys.*, **A131:**623. (17.4)
Yafaev, D. R. (1972): *Dokl. Akad. Nauk. SSSR*, **206:**68 (transl. *Soviet Phys. Dokl.*, **17:**849). (12.1.4, 17.4)
— (1974): *Mat. Sbornik*, **94:**567 (transl. *Math. USSR Sbornik*, **23:**535). (17.4.8)
— (1976): *Teor. i Mat. Fiz.*, **27:**55 (transl. *Theor. Math. Phys.*, **27:**328). (12.1.4, 17.4)

— (1978): *Teor. i Mat. Fiz.*, **37:**48 (transl. *Theor. Math. Phys.*, **37:**867). (16.2.3, 17.4.7)
Yajima, K. (1976): *J. Fac. Sci. Univ. Tokyo IA*, **23:**517. (6)
— (1977): *J. Math. Soc. Japan*, **29:**729. (6.3)
— (1978): *J. Fac. Sci. Univ. Tokyo IA*, **25:**109. (17.4)
— (1979a): *Commun. Math. Phys.*, **69:**101. (18.3)
— (1979b): *J. Fac. Sci. Univ. Tokyo IA*, **26:**377. (6.3)
— (1981a): *Duke Math. J.*, **48:**1. (18.3)
— (1981b): preprint, Princeton. (18.7)
— (1981c): preprint, Princeton. (18.3)
— (1981d): *J. Fac. Sci. Univ. Tokyo IA*, **28:**1. (6.3)
— (1981e): preprint, Princeton . (Ch. 5)
Yakubovskii, O. A. (1967): *Yad. Fiz.*, **5:**1312 (transl. *Soviet J. Nucl. Phys.*, **5:**937). (17.4)
— (1977): *Yad. Fiz.*, **25:**1311 (transl. *Soviet J. Nucl. Phys.*, **25:**694). (17.4)
Yamaguchi, Y. (1959): *Progr. Theor. Phys. Suppl.*, **7:**1. (16.3, 17.2)
Yang, C. N. (1948): *Phys. Rev.*, **74:**764. (15.1)
— (1961): "Elementary Particles," Princeton, Princeton, N.J. (3.7)
— (1968): *Phys. Rev.*, **168:**1920. (17.3)
Yao, T. H. (1973): *J. Math. Phys.*, **14:**1141. (10.3.4)
— (1974): *J. Math. Phys.*, **15:**1211, 1225, 1227, 1232. (10.3.4)
— (1975): *J. Math. Phys.*, **16:**68. (10.3.4)
— (1976): *J. Math. Phys.*, **17:**13. (10.3.4)
Yates, A. C. (1979): *Phys. Rev.*, **A19:**1550. (9.1.2)
Yennie, D. R., F. L. Boos, and D. G. Ravenhall (1965): *Phys. Rev.*, **137:**B882. (18.2)
Yosida, K. (1965): "Functional Analysis," Springer, Berlin. (7.3)
Young, L. A. (1931): *Phys. Rev.*, **38:**1612. (12.1.2)
Yukawa, H. (1935): *Proc. Phys.-Math. Soc. Japan*, **17:**48. (10.3.5)
Zachary, W. W. (1969): *J. Math. Phys.*, **10:**1098. (17.4)
— (1971): *J. Math. Phys.*, **12:**1379 and **14:**2018. (14.6)
— (1976): *J. Math. Phys.*, **17:**1056 and **18:**536. (16.2)
Zakhariev, B. N. (1976): *Nucleonica*, **21:**709. (20)
—, E. B. Plekhanov, and A. A. Suzko (1981): preprint . (14.7)
Zeldovich, Ya. B. (1960): *Z. Eksp. i Teor. Fiz.*, **39:**779 (transl. *Soviet Phys. JETP*, 12:542). (19.2)
Zemach, C. (1964a): *Nuovo Cimento*, **33:**939. (12.1.2)
— (1964b): *Phys. Rev.*, **133:**B1201. (17.4)
— and R. J. Glauber (1956): *Phys. Rev.*, **101:**118. (16)
— and A. Klein (1958): *Nuovo Cimento*, **10:**1078. (9.1.1, 10.3.1)
— and F. Odeh (1960): *Arch. Rat. Mech. Anal.*, **5:**226. (7.3)
Zepalova, M. L. (1979): *Ukr. Fiz. Zh.*, **24:**1512. (17.4)
Zhigunov, V. P. (1979): preprint . (20)
Zhislin, G. M. (1960): *Moskow Mat. Obs.*, **9:**81. (16.1)
— (1972): *Dokl. Akad. Nauk SSSR*, **207:**25 (transl. *Soviet Math. Dokl.*, **13:**1445). (12.1.4, 17.4)
Zickendraht, W. (1964): *Proc. Natl. Acad. Sci. (U.S.)*, **52:**1565. (17.4.6)
— (1965): *Ann. Phys. (N.Y.)*, **35:**18. (17.4.6)
— (1967): *Phys. Rev.*, **159:**1448. (17.4)
Ziegelmann, H. (1972): *Nucl. Phys.*, **A192:**426. (17.4)
Zimm, B. H., and W. B. Dandliker (1954): *J. Phys. Chem.*, **58:**644. (1.5)

Zingl, H., M. Haftel, and H. Zankel (1978): eds., *Few Body Systems and Nuclear Forces, I and II*, Springer Verlag, Berlin. (17.4)
Zirilli, F. (1976): *Nuovo Cim.*, **34A:** 385. (12.1.2)
Zorbas, J. (1974a): *J. Phys. A*, **7:** 1557. (14.6)
— (1974b): *Lett. Nuovo Cim.*, **10:** 121. (16.2)
— (1976a): *Rep Math. Phys.*, **9:** 309. (7.2.5)
— (1976b): *J. Math. Phys.*, **17:** 498. (16.2)
— (1977): *J. Math. Phys.*, **18:** 1112. (14.6, 16.2)
— (1978): *J. Math. Phys.*, **19:** 2426. (14.6)
— (1979): *Rep. Math. Phys.*, **16:** 409. (17.4)
Zubarev, A. L. (1976): *Yad. Fiz.*, **23:** 77 (transl. *Soviet J. Nucl. Phys.*, **23:** 40). (16.4)
Zumino, B. (1956): *N.Y. Univ. Inst. Math. Sci. Research Rept.* CX-23. (16.5)
Zupančič, C. (1965): *Revs. Modern Phys.*, **37:** 330. (17.4.3)
Zwanziger, D. (1964): *Phys. Rev.*, **136:** B558. (8.3.2)
— (1967): *J. Math. Phys.*, **8:** 1858. (14.6)

Index

Texts and Monographs in Physics

Edited by **W. Beiglböck, E. Lieb,** and **W. Thirring**

Polarized Electrons
by **J. Kessler**
1976. ix, 223p. 104 illus.

The Theory of Photons and Electrons
The Relativistic Quantum Field Theory of Charged Particles with Spin One-half
Second Expanded Edition
By **J. M. Jauch** and **F. Rohrlich**
1980. xix, 553p. 59 illus.

Essential Relativity
Special, General, and Cosmological
Revised Second Edition
By **W. Rindler**
1977. xiv, 284p. 44 illus.

Inverse Problems in Quantum Scattering Theory
By **K. Chadan** and **P. C. Sabatier**
1977. xxii, 344p. 34 illus.

Quantum Mechanics
By. **A. Böhm**
1979. xvii, 522p. 105 illus.

Relativistic Particle Physics
By **H. Pilkuhn**
1979. xii, 427p. 85 illus.

The Concepts and Logic of Classical Thermodynamics as a Theory of Heat Engines
Rigourously Constructed upon the Foundation Laid by S. Carnot and F. Reech
By **C. Truesdell** and **S. Bharatha**
1977. xxii, 154p. 15 illus.

Principles of Advanced Mathematical Physics
By **R. D. Richtmyer**
Volume 1
1978. xv, 422p. 45 illus.
Volume II
1981. ix, 322p. 60 illus.

Foundation of Theoretical Mechanics
By **R. M. Santilli**
Vol. I: The Inverse Problem in Newtonian Mechanics
1978. xix, 266p. 5 illus.
Vol. II: Birkhoffian Generalization of Hamiltonian Mechanics
In preparation

Advanced Quantum Theory
And Its Applications Through Feynman Diagrams
By **M. D. Scadron**
1979. xiv, 386p. 78 illus.

Operator Algebras and Quantum Statistical Mechanics
By **O. Bratteli** and **D. W. Robinson**
Vol. I: C*- and W*-Algebras. Symmetry Groups. Decomposition of States
1979. xii, 500p.
Vol. II: Equilibrium States. Models in Quantum Statistical Mechanics
1981. xi, 505p.

The Nuclear Many-Body Problem
By **P. Ring** and **P. Schuck**
1980. xvii, 716p. 171 illus.

Nuclear Reactions with Heavy Ions
By **R. Bass**
1980. viii, 410p. 176 illus.

Scattering Theory of Waves and Particles
Second Edition
By **R. G. Newton**
1982. xvii, 742p. 35 illus.